PRIMES OF THE FORM $x^2 + ny^2$

FERMAT, CLASS FIELD THEORY, AND COMPLEX MULTIPLICATION. THIRD EDITION WITH SOLUTIONS

Primes of the Form $x^2 + ny^2$

Fermat, Class Field Theory, and Complex Multiplication. Third Edition with Solutions

David A. Cox

With contributions by Roger Lipsett

AMS Chelsea Publishing

2020 *Mathematics Subject Classification.* Primary 11A41; Secondary 11F11, 11R11, 11R16, 11R18, 11R37, 11Y11.

For additional information and updates on this book, visit
www.ams.org/bookpages/chel-387

Library of Congress Cataloging-in-Publication Data

Names: Cox, David A., author. | Lipsett, Roger, 1950- other.
Title: Primes of the form $x^2 + ny^2$: Fermat, class field theory, and complex multiplication. Third Edition with Solutions / David A. Cox ; with contributions by Roger Lipsett.
Other titles: Primes of the form p equals x^2 plus ny^2
Description: Third edition. | Providence, Rhode Island : AMS Chelsea Publishing/American Mathematical Society, [2022] | Includes bibliographical references and index. | Summary: "The goal of the new edition of Primes of the Form $x^2 + ny^2$ is to make this wonderful part of number theory available to readers in a form especially suited to self-study, mainly because complete solutions to all exercises are included"– Provided by publisher.
Identifiers: LCCN 2022025796 | ISBN 9781470470289 (paperback acid-free paper) | 9781470471835 (ebook)
Subjects: LCSH: Numbers, Prime. | Mathematics. | AMS: Number theory – Elementary number theory – Primes. | Number theory – Discontinuous groups and automorphic forms – Holomorphic modular forms of integral weight. | Number theory – Algebraic number theory: global fields – Quadratic extensions. | Number theory – Algebraic number theory: global fields – Cubic and quartic extensions. | Number theory – Algebraic number theory: global fields – Cyclotomic extensions. | Number theory – Algebraic number theory: global fields – Class field theory. | Number theory – Computational number theory – Primality.
Classification: LCC QA246 .C69 2022 | DDC 512.7/23–dc23/eng20220829
LC record available at https://lccn.loc.gov/2022025796

∞ The paper used in this book is acid-free and falls within the guidelines
established to ensure permanence and durability.
Visit the AMS home page at `https://www.ams.org/`

10 9 8 7 6 5 4 3 2 1 27 26 25 24 23 22

Contents

Preface

First Edition

Several years ago, while reading Weil's *Number Theory: An Approach Through History*, I noticed a conjecture of Euler concerning primes of the form $x^2 + 14y^2$. That same week I picked up Cohn's *A Classical Invitation to Algebraic Numbers and Class Fields* and saw the same example treated from the point of view of the Hilbert class field. The coincidence made it clear that something interesting was going on, and this book is my attempt to tell the story of this wonderful part of mathematics.

I am an algebraic geometer by training, and number theory has always been more of an avocation than a profession for me. This will help explain some of the curious omissions in the book. There may also be errors of history or attribution (for which I take full responsibility), and doubtless some of the proofs can be improved. Corrections and comments are welcome!

I would like to thank my colleagues in the number theory seminars of Oklahoma State University and the Five Colleges (Amherst College, Hampshire College, Mount Holyoke College, Smith College and the University of Massachusetts) for the opportunity to present material from this book in preliminary form. Special thanks go to Dan Flath and Peter Norman for their comments on earlier versions of the manuscript. I also thank the reference librarians at Amherst College and Oklahoma Slate University for their help in obtaining books through interlibrary loan.

Amherst, Massachusetts
August 1989

Second Edition

The philosophy of the second edition is to preserve as much of the original text as possible. The major changes are:

- A new §15 on Shimura Reciprocity has been added, based on work of Peter Stevenhagen and Alice Gee [**53, 54, 126**] and Bumkyo Cho [**22**].
- The fifteen sections are now organized into four chapters:
  - The original §§1–13, which present a complete solution of $p = x^2 + ny^2$, now constitute Chapters 1, 2 and 3.
  - The new Chapter 4 consists of the original §14 (on elliptic curves) and the new §15 (on Shimura Reciprocity).
- An "Additional References" section has been added to supplement the original references [1]–[112]. This section is divided into five parts:
  - The first part consists of references [A1]–[A24] that are cited in the text. These references (by no means complete) provide updates to the book.

  - The remaining four parts give some references (also not complete) for further reading that are relevant to the topics covered in Chapters 1, 2, 3 and 4.
- The expanded Notation section now includes all notation used in the book. Specialized notation is listed according to the page where it first appears.

The other changes to the text are very minor, mostly to enhance clarity, improve formatting, and simplify some of the proofs. One exception is the addition of new exercises: at the end of §12, Exercise 12.31 shows how Ramanujan could have derived Weber's formula for $\mathfrak{f}_1(\sqrt{-14})^2$ (thanks to Heng Huat Chan), and at the end of §14, Exercise 14.24 gives an elliptic curve primality test for Mersenne numbers due to Dick Gross [**59**] (thanks to Alice Silverberg).

I would like to thank the following people for the errors they found in the first edition and for the suggestions they made: Michael Baake, Dominique Bernardi, Jeff Beyerl, Reinier Bröker, Tony Feng, Nicholas Gavrielides, Lee Goswik, Christian Guenther, Shiv Gupta, Kazuo Hata, Yves Hellegouarach, Norm Hurt, Tim Hutchinson, Trevor Hyde, Maurice Kostas, Susumu Kuninaga, Franz Lemmermeyer, Joseph Lipman, Mario Magioladitis, David May, Stephen Mildenhall, Takashi Ono, Frans Oort, Alf van der Poorten, Jerry Shurman, Alice Silverberg, Neil Sloane, Steve Swanson, Cihangir Tezcan, Satoshi Tomabechi, Fan Xingyuan and Noriko Yui.

My hope is that the second edition of *Primes of the Form* $x^2 + ny^2$ will help bring this wonderful part of number theory to a new audience of students and researchers.

Amherst, Massachusetts
November 2012

Third Edition with Solutions

The goal of the new edition of *Primes of the Form* x^2+ny^2 is to make this wonderful part of number theory available to readers in a form especially suited to self-study, mainly because complete solutions to all exercises are included. The inspiration was an email correspondence with Roger Lipsett. The questions Roger asked about the exercises and the solutions he wrote for most of them led to the two major changes in this edition:

- All exercises in the book have now been carefully checked. Small errors have been fixed and many hints have been clarified and/or expanded. For some exercises, the changes are more substantial, including complete replacement in a few cases. There are also a small number of new exercises.
- Roger and I completed and revised the solutions he wrote. These now appear at the end of the book, along with some suggestions for how to use the solutions. (Briefly, our advice is that rather than just passively reading the solutions, you should actively engage with the exercises and use our solutions as extended hints for what to do.).

This explains why the new edition is labeled "Third Edition with Solutions."

Apart from these changes and typographical corrections, the text is largely the same. Here are two exceptions:

- The second edition added §15 on Shimura Reciprocity, with many details left to the exercises. The process of writing solutions for these exercises revealed some problems, including an error in the statement of Theorem 15.17 in the second edition. The corrected statement is now Theorem 15.22. Revisions to §15 include

fixing this theorem and its proof, together with a thorough rewrite of the entire section. This resulted in significant changes to the exercises.
- There is now a unified bibliography. The 112 references of the first edition were supplemented in the second edition with references A1–A24 that were listed separately. These have now been combined into a single bibliography with the addition of a few new references.

The web site for the book is

`https://dacox.people.amherst.edu/primes.html`

This website includes typographical errors for all editions and a link to supplementary exercises for §§1–3 written by Jeffrey Stopple.

I am especially grateful to the AMS for making this edition of *Primes of the Form* $x^2 + ny^2$ possible. I thank Ina Mette for helpful suggestions and the staff of AMS author support for their help with numerous questions about LaTeX. And, of course, huge thanks to Roger.

Amherst, Massachusetts
March 2022

DAVID A. COX

Notation

The following standard notation will be used throughout the book.

$\mathbb{Z}$, $\mathbb{Q}$, $\mathbb{R}$, $\mathbb{C}$	Integers, rational numbers, real numbers, complex numbers
$\mathrm{Re}(z)$, $\mathrm{Im}(z)$	Real and imaginary parts of $z \in \mathbb{C}$
$\mathfrak{h}$	Upper half plane $\{x + iy \in \mathbb{C} : y > 0\}$
$\mathbb{F}_q$	Finite field with q elements
$\mathbb{Z}_p$	Ring of p-adic integers
$\mathbb{Z}/n\mathbb{Z}$	Ring of integers modulo n
$[a] \in A/B$	Coset of $a \in A$ in the quotient A/B
R^*	Group of units in a commutative ring R with identity
$\mathrm{GL}(2, R)$	Group of invertible matrices $\left(\begin{smallmatrix} a & b \\ c & d \end{smallmatrix}\right)$, $a, b, c, d \in R$
$\mathrm{SL}(2, R)$	Subgroup of $\mathrm{GL}(2, R)$ of matrices with determinant 1
I	2×2 identity matrix $\left(\begin{smallmatrix} 1 & 0 \\ 0 & 1 \end{smallmatrix}\right)$
$\mathrm{Gal}(L/K)$	Galois group of the finite extension $K \subset L$
$[L : K]$	Degree of a the finite extension $K \subset L$
$\mathcal{O}_K$	Ring of algebraic integers in a finite extension K of $\mathbb{Q}$
$\zeta_n = e^{2\pi i/n}$	Standard primitive nth root of unity
$[a, b]$	The set $\{ma + nb : m, n \in \mathbb{Z}\}$
$\gcd(a, b)$	Greatest common divisor of the integers a and b
$\phi(n)$	Euler ϕ-function
$\log(x)$	Logarithm to the base e of $x \in \mathbb{R}$
$[x]$	Greatest integer $\leq x$ for $x \in \mathbb{R}$
$\|S\|$	Number of elements in a finite set S
$G \rtimes H$	Semidirect product, where H acts on G
$\ker(\varphi)$, $\mathrm{im}(\varphi)$	Kernel and image of a homomorphism φ
Q.E.D.	End of a proof or the absence of a proof

Notation for Chapter 1 (§§1–4)

(a/p)	Legendre symbol	11
(a/m)	Jacobi symbol	14
$h(D)$	Class number	24
$C(D)$	Class group	41
$\chi_i(a)$, $\delta(a)$, $\epsilon(a)$	Assigned characters	44
(P/Q)	Extended Jacobi symbol	59
$\mathbb{Z}[\omega]$, $\omega = e^{2\pi i/3}$	Ring for cubic reciprocity	60
$\mathbb{Z}[i]$, $i = \sqrt{-1}$	Ring of Gaussian integers	60
$N(\alpha)$	Norm of α	60
$(\alpha/\pi)_3$, $(\alpha/\pi)_4$	Cubic and biquadratic Legendre symbols	63, 65
(f, λ)	Gaussian period	68

Notation for Chapter 2 (§§5–9)

$N(\mathfrak{a})$	Norm of an ideal	78
I_K, P_K	Groups of ideals and principal ideals of $\mathcal{O}_K$	79
$C(\mathcal{O}_K)$	Ideal class group of $\mathcal{O}_K$	79

$e_{\mathfrak{P}\mid\mathfrak{p}}$, $f_{\mathfrak{P}\mid\mathfrak{p}}$	Ramification and inertial degrees	79
$D_{\mathfrak{P}}$, $I_{\mathfrak{P}}$	Decomposition and inertia groups	80
d_K	Discriminant of K	81
$(D/2)$	Kronecker symbol	82
$((L/K)/\mathfrak{P})$	Artin symbol of $\mathfrak{P} \subset \mathcal{O}_L$	84
$((L/K)/\mathfrak{p})$	Artin symbol of $\mathfrak{p} \subset \mathcal{O}_K$ (Abelian case)	85
$((L/K)/\cdot)$	Artin map	85
$T(\alpha)$, $N(\alpha)$	Trace and norm of α	92
$\mathcal{O}$	Order in a quadratic field	105
$f = [\mathcal{O}_K : \mathcal{O}]$	Conductor of $\mathcal{O}$	105
$I(\mathcal{O})$, $P(\mathcal{O})$	Groups of ideals and principal ideals of $\mathcal{O}$	108
$C(\mathcal{O})$	Ideal class group of $\mathcal{O}$	108
$h(\mathcal{O})$	Class number of $\mathcal{O}$	108
$C^+(\mathcal{O})$	Narrow (or strict) ideal class group	112
$C_s(\mathcal{O})$	Signed ideal class group	112
$I(\mathcal{O}, f), P(\mathcal{O}, f)$	$\mathcal{O}$-ideals and principal $\mathcal{O}$-ideals prime to f	113
$I_K(m)$	$\mathcal{O}_K$-ideals relatively prime to m	114
$P_{K,\mathbb{Z}}(f)$	Subgroup of $I_K(f)$ satisfying $I_K(f)/P_{K,\mathbb{Z}}(f) \simeq C(\mathcal{O})$	115
$\mathfrak{m}$	Modulus in the sense of class field theory	126
$P_{K,1}(\mathfrak{m})$	Important subgroup of $I_K(\mathfrak{m})$	126
$\Phi_{\mathfrak{m}} = \Phi_{L/K,\mathfrak{m}}$	Artin map for the modulus $\mathfrak{m}$	127
$\mathfrak{f}(L/K)$	Class field theory conductor of $K \subset L$	128
$(\alpha/\mathfrak{p})_n$, $(\alpha/\mathfrak{a})_n$	nth power Legendre symbols	130
$(\alpha, \beta/\mathfrak{p})_n$	nth power Hilbert symbol	132
$\mathcal{P}_K$	Set of prime ideals of K	133
$\delta(S)$	Dirichlet density of $S \subset \mathcal{P}_K$	133
$S \dot{\subset} T$	$S \subset T \cup$ finite set	135
$S_{L/K}$, $\tilde{S}_{L/K}$	Primes in K splitting completely in L and variant	135
$N_{L/K}$	Norm map on ideals and ideles	136, 137
$K_{\mathfrak{p}}$	Completion of K at $\mathfrak{p}$	136
$\mathbf{I}_K$, $\mathbf{C}_K$	Idele group and idele class group of K	137
$\Phi_{L/K}$	Idelic Artin map	137

Notation for Chapter 3 (§§10–13)

$L = [\omega_1, \omega_2]$	Lattice in $\mathbb{C}$	157
$\wp(z) = \wp(z; L)$	Weierstrass $\wp$-function	158
$g_2(L)$, $g_3(L)$	Coefficients in the differential equation for $\wp$	158
$G_r(L)$	The sum $\sum_{\omega \in L - \{0\}} 1/\omega^r$	158
$\Delta(L)$	The discriminant $g_2(L)^3 - 27g_3(L)^2$	162
e_1, e_2, e_3	Roots of $4x^3 - g_2(L)x - g_3(L)$	162
$j(L)$	j-invariant of L	162
$j(\tau) = j([1, \tau])$	j-function of $\tau \in \mathfrak{h}$	164
$j(\mathfrak{a})$	j-invariant of $\mathfrak{a} \subset \mathcal{O}$	164
$g_2(\tau)$, $g_3(\tau)$	$g_2(L)$, $g_3(L)$ for the lattice $L = [1, \tau]$	173
$\Delta(\tau)$	$\Delta(L)$ for the lattice $L = [1, \tau]$	173
$q = q(\tau)$	$e^{2\pi i \tau}$, $\tau \in \mathfrak{h}$	177
$\Gamma_0(m)$	Group $\left\{\left(\begin{smallmatrix} a & b \\ c & d \end{smallmatrix}\right) \in \mathrm{SL}(2, \mathbb{Z}) : c \equiv 0 \bmod m\right\}$	177
$C(m)$	Matrices that give the cosets of $\Gamma_0(m) \subset \mathrm{SL}(2, \mathbb{Z})$	179
$\Phi_m(X, Y)$	Modular equation	180
$h(z; L)$	Weber function used to generate ray class fields	189
$\Psi(m)$	Cardinality of $C(m)$	192
$\gamma_2(\tau)$	Cube root of the j-function	195
$\gamma_3(\tau)$	Square root of $j(\tau) - 1728$	200

$\eta(\tau)$	Dedekind η-function	201
$\mathfrak{f}(\tau)$, $\mathfrak{f}_1(\tau)$, $\mathfrak{f}_2(\tau)$	Weber functions	201
$\sigma(z;\tau)$	Weierstrass σ-function	202
$\Gamma_0(2)^t$	Transpose of $\Gamma_0(2)$	208
$\zeta(z)$	Weierstrass ζ-function	218
$H_{\mathcal{O}}(X)$, $H_D(X)$	Class equation	225
$r(\mathcal{O},m)$	$\lvert\{\alpha\in\mathcal{O} : \alpha \text{ primitive, } N(\alpha)=m\}/\mathcal{O}^*\rvert$	226
$\Phi_{m,1}(X,X)$	Product of the multiplicity one factors of $\Phi_m(X,X)$	229
$J(d_1,d_2)$, $F(n)$	Notation for the Gross–Zagier theorem	236

Notation for Chapter 4 (§§14 and 15)

E, $E(K)$	Elliptic curve and its group of points over K	243
$\mathbb{P}^2(K)$	Projective plane over the field K	244, 261
$j(E)$	j-invariant of E	244
$\mathrm{End}_K(E)$	Endomorphism ring of E over K	246, 248
$\deg(\alpha)$	Degree of an isogeny α	247
Frob_q	Frobenius endomorphism of E over $\mathbb{F}_q$	248
$\overline{E}$	Reduction of E modulo a prime	250
$H(D)$	Hurwitz class number	251
$E_0(R)$	Points of E over a ring R	255
$\Gamma(m)$	Congruence subgroup $\{\gamma\in\mathrm{SL}(2,\mathbb{Z}) : \gamma\equiv I \bmod m\}$	265
F_m, F	Fields of modular functions of level m and of all levels	266, 267
$f^\gamma(\tau)$	Action of the matrix γ on $f(\tau)$	266, 269
$\mathbb{Z}_p$	Ring of p-adic integers	267
$\widehat{\mathbb{Z}}$	Product ring $\prod_p \mathbb{Z}_p$	268
$\mathrm{GL}(2,\mathbb{Q})^+$	Elements of $\mathrm{GL}(2,\mathbb{Q})$ with positive determinant	268
$\widehat{\mathbb{Q}}$	Tensor product $\mathbb{Q}\otimes_{\mathbb{Z}}\widehat{\mathbb{Z}}$	269
K^{ab}	Maximal Abelian extension of K	269
$\widehat{K}$, $\widehat{K}^*$	Ring of K-adeles and group of K-ideles	269
$g_{\tau_0}(x)$	Matrix in $\mathrm{GL}(2,\widehat{\mathbb{Q}})$ associated to an idele $x\in\widehat{K}^*$	270
$\widehat{\mathcal{O}}_K$, $\widehat{\mathcal{O}}_K^*$	Ring of $\mathcal{O}_K$-adeles and group of $\mathcal{O}_K$-ideles	270
σ_x	Element of $\mathrm{Gal}(K^{ab}/K)$ associated to $x\in\widehat{K}^*$	270, 271
J_m^1	Subgroup of $\widehat{\mathcal{O}}_K^*$ used in Lemma 15.14	271
$x\mathcal{O}_K$	Fractional $\mathcal{O}_K$-ideal associated to an idele $x\in\widehat{K}^*$	271
$L_{\mathcal{O}}$, $L_{\mathcal{O},m}$	Ring class field and extended ring class field	272
$P_{K,\mathbb{Z},m}(fm)$	Subgroup of $I_K(fm)$ used to define $L_{\mathcal{O},m}$	272
$\widehat{\mathcal{O}}$, $\widehat{\mathcal{O}}^*$	Ring of $\mathcal{O}$-adeles and group of $\mathcal{O}$-ideles	274
$J_{\mathcal{O},m}^1$	Subgroup of $\widehat{\mathcal{O}}_K^*$ used in Lemma 15.20	274
$r(\tau)$	Rogers–Ramanujan continued fraction	275
$\overline{g}_{\tau_0}(u)$	Matrix in $\mathrm{GL}(2,\mathbb{Z}/m\mathbb{Z})$ associated to $u\in(\mathcal{O}/m\mathcal{O})^*$	276
$K_{\mathfrak{p}}$, $\mathcal{O}_{K_{\mathfrak{p}}}$	Completions of K and $\mathcal{O}_K$ at $\mathfrak{p}$	284
$\mathbf{A}_K$, $\mathbf{I}_K$	K-adeles and K-ideles from class field theory	284
$\mathbf{A}_K^{\mathsf{fin}}$, $\mathbf{I}_K^{\mathsf{fin}}$	Finite K-adeles and finite K-ideles	284
$\nu_{\mathfrak{p}}$	$\mathfrak{p}$-adic valuation on K^* and $K_{\mathfrak{p}}^*$	285

Introduction

Most first courses in number theory or abstract algebra prove a theorem of Fermat which states that for an odd prime p,

$$p = x^2 + y^2, \ x, y \in \mathbb{Z} \iff p \equiv 1 \bmod 4.$$

This is only the first of many related results that appear in Fermat's works. For example, Fermat also states that if p is an odd prime, then

$$\begin{aligned} p = x^2 + 2y^2, \ x, y \in \mathbb{Z} &\iff p \equiv 1, 3 \bmod 8 \\ p = x^2 + 3y^2, \ x, y \in \mathbb{Z} &\iff p = 3 \text{ or } p \equiv 1 \bmod 3. \end{aligned}$$

These facts are lovely in their own right, but they also make one curious to know what happens for primes of the form $x^2 + 5y^2$, $x^2 + 6y^2$, etc. This leads to the basic question of the whole book, which we formulate as follows:

BASIC QUESTION 0.1. *Given a positive integer n, which primes p can be expressed in the form*

$$p = x^2 + ny^2$$

where x and y are integers?

We will answer this question completely, and along the way we will encounter some remarkably rich areas of number theory. The first steps will be easy, involving only quadratic reciprocity and the elementary theory of quadratic forms in two variables over $\mathbb{Z}$. These methods work nicely in the special cases considered above by Fermat. Using genus theory and cubic and biquadratic reciprocity, we can treat some more cases, but elementary methods fail to solve the problem in general. To proceed further, we need class field theory. This provides an abstract solution to the problem, but doesn't give explicit criteria for a particular choice of n in $x^2 + ny^2$. The final step uses modular functions and complex multiplication to show that for a given n, there is an algorithm for answering our question of when $p = x^2 + ny^2$.

This book has several goals. The first, to answer the basic question, has already been stated. A second goal is to bridge the gap between elementary number theory and class field theory. Although our basic question is simple enough to be stated in any beginning course in number theory, we will see that its solution is intimately bound up with higher reciprocity laws and class field theory. A related goal is to provide a well-motivated introduction to the classical formulation of class field theory. This will be done by carefully stating the basic theorems and illustrating their power in various concrete situations.

Let us summarize the contents of the book in more detail. We begin in Chapter 1 with the more elementary approaches to the problem, using the works of Fermat, Euler, Lagrange, Legendre and Gauss as a guide. In §1, we will give Euler's proofs of the above theorems of Fermat for primes of the form x^2+y^2, x^2+2y^2

and $x^2 + 3y^2$, and we will see what led Euler to discover quadratic reciprocity. We will also discuss the conjectures Euler made concerning $p = x^2 + ny^2$ for $n > 3$. Some of these conjectures, such as

$$p = x^2 + 5y^2 \iff p \equiv 1, 9 \bmod 20, \tag{0.2}$$

are similar to Fermat's theorems, while others, like

$$p = x^2 + 27y^2 \iff \begin{cases} p \equiv 1 \bmod 3 \text{ and 2 is a} \\ \text{cubic residue modulo } p, \end{cases}$$

are quite unexpected. For later purposes, note that this conjecture can be written in the following form:

$$p = x^2 + 27y^2 \iff \begin{cases} p \equiv 1 \bmod 3 \text{ and } x^3 \equiv 2 \bmod p \\ \text{has an integer solution.} \end{cases} \tag{0.3}$$

In §2, we will study Lagrange's theory of positive definite quadratic forms. After introducing the basic concepts of reduced form and class number, we will develop an elementary form of genus theory which will enable us to prove (0.2) and similar theorems. Unfortunately, for cases like (0.3), genus theory can only prove the partial result that

$$p = \left\{ \begin{array}{c} x^2 + 27y^2 \\ \text{or} \\ 4x^2 + 2xy + 7y^2 \end{array} \right\} \iff p \equiv 1 \bmod 3. \tag{0.4}$$

The problem is that $x^2 + 27y^2$ and $4x^2 + 2xy + 7y^2$ lie in the same genus and hence can't be separated by simple congruences. We will also discuss Legendre's tentative attempts at a theory of composition.

While the ideas of genus theory and composition were already present in the works of Lagrange and Legendre, the real depth of these theories wasn't revealed until Gauss came along. In §3 we will present some basic results in Gauss' *Disquisitiones Arithmeticae*, and in particular we will study the remarkable relationship between genus theory and composition. But for our purposes, the real breakthrough came when Gauss used cubic reciprocity to prove Euler's conjecture (0.3) concerning $p = x^2 + 27y^2$. In §4 we will give a careful statement of cubic reciprocity, and we will explain how it can be used to prove (0.3). Similarly, biquadratic reciprocity can be used to answer our question for $x^2 + 64y^2$. We will see that Gauss clearly recognized the role of higher reciprocity laws in separating forms of the same genus. This section will also begin our study of algebraic integers, for in order to state cubic and biquadratic reciprocity, we must first understand the arithmetic of the rings $\mathbb{Z}[e^{2\pi i/3}]$ and $\mathbb{Z}[i]$.

To go further requires class field theory, which is the topic of Chapter 2. We will begin in §5 with the Hilbert class field, which is the maximal unramified Abelian extension of a given number field. This will enable us to prove the following general result:

THEOREM 0.5. *Let $n \equiv 1, 2 \bmod 4$ be a positive squarefree integer. Then there is an irreducible polynomial $f_n(x) \in \mathbb{Z}[x]$ such that for a prime p dividing neither* *n nor the discriminant of* $f_n(x)$,

$$p = x^2 + ny^2 \iff \begin{cases} (-n/p) = 1 \text{ and } f_n(x) \equiv 0 \bmod p \\ \text{has an integer solution.} \end{cases}$$

While the statement of this theorem is elementary, the polynomial $f_n(x)$ is quite sophisticated: it is the minimal polynomial of a primitive element of the Hilbert class field L of $K = \mathbb{Q}(\sqrt{-n})$.

As an example of Theorem 0.5, we will study the case $n = 14$. We will show that the Hilbert class field of $K = \mathbb{Q}(\sqrt{-14})$ is $L = K(\alpha)$, where $\alpha = \sqrt{2\sqrt{2}-1}$. By Theorem 0.5, this will show that for an odd prime p,

$$(0.6) \qquad p = x^2 + 14y^2 \iff \begin{cases} (-14/p) = 1 \text{ and } (x^2+1)^2 \equiv 8 \bmod p \\ \text{has an integer solution,} \end{cases}$$

which answers our basic question for $x^2 + 14y^2$. The Hilbert class field will also enable us in §6 to give new proofs of the main theorems of genus theory.

The theory sketched so far is very nice, but there are some gaps in it. The most obvious is that the above results for $x^2 + 27y^2$ and $x^2 + 14y^2$ ((0.3) and (0.6) respectively) both follow the same format, but (0.3) does *not* follow from Theorem 0.5, for $n = 27$ is *not* squarefree. There should be a unified theorem that works for *all* positive n, yet the proof of Theorem 0.5 breaks down for general n because $\mathbb{Z}[\sqrt{-n}]$ is not in general the full ring of integers in $\mathbb{Q}(\sqrt{-n})$.

The goal of §§7–9 is to show that Theorem 0.5 holds for *all* positive integers n. This, in fact, is the main theorem of the whole book. In §7 we will study the rings $\mathbb{Z}[\sqrt{-n}]$ for general n, which leads to the concept of an *order* in an imaginary quadratic field. In §8 we will summarize the main theorems of class field theory and the Čebotarev Density Theorem, and in §9 we will introduce a generalization of the Hilbert class field called the ring class field, which is a certain (possibly ramified) Abelian extension of $\mathbb{Q}(\sqrt{-n})$ determined by the order $\mathbb{Z}[\sqrt{-n}]$. Then, in Theorem 9.2, we will use the Artin Reciprocity Theorem to show that Theorem 0.5 holds for *all* $n > 0$, where the polynomial $f_n(x)$ is now the minimal polynomial of a primitive element of the above ring class field. To give a concrete example of what this means, we will apply Theorem 9.2 to the case $x^2 + 27y^2$, which will give us a class field theory proof of (0.3). In §§8 and 9 we will also discuss how class field theory is related to higher reciprocity theorems.

The major drawback to the theory presented in §9 is that it is not constructive: for a given $n > 0$, we have no idea how to find the polynomial $f_n(x)$. From (0.3) and (0.6), we know $f_{27}(x)$ and $f_{14}(x)$, but the methods used in these examples hardly generalize. Chapter 3 will use the theory of complex multiplication to remedy this situation. In §10 we will study elliptic functions and introduce the idea of complex multiplication, and then in §11 we will discuss modular functions for the group $\Gamma_0(m)$ and show that the j-function can be used to generate ring class fields. As an example of the wonderful formulas that can be proved, in §12 we will give Weber's computation that

$$j(\sqrt{-14}) = 2^3\left(323 + 228\sqrt{2} + \left(231 + 161\sqrt{2}\right)\sqrt{2\sqrt{2}-1}\right)^3.$$

These methods will enable us to prove the Baker–Heegner–Stark Theorem on imaginary quadratic fields of class number 1. In §13 of the book we will discuss the class equation, which is the minimal polynomial of $j(\sqrt{-n})$. We will learn how to compute the class equation, which will lead to a constructive solution of $p = x^2 + ny^2$. We will then describe some work by Deuring and by Gross and Zagier. In 1946 Deuring proved a result about the difference of singular j-invariants, which implies an especially elegant version of our main theorem, and drawing on Deuring's work, Gross and Zagier discovered yet more remarkable properties of the class equation.

The first three chapters of the book present a complete solution to the problem of when $p = x^2 + ny^2$. In Chapter 4, we pursue two additional topics, elliptic curves in §14 and Shimura Reciprocity in §15, that give a more modern approach to the study of complex multiplication. We also include applications to primality testing in §14. Then §15 discusses ideles and the field of modular functions, and replaces certain pretty but ad-hoc arguments used in §12 with a more systematic treatment based on Shimura Reciprocity. We also give an unexpected application to $p = x^2 + ny^2$.

Number theory is usually taught at three levels, as an undergraduate course, a beginning graduate course, or a more advanced graduate course. These levels correspond roughly to the first three chapters of the book. Chapter 1 requires only beginning number theory (up to quadratic reciprocity) and a semester of abstract algebra. Since the proofs of quadratic, cubic and biquadratic reciprocity are omitted, this book would be best suited as a supplementary text in a beginning course. For Chapter 2, the reader should know Galois theory and some basic facts about algebraic number theory (these are reviewed in §5), but no previous exposure to class field theory is assumed. The theorems of class field theory are stated without proof, so that this book would be most useful as a supplement to the topics covered in a first graduate course. Chapter 3 assumes the same background as Chapter 2 plus a knowledge of complex analysis, but otherwise it is self-contained. (Brief but complete accounts of the Weierstrass $\wp$-function and modular functions are included in §§10 and 11.) The same is true for §14 in Chapter 4. In §15, the mathematics is more sophisticated, involving topics such as the p-adic integers, tensor products, and ideles. The final two chapters of the book should be suitable for use in a graduate seminar.

There are exercises at the end of each section, many of which consist of working out the details of arguments sketched in the text. Readers learning this material for the first time should find the exercises to be useful, while more sophisticated readers may skip them without loss of continuity. Complete solutions are included at the end of the book.

Many important (and relevant) topics are not covered in the book. An obvious omission in Chapter 1 concerns forms such as $x^2 - 2y^2$, which were certainly considered by Fermat and Euler. Questions of this sort lead to Pell's equation and the class field theory of real quadratic fields. We have also ignored the problem of representing arbitrary integers, not just primes, by quadratic forms, and there are interesting questions to ask about the *number* of such representations (this material is covered in Grosswald's book [**62**]). In Chapter 2 we give a classical formulation of class field theory, with only a brief mention of idele groups. A more modern treatment can be found in Gras [**57**], Neukirch [**100**] or Weil [**133**] (see also §15). We also do not do justice to the use of analytic methods in number theory. For a nice introduction in the case of quadratic fields, see Zagier [**140**]. Our treatment of elliptic curves in Chapter 4 is rather incomplete. See Husemöller [**74**], Knapp [**83**] or Silverman [**118**] for the basic theory, while more advanced topics are covered by Lang [**90**], Shimura [**115**] and Silverman [**119**]. At a more elementary level, there is the wonderful book [**120**] by Silverman and Tate.

There are many books which touch on the number theory encountered in studying the problem of representing primes by $x^2 + ny^2$. Four books that we particularly recommend are Cohn's *A Classical Invitation to Algebraic Numbers and Class Fields* [**25**], Lang's *Elliptic Functions* [**90**], Scharlau and Opolka's *From Fermat to Minkowski* [**108**], and Weil's *Number Theory: An Approach Through History* [**135**]. These books, as well as others to be found in the References, open up an extraordinarily rich area of mathematics. The purpose of this book is to reveal some of this richness and to encourage the reader to learn more about it.

CHAPTER 1

From Fermat to Gauss

1. Fermat, Euler and Quadratic Reciprocity

In this section we will discuss primes of the form x^2+ny^2, where n is a fixed positive integer. Our starting point will be the three theorems of Fermat for odd primes p

$$\begin{aligned} p &= x^2 + y^2, \quad x, y \in \mathbb{Z} \iff p \equiv 1 \bmod 4 \\ p &= x^2 + 2y^2, \quad x, y \in \mathbb{Z} \iff p \equiv 1 \text{ or } 3 \bmod 8 \\ p &= x^2 + 3y^2, \quad x, y \in \mathbb{Z} \iff p = 3 \text{ or } p \equiv 1 \bmod 3 \end{aligned} \tag{1.1}$$

mentioned in the introduction. The goals of §1 are to prove (1.1) and, more importantly, to get a sense of what's involved in studying the equation $p = x^2+ny^2$ when $n > 0$ is arbitrary. This last question was best answered by Euler, who spent 40 years proving Fermat's theorems and thinking about how they can be generalized. Our exposition will follow some of Euler's papers closely, both in the theorems proved and in the examples studied. We will see that Euler's strategy for proving (1.1) was one of the primary things that led him to discover quadratic reciprocity, and we will also discuss some of his remarkable conjectures concerning $p = x^2+ny^2$ for $n > 3$. These conjectures touch on quadratic forms, composition, genus theory, cubic and biquadratic reciprocity, and will keep us busy for the rest of the chapter.

A. Fermat

Fermat's first mention of $p = x^2+y^2$ occurs in a 1640 letter to Mersenne [**45**, Vol. II, p. 212], while $p = x^2 + 2y^2$ and $p = x^2 + 3y^2$ come later, first appearing in a 1654 letter to Pascal [**45**, Vol. II, pp. 310–314]. Although no proofs are given in these letters, Fermat states the results as theorems. Writing to Digby in 1658, he repeats these assertions in the following form:

> Every prime number which surpasses by one a multiple of four is composed of two squares. Examples are 5, 13, 17, 29, 37, 41, etc.
>
> Every prime number which surpasses by one a multiple of three is composed of a square and the triple of another square. Examples are 7, 13, 19, 31, 37, 43, etc.
>
> Every prime number which surpasses by one or three a multiple of eight is composed of a square and the double of another square. Examples are 3, 11, 17, 19, 41, 43, etc.

Fermat adds that he has solid proofs—"firmissimis demonstralibus" [**45**, Vol. II, pp. 402–408 (Latin), Vol. III, pp. 314–319 (French)].

The theorems (1.1) are only part of the work that Fermat did with $x^2 + ny^2$. For example, concerning $x^2 + y^2$, Fermat knew that a positive integer N is the sum of two squares if and only if the quotient of N by its largest square factor is a product of primes congruent to 1 modulo 4 [**45**, Vol. III, Obs. 26, pp. 256–257], and he knew the number of different ways N can be so represented [**45**, Vol. III, Obs. 7, pp. 243–246]. Fermat also studied forms beyond x^2+y^2, x^2+2y^2 and x^2+3y^2. For example, in the 1658 letter to Digby quoted above, Fermat makes the following conjecture about x^2+5y^2, which he admits he can't prove:

> If two primes, which end in 3 or 7 and surpass by three a multiple of four, are multiplied, then their product will be composed of a square and the quintuple of another square.
>
> Examples are the numbers 3, 7, 23, 43, 47, 67, etc. Take two of them, for example 7 and 23; their product 161 is composed of a square and the quintuple of another square. Namely 81, a square, and the quintuple of 16 equal 161.

Fermat's condition on the primes is simply that they be congruent to 3 or 7 modulo 20. In §2 we will present Lagrange's proof of this conjecture, which uses ideas from genus theory and the composition of forms.

Fermat's proofs used the method of infinite descent, but that's often all he said. As an example, here is Fermat's description of his proof for $p = x^2 + y^2$ [**45**, Vol. II, p. 432]:

> If an arbitrarily chosen prime number, which surpasses by one a multiple of four, is not a sum of two squares, then there is a prime number of the same form, less than the given one, and then yet a third still less, etc., descending infinitely until you arrive at the number 5, which is the least of all of this nature, from which it would follow was not the sum of two squares. From this one must infer, by deduction of the impossible, that all numbers of this form are consequently composed of two squares.

This explains the philosophy of infinite descent, but doesn't tell us how to produce the required lesser prime. We have only one complete proof by Fermat. It occurs in one of his marginal notes (the area of a right triangle with integral sides cannot be an integral square [**45**, Vol. III, Obs. 45, pp. 271–272]—for once the margin was big enough!). The methods of this proof (see Weil [**135**, p. 77] or Edwards [**41**, pp. 10–14] for modern expositions) do not apply to our case, so that we are still in the dark. An analysis of Fermat's approach to infinite descent appears in Bussotti [**19**]. Weil's book [**135**] makes a careful study of Fermat's letters and marginal notes, and with some hints from Euler, he reconstructs some of Fermat's proofs. Weil's arguments are quite convincing, but we won't go into them here. For the present, we prefer to leave things as Euler found them, i.e., wonderful theorems but no proofs.

B. Euler

Euler first heard of Fermat's results through his correspondence with Goldbach. In fact, Goldbach's first letter to Euler, written in December 1729, mentions Fermat's conjecture that $2^{2^n}+1$ is always prime [**50**, p. 10]. Shortly thereafter, Euler read some of Fermat's letters that had been printed in Wallis' *Opera* [**129**] (which included the one to Digby quoted above). Euler was intrigued by what he found. For example, writing to Goldbach in June 1730, Euler comments that Fermat's four-square theorem (every positive integer is a sum of four or fewer squares) is a "non inelegans theorema" [**50**, p. 24]. For Euler, Fermat's assertions were serious theorems deserving of proof, and finding the proofs became a life-long project. Euler's first paper on number theory, written in 1732 at age 25, disproves Fermat's claim about $2^{2^n}+1$ by showing that 641 is a factor of $2^{32}+1$ [**43**, Vol. II, pp. 1–5]. Euler's interest in number theory continued unabated for the next 51 years—there was a steady stream of papers introducing many of the fundamental concepts of number theory, and even after his death in 1783, his papers continued to appear until 1830 (see [**43**, Vol. IV–V]). Weil's book [**135**] gives a survey of Euler's work on number theory (other references are Burkhardt [**18**], Edwards [**41**, Chapter 2], Scharlau and Opolka [**108**, Chapter 3], and the introductions to Volumes II–V of Euler's collected works [**43**]).

We can now present Euler's proof of the first of Fermat's theorems from (1.1):

THEOREM 1.2. *An odd prime p can be written as x^2+y^2 if and only if $p \equiv 1 \bmod 4$.*

PROOF. If $p = x^2+y^2$, then congruences modulo 4 easily imply that $p \equiv 1 \bmod 4$. The hard work is proving the converse. We will give a modern version of Euler's proof. Given an odd prime p, there are two basic steps to be proved:

$$\begin{aligned} \textit{Descent Step}: &\text{ If } p \mid x^2+y^2,\ \gcd(x,y)=1, \text{ then } p \text{ can be written}\\ &\text{ as } x^2+y^2 \text{ for some possibly different } x,y.\\ \textit{Reciprocity Step}: &\text{ If } p \equiv 1 \bmod 4, \text{ then } p \mid x^2+y^2,\ \gcd(x,y)=1. \end{aligned}$$

It will soon become clear why we use the names "Descent" and "Reciprocity."

We'll do the Descent Step first since that's what happened historically. The argument below is taken from a 1747 letter to Goldbach [**50**, pp. 416–419] (see also [**43**, Vol. II, pp. 295–327]). We begin with the classical identity

$$(x^2+y^2)(z^2+w^2) = (xz \pm yw)^2 + (xw \mp yz)^2 \tag{1.3}$$

(see Exercise 1.1) which enables one to express composite numbers as sums of squares. The key observation is the following lemma:

LEMMA 1.4. *Suppose that N is a sum of two relatively prime squares, and that $q = x^2+y^2$ is a prime divisor of N. Then N/q is also a sum of two relatively prime squares.*

PROOF. Write $N = a^2+b^2$, where a and b are relatively prime. We also have $q = x^2+y^2$, and thus q divides

$$\begin{aligned} x^2N - a^2q &= x^2(a^2+b^2) - a^2(x^2+y^2)\\ &= x^2b^2 - a^2y^2 = (xb-ay)(xb+ay). \end{aligned}$$

Since q is prime, it divides one of these two factors, and changing the sign of a if necessary, we can assume that $q \mid xb-ay$. Thus $xb-ay = dq$ for some integer d.

We claim that $x \mid a+dy$. Since x and y are relatively prime, this is equivalent to $x \mid (a+dy)y$. However,

$$\begin{aligned} (a+dy)y &= ay+dy^2 = xb-dq+dy^2\\ &= xb - d(x^2+y^2)+dy^2 = xb-dx^2, \end{aligned}$$

which is obviously divisible by x. Furthermore, if we set $a + dy = cx$, then the above equation implies that $b = dx + cy$. Thus we have

$$\begin{aligned} a &= cx - dy \\ b &= dx + cy. \end{aligned} \tag{1.5}$$

Then, using (1.3), we obtain

$$\begin{aligned} N = a^2 + b^2 &= (cx - dy)^2 + (dx + cy)^2 \\ &= (x^2 + y^2)(c^2 + d^2) = q(c^2 + d^2). \end{aligned}$$

Thus $N/q = c^2 + d^2$ is a sum of squares, and (1.5) shows that c and d must be relatively prime since a and b are. This proves the lemma. Q.E.D.

To complete the proof of the Descent Step, let p be an odd prime dividing $N = a^2 + b^2$, where a and b are relatively prime. If a and b are changed by multiples of p, we still have $p \mid a^2 + b^2$. We may thus assume that $|a| < p/2$ and $|b| < p/2$, which in turn implies that $N < p^2/2$. The new a and b may have a greatest common divisor $d > 1$, but p doesn't divide d, so that dividing a and b by d, we may assume that $p \mid N$, $N < p^2/2$, and $N = a^2 + b^2$ where $\gcd(a, b) = 1$. Then all prime divisors $q \neq p$ of N are less than p. If q were a sum of two squares, then Lemma 1.4 would show that N/q would be a multiple of p that is again a sum of two squares. If all such q's were sums of two squares, then repeatedly applying Lemma 1.4 would imply that p itself was of the same form. So if p is not a sum of two squares, there must be a smaller prime q with the same property. Since there is nothing to prevent us from repeating this process indefinitely, we get an infinite decreasing sequence of prime numbers. This contradiction finishes the Descent Step.

This is a classical descent argument, and as Weil argues [**135**, pp. 68–69], it is probably similar to what Fermat did. In §2 we will take another approach to the Descent Step, using the reduction theory of positive definite quadratic forms.

The Reciprocity Step caused Euler a lot more trouble, taking him until 1749. Euler was clearly relieved when he could write to Goldbach "Now have I finally found a valid proof" [**50**, pp. 493–495]. The basic idea is quite simple: since $p \equiv 1 \bmod 4$, we can write $p = 4k + 1$. Then Fermat's Little Theorem implies that

$$(x^{2k} - 1)(x^{2k} + 1) \equiv x^{4k} - 1 \equiv 0 \bmod p$$

for all $x \not\equiv 0 \bmod p$. If $x^{2k} - 1 \not\equiv 0 \bmod p$ for *one* such x, then $p \mid x^{2k} + 1$, so that p divides a sum of relatively prime squares, as desired. For us, the required x is easy to find, since $x^{2k} - 1$ is a polynomial over the field $\mathbb{Z}/p\mathbb{Z}$ and hence has at most $2k < p - 1$ roots. Euler's first proof is quite different, for it uses the calculus of finite differences—see Exercise 1.2 for details. This proves Fermat's claim (1.1) for primes of the form $x^2 + y^2$. Q.E.D.

Euler used the same two-step strategy in his proofs for $x^2 + 2y^2$ and $x^2 + 3y^2$. The Descent Steps are

If $p \mid x^2 + 2y^2$, $\gcd(x, y) = 1$, then p is of the form $x^2 + 2y^2$ for some possibly different x, y

If $p \mid x^2 + 3y^2$, $\gcd(x, y) = 1$, then p is of the form $x^2 + 3y^2$ for some possibly different x, y, and the Reciprocity Steps are

$$\text{If } p \equiv 1, 3 \bmod 8, \text{ then } p \mid x^2 + 2y^2, \ \gcd(x, y) = 1$$
$$\text{If } p \equiv 1 \bmod 3, \text{ then } p \mid x^2 + 3y^2, \ \gcd(x, y) = 1,$$

where p is always an odd prime. In each case, the Reciprocity Step was harder to prove than the Descent Step, and Euler didn't succeed in giving complete proofs of Fermat's theorems (1.1) until 1772, 40 years after he first read about them. Weil discusses the proofs for $x^2 + 2y^2$ and $x^2 + 3y^2$ in [**135**, pp. 178–179, 191, and 210–212], and in Exercises 1.4 and 1.5 we will present a version of Euler's argument for $x^2 + 3y^2$.

C. $p = x^2 + ny^2$ and Quadratic Reciprocity

Let's turn to the general case of $p = x^2 + ny^2$, where n is now any positive integer. To study this problem, it makes sense to start with Euler's two-step strategy. This won't lead to a proof, but the Descent and Reciprocity Steps will both suggest some very interesting questions for us to pursue.

The Descent Step for arbitrary $n > 0$ begins with the identity

$$(x^2 + ny^2)(z^2 + nw^2) = (xz \pm nyw)^2 + n(xw \mp yz)^2 \tag{1.6}$$

(see Exercise 1.1), and Lemma 1.4 generalizes easily for $n > 0$ (see Exercise 1.3). Then suppose that $p \mid x^2 + ny^2$. As in the proof of the Descent Step in Theorem 1.2, we can assume that $|x|, |y| \le p/2$. For $n \le 3$, it follows that $x^2 + ny^2 < p^2$ when p is odd, and then the argument from Theorem 1.2 shows that p is of the form $x^2 + ny^2$ (see Exercise 1.4). One might conjecture that this holds in general, i.e., that $p \mid x^2 + ny^2$ always implies $p = x^2 + ny^2$. Unfortunately this fails even for $n = 5$: for example, $3 \mid 21 = 1^2 + 5 \cdot 2^2$ but $3 \ne x^2 + 5y^2$. Euler knew this, and most likely so did Fermat (remember his speculations about $x^2 + 5y^2$). So the question becomes: how are prime divisors of $x^2 + ny^2$ to be represented? As we will see in §2, the proper language for this is Lagrange's theory of quadratic forms, and a complete solution to the Descent Step will follow from the properties of reduced forms.

Turning to the Reciprocity Step for $n > 0$, the general case asks for congruence conditions on a prime p which will guarantee $p \mid x^2 + ny^2$. To see what kind of congruences we need, note that the conditions of (1.1) can be unified by working modulo $4n$. Thus, given $n > 0$, we're looking for a congruence of the form $p \equiv \alpha, \beta, \ldots \bmod 4n$ which implies $p \mid x^2 + ny^2$, $\gcd(x, y) = 1$. To give a modern formulation of this last condition, we first define the Legendre symbol (a/p). If a is an integer and p an odd prime, then

$$\left(\frac{a}{p}\right) = \begin{cases} 0 & p \mid a \\ 1 & p \nmid a \text{ and } a \text{ is a quadratic residue modulo } p \\ -1 & p \nmid a \text{ and } a \text{ is a quadratic nonresidue modulo } p. \end{cases}$$

We can now restate the condition for $p \mid x^2 + ny^2$ as follows:

LEMMA 1.7. *Let n be a nonzero integer, and let p be an odd prime not dividing n. Then*

$$p \mid x^2 + ny^2, \ \gcd(x, y) = 1 \iff \left(\frac{-n}{p}\right) = 1.$$

PROOF. The basic idea is that if $x^2+ny^2 \equiv 0 \bmod p$ and $\gcd(x,y)=1$, then y must be relatively prime to p and consequently has a multiplicative inverse modulo p. The details are left to the reader (see Exercise 1.6). Q.E.D.

The arguments of the above lemma are quite elementary, but for Euler they were not so easy—he first had to realize that quadratic residues were at the heart of the matter. This took several years, and it's fun to watch his terminology evolve: in 1744, he writes "prime divisors of numbers of the form $aa - Nbb$" [**43**, Vol. II, p. 216]; by 1747 this changes to "residues arising from the division of squares by the prime p" [**43**, Vol. II, p. 313]; and by 1751 the transition is complete—Euler now uses the terms "residua" and "non-residua" freely, with the "quadratic" being understood [**43**, Vol. II, p. 343].

Using Lemma 1.7, the Reciprocity Step can be restated as the following question: is there a congruence $p \equiv \alpha, \beta, \ldots \bmod 4n$ which implies $(-n/p)=1$ when p is prime? This question also makes sense when $n<0$, and in the following discussion n will thus be allowed to be positive or negative. We will see in Corollary 1.19 that the full answer is intimately related to the law of quadratic reciprocity, and in fact the Reciprocity Step was one of the primary things that led Euler to discover quadratic reciprocity.

Euler became intensely interested in this question in the early 1740s, and he mentions numerous examples in his letters to Goldbach. In 1744 Euler collected together his examples and conjectures in the paper *Theoremata circa divisores numerorum in hac forma paa ± qbb contentorum* [**43**, Vol. II, pp. 194–222]. He labels his examples as "theorems," but they are really "theorems found by induction," which is eighteenth-century parlance for conjectures based on working out some particular cases. Here are of some of Euler's conjectures, stated in modern notation:

$$\begin{aligned}
\left(\frac{-3}{p}\right) &= 1 \iff p \equiv 1, 7 \bmod 12\\
\left(\frac{-5}{p}\right) &= 1 \iff p \equiv 1, 3, 7, 9 \bmod 20\\
\left(\frac{-7}{p}\right) &= 1 \iff p \equiv 1, 9, 11, 15, 23, 25 \bmod 28\\
\left(\frac{3}{p}\right) &= 1 \iff p \equiv \pm 1 \bmod 12\\
\left(\frac{5}{p}\right) &= 1 \iff p \equiv \pm 1, \pm 11 \bmod 20\\
\left(\frac{7}{p}\right) &= 1 \iff p \equiv \pm 1, \pm 3, \pm 9 \bmod 28,
\end{aligned} \tag{1.8}$$

where p is an odd prime not dividing n. In looking for a unifying pattern, the bottom three look more promising because of the $\pm$'s. If we rewrite the bottom half of (1.8) using $11 \equiv -9 \bmod 20$ and $3 \equiv -25 \bmod 28$, we obtain

$$\begin{aligned}
\left(\frac{3}{p}\right) &= 1 \iff p \equiv \pm 1 \bmod 12\\
\left(\frac{5}{p}\right) &= 1 \iff p \equiv \pm 1, \pm 9 \bmod 20
\end{aligned}$$

$$\left(\frac{7}{p}\right) = 1 \iff p \equiv \pm 1, \pm 25, \pm 9 \bmod 28.$$

All of the numbers that appear are odd squares!

Before getting carried away, we should note another of Euler's conjectures:

$$\left(\frac{6}{p}\right) = 1 \iff p \equiv \pm 1, \pm 5 \bmod 24.$$

Unfortunately, ± 5 is not a square modulo 24, and the same thing happens for $(10/p)$ and $(14/p)$. But 3, 5 and 7 are prime, while 6, 10 and 14 are composite. Thus it makes sense to make the following conjecture for the prime case:

CONJECTURE 1.9. *If p and q are distinct odd primes, then*

$$\left(\frac{q}{p}\right) = 1 \iff p \equiv \pm \beta^2 \bmod 4q \text{ for some odd integer } \beta.$$

The remarkable fact is that this conjecture is equivalent to the usual statement of quadratic reciprocity:

PROPOSITION 1.10. *If p and q are distinct odd primes, then Conjecture* 1.9 *is equivalent to*

$$\left(\frac{p}{q}\right)\left(\frac{q}{p}\right) = (-1)^{(p-1)(q-1)/4}.$$

PROOF. Let $p^* = (-1)^{(p-1)/2}p$. Then the standard properties

$$(1.11)\qquad \begin{aligned}\left(\frac{-1}{p}\right) &= (-1)^{(p-1)/2}\\ \left(\frac{ab}{p}\right) &= \left(\frac{a}{p}\right)\left(\frac{b}{p}\right)\end{aligned}$$

of the Legendre symbol easily imply that quadratic reciprocity is equivalent to

$$(1.12)\qquad \left(\frac{p^*}{q}\right) = \left(\frac{q}{p}\right)$$

(see Exercise 1.7). Since both sides are ± 1, it follows that quadratic reciprocity can be stated as

$$\left(\frac{q}{p}\right) = 1 \iff \left(\frac{p^*}{q}\right) = 1.$$

Comparing this to Conjecture 1.9, we see that it suffices to show

$$(1.13)\qquad \left(\frac{p^*}{q}\right) = 1 \iff p \equiv \pm \beta^2 \bmod 4q,\ \beta \text{ odd}.$$

The proof of 1.13 is straightforward and is left to the reader (see Exercise 1.8). Q.E.D.

With hindsight, we can see why Euler had trouble with the Reciprocity Steps for $x^2 + 2y^2$ and $x^2 + 3y^2$: he was working out special cases of quadratic reciprocity! Exercise 1.9 will discuss which special cases were involved. We will not prove quadratic reciprocity in this section, but later in §8 we will give a proof using class field theory. Proofs of a more elementary nature can be found in most number theory texts.

The discussion leading up to Conjecture 1.9 is pretty exciting, but was it what Euler did? The answer is yes and no. To explain this, we must look more closely at

Euler's 1744 paper. In addition to conjectures like (1.8), the paper also contained a series of Annotations where Euler speculated on what was happening in general. For simplicity, we will concentrate on the case of (N/p),where $N > 0$. Euler notes in Annotation 13 [**43**, Vol. II, p. 216] that for such N's, all of the conjectures have the form

$$\left(\frac{N}{P}\right) = 1 \iff P \equiv \pm\alpha \bmod 4N$$

for certain odd values of α. Then in Annotation 16 [**43**, Vol. II, pp. 216–217], Euler states that "while 1 is among the values [of the α's], yet likewise any square number, which is prime to $4N$, furnishes a suitable value for α." This is close to what we want, but it doesn't say that the odd squares fill up all possible α's when N is prime. For this, we turn to Annotation 14 [**43**, Vol. II, p. 216], where Euler notes that the number of α's that occur is $(1/2)\phi(N)$. When N is prime, this equals $(N-1)/2$, the number of incongruent squares modulo $4N$ relatively prime to $4N$. Thus what Euler states is fully equivalent to Conjecture 1.9. In 1875, Kronecker identified these Annotations as the first complete statement of quadratic reciprocity [**85**, Vol. II, pp. 3–4].

The problem is that we have to read between the lines to get quadratic reciprocity. Why didn't Euler state it more explicitly? He knew that the prime case was special, for why else would he list the prime cases before the composite ones? The answer to this puzzle, as Weil points out [**135**, pp. 207–209], is that Euler's real goal was to characterize the α's for *all* N, not just primes. To explain this, we need to give a modern description of the $\pm\alpha$'s. The following lemma is at the heart of the matter:

LEMMA 1.14. *If $D \equiv 0, 1 \bmod 4$ is a nonzero integer, then there is a unique homomorphism $\chi : (\mathbb{Z}/D\mathbb{Z})^* \to \{\pm 1\}$ such that $\chi([p]) = (D/p)$ for odd primes p not dividing D. Furthermore,*

$$\chi([-1]) = \begin{cases} 1 & \textit{when } D > 0 \\ -1 & \textit{when } D < 0. \end{cases}$$

PROOF. The proof will make extensive use of the Jacobi symbol. Given $m > 0$ odd and relatively prime to M, recall that the Jacobi symbol (M/m) is defined to be the product

$$\left(\frac{M}{m}\right) = \prod_{i=1}^{r} \left(\frac{M}{p_i}\right)$$

where $m = p_1 \cdots p_r$ is the prime factorization of m. Note that $(M/m) = (N/m)$ when $M \equiv N \bmod m$, and there are the multiplicative identities

$$\begin{aligned} \left(\frac{MN}{m}\right) &= \left(\frac{M}{m}\right)\left(\frac{N}{m}\right) \\ \left(\frac{M}{mn}\right) &= \left(\frac{M}{m}\right)\left(\frac{M}{n}\right) \end{aligned} \tag{1.15}$$

(see Exercise 1.10). The Jacobi symbol satisfies the following version of quadratic reciprocity:

$$\begin{aligned} \left(\frac{-1}{m}\right) &= (-1)^{(m-1)/2} \\ \left(\frac{2}{m}\right) &= (-1)^{(m^2-1)/8} \\ \left(\frac{M}{m}\right) &= (-1)^{(M-1)(m-1)/4}\left(\frac{m}{M}\right) \end{aligned} \tag{1.16}$$

(see Exercise 1.10).

For this lemma, the crucial property of the Jacobi symbol is one usually not mentioned in elementary texts: if $m \equiv n \bmod D$, where m and n are odd and positive and $D \equiv 0, 1 \bmod 4$, then

$$\left(\frac{D}{m}\right) = \left(\frac{D}{n}\right). \tag{1.17}$$

The proof is quite easy when $D \equiv 1 \bmod 4$ and $D > 0$: using quadratic reciprocity (1.16), the two sides of (1.17) become

$$\begin{aligned} &(-1)^{(D-1)(m-1)/4}\left(\frac{m}{D}\right) \\ &(-1)^{(D-1)(n-1)/4}\left(\frac{n}{D}\right). \end{aligned} \tag{1.18}$$

To compare these expressions, first note that the two Jacobi symbols are equal since $m \equiv n \bmod D$. From $D \equiv 1 \bmod 4$ we see that

$$(D-1)(m-1)/4 \equiv (D-1)(n-1)/4 \equiv 0 \bmod 2$$

since m and n are odd. Thus the signs in front of (1.18) are both $+1$, and (1.17) follows. When D is even or negative, a similar argument using the supplementary laws from (1.16) shows that (1.17) still holds (see Exercise 1.11).

It follows from (1.17) that $\chi([m]) = (D/m)$ gives a well-defined homomorphism from $(\mathbb{Z}/D\mathbb{Z})^*$ to $\{\pm 1\}$ (see Exercise 1.12), and the statement concerning $\chi([-1])$ follows from the above properties of the Jacobi symbol (see Exercise 1.12). Finally, the condition that $\chi([p]) = (D/p)$ for odd primes p determines χ uniquely follows because χ is a homomorphism and every class in $(\mathbb{Z}/D\mathbb{Z})^*$ contains a positive odd number (hence a product of odd primes) by part (a) of Exercise 1.12. Q.E.D.

The above proof made heavy use of quadratic reciprocity, which is no accident: Lemma 1.14 is in fact equivalent to quadratic reciprocity and the supplementary laws (see Exercise 1.13). For us, however, the main feature of Lemma 1.14 is that it gives a complete solution of the Reciprocity Step of Euler's strategy:

COROLLARY 1.19. *Let n be a nonzero integer, and let $\chi : (\mathbb{Z}/4n\mathbb{Z})^* \to \{\pm 1\}$ be the homomorphism from Lemma 1.14 when $D = -4n$. If p is an odd prime not dividing n, then the following are equivalent:*

(i) $p \mid x^2 + ny^2$, $\gcd(x, y) = 1$.
(ii) $(-n/p) = 1$.
(iii) $[p] \in \ker(\chi) \subset (\mathbb{Z}/4n\mathbb{Z})^*$.

Proof. (i) and (ii) are equivalent by Lemma 1.7, and since $(-4n/p) = (-n/p)$, (ii) and (iii) are equivalent by Lemma 1.14. Q.E.D.

To see how this solves the Reciprocity Step, write $\ker(\chi) = \{[\alpha], [\beta], [\gamma], \ldots\}$. Then $[p] \in \ker(\chi)$ is equivalent to the congruence $p \equiv \alpha, \beta, \gamma, \ldots \bmod 4n$, which is exactly the kind of condition we were looking for. Actually, Lemma 1.14 allows us to refine this a bit: when $n \equiv 3 \bmod 4$, then congruence can be taken to be of the form $p \equiv \alpha, \beta, \gamma, \ldots \bmod n$ (see Exercise 1.14). We should also note that in all cases, the usual statement of quadratic reciprocity makes it easy to compute the classes in question (see Exercise 1.15 for an example).

To see how this relates to what Euler did in 1744, let N be as in our discussion of Euler's Annotations, and let $D = 4N$ in Lemma 1.14. Then $\ker(\chi)$ consists *exactly* of Euler's $\pm\alpha$'s (when $N > 0$, the lemma also implies that $-1 \in \ker(\chi)$, which explains the $\pm$ signs). The second thing to note is that when N is odd and squarefree, $K = \ker(\chi)$ is uniquely characterized by the following four properties:

(i) K is a subgroup of index 2 in $(\mathbb{Z}/4N\mathbb{Z})^*$.
(ii) $-1 \in K$ when $N > 0$ and $-1 \notin K$ when $N < 0$.
(iii) K has period N if $N \equiv 1 \bmod 4$ and period $4N$ otherwise. (Having period $P > 0$ means that if $[a], [b] \in (\mathbb{Z}/4N\mathbb{Z})^*$, $[a] \in K$ and $a \equiv b \bmod P$, then $[b] \in K$.)
(iv) K does not have any smaller period.

For a proof of this characterization, see Weil [**135**, pp. 287–291]. In the Annotations to his 1744 paper, Euler gives very clear statements of (i)–(iii) (see Annotations 13–16 in [**43**, Vol. II, pp. 216–217]), and as for (iv), he notes that N is not a period when $N \not\equiv 1 \bmod 4$, but says nothing about the possibility of smaller periods (see Annotation 20 in [**43**, Vol. II, p. 219]). So Euler doesn't quite give a complete characterization of $\ker(\chi)$, but he comes incredibly close. It is a tribute to Euler's insight that he could deduce this underlying structure on the basis of examples like (1.8).

D. Beyond Quadratic Reciprocity

We will next discuss some of Euler's conjectures concerning primes of the form $x^2 + ny^2$ for $n > 3$. We start with the cases $n = 5$ and 14 (taken from his 1744 paper), for each will have something unexpected to offer us.

When $n = 5$, Euler conjectured that for odd primes $p \neq 5$,

$$\begin{aligned} p = x^2 + 5y^2 &\iff p \equiv 1, 9 \bmod 20 \\ 2p = x^2 + 5y^2 &\iff p \equiv 3, 7 \bmod 20. \end{aligned} \tag{1.20}$$

Recall from (1.8) that $p \mid x^2 + 5y^2$ is equivalent to $p \equiv 1, 3, 7, 9 \bmod 20$. Hence these four congruence classes break up into two groups $\{1, 9\}$ and $\{3, 7\}$ which have quite different representability properties. This is a new phenomenon, not encountered for $x^2 + ny^2$ when $n \leq 3$. Note also that the classes 3, 7 modulo 20 are the ones that entered into Fermat's speculations on $x^2 + 5y^2$, so something interesting is going on here. In §2 we will see that this is one of the examples that led Lagrange to discover genus theory.

The case $n = 14$ is yet more complicated. Here, Euler makes the following conjecture for odd primes $\neq 7$:

$$(1.21)\qquad \begin{aligned} p = \left\{\begin{matrix} x^2 + 14y^2 \\ 2x^2 + 7y^2 \end{matrix}\right\} &\Longleftrightarrow p \equiv 1, 9, 15, 23, 25, 39 \bmod 56 \\ 3p = x^2 + 14y^2 &\Longleftrightarrow p \equiv 3, 5, 13, 19, 27, 45 \bmod 56. \end{aligned}$$

As with (1.20), the union of the two groups of congruence classes in (1.21) describes those primes for which $(-14/p) = 1$. The new puzzle here is that we don't seem to be able to separate $x^2 + 14y^2$ from $2x^2 + 7y^2$. In §2, we will see that this is not an oversight on Euler's part, for the two quadratic forms $x^2 + 14y^2$ and $2x^2 + 7y^2$ are in the same genus and hence can't be separated by congruence classes. Another puzzle is why (1.20) uses $2p$ while (1.21) uses $3p$. In §2 we will use composition to explain these facts. One could also ask what extra condition is needed to insure $p = x^2 + 14y^2$. This lies much deeper, for as we will see in §5, it involves the Hilbert class field of $\mathbb{Q}(\sqrt{-14})$.

The final examples we want to discuss come from quite a different source, the *Tractatus de numerorum doctrina capita sedecim quae supersunt,* which Euler wrote in the period 1748–1750 [**43**, Vol. V, pp. 182–283]. Euler intended this work to be a basic text for number theory, in the same way that his *Introductio in analysin infinitorum* [**43**, Vol. VIII–IX] was the first real textbook in analysis. Unfortunately, Euler never completed the *Tractatus,* and it was first published only in 1849. Weil [**135**, pp. 192–196] gives a description of what's in the *Tractatus* (see also [**43**, Vol. V, pp. XIX–XXVI]). For us, the most interesting chapters are the two that deal with cubic and biquadratic residues. Recall that a number a is a cubic (resp. biquadratic) residue modulo p if the congruence $x^3 \equiv a \bmod p$ (resp. $x^4 \equiv a \bmod p$) has an integer solution. Euler makes the following conjectures about when 2 is a cubic or biquadratic residue modulo an odd prime p:

$$(1.22)\qquad p = x^2 + 27y^2 \Longleftrightarrow \left\{\begin{array}{l} p \equiv 1 \bmod 3 \text{ and } 2 \text{ is a} \\ \text{cubic residue modulo } p \end{array}\right.$$

$$(1.23)\qquad p = x^2 + 64y^2 \Longleftrightarrow \left\{\begin{array}{l} p \equiv 1 \bmod 4 \text{ and } 2 \text{ is a} \\ \text{biquadratic residue modulo } p \end{array}\right.$$

(see [**43**, Vol. V, pp. 250 and 258]). In §4, we will see that both of these conjectures were proved by Gauss as consequences of his work on cubic and biquadratic reciprocity.

The importance of the examples (1.20)–(1.23) is hard to overestimate. Thanks to Euler's amazing ability to find patterns, we now see some of the serious problems to be tackled (in (1.20) and (1.21)), and we have our first hint of what the final solution will look like (in (1.22) and (1.23)). Much of the next three sections will be devoted to explaining and proving these conjectures. In particular, it should be clear that we need to learn a lot more about quadratic forms. Euler left us with a magnificent series of examples and conjectures, but it remained for Lagrange to develop the language which would bring the underlying structure to light.

E. Exercises

1.1. In this exercise, we prove some identities used by Euler.

(a) Prove (1.3) and its generalization (1.6).

(b) Generalize (1.6) to find an identity of the form

$$(ax^2 + cy^2)(az^2 + cw^2) = (?)^2 + ac(?)^2.$$

This is due to Euler [**43**, Vol. I, p. 424].

1.2. Let p be prime, and let $f(x)$ be a monic polynomial of degree $d < p$. This exercise will describe Euler's proof that the congruence $f(x) \not\equiv 0 \bmod p$ has a solution. Let $\Delta f(x) = f(x+1) - f(x)$ be the difference operator.

(a) For any $k \geq 1$, show that $\Delta^k f(x)$ is an integral linear combination of $f(x), f(x+1), \ldots, f(x+k)$.

(b) Show that $\Delta^d f(x) = d!$.

(c) Euler's argument is now easy to state: if $f(x) \not\equiv 0 \bmod p$ has no solutions, then $p \mid \Delta^d f(x)$ follows from part (a). By part (b), this is impossible.

1.3. Let n be a positive integer.

(a) Formulate and prove a version of Lemma 1.4 when a prime $q = x^2 + ny^2$ divides a number $N = a^2 + nb^2$.

(b) Show that your proof of part (a) works when $n = 3$ and $q = 4$.

1.4. In this exercise, we will prove the Descent Steps for $x^2 + 2y^2$ and $x^2 + 3y^2$.

(a) If a prime p divides $x^2 + 2y^2$, $\gcd(x, y) = 1$, then adapt the argument of Theorem 1.2 to show that $p = x^2 + 2y^2$. Hint: use Exercise 1.3.

(b) Prove that if an odd prime p divides $x^2 + 3y^2$, $\gcd(x, y) = 1$, then $p = x^2 + 3y^2$. The argument is more complicated because the Descent Step fails for $p = 2$. Thus, if it fails for some odd prime p, you have to produce an *odd* prime $q < p$ where it also fails. Hint: part (b) of Exercise 1.3 will be useful.

1.5. If $p = 3k + 1$ is prime, prove that $(-3/p) = 1$. Hint:

$$\begin{aligned} 4(x^{3k} - 1) &= (x^k - 1) \cdot 4(x^{2k} + x^k + 1) \\ &= (x^k - 1)((2x^k + 1)^2 + 3). \end{aligned}$$

Note that Fermat's theorem for $x^2 + 3y^2$ follows from part (b) of Exercise 1.4 and Exercise 1.5.

1.6. Prove Lemma 1.7.

1.7. Use the properties (1.11) of the Legendre symbol to prove that quadratic reciprocity is equivalent to (1.12).

1.8. Prove (1.13).

1.9. In this exercise we will see how the Reciprocity Steps for $x^2 + y^2$, $x^2 + 2y^2$ and $x^2 + 3y^2$ relate to quadratic reciprocity.

(a) Use Lemma 1.7 to show that for a prime $p > 3$,

$$p \mid x^2 + 3y^2,\ \gcd(x, y) = 1 \iff p \equiv 1 \bmod 3$$

is equivalent to

$$\left(\frac{-3}{p}\right) = \left(\frac{p}{3}\right).$$

By (1.12), we recognize this as part of quadratic reciprocity.

(b) Use Lemma 1.7 and the bottom line of (1.11) to show that the statements

$$p \mid x^2 + y^2,\ \gcd(x, y) = 1 \iff p \equiv 1 \bmod 4$$
$$p \mid x^2 + 2y^2,\ \gcd(x, y) = 1 \iff p \equiv 1, 3 \bmod 8$$

are equivalent to the statements

$$\left(\frac{-1}{p}\right) = (-1)^{(p-1)/2}$$
$$\left(\frac{2}{p}\right) = (-1)^{(p^2-1)/8}.$$

1.10. This exercise is concerned with the properties of the Jacobi symbol (M/m) defined in the proof of Lemma 1.14.

(a) Prove that $(M/m) = (N/m)$ when $M = N \bmod m$.
(b) Prove (1.15).
(c) Prove (1.16) using quadratic reciprocity and the two supplementary laws $(-1/p) = (-1)^{(p-1)/2}$ and $(2/p) = (-1)^{(p^2-1)/8}$. Hint: if r and s are odd, show that

$$(rs - 1)/2 \equiv (r - 1)/2 + (s - 1)/2 \bmod 2$$
$$(r^2s^2 - 1)/8 \equiv (r^2 - 1)/8 + (s^2 - 1)/8 \bmod 2.$$

(d) If M is a quadratic residue modulo m, show that $(M/m) = 1$. Give an example to show that the converse is not true.

1.11. Use (1.15) and (1.16) to complete the proof of (1.17) begun in the text.

1.12. This exercise is concerned with the map $\chi : (\mathbb{Z}/D\mathbb{Z})^* \to \{\pm 1\}$ of Lemma 1.14. When m is odd and positive, we define $\chi([m])$ to be the Jacobi symbol (D/m).

(a) Show that any class in $(\mathbb{Z}/D\mathbb{Z})^*$ may be written as $[m]$, where m is odd and positive, and then use (1.17) to show that χ is a well-defined homomorphism on $(\mathbb{Z}/D\mathbb{Z})^*$.
(b) Show that

$$\chi([-1]) = \begin{cases} 1 & \text{if } D > 0 \\ -1 & \text{if } D < 0. \end{cases}$$

(c) If $D \equiv 1 \bmod 4$, show that

$$\chi([2]) = \begin{cases} 1 & \text{if } D \equiv 1 \bmod 8 \\ -1 & \text{if } D \equiv 5 \bmod 8. \end{cases}$$

1.13. In this exercise, we will assume that Lemma 1.14 holds for all nonzero integers $D \equiv 0, 1 \bmod 4$, and we will prove quadratic reciprocity and the supplementary laws.

(a) Let p and q be distinct odd primes, and let $q^* = (-1)^{(q-1)/2}q$. By applying the lemma with $D = q^*$, show that $(q^*/\cdot)$ induces a homomorphism from $(\mathbb{Z}/q\mathbb{Z})^*$ to $\{\pm 1\}$. Since $(\cdot/q)$ can be regarded as a homomorphism between the same two groups and $(\mathbb{Z}/q\mathbb{Z})^*$ is cyclic, conclude that the two are equal.
(b) Use similar arguments to prove the supplementary laws. Hint: apply the lemma with $D = -4$ and 8 respectively.

1.14. Use Lemma 1.14 to prove that when $n \equiv 3 \bmod 4$, there are integers $\alpha, \beta, \gamma, \ldots$ such that for an odd prime p not dividing n, $p \mid x^2 + ny^2$, $\gcd(x, y) = 1$ if and only if $p \equiv \alpha, \beta, \gamma, \ldots \bmod n$.

1.15. Use quadratic reciprocity to determine those classes in $(\mathbb{Z}/84\mathbb{Z})^*$ that satisfy $(-21/p) = 1$. This tells us when $p \mid x^2 + 21y^2$, and thus solves Reciprocity Step when $n = 21$.

1.16. In the discussion following the proof of Lemma 1.14, we stated that $K = \ker(\chi)$ is characterized by the four properties (i)–(iv). When $D = 4q$, where q is an odd prime, prove that (i) and (ii) suffice to determine K uniquely.

2. Lagrange, Legendre and Quadratic Forms

The study of integral quadratic forms in two variables

$$f(x, y) = ax^2 + bxy + cy^2, \quad a, b, c \in \mathbb{Z}$$

began with Lagrange, who introduced the concepts of discriminant, equivalence and reduced form. When these are combined with Gauss' notion of proper equivalence, one has all of the ingredients necessary to develop the basic theory of quadratic forms. We will concentrate on the special case of positive definite forms. Here, Lagrange's theory of reduced forms is especially nice, and in particular we will get a complete solution of the Descent Step from §1. When this is combined with the solution of the Reciprocity Step given by quadratic reciprocity, we will get immediate proofs of Fermat's theorems (1.1) as well as several new results. We will then describe an elementary form of genus theory due to Lagrange, which will enable us to prove some of Euler's conjectures from §1, and we will also be able to solve our basic question of $p = x^2 + ny^2$ for quite a few n. The section will end with some historical remarks concerning Lagrange and Legendre.

A. Quadratic Forms

Our treatment of quadratic forms is taken primarily from Lagrange's "Recherches d'Arithmétique" of 1773–1775 [**86**, pp. 695–795] and Gauss' *Disquisitiones Arithmeticae* of 1801 [**51**, §§153–226]. Most of the terminology is due to Gauss, though many of the terms he introduced refer to concepts used implicitly by Lagrange (with some important exceptions).

A first definition is that a form $ax^2 + bxy + cy^2$ is *primitive* if its coefficients a, b and c are relatively prime. Note that any form is an integer multiple of a primitive form. We will deal exclusively with primitive forms.

An integer m is *represented* by a form $f(x, y)$ if the equation

$$m = f(x, y) \tag{2.1}$$

has an integer solution in x and y. If the x and y in (2.1) are relatively prime, we say that m is *properly represented* by $f(x, y)$. Note that the basic question of the book can be restated as: which primes are represented by the quadratic form $x^2 + ny^2$?

Next, we say that two forms $f(x, y)$ and $g(x, y)$ are *equivalent* if there are integers p, q, r and s such that

$$f(x, y) = g(px + qy, rx + sy) \quad \text{and} \quad ps - qr = \pm 1. \tag{2.2}$$

Since $\det\left(\begin{smallmatrix} p & q \\ r & s \end{smallmatrix}\right) = ps - qr = \pm 1$, this means that $\left(\begin{smallmatrix} p & q \\ r & s \end{smallmatrix}\right)$ is in the group of 2×2 invertible integer matrices $\mathrm{GL}(2, \mathbb{Z})$, and it follows easily that the equivalence of forms is an equivalence relation (see Exercise 2.2). An important observation is that equivalent forms represent the same numbers, and the same is true for proper representations (see Exercise 2.2). Note also that any form equivalent to a primitive form is itself primitive (see Exercise 2.2). Following Gauss, we say that an equivalence is a *proper equivalence* if $ps - qr = 1$, i.e., $\left(\begin{smallmatrix} p & q \\ r & s \end{smallmatrix}\right) \in \mathrm{SL}(2, \mathbb{Z})$, and it is an *improper equivalence* if $ps - qr = -1$ [**51**, §158]. Since $\mathrm{SL}(2, \mathbb{Z})$ is a subgroup of $\mathrm{GL}(2, \mathbb{Z})$, it follows that proper equivalence is also an equivalence relation (see Exercise 2.2).

The notion of equivalence is due to Lagrange, though he simply said that one form "can be transformed into another of the same kind" [**86**, p. 723]. Neither Lagrange nor Legendre made use of proper equivalence. The terms "equivalence" and "proper equivalence" are due to Gauss [**51**, §157], and after stating their definitions, Gauss promises that "the usefulness of these distinctions will soon be made clear" [**51**, §158]. In §3 we will see that he was true to his word.

As an example of these concepts, note that the forms $ax^2 + bxy + cy^2$ and $ax^2 - bxy + cy^2$ are always improperly equivalent via the substitution $(x, y) \mapsto (x, -y)$. But are they properly equivalent? This is not obvious. We will see below that the answer is sometimes yes (for $2x^2 \pm 2xy + 3y^2$) and sometimes no (for $3x^2 \pm 2xy + 5y^2$).

There is a very nice relation between proper representations and proper equivalence:

LEMMA 2.3. *A form $f(x, y)$ properly represents an integer m if and only if $f(x, y)$ is properly equivalent to the form $mx^2 + Bxy + Cy^2$ for some $B, C \in \mathbb{Z}$.*

PROOF. First, suppose that $f(p, q) = m$, where p and q are relatively prime. We can find integers r and s so that $ps - qr = 1$. If $f(x, y) = ax^2 + bxy + cy^2$, then

$$\begin{aligned} f(px + ry, qx + sy) &= f(p, q)x^2 + (2apr + bps + brq + 2cqs)xy + f(r, s)y^2 \\ &= mx^2 + Bxy + Cy^2 \end{aligned}$$

is of the desired form. To prove the converse, note that $mx^2 + Bxy + Cy^2$ represents m properly by taking $(x, y) = (1, 0)$, and the lemma is proved. Q.E.D.

We define the *discriminant* of $ax^2 + bxy + cy^2$ to be $D = b^2 - 4ac$. To see how this definition relates to equivalence, suppose that the forms $f(x, y)$ and $g(x, y)$ have discriminants D and D' respectively, and that

$$f(x, y) = g(px + qy, rx + sy), \qquad p, q, r, s \in \mathbb{Z}.$$

Then a straightforward calculation shows that

$$D = (ps - qr)^2 D'$$

(see Exercise 2.3), so that the two forms have the same discriminant whenever $ps - qr = \pm 1$. Thus equivalent forms have the same discriminant.

The sign of the discriminant D has a strong effect on the behavior of the form. If $f(x, y) = ax^2 + bxy + cy^2$, then we have the identity

$$4af(x, y) = (2ax + by)^2 - Dy^2. \tag{2.4}$$

If $D > 0$, then $f(x, y)$ represents both positive and negative integers, and we call the form *indefinite*, while if $D < 0$, then the form represents only positive integers or only negative ones, depending on the sign of a, and $f(x, y)$ is accordingly called *positive definite* or *negative definite* (see Exercise 2.4). Note that all of these notions are invariant under equivalence.

The discriminant D influences the form in one other way: since $D = b^2 - 4ac$, we have $D \equiv b^2 \bmod 4$, and it follows that the middle coefficient b is even (resp. odd) if and only if $D \equiv 0$ (resp. 1) mod 4.

We have the following necessary and sufficient condition for a number m to be represented by a form of discriminant D:

LEMMA 2.5. *Let $D \equiv 0, 1 \bmod 4$ be an integer and m be an odd integer relatively prime to D. Then m is properly represented by a primitive form of discriminant D if and only if D is a quadratic residue modulo m.*

PROOF. If $f(x, y)$ properly represents m, then by Lemma 2.3, we may assume that $f(x, y) = mx^2 + bxy + cy^2$. Thus $D = b^2 - 4mc$, and $D \equiv b^2 \bmod m$ follows easily.

Conversely, suppose that $D \equiv b^2 \bmod m$. Since m is odd, we can assume that D and b have the same parity (replace b by $b + m$ if necessary), and then $D \equiv 0, 1 \bmod 4$ implies that $D \equiv b^2 \bmod 4m$. This means that $D = b^2 - 4mc$ for some c. Then $mx^2 + bxy + cy^2$ represents m properly and has discriminant D, and the coefficients are relatively prime since m is relatively prime to D. Q.E.D.

For our purposes, the most useful version of Lemma 2.5 will be the following corollary:

COROLLARY 2.6. *Let n be an integer and let p be an odd prime not dividing n. Then $(-n/p) = 1$ if and only if p is represented by a primitive form of discriminant $-4n$.*

PROOF. This follows immediately from Lemma 2.5 since $-4n$ is a quadratic residue modulo p if and only if $(-4n/p) = (-n/p) = 1$. Q.E.D.

This corollary is relevant to the question raised in §1 when we tried to generalize the Descent Step of Euler's strategy. Recall that we asked how to represent prime divisors of $x^2 + ny^2$, $\gcd(x, y) = 1$. Note that Corollary 2.6 gives a first answer to this question, for such primes satisfy $(-n/p) = 1$ by Lemma 1.7, and hence are represented by forms of discriminant $-4n$. The problem is that there are too many quadratic forms of a given discriminant. For example, if the proof of Lemma 2.5 is applied to $(-3/13) = 1$, then we see that 13 is represented by the form $13x^2 + 12xy + 3y^2$ of discriminant -12. This is not very enlightening. So to improve Corollary 2.6, we need to show that every form is equivalent to an especially simple one. Lagrange's theory of reduced forms does this and a lot more.

So far, we've dealt with arbitrary quadratic forms, but from this point on, we will specialize to the positive definite case. These forms include the ones we're most interested in (namely, $x^2 + ny^2$ for $n > 0$), and their theory has a classical simplicity and elegance. In particular, there is an especially nice notion of reduced form.

A primitive positive definite form $ax^2 + bxy + cy^2$ is said to be *reduced* if

$$|b| \leq a \leq c, \text{ and } b \geq 0 \text{ if either } |b| = a \text{ or } a = c. \tag{2.7}$$

(Note that a and c are positive since the form is positive definite.) The basic theorem is the following:

THEOREM 2.8. *Every primitive positive definite form is properly equivalent to a unique reduced form.*

PROOF. The first step is to show that a given form is properly equivalent to one satisfying $|b| \le a \le c$. Among all forms properly equivalent to the given one, pick $f(x, y) = ax^2 + bxy + cy^2$ so that $|b|$ is as small as possible. If $a < |b|$, then

$$g(x, y) = f(x + my, y) = ax^2 + (2am + b)xy + c'y^2$$

is properly equivalent to $f(x, y)$ for any integer m. Since $a < |b|$, we can choose m so that $|2am + b| < |b|$, which contradicts our choice of $f(x, y)$. Thus $a \ge |b|$, and $c \ge |b|$ follows similarly. If $a > c$, we need to interchange the outer coefficients, which is accomplished by the proper equivalence $(x, y) \mapsto (-y, x)$. The resulting form satisfies $|b| \le a \le c$.

The next step is to show that such a form is properly equivalent to a reduced one. By definition (2.7), the form is already reduced unless $b < 0$ and $a = -b$ or $a = c$. In these exceptional cases, $ax^2 - bxy + cy^2$ is reduced, so that we need only show that the two forms $ax^2 \pm bxy + cy^2$ are properly equivalent. This is done as follows:

$$\begin{aligned} a = -b &: (x, y) \mapsto (x + y, y) \text{ takes } ax^2 - axy + cy^2 \text{ to } ax^2 + axy + cy^2. \\ a = c &: (x, y) \mapsto (-y, x) \text{ takes } ax^2 + bxy + ay^2 \text{ to } ax^2 - bxy + ay^2. \end{aligned}$$

The final step in the proof is to show that different reduced forms cannot be properly equivalent. This is the uniqueness part of the theorem. If $f(x, y) = ax^2 + bxy + cy^2$ satisfies $|b| \le a \le c$, then one easily shows that

$$f(x, y) \ge (a - |b| + c)\min(x^2, y^2) \tag{2.9}$$

(see Exercise 2.7). Thus $f(x, y) \ge a - |b| + c$ whenever $xy \ne 0$, and it follows that a is the smallest nonzero value of $f(x, y)$. Furthermore, if $c > a$, then c is the next smallest number represented properly by $f(x, y)$, so that in this case the outer coefficients of a reduced form give the minimum values properly represented by any equivalent form. These observations are due to Legendre [**91**, Vol. I, pp. 77–78].

We now prove uniqueness. For simplicity, assume that $f(x, y) = ax^2 + bxy + cy^2$ is a reduced form that satisfies the strict inequalities $|b| < a < c$. Then

$$a < c < a - |b| + c, \tag{2.10}$$

and by the above considerations, these are the three smallest numbers properly represented by $f(x, y)$. Using (2.9) and (2.10), it follows that

$$\begin{aligned} f(x, y) = a, \ \gcd(x, y) = 1 &\iff (x, y) = \pm(1, 0) \\ f(x, y) = c, \ \gcd(x, y) = 1 &\iff (x, y) = \pm(0, 1) \end{aligned} \tag{2.11}$$

(see Exercise 2.8). Now let $g(x, y)$ be a reduced form equivalent to $f(x, y)$. Since these forms represent the same numbers and are reduced, they must have the same first coefficient a by Legendre's observation. Now consider the third coefficient c' of $g(x, y)$. We know that $a \le c'$ since $g(x, y)$ is reduced. If equality occurred, then the equation $g(x, y) = a$ would have four proper solutions $\pm(1, 0)$ and $\pm(0, 1)$. Since $f(x, y)$ is equivalent to $g(x, y)$, this would contradict (2.11). Thus $a < c'$, and then Legendre's observation shows that $c = c'$. Hence the outer coefficients of $f(x, y)$ and $g(x, y)$ are the same, and since they have the same discriminant, it follows that $g(x, y) = ax^2 \pm bxy + cy^2$.

It remains to show that $f(x,y) = g(x,y)$ when we make the stronger assumption that the forms are properly equivalent. If we assume that

$$g(x,y) = f(px+qy, rx+sy), \qquad ps - qr = 1,$$

then $a = g(1,0) = f(p,r)$ and $c = g(0,1) = f(q,s)$ are proper representations. By (2.11), it follows that $(p,r) = \pm(1,0)$ and $(q,s) = \pm(0,1)$. Then $ps - qr = 1$ implies $\left(\begin{smallmatrix} p & q \\ r & s \end{smallmatrix}\right) = \pm\left(\begin{smallmatrix} 1 & 0 \\ 0 & 1 \end{smallmatrix}\right)$, and $f(x,y) = g(x,y)$ follows easily.

When $a = |b|$ or $a = c$, the above argument breaks down, because the values in (2.10) are no longer distinct. Nevertheless, one can still show that $f(x,y)$ and $g(x,y)$ reduce to $ax^2 \pm bxy + cy^2$, and then the restriction $b \geq 0$ in definition (2.7) implies equality. (See Exercise 2.8, or for the complete details, Scharlau and Opolka [**108**, pp. 36–38].) Q.E.D.

Note that we can now answer our earlier question about equivalence versus proper equivalence. Namely, the forms $3x^2 \pm 2xy + 5y^2$ are clearly equivalent, but since they are both reduced, Theorem 2.8 implies that they are not properly equivalent. On the other hand, of $2x^2 \pm 2xy + 3y^2$, only $2x^2 + 2xy + 3y^2$ is reduced (because $a = |b|$), and by the proof of Theorem 2.8, it is properly equivalent to $2x^2 - 2xy + 3y^2$.

In order to complete the elementary theory of reduced forms, we need one more observation. Suppose that $ax^2 + bxy + cy^2$ is a reduced form of discriminant $D < 0$. Then $b^2 \leq a^2$ and $a \leq c$, so that

$$-D = 4ac - b^2 \geq 4a^2 - a^2 = 3a^2$$

and thus

$$a \leq \frac{\sqrt{(-D)}}{3}. \tag{2.12}$$

If D is fixed, then $|b| \leq a$ and (2.12) imply that there are only finitely many choices for a and b. Since $b^2 - 4ac = D$, the same is true for c, so that there are only a finite number of reduced forms of discriminant D. Then Theorem 2.8 implies that the number of proper equivalence classes is also finite. Following Gauss [**51**, §223], we say that two forms are in the same *class* if they are properly equivalent. We will let $h(D)$ denote the number of classes of primitive positive definite forms of discriminant D, which by Theorem 2.8 is just the number of reduced forms. We have thus proved the following theorem:

THEOREM 2.13. *Let $D < 0$ be fixed. Then the number $h(D)$ of classes of primitive positive definite forms of discriminant D is finite, and furthermore $h(D)$ is equal to the number of reduced forms of discriminant D.* Q.E.D.

The above discussion shows that there is an algorithm for computing reduced forms and class numbers which, for small discriminants, is easily implemented on a computer (see Exercise 2.9). Here are some examples that will prove useful later on:

(2.14)

D	$h(D)$	Reduced Forms of Discriminant D
-4	1	$x^2 + y^2$
-8	1	$x^2 + 2y^2$
-12	1	$x^2 + 3y^2$
-20	2	$x^2 + 5y^2$, $2x^2 + 2xy + 3y^2$
-28	1	$x^2 + 7y^2$
-56	4	$x^2 + 14y^2$, $2x^2 + 7y^2$, $3x^2 \pm 2xy + 5y^2$
-108	3	$x^2 + 27y^2$, $4x^2 \pm 2xy + 7y^2$
-256	4	$x^2 + 64y^2$, $4x^2 + 4xy + 17y^2$, $5x^2 \pm 2xy + 13y^2$

Note, by the way, that $x^2 + ny^2$ is always a reduced form! For a further discussion of the computational aspects of class numbers, see Buell [**16**] and Shanks [**114**] (the algorithm described in [**114**] makes nice use of the theory to be described in §3).

This completes our discussion of positive definite forms. We should mention that there is a corresponding theory for indefinite forms. Its roots go back to Fermat and Euler (both considered special cases, such as $x^2 - 2y^2$), and Lagrange and Gauss each developed a general theory of such forms. There are notions of reduced form, class number, etc., but the uniqueness problem is much more complicated. As Gauss notes, "it can happen that many reduced forms are properly equivalent among themselves" [**51**, §184]. Determining exactly which reduced forms are properly equivalent is not easy (see Lagrange [**86**, pp. 728–740] and Gauss [**51**, §§183–193]). There are also connections with continued fractions and Pell's equation (see [**51**, §§183–205]), so that the indefinite case has a very different flavor. Two modern references are Flath [**46**, Chapter IV] and Zagier [**140**, §§8, 13 and 14].

B. $p = x^2 + ny^2$ and Quadratic Forms

We can now apply the theory of positive definite quadratic forms to solve some of the problems encountered in §1. The Descent Step of Euler's strategy begins with $p \mid x^2 + ny^2$, which by Lemma 1.7 is equivalent to $(-n/p) = 1$. We start by giving a complete solution of the Descent Step:

PROPOSITION 2.15. *Let n be a positive integer and p be an odd prime not dividing n. Then $(-n/p) = 1$ if and only if p is represented by one of the $h(-4n)$ reduced forms of discriminant $-4n$.*

PROOF. This follows immediately from Corollary 2.6 and Theorem 2.8. Q.E.D.

In §1 we showed how quadratic reciprocity gives a general solution of the Reciprocity Step of Euler's strategy. Having just solved the Descent Step, it makes sense to put the two together and see what we get. But rather than just treat the case of forms of discriminant $-4n$, we will state a result that applies to *all* negative discriminants $D < 0$. Recall from Lemma 1.14 that there is a homomorphism $\chi : (\mathbb{Z}/D\mathbb{Z})^* \to \{\pm 1\}$ such that $\chi([p]) = (D/p)$ for odd primes not dividing D. Note that $\ker(\chi) \subset (\mathbb{Z}/D\mathbb{Z})^*$ is a subgroup of index 2. We then have the following general theorem:

THEOREM 2.16. *Let $D \equiv 0, 1 \bmod 4$ be negative, and let $\chi : (\mathbb{Z}/D\mathbb{Z})^* \to \{\pm 1\}$ be the homomorphism from Lemma 1.14. Then, for an odd prime p not dividing D, $[p] \in \ker(\chi)$ if and only if p is represented by one of the $h(D)$ reduced forms of discriminant D.*

PROOF. The definition of χ tells us that $[p] \in \ker(\chi)$ if and only if $(D/p) = 1$. By Lemma 2.5, this last condition is equivalent to being represented by a primitive positive definite form of discriminant D, and then we are done by Theorem 2.8. Q.E.D.

The basic content of this theorem is that there is a congruence $p \equiv \alpha, \beta, \gamma, \ldots$ mod D which gives necessary and sufficient conditions for an odd prime p to be represented by a reduced form of discriminant D. This result is very computational, for we know how to find the reduced forms, and quadratic reciprocity makes it easy to find the congruence classes $\alpha, \beta, \gamma, \ldots$ mod D such that $(D/p) = 1$.

For an example of how Theorem 2.16 works, note that $x^2 + y^2$, $x^2 + 2y^2$ and $x^2 + 3y^2$ are the only reduced forms of discriminants -4, -8 and -12 respectively (this is from (2.14)). Using quadratic reciprocity to find the congruence classes for which $(-1/p)$, $(-2/p)$ and $(-3/p)$ equal 1, we get immediate proofs of Fermat's three theorems (1.1) (see Exercise 2.11). This shows just how powerful a theory we have: Fermat's theorems are now reduced to the status of an exercise. We can also go beyond Fermat, for notice that by (2.14), $x^2 + 7y^2$ is the only reduced form of discriminant -28, and it follows easily that

$$p = x^2 + 7y^2 \iff p \equiv 1, 9, 11, 15, 23, 25 \bmod 28 \tag{2.17}$$

for primes $p \neq 7$ (see Exercise 2.11). Thus we have made significant progress in answering our basic question of when $p = x^2 + ny^2$.

Unfortunately, this method for characterizing $p = x^2 + ny^2$ works only when $h(-4n) = 1$. In 1903, Landau proved a conjecture of Gauss that there are very few n's with this property:

THEOREM 2.18. *Let n be a positive integer. Then*

$$h(-4n) = 1 \iff n = 1, 2, 3, 4 \text{ or } 7.$$

PROOF. We follow Landau's proof [**87**]. The basic idea is simple: $x^2 + ny^2$ is a reduced form, and for $n \notin \{1, 2, 3, 4, 7\}$, we will produce a second reduced form of the same discriminant, showing that $h(-4n) > 1$. We may assume $n > 1$.

First suppose that n is not a prime power. Then n can be written $n = ac$, where $1 < a < c$ and $\gcd(a, c) = 1$ (see Exercise 2.12), and the form

$$ax^2 + cy^2$$

is reduced of disciminant $-4ac = -4n$. Thus $h(-4n) > 1$ when n is not a prime power.

Next suppose that $n = 2^r$. If $r \geq 4$, then

$$4x^2 + 4xy + (2^{r-2} + 1)y^2$$

has relatively prime coefficients and is reduced since $4 \leq 2^{r-2} + 1$. Furthermore, it has discriminant $4^2 - 4 \cdot 4(2^{r-2} + 1) = -16 \cdot 2^{r-2} = -4n$. Thus $h(-4n) > 1$ when $n = 2^r$, $r \geq 4$. One computes directly that $h(-4 \cdot 8) = 2$ (see Exercise 2.12), which leaves us with the known cases $n = 2$ and 4.

Finally, assume that $n = p^r$, where p is an odd prime. If $n + 1$ can be written $n + 1 = ac$, where $2 \leq a < c$ and $\gcd(a, c) = 1$, then

$$ax^2 + 2xy + cy^2$$

is reduced of discriminant $2^2 - 4ac = 4 - 4(n+1) = -4n$. Thus $h(-4n) > 1$ when $n+1$ is not a prime power. But $n = p^r$ is odd, so that $n+1$ is even, and hence it remains to consider the case $n+1 = 2^s$. If $s \geq 6$, then

$$8x^2 + 6xy + (2^{s-3}+1)y^2$$

has relatively prime coefficients and is reduced since $8 \leq 2^{s-3}+1$. Furthermore, it has discriminant $6^2 - 4 \cdot 8(2^{s-3}+1) = 4 - 4 \cdot 2^s = 4 - 4(n+1) = -4n$, and hence $h(-4n) > 1$ when $s \geq 6$. The cases $s = 1, 2, 3, 4$ and 5 correspond to $n = 1, 3, 7, 15$ and 31 respectively. Now $n = 15$ is not a prime power, and one easily computes that $h(-4 \cdot 31) = 3$ (see Exercise 2.12). This leaves us with the three known cases $n = 1, 3$ and 7, and completes the proof of the theorem. Q.E.D.

Note that we've already discussed the cases $n = 1$, 2, 3 and 7, and the case $n = 4$ was omitted since $p = x^2 + 4y^2$ is a trivial corollary of $p = x^2 + y^2$ (p is odd, so that one of x or y must be even). One could also ask if there is a similar finite list of *odd* discriminants $D < 0$ with $h(D) = 1$. The answer is yes, but the proof is *much* more difficult. We will discuss this problem in §7 and give a proof in §12.

C. Elementary Genus Theory

One consequence of Theorem 2.18 is that we need some new ideas to characterize $p = x^2 + ny^2$ when $h(-4n) > 1$. To get a sense of what's involved, consider the example $n = 5$. Here, Theorem 2.16, quadratic reciprocity and (2.14) tell us that

$$\text{(2.19)} \qquad \begin{aligned} p \equiv 1, 3, 7, 9 \mod 20 &\iff \left(\frac{-5}{p}\right) = 1 \\ &\iff p = x^2 + 5y^2 \text{ or } 2x^2 + 2xy + 3y^2. \end{aligned}$$

We need a method of separating reduced forms of the same discriminant, and this is where genus theory comes in. The basic idea is due to Lagrange, who, like us, used quadratic forms to prove conjectures of Fermat and Euler. But rather than working with reduced forms collectively, as we did in Theorem 2.16, Lagrange considers the congruence classes represented in $(\mathbb{Z}/D\mathbb{Z})^*$ by a single form, and he groups together forms that represent the same classes. This turns out to be the basic idea of genus theory!

Let's work out some examples to see how this grouping works. When $D = -20$, one easily computes that

$$\text{(2.20)} \qquad \begin{aligned} x^2 + 5y^2 &\quad \text{represents} \quad 1, 9 \quad \text{in} \quad (\mathbb{Z}/20\mathbb{Z})^* \\ 2x^2 + 2xy + 3y^2 &\quad \text{represents} \quad 3, 7 \quad \text{in} \quad (\mathbb{Z}/20\mathbb{Z})^* \end{aligned}$$

while for $D = -56$ one has

$$\text{(2.21)} \qquad \begin{aligned} x^2 + 14y^2, 2x^2 + 7y^2 &\quad \text{represent} \quad 1, 9, 15, 23, 25, 39 \quad \text{in} \quad (\mathbb{Z}/56\mathbb{Z})^* \\ 3x^2 \pm 2xy + 5y^2 &\quad \text{represent} \quad 3, 5, 13, 19, 27, 45 \quad \text{in} \quad (\mathbb{Z}/56\mathbb{Z})^* \end{aligned}$$

(see Exercise 2.14—the reduced forms are taken from (2.14)). In his memoir on quadratic forms, Lagrange gives a systematic procedure for determining the congruence classes in $(\mathbb{Z}/D\mathbb{Z})^*$ represented by a form of discriminant D [**86**, pp. 759–765], and he includes a table listing various reduced forms together with the corresponding congruence classes [**86**, pp. 766–767]. The examples in Lagrange's table show that this is a very natural way to group forms of the same discriminant.

In general, we say that two primitive positive definite forms of discriminant D are in the same *genus* if they represent the same values in $(\mathbb{Z}/D\mathbb{Z})^*$. Note that equivalent forms represent the same numbers and hence are in the same genus. In particular, each genus consists of a finite number of classes of forms. The above examples show that when $D = -20$, there are two genera, each consisting of a single class, and when $D = -56$, there are again two genera, but this time each genus consists of two classes.

The real impact of this theory becomes clear when we combine it with Theorem 2.16. The basic idea is that genus theory refines our earlier correspondence between congruence classes and representations by reduced forms. For example, when $D = -20$, (2.19) tells us that $p \equiv 1, 3, 7, 9 \bmod 20 \Longleftrightarrow x^2 + 5y^2$ or $2x^2 + 2xy + 3y^2$. If we combine this with (2.20), we obtain

$$(2.22) \qquad \begin{aligned} p = x^2 + 5y^2 &\iff p \equiv 1, 9 \bmod 20 \\ p = 2x^2 + 2xy + 3y^2 &\iff p \equiv 3, 7 \bmod 20 \end{aligned}$$

when $p \neq 5$ is odd. Note that the top line of (2.22) solves Euler's conjecture (1.20) for when $p = x^2 + 5y^2$! The thing that makes this work is that the two genera represent disjoint values in $(\mathbb{Z}/20\mathbb{Z})^*$. Looking at (2.21), we see that the same thing happens when $D = -56$, and then using Theorem 2.16 it is straightforward to prove that

$$(2.23) \qquad \begin{aligned} p = x^2 + 14y^2 \text{ or } 2x^2 + 7y^2 &\iff p \equiv 1, 9, 15, 23, 25, 39 \bmod 56 \\ p = 3x^2 \pm 2xy + 5y^2 &\iff p \equiv 3, 5, 13, 19, 27, 45 \bmod 56 \end{aligned}$$

when $p \neq 7$ is odd (see Exercise 2.15). Note that the top line proves part of Euler's conjecture (1.21) concerning $x^2 + 14y^2$.

In order to combine Theorem 2.16 and genus theory into a general theorem, we must show that the above examples reflect the general case. We first introduce some terminology. Given a negative integer $D \equiv 0, 1 \bmod 4$, the *principal form* is defined to be

$$\begin{aligned} x^2 - \frac{D}{4}y^2, &\qquad D \equiv 0 \bmod 4 \\ x^2 + xy + \frac{1-D}{4}y^2, &\qquad D \equiv 1 \bmod 4. \end{aligned}$$

It is easy to check that the principal form has discriminant D and is reduced (see Exercise 2.16). Note that when $D = -4n$, we get our friend $x^2 + ny^2$. Using the principal form, we can characterize the congruence classes in $(\mathbb{Z}/D\mathbb{Z})^*$ represented by a form of discriminant D:

LEMMA 2.24. *Given a negative integer $D \equiv 0, 1 \bmod 4$, let $\ker(\chi) \subset (\mathbb{Z}/D\mathbb{Z})^*$ be as in Theorem 2.16, and let $f(x, y)$ be a form of discriminant D.*

(i) *The values in $(\mathbb{Z}/D\mathbb{Z})^*$ represented by the principal form of discriminant D form a subgroup $H \subset \ker(\chi)$.*
(ii) *The values in $(\mathbb{Z}/D\mathbb{Z})^*$ represented by $f(x, y)$ form a coset of H in $\ker(\chi)$.*

PROOF. We first show that if a number m is prime to D and is represented by a form of discriminant D, then $[m] \in \ker(\chi)$. By Exercise 2.1, we can write $m = d^2m'$, where m' is properly represented by $f(x, y)$. Then $\chi([m]) = \chi([d^2m']) = \chi([d])^2\chi([m']) = \chi([m'])$. Thus we may assume that m is properly represented by

$f(x, y)$, and then Lemma 2.5 implies that D is a quadratic residue modulo m, i.e., $D = b^2 - km$ for some b and k. When m is odd, the properties of the Jacobi symbol (see Lemma 1.14) imply that

$$\chi([m]) = \left(\frac{D}{m}\right) = \left(\frac{b^2 - km}{m}\right) = \left(\frac{b^2}{m}\right) = \left(\frac{b}{m}\right)^2 = 1$$

and our claim is proved. The case when m is even is covered in Exercise 2.17.

We now turn to statements (i) and (ii) of the lemma. Concerning (i), the above paragraph shows that $H \subset \ker(\chi)$. When $D = -4n$, the identity (1.6) shows that H is closed under multiplication, and hence H is a subgroup. When $D \equiv 1 \bmod 4$, the argument is slightly different: here, notice that

$$4\left(x^2 + xy + \frac{1-D}{4}y^2\right) \equiv (2x + y)^2 \bmod D,$$

which makes it easy to show that H is in fact the subgroup of squares in $(\mathbb{Z}/D\mathbb{Z})^*$ (see Exercise 2.17).

To prove (ii), we need the following observation of Gauss [**51**, §228]:

LEMMA 2.25. *Given a form $f(x, y)$ and an integer M, then $f(x, y)$ properly represents at least one number relatively prime to M.*

PROOF. See Exercise 2.18. Q.E.D.

Now suppose that $D = -4n$. If we apply Lemma 2.25 with $M = 4n$ and then use Lemma 2.3, we may assume that $f(x, y) = ax^2 + bxy + cy^2$, where a is prime to $4n$. Since $f(x, y)$ has discriminant $-4n$, b is even and can be written as $2b'$, and then (2.4) implies that

$$af(x, y) = (ax + b'y)^2 + ny^2.$$

Since a is relatively prime to $4n$, it follows that the values of $f(x, y)$ in $(\mathbb{Z}/4n\mathbb{Z})^*$ lie in the coset $[a]^{-1}H$. Conversely, if $[c] \in [a]^{-1}H$, then $ac \equiv z^2 + nw^2 \bmod 4n$ for some z and w. Using the above identity, it is easy to solve the congruence $f(x, y) \equiv c \bmod 4n$, and thus the coset $[a]^{-1}H$ consists exactly of the values represented in $(\mathbb{Z}/D\mathbb{Z})^*$ by $f(x, y)$. The case $D \equiv 1 \bmod 4$ is similar (see Exercise 2.17), and Lemma 2.24 is proved. Q.E.D.

Since distinct cosets of H are disjoint, Lemma 2.24 implies that different genera represent disjoint values in $(\mathbb{Z}/D\mathbb{Z})^*$. This allows us to describe genera by cosets H' of H in $\ker(\chi)$. We define the *genus of* H' to consist of all forms of discriminant D which represent the values of H' modulo D. Then Lemma 2.24 immediately implies the following refinement of Theorem 2.16:

THEOREM 2.26. *Assume that $D \equiv 0, 1 \bmod 4$ is negative, and let $H \subset \ker(\chi)$ be as in Lemma 2.24. If H' is a coset of H in $\ker(\chi)$ and p is an odd prime not dividing D, then $[p] \in H'$ if and only if p is represented by a reduced form of discriminant D in the genus of H'.* Q.E.D.

This theorem is the main result of our elementary genus theory. It generalizes examples (2.22) and (2.23), and it shows that there are always congruence conditions which characterize when a prime is represented by some form in a given genus.

For us, the most interesting genus is the one containing the principal form, which following Gauss, we call the *principal genus*. When $D = -4n$, the principal form is $x^2 + ny^2$, and since $x^2 + ny^2$ is congruent modulo $4n$ to x^2 or $x^2 + n$, depending on whether y is even or odd, we get the following explicit congruence conditions for this case:

COROLLARY 2.27. *Let n be a positive integer and p an odd prime not dividing n. Then p is represented by a form of discriminant $-4n$ in the principal genus if and only if for some integer β,*

$$p \equiv \beta^2 \text{ or } \beta^2 + n \bmod 4n. \qquad \text{Q.E.D.}$$

There is also a version of this for discriminants $D \equiv 1 \bmod 4$—see Exercise 2.20.

The nicest case of Corollary 2.27 is when the principal genus consists of a single class, for then we get congruence conditions that characterize $p = x^2 + ny^2$. This is what happened when $n = 5$ (see (2.22)), and this isn't the only case. For example, the table of reduced forms in Lagrange's memoir [**86**, pp. 766–767] shows that the same thing happens for $n = 6$, 10, 13, 15, 21, 22 and 30—for each of these n's, the principal genus consists of only one class (see Exercise 2.21). Corollary 2.27 then gives us the following theorems for primes p:

$$\begin{aligned} p = x^2 + 6y^2 &\iff p \equiv 1, 7 \bmod 24 \\ p = x^2 + 10y^2 &\iff p \equiv 1, 9, 11, 19 \bmod 40 \\ p = x^2 + 13y^2 &\iff p \equiv 1, 9, 17, 25, 29, 49 \bmod 52 \\ p = x^2 + 15y^2 &\iff p \equiv 1, 19, 31, 49 \bmod 60 \\ p = x^2 + 21y^2 &\iff p \equiv 1, 25, 37 \bmod 84 \\ p = x^2 + 22y^2 &\iff p \equiv 1, 9, 15, 23, 25, 31, 47, 49, 71, 81 \bmod 88 \\ p = x^2 + 30y^2 &\iff p \equiv 1, 31, 49, 79 \bmod 120. \end{aligned} \tag{2.28}$$

It should be clear that this is a powerful theory! A natural question to ask is how often does the principal genus consist of only one class, i.e., how many theorems like (2.28) do we get? We will explore this question in more detail in §3.

The genus theory just discussed has been very successful, but it hasn't solved all of the problems posed in §1. In particular, we have yet to prove Fermat's conjecture concerning $pq = x^2 + 5y^2$, and we've only done parts of Euler's conjectures (1.20) and (1.21) concerning $x^2 + 5y^2$ and $x^2 + 14y^2$. To complete the proofs, we again turn to Lagrange for help.

Let's begin with $x^2 + 5y^2$. We've already proved the part concerning when a prime p can equal $x^2 + 5y^2$ (see (2.22)), but it remains to show that for primes p and q, we have

$$\begin{aligned} p, q \equiv 3, 7 \bmod 20 &\implies pq = x^2 + 5y^2 \quad \text{(Fermat)} \\ p \equiv 3, 7 \bmod 20 &\implies 2p = x^2 + 5y^2 \quad \text{(Euler)}. \end{aligned} \tag{2.29}$$

Lagrange's argument [**86**, pp. 788–789] is as follows. He first notes that primes congruent to 3 or 7 modulo 20 can be written as $2x^2 + 2xy + 3y^2$ (this is (2.22)), so that both parts of (2.29) can be proved by showing that the product of two numbers represented by $2x^2 + 2xy + 3y^2$ is of the form $x^2 + 5y^2$. He then states the identity

$$\begin{aligned} &(2x^2 + 2xy + 3y^2)(2z^2 + 2zw + 3w^2) \\ &\qquad = (2xz + xw + yz + 3yw)^2 + 5(xw - yz)^2 \end{aligned} \tag{2.30}$$

(see Exercise 2.22), and everything is proved!

Turning to Euler's conjecture (1.21) for x^2+14y^2, we proved part of it in (2.23), but we still need to show that

$$p \equiv 3, 5, 13, 19, 27, 45 \bmod 56 \iff 3p = x^2 + 14y^2.$$

Using (2.23), it suffices to show that 3 times a number represented by $3x^2 \pm 2xy + 5y^2$, or more generally the product of any two such numbers, is of the form $x^2 + 14y^2$. So what we need is another identity of the form (2.30), and in fact there is a version of (2.30) that holds for any form of discriminant $-4n$:

$$\begin{aligned}(ax^2 + 2bxy + cy^2)&(az^2 + 2bzw + cw^2)\\ &= (axz + bxw + byz + cyw)^2 + n(xw - yz)^2\end{aligned} \tag{2.31}$$

(see Exercise 2.21). Applying this to $3x^2 + 2xy + 5y^2$ and $n = 14$, we are done.

We can also explain one other aspect of Euler's conjectures (1.20) and (1.21), for recall that we wondered why (1.20) used $2p$ while (1.21) used $3p$. The answer again involves the identities (2.30) and (2.31): they show that 2 (resp. 3) can be replaced by any value represented by $2x^2 + 2xy + 3y^2$ (resp. $3x^2 \pm 2xy + 5y^2$). But Legendre's observation from the proof of Theorem 2.8 shows that 2 (resp. 3) is the best choice because it's the smallest nonzero value represented by the form in question. We will see below and in §3 that identities like (2.30) and (2.31) are special cases of the composition of quadratic forms.

We now have complete proofs of Euler's conjectures (1.20) and (1.21) for $x^2 + 5y^2$ and $x^2 + 14y^2$. Notice that we've used a lot of mathematics: quadratic reciprocity, reduced quadratic forms, genus theory and the composition of quadratic forms. This amply justifies the high estimate of Euler's insight that was made in §1, and Lagrange is equally impressive for providing the proper tools to understand what lay behind Euler's conjectures.

D. Lagrange and Legendre

We've already described parts of Lagrange's memoir "Recherches d'Arithmétique," but there are some further comments we'd like to add. First, although we credit Lagrange with the discovery of genus theory, it appears only implicitly in his work. The groupings that appear in his tables of reduced forms are striking, but Lagrange's comments on genus theory are a different matter. On the page before the tables begin, Lagrange explains his grouping of forms as follows: "when two different [forms] give the same values of b [in $(\mathbb{Z}/4n\mathbb{Z})^*$], one combines these [forms] into the same case" [**86**, p. 765]. This is the sum total of what Lagrange says about genus theory!

After completing the basic theory of quadratic forms (both definite and indefinite), Lagrange gives some applications to number theory. To motivate his results, he turns to Fermat and Euler, and he quotes from two of our main sources of inspiration: Fermat's 1658 letter to Digby and Euler's 1744 paper on prime divisors of $paa \pm qyy$. Lagrange explicitly states Fermat's results (1.1) on primes of the form $x^2 + ny^2$, $n = 1, 2$ or 3, and he notes Fermat's speculation that $pq = x^2 + 5y^2$ whenever p and q are primes congruent to 3 or 7 modulo 20. Lagrange also mentions several of Euler's conjectures, including (1.20), and he adds "one finds a very large number of similar theorems in Volume XIV of the old *Commentaires de Pétersbourg* [where Euler's 1744 paper appeared], but none of them have been demonstrated until now" [**86**, pp. 775–776].

The last section of Lagrange's memoir is titled "Prime numbers of the form $4nm + b$ which are at the same time of the form $x^2 \pm ny^2$" [**86**, p. 775]. It's clear that Lagrange wanted to prove Theorem 2.26, so that he could read off corollaries like (2.17), (2.22), (2.23) and (2.28). The problem is that these proofs depend on quadratic reciprocity, which Lagrange didn't know in general—he could only prove some special cases. For example, he was able to determine $(\pm 2/p)$, $(\pm 3/p)$ and $(\pm 5/p)$, but he had only partial results for $(\pm 7/p)$. Thus, he could prove all of (2.22) but only parts of the others (see [**86**, pp. 784–793] for the full list of his results). To get the flavor of Lagrange's arguments, the reader should see Exercise 2.23 or Scharlau and Opolka [**108**, pp. 41–43]. At the end of the memoir, Lagrange summarizes what he could prove about quadratic reciprocity, stating his results in terms of Euler's criterion

$$a^{(p-1)/2} \equiv \left(\frac{a}{p}\right) \bmod p.$$

For example, for $(2/p)$, Lagrange states [**86**, p. 794]:

> Thus, if p is a prime number of one of the forms $8n \pm 1$, $2^{(p-1)/2} - 1$ will be divisible by p, and if p is of the form $8n \pm 3$, $2^{(p-1)/2} + 1$ will thus be divisible by p.

We next turn to Legendre. In his 1785 memoir "Recherches d'Analyse Indéterminée" [**92**], the two major results are first, a necessary and sufficient criterion for the equation

$$ax^2 + by^2 + cz^2 = 0, \qquad a, b, c \in \mathbb{Z}$$

to have a nontrivial integral solution, and second, a proof of quadratic reciprocity. Legendre was influenced by Lagrange, but he replaces Lagrange's "$2^{(p-1)/2} - 1$ will be divisible by p" by the simpler phrase "$2^{(p-1)/2} = 1$," where, as he warns the reader, "one has thrown out the multiples of p in the first member" [**92**, p. 516]. He then goes on to state quadratic reciprocity in the following form [**92**, p. 517]:

> c and d being two [odd] prime numbers, the expressions $c^{(d-1)/2}$, $d^{(c-1)/2}$ do not have different signs except when c & d are both of the form $4n - 1$; in all other cases, these expressions will always have the same sign.

Except for the notation, this is a thoroughly modern statement of quadratic reciprocity. Legendre's proof is a different matter, for it is quite incomplete. We won't examine the proof in detail—this is done in Weil [**135**, pp. 328–330 and 344–345]. Suffice it to say that some of the cases are proved rigorously (see Exercise 2.24), some depend on Dirichlet's theorem on primes in arithmetic progressions, and some are a tangle of circular reasoning.

In 1798 Legendre published a more ambitious work, the *Essai sur la Théorie des Nombres*. (The third edition [**91**], published 1830, was titled *Théorie des Nombres*, and all of our references will be to this edition.) Legendre must have been dissatisfied with the notation of the "Recherches", for in the *Essai* he introduces the Legendre symbol (a/p). Then, in a section titled "Theorem containing a law of reciprocity which exists between two arbitrary prime numbers," Legendre states that if n and m are distinct odd primes, then

$$\left(\frac{n}{m}\right) = (-1)^{(n-1)/2\cdot(m-1)/2}\left(\frac{m}{n}\right)$$

(see [**91**, Vol. I, p. 230]). This is where our notation and terminology for quadratic reciprocity come from. Unfortunately, the *Essai* repeats Legendre's incomplete proof from 1785, although by the 1830 edition there had been enough criticism of this proof that Legendre added Gauss' third proof of reciprocity as well as one communicated to him by Jacobi (still maintaining that his original proof was valid).

The *Essai* also contains a treatment of quadratic forms. Like Lagrange, one of Legendre's goals was to prove theorems in number theory using quadratic forms. The difference is that Legendre knows quadratic reciprocity (or at least he thinks he does), and this allows him to state a version of our main result, Theorem 2.26. Legendre calls it his "Théorème General" [**91**, Vol. I, p. 299], and it goes as follows: if $[a]$ is a congruence class lying in $\ker(\chi)$, then

> every prime number comprised of the form $4nx + a \ldots$ will consequently be given by one of the quadratic forms $py^2+2qyz\pm rz^2$ which correspond to the linear form $4nx + a$.

The terminology here is interesting. Euler and Lagrange would speak of numbers "of the form" $4nx+a$ or "of the form" $ax^2+bxy+cy^2$. As the above quote indicates, Legendre distinguished these two by calling them linear forms and quadratic forms respectively. This is where we get the term "quadratic form."

While Legendre's "Théorème" makes no explicit reference to genus theory, the context shows that it's there implicitly. Namely, Legendre's book has tables similar to Lagrange's, with the forms grouped according to the values they represent in $(\mathbb{Z}/D\mathbb{Z})^*$. Since the explanation of the tables immediately precedes the statement of the "Théorème" [**91**, Vol. I, pp. 286–298], it's clear that Legendre's correspondence between linear forms and quadratic forms is exactly that given by Theorem 2.26.

To Legendre, this theorem "is, without contradiction, one of the most general and most important in the theory of numbers" [**91**, Vol. I, p. 302]. Its main consequence is that every entry in his tables becomes a theorem, and Legendre gives several pages of explicit examples [**91**, Vol. I, pp. 305–307]. This is a big advance over what Lagrange could do, and Legendre notes that quadratic reciprocity was the key to his success [**91**, Vol. I, p. 307]:

> Lagrange is the first who opened the way for the study of these sorts of theorems. ... But the methods which served the great geometer are not applicable ... except in very few cases; and the difficulty in this regard could not be completely resolved without the aid of the law of reciprocity.

Besides completing Lagrange's program, Legendre also tried to understand some of the other ideas implicit in Lagrange's memoir. We will discuss one of Legendre's attempts that is particularly relevant to our purposes: his theory of composition. Legendre's basic idea was to generalize the identity (2.30)

$$\begin{aligned}(2x^2+2xy+3y^2)&(2z^2+2zw+3w^2)\\ &= (2xz+xw+yz+3yw)^2+5(xw-yz)^2\end{aligned}$$

used by Lagrange in proving the conjectures of Fermat and Euler concerning $x^2 + 5y^2$. We gave one generalization in (2.31), but Legendre saw that something more general was going on. More precisely, let $f(x, y)$ and $g(x, y)$ be forms of discriminant D. Then a form $F(x, y)$ of the same discriminant is their *composition* provided that

$$f(x,y)g(z,w) = F(B_1(x,y;z,w), B_2(x,y;z,w))$$

where

$$B_i(x,y;z,w) = a_i xz + b_i xw + c_i yz + d_i yw, \qquad i = 1,2$$

are bilinear forms in x, y and z, w. Thus Lagrange's identity shows that $x^2 + 5y^2$ is the composition of $2x^2 + 2xy + 3y^2$ with itself. And this is not the only example we've seen—the reader can check that (1.3), (1.6) and (2.31) are also examples of the composition of forms.

A useful consequence of composition is that whenever $F(x,y)$ is composed of $f(x,y)$ and $g(x,y)$, then the product of numbers represented by $f(x,y)$ and $g(x,y)$ will be represented by $F(x,y)$. This was the idea that enabled us to complete the conjectures of Fermat and Euler for $x^2 + 5y^2$ and $x^2 + 14y^2$.

The basic question is whether any two forms of the same discriminant can be composed, and Legendre showed that the answer is yes [**91**, Vol. II, pp. 27–30]. For simplicity, let's discuss the case where the forms $f(x,y) = ax^2 + 2bxy + cy^2$ and $g(x,y) = a'x^2 + 2b'xy + c'y^2$ have discriminant $-4n$, and a and a' are relatively prime (we can always arrange the last condition by changing the forms by a proper equivalence). Then the Chinese Remainder Theorem shows that there is a number B such that

$$\begin{aligned} B &= \pm b \bmod a \\ B &= \pm b' \bmod a'. \end{aligned} \tag{2.32}$$

It follows that $B^2 + n \equiv b^2 + (ac - b^2) \equiv 0 \bmod a$, so that $a \mid B^2 + n$. The same holds for a', and thus $aa' \mid B^2 + n$. Then Legendre shows that the form

$$F(x,y) = aa'x^2 + 2Bxy + \frac{B^2+n}{aa'}y^2$$

is the composition of $f(x,y)$ and $g(x,y)$. A modern account of Legendre's argument may be found in Weil [**135**, pp. 332–335]), and we will consider this problem (from a slightly different point of view) in §3 when we discuss composition in more detail.

Because of the $\pm$ signs in (2.32), two forms in general may be composed in four different ways. For example, the forms $14x^2 + 10xy + 21y^2$ and $9x^2 + 2xy + 30y^2$ compose to the four forms

$$126x^2 \pm 38xy + 5y^2, \qquad 126x^2 \pm 74xy + 13y^2,$$

and it is easy to show that these forms all lie in different classes (see Exercise 2.26). Since Legendre used equivalence rather than proper equivalence, he sees two rather than four forms here—for him, this operation "leads in general to two solutions" [**91**, Vol. II, p. 28].

One of Legendre's important ideas is that since every form is equivalent to a reduced one, it suffices to work out the compositions of reduced forms. The resulting table would then give the compositions of all possible forms of that discriminant. Let's look at the case $n = 41$, which Legendre does in detail in [**51**, Vol. II, pp. 39–40]. He labels the reduced forms as follows:

$$\begin{aligned} A &= x^2 + 41y^2 \\ B &= 2x^2 + 2xy + 21y^2 \\ C &= 5x^2 + 4xy + 9y^2 \\ D &= 3x^2 + 2xy + 14y^2 \\ E &= 6x^2 + 2xy + 7y^2. \end{aligned} \tag{2.33}$$

(Legendre writes the forms slightly differently, but it's more convenient to work with reduced forms.) He then gives the following table of compositions:

$$
\begin{array}{r|r|r|r|r}
AA = A & BB = A & CC = A \text{ or } B & DD = A \text{ or } C & EE = A \text{ or } C \\
AB = B & BC = C & CD = D \text{ or } E & DE = B \text{ or } C & \\
AC = C & BD = E & CE = D \text{ or } E & & \\
AD = D & BE = D & & & \\
AE = E & & & &
\end{array} \tag{2.34}
$$

This almost looks like the multiplication table for a group, but the binary operation isn't single-valued. To the modern reader, it's clear that Legendre must be doing something slightly wrong.

One problem is that (2.33) lists 5 forms, while the class number is 8. (C, D and E each give two reduced forms, while A and B each give only one.) This is closely related to the ambiguity in Legendre's operation: as long as we work with equivalence rather than proper equivalence, we can't fix the sign of the middle coefficient $2b$ of a reduced form, so that the $\pm$ signs in (2.32) are forced upon us.

This suggests that composition might give a group operation on the *classes* of forms of discriminant D. However, there remain serious problems to be solved. Composition, as defined above, is still a multiple-valued operation. Thus one has to show that the signs in (2.32) can be chosen *uniformly* so that as we vary $f(x, y)$ and $g(x, y)$ within their proper equivalence classes, the resulting compositions are all properly equivalent. Then one has to worry about associativity, inverses, etc. There's a lot of work to be done!

This concludes our discussion of Lagrange and Legendre. While the last few pages have raised more questions than answers, the reader should still be convinced of the richness of the theory of quadratic forms. The surprising fact is that we have barely reached the really interesting part of the theory, for we have yet to consider the work of Gauss.

E. Exercises

2.1. If a form $f(x, y)$ represents an integer m, show that m can be written $m = d^2m'$, where $f(x, y)$ properly represents m'.

2.2. In this exercise we study equivalence and proper equivalence.

(a) Show that equivalence and proper equivalence are equivalence relations.
(b) Show that improper equivalence is not an equivalence relation.
(c) Show that equivalent forms represent the same numbers, and show that the same holds for proper representations.
(d) Show that any form equivalent to a primitive form is itself primitive. Hint: use part (c).

2.3. Let $f(x, y)$ and $g(x, y)$ be forms of discriminants D and D' respectively, and assume that there are integers p, q, r and s such that

$$f(x, y) = g(px + qy, rx + sy).$$

Prove that $D = (ps - qr)^2 D'$.

2.4. Let $f(x, y)$ be a form of discriminant $D \neq 0$.

(a) If $D > 0$, then use (2.4) to prove that $f(x, y)$ represents both positive and negative numbers.

(b) If $D < 0$, then show that $f(x, y)$ represents only positive or only negative numbers, depending on the sign of the coefficient of x^2.

2.5. Formulate and prove a version of Corollary 2.6 which holds for arbitrary discriminants.

2.6. Find a reduced form that is properly equivalent to $126x^2 + 74xy + 13y^2$. Hint: make the middle coefficient small—see the proof of Theorem 2.8.

2.7. Prove (2.9) for forms that satisfy $|b| \leq a \leq c$.

2.8. This exercise is concerned with the uniqueness part of Theorem 2.8.

(a) Prove (2.11).

(b) Prove a version of (2.11) that holds in the exceptional cases $|b| = a$ or $a = c$, and use this to complete the uniqueness part of the proof of Theorem 2.8.

2.9. Use a computer algebra system (such as *Maple*™ or *Mathematica*®) to write a procedure that computes the class number and all reduced forms of a given discriminant $D < 0$. For example, one finds that $h(-32767) = 52$. If you don't use a computer, then you should check the following examples by hand.

(a) Verify the entries in table (2.14).

(b) Compute all reduced forms of discriminants -3, -15, -24, -31 and -52.

2.10. This exercise is concerned with indefinite forms of discriminant $D > 0$, D not a perfect square. The last condition implies that the outer coefficients of a form with discriminant D are nonzero.

(a) Adapt the proof of Theorem 2.8 to show that any form of discriminant D is properly equivalent to $ax^2 + bxy + cy^2$, where

$$|b| \leq |a| \leq |c|.$$

(b) If $ax^2 + bxy + cy^2$ satisfies the above inequalities, prove that

$$|a| \leq \frac{\sqrt{D}}{2}.$$

(c) Conclude that there are only finitely many proper equivalence classes of forms of discriminant D. This proves that the class number $h(D)$ is finite.

2.11. Use Theorem 2.16, quadratic reciprocity and table (2.14) to prove Fermat's three theorems (1.1) and the new result (2.17) for $x^2 + 7y^2$.

2.12. This exercise is concerned with the proof of Theorem 2.18.

(a) If $m > 1$ is an integer which is not a prime power, prove that m can be written $m = ac$ where $1 < a < c$ and $\gcd(a, c) = 1$.

(b) Show that $h(-32) = 2$ and $h(-124) = 3$.

2.13. Use Theorem 2.16, quadratic reciprocity and table (2.14) to prove (2.19), and work out similar results for discriminants -3, -15, -24, -31 and -52.

2.14. Prove (2.20) and (2.21). Hint: use Lemma 2.24.

2.15. Prove (2.23).

2.16. Let D be a number congruent to 1 modulo 4. Show that the form $x^2 + xy + ((1 - D)/4)y^2$ has discriminant D, and show that it is reduced when $D < 0$.

2.17. In this exercise, we will complete the proof of Lemma 2.24 for discriminants $D \equiv 1 \bmod 4$. Let $\chi : (\mathbb{Z}/D\mathbb{Z})^* \to \{\pm 1\}$ be as in Lemma 1.14.

(a) If an even number is properly represented by a form of discriminant D, then show that $D \equiv 1 \bmod 8$. Hint: use Lemma 2.3.

(b) If m is relatively prime to D and is represented by a form of discriminant D, then show that $[m] \in \ker(\chi)$. Hint: use Lemma 2.5 and, when m is even, part (a) and part (c) of Exercise 1.12.

(c) Let $H \subset (\mathbb{Z}/D\mathbb{Z})^*$ be the subgroup of squares. Show that H consists of the values represented by $x^2 + xy + ((1-D)/4)y^2$. Hint: use

$$4\left(x^2 + xy + \frac{1-D}{4}y^2\right) \equiv (2x+y)^2 \bmod D.$$

(d) Let $f(x,y)$ be a form of discriminant D. Show that the values in $(\mathbb{Z}/D\mathbb{Z})^*$ represented by $f(x,y)$ form a coset of H in $\ker(\chi)$. Hint: use (2.4).

2.18. Let $f(x,y) = ax^2 + bxy + cy^2$, where as usual we assume $\gcd(a,b,c) = 1$.

(a) Given a prime p, prove that at least one of $f(1,0)$, $f(0,1)$ and $f(1,1)$ is relatively prime to p.

(b) Prove Lemma 2.25 using part (a) and the Chinese Remainder Theorem.

2.19. Work out the genus theory of Theorem 2.26 for discriminants -15, -24, -31 and -52. Your answers should be similar to (2.22) and (2.23).

2.20. Formulate and prove a version of Corollary 2.27 for negative discriminants $D \equiv 1 \bmod 4$. Hint: by part (c) of Exercise 2.17, H is the subgroup of squares.

2.21. Prove (2.28). Hint: for each n, find the reduced forms and use Lemma 2.24.

2.22. Prove (2.30) and its generalization (2.31).

2.23. The goal of this exercise is to prove that $(-2/p) = 1$ when $p \equiv 1, 3 \bmod 8$. The argument below is due to Lagrange, and is similar to the one used by Euler in his proof of the Reciprocity Step for $x^2 + 2y^2$ [**43**, Vol. II, pp. 240–281].

(a) When $p \equiv 1 \bmod 8$, write $p = 8k+1$, and then use the identity

$$x^{8k} - 1 = ((x^{2k}-1)^2 + 2x^{2k})(x^{4k} - 1)$$

to show that $(-2/p) = 1$.

(b) When $p \equiv 3 \bmod 8$, assume that $(-2/p) = -1$. Show that $(2/p) = 1$, and thus by Corollary 2.6, p is represented by a form of discriminant 8.

(c) Use part (a) of Exercise 2.10 to show that any form of discriminant 8 is properly equivalent to $\pm(x^2 - 2y^2)$.

(d) Show that an odd prime $p = \pm(x^2 - 2y^2)$ is congruent to ± 1 modulo 8.

From parts (a)–(d), it follows easily that $(-2/p) = 1$ when $p \equiv 1, 3 \bmod 8$.

2.24. One of the main theorems in Legendre's 1785 memoir [**91**, pp. 509–513] states that the equation

$$ax^2 + by^2 + cz^2 = 0,$$

where abc is squarefree, has a nontrivial integral solution if and only if

(i) a, b and c are not all of the same sign, and

(ii) $-bc$, $-ac$ and $-ab$ are quadratic residues modulo $|a|$, $|b|$ and $|c|$ respectively.

As we've already noted, Legendre tried to use this result to prove quadratic reciprocity. In this problem, we will treat one of the cases where he succeeded. Let p and q be primes which satisfy $p \equiv 1 \bmod 4$ and $q \equiv 3 \bmod 4$, and assume that $(p/q) = -1$ and $(q/p) = 1$. We will derive a contradiction as follows:

(a) Use Legendre's theorem to show that $x^2 + py^2 - qz^2 = 0$ has a nontrivial integral solution.

(b) Working modulo 4, show that $x^2 + py^2 - qz^2 = 0$ has no nontrivial integral solutions.

In [**135**, pp. 339–345], Weil explains why this argument works.

2.25. The opposite of the form $ax^2+bxy+cy^2$ is the form $ax^2-bxy+cy^2$. Prove that two forms are properly equivalent if and only if their opposites are.

2.26. Verify that $14x^2 + 10xy + 21y^2$ and $9x^2 + 2xy + 30y^2$ compose to the four forms $126x^2 \pm 74xy + 13y^2$ and $126x^2 \pm 38xy + 5y^2$, and show that they all lie in different classes. Hint: use Exercises 2.6 and 2.25.

2.27. Let p be a prime number which is represented by forms $f(x, y)$ and $g(x, y)$ of the same discriminant.

(a) Show that $f(x, y)$ and $g(x, y)$ are equivalent. Hint: use Lemma 2.3, and examine the middle coefficient modulo p.

(b) If $f(x, y) = x^2+ny^2$, and $g(x, y)$ is reduced, then show that $f(x, y)$ and $g(x, y)$ are equal.

3. Gauss, Composition and Genera

While genus theory and composition were implicit in Lagrange's work, these concepts are still primarily linked to Gauss, and for good reason: he may not have been the first to use them, but he was the first to understand their astonishing depth and interconnection. In this section we will prove Gauss' major results on composition and genus theory for the special case of positive definite forms. We will then apply this theory to our question concerning primes of the form $x^2 + ny^2$, and we will also discuss Euler's convenient numbers. These turn out to be those n's for which each genus consists of a single class, and it is still not known exactly how many there are. The section will end with a discussion of Gauss' *Disquisitiones Arithmeticae.*

A. Composition and the Class Group

The basic definition of composition was given in §2: if $f(x, y)$ and $g(x, y)$ are primitive positive definite forms of discriminant D, then a form $F(x, y)$ of the same type is their *composition* provided that

$$f(x, y)g(z, w) = F(B_1(x, y; z, w), B_2(x, y; z, w)),$$

where

$$B_i(x, y; z, w) = a_i xz + b_i xw + c_i yz + d_i yw, \qquad i = 1, 2$$

are integral bilinear forms. Two forms can be composed in many different ways, and the resulting forms need not be properly equivalent. In §2 we gave an example of two forms whose compositions lay in four distinct classes. So if we want a well-defined operation on classes of forms, we must somehow *restrict* the notion of composition. Gauss does this as follows: given the above composition data, he proves that

$$a_1b_2 - a_2b_1 = \pm f(1,0), \qquad a_1c_2 - a_2c_1 = \pm g(1,0) \tag{3.1}$$

(see [**51**, §235] or Exercise 3.1), and then he defines the composition to be a *direct composition* provided that both of the signs in (3.1) are +.

The main result of Gauss' theory of composition is that for a fixed discriminant, direct composition makes the set of classes of forms into a finite Abelian group [**51**, §§236–240, 245 and 249]. Unfortunately, direct composition is an awkward concept to work with, and Gauss' proof of the group structure is long and complicated. So rather than follow Gauss, we will take a different approach to the study of composition. The basic idea is due to Dirichlet [**36**, Supplement X], though his treatment was clearly influenced by Legendre. Before giving Dirichlet's definition, we will need the following lemma:

LEMMA 3.2. *Assume that $f(x,y) = ax^2+bxy+cy^2$ and $g(x,y) = a'x^2+b'xy+c'y^2$ have discriminant D and satisfy $\gcd(a,a',(b+b')/2) = 1$ (since b and b' have the same parity, $(b+b')/2$ is an integer). Then there is a unique integer B modulo $2aa'$ such that*

$$\begin{aligned} B &\equiv b \bmod 2a \\ B &\equiv b' \bmod 2a' \\ B^2 &\equiv D \bmod 4aa'. \end{aligned}$$

PROOF. The first step is to put these congruences into a standard form. If a number B satisfies the first two, then

$$B^2 - (b+b')B + bb' \equiv (B-b)(B-b') \equiv 0 \bmod 4aa',$$

so that the third congruence can be written as

$$(b+b')B \equiv bb' + D \bmod 4aa'.$$

Dividing by 2, this becomes

$$(b+b')/2 \cdot B \equiv (bb'+D)/2 \bmod 2aa'. \tag{3.3}$$

If we multiply the first two congruences of the lemma by a' and a respectively and combine them with (3.3), we see that the three congruences in the statement of the lemma are equivalent to

$$\begin{aligned} a' \cdot B &\equiv a'b \bmod 2aa' \\ a \cdot B &\equiv ab' \bmod 2aa' \\ (b+b')/2 \cdot B &\equiv (bb'+D)/2 \bmod 2aa'. \end{aligned} \tag{3.4}$$

The following lemma tells us about the solvability of these congruences:

LEMMA 3.5. *Let $p_1, q_1, \ldots, p_r, q_r, m$ be numbers with $\gcd(p_1,\ldots,p_r,m) = 1$. Then the congruences*

$$p_iB \equiv q_i \bmod m, \qquad i = 1,\ldots,r$$

have a unique solution modulo m if and only if for all $i,j = 1,\ldots,r$ we have

$$p_iq_j \equiv p_jq_i \bmod m. \tag{3.6}$$

PROOF. See Exercise 3.3. Q.E.D.

Since we are assuming $\gcd(a, a', (b+b')/2) = 1$, the congruences (3.4) satisfy the gcd condition of the above lemma, and the compatibility conditions (3.6) are easy to verify (see Exercise 3.4). The existence and uniqueness of the desired B follow immediately. Q.E.D.

We now give Dirichlet's definition of composition. Let $f(x, y) = ax^2 + bxy + cy^2$ and $g(x, y) = a'x^2 + b'xy + c'y^2$ be primitive positive definite forms of discriminant $D < 0$ which satisfy $\gcd(a, a', (b+b')/2) = 1$. Then the *Dirichlet composition* of $f(x, y)$ and $g(x, y)$ is the form

$$F(x,y) = aa'x^2 + Bxy + \frac{B^2 - D}{4aa'}y^2, \tag{3.7}$$

where B is the integer determined by Lemma 3.2. The basic properties of $F(x, y)$ are:

PROPOSITION 3.8. *Let $f(x, y)$ and $g(x, y)$ be as above. Then the Dirichlet composition $F(x, y)$ defined in* (3.7) *is a primitive positive definite form of discriminant D, and $F(x, y)$ is the direct composition of $f(x, y)$ and $g(x, y)$ in the sense of* (3.1).

PROOF. An easy calculation shows that $F(x, y)$ has discriminant D, and the form is consequently positive definite.

The next step is to prove that $F(x, y)$ is the composition of $f(x, y)$ and $g(x, y)$. We will sketch here the argument and leave the details to the reader. To begin, let $C = (B^2 - D)/4aa'$, so that $F(x, y) = aa'x^2 + Bxy + Cy^2$. Then, using the first two congruences of Lemma 3.2, it is easy to prove that $f(x, y)$ and $g(x, y)$ are properly equivalent to the forms $ax^2 + Bxy + a'Cy^2$ and $a'x^2 + Bxy + aCy^2$ respectively. However, for these last two forms one has the composition identity

$$(ax^2 + Bxy + a'Cy^2)(a'z^2 + Bzw + aCw^2) = aa'X^2 + BXY + CY^2,$$

where $X = xz - Cyw$ and $Y = axw + a'yz + Byw$. It follows easily that $F(x, y)$ is the composition of $f(x, y)$ and $g(x, y)$. With a little more effort, it can be checked that this is a direct composition in Gauss' sense (3.1). The details of these arguments are covered in Exercise 3.5.

It remains to show that $F(x, y)$ is primitive, i.e., that its coefficients are relatively prime. Suppose that some prime p divided all of the coefficients. This would imply that p divided all numbers represented by $F(x, y)$. Since $F(x, y)$ is the composition of $f(x, y)$ and $g(x, y)$, this implies that p divides all numbers of the form $f(x, y)g(z, w)$. But $f(x, y)$ and $g(x, y)$ are primitive, so that by Lemma 2.25, they represent numbers relatively prime to p. Hence $f(x, y)g(z, w)$ also represents a number relatively prime to p. This contradiction completes the proof of the proposition. Q.E.D.

While Dirichlet composition is not as general as direct composition (not all direct compositions satisfy $\gcd(a, a', (b+b')/2) = 1$), it is easier to use in practice since there is an explicit formula (3.7) for the composition. Notice also that the congruence conditions in Lemma 3.2 are similar to the ones (2.32) used by Legendre. This is no accident, for when $D = -4n$ and $\gcd(a, a') = 1$, Dirichlet's formula reduces exactly to the one given by Legendre (see Exercise 3.6).

We can now state our main result on composition:

THEOREM 3.9. *Let $D \equiv 0, 1 \bmod 4$ be negative, and let $C(D)$ be the set of classes of primitive positive definite forms of discriminant D. Then Dirichlet composition induces a well-defined binary operation on $C(D)$ which makes $C(D)$ into a finite Abelian group whose order is the class number $h(D)$.*

Furthermore, the identity element of $C(D)$ is the class containing the principal form

$$x^2 - \frac{D}{4}y^2 \qquad \text{if } D \equiv 0 \bmod 4$$

$$x^2 + xy + \frac{1-D}{4}y^2 \qquad \text{if } D \equiv 1 \bmod 4,$$

and the inverse of the class containing the for $ax^2+bxy+cy^2$ is the class containing $ax^2 - bxy + cy^2$.

REMARKS. Some terminology is in order here.

(i) The group $C(D)$ is called the *class group*, though we will sometimes refer to $C(D)$ as the *form class group* to distinguish it from the ideal class group to be defined later.
(ii) The principal form of discriminant D was introduced in §2. The class it lies in is called the *principal class*. When $D = -4n$, the principal form is $x^2 + ny^2$.
(iii) The form $ax^2 - bxy + cy^2$ is called the *opposite* of $ax^2 + bxy + cy^2$, so that the opposite form gives the inverse under Dirichlet composition.

PROOF. Let $f(x, y) = ax^2 + bxy + cy^2$ and $g(x, y)$ be forms of the given type. Using Lemmas 2.3 and 2.25, we can replace $g(x, y)$ by a properly equivalent form $a'x^2 + b'xy + c'y^2$ where $\gcd(a, a') = 1$. Then the Dirichlet composition of these forms is defined, which proves that Dirichlet composition is defined for any pair of classes in $C(D)$. To get a group structure out of this, we must then prove that:

(i) This operation is well-defined on the level of classes, and
(ii) The induced binary operation makes $C(D)$ into an Abelian group.

The proofs of (i) and (ii) can be done directly using the definition of Dirichlet composition (see Dirichlet [**36**, Supplement X] or Flath [**46**, §V.2]), but the argument is much easier using ideal class groups (to be studied in §7). We will therefore postpone this part of the proof until then. For now, we will assume that (i) and (ii) are true.

Let's next show that the principal class is the identity element of $C(D)$. To compose the principal form with $f(x, y) = ax^2 + bxy + cy^2$, first note that the gcd condition is clearly met, and thus the Dirichlet composition is defined. Then observe that $B = b$ satisfies the conditions of Lemma 3.2, so that by formula (3.7), the Dirichlet composition $F(x, y)$ reduces to the given form $f(x, y)$. This proves that the principal class is the identity.

Finally, given $f(x, y) = ax^2+bxy+cy^2$, its opposite is $f'(x, y) = ax^2-bxy+cy^2$. Since $\gcd(a, a, (b+(-b))/2) = a$ may be > 1, we can't apply Dirichlet composition directly. But if we use the proper equivalence $(x, y) \mapsto (-y, x)$, then we can replace $f'(x, y)$ by $g(x, y) = cx^2 + bxy + ay^2$. Since $\gcd(a, c, (b + b)/2) = \gcd(a, c, b) = 1$, we can apply Dirichlet's formulas to $f(x, y)$ and $g(x, y)$. One checks easily that $B = b$ satisfies the conditions of Lemma 3.2, so that the Dirichlet composition is $acx^2 + bxy + y^2$. We leave it to the reader to show that this form is properly equivalent to the principal form (see Exercise 3.7). This completes the proof of the theorem. Q.E.D.

We can now complete the discussion (begun in §2) of Legendre's theory of composition. To prevent confusion, we will distinguish between a *class* (all forms properly equivalent to a given form) and a *Lagrangian class* (all forms equivalent to a given one). In Theorem 3.9, we studied the composition of classes, while Legendre was concerned with the composition of Lagrangian classes. It is an easy exercise to show that the Lagrangian class of a form is the union of its class and the class of its opposite (see Exercise 3.8). Theorem 3.9 implies that a Lagrangian class is the union of a class and its inverse in the class group $C(D)$. Thus Legendre's "operation" is the multiple-valued operation that multiplication induces on the set $C(D)/\sim$, where $\sim$ is the equivalence relation that identifies $x \in C(D)$ with x^{-1} (see Exercise 3.9). In Legendre's example (2.33), which dealt with forms of discriminant -164, we will see shortly that $C(-164) \simeq \mathbb{Z}/8\mathbb{Z}$, and it is then an easy exercise to show that $C(-164)/\sim$ is isomorphic to the structure given in (2.34) (see Exercise 3.9).

Elements of order ≤ 2 in the class group $C(D)$ play a special role in composition and genus theory. The reduced forms that lie in such classes are easy to find:

LEMMA 3.10. *A reduced form $f(x, y) = ax^2 + bxy + cy^2$ of discriminant D has order ≤ 2 in the class group $C(D)$ if and only if $b = 0$, $a = b$ or $a = c$.*

PROOF. Let $f'(x, y)$ be the opposite of $f(x, y)$. By Theorem 3.9, the class of $f(x, y)$ has order ≤ 2 if and only if the forms $f(x, y)$ and $f'(x, y)$ are properly equivalent. There are two cases to consider:

$|b| < a < c$: Here, $f'(x, y)$ is also reduced, so that by Theorem 2.8, the two forms are properly equivalent $\iff b = 0$.

$a = b$ or $a = c$: In these cases, the proof of Theorem 2.8 shows that the two forms are always properly equivalent.

The lemma now follows immediately. Q.E.D.

For an example of how this works, consider Legendre's example from §2 of forms of discriminant -164. The reduced forms are listed in (2.33), and Lemma 3.10 shows that only $2x^2 + 2xy + 21y^2$ has order 2. Since the class number is 8, the structure theorem for finite Abelian groups shows that the class group $C(-164)$ is $\mathbb{Z}/8\mathbb{Z}$.

A surprising fact is that one doesn't need to list the reduced forms in order to determine the number of elements of order 2 in the class group:

PROPOSITION 3.11. *Let $D \equiv 0, 1 \bmod 4$ be negative, and let r be the number of odd primes dividing D. Define the number μ as follows: if $D \equiv 1 \bmod 4$, then $\mu = r$, and if $D \equiv 0 \bmod 4$, then $D = -4n$, where $n > 0$, and μ is determined by the following table:*

n	μ
$n \equiv 3 \bmod 4$	r
$n \equiv 1, 2 \bmod 4$	$r + 1$
$n \equiv 4 \bmod 8$	$r + 1$
$n \equiv 0 \bmod 8$	$r + 2$

Then the class group $C(D)$ has exactly $2^{\mu-1}$ elements of order ≤ 2.

PROOF. For simplicity, we will treat only the case $D = -4n$ where $n \equiv 1 \bmod 4$. Recall that a form of discriminant $-4n$ may be written as $ax^2+2bxy+cy^2$. The basic idea of the proof is to count the number of reduced forms that satisfy $2b = 0, a = 2b$ or $a = c$, for by Lemma 3.10, this gives the number of classes of order ≤ 2 in $C(-4n)$. Since n is odd, note that r is the number of prime divisors of n.

First, consider forms with $2b = 0$, i.e., the forms ax^2+cy^2, where $ac = n$. Since a and c must be relatively prime and positive, there are 2^r choices for a. To be reduced, we must also have $a < c$, so that we get 2^{r-1} reduced forms of this type.

Next consider forms with $a = 2b$ or $a = c$. Write $n = bk$, where b and k are relatively prime and $0 < b < k$. As above, there are 2^{r-1} such b's. Set $c = (b+k)/2$, and consider the form $2bx^2 + 2bxy + cy^2$. One computes that it has discriminant $-4n$, and since $n \equiv 1 \bmod 4$, its coefficients are relatively prime. We then get 2^{r-1} reduced forms as follows:

$2b < c$: Here, $2bx^2 + 2bxy + cy^2$ is a reduced form.

$2b > c$: Here, $2bx^2 + 2bxy + cy^2$ is properly equivalent to $cx^2 + 2(c-b)xy + cy^2$ via $(x, y) \mapsto (-y, x + y)$. Since $2b > c \Rightarrow 2(c-b) < c$, the latter is reduced.

The next step is to check that this process gives *all* reduced forms with $a = 2b$ or $a = c$. We leave this to the reader (see Exercise 3.10).

We thus have $2^{r-1} + 2^{r-1} = 2^r$ elements of order ≤ 2, which shows that $\mu = r + 1$ in this case. The remaining cases are similar and are left to the reader (see Exercise 3.10, Flath [**46**, §V.5], Gauss [**51**, §§257–258] or Mathews [**97**, pp. 171–173]). Q.E.D.

This is not the last we will see of the number μ, for it also plays an important role in genus theory.

B. Genus Theory

As in §2, we define two forms of discriminant D to be in the same genus if they represent the same values in $(\mathbb{Z}/D\mathbb{Z})^*$. Let's recall the classification of genera given in §2. Consider the subgroups $H \subset \ker(\chi) \subset (\mathbb{Z}/D\mathbb{Z})^*$, where H consists of the values represented by the principal form, and $\chi : (\mathbb{Z}/D\mathbb{Z})^* \longrightarrow \{\pm 1\}$ is defined by $\chi([p]) = (D/p)$ for $p \nmid D$ prime. Then the key result was Lemma 2.24, where we proved that the values represented in $(\mathbb{Z}/D\mathbb{Z})^*$ by a given form $f(x, y)$ are a coset of H in $\ker(\chi)$. This coset determines which genus $f(x, y)$ is in.

Our first step is to relate this theory to the class group $C(D)$. Since all forms in a given class represent the same numbers, sending the class to the coset of $H \subset \ker(\chi)$ it represents defines a map

$$\Phi : C(D) \longrightarrow \ker(\chi)/H. \tag{3.12}$$

Note that a given fiber $\Phi^{-1}(H')$, $H' \in \ker(\chi)/H$, consists of all classes in a given genus (this is what we called the *genus of* H' in Theorem 2.26), and the image of Φ may thus be identified with the set of genera. A crucial observation is that Φ is a group homomorphism:

LEMMA 3.13. *The map Φ which maps a class in $C(D)$ to the coset of values represented in $\ker(\chi)/H$ is a group homomorphism.*

PROOF. Let $f(x, y)$ and $g(x, y)$ be two forms of discriminant D taking values in the cosets H' and H'' respectively. We can assume that their Dirichlet composition $F(x, y)$ is defined, so that a product of values represented by $f(x, y)$ and $g(x, y)$ is represented by $F(x, y)$. Then $F(x, y)$ represents values in $H'H''$, which proves that $H'H''$ is the coset associated to the composition of $f(x, y)$ and $g(x, y)$. Thus Φ is a homomorphism. Q.E.D.

This lemma has the following consequences:

COROLLARY 3.14. *Let $D \equiv 0, 1 \bmod 4$ be negative. Then:*

(i) *All genera of forms of discriminant D consist of the same number of classes.*
(ii) *The number of genera of forms of discriminant D is a power of two.*

PROOF. The first statement follows since all fibers of a homomorphism have the same number of elements. To prove the second, first note that the subgroup H contains all squares in $(\mathbb{Z}/D\mathbb{Z})^*$. This is obvious because if $f(x, y)$ is the principal form, then $f(x, 0) = x^2$. Thus every element in $\ker(\chi)/H$ has order ≤ 2, and it follows from the structure theorem for finite Abelian groups that $\ker(\chi)/H \simeq \{\pm 1\}^m$ for some m. Thus the image of Φ, being a subgroup of $\ker(\chi)/H$, has order 2^k for some k. Since $\Phi(C(D))$ tells us the number of genera, we are done. Q.E.D.

Note also that $\Phi(C(D))$ gives a natural group structure on the set of genera, or as Gauss would say, one can define the composition of genera [**51**, §§246–247].

These elementary facts are nice, but they aren't the whole story. The real depth of the relation between composition and genera is indicated by the following theorem:

THEOREM 3.15. *Let $D \equiv 0, 1 \bmod 4$ be negative. Then:*

(i) *There are $2^{\mu - 1}$ genera of forms of discriminant D, where μ is the number defined in Proposition 3.11.*
(ii) *The principal genus (the genus containing the principal form) consists of the classes in $C(D)^2$, the subgroup of squares in the class group $C(D)$. Thus every form in the principal genus arises by duplication.*

PROOF. We first need to give a more efficient method for determining when two forms are in the same genus. The basic idea is to use certain *assigned characters*, which are defined as follows. Let $p_1, \ldots, p_r$ be the distinct odd primes dividing D. Then consider the functions:

$$
\begin{aligned}
\chi_i(a) &= \left(\frac{a}{p_i}\right) && \text{defined for } a \text{ prime to } p_i,\ i = 1, \ldots, r\\
\delta(a) &= (-1)^{(a-1)/2} && \text{defined for } a \text{ odd}\\
\epsilon(a) &= (-1)^{(a^2-1)/8} && \text{defined for } a \text{ odd.}
\end{aligned}
$$

Rather than using all of these functions, we assign only certain ones, depending on the discriminant D. When $D \equiv 1 \bmod 4$, we define $\chi_1, \ldots, \chi_r$ to be the *assigned characters*, and when $D \equiv 0 \bmod 4$, we write $D = -4n$, and then the *assigned characters* are defined by the following table:

n	assigned characters
$n \equiv 3 \bmod 4$	$\chi_1, \ldots, \chi_r$
$n \equiv 1 \bmod 4$	$\chi_1, \ldots, \chi_r, \delta$
$n \equiv 2 \bmod 8$	$\chi_1, \ldots, \chi_r, \delta\epsilon$
$n \equiv 6 \bmod 8$	$\chi_1, \ldots, \chi_r, \epsilon$
$n \equiv 4 \bmod 8$	$\chi_1, \ldots, \chi_r, \delta$
$n \equiv 0 \bmod 8$	$\chi_1, \ldots, \chi_r, \delta, \epsilon$

Note that the number of assigned characters is exactly the number μ given in Proposition 3.11. It is easy to see that the assigned characters give a homomorphism

$$\Psi : (\mathbb{Z}/D\mathbb{Z})^* \longrightarrow \{\pm 1\}^\mu. \tag{3.16}$$

The crucial property of Ψ is the following:

LEMMA 3.17. *The homomorphism $\Psi : (\mathbb{Z}/D\mathbb{Z})^* \longrightarrow \{\pm 1\}^\mu$ of (3.16) is surjective and its kernel is the subgroup H of values represented by the principal form. Thus Ψ induces an isomorphism*

$$(\mathbb{Z}/D\mathbb{Z})^*/H \xrightarrow{\sim} \{\pm 1\}^\mu.$$

PROOF. When $D \equiv 1 \bmod 4$, the proof is quite easy. First note that if p is an odd prime, then for any $m \geq 1$, the Legendre symbol (a/p) induces a surjective homomorphism

$$(\cdot/p) : (\mathbb{Z}/p^m\mathbb{Z})^* \longrightarrow \{\pm 1\} \tag{3.18}$$

whose kernel is exactly the subgroup of squares of $(\mathbb{Z}/p^m\mathbb{Z})^*$ (see Exercise 3.11). Now let $D = -\prod_{i=1}^{\mu} p_i^{m_i}$ be the prime factorization of D. The Chinese Remainder Theorem tells us that

$$(\mathbb{Z}/D\mathbb{Z})^* \xrightarrow{\sim} \prod_{i=1}^{\mu} (\mathbb{Z}/p_i^{m_i}\mathbb{Z})^*,$$

so that the map Ψ can be interpreted as the map

$$\prod_{i=1}^{\mu} (\mathbb{Z}/p_i^{m_i}\mathbb{Z})^* \longrightarrow \{\pm 1\}^\mu$$

given by $([a_1], \ldots, [a_\mu]) \mapsto ((a_1/p_1), \ldots, (a_\mu/p_\mu))$. By the analysis of (3.18), it follows that Ψ is surjective and its kernel is exactly the subgroup of squares of $(\mathbb{Z}/D\mathbb{Z})^*$. By part (c) of Exercise 2.17, this equals the subgroup H of values represented by the principal form $x^2 + xy + ((1-D)/4)y^2$, and we are done.

The proof is more complicated when $D = -4n$, mainly because the subgroup H represented by $x^2 + ny^2$ may be slightly larger than the subgroup of squares. However, the above argument using the Chinese Remainder Theorem can be adapted to this case. The odd primes dividing n are no problem, but 2 causes considerable difficulty (see Exercise 3.11 for the details). Q.E.D.

We can now prove Theorem 3.15. To prove (i), recall that $\ker(\chi)$ has index 2 in $(\mathbb{Z}/D\mathbb{Z})^*$. By Lemma 3.17, it follows that $\ker(\chi)/H$ has order $2^{\mu-1}$. We know that the number of genera is the order of $\Phi(C(D)) \subset \ker(\chi)/H$, so that it suffices to show $\Phi(C(D)) = \ker(\chi)/H$. Since Φ maps a class to the coset of values it represents, we need to show that every congruence class in $\ker(\chi)$ contains a number represented by a form of discriminant D. This is easy: Dirichlet's theorem on primes in arithmetic progressions tells us that any class in $\ker(\chi)$ contains an odd prime p. But $[p] \in \ker(\chi)$ means that $\chi([p]) = (D/p) = 1$, so that by Lemma 2.5, p is represented by a form of discriminant D, and (i) is proved.

To prove (ii), let C denote the class group $C(D)$. Since $\Phi : C \to \ker(\chi)/H \simeq \{\pm 1\}^{\mu-1}$ is a homomorphism, it follows that $C^2 \subset \ker(\Phi)$, and we get an induced map

$$C/C^2 \longrightarrow \{\pm 1\}^{\mu-1}. \tag{3.19}$$

We compute the order of C/C^2 as follows. The squaring map from C to itself gives a short exact sequence

$$0 \to C_0 \to C \to C^2 \to 0$$

where C_0 is the subgroup of C of elements of order ≤ 2. It follows that the index $[C : C^2]$ equals the order of C_0, which is $2^{\mu-1}$ by Proposition 3.11.

Thus, in the map given in (3.19), both the domain and the range have the same order. But from (i) we know that the map is surjective, so that it must be an isomorphism. Hence C^2 is exactly the kernel of the map Φ. Since $\ker(\Phi)$ consists of the classes in the principal genus, the theorem is proved. Q.E.D.

We have now proved the main theorems of genus theory for primitive positive definite forms. These results are due to Gauss and appear in the fifth section of *Disquisitiones Arithmeticae* [**51**, §§229–287]. Gauss' treatment is more general than ours, for he considers both the definite and indefinite forms, and in particular, he shows that Proposition 3.11 and Theorem 3.15 are true for *any* nonsquare discriminant, positive or negative. His proofs are quite difficult, and at the end of this long series of arguments, Gauss makes the following comment about genus theory [**51**, §287]:

> these theorems are among the most beautiful in the theory of binary forms, especially because, despite their extreme simplicity, they are so profound that a rigorous demonstration requires the help of many other investigations.

Besides these theorems, there is another component to Gauss' genus theory not mentioned so far: Gauss' second proof of quadratic reciprocity [**51**, §262], which uses the genus theory developed above. We will not discuss Gauss' proof since it uses forms of positive discriminant, though the main ideas of the proof are outlined in Exercises 3.12 and 3.13. Many people regard this as the deepest of Gauss' many proofs of quadratic reciprocity.

Gauss' approach to genus theory is somewhat different from ours. In *Disquisitiones*, genera are defined in terms of the *assigned characters* introduced in the proof of Theorem 3.15. Given a form $f(x, y)$ of discriminant D, let $f(x, y)$ represent a number a relatively prime to D. If the μ assigned characters are evaluated at a, then Gauss calls the resulting μ-tuple the *complete character* of $f(x, y)$, and he defines two forms of discriminant D to be in the same genus if they have the same complete character [**51**, §231]. The following lemma shows that this is equivalent to our previous definition of genus:

LEMMA 3.20. *The complete character depends only on the form $f(x, y)$, and two forms of discriminant D lie in the same genus (as defined in §2) if and only if they have the same complete character.*

PROOF. Suppose that $f(x,y)$ represents a, where a is relatively prime to D. Then Gauss' complete character is nothing other than $\Psi([a])$, where Ψ is the map defined in (3.16). By Lemma 2.24, the possible a's lie in a coset $H'$of H in $(\mathbb{Z}/D\mathbb{Z})^*$, and this coset determines the genus of $f(x,y)$. Using Lemma 3.17, it follows that the complete character is uniquely determined by H', and Lemma 3.20 is proved. Q.E.D.

We should mention that Gauss' use of the word "character" is where the modern term "group character" comes from. Also, it is interesting to note that Gauss never mentions the connection between his characters and Lagrange's implicit genus theory. While Gauss' characters make it easy to decide when two forms belong to the same genus (see Exercise 3.14 for an example), they are not very intuitive. Unfortunately, most of Gauss' successors followed his presentation of genus theory, so that readers were presented with long lists of characters and no motivation whatsoever. The simple idea of grouping forms according to the congruence classes they represent was usually not mentioned. This happens in Dirichlet [**36**, pp. 313–316] and in Mathews [**97**, pp. 132–136], although Smith [**122**, pp. 202–207] does discuss congruence classes.

So far we have discussed two ways to formulate genera, Lagrange's and Gauss'. There are many other ways to state the definition, but before we can discuss them, we need some terminology. We say that two forms $f(x,y)$ and $g(x,y)$ are *equivalent over a ring* R if there is a matrix $\left(\begin{smallmatrix} p & q \\ r & s \end{smallmatrix}\right) \in \mathrm{GL}(2,R)$ such that $f(x,y) = g(px+qy, rx+sy)$. If $R = \mathbb{Z}/m\mathbb{Z}$, we say that $f(x,y)$ and $g(x,y)$ are *equivalent modulo* m. We then have the following theorem:

THEOREM 3.21. *Let $f(x,y)$ and $g(x,y)$ be primitive forms of discriminant $D \neq 0$, positive definite if $D < 0$. Then the following statements are equivalent:*

(i) *$f(x,y)$ and $g(x,y)$ are in the same genus, i.e., they represent the same values in $(\mathbb{Z}/D\mathbb{Z})^*$.*
(ii) *$f(x,y)$ and $g(x,y)$ represent the same values in $(\mathbb{Z}/m\mathbb{Z})^*$ for all nonzero integers m.*
(iii) *$f(x,y)$ and $g(x,y)$ are equivalent modulo m for all nonzero integers m.*
(iv) *$f(x,y)$ and $g(x,y)$ are equivalent over the ring of p-adic integers $\mathbb{Z}_p$ for all primes p.*
(v) *$f(x,y)$ and $g(x,y)$ are equivalent over $\mathbb{Q}$ via a matrix in $\mathrm{GL}(2,\mathbb{Q})$ whose entries have denominators prime to $2D$.*
(vi) *$f(x,y)$ and $g(x,y)$ are equivalent over $\mathbb{Q}$ without essential denominator, i.e., given any nonzero m, a matrix in $\mathrm{GL}(2,\mathbb{Q})$ can be found which takes one form to the other and whose entries have denominators prime to m.*

PROOF. It is easy to prove (vi) $\Rightarrow$ (iii) $\Rightarrow$ (ii) $\Rightarrow$ (i) and (vi) $\Rightarrow$ (v) $\Rightarrow$ (i) (see Exercise 3.15), and (iii) $\Leftrightarrow$ (iv) is a standard argument using the compactness of $\mathbb{Z}_p$ (see Borevich and Shafarevich [**11**, p. 41] for an analogous case). A proof of (i) $\Rightarrow$ (iii) appears in Hua [**73**, §12.5, Exercise 4], and (i) $\Rightarrow$ (iv) is in Jones [**79**, pp. 103–104]. Finally, the implication (iv) $\Rightarrow$ (vi) uses the Hasse principle for the equivalence of forms over $\mathbb{Q}$ and may be found in Jones [**79**, Theorem 40] or Siegel [**116**]. Q.E.D.

Some modern texts give yet a different definition, saying that two forms are in the same genus if and only if they are equivalent over $\mathbb{Q}$ (see, for example, Borevich and Shafarevich [**11**, p. 241]). This characterization doesn't hold in general $x^2 + 18y^2$ and $2x^2 + 9y^2$ are rationally equivalent but belong to different genera—see Exercise 3.16), but it does work for *field discriminants*, which means that $D \equiv 1 \bmod 4$, D squarefree, or $D = 4k$, $k \not\equiv 1 \bmod 4$, k squarefree (see Exercise 3.17—we will study such discriminants in more detail in §5). According to Dickson [**34**, Vol. III, pp. 216 and 236], Eisenstein suggested in 1852 that genera could be defined using rational equivalence, and only later, in 1867, did Smith point out that extra assumptions are needed on the denominators.

C. $p = x^2 + ny^2$ and Euler's Convenient Numbers

Our discussion of genus theory has distracted us from our problem of determining when a prime p can be written as $x^2 + ny^2$. Recall from Corollary 2.27 that genus theory gives us congruence conditions for p to be represented by a reduced form in the principal genus. The nicest case is when every genus of discriminant $-4n$ consists of a single class, for then we get congruence conditions that characterize $p = x^2 + ny^2$ (this is what made the examples in (2.28) work). Let's see if the genus theory developed in this section can shed any light on this special case. We have the following result:

THEOREM 3.22. *Let n be a positive integer. Then the following statements are equivalent:*

(i) *Every genus of forms of discriminant $-4n$ consists of a single class.*
(ii) *If $ax^2 + bxy + cy^2$ is a reduced form of discriminant $-4n$, then either $b = 0$, $a = b$ or $a = c$.*
(iii) *Two forms of discriminant $-4n$ are equivalent if and only if they are properly equivalent.*
(iv) *The class group $C(-4n)$ is isomorphic to $(\mathbb{Z}/2\mathbb{Z})^m$ for some integer m.*
(v) *The class number $h(-4n)$ equals $2^{\mu-1}$, where μ is as in Proposition 3.11.*

PROOF. We will prove (i) $\Rightarrow$ (ii) $\Rightarrow$ (iii) $\Rightarrow$ (iv) $\Rightarrow$ (v) $\Rightarrow$ (i). Let C denote the class group $C(-4n)$.

Since the principal genus is C^2 by Theorem 3.15, (i) implies that $C^2 = \{1\}$, so that every element of C has order ≤ 2. Then Lemma 3.10 shows that (i) $\Rightarrow$ (ii).

Next assume (ii), and suppose that two forms of discriminant $-4n$ are equivalent. By Exercise 3.8, we know that one is properly equivalent to the other or its opposite. We may assume that the forms are reduced, so that by assumption $b = 0$, $a = b$ or $a = c$. The proof of Theorem 2.8 shows that forms of this type are always properly equivalent to their opposites, so that the forms are properly equivalent. This proves (ii) $\Rightarrow$ (iii).

Recall that any form is equivalent to its opposite via $(x, y) \mapsto (x, -y)$. Thus (iii) implies that any form and its opposite lie in the same class in C. Since the opposite gives the inverse in C by Theorem 3.9, we see that every class is its own inverse. The structure theorem for finite Abelian groups shows that the only groups with this property are $(\mathbb{Z}/2\mathbb{Z})^m$, and (iii) $\Rightarrow$ (iv) is proved.

Next, Theorem 3.15 implies that the number of genera is $[C : C^2] = 2^{\mu-1}$, so that

$$h(-4n) = |C| = [C : C^2]|C^2| = 2^{\mu-1}|C^2|. \tag{3.23}$$

If (iv) holds, then $C^2 = \{1\}$, and then (v) follows immediately from (3.23). Finally, given (v), (3.23) implies that $C^2 = \{1\}$, so that by Theorem 3.15, the principal genus consists of a single class. Since every genus consists of the same number of classes, (i) follows, and the theorem is proved. Q.E.D.

Notice how this theorem runs the full gamut of what we've done so far: the conditions of Theorem 3.22 involve genera, reduced forms, the class number, the structure of the class group and the relation between equivalence and proper equivalence. For computational purposes, the last condition (v) is especially useful, for it only requires knowing the class number. This makes it much easier to verify that the examples in (2.28) have only one class per genus.

Near the end of the fifth section of *Disquisitiones*, Gauss lists 65 discriminants that satisfy this theorem [**51**, §303]. Grouped according to class number, they are:

$h(-4n)$	n's with one class per genus
1	$1, 2, 3, 4, 7$
2	$5, 6, 8, 9, 10, 12, 13, 15, 16, 18, 22, 25, 28, 37, 58$
4	$21, 24, 30, 33, 40, 42, 45, 48, 57, 60, 70, 72, 78, 85, 88, 93, 102, 112$ $130, 133, 177, 190, 232, 253$
8	$105, 120, 165, 168, 210, 240, 273, 280, 312, 330, 345, 357, 385$ $408, 462, 520, 760$
16	$840, 1320, 1365, 1848$

Gauss was interested in these 65 n's not for their relation to the question of when $p = x^2 + ny^2$, but rather because they had been discovered earlier by Euler in a different context. Euler called a number n a *convenient number* (numerus idoneus) if it satisfies the following criterion:

> Let m be an odd number relatively prime to n which is properly represented by $x^2 + ny^2$. If the equation $m = x^2 + ny^2$ has only one solution with $x, y \geq 0$, then m is a prime number.

Euler was interested in convenient numbers because they helped him find large primes. For example, working with $n = 1848$, he was able to show that

$$18{,}518{,}809 = 197^2 + 1848 \cdot 100^2$$

is prime, a large one for Euler's time. Convenient numbers are a fascinating topic, and the reader should consult Frei [**48**] or Weil [**135**, pp. 219–226] for a fuller discussion. We will confine ourselves to the following remarkable observation of Gauss:

PROPOSITION 3.24. *A positive integer n is a convenient number if and only if for forms of discriminant $-4n$, every genus consists of a single class.*

PROOF. We begin with a lemma:

LEMMA 3.25. *Let m be a positive odd number relatively prime to $n > 1$. Then the number of ways that m is properly represented by a reduced form of discriminant $-4n$ is*

$$2\prod_{p|m}\left(1 + \left(\frac{-n}{p}\right)\right).$$

PROOF. See Exercise 3.20 or Landau [**88**, Vol. 1, p. 144]. Q.E.D.

This classical lemma belongs to an area of quadratic forms that we have ignored, namely the study of the *number* of representations of a number by a form. To see what this has to do with genus theory, note that two forms representing m must lie in the same genus, for the values they represent in $(\mathbb{Z}/4n\mathbb{Z})^*$ are not disjoint. We thus get the following corollary of Lemma 3.25:

COROLLARY 3.26. *Let m be properly represented by a primitive positive definite form $f(x,y)$ of discriminant $-4n$, $n > 1$, and assume that m is odd and relatively prime to n. If r denotes the number of prime divisors of m, then m is properly represented in exactly 2^{r+1} ways by a reduced form in the genus of $f(x,y)$.* Q.E.D.

Now we can prove the proposition. First, assume that there is only one class per genus. If m is properly represented by x^2+ny^2 and $m = x^2+ny^2$ has a unique solution when $x, y \geq 0$, then we need to prove that m is prime. The above corollary shows that m is properly represented by $x^2 + ny^2$ in 2^{r+1} ways since $x^2 + ny^2$ is the only reduced form in its genus. At least 2^{r-1} of these representations satisfy $x, y \geq 0$, and then our assumption on m implies that $r = 1$, i.e., m is a prime power p^a. If $a \geq 2$, then Lemma 3.25 shows that p^{a-2} also has a proper representation, and it follows easily that m has at least two representations in nonnegative integers. This contradiction proves that m is prime, and hence n is a convenient number.

Conversely, assume that n is convenient. Let $f(x,y)$ be a form of discriminant $-4n$, and let $g(x,y)$ be the composition of $f(x,y)$ with itself. We can assume that $g(x,y)$ is reduced, and it suffices to show that $g(x,y) = x^2 + ny^2$ (for then every element in the class group has order ≤ 2, which by Theorem 3.22 implies that there is one class per genus).

Assume that $g(x,y) \neq x^2 + ny^2$, and let p and q be distinct odd primes not dividing n which are represented by $f(x,y)$. (In §9 we will prove that $f(x,y)$ represents infinitely many primes.) Then $g(x,y)$ represents pq, and formula (2.31) shows that $x^2 + ny^2$ does too. By Corollary 3.26, pq has only 8 proper representations by reduced forms of discriminant $-4n$. At least one comes from $g(x,y)$, leaving at most 7 for $x^2 + ny^2$. It follows that pq is uniquely represented by $x^2 + ny^2$ when we restrict to nonnegative integers. This contradicts our assumption that n is convenient. Q.E.D.

Gauss never states Proposition 3.24 formally, but it is implicit in the methods he discusses for factoring large numbers [**51**, §§329–334].

In §2 we asked how many such n's there were. Gauss suggests [**51**, §303] that the 65 given by Euler are the only ones. In 1934 Chowla [**23**] proved that the number of such n's is finite, and by 1973 it was known that Euler's list is complete except for possibly one more n (see Weinberger [**137**]). Whether or not this last n actually exists is still an open question.

From our point of view, the upshot is that there are only finitely many theorems like (2.28) where $p = x^2 + ny^2$ is characterized by simple congruences modulo $4n$. Thus genus theory cannot solve our basic question for all n. In some cases, such as $D = -108$, it's completely useless (all three reduced forms $x^2 + 27y^2$ and $4x^2 \pm 2xy + 7y^2$ lie in the same genus), and even when it's a partial help, such as $D = -56$, we're still stuck (we can separate $x^2 + 14y^2$ and $2x^2 + 7y^2$ from $3x^2 \pm 2xy + 5y^2$, but we can't distinguish between the first two). And notice that by part (iii) of Theorem 3.21, forms in the same genus are equivalent modulo m for *all* $m \neq 0$, so that no matter how m is chosen, there are no congruences

$p \equiv a, b, c, \ldots \bmod m$ which can separate forms in the same genus. Something new is needed. In 1833, Dirichlet described the situation as follows [**35**, Vol. I, p. 201]:

> there lies in the mentioned [genus] theory an incompleteness, in that it certainly shows that a prime number, as soon as it is contained in a linear form [congruence class], necessarily must assume one of the corresponding quadratic forms, only without giving any *a priori* method for deciding which quadratic form it will be. ... It becomes clear that the characteristic property of a single quadratic form belonging to a group [genus] cannot be expressed through the prime numbers in the corresponding linear forms, but necessarily must be expressed by another theory not depending on the elements at hand.

As we already know from Euler's conjectures concerning $x^2 + 27y^2$ and $x^2 + 64y^2$ (see (1.22) and (1.23)), the new theory we're seeking involves residues of higher powers. Gauss rediscovered Euler's conjectures in 1805, and he proved them in the course of his work on cubic and biquadratic reciprocity. In §4 we will give careful statements of these reciprocity theorems and show how they can be used to prove Euler's conjectures.

D. Disquisitiones Arithmeticae

Gauss' *Disquisitiones Arithmeticae* covers a wide range of topics in number theory, including congruences, quadratic reciprocity, quadratic forms (in two and three variables), and the cyclotomic fields $\mathbb{Q}(\zeta_n)$, $\zeta_n = e^{2\pi i/n}$. There are several excellent accounts of what's in *Disquisitiones*, notably Bühler [**17**, Chapter 3], Bachmann [**52**, Vol. X.2.1, pp. 8–40] and Rieger [**105**], and translations into English and German are available (see item [**51**] in the references). Rather than try to survey the whole book, we will instead make some comments on Gauss' treatment of quadratic reciprocity and quadratic forms, for in each case he does things slightly differently from the theory presented in §§2 and 3.

Disquisitiones contains the first published (valid) proof of the law of quadratic reciprocity. One surprise is that Gauss never uses the term "quadratic reciprocity." Instead, Gauss uses the phrase "fundamental theorem," which he explains as follows [**51**, §131]:

> Since almost everything that can be said about quadratic residues depends on this theorem, the term *fundamental theorem* which we will use from now on should be acceptable.

In the more informal setting of his mathematical diary, Gauss uses the term "golden theorem" to describe his high regard for quadratic reciprocity [**52**, Vol. X.1, entries 16, 23 and 30 on pp. 496–501] (see Gray [**58**] for an English translation). Likewise absent from *Disquisitiones* is the Legendre symbol, for Gauss uses the notation aRb or aNb to indicate whether or not a was a quadratic residue modulo b [**51**, §131]. (The Legendre symbol does appear in some of his handwritten notes—see [**52**, Vol. X.1, p. 53]—but this doesn't happen very often.)

One reason why Gauss ignored Legendre's terminology is that Gauss discovered quadratic reciprocity independently of his predecessors. In a marginal note in his copy of *Disquisitiones*, Gauss states that "we discovered the fundamental theorem by induction in March 1795. We found our first proof, the one contained in this section, April 1796" [**51**, p. 468, English editions] or [**52**, Vol. I, p. 476]. In 1795 Gauss was still a student at the Collegium Carolinum in Brunswick, and only later, while at Göttingen, did he discover the earlier work of Euler and Legendre on reciprocity.

Gauss' proof from April 1796 appears in §§135–144 of *Disquisitiones*. The theorem is stated in two forms: the usual version of quadratic reciprocity appears in [**51**, §131], and the more general version that holds for the Jacobi symbol (which we used in the proof of Lemma 1.14) is given in [**51**, §133]. The proof uses complete induction on the prime p, and there are many cases to consider, some of which use reciprocity for the Jacobi symbol (which would hold for numbers smaller than p). As Gauss wrote in 1808, the proof "proceeds by laborious steps and is burdened by detailed calculations" [**52**, Vol. II, p. 4]. In 1857, Dirichlet used the Jacobi symbol to simplify the proof and reduce the number of cases to just two [**35**, Vol. II, pp. 121–138]. It is interesting to note that what Gauss proves in *Disquisitiones* is actually a bit more general than the usual statement of quadratic reciprocity for the Jacobi symbol (see Exercise 3.24). Thus, when Jacobi introduced the Jacobi symbol in 1837 [**77**, Vol. VI, p. 262], he was simply giving a nicer but less general formulation of what was already in *Disquisitiones.*

As we mentioned in our discussion of genus theory, *Disquisitiones* also contains a second proof of reciprocity that is quite different in nature. The first proof is awkward but elementary, while the second uses Gauss' genus theory and is much more sophisticated.

Gauss' treatment of quadratic forms occupies the fifth (and longest) section of *Disquisitiones.* It is not easy reading, for many of the arguments are very complicated. Fortunately, there are more modern texts that cover pretty much the same material (in particular, see either Flath [**46**] or Mathews [**97**]). Gauss starts with the case of positive definite forms, and the theory he develops is similar to the first part of §2. Then, in [**51**, §182], he gives some applications to number theory, which are introduced as follows:

> Let us now consider certain particular cases both because of their remarkable elegance and because of the painstaking work done on them by Euler, who endowed them with an almost classical distinction.

As might be expected, Gauss first proves Fermat's three theorems (1.1), and then he proves Euler's conjecture for $p = x^2 + 5y^2$ using Lagrange's implicit genus theory (his proof is similar to what we did in (2.19), (2.20) and (2.22)). Interestingly enough, Gauss never mentions the relation between this example and genus theory. In contrast to Lagrange and Legendre, Gauss works out few examples. His one comment is that "the reader can derive this proposition [concerning $x^2 + 5y^2$] and an infinite number of other particular ones from the preceding and the following discussions" [**51**, §182].

Gauss always assumed that the middle coefficient was even, so that his forms were written $f(x, y) = ax^2 + 2bxy + cy^2$. He used the ordered triple (a, b, c) to denote $f(x, y)$ [**51**, §153], and he defined its *determinant* to be $b^2 - ac$ [**51**, §154]. Note that the discriminant of $ax^2 + 2bxy + cy^2$ is just 4 times Gauss' determinant.

Gauss did not assume that the coefficients of his forms were relatively prime, and he organized forms into *orders* according to the common divisors of the coefficients. More precisely, the forms $ax^2 + 2bxy + cy^2$ and $a'x^2 + 2b'xy + c'y^2$ are in the same *order* provided that $\gcd(a,b,c) = \gcd(a',b',c')$ and $\gcd(a,2b,c) = \gcd(a',2b',c')$ [**51**, §226]. To get a better idea of how this works, consider a primitive quadratic form $ax^2+bxy+cy^2$. Here, a, b and c are relatively prime integers, and b may be even or odd. We can fit this form into Gauss' scheme as follows:

b even: Then $b = 2b'$, and $ax^2+2b'xy+cy^2$ satisfies $\gcd(a,b',c) = \gcd(a,2b',c) = 1$. Gauss called forms in this order *properly primitive.*

b odd: Then $2ax^2+2bxy+2cy^2$ satisfies $\gcd(2a,b,2c) = 1$ and $\gcd(2a,2b,2c) = 2$. He called forms in this order *improperly primitive.*

All primitive forms are present, though the ones with b odd appear in disguised form. This doesn't affect the class number but does cause problems with composition.

Gauss' classification of forms thus consists of orders, which are made up of genera, which are in turn made up of classes. This is reminiscent of the Linnean classification in biology, where the categories are class, order, family, genus and species. Gauss' terms all appear on Linneaus' list, and it is thus likely that this is where Gauss got his terminology. Since our current term "equivalence class" comes from Gauss' example of *classes* of properly *equivalent* forms, we see that there is an unexpected link between modern set theory and eighteenth-century biology.

Finally, let's make one comment about composition. Gauss' theory of composition has always been one of the more difficult parts of *Disquisitiones* to read, and part of the reason is the complexity of Gauss' presentation. For example, the proof that composition is associative involves checking that 28 equations are satisfied [**51**, §240]. But a multiplicity of equations is not the only difficulty here—there is also an interesting conceptual issue. Namely, in order to define the class group, notice that Gauss has to put the structure of an abstract Abelian group on a set of equivalence classes. Considering that we're talking about the year 1801, this is an amazing level of abstraction. But then, *Disquisitiones* is an amazing book.

E. Exercises

3.1. Assume that $F(x,y) = Ax^2+Bxy+Cy^2$ is the composition of the two forms $f(x,y) = ax^2+bxy+cy^2$ and $g(x,y) = a'x^2+b'xy+c'y^2$ via

$$\begin{aligned} f(x,y)g(z,w) = F(&a_1xz+b_1xw+c_1yz+d_1yw, a_2xz\\ &+b_2xw+c_2yz+d_2yw), \end{aligned}$$

and suppose that all three forms have discriminant $D \neq 0$. The goal of this exercise is to prove Gauss' formulas (3.1).

(a) By specializing the variables x, y, z and w, prove that

$$\begin{aligned} aa' &= Aa_1^2+Ba_1a_2+Ca_2^2\\ ac' &= Ab_1^2+Bb_1b_2+Cb_2^2\\ ab' &= 2Aa_1b_1+B(a_1b_2+a_2b_1)+2Ca_2b_2. \end{aligned}$$

Hint: for the first one, try $x = z = 1$ and $y = w = 0$.

(b) Prove that $a = \pm(a_1b_2 - a_2b_1)$. Hint: prove that

$$a^2(b'^2-4a'c') = (a_1b_2-a_2b_1)^2(B^2-4AC).$$

(c) Prove that $a' = \pm(a_1c_2 - a_2c_1)$.

3.2. Show that the compositions given in (2.30) and (2.31) are not direct compositions.

3.3. Prove Lemma 3.5. Hint: there are $a, a_1, \ldots, a_r$ with $am + \Sigma_{i=1}^r a_i p_i = 1$.

3.4. Verify that the congruences (3.4) satisfy the compatibility conditions stated in Lemma 3.5.

3.5. Assume that $f(x, y) = ax^2 + bxy + cy^2$, $g(x, y) = a'x^2 + b'xy + c'y^2$ and B are as in Lemma 3.2. We want to show that $aa'x^2 + Bxy + Cy^2$, $C = (B^2 - D)/4aa'$, is the direct composition of $f(x, y)$ and $g(x, y)$.

(a) Show that $f(x, y)$ and $g(x, y)$ are properly equivalent to $ax^2+Bxy+a'Cy^2$ and $a'x^2 + Bxy + aCy^2$ respectively. Hint: use $B \equiv b \bmod 2a$ for $f(x, y)$.

(b) Let $X = xz - Cyw$ and $Y = axw + a'yz + Byw$. Then show that

$$\begin{aligned}(ax^2 + Bxy + a'Cy^2)(a'z^2 + Bzw + aCw^2) \\ = aa'X^2 + BXY + CY^2.\end{aligned}$$

Furthermore, show that this is a direct composition in the sense of (3.1). Hint: first show that

$$\begin{aligned}(ax + (B + \sqrt{D})y/2)(a'z + (B + \sqrt{D})w/2) \\ = aa'X + (B + \sqrt{D})Y/2.\end{aligned}$$

(c) Suppose that a form $G(x, y)$ is the direct composition of forms $h(x, y)$ and $k(x, y)$. If $\tilde{h}(x, y)$ is properly equivalent to $h(x, y)$, then show that $G(x, y)$ is also the direct composition of $\tilde{h}(x, y)$ and $k(x, y)$. Hint: for $\tilde{h}(1, 0)$, start with the formula for $\pm\tilde{h}(1, 0)$ given by (3.1).

(d) Use parts (a)–(c) to show that Dirichlet composition is indeed a direct composition.

3.6. This problem studies the relation between Legendre's and Dirichlet's formulas for composition.

(a) Suppose that $f(x, y) = ax^2 + 2bxy + cy^2$ and $g(x, y) = a'x^2 + 2b'xy + c'y^2$ have the same discriminant and satisfy $\gcd(a, a') = 1$. Show that the Dirichlet composition of these forms is the one given by Legendre's formula with both signs + in (2.32).

(b) In Exercise 2.26, we saw that $14x^2 + 10xy + 21y^2$ and $9x^2 + 2xy + 30y^2$ can be composed to obtain $126x^2 \pm 74xy + 13y^2$ and $126x^2 \pm 38xy + 5y^2$. Which one of these four is the direct composition of the original two forms?

3.7. Show that $acx^2 + bxy + y^2$ is properly equivalent to the principal form.

3.8. For us, a *class* consists of all forms properly equivalent to a given form. Let a *Lagrangian class* (this terminology is due to Weil [**135**, p. 319]) consist of all forms equivalent (properly or improperly) to a given form.

(a) Prove that the Lagrangian class of a form is the union of the class of the form and the class of its opposite.

(b) Show that the following statements are equivalent:

(i) The Lagrangian class of $f(x, y)$ equals the class of $f(x, y)$.
(ii) $f(x, y)$ is properly equivalent to its opposite.
(iii) $f(x, y)$ is properly and improperly equivalent to itself.
(iv) The class of $f(x, y)$ has order ≤ 2 in the class group.

3.9. In this problem we will describe the "almost" group structure given by Legendre's theory of composition. Let G be an Abelian group and let $\sim$ be the equivalence relation which identifies a^{-1} and a for all $a \in G$.

(a) Show that multiplication on G induces an operation on $G/\sim$ which takes either one or two values. Furthermore, if a, $b \in G$ and $[a]$, $[b]$ are their classes in $G/\sim$, then show that $[a] \cdot [b]$ takes on only one value if and only if a or b has order ≤ 2 in G.

(b) If G is cyclic of order 8, show that $G/\sim$ is isomorphic (in the obvious sense) to the structure given by (2.33) and (2.34).

(c) If $C(D)$ is the class group of forms of discriminant D, show that $C(D)/\sim$ can be naturally identified with the set of Lagrangian classes of forms of discriminant D (see Exercise 3.8).

3.10. Complete the proof of Proposition 3.11 for the case $D = -4n$, $n \equiv 1 \bmod 4$, and prove all of the remaining cases.

3.11. This exercise is concerned with the proof of Lemma 3.17.

(a) Prove that the map (3.18) is surjective and its kernel is the subgroup of squares.

(b) We next want to prove the lemma when $D = -4n$, $n > 0$. Write $n = 2^a m$ where m is odd, so that we have an isomorphism

$$\varphi : (\mathbb{Z}/D\mathbb{Z})^* \xrightarrow{\sim} (\mathbb{Z}/2^{a+2}\mathbb{Z})^* \times (\mathbb{Z}/m\mathbb{Z})^*.$$

Let $H \subset (\mathbb{Z}/D\mathbb{Z})^*$ denote the subgroup of values represented by $x^2 + ny^2$.

(i) For $H_1 = \{u \in (\mathbb{Z}/2^{a+2}\mathbb{Z})^* : (u, 1) \in \varphi(H)\}$, show that H_1 is the projection of $\varphi(H) \subset (\mathbb{Z}/2^{a+2}\mathbb{Z})^* \times (\mathbb{Z}/m\mathbb{Z})^*$ onto the first factor.

(ii) Show that $\varphi(H) = H_1 \times (\mathbb{Z}/m\mathbb{Z})^{*2}$, so that $H \simeq H_1 \times (\mathbb{Z}/m\mathbb{Z})^{*2}$.

(iii) When $a \geq 4$, show that $H_1 = (\mathbb{Z}/2^{a+2}\mathbb{Z})^{*2}$, where H_1 is as in (i).

(iv) Prove Lemma 3.17 when $D \equiv 0 \bmod 4$. Hint: treat the cases $a = 0$, 1, 2, 3 and ≥ 4 separately. See the description of $(\mathbb{Z}/2^{a+2}\mathbb{Z})^*$ given in Ireland and Rosen [**75**, §4.1].

3.12. In Exercises 3.12 and 3.13 we will sketch Gauss' second proof of quadratic reciprocity. There are two parts to the proof: first, one shows, without using quadratic reciprocity, that for any nonsquare discriminant D,

(*) the number of genera of forms of discriminant D is $\leq 2^{\mu-1}$,

where μ is defined in Proposition 3.11, and second, one shows that (*) implies quadratic reciprocity. This exercise will do the first step, and Exercise 3.13 will take care of the second.

We proved in Exercise 2.10 that when $D > 0$ is not a perfect square, there are only finitely many proper equivalence classes of primitive forms of discriminant D. The set of equivalence classes will be denoted $C(D)$, and as in the positive definite case, $C(D)$ becomes a finite Abelian group under Dirichlet composition (we will prove this in the exercises to §7). We will assume that Proposition 3.11 holds for all nonsquare discriminants D. This is where we pay the price for restricting ourselves to positive definite forms—the proofs in the text only work for $D < 0$. For proofs of these theorems when $D > 0$, see Flath [**46**, Chapter V], Gauss [**51**, §§257–258] or Mathews [**97**, pp. 171–173].

To prove (*), let D be any nonsquare discriminant, and let C denote the class group $C(D)$. Let $H \subset (\mathbb{Z}/D\mathbb{Z})^*$ be the subgroup of values represented by the principal form.

(a) Show that genera can be classified by cosets of H in $(\mathbb{Z}/D\mathbb{Z})^*$. Thus, instead of the map Φ of (3.12), we can use the map

$$\Phi' : C \longrightarrow (\mathbb{Z}/D\mathbb{Z})^*/H,$$

so that $\ker(\Phi')$ is the principal genus and $\Phi'(C)$ is the set of genera. Note that this argument does not use quadratic reciprocity.

(b) Since H contains all squares in $(\mathbb{Z}/D\mathbb{Z})^*$, it follows that $C^2 \subset \ker(\Phi')$. Now adapt the proof of Theorem 3.15 to show that

$$\text{the number of genera is } \leq [C : C^2] = 2^{\mu-1},$$

where the last equality uses Proposition 3.11. This proves (*).

3.13. In this exercise we will show that quadratic reciprocity follows from statement (*) of Exercise 3.12. As we saw in §1, it suffices to show

$$\left(\frac{p^*}{q}\right) = 1 \iff \left(\frac{q}{p}\right) = 1,$$

where p and q are distinct odd primes and $p^* = (-1)^{(p-1)/2}p$.

(a) Show that Lemma 3.17 holds for all nonsquare discriminants D, so that we can use the assigned characters (where $n < 0$ when $D = -4n$ is positive) to distinguish genera.

(b) Assume that $(p^*/q) = 1$. Applying Lemma 2.5 with $D = p^*$ shows that q is represented by a form $f(x, y)$ of discriminant p^*. The number μ from Proposition 3.11 is 1, so that by (*), there is only one genus. Hence the assigned character (there is only one in this case) must equal 1 on any number represented by $f(x, y)$, in particular q. Use this to prove that $(q/p) = 1$. This proves that $(p^*/q) = 1 \Rightarrow (q/p) = 1$.

(c) Next, assume that $(q/p) = 1$ and that either $p \equiv 1 \bmod 4$ or $q \equiv 1 \bmod 4$. Use part (b) to show that $(p^*/q) = 1$.

(d) Finally, assume that $(q/p) = 1$ and that $p \equiv q \equiv 3 \bmod 4$. This time we will consider forms of discriminant pq. Proposition 3.11 shows that $\mu = 2$, so that by (*), there are at most two genera. Furthermore, the assigned characters are $\chi_1(a) = (a/p)$ and $\chi_2(a) = (a/q)$. Now consider the form $f(x, y) = px^2 + pxy + ((p - q)/4)y^2$, which is easily seen to have discriminant pq. Letting $(x, y) = (0, 2)$, it represents $p - q$. Use this to compute the complete character of the forms $f(x, y)$ and $-f(x, y)$, and show that one of these must lie in the principal genus since there are at most two genera. Then show that $(-p/q) = 1$. Note that parts (c) and (d) imply that $(q/p) = 1 \Rightarrow (p^*/q) = 1$, which completes the proof of quadratic reciprocity.

(e) Gauss also used (*) to show that $(2/p) = (-1)^{(p^2-1)/8}$. Adapt the argument given above to prove this. Hint: prove $(2/p) = 1$ if and only if $p \equiv 1, 7 \bmod 8$. For $\Rightarrow$, show that p is properly represented by a form of discriminant 8 and use $-2 \equiv 6 \bmod 8$. For $\Leftarrow$, use the forms $2x^2 + xy + ((1 - p)/8)y^2$ of discriminant p (when $p \equiv 1 \bmod 8$) and

$2x^2 + xy + ((1+p)/8)y^2$ of discriminant $-p$ (when $p \equiv 7 \bmod 8$). Both represent 2.

3.14. Use Gauss' definition of genus to divide the forms of discriminant -164 into genera. Hint: the forms are given in (2.33). Notice that this is much easier than working with our original definition!

3.15. Prove (vi) $\Rightarrow$ (iii) $\Rightarrow$ (ii) $\Rightarrow$ (i) and (vi) $\Rightarrow$ (v) $\Rightarrow$ (i) of Theorem 3.21.

3.16. Prove that the forms $x^2 + 18y^2$ and $2x^2 + 9y^2$ are rationally equivalent but belong to different genera. Hint: if they represent the same values in $(\mathbb{Z}/72\mathbb{Z})^*$, then the same is true for any divisor of 72.

3.17. Let D be a field discriminant, i.e., $D \equiv 1 \bmod 4$, D squarefree, or $D = 4k$, $k \not\equiv 1 \bmod 4$, k squarefree. Let $f(x, y)$ and $g(x, y)$ be rationally equivalent forms of discriminant D. We want to prove that they lie in the same genus.

(a) Let m be prime to D and represented by $g(x, y)$. Show that $f(x, y)$ represents d^2m for some nonzero integer d.

(b) Show that $f(x, y)$ and $g(x, y)$ lie in the same genus. Hint: by Exercise 2.1, $f(x, y)$ properly represents m' where $d'^2m' = d^2m$ for some integer d'. Show that m' is relatively prime to D. To do this, use Lemma 2.3 to write $f(x, y) = m'x^2 + bxy + cy^2$.

3.18. When $D = -4n$ is a field discriminant, we can use Theorem 3.21 to give a different proof that every form in the principal genus is a square (this is part (ii) of Theorem 3.15). Let $f(x, y)$ be a form of discriminant $-4n$ which lies in the principal genus.

(a) Show that $f(x, y)$ properly represents a number of the form a^2, where a is odd and relatively prime to n. Hint: use part (v) of Theorem 3.21.

(b) By (a), we may assume that $f(x, y) = a^2x^2 + 2bxy + cy^2$. Show that $\gcd(a, 2b) = 1$, and conclude that $g(x, y) = ax^2 + 2bxy + acy^2$ has relatively prime coefficients and discriminant $-4n$.

(c) Show that $f(x, y)$ is the Dirichlet composition of $g(x, y)$ with itself.

This argument is due to Arndt (see Smith [**122**, pp. 254–256]), though Arndt proved part (a) using the theorem of Legendre discussed in Exercise 2.24. Note that part (a) can be restated in terms of ternary forms: if $f(x, y)$ is in the principal genus, then part (a) proves that the ternary form $f(x, y) - z^2$ has a nontrivial zero. This result shows that there is a connection between ternary forms and genus theory. It is therefore not surprising that Gauss used ternary forms in his proof of Theorem 3.15.

3.19. Let $C(D)$ be the class group of forms of discriminant $D < 0$. Prove that the following statements are equivalent:

(i) Every genus of discriminant D consists of a single class.

(ii) $C(D) \simeq \{\pm 1\}^{\mu - 1}$, where μ is as in Proposition 3.11.

(iii) Every genus of discriminant D consists of equivalent forms.

3.20. In this exercise we will prove Lemma 3.25. Let $m > 0$ be odd and prime to $n > 1$.

(a) Show that the number of solutions modulo m of the congruence

$$x^2 \equiv -n \bmod m$$

is given by the formula

$$\prod_{p|m}\left(1+\left(\frac{-n}{p}\right)\right).$$

(b) Consider forms $g(x, y)$ of discriminant $-4n$ of the form

$$g(x, y) = mx^2 + 2bxy + cy^2, \quad 0 \le b < m.$$

Show that the map sending $g(x, y)$ to $[b] \in (\mathbb{Z}/m\mathbb{Z})^*$ induces a bijection between the $g(x, y)$'s and the solutions modulo m of $x^2 \equiv -n \bmod m$.

(c) Let $f(x, y)$ have discriminant $-4n$ and let $f(u, v) = m$ be a proper representation. Pick r_0, s_0, so that $us_0 - vr_0 = 1$, and set $r = r_0 + uk$, $s = s_0 + vk$. Note that as $k \in \mathbb{Z}$ varies, we get all solutions of $us - vr = 1$. Then set

$$g(x, y) = f(ux + ry, vx + sy)$$

and show that there is a unique $k \in \mathbb{Z}$ such that $g(x, y)$ satisfies the condition of part (b). This form is denoted $g_{u,v}(x, y)$.

(d) Show that the map sending a proper representation $f(u, v) = m$ to the form $g_{u,v}(x, y)$ is onto, i.e., every form $g(x, y)$ as in part (b) is equal to some $g_{u,v}(x, y)$.

(e) If $g_{u',v'}(x, y) = g_{u,v}(x, y)$, let

$$\begin{pmatrix}\alpha & \beta\\ \gamma & \delta\end{pmatrix} = \begin{pmatrix}u' & v'\\ r' & s'\end{pmatrix}^{-1}\begin{pmatrix}u & v\\ r & s\end{pmatrix}.$$

Show that $f(\alpha x + \gamma y, \beta x + \delta y) = f(x, y)$ and, since $n > 1$, show that $\left(\begin{smallmatrix}\alpha & \beta\\ \gamma & \delta\end{smallmatrix}\right) = \pm\left(\begin{smallmatrix}1 & 0\\ 0 & 1\end{smallmatrix}\right)$. Hint: assume that $f(x, y)$ is reduced, and use the arguments from the uniqueness part of the proof of Theorem 2.8.

(f) Conclude that $g_{u',v'}(x, y) = g_{u,v}(x, y)$ if and only if $(u', v') = \pm(u, v)$, so that the map of part (d) is exactly two-to-one. Combining this with parts (a) and (b), we get a proof of Lemma 3.25.

3.21. This exercise will use Lemma 3.25 to study the equation $m^3 = a^2 + 2b^2$.

(a) If m is odd, use Lemma 3.25 to show that the equations $m = x^2 + 2y^2$ and $m^3 = x^2 + 2y^2$ have the same number of proper solutions.

(b) If $m = a^2 + 2b^2$ is a proper representation, then show that

$$m^3 = (a^3 - 6ab^2)^2 + 2(3a^2b - 2b^3)^2$$

is a proper representation.

(c) Show that the map sending (a, b) to $(a^3 - 6ab^2,\ 3a^2b - 2b^3)$ is injective. Hint: note that

$$(a + b\sqrt{-2})^3 = (a^3 - 6ab^2) + (3a^2b - 2b^3)\sqrt{-2}.$$

(d) Combine parts (a) and (c) to show that all proper representations of $m^3 = x^2 + 2y^2$, m odd, arise from part (b).

3.22. Use Exercise 3.21 to prove Fermat's famous result that $(x, y) = (3, \pm 5)$ are the only integral solutions of the equation $x^3 = y^2 + 2$. Hint: first show that x must be odd, and then apply Exercise 3.21 to the proper representation $x^3 = y^2 + 2\cdot 1^2$. It's likely that Fermat's original proof of this result was similar to the argument presented here, though he would have used a version of Lemma 1.4 to prove part (c) of Exercise 3.21. See Weil [**135**, pp. 68–69 and 71–73] for more details.

3.23. Let p be an odd prime of the form x^2+ny^2, $n > 1$. Use Lemma 3.25 to show that the equation

$$p = x^2+ny^2$$

has a unique solution once we require x and y to be nonnegative. Note also that Lemma 3.25 gives a quick proof of part (a) of Exercise 2.27 for forms of discriminant $-4n$ when $p \nmid n$.

3.24. This exercise will examine a generalization of the Jacobi symbol. Let P and Q be relatively prime nonzero integers, where Q is odd but possibly negative. Then define the extended Jacobi symbol (P/Q) via

$$\left(\frac{P}{Q}\right) = \begin{cases} (P/|Q|) & \text{when } |Q| > 1 \\ 1 & \text{when } |Q| = 1. \end{cases}$$

(a) Prove that when P and Q are odd and relatively prime, then

$$\left(\frac{P}{Q}\right)\left(\frac{Q}{P}\right) = (-1)^{(P-1)(Q-1)/4+(\mathrm{sgn}(P)-1)(\mathrm{sgn}(Q)-1)/4}$$

where $\mathrm{sgn}(P) = P/|P|$.

(b) Gauss' version of part (a) is more complicated to state. First, given P and Q as above, he lets p denote the number of prime factors of Q (counted with multiplicity) for which P is not a quadratic residue. This relates to (P/Q) by the formula

$$\left(\frac{P}{Q}\right) = (-1)^p.$$

Interchanging P and Q, we get a similarly defined number q. To relate the parity of p and q, Gauss states a rule in [**51**, §133] which breaks up into 10 separate cases. Verify that the rule proved in part (a) covers all 10 of Gauss' cases.

(c) Prove the supplementary laws for P odd:

$$\left(\frac{-1}{P}\right) = \mathrm{sgn}(P)(-1)^{(P-1)/2}$$

$$\left(\frac{2}{P}\right) = (-1)^{(P^2-1)/8}.$$

3.25. Let $p \equiv 1 \bmod 8$ be prime.

(a) If $C(-4p)$ is the class group of forms of discriminant $-4p$, then use genus theory to prove that

$$C(-4p) \simeq (\mathbb{Z}/2^a\mathbb{Z}) \times G$$

where $a \geq 1$ and G has odd order. Thus $2 \mid h(-4p)$.

(b) Let $f(x,y) = 2x^2+2xy+((p+1)/2)y^2$. Use Gauss' definition of genus to show that $f(x,y)$ is in the principal genus.

(c) Use Theorem 3.15 to show that $C(-4p)$ has an element of order 4. Thus $4 \mid h(-4p)$.

4. Cubic and Biquadratic Reciprocity

In this section we will study cubic and biquadratic reciprocity and use them to prove Euler's conjectures for $p = x^2 + 27y^2$ and $p = x^2 + 64y^2$ (see (1.22) and (1.23)). An interesting feature of these reciprocity theorems is that each one requires that we extend the notion of integer: for cubic reciprocity we will use the ring

$$\text{(4.1)} \qquad \mathbb{Z}[\omega] = \{a + b\omega : a, b \in \mathbb{Z}\}, \qquad \omega = e^{2\pi i/3} = (-1 + \sqrt{-3})/2,$$

and for biquadratic reciprocity we will use the Gaussian integers

$$\text{(4.2)} \qquad \mathbb{Z}[i] = \{a + bi : a, b \in \mathbb{Z}\}, \qquad i = \sqrt{-1}.$$

Both $\mathbb{Z}[\omega]$ and $\mathbb{Z}[i]$ are subrings of the complex numbers (see Exercise 4.1). Our first task will be to describe the arithmetic properties of these rings and determine their units and primes. We will then define the generalized Legendre symbols $(\alpha/\pi)_3$ and $(\alpha/\pi)_4$ and state the laws of cubic and biquadratic reciprocity. The proofs will be omitted since excellent proofs are already available in print (see especially Ireland and Rosen [**75**, Chapter 9]). At the end of the section we will discuss Gauss' work on reciprocity and say a few words about the origins of class field theory.

A. $\mathbb{Z}[\omega]$ and Cubic Reciprocity

The law of cubic reciprocity is intimately bound up with the ring $\mathbb{Z}[\omega]$ of (4.1). The main tool used to study the arithmetic of $\mathbb{Z}[\omega]$ is the norm function: if $\alpha = a + b\omega$ is in $\mathbb{Z}[\omega]$, then its *norm* $N(\alpha)$ is the nonnegative integer

$$N(\alpha) = \alpha\overline{\alpha} = a^2 - ab + b^2,$$

where $\overline{\alpha}$ is the complex conjugate of α (in Exercise 4.1 we will see that $\overline{\alpha} \in \mathbb{Z}[\omega]$). Note that the norm is multiplicative, i.e., for $\alpha, \beta \in \mathbb{Z}[\omega]$, we have

$$N(\alpha\beta) = N(\alpha)N(\beta)$$

(see Exercise 4.2). Using the norm, one can prove that $\mathbb{Z}[\omega]$ is a Euclidean ring:

PROPOSITION 4.3. *Given $\alpha, \beta \in \mathbb{Z}[\omega]$, $\beta \neq 0$, there are γ, $\delta \in \mathbb{Z}[\omega]$ such that*

$$\alpha = \gamma\beta + \delta \qquad \textit{and} \qquad N(\delta) < N(\beta).$$

Thus $\mathbb{Z}[\omega]$ is a Euclidean ring.

PROOF. The norm function $N(\alpha) = \alpha\overline{\alpha}$ is defined on $\mathbb{Q}(\omega) = \{r + s\omega : r, s \in \mathbb{Q}\}$ and satisfies $N(uv) = N(u)N(v)$ for $u, v \in \mathbb{Q}(\omega)$ (see Exercise 4.2). Then

$$\frac{\alpha}{\beta} = \frac{\alpha\overline{\beta}}{\beta\overline{\beta}} = \frac{\alpha\overline{\beta}}{N(\beta)} \in \mathbb{Q}(\omega),$$

so that $\alpha/\beta = r + s\omega$ for some $r, s \in \mathbb{Q}$. Let r_1, s_1 be integers such that $|r - r_1| \leq 1/2$ and $|s - s_1| \leq 1/2$, and then set $\gamma = r_1 + s_1\omega$ and $\delta = \alpha - \gamma\beta$. Note that $\gamma, \delta \in \mathbb{Z}[\omega]$ and $\alpha = \gamma\beta + \delta$. It remains to show that $N(\delta) < N(\beta)$. To see this, let $\epsilon = \alpha/\beta - \gamma = (r - r_1) + (s - s_1)\omega$, and note that

$$\delta = \alpha - \gamma\beta = \beta(\alpha/\beta - \gamma) = \beta\epsilon.$$

Since the norm is multiplicative, it suffices to prove that $N(\epsilon) < 1$. But

$$N(\epsilon) = N((r - r_1) + (s - s_1)\omega) = (r - r_1)^2 - (r - r_1)(s - s_1) + (s - s_1)^2,$$

and the desired inequality follows from $|r - r_1|$, $|s - s_1| \leq 1/2$. By the standard definition of a Euclidean ring (see, for example, Herstein [**70**, §3.7]), we are done. Q.E.D.

COROLLARY 4.4. *$\mathbb{Z}[\omega]$ is a PID (principal ideal domain) and a UFD (unique factorization domain).*

PROOF. It is well-known that any Euclidean ring is a PID and a UFD—see, for example, Herstein [**70**, Theorems 3.7.1 and 3.7.2]. Q.E.D.

For completeness, let's recall the definitions of PID and UFD. Let R be an integral domain. An ideal of R is *principal* if it can be written in the form $\alpha R = \{\alpha\beta : \beta \in R\}$ for some $\alpha \in R$, and R is a PID if every ideal of R is principal. To explain what a UFD is, we first need to define units, associates and irreducibles:

(i) $\alpha \in R$ is a *unit* if $\alpha\beta = 1$ for some $\beta \in R$.
(ii) α, $\beta \in R$ are *associates* if α is a unit times β. This is equivalent to $\alpha R = \beta R$.
(iii) A nonunit $\alpha \in R$ is *irreducible* if $\alpha = \beta\gamma$ in R implies that β or γ is a unit.

Then R is a UFD if every nonunit $\alpha \neq 0$ can be written as a product of irreducibles, and given two such factorizations of α, each irreducible in the first factorization can be matched up in a one-to-one manner with an associate irreducible in the second. Thus factorization is unique up to order and associates.

It turns out that being a PID is the stronger property: every PID is a UFD (see Ireland and Rosen [**75**, §1.3]), but the converse is not true (see Exercise 4.3). Given a nonunit $\alpha \neq 0$ in a PID R, the following statements are equivalent:

(i) α is irreducible.
(ii) α is prime (an element α of R is *prime* if $\alpha \mid \beta\gamma$ implies $\alpha \mid \beta$ or $\alpha \mid \gamma$).
(iii) αR is a prime ideal (an ideal $\mathfrak{p}$ of R is *prime* if $\beta\gamma \in \mathfrak{p}$ implies $\beta \in \mathfrak{p}$ or $\gamma \in \mathfrak{p}$).
(iv) αR is a maximal ideal.

(See Exercise 4.4 for the proof.)

Since $\mathbb{Z}[\omega]$ is a PID and a UFD, the next step is to determine the units and primes of $\mathbb{Z}[\omega]$. Let's start with the units:

LEMMA 4.5.

(i) *An element $\alpha \in \mathbb{Z}[\omega]$ is a unit if and only if $N(\alpha) = 1$.*
(ii) *The units of $\mathbb{Z}[\omega]$ are $\mathbb{Z}[\omega]^* = \{\pm 1, \pm\omega, \pm\omega^2\}$.*

PROOF. See Exercise 4.5. Q.E.D.

The next step is to describe the primes of $\mathbb{Z}[\omega]$. The following lemma will be useful:

LEMMA 4.6. *If $\alpha \in \mathbb{Z}[\omega]$ and $N(\alpha)$ is a prime in $\mathbb{Z}$, then α is prime in $\mathbb{Z}[\omega]$.*

PROOF. Since $\mathbb{Z}[\omega]$ is a PID, it suffices to prove that α is irreducible. So suppose that $\alpha = \beta\gamma$ in $\mathbb{Z}[\omega]$. Taking norms, we obtain the integer equation

$$N(\alpha) = N(\beta\gamma) = N(\beta)N(\gamma)$$

(recall that the norm is multiplicative). Since $N(\alpha)$ is prime by assumption, this implies that $N(\beta)$ or $N(\gamma)$ is 1, so that β or γ is a unit by Lemma 4.5. Q.E.D.

We can now determine all primes in $\mathbb{Z}[\omega]$:

PROPOSITION 4.7. *Let p be a prime in $\mathbb{Z}$. Then:*

(i) *If $p = 3$, then $1-\omega$ is prime in $\mathbb{Z}[\omega]$ and $3 = -\omega^2(1-\omega)^2$.*
(ii) *If $p \equiv 1 \bmod 3$, then there is a prime $\pi \in \mathbb{Z}[\omega]$ such that $p = \pi\overline{\pi}$, and the primes π and $\overline{\pi}$ are nonassociate in $\mathbb{Z}[\omega]$.*
(iii) *If $p \equiv 2 \bmod 3$, then p remains prime in $\mathbb{Z}[\omega]$.*

Furthermore, every prime in $\mathbb{Z}[\omega]$ is associate to one of the primes listed in (i)–(iii) *above.*

PROOF. Since $N(1-\omega) = 3$, Lemma 4.6 implies that $1-\omega$ is prime in $\mathbb{Z}[\omega]$, and (i) follows. To prove (ii), suppose that $p \equiv 1 \bmod 3$. Then $(-3/p) = 1$, so that p is represented by a reduced form of discriminant -3 (this is Theorem 2.16). The only such form is $x^2 + xy + y^2$, so that p can be written as $a^2 - ab + b^2$. Then $\pi = a + b\omega$ and $\overline{\pi} = a + b\omega^2$ have norms $N(\pi) = N(\overline{\pi}) = p$ and hence are prime in $\mathbb{Z}[\omega]$ by Lemma 4.6. In Exercise 4.7 we will prove that π and $\overline{\pi}$ are nonassociate. The proof of (iii) is left to the reader (see Exercise 4.7).

It remains to show that all primes in $\mathbb{Z}[\omega]$ are associate to one of the above. Let's temporarily call the primes given in (i)–(iii) the *known primes* of $\mathbb{Z}[\omega]$, and let α be any prime of $\mathbb{Z}[\omega]$. Then $N(\alpha) = \alpha\overline{\alpha}$ is an ordinary integer and may be factored into integer primes. But (i)–(iii) imply that any integer prime is a product of known primes in $\mathbb{Z}[\omega]$, and consequently $\alpha\overline{\alpha} = N(\alpha)$ is also a product of known primes. The proposition then follows since $\mathbb{Z}[\omega]$ is a UFD. Q.E.D.

Given a prime π of $\mathbb{Z}[\omega]$, we get the maximal ideal $\pi\mathbb{Z}[\omega]$ of $\mathbb{Z}[\omega]$. The quotient ring $\mathbb{Z}[\omega]/\pi\mathbb{Z}[\omega]$ is a thus a field. We can describe this field more carefully as follows:

LEMMA 4.8. *If π is a prime of $\mathbb{Z}[\omega]$, then the quotient field $\mathbb{Z}[\omega]/\pi\mathbb{Z}[\omega]$ is a finite field with $N(\pi)$ elements. Furthermore, $N(\pi) = p$ or p^2 for some integer prime p, and:*

(i) *If $p = 3$ or $p \equiv 1 \bmod 3$, then $N(\pi) = p$ and $\mathbb{Z}/p\mathbb{Z} \simeq \mathbb{Z}[\omega]/\pi\mathbb{Z}[\omega]$.*
(ii) *If $p \equiv 2 \bmod 3$, then $N(\pi) = p^2$ and $\mathbb{Z}/p\mathbb{Z}$ is the unique subfield of order p of the field $\mathbb{Z}[\omega]/\pi\mathbb{Z}[\omega]$ of p^2 elements.*

PROOF. In §7 we will prove that if π is a nonzero element of $\mathbb{Z}[\omega]$, then $\mathbb{Z}[\omega]/\pi\mathbb{Z}[\omega]$ is a finite ring with $N(\pi)$ elements (see Lemma 7.14 or Ireland and Rosen [**75**, §§9.2 and 14.1]). Then (i) and (ii) follow easily (see Exercise 4.8). Q.E.D.

Given α, β and π in $\mathbb{Z}[\omega]$, we will write $\alpha \equiv \beta \bmod \pi$ to indicate that α and β differ by a multiple of π, i.e., that they give the same element in $\mathbb{Z}[\omega]/\pi\mathbb{Z}[\omega]$. Using this notation, Lemma 4.8 gives us the following analog of Fermat's Little Theorem:

COROLLARY 4.9. *If π is prime in $\mathbb{Z}[\omega]$ and doesn't divide $\alpha \in \mathbb{Z}[\omega]$, then*

$$\alpha^{N(\pi)-1} \equiv 1 \bmod \pi.$$

PROOF. This follows because $(\mathbb{Z}[\omega]/\pi\mathbb{Z}[\omega])^*$ is a finite group with $N(\pi) - 1$ elements. Q.E.D.

Given these properties of $\mathbb{Z}[\omega]$, we can now define the generalized Legendre symbol $(\alpha/\pi)_3$. Let π be a prime of $\mathbb{Z}[\omega]$ not dividing 3 (i.e., not associate to $1-\omega$). It is straightforward to check that $3 \mid N(\pi) - 1$ (see Exercise 4.9). Now suppose that $\alpha \in \mathbb{Z}[\omega]$ is not divisible by π. It follows from Corollary 4.9 that $x = \alpha^{(N(\pi)-1)/3}$ is a root of $x^3 \equiv 1 \bmod \pi$. Since

$$x^3 - 1 \equiv (x-1)(x-\omega)(x-\omega^2) \bmod \pi$$

and π is prime, it follows that

$$\alpha^{(N(\pi)-1)/3} \equiv 1, \omega, \omega^2 \bmod \pi.$$

However, the cube roots of unity $1, \omega, \omega^2$ are incongruent modulo π. To see this, note that if any two were congruent, then we would have $1 \equiv \omega \bmod \pi$, which would contradict π not associate to $1-\omega$ (see Exercise 4.9 for the details). Then we define the *Legendre symbol* $(\alpha/\pi)_3$ to be the unique cube root of unity such that

$$\alpha^{(N(\pi)-1)/3} \equiv \left(\frac{\alpha}{\pi}\right)_3 \bmod \pi. \tag{4.10}$$

The basic properties of the Legendre symbol are easy to work out. First, from (4.10), one can show

$$\left(\frac{\alpha\beta}{\pi}\right)_3 = \left(\frac{\alpha}{\pi}\right)_3 \left(\frac{\beta}{\pi}\right)_3,$$

and second, $\alpha \equiv \beta \bmod \pi$ implies that

$$\left(\frac{\alpha}{\pi}\right)_3 = \left(\frac{\beta}{\pi}\right)_3$$

(see Exercise 4.10). The Legendre symbol may thus be regarded as a group homomorphism from $(\mathbb{Z}[\omega]/\pi\mathbb{Z}[\omega])^*$ to $\mathbb{C}^*$.

An important fact is that the multiplicative group of any finite field is cyclic (see Ireland and Rosen [**75**, §7.1]). In particular, $(\mathbb{Z}[\omega]/\pi\mathbb{Z}[\omega])^*$ is cyclic, which implies that

$$\begin{aligned} \left(\frac{\alpha}{\pi}\right)_3 = 1 &\iff \alpha^{(N(\pi)-1)/3} \equiv 1 \bmod \pi \\ &\iff x^3 \equiv \alpha \bmod \pi \ \text{ has a solution in } \mathbb{Z}[\omega] \end{aligned} \tag{4.11}$$

(see Exercise 4.11). This establishes the link between the Legendre symbol and cubic residues. Note that one-third of $(\mathbb{Z}[\omega]/\pi\mathbb{Z}[\omega])^*$ consists of cubic residues (where the Legendre symbol equals 1), and the remaining two-thirds consist of nonresidues (where the symbol equals ω or ω^2). Later on we will explain how this relates to the more elementary notion of cubic residues of integers.

To state the law of cubic reciprocity, we need one final definition: a prime π is called *primary* if $\pi \equiv \pm 1 \bmod 3$. Given any prime π not dividing 3, one can show that exactly two of the six associates $\pm\pi$, $\pm\omega\pi$ and $\pm\omega^2\pi$ are primary (see Exercise 4.12). Then the law of cubic reciprocity states the following:

THEOREM 4.12. *If π and θ are primary primes in $\mathbb{Z}[\omega]$ of unequal norm, then*

$$\left(\frac{\theta}{\pi}\right)_3 = \left(\frac{\pi}{\theta}\right)_3.$$

PROOF. See Ireland and Rosen [**75**, §§9.4–9.5] or Smith [**122**, pp. 89–91]. Q.E.D.

Notice how simple the statement of the theorem is—it's among the most elegant of all reciprocity theorems (biquadratic reciprocity, to be stated below, is a bit more complicated). The restriction to primary primes is a normalization analogous to the normalization $p > 0$ that we make for ordinary primes. Some books (such as Ireland and Rosen [**75**]) define primary to mean $\pi \equiv -1 \bmod 3$. Since $(-1/\pi)_3 = 1$, this doesn't affect the statement of cubic reciprocity.

There are also supplementary formulas for $(\omega/\pi)_3$ and $(1-\omega/\pi)_3$. Let π be prime and not associate to $1-\omega$. Then we may assume that $\pi \equiv -1 \bmod 3$ (if π is primary, one of $\pm\pi$ satisfies this condition). Writing $\pi = -1 + 3m + 3n\omega$, it can be shown that

$$\begin{aligned} \left(\frac{\omega}{\pi}\right)_3 &= \omega^{m+n} \\ \left(\frac{1-\omega}{\pi}\right)_3 &= \omega^{2m}. \end{aligned} \tag{4.13}$$

The first line of (4.13) is easy to prove (see Exercise 4.13), while the second is more difficult (see Ireland and Rosen [**75**, p. 114] or Exercise 9.13).

Let's next discuss cubic residues of integers. If p is a prime, the basic question is: when does $x^3 \equiv a \bmod p$ have an integer solution? If $p = 3$, then Fermat's Little Theorem tells us that $a^3 \equiv a \bmod 3$ for all a, so that we always have a solution. If $p \equiv 2 \bmod 3$, then the map $a \mapsto a^3$ induces an automorphism of $(\mathbb{Z}/p\mathbb{Z})^*$ since $3 \nmid p-1$ (see Exercise 4.14), and consequently $x^3 \equiv a \bmod p$ is again always solvable. If $p \equiv 1 \bmod 3$, things are more interesting. In this case, $p = \pi\overline{\pi}$ in $\mathbb{Z}[\omega]$, and there is a natural isomorphism $\mathbb{Z}/p\mathbb{Z} \simeq \mathbb{Z}[\omega]/\pi\mathbb{Z}[\omega]$ by Lemma 4.8. Thus, for $p \nmid a$, (4.11) implies that

$$x^3 \equiv a \bmod p \ \text{ is solvable in } \mathbb{Z} \iff \left(\frac{a}{\pi}\right)_3 = 1. \tag{4.14}$$

Furthermore, $(\mathbb{Z}/p\mathbb{Z})^*$ breaks up into three pieces of equal size, one of cubic residues and two of nonresidues.

We can now use cubic reciprocity to prove Euler's conjecture for primes of the form $x^2 + 27y^2$:

THEOREM 4.15. *Let p be a prime. Then $p = x^2+27y^2$ if and only if $p \equiv 1 \bmod 3$ and 2 is a cubic residue modulo p.*

PROOF. First, suppose that $p = x^2 + 27y^2$. This clearly implies that $p \equiv 1 \bmod 3$, so that we need only show that 2 is a cubic residue modulo p. Let $\pi = x + 3\sqrt{-3}y$, so that $p = \pi\overline{\pi}$ in $\mathbb{Z}[\omega]$. It follows that π is prime, and then by (4.14), 2 is a cubic residue modulo p if and only if $(2/\pi)_3 = 1$. However, both 2 and $\pi = x + 3\sqrt{-3}y$ are primary primes, so that cubic reciprocity implies

$$\left(\frac{2}{\pi}\right)_3 = \left(\frac{\pi}{2}\right)_3. \tag{4.16}$$

It thus suffices to prove that $(\pi/2)_3 = 1$. However, from (4.10), we know that

$$\left(\frac{\pi}{2}\right)_3 \equiv \pi \bmod 2 \tag{4.17}$$

since $(N(2) - 1)/3 = 1$. So we need only show that $\pi \equiv 1 \bmod 2$. Since $\sqrt{-3} = 1 + 2\omega$, $\pi = x + 3\sqrt{-3}y = x + 3y + 6y\omega$, so that $\pi \equiv x + 3y \equiv x + y \bmod 2$. But x and y must have opposite parity since $p = x^2 + 27y^2$, and we are done.

Conversely, suppose that $p \equiv 1 \bmod 3$ is prime and 2 is a cubic residue modulo p. We can write p as $p = \pi\overline{\pi}$, and we can assume that π is a primary prime in $\mathbb{Z}[\omega]$. This means that $\pi = a + 3b\omega$ for some integers a and b. Thus

$$4p = 4\pi\overline{\pi} = 4(a^2 - 3ab + 9b^2) = (2a - 3b)^2 + 27b^2.$$

Once we show b is even, it will follow immediately that p is of the form $x^2 + 27y^2$.

We now can use our assumption that 2 is a cubic residue modulo p. From (4.14) we know that $(2/\pi)_3 = 1$, and then cubic reciprocity (4.16) tells us that $(\pi/2)_3 = 1$. But by (4.17), this implies $\pi \equiv 1 \bmod 2$, which we can write as $a + 3b\omega \equiv 1 \bmod 2$. This easily implies that a is odd and b is even, and $p = x^2 + 27y^2$ follows. The theorem is proved. Q.E.D.

B. $\mathbb{Z}[i]$ and Biquadratic Reciprocity

Our treatment of biquadratic reciprocity will be brief since the basic ideas are similar to what we did for cubic residues (for a complete discussion, see Ireland and Rosen [**75**, §§9.7–9.9]). Here, the appropriate ring is the ring of Gaussian integers $\mathbb{Z}[i]$ as defined in (4.2). The norm function $N(a + bi) = a^2 + b^2$ makes $\mathbb{Z}[i]$ into a Euclidean ring, and hence $\mathbb{Z}[i]$ is also a PID and a UFD. The analogs of Lemma 4.5 and 4.6 hold for $\mathbb{Z}[i]$, and it is easy to check that its units are $\pm$ 1 and $\pm i$ (see Exercise 4.16). The primes of $\mathbb{Z}[i]$ are described as follows:

PROPOSITION 4.18. *Let p be a prime in $\mathbb{Z}$. Then:*

(i) *If $p = 2$, then $1 + i$ is prime in $\mathbb{Z}[i]$ and $2 = i^3(1 + i)^2$.*
(ii) *If $p \equiv 1 \bmod 4$, then there is a prime $\pi \in \mathbb{Z}[i]$ such that $p = \pi\overline{\pi}$ and the primes π and $\overline{\pi}$ are nonassociate in $\mathbb{Z}[i]$.*
(iii) *If $p \equiv 3 \bmod 4$, then p remains prime in $\mathbb{Z}[i]$.*

Furthermore, every prime in $\mathbb{Z}[i]$ is associate to one of the primes listed in (i)–(iii) *above.*

PROOF. See Exercise 4.16. Q.E.D.

We also have the following version of Fermat's Little Theorem: if π is prime in $\mathbb{Z}[i]$ and doesn't divide $\alpha \in \mathbb{Z}[i]$, then

$$\alpha^{N(\pi)-1} \equiv 1 \bmod \pi \tag{4.19}$$

(see Exercise 4.16).

These basic facts about the Gaussian integers appear in many elementary texts (e.g., Herstein [54, §3.8]), but such books rarely mention that the whole reason Gauss introduced the Gaussian integers was so that he could state biquadratic reciprocity. We will have more to say about this later.

We can now define the Legendre symbol $(\alpha/\pi)_4$. Given a prime $\pi \in \mathbb{Z}[i]$ not associate to $1 + i$, it can be proved that ± 1, $\pm i$ are distinct modulo π and that $4 \mid N(\pi) - 1$ (see Exercise 4.17). Then, for α not divisible by π, the *Legendre symbol* $(\alpha/\pi)_4$ is defined to be the unique fourth root of unity such that

$$\alpha^{(N(\pi)-1)/4} \equiv \left(\frac{\alpha}{\pi}\right)_4 \bmod \pi. \tag{4.20}$$

As in the cubic case, we see that

$$\left(\frac{\alpha}{\pi}\right)_4 = 1 \iff x^4 \equiv \alpha \bmod \pi \quad \text{is solvable in } \mathbb{Z}[i],$$

and furthermore, the Legendre symbol gives a character from $(\mathbb{Z}[i]/\pi\mathbb{Z}[i])^*$ to $\mathbb{C}^*$, so that $(\mathbb{Z}[i]/\pi\mathbb{Z}[i])^*$ is divided into four equal parts (see Exercise 4.18). When $p \equiv 1 \bmod 4$, we have $p = \pi\overline{\pi}$ with $(\mathbb{Z}[i]/\pi\mathbb{Z}[i])^* \simeq (\mathbb{Z}/p\mathbb{Z})^*$, and the partition can be described as follows: one part consists of biquadratic residues (where the symbol equals 1), another consists of quadratic residues which aren't biquadratic residues (where the symbol equals -1), and the final two parts consist of quadratic nonresidues (where the symbol equals $\pm i$)—see Exercise 4.19.

A prime π of $\mathbb{Z}[i]$ is *primary* if $\pi \equiv 1 \bmod 2+2i$. Any prime not associate to $1+i$ has a unique associate which is primary (see Exercise 4.21). With this normalization, the law of biquadratic reciprocity can be stated as follows:

THEOREM 4.21. *If π and θ are distinct primary primes in $\mathbb{Z}[i]$, then*

$$\left(\frac{\theta}{\pi}\right)_4 = \left(\frac{\pi}{\theta}\right)_4 (-1)^{(N(\theta)-1)(N(\pi)-1)/16}.$$

PROOF. See Ireland and Rosen [**75**, §9.9] or Smith [**122**, pp. 76–37]. Q.E.D.

There are also supplementary laws which state that

$$\begin{aligned} \left(\frac{i}{\pi}\right)_4 &= i^{-(a-1)/2} \\ \left(\frac{1+i}{\pi}\right)_4 &= i^{(a-b-1-b^2)/4} \end{aligned} \tag{4.22}$$

where $\pi = a+bi$ is a primary prime. As in the cubic case, the first line of (4.22) is easy to prove (see Exercise 4.22), while the second is more difficult (see Ireland and Rosen [**75**, Exercises 32–37, p. 136]).

We can now prove Euler's conjecture about $p = x^2 + 64y^2$:

THEOREM 4.23.

(i) *If $\pi = a+bi$ is a primary prime in $\mathbb{Z}[i]$, then*

$$\left(\frac{2}{\pi}\right)_4 = i^{ab/2}.$$

(ii) *If p is prime, then $p = x^2 + 64y^2$ if and only if $p \equiv 1 \bmod 4$ and 2 is a biquadratic residue modulo p.*

PROOF. First note that (i) implies (ii). To see this, let $p \equiv 1 \bmod 4$ be prime. We can write $p = a^2 + b^2 = \pi\overline{\pi}$, where $\pi = a+bi$ is primary. Note that a is odd and b is even. Since $\mathbb{Z}/p\mathbb{Z} \simeq \mathbb{Z}[i]/\pi\mathbb{Z}[i]$, (i) shows that 2 is a biquadratic residue modulo p if and only if b is divisible by 8, and (ii) follows easily.

One way to prove (i) is via the supplementary laws (4.22) since $2 = i^3(1+i)^2$ (see Exercise 4.23). However, in 1857, Dirichlet found a proof of (i) that uses only quadratic reciprocity [**35**, Vol. II, pp. 261–262]. A version of this proof is given in Exercise 4.24 (see also Ireland and Rosen [**75**, Exercises 26–28, p. 64]). Q.E.D.

C. Gauss and Higher Reciprocity

Most of the above theorems were discovered by Gauss in the period 1805–1814, though the bulk of what he knew was never published. Only in 1828 and 1832, long after the research was completed, did Gauss publish his two memoirs on biquadratic residues [**52**, Vol. II, pp. 65–148] (see also [**51**, pp. 511–586, German editions] for a German translation). The first memoir treats the elementary theory of biquadratic residues of integers, and it includes a proof of Euler's conjecture for $x^2 + 64y^2$. In the second memoir, Gauss begins with a careful discussion of the Gaussian integers, and he explains their relevance to biquadratic reciprocity as follows [**52**, Vol. II, §30, p. 102]:

> the theorems on biquadratic residues gleam with the greatest simplicity and genuine beauty only when the field of arithmetic is extended to **imaginary** numbers, so that without restriction, the numbers of the form $a + bi$ constitute the object [of study], where as usual i denotes $\sqrt{-1}$ and the indeterminates a, b denote integral real numbers between $-\infty$ and $+\infty$. We will call such numbers **integral complex numbers** (numeros integros complexos) ...

Gauss' treatment of $\mathbb{Z}[i]$ includes most of what we did above, and in particular the terms norm, associate and primary are due to Gauss.

Gauss' statement of biquadratic reciprocity differs slightly from Theorem 4.21. In terms of the Legendre symbol, his version goes as follows: given distinct primary primes π and θ of $\mathbb{Z}[i]$,

If either π or θ is congruent to 1 modulo 4, then $(\pi/\theta)_4 = (\theta/\pi)_4$.

If both π and θ are congruent to $3 + 2i$ modulo 4, then $(\pi/\theta)_4 = -(\theta/\pi)_4$.

In Exercise 4.25 we will see that this is equivalent to Theorem 4.21. As might be expected, Gauss doesn't use the Legendre symbol in his memoir. Rather, he defines the *biquadratic character* of α with respect to π to be the number $\lambda \in \{0, 1, 2, 3\}$ satisfying $\alpha^{(N(\pi)-1)/4} \equiv i^\lambda \bmod \pi$ (so that $(\alpha/\pi)_4 = i^\lambda$), and he states biquadratic reciprocity using the biquadratic character. For Gauss, this theorem is "the Fundamental Theorem of biquadratic residues" [**52**, Vol. II, §67, p. 138], but instead of giving a proof, Gauss comments that

> In spite of the great simplicity of this theorem, the proof belongs to the most hidden mysteries of higher arithmetic, and at least as things now stand, [the proof] can be explained only by the most subtle investigations, which would greatly exceed the limits of the present memoir.

Later on, we will have more to say about Gauss' proof.

In the second memoir, Gauss also makes his only published reference to cubic reciprocity [**52**, Vol. II, §30, p. 102]:

> The theory of cubic residues must be based in a similar way on a consideration of numbers of the form $a + bh$, where h is an imaginary root of the equation $h^3 - 1 = 0$, say $h = (-1 + \sqrt{-3})/2$, and similarly the theory of residues of higher powers leads to the introduction of other imaginary quantities.

So Gauss was clearly aware of the properties of $\mathbb{Z}[\omega]$, even if he never made them public.

Turning to Gauss' unpublished material, we find that one of the earliest fragments on higher reciprocity, dated around 1805, is the following "Beautiful Observation Made By Induction" [**52**, Vol. VIII, pp. 5 and 11]:

> 2 is a cubic residue or nonresidue of a prime number p of the form $3n+1$, according to whether p is representable by the form $xx + 27yy$ or $4xx + 2xy + 7yy$.

This shows that Euler's conjecture for $x^2 + 27y^2$ was one of Gauss' starting points. And notice that Gauss was aware that he was separating forms in the same genus—the very problem we discussed in §3.

Around the same time, Gauss also rediscovered Euler's conjecture for $x^2 + 64y^2$ [**52**, Vol. X.1, p. 37]. But how did he come to make these conjectures? There are two aspects of Gauss' work that bear on this question. The first has to do with quadratic forms. Let's follow the treatment in Gauss' first memoir on biquadratic residues [**52**, Vol. II, §§12–14, pp. 75–78]. Let $p \equiv 1 \bmod 4$ be prime. If 2 is to be a biquadratic residue modulo p, it follows by quadratic reciprocity that $p \equiv 1 \bmod 8$ (see Exercise 4.26). By Fermat's theorem for $x^2 + 2y^2$, p can be written as $p = a^2 + 2b^2$, and Gauss proves the lovely result that 2 is a biquadratic residue modulo p if and only if $a \equiv \pm 1 \bmod 8$ (see Exercise 4.27). This is nice, but Gauss isn't satisfied:

> Since the decomposition of the number p into a single and double square is bound up so prominently with the classification of the number 2, it would be worth the effort to understand whether the decomposition into two squares, to which the number p is equally liable, perhaps promises a similar success.

Gauss then computes some numerical examples, and they show that when p is written as $a^2 + b^2$, 2 is a biquadratic residue exactly when b is divisible by 8. This could be how Gauss was led to the conjecture in the first place, and the same thing could have happened in the cubic case, where primes $p = 1 \bmod 3$ can be written as $a^2 + 3b^2$.

The cubic case most likely came first, for it turns out that Gauss describes a relation between $x^2 + 27y^2$ and cubic residues in the last section of *Disquisitiones*. This is where Gauss discusses the cyclotomic equation $x^p - 1 = 0$ and proves his celebrated theorem on the constructibility of regular polygons. To see what this has to do with cubic residues, let's describe a little of what he does. Given an odd prime p, let $\zeta_p = e^{2\pi i/p}$ be a primitive pth root of unity, and let g be a primitive root modulo p, i.e., g is an integer such that $[g]$ generates the cyclic group $(\mathbb{Z}/p\mathbb{Z})^*$. Now suppose that $p - 1 = ef$, and let λ be an integer. Gauss then defines [**51**, §343] the *period* (f, λ) to be the sum

$$(f, \lambda) = \sum_{j=0}^{f-1} \zeta_p^{\lambda g^{ej}}.$$

These periods are the key to Gauss' study of the cyclotomic field $\mathbb{Q}(\zeta_p)$. If we fix f, then the periods $(f, 1), (f, g), (f, g^2), \ldots, (f, g^{e-1})$ are the roots of an irreducible integer polynomial of degree e, so that these periods are primitive elements of the unique subfield $\mathbb{Q} \subset K \subset \mathbb{Q}(\zeta_p)$ of degree e over $\mathbb{Q}$. See Cox [**29**, Section 9.2] for more on Gauss' theory of periods.

When $p \equiv 1 \bmod 3$, we can write $p-1 = 3f$, and then the three above periods are $(f,1)$, (f,g) and (f,g^2). Gauss studies this case in [**51**, §358], and by analyzing the products of the periods, he deduces the amazing result that

(4.24) If $4p = a^2 + 27b^2$ and $a \equiv 1 \bmod 3$, then $N = p + a - 2$, where N is the number of solutions modulo p of $x^3 - y^3 \equiv 1 \bmod p$.

To see how cubic residues enter into (4.24), note that $N = 9M + 6$, where M is the number of nonzero cubic residues which, when increased by one, remain a nonzero cubic residue (see Exercise 4.29). Gauss conjectured this result in October 1796 and proved it in July 1797 [**52**, Vol. X.1, entries 39 and 67, pp. 505–506 and 519]. So Gauss was aware of cubic residues and quadratic forms in 1796. Gauss' proof of (4.24) is sketched in Exercise 4.29.

Statement (4.24) is similar to the famous last entry in Gauss' mathematical diary. In this entry, Gauss gives the following analog of (4.24) for the decomposition $p = a^2 + b^2$ of a prime $p \equiv 1 \bmod 4$:

If $p = a^2 + b^2$ and $a + bi$ is primary, then $N = p - 2a - 3$, where N is the number of solutions modulo p of $x^2 + y^2 + x^2y^2 \equiv 1 \bmod p$

(see [**52**, Vol. X.1, entry 146, pp. 571–572]). In general, the study of the solutions of equations modulo p leads to the zeta function of a variety over a finite field. For an introduction to this rich topic, see Ireland and Rosen [**75**, Chapter 11]. In §14 we will see how Gauss' results relate to elliptic curves with complex multiplication.

Going back to the cubic case, there is a footnote in [**51**, §358] which gives another interesting property of the periods $(f,1)$, (f,g) and (f,g^2):

$$\begin{aligned}&((f,1) + \omega(f,g) + \omega^2(f,g^2))^3 = p(a + b\sqrt{-27})/2,\\ &\text{where } 4p = a^2 + 27b^2.\end{aligned} \tag{4.25}$$

The right-hand side is an integer in the ring $\mathbb{Z}[\omega]$, and one can show that $\pi = (a + b\sqrt{-27})/2$ is a primary prime in $\mathbb{Z}[\omega]$ and that $p = \pi\overline{\pi}$. This is how Gauss first encountered $\mathbb{Z}[\omega]$ in connection with cubic residues. Notice also that if we set $\chi(a) = (a/\pi)_3$ and pick the primitive root g so that $\chi(g) = \omega$, then

$$(f,1) + \omega(f,g) + \omega^2(f,g^2) = \sum_{a=1}^{p-1} \chi(a)\zeta_p^a. \tag{4.26}$$

This is an example of what we now call a cubic Gauss sum. See Ireland and Rosen [**75**, §§8.2–8.3] for the basic properties of Gauss sums and a modern treatment of (4.24) and (4.25).

The above discussion shows that Gauss was aware of cubic residues and $\mathbb{Z}[\omega]$ when he made his "Beautiful Observation" of 1805, and it's not surprising that two years later he was able to prove a version of cubic reciprocity [**52**, Vol. VIII, pp. 9–13]. The biquadratic case was harder, taking him until sometime around 1813 or 1814 to find a complete proof. We know this from a letter Gauss wrote Dirichlet in 1828, where Gauss mentions that he has possessed a proof of the "Main Theorem" for around 14 years [**52**, Vol. II, p. 516]. Exact dates are hard to come by, for most of the fragments Gauss left are undated, and it's not easy to match them up with his diary entries. For a fuller discussion of Gauss' work on biquadratic reciprocity, see Bachmann [**52**, Vol. X.2.1, pp. 52–60] or Rieger [**105**].

Gauss' proofs of cubic and biquadratic reciprocity probably used Gauss sums similar to (4.26), and many modern proofs run along the same lines (see Ireland and Rosen [**75**, Chapter 9]). Gauss sums were first used in Gauss' sixth proof of quadratic reciprocity (see [**52**, Vol. II, pp. 55–59] or [**51**, pp. 501–505, German editions]). This is no accident, for as Gauss explained in 1818:

> From 1805 onwards I have investigated the theory of cubic and biquadratic residues ... Theorems were found by induction ... which had a wonderful analogy with the theorems for quadratic residues. On the other hand, for a long time all attempts at complete proofs have been futile. This was the motive for endeavoring to add yet more proofs to those already known for quadratic residues, in the hope that of the many different methods given, one or the other would contribute to the illumination of the related arguments [for cubic and biquadratic residues]. This hope was in no way in vain, for at last tireless labor has led to favorable success. Soon the fruit of this vigilance will be permitted to come to public light ...

(see [**52**, Vol. II, p. 50] or [**51**, p. 497, German editions]). The irony is that Gauss never did publish his proofs, and it was left to Eisenstein and Jacobi to give us the first complete treatments of cubic and biquadratic reciprocity (see Collinson [**28**] or Smith [**122**, pp. 76–92] for more on the history of these reciprocity theorems).

We will conclude this section with some remarks about what happened after Gauss. Number theory was becoming a much larger area of mathematics, and the study of quadratic forms and reciprocity laws began to diverge. In the 1830s and 1840s, Dirichlet introduced L-series and began the analytic study of quadratic forms, and simultaneously, Eisenstein and Jacobi worked out cubic and biquadratic reciprocity. Jacobi studied reciprocity for 5th, 8th and 12th powers, and Eisenstein proved octic reciprocity. Kummer was also studying higher reciprocity, and he introduced his "ideal numbers" to make up for the lack of unique factorization in $\mathbb{Z}(e^{2\pi i/p})$. Both he and Eisenstein were able to prove generalized reciprocity laws using these "ideal numbers" (see Ireland and Rosen [**75**, Chapter 14] and Smith [**122**, pp. 93–126]). In 1871 Dedekind made the transition from "ideal numbers" to ideals in rings of algebraic integers, laying the foundation for modern algebraic number theory and class field theory. Lemmermeyer's book [**93**] contains a wealth of information about reciprocity in the nineteenth century. See also Chapter 8 of the book [**6**] by Berndt, Evans and Williams.

But reciprocity was not the only force leading to class field theory: there was also complex multiplication. Euler, Lagrange, Legendre and others studied transformations of the elliptic integrals

$$\int \frac{dx}{\sqrt{(1-x^2)(1-k^2x^2)}},$$

and they discovered that certain values of k, called *singular moduli*, gave elliptic integrals that could be transformed into complex multiples of themselves. This phenomenon came to be called *complex multiplication*. In working with complex multiplication, Abel observed that singular moduli and the roots of the corresponding transformation equations have remarkable algebraic properties. In modern terms, they generate *Abelian extensions* of $\mathbb{Q}(\sqrt{-n})$, i.e., Galois extensions of $\mathbb{Q}(\sqrt{-n})$ whose Galois group is Abelian. These topics will be discussed in more detail in Chapter 3.

Kronecker extended and completed Abel's work on complex multiplication, and in so doing he made the amazing conjecture that every Abelian extension of $\mathbb{Q}(\sqrt{-n})$ lies in one of the fields described above. Kronecker had earlier conjectured that every Abelian extension of $\mathbb{Q}$ lies in one of the cyclotomic fields $\mathbb{Q}(e^{2\pi i/n})$ (this is the famous Kronecker–Weber Theorem, to be proved in §8). Abelian extensions may seem far removed from reciprocity theorems, but Kronecker also noticed relations between singular moduli and quadratic forms. For example, his results on complex multiplication by $\sqrt{-31}$ led to the following corollary which he was fond of quoting:

$$p = x^2 + 31y^2 \iff \begin{cases} (x^3 - 10x)^2 + 31(x^2 - 1)^2 \equiv 0 \bmod p \\ \text{has an integral solution} \end{cases}$$

(see [**85**, Vol. II, pp. 93 and 99–100, Vol. IV, pp. 123–129]). This is similar to what we just proved for $x^2 + 27y^2$ and $x^2 + 64y^2$ using cubic and biquadratic reciprocity. So something interesting is going on here.

We thus have two interrelated questions of interest:

(i) Is there a general reciprocity law that subsumes the known ones?
(ii) Is there a general method for describing all Abelian extensions of a number field?

The crowning achievement of class field theory is that it solves both of these problems simultaneously: an Abelian extension L of a number field K is classified in terms of data intrinsic to K, and the key ingredient linking L to this data is the Artin Reciprocity Theorem. Complete statements of the theorems of class field theory will be given in Chapter 2, and in Chapter 3 we will explain how complex multiplication is related to the class field theory of imaginary quadratic fields.

For a fuller account of the history of class field theory, see the article by W. and F. Ellison [**42**, §§III–IV] in Dieudonné's *Abrégé d'Histoire des Mathématiques 1700–1900.* Weil has a nice discussion of reciprocity and cyclotomic fields in [**134**] and [**136**], and Edwards describes Kummer's "ideal numbers" in [**41**, Chapter 4]. See also Part I of Vlăduţ's book [**127**] on Kronecker's Jugentraum.

D. Exercises

4.1. Prove that $\mathbb{Z}[\omega]$ and $\mathbb{Z}[i]$ are subrings of the complex numbers and are closed under complex conjugation.

4.2. Let $\mathbb{Q}(\omega) = \{r+s\omega : r, s \in \mathbb{Q}\}$, and define the norm of $r+s\omega$ to be $N(r+s\omega) = (r+s\omega)\overline{(r+s\omega)}$.

(a) Show that $N(r+s\omega) = r^2 - rs + s^2$.
(b) Show that $N(uv) = N(u)N(v)$ for $u, v \in \mathbb{Q}(\omega)$.

4.3. It is well-known that $R = \mathbb{C}[x, y]$ is a UFD (see Herstein [**70**, Corollary 2 to Theorem 3.11.1]). Prove that $I = \{f(x, y) \in R : f(0, 0) = 0\}$ is an ideal of R which is not principal, so that R is not a PID. Hint: $x, y \in I$.

4.4. Given a nonunit $\alpha \neq 0$ in a PID R, prove that α is irreducible $\iff$ α is prime $\iff$ αR is a prime ideal $\iff$ αR is a maximal ideal.

4.5. Prove Lemma 4.5. Hint for (ii): use (i) and (2.4).

4.6. While $\mathbb{Z}[\omega]$ is a PID and a UFD, this exercise will show that the closely related ring $\mathbb{Z}[\sqrt{-3}]$ has neither property.

(a) Show that ± 1 are the only units of $\mathbb{Z}[\sqrt{-3}]$.

(b) Show that 2, $1+\sqrt{-3}$ and $1-\sqrt{-3}$ are nonassociate and irreducible in $\mathbb{Z}[\sqrt{-3}]$. Since $4 = 2 \cdot 2 = (1+\sqrt{-3})(1-\sqrt{-3})$, these elements are not prime and thus $\mathbb{Z}[\sqrt{-3}]$ is not a UFD.

(c) Show that the ideal in $\mathbb{Z}[\sqrt{-3}]$ generated by 2 and $1+\sqrt{-3}$ is not principal. Thus $\mathbb{Z}[\sqrt{-3}]$ is not a PID.

4.7. This exercise is concerned with the proof of Proposition 4.7. Let p be a prime number.

(a) When $p \equiv 1 \bmod 3$, we showed that $p = \pi\overline{\pi}$ where π and $\overline{\pi}$ are prime in $\mathbb{Z}[\omega]$. Prove that π and $\overline{\pi}$ are nonassociate in $\mathbb{Z}[\omega]$.

(b) When $p \equiv 2 \bmod 3$, prove that p is prime in $\mathbb{Z}[\omega]$. Hint: show that p is irreducible. Note that by Lemma 2.5, the equation $p = N(\alpha)$ has no solutions.

4.8. Complete the proof of Lemma 4.8.

4.9. Let π be a prime of $\mathbb{Z}[\omega]$ not associate to $1-\omega$.

(a) Show that $3 \mid N(\pi) - 1$.

(b) If two of $1, \omega, \omega^2$ are congruent modulo π, then show that $1 \equiv \omega \bmod \pi$, and explain why this contradicts our assumption on π. This proves that 1, ω and ω^2 are distinct modulo π.

4.10. Let π be a prime of $\mathbb{Z}[\omega]$ not associate to $1-\omega$, and let $\alpha, \beta \in \mathbb{Z}[\omega]$ be not divisible by π. Verify the following properties of the Legendre symbol.

(a) $(\alpha\beta/\pi)_3 = (\alpha/\pi)_3(\beta/\pi)_3$.

(b) $(\alpha/\pi)_3 = (\beta/\pi)_3$ when $\alpha \equiv \beta \bmod \pi$.

4.11. Let π be prime in $\mathbb{Z}[\omega]$. Assuming that $(\mathbb{Z}[\omega]/\pi\mathbb{Z}[\omega])^*$ is cyclic, prove (4.11).

4.12. Let π be a prime of $\mathbb{Z}[\omega]$ which is not associate to $1-\omega$. Prove that exactly two of the six associates of π are primary.

4.13. Prove the top line of (4.13).

4.14. Use the hints in the text to prove that the congruence $x^3 \equiv a \bmod p$ is always solvable when p is a prime congruent to 2 modulo 3.

4.15. In this problem we will give an application of cubic reciprocity which is similar to Theorem 4.15. Let $p \equiv 1 \bmod 3$ be a prime.

(a) Use the proof of Theorem 4.15 to show that $4p$ can be written in the form $4p = a^2 + 27b^2$, where $a \equiv 1 \bmod 3$. Conclude that $\pi = (a + 3\sqrt{-3}b)/2$ is a primary prime of $\mathbb{Z}[\omega]$ and that $p = \pi\overline{\pi}$.

(b) Show that the supplementary laws (4.13) can be written

$$\begin{aligned}\left(\frac{\omega}{\pi}\right)_3 &= \omega^{2(a+2)/3}\\ \left(\frac{1-\omega}{\pi}\right)_3 &= \omega^{(a+2)/3+b}\end{aligned}$$

where π is as in part (a).

(c) Use part (b) to show that $(3/\pi)_3 = \omega^{2b}$.
(d) Use part (c) and (4.14) to prove that for a prime p,

$$4p = x^2 + 243y^2 \iff \begin{cases} p \equiv 1 \bmod 3 \text{ and } 3 \text{ is a} \\ \text{cubic residue modulo } p. \end{cases}$$

Euler conjectured the result of part (d) (in a slightly different form) in his *Tractatus* [**43**, Vol. V, pp. XXII and 250].

4.16. In this exercise we will discuss the properties of the Gaussian integers $\mathbb{Z}[i]$.
(a) Use the norm function to prove that $\mathbb{Z}[i]$ is Euclidean.
(b) Prove the analogs of Lemmas 4.5 and 4.6 for $\mathbb{Z}[i]$.
(c) Prove Proposition 4.18.
(d) Formulate and prove the analog of Lemma 4.8 for $\mathbb{Z}[i]$.
(e) Prove (4.19).

4.17. If π is a prime of $\mathbb{Z}[i]$ not associate to $1+i$, show that $4 \mid N(\pi) - 1$ and that ± 1 and $\pm i$ are all distinct modulo π.

4.18. This exercise is devoted to the properties of the Legendre symbol $(\alpha/\pi)_4$, where π is a prime of $\mathbb{Z}[i]$ not associate to $1+i$ and α is not divisible by π.
(a) Show that $\alpha^{(N(\pi)-1)/4}$ is congruent to a unique fourth root of unity modulo π. This shows that the Legendre symbol, as given in (4.20), is well-defined. Hint: use Exercise 4.17.
(b) Prove that the analogs of the properties given in Exercise 4.10 hold for $(\alpha/\pi)_4$.
(c) Prove that

$$\left(\frac{\alpha}{\pi}\right)_4 = 1 \iff x^4 \equiv \alpha \bmod \pi \text{ is solvable in } \mathbb{Z}[i].$$

4.19. In this exercise we will study the integer congruence $x^4 \equiv a \bmod p$, where $p \equiv 1 \bmod 4$ is prime and a is an integer not divisible by p.
(a) Write $p = \pi\overline{\pi}$ in $\mathbb{Z}[i]$. Then use (4.20) to show that $(a/\pi)_4^2 = (a/p)$, and conclude that $(a/\pi)_4 = \pm 1$ if and only if $(a/p) = 1$.
(b) Verify the partition of $(\mathbb{Z}/p\mathbb{Z})^*$ described in the discussion following (4.20).

4.20. Here we will study the congruence $x^4 \equiv a \bmod p$ when $p \equiv 3 \bmod 4$ is prime and a is an integer not divisible by p.
(a) Use (4.20) to show that $(a/p)_4 = 1$. Thus a is a fourth power modulo p in the ring $\mathbb{Z}[i]$.
(b) Show that a is the biquadratic residue of an *integer* modulo p if and only if $(a/p) = 1$. Hint: study the squaring map on the subgroup of $(\mathbb{Z}/p\mathbb{Z})^*$ consisting of quadratic residues.

4.21. If a prime π of $\mathbb{Z}[i]$ is not associate to $1+i$, then show that a unique associate of π is primary.

4.22. Prove the top formula of (4.22).

4.23. Use the supplementary laws (4.22) to prove part (i) of Theorem 4.23.

4.24. Let $p \equiv 1 \bmod 4$ be prime, and write $p = a^2 + b^2$, where a is odd and b is even. The goal of this exercise is to present Dirichlet's elementary proof that $(2/\pi)_4 = i^{ab/2}$, where $\pi = a + bi$.

(a) Use quadratic reciprocity for the Jacobi symbol to show that $(a/p) = 1$.
(b) Use $2p = (a+b)^2 + (a-b)^2$ and quadratic reciprocity to show that

$$\left(\frac{a+b}{p}\right) = (-1)^{((a+b)^2-1)/8}.$$

(c) Use part (b) and (4.20) to show that

$$\left(\frac{a+b}{p}\right) = \left(\frac{i}{\pi}\right)_4 i^{ab/2}.$$

Hint: $-1 = i^2$.
(d) From $(a+b)^2 \equiv 2ab \bmod p$, deduce that
  (i) $(a+b)^{(p-1)/2} \equiv (2ab)^{(p-1)/4} \bmod p$.
  (ii) $(a+b/p) = (2ab/\pi)_4$.
(e) Show that $2ab \equiv 2a^2 i \bmod \pi$, and then use part (a) and Exercise 4.19 to show that

$$\left(\frac{2ab}{\pi}\right)_4 = \left(\frac{2i}{\pi}\right)_4.$$

(f) Combine parts (c), (d) and (e) to show that $(2/\pi)_4 = i^{ab/2}$.

4.25. In this exercise we will study Gauss' statement of biquadratic reciprocity.
(a) If π is a primary prime of $\mathbb{Z}[i]$, then show that either $\pi \equiv 1 \bmod 4$ or $\pi \equiv 3 + 2i \bmod 4$.
(b) Let π and θ be distinct primary primes in $\mathbb{Z}[i]$. Show that biquadratic reciprocity is equivalent to the following two statements:

If either π or θ is congruent to 1 modulo 4, then $(\pi/\theta)_4 = (\theta/\pi)_4$.
If π and θ are both congruent to $3+2i$ modulo 4, then $(\pi/\theta)_4 = -(\theta/\pi)_4$.

This is how Gauss states biquadratic reciprocity in [**52**, Vol. II, §67, p. 138].

4.26. If 2 is a biquadratic residue modulo an odd prime p, prove that $p \equiv \pm 1 \bmod 8$.

4.27. In this exercise, we will present Gauss' proof that for a prime $p \equiv 1 \bmod 8$, the biquadratic character of 2 is determined by the decomposition $p = a^2 + 2b^2$. As usual, we write $p = \pi\overline{\pi}$ in $\mathbb{Z}[i]$.
(a) Show that $(-1/\pi)_4 = 1$ when $p \equiv 1 \bmod 8$.
(b) Use the properties of the Jacobi symbol to show that

$$\left(\frac{a}{p}\right) = (-1)^{(a^2-1)/8}.$$

(c) Use the Jacobi symbol to show that $(b/p) = 1$. Hint: write $b = 2^m c$, c odd, and first show that $(c/p) = 1$.
(d) Show that

$$\left(\frac{2}{\pi}\right)_4 = \left(\frac{-2b^2}{\pi}\right)_4 = \left(\frac{a^2}{\pi}\right)_4 = \left(\frac{a}{p}\right).$$

Hint: use Exercise 4.19.

Combining parts (b) and (d), we see that $(2/\pi)_4 = (-1)^{(a^2-1)/8}$, and Gauss' claim follows. If you read Gauss' original argument [**52**, Vol. II, §13], you'll appreciate how much the Jacobi symbol simplifies things.

4.28. Let (f,λ) and (f,μ) be periods, and write $(f,\mu) = \zeta^{\mu_1} + \cdots + \zeta^{\mu_f}$. Then prove that

$$(f,\lambda)\cdot(f,\mu) = \sum_{j=1}^{f}(f,\lambda+\mu_j).$$

4.29. Let $p \equiv 1 \bmod 3$ be prime, and set $p-1=3f$. Let $(f,1)$, (f,g) and (f,g^2) be the periods as in the text. Recall that g is a primitive root modulo p. In this problem we will describe Gauss' proof of (4.24) (see [**51**, §358]). For $i,j \in \{0,1,2\}$, let (ij) be the number of pairs (m,n), $0 \le m,n \le f-1$, such that

$$1+g^{3m+i} \equiv g^{3n+j} \bmod p.$$

(a) Show that the number of solutions modulo p of the equation $x^3 - y^3 \equiv 1 \bmod p$ is $N = 9(00)+6$.
(b) Use Exercise 4.28 to show that

$$\begin{aligned}(f,1)\cdot(f,1) &= f + (00)(f,1) + (01)(f,g) + (02)(f,g^2)\\ (f,1)\cdot(f,g) &= (10)(f,1) + (11)(f,g) + (12)(f,g^2)\end{aligned}$$

and conclude that $(00)+(01)+(02) = f-1$ and $(10)+(11)+(12) = f$. Hint: $(f,0)=f$ and $-1=(-1)^3$.
(c) Show that $(10)=(22), (11)=(20)$ and $(12)=(21)$. Hint: expand $(f,g)\cdot(f,1)$ and compare it to what you got in part (b).
(d) Show that the 9 quantities (ij) reduce to three:

$$\begin{aligned}\alpha &= (12) = (21) = (00)+1\\ \beta &= (01) = (10) = (22)\\ \gamma &= (02) = (20) = (11).\end{aligned}$$

Hint: to prove $(01)=(10)$ and $(02)=(20)$, show that if (m,n) is counted by (10), then the pair (n',m') defined by

$$\begin{aligned}n + f/2 &\equiv n' \bmod f,\ 0 \le n' \le f-1\\ m + f/2 &\equiv m' \bmod f,\ 0 \le m' \le f-1.\end{aligned}$$

is counted by (01). Now use parts (b) and (c).
(e) Note that $(f,1)\cdot(f,g)\cdot(f,g^2)$ is an integer. By expanding this quantity in terms of α, β and γ, show that

$$\alpha^2+\beta^2+\gamma^2-\alpha = \alpha\beta+\beta\gamma+\alpha\gamma.$$

(f) Using part (e), show that

$$(6\alpha-3\beta-3\gamma-2)^2 + 27(\beta-\gamma)^2 = 12(\alpha+\beta+\gamma)+4.$$

(g) Recall that $\alpha+\beta+\gamma = f$ (this was proved in (b)) and that $p-1=3f$. Then use part (f) to show that

$$4p = a^2 + 27b^2,$$

where $a = 6\alpha-3\beta-3\gamma-2$ and $b = \beta-\gamma$.
(h) Let a be as in part (g). Show that

$$a = 9\alpha - 3(\alpha+\beta+\gamma) - 2 = 9\alpha - p - 1.$$

Then use $\alpha = (00) + 1$ and part (a) to conclude that

$$a = N - p + 2.$$

This proves (4.24).

In his first memoir on biquadratic residues [**52**, Vol. II, §§15–20, pp. 78–89], Gauss used a biquadratic analog of the (ij)'s (without any mention of periods) to determine the biquadratic character of 2.

CHAPTER 2

Class Field Theory

5. The Hilbert Class Field and $p = x^2 + ny^2$

In Chapter 1, we used elementary techniques to study the primes represented by $x^2 + ny^2$, $n > 0$. Genus theory told us when $p = x^2 + ny^2$ for a large but finite number of n's, and cubic and biquadratic reciprocity enabled us to treat two cases where genus theory failed. These methods are lovely but limited in scope. To solve $p = x^2 + ny^2$ when $n > 0$ is arbitrary, we will need class field theory, and this is the main task of Chapter 2. But rather than go directly to the general theorems of class field theory, in §5 we will first study the special case of the Hilbert class field. Theorem 5.1 below will use Artin Reciprocity for the Hilbert class field to solve our problem for infinitely many (but not all) $n > 0$. We will then study the case $p = x^2 + 14y^2$ in detail. This is a case where our previous methods failed, but once we determine the Hilbert class field of $\mathbb{Q}(\sqrt{-14})$, Theorem 5.1 will immediately give us a criterion for when $p = x^2 + 14y^2$.

The central notion of this section is the Hilbert class field of a number field K. We do not assume any previous acquaintance with this topic, for one of our goals is to introduce the reader to this more accessible part of class field theory. To see what the Hilbert class field has to do with the problem of representing primes by $x^2 + ny^2$, let's state the main theorem we intend to prove:

THEOREM 5.1. *Let $n > 0$ be an integer satisfying the following condition:*

$$n \text{ squarefree}, \ n \not\equiv 3 \bmod 4. \tag{5.2}$$

Then there is a monic irreducible polynomial $f_n(x) \in \mathbb{Z}[x]$ of degree $h(-4n)$ such that if an odd prime p divides neither n nor the discriminant of $f_n(x)$, then

$$p = x^2 + ny^2 \iff \begin{cases} (-n/p) = 1 \text{ and } f_n(x) \equiv 0 \bmod p \\ \text{has an integer solution.} \end{cases}$$

Furthermore, $f_n(x)$ may be taken to be the minimal polynomial of a real algebraic integer α for which $L = K(\alpha)$ is the Hilbert class field of $K = \mathbb{Q}(\sqrt{-n})$.

While (5.2) does not give all integers $n > 0$, it gives infinitely many, so that Theorem 5.1 represents some real progress. In §9 we will use the full power of class field theory to prove a version of Theorem 5.1 that holds for *all* positive integers n.

A. Number Fields

We will review some basic facts from algebraic number theory, including Dedekind domains, factorization of ideals, and ramification. Most proofs will be omitted, though references will be given. Readers looking for a more complete treatment should consult Borevich and Shafarevich [**11**], Lang [**89**] or Marcus [**96**]. For a compact presentation of this material, see Ireland and Rosen [**75**, Chapter 12].

To begin, we define a *number field* K to be a subfield of the complex numbers $\mathbb{C}$ which has finite degree over $\mathbb{Q}$. The degree of K over $\mathbb{Q}$ is denoted $[K : \mathbb{Q}]$. Given such a field K, we let $\mathcal{O}_K$ denote the algebraic integers of K, i.e., the set of all $\alpha \in K$ which are roots of a monic integer polynomial. The basic structure of $\mathcal{O}_K$ is given in the following proposition:

PROPOSITION 5.3. *Let K be a number field.*

(i) *$\mathcal{O}_K$ is a subring of $\mathbb{C}$ whose field of fractions is K.*
(ii) *$\mathcal{O}_K$ is a free $\mathbb{Z}$-module of rank $[K : \mathbb{Q}]$.*

PROOF. See Borevich and Shafarevich [**11**, §2.2] or Marcus [**96**, Corollaries to Theorems 2 and 9]. Q.E.D.

We will often call $\mathcal{O}_K$ the *ring of integers* of K. To begin our study of $\mathcal{O}_K$, we note that part (ii) of Proposition 5.3 has the following useful consequence concerning the ideals of $\mathcal{O}_K$:

COROLLARY 5.4. *If K is a number field and $\mathfrak{a}$ is a nonzero ideal of $\mathcal{O}_K$, then the quotient ring $\mathcal{O}_K/\mathfrak{a}$ is finite.*

PROOF. See Exercise 5.1. Q.E.D.

Given a nonzero ideal $\mathfrak{a}$ of the ring $\mathcal{O}_K$, its *norm* is defined to be $N(\mathfrak{a}) = |\mathcal{O}_K/\mathfrak{a}|$. Corollary 5.4 guarantees that $N(\mathfrak{a})$ is finite.

When we studied the rings $\mathbb{Z}[\omega]$ and $\mathbb{Z}[i]$ in §4, we used the fact that they were unique factorization domains. In general, the rings $\mathcal{O}_K$ are not UFDs, but they have another property which is almost as good: they are Dedekind domains. This means the following:

THEOREM 5.5. *Let $\mathcal{O}_K$ be the ring of integers in a number field K. Then $\mathcal{O}_K$ is a Dedekind domain, which means that*

(i) *$\mathcal{O}_K$ is integrally closed in K, i.e., if $\alpha \in K$ satisfies a monic polynomial with coefficients in $\mathcal{O}_K$, then $\alpha \in \mathcal{O}_K$.*
(ii) *$\mathcal{O}_K$ is Noetherian, i.e., given any chain of ideals $\mathfrak{a}_1 \subset \mathfrak{a}_2 \subset \cdots$, there is an integer n such that $\mathfrak{a}_n = \mathfrak{a}_{n+1} = \cdots$.*
(iii) *Every nonzero prime ideal of $\mathcal{O}_K$ is maximal.*

PROOF. The proof of (i) follows easily from the properties of algebraic integers (see Lang [**89**, §1.2] or Marcus [**96**, Exercise 4 to Chapter 2]), while (ii) and (iii) are straightforward consequences of Corollary 5.4 (see Exercise 5.1). Q.E.D.

The most important property of a Dedekind domain is that it has unique factorization at the level of ideals. More precisely:

COROLLARY 5.6. *If K is a number field, then any nonzero ideal $\mathfrak{a}$ in $\mathcal{O}_K$ can be written as a product*

$$\mathfrak{a} = \mathfrak{p}_1 \cdots \mathfrak{p}_r$$

of prime ideals, and the decomposition is unique up to order. Furthermore, the $\mathfrak{p}_i$'s are exactly the prime ideals of $\mathcal{O}_K$ containing $\mathfrak{a}$.

PROOF. This corollary holds for any Dedekind domain. For a proof, see Lang [**89**, §1.6] or Marcus [**96**, Chapter 3, Theorem 16]. In Ireland and Rosen [**75**, §12.2] there is a nice proof (due to Hurwitz) that is special to the number field case. Q.E.D.

Prime ideals play an especially important role in algebraic number theory. We will often say "prime" rather than "nonzero prime ideal," and the terms "prime of K" and "nonzero prime ideal of $\mathcal{O}_K$" will be used interchangeably. Notice that when $\mathfrak{p}$ is a prime of K, the quotient ring $\mathcal{O}_K/\mathfrak{p}$ is a finite field by Corollary 5.4 and Theorem 5.5. This field is called the *residue field* of $\mathfrak{p}$.

Besides ideals of $\mathcal{O}_K$, we will also use *fractional ideals*, which are the nonzero finitely generated $\mathcal{O}_K$-submodules of K. The name "fractional ideal" comes from the fact that such an ideal can be written in the form $\alpha\mathfrak{a}$, where $\alpha \in K$ and $\mathfrak{a}$ is an ideal of $\mathcal{O}_K$ (see Exercise 5.2). Readers unfamiliar with fractional ideals should consult Marcus [**96**, Exercise 31 of Chapter 3]. The basic properties of fractional ideals are:

PROPOSITION 5.7. *Let $\mathfrak{a}$ be a fractional $\mathcal{O}_K$-ideal.*

(i) *$\mathfrak{a}$ is invertible, i.e., there is a fractional $\mathcal{O}_K$-ideal $\mathfrak{b}$ such that $\mathfrak{a}\mathfrak{b} = \mathcal{O}_K$. The ideal $\mathfrak{b}$ will be denoted $\mathfrak{a}^{-1}$.*

(ii) *$\mathfrak{a}$ can be written uniquely as a product $\mathfrak{a} = \prod_{i=1}^{r} \mathfrak{p}_i^{r_i}$, $r_i \in \mathbb{Z}$, where the $\mathfrak{p}_i$'s are distinct prime ideals of $\mathcal{O}_K$.*

PROOF. See Lang [**89**, §1.6] or Marcus [**96**, Exercise 31 of Chapter 3]. Q.E.D.

We will let I_K denote the set of all fractional ideals of K. Then I_K is closed under multiplication of ideals (see Exercise 5.2), and part (i) of Proposition 5.7 shows that I_K is a group. The most important subgroup of I_K is the subgroup P_K of *principal fractional ideals*, i.e., those of the form $\alpha\mathcal{O}_K$ for some $\alpha \in K^*$. The quotient I_K/P_K is the *ideal class group* and is denoted by $C(\mathcal{O}_K)$. A basic fact is that $C(\mathcal{O}_K)$ is a finite group (see Borevich and Shafarevich [**11**, §3.7] or Marcus [**96**, Corollary 2 to Theorem 35]). In the case of imaginary quadratic fields, we will see in Theorem 5.30 that the ideal class group is closely related to the form class group defined in §3.

We will next introduce the idea of ramification, which is concerned with the behavior of primes in finite extensions. Suppose that K is a number field, and let L be a finite extension of K. If $\mathfrak{p}$ is a prime ideal of $\mathcal{O}_K$, then $\mathfrak{p}\mathcal{O}_L$ is an ideal of $\mathcal{O}_L$, and hence has a prime factorization

$$\mathfrak{p}\mathcal{O}_L = \mathfrak{P}_1^{e_1} \cdots \mathfrak{P}_g^{e_g}$$

where the $\mathfrak{P}_i$'s are the distinct primes of L containing $\mathfrak{p}$. The integer e_i, also written $e_{\mathfrak{P}_i|\mathfrak{p}}$, is called the *ramification index* of $\mathfrak{p}$ in $\mathfrak{P}_i$. Each prime $\mathfrak{P}_i$ containing $\mathfrak{p}$ also gives a residue field extension $\mathcal{O}_K/\mathfrak{p} \subset \mathcal{O}_L/\mathfrak{P}_i$, and its degree, written f_i or $f_{\mathfrak{P}_i|\mathfrak{p}}$, is the *inertial degree* of $\mathfrak{p}$ in $\mathfrak{P}_i$. The basic relation between the e_i's and f_i's is given by:

THEOREM 5.8. *Let $K \subset L$ be number fields, and let $\mathfrak{p}$ be a prime of K. If e_i (resp. f_i), $i = 1, \ldots, g$ are the ramification indices (resp. inertial degrees) defined above, then*

$$\sum_{i=1}^{g} e_i f_i = [L : K].$$

PROOF. See Borevich and Shafarevich [**11**, §3.5] or Marcus [**96**, Theorem 21]. Q.E.D.

In the situation of Theorem 5.8, we say that a prime $\mathfrak{p}$ of K *ramifies* in L if any of the ramification indices e_i are greater than 1. It can be proved that only a finite number of primes of K ramify in L (see Lang [**89**, §III.2] or Marcus [**96**, Corollary 3 to Theorem 24]).

Most of the extensions $K \subset L$ we will deal with will be Galois extensions, and in this case the above description can be simplified as follows:

THEOREM 5.9. *Let $K \subset L$ be a Galois extension, and let $\mathfrak{p}$ be prime in K.*

(i) *The Galois group $\mathrm{Gal}(L/K)$ acts transitively on the primes of L containing $\mathfrak{p}$, i.e., if $\mathfrak{P}$ and $\mathfrak{P}'$ are primes of L containing $\mathfrak{p}$, then there is $\sigma \in \mathrm{Gal}(L/K)$ such that $\sigma(\mathfrak{P}) = \mathfrak{P}'$.*

(ii) *The primes $\mathfrak{P}_1, \ldots, \mathfrak{P}_g$ of L containing $\mathfrak{p}$ all have the same ramification index e and the same inertial degree f, and the formula of Theorem 5.8 becomes*

$$efg = [L : K].$$

PROOF. For a proof of (i), see Lang [**89**, §1.7] or Marcus [**96**, Theorem 23]. The proof of (ii) follows easily from (i) and is left to the reader (see Exercise 5.3). Q.E.D.

Given a Galois extension $K \subset L$, an ideal $\mathfrak{p}$ of K ramifies if $e > 1$, and is unramified if $e = 1$. If $\mathfrak{p}$ satisfies the stronger condition $e = f = 1$, we say that $\mathfrak{p}$ *splits completely* in L. Such a prime is unramified, and in addition $\mathfrak{p}\mathcal{O}_L$ is the product of $[L : K]$ distinct primes, the maximum number allowed by Theorem 5.9. In §8 we will show that L is determined uniquely by the primes of K that split completely in L.

We will also need some facts concerning decomposition and inertia groups. Let $K \subset L$ be Galois, and let $\mathfrak{P}$ be a prime of L. Then the *decomposition group* and *inertia group* of $\mathfrak{P}$ are defined by

$$\begin{aligned} D_{\mathfrak{P}} &= \{\sigma \in \mathrm{Gal}(L/K) : \sigma(\mathfrak{P}) = \mathfrak{P}\} \\ I_{\mathfrak{P}} &= \{\sigma \in \mathrm{Gal}(L/K) : \sigma(\alpha) \equiv \alpha \bmod \mathfrak{P} \text{ for all } \alpha \in \mathcal{O}_L\}. \end{aligned}$$

It is easy to show that $I_{\mathfrak{P}} \subset D_{\mathfrak{P}}$ and that an element $\sigma \in D_{\mathfrak{P}}$ induces an automorphism $\tilde{\sigma}$ of $\mathcal{O}_L/\mathfrak{P}$ which is the identity on $\mathcal{O}_K/\mathfrak{p}$, $\mathfrak{p} = \mathfrak{P} \cap \mathcal{O}_K$ (see Exercise 5.4). If $\widetilde{G}$ denotes the Galois group of $\mathcal{O}_K/\mathfrak{p} \subset \mathcal{O}_L/\mathfrak{P}$, it follows that $\tilde{\sigma} \in \widetilde{G}$. Thus the map $\sigma \mapsto \tilde{\sigma}$ defines a homomorphism $D_{\mathfrak{P}} \to \widetilde{G}$ whose kernel is exactly the inertia group $I_{\mathfrak{P}}$ (see Exercise 5.4). Then we have:

PROPOSITION 5.10. *Let $D_{\mathfrak{P}}$, $I_{\mathfrak{P}}$ and $\widetilde{G}$ be as above.*

(i) *The homomorphism $D_{\mathfrak{P}} \to \widetilde{G}$ is surjective. Thus $D_{\mathfrak{P}}/I_{\mathfrak{P}} \simeq \widetilde{G}$.*

(ii) *$|I_{\mathfrak{P}}| = e_{\mathfrak{P}|\mathfrak{p}}$ and $|D_{\mathfrak{P}}| = e_{\mathfrak{P}|\mathfrak{p}} f_{\mathfrak{P}|\mathfrak{p}}$.*

PROOF. See Lang [**89**, §1.7] or Marcus [**96**, Theorem 28]. Q.E.D.

The following proposition will help us decide when a prime is unramified or split completely in a Galois extension:

PROPOSITION 5.11. *Let $K \subset L$ be a Galois extension, where $L = K(\alpha)$ for some $\alpha \in \mathcal{O}_L$. Let $f(x)$ be the monic minimal polynomial of α over K, so that $f(x) \in \mathcal{O}_K[x]$. If $\mathfrak{p}$ is prime in $\mathcal{O}_K$ and $f(x)$ is separable modulo $\mathfrak{p}$, then:*

(i) *$\mathfrak{p}$ is unramified in L.*

(ii) *If* $f(x) \equiv f_1(x) \cdots f_g(x) \bmod \mathfrak{p}$, *where the* $f_i(x)$ *are monic, and distinct and irreducible modulo* $\mathfrak{p}$, *then* $\mathfrak{P}_i = \mathfrak{p}\mathcal{O}_L + f_i(\alpha)\mathcal{O}_L$ *is a prime ideal of* $\mathcal{O}_L$, $\mathfrak{P}_i \neq \mathfrak{P}_j$ *for* $i \neq j$, *and*

$$\mathfrak{p}\mathcal{O}_L = \mathfrak{P}_1 \cdots \mathfrak{P}_g.$$

Furthermore, the $f_i(x)$ *all have the same degree, which is equal to the inertial degree* f.

(iii) $\mathfrak{p}$ *splits completely in* L *if and only if* $f(x) \equiv 0 \bmod \mathfrak{p}$ *has a solution in* $\mathcal{O}_K$.

PROOF. Note that (i) and (iii) are immediate consequences of (ii) (see Exercise 5.5). To prove (ii), note that $f(x)$ separable modulo $\mathfrak{p}$ implies that

$$f(x) \equiv f_1(x) \cdots f_g(x) \bmod \mathfrak{p},$$

where the $f_i(x)$ are distinct and irreducible modulo $\mathfrak{p}$. The fact that the above congruence governs the splitting of $\mathfrak{p}$ in $\mathcal{O}_L$ is a general fact that holds for arbitrary finite extensions (see Marcus [**96**, Theorem 27]). However, the decomposition group from Proposition 5.10 makes the proof in the Galois case especially easy. You will work this out in Exercise 5.6. Q.E.D.

B. Quadratic Fields

To better understand the theory just sketched, let's apply it to the case of quadratic number fields. Such a field can be written uniquely in the form $K = \mathbb{Q}(\sqrt{N})$, where $N \neq 0, 1$ is a squarefree integer. The basic invariant of K is its *discriminant* d_K, which is defined to be

$$d_K = \begin{cases} N & \text{if } N \equiv 1 \bmod 4 \\ 4N & \text{otherwise.} \end{cases} \tag{5.12}$$

Note that $d_K \equiv 0, 1 \bmod 4$ and $K = \mathbb{Q}(\sqrt{d_K})$, so that a quadratic field is determined by its discriminant.

The next step is to describe the integers $\mathcal{O}_K$ of K. Writing $K = \mathbb{Q}(\sqrt{N})$, N squarefree, one can show that

$$\mathcal{O}_K = \begin{cases} \mathbb{Z}[\sqrt{N}] & N \not\equiv 1 \bmod 4 \\ \mathbb{Z}\left[\dfrac{1+\sqrt{N}}{2}\right] & N \equiv 1 \bmod 4 \end{cases} \tag{5.13}$$

(see Exercise 5.7 or Marcus [**96**, Corollary 2 to Theorem 1]). Hence the rings $\mathbb{Z}[\omega]$ and $\mathbb{Z}[i]$ from §4 are the full rings of integers in their respective fields. Using the discriminant, this description of $\mathcal{O}_K$ may be written more elegantly as follows:

$$\mathcal{O}_K = \mathbb{Z}\left[\frac{d_K + \sqrt{d_K}}{2}\right] \tag{5.14}$$

(see Exercise 5.7).

We can now explain the restriction (5.2) made on n in Theorem 5.1. Namely, given $n > 0$, let K be the imaginary quadratic field $\mathbb{Q}(\sqrt{-n})$. Then (5.12) and (5.13) imply that

$$d_K = -4n \iff \mathcal{O}_K = \mathbb{Z}[\sqrt{-n}] \iff n \text{ satisfies (5.2)} \tag{5.15}$$

(see Exercise 5.8). Thus the condition (5.2) on n is equivalent to $\mathbb{Z}[\sqrt{-n}]$ being the full ring of integers in K. For other n's, we will see in §7 that $\mathbb{Z}[\sqrt{-n}]$ is no longer a Dedekind domain but still has a lot of interesting structure.

We next want to discuss the arithmetic of a quadratic field K. As in §4, this means describing units and primes, the difference being that "prime" now means "prime ideal". Let's first consider units. Quadratic fields come in two flavors, real ($d_K > 0$) and imaginary ($d_K < 0$), and the units $\mathcal{O}_K^*$ behave quite differently in the two cases. In the imaginary case, there are only finitely many units. In §4 we computed $\mathcal{O}_K^*$ for $K = \mathbb{Q}(\sqrt{-3})$ or $\mathbb{Q}(i)$, and for all other imaginary quadratic fields it turns out that $\mathcal{O}_K^* = \{\pm 1\}$ (see Exercise 5.9). On the other hand, real quadratic fields always have infinitely many units, and determining them is related to Pell's equation and continued fractions (see Borevich and Shafarevich [**11**, §2.7]).

Before describing the primes of $\mathcal{O}_K$, we will need one useful bit of notation: if $D \equiv 0, 1 \bmod 4$, then the *Kronecker symbol* $(D/2)$ is defined by

$$\left(\frac{D}{2}\right) = \begin{cases} 0 & \text{if } D \equiv 0 \bmod 4 \\ 1 & \text{if } D \equiv 1 \bmod 8 \\ -1 & \text{if } D \equiv 5 \bmod 8. \end{cases}$$

We will most often apply this when $D = d_K$ is the discriminant of a quadratic field K. The following proposition tells us about the primes of quadratic fields:

PROPOSITION 5.16. *Let K be a quadratic field of discriminant d_K, and let the nontrivial automorphism of K be denoted $\alpha \mapsto \alpha'$. Let p be prime in $\mathbb{Z}$.*

(i) *If $(d_K/p) = 0$ (i.e., $p \mid d_K$), then $p\mathcal{O}_K = \mathfrak{p}^2$, where $\mathfrak{p}$ is prime in $\mathcal{O}_K$.*
(ii) *If $(d_K/p) = 1$, then $p\mathcal{O}_K = \mathfrak{p}\mathfrak{p}'$, where $\mathfrak{p} \neq \mathfrak{p}'$ are prime in $\mathcal{O}_K$.*
(iii) *If $(d_K/p) = -1$, then $p\mathcal{O}_K$ is prime in $\mathcal{O}_K$.*

Furthermore, the primes in (i)–(iii) *above give all nonzero primes of $\mathcal{O}_K$.*

PROOF. To prove (i), suppose that p is an odd prime dividing d_K, and let $\mathfrak{p}$ be the ideal

$$\mathfrak{p} = p\mathcal{O}_K + \sqrt{d_K}\mathcal{O}_K.$$

Squaring, one obtains

$$\mathfrak{p}^2 = p^2\mathcal{O}_K + p\sqrt{d_K}\mathcal{O}_K + d_K\mathcal{O}_K.$$

However, d_K is squarefree (except for a possible factor of 4) and p is an odd divisor, so that $\gcd(p^2, d_K) = p$. It follows easily that $\mathfrak{p}^2 = p\mathcal{O}_K$, and then the relation $efg = [K : \mathbb{Q}] = 2$ from Theorem 5.9 implies that $\mathfrak{p}$ is a prime ideal. The case when $p = 2$ is similar and is left as part of Exercise 5.10.

Let's next prove (ii) and (iii) for an odd prime p not dividing d_K. The key tool will be Proposition 5.11. Note that $f(x) = x^2 - d_K$ is the minimal polynomial of the primitive element $\sqrt{d_K}$ of K over $\mathbb{Q}$, and since $p \nmid 4d_K$, $f(x)$ is separable modulo p. Then Proposition 5.11 shows that p is unramified in K.

If $(d_K/p) = 1$, then the congruence $x^2 \equiv d_K \bmod p$ has a solution, and consequently p splits completely in K by part (iii) of Proposition 5.11, i.e., $p\mathcal{O}_K = \mathfrak{p}_1\mathfrak{p}_2$ for distinct primes $\mathfrak{p}_1$ and $\mathfrak{p}_2$ of $\mathcal{O}_K$. Since $\mathrm{Gal}(K/\mathbb{Q})$ acts transitively on the primes of K containing p (Theorem 5.9), we must have $\mathfrak{p}_1' = \mathfrak{p}_2$, and it follows that $p\mathcal{O}_K$ factors as claimed. If $(d_K/p) = -1$, then $f(x) = x^2 - d_K$ is irreducible modulo p, and hence by part (ii) of Proposition 5.11, $p\mathcal{O}_K$ is prime in K.

The proof of (ii) and (iii) for $p = 2$ is similar and is left as an exercise (see Exercise 5.10). It remains to prove that the prime ideals listed so far are *all* nonzero primes in $\mathcal{O}_K$. The argument is analogous to what we did in Proposition 4.7, and the details are left to the reader (see Exercise 5.10). Q.E.D.

From this proposition, we get the following immediate corollary which tells us how primes of $\mathbb{Z}$ behave in a quadratic extension:

COROLLARY 5.17. *Let K be a quadratic field of discriminant d_K, and let p be an integer prime. Then:*

(i) *p ramifies in K if and only if p divides d_K.*
(ii) *p splits completely in K if and only if $(d_K/p) = 1$.* Q.E.D.

C. The Hilbert Class Field

The Hilbert class field of a number field K is defined in terms of the unramified Abelian extensions of K. To see what these terms mean, we begin with the "Abelian" part. This is easy, for an extension $K \subset L$ is *Abelian* if it is Galois and $\mathrm{Gal}(L/K)$ is an Abelian group. But we aren't quite ready to define "unramified," for we first need to discuss the ramification of infinite primes.

Prime ideals of $\mathcal{O}_K$ are often called *finite primes* to distinguish them from the *infinite primes*, which are determined by the embeddings of K into $\mathbb{C}$. A *real infinite prime* is an embedding $\sigma : K \to \mathbb{R}$, while a *complex infinite prime* is a pair of complex conjugate embeddings $\sigma, \overline{\sigma} : K \to \mathbb{C}$, $\sigma \neq \overline{\sigma}$. Given an extension $K \subset L$, an infinite prime σ of K *ramifies* in L provided that σ is real but it has an extension to L which is complex. For example, the infinite prime of $\mathbb{Q}$ is unramified in $\mathbb{Q}(\sqrt{2})$ but ramified in $\mathbb{Q}(\sqrt{-2})$.

An extension $K \subset L$ is *unramified* if it is unramified at *all* primes, finite or infinite. While this is a very strong restriction, it can happen that a given field has unramified extensions of arbitrarily high degree (an example is $K = \mathbb{Q}(\sqrt{-2 \cdot 3 \cdot 5 \cdot 7 \cdot 11 \cdot 13})$, a consequence of the work of Golod and Shafarevich on class field towers—see Roquette [**106**]). But if we ask for unramified *Abelian* extensions, a much nicer picture emerges. In §8 we will use class field theory to prove the following result:

THEOREM 5.18. *Given a number field K, there is a finite Galois extension L of K such that:*

(i) *L is an unramified Abelian extension of K.*
(ii) *Any unramified Abelian extension of K lies in L.* Q.E.D.

The field L of Theorem 5.18 is called the *Hilbert class field* of K. It is the maximal unramified Abelian extension of K and is clearly unique.

To unlock the full power of the Hilbert class field L of K, we will use the Artin symbol to link L to the ideal structure of $\mathcal{O}_K$. The following lemma is needed to define the Artin symbol:

LEMMA 5.19. *Let $K \subset L$ be a Galois extension, and let $\mathfrak{p}$ be a prime of $\mathcal{O}_K$ which is unramified in L. If $\mathfrak{P}$ is a prime of $\mathcal{O}_L$ containing $\mathfrak{p}$, then there is a unique element $\sigma \in \mathrm{Gal}(L/K)$ such that for all $\alpha \in \mathcal{O}_L$,*

$$\sigma(\alpha) \equiv \alpha^{N(\mathfrak{p})} \bmod \mathfrak{P},$$

where $N(\mathfrak{p}) = |\mathcal{O}_K/\mathfrak{p}|$ is the norm of $\mathfrak{p}$.

PROOF. As in Proposition 5.10, let $D_{\mathfrak{P}}$ and $I_{\mathfrak{P}}$ be the decomposition and inertia groups of $\mathfrak{P}$. Recall that $\sigma \in D_{\mathfrak{P}}$ induces an element $\tilde{\sigma} \in \widetilde{G}$, where $\widetilde{G}$ is the Galois group of $\mathcal{O}_L/\mathfrak{P}$ over $\mathcal{O}_K/\mathfrak{p}$. Since $\mathfrak{p}$ is unramified in L, part (ii) of Proposition 5.10 tells us that $|I_{\mathfrak{P}}| = e_{\mathfrak{P}|\mathfrak{p}} = 1$, and then the first part of the proposition implies that $\sigma \mapsto \tilde{\sigma}$ defines an isomorphism

$$D_{\mathfrak{P}} \xrightarrow{\sim} \widetilde{G}.$$

The structure of the Galois group $\widetilde{G}$ is well-known: if $\mathcal{O}_K/\mathfrak{p}$ has q elements, then $\widetilde{G}$ is a cyclic group with canonical generator given by the *Frobenius automorphism* $x \mapsto x^q$ (see Hasse [**66**, pp. 40–41]). Thus there is a *unique* $\sigma \in D_{\mathfrak{P}}$ which maps to the Frobenius element. Since $q = N(\mathfrak{p})$ by definition, σ satisfies our desired condition

$$\sigma(\alpha) \equiv \alpha^{N(\mathfrak{p})} \bmod \mathfrak{P} \quad \text{for all } \alpha \in \mathcal{O}_L.$$

To prove uniqueness, note that any σ satisfying this condition must lie in $D_{\mathfrak{P}}$, and then we are done. Q.E.D.

The unique element σ of Lemma 5.19 is called the *Artin symbol* and is denoted $((L/K)/\mathfrak{P})$ since it depends on the prime $\mathfrak{P}$ of L. Its crucial property is that for any $\alpha \in \mathcal{O}_L$, we have

$$\left(\frac{L/K}{\mathfrak{P}}\right)(\alpha) \equiv \alpha^{N(\mathfrak{p})} \bmod \mathfrak{P}, \tag{5.20}$$

where $\mathfrak{p} = \mathfrak{P} \cap \mathcal{O}_K$. The Artin symbol $((L/K)/\mathfrak{P})$ has the following useful properties:

COROLLARY 5.21. *Let $K \subset L$ be a Galois extension, and let $\mathfrak{p}$ be an unramified prime of K. Given a prime $\mathfrak{P}$ of L containing $\mathfrak{p}$, we have*:

(i) *If $\sigma \in \mathrm{Gal}(L/K)$, then*

$$\left(\frac{L/K}{\sigma(\mathfrak{P})}\right) = \sigma \left(\frac{L/K}{\mathfrak{P}}\right) \sigma^{-1}.$$

(ii) *The order of $((L/K)/\mathfrak{P})$ is the inertial degree $f = f_{\mathfrak{P}|\mathfrak{p}}$.*
(iii) *$\mathfrak{p}$ splits completely in L if and only if $((L/K)/\mathfrak{P}) = 1$.*

PROOF. The proof of (i) is a direct consequence of the uniqueness of the Artin symbol. The details are left to the reader (see Exercise 5.12).

To prove (ii), recall from the proof of Lemma 5.19 that since $\mathfrak{p}$ is unramified, the decomposition group $D_{\mathfrak{P}}$ is isomorphic to the Galois group of the finite extension $\mathcal{O}_K/\mathfrak{p} \subset \mathcal{O}_L/\mathfrak{P}$ whose degree is the inertial degree f. By definition, the Artin symbol maps to a generator of the Galois group, so that the Artin symbol has order f as desired.

To prove (iii), recall that $\mathfrak{p}$ splits completely in L if and only if $e = f = 1$. Since we're already assuming that $e = 1$, (iii) follows immediately from (ii). Q.E.D.

When $K \subset L$ is an Abelian extension, the Artin symbol $((L/K)/\mathfrak{P})$ depends only on the underlying prime $\mathfrak{p} = \mathfrak{P} \cap \mathcal{O}_K$. To see this, let be $\mathfrak{P}'$ be another prime containing $\mathfrak{p}$. We've seen that $\mathfrak{P}' = \sigma(\mathfrak{P})$ for some $\sigma \in \mathrm{Gal}(L/K)$. Then Corollary 5.21 implies that

$$\left(\frac{L/K}{\mathfrak{P}'}\right) = \left(\frac{L/K}{\sigma(\mathfrak{P})}\right) = \sigma \left(\frac{L/K}{\mathfrak{P}}\right) \sigma^{-1} = \left(\frac{L/K}{\mathfrak{P}}\right)$$

since $\mathrm{Gal}(L/K)$ is Abelian. It follows that whenever $K \subset L$ is Abelian, the Artin symbol can be written as $((L/K)/\mathfrak{p})$.

To see the relevance of the Artin symbol to reciprocity, let's work out an example. Let $K = \mathbb{Q}(\sqrt{-3})$ and $L = K(\sqrt[3]{2})$. Since $\mathcal{O}_K$ is the ring $\mathbb{Z}[\omega]$ of §4, it's a PID, and consequently a prime ideal $\mathfrak{p}$ can be written as $\pi\mathbb{Z}[\omega]$, where π is prime in $\mathbb{Z}[\omega]$. If π doesn't divide 6, it follows from Proposition 5.11 that π is unramified in L (see part (a) of Exercise 5.14). Since $\mathrm{Gal}(L/K) \simeq \mathbb{Z}/3\mathbb{Z}$ is Abelian, we see that $((L/K)/\pi)$ is defined. To determine which automorphism it is, we need only evaluate it on $\sqrt[3]{2}$. The answer is very nice:

$$\left(\frac{L/K}{\pi}\right)(\sqrt[3]{2}) = \left(\frac{2}{\pi}\right)_3 \sqrt[3]{2}. \tag{5.22}$$

So the Artin symbol generalizes the Legendre symbol! To prove this, let $\mathfrak{P}$ be a prime of $\mathcal{O}_L$ containing π. Then, by (5.20),

$$\begin{aligned}\left(\frac{L/K}{\pi}\right)(\sqrt[3]{2}) &\equiv \sqrt[3]{2}^{\,N(\pi)} \mod \mathfrak{P}\\ &\equiv 2^{(N(\pi)-1)/3}\cdot\sqrt[3]{2} \mod \mathfrak{P}.\end{aligned}$$

However, we know from (4.10) that

$$2^{(N(\pi)-1)/3} \equiv \left(\frac{2}{\pi}\right)_3 \mod \pi,$$

and then $\pi \in \mathfrak{P}$ implies

$$\left(\frac{L/K}{\pi}\right)(\sqrt[3]{2}) \equiv \left(\frac{2}{\pi}\right)_3 \sqrt[3]{2} \mod \mathfrak{P}.$$

Since $((L/K)/\pi)(\sqrt[3]{2})$ equals $\sqrt[3]{2}$ times a cube root of unity (which are distinct modulo $\mathfrak{P}$—see part (a) of Exercise 5.13), (5.22) is proved. In Exercise 5.14, we will generalize (5.22) to the case of the nth power Legendre symbol.

When $K \subset L$ is an unramified Abelian extension, things are especially nice because $((L/K)/\mathfrak{p})$ is defined for *all* primes $\mathfrak{p}$ of $\mathcal{O}_K$. To exploit this, let I_K be the group of all fractional ideals of $\mathcal{O}_K$. As we saw in Proposition 5.7, any fractional ideal $\mathfrak{a} \in I_K$ has a prime factorization

$$\mathfrak{a} = \prod_{i=1}^{r} \mathfrak{p}_i^{r_i}, \quad r_i \in \mathbb{Z},$$

and then we define the Artin symbol $((L/K)/\mathfrak{a})$ to be the product

$$\left(\frac{L/K}{\mathfrak{a}}\right) = \prod_{i=1}^{r} \left(\frac{L/K}{\mathfrak{p}_i}\right)^{r_i}.$$

The Artin symbol thus defines a homomorphism, called the *Artin map*,

$$\left(\frac{L/K}{\cdot}\right) : I_K \longrightarrow \mathrm{Gal}(L/K).$$

Notice that when $K \subset L$ is ramified, the Artin map is not defined on all of I_K. This is one reason why the general theorems of class field theory are complicated to state.

The *Artin Reciprocity Theorem for the Hilbert Class Field* relates the Hilbert class field to the ideal class group $C(\mathcal{O}_K)$ as follows:

THEOREM 5.23. *If L is the Hilbert class field of a number field K, then the Artin map*

$$\left(\frac{L/K}{\cdot}\right) : I_K \longrightarrow \mathrm{Gal}(L/K)$$

is surjective, and its kernel is exactly the subgroup P_K of principal fractional ideals. Thus the Artin map induces an isomorphism

$$C(\mathcal{O}_K) \xrightarrow{\sim} \mathrm{Gal}(L/K). \qquad \text{Q.E.D.}$$

This theorem will follow from the results of §8. The appearance of the class group $C(\mathcal{O}_K)$ explains why L is called a "class field."

If we apply Galois theory to Theorems 5.18 and 5.23, we get the following classification of unramified Abelian extensions of K (see Exercise 5.17):

COROLLARY 5.24. *Given a number field K, there is a one-to-one correspondence between unramified Abelian extensions M of K and subgroups H of the ideal class group $C(\mathcal{O}_K)$. Furthermore, if the extension $K \subset M$ corresponds to the subgroup $H \subset C(\mathcal{O}_K)$, then the Artin map induces an isomorphism*

$$C(\mathcal{O}_K)/H \xrightarrow{\sim} \mathrm{Gal}(M/K). \qquad \text{Q.E.D.}$$

This corollary is *class field theory for unramified Abelian extensions*, and it illustrates one of the main themes of class field theory: a certain class of extensions of K (unramified Abelian extensions) are classified in terms of data intrinsic to K (subgroups of the ideal class group). The theorems we encounter in §8 will follow the same format.

Theorem 5.23 also allows us to characterize the primes of K which split completely in the Hilbert class field:

COROLLARY 5.25. *Let L be the Hilbert class field of a number field K, and let $\mathfrak{p}$ be a prime ideal of K. Then*

$$\mathfrak{p} \text{ splits completely in } L \iff \mathfrak{p} \text{ is a principal ideal.}$$

PROOF. Corollary 5.21 implies that the prime $\mathfrak{p}$ splits completely in L if and only if $((L/K)/\mathfrak{p}) = 1$. Since the Artin map induces an isomorphism $C(\mathcal{O}_K) \simeq \mathrm{Gal}(L/K)$, we see that $((L/K)/\mathfrak{p}) = 1$ if and only if $\mathfrak{p}$ determines the trivial class of $C(\mathcal{O}_K)$. By the definition of the ideal class group, this means that $\mathfrak{p}$ is principal, and the corollary is proved. Q.E.D.

In §8, we will see that the Hilbert class field is characterized by the property that the primes that split completely are exactly the principal prime ideals.

D. Solution of $p = x^2 + ny^2$ for Infinitely Many n

Now that we know about the Hilbert class field, we can prove Theorem 5.1:

PROOF OF THEOREM 5.1. The first step is to relate $p = x^2 + ny^2$ to the behavior of p in the Hilbert class field L. This result is sufficiently interesting to be a theorem in its own right:

THEOREM 5.26. *Let L be the Hilbert class field of $K = \mathbb{Q}(\sqrt{-n})$. Assume that n satisfies (5.2), so that $\mathcal{O}_K = \mathbb{Z}[\sqrt{-n}]$. If p is an odd prime not dividing n, then*

$$p = x^2 + ny^2 \iff p \text{ splits completely in } L.$$

PROOF. Since n satisfies (5.2), (5.15) tells us that $d_K = -4n$ and $\mathcal{O}_K = \mathbb{Z}[\sqrt{-n}]$. Let p be an odd prime not dividing n. Then $p \nmid d_K$, so that p is unramified in K by Corollary 5.17. We will prove the following equivalences:

$$\begin{aligned} p = x^2 + ny^2 &\iff p\mathcal{O}_K = \mathfrak{p}\bar{\mathfrak{p}},\ \mathfrak{p} \neq \bar{\mathfrak{p}},\ \text{and } \mathfrak{p} \text{ is principal in } \mathcal{O}_K \\ &\iff p\mathcal{O}_K = \mathfrak{p}\bar{\mathfrak{p}},\ \mathfrak{p} \neq \bar{\mathfrak{p}},\ \text{and } \mathfrak{p} \text{ splits completely in } L \\ &\iff p \text{ splits completely in } L, \end{aligned} \tag{5.27}$$

and Theorem 5.26 will follow.

For the first equivalence, suppose that $p = x^2 + ny^2 = (x + \sqrt{-n}y)(x - \sqrt{-n}y)$. Setting $\mathfrak{p} = (x + \sqrt{-n}y)\mathcal{O}_K$, then $p\mathcal{O}_K = \mathfrak{p}\bar{\mathfrak{p}}$ must be the prime factorization of $p\mathcal{O}_K$ in $\mathcal{O}_K$. Note that $\mathfrak{p} \neq \bar{\mathfrak{p}}$ since p is unramified in K. Conversely, suppose that $p\mathcal{O}_K = \mathfrak{p}\bar{\mathfrak{p}}$, where $\mathfrak{p}$ is principal. Since $\mathcal{O}_K = \mathbb{Z}[\sqrt{-n}]$, we can write $\mathfrak{p} = (x + \sqrt{-n}y)\mathcal{O}_K$. This implies that $p\mathcal{O}_K = (x^2 + ny^2)\mathcal{O}_K$, and it follows that $p = x^2 + ny^2$.

The second equivalence of (5.27) follows immediately from Corollary 5.25. To prove the final equivalence, we will use the following lemma:

LEMMA 5.28. *Let L be the Hilbert class field of an imaginary quadratic field K, and let τ denote complex conjugation. Then $\tau(L) = L$, and hence L is Galois over $\mathbb{Q}$.*

PROOF. It is easy to see that $\tau(L)$ is an unramified Abelian extension of $\tau(K) = K$. Since L is the maximal such extension, we have $\tau(L) \subset L$, and then $\tau(L) = L$ since they have the same degree over K. Hence $\tau \in \mathrm{Gal}(L/\mathbb{Q})$, which implies that L is Galois over $\mathbb{Q}$ (see Exercise 5.19). Q.E.D.

To finish the proof of (5.27), note that the condition

$$p\mathcal{O}_K = \mathfrak{p}\bar{\mathfrak{p}},\ \mathfrak{p} \neq \bar{\mathfrak{p}},\ \text{and } \mathfrak{p} \text{ splits completely in } L$$

says that p splits completely in K and that some prime of K containing p splits completely in L. Since L is Galois over $\mathbb{Q}$, this is easily seen to be equivalent to p splitting completely in L (see Exercise 5.18), and Theorem 5.26 is proved. Q.E.D.

The next step in the proof of Theorem 5.1 is to give a more elementary way of saying that p splits completely in L. We have the following criterion:

PROPOSITION 5.29. *Let K be an imaginary quadratic field, and let L be a finite extension of K which is Galois over $\mathbb{Q}$. Then:*

(i) *There is a real algebraic integer α such that $L = K(\alpha)$.*
(ii) *Given α as in (i), let $f(x) \in \mathbb{Z}[x]$ denote its monic minimal polynomial. If p is a prime not dividing the discriminant of $f(x)$, then*

$$p \text{ splits completely in } L \iff \begin{cases} (d_K/p) = 1 \text{ and } f(x) \equiv 0 \bmod p \\ \text{has an integer solution.} \end{cases}$$

PROOF. By hypothesis, L is Galois over $\mathbb{Q}$, and thus $[L \cap \mathbb{R} : \mathbb{Q}] = [L : K]$ since $L \cap \mathbb{R}$ is the fixed field of complex conjugation. This implies that for $\alpha \in L \cap \mathbb{R}$,

$$L \cap \mathbb{R} = \mathbb{Q}(\alpha) \iff L = K(\alpha)$$

(see Exercise 5.19). Hence, if $\alpha \in \mathcal{O}_L \cap \mathbb{R}$ satisfies $L \cap \mathbb{R} = \mathbb{Q}(\alpha)$, then α is a real integral primitive element of L over K, and (i) is proved. Furthermore, given such an α, let $f(x)$ be its monic minimal polynomial over $\mathbb{Q}$. Then $f(x) \in \mathbb{Z}[x]$, and since $[L \cap \mathbb{R} : \mathbb{Q}] = [L : K]$, $f(x)$ is also the minimal polynomial of α over K.

To prove the final part of (ii), let p be a prime not dividing the discriminant of $f(x)$. This tells us that $f(x)$ is separable modulo p. By Corollary 5.17 we have

$$p\mathcal{O}_K = \mathfrak{p}\overline{\mathfrak{p}},\ \mathfrak{p} \neq \overline{\mathfrak{p}} \iff \left(\frac{d_K}{p}\right) = 1.$$

We may assume that p splits completely in K, so that $\mathbb{Z}/p\mathbb{Z} \simeq \mathcal{O}_K/\mathfrak{p}$. Since $f(x)$ is separable over $\mathbb{Z}/p\mathbb{Z}$, it is separable over $\mathcal{O}_K/\mathfrak{p}$, and then Proposition 5.11 shows that

$$\begin{aligned} \mathfrak{p} \text{ splits completely in } L &\iff f(x) \equiv 0 \bmod \mathfrak{p} \text{ is solvable in } \mathcal{O}_K \\ &\iff f(x) \equiv 0 \bmod p \text{ is solvable in } \mathbb{Z}, \end{aligned}$$

where the last equivalence again uses $\mathbb{Z}/p\mathbb{Z} \simeq \mathcal{O}_K/\mathfrak{p}$. The proposition now follows from the last equivalence of (5.27). Q.E.D.

We can now prove the main equivalence of Theorem 5.1. Since the Hilbert class field L of $K = \mathbb{Q}(\sqrt{-n})$ is Galois over $\mathbb{Q}$, Proposition 5.29 implies that there is a real algebraic integer α which is a primitive element of L over K. Let $f_n(x)$ be the monic minimal polynomial of α, and let p be an odd prime dividing neither n nor the discriminant of $f_n(x)$. Then Theorem 5.26 and Proposition 5.29 imply that

$$\begin{aligned} p = x^2 + ny^2 &\iff p \text{ splits completely in } L \\ &\iff \begin{cases} (-n/p) = 1 \text{ and } f_n(x) \equiv 0 \bmod p \\ \text{has an integer solution.} \end{cases} \end{aligned}$$

In the second equivalence, recall that n satisfies (5.2), so that $d_k = -4n$, and hence $(d_k/p) = (-n/p)$.

It remains to show that the degree of $f_n(x)$ is the class number $h(-4n)$. Using Galois theory and Theorem 5.23, it follows that $f_n(x)$ has degree

$$[L : K] = |\mathrm{Gal}(L/K)| = |C(\mathcal{O}_K)|.$$

In Theorem 5.30 below we will see that when $d_K < 0$, there is a natural isomorphism

$$C(\mathcal{O}_K) \simeq C(d_K)$$

between the ideal class group $C(\mathcal{O}_K)$ and the form class group $C(d_K)$ from §3. Since $d_K = -4n$ in our case, we have $|C(\mathcal{O}_K)| = |C(-4n)| = h(-4n)$, which completes the proof of Theorem 5.1. Q.E.D.

The polynomial $f_n(x)$ of Theorem 5.1 is not unique—there are lots of primitive elements. However, we can at least predict its degree in advance by computing the class number $h(-4n)$. In §8 we will see that knowing $f_n(x)$ is *equivalent* to knowing the Hilbert class field.

We have now answered our basic question of when $p = x^2 + ny^2$, at least for those n satisfying (5.2). Notice that quadratic forms have almost completely disappeared! We used x^2+ny^2 in Theorem 5.26, but otherwise all of the action took place using *ideals* rather than *forms*. This is typical of what happens in modern algebraic number theory—ideals are the dominant language. At the same time, we don't want to waste the work done on quadratic forms in §§2–3. So can we translate quadratic forms into ideals? In §7 we will study this question in detail. The full story is somewhat complicated, but the case of negative field discriminants is rather nice: here, the form class group $C(d_K)$ from §3 is isomorphic to the ideal class group $C(\mathcal{O}_K)$. More precisely, we get the following theorem, which is a special case of the results of §7:

THEOREM 5.30. *Let K be an imaginary quadratic field of discriminant $d_K < 0$. Then:*

(i) *If $f(x, y) = ax^2 + bxy + cy^2$ is a primitive positive definite quadratic form of discriminant d_K, then*

$$[a, (-b + \sqrt{d_K})/2] = \{ma + n(-b + \sqrt{d_K})/2 : m, n \in \mathbb{Z}\}$$

is an ideal of $\mathcal{O}_K$.

(ii) *The map sending $f(x, y)$ to $[a, (-b + \sqrt{d_K})/2]$ induces an isomorphism between the form class group $C(d_K)$ of §3 and the ideal class group $C(\mathcal{O}_K)$. Hence the order of $C(\mathcal{O}_K)$ is the class number $h(d_K)$.* Q.E.D.

If we combine Theorems 5.30 and 5.23, we see that the Galois group $\mathrm{Gal}(L/K)$ of the Hilbert class field of an imaginary quadratic field K is canonically isomorphic to the form class group $C(d_K)$. Thus the "class" in "Hilbert class field" refers to Gauss' classes of properly equivalent quadratic forms.

This theorem allows us to compute ideal class groups using what we know about quadratic forms. For example, consider the quadratic field $K = \mathbb{Q}(\sqrt{-14})$ of discriminant -56. In §2 we saw that the reduced forms of discriminant -56 are x^2+14y^2, $2x^2+7y^2$ and $3x^2 \pm 2xy + 5y^2$. The form class group $C(-56)$ is thus cyclic of order 4 since only $x^2 + 14y^2$ and $2x^2 + 7y^2$ give classes of order ≤ 2. Then, using Theorem 5.30, we see that the ideal class group $C(\mathcal{O}_K)$ is isomorphic to $\mathbb{Z}/4\mathbb{Z}$, and furthermore, ideal class representatives are given by $[1, \sqrt{-14}] = \mathcal{O}_K$, $[2, \sqrt{-14}]$ and $[3, 1 \pm \sqrt{-14}]$. See Exercises 5.20–5.22 for some other applications of Theorem 5.30.

The final task of §5 is to work out an explicit example of Theorem 5.1. We will discuss the case $p = x^2 + 14y^2$, which was left unresolved at the end of §3. Of course, we know from Theorem 5.1 that there is *some* polynomial $f_{14}(x)$ such that

$$p = x^2 + 14y^2 \iff \begin{cases} (-14/p) = 1 \text{ and } f_{14}(x) \equiv 0 \bmod p \\ \text{has an integer solution,} \end{cases}$$

but so far all we know about $f_{14}(x)$ is that it has degree 4 since $h(-56) = 4$. This illustrates one weakness of Theorem 5.1: it tells us that $f_{14}(x)$ exists, but doesn't tell us how to find it. To determine $f_{14}(x)$ we need to know the Hilbert class field of $\mathbb{Q}(\sqrt{-14})$. The answer is as follows:

PROPOSITION 5.31. *The Hilbert class field of $K = \mathbb{Q}(\sqrt{-14})$ is $L = K(\alpha)$, where $\alpha = \sqrt{2\sqrt{2} - 1}$.*

PROOF. Since $h(-56) = 4$, the Hilbert class field has degree 4 over K. Then $L = K(\alpha)$ will be the Hilbert class field once we show that $K \subset L$ is an unramified Abelian extension of degree 4. It's easy to see that $K \subset L$ is Abelian of degree 4, so that we need only show that it is unramified. Furthermore, since K is imaginary quadratic, the infinite primes are automatically unramified.

Note that $\alpha^2 = 2\sqrt{2} - 1$, so that $\sqrt{2} \in L$. If we let $K_1 = K(\sqrt{2})$, then we have the extensions

$$K \subset K_1 \subset L,$$

and it suffices to show that $K \subset K_1$ and $K_1 \subset L$ are unramified (see Exercise 5.15). Since each of these extensions is obtained by adjoining a square root ($K_1 = K(\sqrt{2})$

and $L = K_1(\sqrt{\mu})$, $\mu = 2\sqrt{2} - 1$), let's first prove a general lemma about this situation:

LEMMA 5.32. *Let $L = K(\sqrt{u})$ be a quadratic extension with $u \in \mathcal{O}_K$, and let $\mathfrak{p}$ be prime in $\mathcal{O}_K$.*

(i) *If $2u \notin \mathfrak{p}$, then $\mathfrak{p}$ is unramified in L.*
(ii) *If $2 \in \mathfrak{p}$, $u \notin \mathfrak{p}$ and $u = b^2 - 4c$ for some $b, c \in \mathcal{O}_K$, then $\mathfrak{p}$ is unramified in L.*

PROOF. (i) Note that $2u \notin \mathfrak{p}$ implies $2 \notin \mathfrak{p}$ and $u \notin \mathfrak{p}$, and then $4u \notin \mathfrak{p}$ since $\mathfrak{p}$ is prime. Thus the discriminant of $x^2 - u$ is $4u \notin \mathfrak{p}$, so that $x^2 - u$ is separable modulo $\mathfrak{p}$. Then $\mathfrak{p}$ is unramified by Proposition 5.11.

(ii) Note that $L = K(\beta)$, where $\beta = (-b + \sqrt{u})/2$ is a root of $x^2 + bx + c$. The discriminant is $b^2 - 4c = u \notin \mathfrak{p}$, so again $\mathfrak{p}$ is unramified by Proposition 5.11. Q.E.D.

Now we can prove Proposition 5.31. To study $K \subset K_1$, let $\mathfrak{p}$ be prime in $\mathcal{O}_K$. Since $K_1 = K(\sqrt{2})$, part (i) of Lemma 5.32 implies that $\mathfrak{p}$ is unramified whenever $2 \notin \mathfrak{p}$. It remains to study the case $2 \in \mathfrak{p}$. Since $\sqrt{-14} \in K$ and $\sqrt{2} \in K_1$, we also have $\sqrt{-7} \in K_1$, i.e., $K_1 = K(\sqrt{-7})$. Since $-7 \notin \mathfrak{p}$ and $-7 = 1^2 - 4 \cdot 2$, $\mathfrak{p}$ is unramified by part (ii) of Lemma 5.32.

The extension $K_1 \subset L$ is almost as easy. We know that $L = K_1(\sqrt{\mu})$, $\mu = 2\sqrt{2} - 1$. Let $\mu' = -2\sqrt{2} - 1$. Since $\sqrt{\mu\mu'} = \sqrt{-7} \in K_1$, it follows that $\sqrt{\mu'} \in L$, and in fact

$$L = K_1(\sqrt{\mu}) = K_1(\sqrt{\mu'}).$$

Now let $\mathfrak{p}$ be prime in K_1. If $2 \notin \mathfrak{p}$, then $\mu + \mu' = -2$ shows that $\mu \notin \mathfrak{p}$ or $\mu' \notin \mathfrak{p}$, and $\mathfrak{p}$ is unramified by part (i) of Lemma 5.32. If $2 \in \mathfrak{p}$, then $\mu \notin \mathfrak{p}$ since $\mu = 2\sqrt{2} - 1$. We also have $\mu = (1 + \sqrt{2})^2 - 4$, and then part (ii) of Lemma 5.32 shows that $\mathfrak{p}$ is unramified. Q.E.D.

We can now characterize when a prime p is represented by $x^2 + 14y^2$:

THEOREM 5.33. *If $p \neq 7$ is an odd prime, then*

$$p = x^2 + 14y^2 \iff \begin{cases} (-14/p) = 1 \text{ and } (x^2+1)^2 \equiv 8 \bmod p \\ \text{has an integer solution.} \end{cases}$$

PROOF. Since $\alpha = \sqrt{2\sqrt{2} - 1}$ is a real integral primitive element of the Hilbert class field of $K = \mathbb{Q}(\sqrt{-14})$, its minimal polynomial $x^4 + 2x^2 - 7 = (x^2 + 1)^2 - 8$ can be chosen to be the polynomial $f_{14}(x)$ of Theorem 5.1. Its discriminant is $-2^{14} \cdot 7$ (see Exercise 5.24), so that the only excluded primes are 2 and 7. Then Theorem 5.33 follows immediately from Theorem 5.1. Q.E.D.

These methods can be used to compute the Hilbert class field in other cases (see Herz [**72**]). For example, in Exercise 5.25, we will see that the Hilbert class field of $K = \mathbb{Q}(\sqrt{-17})$ is $L = K(\alpha)$, where $\alpha = \sqrt{(1 + \sqrt{17})/2}$. This gives us an explicit criterion for a prime to be of the form $x^2 + 17y^2$ (see Exercise 5.26).

One unsatisfactory aspect of these examples is that they don't explain how the primitive element α of the Hilbert class field was found. In general, the Hilbert class field is difficult to describe explicitly, though this can be done for class numbers ≤ 4 (see Herz [**72**]). In §6 we will use genus theory to discover the above primitive elements when $K = \mathbb{Q}(\sqrt{-14})$ or $\mathbb{Q}(\sqrt{-17})$, and in Chapter 3 we will use complex multiplication to give a general method for finding the Hilbert class field of any imaginary quadratic field.

E. Exercises

5.1. Let $\mathcal{O}_K$ be the algebraic integers in a number field K.

(a) Show that a nonzero ideal $\mathfrak{a}$ of $\mathcal{O}_K$ contains a nonzero integer m. Hint: if $\alpha \neq 0$ is in $\mathfrak{a}$, let $x^n + a_1x^{n-1} + \cdots + a_n$ be its minimal polynomial. Show that $m = a_n$ is what we want.

(b) Show that $\mathcal{O}_K/\mathfrak{a}$ is finite whenever $\mathfrak{a}$ is a nonzero ideal of $\mathcal{O}_K$. Hint: if m is the integer from part (a), consider the surjection $\mathcal{O}_K/m\mathcal{O}_K \to \mathcal{O}_K/\mathfrak{a}$. Use part (ii) of Proposition 5.3 to compute the order of $\mathcal{O}_K/m\mathcal{O}_K$.

(c) Use part (b) to show that every nonzero ideal of $\mathcal{O}_K$ is a free $\mathbb{Z}$-module of rank $[K : \mathbb{Q}]$.

(d) If we have ideals $\mathfrak{a}_1 \subset \mathfrak{a}_2 \subset \cdots$, show that there is an integer n such that $\mathfrak{a}_n = \mathfrak{a}_{n+1} = \cdots$. Hint: consider the surjections $\mathcal{O}_K/\mathfrak{a}_1 \to \mathcal{O}_K/\mathfrak{a}_2 \to \cdots$, and use part (b).

(e) Use part (b) to show that a nonzero prime ideal of $\mathcal{O}_K$ is maximal.

5.2. We will study the elementary properties of fractional ideals in a number field K. Recall that $\mathfrak{a} \subset K$ is a fractional ideal if, under ordinary addition and multiplication, it is a nonzero finitely generated $\mathcal{O}_K$-module.

(a) Show that $\mathfrak{a}$ is a fractional ideal if and only if $\mathfrak{a} = \alpha\mathfrak{b}$, where $\alpha \in K$ and $\mathfrak{b}$ is an ideal of $\mathcal{O}_K$. Hint: write each generator of $\mathfrak{a}$ in the form α/β, $\alpha, \beta \in \mathcal{O}_K$. Going the other way, use part (c) of Exercise 5.1 to show that $\alpha\mathfrak{b}$ is a finitely generated $\mathcal{O}_K$-module.

(b) Show that a nonzero fractional ideal is a free $\mathbb{Z}$-module of rank $[K : \mathbb{Q}]$. Hint: use part (a) together with part (c) of Exercise 5.1.

(c) Show that the product of two fractional ideals is a fractional ideal.

5.3. Let $K \subset L$ be a Galois extension, and let $\mathfrak{p} \subset \mathfrak{P}$ be prime ideals of K and L respectively.

(a) If $\sigma \in \mathrm{Gal}(L/K)$, then prove that $e_{\sigma(\mathfrak{P})|\mathfrak{p}} = e_{\mathfrak{P}|\mathfrak{p}}$ and $f_{\sigma(\mathfrak{P})|\mathfrak{p}} = f_{\mathfrak{P}|\mathfrak{p}}$.

(b) Prove part (ii) of Theorem 5.9.

5.4. Let $K \subset L$ be a Galois extension, and let $\mathfrak{P}$ be prime in L. Then we have the decomposition group $D_{\mathfrak{P}} = \{\sigma \in \mathrm{Gal}(L/K) : \sigma(\mathfrak{P}) = \mathfrak{P}\}$ and the inertia group $I_{\mathfrak{P}} = \{\sigma \in \mathrm{Gal}(L/K) : \sigma(\alpha) \equiv \alpha \bmod \mathfrak{P}$ for all $\alpha \in \mathcal{O}_L\}$.

(a) Show that $I_{\mathfrak{P}} \subset D_{\mathfrak{P}}$.

(b) Show that $\sigma \in D_{\mathfrak{P}}$ induces an automorphism $\tilde{\sigma}$ of $\mathcal{O}_L/\mathfrak{P}$ which is the identity on $\mathcal{O}_K/\mathfrak{p}$, $\mathfrak{p} = \mathfrak{P} \cap \mathcal{O}_K$.

(c) Let $\sigma \in D_{\mathfrak{P}}$. Then show that $\sigma \in I_{\mathfrak{P}}$ if and only if the automorphism $\tilde{\sigma}$ from part (b) is the identity.

5.5. In Proposition 5.11, prove that parts (i) and (iii) follow from part (ii).

5.6. In this exercise, we will prove part (ii) of Proposition 5.11. Let $\mathfrak{P}$ be a prime of $\mathcal{O}_L$ containing $\mathfrak{p}$, and let $D_{\mathfrak{P}} = \{\sigma \in \mathrm{Gal}(L/K) : \sigma(\mathfrak{P}) = \mathfrak{P}\}$ be the decomposition group. In Proposition 5.10 we observed that the order of $D_{\mathfrak{P}}$ is ef, where $e = e_{\mathfrak{P}|\mathfrak{p}}$ and $f = f_{\mathfrak{P}|\mathfrak{p}}$.

(a) Since $f(x) \equiv f_1(x) \cdots f_g(x) \bmod \mathfrak{p}$, show that $f_i(\alpha) \in \mathfrak{P}$ for some i. We can assume that $f_1(\alpha) \in \mathfrak{P}$.
(b) Using $f = [\mathcal{O}_L/\mathfrak{P} : \mathcal{O}_K/\mathfrak{p}]$, prove that $f \geq \deg(f_1(x))$.
(c) Since $f_1(\sigma(\alpha)) \in \mathfrak{P}$ for all $\sigma \in D_{\mathfrak{P}}$, show that $\deg(f_1(x)) \geq |D_{\mathfrak{P}}| = ef$. Hint: this is where separability is used.
(d) From parts (b) and (c) conclude that $e = 1$ and $f = \deg(f_1(x))$. Thus $\mathfrak{p}$ is unramified in L.
(e) Show that $\mathfrak{p}\mathcal{O}_L = \mathfrak{P}_1 \cdots \mathfrak{P}_g$ where $\mathfrak{P}_i$ is prime in $\mathcal{O}_L$ and $f_i(\alpha) \in \mathfrak{P}_i$. This shows that all of the $f_i(x)$'s have the same degree.
(f) Show that $\mathfrak{P}_i$ is generated by $\mathfrak{p}$ and $f_i(\alpha)$. Hint: let $I_i = \mathfrak{p}\mathcal{O}_L + f_i(\alpha)\mathcal{O}_L$. Show that ideals $\mathfrak{a}, \mathfrak{b}$ of $\mathcal{O}_L$ satisfy $\mathfrak{a} \subset \mathfrak{b}$ if and only if $\mathfrak{a} = \mathfrak{b}\mathfrak{c}$ for some ideal $\mathfrak{c}$. Then apply this to $\mathfrak{p}\mathcal{O}_L \subset I_i \subset \mathfrak{P}_i$.

5.7. In this problem we will determine the integers in the quadratic field $K = \mathbb{Q}(\sqrt{N})$, where N is a squarefree integer. Let $\alpha \mapsto \alpha'$ denote the nontrivial automorphism of K.

(a) Given $\alpha = r + s\sqrt{N} \in K$, define the *trace* and *norm* of α to be

$$T(\alpha) = \alpha + \alpha' = 2r$$
$$N(\alpha) = \alpha\alpha' = r^2 - s^2N.$$

Then prove that for $\alpha, \beta \in K$,

$$T(\alpha + \beta) = T(\alpha) + T(\beta)$$
$$N(\alpha\beta) = N(\alpha)N(\beta).$$

(b) Given $\alpha \in K$, prove that $\alpha \in \mathcal{O}_K$ if and only if $T(\alpha), N(\alpha) \in \mathbb{Z}$.
(c) Use part (b) to prove the description of $\mathcal{O}_K$ given in (5.13).
(d) Prove the description of $\mathcal{O}_K$ given in (5.14).

5.8. Use (5.12) and (5.13) to prove (5.15).

5.9. In this exercise we will study the units in an imaginary quadratic field K. Let $N(\alpha)$ be the norm of $\alpha \in K$ from Exercise 5.7.

(a) Prove that $\alpha \in \mathcal{O}_K$ is a unit if and only if $N(\alpha) = 1$.
(b) Show that $\mathcal{O}_K^* = \{\pm 1\}$ unless $K = \mathbb{Q}(i)$ or $\mathbb{Q}(\omega)$, in which case $\mathcal{O}_K^* = \{\pm 1, \pm i\}$ or $\{\pm 1, \pm\omega, \pm\omega^2\}$ respectively. Hint: use part (a) and (5.13). Exercises 4.5 and 4.16 will also be useful.

5.10. Let K be a quadratic field of discriminant d_K, and let the nontrivial automorphism of K be $(a + b\sqrt{d_K})' = a - b\sqrt{d_K}$. We want to complete the description of the prime ideals $\mathfrak{p}$ of $\mathcal{O}_K$ begun in Proposition 5.16. Our basic tools will be Proposition 5.11 and the formula $efg = 2$ from Theorem 5.9.

(a) If $2 \mid d_K$, then show that $2\mathcal{O}_K = \mathfrak{p}^2$, $\mathfrak{p} = \mathfrak{p}'$ prime. Hint: write $d_K = 4N$ and set

$$\mathfrak{p} = \begin{cases} 2\mathcal{O}_K + (1 + \sqrt{N})\mathcal{O}_K & N \text{ odd} \\ 2\mathcal{O}_K + \sqrt{N}\mathcal{O}_K & N \text{ even.} \end{cases}$$

(b) If $2 \nmid d_K$, then show that

$$d_K \equiv 1 \bmod 8 \iff 2\mathcal{O}_K = \mathfrak{p}\mathfrak{p}',\ \mathfrak{p} \neq \mathfrak{p}' \text{ prime}$$
$$d_K \equiv 5 \bmod 8 \iff 2\mathcal{O}_K \text{ is prime in } \mathcal{O}_K.$$

Hint: apply Proposition 5.11 to $K = \mathbb{Q}(\alpha)$, $\alpha = (1 + \sqrt{d_K})/2$.

(c) Show that the ideals described in parts (i)–(iii) of Proposition 5.16 give all prime ideals of $\mathcal{O}_K$. Hint: by part (a) of Exercise 5.1, $\mathfrak{p}$ contains a nonzero integer m, which can be assumed to be positive. Thus $\mathfrak{p} \mid m\mathcal{O}_K$, and we are done by unique factorization.

Notice how these results generalize the descriptions given in Propositions 4.7 and 4.18 of the primes in $\mathbb{Z}[\omega]$ and $\mathbb{Z}[i]$.

5.11. This problem will study the norm of a prime $\mathfrak{p}$ in a number field K. Recall that the norm $N(\mathfrak{p})$ is defined by $N(\mathfrak{p}) = |\mathcal{O}_K/\mathfrak{p}|$. Let p be the unique prime of $\mathbb{Z}$ contained in $\mathfrak{p}$.

(a) Show that $N(\mathfrak{p}) = p^f$, where f is the inertial degree of $\mathfrak{p}$ over p.
(b) Now assume that $\mathfrak{p}$ is prime in a quadratic field K. Show that

$$p \mid d_K : \ N(\mathfrak{p}) = p$$
$$p \nmid d_K : \ N(\mathfrak{p}) = \begin{cases} p & p \text{ splits completely in } K \\ p^2 & p\mathcal{O}_K \text{ is prime in } \mathcal{O}_K. \end{cases}$$

Hint: use $efg = 2$.

5.12. This exercise is concerned with the Artin symbol $((L/K)/\mathfrak{P})$.

(a) Prove part (i) of Corollary 5.21.
(b) Let $K \subset L$ be a Galois extension and let $\mathfrak{p}$ be a prime of K unramified in L. Prove that the set $\{((L/K)/\mathfrak{P}) : \mathfrak{P} \text{ is a prime of } L \text{ containing } \mathfrak{p}\}$ is a conjugacy class of $\mathrm{Gal}(L/K)$. This conjugacy class is defined to be the Artin symbol $((L/K)/\mathfrak{p})$ of $\mathfrak{p}$.

5.13. Assume that the number field K contains a primitive nth root of unity ζ. In this problem we will discuss a generalization of the Legendre symbol. Let $a \in \mathcal{O}_K$ and let $\mathfrak{p}$ be a prime ideal of $\mathcal{O}_K$ such that $na \notin \mathfrak{p}$.

(a) Prove that $1, \zeta, \ldots, \zeta^{n-1}$ are distinct modulo $\mathfrak{p}$. Hint: show that $x^n - 1$ is separable modulo $\mathfrak{p}$.
(b) Use part (a) to prove that $n \mid N(\mathfrak{p}) - 1$.
(c) Show that $a^{(N(\mathfrak{p})-1)/n}$ congruent to a unique nth root of unity modulo $\mathfrak{p}$. This allows us to define the nth power Legendre symbol $(\mathfrak{a}/\mathfrak{p})_n$ to be the unique nth root of unity such that

$$a^{(N(\mathfrak{p})-1)/n} \equiv \left(\frac{a}{\mathfrak{p}}\right)_n \mod \mathfrak{p}.$$

(d) Prove that $(\mathfrak{a}/\mathfrak{p})_n = 1 \iff a$ is an nth power residue modulo $\mathfrak{p}$.

5.14. Let K, n, a and $\mathfrak{p}$ be as in the previous exercise, and let $L = K(\sqrt[n]{a})$. Note that L is an Abelian extension of K. In this problem we will relate the Legendre symbol $(a/\mathfrak{p})_n$ to the Artin symbol $((L/K)/\mathfrak{p})$.

(a) Show that $\mathfrak{p}$ is unramified in L. Hint: show that $x^n - a$ is separable modulo $\mathfrak{p}$ and use Proposition 5.11.
(b) Generalize the argument of (5.22) to show that

$$\left(\frac{L/K}{\mathfrak{p}}\right)(\sqrt[n]{a}) = \left(\frac{a}{\mathfrak{p}}\right)_n \sqrt[n]{a}.$$

5.15. Suppose that $K \subset M \subset L$ are number fields.

(a) Let $\mathfrak{p}$ be prime in $\mathcal{O}_K$, and assume that $\mathfrak{p} \subset \mathfrak{P} \subset \mathfrak{P}'$, where $\mathfrak{P}$ (resp. $\mathfrak{P}'$) is prime in $\mathcal{O}_M$ (resp. $\mathcal{O}_L$). Then show that $e_{\mathfrak{P}'|\mathfrak{p}} = e_{\mathfrak{P}'|\mathfrak{P}} e_{\mathfrak{P}|\mathfrak{p}}$.

(b) Prove that a prime $\mathfrak{p}$ of $\mathcal{O}_K$ is unramified in L if and only if $\mathfrak{p}$ is unramified in M and every prime of $\mathcal{O}_M$ lying over $\mathfrak{p}$ is unramified in L.

(c) Prove that L is an unramified extension of K if and only if L is unramified over M and M is unramified over K.

5.16. Let $K \subset L$ be an unramified Abelian extension, and assume that $K \subset M \subset L$. By the previous exercise, $K \subset M$ is unramified, and it is clearly Abelian. We thus have Artin maps

$$\left(\frac{L/K}{\cdot}\right) : I_K \longrightarrow \mathrm{Gal}(L/K)$$

$$\left(\frac{M/K}{\cdot}\right) : I_K \longrightarrow \mathrm{Gal}(M/K)$$

and we also have the restriction map $r : \mathrm{Gal}(L/K) \to \mathrm{Gal}(M/K)$. Then use Lemma 5.19 to prove that

$$\left(\frac{M/K}{\cdot}\right) = r \circ \left(\frac{L/K}{\cdot}\right).$$

5.17. Prove Corollary 5.24. Hint: besides Galois theory and Theorems 5.18 and 5.23, you will also need Exercises 5.15 and 5.16.

5.18. If $K \subset M \subset L$, where L is Galois over K, then prove that a prime $\mathfrak{p}$ of $\mathcal{O}_K$ splits completely in L if and only if it splits completely in M and some prime of $\mathcal{O}_M$ containing $\mathfrak{p}$ splits completely in L.

5.19. Let K be an imaginary quadratic field, and let $K \subset L$ be a Galois extension. As usual, τ will denote complex conjugation.

(a) Show that L is Galois over $\mathbb{Q}$ if and only if $\tau(L) = L$.

(b) If L is Galois over $\mathbb{Q}$, then prove that

 (i) $[L \cap \mathbb{R} : \mathbb{Q}] = [L : K]$.

 (ii) For $\alpha \in L \cap \mathbb{R}$, $L \cap \mathbb{R} = \mathbb{Q}(\alpha) \iff L = K(\alpha)$.

5.20. Show that $\mathbb{Z}[(1+\sqrt{-19})/2]$ is a UFD. Hint: every PID is a UFD (see Ireland and Rosen [**75**, §1.3] or Marcus [**96**, pp. 255–256]). Thus, by Theorem 5.30, it suffices to show that $h(-19) = 1$.

5.21. In this exercise we will study the ring $\mathbb{Z}[\sqrt{-2}]$.

(a) Use Theorem 5.30 to show that $\mathbb{Z}[\sqrt{-2}]$ is a UFD.

(b) Show that $\sqrt{-2}$ is a prime in $\mathbb{Z}[\sqrt{-2}]$.

(c) If $ab = u^3$ in $\mathbb{Z}[\sqrt{-2}]$ and a and b are relatively prime, then prove that a and b are cubes in $\mathbb{Z}[\sqrt{-2}]$.

5.22. We can now give a second proof of Fermat's theorem that $(x, y) = (3, \pm 5)$ are the only integer solutions of the equation $x^3 = y^2 + 2$.

(a) If $x^3 = y^2 + 2$, show that $y + \sqrt{-2}$ and $y - \sqrt{-2}$ are relatively prime in $\mathbb{Z}[\sqrt{-2}]$. Hint: use part (b) of Exercise 5.21.

(b) Use part (c) of Exercise 5.21 to show that $(x, y) = (3, \pm 5)$.

This argument is due to Euler [**43**, Vol. I, Chapter XII, §§191–193], though he assumed (without proof) that Exercise 5.21 was true.

5.23. If $D \equiv 1 \bmod 4$ is negative and squarefree, prove a version of Theorems 5.1 and 5.26 for primes of the form $x^2 + xy + ((1-D)/4)y^2$.

5.24. Prove that the discriminant of $x^4 + bx^2 + c$ equals $2^4c(b^2 - 4c)^2$. Hint: write down the roots explicitly.

5.25. Let $K = \mathbb{Q}(\sqrt{-17})$.

(a) Show that $C(\mathcal{O}_K) \simeq \mathbb{Z}/4\mathbb{Z}$.

(b) Show that the Hilbert class field of K is given by $L = K(\alpha)$, where $\alpha = \sqrt{(1+\sqrt{17})/2}$. Hint: use the methods of Proposition 5.31. The only tricky part concerns primes of $K(\sqrt{17})$ which contain 2. Setting $u = (1+\sqrt{17})/2$ and $u' = (1-\sqrt{17})/2$, note that u and u' satisfy $x = x^2 - 4$.

5.26. Prove an analog of Theorem 5.33 for primes of the form $x^2 + 17y^2$.

6. The Hilbert Class Field and Genus Theory

In Chapter 1 we studied the genus theory of primitive positive definite quadratic forms, and our main result (Theorem 3.15) was that for a fixed discriminant D:

(i) There are $2^{\mu-1}$ genera, where μ is the number defined in Proposition 3.11.

(ii) The principal genus consists of squares of classes.

In this section, we will use Artin Reciprocity for the Hilbert class field of an imaginary quadratic field K to prove (i) and (ii) when D is the discriminant d_K of K. This result is less general than what we proved in §3, but the proof is such a nice application of the Hilbert class field that we couldn't resist including it. Readers more interested in $p = x^2 + ny^2$ may skip to §7 without loss of continuity.

The key to the class field theory interpretation of genus theory is the concept of the *genus field*. Given an imaginary quadratic field K of discriminant d_K, Theorem 5.30 tells us that the form class group $C(d_K)$ is isomorphic to the ideal class group $C(\mathcal{O}_K)$. The principal genus is a subgroup of $C(d_K)$ and hence maps to a subgroup of $C(\mathcal{O}_K)$. By Corollary 5.24, this subgroup determines an unramified Abelian extension of K which is called the *genus field* of K. Theorem 6.1 below will describe the genus field explicitly and show that the characters used in Gauss' definition of genus appear in the Artin map of the genus field. This will take a fair amount of work, but once done, (i) and (ii) above will follow easily by Artin Reciprocity. We will then discuss how the genus field can help in the harder problem of determining the Hilbert class field.

A. Genus Theory for Field Discriminants

Here is the main result of this section:

THEOREM 6.1. *Let K be an imaginary quadratic field of discriminant d_K. Let μ be the number of primes dividing d_K, and let $p_1, \ldots, p_r$ be the odd primes dividing d_K (so that $\mu = r$ or $r+1$ according to whether $d_K \equiv 0$ or $1 \bmod 4$). In addition, set $p_i^* = (-1)^{(p_i-1)/2}p_i$. Then:*

(i) *The genus field of K is the maximal unramified extension of K which is an Abelian extension of $\mathbb{Q}$.*

(ii) *The genus field of K is $K(\sqrt{p_1^*}, \ldots, \sqrt{p_r^*})$.*

(iii) *The number of genera of primitive positive definite forms of discriminant d_K is $2^{\mu-1}$.*

(iv) *The principal genus of primitive positive definite forms of discriminant d_K consists of squares of classes.*

PROOF. First, note that for field discriminants d_K, the number μ defined in the statement of the theorem agrees with the one defined in Proposition 3.11 (see Exercise 6.1). Note also that (iii) and (iv) of the theorem are the facts about genus theory that we want to prove.

To start the proof, let L be the Hilbert class field of K, and let M be the unramified Abelian extension of K corresponding to the subgroup $C(\mathcal{O}_K)^2 \subset C(\mathcal{O}_K)$ via Corollary 5.24. We claim that

(6.2) $\quad M$ is the maximal unramified extension of K Abelian over $\mathbb{Q}$.

To prove this, consider an unramified extension $\tilde{M}$ of K which is Abelian over $\mathbb{Q}$. Then $\tilde{M}$ is also Abelian over K, so that $\tilde{M} \subset L$, and we thus have the following diagram of fields:

$$
(6.3) \qquad \begin{array}{c} L \\ | \\ \tilde{M} \\ | \\ K \\ | \\ \mathbb{Q} \end{array} \left.\begin{array}{c} \\ \\ \\ \\ \\ \end{array}\right] \text{Abelian}
$$

We want the maximal such $\tilde{M}$. Since L is Galois over $\mathbb{Q}$ (see Lemma 5.28), we can interpret (6.3) via Galois theory. Let $G = \mathrm{Gal}(L/\mathbb{Q})$. Then $\tilde{M}$ being Abelian over $\mathbb{Q}$ is equivalent to $[G, G] \subset \mathrm{Gal}(L/\tilde{M})$, where $[G, G]$ is the commutator subgroup of G (see Exercise 6.2). Note also that $[G, G] \subset \mathrm{Gal}(L/K)$ since the latter has index two in G. Thus $\tilde{M}$ satisfies (6.3) if and only if

$$[G, G] \subset \mathrm{Gal}(L/\tilde{M}) \subset \mathrm{Gal}(L/K).$$

It follows by Galois theory that the maximal unramified extension of K Abelian over $\mathbb{Q}$ is the one that corresponds to $[G, G]$. By Theorem 5.23, $\mathrm{Gal}(L/K)$ can be identified with $C(\mathcal{O}_K)$ via the Artin map. If we can show that $[G, G] \subset \mathrm{Gal}(L/K)$ maps to $C(\mathcal{O}_K)^2 \subset C(\mathcal{O}_K)$, then (6.2) will follow from Corollary 5.24.

We first compute $G = \mathrm{Gal}(L/\mathbb{Q})$. We have a short exact sequence

$$1 \longrightarrow \mathrm{Gal}(L/K) \longrightarrow G \longrightarrow \mathrm{Gal}(K/\mathbb{Q}) \longrightarrow 1$$

which splits because complex conjugation τ is in G by Lemma 5.28. Thus G is the semidirect product $\mathrm{Gal}(L/K) \rtimes (\mathbb{Z}/2\mathbb{Z})$, where $\mathbb{Z}/2\mathbb{Z}$ acts by conjugation by τ.

Under the isomorphism $\mathrm{Gal}(L/K) \simeq C(\mathcal{O}_K)$, conjugation by τ operates on $C(\mathcal{O}_K)$ by sending an ideal to its conjugate. To see this, let $\mathfrak{p}$ be a prime ideal of $\mathcal{O}_K$. Then the uniqueness part of Lemma 5.19 shows that

$$(6.4) \qquad \tau \left(\frac{L/K}{\mathfrak{p}} \right) \tau^{-1} = \left(\frac{L/K}{\tau(\mathfrak{p})} \right)$$

(see Exercise 6.3), and our claim follows. However, for any ideal $\mathfrak{a}$ of $\mathcal{O}_K$, we will prove in Lemma 7.14 that the product $\mathfrak{a}\bar{\mathfrak{a}}$ is always a principal ideal, and it follows that the class of $\bar{\mathfrak{a}}$ is the inverse of the class of $\mathfrak{a}$ in $C(\mathcal{O}_K)$. Hence G may be identified with the semidirect product $C(\mathcal{O}_K) \rtimes (\mathbb{Z}/2\mathbb{Z})$, where the nontrivial element of $\mathbb{Z}/2\mathbb{Z}$ acts by sending an element of $C(\mathcal{O}_K)$ to its inverse.

It is now easy to show that $[G, G] = C(\mathcal{O}_K)^2$. First, note that $C(\mathcal{O}_K)^2$ is normal in G (any subgroup of $C(\mathcal{O}_K)$ is, which has unexpected consequences—see Exercise 6.4), and since $\mathbb{Z}/2\mathbb{Z}$ acts trivially on $C(\mathcal{O}_K)/C(\mathcal{O}_K)^2$ (every element is its own inverse), we have

$$\begin{aligned} G/C(\mathcal{O}_K)^2 &\simeq \big(C(\mathcal{O}_K) \rtimes (\mathbb{Z}/2\mathbb{Z})\big)/C(\mathcal{O}_K)^2 \\ &\simeq \big(C(\mathcal{O}_K)/C(\mathcal{O}_K)^2\big) \times (\mathbb{Z}/2\mathbb{Z}), \end{aligned} \tag{6.5}$$

so that $G/C(\mathcal{O}_K)^2$ is Abelian (see Exercise 6.5). It follows that $[G, G] \subset C(\mathcal{O}_K)^2$. For the opposite inclusion, note that if $a \in C(\mathcal{O}_K)$, then $(a, 1) \in C(\mathcal{O}_K) \rtimes (\mathbb{Z}/2\mathbb{Z})$. If τ is the nontrivial element of $\mathbb{Z}/2\mathbb{Z}$, then $(1, \tau)(a, 1)(1, \tau)^{-1} = (a^{-1}, 1)$, so that

$$(a, 1)(1, \tau)(a, 1)^{-1}(1, \tau)^{-1} = (a^2, 1).$$

This proves that $[G, G] = C(\mathcal{O}_K)^2$, and (6.2) is proved.

We will next show that

$$M = K\big(\sqrt{p_1^*}, \dots, \sqrt{p_r^*}\big), \tag{6.6}$$

where p_i^*'s are as in the statement of the theorem. We begin with two preliminary lemmas. The first concerns some general facts about ramification and the Artin symbol:

LEMMA 6.7. *Let L and M be Abelian extensions of a number field K, and let $\mathfrak{p}$ be prime in $\mathcal{O}_K$.*

(i) *$\mathfrak{p}$ is unramified in LM if and only if it is unramified in both L and M.*

(ii) *If $\mathfrak{p}$ is unramified in LM, then under the natural injection*

$$\mathrm{Gal}(LM/K) \longrightarrow \mathrm{Gal}(L/K) \times \mathrm{Gal}(M/K),$$

the Artin symbol $((LM/K)/\mathfrak{p})$ maps to $(((L/K)/\mathfrak{p}), ((M/K)/\mathfrak{p}))$.

PROOF. See Exercise 6.6, or, for a more general version of these facts, Marcus [**96**, Exercises 10–11, pp. 117–118]. Q.E.D.

The second lemma tells us when a quadratic extension $K \subset K(\sqrt{a})$, $a \in \mathbb{Z}$, is unramified:

LEMMA 6.8. *Let K be an imaginary quadratic field of discriminant d_K, and let $K(\sqrt{a})$ be a quadratic extension where $a \in \mathbb{Z}$. Then $K \subset K(\sqrt{a})$ is unramified if and only if a can be chosen so that $a \mid d_K$ and $a \equiv 1 \bmod 4$.*

PROOF. For the most part, the proof is a straightforward application of the techniques used in the proof of Proposition 5.31. See Exercises 6.7, 6.8 and 6.9 for the details. Q.E.D.

We can now prove (6.6). Let $M^* = K(\sqrt{p_1^*}, \dots, \sqrt{p_r^*})$. Since p_i^* divides d_K and satisfies $p_i^* \equiv 1 \bmod 4$, $K \subset K(\sqrt{p_i^*})$ is unramified by Lemma 6.8, and consequently $K \subset M^*$ is unramified by Lemma 6.7. But $M^* = \mathbb{Q}(\sqrt{d_K}, \sqrt{p_1^*}, \dots, \sqrt{p_r^*})$ is clearly Abelian over $\mathbb{Q}$, so that $M^* \subset M$ by the maximality of M.

To prove the opposite inclusion, we first study $\mathrm{Gal}(M/\mathbb{Q})$. Since $\mathbb{Q} \subset M \subset L$ corresponds to $G \supset C(\mathcal{O}_K)^2 \supset \{1\}$ under the Galois correspondence, we have

$$\mathrm{Gal}(M/\mathbb{Q}) \simeq \mathrm{Gal}(L/\mathbb{Q})/\mathrm{Gal}(L/M) = G/C(\mathcal{O}_K)^2,$$

so that by (6.5), $\mathrm{Gal}(M/\mathbb{Q}) \simeq (\mathbb{Z}/2\mathbb{Z})^m$ for some m. Then Galois theory shows that $M = \mathbb{Q}(\sqrt{a_1}, \ldots, \sqrt{a_m})$ where $a_1, \ldots, a_m \in \mathbb{Z}$ (see Exercise 6.10). Thus M is the compositum of quadratic extensions $K \subset K(\sqrt{a_i})$, $a_i \in \mathbb{Z}$, and by Lemma 6.7, each of these is unramified.

It suffices to show that M^* contains all unramified extensions $K \subset K(\sqrt{a})$, $a \in \mathbb{Z}$. By Lemma 6.8, we may assume that $a \equiv 1 \bmod 4$ and that $a \mid d_K$. It follows that a must be of the form $p_{i_1}^* \cdots p_{i_s}^*$, $1 \le i_1 < \cdots < i_s \le r$, so that $K(\sqrt{a})$ is clearly contained in M^*. This completes the proof of (6.6).

We will next show that $[M : \mathbb{Q}] = 2^\mu$. Note that $M = \mathbb{Q}(\sqrt{d_k}, \sqrt{p_1^*}, \ldots, \sqrt{p_r^*})$. When $d_K \equiv 1 \bmod 4$, we have $d_K = p_1^* \cdots p_r^*$, so that $[M : \mathbb{Q}] = 2^r = 2^\mu$ since $\mu = r$ in this case. When $d_K \equiv 0 \bmod 4$, we can write $d_K = -4n$, $n > 0$, and then we have

$$(6.9) \qquad M = \begin{cases} \mathbb{Q}(i, \sqrt{p_i^*}, \ldots, \sqrt{p_r^*}) & n \equiv 1 \bmod 4 \\ \mathbb{Q}(\sqrt{2}, \sqrt{p_1^*}, \ldots, \sqrt{p_r^*}) & n \equiv 6 \bmod 8 \\ \mathbb{Q}(\sqrt{-2}, \sqrt{p_i^*}, \ldots, \sqrt{p_r^*}) & n \equiv 2 \bmod 8 \end{cases}$$

(see Exercise 6.11). Thus $[M : \mathbb{Q}] = 2^{r+1} = 2^\mu$. Since $[C(\mathcal{O}_K) : C(\mathcal{O}_K)^2]$ equals half of $[G : C(\mathcal{O}_K)^2] = [M : \mathbb{Q}] = 2^\mu$, we have proved that

$$(6.10) \qquad [C(\mathcal{O}_K) : C(\mathcal{O}_K)^2] = 2^{\mu-1}.$$

We can now compute the Artin map $((M/K)/\cdot) : I_K \to \mathrm{Gal}(M/K)$. If we set $K_i = K(\sqrt{p_i^*})$, then M is the compositum $K_1 \cdots K_r$, and we have a natural injection

$$(6.11) \qquad \mathrm{Gal}(M/K) \longrightarrow \prod_{i=1}^{r} \mathrm{Gal}(K_i/K).$$

Furthermore, we may identify $\mathrm{Gal}(K_i/K)$ with $\{\pm 1\}$, so that composing the Artin map with (6.11) gives us a homomorphism

$$\Phi_K : I_K \longrightarrow \{\pm 1\}^r.$$

We claim that if $\mathfrak{a}$ is an ideal of $\mathcal{O}_K$ prime to $2d_K$, then $\Phi_K(\mathfrak{a})$ can be computed in terms of Legendre symbols as follows:

$$(6.12) \qquad \Phi_K(\mathfrak{a}) = \left(\left(\frac{N(\mathfrak{a})}{p_1} \right), \ldots, \left(\frac{N(\mathfrak{a})}{p_r} \right) \right),$$

where $N(\mathfrak{a}) = |\mathcal{O}_K/\mathfrak{a}|$ is the norm of $\mathfrak{a}$.

To prove (6.12), we will need one basic fact about norms: if $\mathfrak{a}$ and $\mathfrak{b}$ are ideals of $\mathcal{O}_K$, then $N(\mathfrak{a}\mathfrak{b}) = N(\mathfrak{a})N(\mathfrak{b})$ (see Lemma 7.14 or Marcus [**96**, Theorem 22]). It follows that both sides of (6.12) are multiplicative in $\mathfrak{a}$, so that we may assume that $\mathfrak{a}$ is a prime ideal $\mathfrak{p}$ of $\mathcal{O}_K$. Then Lemma 6.7, applied to (6.11), shows that $((M/K)/\mathfrak{p})$ maps to the r-tuple

$$\left(\left(\frac{K_1/K}{\mathfrak{p}} \right), \ldots, \left(\frac{K_r/K}{\mathfrak{p}} \right) \right).$$

If we can show that

$$(6.13) \qquad \left(\frac{K_i/K}{\mathfrak{p}} \right)(\sqrt{p_i^*}) = \left(\frac{N(\mathfrak{p})}{p_i} \right) \sqrt{p_i^*},$$

then (6.12) will follow immediately.

To prove (6.13), let $\mathfrak{P}$ be a prime of $\mathcal{O}_{K_i}$ containing $\mathfrak{p}$, and set $\sigma = ((K_i/K)/\mathfrak{p})$. By Lemma 5.19 we see that

$$\sigma\big(\sqrt{p_i^*}\big) \equiv \sqrt{p_i^*}^{\,N(\mathfrak{p})} \equiv (p_i^*)^{(N(\mathfrak{p})-1)/2}\sqrt{p_i^*} \bmod \mathfrak{P}. \tag{6.14}$$

Since K is a quadratic field, it follows that $N(\mathfrak{p}) = p$ or p^2 (see Exercise 5.11), and thus there are two cases to consider.

If $N(\mathfrak{p}) = p$, then we know that

$$(p_i^*)^{(p-1)/2} \equiv \left(\frac{p_i^*}{p}\right) \bmod p.$$

Since $p \in \mathfrak{P}$ and $(p_i^*/p) = (p/p_i)$ by quadratic reciprocity, (6.14) reduces to

$$\sigma\big(\sqrt{p_i^*}\big) \equiv \left(\frac{p}{p_i}\right)\sqrt{p_i^*} \equiv \left(\frac{N(\mathfrak{p})}{p_i}\right)\sqrt{p_i^*} \bmod \mathfrak{P},$$

and we are done. If $N(\mathfrak{p}) = p^2$, then by Fermat's Little Theorem,

$$(p_i^*)^{(p^2-1)/2} \equiv \big((p_i^*)^{(p+1)/2}\big)^{p-1} \equiv 1 \bmod p,$$

so that (6.14) becomes

$$\sigma\big(\sqrt{p_i^*}\big) \equiv \sqrt{p_i^*} \equiv \left(\frac{p^2}{p_i}\right)\sqrt{p_i^*} \equiv \left(\frac{N(\mathfrak{p})}{p_i}\right)\sqrt{p_i^*} \bmod \mathfrak{P},$$

and (6.13) is proved. This proves (6.12).

For the rest of the proof, we will assume that $d_K = -4n$, $n > 0$ (see Exercise 6.12 for the case $d_K \equiv 1 \bmod 4$). Here, it is easily checked that the map (6.11) is an isomorphism, and then Artin Reciprocity (Corollary 5.24) for $K \subset M$ means that the map $\Phi_K : I_K \to \{\pm 1\}^r$ of (6.12) induces an isomorphism

$$A : C(\mathcal{O}_K)/C(\mathcal{O}_K)^2 \xrightarrow{\sim} \{\pm 1\}^r,$$

where the A stands for Artin.

It's now time to bring in quadratic forms. Let $C(d_K)$ be the class group of primitive positive definite forms of discriminant $d_K = -4n$, and let P be the principal genus. Recall from the proof of Theorem 3.15 that we have the $\mu = r+1$ assigned characters $\chi_0, \chi_1, \ldots, \chi_r$, where χ_0 is one of δ, ϵ or $\delta\epsilon$, and $\chi_i(a) = (a/p_i)$ for $i = 1, \ldots, r$. In Lemma 3.20, we proved that if $f(x,y)$ represents a number a prime to $4n$, then the genus of $f(x,y)$ is determined by the $(r+1)$-tuple $(\chi_0(a), \chi_1(a), \ldots, \chi_r(a))$. Thus we have an injective map

$$G : C(d_K)/P \longrightarrow \{\pm 1\}^{r+1},$$

where the G stands for Gauss.

To relate the two maps A and G, we will use the isomorphism $C(d_K) \simeq C(\mathcal{O}_K)$ of Theorem 5.30. Since $C(d_K)^2 \subset P$, we get the following diagram:

$$\begin{array}{ccccc} C(d_K)/C(d_K)^2 & \xrightarrow{\ \alpha\ } & C(d_K)/P & \xrightarrow{\ G\ } & \{\pm 1\}^{r+1} \\ \downarrow \wr & & & & \downarrow \pi \\ C(\mathcal{O}_K)/C(\mathcal{O}_K)^2 & & \xrightarrow[A]{\ \sim\ } & & \{\pm 1\}^r \end{array} \tag{6.15}$$

where $\alpha : C(d_K)/C(d_K)^2 \to C(d_K)/P$ is the natural surjection and π is the projection onto the last r factors.

We claim that this diagram commutes, which means that Gauss' definition of genus is amazingly close to the Artin map of the genus field. (The full story of the relation is worked out in Exercise 6.13.)

To prove that (6.15) commutes, let $f(x, y) = ax^2 + 2bxy + cy^2$ be a form of discriminant $-4n$. We can assume that a is relatively prime to $4n$. Then, in (6.15), if we first go across and then down, we see that the class of $f(x, y)$ maps to

$$(6.16) \qquad (\chi_1(a), \ldots, \chi_r(a)) = \left(\left(\frac{a}{p_1}\right), \ldots, \left(\frac{a}{p_r}\right)\right).$$

Let's see what happens when we go the other way. By Theorem 5.30, $f(x, y)$ corresponds to the ideal $\mathfrak{a} = [a, -b + \sqrt{-n}]$ of $\mathcal{O}_K$. However, it is easy to see that the natural map

$$(6.17) \qquad \mathbb{Z}/a\mathbb{Z} \longrightarrow \mathcal{O}_K/\mathfrak{a}$$

is an isomorphism (see Exercise 6.14). Thus $\mathfrak{a}$ has norm $N(\mathfrak{a}) = a$, and our description of the Artin map from (6.12) shows that $f(x, y)$ maps to

$$(6.18) \qquad \left(\left(\frac{N(\mathfrak{a})}{p_1}\right), \ldots, \left(\frac{N(\mathfrak{a})}{p_r}\right)\right) = \left(\left(\frac{a}{p_1}\right), \ldots, \left(\frac{a}{p_r}\right)\right).$$

Comparing (6.16) and (6.18), we see that (6.15) commutes as claimed.

Now everything is easy to prove. If we go down and across in (6.15), the resulting map is injective. By commutativity, it follows that $\alpha : C(d_K)/C(d_K)^2 \to C(d_K)/P$ must be injective, which proves that $C(d_K)^2 = P$, and part (iv) of the theorem is done. The number of genera is thus $[C(d_K) : P] = [C(d_K) : C(d_K)^2] = [C(\mathcal{O}_K) : C(\mathcal{O}_K)^2] = 2^{\mu-1}$ (the last equality is (6.10)), and (iii) follows. Finally, since $P = C(d_K)^2$ corresponds to $C(\mathcal{O}_K)^2$, we see that M is the genus field of K, and then (i) and (ii) follow from (6.2) and (6.6). Theorem 6.1 is proved. Q.E.D.

B. Applications to the Hilbert Class Field

Theorem 6.1 makes it easy to compute the genus field. So let's see if the genus field can help us find the Hilbert class field, which in general is more difficult to compute. The nicest case is when the genus field *equals* the Hilbert class field, which happens for field discriminants where every genus consists of a single class (see Exercise 6.15). In particular, if $d_K = -4n$, then this means that n is one of Euler's convenient numbers (see Proposition 3.24). Of the 65 convenient numbers on Gauss' list in §3, 34 satisfy the additional condition that $d_K = -4n$ (see Exercise 6.15), so that we can determine lots of Hilbert class fields. For example, when $K = \mathbb{Q}(\sqrt{-5})$, Theorem 6.1 tells us that the Hilbert class field is $K(\sqrt{5}) = K(i)$. Other examples are just as easy to work out (see Exercise 6.16).

The more typical situation is when the Hilbert class field is strictly bigger than the genus field. It turns out that the genus field can still provide us with useful information about the Hilbert class field. Let's consider the case $K = \mathbb{Q}(\sqrt{-14})$. Here, the genus field is $M = K(\sqrt{-7}) = K(\sqrt{2})$ by Theorem 6.1. Since the class number is 4, we know that the Hilbert class field is a quadratic extension of M, so that $L = M(\sqrt{u})$ for some $u \in M$. This is already useful information, but we can do better. In Theorem 5.1, we saw the importance of a *real* primitive element of the Hilbert class field. So let's intersect everything with the real numbers. This gives us the quadratic extension $M \cap \mathbb{R} \subset L \cap \mathbb{R}$. Since $M = K(\sqrt{2}) = \mathbb{Q}(\sqrt{-14}, \sqrt{2})$, it follows that $M \cap \mathbb{R} = \mathbb{Q}(\sqrt{2})$. Thus we can write $L \cap \mathbb{R} = \mathbb{Q}(\sqrt{2}, \sqrt{u})$, where $u > 0$ is in $\mathbb{Q}(\sqrt{2})$, and from this it is easy to prove that

$$L = K(\sqrt{u}), \quad u = a + b\sqrt{2} > 0, \quad a, b \in \mathbb{Z}$$

(see Exercise 6.17). Hence genus theory explains the form of the primitive element $\alpha = \sqrt{2\sqrt{2} - 1}$ of Proposition 5.31. In Exercise 6.18, we will continue this discussion and show how one can take $u = a + b\sqrt{2}$ and discover the precise form $u = 2\sqrt{2} - 1$ of the primitive element of the Hilbert class field.

It's interesting to compare this discussion of $x^2 + 14y^2$ to what we did in §3. The genus theory developed in §3 told us when p was represented by $x^2 + 14y^2$ or $2x^2 + 7y^2$, but this partial information didn't help in deciding when $p = x^2 + 14y^2$. In contrast, the genus theory of Theorem 6.1 determines the genus field, which helps us understand the Hilbert class field. The field-theoretic approach seems to have more useful information.

This ends our discussion of genus theory, but it by no means exhausts the topic. For more complete treatments of genus theory from the point of view of class field theory, see Hasse [**67**], Janusz [**78**, §VI.3] and Cohn's two books [**25**, Chapters 14 and 18] and [**27**, Chapter 8]. Genus theory can also be studied by standard methods of algebraic number theory, with no reference to class field theory. Both Cohn [**26**, Chapter XIII] and Hasse [**66**, §§26.8 and 29.3] use the Hilbert symbol in their discussion of genera. For a more elementary approach, see Zagier [**140**]. Genus theory can also be generalized in several ways. It is possible to define the genus field of an arbitrary number field (see Ishida [**76**]), and in another direction, one can formulate genus theory from the point of view of algebraic groups and Tamagawa numbers (Ono [**103**] has a nice introduction to this subject). For a survey of all these aspects of genus theory, see Frei [**49**].

C. Exercises

6.1. Let d_K be the discriminant of a quadratic field. When considering forms of discriminant d_K, show that the number μ from Proposition 3.11 is just the number of primes dividing d_K.

6.2. Suppose that we have fields $K \subset M \subset L$, where L is Galois over K with group $G = \mathrm{Gal}(L/K)$. Prove that M is Abelian over K if and only if $[G, G] \subset \mathrm{Gal}(L/M)$.

6.3. Prove statement (6.4).

6.4. If K is an imaginary quadratic field and M is an unramified Abelian extension of K, then prove that M is Galois over $\mathbb{Q}$. Hint: use the description of $\mathrm{Gal}(L/\mathbb{Q})$, where L is the Hilbert class field of K.

6.5. Prove statement (6.5).

6.6. In this problem we will prove Lemma 6.7. Let $\mathfrak{p}$ be a prime of $\mathcal{O}_K$.

(a) If $\mathfrak{p}$ is unramified in LM, then use Exercise 5.15 to show that it's unramified in both L and M.

(b) Prove the converse of (a). Hint: use Proposition 5.10. For a prime $\mathfrak{P}$ of $\mathcal{O}_{LM}$ containing $\mathfrak{p}$, show that if $\sigma \in I_{\mathfrak{P}}$, then the restrictions $\sigma|_L$ and $\sigma|_M$ lie in the inertia groups of $\mathfrak{P} \cap \mathcal{O}_L$ and $\mathfrak{P} \cap \mathcal{O}_M$ respectively. Argue that the restrictions $\sigma|_L$ and $\sigma|_M$ are the identity. Note that parts (a) and (b) prove part (i) of Lemma 6.7.

(c) Use Exercise 5.16 to prove part (ii) of Lemma 6.7.

(d) With the same hypothesis as Lemma 6.7, show that $\mathfrak{p}$ splits completely in LM if and only if it splits completely in both L and M. In Exercise 8.14 we will see that this result can be proved without assuming that L and M are Galois over K.

6.7. Let $K = \mathbb{Q}(i, \sqrt{2m})$, where $m \in \mathbb{Z}$ is odd and squarefree.

(a) Let $\alpha = (1+i)\sqrt{2m}/2$. Show that $a^2 = im$, and conclude that $\alpha \in \mathcal{O}_K$. (It turns out that 1, $i, \sqrt{2m}$ and α form an integral basis of $\mathcal{O}_K$—see Marcus [**96**, Exercise 42 of Chapter 2].)

(b) Let $\mathfrak{P}$ be the ideal of $\mathcal{O}_K$ generated by $1+i$ and $1+\alpha$. Show that $2\mathcal{O}_K = \mathfrak{P}^4$, and conclude that $\mathfrak{P}$ is prime. Hint: compute $\mathfrak{P}^2$.

6.8. Let K be an imaginary quadratic field. We want to show that if $K \subset K(i)$ is unramified, then $d_K \equiv 12 \bmod 16$.

(a) Show that $K \subset K(i)$ is ramified when $d_K \equiv 1 \bmod 4$. Hint: consider the diagram of fields

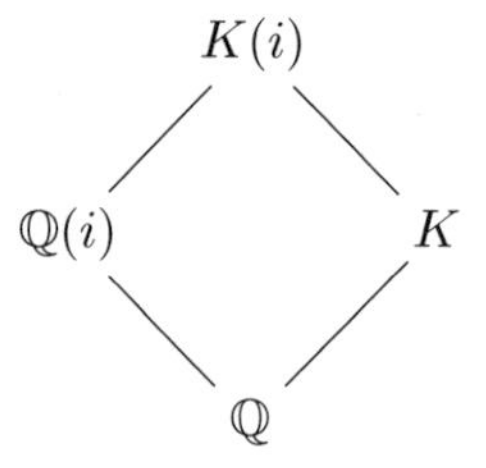

If $K \subset K(i)$ is unramified, show that 2 is unramified in $K(i)$. But 2 ramifies in $\mathbb{Q}(i)$. Exercise 5.15 will be useful.

(b) Show that the extension is ramified when $d_K \equiv 0 \bmod 8$. Hint: if it's unramified, show that the ramification index of 2 in $K(i)$ is at most 2. Then use Exercise 6.7.

Since an even discriminant is of the form $4N$, where $N \equiv 2, 3 \bmod 4$, it follows from parts (a) and (b) that $d_K \equiv 12 \bmod 16$ when $K \subset K(i)$ is unramified.

6.9. In this exercise we will prove Lemma 6.8.

(a) Prove that $K \subset K(\sqrt{a})$ is unramified when $a \mid d_K$ and $a \equiv 1 \bmod 4$. Hint: when $2 \notin \mathfrak{p}$, note that $d_K = ab$, where $K(\sqrt{a}) = K(\sqrt{b})$.

(b) Assume that $K \subset K(\sqrt{a})$ is unramified with a squarefree. Show that $a \mid d_K$. Hint: if p is a prime such that $p \mid a$, $p \nmid d_K$, then analyze p in the fields

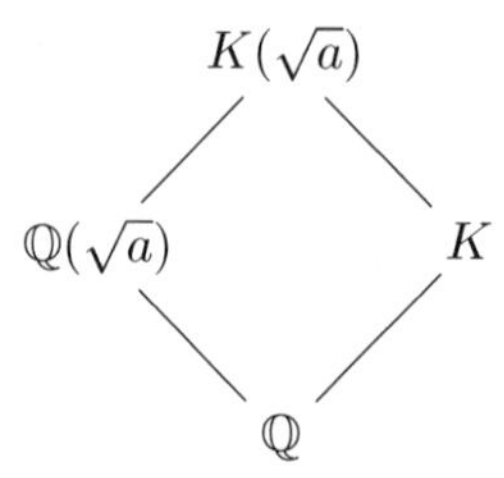

(c) Show that a may be chosen to be odd when d_K is even. Hint: by Proposition 5.16, $2\mathcal{O}_K = \mathfrak{p}^2$, $\mathfrak{p}$ prime in $\mathcal{O}_K$. Set $L = K(\sqrt{a})$ and let $\mathfrak{P}$ be a prime of $\mathcal{O}_L$ containing $\mathfrak{p}$. Then let K' be the fixed field of the inertia group

$I_{\mathfrak{P}} \subset \mathrm{Gal}(L/\mathbb{Q})$. Show that 2 is not ramified in K', so that $K' = \mathbb{Q}(\sqrt{a'})$ for a' odd. Proposition 5.10 will be useful.

(d) Let $K \subset K(\sqrt{a})$ be unramified, where $a \mid d_K$ is odd.

(i) If $a \equiv 3 \bmod 4$, show that $d_K \equiv 12 \bmod 16$. Hint: apply part (a) to $-a$, and then use Exercise 6.8.

(ii) If $d_K \equiv 12 \bmod 16$, show that $K(\sqrt{a}) = K(\sqrt{b})$, where $b \mid d_K$ and $b \equiv 1 \bmod 4$. Hint: factor d_K.

Lemma 6.8 follows easily from parts (a)–(d).

6.10. If M is a Galois extension of $\mathbb{Q}$ and $\mathrm{Gal}(M/\mathbb{Q}) \simeq (\mathbb{Z}/2\mathbb{Z})^m$, then show that $M = \mathbb{Q}(\sqrt{a_1}, \ldots, \sqrt{a_m})$, $a_i \in \mathbb{Z}$ squarefree.

6.11. Prove the description of the genus field M given in (6.9).

6.12. Complete the proof of Theorem 6.1 when $d_K \equiv 1 \bmod 4$, $d_K < 0$.

6.13. Let K be an imaginary quadratic field of discriminant $-4n$. The description of the genus field M given in (6.9) gives us an isomorphism

$$\mathrm{Gal}(M/\mathbb{Q}) \xrightarrow{\sim} \{\pm 1\}^{\mu}.$$

However, we also have maps

$$C(-4n) \longrightarrow C(\mathcal{O}_K) \longrightarrow \mathrm{Gal}(M/K).$$

If we combine these with the natural inclusion $\mathrm{Gal}(M/K) \subset \mathrm{Gal}(M/\mathbb{Q})$, then we get a map

$$C(-4n) \longrightarrow \{\pm 1\}^{\mu}.$$

Show that this map is exactly what Gauss used in his definition of genus. Hint: it's fun to see the characters ϵ, δ and $\epsilon\delta$ from §3 reappear. For example, when n is odd, the key step is to show that

$$\left(\frac{M/K}{\mathfrak{a}}\right)(i) = \delta(N(\mathfrak{a}))i$$

for ideals $\mathfrak{a}$ prime to $4n$. The proof is similar to the proof of (6.13).

6.14. Prove that the map (6.17) is an isomorphism.

6.15. In this exercise we will study when the genus field equals the Hilbert class field.

(a) Prove that the genus field of an imaginary quadratic field K equals its Hilbert class field if and only if for primitive positive forms of discriminant d_K, there is only one class per genus.

(b) Of Gauss' list of 65 convenient numbers n in §3, which satisfy the condition (5.2) that guarantees that $-4n$ is a field discriminant? This gives us a list of fields where we know the Hilbert class field.

6.16. Compute the Hilbert class fields of $\mathbb{Q}(\sqrt{-6})$, $\mathbb{Q}(\sqrt{-10})$ and $\mathbb{Q}(\sqrt{-35})$.

6.17. Let $K = \mathbb{Q}(\sqrt{-14})$, and let L be the Hilbert class field of K. The genus field M of K is $K(\sqrt{-7}) = K(\sqrt{2})$, so that L is a degree 2 extension of M. Use the hints in the text to show that $L = K(\sqrt{u})$, where $u = a + b\sqrt{2} > 0$, $a, b \in \mathbb{Z}$.

6.18. In this exercise we will discover a primitive element for the Hilbert class field L of $K = \mathbb{Q}(\sqrt{-14})$. From the previous exercise, we know that $L = K(\sqrt{u})$, where $u = a + b\sqrt{2} > 0$, $a, b \in \mathbb{Z}$. Let $u' = a - b\sqrt{2}$.

(a) Show that $\text{Gal}(L/\mathbb{Q})$ is the dihedral group

$$\langle \sigma, \tau : \sigma^4 = 1,\ \tau^2 = 1,\ \sigma\tau = \tau\sigma^3 \rangle$$

of order 8, where $\sigma(\sqrt{u}) = \sqrt{u'}$ and τ is complex conjugation. Conclude that $\sigma^2(\sqrt{u}) = -\sqrt{u}$ and $\tau(\sqrt{u'}) = -\sqrt{u'}$.

(b) Show that $\mathbb{Q}(\sqrt{-7})$ is the fixed field of σ^2 and $\sigma\tau$.

(c) Show that $\sqrt{uu'}$ is fixed by σ^2 and $\sigma\tau$, and then using τ, conclude that $\sqrt{uu'} = m\sqrt{-7}$, $m \in \mathbb{Z}$.

(d) Let N be the norm function on $\mathbb{Q}(\sqrt{2})$, and let $\pi = 2\sqrt{2} - 1$. Note that $N(\pi) = -7$. Show that $u = \pi\alpha$, where $N(\alpha) = m^2$. Hint: $\mathbb{Z}[\sqrt{2}]$ is a UFD. You may have to switch u and u'.

(e) Assume that u has no square factors in $\mathbb{Z}[\sqrt{2}]$. Then show that $u = \epsilon\pi n$, where ϵ is a unit and n is a squarefree integer prime to 14. Hint: use Proposition 5.16 to describe the primes in $\mathbb{Z}[\sqrt{2}]$.

(f) Show that n must be ± 1. Hint: note that $u\mathcal{O}_L = \pi\mathcal{O}_L \cdot n\mathcal{O}_L$ is a square, and conclude that any prime dividing n ramifies in L.

(g) Thus $u = \epsilon\pi$ by (f). All units of $\mathbb{Z}[\sqrt{2}]$ are of the form $\pm(\sqrt{2} - 1)^m$ (see Hasse [**66**, pp. 554–556]). Since $N(u) = -7$ and $N(\sqrt{2} - 1) = -1$, we can assume $u = \pi$ since $u > 0$.

This proves that $\sqrt{u} = \sqrt{\pi} = \sqrt{2\sqrt{2} - 1}$ is the desired primitive element.

6.19. Adapt Exercises 6.17 and 6.18 to discover a primitive element for the Hilbert class field of $\mathbb{Q}(\sqrt{-17})$. Hint: see Exercise 5.25. You may assume that the ring of integers in $\mathbb{Q}(\sqrt{17})$ is a UFD and that the units in this ring are all of the form $\pm(4 + \sqrt{17})^m$, $m \in \mathbb{Z}$ (see Borevich and Shafarevich [**11**, p. 422]). One difference is that in the analog of part (f) of the previous exercise, one gets $n = \pm 1$ or ± 2 since 2 ramifies in $\mathbb{Q}(\sqrt{-17})$. Then the analog of part (g) leads to $u = 4 + \sqrt{17}$ or $2(4 + \sqrt{17})$. Rule out the latter by showing that it would violate Theorem 5.1. Note that $u = 4 + \sqrt{17}$ is related to our earlier choice $(1 + \sqrt{17})/2$ via

$$(4 + \sqrt{17}) \cdot (1 + \sqrt{17})/2 = ((5 + \sqrt{17})/2)^2.$$

This problem can be done without using the fact that $\mathbb{Q}(\sqrt{17})$ has class number 1 (see Herz [**72**]).

6.20. Let $K = \mathbb{Q}(\sqrt{-55})$.

(a) Show that $C(\mathcal{O}_K) \simeq \mathbb{Z}/4\mathbb{Z}$.

(b) Determine the Hilbert class field of K. Hint: use the methods of Proposition 5.31. Exercises 6.17 and 6.18 will show you what to look for.

(c) Prove an analog of Theorem 5.33 for primes of the form $x^2 + 55y^2$. Hint: show that $x^2 + xy + 14y^2$ and $x^2 + 55y^2$ represent the same odd numbers.

7. Orders in Imaginary Quadratic Fields

In §5, we solved our basic question of $p = x^2 + ny^2$ for those n's where $\mathbb{Z}[\sqrt{-n}]$ is the full ring of integers $\mathcal{O}_K$ in $K = \mathbb{Q}(\sqrt{-n})$ (see (5.15)). This holds for infinitely many n's, but it also leaves out infinitely many. The full story of what happens for these other n's will be told in §9, and we will see that the answer involves the ring $\mathbb{Z}[\sqrt{-n}]$. Such a ring is an example of an *order* in an imaginary quadratic field, which brings us to the main topic of §7.

We begin this section with a study of orders in a quadratic field K. Unlike $\mathcal{O}_K$, an order $\mathcal{O}$ is usually *not* a Dedekind domain, so that the ideal theory of $\mathcal{O}$ is more complicated. This will lead us to restrict the class of ideals under consideration. In the case of imaginary quadratic fields, there is a nice relation between ideals in orders and quadratic forms. In particular, an order $\mathcal{O}$ has an ideal class group $C(\mathcal{O})$, and we will show that for any discriminant $D < 0$, the form class group $C(D)$ from §3 is naturally isomorphic to $C(\mathcal{O})$ for a suitable order $\mathcal{O}$. Then, to prepare the way for class field theory, we will show how to translate ideals for an order $\mathcal{O}$ in K into terms of the maximal order $\mathcal{O}_K$. The section will conclude with a discussion of class numbers.

A. Orders in Quadratic Fields

An *order* $\mathcal{O}$ in a quadratic field K is a subset $\mathcal{O} \subset K$ such that

(i) $\mathcal{O}$ is a subring of K containing 1.
(ii) $\mathcal{O}$ is a finitely generated $\mathbb{Z}$-module.
(iii) $\mathcal{O}$ contains a $\mathbb{Q}$-basis of K.

Since $\mathcal{O}$ is clearly torsion-free, (ii) and (iii) are equivalent to $\mathcal{O}$ being a free $\mathbb{Z}$-module of rank 2 (see Exercise 7.1). Note also that by (iii), K is the field of fractions of $\mathcal{O}$.

The ring $\mathcal{O}_K$ of integers in K is always an order in K—this follows from the description (5.13) of $\mathcal{O}_K$ given in §5. More importantly, (i) and (ii) above imply that for *any* order $\mathcal{O}$ of K, we have $\mathcal{O} \subset \mathcal{O}_K$ (see Exercise 7.2), so that $\mathcal{O}_K$ is the *maximal order* of K.

To describe orders in quadratic fields more explicitly, first note that by (5.14), the maximal order $\mathcal{O}_K$ can be written as follows:

$$\mathcal{O}_K = [1, w_K], \qquad w_K = \frac{d_K + \sqrt{d_K}}{2}, \tag{7.1}$$

where d_K is the discriminant of K. We now describe all orders in quadratic fields:

LEMMA 7.2. *Let $\mathcal{O}$ be an order in a quadratic field K of discriminant d_K. Then $\mathcal{O}$ has finite index in $\mathcal{O}_K$, and if we set $f = [\mathcal{O}_K : \mathcal{O}]$, then*

$$\mathcal{O} = \mathbb{Z} + f\mathcal{O}_K = [1, fw_K],$$

where w_K is as in (7.1).

PROOF. First note that $[\mathcal{O}_K : \mathcal{O}] < \infty$ since $\mathcal{O}$ and $\mathcal{O}_K$ are free $\mathbb{Z}$-modules of rank 2. Setting $f = [\mathcal{O}_K : \mathcal{O}]$, we have $f\mathcal{O}_K \subset \mathcal{O}$, and then $\mathbb{Z} + f\mathcal{O}_K \subset \mathcal{O}$ follows. However, (7.1) implies $\mathbb{Z} + f\mathcal{O}_K = [1, fw_K]$, so that to prove the lemma, we need only show that $[1, fw_K]$ has index f in $\mathcal{O}_K = [1, w_K]$. This is obvious, and we are done. Q.E.D.

Given an order $\mathcal{O}$ in a quadratic field K, the index $f = [\mathcal{O}_K : \mathcal{O}]$ is called the *conductor* of the order. Another important invariant of $\mathcal{O}$ is its *discriminant*, which is defined as follows. Let $\alpha \mapsto \alpha'$ be the nontrivial automorphism of K, and suppose that $\mathcal{O} = [\alpha, \beta]$. Then the *discriminant* of $\mathcal{O}$ is the number

$$D = \left(\det \begin{pmatrix} \alpha & \beta \\ \alpha' & \beta' \end{pmatrix}\right)^2.$$

The discriminant is independent of the integral basis used, and if we compute D using the basis $\mathcal{O} = [1, fw_k]$ from Lemma 7.2, then we obtain the formula

$$D = f^2 d_K. \tag{7.3}$$

Thus the discriminant satisfies $D \equiv 0, 1 \bmod 4$. From (7.3) we also see that $K = \mathbb{Q}(\sqrt{D})$, so that K is real or imaginary according to whether $D > 0$ or $D < 0$. In fact, one can show that D determines $\mathcal{O}$ uniquely and that any nonsquare integer $D \equiv 0, 1 \bmod 4$ is the discriminant of an order in a quadratic field. See Exercise 7.3 for proofs of these elementary facts. Note that by (7.3), the discriminant of the maximal order $\mathcal{O}_K$ is d_K, which agrees with the definition given in §5.

For an example of an order, consider $\mathbb{Z}[\sqrt{-n}] \subset K = \mathbb{Q}(\sqrt{-n})$. The discriminant of $\mathbb{Z}[\sqrt{-n}]$ is easily computed to be $-4n$, and then (7.3) shows that

$$-4n = f^2 d_K.$$

This makes it easy to compute the conductor of $\mathbb{Z}[\sqrt{-n}]$. This order will be used in §9 when we give the general solution of $p = x^2 + ny^2$.

Now let's study the ideals of an order $\mathcal{O}$. If $\mathfrak{a}$ is a nonzero ideal of $\mathcal{O}$, then the proof of Corollary 5.4 adapts easily to show that $\mathcal{O}/\mathfrak{a}$ is finite (see Exercise 7.4). Thus we can define the *norm* of $\mathfrak{a}$ to be $N(\mathfrak{a}) = |\mathcal{O}/\mathfrak{a}|$. Furthermore, as in the proof of Theorem 5.5, it follows that $\mathcal{O}$ is Noetherian and that every nonzero prime ideal of $\mathcal{O}$ is maximal (see Exercise 7.4). However, it is equally obvious that if the conductor f of $\mathcal{O}$ is greater than 1, then $\mathcal{O}$ is *not* integrally closed in K, so that $\mathcal{O}$ is *not* a Dedekind domain when $f > 1$. Thus we may not assume that the ideals of $\mathcal{O}$ have unique factorization.

To remedy this situation, we will introduce the concept of a *proper ideal* of an order. Namely, given any ideal $\mathfrak{a}$ of $\mathcal{O}$, notice that

$$\mathcal{O} \subset \{\beta \in K : \beta\mathfrak{a} \subset \mathfrak{a}\}$$

since $\mathfrak{a}$ is an ideal of $\mathcal{O}$. However, equality need not occur. For example, if $\mathcal{O} = \mathbb{Z}[\sqrt{-3}]$ is the order of conductor 2 in $K = \mathbb{Q}(\sqrt{-3})$, and $\mathfrak{a}$ is the ideal of $\mathcal{O}$ generated by 2 and $1 + \sqrt{-3}$, then one sees easily that

$$\mathcal{O} \neq \{\beta \in K : \beta\mathfrak{a} \subset \mathfrak{a}\} = \mathcal{O}_K$$

(see Exercise 7.5). In general, we say that an ideal $\mathfrak{a}$ of $\mathcal{O}$ is *proper* whenever equality holds, i.e., when

$$\mathcal{O} = \{\beta \in K : \beta\mathfrak{a} \subset \mathfrak{a}\}.$$

For example, nonzero principal ideals are always proper, and for the maximal order, *all* nonzero ideals are proper (see Exercise 7.6).

We can also extend this terminology to fractional ideals. A *fractional ideal* of $\mathcal{O}$ is a subset of K which is a nonzero finitely generated $\mathcal{O}$-module. One can show that every fractional ideal is of the form $\alpha\mathfrak{a}$, where $\alpha \in K^*$ and $\mathfrak{a}$ is an $\mathcal{O}$-ideal (see Exercise 7.7). Then a fractional $\mathcal{O}$-ideal $\mathfrak{b}$ is *proper* provided that

$$\mathcal{O} = \{\beta \in K : \beta\mathfrak{b} \subset \mathfrak{b}\}.$$

Once we have fractional ideals, we can talk about invertible ideals: a fractional $\mathcal{O}$-ideal $\mathfrak{a}$ is *invertible* if there is another fractional $\mathcal{O}$-ideal $\mathfrak{b}$ such that $\mathfrak{a}\mathfrak{b} = \mathcal{O}$. Note that principal fractional ideals (those of the form $\alpha\mathcal{O}$, $\alpha \in K^*$) are obviously invertible. The basic result is that for orders in quadratic fields, the notions of proper and invertible coincide:

PROPOSITION 7.4. *Let $\mathcal{O}$ be an order in a quadratic field K, and let $\mathfrak{a}$ be a fractional $\mathcal{O}$-ideal. Then $\mathfrak{a}$ is proper if and only if $\mathfrak{a}$ is invertible.*

PROOF. If $\mathfrak{a}$ is invertible, then $\mathfrak{a}\mathfrak{b} = \mathcal{O}$ for some fractional $\mathcal{O}$-ideal $\mathfrak{b}$. If $\beta \in K$ and $\beta\mathfrak{a} \subset \mathfrak{a}$, then we have

$$\beta\mathcal{O} = \beta(\mathfrak{a}\mathfrak{b}) = (\beta\mathfrak{a})\mathfrak{b} \subset \mathfrak{a}\mathfrak{b} = \mathcal{O},$$

and $\beta \in \mathcal{O}$ follows, proving that $\mathfrak{a}$ is proper.

To argue the other way, we will need the following lemma:

LEMMA 7.5. *Let $K = \mathbb{Q}(\tau)$ be a quadratic field, and let $ax^2 + bx + c$ be the minimal polynomial of τ, where a, b and c are relatively prime integers. Then $[1, \tau]$ is a proper fractional ideal for the order $[1, a\tau]$ of K.*

PROOF. First, $[1,\ a\tau]$ is an order since $a\tau$ is an algebraic integer. Then, given $\beta \in K$, note that $\beta[1, \tau] \subset [1, \tau]$ is equivalent to

$$\beta \cdot 1 \in [1, \tau]$$
$$\beta \cdot \tau \in [1, \tau].$$

The first line says $\beta = m + n\tau$, $m, n \in \mathbb{Z}$. To understand the second, note that

$$\beta\tau = m\tau + n\tau^2 = m\tau + \frac{n}{a}(-b\tau - c)$$
$$= \frac{-cn}{a} + \left(\frac{-bn}{a} + m\right)\tau.$$

Since $\gcd(a, b, c) = 1$, we see that $\beta\tau \in [1, \tau]$ if and only if $a \mid n$. It follows that

$$\{\beta \in K : \beta[1, \tau] \subset [1, \tau]\} = [1, a\tau],$$

which proves the lemma. Q.E.D.

Now we are ready to prove that proper fractional ideals are invertible. First note that $\mathfrak{a}$ is a $\mathbb{Z}$-module of rank 2 (see Exercise 7.8), so that $\mathfrak{a} = [\alpha, \beta]$ for some $\alpha, \beta \in K$. Then $\mathfrak{a} = \alpha[1, \tau]$, where $\tau = \beta/\alpha$. If $ax^2 + bx + c$, $\gcd(a, b, c,) = 1$, is the minimal polynomial of τ, then Lemma 7.5 implies that $\mathcal{O} = [1, a\tau]$. Let $\beta \mapsto \beta'$ denote the nontrivial automorphism of K. Since τ' is the other root of $ax^2 + bx + c$, using Lemma 7.5 again shows that $\mathfrak{a}' = \alpha'[1, \tau']$ is a fractional ideal for $[1, a\tau] = [1, a\tau'] = \mathcal{O}$. We claim that

$$\mathfrak{a}\mathfrak{a}' = \frac{N(\alpha)}{a}\mathcal{O}. \tag{7.6}$$

To see why, note that

$$a\mathfrak{a}\mathfrak{a}' = a\alpha\alpha'[1, \tau][1, \tau'] = N(\alpha)[a, a\tau, a\tau', a\tau\tau'].$$

Since $\tau + \tau' = -b/a$ and $\tau\tau' = c/a$, this becomes

$$a\mathfrak{a}\mathfrak{a}' = N(\alpha)[a, a\tau, -b, c] = N(\alpha)[1, a\tau] = N(\alpha)\mathcal{O}$$

since $\gcd(a, b, c) = 1$. This implies (7.6), which shows that $\mathfrak{a}$ is invertible. Q.E.D.

Unfortunately, Proposition 7.4 is not strong enough to prove unique factorization for proper ideals (see Exercise 7.9 for a counterexample). Later we will see that unique factorization holds for a slightly smaller class of ideals, those prime to the conductor.

Given an order $\mathcal{O}$, let $I(\mathcal{O})$ denote the set of proper fractional $\mathcal{O}$-ideals. By Proposition 7.4, $I(\mathcal{O})$ is a group under multiplication: the crucial issues are closure and the existence of inverses, both of which follow from the invertibility of proper ideals (see Exercise 7.10). The principal $\mathcal{O}$-ideals give a subgroup $P(\mathcal{O}) \subset I(\mathcal{O})$, and thus we can form the quotient

$$C(\mathcal{O}) = I(\mathcal{O})/P(\mathcal{O}),$$

which is the *ideal class group* of the order $\mathcal{O}$. When $\mathcal{O}$ is the maximal order $\mathcal{O}_K, I(\mathcal{O}_K)$ and $P(\mathcal{O}_K)$ will be denoted I_K and P_K, respectively. This is the notation used in §5, and in general we reserve the subscript K exclusively for the maximal order. Then the above definition of $C(\mathcal{O}_K)$ agrees with the one given in §5.

B. Orders and Quadratic Forms

We can relate the ideal class group $C(\mathcal{O})$ to the form class group $C(D)$ defined in §3 as follows:

THEOREM 7.7. *Let $\mathcal{O}$ be the order of discriminant D in an imaginary quadratic field K. Then:*

(i) *If $f(x, y) = ax^2 + bxy + cy^2$ is a primitive positive definite quadratic form of discriminant D, then $[a, (-b + \sqrt{D})/2]$ is a proper ideal of $\mathcal{O}$.*
(ii) *The map sending $f(x, y)$ to $[a, (-b + \sqrt{D})/2]$ induces an isomorphism between the form class group $C(D)$ and the ideal class group $C(\mathcal{O})$. Hence the order of $C(\mathcal{O})$ is the class number $h(D)$.*
(iii) *A positive integer m is represented by a form $f(x, y)$ if and only if m is the norm $N(\mathfrak{a})$ of some ideal $\mathfrak{a}$ in the corresponding ideal class in $C(\mathcal{O})$ (recall that $N(\mathfrak{a}) = |\mathcal{O}/\mathfrak{a}|$).*

REMARK. Because of the isomorphism $C(D) \simeq C(\mathcal{O})$, we will sometimes write the class number as $h(\mathcal{O})$ instead of $h(D)$.

PROOF. Let $f(x, y) = ax^2 + bxy + cy^2$ be a primitive positive definite form of discriminant $D < 0$. The roots of $f(x, 1) = ax^2 + bx + c$ are complex, so that there is a unique $\tau \in \mathfrak{h}$ ($\mathfrak{h}$ is the upper half plane) such that $f(\tau, 1) = 0$. We call τ the *root* of $f(x, y)$. Since $a > 0$, it follows that $\tau = (-b + \sqrt{D})/2a$. Thus

$$[a, (-b + \sqrt{D})/2] = [a, a\tau] = a[1, \tau].$$

Note also that $\tau \in K$.

To prove (i), note that by Lemma 7.5, $a[1, \tau]$ is a proper ideal for the order $[1, a\tau]$. However, if f is the conductor of $\mathcal{O}$, then $D = f^2 d_K$ by (7.3), and thus

$$\begin{aligned} a\tau &= \frac{-b + \sqrt{D}}{2} = \frac{-b + f\sqrt{d_K}}{2} \\ &= -\frac{b + fd_K}{2} + f\left(\frac{d_K + \sqrt{d_K}}{2}\right) = -\frac{b + fd_K}{2} + fw_K. \end{aligned}$$

Since $D = b^2 - 4ac$, fd_K and b have the same parity, so that $(b + fd_K)/2 \in \mathbb{Z}$. It follows that $[1, a\tau] = [1, fw_K]$, so that $[1, a\tau] = \mathcal{O}$ by Lemma 7.2. This proves that $a[1, \tau]$ is a proper $\mathcal{O}$-ideal.

To prove (ii), let $f(x,y)$ and $g(x,y)$ be forms of discriminant D, and let τ and τ' be their respective roots. We will prove the following equivalences:

$$\begin{aligned} & f(x,y),\ g(x,y) \text{ are properly equivalent} \\ \iff & \tau' = \frac{p\tau+q}{r\tau+s},\ \begin{pmatrix} p & q \\ r & s \end{pmatrix} \in \mathrm{SL}(2,\mathbb{Z}) \\ \iff & [1,\tau] = \lambda[1,\tau'],\ \lambda \in K^*. \end{aligned} \tag{7.8}$$

To see why this is true, assume that $f(x,y) = g(px+qy, rx+sy)$, where $\left(\begin{smallmatrix} p & q \\ r & s \end{smallmatrix}\right) \in \mathrm{SL}(2,\mathbb{Z})$. Then

$$0 = f(\tau,1) = g(p\tau+q, r\tau+s) = (r\tau+s)^2\, g\!\left(\frac{p\tau+q}{r\tau+s}, 1\right), \tag{7.9}$$

so that $g((p\tau+q)/(r\tau+s),1) = 0$. By an easy computation (see Exercise 7.11),

$$\mathrm{Im}\left(\frac{p\tau+q}{r\tau+s}\right) = \det\begin{pmatrix} p & q \\ r & s \end{pmatrix} |r\tau+s|^{-2}\, \mathrm{Im}(\tau). \tag{7.10}$$

This implies $(p\tau+q)/(r\tau+s) \in \mathfrak{h}$, and thus $\tau' = (p\tau+q)/(r\tau+s)$ by the uniqueness of the root τ'. Conversely, if $\tau' = (p\tau+q)/(r\tau+s)$, then (7.9) shows that $f(x,y)$ and $g(px+qy, rx+sy)$ have the same root, and it follows easily that they must be equal (see Exercise 7.12). This proves the first equivalence of (7.8).

Next, if $\tau' = (p\tau+q)/(r\tau+s)$, let $\lambda = r\tau+s \in K^*$. Then

$$\begin{aligned} \lambda[1,\tau'] &= (r\tau+s)\left[1, \frac{p\tau+q}{r\tau+s}\right] \\ &= [r\tau+s, p\tau+q] = [1,\tau] \end{aligned}$$

since $\left(\begin{smallmatrix} p & q \\ r & s \end{smallmatrix}\right) \in \mathrm{SL}(2,\mathbb{Z})$. Conversely, if $[1,\tau] = \lambda[1,\tau']$ for some $\lambda \in K^*$, then $[1,\tau] = [\lambda, \lambda\tau']$, which implies

$$\begin{aligned} \lambda\tau' &= p\tau+q \\ \lambda &= r\tau+s \end{aligned}$$

for some $\left(\begin{smallmatrix} p & q \\ r & s \end{smallmatrix}\right) \in \mathrm{GL}(2,\mathbb{Z})$. This gives us

$$\tau' = \frac{p\tau+q}{r\tau+s},$$

and then (7.10) shows that $\left(\begin{smallmatrix} p & q \\ r & s \end{smallmatrix}\right) \in \mathrm{SL}(2,\mathbb{Z})$ since τ and τ' are both in $\mathfrak{h}$. This completes the proof of (7.8).

Using (7.8), one easily sees that the map sending $f(x,y)$ to $a[1,\tau]$ induces an *injection*

$$C(D) \longrightarrow C(\mathcal{O}).$$

To show that the map is surjective, let $\mathfrak{a}$ be a proper fractional $\mathcal{O}$-ideal. As in the proof of Proposition 7.4, we can write $\mathfrak{a} = [\alpha,\beta]$ for some α, $\beta \in K$. Switching α and β if necessary, we can assume that $\tau = \beta/\alpha$ lies in $\mathfrak{h}$. Let ax^2+bx+c be the minimal polynomial of τ. We may assume that $\gcd(a,b,c) = 1$ and $a > 0$. Then $f(x,y) = ax^2+bxy+cy^2$ is positive definite of discriminant D (see Exercise 7.12), and $f(x,y)$ maps to $a[1,\tau]$. This ideal lies in the class of $\mathfrak{a} = [\alpha,\beta] = \alpha[1,\tau]$ in $C(\mathcal{O})$, and surjectivity is proved.

We thus have a bijection of sets

$$C(D) \longrightarrow C(\mathcal{O}). \tag{7.11}$$

We next want to see what happens to the group structure, but we first need to review the formulas for Dirichlet composition from §3. Given two primitive positive definite forms $f(x,y) = ax^2 + bxy + cy^2$ and $g(x,y) = a'x^2 + b'xy + c'y^2$ of discriminant D, suppose that $\gcd(a, a', (b+b')/2) = 1$. Then the *Dirichlet composition* of $f(x,y)$ and $g(x,y)$ was defined to be the form

$$F(x,y) = aa'x^2 + Bxy + \frac{B^2 - D}{4aa'}y^2,$$

where B is the unique number modulo $2aa'$ such that

$$\begin{aligned} B &\equiv b \bmod 2a \\ B &\equiv b' \bmod 2a' \\ B^2 &\equiv D \bmod 4aa' \end{aligned} \tag{7.12}$$

(see Lemma 3.2 and (3.7)). In Theorem 3.9 we asserted that Dirichlet composition made $C(D)$ into an Abelian group, but the proof given in §3 was not complete. So our first task is to use the bijection (7.11) to finish the proof of Theorem 3.9.

Given $f(x,y)$, $g(x,y)$ and $F(x,y)$ as above, we get three proper ideals of $\mathcal{O}$:

$$[a, (-b + f\sqrt{d_K})/2],\ [a', (-b' + f\sqrt{d_K})/2] \text{ and } [aa', (-B + f\sqrt{d_K})/2].$$

If we set $\Delta = (-B + f\sqrt{d_K})/2$ and use the top two lines of (7.12), then these ideals can be written as

$$[a, \Delta],\ [a', \Delta] \text{ and } [aa', \Delta].$$

We claim that

$$[a, \Delta][a', \Delta] = [aa', \Delta]. \tag{7.13}$$

To see this, note that $\Delta^2 \equiv -B\Delta \bmod aa'$ by the last line of (7.12). Thus

$$[a, \Delta][a', \Delta] = [aa', a\Delta, a'\Delta, \Delta^2] = [aa', a\Delta, a'\Delta, -B\Delta].$$

However, from (7.12) one easily proves that $\gcd(a, a', B) = 1$ (see Exercise 7.13), and then (7.13) follows immediately.

By (7.11) and (7.13), we see that the Dirichlet composition of $f(x,y)$ and $g(x,y)$ corresponds to the product of their corresponding ideal classes, which proves that Dirichlet composition induces a well-defined binary operation on $C(D)$. Furthermore, since the product of ideals makes $C(\mathcal{O})$ into a group, it follows immediately that $C(D)$ is a group under Dirichlet composition. This completes the proof of Theorem 3.9, and it is now obvious that (7.11) is an isomorphism of groups.

Before we can prove part (iii) of the theorem, we need to learn more about the norm $N(\mathfrak{a}) = |\mathcal{O}/\mathfrak{a}|$ of a proper $\mathcal{O}$-ideal $\mathfrak{a}$. The basic properties of $N(\mathfrak{a})$ are:

LEMMA 7.14. *Let $\mathcal{O}$ be an order in an imaginary quadratic field. Then:*

(i) $N(\alpha\mathcal{O}) = N(\alpha)$ *for* $\alpha \in \mathcal{O}$, $\alpha \neq 0$.
(ii) $N(\mathfrak{a}\mathfrak{b}) = N(\mathfrak{a})N(\mathfrak{b})$ *for proper* $\mathcal{O}$*-ideals* $\mathfrak{a}$ *and* $\mathfrak{b}$.
(iii) $\mathfrak{a}\bar{\mathfrak{a}} = N(\mathfrak{a})\mathcal{O}$ *for a proper* $\mathcal{O}$*-ideal* $\mathfrak{a}$.

PROOF. The proof of (i) is covered in Exercises 7.14 and 7.15. We will next prove a special case of (ii): if $\alpha \neq 0$ in $\mathcal{O}$, we claim that

$$N(\alpha\mathfrak{a}) = N(\alpha)N(\mathfrak{a}). \tag{7.15}$$

To prove this, note that the inclusions $\alpha\mathfrak{a} \subset \alpha\mathcal{O} \subset \mathcal{O}$ give us the short exact sequence

$$0 \to \alpha\mathcal{O}/\alpha\mathfrak{a} \to \mathcal{O}/\alpha\mathfrak{a} \to \mathcal{O}/\alpha\mathcal{O} \to 0,$$

which implies that $|\mathcal{O}/\alpha\mathfrak{a}| = |\mathcal{O}/\alpha\mathcal{O}||\alpha\mathcal{O}/\alpha\mathfrak{a}|$. Since multiplication by α induces an isomorphism $\mathcal{O}/\mathfrak{a} \xrightarrow{\sim} \alpha\mathcal{O}/\alpha\mathfrak{a}$, we get $N(\alpha\mathfrak{a}) = N(\alpha\mathcal{O})N(\mathfrak{a})$, and then (7.15) follows from (i).

Before proving (ii) and (iii), we need to study $N(\mathfrak{a})$. If we write $\mathfrak{a}$ in the form $\mathfrak{a} = \alpha[1, \tau]$, then Lemma 7.5 implies that $\mathcal{O} = [1, a\tau]$. Since $[a, a\tau]$ obviously has index a in $[1, a\tau]$, we obtain

$$N(a[1,\tau]) = a.$$

Then $a \cdot \ \mathfrak{a} \ = \alpha \cdot a[1, \tau]$ and (7.15) imply that

$$N(\mathfrak{a}) = \frac{N(\alpha)}{a}. \tag{7.16}$$

Now (iii) follows immediately by combining (7.16) with the equation

$$\mathfrak{a}\bar{\mathfrak{a}} = \frac{N(\alpha)}{a}\mathcal{O}$$

proved in (7.6). Turning to (ii), note that (iii) implies that

$$N(\mathfrak{a}\mathfrak{b})\mathcal{O} = \mathfrak{a}\mathfrak{b} \cdot \overline{\mathfrak{a}\mathfrak{b}} \ = \ \mathfrak{a}\bar{\mathfrak{a}} \cdot \mathfrak{b}\bar{\mathfrak{b}} = N(\mathfrak{a})\mathcal{O} \cdot N(\mathfrak{b})\mathcal{O} = N(\mathfrak{a})N(\mathfrak{b})\mathcal{O},$$

and then $N(\mathfrak{a}\mathfrak{b}) = N(\mathfrak{a})N(\mathfrak{b})$ follows. Q.E.D.

A useful consequence of this lemma is that if $\mathfrak{a}$ is a proper $\mathcal{O}$-ideal, then $\bar{\mathfrak{a}}$ gives the inverse of $\mathfrak{a}$ in $C(\mathcal{O})$. This follows immediately from $\mathfrak{a}\bar{\mathfrak{a}} = N(\mathfrak{a})\mathcal{O}$. In Exercise 7.16 we will use the isomorphism $C(D) \simeq C(\mathcal{O})$ to give a second proof of this fact.

We can now prove part (iii) of the theorem. If m is represented by $f(x, y)$, then $m = d^2a$, where a is properly represented by $f(x, y)$. We may assume that $f(x, y) = ax^2 + bxy + cy^2$. Then $f(x, y)$ maps to $\mathfrak{a} = a[1, \tau]$, so that $N(\mathfrak{a}) = a$ by (7.16). It follows that $N(d\mathfrak{a}) = d^2a = m$, so that m is the norm of an ideal in the class of $\mathfrak{a}$.

Conversely, assume that $N(\mathfrak{a}) = m$. We know that $\mathfrak{a} = \alpha[1, \tau]$, where $\mathrm{Im}(\tau) > 0$ and $a\tau^2 + b\tau + c = 0$, $\gcd(a, b, c) = 1$ and $a > 0$. Then $f(x, y) = ax^2 + bxy + cy^2$ maps to the class of $\mathfrak{a}$, so that we need only show that $f(x, y)$ represents m.

By (7.16), we know that

$$m = N(\mathfrak{a}) = \frac{N(\alpha)}{a}.$$

However, $\alpha[1, \tau] = \mathfrak{a} \subset \mathcal{O} = [1, a\tau]$, so that $\alpha = p + qa\tau$ and $\alpha\tau = r + sa\tau$ for some integers $p, q, r, s \in \mathbb{Z}$. Thus $(p + qa\tau)\tau = r + sa\tau$, and since $a\tau^2 = -b\tau - c$, comparing coefficients shows that $p = as + bq$. Hence

$$\begin{aligned} m = \frac{N(\alpha)}{a} &= \frac{1}{a}(p^2 - bpq + acq^2) \\ &= \frac{1}{a}((as + bq)^2 - b(as + bq)q + acq^2) \\ &= \frac{1}{a}(a^2s^2 + absq + acq^2) \\ &= as^2 + bsq + cq^2 = f(s, q). \end{aligned}$$

This proves (iii) and completes the proof of Theorem 7.7. Q.E.D.

Notice that Theorem 5.30 is an immediate corollary of Theorem 7.7.

The map $f(x,y) \mapsto \mathfrak{a} = [a, (-b+\sqrt{D})/2]$ of Theorem 7.7 has a natural inverse which is defined as follows. If $\mathfrak{a} = [\alpha, \beta]$ is a proper $\mathcal{O}$-ideal with $\mathrm{Im}(\beta/\alpha) > 0$, then

$$f(x,y) = \frac{N(\alpha x - \beta y)}{N(\mathfrak{a})}$$

is a positive definite form of discriminant D. On the level of classes, this map is the inverse to the map of Theorem 7.7 (see Exercise 7.17).

Theorem 7.7 allows us to translate what we know about quadratic forms into facts about ideal classes. Here is an example that will be useful later on:

COROLLARY 7.17. *Let $\mathcal{O}$ be an order in an imaginary quadratic field. Given a nonzero integer M, then every ideal class in $C(\mathcal{O})$ contains a proper $\mathcal{O}$-ideal whose norm is relatively prime to M.*

PROOF. In Lemma 2.25 we learned that any primitive form represents at least one number relatively prime to M, and the corollary then follows from part (iii) of Theorem 7.7. Q.E.D.

The reader may wonder if Theorem 7.7 holds for real quadratic fields. Simple examples show that this isn't true in general. For instance, when $K = \mathbb{Q}(\sqrt{3})$, the maximal order $\mathcal{O}_K = \mathbb{Z}[\sqrt{3}]$ is a UFD, which implies that $C(\mathcal{O}_K) \simeq \{1\}$. Yet the forms $\pm(x^2 - 3y^2)$ of discriminant $d_K = 12$ are not properly equivalent, so that $C(d_K) \not\simeq \{1\}$ (see Exercise 7.18 for the details). In order to make a version of Theorem 7.7 that holds for real quadratic fields, we need to change the notion of equivalence. In Exercises 7.19–7.24 we will explore two ways of doing this:

1. Change the notion of equivalence of ideals. Instead of using all principal ideals $P(\mathcal{O})$, use only $P^+(\mathcal{O})$, which consists of all principal ideals $\alpha\mathcal{O}$ where $N(\alpha) > 0$. The quotient $I(\mathcal{O})/P^+(\mathcal{O})$ is the *narrow* (or *strict*) *ideal class group* and is denoted by $C^+(\mathcal{O})$. In Exercise 7.21 we will construct a natural isomorphism $C(D) \simeq C^+(\mathcal{O})$ which holds for any order in any quadratic field K. We also have $C^+(\mathcal{O}) = C(\mathcal{O})$ when K is imaginary, and the same is true when K is real and $\mathcal{O}$ has a unit ϵ with $N(\epsilon) = -1$. If K has no such unit, then $|C^+(\mathcal{O})| = 2|C(\mathcal{O})|$.
2. Change the notion of equivalence of forms. Instead of using proper equivalence, use the notion of *signed equivalence*, where $f(x,y)$ and $g(x,y)$ are *signed equivalent* if there is a matrix $\left(\begin{smallmatrix} p & q \\ r & s \end{smallmatrix}\right) \in \mathrm{GL}(2,\mathbb{Z})$ such that

$$f(x,y) = \det\begin{pmatrix} p & q \\ r & s \end{pmatrix} g(px+qy, rx+sy).$$

The set of signed equivalence classes is denoted $C_s(D)$, and in Exercise 7.22 we will see that there is a natural isomorphism $C_s(D) \simeq C(\mathcal{O})$. The criteria for when $C_s(D) = C(D)$ are the same as above.

For other treatments of the relation between forms and ideals, see Borevich and Shafarevich [**11**, Chapter 2, §7.5], Cohn [**25**, §§14.A–C] and Zagier [**140**, §§8 and 10].

C. Ideals Prime to the Conductor

The theory described so far does not interact well with the usual formulation of class field theory. The reason is that class field theory is always stated in terms of the maximal order $\mathcal{O}_K$. So given an order $\mathcal{O}$ in a quadratic field K, we will need to translate proper $\mathcal{O}$-ideals into terms of $\mathcal{O}_K$-ideals. This is difficult to do directly, but becomes much easier once we study $\mathcal{O}$-ideals prime to the conductor.

Given an order $\mathcal{O}$ of conductor f, we say that a nonzero $\mathcal{O}$-ideal $\mathfrak{a}$ is *prime to* f provided that $\mathfrak{a} + f\mathcal{O} = \mathcal{O}$. The following lemma gives the basic properties of $\mathcal{O}$-ideals prime to the conductor:

LEMMA 7.18. *Let $\mathcal{O}$ be an order of conductor f.*

(i) *An $\mathcal{O}$-ideal $\mathfrak{a}$ is prime to f if and only if its norm $N(\mathfrak{a})$ is relatively prime to f.*

(ii) *Every $\mathcal{O}$-ideal prime to f is proper.*

PROOF. To prove (i), let $m_f : \mathcal{O}/\mathfrak{a} \to \mathcal{O}/\mathfrak{a}$ be multiplication by f. Then

$$\mathfrak{a} + f\mathcal{O} = \mathcal{O} \iff m_f \text{ is surjective} \iff m_f \text{ is an isomorphism.}$$

By the structure theorem for finite Abelian groups, m_f is an isomorphism if and only if f is relatively prime to the order $N(\mathfrak{a})$ of $\mathcal{O}/\mathfrak{a}$, and (i) is proved.

To show that an $\mathcal{O}$-ideal $\mathfrak{a}$ prime to f is proper, let $\beta \in K$ satisfy $\beta\mathfrak{a} \subset \mathfrak{a}$. Then β is certainly in $\mathcal{O}_K$, and we thus have

$$\beta\mathcal{O} = \beta(\mathfrak{a} + f\mathcal{O}) = \beta\mathfrak{a} + \beta f\mathcal{O} \subset \mathfrak{a} + f\mathcal{O}_K.$$

However, $f\mathcal{O}_K \subset \mathcal{O}$, which proves that $\beta\mathcal{O} \subset \mathcal{O}$. Thus $\beta \in \mathcal{O}$, which proves that $\mathfrak{a}$ is proper. Q.E.D.

It follows that $\mathcal{O}$-ideals prime to f lie naturally in $I(\mathcal{O})$ and are closed under multiplication (since $N(\mathfrak{a}\mathfrak{b}) = N(\mathfrak{a})N(\mathfrak{b})$ will also be prime to f). The subgroup of fractional ideals they generate is denoted $I(\mathcal{O}, f) \subset I(\mathcal{O})$, and inside of $I(\mathcal{O}, f)$ we have the subgroup $P(\mathcal{O}, f)$ generated by the principal ideals $\alpha\mathcal{O}$ where $\alpha \in \mathcal{O}$ has norm $N(\alpha)$ prime to f. We can then describe $C(\mathcal{O})$ in terms of $I(\mathcal{O}, f)$ and $P(\mathcal{O}, f)$ as follows:

PROPOSITION 7.19. *The inclusion $I(\mathcal{O}, f) \subset I(\mathcal{O})$ induces an isomorphism*

$$I(\mathcal{O}, f)/P(\mathcal{O}, f) \simeq I(\mathcal{O})/P(\mathcal{O}) = C(\mathcal{O}).$$

PROOF. The map $I(\mathcal{O}, f) \to C(\mathcal{O})$ is surjective by Corollary 7.17 (any ideal class in $C(\mathcal{O})$ contains an $\mathcal{O}$-ideal prime to f), and the kernel is $I(\mathcal{O}, f) \cap P(\mathcal{O})$. This obviously contains $P(\mathcal{O}, f)$, but the inclusion $I(\mathcal{O}, f) \cap P(\mathcal{O}) \subset P(\mathcal{O}, f)$ needs proof. An element of $I(\mathcal{O}, f) \cap P(\mathcal{O})$ is a fractional ideal $\alpha\mathcal{O} = \mathfrak{a}\mathfrak{b}^{-1}$, where $\alpha \in K$ and $\mathfrak{a}$ and $\mathfrak{b}$ are $\mathcal{O}$-ideals prime to f. Let $m = N(\mathfrak{b})$. Then $m\mathcal{O} = N(\mathfrak{b})\mathcal{O} = \mathfrak{b}\overline{\mathfrak{b}}$ lies in $P(\mathcal{O}, f)$, and we also have $m\mathfrak{b}^{-1} = \overline{\mathfrak{b}}$. Hence

$$m\alpha\mathcal{O} = \mathfrak{a} \cdot m\mathfrak{b}^{-1} = \mathfrak{a}\overline{\mathfrak{b}} \subset \mathcal{O},$$

which proves that $m\alpha\mathcal{O} \in P(\mathcal{O}, f)$. Then $\alpha\mathcal{O} = m\alpha\mathcal{O} \cdot (m\mathcal{O})^{-1}$ is also in $P(\mathcal{O}, f)$, and the proposition is proved. Q.E.D.

For any order $\mathcal{O}$, ideals prime to the conductor relate nicely to ideals for the maximal order $\mathcal{O}_K$. To explain this, we begin with a definition: given a positive integer m, an $\mathcal{O}_K$-ideal $\mathfrak{a}$ is *prime to* m provided that $\mathfrak{a} + m\mathcal{O}_K = \mathcal{O}_K$. As in

Lemma 7.18, this is equivalent to $\gcd(N(\mathfrak{a}), m) = 1$. Thus, inside of the group of fractional $\mathcal{O}_K$-ideals I_K, we have the subgroup $I_K(m) \subset I_K$ generated by $\mathcal{O}_K$-ideals prime to m.

PROPOSITION 7.20. *Let $\mathcal{O}$ be the order of conductor f in an imaginary quadratic field K. Then:*

(i) *If $\mathfrak{a}$ is an $\mathcal{O}_K$-ideal prime to f, then $\mathfrak{a} \cap \mathcal{O}$ is an $\mathcal{O}$-ideal prime to f of the same norm.*
(ii) *If $\mathfrak{a}$ is an $\mathcal{O}$-ideal prime to f, then $\mathfrak{a}\mathcal{O}_K$ is an $\mathcal{O}_K$-ideal prime to f of the same norm.*
(iii) *The map $\mathfrak{a} \mapsto \mathfrak{a} \cap \mathcal{O}$ induces an isomorphism $I_K(f) \xrightarrow{\sim} I(\mathcal{O}, f)$, and the inverse of this map is given by $\mathfrak{a} \mapsto \mathfrak{a}\mathcal{O}_K$.*

PROOF. To prove (i), let $\mathfrak{a}$ be an $\mathcal{O}_K$-ideal prime to f. Since $\mathcal{O}/\mathfrak{a} \cap \mathcal{O}$ injects into $\mathcal{O}_K/\mathfrak{a}$ and $N(\mathfrak{a})$ is prime to f, so is $N(\mathfrak{a} \cap \mathcal{O})$, which proves that $\mathfrak{a} \cap \mathcal{O}$ is prime to f. As for norms, consider the natural injection

$$\mathcal{O}/\mathfrak{a} \cap \mathcal{O} \longrightarrow \mathcal{O}_K/\mathfrak{a}.$$

Since $\mathfrak{a}$ is prime to f, multiplication by f induces an isomorphism of $\mathcal{O}_K/\mathfrak{a}$. But $f\mathcal{O}_K \subset \mathcal{O}$, so that the above injection is also a surjection. This shows that the norms are equal, and (i) is proved.

To prove (ii), let $\mathfrak{a}$ be an $\mathcal{O}$-ideal prime to f. Since

$$\mathfrak{a}\mathcal{O}_K + f\mathcal{O}_K = (\mathfrak{a} + f\mathcal{O})\mathcal{O}_K = \mathcal{O}\mathcal{O}_K = \mathcal{O}_K$$

we see that $\mathfrak{a}\mathcal{O}_K$ is also prime to f. The statement about norms will be proved below.

Turning to (iii), we claim that

$$(7.21) \qquad \begin{aligned} \mathfrak{a}\mathcal{O}_K \cap \mathcal{O} &= \mathfrak{a} \qquad \text{when } \mathfrak{a} \text{ is an } \mathcal{O}\text{-ideal prime to } f \\ (\mathfrak{a} \cap \mathcal{O})\mathcal{O}_K &= \mathfrak{a} \qquad \text{when } \mathfrak{a} \text{ is an } \mathcal{O}_K\text{-ideal prime to } f. \end{aligned}$$

We start with the top line. If $\mathfrak{a}$ is an $\mathcal{O}$-ideal prime to f, then

$$\begin{aligned} \mathfrak{a}\mathcal{O}_K \cap \mathcal{O} &= (\mathfrak{a}\mathcal{O}_K \cap \mathcal{O})\mathcal{O} \\ &= (\mathfrak{a}\mathcal{O}_K \cap \mathcal{O})(\mathfrak{a} + f\mathcal{O}) \\ &\subset \mathfrak{a} + f(\mathfrak{a}\mathcal{O}_K \cap \mathcal{O}) \subset \mathfrak{a} + \mathfrak{a} \cdot f\mathcal{O}_K. \end{aligned}$$

Since $f\mathcal{O}_K \subset \mathcal{O}$, this proves that $\mathfrak{a}\mathcal{O}_K \cap \mathcal{O} \subset \mathfrak{a}$. The other inclusion is obvious, so that equality follows. Turning to the second line of (7.21), let $\mathfrak{a}$ be an $\mathcal{O}_K$-ideal prime to f. Then

$$\mathfrak{a} = \mathfrak{a}\mathcal{O} = \mathfrak{a}(\mathfrak{a} \cap \mathcal{O} + f\mathcal{O}) \subset (\mathfrak{a} \cap \mathcal{O})\mathcal{O}_K + f\mathfrak{a}.$$

However, $f\mathfrak{a} \subset f\mathcal{O}_K \subset \mathcal{O}$, so that $f\mathfrak{a} \subset \mathfrak{a} \cap \mathcal{O} \subset (\mathfrak{a} \cap \mathcal{O})\mathcal{O}_K$, and $\mathfrak{a} \subset (\mathfrak{a} \cap \mathcal{O})\mathcal{O}_K$ follows. The other inclusion is obvious, which finishes the proof of (7.21). Notice that (7.21) and (i) imply the norm statement of (ii).

From (7.21) we get a bijection on the monoids of $\mathcal{O}_K$- and $\mathcal{O}$-ideals prime to f. If we can show that $\mathfrak{a} \mapsto \mathfrak{a} \cap \mathcal{O}$ preserves multiplication, then we get an isomorphism $I_K(f) \simeq I(\mathcal{O}, f)$ (see Exercise 7.25). But multiplicativity is easy, for the inverse map $\mathfrak{a} \mapsto \mathfrak{a}\mathcal{O}_K$ is obviously multiplicative:

$$(\mathfrak{a}\mathfrak{b})\mathcal{O}_K = \mathfrak{a}\mathcal{O}_K \cdot \mathfrak{b}\mathcal{O}_K.$$

This proves the proposition. Q.E.D.

Using this proposition, it follows that every $\mathcal{O}$-ideal prime to f has a unique decomposition as a product of prime $\mathcal{O}$-ideals which are prime to f (see Exercise 7.26).

We can now describe $C(\mathcal{O})$ in terms of the maximal order:

PROPOSITION 7.22. *Let $\mathcal{O}$ be an order of conductor f in an imaginary quadratic field K. Then there are natural isomorphisms*

$$C(\mathcal{O}) \simeq I(\mathcal{O}, f)/P(\mathcal{O}, f) \simeq I_K(f)/P_{K,\mathbb{Z}}(f),$$

where $P_{K,\mathbb{Z}}(f)$ is the subgroup of $I_K(f)$ generated by principal ideals of the form $\alpha\mathcal{O}_K$, where $\alpha \in \mathcal{O}_K$ satisfies $\alpha \equiv a \bmod f\mathcal{O}_K$ for an integer a relatively prime to f.

REMARK. To keep track of the various ideal groups, remember that the subscript K refers to the maximal order $\mathcal{O}_K$ (as in I_K, $I_K(f)$, etc.), while no subscript refers to the order $\mathcal{O}$ (as in $I(\mathcal{O})$, $I(\mathcal{O}, f)$, etc.).

PROOF. The first isomorphism comes from Proposition 7.19. To prove the second, note that $\mathfrak{a} \mapsto \mathfrak{a}\mathcal{O}_K$ induces an isomorphism $I(\mathcal{O}, f) \simeq I_K(f)$ by Proposition 7.20. Under this isomorphism $P(\mathcal{O}, f) \subset I(\mathcal{O}, f)$ maps to a subgroup $\tilde{P} \subset I_K(f)$. It remains to prove $\tilde{P} = P_{K,\mathbb{Z}}(f)$.

We first show that for $\alpha \in \mathcal{O}_K$,

$$\text{(7.23)} \qquad \alpha \equiv a \bmod f\mathcal{O}_K,\ a \in \mathbb{Z},\ \gcd(a, f) = 1 \Longleftrightarrow \alpha \in \mathcal{O},\ \gcd(N(\alpha), f) = 1.$$

Going one way, assume that $\alpha \equiv a \bmod f\mathcal{O}_K$, where $a \in \mathbb{Z}$ is relatively prime to f. Then $N(\alpha) \equiv a^2 \bmod f$ follows easily (see Exercise 7.27), so that $\gcd(N(\alpha), f) = \gcd(a^2, f) = 1$. Since $f\mathcal{O}_K \subset \mathcal{O}$, we also see that $\alpha \in \mathcal{O}$. Conversely, let $\alpha \in \mathcal{O} = [1, fw_K]$ have norm prime to f. Writing $\alpha = a + bfw_K$, we see that $\alpha \equiv a \bmod f\mathcal{O}_K$. Since $\gcd(N(\alpha), f) = 1$ and $N(\alpha) \equiv a^2 \bmod f$, we must have $\gcd(a, f) = 1$. This completes the proof of (7.23).

We know that $P(\mathcal{O}, f)$ is generated by the ideals $\alpha\mathcal{O}$, where $\alpha \in \mathcal{O}$ and $N(\alpha)$ is relatively prime to f. Thus $\tilde{P}$ is generated by the corresponding ideals $\alpha\mathcal{O}_K$, and by (7.23), this implies that $\tilde{P} = P_{K,\mathbb{Z}}(f)$. Q.E.D.

In §9 we will use this proposition to link $C(\mathcal{O})$ to the class field theory of K. For other discussions of the relation between ideals of $\mathcal{O}$ and $\mathcal{O}_K$, see Deuring [**32**, §8] and Lang [**90**, §8.1].

D. The Class Number

One of the nicest applications of Proposition 7.22 is a formula for the class number $h(\mathcal{O})$ in terms of its conductor f and the class number $h(\mathcal{O}_K)$ of the maximal order. Before we can state the formula, we need to recall some terminology from §5. Given an odd prime p, we have the Legendre symbol (d_K/p), and for $p = 2$ we have the Kronecker symbol:

$$\left(\frac{d_K}{2}\right) = \begin{cases} 0 & \text{if } 2 \mid d_K \\ 1 & \text{if } d_K \equiv 1 \bmod 8 \\ -1 & \text{if } d_K \equiv 5 \bmod 8. \end{cases}$$

(Recall that $d_K \equiv 1 \bmod 4$ when d_K is odd.) We can now state our formula for $h(\mathcal{O})$:

THEOREM 7.24. *Let $\mathcal{O}$ be the order of conductor f in an imaginary quadratic field K. Then*

$$h(\mathcal{O}) = \frac{h(\mathcal{O}_K)f}{[\mathcal{O}_K^* : \mathcal{O}^*]} \prod_{p|f} \left(1 - \left(\frac{d_K}{p}\right)\frac{1}{p}\right).$$

Furthermore, $h(\mathcal{O})$ is always an integer multiple of $h(\mathcal{O}_K)$.

PROOF. By Theorem 7.7 and Proposition 7.22, we have

$$h(\mathcal{O}) = |C(\mathcal{O})| = |I_K(f)/P_{K,\mathbb{Z}}(f)|$$
$$h(\mathcal{O}_K) = |C(\mathcal{O}_K)| = |I_K/P_K|.$$

Since $I_K(f) \subset I_K$ and $P_{K,\mathbb{Z}}(f) \subset I_K(f) \cap P_K$, we get an exact sequence

$$(7.25)\qquad \begin{array}{ccccccc} 0 \longrightarrow & I_K(f) \cap P_K/P_{K,\mathbb{Z}}(f) & \longrightarrow & I_K(f)/P_{K,\mathbb{Z}}(f) & \longrightarrow & I_K/P_K \\ & & & \downarrow\wr & & \downarrow\wr \\ & & & C(\mathcal{O}) & \longrightarrow & C(\mathcal{O}_K). \end{array}$$

We know from Corollary 7.17 that every class in $C(\mathcal{O}_K)$ contains an $\mathcal{O}_K$-ideal whose norm is relatively prime to f. This implies that $C(\mathcal{O}) \to C(\mathcal{O}_K)$ is surjective, which proves that $h(\mathcal{O}_K)$ divides $h(\mathcal{O})$. Furthermore, (7.25) then implies that

$$(7.26)\qquad \frac{h(\mathcal{O})}{h(\mathcal{O}_K)} = |I_K(f) \cap P_K/P_{K,\mathbb{Z}}(f)|.$$

It remains to compute the order of $I_K(f) \cap P_K/P_{K,\mathbb{Z}}(f)$. The key idea is to relate this quotient to $(\mathcal{O}_K/f\mathcal{O}_K)^*$.

Given $[\alpha] \in (\mathcal{O}_K/f\mathcal{O}_K)^*$, the ideal $\alpha\mathcal{O}_K$ is prime to f and thus lies in $I_K(f) \cap P_K$. Furthermore, if $\alpha \equiv \beta \bmod f\mathcal{O}_K$, we can choose $u \in \mathcal{O}_K$ with $u\alpha \equiv u\beta \equiv 1 \bmod f\mathcal{O}_K$. Then the ideals $u\alpha\mathcal{O}_K$ and $u\beta\mathcal{O}_K$ lie in $P_{K,\mathbb{Z}}(f)$, and since

$$\alpha\mathcal{O}_K \cdot u\beta\mathcal{O}_K = \beta\mathcal{O}_K \cdot u\alpha\mathcal{O}_K,$$

$\alpha\mathcal{O}_K$ and $\beta\mathcal{O}_K$ lie in the same class in $I_K(f) \cap P_K/P_{K,\mathbb{Z}}(f)$. Consequently, the map

$$\phi : (\mathcal{O}_K/f\mathcal{O}_K)^* \longrightarrow I_K(f) \cap P_K/P_{K,\mathbb{Z}}(f)$$

sending $[\alpha]$ to $[\alpha\mathcal{O}_K]$ is a well-defined homomorphism.

We will first show that ϕ is surjective. An element of $I_K(f) \cap P_K$ can be written as $\alpha\mathcal{O}_K = \mathfrak{a}\mathfrak{b}^{-1}$, where $\alpha \in K$ and $\mathfrak{a}$ and $\mathfrak{b}$ are $\mathcal{O}_K$-ideals prime to f. Letting $m = N(\mathfrak{b})$, we've seen that $\overline{\mathfrak{b}} = m\mathfrak{b}^{-1}$, so that $m\alpha\mathcal{O}_K = \mathfrak{a}\overline{\mathfrak{b}}$, which implies that $m\alpha \in \mathcal{O}_K$. Note also that $m\alpha\mathcal{O}_K$ is prime to f. Since $m\mathcal{O}_K \in P_{K,\mathbb{Z}}(f)$, it follows $[\alpha\mathcal{O}_K] = [m\alpha\mathcal{O}_K] = \phi([m\alpha])$, proving that ϕ is surjective.

To determine the kernel of ϕ, we will assume that $\mathcal{O}_K^* = \{\pm 1\}$ (by Exercise 5.9, this means that $K \neq \mathbb{Q}(\sqrt{-3})$ or $\mathbb{Q}(i)$). In this case we will show that there is an exact sequence

$$(7.27)\qquad 1 \longrightarrow (\mathbb{Z}/f\mathbb{Z})^* \stackrel{\psi}{\longrightarrow} (\mathcal{O}_K/f\mathcal{O}_K)^* \stackrel{\phi}{\longrightarrow} I_K(f) \cap P_K/P_{K,\mathbb{Z}}(f) \longrightarrow 1$$

where ψ is the obvious injection. The definition of $P_{K,\mathbb{Z}}(f)$ makes it clear that $\mathrm{im}(\psi) \subset \ker(\phi)$. Going the other way, let $[\alpha] \in \ker(\phi)$. Then $\alpha\mathcal{O}_K \in P_{K,\mathbb{Z}}(f)$, i.e., $\alpha\mathcal{O}_K = \beta\mathcal{O}_K \cdot \gamma^{-1}\mathcal{O}_K$, where β and γ satisfy $\beta \equiv b \bmod f\mathcal{O}_K$ and $\gamma \equiv c \bmod f\mathcal{O}_K$ for some $[b]$ and $[c]$ in $(\mathbb{Z}/f\mathbb{Z})^*$. Since $\mathcal{O}_K^* = \{\pm 1\}$, it follows that $\alpha = \pm\beta\gamma^{-1}$, and one then easily sees that $\pm[b][c]^{-1} \in (\mathbb{Z}/f\mathbb{Z})^*$ maps to $[\alpha] \in (\mathcal{O}_K/f\mathcal{O}_K)^*$. This proves exactness.

It is well-known that

$$|(\mathbb{Z}/f\mathbb{Z})^*| = f\prod_{p|f}\left(1-\frac{1}{p}\right),$$

and in Exercises 7.28 and 7.29 we will show that

$$|(\mathcal{O}_K/f\mathcal{O}_K)^*| = f^2\prod_{p|f}\left(1-\frac{1}{p}\right)\left(1-\left(\frac{d_K}{p}\right)\frac{1}{p}\right).$$

Using these formulas and (7.26), we obtain

$$\frac{h(f^2d_K)}{h(d_K)} = |I_K(f)\cap P_K/P_{K,\mathbb{Z}}(f)| = f\prod_{p|f}\left(1-\left(\frac{d_K}{p}\right)\frac{1}{p}\right),$$

which proves the desired formula since $|\mathcal{O}_K^*| = |\mathcal{O}^*| = 2$. In Exercise 7.30 we will indicate how to modify this argument when $\mathcal{O}_K^* \neq \{\pm 1\}$. Q.E.D.

This theorem may also be proved by analytic methods—see, for example, Zagier [**140**, §8, Exercise 8].

By Theorem 7.24, we can relate the class numbers $h(m^2D)$ and $h(D)$ as follows:

COROLLARY 7.28. *Let $D \equiv 0, 1 \bmod 4$ be negative, and let m be a positive integer. Then*

$$h(m^2D) = \frac{h(D)m}{[\mathcal{O}^* : \mathcal{O}'^*]}\prod_{p|m}\left(1-\left(\frac{D}{p}\right)\frac{1}{p}\right),$$

where $\mathcal{O}$ and $\mathcal{O}'$ are the orders of discriminant D and m^2D, respectively (and $\mathcal{O}'$ has index m in $\mathcal{O}$).

PROOF. Suppose that the order $\mathcal{O}$ has discriminant D and conductor f. Then the order $\mathcal{O}' \subset \mathcal{O}$ of index m has discriminant m^2D and conductor mf, and the corollary follows from Theorem 7.24 (see Exercise 7.31). This corollary is due to Gauss, and his proof may be found in *Disquisitiones* [**51**, §§254–256]. Q.E.D.

The only method we learned in §2 for computing class numbers $h(D)$ for $D < 0$ was to count reduced forms. This becomes awkward as $|D|$ gets large, but other methods are available. By Theorem 7.24, we are reduced to computing $h(d_K)$, and here one has the classical formula

$$h(d_K) = -\frac{|\mathcal{O}_K^*|}{2|d_K|}\sum_{n=1}^{|d_K|-1}\left(\frac{d_K}{n}\right)n, \tag{7.29}$$

where (d_K/n) is defined for $n = p_1\cdots p_r$, p_i prime, by $(d_K/n) = \prod_{i=1}^r(d_K/p_i)$. This formula is usually proved by analytic methods (see Borevich and Shafarevich [**11**, Chapter 5, Section 4], or Zagier [**140**, §9]), but there is also a purely algebraic proof (see Orde [**104**]).

While (7.29) enables us to compute $h(d_K)$ for a given imaginary quadratic field, it doesn't reveal the way $h(d_K)$ grows as $|d_K|$ gets large. Gauss noticed this growth empirically in *Disquisitiones* [**51**, §302], but there were no complete proofs until the 1930s. The best result is due to Siegel [**117**], who proved in 1935 that

$$\lim_{d_K\to-\infty}\frac{\log h(d_K)}{\log|d_K|} = \frac{1}{2}.$$

This implies that given any $\epsilon > 0$, there is a constant $C(\epsilon)$ such that

$$h(d_K) > C(\epsilon)|d_K|^{(1/2)-\epsilon}$$

for all field discriminants $d_K < 0$. Unfortunately, the constant $C(\epsilon)$ in Siegel's proof is not effectively computable given what we currently know about L-series (these difficulties are related to the Riemann Hypothesis). However, work by Goldfeld, Gross, Zagier and Oesterlé in the 1980s led to the weaker formula

$$h(d_K) > \frac{\log|d_K|}{55} \prod_{p|d_K,\, p<d_K} \left(1 - \frac{[2\sqrt{p}]}{p+1}\right),$$

where [ ] is the greatest integer function and log is to the base e. For a fuller discussion of this result and its implications, see Oesterlé [**102**, A17] and Zagier [**141**].

These results on the growth of $h(d_K)$ imply that there are only finitely many orders with given class number h (see Exercise 7.32). Nevertheless, even when h is small, determining exactly which orders have class number h remains a difficult problem. For the case of class number 1, the answer is given by the following theorem due independently to Baker [**4**], Heegner [**68**] and Stark [**123**]:

THEOREM 7.30.

(i) *If K is an imaginary quadratic field of discriminant d_K, then*

$$h(d_K) = 1 \iff d_K = -3, -4, -7, -8, -11, -19, -43, -67, -163.$$

(ii) *If $D \equiv 0, 1 \bmod 4$ is negative, then*

$$\begin{aligned} h(D) = 1 \iff D = &-3, -4, -7, -8, -11, -12, -16, \\ &-19, -27, -28, -43, -67, -163. \end{aligned}$$

PROOF. First note that (i) $\Rightarrow$ (ii). To see why, assume that $h(D) = 1$. If we write $D = f^2 d_K$, then Theorem 7.24 tells us that $h(d_K)|h(D)$, and thus $h(d_K) = 1$. By (i), this determines the possibilities for d_K, but we still need to see which conductors $f > 1$ can occur. First, suppose that $\mathcal{O}_K^* = \{\pm 1\}$. If $f > 2$, then

$$f \prod_{p|f} \left(1 - \left(\frac{d_K}{p}\right)\frac{1}{p}\right) > 1,$$

so that by Theorem 7.24, this case can be excluded. One then calculates directly (using (i) and Theorem 7.24) that $f = 2$ happens only when $d_K = -7$, i.e., $D = -28$. The argument when $\mathcal{O}_K^* \neq \{\pm 1\}$ is similar and is left to the reader (see Exercise 7.33).

The proof of (i) is a different matter. When the discriminant is even, the theorem was proved in §2 by an elementary argument due to Landau (see Theorem 2.18. But when the discriminant is odd, the proof is *much* more difficult. In §12 we will give a complete proof of (i) using modular functions and complex multiplication. Q.E.D.

E. Exercises

7.1. Let K be a finite extension of $\mathbb{Q}$ of degree n, and let $M \subset K$ be a finitely generated $\mathbb{Z}$-module.

(a) Prove that M is a free $\mathbb{Z}$-module.

(b) Prove that M has rank n if and only if M contains a $\mathbb{Q}$-basis of K.

7.2. Let $\mathcal{O}$ be an order in a quadratic field K. Prove that $\mathcal{O} \subset \mathcal{O}_K$.

7.3. This exercise is concerned with the conductor and discriminant of an order $\mathcal{O}$ in a quadratic field K. Let $\alpha \mapsto \alpha'$ be the nontrivial automorphism of K.

(a) If $\mathcal{O} = [\alpha, \beta]$, then the discriminant is defined to be

$$D = \left(\det \begin{pmatrix} \alpha & \beta \\ \alpha' & \beta' \end{pmatrix}\right)^2$$

Prove that the D is independent of the basis used and hence depends only on $\mathcal{O}$.

(b) Use the basis $\mathcal{O} = [1, fw_K]$ from Lemma 7.2 to prove that $D = f^2 d_K$.

(c) Use part (b) and Lemma 7.2 to prove that an order in a quadratic field is uniquely determined by its discriminant.

(d) If $D \equiv 0, 1 \bmod 4$ is nonsquare, then show that there is an order in a quadratic field whose discriminant is D.

7.4. Let $\mathcal{O}$ be an order in a quadratic field K.

(a) If $\mathfrak{a}$ is a nonzero ideal of $\mathcal{O}$, prove that $\mathfrak{a}$ contains a nonzero integer m. Hint: take $\alpha \in \mathfrak{a}$, and use Lemma 7.2 to show that $\alpha' \in \mathcal{O}$, where $\alpha \mapsto \alpha'$ is the nontrivial automorphism of K.

(b) If $\mathfrak{a}$ is a nonzero ideal of $\mathcal{O}$, show that $\mathcal{O}/\mathfrak{a}$ is finite. Hint: take the integer m from part (a) and show that $\mathcal{O}/m\mathcal{O}$ is finite.

(c) Use part (b) to show that every nonzero prime ideal of $\mathcal{O}$ is maximal.

(d) Use part (b) to show that $\mathcal{O}$ is Noetherian.

7.5. Let $K = \mathbb{Q}(\sqrt{-3})$, and let $\mathfrak{a}$ be the ideal of $\mathcal{O} = \mathbb{Z}[\sqrt{-3}]$ generated by 2 and $1 + \sqrt{-3}$. Show that

$$\{\beta \in K : \beta\mathfrak{a} \subset \mathfrak{a}\} = \mathcal{O}_K \neq \mathcal{O}.$$

7.6. Let K be a quadratic field.

(a) Show that for any order of K, nonzero principal ideals are proper.

(b) Show that for the maximal order $\mathcal{O}_K$, nonzero ideals are proper.

7.7. Let $\mathcal{O}$ be an order of K, and let $\mathfrak{b} \subset K$ be an $\mathcal{O}$-module (note that $\mathfrak{b}$ need not be contained in $\mathcal{O}$). Show that $\mathfrak{b}$ is finitely generated as an $\mathcal{O}$-module if and only if $\mathfrak{b}$ is of the form $\alpha\mathfrak{a}$, where $\alpha \in K$ and $\mathfrak{a}$ is an $\mathcal{O}$-ideal.

7.8. Show that a nonzero fractional $\mathcal{O}$-ideal $\mathfrak{a}$ is a free $\mathbb{Z}$-module of rank 2 when K is a quadratic field. Hint: use the previous exercise and part (b) of Exercise 7.4.

7.9. Let $\mathcal{O} = \mathbb{Z}[\sqrt{-3}]$, which is an order of conductor 2 in the imaginary quadratic field $K = \mathbb{Q}(\sqrt{-3})$.

(a) Show that $C(\mathcal{O}) \simeq \{1\}$, so that the proper ideals of $\mathcal{O}$ are exactly the principal ideals. Hint: use Theorem 7.7 and what we know from §2.

(b) Show that if unique factorization holds for proper ideals of $\mathcal{O}$, then $\mathcal{O}$ is a UFD.

(c) Show that 2, $1 + \sqrt{-3}$ and $1 - \sqrt{-3}$ are irreducible (in the sense of §4) in $\mathcal{O}$. Since $4 = 2 \cdot 2 = (1 + \sqrt{-3})(1 - \sqrt{-3})$, this shows that $\mathcal{O}$ is not a UFD.

This example shows that unique factorization can fail for proper ideals.

7.10. If $\mathfrak{a}$ and $\mathfrak{b}$ are invertible fractional ideals for an order $\mathcal{O}$, then prove that $\mathfrak{ab}$ and $\mathfrak{a}^{-1}$ (where $\mathfrak{a}^{-1}$ is the fractional $\mathcal{O}$-ideal such that $\mathfrak{a}\mathfrak{a}^{-1} = \mathcal{O}$) are also invertible fractional $\mathcal{O}$-ideals.

7.11. Prove (7.10).

7.12. Let $f(x, y) = ax^2 + bxy + cy^2$ be a quadratic form with integer coefficients, and let τ be a root of $ax^2 + bx + c = 0$.

(a) Prove that $f(x, y)$ is positive definite if and only if $a > 0$ and $\tau \notin \mathbb{R}$.
(b) When $f(x, y)$ is positive definite and $\gcd(a, b, c) = 1$, prove that the discriminant of $f(x, y)$ is D, where D is the discriminant of the order $\mathcal{O} = [1, a\tau]$.
(c) Prove that two primitive positive definite forms which have the same root τ must be equal.

7.13. Let $ax^2 + bxy + cy^2$ and $a'x^2 + b'xy + c'y^2$ be two primitive positive definite forms of the same discriminant. Assume that $\gcd(a, a', (b + b')/2) = 1$, and let B be the unique integer modulo $2aa'$ which satisfies the three conditions of (7.12). Prove that $\gcd(a, a', B) = 1$.

7.14. Let $\mathcal{O} = [1, u]$ be an order in a quadratic field, and pick $\alpha = a + bu \in \mathcal{O}$, $\alpha \neq 0$. Since $\mathcal{O}$ is a ring, αu can be written $\alpha u = c + du$.

(a) Show that $N(\alpha) = ad - bc \neq 0$.
(b) Since $\alpha\mathcal{O} = [\alpha, \alpha u] = [a + bu, c + du] \subset \mathcal{O} = [1, u]$ and $ad - bc \neq 0$, it is a standard fact (proved in Exercise 7.15) that $|\mathcal{O}/\alpha\mathcal{O}| = |ad - bc|$. Thus part (a) proves the general relation that $N(\alpha\mathcal{O}) = |N(\alpha)|$.

7.15. Let $M = \mathbb{Z}^2$, and suppose that $A = \left(\begin{smallmatrix} a & b \\ c & d \end{smallmatrix}\right)$ is an integer matrix with $\det(A) = ad - bc \neq 0$. Writing $M = [e_1, e_2]$, note that $MA = [ae_1 + ce_2, be_1 + de_2]$. Our goal is to prove that $|M/MA| = |\det(A)|$.

(a) Show that the result is true when $c = 0$. Hint: use the division algorithm to write an element of M as $ue_1 + ve_2 + w(be_1 + de_2)$ where $u, v, w \in \mathbb{Z}$ and $0 \leq v < |d|$.
(b) Let $B \in \mathrm{GL}(2, \mathbb{Z})$. Show that the result is true for A if and only if it is true for BA. Hint: use the automorphism of M induced by B.
(c) Explain how to find $B \in \mathrm{GL}(2, \mathbb{Z})$ such that $BA = \left(\begin{smallmatrix} a' & b' \\ 0 & d' \end{smallmatrix}\right)$. Hint: if $c \neq 0$, prove that there exists $B \in \mathrm{GL}(2, \mathbb{Z})$ such that $BA = \left(\begin{smallmatrix} a' & b' \\ c' & d' \end{smallmatrix}\right)$ with $|c'| < |c|$. This is easy to do when $|a| < |c|$ (swap rows) and not difficult when $|a| \geq |c|$ (dividing a by c tells you which row operation to use).
(d) Conclude that $|M/MA| = |\det(A)|$.

7.16. Let $\mathcal{O}$ be the order of discriminant D in an imaginary quadratic field K, and let $\mathfrak{a}$ be a proper $\mathcal{O}$-ideal. In this exercise we will give two proofs that the class of $\overline{\mathfrak{a}}$ is the inverse of the class of $\mathfrak{a}$ in $C(\mathcal{O})$.

(a) Prove this assertion using part (iii) of Lemma 7.14.
(b) In §3, we proved that the class of the opposite form $ax^2 - bxy + cy^2$ is the inverse of the class of the form $ax^2 + bxy + cy^2$. Using the isomorphism $C(D) \simeq C(\mathcal{O})$ from Theorem 7.7, show that the class of $\overline{\mathfrak{a}}$ is the inverse of the class of $\mathfrak{a}$ in $C(\mathcal{O})$.

7.17. Let $\mathcal{O}$ be the order of discriminant D in the imaginary quadratic field K.

(a) Show that the map sending a proper $\mathcal{O}$-ideal $\mathfrak{a} = [\alpha, \beta]$, $\text{Im}(\beta/\alpha) > 0$, to the quadratic form

$$f(x,y) = \frac{N(\alpha x - \beta y)}{N(\mathfrak{a})}$$

induces a well-defined map $C(\mathcal{O}) \to C(D)$ which is the inverse of the map $ax^2+bxy+cy^2 \mapsto [a, (-b+\sqrt{D})/2]$ of Theorem 7.7. Hint: use (7.16) and Exercise 7.12.

(b) Give examples to show that the map $ax^2+bxy+cy^2 \mapsto [a, (-b+\sqrt{D})/2]$ of Theorem 7.7 is neither injective nor surjective on the level of forms and ideals.

7.18. The field $K = \mathbb{Q}(\sqrt{3})$ has discriminant $d_K = 12$, and $\mathcal{O}_K = \mathbb{Z}[\sqrt{3}]$ by (5.13).

(a) Use the absolute value of the norm function to show that $\mathcal{O}_K$ is Euclidean, and conclude that $C(\mathcal{O}_K) \simeq \{1\}$.

(b) Show that the form class group $C(d_K) = C(12)$ is nontrivial. Hint: show that the forms $\pm(x^2 - 3y^2)$ are not properly equivalent. You will need to show that the equation $a^2 - 3c^2 = 1$ has no integer solutions.

This shows that $C(d_K) \not\simeq C(\mathcal{O}_K)$ for $K = \mathbb{Q}(\sqrt{3})$.

7.19. In Exercises 7.19–7.24 we will explore two versions of Theorem 7.7 that hold for real quadratic fields K. To begin, we will study the orientation of a basis α, β of a proper ideal $\mathfrak{a} = [\alpha, \beta]$ of an order $\mathcal{O}$ in K. Let $\alpha \mapsto \alpha'$ denote the nontrivial automorphism of K.

(a) Prove that $\alpha'\beta - \alpha\beta' \in \mathbb{R}^*$. We then define $\text{sgn}(\alpha, \beta)$ to be the sign of the nonzero real number $\alpha'\beta - \alpha\beta'$.

(b) Let $\left(\begin{smallmatrix} p & q \\ r & s \end{smallmatrix}\right) \in \text{GL}(2, \mathbb{Z})$, and set $\tilde{\alpha} = p\alpha + q\beta$, $\tilde{\beta} = r\alpha + s\beta$. Note that $\mathfrak{a} = [\alpha, \beta] = [\tilde{\alpha}, \tilde{\beta}]$. Prove that

$$\text{sgn}(\tilde{\alpha}, \tilde{\beta}) = \det \begin{pmatrix} p & q \\ r & s \end{pmatrix} \text{sgn}(\alpha, \beta).$$

We say that α, β are positively oriented if $\text{sgn}(\alpha, \beta) > 0$ and negatively oriented otherwise. By part (b), two bases of $\mathfrak{a}$ have the same orientation if and only if their transition matrix is in $\text{SL}(2, \mathbb{Z})$.

7.20. Theorem 7.7 was proved using a map from quadratic forms to ideals. In the real quadratic case, such a map is harder to describe (see Exercise 7.24), but it is relatively easy to go from ideals to forms. The goal of this exercise is to show how this is done. Let $\mathcal{O}$ be an order in a real quadratic field K, and let $\mathfrak{a} = [\alpha, \beta]$ be a proper $\mathcal{O}$-ideal. Then define the quadratic form $f(x, y)$ by the formula

$$f(x,y) = \frac{N(\alpha x - \beta y)}{N(\mathfrak{a})}.$$

At this point, all we know is that $f(x, y)$ has rational coefficients. Let $\tau = \beta/\alpha$, and let $ax^2 + bx + c$ be the minimal polynomial of τ. We can assume that $a, b, c \in \mathbb{Z}$, $a > 0$ and $\gcd(a, b, c) = 1$.

(a) Prove that $N(\mathfrak{a}) = |N(\alpha)|/a$. Hint: adapt the proof of (7.16) to the real quadratic case. Exercise 7.14 will be useful.

(b) Use part (a) to prove that $f(x, y) = \text{sgn}(N(\alpha))(ax^2 + bxy + cy^2)$. Thus $f(x, y)$ has relatively prime integer coefficients.

(c) Prove that the discriminant of $f(x,y)$ is D, where D is the discriminant of $\mathcal{O}$. Hint: see Exercise 7.12.

7.21. In this exercise we will construct a bijection $C^+(\mathcal{O}) \simeq C(D)$, where $C^+(\mathcal{O})$ is defined in the text.

(a) Let $\mathfrak{a}$ be a proper $\mathcal{O}$-ideal, and write $\mathfrak{a} = [\alpha, \beta]$ where $\operatorname{sgn}(\alpha, \beta) > 0$ (see Exercise 7.19). Then let $f(x,y)$ be the corresponding quadratic form defined in Exercise 7.20. If $\tilde{\alpha}, \tilde{\beta}$ is another positively oriented basis of $\mathfrak{a}$, then show that the corresponding form $g(x,y)$ from Exercise 7.20 is properly equivalent to $f(x,y)$. Furthermore, show that *all* forms properly equivalent to $f(x,y)$ arise in this way.

(b) If $\lambda \in \mathcal{O}$ and $N(\lambda) > 0$, then show that $\lambda\mathfrak{a}$ gives the same class of forms as $\mathfrak{a}$. Hint: show that $\operatorname{sgn}(\lambda\alpha, \lambda\beta) = \operatorname{sgn}(N(\lambda))\operatorname{sgn}(\alpha, \beta)$.

(c) From parts (a), (b) and Exercise 7.20 we get a well-defined map $C^+(\mathcal{O}) \to C(D)$. To show that the map is injective, suppose that $\mathfrak{a}$ and $\tilde{\mathfrak{a}}$ give the same class in $C(D)$. By part (a), we can choose positively oriented bases $\mathfrak{a} = [\alpha, \beta]$ and $\tilde{\mathfrak{a}} = [\tilde{\alpha}, \tilde{\beta}]$ which give the same form $f(x,y)$.

  (i) Using Exercise 7.19, show that $\operatorname{sgn}(N(\alpha)) = \operatorname{sgn}(N(\tilde{\alpha}))$. Hence $N(\alpha\tilde{\alpha}) > 0$. Then replacing $\mathfrak{a}$ and $\tilde{\mathfrak{a}}$ by $\alpha\tilde{\alpha}\mathfrak{a}$ and $\alpha^2\tilde{\mathfrak{a}}$ respectively allows us to assume that $\alpha = \tilde{\alpha}$, i.e., $\mathfrak{a} = [\alpha, \beta]$ and $\tilde{\mathfrak{a}} = [\alpha, \tilde{\beta}]$.

  (ii) Let $\tau = \beta/\alpha$ and $\tilde{\tau} = \tilde{\beta}/\alpha$. Show that $f(\tau, 1) = f(\tilde{\tau}, 1) = 0$, so that $\tilde{\tau} = \tau$ or τ'. Then show that $\tilde{\tau} = \tau'$ contradicts $\operatorname{sgn}(\alpha, \tilde{\beta}) > 0$, which proves that $\beta = \tilde{\beta}$.

(d) To prove surjectivity, let $f(x,y) = ax^2+bxy+cy^2$ be a form of discriminant D, and let τ be either of the roots of $ax^2 + bx + c = 0$. First show that $a\tau \in \mathcal{O}$. Then define an $\mathcal{O}$-ideal $\mathfrak{a}$ as follows: if $a > 0$, then

$$\mathfrak{a} = [a, a\tau] \qquad \text{where } f(\tau, 1) = 0,\ \operatorname{sgn}(1, \tau) > 0,$$

and if $a < 0$, then

$$\mathfrak{a} = \sqrt{D}\,[a, a\tau] \qquad \text{where } f(\tau, 1) = 0,\ \operatorname{sgn}(1, \tau) < 0.$$

Show that $\mathfrak{a}$ is a proper $\mathcal{O}$-ideal and that the form corresponding to $\mathfrak{a}$ from Exercise 7.20 is exactly $f(x,y)$.

This completes the proof that $C^+(\mathcal{O}) \to C(D)$ is a bijection.

7.22. This exercise will construct a bijection $C(\mathcal{O}) \simeq C_s(D)$, where $C_s(D)$ is defined in the text. Our treatment of $C_s(D)$ is based on Zagier [**140**, §8].

(a) Let $\mathfrak{a} = [\alpha, \beta]$ be a proper $\mathcal{O}$-ideal, where this time we make no assumptions about $\operatorname{sgn}(\alpha, \beta)$. Define $f(x,y)$ to be the quadratic form

$$f(x,y) = \operatorname{sgn}(\alpha, \beta)\frac{N(\alpha x - \beta y)}{N(\mathfrak{a})},$$

which by Exercise 7.20 has relatively prime integer coefficients and discriminant D. Show that as we vary over all bases of $\mathfrak{a}$, the corresponding forms vary over all forms signed equivalent to $f(x,y)$.

(b) Show that the map $\mathfrak{a} \mapsto f(x,y)$ from part (a) induces a well-defined bijection $C(\mathcal{O}) \simeq C_s(D)$. Hint: adapt parts (b)–(d) of Exercise 7.21.

7.23. This exercise will study the relations between $C(\mathcal{O})$, $C^+(\mathcal{O})$, $C(D)$ and $C_s(D)$.

(a) Let K be an imaginary quadratic field.

(i) Show that $P^+(\mathcal{O}) = P(\mathcal{O})$, so that $C^+(\mathcal{O})$ always equals $C(\mathcal{O})$.

(ii) The relation between $C(D)$ and $C_s(D)$ is more interesting. Namely, in $C(D)$, we had to explicitly assume that we were only dealing with positive definite forms. However, in $C_s(D)$, one uses both positive definite and negative definite forms. Show that any negative definite form is signed equivalent to a positive definite one, and conclude that $C(D) \simeq C_s(D)$.

(b) Now assume that K is a real quadratic field.

(i) Show that there are natural surjections

$$\begin{aligned} C^+(\mathcal{O}) &\longrightarrow C(\mathcal{O}) \\ C(D) &\longrightarrow C_s(D) \end{aligned}$$

which fit together with the bijections of Exercises 7.21 and 7.22 to give a commutative diagram

$$\begin{array}{ccc} C^+(\mathcal{O}) & \xrightarrow{\sim} & C(D) \\ \downarrow & & \downarrow \\ C(\mathcal{O}) & \xrightarrow{\sim} & C_s(D). \end{array}$$

(ii) Show that the kernel of $C^+(\mathcal{O}) \to C(\mathcal{O})$ is $P(\mathcal{O})/P^+(\mathcal{O})$ and that $P(\mathcal{O}) = P^+(\mathcal{O}) \cup \sqrt{d_K}P^+(\mathcal{O})$. Then conclude that

$$\frac{|C^+(\mathcal{O})|}{|C(\mathcal{O})|} = \begin{cases} 1 & \text{if } \mathcal{O} \text{ has a unit of norm } -1 \\ 2 & \text{otherwise.} \end{cases}$$

(iii) From (i) and (ii), conclude that

$$\frac{|C(D)|}{|C_s(D)|} = \begin{cases} 1 & \text{if } \mathcal{O} \text{ has a unit of norm } -1 \\ 2 & \text{otherwise.} \end{cases}$$

7.24. Write down inverses to the bijections $C^+(\mathcal{O}) \xrightarrow{\sim} C(D)$ and $C(\mathcal{O}) \xrightarrow{\sim} C_s(D)$ of Exercises 7.21 and 7.22. Hint: see part (d) of Exercise 7.21. Note that the answer is more complicated than the map $ax^2 + bxy + cy^2 \to [a, (-b + \sqrt{D})/2]$ of Theorem 7.7.

7.25. Let $\phi : \{\mathcal{O}_K\text{-ideals prime to } f\} \to \{\mathcal{O}\text{-ideals prime to } f\}$ be a bijection which preserves multiplication. Show that we can extend ϕ to an isomorphism

$$\overline{\phi} : I_K(f) \xrightarrow{\sim} I(\mathcal{O}, f).$$

7.26. Let $\mathcal{O}$ be an order of conductor f.

(a) Let $\mathfrak{a}$ be an ideal of $\mathcal{O}$ which is relatively prime to f. Prove that $\mathfrak{a}$ is a prime $\mathcal{O}$-ideal if and only if $\mathfrak{a}\mathcal{O}_K$ is a prime $\mathcal{O}_K$-ideal. Hint: use Proposition 7.20 to show that $\mathcal{O}/\mathfrak{a} \simeq \mathcal{O}_K/\mathfrak{a}\mathcal{O}_K$.

(b) Use part (a) and the unique factorization of ideals in $\mathcal{O}_K$ to show that $\mathcal{O}$-ideals relatively prime to the conductor can be factored uniquely into prime $\mathcal{O}$-ideals (which are also relatively prime to f).

7.27. If $\alpha, \beta \in \mathcal{O}_K$ and $\alpha \equiv \beta \bmod m\mathcal{O}_K$ for some integer m, then prove that $N(\alpha) \equiv N(\beta) \bmod m$.

7.28. Let K be a quadratic field, and let $\mathfrak{p}$ be prime in $\mathcal{O}_K$. The goal of this exercise is to prove that

$$|(\mathcal{O}_K/\mathfrak{p}^n)^*| = N(\mathfrak{p})^{n-1}(N(\mathfrak{p})-1).$$

The formula is true if $n = 1$, and the general case follows easily by induction once we prove that there is an exact sequence

$$1 \longrightarrow \mathcal{O}_K/\mathfrak{p} \stackrel{\phi}{\longrightarrow} (\mathcal{O}_K/\mathfrak{p}^n)^* \longrightarrow (\mathcal{O}_K/\mathfrak{p}^{n-1})^* \longrightarrow 1$$

for $n \geq 2$. For the rest of the exercise fix an integer $n \geq 2$.

(a) Show that $(\mathcal{O}_K/\mathfrak{p}^n)^* \to (\mathcal{O}_K/\mathfrak{p}^{n-1})^*$ is onto. Hint: take an element $[\alpha] \in (\mathcal{O}_K/\mathfrak{p}^{n-1})^*$, which means that $\alpha\beta = 1+\gamma$, where $\beta \in \mathcal{O}_K$ and $\gamma \in \mathfrak{p}^{n-1}$. Then show that $\alpha(\beta+\gamma\delta) - 1 \in \mathfrak{p}^n$ for some $\delta \in \mathcal{O}_K$.

(b) By unique factorization, we know that $\mathfrak{p}^n$ is a proper subset of $\mathfrak{p}^{n-1}$. Pick $u \in \mathfrak{p}^{n-1}$ such that $u \notin \mathfrak{p}^n$.
  (i) Given $\alpha \in \mathcal{O}_K$, show that $[1+\alpha u] \in (\mathcal{O}_K/\mathfrak{p}^n)^*$.
  (ii) From (i), it is easy to define a map $\phi : \mathcal{O}_K/\mathfrak{p} \to (\mathcal{O}_K/\mathfrak{p}^n)^*$. With this definition of ϕ, show that the above sequence is exact.

7.29. Let K be an imaginary quadratic field.

(a) Let $\mathfrak{a} = \prod_{i=1}^r \mathfrak{p}_i^{n_i}$ be the factorization of $\mathfrak{a}$ into powers of distinct primes. Show that there is a natural isomorphism

$$\mathcal{O}_K/\mathfrak{a} \simeq \prod_{i=1}^r (\mathcal{O}_K/\mathfrak{p}_i^{n_i}).$$

This is the Chinese Remainder Theorem for $\mathcal{O}_K$. Hint: it is easy to construct a map and show it is injective. Then use part (ii) of Lemma 7.14.

(b) Use part (a) and the previous exercise to show that if $\mathfrak{a}$ is a nonzero ideal of $\mathcal{O}_K$, then

$$|(\mathcal{O}_K/\mathfrak{a})^*| = N(\mathfrak{a}) \prod_{\mathfrak{p}|\mathfrak{a}} \left(1 - \frac{1}{N(\mathfrak{p})}\right).$$

Notice the similarity to the usual formula for $\phi(n) = |(\mathbb{Z}/n\mathbb{Z})^*|$.

(c) If m is a positive integer, conclude that

$$|(\mathcal{O}_K/m\mathcal{O}_K)^*| = m^2 \prod_{p|m} \left(1 - \frac{1}{p}\right)\left(1 - \left(\frac{d_K}{p}\right)\frac{1}{p}\right),$$

where (d_K/p) is the Kronecker symbol when $p = 2$.

7.30. Let K be an imaginary quadratic field, and let $f > 1$ be a positive integer.

(a) Use the obvious maps

$$\{\pm 1\} \longrightarrow (\mathbb{Z}/f\mathbb{Z})^* \times \mathcal{O}_K^*$$
$$(\mathbb{Z}/f\mathbb{Z})^* \times \mathcal{O}_K^* \longrightarrow (\mathcal{O}_K/f\mathcal{O}_K)^*$$

and the maps from (7.27) to prove that there is an exact sequence

$$\begin{aligned} 1 \longrightarrow \{\pm 1\} \longrightarrow (\mathbb{Z}/f\mathbb{Z})^* \times \mathcal{O}_K^* &\longrightarrow (\mathcal{O}_K/f\mathcal{O}_K)^* \\ &\longrightarrow I_K(f) \cap P_K/P_{K,\mathbb{Z}}(f) \longrightarrow 1. \end{aligned}$$

Notice that when $\mathcal{O}_K^* = \{\pm 1\}$, this sequence is equivalent to (7.27).

(b) Use the exact sequence of part (a) to prove Theorem 7.24 for all imaginary quadratic fields.

7.31. Prove Corollary 7.28.

7.32. In this exercise we will use the inequality

$$(*)\qquad h(d_K) > \frac{\log|d_K|}{55}\prod_{p|d_K,p<d_K}\left(1-\frac{[2\sqrt{p}]}{p+1}\right)$$

to study the equation $h(d_K) = h$, where $h > 0$ is a fixed integer and d_K varies over all negative discriminants.

(a) Show that $1 - [2\sqrt{p}]/(p+1) \geq 1/2$ when $p \geq 11$.

(b) If $h(d_K) = h$, then use part (a) and genus theory to conclude that

$$\prod_{p|d_K,p<d_K}\left(1-\frac{[2\sqrt{p}]}{p+1}\right) \geq \frac{1}{3\cdot 2^{\nu_2(h)+2}},$$

where $\nu_2(h)$ is the highest power of 2 dividing h. Hint: use Theorems 3.15 or 6.1 to show that d_K is divisible by at most $\nu_2(h)+1$ distinct primes.

(c) If $h(d_K) = h$, then show that (*) gives us the following estimate for $|d_K|$:

$$|d_K| \leq e^{165\cdot 2^{\nu_2(h)+2}h}.$$

It follows that there are only finitely many negative discriminants with class number at most h. Unfortunately, this bound is rarely useful in practice. On the other hand, a more careful analysis of $(*)$ can be used to prove $|d_K| \leq e^{165}$ when $h = 3$ (see Oesterlé [**101**]). As Oesterlé explains in [**102**], this is strong enough to solve the class number 3 problem.

(d) If h is fixed and $D \equiv 0, 1 \bmod 4$ varies over all negative integers, show that the equation $h(D) = h$ has only finitely many solutions. Hint: use genus theory to bound the number of primes dividing D, and then use Theorem 7.24.

7.33. In Theorem 7.30, complete the proof of (i) $\Rightarrow$ (ii) sketched in the text.

8. Class Field Theory and the Čebotarev Density Theorem

In this section we will present a classical formulation of class field theory, where Abelian extensions of a number field are described in terms of certain *generalized ideal class groups*. After stating the main theorems (without proof), we will illustrate their use by proving the Kronecker–Weber Theorem and the existence of the Hilbert class field. We will then discuss generalized reciprocity theorems for the nth power Legendre symbol $(\alpha/\mathfrak{p})_n$ and show how quadratic reciprocity follows from class field theory.

The Čebotarev Density Theorem hasn't been mentioned before, but it provides some important information about the behavior of the Artin map. One of its classic applications is Dirichlet's theorem on primes in arithmetic progressions, and in §9 we will use the same methods to study primes represented by a given quadratic form. Another consequence of the Density Theorem is that a Galois extension of a number field is determined uniquely by the primes in the base field that split completely in the extension. As we will see, this is closely related to our basic problem of characterizing the primes represented by $x^2 + ny^2$.

Our account of class field theory will be incomplete in several ways. At the end of the section we will discuss two of the most obvious omissions, norms and ideles.

A. The Theorems of Class Field Theory

We begin our treatment of class field theory with the notion of a *modulus*. Given a number field K, a *modulus* in K is a formal product

$$\mathfrak{m} = \prod_{\mathfrak{p}} \mathfrak{p}^{n_{\mathfrak{p}}}$$

over all primes $\mathfrak{p}$, finite or infinite, of K (see §5), where the exponents must satisfy:

(i) $n_{\mathfrak{p}} \geq 0$, and at most finitely many are nonzero.
(ii) $n_{\mathfrak{p}} = 0$ wherever $\mathfrak{p}$ is a complex infinite prime.
(iii) $n_{\mathfrak{p}} \leq 1$ whenever $\mathfrak{p}$ is a real infinite prime.

A modulus $\mathfrak{m}$ may thus be written $\mathfrak{m}_0\mathfrak{m}_\infty$, where $\mathfrak{m}_0$ is an $\mathcal{O}_K$-ideal and $\mathfrak{m}_\infty$ is a product of distinct real infinite primes of K. When all of the exponents $n_{\mathfrak{p}}$ are zero, we set $\mathfrak{m} = 1$. Note that for a purely imaginary field K (the case we're most interested in), a modulus may be regarded simply as an ideal of $\mathcal{O}_K$.

Given a modulus $\mathfrak{m}$, let $I_K(\mathfrak{m})$ be the group of all fractional $\mathcal{O}_K$-ideals relatively prime to $\mathfrak{m}$ (which means relatively prime to $\mathfrak{m}_0$), and let $P_{K,1}(\mathfrak{m})$ be the subgroup of $I_K(\mathfrak{m})$ generated by the principal ideals $\alpha\mathcal{O}_K$, where $\alpha \in \mathcal{O}_K$ satisfies

$$\alpha \equiv 1 \bmod \mathfrak{m}_0 \quad \text{and} \quad \sigma(\alpha) > 0 \text{ for every real infinite prime } \sigma \text{ dividing } \mathfrak{m}_\infty.$$

A basic result is that $P_{K,1}(\mathfrak{m})$ has finite index in $I_K(\mathfrak{m})$. When K is imaginary quadratic, this is proved in Exercise 8.1, while the general case may be found in Janusz [**78**, Chapter IV.1]. A subgroup $H \subset I_K(\mathfrak{m})$ is called a *congruence subgroup* for $\mathfrak{m}$ if it satisfies

$$P_{K,1}(\mathfrak{m}) \subset H \subset I_K(\mathfrak{m}),$$

and the quotient

$$I_K(\mathfrak{m})/H$$

is called a *generalized ideal class group* for $\mathfrak{m}$.

For an example of these concepts, consider the modulus $\mathfrak{m} = 1$. Then $P_K = P_{K,1}(1)$ is a congruence subgroup, so that the ideal class group $C(\mathcal{O}_K) = I_K/P_K$ is a generalized ideal class group. We also get some interesting examples from §7. Let $\mathcal{O}$ be an order of conductor f in an imaginary quadratic field K. In Proposition 7.22 we proved that the ideal class group $C(\mathcal{O})$ can be written

$$C(\mathcal{O}) \simeq I_K(f)/P_{K,\mathbb{Z}}(f),$$

where $P_{K,\mathbb{Z}}(f)$ is generated by the principal ideals $\alpha\mathcal{O}_K$ for $\alpha \equiv a \bmod f\mathcal{O}_K$, $a \in \mathbb{Z}$ and $\gcd(a, f) = 1$. If we use the modulus $f\mathcal{O}_K$, then the definition of $P_{K,1}(f\mathcal{O}_K)$ shows that

$$P_{K,1}(f\mathcal{O}_K) \subset P_{K,\mathbb{Z}}(f) \subset I_K(f) = I_K(f\mathcal{O}_K), \tag{8.1}$$

and thus $P_{K,\mathbb{Z}}(f)$ is a congruence subgroup for $f\mathcal{O}_K$. This proves that $C(\mathcal{O})$ is a generalized ideal class group of K for the modulus $f\mathcal{O}_K$. In §7, the group $P_{K,\mathbb{Z}}(f)$ seemed awkward, but it's a very natural object from the point of view of class field theory. (In what follows, $P_{K,1}(f\mathcal{O}_K)$ will be written $P_{K,1}(f)$ for simplicity.)

The basic idea of class field theory is that the generalized ideal class groups are the Galois groups of all Abelian extensions of K, and the link between these two is provided by the Artin map. To make this precise, we need to define the *Artin map* of an Abelian extension of K.

Let $\mathfrak{m}$ be a modulus divisible by all ramified primes of an Abelian extension $K \subset L$. Given a prime $\mathfrak{p}$ not dividing $\mathfrak{m}$, we have the Artin symbol

$$\left(\frac{L/K}{\mathfrak{p}}\right) \in \mathrm{Gal}(L/K)$$

from §5. As in the discussion preceding Theorem 5.23, the Artin symbol extends by multiplicativity to give us a homomorphism

$$\Phi_{\mathfrak{m}} : I_K(\mathfrak{m}) \longrightarrow \mathrm{Gal}(L/K)$$

which is called the *Artin map* for $K \subset L$ and $\mathfrak{m}$. When we want to refer explicitly to the extension involved, we will write $\Phi_{L/K,\mathfrak{m}}$ instead of $\Phi_{\mathfrak{m}}$.

The first theorem of class field theory tells us that $\mathrm{Gal}(L/K)$ is a generalized ideal class group for some modulus:

THEOREM 8.2. *Let $K \subset L$ be an Abelian extension, and let $\mathfrak{m}$ be a modulus divisible by all primes of K, finite or infinite, that ramify in L. Then:*

(i) *The Artin map $\Phi_{\mathfrak{m}}$ is surjective.*

(ii) *If the exponents of the finite primes $\mathfrak{m}$ are sufficiently large, then $\ker(\Phi_{\mathfrak{m}})$ is a congruence subgroup for $\mathfrak{m}$, i.e.,*

$$P_{K,1}(\mathfrak{m}) \subset \ker(\Phi_{\mathfrak{m}}) \subset I_K(\mathfrak{m}),$$

and consequently the isomorphism

$$I_K(\mathfrak{m})/\ker(\Phi_{\mathfrak{m}}) \xrightarrow{\sim} \mathrm{Gal}(L/K)$$

shows that $\mathrm{Gal}(L/K)$ is a generalized ideal class group for the modulus $\mathfrak{m}$.

PROOF. See Janusz [**78**, Chapter V, Theorem 5.7]. Q.E.D.

This theorem is sometimes called the *Artin Reciprocity Theorem*. The key ingredient is the condition $P_{K,1}(\mathfrak{m}) \subset \ker(\Phi_{\mathfrak{m}})$, for it says (roughly) that the Artin symbol $((L/K)/\mathfrak{p})$ depends only on $\mathfrak{p}$ up to multiplication by α, $\alpha \equiv 1 \bmod \mathfrak{m}$. Later in this section we will see how Artin Reciprocity relates to quadratic, cubic and biquadratic reciprocity.

Let's work out an example of Theorem 8.2. Consider the extension $\mathbb{Q} \subset \mathbb{Q}(\zeta_m)$, where $\zeta_m = e^{2\pi i/m}$ is a primitive mth root of unity, and let $\mathfrak{m}$ be the modulus $m\infty$, where ∞ is the real infinite prime of $\mathbb{Q}$. Using Proposition 5.11, one sees that any prime not dividing $\mathfrak{m}$ is unramified in $\mathbb{Q}(\zeta_m)$ (see Exercise 8.2), and it follows that the Artin map

$$\Phi_{\mathfrak{m}} : I_{\mathbb{Q}}(\mathfrak{m}) \longrightarrow \mathrm{Gal}(\mathbb{Q}(\zeta_m)/\mathbb{Q}) \simeq (\mathbb{Z}/m\mathbb{Z})^*$$

is defined. The map $\Phi_{\mathfrak{m}}$ can be described as follows: given $(a/b)\mathbb{Z} \in I_{\mathbb{Q}}(\mathfrak{m})$, where $a/b > 0$ and $\gcd(a,m) = \gcd(b,m) = 1$, then

$$\Phi_{\mathfrak{m}}\left(\frac{a}{b}\mathbb{Z}\right) = [a][b]^{-1} \in (\mathbb{Z}/m\mathbb{Z})^*. \tag{8.3}$$

It follows easily that

$$\ker(\Phi_{\mathfrak{m}}) = P_{\mathbb{Q},1}(\mathfrak{m}) \tag{8.4}$$

(see Exercise 8.2). The importance of this computation will soon become clear.

One difficulty with Theorem 8.2 is that the $\mathfrak{m}$ for which $\ker(\Phi_\mathfrak{m})$ is a congruence subgroup is not unique. In fact, if $P_{K,1}(\mathfrak{m}) \subset \ker(\Phi_\mathfrak{m})$ and $\mathfrak{n}$ is any modulus divisible by $\mathfrak{m}$ (it's clear what this means), then

$$P_{K,1}(\mathfrak{m}) \subset \ker(\Phi_\mathfrak{m}) \implies P_{K,1}(\mathfrak{n}) \subset \ker(\Phi_\mathfrak{n})$$

(see Exercise 8.4), so that $\mathrm{Gal}(L/K)$ is a generalized ideal class group for infinitely many moduli. However, there is one modulus which is better than the others:

THEOREM 8.5. *Let $K \subset L$ be an Abelian extension. Then there is a modulus $\mathfrak{f} = \mathfrak{f}(L/K)$ such that*

(i) *A prime of K, finite or infinite, ramifies in L if and only if it divides $\mathfrak{f}$.*
(ii) *Let $\mathfrak{m}$ be a modulus divisible by all primes of K which ramify in L. Then $\ker(\Phi_\mathfrak{m})$ is a congruence subgroup for $\mathfrak{m}$ if and only if $\mathfrak{f} \mid \mathfrak{m}$.*

PROOF. See Janusz [**78**, Chapter V, §6 and Theorem 11.11]. Q.E.D.

The modulus $\mathfrak{f}(L/K)$ is uniquely determined by $K \subset L$ and is called the *conductor* of the extension, and for this reason Theorem 8.5 is often called the *Conductor Theorem.* In Exercise 8.5 we will compute the conductor of $\mathbb{Q} \subset \mathbb{Q}(\zeta_m)$ (it need not be m), and in §9 we will compute the conductor of a ring class field.

The final theorem of class field theory is the *Existence Theorem*, which asserts that *every* generalized ideal class group is the Galois group of some Abelian extension $K \subset L$. More precisely:

THEOREM 8.6. *Let $\mathfrak{m}$ be a modulus of K, and let H be a congruence subgroup for $\mathfrak{m}$, i.e.,*

$$P_{K,1}(\mathfrak{m}) \subset H \subset I_K(\mathfrak{m}).$$

Then there is a unique Abelian extension L of K, all of whose ramified primes, finite or infinite, divide $\mathfrak{m}$, such that if

$$\Phi_\mathfrak{m} : I_K(\mathfrak{m}) \longrightarrow \mathrm{Gal}(L/K)$$

is the Artin map of $K \subset L$, then

$$H = \ker(\Phi_\mathfrak{m}).$$

PROOF. See Janusz [**78**, Chapter V, Theorem 9.16]. Q.E.D.

The importance of this theorem is that it allows us to construct Abelian extensions of K with specified Galois group and restricted ramification. This will be very useful in the applications that follow.

Now that we've stated the basic theorems of class field theory, the next step is to indicate how they are used. We will start with two of the nicest applications: proofs of the Kronecker–Weber Theorem and the existence of the Hilbert class field. A key tool in both proofs is the following corollary of the uniqueness part of Theorem 8.6:

COROLLARY 8.7. *Let L and M be Abelian extensions of K. Then $L \subset M$ if and only if there is a modulus $\mathfrak{m}$, divisible by all primes of K ramified in either L or M, such that*

$$P_{K,1}(\mathfrak{m}) \subset \ker(\Phi_{M/K,\mathfrak{m}}) \subset \ker(\Phi_{L/K,\mathfrak{m}}).$$

PROOF. First, assume that $L \subset M$, and let $r : \mathrm{Gal}(M/K) \to \mathrm{Gal}(L/K)$ be the restriction map. By Theorem 8.2 and Exercise 8.4, there is a modulus $\mathfrak{m}$ for which $\ker(\Phi_{L/K,\mathfrak{m}})$ and $\ker(\Phi_{M/K,\mathfrak{m}})$ are both congruence subgroups for $\mathfrak{m}$. The proof of Exercise 5.16 shows that $r \circ \Phi_{M/K,\mathfrak{m}} = \Phi_{L/K,\mathfrak{m}}$, and then $\ker(\Phi_{M/K,\mathfrak{m}}) \subset \ker(\Phi_{L/K,\mathfrak{m}})$ follows immediately.

Going the other way, assume that $P_{K,1}(\mathfrak{m}) \subset \ker(\Phi_{M/K,\mathfrak{m}}) \subset \ker(\Phi_{L/K,\mathfrak{m}})$. Then, under the map $\Phi_{M/K,\mathfrak{m}}\colon I_K(\mathfrak{m}) \to \mathrm{Gal}(M/K)$, the subgroup $\ker(\Phi_{L/K,\mathfrak{m}}) \subset I_K(\mathfrak{m})$ maps to a subgroup $H \subset \mathrm{Gal}(M/K)$. By Galois theory, H corresponds to an intermediate field $K \subset \tilde{L} \subset M$. The first part of the proof, applied to $\tilde{L} \subset M$, shows that $\ker(\Phi_{\tilde{L}/K,\mathfrak{m}}) = \ker(\Phi_{L/K,\mathfrak{m}})$. Then the uniqueness part of Theorem 8.6 shows that $L = \tilde{L} \subset M$, and we are done. Q.E.D.

We can now prove the Kronecker–Weber Theorem, which classifies all Abelian extensions of $\mathbb{Q}$:

THEOREM 8.8. *Let L be an Abelian extension of $\mathbb{Q}$. Then there is a positive integer m such that $L \subset \mathbb{Q}(\zeta_m)$, $\zeta_m = e^{2\pi i/m}$.*

PROOF. By the Artin Reciprocity Theorem (Theorem 8.2), there is a modulus $\mathfrak{m}$ such that $P_{\mathbb{Q},1}(\mathfrak{m}) \subset \ker(\Phi_{L/\mathbb{Q},\mathfrak{m}})$, and by Exercise 8.4, we may assume that $\mathfrak{m} = m\infty$. By (8.4) we know that $P_{\mathbb{Q},1}(\mathfrak{m}) = \ker(\Phi_{\mathbb{Q}(\zeta_m)/\mathbb{Q},\mathfrak{m}})$, so that

$$P_{\mathbb{Q},1}(\mathfrak{m}) = \ker(\Phi_{\mathbb{Q}(\zeta_m)/\mathbb{Q},\mathfrak{m}}) \subset \ker(\Phi_{L/K,\mathfrak{m}}).$$

Then $L \subset \mathbb{Q}(\zeta_m)$ follows from Corollary 8.7. Q.E.D.

We should mention that the Kronecker–Weber Theorem can be proved without using class field theory (see Marcus [**96**, Chapter 4, Exercises 29–36]).

Next, let's discuss the Hilbert class field. To define it, apply the Existence Theorem (Theorem 8.6) to the modulus $\mathfrak{m} = 1$ and the subgroup $P_K \subset I_K$ (note that $P_K = P_{K,1}(\mathfrak{m})$ in this case). Thus there is a unique Abelian extension L of K, unramified since $\mathfrak{m} = 1$, such that the Artin map induces an isomorphism

$$C(\mathcal{O}_K) = I_K/P_K \xrightarrow{\sim} \mathrm{Gal}(L/K). \tag{8.9}$$

L is the *Hilbert class field* of K, and its main property is the following:

THEOREM 8.10. *The Hilbert class field L is the maximal unramified Abelian extension of K.*

PROOF. We already know that L is an unramified extension. Let M be another unramified extension. The first part of the Conductor Theorem (Theorem 8.5) implies that $\mathfrak{f}(M/K) = 1$ since a prime ramifies if and only if it divides the conductor, and then the second part tells us that $\ker(\Phi_{M/K,1})$ is a congruence subgroup for the modulus 1, so that

$$P_K \subset \ker(\Phi_{M/K,1}).$$

By the definition of the Hilbert class field, this becomes

$$P_K = \ker(\Phi_{L/K,1}) \subset \ker(\Phi_{M/K,1}),$$

and then $M \subset L$ follows from Corollary 8.7. Q.E.D.

Notice that Theorems 5.18 and 5.23 from §5 are immediate consequences of (8.9) and Theorem 8.10.

There is a generalization of the Hilbert class field called the *ray class field*. Given any modulus $\mathfrak{m}$, the Existence Theorem shows that there is a unique Abelian extension $K_{\mathfrak{m}}$ of K such that

$$P_{K,1}(\mathfrak{m}) = \ker(\Phi_{K_{\mathfrak{m}}/K,\mathfrak{m}}).$$

$K_{\mathfrak{m}}$ is called the *ray class field* for the modulus $\mathfrak{m}$, and when $\mathfrak{m} = 1$, this reduces to the Hilbert class field. Another example is given by the cyclotomic field $\mathbb{Q}(\zeta_m)$: here, (8.4) shows that $\mathbb{Q}(\zeta_m)$ is the ray class field of $\mathbb{Q}$ for the modulus $m\infty$. We also get a nice interpretation of the conductor $\mathfrak{f}(L/K)$ of an arbitrary Abelian extension L of K: it's the smallest modulus $\mathfrak{m}$ for which L is contained in the ray class field $K_{\mathfrak{m}}$ (see Exercise 8.6).

Besides proving these classical results, class field theory is also the source of most reciprocity theorems. In particular, we will discuss some reciprocity theorems for the nth power Legendre symbol $(\alpha/\mathfrak{p})_n$ mentioned in §5. To define this symbol, let K be a number field containing a primitive nth root of unity ζ, and let $\mathfrak{p}$ be a prime ideal of $\mathcal{O}_K$. Then, for $\alpha \in \mathcal{O}_K$ prime to $\mathfrak{p}$, we have Fermat's Little Theorem

$$\alpha^{N(\mathfrak{p})-1} \equiv 1 \bmod \mathfrak{p}.$$

Suppose that in addition $\mathfrak{p}$ is prime to n. It can be shown that $n \mid N(\mathfrak{p}) - 1$ (see Exercise 5.13), and it follows that $x = \alpha^{(N(\mathfrak{p})-1)/n}$ is a solution of the congruence $x^n \equiv 1 \bmod \mathfrak{p}$. Consequently

$$\alpha^{(N(\mathfrak{p})-1)/n} \equiv 1, \zeta, \ldots, \zeta^{n-1} \bmod \mathfrak{p}.$$

Since the nth roots of unity are distinct modulo $\mathfrak{p}$ (see Exercise 5.13), $\alpha^{(N(\mathfrak{p})-1)/n}$ is congruent modulo $\mathfrak{p}$ to a *unique* nth root of unity. This root of unity is defined to be the nth *power Legendre symbol* $(\alpha/\mathfrak{p})_n$, so that $(\alpha/\mathfrak{p})_n$ satisfies the congruence

$$\alpha^{(N(\mathfrak{p})-1)/n} \equiv \left(\frac{\alpha}{\mathfrak{p}}\right)_n \bmod \mathfrak{p}.$$

This symbol is a natural generalization of the Legendre symbols $(\alpha/\pi)_3$ and $(\alpha/\pi)_4$ from cubic and biquadratic reciprocity.

The nth power Legendre symbol can be defined for more general ideals as follows: given an ideal $\mathfrak{a}$ of $\mathcal{O}_K$ which is prime to n and α, we set $(\alpha/\mathfrak{a})_n$ to be the product

$$\left(\frac{\alpha}{\mathfrak{a}}\right)_n = \prod_{i=1}^{r} \left(\frac{\alpha}{\mathfrak{p}_i}\right)_n,$$

where $\mathfrak{a} = \mathfrak{p}_1 \cdots \mathfrak{p}_r$ is the prime factorization of $\mathfrak{a}$. Thus, if $\mathfrak{m}$ is a modulus of K such that every prime containing $n\alpha$ divides $\mathfrak{m}$, then the nth power Legendre symbol gives a homomorphism

$$\left(\frac{\alpha}{\cdot}\right)_n : I_K(\mathfrak{m}) \longrightarrow \mu_n,$$

where $\mu_n \subset \mathbb{C}^*$ is the group of nth roots of unity.

We will prove two reciprocity theorems for the nth power Legendre symbol, but first we need to recall a fact from Galois theory. If K has a primitive nth root of unity, then for $\alpha \in K$, the extension $K \subset L = K(\sqrt[n]{\alpha})$ is Galois, and if $\sigma \in \mathrm{Gal}(L/K)$, then $\sigma(\sqrt[n]{\alpha}) = \zeta \sqrt[n]{\alpha}$ for some nth root of unity ζ. This gives us a map $\sigma \mapsto \zeta$, which defines an injective homomorphism

$$\mathrm{Gal}(L/K) \hookrightarrow \mu_n.$$

We can now state our first reciprocity theorem for $(\alpha/\mathfrak{a})_n$:

THEOREM 8.11 (**Weak Reciprocity**). *Let K be a number field containing a primitive nth root of unity, and let $L = K(\sqrt[n]{\alpha})$, where $\alpha \in \mathcal{O}_K$ is nonzero. Assume that $\mathfrak{m}$ is a modulus divisible by all primes of K containing $n\alpha$, and assume in addition that $\ker(\Phi_{L/K,\mathfrak{m}})$ is a congruence subgroup for $\mathfrak{m}$. Then there is a commutative diagram*

$$\begin{array}{ccc} I_K(\mathfrak{m}) & \xrightarrow{\Phi_{L/K,\mathfrak{m}}} & \mathrm{Gal}(L/K) \\ & \searrow^{(\alpha/\cdot)_n} & \downarrow \\ & & \mu_n, \end{array}$$

where $\mathrm{Gal}(L/K) \hookrightarrow \mu_n$ is the natural injection. Thus, if G is the image of $\mathrm{Gal}(L/K)$ in μ_n, then the nth power Legendre symbol $(\alpha/\mathfrak{a})_n$ induces a surjective homomorphism

$$\left(\frac{\alpha}{\cdot}\right)_n : I_K(\mathfrak{m})/P_{K,1}(\mathfrak{m}) \longrightarrow G \subset \mu_n.$$

PROOF. To prove that the diagram commutes, it suffices to show

$$\left(\frac{L/K}{\mathfrak{p}}\right)(\sqrt[n]{\alpha}) = \left(\frac{\alpha}{\mathfrak{p}}\right)_n \sqrt[n]{\alpha}.$$

This is an easy consequence of the definition of the Artin symbol (from Lemma 5.19). The case $n = 3$ was proved in (5.22), and for general n, see Exercise 5.14.

Turning to the final statement of the theorem, recall that $\ker(\Phi_{L/K,\mathfrak{m}})$ is a congruence subgroup for $\mathfrak{m}$. Thus $P_{K,1}(\mathfrak{m}) \subset \ker(\Phi_{L/K,\mathfrak{m}}) \subset I_K(\mathfrak{m})$, so that the Artin map $\Phi_{L/K,\mathfrak{m}}$ induces a surjective homomorphism

$$I_K(\mathfrak{m})/P_{K,1}(\mathfrak{m}) \longrightarrow I_K(\mathfrak{m})/\ker(\Phi_{L/K,\mathfrak{m}}) \xrightarrow{\sim} \mathrm{Gal}(L/K).$$

Using the above commutative diagram, the theorem follows immediately. Q.E.D.

This result is called "Weak Reciprocity" because rather than giving formulas for computing $(\alpha/\mathfrak{a})_n$, the theorem simply asserts that the symbol is a homomorphism on an appropriate group. Nevertheless, Weak Reciprocity is a powerful result. For example, let's use it to prove quadratic reciprocity:

THEOREM 8.12. *Let p and q be distinct odd primes. Then*

$$\left(\frac{p}{q}\right)\left(\frac{q}{p}\right) = (-1)^{(p-1)(q-1)/4}.$$

PROOF. Recall from §1 that quadratic reciprocity can be written in the form

$$\left(\frac{p^*}{q}\right) = \left(\frac{q}{p}\right)$$

where $p^* = (-1)^{(p-1)/2}p$.

The first step is to study $\mathbb{Q} \subset \mathbb{Q}(\sqrt{p^*})$. By (8.3) and (8.4), $\mathrm{Gal}(\mathbb{Q}(\zeta_p)/\mathbb{Q})$ is a generalized ideal class group for the modulus $p\infty$, which implies that the same is true for any subfield of $\mathbb{Q}(\zeta_p)$ (see Exercise 8.7). Since $\mathrm{Gal}(\mathbb{Q}(\zeta_p)/\mathbb{Q})$ is cyclic of order $p-1$, there is a unique subfield $\mathbb{Q} \subset K \subset \mathbb{Q}(\zeta_p)$ which is quadratic over $\mathbb{Q}$. Then $\mathrm{Gal}(K/\mathbb{Q})$ is a generalized ideal class group for $p\infty$, which implies that p is the only finite prime of $\mathbb{Q}$ that ramifies in K. If we write $K = \mathbb{Q}(\sqrt{m})$, m

squarefree, then Corollary 5.17 implies that $m = p^*$, and hence $K = \mathbb{Q}(\sqrt{p^*})$ (see Exercise 8.7).

To apply Theorem 8.11 when $n = 2$, the modulus must be divisible by 2. Since p is odd, $\zeta_{2p} = -\zeta_p$, so $\mathbb{Q}(\zeta_{2p}) = \mathbb{Q}(\zeta_p)$, and by (8.3) and (8.4), $\mathrm{Gal}(\mathbb{Q}(\zeta_{2p})/\mathbb{Q})$ is a generalized ideal class group for the modulus $2p\infty$. It follows that Weak Reciprocity applies to $K/\mathbb{Q}$ for this modulus. This gives a surjective homomorphism

$$I_{\mathbb{Q}}(2p\infty)/P_{\mathbb{Q},1}(2p\infty) \longrightarrow \{\pm 1\}. \tag{8.13}$$

However, we also have isomorphisms

$$(\mathbb{Z}/p\mathbb{Z})^* \xrightarrow{\sim} (\mathbb{Z}/2p\mathbb{Z})^* \xrightarrow{\sim} I_{\mathbb{Q}}(2p\infty)/P_{\mathbb{Q},1}(2p\infty),$$

where the first isomorphism follows since p is odd (a even $\Rightarrow a + p$ is odd) and the second sends $[a] \in (\mathbb{Z}/2p\mathbb{Z})^*$ to $[a\mathbb{Z}] \in I_{\mathbb{Q}}(2p\infty)/P_{\mathbb{Q},1}(2p\infty)$ when $a > 0$ (see Exercise 8.7). Composing this map with (8.13) shows that $(p^*/\cdot)$ induces a surjective homomorphism from $(\mathbb{Z}/p\mathbb{Z})^*$ to $\{\pm 1\}$. But the Legendre symbol $(\cdot/p)$ is also a surjective homomorphism between the same two groups, and since $(\mathbb{Z}/p\mathbb{Z})^*$ is cyclic, there is only one such homomorphism. This proves that

$$\left(\frac{p^*}{q}\right) = \left(\frac{q}{p}\right),$$

and we are done. Q.E.D.

The proof just given is closely related to the discussion of quadratic reciprocity from §1. Recall that a key result implicit in Euler's work was Lemma 1.14, which showed that $(D/\cdot)$ gives a well-defined homomorphism defined on $(\mathbb{Z}/D\mathbb{Z})^*$ when $D \equiv 0, 1 \bmod 4$. The above argument uses Weak Reciprocity to prove this when $D = p^*$. In this way Weak Reciprocity (or more generally, Artin Reciprocity) may be regarded as a far-reaching generalization of Lemma 1.14.

Before we can state our second reciprocity theorem for the nth power Legendre symbol, we need some notation: if α and β are in $\mathcal{O}_K$, then $(\alpha/\beta\mathcal{O}_K)_n$ is written simply $(\alpha/\beta)_n$, when defined. Then we have the following reciprocity theorem for $(\alpha/\beta)_n$:

THEOREM 8.14 (**Strong Reciprocity**). *Let K be a number field containing a primitive nth root of unity, and suppose that $\alpha, \beta \in \mathcal{O}_K$ are relatively prime to each other and to n. Then*

$$\left(\frac{\alpha}{\beta}\right)_n \left(\frac{\beta}{\alpha}\right)_n^{-1} = \prod_{\mathfrak{p} \mid n\infty} \left(\frac{\alpha, \beta}{\mathfrak{p}}\right)_n,$$

where $(\alpha, \beta/\mathfrak{p})_n$ is the nth power Hilbert symbol (to be discussed below) and ∞ is the product of the real infinite primes of K (which can occur only when $n = 2$).

PROOF. While Weak Reciprocity was an immediate consequence of Artin Reciprocity, Strong Reciprocity is a different matter, for here one must first study the nth power Hilbert symbol

$$\left(\frac{\alpha, \beta}{\mathfrak{p}}\right)_n.$$

This symbol is an nth root of unity defined using the local class field theory of the completion $K_{\mathfrak{p}}$ of K at the prime $\mathfrak{p}$. Since we haven't discussed local methods, we can't even give a precise definition. A full discussion of the Hilbert symbol is given in Hasse [**65**, Part II, §§11–12, pp. 53–64] and Neukirch [**100**, §§III.5 and IV.9, pp.

50–55 and 110–112], and both references present a complete proof of the Strong Reciprocity theorem. In Exercise 8.9 we will list the main properties of the Hilbert symbol. Q.E.D.

To get a better idea of how Strong Reciprocity works, let's apply it to cubic reciprocity. Here, $n = 3$ and $K = \mathbb{Q}(\omega)$, $\omega = e^{2\pi i/3}$, and the only prime of $\mathcal{O}_K$ dividing 3 is $\lambda = 1 - \omega$. Thus, given nonassociate primes π and θ in $\mathcal{O}_K$, Strong Reciprocity tells us that

$$\left(\frac{\pi}{\theta}\right)_3 \left(\frac{\theta}{\pi}\right)_3^{-1} = \left(\frac{\pi, \theta}{\lambda}\right)_3.$$

Hence, to prove cubic reciprocity, it suffices to show that

$$\pi,\ \theta \text{ primary} \implies \left(\frac{\pi, \theta}{\lambda}\right)_3 = 1. \tag{8.15}$$

The proof of cubic reciprocity is thus reduced to a purely local computation in the completion K_λ of K at λ. Given the properties of the Hilbert symbol, (8.15) is not difficult to prove (see Exercise 8.9). Biquadratic reciprocity can be proved similarly, though the proof is a bit more complicated (see Hasse [**65**, Part II, §20, pp. 105–106]). This shows that class field theory encompasses all of the reciprocity theorems we've seen so far.

B. The Čebotarev Density Theorem

The Čebotarev Density Theorem will provide some very useful information about the Artin map. But first, we need to define the notion of *Dirichlet density.*

Let K be a number field, and let $\mathcal{P}_K$ be the set of all finite primes of K. Given a subset $\mathcal{S} \subset \mathcal{P}_K$, the *Dirichlet density* of $\mathcal{S}$ is defined to be

$$\delta(\mathcal{S}) = \lim_{s \to 1^+} \frac{\sum_{\mathfrak{p} \in \mathcal{S}} N(\mathfrak{p})^{-s}}{-\log(s-1)},$$

provided the limit exists. The basic properties of the Dirichlet density are:

(i) $\delta(\mathcal{P}_K) = 1$.
(ii) If $\mathcal{S} \subset \mathcal{T}$ and $\delta(\mathcal{S})$ and $\delta(\mathcal{T})$ exist, then $\delta(\mathcal{S}) \leq \delta(\mathcal{T})$.
(iii) If $\delta(\mathcal{S})$ exists, then $0 \leq \delta(\mathcal{S}) \leq 1$.
(iv) If $\mathcal{S}$ and $\mathcal{T}$ are disjoint and $\delta(\mathcal{S})$ and $\delta(\mathcal{T})$ exist, then $\delta(\mathcal{S} \cup \mathcal{T}) = \delta(\mathcal{S}) + \delta(\mathcal{T})$.
(v) If $\mathcal{S}$ is finite, then $\delta(\mathcal{S}) = 0$.
(vi) If $\delta(\mathcal{S})$ exists and $\mathcal{T}$ differs from $\mathcal{S}$ by finitely many elements, then $\delta(\mathcal{T}) = \delta(\mathcal{S})$.

To prove these properties, one first must study the Dirichlet ζ-function $\zeta_K(s)$ of K. This function is defined by

$$\zeta_K(s) = \sum_{\mathfrak{a} \subset \mathcal{O}_K} N(\mathfrak{a})^{-s} = \prod_{\mathfrak{p} \in \mathcal{P}_K} (1 - N(\mathfrak{p})^{-s})^{-1}.$$

One can prove without difficulty that $\zeta_K(s)$ converges absolutely for $\mathrm{Re}(s) > 1$ (see Janusz [**78**, §IV.4] or Neukirch [**100**, §V.6]). This implies that for any $\mathcal{S} \subset \mathcal{P}_K$, the sum $\sum_{\mathfrak{p} \in \mathcal{S}} N(\mathfrak{p})^{-s}$ converges absolutely for $\mathrm{Re}(s) > 1$ (see Exercise 8.10). A much deeper property of $\zeta_K(s)$ is that it has a simple pole at $s = 1$, which enables one to prove that

$$1 = \lim_{s \to 1^+} \frac{\log(\zeta_K(s))}{-\log(s-1)} = \lim_{s \to 1^+} \frac{\sum_{\mathfrak{p} \in \mathcal{P}_K} N(\mathfrak{p})^{-s}}{-\log(s-1)}$$

(see Janusz [**78**, §IV.4] or Neukirch [**100**, §V.6]). This proves (i), and it is now straightforward to prove (ii)–(vi) (see Exercise 8.10).

There is one more property of the Dirichlet density which is sometimes useful. Let $\mathcal{P}_{K,1} = \{\mathfrak{p} \in \mathcal{P}_K : N(\mathfrak{p}) \text{ is prime}\}$. $\mathcal{P}_{K,1}$ is sometimes called the degree 1 primes in K (recall that in general, $N(\mathfrak{p}) = p^f$, where f is the inertial degree of $p \in \mathfrak{p}$ in the extension $\mathbb{Q} \subset K$). Then one can prove that

$$\delta(\mathcal{S}) = \delta(\mathcal{S} \cap \mathcal{P}_{K,1}) \tag{8.16}$$

whenever $\delta(\mathcal{S})$ exists (see Janusz [**78**, §IV.4] or Neukirch [**100**, §V.6]).

Now let L be a Galois extension of K, possibly non-Abelian. If $\mathfrak{p}$ is a prime of K unramified in L, then different primes $\mathfrak{P}$ of L containing $\mathfrak{p}$ may give us different Artin symbols $((L/K)/\mathfrak{P})$. But all of the $((L/K)/\mathfrak{P})$ are conjugate by Corollary 5.21, and in fact they form a complete conjugacy class in $\mathrm{Gal}(L/K)$ (see Exercise 5.12). Thus we can define the Artin symbol $((L/K)/\mathfrak{p})$ of $\mathfrak{p}$ to be this conjugacy class in $\mathrm{Gal}(L/K)$. We can now state the *Čebotarev Density Theorem*:

THEOREM 8.17. *Let L be a Galois extension of K, and let $\langle\sigma\rangle$ be the conjugacy class of an element $\sigma \in \mathrm{Gal}(L/K)$. Then the set*

$$\mathcal{S} = \{\mathfrak{p} \in \mathcal{P}_K : \mathfrak{p} \textit{ is unramified in } L \textit{ and } ((L/K)/\mathfrak{p}) = \langle\sigma\rangle\}$$

has Dirichlet density

$$\delta(\mathcal{S}) = \frac{|\langle\sigma\rangle|}{|\mathrm{Gal}(L/K)|} = \frac{|\langle\sigma\rangle|}{[L:K]}.$$

PROOF. See Janusz [**78**, Chapter V, Theorem 10.4] or Neukirch [**100**, Chapter V, Theorem 6.4]. Q.E.D.

Notice that the set S of the theorem must be infinite since it has positive density (this follows from property (v) above). In particular, we get the following corollary for Abelian extensions:

COROLLARY 8.18. *Let L be an Abelian extension of K, and let $\mathfrak{m}$ be a modulus divisible by all primes that ramify in L. Then, given any element $\sigma \in \mathrm{Gal}(L/K)$, the set of primes $\mathfrak{p}$ not dividing $\mathfrak{m}$ such that $((L/K)/\mathfrak{p}) = \sigma$ has density $1/[L:K]$ and hence is infinite.*

PROOF. When $\mathrm{Gal}(L/K)$ is Abelian, the conjugacy class $\langle\sigma\rangle$ reduces to $\{\sigma\}$. Q.E.D.

This corollary shows that the Artin map $\phi_{L/K,\mathfrak{m}} : I_K(\mathfrak{m}) \to \mathrm{Gal}(L/K)$ is surjective in a very strong sense.

An especially nice case is when $K = \mathbb{Q}$ and $L = \mathbb{Q}(\zeta_m)$, for here Corollary 8.18 gives a quick proof of Dirichlet's theorem on primes in arithmetic progressions (the details are left to the reader—see Exercise 8.11). In §9 we will apply these same ideas to study the primes represented by a fixed quadratic form $ax^2 + bxy + cy^2$.

Another application of Čebotarev Density concerns primes that split completely in a Galois extension $K \subset L$. Namely, if we apply Theorem 8.17 to the conjugacy class of the identity element, we see that the primes in K for which $((L/K)/\mathfrak{p}) = 1$ have density $1/[L:K]$. However, from Corollary 5.21, we know that

$$\left(\frac{L/K}{\mathfrak{p}}\right) = 1 \iff \mathfrak{p} \text{ splits completely in } L.$$

Thus the primes that split completely in L have density $1/[L:K]$, and in particular there are infinitely many of them. The unexpected fact is that these primes characterize the extension $K \subset L$ uniquely. Before we can prove this, we need to introduce some terminology.

Given two sets $\mathcal{S}$ and $\mathcal{T}$, we say that $\mathcal{S} \mathrel{\dot{\subset}} \mathcal{T}$ if $\mathcal{S} \subset \mathcal{T} \cup \Sigma$ for some finite set Σ, and $\mathcal{S} \doteq \mathcal{T}$ means that $\mathcal{S} \mathrel{\dot{\subset}} \mathcal{T}$ and $\mathcal{T} \mathrel{\dot{\subset}} \mathcal{S}$. Also, given a finite extension $K \subset L$, we set

$$\mathcal{S}_{L/K} = \{\mathfrak{p} \in \mathcal{P}_K : \mathfrak{p} \text{ splits completely in } L\}.$$

We can now state our result:

THEOREM 8.19. *Let L and M be Galois extensions of K. Then*

(i) $L \subset M \iff \mathcal{S}_{M/K} \mathrel{\dot{\subset}} \mathcal{S}_{L/K}$.
(ii) $L = M \iff \mathcal{S}_{M/K} \doteq \mathcal{S}_{L/K}$.

PROOF. Notice that (ii) is an immediate consequence of (i). As for (i), we will prove the following more general result which applies when only one of L or M is Galois over K. This will be useful in §§9 and 11.

PROPOSITION 8.20. *Let L and M be finite extensions of K.*

(i) *If M is Galois over K, then $L \subset M \iff \mathcal{S}_{M/K} \mathrel{\dot{\subset}} \mathcal{S}_{L/K}$.*
(ii) *If L is Galois over K, then $L \subset M \iff \tilde{\mathcal{S}}_{M/K} \mathrel{\dot{\subset}} \mathcal{S}_{L/K}$, where $\tilde{\mathcal{S}}_{M/K}$ is the set*

$$\tilde{\mathcal{S}}_{M/K} = \{\mathfrak{p} \in \mathcal{P}_K : \mathfrak{p} \text{ unramified in } M \text{ and } f_{\mathfrak{P}|\mathfrak{p}} = 1 \text{ for some prime } \mathfrak{P} \text{ of } M\}.$$

REMARK. If M is Galois over K, then $\tilde{\mathcal{S}}_{M/K} = \mathcal{S}_{M/K}$ (see Exercise 8.12), and thus either part of Proposition 8.20 implies Theorem 8.19.

PROOF. We start with the proof of (ii). When $L \subset M$, it is easy to see that $\tilde{\mathcal{S}}_{M/K} \mathrel{\dot{\subset}} \mathcal{S}_{L/K}$ (see Exercise 8.12). Conversely, assume that $\tilde{\mathcal{S}}_{M/K} \mathrel{\dot{\subset}} \mathcal{S}_{L/K}$, and let N be a Galois extension of K containing both L and M. By Galois theory, it suffices to show that $\mathrm{Gal}(N/M) \subset \mathrm{Gal}(N/L)$. Thus, given $\sigma \in \mathrm{Gal}(N/M)$, we need to prove that $\sigma|_L$ is the identity.

By the Čebotarev Density Theorem, there is a prime $\mathfrak{p}$ in K, unramified in N, such that $((N/K)/\mathfrak{p})$ is the conjugacy class of σ. Thus there is some prime $\mathfrak{P}$ of N containing $\mathfrak{p}$ such that $((N/K)/\mathfrak{P}) = \sigma$. We claim that $\mathfrak{p} \in \tilde{\mathcal{S}}_{M/K}$. To see why, let $\mathfrak{P}' = \mathfrak{P} \cap \mathcal{O}_M$. Then, for $\alpha \in \mathcal{O}_M$, we have

$$\alpha \equiv \sigma(\alpha) \equiv \alpha^{N(\mathfrak{p})} \bmod \mathfrak{P}'.$$

The first congruence follows from $\sigma|_M = 1$, and the second follows by the definition of the Artin symbol (see Lemma 5.19). Thus $\mathcal{O}_M/\mathfrak{P}' \simeq \mathcal{O}_K/\mathfrak{p}$, so that $f_{\mathfrak{P}'|\mathfrak{p}} = 1$. This shows that $\mathfrak{p} \in \tilde{\mathcal{S}}_{M/K}$ as claimed.

The Density Theorem implies that there are infinitely many such $\mathfrak{p}$'s. Thus our hypothesis $\tilde{\mathcal{S}}_{M/K} \mathrel{\dot{\subset}} \mathcal{S}_{L/K}$ allows us to assume that $\mathfrak{p} \in \mathcal{S}_{L/K}$, i.e., $((L/K)/\mathfrak{p}) = 1$. But [**96**, Exercise 4.11(b)] tells us that $((L/K)/\mathfrak{p}) = ((N/K)/\mathfrak{P})|_L$. Since $\sigma = ((N/K)/\mathfrak{P})$, we see that $\sigma|_L = 1$ as desired.

To prove (i), first note $L \subset M$ easily implies $\mathcal{S}_{M/K} \dot{\subset} \mathcal{S}_{L/K}$ (see Exercise 8.12). To prove the other direction, let L' be the Galois closure of L over K. It is a standard fact that a prime of K splits completely in L if and only if it splits completely in L' (see Exercises 8.13–8.15 or Marcus [**96**, Corollary to Theorem 31]). This implies that $\mathcal{S}_{L/K} = \mathcal{S}_{L'/K}$. Since M is Galois over K, we've already observed that $\tilde{\mathcal{S}}_{M/K} = \mathcal{S}_{M/K}$. Thus our hypothesis $\mathcal{S}_{M/K} \dot{\subset} \mathcal{S}_{L/K}$ can be written $\tilde{\mathcal{S}}_{M/K} \dot{\subset} \mathcal{S}_{L'/K}$, so that by part (ii) we obtain $L' \subset M$, which obviously implies $L \subset M$. This completes the proofs of Proposition 8.20 and Theorem 8.19. Q.E.D.

Theorem 8.19 is closely related to Corollary 8.7. The reason is that if $K \subset L$ is Abelian, then the set $\mathcal{S}_{L/K}$ of primes that split completely is, up to a finite set, exactly the prime ideals in $\ker(\Phi_{L/K,\mathfrak{m}})$, where $\mathfrak{m}$ is any modulus divisible by all of the ramified primes. Thus we don't need the whole kernel of the Artin map to determine the extension—just the primes in it will suffice! In particular, this shows that Theorem 8.19 is relevant to our question of which primes p are of the form $x^2 + ny^2$. To see why, consider the situation of Theorem 5.1. Here, K is an imaginary quadratic field of discriminant $d_K = -4n$ (which means that n satisfies (5.2)). Then, by Theorem 5.26,

$$p = x^2 + ny^2 \iff p \text{ splits completely in the Hilbert class field of } K$$

whenever p is an odd prime not dividing n. Thus Theorem 8.19 shows that the primes represented by $x^2 + ny^2$ characterize the Hilbert class field of $\mathbb{Q}(\sqrt{-n})$ uniquely. In §9 we will give a version of this result that holds for arbitrary n.

C. Norms and Ideles

Our discussion of class field theory has omitted several important topics. To give the reader a sense of what's been left out, we will say a few words about norms and ideles.

Given a finite extension $K \subset L$, there is the norm map $N_{L/K} : L^* \to K^*$, and $N_{L/K}$ can be extended to a map of ideals

$$N_{L/K} : I_L \longrightarrow I_K$$

(see Janusz [**78**, §1.8]). The importance of the norm map is that it gives a precise description of the kernel of the Artin map. Specifically, let L be an Abelian extension of K, and let $\mathfrak{m}$ be a modulus for which $P_{K,1}(\mathfrak{m}) \subset \ker(\Phi_{L/K,\mathfrak{m}})$. Then an important part of the Artin Reciprocity Theorem states that

$$\ker(\Phi_{L/K,\mathfrak{m}}) = N_{L/K}(I_L(\mathfrak{m}))P_{K,1}(\mathfrak{m}) \tag{8.21}$$

(see Janusz [**78**, Chapter V, Theorem 5.7]). Norms play an essential role in the proofs of the theorems of class field theory.

Class field theory can be presented without reference to ideles (as we have done above), but the idelic approach has some distinct advantages. Before we can see why, we need some definitions. Given a number field K, the *idele group* $\mathbf{I}_K$ is the restricted product

$$\mathbf{I}_K = \prod_{\mathfrak{p}}{}^{*} K_{\mathfrak{p}}^*,$$

where $\mathfrak{p}$ runs over all primes of K, finite and infinite, and $K_{\mathfrak{p}}$ is the completion of K at $\mathfrak{p}$. The symbol $\prod_{\mathfrak{p}}^*$ means that $\mathbf{I}_K$ consists of all tuples $(x_{\mathfrak{p}})$ such that $x_{\mathfrak{p}} \in \mathcal{O}_{K_{\mathfrak{p}}}^*$ for all but finitely many $\mathfrak{p}$. The idele group $\mathbf{I}_K$ is a locally compact topological group, and the multiplicative group K^* embeds naturally in $\mathbf{I}_K$ as a discrete subgroup (see Neukirch [**100**, §IV.2] for all of this). The quotient group

$$\mathbf{C}_K = \mathbf{I}_K/K^*$$

is called the *idele class group*.

We can now restate the theorems of class field theory using ideles. Given an Abelian extension L of K, there is an Artin map

$$\Phi_{L/K} : \mathbf{C}_K \longrightarrow \mathrm{Gal}(L/K) \tag{8.22}$$

which is continuous and surjective. This is the idele-theoretic analog of the Artin Reciprocity Theorem. Note that $\ker(\Phi_{L/K})$ is a closed subgroup of finite index in $\mathbf{C}_K$. There is also an idelic version of the Existence Theorem, which asserts that there is a one-to-one correspondence between the Abelian extensions of K and the closed subgroups of finite index in $\mathbf{C}_K$. The nice feature of this approach is that it always uses the same group $\mathbf{C}_K$, unlike our situation, where we had to vary the modulus $\mathfrak{m}$ in $I_K(\mathfrak{m})$ as we moved from one Abelian extension to the next.

Norms also play an important role in the idelic theory. Given an Abelian extension L of K, there is a norm map

$$N_{L/K} : \mathbf{C}_L \longrightarrow \mathbf{C}_K,$$

and then the idelic analog of (8.21) asserts that the kernel of the Artin map $\Phi_{L/K} : \mathbf{C}_K \to \mathrm{Gal}(L/K)$ is exactly $N_{L/K}(\mathbf{C}_L)$. Thus the subgroups of $\mathbf{C}_K$ of finite index are precisely the norm groups $N_{L/K}(\mathbf{C}_L)$.

Standard references for the idele theoretic formulation of class field theory are Neukirch [**100**] and Weil [**133**]. Neukirch also explains carefully the relation between the two approaches to class field theory. See also Gras [**57**]. We will say more about ideles in §15.

D. Exercises

8.1. Let K be an imaginary quadratic field, and let $\mathfrak{m}$ be a modulus for K (which can be regarded as an ideal of $\mathcal{O}_K$). We want to show that $P_{K,1}(\mathfrak{m})$ has finite index in $I_K(\mathfrak{m})$.

(a) Show that the map $\alpha \mapsto \alpha\mathcal{O}_K$ induces a well-defined homomorphism

$$\phi : (\mathcal{O}_K/\mathfrak{m})^* \longrightarrow I_K(\mathfrak{m}) \cap P_K/P_{K,1}(\mathfrak{m}),$$

and then show that there is an exact sequence

$$\mathcal{O}_K^* \longrightarrow (\mathcal{O}_K/\mathfrak{m})^* \xrightarrow{\phi} I_K(\mathfrak{m}) \cap P_K/P_{K,1}(\mathfrak{m}) \longrightarrow 1.$$

Conclude that $I_K(\mathfrak{m}) \cap P_K/P_{K,1}(\mathfrak{m})$ is finite. Hint: look at the proof of Theorem 7.24.

(b) Adapt the exact sequence (7.25) to show that $I_K(\mathfrak{m})/P_{K,1}(\mathfrak{m})$ is finite (recall that $C(\mathcal{O}_K)$ is finite by §2 and Theorem 7.7).

8.2. This problem is concerned with the Artin map of the cyclotomic extension $\mathbb{Q} \subset \mathbb{Q}(\zeta_m)$, where $\zeta_m = e^{2\pi i/m}$. We will assume that $m > 2$.

(a) Use Proposition 5.11 to prove that all finite ramified primes of this extension divide m. Thus the Artin map $\Phi_{m\infty}$ is defined.

(b) Show that $\Phi_{m\infty} : I_{\mathbb{Q}}(m\infty) \to \mathrm{Gal}(\mathbb{Q}(\zeta_m)/\mathbb{Q}) \simeq (\mathbb{Z}/m\mathbb{Z})^*$ is as described in (8.3). Hint: use Lemma 5.19.

(c) Conclude that $\ker(\Phi_{m\infty}) = P_{\mathbb{Q},1}(m\infty)$.

8.3. Let $\mathbb{Q} \subset \mathbb{Q}(\zeta_m)$ be as in the previous exercise, and assume that $m > 2$.

(a) Show that $\mathbb{R} \cap \mathbb{Q}(\zeta_m) = \mathbb{Q}(\cos(2\pi/m))$, and then use this to conclude that $[\mathbb{Q}(\cos(2\pi/m)) : \mathbb{Q}] = (1/2)\phi(m)$.

(b) Compute the Artin map

$$\Phi_m : I_{\mathbb{Q}}(m) \longrightarrow \mathrm{Gal}(\mathbb{Q}(\cos(2\pi/m))/\mathbb{Q}) \simeq (\mathbb{Z}/m\mathbb{Z})^*/\{\pm 1\}.$$

Hint: use the previous exercise.

(c) Show that $\ker(\Phi_m) = P_{\mathbb{Q},1}(m)$.

8.4. Let $K \subset L$ be an Abelian extension, and let $\mathfrak{m}$ be a modulus for which the Artin map $\Phi_{\mathfrak{m}}$ is defined. If $\mathfrak{n}$ is another modulus and $\mathfrak{m} \mid \mathfrak{n}$, prove that

$$P_{K,1}(\mathfrak{m}) \subset \ker(\Phi_{\mathfrak{m}}) \Longrightarrow P_{K,1}(\mathfrak{n}) \subset \ker(\Phi_{\mathfrak{n}}).$$

Use this to show that if $\mathrm{Gal}(L/K)$ is a generalized ideal class group for $\mathfrak{m}$, then it is also a generalized ideal class group for $\mathfrak{n}$. Hint: use part (i) of Theorem 8.2.

8.5. Prove that the conductor of the cyclotomic extension $\mathbb{Q} \subset \mathbb{Q}(\zeta_m)$ is given by

$$\mathfrak{f}(\mathbb{Q}(\zeta_m)/\mathbb{Q}) = \begin{cases} 1 & m \leq 2 \\ (m/2)\infty & m = 2n,\ n > 1 \text{ odd} \\ m\infty & \text{otherwise.} \end{cases}$$

Hint: when $m > 2$, use Theorem 8.5 and Exercise 8.2 to show that the conductor is of the form $n\infty$ for some n dividing m. Then use Corollary 8.7 to show that $\mathbb{Q}(\zeta_m) \subset \mathbb{Q}(\zeta_n)$, which implies that $\phi(m) \mid \phi(n)$. The formula for $\mathfrak{f}(\mathbb{Q}(\zeta_m)/\mathbb{Q})$ now follows from elementary arguments about the Euler ϕ-function.

8.6. This exercise is concerned with conductors.

(a) Given a modulus $\mathfrak{m}$ for a number field K, let $K_{\mathfrak{m}}$ denote the ray class field defined in the text. If L is an Abelian extension of K, then show that the conductor $\mathfrak{f}(L/K)$ is the greatest common divisor of all moduli $\mathfrak{m}$ for which $L \subset K_{\mathfrak{m}}$.

(b) If L is an Abelian extension of $\mathbb{Q}$, let m be the smallest positive integer for which $L \subset \mathbb{Q}(\zeta_m)$ (note that m exists by the Kronecker–Weber Theorem). Then show that

$$\mathfrak{f}(L/\mathbb{Q}) = \begin{cases} m & \text{if } L \subset \mathbb{R} \\ m\infty & \text{otherwise.} \end{cases}$$

8.7. In this exercise we will fill in some of the details omitted in the proof of quadratic reciprocity given in Theorem 8.12. Let p be an odd prime.

(a) If $K \subset L$ is an Abelian extension such that $\mathrm{Gal}(L/K)$ is a generalized ideal class group for the modulus $\mathfrak{m}$ of K, then prove that the same is true for any intermediate field $K \subset M \subset L$.

(b) If K is a quadratic field which ramifies only at p, then use Corollary 5.17 to show that $K = \mathbb{Q}(\sqrt{p^*})$, $p^* = (-1)^{(p-1)/2}p$.

(c) Verify the isomorphisms

$$(\mathbb{Z}/p\mathbb{Z})^* \xrightarrow{\sim} (\mathbb{Z}/2p\mathbb{Z})^* \xrightarrow{\sim} I_{\mathbb{Q}}(2p\infty)/P_{\mathbb{Q},1}(2p\infty)$$

described in the proof of Theorem 8.12.

8.8. This exercise will adapt the proof of Theorem 8.12 to prove the supplementary formula $(2/p) = (-1)^{(p^2-1)/8}$ from quadratic reciprocity.

(a) Use Exercise 8.2 to construct isomorphisms

$$I_{\mathbb{Q}}(8\infty)/P_{\mathbb{Q},1}(8\infty) \simeq \mathrm{Gal}(\mathbb{Q}(\zeta_8)/\mathbb{Q}) \simeq (\mathbb{Z}/8\mathbb{Z})^*,$$

and conclude that $I_{\mathbb{Q}}(8\infty)/P_{\mathbb{Q},1}(8\infty) = \{[\mathbb{Z}],[3\mathbb{Z}],[5\mathbb{Z}],[7\mathbb{Z}]\}$.

(b) Let $H = \{\mathbb{Z}, 7\mathbb{Z}\}P_{\mathbb{Q},1}(8\infty)$. Show that via the Existence Theorem, H corresponds to $\mathbb{Q}(\sqrt{2})$. Hint: using the arguments of Theorem 8.12 and part (b) of Exercise 8.7, show that H corresponds to one of $\mathbb{Q}(i)$, $\mathbb{Q}(\sqrt{2})$ or $\mathbb{Q}(\sqrt{-2})$. Then observe that $[7\mathbb{Z}] \in I_Q(8\infty)/P_{\mathbb{Q},1}(8\infty)$ maps to $-1 \in (\mathbb{Z}/8\mathbb{Z})^*$. What element of $\mathrm{Gal}(\mathbb{Q}(\zeta_8)/\mathbb{Q})$ does this correspond to?

(c) Use Weak Reciprocity to show that $(2/\cdot)$ induces a well-defined homomorphism on $(\mathbb{Z}/8\mathbb{Z})^*$ whose kernel is $\{\pm 1\}$.

(d) Show that $(2/p) = (-1)^{(p^2-1)/8}$.

8.9. In this exercise we will use Strong Reciprocity and the properties of the Hilbert symbol to prove cubic reciprocity. We will assume that the reader is familiar with p-adic fields (see Gouvêa [**56**]). To list the properties of the Hilbert symbol, let K be a number field containing a primitive nth of unity, and let $\mathfrak{p}$ be a prime of K. The completion of K at $\mathfrak{p}$ will be denoted $K_{\mathfrak{p}}$. Then the Hilbert symbol $(\alpha, \beta/\mathfrak{p})_n$ is defined for $\alpha, \beta \in K_{\mathfrak{p}}^*$ and gives a map

$$\left(\frac{\cdot,\cdot}{\mathfrak{p}}\right)_n : K_{\mathfrak{p}}^* \times K_{\mathfrak{p}}^* \longrightarrow \mu_n,$$

where μ_n is the group of nth roots of unity. The Hilbert symbol has the following properties:

(i) $(\alpha\alpha', \beta/\mathfrak{p})_n = (\alpha, \beta/\mathfrak{p})_n(\alpha', \beta/\mathfrak{p})_n$.

(ii) $(\alpha', \beta\beta'/\mathfrak{p})_n = (\alpha, \beta/\mathfrak{p})_n(\alpha, \beta'/\mathfrak{p})_n$.

(iii) $(\alpha, \beta/\mathfrak{p})_n = (\beta, \alpha/\mathfrak{p})_n^{-1}$.

(iv) $(\alpha, -\alpha/\mathfrak{p})_n = 1$.

(v) $(\alpha, 1-\alpha/\mathfrak{p})_n = 1$.

For proofs of these properties of the Hilbert symbol, see Neukirch [**100**, §III.5].

Now let's specialize to the case $n = 3$ and $K = \mathbb{Q}(\omega)$, $\omega = e^{2\pi i/3}$. As we saw in (8.15), Strong Reciprocity shows that cubic reciprocity is equivalent to the assertion

$$\pi, \theta \text{ primary in } \mathcal{O}_K \Longrightarrow \left(\frac{\pi,\theta}{\lambda}\right)_3 = 1$$

where $\lambda = 1 - \omega$. Recall that π primary means that $\pi \equiv \pm 1 \bmod 3\mathcal{O}_K$. In §4 we saw that replacing π by $-\pi$ doesn't affect the statement of cubic reciprocity, so that we can assume that $\pi \equiv \theta \equiv 1 \bmod \lambda^2\mathcal{O}_K$ (note that λ^2 and 3 differ by a unit in $\mathcal{O}_K$). Let K_λ be the completion of K at λ, and let $\mathcal{O}_\lambda$ be the valuation ring of K_λ. We will use the properties of the cubic Hilbert symbol to show that

$$\alpha, \beta \equiv 1 \bmod \lambda^2\mathcal{O}_\lambda \Longrightarrow \left(\frac{\alpha,\beta}{\lambda}\right)_3 = 1,$$

and then cubic reciprocity will be proved.

(a) If $\alpha \equiv 1 \bmod \lambda^4\mathcal{O}_\lambda$, then prove that $\alpha = u^3$ for some $u \in \mathcal{O}_\lambda$. Hint: if $\alpha = u_n^3 \bmod \lambda^n\mathcal{O}_\lambda$ for $n \geq 4$, then show that $a \in \mathcal{O}_\lambda$ can be chosen so that $\alpha \equiv (u_n + a\lambda^{n-2})^3 \bmod \lambda^{n+1}\mathcal{O}_\lambda$.

(b) If $\alpha \in \mathcal{O}_\lambda^*$ and $\alpha \equiv \alpha' \bmod \lambda^4\mathcal{O}_\lambda$, then prove that

$$\left(\frac{\alpha, \beta}{\lambda}\right)_3 = \left(\frac{\alpha', \beta}{\lambda}\right)_3$$

for any $\beta \in K_\lambda^*$. Hint: use part (a) and property (i) above. Remember that $(\alpha, \beta/\lambda)_3$ is a cube root of unity.

(c) Now assume that $\alpha \equiv \beta \equiv 1 \bmod \lambda^2\mathcal{O}_\lambda$, and write $\alpha = 1 + a\lambda^2, a \in \mathcal{O}_\lambda$. Then first, apply property (v) to $1 + a\beta\lambda^2$, and second, apply part (b) to $1 + a\beta\lambda^2 \equiv 1 + a\lambda^2 \bmod \lambda^4\mathcal{O}_\lambda$. This proves that

$$1 = \left(\frac{1 + a\lambda^2, -a\beta\lambda^2}{\lambda}\right)_3.$$

From here, properties (ii) and (v) easily imply that $(\alpha, \beta/\lambda)_3 = 1$, which completes the proof of cubic reciprocity.

8.10. In this exercise we will study the properties of the Dirichlet density.

(a) Assuming that $\zeta_K(s) = \Sigma_{\mathfrak{a} \subset \mathcal{O}_K} N(\mathfrak{a})^{-s}$ converges absolutely for all s with $\mathrm{Re}(s) > 1$, show that for $\mathcal{S} \subset \mathcal{P}_K$, the sum $\Sigma_{\mathfrak{p} \in \mathcal{S}} N(\mathfrak{p})^{-s}$ also converges absolutely for $\mathrm{Re}(s) > 1$.

(b) Use part (a) to prove that properties (ii)–(vi) of the Dirichlet density follow from (i) and the definition.

8.11. Apply the Čebotarev Density Theorem to the cyclotomic extension $\mathbb{Q} \subset \mathbb{Q}(\zeta_m)$ to show that the primes in a fixed congruence class in $(\mathbb{Z}/m\mathbb{Z})^*$ have Dirichlet density $1/\phi(m)$. This proves Dirichlet's theorem that there are infinitely many primes in an arithmetic progression where the first term and common difference are relatively prime.

8.12. Let M be a finite extension of a number field K, and let $\tilde{\mathcal{S}}_{M/K}$ be the set of primes of $\mathcal{O}_K$ defined in Proposition 8.20.

(a) If M is Galois over K, then show that $\tilde{\mathcal{S}}_{M/K}$ equals the set $\mathcal{S}_{M/K}$ of Theorem 8.19.

(b) If L is a Galois extension of K and $L \subset M$, then show that $\tilde{\mathcal{S}}_{M/K} \subset \mathcal{S}_{L/K}$.

(c) If $L \subset M$ are finite extensions of K, then show that $\mathcal{S}_{M/K} \subset \mathcal{S}_{L/K}$.

8.13. Let $K \subset N$ be a Galois extension, and let $\mathfrak{P}$ be a prime of $\mathcal{O}_N$. Set $\mathfrak{p} = \mathfrak{P} \cap \mathcal{O}_K$, $e = e_{\mathfrak{P}|\mathfrak{p}}$ and $f = f_{\mathfrak{P}|\mathfrak{p}}$. If $D_\mathfrak{P} \subset \mathrm{Gal}(N/K)$ is the decomposition group of $\mathfrak{P}$, we will denote the fixed field of $D_\mathfrak{P}$ by $N_\mathfrak{P}$. From Proposition 5.10, we know that $|D_\mathfrak{P}| = ef$, and Galois theory tells us that $[N : N_\mathfrak{P}] = |D_\mathfrak{P}|$. Let $\mathfrak{P}' = \mathfrak{P} \cap \mathcal{O}_{N_\mathfrak{P}}$.

(a) Prove that $e_{\mathfrak{P}'|\mathfrak{p}} = f_{\mathfrak{P}'|\mathfrak{p}} = 1$. Hint: by Proposition 5.10, the map $D_\mathfrak{P} \to \widetilde{G}$ is surjective, where $\widetilde{G}$ is the Galois group of $\mathcal{O}_K/\mathfrak{p} \subset \mathcal{O}_N/\mathfrak{P}$. Use $\mathcal{O}_K/\mathfrak{p} \subset \mathcal{O}_{N_\mathfrak{P}}/\mathfrak{P}' \subset \mathcal{O}_N/\mathfrak{P}$, and remember that the e's and f's are multiplicative (see Exercise 5.15).

(b) Given an intermediate field $K \subset M \subset N$, let $\mathfrak{P}_M = \mathfrak{P} \cap \mathcal{O}_M$. Prove that

$$e_{\mathfrak{P}_M|\mathfrak{p}} = f_{\mathfrak{P}_M|\mathfrak{p}} = 1 \iff M \subset N_\mathfrak{P}.$$

Hint: if $M \subset N_\mathfrak{P}$, then apply part (a). Conversely, show that the compositum $N_\mathfrak{P}M$ is the fixed field for the decomposition group of $\mathfrak{P}$ in $\mathrm{Gal}(N/M)$. By applying the result of part (a) to $M \subset N_\mathfrak{P}M$ and computing degrees, one sees that $N_\mathfrak{P}M = N_\mathfrak{P}$, which implies $M \subset N_\mathfrak{P}$.

8.14. Let L and M be finite extensions of a number field K, and let $\mathfrak{p}$ be a prime of K that splits completely in L and M. Then prove that $\mathfrak{p}$ splits completely in LM. Hint: let N be a Galois extension of K containing both L and M, and let $\mathfrak{P}$ be a prime of N containing $\mathfrak{p}$. From Exercise 8.13 we get the intermediate field $K \subset N_{\mathfrak{P}} \subset N$. Then use part (b) of that exercise to show that L and M lie in $N_{\mathfrak{P}}$, which implies $LM \subset N_{\mathfrak{P}}$.

8.15. Let L be a finite extension of a number field K, and let L' be the Galois closure of L over K. The goal of this exercise is to prove that $\mathcal{S}_{L/K} = \mathcal{S}_{L'/K}$. By part (c) of Exercise 8.12, we have $\mathcal{S}_{L'/K} \subset \mathcal{S}_{L/K}$, so that it suffices to show that a prime of K that splits completely in L also splits completely in L'.

(a) Let $\sigma : L \to \mathbb{C}$ be an embedding which is the identity on K, and let $\mathfrak{p}$ be an ideal of K which splits completely in L. Then prove that $\mathfrak{p}$ splits completely in $\sigma(L)$.

(b) Since L' is the compositum of the $\sigma(L)$'s, use the previous exercise to show that $\mathfrak{p}$ splits completely in L'.

8.16. Let $K \subset M$ be a finite extension of number fields. Then prove that $K \subset M$ is a Galois extension if and only if $\tilde{\mathcal{S}}_{M/K} = \mathcal{S}_{M/K}$. Hint: one implication is covered in part (a) of Exercise 8.12, and the other implication is an easy consequence of Proposition 8.20 and Exercise 8.15.

8.17. The definition of $P_{K,1}(\mathfrak{m})$ differs from the standard definition, which uses valuations and multiplicative congruences. See, for example, [**78**, Chapter IV]. One can show without difficulty that the equivalence of the two definitions reduces to the following claim: $P_{K,1}(\mathfrak{m})$ contains all principal fractional ideals of the form $(a/b)\mathcal{O}_K$ where $a, b \in \mathcal{O}_K$ are relatively prime to $\mathfrak{m}_0$, $a \equiv b \bmod \mathfrak{m}_0$, and $\sigma(a/b) > 0$ for all real infinite primes dividing $\mathfrak{m}$.

(a) Let $\sigma_1, \ldots, \sigma_r$ be the real infinite primes dividing $\mathfrak{m}$, and for each i, pick $\epsilon_i \in \{\pm 1\}$. Prove that there is $\lambda \in \mathcal{O}_K$ such that $\lambda \equiv 1 \bmod \mathfrak{m}_0$, and $\sigma_i(\epsilon_i \lambda) > 0$ for all i. Hint: by the Approximation Theorem (Theorem 1.1 of [**78**, Chapter IV]), there is $\alpha \in K^*$ such that $\sigma_i(\epsilon_i \alpha) > 0$ for all i. Argue that α can be chosen to lie in $\mathcal{O}_K$. Then let $\lambda = 1 + d\alpha\beta^2$ where d is a sufficiently large positive integer and β is a nonzero element of $\mathfrak{m}_0$.

(b) Prove the claim made at the beginning of the exercise. Hint: multiply a, b by a suitable $c \in \mathcal{O}_K$ to ensure $a, b \equiv 1 \bmod \mathfrak{m}_0$, and then multiply both by λ from part (a) for a suitable choice of ϵ_i to make $\sigma_i(a), \sigma_i(b) > 0$ for all i.

9. Ring Class Fields and $p = x^2 + ny^2$

Theorem 5.1 used the Hilbert class field to characterize $p = x^2 + ny^2$ when n is a positive, squarefree and $n \not\equiv 3 \bmod 4$. In §4, we also proved that for an odd prime p,

$$p = x^2 + 27y^2 \iff \begin{cases} p \equiv 1 \bmod 3 \text{ and } x^3 \equiv 2 \bmod p \\ \text{has an integer solution} \end{cases}$$

$$p = x^2 + 64y^2 \iff \begin{cases} p \equiv 1 \bmod 4 \text{ and } x^4 \equiv 2 \bmod p \\ \text{has an integer solution.} \end{cases}$$

These earlier results follow the format of Theorem 5.1 (note that both exponents are class numbers), yet *neither* is a corollary of the theorem, for 27 and 64 are not squarefree. In §9 we will use the theory developed in §§7 and 8 to overcome this limitation. Specifically, given an order $\mathcal{O}$ in an imaginary quadratic field K, we will construct a generalization of the Hilbert class field called the *ring class field* of $\mathcal{O}$. Then, using the ring class field of the order $\mathbb{Z}[\sqrt{-n}]$, where $n > 0$ is now arbitrary, we will prove a version of Theorem 5.1 that holds for all n (see Theorem 9.2 below). This, of course, is the main theorem of the whole book. The basic idea is that the criterion for $p = x^2 + ny^2$ is determined by a primitive element of the ring class field of $\mathbb{Z}[\sqrt{-n}]$. To see how this works in practice, we will describe the ring class fields of $\mathbb{Z}[\sqrt{-27}]$ and $\mathbb{Z}[\sqrt{-64}]$, and then Theorem 9.2 will give us class field theory proofs of Euler's conjectures for $p = x^2 + 27y^2$ or $x^2 + 64y^2$. To complete the circle of ideas, we will then explain how class field theory implies those portions of cubic and biquadratic reciprocity used in §4 in our earlier discussion of $x^2 + 27y^2$ and $x^2 + 64y^2$.

The remainder of the section will explore two other aspects of ring class fields. We will first use the Čebotarev Density Theorem to prove that a primitive positive definite quadratic form represents infinitely many prime numbers. Then, in a different direction, we will give a purely field-theoretic characterization of ring class fields and their subfields.

A. Solution of $p = x^2 + ny^2$ for All n

Before introducing ring class fields, we need some notation. If K is a number field, an ideal $\mathfrak{m}$ of $\mathcal{O}_K$ can be regarded as a modulus, and in §8 we defined the ideal groups $I_K(\mathfrak{m})$ and $P_{K,1}(\mathfrak{m})$. In this section, $\mathfrak{m}$ will usually be a principal ideal $\alpha\mathcal{O}_K$, and the above groups will be written $I_K(\alpha)$ and $P_{K,1}(\alpha)$.

To define a ring class field, let $\mathcal{O}$ be an order of conductor f in an imaginary quadratic field K. We know from Proposition 7.22 that the ideal class group $C(\mathcal{O})$ can be written

$$(9.1) \qquad C(\mathcal{O}) \simeq I_K(f)/P_{K,\mathbb{Z}}(f)$$

(recall that $P_{K,\mathbb{Z}}(f)$ is generated by the principal ideals $\alpha\mathcal{O}_K$ where $\alpha \equiv a \bmod f\mathcal{O}_K$ for some integer a with $\gcd(a, f) = 1$). Furthermore, in §8 we saw that

$$P_{K,1}(f) \subset P_{K,\mathbb{Z}}(f) \subset I_K(f),$$

so that $C(\mathcal{O})$ is a generalized ideal class group of K for the modulus $f\mathcal{O}_K$ (see (8.1)). By the Existence Theorem (Theorem 8.6), this data determines a unique Abelian extension L of K, which is called the *ring class field* of the order $\mathcal{O}$. The basic properties of the ring class field L are, first, all primes of K ramified in L must divide $f\mathcal{O}_K$, and second, the Artin map and (9.1) give us isomorphisms

$$C(\mathcal{O}) \simeq I_K(f)/P_{K,\mathbb{Z}}(f) \simeq \mathrm{Gal}(L/K).$$

In particular the degree of L over K is the class number, i.e., $[L : K] = h(\mathcal{O})$. For an example of a ring class field, note that the ring class field of the maximal order $\mathcal{O}_K$ is the Hilbert class field of K (see Exercise 9.1). Later in this section we will give other examples of ring class fields.

We can now state the main theorem of the book:

THEOREM 9.2. *Let $n > 0$ be an integer. Then there is a monic irreducible polynomial $f_n(x) \in \mathbb{Z}[x]$ of degree $h(-4n)$ such that if an odd prime p divides* *neither n nor the discriminant of $f_n(x)$, then*

$$p = x^2 + ny^2 \iff \begin{cases} (-n/p) = 1 \text{ and } f_n(x) \equiv 0 \bmod p \\ \text{has an integer solution.} \end{cases}$$

Furthermore, $f_n(x)$ may be taken to be the minimal polynomial of a real algebraic integer α for which $L = K(\alpha)$ is the ring class field of the order $\mathbb{Z}[\sqrt{-n}]$ in the imaginary quadratic field $K = \mathbb{Q}(\sqrt{-n})$.

Finally, if $f_n(x)$ is any monic integer polynomial of degree $h(-4n)$ for which the above equivalence holds, then $f_n(x)$ is irreducible over $\mathbb{Z}$ and is the minimal polynomial of a primitive element of the ring class field L described above.

REMARK. This theorem generalizes Theorem 5.1, and the last part of the theorem shows that knowing $f_n(x)$ is *equivalent* to knowing the ring class field of $\mathbb{Z}[\sqrt{-n}]$.

PROOF. Before proceeding with the proof, we will first prove the following general fact about ring class fields:

LEMMA 9.3. *Let L be the ring class field of an order $\mathcal{O}$ in an imaginary quadratic field K. Then L is a Galois extension of $\mathbb{Q}$, and its Galois group can be written as a semidirect product*

$$\mathrm{Gal}(L/\mathbb{Q}) \simeq \mathrm{Gal}(L/K) \rtimes (\mathbb{Z}/2\mathbb{Z})$$

where the nontrivial element of $\mathbb{Z}/2\mathbb{Z}$ acts on $\mathrm{Gal}(L/K)$ by sending σ to σ^{-1}.

PROOF. In the case of the Hilbert class field, this lemma was proved in §6 (see the discussion following (6.3)). To do the general case, we first need to show that $\tau(L) = L$, where τ denotes complex conjugation. Let $\mathfrak{m}$ denote the modulus $f\mathcal{O}_K$, and note that $\tau(\mathfrak{m}) = \mathfrak{m}$. Since $\ker(\Phi_{L/K,\mathfrak{m}}) = P_{K,\mathbb{Z}}(f)$, an easy computation shows that

$$\ker(\Phi_{\tau(L)/K,\mathfrak{m}}) = \tau(\ker(\Phi_{L/K,\mathfrak{m}})) = \tau(P_{K,\mathbb{Z}}(f)) = P_{K,\mathbb{Z}}(f)$$

(see Exercise 9.2), and thus $\ker(\Phi_{\tau(L)/K,\mathfrak{m}}) = \ker(\Phi_{L/K,\mathfrak{m}})$. Then $\tau(L) = L$ follows from Corollary 8.7.

As we noticed in the proof of Lemma 5.28, this implies that L is Galois over $\mathbb{Q}$, so that we have an exact sequence

$$1 \longrightarrow \mathrm{Gal}(L/K) \longrightarrow \mathrm{Gal}(L/\mathbb{Q}) \longrightarrow \mathrm{Gal}(K/\mathbb{Q})(\simeq \mathbb{Z}/2\mathbb{Z}) \longrightarrow 1.$$

Since $\tau \in \mathrm{Gal}(L/\mathbb{Q})$, $\mathrm{Gal}(L/\mathbb{Q})$ is the semidirect product $\mathrm{Gal}(L/K) \rtimes (\mathbb{Z}/2\mathbb{Z})$, where the nontrivial element of $\mathbb{Z}/2\mathbb{Z}$ acts by conjugation by τ. However, for a prime $\mathfrak{p}$ of K, Lemma 5.19 implies that

$$\tau \left(\frac{L/K}{\mathfrak{p}}\right) \tau^{-1} = \left(\frac{L/K}{\tau(\mathfrak{p})}\right) = \left(\frac{L/K}{\overline{\mathfrak{p}}}\right)$$

(see Exercise 6.3). Thus, under the isomorphism $I_K(f)/P_{K,\mathbb{Z}}(f) \simeq \mathrm{Gal}(L/K)$, conjugation by τ in $\mathrm{Gal}(L/K)$ corresponds to the usual action of τ on $I_K(f)$. But if $\mathfrak{a}$ is any ideal in $I_K(f)$, then $\mathfrak{a}\overline{\mathfrak{a}} = N(\mathfrak{a})\mathcal{O}_K$ lies in $P_{K,\mathbb{Z}}(f)$ since $N(\mathfrak{a})$ is prime to f. Thus $\overline{\mathfrak{a}}$ gives the *inverse* of $\mathfrak{a}$ in the quotient $I_K(f)/P_{K,\mathbb{Z}}(f)$, and the lemma is proved. Q.E.D.

We can now proceed with the proof of Theorem 9.2. Let L be the ring class field of $\mathbb{Z}[\sqrt{-n}]$. We start by relating $p = x^2 + ny^2$ to the behavior of p in L:

THEOREM 9.4. *Let $n > 0$ be an integer, and L be the ring class field of the order $\mathbb{Z}[\sqrt{-n}]$ in the imaginary quadratic field $K = \mathbb{Q}(\sqrt{-n})$. If p is an odd prime not dividing n, then*

$$p = x^2 + ny^2 \iff p \text{ splits completely in } L.$$

PROOF. Let $\mathcal{O} = \mathbb{Z}[\sqrt{-n}]$. The discriminant of $\mathcal{O}$ is $-4n$, and then $-4n = f^2 d_K$ by (7.3), where f is the conductor of $\mathcal{O}$. Let p be an odd prime not dividing n. Then $p \nmid f^2 d_K$, which implies that p is unramified in K. We will prove the following equivalences:

$$\begin{aligned} p = x^2 + ny^2 &\iff p\mathcal{O}_K = \mathfrak{p}\overline{\mathfrak{p}},\ \mathfrak{p} \neq \overline{\mathfrak{p}},\ \text{and } \mathfrak{p} = \alpha\mathcal{O}_K,\ \alpha \in \mathcal{O} \\ &\iff p\mathcal{O}_K = \mathfrak{p}\overline{\mathfrak{p}},\ \mathfrak{p} \neq \overline{\mathfrak{p}},\ \text{and } \mathfrak{p} \in P_{K,\mathbb{Z}}(f) \\ &\iff p\mathcal{O}_K = \mathfrak{p}\overline{\mathfrak{p}},\ \mathfrak{p} \neq \overline{\mathfrak{p}},\ \text{and } ((L/K)/\mathfrak{p}) = 1 \\ &\iff p\mathcal{O}_K = \mathfrak{p}\overline{\mathfrak{p}},\ \mathfrak{p} \neq \overline{\mathfrak{p}},\ \text{and } \mathfrak{p} \text{ splits completely in } L \\ &\iff p \text{ splits completely in } L, \end{aligned}$$

and Theorem 9.4 will follow.

For the first equivalence, suppose that $p = x^2 + ny^2 = (x + \sqrt{-n}y)(x - \sqrt{-n}y)$. If we set $\mathfrak{p} = (x + \sqrt{-n}y)\mathcal{O}_K$, then $p\mathcal{O}_K = \mathfrak{p}\overline{\mathfrak{p}}$ is the prime factorization of $p\mathcal{O}_K$ in $\mathcal{O}_K$. Note that $x + \sqrt{-n}y \in \mathcal{O}$, and $\mathfrak{p} \neq \overline{\mathfrak{p}}$ since p is unramified in K. Conversely, if $p\mathcal{O}_K = \mathfrak{p}\overline{\mathfrak{p}}$, where $\mathfrak{p} = (x + \sqrt{-n}y)\mathcal{O}_K$, then it follows easily that $p = x^2 + ny^2$.

Since $p \nmid f$, the second equivalence follows from Proposition 7.22. The next two equivalences are equally easy: the isomorphism $I_K(f)/P_{K,\mathbb{Z}}(f) \simeq \mathrm{Gal}(L/K)$ given by the Artin map shows that $\mathfrak{p} \in P_{K,\mathbb{Z}}(f)$ if and only if $((L/K)/\mathfrak{p}) = 1$, and then Corollary 5.21 shows that $((L/K)/\mathfrak{p}) = 1$ if and only if $\mathfrak{p}$ splits completely in L. Finally, recall from Lemma 9.3 that L is Galois over $\mathbb{Q}$. Thus, the proof of the last equivalence is identical to the proof of the last equivalence of (5.27). This completes the proof of Theorem 9.4. Q.E.D.

The next step is to prove the main equivalence of Theorem 9.2. By Lemma 9.3, the ring class field L is Galois over $\mathbb{Q}$, and thus Proposition 5.29 enables us to find a real algebraic integer α such that $L = K(\alpha)$. Let $f_n(x) \in \mathbb{Z}[x]$ be the minimal polynomial of α over K. Since $\mathcal{O}$ has discriminant $-4n$, the degree of $f_n(x)$ is $[L : K] = h(\mathcal{O}) = h(-4n)$. Then, combining Theorem 9.4 with the last part of Proposition 5.29, we have

$$\begin{aligned} p = x^2 + ny^2 &\iff p \text{ splits completely in } L \\ &\iff \begin{cases} (-n/p) = 1 \ \textit{textand}\ f_n(x) \equiv 0 \bmod p \\ \text{has an integer solution,} \end{cases} \end{aligned}$$

whenever p is an odd prime dividing neither n nor the discriminant of $f_n(x)$. This proves the main equivalence of Theorem 9.2.

The final part of the theorem is concerned with the "uniqueness" of $f_n(x)$. Of course, there are infinitely many real algebraic integers which are primitive elements of the extension $K \subset L$, and correspondingly there are infinitely many $f_n(x)$'s. So the best we could hope for in the way of uniqueness is that these are *all* of the possible $f_n(x)$'s. This is almost what the last part of the statement of Theorem 9.2 asserts—the $f_n(x)$'s that can occur are exactly the monic integer polynomials which are minimal polynomials of primitive elements (not necessarily real) of L over K.

To prove this assertion, let $f_n(x)$ be a monic integer polynomial of degree $h(-4n)$ which satisfies the equivalence of Theorem 9.2. Then let $g(x) \in K[x]$ be an irreducible factor of $f_n(x)$ over K, and let $M = K(\alpha)$ be the field generated by a root of $g(x)$. Note that α is an algebraic integer. If we can show that $L \subset M$, then

$$h(-4n) = [L : K] \leq [M : K] = \deg(g(x)) \leq \deg(f_n(x)) = h(-4n),$$

which will prove that $L = M = K(\alpha)$ and that $f_n(x)$ is the minimal polynomial of α over K (and hence over $\mathbb{Q}$).

It remains to prove $L \subset M$. Since L is Galois over $\mathbb{Q}$ by Lemma 9.3, Proposition 8.20 tells us that $L \subset M$ if and only if $\tilde{\mathcal{S}}_{M/\mathbb{Q}} \dot{\subset} \mathcal{S}_{L/\mathbb{Q}}$, where:

$$\mathcal{S}_{L/\mathbb{Q}} = \{p \text{ prime} : p \text{ splits completely in } L\}$$
$$\tilde{\mathcal{S}}_{M/\mathbb{Q}} = \{p \text{ prime} : p \text{ is unramified in } L \text{ and } f_{\mathfrak{P}|p} = 1 \text{ for some prime } \mathfrak{P} \text{ of } M\}.$$

Let's first study $\mathcal{S}_{L/\mathbb{Q}}$. By Theorem 9.4, this is the set of primes p represented by $x^2 + ny^2$. Since $f_n(x)$ satisfies the equivalence of Theorem 9.2, it follows that $\mathcal{S}_{L/\mathbb{Q}}$ is (with finitely many exceptions) the set of primes p which split completely in K and for which $f_n(x) \equiv 0 \bmod p$ has a solution.

To prove $\tilde{\mathcal{S}}_{M/\mathbb{Q}} \dot{\subset} \mathcal{S}_{L/\mathbb{Q}}$, suppose that $p \in \tilde{\mathcal{S}}_{M/\mathbb{Q}}$. Then $f_{\mathfrak{P}|p} = 1$ for some prime $\mathfrak{P}$ of M, and if we set $\mathfrak{p} = \mathfrak{P} \cap \mathcal{O}_K$, then $1 = f_{\mathfrak{P}|p} = f_{\mathfrak{P}|\mathfrak{p}} f_{\mathfrak{p}|p}$. Thus $f_{\mathfrak{p}|p} = 1$, which implies that p splits completely in K (since it's unramified). Note also that $f_n(x) \equiv 0 \bmod \mathfrak{P}$ has a solution in $\mathcal{O}_M$ since $\alpha \in \mathcal{O}_M$ and $g(\alpha) = f_n(\alpha) = 0$. But $f_{\mathfrak{P}|p} = 1$ implies that $\mathbb{Z}/p\mathbb{Z} \simeq \mathcal{O}_M/\mathfrak{P}$, and hence $f_n(x) \equiv 0 \bmod p$ has an integer solution. By the above description of $\mathcal{S}_{L/\mathbb{Q}}$, it follows that $p \in \mathcal{S}_{L/\mathbb{Q}}$. This proves $\tilde{\mathcal{S}}_{M/\mathbb{Q}} \dot{\subset} \mathcal{S}_{L/\mathbb{Q}}$ and completes the proof of Theorem 9.2. Q.E.D.

There are also versions of Theorems 9.2 and 9.4 that characterize which primes are represented by the form $x^2 + xy + ((1-D)/4)y^2$, where $D \equiv 1 \bmod 4$ is negative (see Exercise 9.3).

B. The Ring Class Fields of $\mathbb{Z}[\sqrt{-27}]$ and $\mathbb{Z}[\sqrt{-64}]$

Theorem 9.2 shows how the ring class field solves our basic problem of determining when $p = x^2 + ny^2$, and the last part of the theorem points out that our problem is in fact *equivalent* to finding the appropriate ring class field. To see how this works in practice, we will next use Theorem 9.2 to give new proofs of Euler's conjectures for when a prime is represented by x^2+27y^2 or x^2+64y^2 (proved in §4 as Theorems 4.15 and 4.23). The first step, of course, is to determine the ring class fields involved:

PROPOSITION 9.5.

(i) *The ring class field of the order* $\mathbb{Z}[\sqrt{-27}] \subset K = \mathbb{Q}(\sqrt{-3})$ *is* $L = K(\sqrt[3]{2})$.
(ii) *The ring class field of the order* $\mathbb{Z}[\sqrt{-64}] \subset K = \mathbb{Q}(i)$ *is* $L = K(\sqrt[4]{2})$.

PROOF. To prove (i), let L be the ring class field of $\mathbb{Z}[\sqrt{-27}]$. Although L is defined abstractly by class field theory, we still know the following facts about L:

(i) L is a cubic Galois extension of $K = \mathbb{Q}(\sqrt{-3})$ since $[L : K] = h(-4\cdot 27) = 3$.
(ii) L is Galois over $\mathbb{Q}$ with group $\mathrm{Gal}(L/\mathbb{Q})$ isomorphic to the symmetric group S_3. This follows from Lemma 9.3 since S_3 is isomorphic to the semidirect product $(\mathbb{Z}/3\mathbb{Z}) \rtimes (\mathbb{Z}/2\mathbb{Z})$ with $\mathbb{Z}/2\mathbb{Z}$ acting nontrivially.

(iii) All primes of K that ramify in L must divide $6\mathcal{O}_K$. To see this, note that $\mathbb{Z}[\sqrt{-27}] = \mathbb{Z}[3\sqrt{-3}]$ is an order of conductor 6 (this follows from $\mathcal{O}_K = \mathbb{Z}[(-1+\sqrt{-3})/2]$), so that L corresponds to a generalized ideal class group for the modulus $6\mathcal{O}_K$. By the Existence Theorem (Theorem 8.6), the ramification must divide the modulus.

We will show that only four fields satisfy these conditions. To see this, first note that K contains a primitive cube root of unity, and hence any cubic Galois extension of K is of the form $K(\sqrt[3]{u})$ for some $u \in K$. (This is a standard result of Galois theory—see Artin [**2**, Corollary to Theorem 25].) However, the fact that $\mathrm{Gal}(L/\mathbb{Q}) \simeq S_3$ allows us to assume that u is an ordinary integer. More precisely, we have:

LEMMA 9.6. *If M is a cubic extension of $K = \mathbb{Q}(\sqrt{-3})$ with* $\mathrm{Gal}(M/\mathbb{Q}) \simeq S_3$, *then $M = K(\sqrt[3]{m})$ for some cubefree positive integer m.*

PROOF. The idea is to modify the classical proof that $M = K(\sqrt[3]{u})$ for some $u \in K$. We know that M is Galois over $\mathbb{Q}$ and that complex conjugation τ is in $\mathrm{Gal}(M/\mathbb{Q})$. Furthermore, if σ is a generator of $\mathrm{Gal}(M/K) \simeq \mathbb{Z}/3\mathbb{Z}$, then $\mathrm{Gal}(M/\mathbb{Q}) \simeq S_3$ implies that $\tau\sigma\tau = \sigma^{-1}$.

By Proposition 5.29, we can find a real algebraic integer α such that $M = K(\alpha)$. Then define $u_i \in M$ by

$$u_i = \alpha + \omega^i\sigma^{-1}(\alpha) + \omega^{2i}\sigma^{-2}(\alpha), \quad i = 0, 1, 2.$$

The u_i's are algebraic integers satisfying $\sigma(u_i) = \omega^i u_i$, and note that $\tau(u_i) = u_i$, since α is real and $\tau\sigma\tau = \sigma^{-1}$. Thus the u_i's are all real. Then u_0 is fixed by both σ and τ, which implies that $u_0 \in \mathbb{Z}$. Similar arguments show that u_1^3 and u_2^3 are also integers. If $u_1 \neq 0$, we claim that $M = K(u_1)$. This is easy to see, for $[M : K] = 3$, and thus $M \neq K(u_1)$ could only happen when $u_1 \in K$. Since u_1 is real, this would force u_1 to be an integer, which would contradict $\sigma(u_1) = \omega u_1$ and $u_1 \neq 0$. This proves our claim, and if we set $m = u_1^3 \in \mathbb{Z}$, it follows that $M = K(u_1) = K(\sqrt[3]{m})$. We may assume that m is positive and cubefree, and we are done.

If $u_2 \neq 0$, we are done by a similar argument. The remaining case to consider is when $u_1 = u_2 = 0$. However, in this situation a simple application of Cramer's rule shows that our original α would lie in K and hence be rational (since we chose α to be real in the first place). The details of this argument are left to the reader (see Exercise 9.4), and this completes the proof of Lemma 9.6. Q.E.D.

Once we know $L = K(\sqrt[3]{m})$ for some cubefree integer m, the next step is to use the ramification of $K \subset L$ to restrict m. Specifically, it is easy to show that *any* prime of $\mathcal{O}_K$ dividing m ramifies in $K(\sqrt[3]{m})$ (see Exercise 9.5). However, by (iii) above, we know that all ramified primes divide $6\mathcal{O}_K$, and consequently 2 and 3 are the only integer primes that can divide m. Since m is also positive and cubefree, it must be one of the following eight numbers:

$$2,\ 3,\ 4,\ 6,\ 9,\ 12,\ 18,\ 36,$$

and this in turn implies that L must be one of the following four fields:

$$K(\sqrt[3]{2}),\ K(\sqrt[3]{3}),\ K(\sqrt[3]{6}),\ K(\sqrt[3]{12}) \tag{9.7}$$

(see Exercise 9.6). All four of these fields satisfy conditions (i)–(iii) above, so that we will need something else to decide which one is the ring class field L.

Surprisingly, the extra ingredient is none other than Theorem 9.2. More precisely, each field listed in (9.7) gives a different candidate for the polynomial $f_{27}(x)$ that characterizes $p = x^2 + 27y^2$, and then numerical computations can determine which one is the correct field. To illustrate what this means, suppose that L were $K(\sqrt[3]{3})$, the second field in (9.7). This would imply that $f_{27}(x) = x^3 - 3$, which has discriminant -3^5 (see Exercise 9.7). If Theorem 9.2 held with this particular $f_{27}(x)$, then the congruence $x^3 \equiv 3 \bmod 31$ would have a solution since $31 = 2^2 + 27 \cdot 1^2$ is of the form $x^2 + 27y^2$. Using a computer, it is straightforward to show that there are no solutions, so that $K(\sqrt[3]{3})$ can't be the ring class field in question. Similar arguments (also using $p = 31$) suffice to rule out the third and fourth fields given in (9.7) (see Exercise 9.8), and it follows that $L = K(\sqrt[3]{2})$ as claimed.

The second part of the proposition, which concerns the ring class field of the order $\mathbb{Z}[\sqrt{-64}] \subset K = \mathbb{Q}(i)$, is easier to prove than the first, for in this case one can show that $K(\sqrt[4]{2})$ is the *unique* field satisfying the analogs of conditions (i)–(iii) above (see Exercise 9.9). Q.E.D.

Another example of a ring class field is given in Exercise 9.10, where we will show that the field $K(\sqrt[3]{3})$ from (9.7) is the ring class field of the order $\mathbb{Z}[9\omega]$ of conductor 9 in $K = \mathbb{Q}(\sqrt{-3})$.

If we combine Theorem 9.2 with the explicit ring class fields of Proposition 9.5, then we get the following characterizations of when $p = x^2 + 27y^2$ and $p = x^2 + 64y^2$ (proved earlier as Theorems 4.15 and 4.23):

THEOREM 9.8.

(i) *If $p > 3$ is prime, then*

$$p = x^2 + 27y^2 \iff \begin{cases} p \equiv 1 \bmod 3 \text{ and } x^3 \equiv 2 \bmod p \\ \text{has an integer solution.} \end{cases}$$

(ii) *If p is an odd prime, then*

$$p = x^2 + 64y^2 \iff \begin{cases} p \equiv 1 \bmod 4 \text{ and } x^4 \equiv 2 \bmod p \\ \text{has an integer solution.} \end{cases}$$

PROOF. By Proposition 9.5, the ring class field of $\mathbb{Z}[\sqrt{-27}]$ is $L = K(\sqrt[3]{2})$, where $K = \mathbb{Q}(\sqrt{-3})$. Since $\sqrt[3]{2}$ is a real algebraic integer, the polynomial $f_{27}(x)$ of Theorem 9.2 may be taken to be $x^3 - 2$. Then the main equivalence of Theorem 9.2 is exactly what we need, once one checks that the condition $(-27/p) = 1$ is equivalent to the congruence $p \equiv 1 \bmod 3$. The final detail to check is that the discriminant of $x^3 - 2$ is $-2^2 \cdot 3^3$ (see Exercise 9.7), so that the only excluded primes are 2 and 3, and then (i) follows. The proof of (ii) is similar and is left to the reader (see Exercise 9.11). Q.E.D.

Besides allowing us to prove Theorem 9.8, the ring class fields determined in Proposition 9.5 have other uses. For example, if we combine them with Weak Reciprocity from §8, we then get the following partial results concerning cubic and biquadratic reciprocity:

THEOREM 9.9.

(i) *If a primary prime π of $\mathbb{Z}[\omega]$, $\omega = e^{2\pi i/3}$, is relatively prime to 6, then*

$$\left(\frac{2}{\pi}\right)_3 = \left(\frac{\pi}{2}\right)_3.$$

(ii) *If $\pi = a + bi$ is a primary prime of $\mathbb{Z}[i]$, then*

$$\left(\frac{2}{\pi}\right)_4 = i^{ab/2}.$$

REMARK. Notice that these are *exactly* the portions of cubic and biquadratic reciprocity used in our discussion of $p = x^2 + 27y^2$ and $x^2 + 64y^2$ in §4 (see Theorems 4.15 and 4.23).

PROOF. We will prove (i) and leave the proof of (ii) as an exercise (see Exercise 9.12). The basic idea is to combine Weak Reciprocity (Theorem 8.11) with the explicit description of the ring class field given in Proposition 9.5.

If $K = \mathbb{Q}(\omega)$, then $\mathcal{O}_K$ is the ring $\mathbb{Z}[\omega]$ from §4. Thus $L = K(\sqrt[3]{2})$ is the ring class field of the order of conductor 6, and hence corresponds to a subgroup of $I_K(6)$ containing $P_{K,1}(6)$. This shows that the conductor $\mathfrak{f}$ divides $6\mathcal{O}_K$. Then Weak Reciprocity tells us that the cubic Legendre symbol $(2/\cdot)_3$ induces a well-defined homomorphism

$$I_K(6)/P_{K,1}(6) \longrightarrow \mu_3$$

where μ_3 is the group of cube roots of unity. However, the map sending $\alpha \in \mathcal{O}_K$ to the principal ideal $\alpha\mathcal{O}_K$ induces a homomorphism

$$(\mathcal{O}_K/6\mathcal{O}_K)^* \longrightarrow I_K(6)/P_{K,1}(6)$$

(this is similar to what we did in (7.27) and part (c) of Exercise 9.21). Combining these two maps, the Legendre symbol $(2/\cdot)_3$ induces a well-defined homomorphism

$$(\mathcal{O}_K/6\mathcal{O}_K)^* \longrightarrow \mu_3. \tag{9.10}$$

Recall that π is primary by assumption, which means that $\pi \equiv \pm 1 \bmod 3\mathcal{O}_K$. Replacing π by $-\pi$ affects neither $(2/\pi)_3$ nor $(\pi/2)_3$, so that we can assume $\pi \equiv 1 \bmod 3\mathcal{O}_K$. Now consider the isomorphism

$$(\mathcal{O}_K/6\mathcal{O}_K)^* \simeq (\mathcal{O}_K/2\mathcal{O}_K)^* \times (\mathcal{O}_K/3\mathcal{O}_K)^*. \tag{9.11}$$

By (9.10), $(2/\cdot)_3$ is a homomorphism on the group $(\mathcal{O}_K/6\mathcal{O}_K)^*$, and the condition $\pi \equiv 1 \bmod 3\mathcal{O}_K$ means we are restricting this homomorphism to the subgroup $(\mathcal{O}_K/2\mathcal{O}_K)^* \times \{1\}$ relative to (9.11). But the cubic Legendre symbol $(\cdot/2)_3$ can also be regarded as a homomorphism on this subgroup, and we thus need only show that these homomorphisms are equal.

To prove this, first note that $(\mathcal{O}_K/2\mathcal{O}_K)^* \times \{1\}$ is cyclic of order 3 ($\mathcal{O}_K/2\mathcal{O}_K$ is a field with four elements), and the class of $\theta = 1+3\omega$ in $(\mathcal{O}_K/6\mathcal{O}_K)^*$ is a generator. Thus, to show that the two homomorphisms are equal, it suffices to prove that

$$\left(\frac{2}{\theta}\right)_3 = \left(\frac{\theta}{2}\right)_3.$$

Using (4.10), this is straightforward to check—see Exercise 9.12 for the details. Theorem 9.9 is proved. Q.E.D.

C. Primes Represented by Positive Definite Quadratic Forms

As an application of ring class fields, we will prove the classic theorem that a primitive positive definite quadratic form $ax^2+bxy+cy^2$ represents infinitely many prime numbers. The basic idea is to compute the Dirichlet density (in the sense of §8) of the set of primes represented by $ax^2 + bxy + cy^2$, for once we show that the density is positive, there must be infinitely many primes represented. Here is the precise statement of what we will prove:

THEOREM 9.12. *Let $ax^2 + bxy + cy^2$ be a primitive positive definite quadratic form of discriminant $D < 0$, and let $\mathcal{S}$ be the set of primes represented by $ax^2 + bxy + cy^2$. Then the Dirichlet density $\delta(\mathcal{S})$ exists and is given by the formula*

$$\delta(\mathcal{S}) = \begin{cases} \dfrac{1}{2h(D)} & \text{if } ax^2 + bxy + cy^2 \text{ is properly equivalent to its opposite} \\ \dfrac{1}{h(D)} & \text{otherwise.} \end{cases}$$

In particular, $ax^2 + bxy + cy^2$ represents infinitely many prime numbers.

PROOF. Let $\mathcal{O}$ be the order of the discriminant D, and let $K = \mathbb{Q}(\sqrt{D})$. By (7.3), we have $D = f^2 d_K$, where f is the conductor of $\mathcal{O}$. As in the statement of the theorem, let $\mathcal{S} = \{p \text{ prime} : p = ax^2 + bxy + cy^2\}$. The goal is to compute the Dirichlet density of $\mathcal{S}$.

We first relate $\mathcal{S}$ to the generalized ideal class group $I_K(f)/P_{K,\mathbb{Z}}(f)$. By the isomorphism $C(D) \simeq C(\mathcal{O})$ of Theorem 7.7, the class $[ax^2 + bxy + cy^2] \in C(D)$ corresponds to the class $[\mathfrak{a}_0] \in C(\mathcal{O})$ for some proper $\mathcal{O}$-ideal $\mathfrak{a}_0$. Then part (iii) of Theorem 7.7 tells us that

$$\mathcal{S} = \{p \text{ prime} : p = N(\mathfrak{b}),\ \mathfrak{b} \in [\mathfrak{a}_0]\}. \tag{9.13}$$

We need to state this in terms of the maximal order $\mathcal{O}_K$. By Corollary 7.17 we may assume that $\mathfrak{a}_0$ is prime to f, and from here on we will consider only primes p not dividing f. Under the map $\mathfrak{a} \mapsto \mathfrak{a}\mathcal{O}_K$, we know that $\mathfrak{b} \in [\mathfrak{a}_0] \in C(\mathcal{O})$ corresponds to $\mathfrak{b}\mathcal{O}_K \in [\mathfrak{a}_0\mathcal{O}_K] \in I_K(f)/P_{K,\mathbb{Z}}(f)$ (Proposition 7.22). Furthermore, $\mathfrak{b}$ and $\tilde{\mathfrak{b}} = \mathfrak{b}\mathcal{O}_K$ have the same norm when prime to f (Proposition 7.20). Thus (9.13) implies

$$\mathcal{S} \doteq \{p \text{ prime} : p \nmid f,\ p = N(\tilde{\mathfrak{b}}),\ \tilde{\mathfrak{b}} \in [\mathfrak{a}_0\mathcal{O}_K]\}.$$

Since p is prime, the equation $p = N(\tilde{\mathfrak{b}})$ forces $\tilde{\mathfrak{b}}$ to be prime, so that this description of $\mathcal{S}$ can be written

$$\mathcal{S} \doteq \{p \text{ prime} : p \nmid f,\ p = N(\mathfrak{p}),\ \mathfrak{p} \text{ prime},\ \mathfrak{p} \in [\mathfrak{a}_0\mathcal{O}_K]\}. \tag{9.14}$$

If L is the ring class field of $\mathcal{O}$, then Artin Reciprocity gives us an isomorphism

$$I_K(f)/P_{K,\mathbb{Z}}(f) \simeq \mathrm{Gal}(L/K). \tag{9.15}$$

Under this isomorphism, the class of $\mathfrak{a}_0\mathcal{O}_K$ maps to an element $\sigma_0 \in \mathrm{Gal}(L/K)$, which we can regard as an element of $\mathrm{Gal}(L/\mathbb{Q})$. Letting $\langle\sigma_0\rangle$ denote its conjugacy class in $\mathrm{Gal}(L/\mathbb{Q})$, we claim that

$$\mathcal{S} \doteq \left\{p \text{ prime} : p \text{ unramified in } L,\ \left(\frac{L/\mathbb{Q}}{p}\right) = \langle\sigma_0\rangle\right\}. \tag{9.16}$$

The right hand side of (9.16) will be denoted $\mathcal{S}'$, so that we must prove $\mathcal{S} \doteq \mathcal{S}'$.

To show that $\mathcal{S}' \,\dot{\subset}\, \mathcal{S}$, take any $p \in \mathcal{S}'$. Thus $((L/\mathbb{Q})/p) = \langle\sigma_0\rangle$, which means that $((L/\mathbb{Q})/\mathfrak{P}) = \sigma_0$ for some prime $\mathfrak{P}$ of L containing p. Then $\mathfrak{p} = \mathfrak{P} \cap \mathcal{O}_K$ is a prime of K containing p, and we claim that $p = N(\mathfrak{p})$. To see this, note that for any $\alpha \in \mathcal{O}_L$,

$$\sigma_0(\alpha) \equiv \alpha^p \bmod \mathfrak{P} \tag{9.17}$$

since $\sigma_0 = ((L/\mathbb{Q})/\mathfrak{P})$. But we also have $\sigma_0 \in \mathrm{Gal}(L/K)$, so that when $\alpha \in \mathcal{O}_K$, the above congruence reduces to

$$\alpha \equiv \alpha^p \bmod \mathfrak{p}.$$

This implies $\mathcal{O}_K/\mathfrak{p} \simeq \mathbb{Z}/p\mathbb{Z}$, and $N(\mathfrak{p}) = p$ follows. This fact and (9.17) then imply that σ_0 is the Artin symbol $((L/K)/\mathfrak{p})$. Since $[\mathfrak{a}_0\mathcal{O}_K] \in I_K(f)/P_{K,\mathbb{Z}}(f)$ corresponds to $\sigma_0 \in \mathrm{Gal}(L/K)$ under the isomorphism (9.15), it follows that $\mathfrak{p}$ is in the class of $\mathfrak{a}_0\mathcal{O}_K$. Then (9.14) implies that $p \in \mathcal{S}$, at least when $p \nmid f$, and $\mathcal{S}' \dot{\subset} \mathcal{S}$ follows. The opposite inclusion is straightforward and is left to the reader (see Exercise 9.14). This completes the proof of (9.16).

From (9.16), the Čebotarev Density Theorem shows that $\mathcal{S}$ has Dirichlet density

$$\delta(\mathcal{S}) = \frac{|\langle\sigma_0\rangle|}{[L:\mathbb{Q}]}.$$

However, since $\sigma_0 \in \mathrm{Gal}(L/K)$, Lemma 9.3 implies that $\langle\sigma_0\rangle = \{\sigma_0, \sigma_0^{-1}\}$ (see Exercise 9.15). Since $[L:\mathbb{Q}] = 2h(D)$, we see that

$$\delta(\mathcal{S}) = \begin{cases} \dfrac{1}{2h(D)} & \sigma_0 \text{ has order} \leq 2 \\ \dfrac{1}{h(D)} & \text{otherwise.} \end{cases}$$

Now σ_0 has order ≤ 2 if and only if $ax^2+bxy+cy^2$ has order ≤ 2 in $C(D)$, and this last statement means that $ax^2 + bxy + cy^2$ is properly equivalent to its opposite. This completes the proof of Theorem 9.12. Q.E.D.

As an example of what the theorem says, consider forms of discriminant -56. The class number is 4, and we know the reduced forms from §2. Then Theorem 9.12 implies that

$$\begin{aligned} &\delta(\{p \text{ prime} : p = x^2 + 14y^2\}) = \tfrac{1}{8} \\ &\delta(\{p \text{ prime} : p = 2x^2 + 7y^2\}) = \tfrac{1}{8} \\ &\delta(\{p \text{ prime} : p = 3x^2 \pm 2xy + 5y^2\}) = \tfrac{1}{4}. \end{aligned}$$

Notice that these densities sum to 1/2, which is the density of primes for which $(-56/p) = 1$. This example is no accident, for given any negative discriminant, the densities of primes represented by the reduced forms (counted properly) always sum to 1/2 (see Exercise 9.17).

A weaker form of Theorem 9.12, which asserts that $ax^2 + bxy + cy^2$ represents infinitely many primes, was first stated by Dirichlet in 1840, though his proof applied only to a restricted class of discriminants (see [**35**, Vol. I, pp. 497–502]). A complete proof was given by Weber in 1882 [**130**], and in 1954 Briggs [**13**] found an "elementary" proof (in the sense of the "elementary" proofs of the Prime Number Theorem due to Erdös and Selberg).

D. Ring Class Fields and Generalized Dihedral Extensions

We will conclude §9 by asking if there is an intrinsic characterization of ring class fields. We know that they are Abelian extensions of K, but which ones? The remarkable fact is that there is a purely field-theoretic way to characterize ring class fields and their subfields. The key idea is to work with the Galois group over $\mathbb{Q}$. We used this strategy in §6 in dealing with the genus field, and here it will be similarly successful. For the genus field, we wanted $\mathrm{Gal}(L/\mathbb{Q})$ to be Abelian, while in the present case we will allow slightly more complicated Galois groups. The crucial notion is when an extension of K is *generalized dihedral* over $\mathbb{Q}$. To define this, let K be an imaginary quadratic field, and let L be an Abelian extension of K which is Galois over $\mathbb{Q}$. As we saw in the proof of Lemma 9.3, complex conjugation τ is an automorphism of L, and the Galois group $\mathrm{Gal}(L/K)$ can be written as a semidirect product

$$\mathrm{Gal}(L/\mathbb{Q}) \simeq \mathrm{Gal}(L/K) \rtimes (\mathbb{Z}/2\mathbb{Z}),$$

where the nontrivial element of $\mathbb{Z}/2\mathbb{Z}$ acts on $\mathrm{Gal}(L/K)$ via conjugation by τ. We say that L is *generalized dihedral* over $\mathbb{Q}$ if this action sends every element in $\mathrm{Gal}(L/K)$ to its inverse.

In Lemma 9.3 we proved that every ring class field L is generalized dihedral over $\mathbb{Q}$, and it is easy to show that every subfield of L containing K is also generalized dihedral over $\mathbb{Q}$ (see Exercise 9.18). The unexpected result, due to Bruckner [**15**], is that this gives *all* extensions of K which are generalized dihedral over $\mathbb{Q}$:

THEOREM 9.18. *Let K be an imaginary quadratic field. Then an Abelian extension L of K is generalized dihedral over $\mathbb{Q}$ if and only if L is contained in a ring class field of K.*

PROOF. By the above discussion, we know that any extension of K contained in a ring class field is generalized dihedral over $\mathbb{Q}$. To prove the converse, fix an Abelian extension L of K which is generalized dihedral over $\mathbb{Q}$. By Artin Reciprocity, there is an ideal $\mathfrak{m}$ and a subgroup $P_{K,1}(\mathfrak{m}) \subset H \subset I_K(\mathfrak{m})$ such that the Artin map induces an isomorphism

$$I_K(\mathfrak{m})/H \xrightarrow{\sim} \mathrm{Gal}(L/K). \tag{9.19}$$

We saw in §8 that all of this remains true when $\mathfrak{m}$ is enlarged, so that we may assume that $\mathfrak{m} = f\mathcal{O}_K$ for some integer f, and we can also assume that f is divisible by the discriminant d_K of K (this will be useful later in the proof). To prove the theorem, it suffices to show that $P_{K,\mathbb{Z}}(f) \subset H$, for this will imply that L lies in the ring class field of the order of conductor f in $\mathcal{O}_K$. From the definition of $P_{K,\mathbb{Z}}(f)$, this means that we have to prove the following for elements $u \in \mathcal{O}_K$:

$$c \in \mathbb{Z},\ c \text{ prime to } f,\ u \equiv c \bmod f \implies u\mathcal{O}_K \in H. \tag{9.20}$$

The first step is to use the fact that $P_{K,1}(f\mathcal{O}_K) \subset H$: if $\alpha, \beta \in \mathcal{O}_K$ are prime to f, then we claim that

$$\alpha \equiv \beta \bmod f\mathcal{O}_K \implies (\alpha\mathcal{O}_K \in H \text{ if and only if } \beta\mathcal{O}_K \in H). \tag{9.21}$$

To prove this, pick an element $\gamma \in \mathcal{O}_K$ such that $\alpha\gamma \equiv 1 \bmod f\mathcal{O}_K$. Then $\beta\gamma \equiv 1 \bmod f\mathcal{O}_K$ also holds, so that $\alpha\gamma\mathcal{O}_K$ and $\beta\gamma\mathcal{O}_K$ both lie in $P_{K,1}(f\mathcal{O}_K) \subset H$, and (9.21) follows immediately. One consequence of (9.21) is that (9.20) is equivalent to the simpler statement

$$c \in \mathbb{Z},\ c \text{ prime to } f \implies c\mathcal{O}_K \in H. \tag{9.22}$$

So we need to see how (9.22) follows from L being generalized dihedral over $\mathbb{Q}$. Under the isomorphism (9.19), we know that conjugation by τ on $\mathrm{Gal}(L/K)$ corresponds to the usual action of τ on $I_K(f)$. Then L being generalized dihedral over $\mathbb{Q}$ means that for $\mathfrak{a} \in I_K(f)$, the class of $\overline{\mathfrak{a}}$ gives the inverse of $\mathfrak{a}$ in $I_K(f)/H$, which in turn means that $\mathfrak{a}\bar{\mathfrak{a}} \in H$. Since $\mathfrak{a}\bar{\mathfrak{a}} = N(\mathfrak{a})\mathcal{O}_K$ by Lemma 7.14, we see that for any ideal $\mathfrak{a} \in I_K(f)$, we have

$$N(\mathfrak{a})\mathcal{O}_K \in H. \tag{9.23}$$

It remains to prove that (9.23) implies (9.22). Note first that it suffices to prove (9.22) when c is a prime p not dividing f. Recall that $d_K \mid f$, so that p is unramified in K. There are two cases to consider, depending on whether or not p splits in K. If p splits, then $p = N(\mathfrak{p})$, where $\mathfrak{p}$ is a prime factor of $p\mathcal{O}_K$. Then, by (9.23), we have $p\mathcal{O}_K = N(\mathfrak{p})\mathcal{O}_K \in H$, as desired. If p doesn't split, then $(d_K/p) = -1$ by Corollary 5.17. Let q be a prime such that $q \equiv -p \bmod f$ (such primes exist by Dirichlet's theorem). We claim that q splits completely in K. The proof will use the character χ from Lemma 1.14. Recall that this lemma states that the Legendre symbol $(d_K/\cdot)$ induces a well-defined homomorphism $\chi : (\mathbb{Z}/d_K\mathbb{Z})^* \to \{\pm 1\}$, and since $d_K < 0$, we also have $\chi([-1]) = -1$. Since $d_K \mid f$, we have $q \equiv -p \bmod d_K$, and thus

$$\left(\frac{d_K}{q}\right) = \chi([q]) = \chi([-p]) = \chi([-1])\chi([p]) = -\left(\frac{d_K}{p}\right) = 1.$$

Hence q splits completely in K. The argument for the split case implies that $q\mathcal{O}_K \in H$, and then $q \equiv -p \bmod f\mathcal{O}_K$ and (9.21) imply that $(-p)\mathcal{O}_K \in H$. Thus $p\mathcal{O}_K \in H$, which proves (9.22) and completes the proof of Theorem 9.18. Q.E.D.

In Exercises 9.19–9.24, we will explore some other aspects of ring class fields, including a computation of the conductor (in the sense of class field theory) of a ring class field. For further discussion of ring class fields, see Bruckner [**15**], Cohn [**25**, §15.1] and Cohn [**27**, Chapter 8].

E. Exercises

9.1. Prove that the Hilbert class field of an imaginary quadratic field is the ring class field of the maximal order.

9.2. Let $\mathcal{O}$ be the order of conductor f in the imaginary quadratic field K, and let L be the ring class field of $\mathcal{O}$. Let $\mathfrak{m} = f\mathcal{O}_K$ and let τ denote complex conjugation.

(a) Show that $\tau(\mathfrak{m}) = \mathfrak{m}$ and that $\tau(P_{K,\mathbb{Z}}(f)) = P_{K,\mathbb{Z}}(f)$.

(b) Show that $\ker(\Phi_{\tau(L)/K,\mathfrak{m}}) = \tau(\ker(\Phi_{L/K,\mathfrak{m}}))$.

(c) Using $\ker(\Phi_{L/K,\mathfrak{m}}) = P_{K,\mathbb{Z}}(f)$, conclude that

$$\ker(\Phi_{\tau(L)/K,\mathfrak{m}}) = \ker(\Phi_{L/K,\mathfrak{m}}).$$

9.3. Formulate and prove versions of Theorems 9.2 and 9.4 for primes represented by the principal form $x^2 + xy + ((1-D)/4)y^2$ when $D \equiv 1 \bmod 4$ is negative.

9.4. Let u_i, $i = 0, 1, 2$, be as in the proof of Lemma 9.6. If $u_1 = u_2 = 0$, then use Cramer's rule to prove that $\alpha \in K$.

9.5. Let $L = K(\sqrt[3]{m})$ be a cubic extension of K where m is a cubefree integer and K is an imaginary quadratic field. If $\mathfrak{p}$ is any prime of K dividing m, then prove that $\mathfrak{p}$ ramifies in L.

9.6. Verify that if $K = \mathbb{Q}(\sqrt{-3})$ and $L = K(\sqrt[3]{m})$, where m is a cubefree integer of the form 2^a3^b, then L is one of the four fields listed in (9.7).

9.7. Prove that the discriminant of the cubic polynomial $x^3 - a$ is $-27a^2$.

9.8. Use the arguments outlined in the proof of Proposition 9.5 to show that none of the fields $K(\sqrt[3]{3})$, $K(\sqrt[3]{6})$ and $K(\sqrt[3]{12})$ can be the ring class field of the order $\mathbb{Z}[\sqrt{-27}]$. Hint: use $31 = 2^2 + 27 \cdot 1^2$.

9.9. Prove part (ii) of Proposition 9.5 using the hints given in the text.

9.10. This exercise is concerned with the order $\mathbb{Z}[9\omega]$ of conductor 9 in the field $K = \mathbb{Q}(\omega)$, $\omega = e^{2\pi i/3}$.

(a) Prove that $L = K(\sqrt[3]{3})$ is the ring class field of $\mathbb{Z}[9\omega]$. Hint: adapt the proof of Proposition 9.5.

(b) Use Exercise 9.3 to prove that for primes $p \geq 5$, $p = x^2 + xy + 61y^2$ if and only if $p \equiv 1 \bmod 3$ and 3 is a cubic residue modulo p.

(c) Use part (b) to prove that for primes $p \geq 5$, $4p = x^2 + 243y^2$ if and only if $p \equiv 1 \bmod 3$ and 3 is a cubic residue modulo p. Note that this result, conjectured by Euler, was proved earlier in Exercise 4.15 using the supplementary laws of cubic reciprocity.

9.11. Prove part (ii) of Theorem 9.8.

9.12. This exercise is concerned with the proof of Theorem 9.9.

(a) Let $\theta = 1 + 3\omega$. To prove that $(2/\theta)_3 = (\theta/2)_3$, first use (4.10) to show

$$\begin{aligned}
\left(\frac{2}{\theta}\right)_3 &= \left(\frac{2}{1+3\omega}\right)_3 \equiv 2^{(N(1+3\omega)-1)/3} \equiv 4 \bmod (1+3\omega)\mathcal{O}_K \\
\left(\frac{\theta}{2}\right)_3 &= \left(\frac{1+3\omega}{2}\right)_3 = \left(\frac{1+\omega}{2}\right)_3 \equiv (1+\omega)^{(N(2)-1)/3} \bmod 2\mathcal{O}_K \\
&\equiv 1+\omega \bmod 2\mathcal{O}_K,
\end{aligned}$$

and then note that $1 + \omega + \omega^2 = 0$ and $4 - \omega^2 = -(1+2\omega)(1+3\omega)$.

(b) Prove part (ii) of Theorem 9.9.

9.13. Let $K = \mathbb{Q}(\omega)$, $\omega = e^{2\pi i/3}$. In this exercise we will use the ring class field $K(\sqrt[3]{3})$ from Exercise 9.10 to prove the supplementary laws of cubic reciprocity. Let $p \equiv 1 \bmod 3$ be prime. In Exercise 4.15 we saw that $4p = a^2 + 27b^2$, which gave us the factorization $p = \pi\overline{\pi}$ where $\pi = (a + \sqrt{-27}b)/2$ is primary. We can assume that $a \equiv 1 \bmod 3$.

(a) Prove that $(\omega/\pi)_3 = \omega^{2(a+2)/3}$. Hint: use (4.10).

(b) Adapt the proof of Theorem 9.9 to prove that $(3/\pi)_3 = \omega^{2b}$. Hint: use Exercise 9.10.

(c) Use $3 = -\omega^2(1-\omega)^2$ to prove that $(1 - \omega/\pi)_3 = \omega^{b+(a+2)/3}$.

(d) Show that the results of parts (a) and (c) imply the supplementary laws for cubic reciprocity as stated in (4.13).

9.14. Let $\mathcal{S}$ and $\mathcal{S}'$ be the two sets of primes defined in the proof of Theorem 9.12. Prove that $\mathcal{S} \mathrel{\dot{\subset}} \mathcal{S}'$. Hint: use (9.14).

9.15. Let K be an imaginary quadratic field, and let $K \subset L$ be an Abelian extension which is generalized dihedral over $\mathbb{Q}$. If $\sigma \in \mathrm{Gal}(L/K) \subset \mathrm{Gal}(L/\mathbb{Q})$, then prove that the conjugacy class $\langle\sigma\rangle$ of σ in $\mathrm{Gal}(L/\mathbb{Q})$ is the set $\{\sigma, \sigma^{-1}\}$.

9.16. In this exercise we will use (8.16) to give a different proof of Theorem 9.12. We will use the notation of the proof of Theorem 9.12. Thus $\mathcal{O}$ is the order of conductor f in an imaginary quadratic field K, and L is the ring class field of $\mathcal{O}$. Let

$$\mathcal{S} = \{p \text{ prime} : p = ax^2 + bxy + cy^2\}.$$

(a) If $ax^2 + bxy + cy^2$ gives us the class $[\mathfrak{a}_0\mathcal{O}_K] \in I_K(f)/P_{K,\mathbb{Z}}(f)$, show that

$$\mathcal{S} \doteq \{p \text{ prime} : p \nmid f,\, p\mathcal{O}_K = \mathfrak{p}\overline{\mathfrak{p}},\, \mathfrak{p} \in [\mathfrak{a}_0\mathcal{O}_K]\}.$$

Hint: use (9.14).

(b) Use the Čebotarev Density Theorem to show that

$$\mathcal{S}'' = \{\mathfrak{p} \in \mathcal{P}_K : \mathfrak{p} \in [\mathfrak{a}_0\mathcal{O}_K]\}$$

has Dirichlet density $\delta(\mathcal{S}'') = 1/h(D)$. Then use (8.16) to show that $\delta(\mathcal{S}'' \cap \mathcal{P}_{K,1}) = 1/h(D)$. Recall that $\mathcal{P}_{K,1} = \{\mathfrak{p} \in \mathcal{P}_K : N(\mathfrak{p}) \text{ is prime}\}$.

(c) Show that the mapping $\mathfrak{p} \mapsto N(\mathfrak{p})$ from $\mathcal{S}'' \cap \mathcal{P}_{K,1}$ to $\mathcal{S}$ is either two-to-one or one-to-one, depending on whether or not $\mathfrak{a}_0\mathcal{O}_K$ has order ≤ 2 in the class group. Then use part (b) to prove Theorem 9.12.

9.17. Fix a negative discriminant D.

(a) Use Theorem 9.12 to show that the sum of the densities of the primes represented by the reduced forms of discriminant D with middle coefficient $b \geq 0$ is always $1/2$.

(b) To explain the result of (a), first use Lemma 2.5 to show that the primes represented by the forms listed in part (a) are, up to a finite set, exactly the primes for which $(D/p) = 1$. Then use the Čebotarev Density Theorem to show that this set has density $1/2$.

9.18. Let K be an imaginary quadratic field. Use Lemma 9.3 to prove that any intermediate field between K and a ring class field of K is generalized dihedral over $\mathbb{Q}$.

9.19. An imaginary quadratic field K has infinitely many ring class fields associated with it. In this exercise we will work out the relation between the different ring class fields.

(a) If $\mathcal{O}_1$ and $\mathcal{O}_2$ are orders in K, then we get ring class fields L_1 and L_2. Prove that

$$\mathcal{O}_1 \subset \mathcal{O}_2 \Longrightarrow L_2 \subset L_1.$$

(b) If f_i is the conductor of $\mathcal{O}_i$, then prove that $\mathcal{O}_1 \subset \mathcal{O}_2$ if and only if $f_2 \mid f_1$, and conclude that the result of part (a) can be stated in terms of conductors as follows:

$$f_2 \mid f_1 \Longrightarrow L_2 \subset L_1.$$

In Exercise 9.24, we will see that the converse of this implication is false.

(c) Show that the Hilbert class field is contained in the ring class field of any order, and conclude that $h(d_K) \mid h(f^2 d_K)$. This fact was proved earlier in Theorem 7.24.

9.20. Let L be the ring class field of an order $\mathcal{O}$ in an imaginary quadratic field K. Such a field has two "conductors" associated to it: first, there is the conductor f of the order $\mathcal{O}$, and second, there is the class field theory conductor $\mathfrak{f}(L/K)$ of L as an Abelian extension of K. There should be a close relation between these conductors, and the obvious guess would be that

$$\mathfrak{f}(L/K) = f\mathcal{O}_K.$$

In Exercises 9.20–9.23, we will show that the answer is a bit more complicated: the conductor is given by the formula

$$\mathfrak{f}(L/K) = \begin{cases} \mathcal{O}_K & f = 2 \text{ or } 3,\ K = \mathbb{Q}(\sqrt{-3}) \\ \mathcal{O}_K & f = 2,\ K = \mathbb{Q}(i) \\ (f/2)\mathcal{O}_K & f = 2f',\ f' \text{ odd},\ 2 \text{ splits completely in } K \\ f\mathcal{O}_K & \text{otherwise.} \end{cases}$$

To begin the proof, let f be a positive integer, and let K be an imaginary quadratic field. Assume that $f = 2f'$, where f' is odd and 2 splits completely in K. Let L and L' be the ring class fields of K corresponding to the orders of conductor f and f' respectively. Then prove that

$$\mathfrak{f}(L/K) = \mathfrak{f}(L'/K).$$

Hint: first show that $L' \subset L$, and then use Theorem 7.24 to conclude that $L' = L$.

9.21. Let L be the ring class field of the order of conductor f in an imaginary quadratic field K, and assume that $\mathfrak{f}(L/K) \neq f\mathcal{O}_K$.

(a) Show that $f\mathcal{O}_K = \mathfrak{p}\mathfrak{m}$, where $\mathfrak{p}$ is prime and $\mathfrak{f}(L/K) \mid \mathfrak{m}$. We will fix $\mathfrak{p}$ and $\mathfrak{m}$ for the rest of this exercise.

(b) Prove that $I_K(f) \cap P_{K,1}(\mathfrak{m}) \subset P_{K,\mathbb{Z}}(f)$.

(c) Consider the natural maps

$$\pi : (\mathcal{O}_K/f\mathcal{O}_K)^* \longrightarrow (\mathcal{O}_K/\mathfrak{m})^*$$
$$\beta : (\mathbb{Z}/f\mathbb{Z})^* \longrightarrow (\mathcal{O}_K/f\mathcal{O}_K)^*.$$

Show that $\ker(\pi) \subset \mathcal{O}_K^* \cdot \mathrm{im}(\beta)$. Hint: use part (b) and show that elements of $P_{K,\mathbb{Z}}(f)$ can be represented as $(\gamma/\delta)\mathcal{O}_K$ where $\gamma, \delta \in \mathcal{O}_K$ satisfy $\gamma \equiv c \bmod f\mathcal{O}_K$ for $c \in \mathbb{Z}$ with $\gcd(c, f) = 1$ and $\delta \equiv 1 \bmod f\mathcal{O}_K$.

9.22. In this exercise we will assume that $\mathcal{O}_K^* = \{\pm 1\}$ (by Exercise 5.9, this excludes the fields $\mathbb{Q}(\sqrt{-3})$ and $\mathbb{Q}(i)$). Let K, f and L be as in the previous exercise, and assume in addition that if $f = 2f'$, f' odd, then 2 doesn't split completely in K. Our goal is to prove that

$$\mathfrak{f}(L/K) = f\mathcal{O}_K.$$

We will argue by contradiction. Suppose that $\mathfrak{f}(L/K) \neq f\mathcal{O}_K$. Exercise 9.21 implies that $f\mathcal{O}_K = \mathfrak{p}\mathfrak{m}$, where $\mathfrak{p}$ is prime and $\mathfrak{f}(L/K) \mid \mathfrak{m}$. Furthermore, if π and β are the natural maps

$$\pi : (\mathcal{O}_K/f\mathcal{O}_K)^* \longrightarrow (\mathcal{O}_K/\mathfrak{m})^*$$
$$\beta : (\mathbb{Z}/f\mathbb{Z})^* \longrightarrow (\mathcal{O}_K/f\mathcal{O}_K)^*,$$

then Exercise 9.21 also implies that $\ker(\pi) \subset \mathcal{O}_K^* \cdot \mathrm{im}(\beta)$, and since $\mathcal{O}_K^* = \{\pm 1\}$, we see that

$$\ker(\pi) \subset \mathrm{im}(\beta).$$

We will show that this inclusion leads to a contradiction.

(a) Use Exercise 7.29 to prove that

$$|\ker(\pi)| = \begin{cases} N(\mathfrak{p}) & \mathfrak{p} \mid \mathfrak{m} \\ N(\mathfrak{p}) - 1 & \mathfrak{p} \nmid \mathfrak{m}. \end{cases}$$

(b) Note that $N(\mathfrak{p}) = p$ or p^2, where p is the unique integer prime contained in $\mathfrak{p}$. Suppose first that $N(\mathfrak{p}) = p$.

(i) Show that $\mathfrak{m} = m\bar{\mathfrak{p}}$ for some integer m.

(ii) Use (i) to show that the map $(\mathbb{Z}/f\mathbb{Z})^* \to (\mathcal{O}_K/\mathfrak{m})^*$ is injective, and conclude that $\ker(\pi) \cap \mathrm{im}(\beta) = \{1\}$.

(iii) Since $\ker(\pi) \subset \mathrm{im}(\beta)$, (ii) implies that $\ker(\pi) = \{1\}$. Use part (a) to show that $p = 2$, 2 splits completely in K, and $f = 2m$ where m is odd. This contradicts our assumption on f.

(c) It remains to consider the case when $N(\mathfrak{p}) = p^2$. Here, $f = pm$ and $\mathfrak{m} = m\mathcal{O}_K$.

(i) Show that $\ker(\pi) \cap \mathrm{im}(\beta) \simeq \ker(\theta)$, where $\theta : (\mathbb{Z}/f\mathbb{Z})^* \to (\mathbb{Z}/m\mathbb{Z})^*$ is the natural map.

(ii) Since $\ker(\pi) \subset \mathrm{im}(\beta)$, (i) implies that $|\ker(\pi)| \leq |\ker(\theta)|$, and we know $|\ker(\pi)|$ from part (a). Now compute $|\ker(\theta)|$ and use this to show that $|\ker(\pi)| \leq |\ker(\theta)|$ is impossible. Again we have a contradiction.

9.23. Recall the formula for the conductor $\mathfrak{f}(L/K)$ stated in Exercise 9.20.

(a) Using Exercises 9.20 and 9.22, prove that the desired formula holds when $\mathcal{O}_K^* = \{\pm 1\}$.

(b) Adapt the proof of Exercise 9.22 to the case $\mathcal{O}_K^* \neq \{\pm 1\}$, and prove the formula for $\mathfrak{f}(L/K)$ for all K. Hint: show that $\ker(\pi) \subset \mathcal{O}_K^* \cdot \mathrm{im}(\beta)$ leads to a contradiction. Use Exercise 9.22 to reduce to the case where $\ker(\pi) \not\subset \mathrm{im}(\beta)$ and prove that there is an exact sequence

$$1 \longrightarrow \ker(\pi) \cap \mathrm{im}(\beta) \longrightarrow \ker(\pi) \longrightarrow \mathcal{O}_K^*/\big(\mathcal{O}_K^* \cap \mathrm{im}(\beta)\big).$$

9.24. Use the conductor formula from Exercise 9.20 to give infinitely many examples where $\mathfrak{f}(L/K) \neq f\mathcal{O}_K$. Also show that the converse of part (b) of Exercise 9.19 is not true in general (i.e., $L_2 \subset L_1$ need not imply $f_2 \mid f_1$).

CHAPTER 3

Complex Multiplication

10. Elliptic Functions and Complex Multiplication

In Chapter 2 we solved our problem of when a prime p can be written in the form x^2+ny^2. The criterion from Theorem 9.2 states that, with finitely many exceptions,

$$p = x^2 + ny^2 \iff \begin{cases} (-n/p) = 1 \text{ and } f_n(x) \equiv 0 \bmod p \\ \text{has an integer solution.} \end{cases}$$

The key ingredient is the polynomial $f_n(x)$, which we know is the minimal polynomial of a primitive element of the ring class field of $\mathbb{Z}[\sqrt{-n}]$. But the proof of Theorem 9.2 doesn't explain how to find such a primitive element, so that currently we have only an abstract solution of the problem of $p = x^2 + ny^2$. In this chapter, we will use modular functions and the theory of complex multiplication to give a systematic method for finding $f_n(x)$.

In §10 we will study elliptic functions and introduce the idea of complex multiplication. A key role is played by the j-invariant of a lattice, and we will show that if $\mathcal{O}$ is an order in an imaginary quadratic field K, then its j-invariant $j(\mathcal{O})$ is an algebraic number. But before we can get to the real depth of the subject, we need to learn about modular functions. Thus §11 will present a brief but complete account of the main properties of modular functions for $\Gamma_0(m)$, including the modular equation. Then we will prove that $j(\mathcal{O})$ is not only an algebraic integer, but also that it generates (over K) the ring class field of $\mathcal{O}$. This theorem, often called the "First Main Theorem" of complex multiplication, is the main result of §11. In §12 we will compute $j(\mathcal{O})$ in some special cases, and in §13 we will complete our study of $j(\mathcal{O})$ by describing an algorithm for computing its minimal polynomial (the so-called "class equation"). When applied to the order $\mathbb{Z}[\sqrt{-n}]$, this theory will give us an algorithm for constructing the polynomial $f_n(x)$ that solves $p = x^2 + ny^2$. We will then have a complete solution of the basic problem of the book.

Before we can begin our discussion of complex multiplication, we need to learn some basic facts about elliptic functions and j-invariants.

A. Elliptic Functions and the Weierstrass $\wp$-Function

To start, we define a *lattice* to be an additive subgroup L of $\mathbb{C}$ which is generated by two complex numbers ω_1 and ω_2 that are linearly independent over $\mathbb{R}$. We express this by writing $L = [\omega_1, \omega_2]$. Then an *elliptic function* for L is a function $f(z)$ defined on $\mathbb{C}$, except for isolated singularities, which satisfies the following two conditions:

(i) $f(z)$ is meromorphic on $\mathbb{C}$.
(ii) $f(z + \omega) = f(z)$ for all $\omega \in L$.

If $L = [\omega_1, \omega_2]$, note that the second condition is equivalent to

$$f(z+\omega_1) = f(z+\omega_2) = f(z).$$

Thus an elliptic function is a doubly-periodic meromorphic function, and elements of L are often referred to as periods.

One of the most important elliptic functions is the Weierstrass $\wp$-function, which is defined as follows: given a complex number z not in the lattice L, we set

$$\wp(z; L) = \frac{1}{z^2} + \sum_{\omega \in L - \{0\}} \left(\frac{1}{(z-\omega)^2} - \frac{1}{\omega^2} \right).$$

When working with a fixed lattice L, we will usually write $\wp(z)$ instead of $\wp(z; L)$. Here are some basic properties of the $\wp$-function:

THEOREM 10.1. *Let $\wp(z)$ be the Weierstrass $\wp$-function for the lattice L.*

(i) *$\wp(z)$ is an elliptic function for L whose singularities consist of double poles at the points of L.*

(ii) *$\wp(z)$ satisfies the differential equation*

$$\wp'(z)^2 = 4\wp(z)^3 - g_2(L)\wp(z) - g_3(L),$$

where the constants $g_2(L)$ and $g_3(L)$ are defined by

$$g_2(L) = 60 \sum_{\omega \in L - \{0\}} \frac{1}{\omega^4}$$

$$g_3(L) = 140 \sum_{\omega \in L - \{0\}} \frac{1}{\omega^6}.$$

(iii) *$\wp(z)$ satisfies the addition law*

$$\wp(z+w) = -\wp(z) - \wp(w) + \frac{1}{4}\left(\frac{\wp'(z) - \wp'(w)}{\wp(z) - \wp(w)} \right)^2$$

whenever $z, w \notin L$ and $z + w \notin L$.

PROOF. The first step is to prove the following lemma:

LEMMA 10.2. *If L is a lattice and $r > 2$, then the series*

$$G_r(L) = \sum_{\omega \in L - \{0\}} \frac{1}{\omega^r}$$

converges absolutely.

PROOF. If $L = [\omega_1, \omega_2]$, then we need to show that the series

$$\sum_{\omega \in L - \{0\}} \frac{1}{|\omega|^r} = {\sum_{m,n}}' \frac{1}{|m\omega_1 + n\omega_2|^r}$$

converges, where $\sum'_{m,n}$ denotes summation over all ordered pairs $(m, n) \neq (0, 0)$ of integers. If we let $M = \min\{|x\omega_1 + y\omega_2| : x^2 + y^2 = 1\}$, then it is easy to see that for all $x, y \in \mathbb{R}$,

$$|x\omega_1 + y\omega_2| \geq M\sqrt{x^2 + y^2}$$

(see Exercise 10.1), and it follows that

$$\sum_{m,n}{}' \frac{1}{|m\omega_1 + n\omega_2|^r} \leq \frac{1}{M^r} \sum_{m,n}{}' \frac{1}{(m^2+n^2)^{r/2}}.$$

By comparing the sum on the right to the integral

$$\iint_{x^2+y^2\geq 1} \frac{1}{(x^2+y^2)^{r/2}}\, dx\, dy,$$

it is easy to show that the sum in question converges when $r > 2$ (see Exercise 10.1). Q.E.D.

We can now show that $\wp(z)$ is holomorphic outside L. Namely, if Ω is a compact subset of $\mathbb{C}$ missing L, it suffices to show that the sum in

$$\wp(z) = \frac{1}{z^2} + \sum_{\omega \in L - \{0\}} \left(\frac{1}{(z-\omega)^2} - \frac{1}{\omega^2} \right)$$

converges absolutely and uniformly on Ω. Pick a number R such that $|z| \leq R$ for all $z \in \Omega$. Now suppose that $z \in \Omega$ and that $\omega \in L$ satisfies $|\omega| \geq 2R$. Then $|z - \omega| \geq \frac{1}{2}|\omega|$, and one sees that

$$\left| \frac{1}{(z-\omega)^2} - \frac{1}{\omega^2} \right| = \left| \frac{z(2\omega - z)}{\omega^2 (z-\omega)^2} \right| \leq \frac{R(2|\omega| + \frac{1}{2}|\omega|)}{|\omega|^2 (\frac{1}{2}|\omega|^2)} = \frac{10R}{|\omega|^3}.$$

Since the inequality $|\omega| \geq 2R$ holds for all but finitely many elements of L, it follows from Lemma 10.2 that the sum in the $\wp$-function converges absolutely and uniformly on Ω. Thus $\wp(z)$ is holomorphic on $\mathbb{C} - L$ and has a double pole at the origin.

Notice that since $(-z-\omega)^2 = (z-(-\omega))^2$, the identity $\wp(-z) = \wp(z)$ follows immediately from absolute convergence. Thus the $\wp$-function is an even function.

To show that $\wp(z)$ is periodic is a bit trickier. We first differentiate the series for $\wp(z)$ to obtain

$$\wp'(z) = -2 \sum_{\omega \in L} \frac{1}{(z-\omega)^3}.$$

Arguing as above, this series converges absolutely, and it follows easily that $\wp'(z)$ is an elliptic function for L (see Exercise 10.2). Now suppose that $L = [\omega_1, \omega_2]$. The functions $\wp(z)$ and $\wp(z+\omega_i)$ have the same derivative (since $\wp'(z)$ is periodic), and hence they differ by a constant, say $\wp(z) = \wp(z+\omega_i) + C$. Evaluating this at $-\omega_i/2$ (which is not in L), we obtain

$$\wp(-\omega_i/2) = \wp(-\omega_i/2 + \omega_i) + C = \wp(\omega_i/2) + C.$$

Since $\wp(z)$ is an even function, C must be zero, and periodicity is proved. It follows that the poles of $\wp(z)$ are all double poles and lie exactly on the points of L, and (i) is proved.

Turning to (ii), we first compute the Laurent expansion of $\wp(z)$ about the origin:

LEMMA 10.3. *Let $\wp(z)$ be the $\wp$-function for the lattice L, and let $G_r(L)$ be the constants defined in Lemma* 10.2. *Then, in a neighborhood of the origin, we have*

$$\wp(z) = \frac{1}{z^2} + \sum_{n=1}^{\infty} (2n+1) G_{2n+2}(L)\, z^{2n}.$$

PROOF. For $|x| < 1$, we have the series expansion

$$\frac{1}{(1-x)^2} = 1 + \sum_{n=1}^{\infty} (n+1)\, x^n$$

(see Exercise 10.3). Thus, if $|z| < |\omega|$, then we can put $x = z/\omega$ in the above series, and it follows easily that

$$\frac{1}{(z-\omega)^2} - \frac{1}{\omega^2} = \sum_{n=1}^{\infty} \frac{n+1}{\omega^{n+2}}\, z^n.$$

Summing over all $\omega \in L - \{0\}$ and using absolute convergence, we obtain

$$\wp(z) = \frac{1}{z^2} + \sum_{n=1}^{\infty} (n+1) G_{n+2}(L)\, z^n.$$

Since the $\wp(z)$ is an even function, all of the odd coefficients must vanish, giving us the desired Laurent expansion. Q.E.D.

From this lemma, we see that

$$\wp'(z) = \frac{-2}{z^3} + \sum_{n=1}^{\infty} 2n(2n+1) G_{2n+2}(L)\, z^{2n-1},$$

and then one computes the first few terms of $\wp(z)^3$ and $\wp'(z)^2$ as follows:

$$\begin{aligned} \wp(z)^3 &= \frac{1}{z^6} + \frac{9G_4(L)}{z^2} + 15G_6(L) + \cdots \\ \wp'(z)^2 &= \frac{4}{z^6} - \frac{24G_4(L)}{z^2} - 80G_6(L) + \cdots, \end{aligned}$$

where $+\cdots$ indicates terms involving positive powers of z (see Exercise 10.4). Now consider the elliptic function

$$F(z) = \wp'(z)^2 - 4\wp(z)^3 + 60G_4(L)\wp(z) + 140G_6(L).$$

Using the above expansions, it is easy to see that $F(z)$ vanishes at the origin, and then by periodicity, $F(z)$ vanishes at all points of L. But it is also holomorphic on $\mathbb{C} - L$, so that $F(z)$ is holomorphic on all of $\mathbb{C}$. An easy argument using Liouville's Theorem shows that $F(z)$ is constant (see Exercise 10.5), so that $F(z)$ is identically zero. Since $g_2(L)$ and $g_3(L)$ were defined to be $60G_4(L)$ and $140G_6(L)$ respectively, the proof of (ii) is complete.

In order to prove (iii), we will need the following lemma:

LEMMA 10.4. *If $z, w \notin L$, then $\wp(z) = \wp(w)$ if and only if $z \equiv \pm w \bmod L$.*

PROOF. The $\Leftarrow$ direction of the proof is trivial since $\wp(z)$ is an even function. To argue the other way, suppose that $L = [\omega_1, \omega_2]$, and fix a number $-1 < \delta < 0$. Let $\mathbf{P}$ denote the parallelogram $\{s\omega_1 + t\omega_2 : \delta \le s, t \le \delta + 1\}$, and let Γ be its boundary oriented counterclockwise. Note that every complex number is congruent modulo L to a number in $\mathbf{P}$ (see Exercise 10.6).

Fix w and consider the function $f(z) = \wp(z) - \wp(w)$. By adjusting δ, we can arrange that $f(z)$ has no zeros or poles on Γ. Then it is well-known that

$$\frac{1}{2\pi i} \int_{\Gamma} \frac{f'(z)}{f(z)} dz = Z - P,$$

where Z (resp. P) is the number of zeros (resp. poles) of $f(z)$ in $\mathbf{P}$, counting multiplicity. Since $f'(z)/f(z)$ is periodic, the integrals on opposite sides of Γ cancel, and thus $\int_\Gamma (f'(z)/f(z))dz = 0$. This shows that $Z = P$. However, P is easy to compute: from the definition of $\mathbf{P}$, it's obvious that 0 is the only pole of $f(z) = \wp(z) - \wp(w)$ in $\mathbf{P}$. It's a double pole, and thus $Z = P = 2$, so that $f(z)$ has two zeros (counting multiplicity) in $\mathbf{P}$.

There are now two cases to consider. If $w \not\equiv -w \bmod L$, then modulo L, w and $-w$ give rise to two distinct points of $\mathbf{P}$, both of which are zeros of $f(z) = \wp(z) - \wp(w)$. Since $Z = 2$, these are all of the zeros, and their multiplicity is one, i.e., $\wp'(w) \neq 0$. If $w \equiv -w \bmod L$, then $2w \in L$. Since $\wp'(z)$ is an odd function (being the derivative of an even function), we obtain

$$\wp'(w) = \wp'(w - 2w) = \wp'(-w) = -\wp'(w),$$

which forces $\wp'(w) = 0$. Thus modulo L, w gives rise to a zero of $f(z)$ of multiplicity ≥ 2 in $\mathbf{P}$, and again $Z = 2$ implies that these are all. This proves the lemma. Q.E.D.

The proof of Lemma 10.4 yields the following useful corollary:

COROLLARY 10.5. *If $w \notin L$, then $\wp'(w) = 0$ if and only if $2w \in L$.* Q.E.D.

Now we can finally prove the addition theorem. Fix $w \notin L$, and consider the elliptic function

$$G(z) = \wp(z + w) + \wp(z) + \wp(w) - \frac{1}{4}\left(\frac{\wp'(z) - \wp'(w)}{\wp(z) - \wp(w)}\right)^2.$$

If we can show that $G(z)$ is holomorphic on $\mathbb{C}$ and vanishes at the origin, then as in (ii), Liouville's Theorem will imply that $G(z)$ vanishes identically, and the addition theorem will be proved.

Using Lemma 10.4, we see that the possible singularities of $G(z)$ come from three sources: L, $L + \{w\}$ and $L - \{w\}$. By periodicity, it suffices to consider $G(0)$, $G(w)$ and $G(-w)$. Let's begin with $G(0)$. Using the Laurent expansions for $\wp(z)$ and $\wp'(z)$, one sees that

$$\frac{1}{4}\left(\frac{\wp'(z) - \wp'(w)}{\wp(z) - \wp(w)}\right)^2 = \frac{1}{4}\left(\frac{-2/z^3 - \wp'(w) + \cdots}{1/z^2 - \wp(w) + \cdots}\right)^2 = \frac{1}{z^2} + 2\wp(w) + \cdots,$$

where as usual, $+\cdots$ means terms involving positive powers of z. Hence

$$G(z) = \wp(z + w) + \wp(w) + \frac{1}{z^2} + \cdots - \frac{1}{z^2} - 2\,\wp(w) - \cdots,$$

and it follows that $G(0) = 0$.

To simplify the remainder of the argument, we will assume that $2w \notin L$. Turning to $G(w)$, we use L'Hospital's rule to obtain

$$G(w) = \wp(2w) + 2\wp(w) - \frac{1}{4}\left(\frac{\wp''(w)}{\wp'(w)}\right)^2.$$

Since $2w \notin L$, Corollary 10.5 shows that $\wp'(w) \neq 0$, and thus $G(w)$ is defined. It remains to consider $G(-w)$. We begin with some Laurent expansions about $z = -w$:

$$\wp(z + w) = \frac{1}{(z + w)^2} + \cdots$$
$$\wp(z) = \wp(-w) + \wp'(-w)(z + w) + \cdots = \wp(w) - \wp'(w)(z + w) + \cdots,$$

where $+\cdots$ now refers to higher powers of $z+w$. Since $\wp'(w) \neq 0$, these formulas make it easy to show that $G(-w)$ is defined (see Exercise 10.7). This shows that $G(z)$ is holomorphic and vanishes at 0, so that $G(z)$ vanishes everywhere.

To complete the proof, we need to consider the case $2w \in L$. We leave this to the reader (see Exercise 10.7). We have now proved Theorem 10.1. Q.E.D.

There are many more results connected with the Weierstrass $\wp$-function, and we refer the reader to Chandrasekharan [**21**, Chapter III], Lang [**90**, Chapter 1] or Whittaker and Watson [**138**, Chapter XX] for more details.

B. The j-Invariant of a Lattice

Elliptic functions depend on which lattice is being used, but sometimes different lattices can have basically the same elliptic functions. We say that two lattices L and L' are *homothetic* if there is a nonzero complex number λ such that $L' = \lambda L$. Note that homothety is an equivalence relation. It is easy to check how homothety affects elliptic functions: if $f(z)$ is an elliptic function for L, then $f(\lambda^{-1}z)$ is an elliptic function for λL. Furthermore, the $\wp$-function transforms as follows:

$$\wp(\lambda z; \lambda L) = \lambda^{-2}\wp(z; L).$$

Thus we would like to classify lattices up to homothety, and this is where the j-invariant comes in.

Given a lattice L, we have the constants $g_2(L)$ and $g_3(L)$ which appear in the differential equation for $\wp(z)$. It is customary to set

$$\Delta(L) = g_2(L)^3 - 27g_3(L)^2.$$

The number $\Delta(L)$ is closely related to the discriminant of the cubic polynomial $4x^3 - g_2(L)x - g_3(L)$ that appears in the differential equation for $\wp(z)$. In fact, if e_1, e_2, and e_3 are the roots of this polynomial, then one can show that

$$\Delta(L) = 16(e_1 - e_2)^2(e_1 - e_3)^2(e_2 - e_3)^2 \tag{10.6}$$

(see Exercise 10.8). An important fact is that $\Delta(L)$ never vanishes, i.e.,

PROPOSITION 10.7. *If L is a lattice, then $\Delta(L) \neq 0$.*

PROOF. If $w \notin L$ and $2w \in L$, then Corollary 10.5 implies that $\wp'(w) = 0$. Then the differential equation from Theorem 10.1 tells us that

$$0 = \wp'(w)^2 = 4\wp(w)^3 - g_2(L)\wp(w) - g_3(L),$$

so that $\wp(w)$ is a root of $4x^3 - g_2(L)x - g_3(L)$. If $L = [\omega_1, \omega_2]$, this process gives three roots $\wp(\omega_1/2)$, $\wp(\omega_2/2)$ and $\wp((\omega_1+\omega_2)/2)$, which are distinct by Lemma 10.4 since $\pm\omega_1/2$, $\pm\omega_2/2$ and $\pm(\omega_1+\omega_2)/2$ are distinct modulo L. Thus the roots of $4x^3 - g_2(L)x - g_3(L)$ are distinct, and $\Delta(L) \neq 0$ by (10.6). Q.E.D.

The j-invariant $j(L)$ of the lattice L is defined to be the complex number

$$j(L) = 1728\frac{g_2(L)^3}{g_2(L)^3 - 27g_3(L)^2} = 1728\frac{g_2(L)^3}{\Delta(L)}. \tag{10.8}$$

Note that $j(L)$ is always defined since $\Delta(L) \neq 0$. The reason for the factor of 1728 will become clear in §11. The remarkable fact is that the j-invariant $j(L)$ characterizes the lattice L up to homothety:

THEOREM 10.9. *If L and L' are lattices in $\mathbb{C}$, then $j(L) = j(L')$ if and only if L and L' are homothetic.*

PROOF. It is easy to see that homothetic lattices have the same j-invariant. Namely, if $\lambda \in \mathbb{C}^*$, then the definition of $g_2(L)$ and $g_3(L)$ implies that

$$\begin{aligned} g_2(\lambda L) &= \lambda^{-4} g_2(L) \\ g_3(\lambda L) &= \lambda^{-6} g_3(L), \end{aligned} \tag{10.10}$$

and $j(\lambda L) = j(L)$ follows easily.

Now suppose that L and L' are lattices such that $j(L) = j(L')$. We first claim that there is a complex number λ such that

$$\begin{aligned} g_2(L') &= \lambda^{-4} g_2(L) \\ g_3(L') &= \lambda^{-6} g_3(L). \end{aligned} \tag{10.11}$$

When $g_2(L') \neq 0$ and $g_3(L') \neq 0$, we can pick a number λ such that

$$\lambda^4 = \frac{g_2(L)}{g_2(L')}.$$

Since $j(L) = j(L')$, some easy algebra shows that

$$\lambda^{12} = \left(\frac{g_3(L)}{g_3(L')}\right)^2,$$

so that

$$\lambda^6 = \pm\frac{g_3(L)}{g_3(L')}.$$

Replacing λ by $i\lambda$ if necessary, we can assume that the above sign is $+$, and then (10.11) follows. The proof when $g_2(L') = 0$ or $g_3(L') = 0$ is similar and is left to the reader (see Exercise 10.9).

To exploit (10.11), we need to learn more about the Laurent expansion of the $\wp$-function:

LEMMA 10.12. *Let $\wp(z)$ be the Weierstrass $\wp$-function for the lattice L, and as in Lemma 10.3, let*

$$\wp(z) = \frac{1}{z^2} + \sum_{n=1}^{\infty} (2n+1) G_{2n+2}(L)\, z^{2n}$$

be its Laurent expansion. Then for $n \geq 1$, the coefficient $(2n+1)G_{2n+2}(L)$ of z^{2n} is a polynomial with rational coefficients, independent of L, in $g_2(L)$ and $g_3(L)$.

PROOF. For simplicity, we will write the coefficients of the Laurent expansion as $a_n = (2n+1)G_{2n+2}(L)$. To get a relation among the a_n's, we differentiate the equation $\wp'(z)^2 = 4\wp(z)^3 - g_2(L)\wp(z) - g_3(L)$ to obtain

$$\wp''(z) = 6\wp(z)^2 - (1/2)g_2(L).$$

By substituting in the Laurent expansion for $\wp(z)$ and comparing the coefficients of z^{2n-2}, one easily sees that for $n \geq 3$,

$$2n(2n-1)a_n = 6\left(2a_n + \sum_{i=1}^{n-2} a_i\, a_{n-1-i}\right)$$

(see Exercise 10.10), and hence

$$(2n+3)(n-2)a_n = 3\sum_{i=1}^{n-2} a_i\, a_{n-1-i}.$$

Since $g_2(L) = 60G_4(L) = 20a_1$ and $g_3(L) = 140G_6(L) = 28a_2$, an easy induction shows that a_n is a polynomial with rational coefficients in $g_2(L)$ and $g_2(L)$. This proves the lemma. Q.E.D.

Now suppose that we have lattices L and L' such that (10.11) holds for some constant λ. We claim that $L' = \lambda L$. To see this, first note that by (10.10), we have $g_2(L') = g_2(\lambda L)$ and $g_3(L') = g_3(\lambda L)$. This implies that $(2n+1)G_{2n+2}(L') = (2n+1)G_{2n+2}(\lambda L)$ for all $n \geq 1$ by Lemma 10.12, so that $\wp(z; L')$ and $\wp(z; \lambda L)$ have the same Laurent expansion about 0. Hence the two functions agree in a neighborhood of the origin, which implies that $\wp(z; L') = \wp(z; \lambda L)$ everywhere. Since the lattice is the set of poles of the $\wp$-function, this proves that $L' = \lambda L$, and Theorem 10.9 follows. Q.E.D.

Besides the notion of the j-invariant of a lattice, there is another way to think about the j-invariant which will be useful when we study modular functions. Given a complex number τ in the upper half plane $\mathfrak{h} = \{\tau \in \mathbb{C} : \mathrm{Im}(\tau) > 0\}$, we get the lattice $[1, \tau]$, and then the *j-function* $j(\tau)$ is defined by

$$j(\tau) = j([1,\tau]).$$

The analytic properties of $j(\tau)$ play an important role in the theory of complex multiplication and will be studied in detail in §11.

C. Complex Multiplication

We begin with the simple observation that orders in imaginary quadratic fields give rise to a natural class of lattices. Namely, let $\mathcal{O}$ be an order in the imaginary quadratic field K, and let $\mathfrak{a}$ be a proper fractional $\mathcal{O}$-ideal. We know from §7 that $\mathfrak{a} = [\alpha, \beta]$ for some $\alpha, \beta \in K$ (see Exercise 7.8). We can regard K as a subset of $\mathbb{C}$, and since K is imaginary quadratic, α and β are linearly independent over $\mathbb{R}$ (see Exercise 10.11). Thus $\mathfrak{a} = [\alpha, \beta]$ is a lattice in $\mathbb{C}$, and consequently the j-invariant $j(\mathfrak{a})$ is defined. These complex numbers, often called *singular moduli*, have some remarkable properties which will be explored in §11. For now, we have the more modest goal of trying to motivate the idea of complex multiplication.

In order to simplify our discussion of complex multiplication, we will fix the lattice L. As usual, $\wp(z; L)$ is written $\wp(z)$, and to simplify things further, $g_2(L)$ and $g_3(L)$ will be written g_2 and g_3. The basic idea of complex multiplication goes back to the addition law for the $\wp$-function, proved in part (iii) of Theorem 10.1. If we specialize to the case $z = w$, then L'Hospital's rule gives the following duplication formula for the $\wp$-function:

$$\wp(2z) = -2\wp(z) + \frac{1}{4}\left(\frac{\wp''(z)}{\wp'(z)}\right)^2. \tag{10.13}$$

However, the differential equation from Theorem 10.1 implies that

$$\wp'(z)^2 = 4\wp(z)^3 - g_2\wp(z) - g_3$$
$$\wp''(z) = 6\wp(z)^2 - (1/2)g_2,$$

and substituting these expressions into (10.13), we obtain

$$\wp(2z) = -2\wp(z) + \frac{(12\wp(z)^2 - g_2)^2}{16(4\wp(z)^3 - g_2\wp(z) - g_3)}.$$

Thus $\wp(2z)$ is a rational function in $\wp(z)$. More generally, one can show by induction that for any positive integer n, $\wp(nz)$ is a rational function in $\wp(z)$ (see Exercise 10.12). So the natural question to ask is whether there are any other complex numbers α for which $\wp(\alpha z)$ is a rational function in $\wp(z)$. The answer is rather surprising:

THEOREM 10.14. *Let L be a lattice, and let $\wp(z)$ be the $\wp$-function for L. Then, for a number $\alpha \in \mathbb{C} - \mathbb{Z}$, the following statements are equivalent:*

(i) *$\wp(\alpha z)$ is a rational function in $\wp(z)$.*
(ii) *$\alpha L \subset L$.*
(iii) *There is an order $\mathcal{O}$ in an imaginary quadratic field K such that $\alpha \in \mathcal{O}$ and L is homothetic to a proper fractional $\mathcal{O}$-ideal.*

Furthermore, if these conditions are satisfied, then $\wp(\alpha z)$ can be written in the form

$$\wp(\alpha z) = \frac{A(\wp(z))}{B(\wp(z))}$$

where $A(x)$ and $B(x)$ are relatively prime polynomials such that

$$\deg(A(x)) = \deg(B(x)) + 1 = [L : \alpha L] = N(\alpha).$$

PROOF. (i) $\Rightarrow$ (ii). If $\wp(\alpha z)$ is a rational function in $\wp(z)$, then there are polynomials $A(x)$ and $B(x)$ such that

$$B(\wp(z))\wp(\alpha z) = A(\wp(z)). \tag{10.15}$$

Since $\wp(z)$ and $\wp(\alpha z)$ have double poles at the origin, it follows from (10.15) that

$$\deg(A(x)) = \deg(B(x)) + 1. \tag{10.16}$$

Now let $\omega \in L$. Then (10.15) and (10.16) show that $\wp(\alpha z)$ has a pole at ω, which means that $\wp(z)$ has a pole at $\alpha\omega$. Since the poles of $\wp(z)$ are exactly the period lattice L, this implies that $\alpha\omega \in L$, and $\alpha L \subset L$ follows.

(ii) $\Rightarrow$ (i). If $\alpha L \subset L$, it follows that $\wp(\alpha z)$ is meromorphic and has L as a lattice of periods. Furthermore, note that $\wp(\alpha z)$ is an even function since $\wp(z)$ is. Then the following theorem immediately implies that $\wp(\alpha z)$ is a rational function in $\wp(z)$:

LEMMA 10.17. *Any even elliptic function for L is a rational function in $\wp(z)$.*

PROOF. The proof of this assertion is covered in Exercise 10.13. Q.E.D.

(ii) $\Rightarrow$ (iii). Suppose that $\alpha L \subset L$. Replacing L by λL for suitable λ, we can assume that $L = [1, \tau]$ for some $\tau \in \mathbb{C} - \mathbb{R}$. Then $\alpha L \subset L$ means that $\alpha = a + b\tau$ and $\alpha\tau = c + d\tau$ for some integers a, b, c and d. Taking the quotient of the two equations, we obtain

$$\tau = \frac{c + d\tau}{a + b\tau},$$

which gives us the quadratic equation

$$b\tau^2 + (a - d)\tau - c = 0.$$

Since τ is not real, we must have $b \neq 0$, and then $K = \mathbb{Q}(\tau)$ is an imaginary quadratic field. It follows that

$$\mathcal{O} = \{\beta \in K : \beta L \subset L\}$$

is an order of K for which L is a proper fractional $\mathcal{O}$-ideal, and since α is obviously in $\mathcal{O}$, we are done.

(iii) $\Rightarrow$ (ii). This implication is trivial.

Finally, we prove the last statement of the theorem. If (i) holds, then by (10.16), we have

$$\wp(\alpha z) = \frac{A(\wp(z))}{B(\wp(z))}, \tag{10.18}$$

where $\deg(A(x)) = \deg(B(x)) + 1$. Since (ii) also holds, we have $\alpha L \subset L$, and in Corollary 11.27, we will show that $N(\alpha) = [L : \alpha L]$. It remains to prove that the degree of $A(x)$ is the index $[L : \alpha L]$.

Fix $z \in \mathbb{C}$ such that $2z \notin (1/\alpha)L$, and consider the polynomial $A(x) - \wp(\alpha z)B(x)$. This polynomial has the same degree as $A(x)$, and z can be chosen so that it has distinct roots (see Exercise 10.14). Then consider the lattices $L \subset (1/\alpha)L$, and let $\{w_i\}$ be coset representatives of L in $(1/\alpha)L$. We claim that

(10.19) The $\wp(z + w_i)$ are distinct and give all roots of $A(x) - \wp(\alpha z)B(x)$.

This will imply $\deg(A(x)) = [(1/\alpha)L : L] = [L : \alpha L]$, and the theorem will be proved.

To prove (10.19), we first show that the $\wp(z+w_i)$ are distinct. If not, we would have $\wp(z+w_i) = \wp(z+w_j)$ for some $i \neq j$. Then Lemma 10.4 implies that $z+w_i \equiv \pm(z + w_j) \bmod L$. The plus sign implies $w_i \equiv w_j \bmod L$, which contradicts $i \neq j$, and the minus sign implies $2z \equiv -w_j - w_i \bmod L$, which contradicts $2z \notin (1/\alpha)L$. Thus the $\wp(z + w_i)$ are distinct.

From (10.18), we see that $A(\wp(z + w_i)) = \wp(\alpha(z + w_i))B(\wp(z + w_i))$. But $w_i \in (1/\alpha)L$, so that $\alpha(z + w_i) \equiv \alpha z \bmod L$, and hence $\wp(\alpha(z + w_i)) = \wp(\alpha z)$. This shows that the $\wp(z + w_i)$ are roots of $A(x) - \wp(\alpha z)B(x)$. To see that all roots arise this way, let u be another root. Note that $B(u) \neq 0$ since $B(u) = 0$ implies $A(u) = 0$, which is impossible since $A(x)$ and $B(x)$ are relatively prime. By adapting the argument of Lemma 10.4, it is easy to see that $u = \wp(w)$ for some complex number w (see Exercise 10.14). Then

$$\wp(\alpha z) = \frac{A(u)}{B(u)} = \frac{A(\wp(w))}{B(\wp(w))} = \wp(\alpha w),$$

and using Lemma 10.4 again, we see that $\alpha w \equiv \pm\alpha z \bmod L$. Changing w to $-w$ if necessary (which doesn't affect $u = \wp(w) = \wp(-w)$), we can assume that $w \equiv z \bmod (1/\alpha)L$. Working modulo L, this means $w \equiv z + w_i \bmod L$ for some i, and thus $u = \wp(w) = \wp(z + w_i)$ is one of the known roots. This proves (10.19), and we are done with Theorem 10.14. Q.E.D.

This theorem shows that if an elliptic function has multiplication by some $\alpha \in \mathbb{C}-\mathbb{R}$, then it has multiplication by an entire order $\mathcal{O}$ in an imaginary quadratic field. Notice that all of the elements of $\mathcal{O} - \mathbb{Z}$ are genuinely complex, i.e., not real. This accounts for the name *complex multiplication*.

One important consequence of Theorem 10.14 is that complex multiplication is an intrinsic property of the lattice. So rather than talk about elliptic functions with complex multiplication, it makes more sense to talk about lattices with complex multiplication. Since changing the lattice by a constant multiple doesn't affect the complex multiplications, we will work with homothety classes of lattices.

Using Theorem 10.14, we can relate homothety classes of lattices and ideal class groups of orders as follows. Fix an order $\mathcal{O}$ in an imaginary quadratic field, and consider those lattices $L \subset \mathbb{C}$ which have $\mathcal{O}$ as their full ring of complex multiplications. By Theorem 10.14, we can assume that L is a proper fractional $\mathcal{O}$-ideal, and conversely, every proper fractional $\mathcal{O}$-ideal is a lattice with $\mathcal{O}$ as its ring of complex multiplications. Furthermore, two proper fractional $\mathcal{O}$-ideals are homothetic as lattices if and only if they determine the same class in the ideal class group $C(\mathcal{O})$ (see Exercise 10.15). We have thus proved the following:

COROLLARY 10.20. *Let $\mathcal{O}$ be an order in an imaginary quadratic field. Then there is a one-to-one correspondence between the ideal class group $C(\mathcal{O})$ and the homothety classes of lattices with $\mathcal{O}$ as their full ring of complex multiplications.* Q.E.D.

It follows that the class number $h(\mathcal{O})$ tells us the number of homothety classes of lattices having $\mathcal{O}$ as their full ring of complex multiplications.

Here are some examples. First, consider all lattices which have complex multiplication by $\sqrt{-3}$. This means that we are dealing with an order $\mathcal{O}$ containing $\sqrt{-3}$ in the field $K = \mathbb{Q}(\sqrt{-3})$. Then $\mathcal{O}$ must be either $\mathbb{Z}[\sqrt{-3}]$ or $\mathbb{Z}[\omega]$, $\omega = e^{2\pi i/3}$, and since both of these have class number 1, the only lattices are $[1, \sqrt{-3}]$ and $[1, \omega]$. Thus, up to homothety, there are only two lattices with complex multiplication by $\sqrt{-3}$. Next, consider complex multiplication by $\sqrt{-5}$. Here, $K = \mathbb{Q}(\sqrt{-5})$, and the only order containing $\sqrt{-5}$ is the maximal order $\mathcal{O}_K = \mathbb{Z}[\sqrt{-5}]$. The class number is $h(-20) = 2$, and since we know the reduced forms of discriminant -20, the results of §7 show that up to homothety, the only lattices with complex multiplication by $\sqrt{-5}$ are $[1, \sqrt{-5}]$ and $[2, 1+\sqrt{-5}]$ (see Exercise 10.16).

The discussion so far has concentrated on the elliptic functions and their lattices. Since our ultimate goal involves the j-invariant of the lattices, we need to indicate how complex multiplication influences the j-invariant. Let's start with the simplest case, complex multiplication by $i = \sqrt{-1}$. Up to a multiple, the only possible lattice is $L = [1, i]$. To compute $j(L) = j(i)$, note that $iL = L$, so that by the homogeneity (10.11) of $g_3(L)$,

$$g_3(L) = g_3(iL) = i^{-6} g_3(L) = -g_3(L).$$

This implies that $g_3(L) = 0$, and then the formula (10.8) for the j-invariant tells us that $j(i) = 1728$. Similarly, one can show that if $L = [1, \omega]$, $\omega = e^{2\pi i/3}$, then $g_2(L) = 0$, which tells us that $j(\omega) = 0$ (see Exercise 10.17).

A more interesting example is given by complex multiplication by $\sqrt{-2}$. By the above methods, the only lattice involved is $[1, \sqrt{-2}]$, up to homothety. We will follow the exposition of Stark [**124**] and show that

$$j(\sqrt{-2}) = 8000.$$

Since $N(\sqrt{-2}) = 2$, Theorem 10.14 tells us that

$$\wp(\sqrt{-2}\, z) = \frac{A(\wp(z))}{B(\wp(z))}$$

where $A(x)$ is quadratic and $B(x)$ is linear. By dividing $B(x)$ into $A(x)$, we can write this as

$$\wp(\sqrt{-2}\,z) = a\wp(z) + b + \frac{1}{c\wp(z)+d}, \tag{10.21}$$

where a and c are nonzero complex numbers. To exploit this identity, we will use the Laurent expansion of $\wp(z)$ at $z=0$. The differential equation for $\wp(z)$ shows that the first few terms of the Laurent expansion are

$$\wp(z) = \frac{1}{z^2} + \frac{g_2}{20}z^2 + \frac{g_3}{28}z^4 + \frac{g_2^2}{1200}z^6 + \cdots$$

(this follows easily from the proof of Lemma 10.12—see Exercise 10.18). To simplify this expansion, first note that g_2 and g_3 are nonzero, for otherwise there would be complex multiplication by i or ω, which can't happen for $L = [1, \sqrt{-2}]$ (see Exercise 10.19). Then, replacing L by a suitable multiple, the homogeneity of g_2 and g_3 allows us to assume that $g_2 = 20g$ and $g_3 = 28g$ for some number g (see Exercise 10.19). With this choice of lattice, the expansion for $\wp(z)$ can be written

$$\wp(z) = \frac{1}{z^2} + g\,z^2 + g\,z^4 + \frac{g^2}{3}z^6 + \cdots,$$

and it follows that the expansion for $\wp(\sqrt{-2}\,z)$ is

$$\wp(\sqrt{-2}\,z) = \frac{-1}{2z^2} - 2g\,z^2 + 4g\,z^4 - \frac{8g^2}{3}z^6 + \cdots.$$

Now the constants a and b appearing in (10.21) are the unique constants such that $\wp(\sqrt{-2}\,z) - a\wp(z) - b$ is zero when $z=0$. Comparing the above expansions for $\wp(z)$ and $\wp(\sqrt{-2}\,z)$, we see that $a = -1/2$ and $b = 0$. Then (10.21) tells us the remarkable fact that $(\wp(\sqrt{-2}\,z) + \frac{1}{2}\wp(z))^{-1}$ is a linear polynomial in $\wp(z)$. Using the above expansions, one computes that

$$\begin{aligned}\left(\wp(\sqrt{-2}\,z) + \frac{1}{2}\wp(z)\right)^{-1} &= \left(-\frac{3g}{2}z^2 + \frac{9g}{2}z^4 - \frac{5g^2}{2}z^6 + \cdots\right)^{-1}\\ &= -\frac{2}{3g\,z^2} - \frac{2}{g} - \frac{2}{3g}\left(9 - \frac{5g}{3}\right)z^2 + \cdots\end{aligned} \tag{10.22}$$

(see Exercise 10.19). By (10.21), this expression is linear in $\wp(z)$. Looking at the behavior at $z=0$, it follows that the bottom line of (10.22) must equal

$$-\frac{2}{3g}\wp(z) - \frac{2}{g},$$

and then comparing the coefficients of z^2 implies that

$$-\frac{2}{3g}\left(9 - \frac{5g}{3}\right) = -\frac{2}{3g}g.$$

Solving this equation for g yields $g = \frac{27}{8}$, so that

$$\begin{aligned} g_2 &= 20g = \frac{5\cdot 27}{2}\\ g_3 &= 28g = \frac{7\cdot 27}{2},\end{aligned}$$

and thus

$$j(\sqrt{-2}) = 1728\frac{g_2^3}{g_2^3 - 27g_3^2} = 8000 = 20^3.$$

By a similar computation, one can also show that

$$j\left(\frac{1+\sqrt{-7}}{2}\right) = -3375 = (-15)^3$$

(see Exercise 10.20). In §12 we will explain why these numbers are cubes.

Besides allowing us to compute $j(\sqrt{-2})$ and $j((1+\sqrt{-7})/2)$, the Laurent series of the $\wp$-function can be used to give an elementary proof that the j-invariant of a lattice with complex multiplication is an algebraic number:

THEOREM 10.23. *Let $\mathcal{O}$ be an order in an imaginary quadratic field, and let $\mathfrak{a}$ be a proper fractional $\mathcal{O}$-ideal. Then $j(\mathfrak{a})$ is an algebraic number of degree at most $h(\mathcal{O})$.*

PROOF. By Lemma 10.12, the Laurent expansion of $\wp(z)$ can be written

$$\wp(z) = \frac{1}{z^2} + \sum_{n=1}^{\infty} a_n(g_2, g_3)\, z^{2n},$$

where each $a_n(g_2, g_3)$ is a polynomial in g_2 and g_3 with rational coefficients. To emphasize the dependence on g_2 and g_3, we will write $\wp(z)$ as $\wp(z; g_2, g_3)$.

By assumption, for any $\alpha \in \mathcal{O}$, $\wp(\alpha z)$ is a rational function in $\wp(z)$, say

$$\wp(\alpha z; g_2, g_3) = \frac{A(\wp(z; g_2, g_3))}{B(\wp(z; g_2, g_3))}. \tag{10.24}$$

We then have the Laurent expansion

$$\wp(\alpha z; g_2, g_3) = \frac{1}{\alpha^2 z^2} + \sum_{n=1}^{\infty} a_n(g_2, g_3)\alpha^{2n} z^{2n},$$

which means that (10.24) can be regarded as an identity in the field $\mathbb{C}((z))$ of formal meromorphic Laurent series. Recall that $\mathbb{C}((z))$ is the field of fractions of the formal power series ring $\mathbb{C}[[z]]$, so that an element of $\mathbb{C}((z))$ is a series of the form $\sum_{n=-M}^{\infty} b_n z^n$, $b_n \in \mathbb{C}$.

Now let σ be any automorphism of $\mathbb{C}$. Then σ induces an automorphism of $\mathbb{C}((z))$ by acting on the coefficients. Thus, if we apply σ to (10.24), we obtain the identity

$$\wp(\sigma(\alpha) z; \sigma(g_2), \sigma(g_3)) = \frac{A^\sigma(\wp(z; \sigma(g_2), \sigma(g_3)))}{B^\sigma(\wp(z; \sigma(g_2), \sigma(g_3)))}, \tag{10.25}$$

where $A^\sigma(x)$ (resp. $B^\sigma(x)$) is the polynomial obtained by applying σ to the coefficients of $A(x)$ (resp. $B(x)$). This follows because $a_n(g_2, g_3)$ is a polynomial in g_2 and g_3 with rational coefficients. We don't know much about $\sigma(g_2)$ and $\sigma(g_3)$, but $g_2^3 - 27g_3^2 \neq 0$ implies $\sigma(g_2)^3 - 27\sigma(g_3)^2 \neq 0$. In §11, we will prove that this condition on $\sigma(g_2)$ and $\sigma(g_3)$ guarantees that there is a lattice L such that

$$\begin{aligned} g_2(L) &= \sigma(g_2) \\ g_3(L) &= \sigma(g_3) \end{aligned}$$

(see Corollary 11.7). Thus the formal Laurent series $\wp(z; \sigma(g_2), \sigma(g_3))$ is the Laurent series of the $\wp$-function $\wp(z; L)$, and then (10.25) tells us that $\wp(z; L)$ has complex multiplication by $\sigma(\alpha)$. This holds for any $\alpha \in \mathcal{O}$, so that if $\mathcal{O}'$ is the ring of all complex multiplications of L, then we have proved that

$$\mathcal{O} = \sigma(\mathcal{O}) \subset \mathcal{O}'.$$

If we replace σ with σ^{-1} and interchange $\mathfrak{a}$ and L, the above argument shows that $\mathcal{O}' \subset \mathcal{O}$, which shows that $\mathcal{O} = \mathcal{O}'$ is the ring of all complex multiplications of both $\mathfrak{a}$ and L.

Now consider j-invariants. The above formulas for $g_2(L)$ and $g_3(L)$ imply that

$$j(L) = \sigma(j(\mathfrak{a})). \tag{10.26}$$

Since L has $\mathcal{O}$ as its ring of complex multiplications, Corollary 10.20 implies that there are only $h(\mathcal{O})$ possibilities for $j(L)$. By (10.26), there are thus at most $h(\mathcal{O})$ possibilities for $\sigma(j(\mathfrak{a}))$. Since σ was an *arbitrary* automorphism of $\mathbb{C}$, it follows that $j(\mathfrak{a})$ must be an algebraic number, and in fact the degree of its minimal polynomial over $\mathbb{Q}$ is at most $h(\mathcal{O})$. This proves the theorem. Q.E.D.

In §11 we will prove the stronger result that $j(\mathfrak{a})$ is an algebraic *integer* and that the degree of its minimal polynomial *equals* the class number $h(\mathcal{O})$. But we thought it worthwhile to show what can be done by elementary means. Furthermore, the method of proof used above (the action of an automorphism on the coefficients of a Laurent expansion) is similar to some of the arguments to be given in §11.

For a more classical introduction to complex multiplication, the reader should consult the book [**12**] by Borwein and Borwein.

D. Exercises

10.1. This exercise is concerned with the proof of Lemma 10.2.

(a) If $L = [\omega_1, \omega_2]$ is a lattice, let $M = \min\{|x\omega_1 + y\omega_2| : x^2 + y^2 = 1\}$. Show that $M > 0$ and that $|x\omega_1 + y\omega_2| \geq M\sqrt{x^2+y^2}$ for all $x, y \in \mathbb{R}$.

(b) Show that the double integral $\iint_{x^2+y^2\geq 1}(x^2+y^2)^{-r/2}\,dx\,dy$ converges when $r > 2$.

(c) Show that the series $\sum'_{m,n}(m^2+n^2)^{-r/2}$ converges when $r > 2$. Hint: compare the series to the integral in part (b).

10.2. In the proof of Theorem 10.1, we proved that $\wp'(z) = -2\sum_{\omega\in L}(z-\omega)^{-3}$.

(a) Show that this series converges absolutely for $z \notin L$.

(b) Using (a), show that $\wp'(z+\omega) = \wp'(z)$ for $\omega \in L$.

10.3. Show that for $|x| < 1$, $(1-x)^{-2} = \sum_{n=0}^{\infty}(n+1)x^n$. Hint: differentiate the standard identity $(1-x)^{-1} = \sum_{n=0}^{\infty} x^n$.

10.4. Use Lemma 10.3 to show that

$$\begin{aligned}\wp(z)^3 &= \frac{1}{z^6} + \frac{9G_4(L)}{z^2} + 15G_6(L) + \cdots\\ \wp'(z)^2 &= \frac{4}{z^6} - \frac{24G_4(L)}{z^2} - 80G_6(L) + \cdots,\end{aligned}$$

where $+\cdots$ indicates terms involving positive powers of z.

10.5. Use Liouville's Theorem to show that a holomorphic elliptic function $f(z)$ must be constant. Hint: consider the continuous function $|f(z)|$ on the parallelogram $\{s\omega_1 + t\omega_2 : 0 \leq s,t \leq 1\}$. Exercise 10.6 will be useful.

10.6. Let $L = [\omega_1, \omega_2]$ be a lattice. For a fixed $\alpha \in \mathbb{C}$, consider the parallelogram $\mathbf{P} = \{\alpha + s\omega_1 + t\omega_2 : 0 \leq s,t \leq 1\}$. Show that if $z \in \mathbb{C}$, then there is $z' \in \mathbf{P}$ such that $z \equiv z' \bmod L$. Note that the parallelogram used in Lemma 10.4 corresponds to $\alpha = \delta\omega_1 + \delta\omega_2$.

10.7. As in the proof of the addition theorem, let

$$G(z) = \wp(z+w) + \wp(z) + \wp(w) - \frac{1}{4}\left(\frac{\wp'(z) - \wp'(w)}{\wp(z) - \wp(w)}\right)^2.$$

(a) If $2w \notin L$, complete the argument begun in the text to show that $G(-w)$ is defined.

(b) Prove the addition law when $2w \in L$. Hint: take a sequence of points w_i converging to w such that $2w_i \notin L$ for all i.

10.8. Let $4x^3 - g_2x - g_3$ be a cubic polynomial with roots e_1, e_2 and e_3.

(a) Show that $e_1+e_2+e_3 = 0$, $e_1e_2+e_1e_3+e_2e_3 = -g_2/4$ and $e_1e_2e_3 = g_3/4$.

(b) Using (a), show that $g_2^3 - 27g_3^2 = 16(e_1-e_2)^2(e_1-e_3)^2(e_2-e_3)^2$.

10.9. Let L and L' be lattices such that $j(L) = j(L')$. If $g_2(L') = 0$ or $g_3(L') = 0$, prove that there is a complex number λ such that (10.11) holds. Hint: they can't both be zero by Proposition 10.7.

10.10. As in Lemma 10.3, let the Laurent expansion of the $\wp$-function about 0 be $\wp(z) = z^{-2} + \sum_{n=1}^{\infty} a_n\, z^{2n}$, where $a_n = (2n+1)G_{2n+2}$.

(a) Use the differential equation for the $\wp$-function to show that $\wp''(z) = 6\wp(z)^2 - (1/2)g_2(L)$.

(b) Use part (a) to show that for $n \geq 3$,

$$2n(2n-1)a_n = 6\left(2a_n + \sum_{i=1}^{n-2} a_i\, a_{n-1-i}\right).$$

10.11. Let K be an imaginary quadratic field, which we regard as a subfield of $\mathbb{C}$.

(a) If $\mathcal{O}$ is an order in K and $\mathfrak{a} = [\alpha, \beta]$ is a proper fractional $\mathcal{O}$-ideal, then show that α and β are linearly independent over $\mathbb{R}$. Thus $\mathfrak{a} \subset \mathbb{C}$ is a lattice.

(b) Conversely, let $L \subset \mathbb{C}$ be a lattice which is contained in K. Show that L is a proper fractional $\mathcal{O}$-ideal for some order $\mathcal{O}$ of K. Hint: Lemma 7.5.

10.12. Let L be a lattice, and let n be a positive integer.

(a) Prove that $\wp(nz)$ is a rational function in $\wp(z)$. Hint: use the addition law and induction on n. For a quicker proof, use Lemma 10.17.

(b) Adapt the proof of Theorem 10.14 to show that the numerator of the rational function of part (a) has degree n^2 and the denominator has degree $n^2 - 1$.

10.13. In this exercise we will show how to express elliptic functions for a given lattice L in terms of $\wp(z)$ and $\wp'(z)$.

(a) Let $f(z)$ be an even elliptic function which is holomorphic on $\mathbb{C} - L$. Prove that $f(z)$ is a polynomial in $\wp(z)$. Hint: show that there is a polynomial $A(x)$ such that the Laurent expansion of $f(z) - A(\wp(z))$ has only terms of nonnegative degree. Then use Exercise 10.5.

(b) Let $f(z)$ be an even elliptic function that has a pole of order m at $w \in \mathbb{C}$. Assume that $w \notin L$. Prove that $(\wp(z) - \wp(w))^m f(z)$ is holomorphic at w.

(c) Show that an even elliptic function $f(z)$ is a rational function in $\wp(z)$. This will prove Lemma 10.17. Hint: write $L = [\omega_1, \omega_2]$, and consider the parallelogram $\mathbf{P} = \{s\omega_1 + t\omega_2 : 0 \leq s, t \leq 1\}$. Note that only finitely many poles of $f(z)$ lie in $\mathbf{P}$. Now use part (b) to find a polynomial $B(x)$ such that $B(\wp(z))f(z)$ is holomorphic on $\mathbb{C} - L$ (use Exercise 10.6). Then the claim follows easily by part (a).

(d) Show that all elliptic functions for L are rational functions in $\wp(z)$ and $\wp'(z)$. Hint:

$$f(z) = \frac{f(z) + f(-z)}{2} + \left(\frac{f(z) - f(-z)}{2\wp'(z)}\right)\wp'(z).$$

10.14. This exercise is concerned with the proof of Theorem 10.14.

(a) Let $A(x)$ and $B(x)$ be relatively prime polynomials. Prove that there are only finitely many complex numbers λ such that the polynomial $A(x) - \lambda B(x)$ has a multiple root. Hint: show that every multiple root is a root of $A(x)B'(x) - A'(x)B(x)$.

(b) Adapt the proof of Lemma 10.4 to show that for any complex number u, the equation $u = \wp(w)$ always has a solution.

10.15. Let $\mathfrak{a}$ and $\mathfrak{b}$ be two proper fractional $\mathcal{O}$-ideals, where $\mathcal{O}$ is an order in an imaginary quadratic field. Prove that $\mathfrak{a}$ and $\mathfrak{b}$ determine the same class in the ideal class group $C(\mathcal{O})$ if and only if they are homothetic as lattices in $\mathbb{C}$.

10.16. In this exercise we will study lattices with complex multiplication by a fixed complex number $\alpha \in \mathbb{C} - \mathbb{Z}$.

(a) Verify that up to a multiple, the only lattices with complex multiplication by $\sqrt{-5}$ are $[1, \sqrt{-5}]$ and $[2, 1 + \sqrt{-5}]$.

(b) Determine, up to a multiple, all lattices with complex multiplication by $\sqrt{-14}$. Hint: see the example following Theorem 5.30.

(c) Let K be an imaginary quadratic field of discriminant d_K, and let $\alpha \in \mathcal{O}_K - \mathbb{Z}$. Show that up to homothety, the number of lattices with complex multiplication by α is given by

$$\sum_{f \mid [\mathcal{O}_K : \mathbb{Z}[\alpha]]} h(f^2 d_K).$$

10.17. Let $\omega = e^{2\pi i/3}$, and let L be the lattice $[1, \omega]$. Show that $g_2(L) = j(\omega) = 0$.

10.18. Use the proof of Lemma 10.12 to show that in a neighborhood of $z = 0$, the Laurent expansion of the $\wp$-function is

$$\wp(z) = \frac{1}{z^2} + \frac{g_2}{20}z^2 + \frac{g_3}{28}z^4 + \frac{g_2^2}{1200}z^6 + \cdots.$$

10.19. This exercise is concerned with the computation $j(\sqrt{-2}) = 8000$.

(a) If L is a lattice with $g_2(L) = 0$, then prove that L is a multiple of $[1, \omega]$, where $\omega = e^{2\pi i/3}$. Hint: use Theorem 10.9 and Exercise 10.17.

(b) Similarly, show that if $g_3(L) = 0$, then L is a multiple of $[1, i]$.

(c) If L is a lattice with $g_2 g_3 \neq 0$, then show that there is a nonzero complex number λ such that for some $g \in \mathbb{C}$, $\lambda^{-4} g_2 = 20g$ and $\lambda^{-6} g_3 = 28g$.

(d) Verify the computations made in (10.22).

10.20. Show that $j((1 + \sqrt{-7})/2) = -3375$.

11. Modular Functions and Ring Class Fields

In §10 we studied complex multiplication, and we saw that for an order $\mathcal{O}$ in an imaginary quadratic field, the j-invariant $j(\mathfrak{a})$ of a proper fractional $\mathcal{O}$-ideal $\mathfrak{a}$ is an algebraic number. This suggests a strong connection with number theory, and the goal of §11 is to unravel this connection by relating $j(\mathfrak{a})$ to the ring class field of $\mathcal{O}$ introduced in §9. The precise statement of this relation is the "First Main Theorem" of complex multiplication, which is the main result of this section:

THEOREM 11.1. *Let $\mathcal{O}$ be an order in an imaginary quadratic field K, and let $\mathfrak{a}$ be a proper fractional $\mathcal{O}$-ideal. Then the j-invariant $j(\mathfrak{a})$ is an algebraic integer and $K(j(\mathfrak{a}))$ is the ring class field of the order $\mathcal{O}$.*

For a fixed order $\mathcal{O}$, we will prove in §13 that the $j(\mathfrak{a})$'s are all conjugate and hence are roots of the same irreducible polynomial over $\mathbb{Q}$. This polynomial is called the *class equation* of $\mathcal{O}$ and will be studied in detail in §13.

Of special interest is the case when $\mathcal{O} = \mathbb{Z}[\sqrt{-n}]$. Here, Theorem 11.1 implies that $j(\mathcal{O}) = j(\sqrt{-n})$ is an algebraic integer and is a primitive element of the ring class field of $\mathbb{Z}[\sqrt{-n}]$. It is elementary to see that $j(\sqrt{-n})$ is real (see Exercise 11.1), and thus, by Theorem 9.2, the class equation of $\mathbb{Z}[\sqrt{-n}]$ can be used to characterize primes of the form $p = x^2 + ny^2$.

Before we can prove Theorem 11.1, we need to learn about modular functions and the modular equation. The first step is to study the j-function $j(\tau)$ in detail.

A. The j-Function

The j-invariant $j(L)$ of a lattice L was defined in §10 in terms of the constants $g_2(L)$ and $g_3(L)$. Given τ in the upper half plane $\mathfrak{h}$, we get the lattice $[1, \tau]$, and then the j-function $j(\tau)$ is defined by

$$j(\tau) = j([1, \tau]).$$

We also define $g_2(\tau)$ and $g_3(\tau)$ by

$$g_2(\tau) = g_2([1, \tau]) = 60 \sum_{m,n}{}' \frac{1}{(m + n\tau)^4}$$

$$g_3(\tau) = g_3([1, \tau]) = 140 \sum_{m,n}{}' \frac{1}{(m + n\tau)^6},$$

where $\sum'_{m,n}$ denotes summation over all ordered pairs of integers $(m, n) \neq (0, 0)$. By (10.8), it follows that $j(\tau)$ is given by the formula

$$j(\tau) = 1728 \, \frac{g_2(\tau)^3}{\Delta(\tau)},$$

where $\Delta(\tau) = g_2(\tau)^3 - 27 g_3(\tau)^2$.

The properties of $j(\tau)$ are closely related to the action of $\mathrm{SL}(2, \mathbb{Z})$ on the upper half plane $\mathfrak{h}$. This action is defined as follows: if $\tau \in \mathfrak{h}$ and $\gamma = \left(\begin{smallmatrix} a & b \\ c & d \end{smallmatrix}\right) \in \mathrm{SL}(2, \mathbb{Z})$, then

$$\gamma\tau = \frac{a\tau + b}{c\tau + d}.$$

It is easy to check that $\gamma\tau \in \mathfrak{h}$ (see Exercise 11.2), and we say that $\gamma\tau$ and τ are $\mathrm{SL}(2, \mathbb{Z})$-equivalent. Then the j-function has the following properties:

THEOREM 11.2.

(i) *$j(\tau)$ is a holomorphic function on $\mathfrak{h}$.*

(ii) *If τ and τ' lie in $\mathfrak{h}$, then $j(\tau) = j(\tau')$ if and only if $\tau' = \gamma\tau$ for some $\gamma \in \mathrm{SL}(2,\mathbb{Z})$. In particular, $j(\tau)$ is $\mathrm{SL}(2,\mathbb{Z})$-invariant.*

(iii) *$j : \mathfrak{h} \to \mathbb{C}$ is surjective.*

(iv) *For $\tau \in \mathfrak{h}$, $j'(\tau) \neq 0$, except in the following cases:*

 (a) *$\tau = \gamma i$, $\gamma \in \mathrm{SL}(2,\mathbb{Z})$, where $j'(\tau) = 0$, $j''(\tau) \neq 0$.*

 (b) *$\tau = \gamma\omega$, $\omega = e^{2\pi i/3}$, $\gamma \in \mathrm{SL}(2,\mathbb{Z})$, where $j'(\tau) = j''(\tau) = 0$, $j'''(\tau) \neq 0$.*

PROOF. To prove (i), recall from Proposition 10.7 that $\Delta(\tau)$ never vanishes. Thus it suffices to show that $g_2(\tau)$ and $g_3(\tau)$ are holomorphic. For $g_2(\tau)$, this works as follows. By Lemma 10.2, the sum defining $g_2(\tau)$ converges absolutely, but we still must show that the convergence is uniform on compact subsets of $\mathfrak{h}$. To see this, first note that $g_2(\tau + 1) = g_2(\tau)$ (this follows from absolute convergence). Thus it suffices to show that convergence is uniform when τ satisfies $|\mathrm{Re}(\tau)| \leq 1/2$ and $\mathrm{Im}(\tau) \geq \epsilon$, where $\epsilon < 1$ is an arbitrary positive number. In this case it is easy to show that

$$|m + n\tau| \geq \frac{\epsilon}{2}\sqrt{m^2 + n^2}$$

(see Exercise 11.3), and then uniform convergence is immediate. The proof for $g_3(\tau)$ is similar, so that $g_2(\tau)$, $g_3(\tau)$, $\Delta(\tau)$ and $j(\tau)$ are all holomorphic on $\mathfrak{h}$.

Turning to (ii), we need to recall the following fact from §7: if $\tau, \tau' \in \mathfrak{h}$, then

$$[1, \tau] \text{ and } [1, \tau'] \text{ are homothetic} \iff \tau' = \gamma\tau \text{ for some } \gamma \in \mathrm{SL}(2,\mathbb{Z}).$$

See (7.8) for the proof (in §7, we assumed that τ and τ' lay in an imaginary quadratic field, but the proof given for (7.8) holds for arbitrary $\tau, \tau' \in \mathfrak{h}$). From Theorem 10.9, we also know that

$$j(\tau) = j(\tau') \iff [1, \tau] \text{ and } [1, \tau'] \text{ are homothetic.}$$

Combining these two equivalences, (ii) is immediate.

Before we can prove (iii), we need to compute the limits of $g_2(\tau)$ and $g_3(\tau)$ as $\mathrm{Im}(\tau) \to \infty$. To study $g_2(\tau)$, write

$$g_2(\tau) = 60 \sum_{m,n}{}' \frac{1}{(m + n\tau)^4} = 60\left(2\sum_{m=1}^{\infty} \frac{1}{m^4} + \sum_{\substack{m,n=-\infty \\ n \neq 0}}^{\infty} \frac{1}{(m + n\tau)^4}\right).$$

Using the uniform convergence proved in (i), we see that

$$\lim_{\mathrm{Im}(\tau)\to\infty} g_2(\tau) = 120 \sum_{m=1}^{\infty} \frac{1}{m^4},$$

and then the well-known formula $\sum_{m=1}^{\infty} 1/m^4 = \pi^4/90$ (see Serre [**112**, §VII.4.1]) implies that

$$\lim_{\mathrm{Im}(\tau)\to\infty} g_2(\tau) = \frac{4}{3}\pi^4.$$

The case of $g_3(\tau)$ is similar. Here, the key formula is $\sum_{m=1}^{\infty} 1/m^6 = \pi^6/945$ (see Serre [**112**, §VII.4.1]), and one obtains

$$\lim_{\mathrm{Im}(\tau)\to\infty} g_3(\tau) = \frac{8}{27}\pi^6.$$

These limits imply that

$$\lim_{\operatorname{Im}(\tau)\to\infty} \Delta(\tau) = \left(\frac{4}{3}\pi^4\right)^3 - 27\left(\frac{8}{27}\pi^6\right)^2 = 0,$$

and it follows easily that

$$\lim_{\operatorname{Im}(\tau)\to\infty} j(\tau) = \infty. \tag{11.3}$$

We will also need the following lemma:

LEMMA 11.4. *Every $\tau \in \mathfrak{h}$ is $\mathrm{SL}(2,\mathbb{Z})$-equivalent to a point $\tau' \in \mathfrak{h}$ which satisfies $|\operatorname{Re}(\tau')| \le 1/2$ and $\operatorname{Im}(\tau') \ge 1/2$.*

PROOF. If $\operatorname{Im}(\tau) \ge 1/2$, then there is an integer m such that $\tau' = \tau + m$ satisfies $|\operatorname{Re}(\tau')| \le 1/2$ and $\operatorname{Im}(\tau') \ge 1/2$. Since $\tau' = \tau + m = \left(\begin{smallmatrix} 1 & m \\ 0 & 1 \end{smallmatrix}\right)\tau$, we are done in this case.

If $\operatorname{Im}(\tau) < 1/2$, then by the argument of the previous paragraph, we can assume $|\operatorname{Re}(\tau)| \le 1/2$. It follows that $|\tau| < 1/\sqrt{2}$, so that

$$\operatorname{Im}\left(\frac{-1}{\tau}\right) = \frac{\operatorname{Im}(\tau)}{|\tau|^2} > 2\operatorname{Im}(\tau).$$

Since $-1/\tau = \left(\begin{smallmatrix} 0 & -1 \\ 1 & 0 \end{smallmatrix}\right)\tau$, we can more than double the imaginary part of τ by using an element of $\mathrm{SL}(2,\mathbb{Z})$. Repeating this process as often as needed, we eventually obtain a $\mathrm{SL}(2,\mathbb{Z})$-equivalent point $\tau' \in \mathfrak{h}$ which satisfies $\operatorname{Im}(\tau') \ge 1/2$. Q.E.D.

This lemma is related to the idea of finding a *fundamental domain* for the action of $\mathrm{SL}(2,\mathbb{Z})$ on $\mathfrak{h}$. We won't use this concept in the text, but there is an interesting relation between fundamental domains and reduced forms (in the sense of Theorem 2.8). See Exercise 11.4 for the details.

We can now show that the j-function is surjective. Since it's holomorphic and nonconstant, its image is an open subset of $\mathbb{C}$. If we can show that the image is closed, surjectivity will follow. So take a sequence of points $j(\tau_k)$ which converges to some $w \in \mathbb{C}$. We need to show that $w = j(\tau)$ for some $\tau \in \mathfrak{h}$. By Lemma 11.4, we can assume that each τ_k lies in the region $R = \{\tau \in \mathfrak{h} : |\operatorname{Re}(\tau)| \le 1/2,\ \operatorname{Im}(\tau) \ge 1/2\}$. If the imaginary parts of the τ_k's were unbounded, then by the limit (11.3), the $j(\tau_k)$'s would have a subsequence which converged to ∞. This is clearly impossible. But once the imaginary parts are bounded, the τ_k's lie in a compact subset of $\mathfrak{h}$. Then they have a subsequence converging to some $\tau \in \mathfrak{h}$, and it follows by continuity that $j(\tau) = w$, as desired.

The proof of (iv) will use the following lemma:

LEMMA 11.5. *If $\tau, \tau' \in \mathfrak{h}$, then there exist neighborhoods U of τ and V of τ' such that the set $\{\gamma \in \mathrm{SL}(2,\mathbb{Z}) : \gamma(U) \cap V \ne \emptyset\}$ is finite.*

PROOF. This lemma says that $\mathrm{SL}(2,\mathbb{Z})$ acts properly discontinuously on $\mathfrak{h}$. The proof is given in Exercise 11.5. Q.E.D.

COROLLARY 11.6. *If $\tau \in \mathfrak{h}$, then τ has a neighborhood U such that for all $\gamma \in \mathrm{SL}(2,\mathbb{Z})$,*

$$\gamma(U) \cap U \ne \emptyset \iff \gamma\tau = \tau.$$

PROOF. See Exercise 11.5. Q.E.D.

Now suppose that $j'(\tau) = 0$. Then τ has a neighborhood U such that for w sufficiently close to $j(\tau)$, there are $\tau' \neq \tau'' \in U$ such that $j(\tau') = j(\tau'') = w$. By (ii), $\tau'' = \gamma\tau'$ for some $\gamma \neq \pm I$, where $I = \left(\begin{smallmatrix} 1 & 0 \\ 0 & 1 \end{smallmatrix}\right)$. Thus $\gamma(U) \cap U \neq \emptyset$. By shrinking U and using Corollary 11.6, it follows that $\gamma\tau = \tau$, $\gamma \neq \pm I$. This is a very strong restriction on τ. To see why, let $\gamma = \left(\begin{smallmatrix} a & b \\ c & d \end{smallmatrix}\right)$. Then $\gamma\tau = \tau$ implies that

$$[1, \tau] = (c\tau + d)[1, \tau]$$

(see the proof of (7.8)), and since $\gamma \neq \pm I$, an easy argument shows that $c \neq 0$ (see Exercise 11.6). Thus $\alpha = c\tau + d \notin \mathbb{Z}$, so that by Theorem 10.14, the lattice $[1, \tau]$ has complex multiplication by an order $\mathcal{O}$ in an imaginary quadratic field. Furthermore, $\alpha[1, \tau] = [1, \tau]$ implies that $\alpha \in \mathcal{O}^*$. However, we know that $\mathcal{O}^* = \{\pm 1\}$ unless $\mathcal{O} = \mathcal{O}_K$ for $K = \mathbb{Q}(i)$ or $\mathbb{Q}(\omega)$, $\omega = e^{2\pi i/3}$ (see Exercise 11.6). Both of these orders have class number 1, so that $[1, \tau]$ is homothetic to either $[1, i]$ or $[1, \omega]$. Thus $j'(\tau) = 0$ implies that τ is $\mathrm{SL}(2, \mathbb{Z})$-equivalent to either i or ω.

When τ is $\mathrm{SL}(2, \mathbb{Z})$-equivalent to i, we may assume that $\tau = i$, and we need to show that $j'(i) = 0$ and $j''(i) \neq 0$. To prove the former, note that

$$j(\tau) - 1728 = 1728\,\frac{27g_3(\tau)^2}{\Delta(\tau)}.$$

In §10 we proved that $g_3(i) = 0$, and $j'(i) = 0$ follows immediately. Now suppose that $j''(i) = 0$. Then i is at least a triple zero of $j(\tau) - 1728$, so that for w sufficiently near 1728, there are distinct points τ, τ' and τ'' near i such that $j(\tau) = j(\tau') = j(\tau'') = w$. Then $\tau' = \gamma_1\tau$, $\tau'' = \gamma_2\tau$, where $\pm I$, $\pm\gamma_1$ and $\pm\gamma_2$ are all distinct elements of $\mathrm{SL}(2, \mathbb{Z})$. By Corollary 11.6, $\gamma_1 i = \gamma_2 i = i$, so that at least 6 elements of $\mathrm{SL}(2, \mathbb{Z})$ fix i. Since only 4 elements of $\mathrm{SL}(2, \mathbb{Z})$ fix i (see Exercise 11.6), we see that $j''(i) \neq 0$. The case when $\tau = \omega$ is similar and is left to the reader (see Exercise 11.6). Theorem 11.2 is proved. Q.E.D.

The surjectivity of the j-function implies the following result used in §10:

COROLLARY 11.7. *Let g_2 and g_3 be complex numbers such that $g_2^3 - 27g_3^2 \neq 0$. Then there is a lattice L such that $g_2(L) = g_2$ and $g_3(L) = g_3$.*

PROOF. Since the j-function is surjective and $g_2^3 - 27g_3^2 \neq 0$, there is $\tau \in \mathfrak{h}$ such that

$$j(\tau) = 1728\,\frac{g_2^3}{g_2^3 - 27g_3^2}.$$

Arguing as in the proof of (10.11), this equation implies that there is a nonzero complex number λ such that

$$g_2 = \lambda^{-4} g_2(\tau)$$
$$g_3 = \lambda^{-6} g_3(\tau).$$

Using (10.10), it follows that $L = \lambda[1, \tau]$ is the desired lattice. Q.E.D.

Since $j(\tau)$ is invariant under $\mathrm{SL}(2, \mathbb{Z})$, we see that

$$j(\tau + 1) = j\left(\begin{pmatrix} 1 & 1 \\ 0 & 1 \end{pmatrix}\tau\right) = j(\tau).$$

This implies that $j(\tau)$ is a holomorphic function in $q = q(\tau) = e^{2\pi i\tau}$, defined in the region $0 < |q| < 1$. Consequently $j(\tau)$ has a Laurent expansion

$$j(\tau) = \sum_{n=-\infty}^{\infty} c_n\, q^n,$$

which is called the *q-expansion* of $j(\tau)$. The following theorem will be used often in what follows:

THEOREM 11.8. *The q-expansion of $j(\tau)$ is*

$$j(\tau) = \frac{1}{q} + 744 + 196884\, q + \cdots = \frac{1}{q} + \sum_{n=0}^{\infty} c_n\, q^n,$$

where the coefficients c_n are integers for all $n \geq 0$.

PROOF. We will prove this in §12 using the Weber functions and the Weierstrass σ-function. See Apostol [**1**, §1.15] or Lang [**90**, §4.1] for other proofs. Q.E.D.

This theorem is the reason that the factor 1728 appears in the definition of the j-invariant: it's exactly the factor needed to guarantee that all of the coefficients of the q-expansion are integers without any common divisor.

B. Modular Functions for $\Gamma_0(m)$

One can define modular functions for any subgroup of $\mathrm{SL}(2,\mathbb{Z})$, but we will concentrate on the subgroups $\Gamma_0(m)$ of $\mathrm{SL}(2,\mathbb{Z})$, which are defined as follows: if m is a positive integer, then

$$\Gamma_0(m) = \left\{ \begin{pmatrix} a & b \\ c & d \end{pmatrix} \in \mathrm{SL}(2,\mathbb{Z}) : c \equiv 0 \bmod m \right\}.$$

Note that $\Gamma_0(1) = \mathrm{SL}(2,\mathbb{Z})$. Then a *modular function for* $\Gamma_0(m)$ is a complex-valued function $f(\tau)$ defined on the upper half plane $\mathfrak{h}$, except for isolated singularities, which satisfies the following three conditions:

(i) $f(\tau)$ is meromorphic on $\mathfrak{h}$.
(ii) $f(\tau)$ is invariant under $\Gamma_0(m)$.
(iii) $f(\tau)$ is meromorphic at the cusps.

By (ii), we mean that $f(\gamma\tau) = f(\tau)$ for all $\tau \in \mathfrak{h}$ and $\gamma \in \Gamma_0(m)$. To explain (iii), more work is needed. Suppose that $f(\tau)$ satisfies (i) and (ii), and take $\gamma \in \mathrm{SL}(2,\mathbb{Z})$. We claim that $f(\gamma\tau)$ has period m. To see this, note that $\tau + m = U\tau$, where $U = \left(\begin{smallmatrix} 1 & m \\ 0 & 1 \end{smallmatrix}\right)$. An easy calculation shows that $\gamma U \gamma^{-1} \in \Gamma_0(m)$, and we then obtain

$$f(\gamma(\tau + m)) = f(\gamma U \tau) = f(\gamma U \gamma^{-1} \gamma\tau) = f(\gamma\tau)$$

since $f(\tau)$ is $\Gamma_0(m)$-invariant. It follows that if $q = q(\tau) = e^{2\pi i\tau}$, then $f(\gamma\tau)$ is a holomorphic function in $q^{1/m}$, defined for $0 < |q^{1/m}| < 1$. Thus $f(\gamma\tau)$ has a Laurent expansion

$$f(\gamma\tau) = \sum_{n=-\infty}^{\infty} a_n\, q^{n/m},$$

which by abuse of notation we will call the *q-expansion* of $f(\gamma\tau)$. Then $f(\tau)$ is *meromorphic at the cusps* if for all $\gamma \in \mathrm{SL}(2,\mathbb{Z})$, the q-expansion of $f(\gamma\tau)$ has only finitely many nonzero coefficients for negative exponents.

The basic example of such a function is given by $j(\tau)$. It is holomorphic on $\mathfrak{h}$, invariant under $\mathrm{SL}(2,\mathbb{Z})$, and Theorem 11.8 implies that it is meromorphic at the cusps. Thus $j(\tau)$ is a modular function for $\mathrm{SL}(2,\mathbb{Z}) = \Gamma_0(1)$. The remarkable fact is that modular functions for both $\mathrm{SL}(2,\mathbb{Z})$ and $\Gamma_0(m)$ are easily described in terms of the j-function:

THEOREM 11.9. *Let m be a positive integer.*

(i) *$j(\tau)$ is a modular function for $\mathrm{SL}(2,\mathbb{Z})$. Furthermore, every modular function for $\mathrm{SL}(2,\mathbb{Z})$ is a rational function in $j(\tau)$.*

(ii) *$j(\tau)$ and $j(m\tau)$ are modular functions for $\Gamma_0(m)$. Furthermore, every modular function for $\Gamma_0(m)$ is a rational function of $j(\tau)$ and $j(m\tau)$.*

PROOF. Note that (i) is a special case of (ii). It is stated separately not only because of its independent interest, but also because it's what we must prove first.

Before beginning the proof, let's make a comment about q-expansions. Our definition requires checking the q-expansion of $f(\gamma\tau)$ for all $\gamma \in \mathrm{SL}(2,\mathbb{Z})$. Since $f(\tau)$ is $\Gamma_0(m)$-invariant, we actually need only consider the q-expansions of $f(\gamma_i\tau)$, where the γ_i's are right coset representatives of $\Gamma_0(m) \subset \mathrm{SL}(2,\mathbb{Z})$. So there are only finitely many q-expansions to check. The nicest case is when $f(\tau)$ is a modular function for $\mathrm{SL}(2,\mathbb{Z})$, for here we need only consider the q-expansion of $f(\tau)$.

We can now prove (i). We've seen that $j(\tau)$ is a modular function for $\mathrm{SL}(2,\mathbb{Z})$, so we need only show that every modular function $f(\tau)$ for $\mathrm{SL}(2,\mathbb{Z})$ is a rational function in $j(\tau)$. We will begin by studying some special cases. We say that a modular function $f(\tau)$ is *holomorphic at* ∞ if its q-expansion involves only nonnegative powers of q.

LEMMA 11.10.

(i) *A holomorphic modular function for $\mathrm{SL}(2,\mathbb{Z})$ which is holomorphic at ∞ is constant.*

(ii) *A holomorphic modular function for $\mathrm{SL}(2,\mathbb{Z})$ is a polynomial in $j(\tau)$.*

PROOF. To prove (i), let $f(\tau)$ be the modular function in question. Since $f(\tau)$ is holomorphic at ∞, we know that $f(\infty) = \lim_{\mathrm{Im}(\tau)\to\infty} f(\tau)$ exists as a complex number. We will show that $f(\mathfrak{h} \cup \{\infty\})$ is compact. By the maximum modulus principle, this will imply that $f(\tau)$ is constant.

Let $f(\tau_k)$ be a sequence of points in the image. We need to find a subsequence that converges to a point of the form $f(\tau)$ for some $\tau \in \mathfrak{h} \cup \{\infty\}$. Since $f(\tau)$ is invariant under the action of $\mathrm{SL}(2,\mathbb{Z})$, we can assume that the τ_k's lie in the region $R = \{\tau \in \mathfrak{h} : |\mathrm{Re}(\tau)| \leq 1/2,\ \mathrm{Im}(\tau) \geq 1/2\}$ (see Lemma 11.4). If the imaginary parts of the τ_k's are unbounded, then by the above limit, a subsequence converges to $f(\infty)$. If the imaginary parts are bounded, then the τ_k's lie in a compact subset of $\mathfrak{h}$, and the desired subsequence is easily found. This proves (i).

Turning to (ii), let $f(\tau)$ be a holomorphic modular function for $\mathrm{SL}(2,\mathbb{Z})$. Its q-expansion has only finitely many terms with negative powers of q. Since the q-expansion of $j(\tau)$ begins with $1/q$, it is easy to find a polynomial $A(x)$ such that $f(\tau) - A(j(\tau))$ is holomorphic at ∞. Since it is also holomorphic on $\mathfrak{h}$, it is constant by (i). Thus $f(\tau)$ is a polynomial in $j(\tau)$, and the lemma is proved. Q.E.D.

To treat the general case, let $f(\tau)$ be an arbitrary modular function for $\mathrm{SL}(2,\mathbb{Z})$, possibly with poles on $\mathfrak{h}$. If we can find a polynomial $B(x)$ such that $B(j(\tau))f(\tau)$ is holomorphic on $\mathfrak{h}$, then the lemma will imply that $f(\tau)$ is a rational function in $j(\tau)$. Since $f(\tau)$ has a meromorphic q-expansion, it follows that $f(\tau)$ has only finitely many poles in the region $R = \{\tau \in \mathfrak{h} : |\mathrm{Re}(\tau)| \le 1/2,\ \mathrm{Im}(\tau) \ge 1/2\}$, and since $f(\tau)$ is $\mathrm{SL}(2,\mathbb{Z})$-invariant, Lemma 11.4 implies that every pole of $f(\tau)$ is $\mathrm{SL}(2,\mathbb{Z})$-equivalent to one in R. It follows that if $B(j(\tau))f(\tau)$ has no poles in R, then it is holomorphic on $\mathfrak{h}$.

Suppose that $f(\tau)$ has a pole of order m at $\tau_0 \in R$. Then $(j(\tau)-j(\tau_0))^m f(\tau)$ is holomorphic at τ_0. In this way we can find a polynomial $B(x)$ such that $B(j(\tau))f(\tau)$ has no poles in R. By the previous paragraph, we conclude that $f(\tau)$ is a rational function in $j(\tau)$, which completes the proof of part (i) of Theorem 11.9.

To prove part (ii), it is trivial to show that $j(\tau)$ is a modular function for $\Gamma_0(m)$. As for $j(m\tau)$, it is certainly holomorphic, and to check its invariance properties, let $\gamma = \left(\begin{smallmatrix} a & b \\ c & d \end{smallmatrix}\right) \in \Gamma_0(m)$. Then

$$j(m\gamma\tau) = j\left(\frac{m(a\tau+b)}{c\tau+d}\right) = j\left(\frac{a\cdot m\tau + bm}{c/m\cdot m\tau + d}\right).$$

Since $\gamma \in \Gamma_0(m)$, it follows that $\gamma' = \left(\begin{smallmatrix} a & bm \\ c/m & d \end{smallmatrix}\right) \in \mathrm{SL}(2,\mathbb{Z})$. Thus

$$j(m\gamma\tau) = j(\gamma' m\tau) = j(m\tau),$$

which proves that $j(m\tau)$ is $\Gamma_0(m)$-invariant.

In order to show that $j(m\tau)$ is meromorphic at the cusps, we first relate $\Gamma_0(m)$ to the set of matrices

$$C(m) = \left\{\begin{pmatrix} a & b \\ 0 & d \end{pmatrix} : ad = m,\ a > 0,\ 0 \le b < d,\ \gcd(a,b,d) = 1\right\}.$$

The matrix $\sigma_0 = \left(\begin{smallmatrix} m & 0 \\ 0 & 1 \end{smallmatrix}\right) \in C(m)$ has two properties of interest: first, $\sigma_0\tau = m\tau$, and second,

$$\Gamma_0(m) = (\sigma_0^{-1}\mathrm{SL}(2,\mathbb{Z})\,\sigma_0) \cap \mathrm{SL}(2,\mathbb{Z})$$

(see Exercise 11.8). Note that these two properties account for the $\Gamma_0(m)$-invariance of $j(m\tau)$ proved above. More generally, we have the following lemma:

LEMMA 11.11. *For $\sigma \in C(m)$, the set*

$$(\sigma_0^{-1}\mathrm{SL}(2,\mathbb{Z})\,\sigma) \cap \mathrm{SL}(2,\mathbb{Z})$$

is a right coset of $\Gamma_0(m)$ in $\mathrm{SL}(2,\mathbb{Z})$. This induces a one-to-one correspondence between right cosets of $\Gamma_0(m)$ and elements of $C(m)$.

PROOF. See Exercise 11.8. Q.E.D.

This lemma implies that $[\mathrm{SL}(2,\mathbb{Z}) : \Gamma_0(m)] = |C(m)|$. One can also compute the number of elements in $C(m)$; the formula is

$$|C(m)| = m\prod_{p|m}\left(1+\frac{1}{p}\right)$$

(see Exercise 11.9), and thus the index of $\Gamma_0(m)$ is $\mathrm{SL}(2,\mathbb{Z})$ is $m\prod_{p|m}(1+1/p)$.

We can now compute some q-expansions. Fix $\gamma \in \mathrm{SL}(2,\mathbb{Z})$, and choose $\sigma \in C(m)$ so that γ lies in the right coset corresponding to σ in Lemma 11.11. This means that $\sigma_0\gamma = \tilde{\gamma}\sigma$ for some $\tilde{\gamma} \in \mathrm{SL}(2,\mathbb{Z})$, and hence $j(m\gamma\tau) = j(\sigma_0\gamma\tau) = j(\tilde{\gamma}\sigma\tau) = j(\sigma\tau)$ since $j(\tau)$ is $\mathrm{SL}(2,\mathbb{Z})$-invariant. Hence

$$j(m\gamma\tau) = j(\sigma\tau). \tag{11.12}$$

Let $\sigma = \left(\begin{smallmatrix} a & b \\ 0 & d \end{smallmatrix}\right)$. We know from Theorem 11.8 that the q-expansion of $j(\tau)$ is

$$j(\tau) = \frac{1}{q} + \sum_{n=0}^{\infty} c_n\, q^n, \quad c_n \in \mathbb{Z},$$

and since $\sigma\tau = (a\tau + b)/d$, it follows that

$$q(\sigma\tau) = e^{2\pi i(a\tau+b)/d} = e^{2\pi i b/d} q^{a/d}.$$

If we set $\zeta_m = e^{2\pi i/m}$, we can write this as

$$q(\sigma\tau) = \zeta_m^{ab}(q^{1/m})^{a^2}$$

since $ad = m$. This gives us the q-expansion

$$j(m\gamma\tau) = j(\sigma\tau) = \frac{\zeta_m^{-ab}}{(q^{1/m})^{a^2}} + \sum_{n=0}^{\infty} c_n \zeta_m^{abn}(q^{1/m})^{a^2 n}, \quad c_n \in \mathbb{Z}. \tag{11.13}$$

There are only finitely many negative exponents, which shows that $j(m\tau)$ is meromorphic at the cusps, and thus $j(m\tau)$ is a modular function for $\Gamma_0(m)$.

The next step is to introduce the modular equation $\Phi_m(X, Y)$. Let the right cosets of $\Gamma_0(m)$ in $\mathrm{SL}(2, \mathbb{Z})$ be $\Gamma_0(m)\gamma_i$, $i = 1, \ldots, |C(m)|$. Then consider the polynomial in X

$$\Phi_m(X, \tau) = \prod_{i=1}^{|C(m)|} (X - j(m\gamma_i\tau)).$$

We will prove that this expression is a polynomial in X and $j(\tau)$. To see this, consider the coefficients of $\Phi_m(X, \tau)$. Being symmetric polynomials in the $j(m\gamma_i\tau)$'s, they are certainly holomorphic. To check invariance under $\mathrm{SL}(2, \mathbb{Z})$, pick $\gamma \in \mathrm{SL}(2, \mathbb{Z})$. Then the cosets $\Gamma_0(m)\gamma_i\gamma$ are a permutation of the $\Gamma_0(m)\gamma_i$'s, and since $j(m\tau)$ is invariant under $\Gamma_0(m)$, the $j(m\gamma_i\gamma\tau)$'s are a permutation of the $j(m\gamma_i\tau)$'s. This shows that the coefficients of $\Phi_m(X, \tau)$ are invariant under $\mathrm{SL}(2, \mathbb{Z})$.

We next have to show that the coefficients are meromorphic at infinity. Rather than expand in powers of q, it suffices to expand in terms of $q^{1/m} = e^{2\pi i\tau/m}$ and show that only finitely negative exponents appear.

By (11.12), we know that $j(m\gamma_i\tau) = j(\sigma\tau)$ for some $\sigma \in C(m)$, and then (11.13) shows that the q-expansion for $j(m\gamma_i\tau)$ has only finitely many negative exponents. Since the coefficients are polynomials in the $j(m\gamma_i\tau)$'s they clearly are meromorphic at the cusps.

This proves that the coefficients of $\Phi_m(X, \tau)$ are holomorphic modular functions, and thus, by Lemma 11.10, they are polynomials in $j(\tau)$. This means that there is a polynomial

$$\Phi_m(X, Y) \in \mathbb{C}[X, Y]$$

such that

$$\Phi_m(X, j(\tau)) = \prod_{i=1}^{|C(m)|} (X - j(m\gamma_i\tau)). \tag{11.14}$$

The equation $\Phi_m(X, Y) = 0$ is called the *modular equation*, and by abuse of terminology we will call $\Phi_m(X, Y)$ the modular equation. Using some simple field theory, it can be proved that $\Phi_m(X, Y)$ is irreducible as a polynomial in X (see Exercise 11.10).

By (11.12), each $j(m\gamma_i\tau)$ can be written $j(\sigma\tau)$ for a unique $\sigma \in C(m)$. Thus we can also express the modular equation in the form

$$\Phi_m(X, j(\tau)) = \prod_{\sigma \in C(m)} (X - j(\sigma\tau)). \tag{11.15}$$

Note that $j(m\tau)$ is always one of the $j(\sigma\tau)$'s since $\left(\begin{smallmatrix} m & 0 \\ 0 & 1 \end{smallmatrix}\right) \in C(m)$. Hence

$$\Phi_m(j(m\tau), j(\tau)) = 0,$$

which is one of the important properties of the modular equation. Note that the degree of $\Phi_m(X,Y)$ in X is $|C(m)|$, which we know equals $m\prod_{p|m}(1+1/p)$.

Now let $f(\tau)$ be an arbitrary modular function for $\Gamma_0(m)$. To prove that $f(\tau)$ is a rational function in $j(\tau)$ and $j(m\tau)$, consider the function

$$\begin{aligned} G(X,\tau) &= \Phi_m(X, j(\tau)) \sum_{i=1}^{|C(m)|} \frac{f(\gamma_i\tau)}{X - j(m\gamma_i\tau)} \\ &= \sum_{i=1}^{|C(m)|} f(\gamma_i\tau) \prod_{j\neq i}(X - j(m\gamma_j\tau)). \end{aligned} \tag{11.16}$$

This is a polynomial in X, and we claim that its coefficients are modular functions for $\mathrm{SL}(2,\mathbb{Z})$. The proof is similar to what we did for the modular equation, and the details are left to the reader (see Exercise 11.11). But once the coefficients are modular functions for $\mathrm{SL}(2,\mathbb{Z})$, they are rational functions of $j(\tau)$ by what we proved above. Hence $G(X,\tau)$ is a polynomial $G(X, j(\tau)) \in \mathbb{C}(j(\tau))[X]$.

We can assume that γ_1 is the identity matrix. By the product rule, we obtain

$$\frac{\partial \Phi_m}{\partial X}(j(m\tau), j(\tau)) = \prod_{j\neq 1}(j(m\tau)) - j(m\gamma_j\tau)).$$

Thus, substituting $X = j(m\tau)$ in (11.16) gives

$$G(j(m\tau), j(\tau)) = f(\tau)\frac{\partial \Phi_m}{\partial X}(j(m\tau), j(\tau)).$$

Now $\Phi_m(X, j(\tau))$ is irreducible (see Exercise 11.10) and hence separable, so that $(\partial/\partial X)\Phi_m(j(m\tau), j(\tau)) \neq 0$. Thus we can write

$$f(\tau) = \frac{G(j(m\tau), j(\tau))}{\frac{\partial \Phi_m}{\partial X}(j(m\tau), j(\tau))}, \tag{11.17}$$

which proves that $f(\tau)$ is a rational function in $j(\tau)$ and $j(m\tau)$, This completes the proof of Theorem 11.9. Q.E.D.

There is a large literature on modular functions, and the reader may wish to consult Apostol [**1**], Koblitz [**84**], Lang [**90**] or Shimura [**115**] to learn more about these remarkable functions.

C. The Modular Equation $\Phi_m(X, Y)$

The modular equation, as defined by equations (11.14) or (11.15), will play a crucial role in what follows. In particular, we will make heavy use of the arithmetic properties of $\Phi_m(X,Y)$, which are given in the following theorem:

THEOREM 11.18. *Let m be a positive integer.*

(i) $\Phi_m(X,Y) \in \mathbb{Z}[X,Y]$.

(ii) $\Phi_m(X, Y)$ *is irreducible when regarded as a polynomial in* X.
(iii) $\Phi_m(X, Y) = \Phi_m(Y, X)$ *if* $m > 1$.
(iv) *If* m *is not a perfect square, then* $\Phi_m(X, X)$ *is a polynomial of degree* > 1 *whose leading coefficient is* ± 1.
(v) *If* m *is a prime* p*, then* $\Phi_p(X, Y) \equiv (X^p - Y)(X - Y^p) \bmod p\mathbb{Z}[X, Y]$.

PROOF. To prove (i), it suffices to show that an elementary symmetric function $f(\tau)$ in the $j(\sigma\tau)$'s, $\sigma \in C(m)$, is a polynomial in $j(\tau)$ with integer coefficients. We begin by studying the q-expansion of $f(\tau)$ in more detail.

Let $\zeta_m = e^{2\pi i/m}$. By (11.13), each $j(\sigma\tau)$ lies in the field of formal meromorphic Laurent series $\mathbb{Q}(\zeta_m)((q^{1/m}))$, and since $f(\tau)$ is an integer polynomial in the $j(\sigma\tau)$'s, $f(\tau)$ also lies in $\mathbb{Q}(\zeta_m)((q^{1/m}))$.

We claim that $f(\tau)$ is contained in the smaller field $\mathbb{Q}((q^{1/m}))$. To see this, we will use Galois theory. An automorphism $\psi \in \mathrm{Gal}(\mathbb{Q}(\zeta_m)/\mathbb{Q})$ determines an automorphism of $\mathbb{Q}(\zeta_m)((q^{1/m}))$ by acting on the coefficients. Given $\sigma = \left(\begin{smallmatrix} a & b \\ 0 & d \end{smallmatrix}\right) \in C(m)$, let's see how ψ affects $j(\sigma\tau)$. We know that $\psi(\zeta_m) = \zeta_m^k$ for some integer k relatively prime to m, and from (11.13), it follows that

$$\psi(j(\sigma\tau)) = \frac{\zeta_m^{-abk}}{(q^{1/m})^{a^2}} + \sum_{n=0}^{\infty} c_n \zeta_m^{abkn} (q^{1/m})^{a^2 n}$$

since all of the c_n's are integers. Let b' be the unique integer $0 \le b' < d$ such that $b' \equiv bk \bmod d$. Since $ad = m$, we have $\zeta_m^{abk} = \zeta_m^{ab'}$, and consequently the above formula can be written

$$\psi(j(\sigma\tau)) = \frac{\zeta_m^{-ab'}}{(q^{1/m})^{a^2}} + \sum_{n=0}^{\infty} c_n \zeta_m^{ab'n} (q^{1/m})^{a^2 n}.$$

If we let $\sigma' = \left(\begin{smallmatrix} a & b' \\ 0 & d \end{smallmatrix}\right)$, then $\sigma' \in C(m)$, and (11.13) implies that

$$\psi(j(\sigma\tau)) = j(\sigma'\tau).$$

Thus the elements of $\mathrm{Gal}(\mathbb{Q}(\zeta_m)/\mathbb{Q})$ permute the $j(\sigma\tau)$'s. Since $f(\tau)$ is symmetric in the $j(\sigma\tau)$'s, it follows that $f(\tau) \in \mathbb{Q}((q^{1/m}))$.

We conclude that $f(\tau) \in \mathbb{Z}((q))$ since the q-expansion of $f(\tau)$ involves only integral powers of q and the coefficients of the q-expansion are algebraic integers. It remains to show that $f(\tau)$ is an integer polynomial in $j(\tau)$. By Lemma 11.10, we can find $A(X) \in \mathbb{C}[X]$ such that $f(\tau) = A(j(\tau))$. Recall from the proof of Lemma 11.10 that $A(X)$ was chosen so that the q-expansion of $f(\tau) - A(j(\tau))$ has only terms of degree > 0. Since the expansions of $f(\tau)$ and $j(\tau)$ have integer coefficients and $j(\tau) = 1/q + \cdots$, it follows that $A(X) \in \mathbb{Z}[X]$. Thus $f(\tau) = A(j(\tau))$ is an integer polynomial in $j(\tau)$, and (i) is proved.

We should mention that the passage from the coefficients of the q-expansion to the coefficients of the polynomial $A(X)$ is a special case of Hasse's q-expansion principle—see Exercise 11.12 for a precise formulation.

A proof of (ii) is given in Exercise 11.10, and a proof of (iii) may be found in Lang [**90**, §5.2, Theorem 3].

Turning to (iv), assume that m is not a square. We want to study the leading term of the integer polynomial $\Phi_m(X, X)$. Replacing X with $j(\tau)$, it suffices to study the coefficient of the most negative power of q in the q-expansion of

$\Phi_m(j(\tau), j(\tau))$. However, given $\sigma = \left(\begin{smallmatrix} a & b \\ 0 & d \end{smallmatrix}\right) \in C(m)$, (11.13) tells us that

$$j(\tau) - j(\sigma\tau) = \frac{1}{q} - \frac{\zeta_m^{-ab}}{q^{a/d}} + \sum_{n=0}^{\infty} d_n (q^{1/m})^n \tag{11.19}$$

for some coefficients d_n. Since m is not a perfect square, we know that $a \neq d$, i.e., $a/d \neq 1$. Thus the coefficient of the most negative term in (11.19) is a root of unity. By (11.15), $\Phi_m(j(\tau), j(\tau))$ is the product of the factors (11.19), so that the coefficient of the most negative power of q in $\Phi_m(j(\tau), j(\tau))$ is also a root of unity. But this coefficient is an integer, and thus it must be ± 1, as claimed.

Finally, we turn to (v). Here, we are assuming that $m = p$, where p is prime. Let $\zeta_p = e^{2\pi i/p}$. We will use the following notation: given $f(\tau)$ and $g(\tau)$ in $\mathbb{Z}[\zeta_p]((q^{1/p}))$ and $\alpha \in \mathbb{Z}[\zeta_p]$, we will write

$$f(\tau) \equiv g(\tau) \bmod \alpha$$

to indicate that $f(\tau) - g(\tau) \in \alpha\mathbb{Z}[\zeta_p]((q^{1/p}))$.

Since p is prime, the elements of $C(p)$ are easy to write down:

$$\begin{aligned} \sigma_i &= \begin{pmatrix} 1 & i \\ 0 & p \end{pmatrix}, \qquad i = 0, \ldots, p-1 \\ \sigma_p &= \begin{pmatrix} p & 0 \\ 0 & 1 \end{pmatrix}. \end{aligned}$$

If $0 < i \leq p-1$, then (11.13) tells us that

$$j(\sigma_i\tau) = \frac{\zeta_p^{-i}}{q^{1/p}} + \sum_{n=0}^{\infty} c_n \zeta_p^{in} (q^{1/p})^n \equiv \frac{1}{q^{1/p}} + \sum_{n=0}^{\infty} c_n (q^{1/p})^n \bmod 1 - \zeta_p,$$

which implies that

$$j(\sigma_i\tau) \equiv j(\sigma_0\tau) \bmod 1 - \zeta_p \tag{11.20}$$

for $0 \leq i \leq p-1$. Turning to $j(\sigma_p\tau)$, here (11.13) tells us that

$$j(\sigma_p\tau) = \frac{1}{q^p} + \sum_{n=0}^{\infty} c_n\, q^{pn},$$

and since $c_n^p \equiv c_n \bmod p$, it follows easily that

$$j(\sigma_p\tau) \equiv j(\tau)^p \bmod p.$$

Since $1 - \zeta_p$ divides p in $\mathbb{Z}[\zeta_p]$ (see Exercise 11.13), the above congruence can be written

$$j(\sigma_p\tau) \equiv j(\tau)^p \bmod 1 - \zeta_p. \tag{11.21}$$

Then (11.20) and (11.21) imply that

$$\begin{aligned} \Phi_p(X, j(\tau)) &= \prod_{i=0}^{p} (X - j(\sigma_i\tau)) \\ &\equiv (X - j(\sigma_0\tau))^p (X - j(\tau)^p) \bmod 1 - \zeta_p \\ &\equiv (X^p - j(\sigma_0\tau)^p)(X - j(\tau)^p) \bmod 1 - \zeta_p, \end{aligned}$$

where we are now working in the ring $\mathbb{Z}[\zeta_p]((q^{1/p}))[X]$. However, the argument used to prove (11.21) is easily adapted to prove that

$$j(\tau) = j(\sigma_0\tau)^p \bmod 1 - \zeta_p$$

(see Exercise 11.14), and then we obtain

$$\Phi_p(X, j(\tau)) \equiv (X^p - j(\tau))(X - j(\tau)^p) \bmod 1 - \zeta_p.$$

The two sides of this congruence lie in $\mathbb{Z}((q))[X]$, so that the coefficients of the difference are ordinary integers divisible by $1 - \zeta_p$ in the ring $\mathbb{Z}[\zeta_p]$. This implies that all of the coefficients are divisible by p (see Exercise 11.13), and thus

$$\Phi_p(X, j(\tau)) \equiv (X^p - j(\tau))(X - j(\tau)^p) \bmod p\mathbb{Z}((q))[X].$$

Then the Hasse q-expansion principle (used in the proof of (i)) shows that

$$\Phi_p(X, Y) \equiv (X^p - Y)(X - Y^p) \bmod p\mathbb{Z}[X, Y],$$

as desired (see Exercise 11.15). The above congruence was first discovered by Kronecker (in a slightly different context) and is sometimes called *Kronecker's congruence.* This completes the proof of Theorem 11.18. Q.E.D.

The properties of the modular equation are straightforward consequences of the properties of the j-function, which makes the modular equation seem like a reasonable object to deal with. This is true as long as one works at the abstract level, but as soon as one asks for concrete examples, the situation gets surprisingly complicated. For example, when $m = 3$, Smith [**121**] showed that $\Phi_3(X, Y)$ is the polynomial

$$\begin{aligned} &X(X + 2^{15} \cdot 3 \cdot 5^3)^3 + Y(Y + 2^{15} \cdot 3 \cdot 5^3)^3 + 2^3 \cdot 3^2 \cdot 31\, X^2Y^2(X + Y) \\ &- X^3Y^3 - 2^2 \cdot 3^3 \cdot 9907\, XY(X^2 + Y^2) + 2 \cdot 3^4 \cdot 13 \cdot 193 \cdot 6367\, X^2Y^2 \\ &+ 2^{16} \cdot 3^5 \cdot 5^3 \cdot 17 \cdot 263\, XY(X + Y) - 2^{31} \cdot 5^6 \cdot 22973\, XY. \end{aligned} \tag{11.22}$$

The modular equation $\Phi_m(X, Y)$ has been computed for $m = 5$, 7 and 11 (see Hermann [**69**] and Kaltofen and Yui [**82**]), and in §13 we will discuss the problem of computing $\Phi_m(X, Y)$ for general m.

Before we can apply the modular equation to complex multiplication, one task remains: we need to understand the modular equation in terms of j-invariants of lattices. The basic idea is that if L is a lattice, then the roots of $\Phi_m(X, j(L)) = 0$ are given by the j-invariants of those sublattices $L' \subset L$ which satisfy:

(i) L' is a sublattice of index m in L, i.e., $[L : L'] = m$.
(ii) The quotient L/L' is a cyclic group.

In this situation, we say that L' is a *cyclic sublattice* of L of index m. Here is the precise statement of what we want to prove:

THEOREM 11.23. *Let m be a positive integer. If $u, v \in \mathbb{C}$, then $\Phi_m(u, v) = 0$ if and only if there is a lattice L and a cyclic sublattice $L' \subset L$ of index m such that $u = j(L')$ and $v = j(L)$.*

PROOF. We will first study the cyclic sublattices of the lattice $[1, \tau]$, $\tau \in \mathfrak{h}$:

LEMMA 11.24. *Let $\tau \in \mathfrak{h}$, and consider the lattice $[1, \tau]$.*

(i) *Given a cyclic sublattice $L' \subset [1, \tau]$ of index m, there is a unique $\sigma = \left(\begin{smallmatrix} a & b \\ 0 & d \end{smallmatrix}\right) \in C(m)$ such that $L' = d[1, \sigma\tau]$.*
(ii) *Conversely, if $\sigma = \left(\begin{smallmatrix} a & b \\ 0 & d \end{smallmatrix}\right) \in C(m)$, then $d[1, \sigma\tau]$ is a cyclic sublattice of $[1, \tau]$ of index m.*

PROOF. First recall that $C(m)$ is the set of matrices

$$C(m) = \left\{ \left(\begin{smallmatrix} a & b \\ 0 & d \end{smallmatrix}\right) : ad = m,\ a > 0,\ 0 \le b < d,\ \gcd(a,b,d) = 1 \right\}.$$

A sublattice $L' \subset L = [1,\tau]$ can be written $L' = [a\tau + b, c\tau + d]$, and in Exercise 7.15 we proved that $[L : L'] = |ad - bc| = m$. Furthermore, a standard argument using elementary divisors shows that

$$L/L' \text{ is cyclic} \iff \gcd(a,b,c,d) = 1 \tag{11.25}$$

(see, for example, Lang [**90**, pp. 51–52]). Another proof of (11.25) is given in Exercise 11.16.

Now suppose that $L' \subset [1,\tau]$ is cyclic of index m. If d is the smallest positive integer contained in L', then it follows easily that L' is of the form $L' = [d, a\tau + b]$ (see Exercise 11.17). We may assume that $a > 0$, and then $ad = m$. However, if k is any integer, then

$$L' = [d, (a\tau + b) + kd] = [d, a\tau + (b + kd)],$$

so that by choosing k appropriately, we can assume $0 \le b < d$. We also have $\gcd(a,b,d) = 1$ by (11.25), and thus the matrix $\sigma = \left(\begin{smallmatrix} a & b \\ 0 & d \end{smallmatrix}\right)$ lies in $C(m)$. Then

$$L' = [d, a\tau + b] = d[1, (a\tau + b)/d] = d[1, \sigma\tau]$$

shows that L' has the desired form. It is straightforward to prove that $\sigma \in C(m)$ is uniquely determined by L' (see Exercise 11.17), and (i) is proved.

The proof of (ii) follows immediately from (11.25), and we are done. Q.E.D.

By this lemma, the j-invariants of the cyclic sublattices L' of index m of $[1,\tau]$ are given by

$$j(L') = j(d[1,\sigma\tau]) = j([1,\sigma\tau]) = j(\sigma\tau).$$

By (11.15), it follows that the roots of $\Phi_m(X, j(\tau)) = 0$ are exactly the j-invariants of the cyclic sublattices of index m of $[1,\tau]$. It is now easy to complete the proof of Theorem 11.23 (see Exercise 11.18 for the details). Q.E.D.

D. Complex Multiplication and Ring Class Fields

To prove Theorem 11.1, we will apply the modular equation to lattices with complex multiplication. The key point is that such lattices have some especially interesting cyclic sublattices. To construct these sublattices, we will use the notion of a *primitive* ideal. Given an order $\mathcal{O}$, we say that a proper $\mathcal{O}$-ideal is *primitive* if it is not of the form $d\mathfrak{a}$ where $d > 1$ is an integer and $\mathfrak{a}$ is a proper $\mathcal{O}$-ideal. Then primitive ideals give us cyclic sublattices as follows:

LEMMA 11.26. *Let $\mathcal{O}$ be an order in an imaginary quadratic field, and let $\mathfrak{b}$ be a proper fractional $\mathcal{O}$-ideal. Then, given a proper $\mathcal{O}$-ideal $\mathfrak{a}$, $\mathfrak{ab}$ is a sublattice of $\mathfrak{b}$ of index $N(\mathfrak{a})$, and $\mathfrak{ab}$ is a cyclic sublattice if and only if $\mathfrak{a}$ is a primitive ideal.*

PROOF. Replacing $\mathfrak{b}$ by a multiple, we can assume that $\mathfrak{b} \subset \mathcal{O}$. Then the exact sequence

$$0 \longrightarrow \mathfrak{b}/\mathfrak{ab} \longrightarrow \mathcal{O}/\mathfrak{ab} \longrightarrow \mathcal{O}/\mathfrak{b} \longrightarrow 0$$

implies that $[\mathfrak{b} : \mathfrak{ab}]N(\mathfrak{b}) = N(\mathfrak{ab}) = N(\mathfrak{a})N(\mathfrak{b})$, and $[\mathfrak{b} : \mathfrak{ab}] = N(\mathfrak{a})$ follows.

Now assume that $\mathfrak{b}/\mathfrak{ab}$ is not cyclic. By part (a) of Exercise 11.16, it follows that $\mathfrak{b}/\mathfrak{ab}$ contains a subgroup isomorphic to $(\mathbb{Z}/d\mathbb{Z})^2$ for some $d > 1$, so that there is a sublattice $\mathfrak{ab} \subset \mathfrak{b}' \subset \mathfrak{b}$ such that $\mathfrak{b}'/\mathfrak{ab} \simeq (\mathbb{Z}/d\mathbb{Z})^2$. Since $\mathfrak{b}'$ is rank 2, this implies that $\mathfrak{a}\mathfrak{b} = d\mathfrak{b}'$, and then $\mathfrak{a} = d\mathfrak{b}'\mathfrak{b}^{-1}$. But $\mathfrak{b}'\mathfrak{b}^{-1} \subset \mathcal{O}$ since $\mathfrak{b}' \subset \mathfrak{b}$, which shows that $\mathfrak{a}$ is not primitive.

The converse, that $\mathfrak{a}$ not primitive implies that $\mathfrak{b}/\mathfrak{a}\mathfrak{b}$ not cyclic, is even easier to prove, and is left to the reader (see Exercise 11.19). This completes the proof of the lemma. Q.E.D.

When we apply this lemma, $\mathfrak{a}$ will often be a principal ideal $\mathfrak{a} = \alpha\mathcal{O}$, $\alpha \in \mathcal{O}$. In this case, $\alpha\mathcal{O}$ is primitive as an ideal if and only if α is primitive as an element of $\mathcal{O}$ (which means that α is not of the form $d\beta$ where $d > 1$ and $\beta \in \mathcal{O}$). Since $N(\alpha) = N(\alpha\mathcal{O})$ by Lemma 7.14, we get the following corollary of Lemma 11.26:

COROLLARY 11.27. *Let $\mathcal{O}$ and $\mathfrak{b}$ be as above. Then, given $\alpha \in \mathcal{O}$, $\alpha\mathfrak{b}$ is a sublattice of $\mathfrak{b}$ of index $N(\alpha)$, and $\alpha\mathfrak{b}$ is a cyclic sublattice if and only if α is primitive.* Q.E.D.

We are now ready to prove Theorem 11.1, the "First Main Theorem" of complex multiplication.

PROOF OF THEOREM 11.1. Let $\mathfrak{a}$ be a proper fractional $\mathcal{O}$-ideal, where $\mathcal{O}$ is an order in an imaginary quadratic field K. We must prove that $j(\mathfrak{a})$ is an algebraic integer and that $K(j(\mathfrak{a}))$ is the ring class field of $\mathcal{O}$. We will follow the proof given by Deuring in [**32**, §10].

Let's first use the modular equation to prove that $j(\mathfrak{a})$ is an algebraic integer. The basic idea is quite simple: let $\alpha \in \mathcal{O}$ be primitive so that by the above corollary, $\alpha\mathfrak{a}$ is a cyclic sublattice of $\mathfrak{a}$ of index $m = N(\alpha)$. Then, by Theorem 11.23, we know that

$$0 = \Phi_m(j(\alpha\mathfrak{a}), j(\mathfrak{a})) = \Phi_m(j(\mathfrak{a}), j(\mathfrak{a})) = 0$$

since $j(\alpha\mathfrak{a}) = j(\mathfrak{a})$. Thus $j(\mathfrak{a})$ is a root of $\Phi_m(X, X)$. Since $\Phi_m(X, Y)$ has integer coefficients (part (i) of Theorem 11.18), this shows that $j(\mathfrak{a})$ is an algebraic number. Furthermore, if we can pick α so that $m = N(\alpha)$ is not a perfect square, then the leading coefficient of $\Phi_m(X, X)$ is ± 1 (part (iv) of Theorem 11.18), and thus $j(\mathfrak{a})$ will be an algebraic integer. So can we find a primitive $\alpha \in \mathcal{O}$ such that $N(\alpha)$ is not a perfect square? We will see below in (11.28) that $\mathcal{O}$ has lots of α's such that $N(\alpha)$ is prime. Such an α is certainly primitive of nonsquare norm. For a more elementary proof, let f be the conductor of $\mathcal{O}$. By Lemma 7.2, $\mathcal{O} = [1, fw_K]$, $w_K = (d_K + \sqrt{d_K})/2$. Then $\alpha = fw_K$ is primitive in $\mathcal{O}$, and one easily sees that its norm $N(\alpha)$ is not a perfect square (see Exercise 11.20).

Let L denote the ring class field of $\mathcal{O}$. In order to prove $L = K(j(\mathfrak{a}))$, we will study how integer primes decompose in L and $K(j(\mathfrak{a}))$. We will make extensive use of the results of §8, especially Proposition 8.20. As usual, f and D will denote the conductor and discriminant of $\mathcal{O}$.

Let's first study how integer primes behave in the ring class field L. Let $\mathcal{S}_{L/\mathbb{Q}}$ be the set of primes that split completely in L. We claim that

$$\mathcal{S}_{L/\mathbb{Q}} \doteq \{p \text{ prime} : p = N(\alpha) \text{ for some } \alpha \in \mathcal{O}\}. \tag{11.28}$$

(As noted above, this shows that there are α's in $\mathcal{O}$ with $N(\alpha)$ prime.) When $D \equiv 0 \bmod 4$, then $\mathcal{O} = \mathbb{Z}[\sqrt{-n}]$ for some positive integer n. Thus $N(\alpha) = N(x + y\sqrt{-n}) = x^2 + ny^2$, so that (11.28) says, with finitely many exceptions, that the primes splitting completely in L are those represented by $x^2 + ny^2$. This was proved in Theorem 9.4. The case when $D \equiv 1 \bmod 4$ is similar and was covered in Exercise 9.3. Hence we have proved (11.28).

Let $M = K(j(\mathfrak{a}))$. Since L is Galois over $\mathbb{Q}$ by Lemma 9.3, part (i) of Proposition 8.20 shows that $M \subset L$ is equivalent to

$$\mathcal{S}_{L/\mathbb{Q}} \,\dot{\subset}\, \mathcal{S}_{M/\mathbb{Q}}. \tag{11.29}$$

Take $p \in \mathcal{S}_{L/\mathbb{Q}}$, and assume that p is unramified in M (this excludes only finitely many p's). By (11.28), $p = N(\alpha)$ for some $\alpha \in \mathcal{O}$. Then $\alpha\mathfrak{a} \subset \mathfrak{a}$ is a sublattice of index $N(\alpha) = p$, and is cyclic since p is prime. Thus

$$0 = \Phi_p(j(\alpha\mathfrak{a}), j(\mathfrak{a})) = \Phi_p(j(\mathfrak{a}), j(\mathfrak{a})).$$

Using Kronecker's congruence from part (v) of Theorem 11.18, this implies that

$$0 = \Phi_p(j(\mathfrak{a}), j(\mathfrak{a})) = -(j(\mathfrak{a})^p - j(\mathfrak{a}))^2 + p\beta$$

for some $\beta \in \mathcal{O}_M$. Now let $\mathfrak{P}$ be any prime of M containing p. The above equation then implies that

$$j(\mathfrak{a})^p \equiv j(\mathfrak{a}) \bmod \mathfrak{P}. \tag{11.30}$$

We claim the following:

(i) $\mathcal{O}_K[j(\mathfrak{a})] \subset \mathcal{O}_M$ has finite index.

(ii) If $p \nmid [\mathcal{O}_M : \mathcal{O}_K[j(\mathfrak{a})]]$, then (11.30) implies that $\alpha^p \equiv \alpha \bmod \mathfrak{P}$ for all $\alpha \in \mathcal{O}_M$.

The proof of (i) is a direct consequence of $M = K(j(\mathfrak{a}))$ and is left to the reader (see Exercise 11.21). As for (ii), note that p splits completely in L, so that it splits completely in K, and hence $p \in \mathfrak{p} \subset \mathfrak{P}$ for some ideal $\mathfrak{p}$ of norm p. This implies that $\alpha^p \equiv \alpha \bmod \mathfrak{P}$ holds for all $\alpha \in \mathcal{O}_K$, and consequently the congruence holds for all $\alpha \in \mathcal{O}_K[j(\mathfrak{a})]$ by (11.30). Then (ii) follows easily (see Exercise 11.21).

From (ii) it follows that $f_{\mathfrak{P}|p} = 1$, and since this holds for any $\mathfrak{P}$ containing p, we see that p splits completely in M. This proves (11.29), and $M \subset L$ follows.

The inclusion $M = K(j(\mathfrak{a})) \subset L$ shows that the ring class field L contains the j-invariants of *all* proper fractional $\mathcal{O}$-ideals. Let $h = h(\mathcal{O})$, and let $\mathfrak{a}_i$, $i = 1, \ldots, h$ be class representatives for $C(\mathcal{O})$. It follows that any $j(\mathfrak{a})$ equals one of $j(\mathfrak{a}_1), \ldots, j(\mathfrak{a}_h)$, and furthermore $j(\mathfrak{a}_1), \ldots, j(\mathfrak{a}_h)$ are distinct. Thus

$$\Delta = \prod_{i<j} (j(\mathfrak{a}_i) - j(\mathfrak{a}_j)) \tag{11.31}$$

is a nonzero element of $\mathcal{O}_L$.

To prove the opposite inclusion $L \subset M$, we will use the criterion $\tilde{\mathcal{S}}_{M/\mathbb{Q}} \,\dot{\subset}\, \mathcal{S}_{L/\mathbb{Q}}$ from part (ii) of Proposition 8.20. So let $p \in \tilde{\mathcal{S}}_{M/\mathbb{Q}}$, which means that p is unramified in M and $f_{\mathfrak{P}|p} = 1$ for some prime $\mathfrak{P}$ of M containing p. In particular, this implies that p splits completely in K, and thus $p = N(\mathfrak{p})$ for some prime ideal of $\mathcal{O}_K$. Then Proposition 7.20 tells us that $p = N(\mathfrak{p} \cap \mathcal{O})$ (we can assume that p doesn't divide f—this excludes finitely many primes). If we can show that $\mathfrak{p} \cap \mathcal{O}$ is a principal ideal $\alpha\mathcal{O}$, then $p = N(\alpha)$ implies that $p \in \mathcal{S}_{L/\mathbb{Q}}$ by (11.28). We may assume that p is relatively prime to the element Δ of (11.31).

Let $\mathfrak{a}' = (\mathfrak{p} \cap \mathcal{O})\mathfrak{a}$. Since $\mathfrak{p} \cap \mathcal{O}$ has norm p, $\mathfrak{a}' \subset \mathfrak{a}$ is a sublattice of index p by Lemma 11.26, and it is cyclic since p is prime. Thus $\Phi_p(j(\mathfrak{a}'), j(\mathfrak{a})) = 0$. Using Kronecker's congruence again, we can write this as

$$0 = \Phi_p(j(\mathfrak{a}'), j(\mathfrak{a})) = (j(\mathfrak{a}')^p - j(\mathfrak{a}))(j(\mathfrak{a}') - j(\mathfrak{a})^p) + pQ(j(\mathfrak{a}'), j(\mathfrak{a}))$$

for some polynomial $Q(X,Y) \in \mathbb{Z}[X,Y]$. Let $\tilde{\mathfrak{P}}$ be a prime of L containing $\mathfrak{P}$. Since $j(\mathfrak{a}')$ and $j(\mathfrak{a})$ are algebraic integers lying in L, the above equation implies that $pQ(j(\mathfrak{a}'),j(\mathfrak{a})) \in \tilde{\mathfrak{P}}$. Thus

$$(11.32) \qquad j(\mathfrak{a}')^p \equiv j(\mathfrak{a}) \bmod \tilde{\mathfrak{P}} \qquad \text{or} \qquad j(\mathfrak{a}') \equiv j(\mathfrak{a})^p \bmod \tilde{\mathfrak{P}}.$$

However, we also know $f_{\mathfrak{P}|p} = 1$, which tells us that $j(\mathfrak{a})^p \equiv j(\mathfrak{a}) \bmod \mathfrak{P}$, and since $\mathfrak{P} \subset \tilde{\mathfrak{P}}$, we obtain

$$(11.33) \qquad j(\mathfrak{a})^p \equiv j(\mathfrak{a}) \bmod \tilde{\mathfrak{P}}.$$

It is straightforward to show that (11.32) and (11.33) imply

$$j(\mathfrak{a}) \equiv j(\mathfrak{a}') \bmod \tilde{\mathfrak{P}}.$$

If $\mathfrak{a}$ and $\mathfrak{a}'$ lay in distinct ideal classes in $C(\mathcal{O})$, then $j(\mathfrak{a}) - j(\mathfrak{a}')$ would be one of the factors of Δ from (11.31), and p and Δ would not be relatively prime. This contradicts our choice of p, so that $\mathfrak{a}$ and $\mathfrak{a}' = (\mathfrak{p} \cap \mathcal{O})\mathfrak{a}$ must lie in the same ideal class in $C(\mathcal{O})$. This forces $\mathfrak{p} \cap \mathcal{O}$ to be a principal ideal, which as we showed above, implies that $p \in \mathcal{S}_{L/\mathbb{Q}}$. Thus $\tilde{\mathcal{S}}_{M/\mathbb{Q}} \dot{\subset} \mathcal{S}_{L/\mathbb{Q}}$, which completes the proof that $L = M$. Theorem 11.1 is proved. Q.E.D.

As an application of Theorem 11.1, let's see what it tells us about the Abelian extensions of an imaginary quadratic field K. First, we know that the Hilbert class field of K is the ring class field of the maximal order $\mathcal{O}_K$. Thus we get the following corollary of Theorem 11.1:

COROLLARY 11.34. *If K is an imaginary quadratic field, then $K(j(\mathcal{O}_K))$ is the Hilbert class field of K.* Q.E.D.

Besides the Hilbert class field, Theorem 11.1 also allows us to describe other Abelian extensions of K. Recall that in Theorem 9.18 we proved that an Abelian extension of K is generalized dihedral over $\mathbb{Q}$ if and only if it lies in some ring class field of K. Combining this with Theorem 11.1, we get the following result:

COROLLARY 11.35. *Let K be an imaginary quadratic field, and let $K \subset L$ be a finite extension. Then L is an Abelian extension of K which is generalized dihedral over $\mathbb{Q}$ if and only if there is an order $\mathcal{O}$ in K such that $L \subset K(j(\mathcal{O}))$.* Q.E.D.

To complete our discussion of ring class fields and complex multiplication, we need to compute the Artin map of a ring class field using j-invariants. The answer is given by the following theorem:

THEOREM 11.36. *Let $\mathcal{O}$ be an order in an imaginary quadratic field K, and let L be the ring class field of $\mathcal{O}$. If $\mathfrak{a}$ is a proper fractional $\mathcal{O}$-ideal and $\mathfrak{p}$ is a prime ideal of $\mathcal{O}_K$ relatively prime to the conductor of $\mathcal{O}$, then*

$$\left(\frac{L/K}{\mathfrak{p}}\right)(j(\mathfrak{a})) = j\big(\overline{\mathfrak{p} \cap \mathcal{O}}\,\mathfrak{a}\big).$$

PROOF. For analytic proofs, see Deuring [**32**, §15], Lang [**90**, Chapter 12, §3] or Cohn [**27**, §11.2], while algebraic proofs (which use the reduction theory of elliptic curves) may be found in Lang [**90**, Chapter 10, §3] or Shimura [**115**, §5.4]. We will use this theorem (in the guise of Corollary 11.37 below) in §12 when we compute some j-invariants, though our discussion of the class equation in §13 will use only Theorem 11.1. Q.E.D.

In terms of the ideal class group, Theorem 11.36 can be stated as follows:

COROLLARY 11.37. *Let $\mathcal{O}$ be an order in an imaginary quadratic field K, and let L be the ring class field of $\mathcal{O}$. Given proper fractional $\mathcal{O}$-ideals $\mathfrak{a}$ and $\mathfrak{b}$, define $\sigma_{\mathfrak{a}}(j(\mathfrak{b}))$ by the formula*

$$\sigma_{\mathfrak{a}}(j(\mathfrak{b})) = j(\overline{\mathfrak{a}}\mathfrak{b}).$$

Then $\sigma_{\mathfrak{a}}$ is a well-defined element of $\mathrm{Gal}(L/K)$, *and $\mathfrak{a} \mapsto \sigma_{\mathfrak{a}}$ induces an isomorphism*

$$C(\mathcal{O}) \xrightarrow{\sim} \mathrm{Gal}(L/K).$$

PROOF. This is a straightforward consequence of Theorem 11.36 and the isomorphisms

$$C(\mathcal{O}) \simeq I(\mathcal{O}, f)/P(\mathcal{O}, f) \simeq I_K(f)/P_{K,\mathbb{Z}}(f),$$

where f is the conductor of $\mathcal{O}$. See Exercise 11.22 for the details. Q.E.D.

The "First Main Theorem" of complex multiplication allowed us to describe some of the Abelian extensions of K, namely those which are generalized dihedral over $\mathbb{Q}$. The "Second Main Theorem" of complex multiplication answers the question of how to describe *all* Abelian extensions of K. By class field theory, every Abelian extension lies in a ray class field for some modulus $\mathfrak{m}$ of K, so that we need only find generators for the ray class fields of K. Rather than work with an arbitrary modulus $\mathfrak{m}$, we will describe the ray class fields only for moduli of the form $N\mathcal{O}_K$, where N is a positive integer. It is easy to see that any Abelian extension of K lies in such a ray class field (see Exercise 11.23).

The basic idea is that the ray class field of $N\mathcal{O}_K$ is obtained by adjoining, first, the j-invariant $j(L)$ of some lattice L, and second, some values of the Weierstrass $\wp$-function evaluated at N-division points of the lattice L, i.e., if $L = [\alpha, \beta]$, then we use

$$\wp\Big(\frac{m\alpha + n\beta}{N}; L\Big) \tag{11.38}$$

for suitable m and n. The observation that (11.38) generates Abelian extensions of K goes back to Abel. The problem is that these values aren't invariant enough: if we multiply the lattice by a constant, the j-invariant remains the same, but the values (11.38) change. To remedy this problem, we introduce a variant of the Weierstrass $\wp$-function called the *Weber function.* Given the lattice L, the Weber function $h(z; L)$ is defined by

$$h(z;L) = \begin{cases} \dfrac{g_2(L)^2}{\Delta(L)}\wp(z;L)^2 & \text{if } g_3(L) = 0 \\[2ex] \dfrac{g_3(L)}{\Delta(L)}\wp(z;L)^3 & \text{if } g_2(L) = 0 \\[2ex] \dfrac{g_2(L)g_3(L)}{\Delta(L)}\wp(z;L) & \text{otherwise,} \end{cases}$$

where $\Delta(L) = g_2(L)^3 - 27g_3(L)^2$. It is easy to check that $h(\lambda z; \lambda L) = h(z; L)$ for all $\lambda \in \mathbb{C}^*$ (see Exercise 11.24).

We can now state the "Second Main Theorem" of complex multiplication, which uses singular j-invariants and the Weber function to generate ray class fields:

THEOREM 11.39. *Let K be an imaginary quadratic field of discriminant d_K, and let N be a positive integer.*

(i) *$K(j(\mathcal{O}_K), h(1/N; \mathcal{O}_K))$ is the ray class field for the modulus $N\mathcal{O}_K$.*
(ii) *Let $\mathcal{O}$ be the order of conductor N in K. Then $K(j(\mathcal{O}), h(w_K; \mathcal{O}))$, where $w_K = (d_K + \sqrt{d_K})/2$, is the ray class field for the modulus $N\mathcal{O}_K$.*

PROOF. Notice that in each case we obtain the ray class field by adjoining the j-invariant of a lattice and the Weber function of one N-division point. The proof of (i) may be found in Deuring [**32**, §26] or Lang [**90**, §10.3, Corollary to Theorem 7], and the proof of (ii) follows from Satz 1 of Franz [**47**]. These references also explain how to generate the ray class field of an arbitrary modulus $\mathfrak{m}$ of K. Q.E.D.

The theory of complex multiplication, even in the one variable case described here, is an active area of research. See, for example, the books *Elliptic Functions and Rings of Integers* [**20**] by Cassou–Noguès and Taylor and *Arithmetic on Elliptic Curves with Complex Multiplication* [**60**] by Gross.

E. Exercises

11.1. This exercise will study j-invariants and complex conjugation.

(a) Let L be a lattice, and let $\overline{L}$ denote the lattice obtained by complex conjugation. Prove that $g_2(\overline{L}) = \overline{g_2(L)}$, $g_3(\overline{L}) = \overline{g_3(L)}$ and $j(\overline{L}) = \overline{j(L)}$.
(b) Let $\mathfrak{a}$ be a proper fractional $\mathcal{O}$-ideal, where $\mathcal{O}$ is an order in an imaginary quadratic field. Show that $j(\mathfrak{a})$ is a real number if and only if the class of $\mathfrak{a}$ has order ≤ 2 in the ideal class group $C(\mathcal{O})$. Hint: use part (a) and Theorem 10.9.

One consequence of part (b) is that $j(\mathcal{O})$ is real for any order $\mathcal{O}$.

11.2. If $\tau \in \mathfrak{h}$ and $\gamma = \left(\begin{smallmatrix} a & b \\ c & d \end{smallmatrix}\right) \in \mathrm{SL}(2, \mathbb{Z})$, then show that

$$\gamma\tau = \frac{a\tau + b}{c\tau + d}$$

also lies in $\mathfrak{h}$. This shows that $\mathrm{SL}(2, \mathbb{Z})$ acts on $\mathfrak{h}$. Hint: use (7.10).

11.3. Let τ satisfy $|\mathrm{Re}(\tau)| \leq 1/2$ and $\mathrm{Im}(\tau) \geq \epsilon$, where $\epsilon < 1$ is fixed. Our goal is to show that for $x, y \in \mathbb{R}$,

$$|x + y\tau| \geq \frac{\epsilon}{2}\sqrt{x^2 + y^2}.$$

If we let $\tau = a + bi$, then the above is equivalent to

$$(x + ay)^2 + b^2y^2 \geq \frac{\epsilon^2}{4}(x^2 + y^2).$$

(a) Show that the inequality is true when $|x + ay| \geq (\epsilon/2)|x|$.
(b) When $|x + ay| < (\epsilon/2)|x|$, use $|a| \leq 1/2$ and $\epsilon < 1$ to show that $|x| < |y|$.
(c) Using (b), show that the inequality is true when $|x + ay| < (\epsilon/2)|x|$.

11.4. In Lemma 11.4 we showed that every point of $\mathfrak{h}$ is $\mathrm{SL}(2, \mathbb{Z})$-equivalent to a point in the region $\{\tau \in \mathfrak{h} : |\mathrm{Re}(\tau)| \leq 1/2,\ \mathrm{Im}(\tau) \geq 1/2\}$. In this exercise we will study the smaller region

$$\begin{aligned} F = \{\tau \in \mathfrak{h} : |\mathrm{Re}(\tau)| \leq 1/2,\ |\tau| \geq 1, \text{ and} \\ \mathrm{Re}(\tau) \leq 0 \text{ if } |\mathrm{Re}(\tau)| = 1/2 \text{ or } |\tau| = 1\}, \end{aligned}$$

and we will show that every point of $\mathfrak{h}$ is $\mathrm{SL}(2,\mathbb{Z})$-equivalent to a *unique* point of F. This is usually expressed by saying that F is a *fundamental domain* for the action of SL2, $\mathbb{Z}$) on $\mathfrak{h}$. Our basic tool will be positive definite quadratic forms $f(x,y) = ax^2+bxy+cy^2$, where we allow a, b and c to be real numbers. We say that two such forms $f(x,y)$ and $g(x,y)$ are $\mathbb{R}^+$-equivalent if there is $\left(\begin{smallmatrix} p & q \\ r & s \end{smallmatrix}\right) \in \mathrm{SL}(2,\mathbb{Z})$ such that

$$f(x,y) = \lambda g(px+qy, rx+sy)$$

for some $\lambda > 0$ in $\mathbb{R}$. We also say that $f(x,y) = ax^2+bxy+cy^2$ is reduced if

$$a \le |b| \le c, \text{ and } b \ge 0 \text{ if } a = |b| \text{ or } |b| = c.$$

This is consistent with the definition given in §2.

(a) Show that $\mathbb{R}^+$-equivalence of positive definite forms is an equivalence relation.

(b) Show that every positive definite form is $\mathbb{R}^+$-equivalent to a reduced form, and that two reduced forms are $\mathbb{R}^+$-equivalent if and only if one is a constant multiple of the other. Hint: see the proof of Theorem 2.8. Show that among all forms properly equivalent to a given positive definite form with real coefficients, there is one with minimal $|b|$. Equation (2.4) is helpful.

(c) Show that every positive definite form $f(x,y) = ax^2+bxy+cy^2$ can be written uniquely as $f(x,y) = a|x-\tau y|^2$, where $\tau \in \mathfrak{h}$. In this case we say that τ is the *root* of $f(x,y)$ (this is consistent with the terminology used in §7). Furthermore, show that $b = -2a\mathrm{Re}(\tau)$ and $c = a|\tau|^2$.

(d) Show that two positive definite forms are $\mathbb{R}^+$-equivalent if and only if their roots are $\mathrm{SL}(2,\mathbb{Z})$-equivalent. Hint: see the proof of (7.8).

(e) Show that a positive definite form is reduced if and only if its root lies in the fundamental domain F.

(f) Conclude that every $\tau \in \mathfrak{h}$ is $\mathrm{SL}(2,\mathbb{Z})$-equivalent to a unique point of F.

This exercise shows that there is a remarkable relation between reduced forms and fundamental domains. Similar considerations led Gauss (unpublished, of course) to discover the idea of a fundamental domain in the early 1800s. See Cox [**31**] for more details.

11.5. In this exercise we will prove Lemma 11.5 and Corollary 11.6.

(a) Let M and ϵ be positive constants, and define $C \subset \mathfrak{h}$ by

$$C = \{\tau \in \mathfrak{h} : |\mathrm{Re}(\tau)| \le M,\ \epsilon \le \mathrm{Im}(\tau) \le 1/\epsilon\}.$$

We want to show that the set $\Delta(C) = \{\gamma \in \mathrm{SL}(2,\mathbb{Z}) : \gamma(C) \cap C \ne \emptyset\}$ is finite. So take $\gamma = \left(\begin{smallmatrix} a & b \\ c & d \end{smallmatrix}\right) \in \Delta(C)$, which means that there is $\tau \in C$ such that $\gamma\tau \in C$. If we can bound $|a|$, $|b|$, $|c|$ and $|d|$ in terms of M and ϵ, then finiteness will follow.

(i) Use (7.10) to show that $|c\tau + d| \le 1/\epsilon$.

(ii) Use $|c\tau+d|^2 = (c\mathrm{Re}(\tau)+d)^2 + c^2\mathrm{Im}(\tau)^2$ to show that $|c| \le 1/\epsilon^2$ and $|d| \le (\epsilon + M)/\epsilon^2$.

(iii) Show that $\gamma^{-1} \in \Delta(C)$ By (ii), this implies that $|a| \le (\epsilon+M)/\epsilon^2$.

(iv) Show that $|b| \le |c\tau+d||\gamma\tau| + |a||\tau|$. Conclude that $|b|$ is bounded in terms of M and ϵ.

(b) Use part (a) to show that if U is a neighborhood of $\tau \in \mathfrak{h}$ such that $\overline{U} \subset \mathfrak{h}$ is compact, then $\{\gamma \in \mathrm{SL}(2,\mathbb{Z}) : \gamma(U) \cap U \neq \emptyset\}$ is finite. This will prove Lemma 11.5.

(c) Prove Corollary 11.6.

11.6. This exercise is concerned with the proof of part (iv) of Theorem 11.2.

(a) Suppose that $\gamma = \left(\begin{smallmatrix} a & b \\ c & d \end{smallmatrix}\right) \in \mathrm{SL}(2,\mathbb{Z})$, $\gamma \neq \pm I$, and that $\gamma\tau = \tau$ for some $\tau \in \mathfrak{h}$. We saw in the text that this implies $[1,\tau] = (c\tau + d)[1,\tau]$. Prove that $c \neq 0$. Hint: show that $c = 0$ implies $\gamma = \pm\left(\begin{smallmatrix} 1 & m \\ 0 & 1 \end{smallmatrix}\right)$. But such a γ with $m \neq 0$ has no fixed points on $\mathfrak{h}$.

(b) Let $\mathcal{O}$ be an order in an imaginary quadratic field such that $\mathcal{O}^* \neq \{\pm 1\}$. Prove that $\mathcal{O} = \mathcal{O}_K$ for $K = \mathbb{Q}(i)$ or $\mathbb{Q}(\omega)$, $\omega = e^{2\pi i/3}$. Hint: when $\mathcal{O} = \mathcal{O}_K$, see Exercise 5.9.

(c) Show that the only elements of $\mathrm{SL}(2,\mathbb{Z})$ fixing i are $\pm\left(\begin{smallmatrix} 1 & 0 \\ 0 & 1 \end{smallmatrix}\right)$ and $\pm\left(\begin{smallmatrix} 0 & 1 \\ -1 & 0 \end{smallmatrix}\right)$.

(d) If $\omega = e^{2\pi i/3}$, show that $j'(\omega) = j''(\omega) = 0$ but $j'''(\omega) \neq 0$.

11.7. Let $f(\tau)$ be a modular function for $\mathrm{SL}(2,\mathbb{Z})$.

(a) If $f(\tau)$ has a pole of order m at i, then prove that m is even. Hint: write $f(\tau) = g(\tau)/(\tau - i)^m$, where $g(\tau)$ is holomorphic and nonvanishing at i. Note that i is fixed by $\left(\begin{smallmatrix} 0 & 1 \\ -1 & 0 \end{smallmatrix}\right)$.

(b) If $f(\tau)$ has a pole of order m at $\tau = \omega$, $\omega = e^{2\pi i/3}$, then prove that m is divisible by 3. Hint: argue as in part (a). Note that ω is fixed by $\left(\begin{smallmatrix} 1 & 1 \\ -1 & 0 \end{smallmatrix}\right)$.

11.8. As in the proof of Theorem 11.9, let

$$\Gamma_0(m) = \left\{\begin{pmatrix} a & b \\ c & d \end{pmatrix} \in \mathrm{SL}(2,\mathbb{Z}) : c \equiv 0 \bmod m\right\}$$

$$C(m) = \left\{\begin{pmatrix} a & b \\ 0 & d \end{pmatrix} : ad = m,\ a > 0,\ 0 \leq b < d,\ \gcd(a,b,d) = 1\right\},$$

and let $\sigma_0 = \left(\begin{smallmatrix} m & 0 \\ 0 & 1 \end{smallmatrix}\right) \in C(m)$.

(a) Show that $\Gamma_0(m) = (\sigma_0^{-1}\mathrm{SL}(2,\mathbb{Z})\,\sigma_0) \cap \mathrm{SL}(2,\mathbb{Z})$.

(b) If $\sigma \in C(m)$, then show that $(\sigma_0^{-1}\mathrm{SL}(2,\mathbb{Z})\,\sigma) \cap \mathrm{SL}(2,\mathbb{Z})$ is a coset of $\Gamma_0(m)$ in $\mathrm{SL}(2,\mathbb{Z})$. Hint: first show that $(\sigma_0^{-1}\mathrm{SL}(2,\mathbb{Z})\,\sigma) \cap \mathrm{SL}(2,\mathbb{Z}) \neq \emptyset$.

(c) In the construction of part (b), show that different σ's give different cosets, and that *all* cosets of $\Gamma_0(m)$ in $\mathrm{SL}(2,\mathbb{Z})$ arise in this way.

11.9. Let m be a positive integer, and let $\Psi(m)$ denote the number of triples (a,b,d) of integers which satisfy $ad = m$, $a > 0$, $0 \leq b < d$ and $\gcd(a,b,d) = 1$. Thus $\Psi(m) = |C(m)|$, where $C(m)$ is the set of matrices defined in the previous exercise. The goal of this exercise is to prove that

$$\Psi(m) = m\prod_{p \mid m}\left(1 + \frac{1}{p}\right).$$

(a) If we fix a positive divisor d of m, then $a = m/d$ is determined. Show that the number of possible b's for this d is given by

$$\frac{d}{\gcd(d, m/d)}\phi(\gcd(d, m/d)),$$

where ϕ denotes the Euler ϕ-function.

(b) Use the formula of part (a) to prove that $\Psi(m)$ is multiplicative, i.e., that if m_1 and m_2 are relatively prime, then $\Psi(m_1 m_2) = \Psi(m_1)\Psi(m_2)$.

(c) Use the formula of part (a) to prove that if p is a prime, then

$$\Psi(p^r) = p^r + p^{r-1}.$$

(d) Use parts (b) and (c) to prove the desired formula for $\Psi(m)$.

11.10. In this exercise we will show that $\Phi_m(X,Y)$ is irreducible as a polynomial in X (which will prove part (ii) of Theorem 11.18). Let γ_i be coset representatives for $\Gamma_0(m)$ in $\mathrm{SL}(2,\mathbb{Z})$. As we saw in (11.14), we can write

$$\Phi_m(X, j(\tau)) = \prod_{i=1}^{|C(m)|} (X - j(m\gamma_i\tau)).$$

Let $\mathcal{F}_m$ be the field $\mathbb{C}(j(\tau), j(m\tau))$. Since $\Phi_m(X, j(\tau))$ has coefficients in $\mathbb{C}(j(\tau))$ and $j(m\tau)$ is a root, it follows that $[\mathcal{F}_m : \mathbb{C}(j(\tau))] \leq \Psi(m) = |C(m)|$. If we can prove equality, then $\Phi_m(X, j(\tau))$ will be the minimal polynomial of $j(m\tau)$ over $\mathbb{C}(j(\tau))$, and irreducibility will follow.

(a) Let $\mathcal{F}$ be the field of all meromorphic functions on $\mathfrak{h}$, which contains $\mathcal{F}_m$ as a subfield. For $\gamma \in \mathrm{SL}(2,\mathbb{Z})$, show that $f(\tau) \mapsto f(\gamma\tau)$ is an embedding of $\mathcal{F}_m$ into $\mathcal{F}$ which is the identity on $\mathbb{C}(j(\tau))$.

(b) Use (11.13) to prove that $j(m\gamma_i\tau) \neq j(m\gamma_j\tau)$ for $i \neq j$. This shows that the embeddings constructed in part (a) are distinct, which implies that $[\mathcal{F}_m : \mathbb{C}(j(\tau))] \geq \Psi(m)$. This proves the desired equality.

11.11. Show that the coefficients of $G(X,\tau)$ (as defined in (11.16)) are modular functions for $\mathrm{SL}(2,\mathbb{Z})$. Hint: argue as in the case of the modular function. You will use the fact that $f(\gamma_i\tau)$ has a meromorphic q-expansion.

11.12. Let $A \subset \mathbb{C}$ be an additive subgroup, and let $f(\tau)$ be a holomorphic modular function. Suppose that its q-expansion is

$$f(\tau) = \sum_{n=-M}^{\infty} a_n q^n,$$

and that $a_n \in A$ for all $n \leq 0$. Then prove the Hasse q-expansion principle, which states that $f(\tau)$ is a polynomial in $j(\tau)$ with coefficients in A. Hint: since the q-expansion of $j(\tau)$ has integer coefficients and begins with $1/q$, the polynomial $A(x)$ used in part (ii) of Lemma 11.10 must have coefficients in A.

11.13. Let p be a prime, and let $\zeta_p = e^{2\pi i/p}$.

(a) Prove that $p = (1-\zeta_p)(1-\zeta_p^2)\cdots(1-\zeta_p^{p-1})$. Hint: use the factorization of $x^{p-1} + \cdots + x + 1$.

(b) Given $\alpha \in \mathbb{Z}[\zeta_p]$, define the norm $N_{\mathbb{Q}(\zeta_p)/\mathbb{Q}}(\alpha)$ to be the number

$$N_{\mathbb{Q}(\zeta_p)/\mathbb{Q}}(\alpha) = \prod_{\sigma \in \mathrm{Gal}(\mathbb{Q}(\zeta_p)/\mathbb{Q})} \sigma(\alpha).$$

For simplicity, we will write $N(\alpha)$ instead of $N_{\mathbb{Q}(\zeta_p)/\mathbb{Q}}(\alpha)$. Prove that $N(\alpha)$ is an integer, and show that $N(\alpha\beta) = N(\alpha)N(\beta)$ and $N(1-\zeta_p) = p$.

(c) If an integer a can be written $a = (1-\zeta_p)\alpha$ where $\alpha \in \mathbb{Z}[\zeta_p]$, then use part (b) to prove that a is divisible by p.

11.14. Adapt the proof of (11.21) to show that $j(\tau) \equiv j(\sigma_0\tau)^p \bmod p$.

11.15. Let $f(X,Y) \in \mathbb{Z}[X,Y]$ be a polynomial such that $f(X,j(\tau)) \in p\mathbb{Z}((q))[X]$. Prove that $f(X,Y) \in p\mathbb{Z}[X,Y]$. Hint: apply the q-expansion principle (proved in Exercise 11.12) to the coefficients of X.

11.16. Let $M = \mathbb{Z}^2$, thought of as column vectors, and let A be a 2×2 integer matrix with $\det(A) \neq 0$. Exercise 7.15, applied to the transpose of A, implies that M/AM is a finite group of order $|\det(A)|$. The object of this exercise is to prove that M/AM is cyclic if and only if the entries of A are relatively prime.

(a) Let G be a finite Abelian group. Prove that G is not cyclic if and only if G contains a subgroup isomorphic to $(\mathbb{Z}/d\mathbb{Z})^2$ for some integer $d > 1$. Hint: use the structure theorem for finite Abelian groups.

(b) Assume that the entries of A have a common divisor $d > 1$, and prove that M/AM is not cyclic. Hint: write $A = dA'$, where A' is an integer matrix, and note that $A'M/dA'M \subset M/AM$. Then use part (a).

(c) Finally, assume that M/AM is not cyclic, and prove that the entries of A have a common divisor $d > 1$. Hint: by part (a), there is $AM \subset M' \subset M$ such that $M'/AM \simeq (\mathbb{Z}/d\mathbb{Z})^2$ for some $d > 1$. Prove that $AM = dM'$, and conclude that d divides the entries of A.

11.17. This exercise is concerned with the proof of Lemma 11.24.

(a) Let L' be a sublattice of $[1,\tau]$ of finite index, and let d be the smallest positive integer in L'. Then prove that $L' = [d, a\tau + b]$ for some integers a and b.

(b) Let $\tau \in \mathfrak{h}$ and let $C(m)$ be the set of matrices defined in the text. Given $\sigma, \sigma' \in C(m)$ such that $d[1,\sigma\tau] = d'[1,\sigma'\tau]$, prove that $\sigma = \sigma'$.

11.18. In the text, we proved that for $\tau \in \mathfrak{h}$, the roots of $\Phi_m(X,j(\tau)) = 0$ are the j-invariants of the cyclic sublattices of index m of $[1,\tau]$. Use this fact and the surjectivity of the j-function to prove Theorem 11.23.

11.19. Let $\mathcal{O}$ be an order, and let $\mathfrak{b}$ be a proper fractional $\mathcal{O}$-ideal. If $\mathfrak{a}$ is a proper $\mathcal{O}$-ideal which is not primitive, then prove that $\mathfrak{b}/\mathfrak{a}\mathfrak{b}$ is not cyclic. Hint: part (a) of Exercise 11.16 will be useful.

11.20. Let $\mathcal{O}$ be an order in an imaginary quadratic field K of conductor f. Letting $w_K = (d_K + \sqrt{d_k})/2$, we proved in Lemma 7.2 that $\mathcal{O} = [1, fw_K]$. Prove that $\alpha = fw_K$ is a primitive element of $\mathcal{O}$ whose norm is not a perfect square.

11.21. Let $K \subset L$ be an extension of number fields, and let $\alpha \in \mathcal{O}_L$ satisfy $L = K(\alpha)$.

(a) Prove that $\mathcal{O}_K[\alpha]$ has finite index in $\mathcal{O}_L$. Hint: by Proposition 5.3, we know that $\mathcal{O}_L$ is a free $\mathbb{Z}$-module of rank $[L:\mathbb{Q}]$. Then show that $\mathcal{O}_K[\alpha]$ has the same rank.

(b) Let $\mathfrak{P}$ be a prime ideal of $\mathcal{O}_L$, and suppose that $N(\mathfrak{P}) = p^f$, where p is relatively prime to $[\mathcal{O}_L : \mathcal{O}_K[\alpha]]$. If $\beta^p \equiv \beta \bmod \mathfrak{P}$ holds for all $\beta \in \mathcal{O}_K[\alpha]$, then show that the same congruence holds for all $\beta \in \mathcal{O}_L$. Hint: if $N = [\mathcal{O}_L : \mathcal{O}_K[\alpha]]$, then multiplication by N induces an isomorphism of $\mathcal{O}_L/\mathfrak{P}$.

11.22. Complete the proof of Corollary 11.37.

11.23. Let K be an imaginary quadratic field, and let L be an Abelian extension of K. Prove that there is a positive integer N such that L is contained in the ray class field for the modulus $N\mathcal{O}_K$.

11.24. If L is a lattice and $h(z; L)$ is the Weber function defined in the text, then prove that $h(\lambda z; \lambda L) = h(z; L)$ for any $\lambda \in \mathbb{C}^*$.

12. Modular Functions and Singular j-Invariants

The j-invariant $j(L)$ of a lattice with complex multiplication is often called a *singular j-invariant* or *singular modulus.* In §11 we learned about the fields generated by singular moduli, and in this section we will compute some of these remarkable numbers. One of our main tools will be the function $\gamma_2(\tau)$, which is defined by

$$\gamma_2(\tau) = \sqrt[3]{j(\tau)}.$$

We will show that $\gamma_2(3\tau)$ is a modular function for $\Gamma_0(9)$, and we will use $\gamma_2(\tau)$ to generate ring class fields for orders of discriminant not divisible by 3. This will explain why the j-invariants computed in §10 were perfect cubes.

We will then give a careful treatment of some of the results contained in Volume III of Weber's monumental *Lehrbuch der Algebra* [**131**]. There is a wealth of material in this book, far more than we could ever cover here. We will concentrate on some applications of the Dedekind η-function $\eta(\tau)$ and the three Weber functions $\mathfrak{f}(\tau)$, $\mathfrak{f}_1(\tau)$ and $\mathfrak{f}_2(\tau)$. These functions are closely related to $\gamma_2(\tau)$ and $j(\tau)$ and make it easy to compute the j-invariants of most orders of class number 1. The Weber functions also give some interesting modular functions, which will enable us to compute that

$$j(\sqrt{-14}) = 2^3\left(323 + 228\sqrt{2} + (231 + 161\sqrt{2})\sqrt{2\sqrt{2} - 1}\right)^3. \tag{12.1}$$

At the end of the section, we will present Heegner's proof of the Baker–Heegner–Stark Theorem on imaginary quadratic fields of class number 1.

A. The Cube Root of the j-Function

Our first task is to study the cube root $\gamma_2(\tau)$ of the j-function. Recall from §11 that $j(\tau)$ can be written as the quotient

$$j(\tau) = 1728\,\frac{g_2(\tau)^3}{\Delta(\tau)}.$$

The function $\Delta(\tau)$ is nonvanishing and holomorphic on the simply connected domain $\mathfrak{h}$, and hence has a holomorphic cube root $\sqrt[3]{\Delta(\tau)}$. Since $\Delta(\tau)$ is real-valued on the imaginary axis (see Exercise 12.1), we can choose $\sqrt[3]{\Delta(\tau)}$ with the same property. Using this cube root, we define

$$\gamma_2(\tau) = 12\,\frac{g_2(\tau)}{\sqrt[3]{\Delta(\tau)}}.$$

Since $g_2(\tau)$ is also real on the imaginary axis (see Exercise 12.1), it follows that $\gamma_2(\tau)$ is the unique cube root of $j(\tau)$ which is real-valued on the imaginary axis.

For us, the main property of $\gamma_2(\tau)$ is that it can be used to generate all ring class fields of orders of discriminant not divisible by 3. Note that τ needs to be chosen carefully, for replacing τ by $\tau + 1$ doesn't affect $j(\tau)$, but we will see below that $\gamma_2(\tau+1) = \zeta_3^{-1}\gamma_2(\tau)$, where $\zeta_3 = e^{2\pi i/3}$. The necessity to normalize τ leads to the following theorem:

THEOREM 12.2. *Let $\mathcal{O}$ be an order of discriminant D in an imaginary quadratic field K. Assume that $3 \nmid D$, and write $\mathcal{O} = [1, \tau_0]$, where*

$$\tau_0 = \begin{cases} \sqrt{-m} & D = -4m \equiv 0 \bmod 4 \\ \dfrac{3+\sqrt{-m}}{2} & D = -m \equiv 1 \bmod 4. \end{cases}$$

Then $\gamma_2(\tau_0)$ is an algebraic integer and $K(\gamma_2(\tau_0))$ is the ring class field of $\mathcal{O}$. Furthermore, $\mathbb{Q}(\gamma_2(\tau_0)) = \mathbb{Q}(j(\tau_0))$.

Let's first see how this theorem relates to the j-invariants computed in §10. When $\mathcal{O}$ has class number one, we know that $j(\mathcal{O})$ is an integer, so that by Theorem 12.2, $\gamma_2(\tau_0)$ is also an integer when $3 \nmid D$. This explains why

$$j(i) = 12^3$$

$$j(\sqrt{-2}) = 20^3$$

$$j\left(\frac{1+\sqrt{-7}}{2}\right) = -15^3$$

are all perfect cubes. (In the last case, note that $j((1+\sqrt{-7})/2) = j((3+\sqrt{-7})/2)$, so that Theorem 12.2 does apply.)

PROOF OF THEOREM 12.2. By Theorem 11.1, we know that $K(j(\tau_0))$ is the ring class field of $\mathcal{O} = [1, \tau_0]$. Thus, to prove Theorem 12.2, it suffices to prove that

$$\mathbb{Q}(\gamma_2(\tau_0)) = \mathbb{Q}(j(\tau_0))$$

whenever $3 \nmid D$. The first proof of this theorem was due to Weber [**131**, §125], and modern proofs have been given by Birch [**10**] and Schertz [**110**]. Our presentation is based on [**110**], though readers may wish to consult his 2010 book [**109**].

The first step of the proof is to show that $\gamma_2(3\tau)$ is a modular function.

PROPOSITION 12.3. *$\gamma_2(3\tau)$ is a modular function for the group $\Gamma_0(9)$.*

PROOF. We first study how $\gamma_2(\tau)$ transforms under elements of $\mathrm{SL}(2,\mathbb{Z})$. We claim that

$$\begin{aligned} \gamma_2(-1/\tau) &= \gamma_2(\tau) \\ \gamma_2(\tau+1) &= \zeta_3^{-1}\gamma_2(\tau), \end{aligned} \tag{12.4}$$

where $\zeta_3 = e^{2\pi i/3}$. The first line of (12.4) is easy to prove, for $\gamma_2(-1/\tau)$ is a cube root of $j(-1/\tau) = j(\tau)$. But $-1/\tau$ lies on the imaginary axis whenever τ does, so that $\gamma_2(-1/\tau)$ is a cube root of $j(\tau)$ which is real on the imaginary axis. By the definition of $\gamma_2(\tau)$, this implies $\gamma_2(-1/\tau) = \gamma_2(\tau)$.

To prove the second line of (12.4), consider the q-expansion of $\gamma_2(\tau)$. We know that

$$j(\tau) = q^{-1} + \sum_{n=0}^{\infty} c_n q^n = q^{-1}h(q),$$

where $h(q)$ is holomorphic for $|q| < 1$ and $h(0) = 1$. We can therefore write $h(q) = u(q)^3$, where $u(q)$ is holomorphic and $u(0) = 1$. Note also that $u(q)$ has rational coefficients since $h(q)$ does (see Exercise 12.2). Then $q^{-1/3}u(q)$ is a cube root of $j(\tau)$ which is real-valued on the imaginary axis, and it follows that

$$\gamma_2(\tau) = q^{-1/3}u(q) = q^{-1/3}\Big(1 + \sum_{n=1}^{\infty} b_n q^n\Big), \qquad b_n \in \mathbb{Q}. \tag{12.5}$$

It is now trivial to see that $\gamma_2(\tau+1) = \zeta_3^{-1}\gamma_2(\tau)$ and (12.4) is proved.

We next claim that if $\left(\begin{smallmatrix} a & b \\ c & d \end{smallmatrix}\right) \in \mathrm{SL}(2,\mathbb{Z})$, then

$$\gamma_2\left(\frac{a\tau+b}{c\tau+d}\right) = \zeta_3^{ac-ab+a^2cd-cd}\gamma_2(\tau). \tag{12.6}$$

To see this, first note that (12.6) holds for $S = \left(\begin{smallmatrix} 0 & -1 \\ 1 & 0 \end{smallmatrix}\right)$ and $T = \left(\begin{smallmatrix} 1 & 1 \\ 0 & 1 \end{smallmatrix}\right)$ by (12.4). It is well-known that these two matrices generate $\mathrm{SL}(2,\mathbb{Z})$ (see Serre [**112**, §VII.1] or Exercise 12.3). Then (12.6) follows by induction on the length of $\left(\begin{smallmatrix} a & b \\ c & d \end{smallmatrix}\right)$ as a word in S and T (see Exercise 12.5).

Given (12.6), it follows easily that $\gamma_2(\tau)$ is invariant under the group of matrices

$$\tilde{\Gamma}(3) = \left\{\begin{pmatrix} a & b \\ c & d \end{pmatrix} : b \equiv c \equiv 0 \bmod 3\right\}.$$

This group is related to $\Gamma_0(9)$ by the identity

$$\Gamma_0(9) = \begin{pmatrix} 1/3 & 0 \\ 0 & 1 \end{pmatrix}\tilde{\Gamma}(3)\begin{pmatrix} 3 & 0 \\ 0 & 1 \end{pmatrix},$$

and a simple computation then shows that $\gamma_2(3\tau)$ is invariant under $\Gamma_0(9)$ (see Exercise 12.5). The group $\tilde{\Gamma}(3)$ is not the largest subgroup of $\mathrm{SL}(2,\mathbb{Z})$ fixing $\gamma_2(\tau)$, but it's the one that relates most easily to the $\Gamma_0(m)$'s (see Exercise 12.5).

To finish the proof that $\gamma_2(3\tau)$ is a modular function for $\Gamma_0(9)$, we need to check its behavior at the cusps. Let $\gamma \in \mathrm{SL}(2,\mathbb{Z})$. By Theorem 11.9, $j(3\tau)$ is a modular function for $\Gamma_0(3)$, so that $j(3\gamma\tau)$ has a meromorphic expansion in powers of $q^{1/3}$. Taking cube roots, this implies that $\gamma_2(3\gamma\tau)$ has a meromorphic expansion in powers of $q^{1/9}$, which proves that $\gamma_2(3\tau)$ is meromorphic at the cusps. This proves the proposition. Q.E.D.

Once we know that $\gamma_2(3\tau)$ is a modular function for $\Gamma_0(9)$, Theorem 11.9 tells us that it is a rational function in $j(\tau)$ and $j(9\tau)$. The following proposition will give us information about the coefficients of this rational function:

PROPOSITION 12.7. *Let $f(\tau)$ be a modular function for $\Gamma_0(m)$ whose q-expansion has rational coefficients. Then:*

(i) *$f(\tau) \in \mathbb{Q}(j(\tau), j(m\tau))$.*

(ii) *Assume in addition that $f(\tau)$ is holomorphic on $\mathfrak{h}$, and let $\tau_0 \in \mathfrak{h}$. If*

$$\frac{\partial\Phi_m}{\partial X}(j(m\tau_0), j(\tau_0)) \neq 0,$$

then $f(\tau_0) \in \mathbb{Q}(j(\tau_0), j(m\tau_0))$.

REMARK. Note that the hypothesis of the theorem involves only the expansion of $f(\tau)$ in powers of $q^{1/m}$. For general $\gamma \in \mathrm{SL}(2,\mathbb{Z})$, the expansion of $f(\gamma\tau)$ need not have coefficients in $\mathbb{Q}$.

PROOF. To prove (i), we will use the representation

$$f(\tau) = \frac{G(j(m\tau), j(\tau))}{\frac{\partial}{\partial X}\Phi_m(j(m\tau), j(\tau))} \tag{12.8}$$

given by (11.17). Since the denominator clearly lies in $\mathbb{Q}(j(\tau), j(m\tau))$ (part (i) of Theorem 11.18), it suffices to show that the same holds for the numerator. We know that $G(j(m\tau), j(\tau))$ lies in $\mathbb{C}(j(\tau))[j(m\tau)]$, so that

$$G(j(m\tau), j(\tau)) = \frac{P(j(m\tau), j(\tau))}{Q(j(\tau))},$$

where $P(X, Y)$ and $Q(Y) \neq 0$ are polynomials with complex coefficients. Let's write these polynomials as

$$P(X, Y) = \sum_{i=0}^{N}\sum_{k=0}^{M} a_{ik}\, X^i Y^k$$

$$Q(Y) = \sum_{l=0}^{L} b_l Y^l.$$

Then (12.8) implies that

$$P(j(m\tau), j(\tau)) = f(\tau)\frac{\partial\Phi_m}{\partial X}(j(m\tau), j(\tau))Q(j(\tau)),$$

which we can write as

$$\sum_{i=0}^{N}\sum_{k=0}^{M} a_{ik} j(m\tau)^i j(\tau)^k = f(\tau)\frac{\partial\Phi_m}{\partial X}(j(m\tau), j(\tau))\Big(\sum_{l=0}^{L} b_l j(\tau)^l\Big).$$

Substituting in the q-expansions of $f(\tau)$, $j(\tau)$ and $j(m\tau)$ and equating coefficients of powers of $q^{1/m}$, we get an infinite system of homogeneous linear equations with the a_{ik}'s and b_l's as unknowns. The q-expansions of $f(\tau)$, $j(\tau)$ and $j(m\tau)$ all have coefficients in $\mathbb{Q}$, and the coefficients of $(\partial/\partial X)\Phi_m(X, Y)$ are also rational. Thus the coefficients of our system of equations all lie in $\mathbb{Q}$. This system has a solution over $\mathbb{C}$ which is nontrivial in the b_l's (since $Q(j(\tau)) \neq 0$), and hence must have a solution over $\mathbb{Q}$ also nontrivial in the b_l's. This proves that $P(X, Y)$ and $Q(Y) \neq 0$ can be chosen to have rational coefficients, which proves part (i).

To prove (ii), let's go back to the definition of $G(X, j(\tau))$ given in (11.16). Since $f(\tau)$ is holomorphic on $\mathfrak{h}$, the coefficients of $G(X, j(\tau))$ are also holomorphic on $\mathfrak{h}$. As we saw in Lemma 11.10, this means that the coefficients are polynomials in $j(\tau)$. Thus, in the representation of $f(\tau)$ given by (12.8), the numerator $G(j(m\tau), j(\tau))$ is a polynomial in $j(m\tau)$ and $j(\tau)$. By a slight modification of the argument for part (i), we can assume that it has rational coefficients (see Exercise 12.6). Consequently, whenever the denominator doesn't vanish at τ_0, we can evaluate this expression at $\tau = \tau_0$ to conclude that $f(\tau_0)$ lies in $\mathbb{Q}(j(\tau_0), j(m\tau_0))$. Q.E.D.

We want to apply this proposition to $\gamma_2(\tau_0)$, where τ_0 is given in the statement of Theorem 12.2. By (12.5), we see that the q-expansion of $\gamma_2(3\tau)$ has rational coefficients. Since it is a modular function for $\Gamma_0(9)$, Proposition 12.7 tells us that

$$\gamma_2(3\tau) \in \mathbb{Q}(j(\tau), j(9\tau)).$$

Since we're concerned about $\gamma_2(\tau_0)$, we need to evaluate the above expression at $\tau = \tau_0/3$. We will for the moment assume that

$$\frac{\partial \Phi_9}{\partial X}(j(3\tau_0), j(\tau_0/3)) \neq 0. \tag{12.9}$$

Since $\gamma_2(3\tau)$ is holomorphic, the second part of Proposition 12.7 then implies that $\gamma_2(\tau_0) \in \mathbb{Q}(j(\tau_0/3), j(3\tau_0))$, which we can write as

$$\gamma_2(\tau_0) \in \mathbb{Q}(j([1, \tau_0/3]), j([1, 3\tau_0])). \tag{12.10}$$

To see what this says about $\gamma_2(\tau_0)$, recall that $\mathcal{O} = [1, \tau_0]$. Then $\mathcal{O}' = [1, 3\tau_0]$ is the order of index 3 in $\mathcal{O}$, and the special form of τ_0 implies that $[1, \tau_0/3]$ is a proper fractional $\mathcal{O}'$-ideal (this follows from Lemma 7.5 and $3 \nmid D$—see Exercise 12.7). Thus, by Theorem 11.1, both $j(\tau_0/3)$ and $j(3\tau_0)$ generate the ring class field L' of the order $\mathcal{O}'$. Consequently, (12.10) implies that $\gamma_2(\tau_0)$ lies in the ring class field L'.

Let L denote the ring class field of $\mathcal{O}$, so that $L \subset L'$. To compute the degree of this extension, recall that the class number is the degree of the ring class field over K. Since the discriminant of $\mathcal{O}$ is D, this means that $[L' : L] = h(9D)/h(D)$. Corollary (7.28) implies that

$$h(9D) = \frac{3h(D)}{[\mathcal{O}^* : \mathcal{O}'^*]}\left(1 - \left(\frac{D}{3}\right)\frac{1}{3}\right),$$

and since $3 \nmid D$, we see that $L \subset L'$ is an extension of degree 2 or 4. Now consider the following diagram of fields:

$$\begin{array}{ccc} \mathbb{Q}(j(\tau_0)) & \subset & L \\ \cap & & \cap \\ \mathbb{Q}(\gamma_2(\tau_0)) & \subset & L'. \end{array}$$

We know that L has degree 2 over $\mathbb{Q}(j(\tau_0))$, and by the above computation, L' has degree 2 or 4 over L. It follows that the degree of $\mathbb{Q}(\gamma_2(\tau_0))$ over $\mathbb{Q}(j(\tau_0))$ is a power of 2. But recall that $\gamma_2(\tau_0)$ is the real cube root of $j(\tau_0)$, which means that the extension $\mathbb{Q}(j(\tau_0)) \subset \mathbb{Q}(\gamma_2(\tau_0))$ has degree 1 or 3. Hence this degree must be 1, which proves that $\mathbb{Q}(j(\tau_0)) = \mathbb{Q}(\gamma_2(\tau_0))$.

We are not quite done with the theorem, for we still have to verify that (12.9) is satisfied, i.e., that

$$\frac{\partial \Phi_9}{\partial X}(j(3\tau_0), j(\tau_0/3)) \neq 0.$$

For later purposes, we will prove the following general lemma:

LEMMA 12.11. *Let $\mathcal{O}$ be an order in an imaginary quadratic field, and assume that $\mathcal{O}^* = \{\pm 1\}$. Write $\mathcal{O} = [1, \alpha]$, and assume that for some integer s, $s \mid T(\alpha)$ and $\gcd(s^2, N(\alpha))$ is squarefree, where $T(\alpha)$ and $N(\alpha)$ are the trace and norm of α. Then for any positive integer m,*

$$\frac{\partial \Phi_m}{\partial X}(j(m\alpha/s), j(\alpha/s)) \neq 0.$$

PROOF. Since $\Phi_m(j(m\alpha/s), j(\alpha/s)) = 0$, the nonvanishing of the partial derivative means that $j(m\alpha/s)$ is not a multiple root of the polynomial

$$\Phi_m(X, j(\alpha/s)) = \prod_{\sigma \in C(m)} (X - j(\sigma\alpha/s)).$$

Thus we must show that

$$j(m\alpha/s) \neq j(\sigma\alpha/s), \qquad \sigma \in C(m), \quad \sigma \neq \sigma_0 = \begin{pmatrix} m & 0 \\ 0 & 1 \end{pmatrix}.$$

So pick $\sigma = \left(\begin{smallmatrix} a & b \\ 0 & d \end{smallmatrix}\right) \in C(m)$, $\sigma \neq \sigma_0$, and assume that $j(m\alpha/s) = j(\sigma\alpha/s)$. In terms of lattices, this means that there is a complex number λ such that

$$\lambda[1, m\alpha/s] = [d, a\alpha/s + b]. \tag{12.12}$$

We will show that this leads to a contradiction when $\mathcal{O}^* = \{\pm 1\}$.

The idea is to prove that λ is a unit of $\mathcal{O}$. To see this, note that by Lemma 11.24, both $[1, m\alpha/s]$ and $[d, a\alpha/s+b]$ have index m in $[1, \alpha/s]$, so that λ must have norm 1. Furthermore, we have

$$s\lambda \in s[d, a\alpha/s + b] = [sd, a\alpha + sb] \subset [s, \alpha].$$

Writing $s\lambda = us + v\alpha$, u, $v \in \mathbb{Z}$, and taking norms, we obtain

$$s^2 = s^2 N(\lambda) = N(us + v\alpha) = u^2 s^2 + usvT(\alpha) + v^2 N(\alpha).$$

Since $s \mid T(\alpha)$, it follows that $s^2 \mid v^2 N(\alpha)$, and since $\gcd(s^2, N(\alpha))$ is squarefree, we must have $s \mid v$. This shows that $\lambda \in [1, \alpha] = \mathcal{O}$, so that λ is a unit since it has norm 1. Then $\mathcal{O}^* = \{\pm 1\}$ implies that $\lambda = \pm 1$, and hence $[1, m\alpha/s] = [d, a\alpha/s+b]$, which contradicts $\sigma \neq \sigma_0$ by the uniqueness part of Lemma 11.24. The lemma is proved. Q.E.D.

We want to apply this lemma to the case $s = 3$, $m = 9$ and $\alpha = \tau_0$. Using the special form of τ_0, it is easy to see that the norm and trace conditions are satisfied (note that the discriminant of $\mathcal{O} = [1, \tau_0]$ is $D = T(\tau_0)^2 - 4N(\tau_0)$). Thus (12.9) holds except possibly when $\mathcal{O}$ is $\mathbb{Z}[i]$ or $\mathbb{Z}[\zeta_3]$. The latter can't occur since 3 doesn't divide the discriminant, and when $\mathcal{O} = \mathbb{Z}[i]$, a simple argument shows that (12.12) is impossible (see Exercise 12.8). This completes the proof of Theorem 12.2. Q.E.D.

This theorem tells us about the behavior of $\gamma_2(\tau_0)$ when 3 doesn't divide the discriminant D. For completeness, let's record what happens when D is a multiple of 3 (see Schertz [**110**] for a proof):

THEOREM 12.13. *Let $\mathcal{O}$ be an order of discriminant D in an imaginary quadratic field K. Assume $3 \mid D$ and $D < -3$ and write $\mathcal{O} = [1, \tau_0]$, where*

$$\tau_0 = \begin{cases} \sqrt{-m} & D = -4m \equiv 0 \bmod 4 \\ \dfrac{3 + \sqrt{-m}}{2} & D = -m \equiv 1 \bmod 4. \end{cases}$$

Then $K(\gamma_2(\tau_0))$ is the ring class field of the order $\mathcal{O}' = [1, 3\tau_0]$ and is an extension of degree 3 of the ring class field of $\mathcal{O}$. Furthermore, $\mathbb{Q}(\gamma_2(\tau_0)) = \mathbb{Q}(j(3\tau_0))$. Q.E.D.

Besides the cube root $\gamma_2(\tau)$ of $j(\tau)$, there is also an interesting square root to consider. This follows because

$$j(\tau) - 1728 = 1728\left(\frac{g_2(\tau)^3}{\Delta(\tau)} - 1\right) = 6^6\, \frac{g_3(\tau)^2}{\Delta(\tau)}$$

since $\Delta(\tau) = g_2(\tau)^3 - 27g_3(\tau)^2$. Hence we can define

$$\gamma_3(\tau) = \sqrt{j(\tau) - 1728} = 216\, \frac{g_3(\tau)}{\sqrt{\Delta(\tau)}}.$$

Similar to Theorems 12.2 and 12.13, τ_0 can be chosen so that $\gamma_3(\tau_0)$ generates a field closely related to the ring class field of $\mathcal{O} = [1, \tau_0]$. See Schertz [**111**, Theorem 3].

B. The Weber Functions

To work effectively with $\gamma_2(\tau)$, we need good formulas for computing it. This leads us to our next topic, the Dedekind η-function $\eta(\tau)$ and the three Weber functions $\mathfrak{f}(\tau)$, $\mathfrak{f}_1(\tau)$ and $\mathfrak{f}_2(\tau)$. If $\tau \in \mathfrak{h}$, we let $q = e^{2\pi i \tau}$ as usual, and then the *Dedekind η-function* is defined by the formula

$$\eta(\tau) = q^{1/24} \prod_{n=1}^{\infty} (1 - q^n).$$

Note that this product converges (and is nonzero) for $\tau \in \mathfrak{h}$ since $0 < |q| < 1$.

We then define the *Weber functions* $\mathfrak{f}(\tau)$, $\mathfrak{f}_1(\tau)$ and $\mathfrak{f}_2(\tau)$ in terms of the η-function as follows:

$$(12.14)\qquad \begin{aligned} \mathfrak{f}(\tau) &= \zeta_{48}^{-1} \frac{\eta((\tau+1)/2)}{\eta(\tau)} \\ \mathfrak{f}_1(\tau) &= \frac{\eta(\tau/2)}{\eta(\tau)} \\ \mathfrak{f}_2(\tau) &= \sqrt{2}\, \frac{\eta(2\tau)}{\eta(\tau)}, \end{aligned}$$

where $\zeta_{48} = e^{2\pi i/48}$. From these definitions, one gets the following product expansions for the Weber functions:

$$(12.15)\qquad \begin{aligned} \mathfrak{f}(\tau) &= q^{-1/48} \prod_{n=1}^{\infty} (1 + q^{n-1/2}) \\ \mathfrak{f}_1(\tau) &= q^{-1/48} \prod_{n=1}^{\infty} (1 - q^{n-1/2}) \\ \mathfrak{f}_2(\tau) &= \sqrt{2}\, q^{1/24} \prod_{n=1}^{\infty} (1 + q^{n}) \end{aligned}$$

(see Exercise 12.9), and we also get the following useful identities connecting the Weber functions:

$$(12.16)\qquad \begin{aligned} \mathfrak{f}(\tau)\, \mathfrak{f}_1(\tau)\, \mathfrak{f}_2(\tau) &= \sqrt{2} \\ \mathfrak{f}_1(2\tau)\, \mathfrak{f}_2(\tau) &= \sqrt{2} \end{aligned}$$

(see Exercise 12.9).

Much deeper lie the following relations between $\eta(\tau)$, $\mathfrak{f}(\tau)$, $\mathfrak{f}_1(\tau)$ and $\mathfrak{f}_2(\tau)$ and the previously defined functions $j(\tau)$, $\gamma_2(\tau)$ and $\Delta(\tau)$:

THEOREM 12.17. *If $\tau \in \mathfrak{h}$, then $\Delta(\tau) = (2\pi)^{12} \eta(\tau)^{24}$ and*

$$\gamma_2(\tau) = \frac{\mathfrak{f}(\tau)^{24} - 16}{\mathfrak{f}(\tau)^8} = \frac{\mathfrak{f}_1(\tau)^{24} + 16}{\mathfrak{f}_1(\tau)^8} = \frac{\mathfrak{f}_2(\tau)^{24} + 16}{\mathfrak{f}_2(\tau)^8}.$$

REMARK. Since $j(\tau) = \gamma_2(\tau)^3$, this theorem gives us some remarkable formulas for computing the j-function.

PROOF. We need to relate $\eta(\tau)$ and the Weber functions to the Weierstrass $\wp$-function. Let $\wp(z) = \wp(z;\tau)$ denote the $\wp$-function for the lattice $[1,\tau]$, and set

$$e_1 = \wp(\tau/2), \qquad e_2 = \wp(1/2), \qquad e_3 = \wp((\tau+1)/2).$$

We will prove the following formulas for the differences $e_i - e_j$:

$$\begin{aligned} e_2 - e_1 &= \pi^2\,\eta(\tau)^4\,\mathfrak{f}(\tau)^8 \\ e_2 - e_3 &= \pi^2\,\eta(\tau)^4\,\mathfrak{f}_1(\tau)^8 \\ e_3 - e_1 &= \pi^2\,\eta(\tau)^4\,\mathfrak{f}_2(\tau)^8. \end{aligned} \tag{12.18}$$

The basic strategy of the proof is to express $e_i - e_j$ in terms of the Weierstrass σ-function (see below), and then use the product expansion of the σ-function to get product expansions for $e_i - e_j$. Proofs will appear in the exercises.

The Weierstrass σ-function is defined as follows. Let $\tau \in \mathfrak{h}$, and let L be the lattice $[1,\tau]$. Then the Weierstrass σ-function is the product

$$\sigma(z;\tau) = z \prod_{\omega \in L - \{0\}} \left(1 - \frac{z}{\omega}\right) e^{z/\omega + (1/2)(z/\omega)^2}.$$

Note that $\sigma(z;\tau)$ is an odd function in z. We will usually write $\sigma(z;\tau)$ more simply as $\sigma(z)$. The σ-function is not periodic, but there are complex numbers η_1 and η_2, depending only on τ, such that

$$\begin{aligned} \sigma(z+\tau) &= -e^{\eta_1(z+\tau/2)}\,\sigma(z) \\ \sigma(z+1) &= -e^{\eta_2(z+1/2)}\,\sigma(z), \end{aligned}$$

and the numbers η_1 and η_2 satisfy the Legendre relation $\eta_2\tau - \eta_1 = 2\pi i$ (see Exercise (12.10)). The σ-function is related to the $\wp$-function by the formula

$$\wp(z) - \wp(w) = -\frac{\sigma(z+w)\,\sigma(z-w)}{\sigma^2(z)\,\sigma^2(w)}$$

whenever z and w do not lie in L (see Exercise 12.11). Since $e_1 = \wp(\tau/2)$, $e_2 = \wp(1/2)$ and $e_3 = \wp((\tau+1)/2)$, it follows easily that

$$\begin{aligned} e_2 - e_1 &= -e^{-\eta_2\tau/2}\,\frac{\sigma^2\left(\dfrac{\tau+1}{2}\right)}{\sigma^2\left(\dfrac{1}{2}\right)\sigma^2\left(\dfrac{\tau}{2}\right)} \\ e_2 - e_3 &= -e^{\eta_2(\tau+1)/2}\,\frac{\sigma^2\left(\dfrac{\tau}{2}\right)}{\sigma^2\left(\dfrac{1}{2}\right)\sigma^2\left(\dfrac{\tau+1}{2}\right)} \\ e_3 - e_1 &= e^{\eta_1(\tau+1)/2}\,\frac{\sigma^2\left(\dfrac{1}{2}\right)}{\sigma^2\left(\dfrac{\tau+1}{2}\right)\sigma^2\left(\dfrac{\tau}{2}\right)} \end{aligned}$$

(see Exercise 12.12).

There is also the following q-product expansion for the σ-function:

$$\sigma(z;\tau) = \frac{1}{2\pi i}\, e^{\eta_2 z^2/2}\,(q_z^{1/2} - q_z^{-1/2}) \prod_{n=1}^{\infty} \frac{(1 - q_\tau^n q_z)(1 - q_\tau^n/q_z)}{(1 - q_\tau^n)^2},$$

where $q_\tau = e^{2\pi i\tau}$ and $q_z = e^{2\pi i z}$ (see Exercise 12.13). Using this product expansion, we obtain the formulas

$$\begin{aligned}
\sigma\left(\frac{1}{2}\right) &= \frac{1}{2\pi} e^{\eta_2/8} \frac{\mathfrak{f}_2(\tau)^2}{\eta(\tau)^2}\\
\sigma\left(\frac{\tau}{2}\right) &= \frac{i}{2\pi} e^{\eta_2\tau^2/8} q^{-1/8} \frac{\mathfrak{f}_1(\tau)^2}{\eta(\tau)^2}\\
\sigma\left(\frac{\tau+1}{2}\right) &= \frac{1}{2\pi} e^{\eta_2(\tau+1)^2/8} q^{-1/8} \frac{\mathfrak{f}(\tau)^2}{\eta(\tau)^2}
\end{aligned}$$

(see Exercise 12.14). It is now straightforward to derive the desired formulas (12.18) for $e_i - e_j$ (see Exercise 12.14).

To relate this to $\Delta(\tau)$, recall that $\Delta(\tau) = 16(e_2 - e_1)^2(e_2 - e_3)^2(e_3 - e_1)^2$ by (10.6). Using (12.18), it is now easy to express $\Delta(\tau)$ in terms of the η-function:

$$\begin{aligned}
\Delta(\tau) &= 16(e_2 - e_1)^2(e_2 - e_3)^2(e_3 - e_1)^2\\
&= 16\pi^{12}\,\eta(\tau)^{24}\,\mathfrak{f}(\tau)^{16}\,\mathfrak{f}_1(\tau)^{16}\,\mathfrak{f}_2(\tau)^{16}\\
&= (2\pi)^{12}\,\eta(\tau)^{24},
\end{aligned}$$

where the last line follows by (12.16).

Turning to $\gamma_2(\tau)$, we know that

$$\gamma_2(\tau) = \sqrt[3]{j(\tau)} = \frac{12 g_2(\tau)}{\sqrt[3]{\Delta(\tau)}},$$

where the cube root is chosen to be real-valued on the imaginary axis. Using what we just proved about $\Delta(\tau)$, this formula can be written

$$\gamma_2(\tau) = \frac{3 g_2(\tau)}{4\pi^4\,\eta(\tau)^8}$$

since $\eta(\tau)$ is real valued on the imaginary axis. Thus, to express $\gamma_2(\tau)$ in terms of Weber functions, we need to express $g_2(\tau)$ in terms of $\eta(\tau)$, $\mathfrak{f}(\tau)$, $\mathfrak{f}_1(\tau)$ and $\mathfrak{f}_2(\tau)$.

The idea is to write $g_2(\tau)$ in terms of the $e_i - e_j$'s. Recall from the proof of Proposition 10.7 that the e_i's are the roots of $4x^3 - g_2(\tau)x - g_3(\tau)$, which implies that $g_2(\tau) = -4(e_1e_2 + e_1e_3 + e_2e_3)$ (see Exercise 10.8). Then, using $e_1 + e_2 + e_3 = 0$, one obtains

$$3g_2(\tau) = 4((e_2 - e_1)^2 - (e_2 - e_3)(e_3 - e_1))$$

(see Exercise 12.15). Substituting in the formulas from (12.18) yields

$$3g_2(\tau) = 4\pi^4\,\eta(\tau)^8\big(\mathfrak{f}(\tau)^{16} - \mathfrak{f}_1(\tau)^8\,\mathfrak{f}_2(\tau)^8\big),$$

so that

$$\begin{aligned}
\gamma_2(\tau) &= \mathfrak{f}(\tau)^{16} - \mathfrak{f}_1(\tau)^8\,\mathfrak{f}_2(\tau)^8\\
&= \mathfrak{f}(\tau)^{16} - \frac{16}{\mathfrak{f}(\tau)^8}\\
&= \frac{\mathfrak{f}(\tau)^{24} - 16}{\mathfrak{f}(\tau)^8},
\end{aligned}$$

where we have again used the basic identity (12.16). The other two formulas for $\gamma_2(\tau)$ stated in Theorem 12.17 are proved similarly and are left to the reader (see Exercise 12.15). This completes the proof of the theorem. Q.E.D.

Using these formulas it is easy to show that the q-expansions of $\gamma_2(\tau)$ and $j(\tau)$ have integer coefficients (see Exercise 12.16), and this proves Theorem 11.8. We can also use Theorem 12.17 to study the transformation properties of $\eta(\tau)$, $\mathfrak{f}(\tau)$, $\mathfrak{f}_1(\tau)$ and $\mathfrak{f}_2(\tau)$:

COROLLARY 12.19. *For a positive integer n, let $\zeta_n = e^{2\pi i/n}$. Then*

$$\begin{aligned}\eta(\tau+1) &= \zeta_{24}\,\eta(\tau)\\ \eta(-1/\tau) &= \sqrt{-i\tau}\,\eta(\tau),\end{aligned}$$

where the square root is chosen to be positive on the imaginary axis. Furthermore,

$$\begin{aligned}\mathfrak{f}(\tau+1) &= \zeta_{48}^{-1}\mathfrak{f}_1(\tau)\\ \mathfrak{f}_1(\tau+1) &= \zeta_{48}^{-1}\mathfrak{f}(\tau)\\ \mathfrak{f}_2(\tau+1) &= \zeta_{24}\,\mathfrak{f}_2(\tau),\end{aligned}$$

and

$$\begin{aligned}\mathfrak{f}(-1/\tau) &= \mathfrak{f}(\tau)\\ \mathfrak{f}_1(-1/\tau) &= \mathfrak{f}_2(\tau)\\ \mathfrak{f}_2(-1/\tau) &= \mathfrak{f}_1(\tau).\end{aligned}$$

PROOF. The definition of $\eta(\tau)$ makes the formula for $\eta(\tau+1)$ obvious. Turning to $\eta(-1/\tau)$, first consider $\Delta(\tau) = (2\pi)^{12}\eta(\tau)^{24}$. For a lattice L, we know

$$\Delta(L) = g_2(L)^3 - 27g_3(L)^2.$$

In (10.10) we showed that $g_2(\lambda L) = \lambda^{-4}g_2(L)$ and $g_3(\lambda L) = \lambda^{-6}g_3(L)$. Hence

$$\Delta(\lambda L) = \lambda^{-12}\Delta(L).$$

This gives us the formula

$$\Delta(-1/\tau) = \Delta([1,-1/\tau]) = \Delta(\tau^{-1}[1,\tau]) = \tau^{12}\Delta([1,\tau]) = \tau^{12}\Delta(\tau),$$

and taking 24th roots, we obtain

$$\eta(-1/\tau) = \epsilon\sqrt{-i\tau}\,\eta(\tau)$$

for some root of unity ϵ. Both sides take positive real values on the imaginary axis, which forces ϵ to be 1. This proves that $\eta(\tau)$ transforms as desired.

Turning to the Weber functions, their behavior under $\tau \mapsto \tau+1$ and $\tau \mapsto -1/\tau$ is a simple consequence of their definitions and the transformation properties of $\eta(\tau)$ (see Exercise 12.17). Q.E.D.

We will make extensive use of these transformation properties in the latter part of this section.

C. j-Invariants of Orders of Class Number 1

Using the properties of the Weber functions, we can now compute the j-invariants for orders of class number 1. In §7 we saw that there are exactly 13 such orders, with discriminants

$$-3,\ -4,\ -7,\ -8,\ -11,\ -12,\ -16,\ -19,\ -27,\ -28,\ -43,\ -67,\ -163$$

(we will prove this in Theorem 12.34 below). The j-invariants of these orders are integers, and if we restrict ourselves to those where 3 doesn't divide the discriminant (10 of the above 13), then Theorem 12.2 tells us that the j-invariant is a cube. So in these cases we need only compute $\gamma_2(\tau_0)$, where τ_0 is an appropriately chosen element of the order. Rather than compute $\gamma_2(\tau_0)$ directly, we will use the Weber functions to approximate its value to within $\pm.5$. Since $\gamma_2(\tau_0)$ is an integer, this will determine its value uniquely. This scheme for computing these j-invariants is due to Weber [**131**, §125].

The ten j-invariants we want to compute are given in the following table:

(12.20)

d_K	τ_0	$\gamma_2(\tau_0)$	$j(\mathcal{O}) = j(\tau_0)$
-4	i	$12 = 2^2 \cdot 3$	12^3
-7	$(3+\sqrt{-7})/2$	$-15 = -3 \cdot 5$	-15^3
-8	$\sqrt{-2}$	$20 = 2^2 \cdot 5$	20^3
-11	$(3+\sqrt{-11})/2$	$-32 = -2^5$	-32^3
-16	$2i$	$66 = 2 \cdot 3 \cdot 11$	66^3
-19	$(3+\sqrt{-19})/2$	$-96 = -2^5 \cdot 3$	-96^3
-28	$\sqrt{-7}$	$255 = 3 \cdot 5 \cdot 17$	255^3
-43	$(3+\sqrt{-43})/2$	$-960 = -2^6 \cdot 3 \cdot 5$	-960^3
-67	$(3+\sqrt{-67})/2$	$-5280 = -2^5 \cdot 3 \cdot 5 \cdot 11$	-5280^3
-163	$(3+\sqrt{-163})/2$	$-640320 = -2^6 \cdot 3 \cdot 5 \cdot 23 \cdot 29$	-640320^3

For completeness, we also give the j-invariants of the three orders of discriminant divisible by 3:

d_K	τ_0	$j(\mathcal{O}) = j(\tau_0)$
-3	$(1+\sqrt{-3})/2$	0
-12	$\sqrt{-3}$	$54000 = 2^4 \cdot 3^3 \cdot 5^3$
-27	$(1+3\sqrt{-3})/2$	$-12288000 = -2^{15} \cdot 3 \cdot 5^3$

We computed $j((1+\sqrt{-3})/2) = 0$ in §10, and we will prove $j(\sqrt{-3}) = 54000$ in §13. As predicted by Theorem 12.13, the last two entries are not perfect cubes.

To start the computation, first consider the case of even discriminant. Here, $\tau_0 = \sqrt{-m}$ where $m = 1$, 2, 4 or 7. Setting $q = e^{2\pi i\sqrt{-m}} = e^{-2\pi\sqrt{m}}$, we claim that

$$\gamma_2(\sqrt{-m}) = [\![256\, q^{2/3} + q^{-1/3}]\!], \tag{12.21}$$

where $[\![\ ]\!]$ is the nearest integer function (i.e., for a real number $x \notin \mathbb{Z} + \frac{1}{2}$, $[\![x]\!]$ is the integer nearest to x).

To prove this, we will write $\gamma_2(\tau)$ in terms of the Weber function $\mathfrak{f}_2(\tau)$:

$$\gamma_2(\sqrt{-m}) = \mathfrak{f}_2(\sqrt{-m})^{16} + \frac{16}{\mathfrak{f}_2(\sqrt{-m})^8}. \tag{12.22}$$

Using $q = e^{-2\pi\sqrt{m}}$ as above, (12.15) gives us

$$\mathfrak{f}_2(\sqrt{-m}) = \sqrt{2}\, q^{1/24} \prod_{n=1}^{\infty} (1 + q^n),$$

and to estimate the infinite product, we use the inequality $1 + x < e^x$ for $x > 0$. This yields

$$1 < \prod_{n=1}^{\infty} (1 + q^n) < \prod_{n=1}^{\infty} e^{q^n} = e^{q/(1-q)},$$

and we can simplify the exponent by noting that $q/(1-q) \le q/(1-e^{-2\pi}) < 1.002q$ since $q \le e^{-2\pi}$. Thus we have the following inequalities for $\mathfrak{f}_2(\sqrt{-m})$:

$$\sqrt{2}\, q^{1/24} < \mathfrak{f}_2(\sqrt{-m}) < \sqrt{2}\, q^{1/24} e^{1.002q},$$

and applying this to (12.22), we get upper and lower bounds for $\gamma_2(\sqrt{-m})$:

$$256\, q^{2/3} + q^{-1/3} e^{-8.016q} < \gamma_2(\sqrt{-m}) < 256\, q^{2/3} e^{16.032q} + q^{-1/3}. \tag{12.23}$$

To see how sharp these bounds are, consider their difference

$$E = 256\, q^{2/3}(e^{16.032q} - 1) + q^{-1/3}(1 - e^{-8.016q}).$$

Using the inequality

$$1 - e^{-x} < \frac{x}{1-x}, \qquad 0 < x < 1,$$

one sees that

$$\begin{aligned} E &< 256\, q^{2/3}(e^{16.032q} - 1) + q^{-1/3} 8.016\, q/(1 - 8.016q) \\ &= 256\, q^{2/3}(e^{16.032q} - 1) + 8.016\, q^{2/3}/(1 - 8.016\, q). \end{aligned}$$

The last quantity is an increasing function in q, and then $q < e^{-2\pi}$ easily implies that $E < .25$. Since $\gamma_2(\sqrt{-m})$ is an integer, this means that $[\![x]\!] = \gamma_2(\sqrt{-m})$ for any x lying between the upper and lower limits of (12.23). In particular, $256\, q^{2/3} + q^{-1/3}$ lies between these limits, which proves (12.21). Using a hand calculator, it is now trivial to compute the corresponding entries in table (12.20) (see Exercise 12.18).

Turning to the case of odd discriminant, let $\tau_0 = (3 + \sqrt{-m})/2$, $m = 7$, 11, 19, 43, 67 or 163, and we again want to compute

$$\gamma_2(\tau_0) = \mathfrak{f}_2(\tau_0)^{16} + \frac{16}{\mathfrak{f}_2(\tau_0)^8}.$$

Our previous techniques won't work, for $q = e^{2\pi i(3+\sqrt{-m})/2} = -e^{-\pi\sqrt{m}}$ is negative in this case. But Weber uses the following clever trick: from (12.16), we know that

$$\mathfrak{f}_2(\tau_0) = \frac{\sqrt{2}}{\mathfrak{f}_1(2\tau_0)},$$

and then the transformation properties from Corollary 12.19 imply

$$\begin{aligned} \mathfrak{f}_1(2\tau_0) = \mathfrak{f}_1(3 + \sqrt{-m}) &= \zeta_{48}^{-1} \mathfrak{f}(2 + \sqrt{-m}) \\ &= \zeta_{48}^{-2} \mathfrak{f}_1(1 + \sqrt{-m}) = \zeta_{48}^{-3} \mathfrak{f}(\sqrt{-m}). \end{aligned}$$

Combining the above equations implies that

$$\mathfrak{f}_2(\tau_0) = \frac{\sqrt{2}\,\zeta_{16}}{\mathfrak{f}(\sqrt{-m})},$$

and thus

$$\gamma_2(\tau_0) = \frac{256}{\mathfrak{f}(\sqrt{-m})^{16}} - \mathfrak{f}(\sqrt{-m})^8.$$

From here, our previous methods easily imply that if $m = 7, 11, 19, 43, 67$ or 163, and $q = e^{-2\pi\sqrt{m}}$, then

$$\gamma_2((3+\sqrt{-m})/2) = [\![-q^{-1/6} + 256\,q^{1/3}]\!],$$

where $[\![\ ]\!]$ is again the nearest integer function. Using a hand calculator, we can now complete our table (12.20) of singular j-invariants (see Exercise 12.18).

D. Weber's Computation of $j(\sqrt{-14})$

We next want to compute some singular j-invariants when the class number is greater than 1. There are several ways one can proceed. For example, when the class number is 2, the Kronecker Limit Formula gives an elegant method to determine the j-invariant, and this method generalizes to the case of orders with only one class per genus. (Recall from §3 that for discriminants $-4n$, this condition means that n is one of Euler's convenient numbers.) For example, when $n = 105$, Weber [**131**, §143] shows that

$$\mathfrak{f}(\sqrt{-105})^6 = \sqrt{2}^{-13}(1+\sqrt{3})^3(1+\sqrt{5})^3(\sqrt{3}+\sqrt{7})^3(\sqrt{5}+\sqrt{7}),$$

which would then allow us to compute $\gamma_2(\sqrt{-105})$ and hence $j(\sqrt{-105})$. (The radicals appearing in the above formula are not surprising, since in this case the Hilbert class field equals the genus field, which we know by Theorem 6.1—see Exercise 12.19.) Other examples may be found in Weber [**131**, pp. 721–726] or [**132**], and a modern treatment of the Kronecker Limit Formula is in Lang [**90**, Chapter 20].

We will instead take a different route and compute $j(\sqrt{-14})$, an example particularly relevant to earlier sections. Namely, $K(j(\sqrt{-14}))$ is the Hilbert class field of $K = \mathbb{Q}(\sqrt{-14})$ since $\mathcal{O}_K = [1, \sqrt{-14}]$. We determined this field in §5, so that finding $j(\sqrt{-14})$ will give us a second and quite different way of finding the Hilbert class field of $\mathbb{Q}(\sqrt{-14})$. Our exposition will again follow Weber [**131**, §144], using ideas from Schertz [**110**] to fill in the details omitted by Weber.

A key fact we will use is that in many cases, one can generate ring class fields using small powers of the Weber functions. Weber gives a long list of such theorems in [**131**, §§126–127], and modern proofs have been given by Birch [**10**] and Schertz [**109**–**111**]. We will discuss two cases which will be useful to our purposes:

THEOREM 12.24. *Given an integer $m > 0$ not divisible by 3, let $\mathcal{O} = [1, \sqrt{-m}]$, which is an order in $K = \mathbb{Q}(\sqrt{-m})$. Then:*

(i) *For $m \equiv 6 \bmod 8$, $\mathfrak{f}_1(\sqrt{-m})^2$ is an algebraic integer and $K(\mathfrak{f}_1(\sqrt{-m})^2)$ is the ring class field of $\mathcal{O}$.*
(ii) *For $m \equiv 3 \bmod 4$, $\mathfrak{f}(\sqrt{-m})^2$ is an algebraic integer and $K(\mathfrak{f}(\sqrt{-m})^2)$ is the ring class field of $\mathcal{O}$.*

PROOF. We begin with (i). Multiplying out the identity

$$j(\sqrt{-m}) = \gamma_2(\sqrt{-m})^3 = \left(\frac{\mathfrak{f}_1(\sqrt{-m})^{24}+16}{\mathfrak{f}_1(\sqrt{-m})^8}\right)^3,$$

it follows that $\mathfrak{f}_1(\sqrt{-m})^2$ is a root of a monic polynomial with coefficients in the ring $\mathbb{Z}[j(\sqrt{-m})]$. But $j(\sqrt{-m})$ is an algebraic integer, which implies that the same is true for $\mathfrak{f}_1(\sqrt{-m})^2$.

We know that $L = K(j(\sqrt{-m}))$ is the ring class field of $[1, \sqrt{-m}]$, and since $j(\sqrt{-m})$ is a polynomial in $\mathfrak{f}_1(\sqrt{-m})^2$, we need only show that $\mathfrak{f}_1(\sqrt{-m})^2$ lies in L. Actually, it suffices to show that $\mathfrak{f}_1(\sqrt{-m})^6$ lies in L. This is a consequence of Theorems 12.2 and 12.17, for since $3 \nmid m$, we have $\gamma_2(\sqrt{-m}) \in L$. We also know that

$$\gamma_2(\sqrt{-m}) = \frac{\mathfrak{f}_1(\sqrt{-m})^{24} + 16}{\mathfrak{f}_1(\sqrt{-m})^8}.$$

When $\mathfrak{f}_1(\sqrt{-m})^6$ lies in L, so does $\mathfrak{f}_1(\sqrt{-m})^{24}$. The above equation then implies that $\mathfrak{f}_1(\sqrt{-m})^8 \in L$, and $\mathfrak{f}_1(\sqrt{-m})^2 \in L$ follows immediately.

The next step in the proof is to show that $\mathfrak{f}_1(8\tau)^6$ is a modular function:

PROPOSITION 12.25. *$\mathfrak{f}_1(8\tau)^6$ is a modular function for the group $\Gamma_0(32)$.*

PROOF. We first study the transformation properties of $f_1(\tau)^6$. Consider the group

$$\Gamma_0(2)^t = \left\{ \begin{pmatrix} a & b \\ c & d \end{pmatrix} : b \equiv 0 \bmod 2 \right\}.$$

In Exercise 12.4, we will show that the matrices

$$-I = \begin{pmatrix} -1 & 0 \\ 0 & -1 \end{pmatrix}, \quad U = \begin{pmatrix} 1 & 0 \\ 1 & 1 \end{pmatrix}, \quad V = \begin{pmatrix} 1 & 2 \\ 0 & 1 \end{pmatrix}$$

generate $\Gamma_0(2)^t$. Using Corollary 12.19, $\mathfrak{f}_1(\tau)^6$ transforms under U and V as follows:

$$\begin{aligned} \mathfrak{f}_1(U\tau)^6 &= -i\mathfrak{f}_1(\tau)^6 \\ \mathfrak{f}_1(V\tau)^6 &= -i\mathfrak{f}_1(\tau)^6 \end{aligned}$$

(see Exercise 12.20). Then we get the general transformation law for $\mathfrak{f}_1(\tau)^6$:

$$\mathfrak{f}_1(\gamma\tau)^6 = i^{-ac-(1/2)bd+(1/2)b^2c}\mathfrak{f}_1(\tau)^6, \quad \gamma = \begin{pmatrix} a & b \\ c & d \end{pmatrix} \in \Gamma_0(2)^t. \tag{12.26}$$

This can be proved by induction on the length of γ as a word in $-I$, U and V. A more enlightening way to prove (12.26) is sketched in Exercise 12.21. See also §15.

Now consider the function $\mathfrak{f}_1(8\tau)^6$, and let $\gamma \in \Gamma_0(32)$. Then

$$8\gamma\tau = 8\begin{pmatrix} a & b \\ 32c & d \end{pmatrix}\tau = \begin{pmatrix} a & 8b \\ 4c & d \end{pmatrix} 8\tau = \tilde{\gamma}8\tau.$$

Since $\tilde{\gamma} \in \Gamma_0(2)^t$, it follows easily from (12.26) that $\mathfrak{f}_1(8\gamma\tau)^6 = \mathfrak{f}_1(\tilde{\gamma}8\tau)^6 = \mathfrak{f}_1(8\tau)^6$, which proves that $\mathfrak{f}_1(8\tau)^6$ is invariant under $\Gamma_0(32)$. To check the cusps, suppose that $\gamma \in \mathrm{SL}(2, \mathbb{Z})$. Under the correspondence between cosets of $\Gamma_0(8)$ and matrices in $C(8)$ given by Lemma 11.11, there is $\sigma \in C(8)$ and $\tilde{\gamma} \in \mathrm{SL}(2, \mathbb{Z})$ such that

$$8\gamma\tau = \tilde{\gamma}\sigma\tau.$$

Writing $\tilde{\gamma}$ as a product of various powers of $\left(\begin{smallmatrix} 1 & 1 \\ 0 & 1 \end{smallmatrix}\right)$ and $\left(\begin{smallmatrix} 0 & -1 \\ 1 & 0 \end{smallmatrix}\right)$, the transformation properties of Corollary 12.19 imply that

$$\mathfrak{f}_1(8\gamma\tau)^6 = \mathfrak{f}_1(\tilde{\gamma}\sigma\tau)^6 = \epsilon\mathfrak{f}(\sigma\tau)^6, \ \epsilon\mathfrak{f}_1(\sigma\tau)^6, \text{ or } \epsilon\mathfrak{f}_2(\sigma\tau)^6$$

for some root of unity ϵ. Since $\sigma = \left(\begin{smallmatrix} a & b \\ 0 & d \end{smallmatrix}\right)$, where $ad = 8$, we have

$$e^{2\pi i\sigma\tau} = \zeta_d^b q^{a/d} = \zeta_d^b (q^{1/8})^{a^2},$$

and hence, the product expansions for the Weber functions imply that $\mathfrak{f}_1(8\gamma\tau)^6$ is meromorphic in $q^{1/8}$. This proves that $\mathfrak{f}_1(8\tau)^6$ is a modular function for $\Gamma_0(32)$. Q.E.D.

The next step in proving Theorem 12.24 is to determine some field (not necessarily the smallest) containing $\mathfrak{f}_1(\sqrt{-m})^6$. The key point is that $\mathfrak{f}_1(8\tau)^6$ is not only a modular function for $\Gamma_0(32)$, it's also holomorphic and its q-expansion is integral. Thus Proposition 12.7 tells us that $\mathfrak{f}_1(8\tau)^6 = R(j(\tau), j(32\tau))$ for some rational function $R(X,Y) \in \mathbb{Q}(X,Y)$. We will write this in the form

$$\mathfrak{f}_1(\tau)^6 = R(j(\tau/8), j(4\tau)). \tag{12.27}$$

Using Lemma 12.11 with $m = 32$ and $s = 8$, we see that

$$\frac{\partial \Phi_{32}}{\partial X}(j(4\sqrt{-m}), j(\sqrt{-m}/8)) \neq 0,$$

and thus, by Proposition 12.7, we conclude that

$$\mathfrak{f}_1(\sqrt{-m})^6 = R(j(\sqrt{-m}/8), j(4\sqrt{-m})) = R(j([8,\sqrt{-m}]), j([1,4\sqrt{-m}])). \tag{12.28}$$

To identify what field this lies in, let L' denote the ring class field of the order $\mathcal{O}' = [1, 4\sqrt{-m}]$. Since $[8, \sqrt{-m}]$ is a fractional proper ideal for $\mathcal{O}'$ (this uses Lemma 7.5 and $m \equiv 6 \bmod 8$—see Exercise 12.22), it follows that $\mathfrak{f}_1(\sqrt{-m})^6 \in L'$.

We want to prove that $\mathfrak{f}_1(\sqrt{-m})^6$ lies in the smaller field L. This is the situation that occurred in the proof of Theorem 12.2, but here we will need more than just a degree calculation. The crucial new idea will be to relate Galois theory and modular functions.

Let's first study the Galois theory of $L \subset L'$. The orders $\mathcal{O}'$ and $\mathcal{O}$ have discriminants $-64m$ and $-4m$ respectively, so that Corollary 7.28 implies that $h(-64m)) = 4h(-4m)$. Thus L' has degree 4 over L. Furthermore, the isomorphisms $C(\mathcal{O}') \simeq \mathrm{Gal}(L'/K)$ and $C(\mathcal{O}) \simeq \mathrm{Gal}(L/K)$ imply that

$$\mathrm{Gal}(L'/L) \simeq \ker(C(\mathcal{O}') \to C(\mathcal{O})).$$

In Exercise 12.22 we show that $[4, 1+\sqrt{-m}]$ is a proper $\mathcal{O}'$-ideal which lies in the above kernel and has order 4. It follows that $L \subset L'$ is a cyclic extension of degree 4.

The goal of the remainder of the proof will be to compute $\sigma(\mathfrak{f}_1(\sqrt{-m})^6)$ for some generator σ of $\mathrm{Gal}(L'/L)$. At the end of §11 we described an isomorphism

$$C(\mathcal{O}') \simeq \mathrm{Gal}(L'/K)$$

as follows. Given a class $[\mathfrak{a}] \in C(\mathcal{O}')$, let the corresponding automorphism be $\sigma_{\mathfrak{a}} \in \mathrm{Gal}(L'/K)$. If we write $L' = K(j(\mathfrak{b}))$ for some proper fractional $\mathcal{O}'$-ideal $\mathfrak{b}$, then Corollary 11.37 states that

$$\sigma_{\mathfrak{a}}(j(\mathfrak{b})) = j(\overline{\mathfrak{a}}\mathfrak{b}).$$

To exploit this, let $\mathfrak{b} = [8, \sqrt{-m}] = 8[1, \sqrt{-m}/8]$, so that (12.28) can be written

$$\mathfrak{f}_1(\sqrt{-m})^6 = R(j(\mathfrak{b}), j(\mathcal{O}')).$$

Now let $\mathfrak{a} = [4, 1+\sqrt{-m}]$, and let the corresponding automorphism be $\sigma = \sigma_{\mathfrak{a}} \in \mathrm{Gal}(L'/L)$. Note that σ is a generator of $\mathrm{Gal}(L'/L)$. Hence to prove that $\mathfrak{f}_1(\sqrt{-m})^6$ lies in L, we need only prove that it is fixed by σ. Using the above formula for $\mathfrak{f}_1(\sqrt{-m})^6$, we compute

$$\sigma(\mathfrak{f}_1(\sqrt{-m})^6) = R(\sigma(j(\mathfrak{b})), \sigma(j(\mathcal{O}'))) = R(j(\overline{\mathfrak{a}}\mathfrak{b}), j(\overline{\mathfrak{a}})).$$

Since $m \equiv 6 \bmod 8$, one easily sees that

$$\bar{\mathfrak{a}}\mathfrak{b} = [8, -2+\sqrt{-m}], \qquad \bar{\mathfrak{a}} = [4, -1+\sqrt{-m}]$$

(see Exercise 12.22), and hence $\sigma(\mathfrak{f}_1(\sqrt{-m})^6)$ can be written

$$\sigma(\mathfrak{f}_1(\sqrt{-m})^6) = R(j([8,-2+\sqrt{-m}]), j([4,-1+\sqrt{-m}])). \tag{12.29}$$

Now let $\gamma = \left(\begin{smallmatrix} 1 & -2 \\ 1 & -1 \end{smallmatrix}\right) \in \Gamma_0(2)^t$. If we substitute $\gamma\tau$ for τ in (12.27), we get

$$\mathfrak{f}_1(\gamma\tau)^6 = R(j(\gamma\tau/8), j(4\gamma\tau)).$$

Since $\gamma\tau = (\tau-2)/(\tau-1)$, one sees that

$$\begin{aligned} [1,\gamma\tau/8] &\text{ is homothetic to } [8(\tau-1), \tau-2] = [8,-2+\tau] \\ [1, 4\gamma\tau] &\text{ is homothetic to } [\tau-1, 4(\tau-2)] = [4,-1+\tau], \end{aligned}$$

and thus

$$\mathfrak{f}_1(\gamma\tau)^6 = R(j([8,-2+\tau]), j([4,-1+\tau])).$$

Evaluating this at $\tau = \sqrt{-m}$ and using (12.29), we see that

$$\sigma(\mathfrak{f}_1(\sqrt{-m})^6) = \mathfrak{f}_1(\gamma\sqrt{-m})^6.$$

However, (12.26) shows that $\mathfrak{f}_1(\gamma\tau)^6 = \mathfrak{f}_1(\tau)^6$ for all τ, which shows that $\mathfrak{f}_1(\sqrt{-m})^6$ is fixed by σ and hence lies in the ring class field L. This completes the proof of (i).

The proof of (ii) is similar to what we did for (i), though this case is a little more difficult. We will sketch the main steps of the proof in Exercise 12.23. This completes the proof of Theorem 12.24. Q.E.D.

REMARK. The above equation $\sigma(\mathfrak{f}_1(\sqrt{-m})^6) = \mathfrak{f}_1(\gamma\sqrt{-m})^6$ is significant, for it allows us to compute the action of $\sigma \in \mathrm{Gal}(L'/K)$ using the matrix $\gamma \in \mathrm{SL}(2,\mathbb{Z})$. This correspondence between Galois automorphisms and linear fractional transformations is not unexpected, for the $\mathfrak{f}_1(\gamma\tau)^6$'s are the conjugates of $\mathfrak{f}_1(\tau)^6$ over $\mathbb{Q}(j(\tau))$, hence when we specialize to $\tau = \sqrt{-m}$, the conjugates of $\mathfrak{f}_1(\sqrt{-m})^6$ should lie among the $\mathfrak{f}_1(\gamma\sqrt{-m})^6$'s. What's surprising is that there's a systematic way of finding γ. This is the basic content of the *Shimura Reciprocity Law.* We will explain how this works in Theorem 15.23 in §15 when we use Shimura Reciprocity to give a second proof that $\mathfrak{f}_1(\sqrt{-m})^6 \in L$. See also Lang [**90**, Chapter 11] or Shimura [**115**, §6.8].

We can now begin Weber's computation of $j(\sqrt{-14})$ from [**131**, §144]. Let $K = \mathbb{Q}(\sqrt{-14})$. Since $\mathcal{O}_K = [1, \sqrt{-14}]$, $L = K(j(\sqrt{-14}))$ is the Hilbert class field of K. As we saw in §5, $\mathrm{Gal}(L/K) \simeq C(\mathcal{O}_K)$ is cyclic of order 4. Furthermore, we can use the results of §6 to determine part of this extension. Recall that the *genus field* M of K is the intermediate field $K \subset M \subset L$ corresponding to the subgroup of squares. When $K = \mathbb{Q}(\sqrt{-14})$, Theorem 6.1 tells us that $M = K(\sqrt{-7}) = K(\sqrt{2})$. Thus

$$K \subset K(\sqrt{2}) \subset L.$$

We will compute $\mathfrak{f}_1(\sqrt{-14})^2$, which lies in the Hilbert class field L since $m = 14$ satisfies the hypothesis of the first part of Theorem 12.24. Let σ be the unique element of $\mathrm{Gal}(L/K)$ of order 2, so that the fixed field of σ is the genus field $K(\sqrt{2})$. The key step in the computation is to show that

$$\sigma(\mathfrak{f}_1(\sqrt{-14})^2) = \mathfrak{f}_2(\sqrt{-14}/2)^2. \tag{12.30}$$

We start with the equation

$$\mathfrak{f}_1(\sqrt{-m})^6 = R(j(\mathfrak{b}), j(\mathcal{O}'))$$

from Theorem 12.24, where $\mathcal{O}' = [1, 4\sqrt{-14}]$ and $\mathfrak{b} = [8, \sqrt{-14}]$. If $\mathcal{O}'$ and L' are as in the proof of Theorem 12.24, then $\mathfrak{b}$ determines a class in $C(\mathcal{O}')$ and hence an automorphism $\sigma_{\mathfrak{b}} \in \mathrm{Gal}(L'/K)$. It is easy to check that $\mathfrak{b}$ maps to the unique element of order 2 in $C(\mathcal{O}_K)$ (see Exercise 12.24), and consequently, the restriction of $\sigma_{\mathfrak{b}}$ to L is the above automorphism σ. By abuse of notation, we will write $\sigma = \sigma_{\mathfrak{b}}$. Then, using Corollary 11.37, we obtain

$$\sigma(\mathfrak{f}_1(\sqrt{-14})^6) = R(j(\overline{\mathfrak{b}}\mathfrak{b}), j(\overline{\mathfrak{b}})) = R(j(\mathcal{O}'), j(\mathfrak{b}))$$

since $\overline{\mathfrak{b}} = \mathfrak{b}$ and $\overline{\mathfrak{b}}\mathfrak{b} = [2, 8\sqrt{-14})] = 2\mathcal{O}'$. Thus

$$\sigma(\mathfrak{f}_1(\sqrt{-14})^6) = R(j([1, 4\sqrt{-14}]), j([8, \sqrt{-14}])). \tag{12.31}$$

Let $\gamma = \left(\begin{smallmatrix} 0 & -1 \\ 1 & 0 \end{smallmatrix}\right)$, and note that $\mathfrak{f}_2(\tau) = \mathfrak{f}_1(\gamma\tau)$ by Corollary 12.19. Combining this with (12.27), we get

$$\begin{aligned} \mathfrak{f}_2(\tau)^6 = \mathfrak{f}_1(\gamma\tau)^6 &= R(j(\gamma\tau/8), j(4\gamma\tau)) \\ &= R(j([1, 8\tau]), j([4, \tau])). \end{aligned}$$

Evaluating this at $\tau = \sqrt{-14}/2$ and using (12.31), we obtain

$$\sigma(\mathfrak{f}_1(\sqrt{-14})^6) = \mathfrak{f}_2(\sqrt{-14}/2)^6.$$

(See Theorem 15.31 in §15 for a second proof of this that uses Shimura Reciprocity.) If we take the cube root of each side, we see that

$$\sigma(\mathfrak{f}_1(\sqrt{-14})^2) = \zeta_3^i\, \mathfrak{f}_2(\sqrt{-14}/2)^2$$

for some cube root of unity ζ_3^i. It remains to prove that the cube root is 1. From (12.16) we know that $\mathfrak{f}_1(\tau)\mathfrak{f}_2(\tau/2) = \sqrt{2}$, so that

$$\mathfrak{f}_1(\sqrt{-14})^2 \sigma(\mathfrak{f}_1(\sqrt{-14})^2) = \zeta_3^i\, \mathfrak{f}_1(\sqrt{-14})^2\, \mathfrak{f}_2(\sqrt{-14}/2)^2 = 2\zeta_3^i. \tag{12.32}$$

Since $\mathfrak{f}_1(\sqrt{-14})^2 \sigma(\mathfrak{f}_1(\sqrt{-14})^2)$ is fixed by σ, it lies in $K(\sqrt{2})$, and hence $\zeta_3^i \in K(\sqrt{2}) = \mathbb{Q}(\sqrt{2}, \sqrt{-7})$. This forces the root of unity to be 1, and (12.30) is proved.

Now let $\alpha = \mathfrak{f}_1(\sqrt{-14})^2$. From (12.32) we see that $\alpha\sigma(\alpha) = 2$, so that $\alpha + \sigma(\alpha) = \alpha + 2/\alpha$ lies in $K(\sqrt{2})$. But α is clearly real, so that $\alpha + 2/\alpha \in \mathbb{Q}(\sqrt{2})$, and furthermore, α and $2/\alpha = \sigma(\alpha)$ are algebraic integers by Theorem 12.24. It follows that

$$\alpha + \frac{2}{\alpha} = a + b\sqrt{2}, \qquad a, b \in \mathbb{Z}. \tag{12.33}$$

We will use a wonderful argument of Weber to show that a and b are both positive. Namely, (12.33) gives a quadratic equation for α, and since α is real and positive (see the product formula for $\mathfrak{f}_1(\tau)$), the discriminant must be nonnegative, i.e.,

$$(a + b\sqrt{2})^2 \geq 8.$$

Let σ_1 be a generator of $\mathrm{Gal}(L/K)$ (so $\sigma = \sigma_1^2$). Then $\sigma_1(\sqrt{2}) = -\sqrt{2}$, and hence

$$\sigma_1(\alpha) + \frac{2}{\sigma_1(\alpha)} = a - b\sqrt{2}.$$

But $\sigma_1(\alpha)$ cannot be real, for then $L \cap \mathbb{R} = \mathbb{Q}(\alpha)$ would be Galois over $\mathbb{Q}$, which contradicts $\mathrm{Gal}(L/\mathbb{Q}) \simeq D_8$ (see Lemma 9.3). Thus the discriminant of the resulting quadratic equation must be negative, i.e.,

$$(a - b\sqrt{2})^2 < 8.$$

Subtracting these two inequalities gives

$$4ab\sqrt{2} > 0,$$

so that a and b are positive since $\alpha > 0$.

As a and b range over all positive integers, the resulting numbers $a + b\sqrt{2}$ form a discrete subset of $\mathbb{R}$ (by contrast, $\mathbb{Z}[\sqrt{2}]$ is dense in $\mathbb{R}$). Thus we can compute a and b by approximating $\alpha + 2/\alpha$ sufficiently closely. Setting $q = e^{-\pi\sqrt{14}}$, (12.15) implies

$$\frac{2}{\alpha} = \mathfrak{f}_2(\sqrt{-14}/2)^2 = 2q^{1/12}\prod_{n=1}^{\infty}(1+q^n)^2.$$

Applying the methods used in our class number 1 calculations, we see that

$$2q^{1/12} < \frac{2}{\alpha} < 2q^{1/12}e^{2.004q},$$

and thus

$$q^{-1/12}e^{-2.004q} < \alpha < q^{-1/12}.$$

These inequalities imply that

$$\alpha + \frac{2}{\alpha} \approx q^{-1/12} + 2q^{1/12} \approx 2.6633 + .7509 = 3.4142,$$

with an error of at most 10^{-4}. Compare this to the smallest values of $a + b\sqrt{2}$, $a, b > 0$:

$$1 + \sqrt{2} \approx 2.4142 < 2 + \sqrt{2} \approx 3.4142 < 1 + 2\sqrt{2} \approx 3.8284.$$

It follows that $\alpha + 2/\alpha = 2 + \sqrt{2}$, and then the quadratic formula implies

$$\alpha = \frac{2 + \sqrt{2} \pm \sqrt{4\sqrt{2} - 2}}{2} = \frac{\sqrt{2} + 1 \pm \sqrt{2\sqrt{2} - 1}}{\sqrt{2}}.$$

Since $\alpha \approx 2.6633$ is the larger root, we have

$$\alpha = \mathfrak{f}_1(\sqrt{-14})^2 = \frac{\sqrt{2} + 1 + \sqrt{2\sqrt{2} - 1}}{\sqrt{2}},$$

and we can now compute $\gamma_2(\sqrt{-14})$:

$$\begin{aligned}
\gamma_2(\sqrt{-14}) &= \mathfrak{f}_1(\sqrt{-14})^{16} + \frac{16}{\mathfrak{f}_1(\sqrt{-14})^8} \\
&= \alpha^8 + \frac{16}{\alpha^4} = \alpha^8 + \left(\frac{2}{\alpha}\right)^4 \\
&= \left(\frac{\sqrt{2} + 1 + \sqrt{2\sqrt{2} - 1}}{\sqrt{2}}\right)^8 + \left(\frac{\sqrt{2} + 1 - \sqrt{2\sqrt{2} - 1}}{\sqrt{2}}\right)^4 \\
&= 2\left(323 + 228\sqrt{2} + (231 + 161\sqrt{2})\sqrt{2\sqrt{2} - 1}\right),
\end{aligned}$$

where the last step was done using a computer. Cubing this, we get the formula for $j(\sqrt{-14})$ given in (12.1).

An immediate corollary is that $L = K(\sqrt{2\sqrt{2}-1})$ is the Hilbert class field of $K = \mathbb{Q}(\sqrt{-14})$. This method of determining L is more difficult than what we did in §5, but it's worth the effort—the formulas are simply wonderful! These same techniques can be used to determine $j(\sqrt{-46})$ and $j(\sqrt{-142})$ (see Exercise 12.25), and in [**131**, §144] Weber does 7 other cases by similar methods.

The examples done so far are only a small fraction of the singular j-invariants computed by Weber in [**131**]. He uses a wide variety of methods and devotes many sections to computations—the interested reader should consult §§125, 128, 129, 130, 131, 135, 139, 143, and 144 for more examples. We should also mention that in 1927, Berwick [**7**] published the j-invariants (in factored form) of all known orders of class number ≤ 3. For a discussion of how singular moduli were computed in the 1960s, see Herz [**71**].

Another person who thought deeply about singular moduli was Ramanujan. In Exercise 12.31 we will explain how one of his amazing identities leads to an easy proof of the formula

$$\mathfrak{f}_1(\sqrt{-14})^2 = \frac{\sqrt{2}+1+\sqrt{2\sqrt{2}-1}}{\sqrt{2}}$$

used in Weber's computation of $j(\sqrt{-14})$.

E. Imaginary Quadratic Fields of Class Number 1

We will end this section with another application of the Weber functions: the determination of all imaginary quadratic fields of class number 1.

THEOREM 12.34. *Let K be an imaginary quadratic field of discriminant d_K. Then*

$$h(d_K) = 1 \iff d_K = -3,\ -4,\ -7,\ -8,\ -11,\ -19,\ -43,\ -67,\ -163.$$

REMARK. As we saw in Theorem 7.30, this theorem enables us to determine all discriminants D with $h(D) = 1$.

PROOF. This theorem was proved by Heegner [**68**] in 1952, but his proof was not accepted at first, partly because of his heavy reliance on Weber. In 1966 complete proofs were found independently by Baker [**4**] and Stark [**123**], which led people to look back at Heegner's work and realize that he did have a complete proof after all (see Birch [**9**] and Stark [**125**]). We will follow Stark's presentation [**125**] of Heegner's argument.

The first part of the proof is quite elementary. Let d_K be a discriminant such that $h(d_K) = 1$. Recall from Theorem 2.18 that $h(-4n) = 1$ if and only if $-4n = -4, -8, -12, -16$ or -28. Thus, if $d_K \equiv 0 \bmod 4$, then $d_K = -4$ or -8 since d_K is a field discriminant. So we may assume $d_K \equiv 1 \bmod 4$, and then Theorem 3.15 implies that there are $2^{\mu-1}$ genera of forms of discriminant d_K, where μ is the number of primes dividing d_K. Since $h(d_K) = 1$, it follows that $\mu = 1$, so that $d_K = -p$, where $p \equiv 3 \bmod 4$ is prime.

If $p \equiv 7 \bmod 8$, then Theorem 7.24 implies that

$$h(-4p) = 2h(-p)\left(1 - \left(\frac{-p}{2}\right)\frac{1}{2}\right) = h(-p) = 1,$$

and using Theorem 2.18 again, we see that $p = 7$.

We are thus reduced to the case $p \equiv 3 \bmod 8$, and of course we may assume that $p \neq 3$. Then Theorem 7.24 tells us that

$$h(-4p) = 2h(-p)\left(1 - \left(\frac{-p}{2}\right)\frac{1}{2}\right) = 3h(-p) = 3.$$

This implies that $\mathbb{Q}(j(\sqrt{-p}))$ has degree 3 over $\mathbb{Q}$. By the second part of Theorem 12.24, we know that $\mathfrak{f}(\sqrt{-p})^2 \in K(j(\sqrt{-p}))$, and since $\mathfrak{f}(\sqrt{-p})^2$ is real, we see that $\mathfrak{f}(\sqrt{-p})^2$ generates a cubic extension of $\mathbb{Q}$.

Let $\tau_0 = (3+\sqrt{-p})/2$, and set $\alpha = \zeta_8^{-1}\mathfrak{f}_2(\tau_0)^2$. We can relate this to $\mathfrak{f}(\sqrt{-p})^2$ as follows. We know from (12.16) that

$$\mathfrak{f}_1(2\tau_0)\,\mathfrak{f}_2(\tau_0) = \sqrt{2},$$

and Corollary 12.19 tells us that

$$\mathfrak{f}_1(2\tau_0) = \mathfrak{f}_1(3+\sqrt{-p}) = \zeta_{48}^{-3}\mathfrak{f}(\sqrt{-p}) = \zeta_{16}^{-1}\mathfrak{f}(\sqrt{-p}).$$

These formulas imply that $\alpha = 2/\mathfrak{f}(\sqrt{-p})^2$, and hence α generates the cubic extension $\mathbb{Q}(\mathfrak{f}(\sqrt{-p})^2)$. Note also that α^4 generates the same cubic extension.

Let's study the minimal polynomial of α^4. Since $\mathcal{O} = [1, \tau_0]$ and $h(-p) = 1$, we know that $j(\tau_0)$ is an integer, and then $\gamma_2(\tau_0)$ is also an integer by Theorem 12.2. Since

$$\gamma_2(\tau_0) = \frac{\mathfrak{f}_2(\tau_0)^{24} + 16}{\mathfrak{f}_2(\tau_0)^8},$$

it follows that $\alpha^4 = -\mathfrak{f}_2(\tau_0)^8$ is a root of the cubic equation

$$x^3 - \gamma_2(\tau_0)x - 16 = 0. \tag{12.35}$$

This is the minimal polynomial of α^4 over $\mathbb{Q}$.

However, α is also cubic over $\mathbb{Q}$, and thus satisfies an equation of the form

$$x^3 + ax^2 + bx + c = 0,$$

where a, b and c lie in $\mathbb{Z}$ since α is an algebraic integer. Heegner's insight was that this equation put some very strong constraints on the equation satisfied by α^4. In fact, moving the even degree terms to the right and squaring, we get

$$(x^3 + bx)^2 = (-ax^2 - c)^2,$$

so that α satisfies

$$x^6 + (2b - a^2)x^4 + (b^2 - 2ac)x^2 - c^2 = 0.$$

Hence α^2 satisfies the cubic equation

$$x^3 + ex^2 + fx + g = 0, \qquad e = 2b - a^2, \quad f = b^2 - 2ac, \quad g = -c^2,$$

and repeating this process, we see that α^4 satisfies the cubic equation

$$x^3 + (2f - e^2)x^2 + (f^2 - 2eg)x - g^2.$$

By the uniqueness of the minimal polynomial, this equation must equal (12.35). Comparing coefficients, we obtain

$$\begin{aligned} 2f - e^2 &= 0 \\ f^2 - 2eg &= -\gamma_2(\tau_0) \\ g^2 &= 16. \end{aligned} \tag{12.36}$$

The third equation of (12.36) implies $g = \pm 4$, and since $g = -c^2$, we have $g = -4$ and $c = \pm 2$. However, changing α to $-\alpha$ leaves α^4 fixed but takes a, b, c to $-a, b, -c$. Thus we may assume $c = 2$, and it follows that

$$\gamma_2(\tau_0) = -f^2 - 8e = -(b^2 - 4a)^2 - 8(2b - a^2).$$

It remains to determine the possible a's and b's.

The first equation $2f = e^2$ of (12.36) may be written

$$2(b^2 - 4a) = (2b - a^2)^2,$$

which implies that a and b are even. If we set $X = -a/2$ and $Y = (b - a^2)/2$, then a little algebra shows that X and Y are integer solutions of the Diophantine equation

$$2X(X^3 + 1) = Y^2$$

(see Exercise 12.26). This equation has the following integer solutions:

PROPOSITION 12.37. *The only integer solutions of the Diophantine equation*

$$2X(X^3 + 1) = Y^2$$

are $(X, Y) = (0, 0)$, $(-1, 0)$, $(1, \pm 2)$, *and* $(2, \pm 6)$.

PROOF. Let (X, Y) be an integer solution. Since X and $X^3 + 1$ are relatively prime, the equation $2X(X^3+1) = Y^2$ implies that $\pm(X^3+1)$ is a square or twice a square. Thus X, Y gives an integer solution of one of four Diophantine equations. These equations, together with some of their obvious solutions, may be written as follows:

(i) $X^3 + 1 = Z^2$, $(X, Z) = (-1, 0)$, $(0, \pm 1)$, $(2, \pm 3)$.
(ii) $X^3 + 1 = -Z^2$, $(X, Z) = (-1, 0)$.
(iii) $W^6 + 1 = 2Z^2$, $(W, Z) = (1, \pm 1)$.
(iv) $X^3 + 1 = -2Z^2$, $(X, Z) = (-1, 0)$.

To explain (iii), note that if $X^3+1 = 2Z^2$, then $2X(X^3+1) = Y^2$ implies that $X = W^2$ for some W, which by substitution gives us $W^6 + 1 = 2Z^2$. In Exercises 12.27–12.29, we will show that the solutions listed above are *all* integer solutions of these four equations. Once this is done, the proposition follows easily.

The integer solutions of equations (ii)–(iv) are relatively easy to find. We need nothing more than the techniques used when we considered the equation $Y^2 = X^3 - 2$ in Exercises 5.21 and 5.22. See Exercise 12.27 for the details of these three cases.

The integer solutions of equation (i) are more difficult to find, and the elementary methods used in (ii)–(iv) don't suffice. Fortunately, we can turn to Euler for help, for in 1738 he used Fermat's technique of infinite descent to determine all integer (and rational) solutions of (i) (see [**43**, Vol. II, pp. 56–58]). A version of Euler's argument may be found in Exercises 12.28 and 12.29. This completes the proof of the proposition. Q.E.D.

Once we know the solutions of $2X(X^3 + 1) = Y^2$, we can compute a, b and hence $\gamma_2(\tau_0)$. This gives us the following table:

X	Y	$a=-2X$	$b=4X^2+2Y$	$\gamma_2(\tau_0)=-(b^2-4a)^2-8(2b-a^2)$
0	0	0	0	0
-1	0	2	4	-96
1	2	-2	8	-5280
1	-2	-2	0	-32
2	6	-4	28	-640320
2	-6	-4	4	-960

Note that these $\gamma_2(\tau_0)$'s are among those computed earlier in table (12.20). Since $j(\mathcal{O}_K)$ determines K uniquely (see Exercise 12.30), it follows that we now know all imaginary quadratic fields of class number 1. This proves the theorem. Q.E.D.

Note that Heegner's argument is clever but elementary—the hard part is proving that $\mathfrak{f}(\sqrt{-p})^2$ lies in the appropriate ring class field. Thus Weber could have solved the class number 1 problem in 1908! We should also mention that there is a more elementary version of the above argument which makes no use of the Weber functions (see Stark [**125**]).

F. Exercises

12.1. Show that $g_2(\tau)$, $g_3(\tau)$ and $\Delta(\tau)$ are real-valued when τ is purely imaginary. Hint: use Exercise (11.1).

12.2. Let $F(q) = 1 + \sum_{n=1}^{\infty} a_n q^n$ be a power series which converges in a neighborhood of the origin.

(a) Show that for any positive integer m, there is a unique power series $G(q)$, converging in a possibly smaller neighborhood of 0, such that $F(q) = G(q)^m$ and $G(0) = 1$.

(b) If in addition the coefficients of $F(q)$ are rational numbers, show that the power series $G(q)$ from part (a) also has rational coefficients.

12.3. In this exercise we will prove that $S = \begin{pmatrix} 0 & -1 \\ 1 & 0 \end{pmatrix}$ and $T = \begin{pmatrix} 1 & 1 \\ 0 & 1 \end{pmatrix}$ generate $\mathrm{SL}(2,\mathbb{Z})$. Let Γ be the subgroup of $\mathrm{SL}(2,\mathbb{Z})$ generated by S and T.

(a) Show that every element of $\mathrm{SL}(2,\mathbb{Z})$ of the form $\begin{pmatrix} a & b \\ 0 & d \end{pmatrix}$ or $\begin{pmatrix} 0 & b \\ c & d \end{pmatrix}$ lies in Γ.

(b) Fix $\gamma_0 \in \mathrm{SL}(2,\mathbb{Z})$, and pick $\gamma \in \Gamma$ so that $\gamma\gamma_0 = \begin{pmatrix} a & b \\ c & d \end{pmatrix}$ has the minimal $|c|$.

(i) If $a = 0$ or $c = 0$, then use part (a) to show that $\gamma_0 \in \Gamma$.

(ii) If $c \neq 0$, then, of the γ's that give the minimal $|c|$, choose one that has the minimal $|a|$. Use

$$S\begin{pmatrix} a & b \\ c & d \end{pmatrix} = \begin{pmatrix} -c & * \\ a & * \end{pmatrix}$$

to show that $|a| \geq |c|$, and then use

$$T^{\pm 1}\begin{pmatrix} a & b \\ c & d \end{pmatrix} = \begin{pmatrix} a \pm c & * \\ c & * \end{pmatrix}$$

to show that $a = 0$. Conclude that $\gamma_0 \in \Gamma$.

(c) Use parts (a) and (b) to show that S and T generate $\mathrm{SL}(2,\mathbb{Z})$.

12.4. This exercise will give generators for the following subgroups of $\mathrm{SL}(2,\mathbb{Z})$:

$$\Gamma_0(2) = \left\{\begin{pmatrix} a & b \\ c & d \end{pmatrix} \in \mathrm{SL}(2,\mathbb{Z}) : c \equiv 0 \bmod 2\right\}$$

$$\Gamma_0(2)^t = \left\{\begin{pmatrix} a & b \\ c & d \end{pmatrix} \in \mathrm{SL}(2,\mathbb{Z}) : b \equiv 0 \bmod 2\right\}$$

$$\Gamma(2) = \left\{\begin{pmatrix} a & b \\ c & d \end{pmatrix} \in \mathrm{SL}(2,\mathbb{Z}) : b \equiv c \equiv 0 \bmod 2\right\}.$$

Let $I = \left(\begin{smallmatrix} 1 & 0 \\ 0 & 1 \end{smallmatrix}\right)$, $A = \left(\begin{smallmatrix} 1 & 0 \\ 1 & 1 \end{smallmatrix}\right)$ and $B = \left(\begin{smallmatrix} 1 & 1 \\ 0 & 1 \end{smallmatrix}\right)$.

(a) Modify the argument of Exercise 12.3 to show that $-I$, A^2 and B generate $\Gamma_0(2)$. Hint: let Γ be generated by $-I$, A^2 and B. Given $\gamma_0 \in \Gamma_0(2)$, choose $\gamma \in \Gamma$ so that $\gamma\gamma_0 = \left(\begin{smallmatrix} a & b \\ c & d \end{smallmatrix}\right)$ is minimal in the sense of Exercise 12.3. If $c \neq 0$, show that $|a| < |c|$, and then use

$$A^{\pm 2}\begin{pmatrix} a & b \\ c & d \end{pmatrix} = \begin{pmatrix} a & * \\ c \pm 2a & * \end{pmatrix}$$

to prove that $a = 0$, which is impossible in this case.

(b) Show that $-I$, A and B^2 generate $\Gamma_0(2)^t$. In the text, these generators are denoted $-I$, U and V respectively.

(c) Adapt the argument of part (a) to show that $-I$, A^2 and B^2 generate $\Gamma(2)$.

12.5. This exercise is concerned with the properties of $\gamma_2(\tau)$.

(a) Prove (12.6) by induction on the length of $\left(\begin{smallmatrix} a & b \\ c & d \end{smallmatrix}\right)$ as a word in the matrices S and T of Exercise 12.3.

(b) Use (12.6) to show that $\gamma_2(\tau)$ is invariant under the group

$$\tilde{\Gamma}(3) = \left\{\begin{pmatrix} a & b \\ c & d \end{pmatrix} : b \equiv c \equiv 0 \bmod 3\right\}.$$

(c) Show that

$$\Gamma_0(9) = \begin{pmatrix} 1/3 & 0 \\ 0 & 1 \end{pmatrix} \tilde{\Gamma}(3) \begin{pmatrix} 3 & 0 \\ 0 & 1 \end{pmatrix},$$

and conclude that $\gamma_2(3\tau)$ is invariant under $\Gamma_0(9)$.

(d) Use (12.6) to show that the exact subgroup of $\mathrm{SL}(2,\mathbb{Z})$ under which $\gamma_2(\tau)$ is invariant is

$$\left\{\begin{pmatrix} a & b \\ c & d \end{pmatrix} \in \mathrm{SL}(2,\mathbb{Z}) : a \equiv d \equiv 0 \bmod 3 \text{ or } b \equiv c \bmod 3\right\}.$$

12.6. Complete the proof of part (ii) of Proposition 12.7 using the hints given in the text.

12.7. Let $\mathcal{O} = [1, \tau_0]$ be an order of discriminant D in an imaginary quadratic field, and assume that $\tau_0 = \sqrt{-m}$ or $(3 + \sqrt{-m})/2$, depending on whether $D \equiv 0$ or $1 \bmod 4$. Let $\mathcal{O}' = [1, 3\tau_0]$ be the order of index 3 in $\mathcal{O}$. If $3 \nmid D$, then prove that $[1, \tau_0/3]$ is a proper fractional $\mathcal{O}'$-ideal. Hint: use Lemma 7.5.

12.8. Adapt the argument of Lemma 12.11 to show that

$$\frac{\partial \Phi_9}{\partial X}(j(3i), j(i/3)) \neq 0.$$

Hint: it suffices to show that (12.12) cannot hold.

12.9. This exercise is concerned with the elementary properties of the Weber functions.

(a) Prove the product expansions (12.15).
(b) Prove the top line of (12.16). Hint: use the product expansions to show that

$$\eta(\tau)\,\mathfrak{f}(\tau)\,\mathfrak{f}_1(\tau)\,\mathfrak{f}_2(\tau) = \sqrt{2}\,\eta(\tau).$$

(c) Prove the bottom line of (12.16). Hint: use the definitions.

12.10. Exercises 12.10, 12.11 and 12.13 will explore the Weierstrass σ-function. The basic properties of $\sigma(z;\tau)$ will be covered, though we will neglect the details of convergence. For a careful treatment, see Chandrasekharan [**21**, Chapter IV], Lang [**90**, Chapter 18], and Whittaker and Watson [**138**, Chapter XX]. As in the text, the σ-function is defined by

$$\sigma(z;\tau) = z \prod_{\omega \in L-\{0\}} \left(1 - \frac{z}{\omega}\right) e^{z/\omega + (1/2)(z/\omega)^2},$$

where $L = [1, \tau]$. Note that $\sigma(z;\tau)$ is an odd function in z. We will write $\sigma(z)$ instead of $\sigma(z;\tau)$.

(a) Define the Weierstrass ζ-function $\zeta(z)$ (which is different from the famous Riemann ζ-function) by

$$\zeta(z) = \frac{\sigma'(z)}{\sigma(z)}.$$

Using the definition of $\sigma(z)$, show that

$$\zeta(z) = \frac{1}{z} + \sum_{\omega \in L-\{0\}} \left(\frac{1}{z-\omega} + \frac{1}{\omega} + \frac{z}{\omega^2}\right).$$

(b) Show that the ζ-function is related to the $\wp$-function by the formula

$$\wp(z) = -\zeta'(z).$$

(c) By part (b), it follows that if $\omega \in L$, then $\zeta(z+\omega) - \zeta(z)$ is a constant depending only on ω. Since $L = [1, \tau]$, we define η_1 and η_2 by the formulas

$$\begin{aligned} \eta_1 &= \zeta(z+\tau) - \zeta(z) \\ \eta_2 &= \zeta(z+1) - \zeta(z). \end{aligned}$$

Then prove Legendre's relation

$$\eta_2\tau - \eta_1 = 2\pi i.$$

Hint: consider $\int_\Gamma \zeta(z)dz$, where Γ is the boundary, oriented counterclockwise, of the parallelogram $\mathbf{P}$ used in the proof of Lemma 10.4. By standard residue theory, the integral equals $2\pi i$ by part (a). But the defining relations for η_1 and η_2 allow one to compute the integral directly.

(d) We can now show that

$$\begin{aligned} \sigma(z+\tau) &= -e^{\eta_1(z+\frac{\tau}{2})}\sigma(z) \\ \sigma(z+1) &= -e^{\eta_2(z+\frac{1}{2})}\sigma(z). \end{aligned}$$

(i) Show that

$$\frac{d}{dz}\frac{\sigma(z+\tau)}{\sigma(z)} = \eta_1 \frac{\sigma(z+\tau)}{\sigma(z)},$$

and conclude that for some constant C,

$$\sigma(z+\tau) = C\, e^{\eta_1 z}\, \sigma(z).$$

(ii) Determine the constant C in (i) by evaluating the above identity at $z = -\tau/2$. This will prove the desired formula for $\sigma(z+\tau)$. Hint: recall that $\sigma(z)$ is an odd function.

(iii) In a similar way, prove the formula for $\sigma(z+1)$.

12.11. The goal of this exercise is to prove the formula

$$\wp(z) - \wp(w) = -\frac{\sigma(z+w)\,\sigma(z-w)}{\sigma^2(z)\,\sigma^2(w)}.$$

Fix $w \notin L = [1, \tau]$, and consider the function

$$f(z) = -\frac{\sigma(z+w)\,\sigma(z-w)}{\sigma^2(z)\,\sigma^2(w)}.$$

(a) Show that $f(z)$ is an even elliptic function for L. By Lemma 10.17, this implies that $f(z)$ is a rational function in $\wp(z)$.

(b) Show that $f(z)$ is holomorphic on $\mathbb{C} - L$ and that its Laurent expansion at $z = 0$ begins with $1/z^2$.

(c) Conclude from part (b) that $f(z) = \wp(z) + C$ for some constant C, and evaluate the constant by setting $z = w$. This proves the desired formula.

12.12. Use the previous exercise to show that

$$e_2 - e_1 = -e^{-\eta_2\tau/2}\frac{\sigma^2\left(\dfrac{\tau+1}{2}\right)}{\sigma^2\left(\dfrac{1}{2}\right)\sigma^2\left(\dfrac{\tau}{2}\right)}$$

$$e_2 - e_3 = -e^{\eta_2(\tau+1)/2}\frac{\sigma^2\left(\dfrac{\tau}{2}\right)}{\sigma^2\left(\dfrac{1}{2}\right)\sigma^2\left(\dfrac{\tau+1}{2}\right)}$$

$$e_3 - e_1 = e^{\eta_1(\tau+1)/2}\frac{\sigma^2\left(\dfrac{1}{2}\right)}{\sigma^2\left(\dfrac{\tau+1}{2}\right)\sigma^2\left(\dfrac{\tau}{2}\right)}.$$

Hint: for $e_2 - e_1$, use the fact that

$$\sigma\left(\frac{1-\tau}{2}\right) = \sigma\left(-\frac{1+\tau}{2}+1\right) = -e^{\eta_2(-(\tau+1)/2+1/2)}\,\sigma\left(-\frac{1+\tau}{2}\right)$$
$$= e^{-\eta_2\tau/2}\,\sigma\left(\frac{\tau+1}{2}\right).$$

12.13. The final fact we need to know about the σ-function is its q-product expansion

$$\sigma(z;\tau) = \frac{1}{2\pi i} e^{\eta_2 z^2/2}(q_z^{1/2} - q_z^{-1/2}) \prod_{n=1}^{\infty} \frac{(1-q_\tau^n q_z)(1-q_\tau^n/q_z)}{(1-q_\tau^n)^2},$$

where $q_\tau = e^{2\pi i\tau}$ and $q_z = e^{2\pi i z}$. To prove this, let $f(z)$ denote the right-hand side of the above equation.

(a) Show that the zeros of $f(z)$ and $\sigma(z)$ are exactly the points of L. Thus $\sigma(z)/f(z)$ is holomorphic on $\mathbb{C} - L$.
(b) Show that $\sigma(z)/f(z)$ has periods $L = [1,\tau]$.
(c) Show that $\sigma(z)/f(z)$ is holomorphic at $z = 0$ and takes the value 1 there.
(d) Conclude that $\sigma(z) = f(z)$. Hint: use Exercise 10.5.

12.14. This exercise will complete the proof of the formulas (12.18), which express the differences $e_i - e_j$ in terms of $\eta(\tau)$ and the Weber functions.

(a) Use the product expansion from Exercise 12.13 to show that

$$\begin{aligned}\sigma\left(\frac{1}{2}\right) &= \frac{1}{2\pi} e^{\eta_2/8} \frac{\mathfrak{f}_2(\tau)^2}{\eta(\tau)^2}\\ \sigma\left(\frac{\tau}{2}\right) &= \frac{i}{2\pi} e^{\eta_2\tau^2/8} q^{-1/8} \frac{\mathfrak{f}_1(\tau)^2}{\eta(\tau)^2}\\ \sigma\left(\frac{\tau+1}{2}\right) &= \frac{1}{2\pi} e^{\eta_2(\tau+1)^2/8} q^{-1/8} \frac{\mathfrak{f}(\tau)^2}{\eta(\tau)^2}.\end{aligned}$$

(b) Use part (a) and the formulas from Exercise 12.12 to show that

$$\begin{aligned}e_2 - e_1 &= \pi^2 \eta(\tau)^4 \mathfrak{f}(\tau)^8\\ e_2 - e_3 &= \pi^2 \eta(\tau)^4 \mathfrak{f}_1(\tau)^8\\ e_3 - e_1 &= \pi^2 \eta(\tau)^4 \mathfrak{f}_2(\tau)^8.\end{aligned}$$

This proves (12.18). Hint: use (12.16).

12.15. In this exercise we will complete the proof of Theorem 12.17. Recall from Exercise 10.8 that $g_2(\tau) = -4(e_1e_2 + e_1e_3 + e_2e_3)$ and $e_1 + e_2 + e_3 = 0$.

(a) Show that

$$3g_2(\tau) = 4((e_2 - e_1)^2 - (e_2 - e_3)(e_3 - e_1)).$$

(b) The identity of part (a), together with the formulas for $e_i - e_j$, were used in the text to derive a formula for $\gamma_2(\tau)$ in terms of $\mathfrak{f}(\tau)$. Find two other identities for $3g_2(\tau)$ similar to the one given in part (a), and use them to derive formulas for $\gamma_2(\tau)$ in terms of $\mathfrak{f}_1(\tau)$ and $\mathfrak{f}_2(\tau)$.

12.16. Use the formulas for $\gamma_2(\tau)$ from Theorem 12.17 to show that the q-expansion of the j-function has integral coefficients. This proves Theorem 11.8.

12.17. Complete the proof of Corollary 12.19.

12.18. Verify the calculations made in table (12.20).

12.19. Use Theorem 6.1 to determine the Hilbert class field of $K = \mathbb{Q}(\sqrt{-105})$, and show that its maximal real subfield is $\mathbb{Q}(\sqrt{3}, \sqrt{5}, \sqrt{7})$. Hint: use Theorem 3.22 to show that the genus field equals the Hilbert class field in this case.

12.20. This exercise is concerned with the properties of the Weber function $\mathfrak{f}_1(\tau)$. Let $I = \left(\begin{smallmatrix} 1 & 0 \\ 0 & 1 \end{smallmatrix}\right)$, $U = \left(\begin{smallmatrix} 1 & 0 \\ 1 & 1 \end{smallmatrix}\right)$ and $V = \left(\begin{smallmatrix} 1 & 2 \\ 0 & 1 \end{smallmatrix}\right)$.

(a) Use Corollary 12.19 to show $\mathfrak{f}_1(U\tau)^6 = \mathfrak{f}_1(V\tau)^6 = -i\,\mathfrak{f}_1(\tau)^6$.

(b) In Exercise 12.4 we proved that $-I$, U and V generate $\Gamma_0(2)^t$. Use induction on the length of $\gamma = \left(\begin{smallmatrix} a & b \\ c & d \end{smallmatrix}\right) \in \Gamma_0(2)^t$ as a word in $-I$, U and V to show that

$$\mathfrak{f}_1(\gamma\tau)^6 = i^{-ac-(1/2)bd+(1/2)b^2c}\,\mathfrak{f}_1(\tau)^6.$$

12.21. In this exercise we will show how to discover the transformation law for $\mathfrak{f}_1(\tau)$ proved in part (b) of Exercise 12.20. Let $-I$, U and V be as in Exercise12.20. We will be using the groups

$$\Gamma(2) = \left\{\begin{pmatrix} a & b \\ c & d \end{pmatrix} \in \mathrm{SL}(2,\mathbb{Z}) : b \equiv c \equiv 0 \bmod 2\right\}$$
$$\tilde{\Gamma}(8) = \left\{\begin{pmatrix} a & b \\ c & d \end{pmatrix} \in \mathrm{SL}(2,\mathbb{Z}) : b \equiv c \equiv 0 \bmod 8\right\}.$$

Note that $\tilde{\Gamma}(8) \subset \Gamma(2) \subset \Gamma_0(2)^t$, and recall from Exercise 12.4 that $-I$, U^2 and V generate $\Gamma(2)$.

(a) Show that $\Gamma(2)$ has index 2 in $\Gamma_0(2)^t$ with I and U as coset representatives.

(b) Show that $\tilde{\Gamma}(8)$ is normal in $\mathrm{SL}(2,\mathbb{Z})$ and that the quotient $\Gamma(2)/\tilde{\Gamma}(8)$ is Abelian. Hint: compute $[U^2, V]$.

(c) We can now discover how $\mathfrak{f}_1(\tau)^6$ transforms under $\gamma = \left(\begin{smallmatrix} a & b \\ c & d \end{smallmatrix}\right) \in \Gamma(2)$. Write

$$\gamma = \pm\prod_{i=1}^{s} U^{2a_i}V^{b_i},$$

and set $A = \sum_{i=1}^{s} a_i$ and $B = \sum_{i=1}^{s} b_i$.

(i) Show that $\mathfrak{f}_1(\gamma\tau)^6 = i^{-2A-B}\mathfrak{f}_1(\tau)^6$.

(ii) Use part (b) to show that $\gamma \equiv U^{2A}V^B \bmod \tilde{\Gamma}(8)$, which means that

$$\begin{pmatrix} a & b \\ c & d \end{pmatrix} \equiv \begin{pmatrix} 1 & 2B \\ 2A & 1+4AB \end{pmatrix} \bmod \tilde{\Gamma}(8).$$

(iii) Use (ii) to show that $ac \equiv 2A \bmod 8$ and $bd = 2B \bmod 8$.

(iv) Conclude that for all $\gamma \in \Gamma(2)$,

$$\mathfrak{f}_1(\gamma\tau)^6 = i^{-ac-(1/2)bd}\,\mathfrak{f}_1(\tau)^6.$$

(d) Now take $\gamma = \left(\begin{smallmatrix} a & b \\ c & d \end{smallmatrix}\right) \in \Gamma_0(2)^t$, $\gamma \notin \Gamma(2)$. By part (a), we can write $\gamma = U\tilde{\gamma}$ for some $\tilde{\gamma} \in \Gamma(2)$. Then use part (c) to show that

$$\mathfrak{f}_1(\gamma\tau)^6 = i^{-ac-(1/2)bd+(1/2)b^2}\,\mathfrak{f}_1(\tau)^6.$$

Hint: observe that $a^2 \equiv 1 \bmod 4$ in this case.

(e) To unify the formulas of parts (c) and (d), take $\gamma = \left(\begin{smallmatrix} a & b \\ c & d \end{smallmatrix}\right) \in \Gamma_0(2)^t$. Show that

$$\tfrac{1}{2}b^2c \equiv \begin{cases} 0 \bmod 4 & \gamma \in \Gamma(2) \\ \tfrac{1}{2}b^2 \bmod 4 & \gamma \notin \Gamma(2). \end{cases}$$

From here, it follows immediately that

$$\mathfrak{f}_1(\gamma\tau)^6 = i^{-ac-(1/2)bd+(1/2)b^2c}\,\mathfrak{f}_1(\tau)^6$$

for all $\gamma \in \Gamma_0(2)^t$. In §15 we will derive a transformation law for $\mathfrak{f}_1(\gamma\tau)^2$ using a different method.

12.22. Let $\mathcal{O} = [1, \sqrt{-m}]$ and $\mathcal{O}' = [1, 4\sqrt{-m}]$, where $m > 0$ is an integer satisfying $m \equiv 6 \bmod 8$. Note that $\mathcal{O}'$ is the order of index 4 in $\mathcal{O}$. Let $\mathfrak{a} = [4, 1+\sqrt{-m}]$ and $\mathfrak{b} = [8, \sqrt{-m}]$.

(a) Show that $\mathfrak{a}$ and $\mathfrak{b}$ are proper fractional $\mathcal{O}'$-ideals. Hint: use Lemma 7.5.
(b) Show that the class of $\mathfrak{a}$ has order 4 in $C(\mathcal{O}')$ and is in the kernel of the natural map $C(\mathcal{O}') \to C(\mathcal{O})$.
(c) Verify that $\bar{\mathfrak{a}}\mathfrak{b} = [8, -2+\sqrt{-m}]$ and $\bar{\mathfrak{a}} = [4, -1+\sqrt{-m}]$.

12.23. In this exercise we will prove part (ii) of Theorem 12.24. We are thus concerned with $\mathfrak{f}(\sqrt{-m})^2$, where $m \equiv 3 \bmod 4$ is a positive integer not divisible by 3. Let L denote the ring class field of the order $\mathcal{O} = [1, \sqrt{-m}]$.

(a) Show that $\mathfrak{f}(\sqrt{-m})^6 \in L$ implies that $L = K(\mathfrak{f}(\sqrt{-m})^2)$.
(b) By Corollary 12.19, we have $\mathfrak{f}(\tau)^6 = \zeta_8\mathfrak{f}_1(\tau+1)^6$. Use this to prove that $\mathfrak{f}(8\tau)^6$ is a modular function for $\Gamma_0(64)$. Hint: show that $\mathfrak{f}_1(\tau)^6$ is a modular function for the group $\tilde{\Gamma}(8)$ defined in Exercise 12.21. Since $\tilde{\Gamma}(8)$ is normal in $\mathrm{SL}(2,\mathbb{Z})$, this implies that $\mathfrak{f}(\tau)^6$ is also invariant under $\tilde{\Gamma}(8)$.
(c) Use Proposition 12.7 and Lemma 12.11 to show that

$$\mathfrak{f}(\sqrt{-m})^6 = S(j([8, \sqrt{-m}]), j([1, 8\sqrt{-m}]))$$

for some rational function $S(X,Y) \in \mathbb{Q}(X,Y)$.
(d) Let $\mathcal{O}'$ be the order $[1, 8\sqrt{-m}]$. Show that $\mathfrak{a} = [8, 2+\sqrt{-m}]$ and $\mathfrak{b} = [8, \sqrt{-m}]$ are proper fractional $\mathcal{O}'$-ideals. Then use part (c) to conclude that $\mathfrak{f}(\sqrt{-m})^6$ lies in the ring class field L' of $\mathcal{O}'$.
(e) Show that the extension $L \subset L'$ has degree 8 and that under the isomorphism $C(\mathcal{O}') \simeq \mathrm{Gal}(L'/K)$, the classes of the ideals $\mathfrak{a}$ and $\mathfrak{b}$ map to generators σ_1 and σ_2 of $\mathrm{Gal}(L'/L)$. Thus we need to prove that $\mathfrak{f}(\sqrt{-m})^6$ is fixed by both σ_1 and σ_2.
(f) Using part (c) and Corollary 11.37, show that

$$\begin{aligned}\sigma_1(\mathfrak{f}(\sqrt{-m})^6) &= S(j([4, 3+2\sqrt{-m}]), j([8, 6+\sqrt{-m}]))\\ \sigma_2(\mathfrak{f}(\sqrt{-m})^6) &= S(j([1, 8\sqrt{-m}]), j([8, \sqrt{-m}]))\end{aligned}$$

(this is where $m \equiv 3 \bmod 4$ is used).
(g) Let $\gamma_1 = \left(\begin{smallmatrix} 2 & 11 \\ 1 & 6 \end{smallmatrix}\right)$ and $\gamma_2 = \left(\begin{smallmatrix} 0 & -1 \\ 1 & 0 \end{smallmatrix}\right)$. Then show that

$$\begin{aligned}\mathfrak{f}(\gamma_1\tau)^6 &= S(j([4, 3+2\tau]), j([8, 6+\tau]))\\ \mathfrak{f}(\gamma_2\tau)^6 &= S(j([1, 8\tau]), j([8, \tau])).\end{aligned}$$

(h) Use Corollary 12.19 to show that $\mathfrak{f}(\tau)^6$ is invariant under both γ_1 and γ_2. Then parts (f) and (g) imply that $\mathfrak{f}(\sqrt{-m})^6$ is fixed by σ_1 and σ_2 which completes the proof.

12.24. Consider the orders $\mathcal{O} = [1, \sqrt{-14}]$ and $\mathcal{O}' = [1, 4\sqrt{-14}]$. By part (a) of Exercise 12.22, we know that $\mathfrak{b} = [8, \sqrt{-14}]$ is a proper fractional $\mathcal{O}'$-ideal. Under the natural map $C(\mathcal{O}') \to C(\mathcal{O})$, show that $\mathfrak{b}$ maps to the unique element of order 2 of $C(\mathcal{O})$.

12.25. Compute $j(\sqrt{-46})$ and $j(\sqrt{-142})$. Hint: in each case the class number is 4. Note also that $46 \equiv 142 \equiv 6 \bmod 8$, so that part (i) of Theorem 12.24 applies.

12.26. Let (a, b) be a solution of the Diophantine equation $2(b^2 - 4a) = (2b - a^2)^2$.

(a) Show that a and b must be even.

(b) If we set $X = -a/2$ and $Y = (b - a^2)/2$, then show that X and Y are integer solutions of the Diophantine equation $2X(X^3 + 1) = Y^2$.

12.27. This exercise will discuss three of the Diophantine equations that arose in the proof of Proposition 12.37. In each case, the methods used in Exercises 5.21 and 5.22 are sufficient to determine the integer solutions.

(a) Show that the only integer solution of $X^3 + 1 = -Z^2$ is $(X, Z) = (-1, 0)$. Hint: work in the ring $\mathbb{Z}[i]$.

(b) Prove that the only integer solutions of $W^6 + 1 = 2Z^2$ are $(W, Z) = (\pm 1, \pm 1)$. Hint: work in $\mathbb{Z}[\omega]$, $\omega = e^{2\pi i/3}$. The fact that $3 \nmid W^2 + 1$ will be useful.

(c) Show that the only integer solutions of $X^3 + 1 = -2Z^2$ are $(X, Z) = (-1, 0)$. Hint: work in the ring $\mathbb{Z}[\sqrt{-2}]$.

12.28. Exercises 12.28 and 12.29 will present Euler's proof [**43**, Vol. II, pp. 56–58] that the only rational solutions of $X^3 + 1 = Z^2$ are $(X, Y) = (-1, 0)$, $(0, \pm 1)$ and $(2, \pm 3)$. In this exercise we will show that there are no relatively prime positive integers c and b such that $bc(c^2 - 3bc + 3b^2)$ is a perfect square when $c \neq b$ and $3 \nmid c$. The proof will use infinite descent. Then Exercise 12.29 will use this result to study $X^3 + 1 = Z^2$.

(a) Let c and b be positive relatively prime integers such that the product $bc(c^2 - 3bc + 3b^2)$ is a perfect square, and assume also that $c \neq b$ and $3 \nmid c$. Show that b, c and $c^2 - 3bc + 3b^2$ are relatively prime, and conclude that each is a perfect square. Then write $c^2 - 3bc + 3b^2 = (\frac{m}{n}b - c)^2$, where $n > 0$, $m > 0$ and $\gcd(m, n) = 1$. Show that this implies

$$\frac{b}{c} = \frac{2mn - 3n^2}{m^2 - 3n^2}.$$

Also show that $2mn - 3n^2 > 0$ and $m^2 - 3n^2 > 0$. There are two cases to consider, depending on whether $3 \nmid m$ or $3 \mid m$.

(b) Preserving the notation of part (a), let's consider the case $3 \nmid m$.

(i) Show that $b = 2mn - 3n^2$ and $c = m^2 - 3n^2$.

(ii) Since c is a perfect square, we can write $m^2 - 3n^2 = (\frac{p}{q}n - m)^2$, where $p > 0$, $q > 0$ and $\gcd(p, q) = 1$. Show that p and q may be chosen so that $3 \nmid p$, and show also that

$$\frac{m}{n} = \frac{p^2 + 3q^2}{2pq}.$$

(iii) Prove that

$$\frac{b}{n^2} = \frac{p^2 - 3pq + 3q^2}{pq},$$

and conclude that $pq(p^2 - 3pq + 3q^2)$ is a perfect square. Show also that $p \neq q$. Hint: use (i) and (ii) to show that $p = q$ implies $b = c = 1$.

(iv) By (ii) and (iii) we see that p and q satisfy the same conditions as c and b. Now prove that $q < b$, which shows that the new solution is "smaller." Hint: note that $q \mid b$, so that $q < b$ unless $q = n = b$. Use (i) and (ii) to show that $b = c = 1$ in this case.

(c) With the same notation as (a), we will now consider the case $3 \mid m$. Then $m = 3k$, so that by (a),

$$\frac{b}{c} = \frac{n^2 - 2nk}{n^2 - 3k^2}.$$

Furthermore, $\gcd(m, n) = 1$ implies $\gcd(k, n) = 1$ and $3 \nmid n$.

(i) Show that $b = 2kn - n^2$ and $c = 3k^2 - n^2$.

(ii) Prove that $c = 3k^2 - n^2$ is impossible since c is a perfect square. Hint: work modulo 3.

Thus, given c and b satisfying the above conditions, we can always produce a pair of integers satisfying the same conditions, but with strictly smaller b. By infinite descent, no such c and b can exist.

12.29. We can now show that the only rational solutions of $X^3 + 1 = Z^2$ are $(X, Y) = (-1, 0)$, $(0, \pm 1)$ and $(2, \pm 3)$. Let (X, Y) be a rational solution, and write $X = a/b$, where $b > 0$ and $\gcd(a, b) = 1$. Assume in addition that $a/b \neq -1$, 0 or 2, and set $c = a + b$. Our goal is to derive a contradiction.

(a) Show that $b(a^3 + b^3) = bc(c^2 - 3bc + 3b^2)$ is a perfect square and that b and c are relatively prime, positive, and unequal.

(b) It follows from Exercise 12.28 that $3 \mid c$. Then $c = 3d$ and $3 \nmid b$. Show that $bd(b^2 - 3bd + 3d^2)$ is a perfect square, and use Exercise 12.28 to show that $b = d$. This implies $b = d = 1$, and hence $c = 3$. Then $a/b = 2$, which contradicts our initial assumption.

12.30. If K and K' are imaginary quadratic fields and $j(\mathcal{O}_K) = j(\mathcal{O}_{K'})$, then prove that $K = K'$. Hint: use Theorem 10.9.

12.31. It turns out that Ramanujan could have proved the formula

$$\mathfrak{f}_1(\sqrt{-14})^2 = \frac{\sqrt{2} + 1 + \sqrt{2\sqrt{2} - 1}}{\sqrt{2}}$$

used in Weber's computation of $j(\sqrt{-14})$. In his notebooks, Ramanujan used

$$f(-q) = \prod_{n=1}^{\infty} (1 - q^n), \quad q = e^{2\pi i \tau},$$

which is related to the Dedekind η-function by $f(-q) = q^{-1/24}\eta(\tau)$. Entry 55 from Chapter 25 of Berndt [**5**, p. 209] states that if we define

$$P = \frac{f^2(-q)}{q^{1/2} f^2(-q^7)}, \quad Q = \frac{f^2(-q^2)}{q f^2(-q^{14})},$$

then

$$PQ + \frac{49}{PQ} = \left(\frac{Q}{P}\right)^3 - 8\frac{Q}{P} - 8\frac{P}{Q} + \left(\frac{P}{Q}\right)^3.$$

This is one of 23 "P-Q modular equations" given in [**5**], which were stated by Ramanujan without proof in his notebooks. Full proofs appear in [**5**].

In this exercise you will show that Weber's formula for $\mathfrak{f}_1(\sqrt{-14})^2$ follows from the above P-Q modular equation by setting $q = e^{-\pi\sqrt{2/7}}$. As in the text, let $\alpha = \mathfrak{f}_1(\sqrt{-14})^2$.

(a) Prove that $P = \sqrt{14}/\alpha$ and $Q = \sqrt{7/2}\alpha$ when $q = e^{-\pi\sqrt{2/7}}$, and conclude that $PQ = 7$ and $Q/P = \alpha^2/2$. Hint: the formula for $\eta(-1/\tau)$ from Corollary 12.19 will be useful.

(b) Use Ramanujan's P-Q modular equation to derive Weber's formula for $\alpha = \mathfrak{f}_1(\sqrt{-14})^2$. Hint: use the modular equation to find a cubic equation with $\beta = \alpha^2/2 + 2/\alpha^2$ as root. Factoring this cubic will give a quadratic equation satisfied by β.

This exercise is due to Heng Huat Chan.

13. The Class Equation

Now that we have discussed singular j-invariants and computed some examples, it is time to turn our attention to their minimal polynomials. Given an order $\mathcal{O}$ in an imaginary quadratic field K, $H_{\mathcal{O}}(X)$ will denote the monic minimal polynomial of $j(\mathcal{O})$ over $\mathbb{Q}$. Note that $H_{\mathcal{O}}(X)$ has integer coefficients since $j(\mathcal{O})$ is an algebraic integer. The equation $H_{\mathcal{O}}(X) = 0$ is called the *class equation*, and by abuse of terminology we will refer to $H_{\mathcal{O}}(X)$ as the class equation. Since $\mathcal{O}$ is uniquely determined by its discriminant D, we will often write $H_D(X)$ instead of $H_{\mathcal{O}}(X)$.

For an example of a class equation, consider the order $\mathbb{Z}[\sqrt{-14}]$ of discriminant -56. Its j-invariant is $j(\sqrt{-14})$, which we computed in (12.1). Thus the minimal polynomial of $j(\sqrt{-14})$ is

$$\begin{aligned} H_{-56}(X) = X^4 &- 2^8 \cdot 19 \cdot 937 \cdot 3559\, X^3 + 2^{13} \cdot 251421776987\, X^2 \\ &+ 2^{20} \cdot 3 \cdot 11^6 \cdot 19 \cdot 21323\, X + (2^8 \cdot 11^2 \cdot 17 \cdot 41)^3, \end{aligned} \tag{13.1}$$

where the coefficients have been factored into primes. Note that the constant term, being the norm of $j(\sqrt{-14}) = \gamma_2(\sqrt{-14})^3$, is a cube by Theorem 12.2.

The first part of §13 will describe an algorithm for computing the class equation $H_D(X)$ for *any* discriminant D. We have a special reason to be interested in this question, for by Theorem 9.2, the polynomial $H_{-4n}(X)$ gives us the criterion for when a prime is of the form x^2+ny^2. Thus our algorithm will provide a constructive version of Theorem 9.2. In the second part of §13, we will discuss work of Deuring, Gross and Zagier on the class equation. We will see that there are strong restrictions on primes dividing the discriminant and constant term of the class equation. The small size of the primes appearing in the constant term of (13.1) is thus no accident.

A. Computing the Class Equation

We will begin by giving a more precise description of the class equation:

PROPOSITION 13.2. *Let $\mathcal{O}$ be an order in an imaginary quadratic field K, and let $\mathfrak{a}_i$, $i = 1, \ldots, h$, be ideal class representatives (so that h is the class number). Then the class equation is given by the formula*

$$H_{\mathcal{O}}(X) = \prod_{i=1}^{h} (X - j(\mathfrak{a}_i)).$$

PROOF. This result follows easily from Corollary 11.37 (see Exercise 13.1), but there is a more elementary argument which we will now give.

By Theorem 11.1, $K(j(\mathcal{O}))$ is the ring class field of $\mathcal{O}$. Thus $[K(j(\mathcal{O})) : K] = h$, and since $j(\mathcal{O})$ is real by Exercise 11.1, it follows that $[\mathbb{Q}(j(\mathcal{O})) : \mathbb{Q}] = h$. This shows that $H_{\mathcal{O}}(X)$ has degree h. Now let α be a root of $H_{\mathcal{O}}(X)$, and let σ be an automorphism of $\mathbb{C}$ that takes $j(\mathcal{O})$ to α. In the proof of Theorem 10.23 we showed that $\sigma(j(\mathcal{O})) = j(\mathfrak{a})$ for some proper fractional $\mathcal{O}$-ideal $\mathfrak{a}$ (see (10.26)). Hence every root of $H_\mathcal{O}(X)$ is also a root of $\prod_{i=1}^h (X - j(\mathfrak{a}_i))$, and since both polynomials are monic of degree h, they must be equal. Q.E.D.

An important consequence of this proposition is that $H_\mathcal{O}(X)$ is the minimal polynomial of $j(\mathfrak{a})$, where $\mathfrak{a}$ is *any* proper fractional $\mathcal{O}$-ideal.

The algorithm we will present for computing $H_\mathcal{O}(X)$ uses the theory of complex multiplication, and in particular, the polynomial $\Phi_m(X, X)$ obtained by setting $X = Y$ in the modular equation plays an important role. Since $\Phi_1(X, Y) = X - Y$ implies that $\Phi_1(X, X)$ is identically zero, we focus on the case $m > 1$. Then:

LEMMA 13.3. *Let $m > 1$. If $\mathcal{O}$ has a primitive element of norm m, then the class equation $H_\mathcal{O}(X)$ is an irreducible factor of $\Phi_m(X, X)$. Furthermore, every irreducible factor of $\Phi_m(X, X)$ arises in this way.*

PROOF. Let $\alpha \in \mathcal{O}$ be a primitive element of norm m. Corollary 11.27 tells us that $\alpha\mathcal{O} \subset \mathcal{O}$ is a cyclic sublattice of index m, and it follows that

$$0 = \Phi_m(j(\alpha\mathcal{O}), j(\mathcal{O})) = \Phi_m(j(\mathcal{O}), j(\mathcal{O})).$$

Thus $j(\mathcal{O})$ is a root of $\Phi_m(X, X)$, which implies that its minimal polynomial $H_\mathcal{O}(X)$ is a factor of $\Phi_m(X, X)$.

To show that every irreducible factor of $\Phi_m(X, X)$ is a class equation, suppose that $\Phi_m(\beta, \beta) = 0$. Then Theorem 11.23 implies that $\beta = j(L) = j(L')$, where $L' \subset L$ is a cyclic sublattice of index m. By Theorem 10.9, $L' = \alpha L$ for some complex number α, and then α is primitive of norm m by Corollary 11.27. Thus $\alpha \notin \mathbb{Z}$ since $m > 1$, so that L has complex multiplication by α. By Theorem 10.14, this means that up to homothety, L is a proper fractional $\mathcal{O}$-ideal for some order $\mathcal{O}$ in an imaginary quadratic field. Then $\beta = j(L)$ has $H_\mathcal{O}(X)$ as its minimal polynomial, and hence $H_\mathcal{O}(X)$ is the corresponding irreducible factor of $\Phi_m(X, X)$. Q.E.D.

Our next task is to determine what power of $H_\mathcal{O}(X)$ appears in the factorization of $\Phi_m(X, X)$. The answer involves the number $r(\mathcal{O}, m)$, which is defined as follows. Given an order $\mathcal{O}$ in an imaginary quadratic field and an integer $m > 1$, set

$$r(\mathcal{O}, m) = \left|\{\alpha \in \mathcal{O} : \alpha \text{ is primitive}, \ N(\alpha) = m\} / \mathcal{O}^*\right|,$$

where the units $\mathcal{O}^*$ act by sending α to $\epsilon\alpha$ for $\epsilon \in \mathcal{O}^*$. It is easy to see that $r(\mathcal{O}, m)$ is finite, and for a given m, there are only finitely many orders with $r(\mathcal{O}, m) > 0$ (see Exercise 13.2). Then the following theorem tells us how to factor $\Phi_m(X, X)$:

THEOREM 13.4. *If $m > 1$, there is a constant $c_m \in \mathbb{C}^*$ such that*

$$\Phi_m(X, X) = c_m \prod_\mathcal{O} H_\mathcal{O}(X)^{r(\mathcal{O}, m)}.$$

PROOF. Fix an order $\mathcal{O}$, and pick a number τ_0 in the upper half plane such that $\mathcal{O} = [1, \tau_0]$. To prove the theorem, it suffices to show that $j(\mathcal{O}) = j(\tau_0)$ is a root of $\Phi_m(X, X)$ of multiplicity $r(\mathcal{O}, m)$.

We begin by studying the multiplicity of $j(\tau_0)$ as a root of $\Phi_m(X, j(\tau_0))$. Using the standard factorization

$$\Phi_m(X, j(\tau_0)) = \prod_{\sigma \in C(m)} (X - j(\sigma\tau_0)),$$

we see that

$$\Phi_m(X, j(\tau_0)) = (X - j(\tau_0))^r \prod_{j(\sigma\tau_0)\neq j(\tau_0)} (X - j(\sigma\tau_0)),$$

where

$$r = |\{\sigma \in C(m) : j(\sigma\tau_0) = j(\tau_0)\}|. \tag{13.5}$$

Thus $j(\tau_0)$ is a root of multiplicity r of $\Phi_m(X, j(\tau_0))$.

We will next show that the number r of (13.5) is the multiplicity of $j(\tau_0)$ as a root of $\Phi_m(X, X)$. To see what's involved, suppose that we have a polynomial $F(X, Y)$ and a number X_0 such that $F(X_0, X_0) = 0$. Then X_0 is a root of both $F(X, X)$ and $F(X, X_0)$, but in general, the multiplicities of these roots are different (see Exercise 13.3 for an example). So it will take a special argument to show that $j(\tau_0)$ has the same multiplicity for both $\Phi_m(X, X)$ and $\Phi_m(X, j(\tau_0))$. The basic idea is to show that

$$\lim_{u\to j(\tau_0)} \frac{\Phi_m(u, u)}{\Phi_m(u, j(\tau_0))}$$

is nonzero, which will force the multiplicities to be equal (see Exercise 13.3). To study this limit, note that

$$\begin{aligned}\lim_{u\to j(\tau_0)} \frac{\Phi_m(u, u)}{\Phi_m(u, j(\tau_0))} &= \lim_{\tau\to\tau_0} \frac{\Phi_m(j(\tau), j(\tau))}{\Phi_m(j(\tau), j(\tau_0))}\\ &= \lim_{\tau\to\tau_0} \prod_{\sigma\in C(m)} \frac{j(\tau) - j(\sigma\tau)}{j(\tau) - j(\sigma\tau_0)}.\end{aligned}$$

It suffices to compute the limit of each factor individually. Note that if $j(\tau_0) \neq j(\sigma\tau_0)$, then the limit of the corresponding factor is 1. Hence we need to study the limit

$$\lim_{\tau\to\tau_0} \frac{j(\tau) - j(\sigma\tau)}{j(\tau) - j(\sigma\tau_0)} \tag{13.6}$$

when $\sigma \in C(m)$ satisfies $j(\tau_0) = j(\sigma\tau_0)$.

The equality $j(\tau_0) = j(\sigma\tau_0)$ implies that there is some $\gamma \in \mathrm{SL}(2, \mathbb{Z})$ such that $\sigma\tau_0 = \gamma\tau_0$. If we set $\tilde{\sigma} = \gamma^{-1}\sigma$, then $\tilde{\sigma}$ fixes τ_0. Note also that $\det(\tilde{\sigma}) = m$ and that the entries of $\tilde{\sigma}$ are relatively prime. Using $\tilde{\sigma}$, the limit (13.6) can be written

$$\lim_{\tau\to\tau_0} \frac{j(\tau) - j(\tilde{\sigma}\tau)}{j(\tau) - j(\tau_0)}.$$

Consider the Taylor expansion of $j(\tau)$ about $\tau = \tau_0$:

$$j(\tau) = j(\tau_0) + a_k(\tau - \tau_0)^k + \cdots, \qquad a_k \neq 0.$$

Substituting $\tilde{\sigma}\tau$ for τ, we get the series

$$j(\tilde{\sigma}\tau) = j(\tau_0) + a_k(\tilde{\sigma}\tau - \tau_0)^k + \cdots,$$

and then one computes that

$$\begin{aligned}\frac{j(\tau) - j(\tilde{\sigma}\tau)}{j(\tau) - j(\tau_0)} &= \frac{a_k((\tau - \tau_0)^k - (\tilde{\sigma}\tau - \tau_0)^k) + \cdots}{a_k(\tau - \tau_0)^k + \cdots}\\ &= 1 - \left(\frac{\tilde{\sigma}\tau - \tau_0}{\tau - \tau_0}\right)^k + \cdots.\end{aligned}$$

Since

$$\lim_{\tau\to\tau_0} \frac{\tilde{\sigma}\tau - \tau_0}{\tau - \tau_0} = \lim_{\tau\to\tau_0} \frac{\tilde{\sigma}\tau - \tilde{\sigma}\tau_0}{\tau - \tau_0} = \tilde{\sigma}'(\tau_0),$$

it follows that the limit (13.6) equals $1 - \tilde{\sigma}'(\tau_0)^k$, and thus we need to prove that

$$\tilde{\sigma}'(\tau_0)^k \neq 1, \tag{13.7}$$

where k is the order of vanishing of $j(\tau) - j(\tau_0)$ at τ_0.

If we write $\tilde{\sigma} = \left(\begin{smallmatrix} a & b \\ c & d \end{smallmatrix}\right)$, then an easy computation shows that

$$\tilde{\sigma}'(\tau_0) = \frac{m}{(c\tau_0 + d)^2}.$$

Note also that $c \neq 0$ since $\tilde{\sigma}$ fixes τ_0 (see Exercise 13.4). Now suppose that $j(\tau_0) \neq 0$ or 1728. Then, by part (iv) of Theorem 11.2, it follows that $k = 1$, so that (13.7) reduces to

$$\frac{m}{(c\tau_0 + d)^2} \neq 1,$$

which is obvious since $c \neq 0$ and τ_0 is not a real number. When $j(\tau_0) = 1728$, we can assume that $\tau_0 = i$ (recall that $j(i) = 1728$), and then Theorem 11.2 tells us that $k = 2$. Thus if (13.7) failed to hold, we would have

$$\frac{m^2}{(ci + d)^4} = 1,$$

which implies that $c = \pm\sqrt{m}$ and $d = 0$ (see Exercise 13.4). Then $\tilde{\sigma}(i) = i$ tells us that $a = 0$ and $b = \mp\sqrt{m}$. So either $\tilde{\sigma}$ doesn't have integer entries (when m is not a perfect square), or the entries are integers with a common divisor (since $m > 1$). Both cases contradict what we know about $\tilde{\sigma}$, so that (13.7) holds when $j(\tau_0) = 1728$. The case when $j(\tau_0) = 0$ is similar and is left to the reader (see Exercise 13.4).

We should mention that the standard treatment of (13.6) in the literature (see Deuring [**32**, §12] or Lang [**90**, Appendix to §10]) seems to be incomplete.

We have thus shown that the multiplicity of $j(\tau_0)$ as a root of $\Phi_m(X, X)$ is

$$r = |\{\sigma \in C(m) : j(\sigma\tau_0) = j(\tau_0)\}|,$$

and it remains to show that $r = r(\mathcal{O}, m)$, where

$$r(\mathcal{O}, m) = |\{\alpha \in \mathcal{O} : \alpha \text{ is primitive},\ N(\alpha) = m\}/\mathcal{O}^*|.$$

To prove the desired equality, we will construct a map $\alpha \mapsto \sigma$. Namely, if $\alpha \in \mathcal{O}$ is primitive of norm m, then by Corollary 11.27, $\alpha\mathcal{O}$ is a cyclic sublattice of $\mathcal{O}$ of index m, and since $\mathcal{O} = [1, \tau_0]$, Lemma 11.24 implies that there is a unique $\sigma = \left(\begin{smallmatrix} a & b \\ 0 & d \end{smallmatrix}\right) \in C(m)$ such that $\alpha\mathcal{O} = d[1, \sigma\tau_0]$. Then σ satisfies $j(\sigma\tau_0) = j(\tau_0)$, and note also that if $\epsilon \in \mathcal{O}^*$, then $\epsilon\alpha$ maps to the same σ that α does. Thus we have constructed a well-defined map

$$\{\alpha \in \mathcal{O} : \alpha \text{ is primitive and } N(\alpha) = m\}/\mathcal{O}^* \longrightarrow \{\sigma \in C(m) : j(\sigma\tau_0) = j(\tau_0)\}.$$

This map is easily seen to be bijective (see Exercise 13.5), which proves that $r = r(\mathcal{O}, m)$. This completes the proof of Theorem 13.4. Q.E.D.

Besides knowing the factorization of $\Phi_m(X, X)$, its degree is easy to compute:

PROPOSITION 13.8. *If $m > 1$, then the degree of $\Phi_m(X,X)$ is*

$$2\sum_{\substack{a|m\\ a>\sqrt{m}}} \frac{a}{\gcd(a,m/a)}\,\phi(\gcd(a,m/a)) + \phi(\sqrt{m}),$$

where ϕ is the Euler ϕ-function and $\phi(\sqrt{m}) = 0$ when m is not a perfect square.

PROOF. The proof of this proposition is given in Exercise 13.6. Q.E.D.

If we write $r(\mathcal{O}, m)$ as $r(D,m)$, where D is the discriminant of $\mathcal{O}$, then Proposition 13.8 and Theorem 13.4 allow us to express the degree of $\Phi_m(X,X)$ in two ways. This gives us the following corollary, which is one of Kronecker's class number relations:

COROLLARY 13.9. *If $m > 1$, then*

$$\sum_D r(D,m)\,h(D) = 2\sum_{\substack{a|m\\ a>\sqrt{m}}} \frac{a}{\gcd(a,m/a)}\,\phi(\gcd(a,m/a)) + \phi(\sqrt{m}). \quad \text{Q.E.D.}$$

To illustrate the above theorems, let's study the case $m = 3$. There are only four orders with primitive elements of norm 3, namely $\mathbb{Z}[\omega]$, $\mathbb{Z}[\sqrt{-3}]$, $\mathbb{Z}[\sqrt{-2}]$ and $\mathbb{Z}[(1+\sqrt{-11})/2]$, and the corresponding $r(D,3)$'s are 1, 1, 2 and 2 respectively (see Exercise 13.7). Then Theorem 13.4 tells us that

$$\Phi_3(X,X) = \pm H_{-3}(X)H_{-12}(X)H_{-8}(X)^2 H_{11}(X)^2, \tag{13.10}$$

and since $\Phi_3(X,X)$ has degree 6 by Proposition 13.8, we get the following class number relation:

$$6 = h(-3) + h(-12) + 2h(-8) + 2h(-11).$$

This equation implies that all four class numbers must be one.

We can work out (13.10) more explicitly, for we know $\Phi_3(X,Y)$ from (11.22). Setting $X = Y$ gives us

$$\begin{aligned}\Phi_3(X,X) = &- X^6 + 4464\ X^5 + 2585778176\ X^4 + 17800519680000\ X^3\\ &- 769939996672000000\ X^2 + 37108517437440000000000,\end{aligned}$$

and factoring this over $\mathbb{Q}$, we obtain

$$\Phi_3(X,X) = -X(X-54000)(X-8000)^2(X+32768)^2.$$

In §§10 and 12, we computed the j-invariants $j((1+\sqrt{-3})/2) = 0$, $j(\sqrt{-2}) = 8000$ and $j((1+\sqrt{-11})/2) = -32768$. Thus we recognize three of the above four factors, and it follows that the fourth must be $H_{-12}(X)$, i.e.,

$$H_{-12}(X) = X - j(\sqrt{-3}) = X - 54000.$$

This proves that $j(\sqrt{-3}) = 54000$.

Let's now turn to the general problem of computing a given class equation $H_D(X)$. Since $\Phi_m(X,X)$ will have many factors, we need to know which one is the particular $H_D(X)$ we're interested in. The basic idea is to use *multiplicities* to distinguish the factors we seek. In particular, the factors of multiplicity one play an especially important role. Let's define the polynomial

$$\Phi_{m,1}(X,X) = \prod_{r(D,m)=1} H_D(X).$$

By Theorem 13.4, we know that $\Phi_{m,1}(X,X)$ is the product of the multiplicity one factors of $\Phi_m(X,X)$. We can describe $\Phi_{m,1}(X,X)$ as follows:

PROPOSITION 13.11. *If $m>1$, then $\Phi_{m,1}(X,X)$ equals*

$$\begin{array}{ll} H_{-4}(X)H_{-8}(X) & \text{if } m=2 \\ H_{-m}(X)H_{-4m}(X) & \text{if } m\equiv 3 \bmod 4 \text{ and } m\neq 3k^2,\ k>1 \\ H_{-4m}(X) & \text{if } m>2,\ m\not\equiv 3 \bmod 4 \text{ or } m=3k^2,\ k>1. \end{array}$$

PROOF. Let's first show that the $H_D(X)$'s listed above are factors of multiplicity one of $\Phi_m(X,X)$. Since $\pm\sqrt{-m}$ are the only primitive norm m elements of $\mathbb{Z}[\sqrt{-m}]$, it follows that $H_{-4m}(X)$ is a factor of multiplicity 1. When $m=2$, the elements of norm 2 in $\mathbb{Z}[i]$ are $\pm 1\pm i$, which are all associate under $\mathbb{Z}[i]^*$. Thus $H_{-4}(X)$ is also a factor of multiplicity 1. Finally, when $m\equiv 3 \bmod 4$ and $m\neq 3k^2$, $k>1$, we need to consider the multiplicity of $H_{-m}(X)$. The order $\mathbb{Z}[(1+\sqrt{-m})/2]$ has at least two primitive norm m elements, namely $\pm\sqrt{-m}$. To see if there are any others, suppose that $a+b(1+\sqrt{-m})/2$ is also primitive of norm m. Then $b\neq 0$ and, taking norms,

$$4m=(2a+b)^2+mb^2.$$

Thus $b=\pm 1$ or ± 2, and $b=\pm 2$ leads to the solutions we already know. So what happens if $b=\pm 1$? This clearly implies $3m=(2a+b)^2$, so that $m=3k^2$, and since $k>1$ is excluded by hypothesis, we see that $m=3$. Here, $b=\pm 1$ leads to 4 more solutions, but since $|\mathbb{Z}[\zeta_3]^*|=6$, we still get a multiplicity 1 factor.

The next step is to show that these are the *only* factors of multiplicity one. So suppose that $r(\mathcal{O},m)=1$ for some order $\mathcal{O}$. For simplicity, let's also assume that $\mathcal{O}^*=\{\pm 1\}$. Given $\alpha\in\mathcal{O}$ primitive of norm m, note that $\pm\alpha$ and $\pm\overline{\alpha}$ are also primitive of the same norm. Then $r(\mathcal{O},m)=1$ implies that $\overline{\alpha}=\pm\alpha$. But $\overline{\alpha}=\alpha$ is easily seen to be impossible (α is primitive and $m>1$), so that $\overline{\alpha}=-\alpha$. This means that α is a rational multiple of $\sqrt{D}$, where D is the discriminant of $\mathcal{O}$. The argument now breaks up into two cases.

If $D\equiv 0 \bmod 4$, then $\mathcal{O}=[1,\sqrt{D}/2]$, so that α, being primitive, must be $\pm\sqrt{D}/2$. This implies that $m=N(\alpha)=-D/4$, hence $D=-4m$. The corresponding factor is thus $H_{-4m}(X)$, which is one of the ones we know.

If $D\equiv 1 \bmod 4$, then $\mathcal{O}=[1,(1+\sqrt{D})/2]$, so that $\alpha=a+b(1+\sqrt{D})/2$. Since α is a multiple of $\sqrt{D}$, we have $2a+b=0$, and since a and b are relatively prime (α is primitive), we have $b=\pm 2$. This means that $\alpha=\pm\sqrt{D}$, so that $m=N(\alpha)=-D$. Thus $D=-m$, and this will be the other case we know once we prove that $m\neq 3k^2$, $k>1$. So suppose that m has this form. Then $D=-3k^2$, which means that $\mathcal{O}$ is the order of conductor k in $\mathbb{Z}[\zeta_3]$. One easily computes that $\pm k\sqrt{-3}$ and $\pm k(1-\zeta_3)$ are primitive elements of $\mathcal{O}$ of norm $3k^2=m$. Since we are assuming $\mathcal{O}^*=\{\pm 1\}$, this contradicts our assumption that $r(\mathcal{O},m)=1$.

It remains to consider the case when $\mathcal{O}^*\neq\{\pm 1\}$. We leave it to the reader to check that when $\mathcal{O}=\mathbb{Z}[\zeta_3]$ (resp. $\mathcal{O}=\mathbb{Z}[i]$), $r(\mathcal{O},m)=1$ implies that $m=3$ (resp. $m=2$) (see Exercise 13.8). This completes the proof of Proposition 13.11. Q.E.D.

It is now fairly easy to compute $H_D(X)$ for most D using the $\Phi_m(X,X)$'s. In the discussion that follows, m will denote a positive integer, and for simplicity we will assume $m>3$. It turns out that there are three cases to consider.

If $m\not\equiv 3 \bmod 4$ or $m=3k^2$, then Proposition 13.11 tells us that

$$H_{-4m}(X)=\Phi_{m,1}(X,X),$$

so that once we factor $\Phi_m(X,X)$ into irreducibles, we know $H_{-4m}(X)$.

Next, if $m \equiv 3 \bmod 8$ and $m \neq 3k^2$, then Proposition 13.11 tells us that

$$H_{-m}(X)H_{-4m}(X) = \Phi_{m,1}(X,X). \tag{13.12}$$

But since $m > 3$ and $m \equiv 3 \bmod 8$, it follows from Corollary 7.28 that $h(-4m) = 3h(-m)$, so that $H_{-4m}(X)$ has greater degree than $H_{-m}(X)$. It follows that factoring $\Phi_m(X,X)$ determines both $H_{-m}(X)$ and $H_{-4m}(X)$.

Finally, if $m \equiv 7 \bmod 8$, then (13.12) still holds, but this time more work is needed since $H_{-m}(X)$ and $H_{-4m}(X)$ have the same degree by Corollary 7.28. We claim that

$$H_{-m}(X) = \gcd(\Phi_{m,1}(X,X), \Phi_{(m+1)/4}(X,X)). \tag{13.13}$$

To see this, first note that $H_{-m}(X)$ divides $\Phi_{(m+1)/4}(X,X)$ since in the order of discriminant $-m$, $(1+\sqrt{-m})/2$ is primitive of norm $(m+1)/4$. If we turn to the order of discriminant $-4m$, there are *no* primitive elements of norm $(m+1)/4$ (see Exercise 13.9), and (13.13) follows. Thus, to determine $H_{-m}(X)$ and $H_{-4m}(X)$, we need to factor both $\Phi_m(X,X)$ and $\Phi_{(m+1)/4}(X,X)$ into irreducibles.

Using the above process, it is now easy to compute $H_D(X)$ for any $D \neq -3k^2$, k odd, assuming that we know the requisite modular equation (or equations). Some simple examples are given in Exercise 13.10.

B. Computing the Modular Equation

To complete our algorithm for finding the class equation, we need to know how to compute the modular equation $\Phi_m(X,Y) = 0$. This turns out to be the weak link in our theory, for while such an algorithm exists, it is so cumbersome that it can be implemented only for very small m.

The first step in computing $\Phi_m(X,Y)$ is to reduce to the case when m is prime. This is done by means of the following proposition:

PROPOSITION 13.14. *Fix an integer $m > 1$ and set $\Psi(m) = m\prod_{p|m}(1+1/p)$, which is the degree of $\Phi_m(X,Y)$ as a polynomial in X.*

(i) *If $m = m_1m_2$, where m_1 and m_2 are relatively prime, then*

$$\Phi_m(X,Y) = \prod_{i=1}^{\Psi(m_2)} \Phi_{m_1}(X,\xi_i),$$

where $X = \xi_i$ are the roots of $\Phi_{m_2}(X,Y) = 0$.

(ii) *If $m = p^a$, where p is prime and $a > 1$, then*

$$\Phi_m(X,Y) = \begin{cases} \dfrac{\prod_{i=1}^{\Psi(p^{a-1})} \Phi_p(X,\xi_i)}{\Phi_{p^{a-2}}(X,Y)^p} & a > 2 \\[2ex] \dfrac{\prod_{i=1}^{p+1} \Phi_p(X,\xi_i)}{(X-Y)^{p+1}} & a = 2, \end{cases}$$

where $X = \xi_i$ are the roots of $\Phi_{p^{a-1}}(X,Y) = 0$.

PROOF. See Weber [**131**, §69]. Q.E.D.

Now let p be a prime. To compute $\Phi_p(X,Y)$, we will follow Kaltofen and Yui [**82**] and Yui [**139**]. First note that by parts (iii) and (v) of Theorem 11.18, we have

$$\Phi_p(X,Y) = \Phi_p(Y,X), \quad \Phi_p(X,Y) \equiv (X^p - Y)(X - Y^p) \bmod p\mathbb{Z}[X,Y],$$

and we also know that $\Phi_p(X,Y)$ is monic of degree $\Psi(p) = p+1$ as a polynomial in X. Thus we can write $\Phi_p(X,Y)$ in the following form:

$$(X^p - Y)(X - Y^p) + p \sum_{0 \le i \le p} c_{ii}\, X^i Y^i + p \sum_{0 \le i < j \le p} c_{ij}(X^i Y^j + X^j Y^i), \tag{13.15}$$

where the coefficients c_{ij}'s are integers. We will use the q-expansion of the j-function to obtain a finite system of equations that can be solved uniquely for the c_{ij}'s.

By the definition of the modular equation, we have the identity

$$\Phi_p(j(p\tau), j(\tau)) = 0.$$

Substituting the q-expansions for $j(\tau)$ and $j(p\tau)$ into this equation and using (13.15), we obtain

$$\begin{aligned} 0 &= (j(p\tau)^p - j(\tau))(j(p\tau) - j(\tau)^p) \\ &\quad + p \sum_{0 \le i \le p} c_{ii} j(p\tau)^i j(\tau)^i + p \sum_{0 \le i < j \le p} c_{ij}(j(p\tau)^i j(\tau)^j + j(p\tau)^j j(\tau)^i). \end{aligned} \tag{13.16}$$

If we equate the coefficients of the different powers of q, then we get an infinite number of linear equations in the variables c_{ij}. We can reduce to a finite number of equations as follows:

PROPOSITION 13.17. *The finite system of linear equations obtained by equating coefficients of nonpositive powers of q in* (13.16) *has a unique solution given by the coefficients c_{ij} of the modular equation.*

PROOF. Since the modular equation provides one solution, it suffices to prove uniqueness. Using (13.15), a solution of these equations gives a polynomial $F(X,Y)$ with the following three properties:

(i) $F(X,Y)$ is monic of degree $p+1$ in X.
(ii) $F(X,Y) = F(Y,X)$.
(iii) $\lim_{\mathrm{Im}(\tau)\to\infty} F(j(p\tau), j(\tau)) = 0$.

To explain the last property, note that the q-expansion of $F(j(p\tau), j(\tau))$ contains no nonpositive powers of q since $F(X,Y)$ comes from a solution of our finite system of equations. Since $q \to 0$ as $\mathrm{Im}(\tau) \to \infty$, (iii) follows.

We claim these properties force $F(X,Y) = \Phi_p(X,Y)$, which will prove uniqueness. The idea is to study $F(j(p\tau), j(\tau))$, which is a modular function for $\Gamma_0(p)$. We will first show that $F(j(p\tau), j(\tau))$ vanishes at the cusps, which means that

$$\lim_{\mathrm{Im}(\tau)\to\infty} F(j(p\gamma\tau), j(\gamma\tau)) = 0 \qquad \text{for all } \gamma \in \mathrm{SL}(2,\mathbb{Z}). \tag{13.18}$$

Using (11.12), this is equivalent to showing

$$\lim_{\mathrm{Im}(\tau)\to\infty} F(j(\sigma\tau), j(\tau)) = 0 \qquad \text{for all } \sigma \in C(p).$$

When $\sigma = \left(\begin{smallmatrix} p & 0 \\ 0 & 1 \end{smallmatrix}\right)$, we're done by (iii), and when $\sigma \neq \left(\begin{smallmatrix} p & 0 \\ 0 & 1 \end{smallmatrix}\right)$, σ must be of the form $\left(\begin{smallmatrix} 1 & i \\ 0 & p \end{smallmatrix}\right)$ since p is prime. If we set $u = \sigma\tau = (\tau + i)/p$, then $\tau = pu - i$, and

$$\begin{aligned} \lim_{\mathrm{Im}(\tau)\to\infty} F(j(\sigma\tau), j(\tau)) &= \lim_{\mathrm{Im}(\tau)\to\infty} F(j(u), j(pu - i)) \\ &= \lim_{\mathrm{Im}(u)\to\infty} F(j(u), j(pu)) \\ &= \lim_{\mathrm{Im}(u)\to\infty} F(j(pu), j(u)) = 0, \end{aligned}$$

where we used (ii) and (iii) above. This proves (13.18).

Thus $F(j(p\tau), j(\tau))$ is a holomorphic modular function for $\Gamma_0(p)$ which vanishes at the cusps. For modular functions for $\mathrm{SL}(2,\mathbb{Z})$, we proved in Lemma (11.10) that such a function is zero, and the proof extends easily to the case of $\Gamma_0(p)$ (see Exercise 13.11). This shows that $F(j(p\tau), j(\tau)) = 0$, so that $j(p\tau)$ is a root of $F(X, j(\tau))$ and $\Phi_p(X, j(\tau))$. Since the latter is irreducible over $\mathbb{C}(j(\tau))$, it must divide $F(X, J(\tau))$. Both $F(X,Y)$ and $\Phi_p(X,Y)$ are monic of the same degree, and hence they must be equal. Q.E.D.

Looking at the q-expansions for $j(\tau)$ and $j(p\tau)$, the most negative power of q in (13.16) is q^{-p^2-p}, and it follows that the system of equations described in Proposition 13.17 has p^2+p+1 equations in the $(p^2+3p+2)/2$ unknowns c_{ij}. With some cleverness, one can reduce to p^2+p equations in $(p^2+3p)/2$ unknowns (see Yui [**139**]). These equations have been written down explicitly by Yui [**139**], though the resulting expressions are *extremely* complicated. For a discussion of the computational aspects of these equations, see Kaltofen and Yui [**82**].

We are not quite done, for our equations for $\Phi_p(X,Y)$ involve the q-expansions of $j(\tau)$ and $j(p\tau)$. Hence we need to calculate those coefficients of the q-expansions which contribute to negative powers of q in (13.16). It suffices to do this for $j(\tau)$, and because the most negative power of q in (13.16) is q^{-p^2-p}, we need only the first p^2+p coefficients of the q-expansion of the j-function. In §12 we found some nice formulas for

$$j(\tau) = 1728\,\frac{g_2(\tau)^3}{\Delta(\tau)},$$

but to get the q-expansion, we need *series* expansions of the numerator and denominator. For $g_2(\tau)$, we use the classical formula

$$g_2(\tau) = \frac{(2\pi)^4}{12}\left(1 + 240\sum_{n=1}^{\infty}\sigma_3(n)q^n\right),$$

where $\sigma_3(n) = \sum_{d|n} d^3$ (see Lang [**90**, §4.1] or Serre [**112**, §VII.4.2]), and for $\Delta(\tau)$, we know from Theorem 12.17 that

$$\Delta(\tau) = (2\pi)^{12} q \prod_{n=1}^{\infty}(1-q^n)^{24}.$$

This is still not a series, but if we use Euler's famous identity

$$\prod_{n=1}^{\infty}(1-q^n) = \sum_{n=-\infty}^{\infty} q^{n(3n+1)/2}$$

(see Hardy and Wright [**63**, §19.9], then it becomes straightforward to write a program to compute the q-expansion of $j(\tau)$. A description of how to do this is in Hermann [**69**] (he also gives an alternate approach to calculating the modular equation), and one finds that the first few terms of the q-expansion are

$$\begin{aligned} j(\tau) =& \frac{1}{q} + 744 + 196884\,q + 21493760\,q^2 + 864299970\,q^3 \\ &+ 20245856256\,q^4 + 333202640600\,q^5 + \cdots . \end{aligned}$$

These formulas also give a second proof that the q-expansion of $j(\tau)$ has integer coefficients (see Exercise 13.12).

The conclusion of this rather long discussion is that for any integer $m > 0$, we can compute $\Phi_m(X, Y)$, which then gives us $\Phi_m(X, X)$ by setting $X = Y$. There are known algorithms for factoring $\Phi_m(X, X)$ into irreducibles, and then the discussion following Proposition 13.11 shows how to compute $H_{\mathcal{O}}(X)$. We have thus proved the following theorem:

THEOREM 13.19. *Given an order $\mathcal{O}$ in an imaginary quadratic field, there is an algorithm for computing the class equation $H_{\mathcal{O}}(X)$.* Q.E.D.

The problem with this theorem is that our algorithm for computing $H_{\mathcal{O}}(X)$ requires knowing $\Phi_m(X, Y)$. Modular equations are extremely complicated polynomials and are difficult to compute. We saw in (11.22) that $\Phi_3(X, Y)$ is very large, and things get worse as m increases. For example, the printout of $\Phi_{11}(X, Y)$ takes over two single-spaced pages, and some of the coefficients have over 120 digits (see Kaltofen and Yui [**82**]). In general, Cohen [**24**] proved that the maximum of the absolute values of the coefficients of $\Phi_m(X, Y)$ is asymptotic to $\exp(6\Psi(m)\log(m))$, where $\Psi(m) = m\prod_{p|m}(1 + 1/p)$, so that the growth is exponential in m. Hence the above algorithm is not a practical way to compute class equations.

A more efficient approach to computing $H_D(X)$ has been developed by Kaltofen and Yui [**81**]. The basic idea is to compute $H_D(X)$ directly from the formula

$$H_D(X) = \prod_{i=1}^{h}(X - j(\mathfrak{a}_i)).$$

We know how to find the $h = h(D)$ reduced forms of discriminant D, and then the $\mathfrak{a}_i$'s can be taken to be the proper $\mathcal{O}$-ideals corresponding to the reduced forms via Theorem 7.7. Since $H_D(X)$ has integral coefficients, we need only compute $j(\mathfrak{a}_i)$ numerically to a sufficiently high degree of precision, and the formulas for $j(\tau)$ given in §12 are ideal for this purpose. For an example of how this works, consider the case of discriminant $D = -71$. Here, the class number is $h(-71) = 7$, and the above process shows that the minimal polynomial of $j((1 + \sqrt{-71})/2)$ is

$$\begin{aligned} H_{-71}(X) = X^7 &+ 5 \cdot 7 \cdot 31 \cdot 127 \cdot 233 \cdot 9769\, X^6 \\ &- 2 \cdot 5 \cdot 7 \cdot 44171287694351\, X^5 \\ &+ 2 \cdot 3 \cdot 7 \cdot 2342715209763043144031\, X^4 \\ &- 3 \cdot 7 \cdot 31 \cdot 1265029590537220860166039\, X^3 \\ &+ 2 \cdot 7 \cdot 11^3 \cdot 67 \cdot 229 \cdot 17974026192471785192633\, X^2 \\ &- 7 \cdot 11^6 \cdot 17^6 \cdot 1420913330979618293\, X \\ &+ (11^3 \cdot 17^2 \cdot 23 \cdot 41 \cdot 47 \cdot 53)^3. \end{aligned} \tag{13.20}$$

(This example was taken from the preliminary version of [**81**]—all primes < 1000 were factored out of the coefficients.) Note that the constant term is a cube, as predicted by Theorem 12.2.

We can apply the algorithm of Theorem 13.19 to give a constructive version of Theorem 9.2, but before we do this, we need to learn about the work of Deuring, Gross and Zagier on the class equation.

C. Theorems of Deuring, Gross and Zagier

In 1946 Deuring [**33**] proved a remarkable result concerning prime divisors of the difference of two singular moduli. To state Deuring's theorem precisely, let $\mathcal{O}_1$ and $\mathcal{O}_2$ be orders in imaginary quadratic fields K_1 and K_2 respectively, and for $i = 1, 2$, let $\mathfrak{a}_i$ be a proper fractional $\mathcal{O}_i$-ideal. Then we have:

THEOREM 13.21. *Let L be a number field containing $j(\mathfrak{a}_1)$ and $j(\mathfrak{a}_2)$, and let $\mathfrak{P}$ be a prime of L lying over the prime number p. When $K_1 = K_2$, assume in addition that p divides neither of the conductors of $\mathcal{O}_1$ and $\mathcal{O}_2$. If $j(\mathfrak{a}_1) \neq j(\mathfrak{a}_2)$, then*

$$j(\mathfrak{a}_1) \equiv j(\mathfrak{a}_2) \bmod \mathfrak{P} \Longrightarrow \begin{cases} p \text{ splits completely} \\ \text{in neither } K_1 \text{ nor } K_2. \end{cases}$$

PROOF. The proof uses reduction theory of elliptic curves. See Deuring [**33**] or Lang [**90**, §13.4]. Q.E.D.

We can use this theorem to study the constant term and discriminant of the class equation:

COROLLARY 13.22. *Let $D < 0$ be a discriminant, and let p be prime.*

(i) *If p divides the constant term of $H_D(X)$ and $\mathbb{Q}(\sqrt{D}) \neq \mathbb{Q}(\sqrt{-3})$, then either $p = 3$ or $p \equiv 2 \bmod 3$, and $(D/p) \neq 1$.*

(ii) *If p divides the discriminant of $H_D(X)$, then $(D/p) \neq 1$.*

PROOF. Let $\mathfrak{a}_1, \ldots, \mathfrak{a}_h$, $h = h(D)$, be ideal class representatives for the order of discriminant D. To prove (i), note that the constant term of the class equation is

$$C = \pm \prod_{i=1}^{h} j(\mathfrak{a}_i).$$

If $p \mid C$, then in some number field L, there is a prime $\mathfrak{P}$ containing p that divides some $j(\mathfrak{a}_i)$. Since

$$j(\mathfrak{a}_i) = j(\mathfrak{a}_i) - 0 = j(\mathfrak{a}_i) - j((1+\sqrt{-3})/2),$$

we know by Theorem 13.21 that p splits in neither $\mathbb{Q}(\sqrt{D})$ nor $\mathbb{Q}(\sqrt{-3})$, and (i) follows immediately.

To prove (ii), note that the discriminant of $H_D(X)$ is

$$\operatorname{disc}(H_D(X)) = \prod_{i<j} \big(j(\mathfrak{a}_i) - j(\mathfrak{a}_j)\big)^2.$$

Thus, if $p \mid \operatorname{disc}(H_D(X))$, then some $\mathfrak{P}$ lying over p divides some $j(\mathfrak{a}_i) - j(\mathfrak{a}_j)$. If $p \nmid D$, then Theorem 13.21 implies that p doesn't split in $\mathbb{Q}(\sqrt{D})$, and $(D/p) = -1$ follows. If $p \mid D$, then $(D/p) = 0$, so that $(D/p) \neq 1$ in either case. Q.E.D.

One of our original motivations for studying complex multiplication came from the question of when a prime can be written in the form $x^2 + ny^2$. Using the class equation, we can now prove a constructive version of the main result of this book, Theorem 9.2:

THEOREM 13.23. *Let n be a positive integer. Then there is a monic irreducible polynomial $f_n(X)$ of degree $h(-4n)$ such that for an odd prime p not dividing n,*

$$p = x^2 + ny^2 \Longleftrightarrow \begin{cases} (-n/p) = 1 \text{ and } f_n(X) \equiv 0 \bmod p \\ \text{has an integer solution.} \end{cases}$$

Furthermore, there is an algorithm for finding $f_n(X)$.

PROOF. The order of discriminant $-4n$ is $\mathcal{O} = [1, \sqrt{-n}]$, so that by Theorem 11.1, $j(\sqrt{-n})$ is a real algebraic integer and is a primitive element of the ring class field of $\mathcal{O}$. Since $H_{-4n}(X)$ is the minimal polynomial of $j(\sqrt{-n})$, we can set $f_n(X) = H_{-4n}(X)$ in Theorem 9.2, and then the desired equivalence holds for primes dividing neither $-4n$ nor the discriminant of $H_{-4n}(X)$. But when a prime divides the discriminant, Corollary 13.22 tells us that $(-4n/p) \neq 1$. Since both sides of the desired equivalence imply $(-n/p) = 1$, the discriminant condition is superfluous. Finally, by Theorem 13.19, there is an algorithm for finding $H_{-4n}(X)$, and the theorem is proved. Q.E.D.

From a computational point of view, this result is not ideal. The polynomials $H_{-4n}(X)$ are difficult to compute, and as indicated by $H_{-56}(X)$ and $H_{-71}(X)$ (see (13.1) and (13.20)), they are excessively complicated. The real value of Theorem 13.23 is the way it links the ideas of class field theory and complex multiplication to the elementary question of when a prime can be written in the form $x^2 + ny^2$.

Deuring's study of $j(\mathfrak{a}_1) - j(\mathfrak{a}_2)$ prompted the work of Gross and Zagier [**61**] which determines *exactly* which primes divide such a difference. Their results apply only to field discriminants, but one gets very complete information in this case. Let d_1 and d_2 be the discriminants of imaginary quadratic fields K_1 and K_2 respectively. We will assume that d_1 and d_2 are relatively prime. Then set

$$J(d_1, d_2) = \left(\prod_{i=1}^{h_1}\prod_{j=1}^{h_2}\big(j(\mathfrak{a}_i) - j(\mathfrak{b}_j)\big)\right)^{4/w_1w_2},$$

where $\mathfrak{a}_1, \ldots, \mathfrak{a}_{h_1}$ are ideal class representatives of $\mathcal{O}_{K_1}$, $\mathfrak{b}_1, \ldots, \mathfrak{b}_{h_2}$ are ideal class representatives of $\mathcal{O}_{K_2}$, and $w_1 = |\mathcal{O}_{K_1}^*|$, $w_2 = |\mathcal{O}_{K_2}^*|$. Note that $J(d_1, d_2)$ is an integer when $d_1, d_2 < -4$, and that $J(d_1, d_2)^2$ is always an integer (see Exercise 13.13).

To state Gross and Zagier's formula for $J(d_1, d_2)^2$, we will need functions $\epsilon(n)$ and $F(m)$, which are defined as follows. First, if p is a prime, we set

$$\epsilon(p) = \begin{cases} (d_1/p) & \text{if } p \nmid d_1 \\ (d_2/p) & \text{if } p \nmid d_2. \end{cases}$$

The reader can easily check that this is well-defined whenever $(d_1d_2/p) \neq -1$ (see Exercise 13.14). Then, if $n = \prod_{i=1}^r p_i^{a_i}$, we set

$$\epsilon(n) = \prod_{i=1}^{r} \epsilon(p_i)^{a_i},$$

where we assume that $(d_1d_2/p_i) \neq -1$ for all i. Finally, $F(m)$ is defined by the formula

$$F(m) = \prod_{\substack{nn'=m \\ n,n'>0}} n^{\epsilon(n')}.$$

This is well-defined when all primes p dividing m satisfy $(d_1d_2/p) \neq -1$.

We can now state the main theorem of Gross and Zagier [**61**]:

THEOREM 13.24. *With the above notation,*

$$J(d_1, d_2)^2 = \pm \prod_{\substack{x^2 < d_1 d_2 \\ x^2 \equiv d_1 d_2 \bmod 4}} F\Big(\frac{d_1 d_2 - x^2}{4}\Big).$$

PROOF. First note that $F((d_1d_2 - x^2)/4)$ is always defined since any prime p dividing $(d_1d_2 - x^2)/4$ satisfies $(d_1d_2/p) \neq -1$ (see Exercise 13.14). The paper [**61**] contains two proofs of this theorem, one algebraic and one analytic. The algebraic proof, which uses reduction theory of elliptic curves, is given only for prime discriminants. A general version of this proof appears in Dorman [**38**]. Q.E.D.

This theorem gives the following corollary:

COROLLARY 13.25. *Let p be a prime dividing $J(d_1, d_2)^2$. Then:*

(i) $(d_1/p) \neq 1$ *and* $(d_2/p) \neq 1$.
(ii) *p divides a positive integer of the form $(d_1d_2 - x^2)/4$.*
(iii) $p \leq d_1d_2/4$.

PROOF. If p divides $J(d_1, d_2)^2$, it must divide some $F((d_1d_2 - x^2)/4)$, and the formula for $F(m)$ then shows that p divides $(d_1d_2 - x^2)/4$. This easily implies parts (ii) and (iii) of the corollary.

It remains to prove part (i). We will first consider the following lemma which tells us how to compute $F(m)$:

LEMMA 13.26. *Let m be a positive integer of the form $(d_1d_2 - x^2)/4$. Then $F(m) = 1$ unless m can be written in the form*

$$m = p^{2a+1} p_1^{2a_1} \cdots p_r^{2a_r} q_1^{b_1} \cdots q_s^{b_s},$$

where $\epsilon(p) = \epsilon(p_1) = \cdots = \epsilon(p_r) = -1$ and $\epsilon(q_1) = \cdots = \epsilon(q_s) = 1$. In this case,

$$F(m) = p^{(a+1)(b_1+1)\cdots(b_s+1)}.$$

In particular, $p \mid F(m)$ means that p is the only prime dividing m with an odd exponent and $\epsilon(p) = -1$.

PROOF. See Exercises 13.15 and 13.16. Q.E.D.

We can now complete the proof of Corollary 13.25. The above lemma shows that $\epsilon(p) = -1$ for any prime p dividing $F(m)$. It is easy to see that $\epsilon(p) = -1$ implies $(d_1/p) \neq 1$ and $(d_2/p) \neq 1$ (see Exercise 13.14), and the corollary follows. Q.E.D.

Note that this corollary implies Deuring's theorem in the case of relatively prime field discriminants. We should also mention that when $d_1d_2 \equiv 1 \bmod 8$, one gets better upper bounds on p (see Exercise 13.17).

If we apply Corollary 13.25 when $d_2 = -3$, then we can strengthen Deuring's result about the constant term of the class equation:

COROLLARY 13.27. *Let d_K be the discriminant of an imaginary quadratic field K, and assume that $3 \nmid d_K$. If p is a prime dividing the constant term of $H_{d_K}(X)$, then $(d_K/p) \neq 1$ and either $p = 3$ or $p \equiv 2 \bmod 3$. Furthermore, $p \leq 3|d_K|/4$.*

PROOF. If $\mathfrak{a}_1 \ldots, \mathfrak{a}_h$ are ideal class representatives of $\mathcal{O}_K$, then

$$J(d_K, -3)^2 = \left(\prod_{i=1}^{h} j(\mathfrak{a}_i)\right)^{4/3w},$$

where $w = |\mathcal{O}_K^*|$. Thus the primes dividing $J(d_K, -3)^2$ are the same as the primes dividing the constant term of $H_{d_K}(X)$, and we are done by the previous corollary. Q.E.D.

For an example of how good these estimates are, consider $H_{-56}(X)$. We know from (13.1) that the constant term is

$$(2^8 \cdot 11^2 \cdot 17 \cdot 41)^3.$$

Corollary 13.27 gives us the estimate $p \leq 3|-56|/4 = 42$, which is as good as one can get. The reader should also check the constant term of $H_{-71}(X)$ given in (13.20)—the estimate is again as good as possible. Of course, one could use Theorem 13.24 to compute these constant terms directly (see Exercise 13.18).

Gross and Zagier also have similar theorems for primes dividing the discriminant of the class equation. Rather than give the formula for the multiplicities of the primes, we will just state the following corollary of their result:

THEOREM 13.28. *Let d_K be the discriminant of an imaginary quadratic field K, and let p be a prime dividing the discriminant of $H_{d_K}(X)$. Then $(d_K/p) \neq 1$ and $p \leq |d_K|$.*

PROOF. For prime discriminants, this is proved by Gross and Zagier in [**61**], and the general case is in Dorman [**37**]. Q.E.D.

This theorem strengthens Deuring's result about the discriminant of the class equation. For an example of the bound $p \leq |d_K|$, consider $H_{-56}(X)$. One computes that its discriminant is

$$-2^{116} \cdot 7^{13} \cdot 11^{10} \cdot 17^6 \cdot 29^4 \cdot 31^2 \cdot 37^2 \cdot 41^2 \cdot 43^2 \cdot 47^2 \cdot 53^2.$$

Theorem 13.28 gives the bound $p \leq 56$ on the primes that can appear, which again is the best possible.

D. Exercises

13.1. Use Corollary 11.37 to prove Proposition 13.2.

13.2. If $\mathcal{O}$ is an order in an imaginary quadratic field and m is a positive integer, then we define $r(\mathcal{O}, m) = |\{\alpha \in \mathcal{O} : \alpha \text{ is primitive and } N(\alpha) = m\}/\mathcal{O}^*|$, where $\mathcal{O}^*$ acts by multiplication.

(a) Prove that $r(\mathcal{O}, m)$ is finite.
(b) For fixed $m > 1$, prove that there are only finitely many orders $\mathcal{O}$ such that $r(\mathcal{O}, m) > 0$.

13.3. Let $F(X, Y) \in \mathbb{C}[X, Y]$ and suppose that $F(X_0, X_0) = 0$. Then X_0 is a root of both $F(X, X_0)$ and $F(X, X)$.

(a) If $F(X, Y) = X^3 + Y^3 + XY$, then show that 0 is a root of $F(X, 0)$ and $F(X, X)$ of different multiplicities. Note that the polynomial $F(X, Y)$ is symmetric.

(b) If $F(X,Y)$ and X_0 satisfy the additional condition that

$$\lim_{X\to X_0} \frac{F(X,X)}{F(X,X_0)}$$

exists and is nonzero, then show that X_0 is a root of $F(X,X_0)$ and $F(X,X)$ of the same multiplicity.

13.4. This exercise is concerned with the proof of (13.7). Recall that $\tilde{\sigma}(\tau_0) = \tau_0$, where $\tilde{\sigma} = \left(\begin{smallmatrix} a & b \\ c & d \end{smallmatrix}\right)$ has relatively prime entries and determinant $m > 1$.

(a) Prove that $c \neq 0$.

(b) When $j(\tau_0) = 1728$, we can assume that $\tau_0 = i$. Then show that $m^2 = (ci+d)^4$ implies $d = 0$. Since $\tilde{\sigma}(i) = i$, conclude that $a = 0$, and derive a contradiction.

(c) When $j(\tau_0) = 0$, argue as in part (b) to complete the proof of (13.7).

13.5. Let $m > 1$, and let $\mathcal{O} = [1, \tau_0]$ be an order in an imaginary quadratic field. Consider the sets

$$\begin{aligned} \mathcal{A} &= \{\alpha \in \mathcal{O} : \alpha \text{ is primitive and } N(\alpha) = m\}/\mathcal{O}^* \\ \mathcal{B} &= \{\sigma \in C(m) : j(\sigma\tau_0) = j(\tau_0)\}. \end{aligned}$$

In the proof of Theorem 13.4, we showed how an element $[\alpha] \in \mathcal{A}$ determines a unique $\sigma \in \mathcal{B}$. Prove that the map $[\alpha] \mapsto \sigma$ defines a bijection $\mathcal{A} \xrightarrow{\sim} \mathcal{B}$.

13.6. The goal of this exercise is to prove the formula for the degree N of $\Phi_m(X,X)$ given in Proposition 13.8.

(a) Prove that q^{-N} is the most negative power of q in the q-expansion of $\Phi_m(j(\tau), j(\tau))$.

(b) If $\sigma = \left(\begin{smallmatrix} a & b \\ 0 & d \end{smallmatrix}\right) \in C(m)$, then use (11.19) to show that the q-expansion of $j(\tau) - j(\sigma\tau)$ is

$$\begin{array}{rl} q^{-1} - \zeta_m^{-ab} q^{-a/d} + \cdots & \text{when } a < d \\ -\zeta_m^{-ab} q^{-a/d} + q^{-1} + \cdots & \text{when } a > d \\ (1 - \zeta_m^{-ab}) q^{-1} + \cdots & \text{when } a = d, \end{array}$$

where $\zeta_m = e^{2\pi i/m}$. The last possibility can occur only when m is a perfect square, and in this case, $\zeta_m^{-ab} \neq 1$ since $\sigma \in C(m)$.

(c) Given a, we know that $d = m/a$. In part (a) of Exercise 11.9 we showed that the number of possible $\sigma \in C(m)$ with this a and d was

$$\frac{d}{e}\phi(e),$$

where $e = \gcd(a,d)$. Use this formula and part (b) to show that the degree N of $\Phi_m(X,X)$ equals

$$\sum_{\substack{a|m \\ a<\sqrt{m}}} \frac{d}{e}\phi(e) + \sum_{\substack{a|m \\ a>\sqrt{m}}} \frac{a}{d}\cdot\frac{d}{e}\phi(e) + \phi(\sqrt{m}).$$

(d) Show that the first two sums in the above expression are equal. This proves the formula for N given in Proposition 13.8.

13.7. This exercise is concerned with some examples of Theorem 13.4.

(a) Verify that $r(-3,3) = r(-12,3) = 1$, $r(-8,3) = r(-11,3) = 2$, and also show that $r(D,3) = 0$ for all other discriminants. This proves that

$$\Phi_3(X,X) = \pm H_{-3}(X)H_{-12}(X)H_{-8}(X)^2H_{-11}(X)^2.$$

(b) Use the method of part (a) to write down the factorization of $\Phi_5(X,X)$.

13.8. The proof of Proposition 13.11 requires the following facts about the orders of discriminant -3 and -4 ($\mathbb{Z}[\omega]$ and $\mathbb{Z}[i]$ respectively).

(a) If $m > 1$, show that $r(-3,m) = 1$ if and only if $m = 3$.
(b) If $m > 1$, show that $r(-4,m) = 1$ if and only if $m = 2$.

13.9. Let $m \equiv 3 \bmod 4$ be an integer > 3. Show that the order $\mathbb{Z}[\sqrt{-m}]$ of discriminant $-4m$ has no primitive elements of norm $(m+1)/4$.

13.10. In this exercise we will illustrate the algorithm given in the text for computing $H_D(X)$.

(a) Show that $H_{-56}(X)$ is determined by $\Phi_{14}(X,X)$.
(b) Show that $H_{-11}(X)$ and $H_{-44}(X)$ are determined by $\Phi_{11}(X,X)$.
(c) Show that $H_{-7}(X)$, $H_{-28}(X)$ are determined by $\Phi_7(X,X)$, $\Phi_2(X,X)$.

13.11. Let $f(\tau)$ be a modular function for $\Gamma_0(m)$ which vanishes at the cusps.

(a) If γ_i, $i = 1,\ldots,|C(m)|$, are coset representatives for $\Gamma_0(m) \subset \mathrm{SL}(2,\mathbb{Z})$, then show that

$$\prod_{i=1}^{|C(m)|} f(\gamma_i\tau)$$

is a modular function for $\mathrm{SL}(2,\mathbb{Z})$ which vanishes at infinity.

(b) If in addition $f(\tau)$ is holomorphic on $\mathfrak{h}$, then show that $f(\tau)$ is identically zero. Hint: use part (a) and Lemma 11.10.

13.12. Use the formulas

$$g_2(\tau) = \frac{(2\pi)^4}{12}\left(1 + 240\sum_{n=1}^{\infty}\sigma_3(n)q^n\right)$$

$$\Delta(\tau) = (2\pi)^{12}q\prod_{n=1}^{\infty}(1-q^n)^{24}$$

to show that the coefficients of the q-expansion of $j(\tau)$ are integral. This is the classical method used to prove Theorem 11.8.

13.13. Let $J(d_1,d_2)$ be as defined in the text.

(a) If d_1, $d_2 < -4$, then show that $J(d_1,d_2)$ is an integer. Hint: use Galois theory.
(b) Show that $J(d_1,d_2)^2$ is always an integer. Hint: when d_1 or d_2 is -3, recall that $j((1+\sqrt{-3})/2) = 0$. Theorem 12.2 will be useful.

13.14. Let $\epsilon(m)$ and $F(n)$ be as defined in the text, and let p be a prime number.

(a) Show that $\epsilon(p)$ is defined whenever $(d_1d_2/p) \neq -1$.
(b) Assume that p divides a number of the form $(d_1d_2 - x^2)/4$. Then prove that $(d_1d_2/p) \neq -1$.
(c) Show that $\epsilon(p) = -1$ implies that $(d_1/p) \neq 1$ and $(d_2/p) \neq 1$.

13.15. Exercises 13.15 and 13.16 will prove Lemma 13.26. In this exercise we will show that any positive integer of the form $m = (d_1d_2 - x^2)/4$ satisfies $\epsilon(m) = -1$. We will need the following extension of the Legendre symbol. Let $D \equiv 0, 1 \bmod 4$, and let $\chi : (\mathbb{Z}/D\mathbb{Z})^* \to \{\pm 1\}$ be the homomorphism from Lemma 1.14 (so that $\chi([p]) = (D/p)$ when p is a prime not dividing D). Then for any integer m relatively prime to D, set

$$\left(\frac{D}{m}\right) = \chi([m]).$$

(a) Show that (D/m) is multiplicative in D and m and depends only on the congruence class of m modulo D. Also, when $m = p_1^{a_1} \cdots p_r^{a_r}$ is positive, show that

$$\left(\frac{D}{m}\right) = \prod_{i=1}^{r} \left(\frac{D}{p_i}\right)^{a_i},$$

where (D/p_i) is the usual Kronecker symbol. Thus, when m is odd and positive, (D/m) is just the Jacobi symbol. Finally, show that $(D/-1) = \operatorname{sgn}(D)$. Hint: see Lemma (1.14).

(b) We will need the following limited version of quadratic reciprocity for (D/m). Namely, if $D \equiv 1 \bmod 4$ is relatively prime to $m \equiv 0, 1 \bmod 4$, then prove that $(D/m) = (m/|D|)$. Furthermore, if D and m have opposite signs, then prove that $(D/m) = (m/D)$.

(c) Let m be a positive integer such that $\epsilon(m)$ is defined. If m is relatively prime to d_1, then show that $\epsilon(m) = (d_1/m)$.

(d) Now we can prove that $\epsilon(m) = -1$ when $m = (d_1d_2 - x^2)/4$. We can assume $d_1 \equiv 1 \bmod 4$, and write $m = ab$, where $a \mid d_1$, $a, b > 0$ and $\gcd(d_1, b) = 1$.
   - (i) Show that $\epsilon(m) = (d_2/a)(d_1/b)$.
   - (ii) Show that $(d_1/b) = -(\varepsilon a/d_2)$, where $\varepsilon = (-1)^{(a-1)/2}$. Hint: write $d_1 = \varepsilon a d$ and note that $(d_1/b) = (d_1/4b) = (\varepsilon a/4b)(d/4b)$. Use $4ab = d_1d_2 - x^2$ to show that $4b \equiv \varepsilon d d_2 \bmod a$ (remember that a has no square factors) and then apply quadratic reciprocity to $(\varepsilon a/d)$.
   - (iii) Use quadratic reciprocity to prove that $\epsilon(m) = -1$. Hint: remember that $d_2 < 0$.

13.16. Let m be a positive integer such that $\epsilon(m) = -1$. The goal of this exercise is to compute $F(m)$. We will use the function $s(m)$ defined by

$$s(m) = \sum_{\substack{n \mid m \\ n > 0}} \epsilon(n).$$

Note that $s(m)$ is defined whenever $\epsilon(m)$ is. Given a prime p, let $\nu_p(m)$ be the highest power of p dividing m.

(a) If m_1 and m_2 are relatively prime integers such that $\epsilon(m_1)$ and $\epsilon(m_2)$ are defined, then prove that

$$F(m_1m_2) = F(m_1)^{s(m_2)} F(m_2)^{s(m_1)}.$$

(b) Suppose that $m = p_1^{a_1} \cdots p_r^{a_r} q_1^{b_1} \cdots q_s^{b_s}$, where $\epsilon(p_i) = -1$ and $\epsilon(q_i) = 1$ for all i. Prove that

$$s(m) = \begin{cases} 0 & \text{some } a_i \text{ is odd} \\ \prod_{i=1}^{s}(b_i + 1) & \text{all } a_i\text{'s are even.} \end{cases}$$

(c) If $\epsilon(m) = -1$, show that there is at least one prime p with $\epsilon(p) = -1$ and $\nu_p(m)$ odd. Conclude that $s(m) = 0$.

(d) Suppose that $\epsilon(m) = -1$, and that m is divisible by two primes p and q with $\epsilon(p) = \epsilon(q) = -1$ and $\nu_p(m)$ and $\nu_q(m)$ odd. Prove that $F(m) = 1$. Hint: write $m = p^{2a+1}q^{2b+1}m'$, and use parts (a) and (b).

(e) Finally, suppose that m is divisible by a unique prime p with $\epsilon(p) = -1$ and $\nu_p(m)$ odd. Then m can be written $m = p^{2a+1}p_1^{a_1}\cdots p_r^{a_r}q_1^{b_1}\cdots q_s^{b_s}$, where $\epsilon(p) = \epsilon(p_i) = -1$ and $\epsilon(q_i) = 1$ for all i. Prove that

$$F(m) = p^{(a+1)(b_1+1)\cdots(b_s+1)}.$$

Hint: show that $F(p^{2a+1}) = p^{a+1}$, and use parts (a) and (b).

By parts (c)–(e), we see that when $\epsilon(m) = -1$, $F(m)$ is computed by the formulas given in Lemma 13.26. Thus Lemma 13.26 is an immediate corollary of this exercise and the previous one.

13.17. Let p be a prime dividing $J(d_1, d_2)^2$. In Corollary 13.25, we showed that $p \le d_1d_2/4$. In some cases, this estimate can be improved.

(a) If $d_1d_2 \equiv 1 \bmod 8$, then prove that $p < d_1d_2/8$. Hint: $p \mid (d_1d_2 - x^2)/4$. When $p = 2$, note that $d_1d_2 \equiv 1 \bmod 8$ implies $d_1d_2 \ge 33$.

(b) If $d_1 \equiv d_2 \equiv 5 \bmod 8$, then prove that $p < d_1d_2/16$. Hint: when p is odd, we have $p \mid (d_1d_2 - x^2)/8$. To rule out the case $2p = (d_1d_2 - x^2)/4$, use Exercise 13.15 and Lemma 13.26. When $p = 2$, see part (a).

13.18. Apply Theorem 13.24 to determine the constant terms of $H_{-56}(X)$ and $H_{-71}(X)$, and compare your results with (13.1) and (13.20). Hint: use Lemma 13.26 to compute $F(m)$.

CHAPTER 4

Additional Topics

14. Elliptic Curves

In the first three chapters of the book, we solved our basic question concerning primes of the form $x^2 + ny^2$. But the classical version of complex multiplication presented in Chapter 3 does not do justice to more recent developments. In this final chapter of the book, we will discuss two additional topics, elliptic curves and Shimura Reciprocity.

In the modern study of complex multiplication, elliptic functions are replaced with elliptic curves. In §14, we will give some of the basic definitions and theorems concerning elliptic curves, and we will discuss complex multiplication and elliptic curves over finite fields. Then, to illustrate the power of what we've done, we will examine two primality tests from the late 1980s that involve elliptic curves, one of which makes use of the class equation.

In §15, we turn our attention to a quite different topic, Shimura Reciprocity. This concerns the deep interaction between Galois theory and special values of modular functions. We saw hints of this in §12 when we gave Weber's computation of $j(\sqrt{-14})$. Using papers of Alice Gee and Peter Stevenhagen [**53, 54, 126**] and Bumkyo Cho [**22**] as a guide, we will revisit parts of §12 from this point of view and give an interesting twist on the question of $p = x^2 + ny^2$.

The two sections of this chapter can be read independently of each other.

A. Elliptic Curves and Weierstrass Equations

Our treatment of elliptic curves will not be self-contained, for our purpose is to entice the reader into learning more about this lovely subject. Excellent introductions to elliptic curves are available, notably the books by Husemöller [**74**], Knapp [**83**], Koblitz [**84**], Silverman [**118**] and Silverman and Tate [**120**], and more advanced topics are discussed in the books by Lang [**90**], Shimura [**115**] and Silverman [**119**].

Given a field K of characteristic different from 2 or 3, an *elliptic curve E over K* is an equation of the form

$$y^2 = 4x^3 - g_2x - g_3, \tag{14.1}$$

where

$$g_2, g_3 \in K \quad \text{and} \quad \Delta = g_2^3 - 27g_3^2 \neq 0.$$

For reasons that will soon become clear, this equation is called the *Weierstrass equation* of E. When K has characteristic 2 or 3, a more complicated defining equation is needed (see Silverman [**118**, Appendix A]).

Given an elliptic curve E over K, we define $E(K)$ to be the set of solutions

$$E(K) = \{(x, y) \in K \times K : y^2 = 4x^3 - g_2x - g_3\} \cup \{\infty\}.$$

The symbol ∞ appears because in algebraic geometry, it is best to work with homogeneous equations in projective space. Equation (14.1) defines a curve in the affine space K^2, but in the projective space $\mathbb{P}^2(K)$ there is an extra "point at infinity" (see Exercise 14.1 for the details). Given a field extension $K \subset L$, we can also define $E(K) \subset E(L)$ in an obvious way.

Over the complex numbers $\mathbb{C}$, the Weierstrass $\wp$-function gives us elliptic curves as follows. Let $L \subset \mathbb{C}$ be a lattice, and let $\wp(z) = \wp(z; L)$ be the corresponding $\wp$-function. Then we have the differential equation

$$\wp'(z)^2 = 4\wp(z)^3 - g_2(L)\wp(z) - g_3(L)$$

of Theorem 10.1, which gives us the elliptic curve E defined by

$$y^2 = 4x^3 - g_2(L)x - g_3(L).$$

If $z \notin L$, then $\wp(z)$ and $\wp'(z)$ are defined, and the differential equation shows that $(\wp(z), \wp'(z))$ is in $E(\mathbb{C})$. Since $\wp(z)$ and $\wp'(z)$ are also periodic for L, we get a well-defined mapping

$$(\mathbb{C} - L)/L \longrightarrow E(\mathbb{C}) - \{\infty\}.$$

It is easy to show that this map is a bijection (see Exercise 14.2), and consequently we get a bijection

$$\mathbb{C}/L \simeq E(\mathbb{C}) \tag{14.2}$$

by sending $0 \in \mathbb{C}$ to $\infty \in E(\mathbb{C})$. Both $\mathbb{C}/L$ and $E(\mathbb{C})$ have natural structures as Riemann surfaces, and it can be shown that the above map is biholomorphic.

The unexpected fact is that *every* elliptic curve over $\mathbb{C}$ arises from a *unique* Weierstrass $\wp$-function. More precisely, we have the following result:

PROPOSITION 14.3. *Let E be an elliptic curve over $\mathbb{C}$ given by the Weierstrass equation*

$$y^2 = 4x^3 - g_2x - g_3, \quad g_2, g_3 \in \mathbb{C},\ g_2^3 - 27g_3^2 \neq 0.$$

Then there is a unique lattice $L \subset \mathbb{C}$ such that

$$\begin{aligned} g_2 &= g_2(L) \\ g_3 &= g_3(L). \end{aligned}$$

PROOF. The existence of L was proved in Corollary 11.7, and the uniqueness follows from the proof of Theorem 10.9 (see Exercise 14.3). Q.E.D.

Proposition 14.3 is often called the uniformization theorem for elliptic curves. Note that it is a consequence of the properties of the j-function.

The mention of the j-function prompts our next definition: if an elliptic curve E over a field K is defined by the Weierstrass equation (14.1), then the *j-invariant* $j(E)$ is defined to be the number

$$j(E) = 1728\frac{g_2^3}{g_2^3 - 27g_3^2} = 1728\frac{g_2^3}{\Delta} \in K.$$

Note that $j(E)$ is well-defined since $\Delta \neq 0$, and the factor of 1728 doesn't cause trouble since K has characteristic different from 2 and 3 (the definition of the j-invariant is more complicated in the latter case—see Silverman [**118**, Appendix A]). Over the complex numbers, notice that

$$j(L) = j(E)$$

whenever E is the elliptic curve determined by the lattice $L \subset \mathbb{C}$.

To define isomorphisms of elliptic curves, let E and E' be elliptic curves over K, defined by Weierstrass equations $y^2 = 4x^3 - g_2x - g_3$ and $y^2 = 4x^3 - g_2'x - g_3'$ respectively. Then E and E' are *isomorphic over* K if there is a nonzero $c \in K$ such that

$$\begin{aligned} g_2' &= c^4 g_2 \\ g_3' &= c^6 g_3. \end{aligned}$$

In this case, note that the map sending (x, y) to (c^2x, c^3y) induces a bijection

$$E(K) \simeq E'(K).$$

It is trivial to check that isomorphic elliptic curves have the same j-invariant.

Over the complex numbers, isomorphisms of elliptic curves are related to lattices and j-invariants as follows:

PROPOSITION 14.4. *Let E and E' be elliptic curves corresponding to lattices L and L' respectively. Then the following statements are equivalent:*

(i) *E and E' are isomorphic over $\mathbb{C}$.*
(ii) *L and L' are homothetic.*
(iii) $j(E) = j(E')$.

PROOF. This follows easily from Theorem 10.9. We leave the details to the reader (see Exercise 14.4). Q.E.D.

What is more interesting is that part of this proposition generalizes to any algebraically closed field:

PROPOSITION 14.5. *Let E and E' be elliptic curves over a field K of characteristic different from 2 or 3.*

(i) *E and E' have the same j-invariant if and only if they are isomorphic over a finite extension of K.*
(ii) *If K is algebraically closed, then E and E' have the same j-invariant if and only if they are isomorphic over K.*

PROOF. The proof is basically a transcription of the algebraic part of the proof of Theorem 10.9—see Exercise 14.4. Q.E.D.

Over non-algebraically closed fields, nonisomorphic elliptic curves may have the same j-invariant (see Exercise 14.4 for an example over $\mathbb{Q}$). Later, we will discuss the isomorphism classes of elliptic curves over a finite field.

Finally, we need to discuss the group structure on an elliptic curve. The basic idea is to translate the addition law for the Weierstrass $\wp$-function into algebraic terms. To see how this works, let E be an elliptic curve over K, and let P_1 and P_2 be two points in $E(K)$. Our goal is to define $P_1 + P_2 \in E(K)$. If $P_1 = \infty$, we define

$$P_1 + P_2 = \infty + P_2 = P_2,$$

and the case $P_2 = \infty$ is treated similarly. Thus ∞ will be the identity element of $E(K)$. For the remaining cases, we may write $P_1 = (x_1, y_1)$ and $P_2 = (x_2, y_2)$. If $x_1 \neq x_2$, then we define

$$P_1 + P_2 = (x_3, y_3),$$

where x_3 and y_3 are given by

$$\begin{aligned} x_3 &= -x_1 - x_2 + \frac{1}{4}\left(\frac{y_1 - y_2}{x_1 - x_2}\right)^2 \\ y_3 &= -y_1 - (x_3 - x_1)\left(\frac{y_1 - y_2}{x_1 - x_2}\right). \end{aligned} \tag{14.6}$$

These formulas come from the addition laws for $\wp(z+w)$ and $\wp'(z+w)$ (see Theorem 10.1 and Exercise 14.5).

We still need to consider what happens when $x_1 = x_2$. In this case, the Weierstrass equation implies that $y_1 = \pm y_2$, so that there are two cases to consider. When $y_1 = -y_2$, we define

$$P_1 + P_2 = \infty.$$

This formula tells us that the inverse of $(x, y) \in E(K)$ is $(x, -y)$. Finally, suppose that $P_1 = P_2$, where $y_1 = y_2 \neq 0$. Here, we define

$$P_1 + P_2 = 2P_1 = (x_3, y_3),$$

where x_3 and y_3 are given by

$$\begin{aligned} x_3 &= -2x_1 + \frac{1}{16}\left(\frac{12x_1^2 - g_2}{y_1}\right)^2 \\ y_3 &= -y_1 - (x_3 - x_1)\left(\frac{12x_1^2 - g_2}{2y_1}\right). \end{aligned} \tag{14.7}$$

These formulas come from the duplication laws for $\wp(2z)$ and $\wp'(2z)$ (see (10.13) and Exercise 14.5). The major fact is that we get a group:

THEOREM 14.8. *If E is an elliptic curve over a field K, then $E(K)$ is a group (with ∞ as identity) under the binary operation defined above.*

PROOF. See Husemöller [**74**], Koblitz [**84**] or Silverman [**118**] for a proof. These references also explain a lovely geometric interpretation of the above formulas. Q.E.D.

If E is an elliptic curve over K and $K \subset L$ is a field extension, then it is easy to show that $E(K)$ is a subgroup of $E(L)$.

Over the complex numbers, we saw in (14.2) that there is a bijection $\mathbb{C}/L \simeq E(\mathbb{C})$. Notice that both of these objects are groups: $\mathbb{C}/L$ has a natural group structure induced by addition of complex numbers, and $E(\mathbb{C})$ has the group structure defined in Theorem 14.8. It is immediate that the map $\mathbb{C}/L \simeq E(\mathbb{C})$ is a group isomorphism.

B. Complex Multiplication and Elliptic Curves

The next topic to discuss is the complex multiplication of elliptic curves. The idea is to take the theory developed in §§10 and 11 and translate lattices into elliptic curves. The crucial step is to get an algebraic description of complex multiplication, which can then be used over arbitrary fields.

Let's start by describing the endomorphism ring of an elliptic curve E over $\mathbb{C}$. Namely, if E corresponds to the lattice L, we define

$$\mathrm{End}_{\mathbb{C}}(E) = \{\alpha \in \mathbb{C} : \alpha L \subset L\}.$$

This is clearly a subring of $\mathbb{C}$, and note that $\mathbb{Z} \subset \mathrm{End}_{\mathbb{C}}(E)$. Then we say that E has *complex multiplication* if $\mathbb{Z} \neq \mathrm{End}_{\mathbb{C}}(E)$. From Theorem 10.14, it follows that E has complex multiplication if and only if L does, and in this case, $\mathrm{End}_{\mathbb{C}}(E)$ is an order $\mathcal{O}$ in an imaginary quadratic field.

For $\alpha \in \mathcal{O}$, the inclusion $\alpha L \subset L$ gives a group homomorphism $\alpha : \mathbb{C}/L \to \mathbb{C}/L$. Combined with (14.2), we see that $\alpha \in \mathrm{End}_{\mathbb{C}}(E)$ induces a group homomorphism

$$\alpha : E(\mathbb{C}) \longrightarrow E(\mathbb{C}).$$

In terms of the x and y coordinates of a point in $E(\mathbb{C})$, this map can be described as follows:

PROPOSITION 14.9. *Given $\alpha \neq 0 \in \mathrm{End}_{\mathbb{C}}(E)$, there is a rational function $R(x) \in \mathbb{C}(x)$ such that for $(x, y) \in E(\mathbb{C})$, we have*

$$\alpha(x,y) = \Big(R(x), \frac{1}{\alpha}R'(x)y\Big),$$

where $R'(x) = (d/dx)R(x)$.

PROOF. Given $\alpha L \subset L$, we saw in Theorem 10.14 that there is a rational function $R(x)$ such that $\wp(\alpha z) = R(\wp(z))$. Differentiating with respect to z gives $\wp'(\alpha z)\alpha = R'(\wp(z))\wp'(z)$, and thus $\wp'(\alpha z) = (1/\alpha)R'(\wp(z))\wp'(z)$. Since $\alpha : E(\mathbb{C}) \to E(\mathbb{C})$ comes from $\alpha : \mathbb{C}/L \to \mathbb{C}/L$ via the map $z \mapsto (\wp(z), \wp'(z))$, the proposition follows immediately. Q.E.D.

Because of the algebraic nature of $\alpha \in \mathrm{End}_{\mathbb{C}}(E)$, we write $\alpha : E \to E$ instead of $\alpha : E(\mathbb{C}) \to E(\mathbb{C})$. When $\alpha \neq 0$, we say that α is an *isogeny* from E to itself. The most important invariant of an isogeny is its *degree* $\deg(\alpha)$, which is defined to be the order of its kernel. More precisely, if E corresponds to the lattice L, then it is easy to see that the kernel of $\alpha : E(\mathbb{C}) \to E(\mathbb{C})$ is isomorphic to $L/\alpha L$ (see Exercise 14.6). Thus, by Theorem 10.14, it follows that

$$\deg(\alpha) = |L/\alpha L| = N(\alpha),$$

where $N(\alpha)$ is the norm of $\alpha \in \mathcal{O} = \mathrm{End}_{\mathbb{C}}(E)$.

For an example of complex multiplication, consider the elliptic curve E defined by the Weierstrass equation

$$y^2 = 4x^3 - 30x - 28.$$

We claim that $\mathrm{End}_{\mathbb{C}}(E) = \mathbb{Z}[\sqrt{-2}]$, and that for $(x, y) \in E(\mathbb{C})$, complex multiplication by $\sqrt{-2}$ is an isogeny of degree 2 given by the formula

$$\sqrt{-2}(x,y) = \Big(-\frac{2x^2+4x+9}{4(x+2)}, -\frac{1}{\sqrt{-2}}\frac{2x^2+8x-1}{4(x+2)^2}y\Big). \tag{14.10}$$

It turns out that the major work of this claim was proved in §10 when we considered the lattice $L = [1, \sqrt{-2}]$. Namely, in the discussion surrounding (10.21) and (10.22), we showed that for some λ,

$$g_2(\lambda L) = \frac{5 \cdot 27}{2}$$
$$g_3(\lambda L) = \frac{7 \cdot 27}{2}.$$

If we set $\lambda' = \sqrt{3/2}\,\lambda$, then it follows that

$$\begin{aligned} g_2(\lambda' L) &= 30 \\ g_3(\lambda' L) &= 28, \end{aligned}$$

so that E comes from the lattice $\lambda' L = \lambda'[1, \sqrt{-2}]$ and hence has complex multiplication by $\sqrt{-2}$. Furthermore, the formula for $\wp(\sqrt{-2}z)$ given in (10.21) and (10.22) easily combines with Proposition 14.9 to prove (14.10) (see Exercise 14.7).

For an elliptic curve E over an arbitrary field K (as always, of characteristic $\neq 2, 3$), we can't use lattices to define complex multiplication. But as indicated by Proposition 14.9, there is a purely algebraic definition of the endomorphism ring $\mathrm{End}_K(E)$ that depends only on the defining equation of E (see Silverman [**118**, Chapter III]). Because of the group structure of E, $\mathrm{End}_K(E)$ always contains $\mathbb{Z}$, and if K has characteristic zero, we say that E has *complex multiplication* if $\mathrm{End}_{\overline{K}}(E) \neq \mathbb{Z}$, where $\overline{K}$ is the algebraic closure of K (thus the complex multiplications may only be defined over finite extensions of K). When K is a finite field, we will see that $\mathrm{End}_K(E)$ is always bigger than $\mathbb{Z}$. For this reason, the term "complex multiplication" is rarely used when K has positive characteristic.

When $K \subset \mathbb{C}$, we can describe the endomorphism ring $\mathrm{End}_K(E)$ as follows. Let $\alpha \in \mathrm{End}_{\mathbb{C}}(E)$, and use Proposition 14.9 to write $\alpha(x, y) = (R(x), (1/\alpha)R'(x)y)$ for $(x, y) \in E(\mathbb{C})$. Then

$$\alpha \in \ \mathrm{End}_K(E) \iff R(x),\ \frac{1}{\alpha}R'(x) \in K(x).$$

Another interesting case is when $K = \mathbb{F}_q$ is a finite field. Here, the map sending (x, y) to (x^q, y^q) clearly defines a group homomorphism $E(L) \to E(L)$ for any field L containing K (see Exercise 14.8). This gives an element $Frob_q \in \mathrm{End}_K(E)$, which is called the *Frobenius endomorphism* of E. It will play an important role later on. Notice that this map is *not* of the form $(R(x), (1/\alpha)R'(x)y)$.

In this more abstract setting, one can still define the degree of a nonzero isogeny $\alpha \in \mathrm{End}_K(E)$. When $K \subset \mathbb{C}$, the degree of α is the order of $\ker(\alpha : E(\mathbb{C}) \to E(\mathbb{C}))$, while over a finite field, the degree is more subtle to define. For example, the Frobenius isogeny $Frob_q$ always has degree q even though $Frob_q : E(L) \to E(L)$ is injective for any field $K \subset L$. See Silverman [**118**, §III.4] for a precise definition of the degree of an isogeny.

Besides isogenies from E to itself (which are recorded by $\mathrm{End}_K(E)$), one can also define the notion of an isogeny α between different elliptic curves E and E' over the same field K. For simplicity, we will confine our remarks to the case $K = \mathbb{C}$. In this situation, E and E' correspond to lattices L and L'. If $\alpha \neq 0$ is a complex number such that $\alpha L \subset L'$, then multiplication by α induces a map

$$\alpha : E(\mathbb{C}) \longrightarrow E'(\mathbb{C})$$

with kernel $L'/\alpha L$, and we say that α is an *isogeny* from E to E'. As in Proposition 14.9, one can show that α is essentially algebraic in nature (see Exercise 14.9), so that we can write α as $\alpha : E \to E'$ and we say that α is an *isogeny* from E to E'.

The notion of isogeny has a close relation to the modular equation. We define an isogeny $\alpha : E \to E'$ to be *cyclic* if its kernel $L'/\alpha L$ is cyclic. Then we have:

PROPOSITION 14.11. *Let E and E' be elliptic curves over $\mathbb{C}$. Then there is a cyclic isogeny α from E to E' of degree m if and only if $\Phi_m(j(E), j(E')) = 0$.*

PROOF. This follows easily from the analysis of $\Phi_m(u, v) = 0$ given in Theorem 11.23 (see Exercise 14.10). Q.E.D.

For a more complete treatment of these topics, see Lang [**90**, Chapters 2 and 5] and Silverman [**118**, Chapter III].

C. Elliptic Curves over Finite Fields

So far, we've translated concepts about lattices into concepts about elliptic curves. If this were all that happened, there would be no special reason to study elliptic curves. The important point is that the algebraic formulation allows us to state some fundamentally new results, the most interesting of which involve elliptic curves over a finite field $\mathbb{F}_q$. As usual, we will assume that $\mathbb{F}_q$ has characteristic greater than 3, i.e., $q = p^a$, $p > 3$.

When E is an elliptic curve over $\mathbb{F}_q$, the group of solutions $E(\mathbb{F}_q)$ is a finite Abelian group, and it is easy to see that its order $|E(\mathbb{F}_q)|$ is at most $2q + 1$ (see Exercise 14.11). In 1934, Hasse proved the following stronger bound conjectured by Artin:

THEOREM 14.12. *If E is an elliptic curve over $\mathbb{F}_q$, then*

$$q + 1 - 2\sqrt{q} \leq |E(\mathbb{F}_q)| \leq q + 1 + 2\sqrt{q}.$$

PROOF. We will discuss some of the ideas used in the proof. The key ingredient is the isogeny $Frob_q \in \mathrm{End}_{\mathbb{F}_q}(E)$ defined by $Frob_q(x, y) = (x^q, y^q)$.

We can form the isogeny $1 - Frob_q$, and it follows easily that if $\overline{\mathbb{F}}_q$ is the algebraic closure of $\mathbb{F}_q$, then

$$E(\mathbb{F}_q) = \ker(1 - Frob_q : E(\overline{\mathbb{F}}_q) \to E(\overline{\mathbb{F}}_q))$$

(see Exercise 14.12). The next step is to show that $1 - Frob_q$ is a separable isogeny, which implies that

$$|E(\mathbb{F}_q)| = \deg(1 - Frob_q). \tag{14.13}$$

From here, the proof is a straightforward consequence of the basic properties of isogenies (see Silverman [**118**, Chapter V, Theorem 1.1]). Q.E.D.

In 1946, Weil proved a similar result for algebraic curves over finite fields, and in 1974, Deligne proved a vast generalization (conjectured by Weil) to higher-dimensional algebraic varieties. For further discussion and references, see Ireland and Rosen [**75**, Chapter 11] and Silverman [**118**, §V.2].

Elliptic curves over finite fields come in two types, *ordinary* and *supersingular*, as determined by their endomorphism rings:

THEOREM 14.14. *If E is an elliptic curve over $\mathbb{F}_q$, then the endomorphism ring $\mathrm{End}_{\overline{\mathbb{F}}_q}(E)$ is either an order in an imaginary quadratic field or an order in a quaternion algebra.*

REMARKS.

(i) We say that E is *ordinary* in the former case and *supersingular* in the latter.

(ii) Notice that for elliptic curves over a finite field K, $\mathrm{End}_{\overline{K}}(E)$ is *always* larger than $\mathbb{Z}$.

PROOF. See Silverman [**118**, Chapter V, Theorem 3.1]. Q.E.D.

There are many known criteria for E to be supersingular (see Husemöller [**74**, p. 258] for an exhaustive list). Over a prime field $\mathbb{F}_p$, there is a special criterion which will be useful later on:

PROPOSITION 14.15. *Let E be an elliptic curve over $\mathbb{F}_p$. If $p > 3$, then E is supersingular if and only if*

$$|E(\mathbb{F}_p)| = p + 1.$$

PROOF. See Silverman [**118**, Chapter V, Exercise 5.10]. Q.E.D.

It is interesting to note that $|E(\mathbb{F}_p)| = p+1$ is exactly in the center of the range $p+1-2\sqrt{p} \leq |E(\mathbb{F}_p)| \leq p+1+2\sqrt{p}$ allowed by Hasse's theorem.

From the point of view of endomorphisms, ordinary elliptic curves over finite fields behave like elliptic curves over $\mathbb{C}$ with complex multiplication, since in each case, the endomorphism ring is an order in an imaginary quadratic field. This suggests a deeper relation between these two classes, which leads to our next topic, reduction of elliptic curves.

The basic idea of reduction is the following. Let K be a number field, and let E be an elliptic curve defined by

$$y^2 = 4x^3 - g_2x - g_3, \qquad g_2, g_3 \in K.$$

If $\mathfrak{p}$ is prime in $\mathcal{O}_K$, we want to "reduce" E modulo $\mathfrak{p}$. This can't be done in general, but suppose that g_2 and g_3 can be written in the form α/β, where $\alpha, \beta \in \mathcal{O}_K$ and $\beta \notin \mathfrak{p}$. Then we can define $[g_2]$ and $[g_3]$ in $\mathcal{O}_K/\mathfrak{p}$. If, in addition, we have

$$\Delta = [g_2]^3 - 27[g_3]^2 \neq 0 \in \mathcal{O}_K/\mathfrak{p},$$

then

$$y^2 = 4x^3 - [g_2]x - [g_3]$$

is an elliptic curve $\overline{E}$ over the finite field $\mathcal{O}_K/\mathfrak{p}$. In this case we call $\overline{E}$ the *reduction* of E modulo $\mathfrak{p}$, and we say that E has *good reduction* modulo $\mathfrak{p}$.

When E has complex multiplication and good reduction, Deuring, drawing on examples of Gauss, discovered an astonishing relation between the complex multiplication of E and the number of points in $\overline{E}(\mathcal{O}_K/\mathfrak{p})$. Rather than state his result in its full generality, we will present a version that concerns only elliptic curves over the prime field $\mathbb{F}_p$.

To set up the situation, let $\mathcal{O}$ be an order in an imaginary quadratic field K, and let L be the ring class field of $\mathcal{O}$. Let p be a prime in $\mathbb{Z}$ which splits completely in L, and we will fix a prime $\mathfrak{P}$ of L lying above p, so that $\mathcal{O}_L/\mathfrak{P} \simeq \mathbb{F}_p$. Finally, let E be an elliptic curve over L which has good reduction at $\mathfrak{P}$. With these hypotheses, the reduction $\overline{E}$ is an elliptic curve over $\mathbb{F}_p$. Then we have the following theorem:

THEOREM 14.16. *Let $\mathcal{O}$, L, p and $\mathfrak{P}$ be as above, and let E be an elliptic curve over L with $\mathrm{End}_{\mathbb{C}}(E) = \mathcal{O}$. If E has good reduction modulo $\mathfrak{P}$, then there is $\pi \in \mathcal{O}$ such that $p = \pi\overline{\pi}$ and*

$$|\overline{E}(\mathbb{F}_p)| = p + 1 - (\pi + \overline{\pi}).$$

Furthermore, $\mathrm{End}_{\overline{\mathbb{F}}_p}(\overline{E}) = \mathcal{O}$, and every elliptic curve over $\mathbb{F}_p$ with endomorphism ring (over $\overline{\mathbb{F}}_p$) equal to $\mathcal{O}$ arises in this way.

PROOF. The basic idea is that when the above hypotheses are fulfilled, reduction induces an isomorphism

$$\mathrm{End}_{\mathbb{C}}(E) \xrightarrow{\sim} \mathrm{End}_{\overline{\mathbb{F}}_p}(\overline{E})$$

that preserves degrees. The proof of this fact is well beyond the scope of this book (see Lang [**90**, Chapter 13, Theorem 12]).

From the above isomorphism, it follows that there is some prime $\pi \in \mathrm{End}_{\mathbb{C}}(E)$ which reduces to $Frob_p \in \mathrm{End}_{\overline{\mathbb{F}}_p}(\overline{E})$. Since $Frob_p$ has degree p, so does π. Over the complex numbers, we know that the degree of $\pi \in \mathcal{O} = \mathrm{End}_{\mathbb{C}}(E)$ is just its norm, so that $N(\pi) = p$. Thus we can write $p = \pi\overline{\pi}$ in $\mathcal{O}$.

It is now trivial to compute the number of points on $\overline{E}$. As we noted in (14.13),

$$|\overline{E}(\mathbb{F}_p)| = \deg(1 - Frob_p).$$

Since the reduction map preserves degrees, it follows that

$$\begin{aligned}\deg(1 - Frob_p) &= \deg(1-\pi) = N(1-\pi) = (1-\pi)(1-\overline{\pi})\\ &= p + 1 - (\pi + \overline{\pi})\end{aligned}$$

since $p = \pi\overline{\pi}$. This proves the desired formula for $|\overline{E}(\mathbb{F}_p)|$.

See Rubin and Silverberg [**107**, Lemma 8.1] for a proof of the final part of the theorem. Q.E.D.

The remarkable fact is that we've already seen two examples of this theorem. First, in (4.24), we stated the following result of Gauss: if $p \equiv 1 \bmod 3$ is prime, then

$$\text{(14.17)} \qquad \begin{array}{l}\text{If } 4p = a^2 + 27b^2 \text{ and } a \equiv 1 \bmod 3, \text{ then } N = p + a - 2, \text{ where}\\ N \text{ is the number of solutions modulo } p \text{ of } x^3 - y^3 \equiv 1 \bmod p.\end{array}$$

We can relate this to Deuring's theorem as follows. The coordinate change $(x, y) \mapsto (3x/(1+y), 9(1-y)/(1+y))$ transforms the curve $x^3 = y^3 + 1$ into the elliptic curve E defined by $y^2 = 4x^3 - 27$ (see Exercise 14.13). Gauss didn't count the three points at infinity that lie on $x^3 = y^3 + 1$, and when these are taken into account, then (14.17) asserts that $|E(\mathbb{F}_p)| = p + 1 + a$. Since $p \equiv 1 \bmod 3$, we can write $p = \pi\overline{\pi}$ in $\mathbb{Z}[\omega]$, $\omega = e^{2\pi i/3}$. In §4, we saw that π may be chosen to be primary, which means $\pi \equiv \pm 1 \bmod 3$. Thus we may assume $\pi \equiv 1 \bmod 3$, so that $\pi = A + 3B\omega$, where $A \equiv 1 \bmod p$. Then an easy calculation shows that

$$4p = (-(2A - 3B))^2 + 27B^2.$$

Since $2A - 3B = \pi + \overline{\pi}$ and $-(2A - 3B) \equiv 1 \bmod 3$, it follows that (14.17) may be stated as follows:

$$\text{If } p = \pi\overline{\pi} \text{ in } \mathbb{Z}[\omega] \text{ and } \pi \equiv 1 \bmod 3, \text{ then } |E(\mathbb{F}_p)| = p + 1 - (\pi + \overline{\pi}).$$

Since E is the reduction of $y^2 = 4x^3 - 27$, which has complex multiplication by $\mathbb{Z}[\omega]$ (see Exercise 14.13), Gauss' observation (14.17) really is a special case of Deuring's theorem.

Similarly, one can check that Gauss' last diary entry, which concerned the number of solutions of $x^2 + y^2 + x^2y^2 \equiv 1 \bmod p$, is also a special case of Deuring's theorem. See the discussion following (4.24) and Exercise 14.14.

As an application of Deuring's theorem, we can give a formula for the number of elliptic curves over $\mathbb{F}_p$ which have a preassigned number of points. We first need some notation. Given an order $\mathcal{O}$ in an imaginary quadratic field K, we define the *Hurwitz class number* $H(\mathcal{O})$ to be the weighted sum of class numbers

$$H(\mathcal{O}) = \sum_{\mathcal{O} \subset \mathcal{O}' \subset \mathcal{O}_K} \frac{2}{|\mathcal{O}'^*|} h(\mathcal{O}')$$

We also write $H(\mathcal{O})$ as $H(D)$, where D is the discriminant of $\mathcal{O}$. Then we have the following theorem of Deuring:

THEOREM 14.18. *Let $p > 3$ be prime, and let $N = p + 1 - a$ be an integer, where $-2\sqrt{p} \le a \le 2\sqrt{p}$. Then the number of elliptic curves E over $\mathbb{F}_p$ which have $|E(\mathbb{F}_p)| = p + 1 - a$ is*

$$\frac{p-1}{2} H(a^2 - 4p).$$

PROOF. Let π be a root of $x^2 - ax + p$. Since $-2\sqrt{p} \le a \le 2\sqrt{p}$, the quadratic formula shows that $\mathcal{O}_a = \mathbb{Z}[\pi]$ is an order in an imaginary quadratic field K. One can also check that $\mathcal{O}_a$ has discriminant $a^2 - 4p$ and p doesn't divide the conductor of $\mathcal{O}_a$ (in fact, it doesn't divide the discriminant when $a \neq 0$), and hence the same is true for any order $\mathcal{O}'$ containing $\mathcal{O}_a$ (see Exercise 14.15).

We will start with the case $a \neq 0$, which by Proposition 14.15 means that all of the elliptic curves involved are ordinary. Given an order $\mathcal{O}'$ containing $\mathcal{O}_a$ and a proper $\mathcal{O}'$-ideal $\mathfrak{a}$, we will produce a collection of elliptic curves E_c with good reduction modulo p. Namely, let L' be the ring class field of $\mathcal{O}'$. Since $p = \pi\overline{\pi}$ in $\mathcal{O}_a \subset \mathcal{O}'$, it follows from Theorem 9.4 and Exercise 9.3 that p splits completely in L'. Thus, if $\mathfrak{P}$ is any prime of L' containing p, then $\mathcal{O}_{L'}/\mathfrak{P} \simeq \mathbb{F}_p$.

First, assume that $\mathcal{O}' \neq \mathbb{Z}(i)$ or $\mathbb{Z}[\omega]$, $\omega = e^{2\pi i/3}$, so that $j(\mathfrak{a}) \neq 0, 1728$. Let

$$k = \frac{27 j(\mathfrak{a})}{j(\mathfrak{a}) - 1728}$$

and define the collection of elliptic curves E_c over L' by the Weierstrass equations

$$y^2 = 4x^3 - kc^2 x - kc^3,$$

where $c \in \mathcal{O}_{L'} - \mathfrak{P}$ is arbitrary. A computation shows that $j(E_c) = j(\mathfrak{a})$. We can reduce k modulo L provided that $j(\mathfrak{a}) - 1728 \notin \mathfrak{P}$. Since $1728 = j(i)$, Theorem 13.21 implies that

$$j(\mathfrak{a}) = 1728 \bmod \mathfrak{P} \implies p \text{ does not split in } K \text{ or } \mathbb{Q}(i)$$

(when $K = \mathbb{Q}(i)$, note that the conductor condition of Theorem 13.21 is satisfied). However, p splits in K, and thus $j(\mathfrak{a}) - 1728 \notin \mathfrak{P}$, as desired.

Then one computes that in $\mathcal{O}_{L'}/\mathfrak{P} \simeq \mathbb{F}_p$,

$$\Delta = [kc^2]^3 - 27[kc^3]^2 = 1728[c^6]\left[\frac{27^3 j(\mathfrak{a})^2}{(j(\mathfrak{a}) - 1728)^3}\right].$$

By the argument used to prove $j(\mathfrak{a}) - 1728 \notin \mathfrak{P}$, Theorem 13.21 and $j(\omega) = 0$ show that $j(\mathfrak{a}) \notin \mathfrak{P}$. It follows that E_c has good reduction modulo $\mathfrak{P}$ since $c \notin \mathfrak{P}$.

If $\mathcal{O}' = \mathbb{Z}[i]$ or $\mathbb{Z}[\omega]$, then $L' = K$. Here, we will use the collection of elliptic curves E_c defined by

$$\begin{aligned} y^2 &= 4x^3 - cx, \quad & c &\notin \pi\mathbb{Z}[i] \\ y^2 &= 4x^3 - c, \quad & c &\notin \pi\mathbb{Z}[\omega]. \end{aligned}$$

One easily checks that these curves have good reduction modulo π and $\mathcal{O}'$ as their endomorphism ring.

Theorem 14.16 assures us that every ordinary elliptic curve $\overline{E}$ over $\mathbb{F}_p$ arises from reduction of some elliptic curve with complex multiplication. Given this, it follows without difficulty that $\overline{E}$ is in fact the reduction of one of the E_c's constructed above (see Exercise 14.16).

Given $\mathcal{O}'$, there are $h(\mathcal{O}')$ distinct j-invariants $j(\mathfrak{a})$, and hence for a fixed a, we have

$$\sum_{\mathcal{O}_a \subset \mathcal{O}'} h(\mathcal{O}')$$

distinct collections of elliptic curves E_c. Furthermore, another application of Theorem 13.21 shows that different collections reduce to elliptic curves with different j-invariants. Since each collection E_c gives us $p-1$ curves over $\mathbb{F}_p$, we get

$$(p-1)\sum_{\mathcal{O}_a \subset \mathcal{O}'} h(\mathcal{O}')$$

elliptic curves over $\mathbb{F}_p$. But which of these have $p+1-a$ points on them? The problem is that Theorem 14.16 implies that $|\overline{E}_c(\mathbb{F}_p)|$ is determined by *some* element of $\mathcal{O}'$ of norm p, but it need not be π! All curves in a given collection have the same j-invariant, but they need not be isomorphic over $\mathbb{F}_p$, and hence they may have different numbers of points. In fact, this is always the case:

PROPOSITION 14.19. *Let E and E' be elliptic curves over $\mathbb{F}_p$. If E is ordinary, then E and E' are isomorphic over $\mathbb{F}_p$ if and only if $j(E) = j(E')$ and $|E(\mathbb{F}_p)| = |E'(\mathbb{F}_p)|$.*

PROOF. One direction of the proof is obvious, but the other requires some more advanced concepts. We will give the details since this result doesn't appear in standard references. The key ingredient is a theorem of Tate, which asserts that curves with the same number of points over a finite field K are isogenous over K (see Husemöller [**74**, §13.8]). Applying this to $|E(\mathbb{F}_p)| = |E'(\mathbb{F}_p)|$, we get an isogeny $\lambda : E \to E'$ defined over $\mathbb{F}_p$. Since E and E' have the same j-invariant, we can also find an isomorphism $\phi : E' \to E$ defined over some extension $\mathbb{F}_{p^a}$ (see Proposition 14.5). Thus $\phi \circ \lambda \in \mathrm{End}_{\overline{\mathbb{F}}_p}(E)$, which is commutative since E is ordinary. It follows that $Frob_p \circ (\phi \circ \lambda) = (\phi \circ \lambda) \circ Frob_p$, so that $\phi \circ \lambda$ is defined over $\mathbb{F}_p$. Then, given $\sigma \in \mathrm{Gal}(\overline{\mathbb{F}}_p/\mathbb{F}_p)$, we have

$$\phi^\sigma \circ \lambda = \phi^\sigma \circ \lambda^\sigma = (\phi \circ \lambda)^\sigma = \phi \circ \lambda,$$

where the last equality holds since $\phi \circ \lambda$ is defined over $\mathbb{F}_p$. Since isogenies are surjective over $\overline{\mathbb{F}}_p$, it follows easily that $\phi^\sigma = \phi$. This is true for all $\sigma \in \mathrm{Gal}(\overline{\mathbb{F}}_p/\mathbb{F}_p)$, which implies that the isomorphism $\phi : E' \to E$ is defined over $\mathbb{F}_p$. Q.E.D.

We claim that the collection $\overline{E}_c$ contains $(p-1)/|\mathcal{O}'^*|$ curves with $p+1-a$ points. This will immediately imply our desired formula. Let's first consider the case when $\overline{E}_c$ corresponds to a j-invariant $j(\mathfrak{a}) \neq 0$ or 1728. Here, the only solutions of $N(\alpha) = p$ in $\mathcal{O}'$ are $\alpha = \pm\pi$ and $\pm\overline{\pi}$ (see Exercise 14.17). Thus, for each c, Deuring's theorem tells us that

$$|\overline{E}_c(\mathbb{F}_p)| = p + 1 \pm a.$$

The curves $\overline{E}_c$ fall into two isomorphism classes, each consisting of $(p-1)/2$ curves, corresponding to whether $[c] \in (\mathbb{F}_p^*)^2$ or not (see Exercise 14.18). By the above proposition, nonisomorphic curves have a different number of elements, and hence we see that exactly half of the E_c's have $p+1-a$ elements. Since $\mathcal{O}'^* = \{\pm 1\}$, we get $(p-1)/2 = (p-1)/|\mathcal{O}'^*|$ curves with $p+1-a$ points.

When $j(\mathfrak{a}) = 1728$, things are more complicated. Here, $\mathcal{O}' = \mathbb{Z}[i]$, and $p = \pi\overline{\pi}$ implies that $p \equiv 1 \bmod 4$. The only solutions of $N(\alpha) = p$ are $\alpha = \pm\pi, \pm\overline{\pi}, \pm i\pi$, and

$\pm i\overline{\pi}$ (see Exercise 14.17), and thus there are at most four possibilities for $|\overline{E}_c(\mathbb{F}_p)|$. But there are four isomorphism classes of curves with $j = 1728$ in this case, each consisting of $(p-1)/4$ curves (see Exercise 14.18). It follows that there are exactly $(p-1)/|\mathcal{O}'^*|$ curves with $p+1-a$ points. The case $j = 0$ is similar and is left to the reader.

It remains to study the case $a = 0$, which concerns the number of supersingular curves over $\mathbb{F}_p$. Since Theorem 14.16 doesn't apply to this case, we will take a more indirect approach. Given any a in the range $-2\sqrt{p} \leq a \leq 2\sqrt{p}$, we just proved that when $a \neq 0$, there are $((p-1)/2)H(a^2-4p)$ elliptic curves over $\mathbb{F}_p$ with $p+1-a$ points. Let SS denote the number of supersingular curves. Since there are $p(p-1)$ elliptic curves over $\mathbb{F}_p$ (see Exercise 14.19), it follows that

$$p(p-1) = SS + \sum_{0<|a|\leq 2\sqrt{p}} \frac{p-1}{2} H(a^2-4p). \tag{14.20}$$

However, we claim that there is a class number formula

$$2p = \sum_{0\leq|a|\leq 2\sqrt{p}} H(a^2-4p). \tag{14.21}$$

Since (14.20) and (14.21) imply that $SS = ((p-1)/2)H(-4p)$, we need only prove the formula (14.21).

To prove this, note that $H(a^2-4p) = H(\mathcal{O}_a)$, so that by definition, the right-hand side of (14.21) equals

$$\sum_{0\leq|a|\leq 2\sqrt{p}} \sum_{\mathcal{O}_a\subset\mathcal{O}'} \frac{2}{|\mathcal{O}'^*|} h(\mathcal{O}').$$

If we define the function $\chi(a)$ by

$$\chi(a) = \begin{cases} 1 & \text{if } \mathcal{O}_a \subset \mathcal{O}' \\ 0 & \text{otherwise,} \end{cases}$$

then the above sum can be written as

$$\sum_{\mathcal{O}'} \left(\frac{2}{|\mathcal{O}'^*|} \sum_{0\leq|a|\leq 2\sqrt{p}} \chi(a) \right) h(\mathcal{O}').$$

It is easy to prove that the quantity in parentheses is $r(\mathcal{O}', p)$, which we defined in §13 to be $|\{\pi \in \mathcal{O}' : N(\pi) = p\}/\mathcal{O}'^*|$ (see Exercise 14.20). Thus the right-hand side of (14.21) becomes

$$\sum_{\mathcal{O}'} r(\mathcal{O}', p)\, h(\mathcal{O}').$$

In Corollary 13.9 we proved that the above sum equals $2p$, and (14.21) is proved. This completes the proof of Theorem 14.18. Q.E.D.

Recall that Corollary 13.9 was part of our study of the polynomial $\Phi_m(X, X)$. It is rather unexpected that the modular equation has a connection with supersingular curves over $\mathbb{F}_p$. This is just more evidence of the amazing richness of the study of elliptic curves. To pursue these topics further, the reader should consult Lang [**90**] and Shimura [**115**]. Also, see the monographs by Cassou–Noguès and Taylor [**20**] and by Gross [**60**] for an introduction to some interesting research questions concerning elliptic curves and complex multiplication.

In some cases, the connection between $\Phi_n(X, X)$ and supersingular curves can be made more explicit by using the class equation and the polynomial $\mathsf{ss}_p(x) \in \mathbb{F}_p[x]$ whose roots are the j-invariants of supersingular curves in characteristic p—this is the *supersingular polynomial.* For simplicity, suppose that $p \equiv 1 \bmod 12$. Then the class equation $H_{-4p}(x)$ is related to $\mathsf{ss}_p(x)$ via the congruence

$$H_{-4p}(x) \equiv \big(\gcd(x \cdot \mathsf{ss}_p(x), (x - 1728)^{(p-1)/2} + 1)\big)^2 \bmod p.$$

See Brillhart and Morton [**14**, Proposition 11], which also gives congruences for other values of p mod 12. These congruences are related to Elkies' 1987 proof that every elliptic curve over $\mathbb{Q}$ has supersingular reduction at infinitely many primes.

D. Elliptic Curve Primality Tests

In the latter part of the twentieth century, some surprising applications of elliptic curves to problems involving factoring and primality were discovered. In 1985, Lenstra announced an elliptic curve factoring method [**94**], and a year later, Goldwasser and Kilian adapted Lenstra's method to obtain an elliptic curve primality test [**55**]. Both methods use the properties of elliptic curves over finite fields. We will concentrate on the Goldwasser–Kilian Test and its variation, the Goldwasser–Kilian–Atkin Test. This last test is especially interesting, for it uses the class equations from §13. Thus, the polynomial $H_{-4n}(X)$, which appears in our criterion for when p is of the form $x^2 + ny^2$, can actually be used to prove that p is prime! Our treatment of these tests will not be complete, and for further details, we refer the reader to the articles by Goldwasser and Kilian [**55**], Lenstra [**94**] and Morain [**99**]. See also the 1993 article [**3**] by Atkin and Morain for a definitive presentation.

Given a potential prime l, the goal of these tests is to prove the primality of l by considering elliptic curves over the field $\mathbb{Z}/l\mathbb{Z}$. Since we don't know that l is prime, we must treat $\mathbb{Z}/l\mathbb{Z}$ as a ring, and thus we need a theory of elliptic curves over *rings*. Fortunately, the basic ideas carry over quite easily. Let R be any commutative ring with identity where 2 and 3 are units. Then an *elliptic curve E over R* is a Weierstrass equation of the usual form

$$y^2 = 4x^3 - g_2x - g_3, \qquad g_2, g_3 \in R,$$

where we now require that

$$\Delta = g_2^3 - 27g_3^2 \in R^*. \tag{14.22}$$

Note that since Δ is a unit in R, the j-invariant

$$j(E) = 1728\frac{g_2^3}{\Delta} \in R$$

is defined.

Given an elliptic curve E over R, we set

$$E_0(R) = \{(x, y) \in R \times R : y^2 = 4x^3 - g_2x - g_3\} \cup \{\infty\}.$$

The reason for the new notation is that $E_0(R)$ may *fail* to be a group! To see this, consider $P_1 = (x_1, y_2)$ and $P_2 = (x_2, y_2)$ in $E_0(R)$. If $x_1 \neq x_2$, then we would like to define

$$P_1 + P_2 = (x_3, y_3),$$

where x_3 and y_3 are given by the formulas (14.6). The problem comes from the denominator $x_1 - x_2$: it is nonzero in R, but it need *not* be invertible! For this reason, the binary operation is only partially defined on $E_0(R)$. Using tools from algebraic geometry, one can define a superset $E(R)$ of $E_0(R)$ which is a group, but we prefer to use $E_0(R)$ because it is easier to work with in practice.

If E is an elliptic curve over $\mathbb{Z}/l\mathbb{Z}$, the potentially incomplete group structure on $E_0(\mathbb{Z}/l\mathbb{Z})$ is not a problem. Namely, if we ever found P_1 and P_2 in $E_0(\mathbb{Z}/l\mathbb{Z})$ such that $P_1 + P_2$ wasn't defined, then it would follow automatically that l must be composite, and the noninvertible denominator would give us a factor of l (just compute the appropriate gcd). This observation is the driving force of Lenstra's elliptic curve factoring algorithm (see [**94**]).

Before discussing the Goldwasser–Kilian Test, let's review some basic ideas concerning primality testing. We regard l as an input of length $[\log_{10} l]$, where $[\ ]$ is the greatest integer function. The length is thus bounded by a constant times $\log l = \log_e l$, which we express by writing $[\log_{10} l] = O(\log l)$. The most interesting question concerning a primality test is its running time: given an input l, how long, as a function of $\log l$, does it take a given algorithm to prove that l is (or is not) prime? The simplest algorithm (divide by all numbers $\leq \sqrt{l}$) requires

$$\sqrt{l} = e^{(1/2)\log l}$$

divisions, and hence runs in *exponential* time. What we really want is an algorithm that runs in *polynomial* time, i.e., where the running time is $O((\log l)^d)$ for some fixed d. A polynomial time algorithm was discovered in 2002 by Agrawal, Kayal and Saxena, and a modified version of their algorithm due to Lenstra and Pomerance [**95**] has a running time of $O\big((\log l)^6(\log\log l)^c\big)$ for some computable constant c.

Another sort of algorithm commonly used is what is called a *probabilistic primality test*. Such a test has two outputs, "prime" and "composite or unluckily prime." In the former case, the algorithm proves the primeness of l, while in the latter case, it says either that l is composite or that l is prime and we were unlucky. A nice discussion of probabilistic primality tests may be found in Wagon's article [**128**]. For our purposes, we will explain this concept by considering the following very special probabilistic primality test.

Let l be our potential prime, relatively prime to 6, and suppose that we have an elliptic curve E over $\mathbb{Z}/l\mathbb{Z}$ over with the following two properties:

(i) $l + 1 - 2\sqrt{l} \leq |E_0(\mathbb{Z}/l\mathbb{Z})| \leq l + 1 + 2\sqrt{l}$.
(ii) $|E_0(\mathbb{Z}/l\mathbb{Z})| = 2q$, where q is an odd prime.

In certain situations, this setup can be used to prove primality:

LEMMA 14.23. *Let l and E be as above, and assume $l > 33$. Let $P \neq \infty$ be in $E_0(\mathbb{Z}/l\mathbb{Z})$. If qP is defined and equal to ∞ in $E_0(\mathbb{Z}/l\mathbb{Z})$, then l is prime.*

PROOF. Assume that l is not prime, and let $p \leq \sqrt{l}$ be a prime divisor of l. Using the natural map $\mathbb{Z}/l\mathbb{Z} \to \mathbb{Z}/p\mathbb{Z} = \mathbb{F}_p$, we can reduce the equation of E modulo p, and by (14.22), we get an elliptic curve $\overline{E}$ over $\mathbb{F}_p$. Furthermore, we get a natural map

$$E_0(\mathbb{Z}/l\mathbb{Z}) \longrightarrow \overline{E}(\mathbb{F}_p)$$

which takes $P = (x, y) \neq \infty$ in $E_0(\mathbb{Z}/l\mathbb{Z})$ to $\overline{P} = (\overline{x}, \overline{y}) \neq \infty$ in $\overline{E}(\mathbb{F}_p)$. Since this map is also clearly a homomorphism (wherever defined), it follows that $q\overline{P} = \infty$ in $\overline{E}(\mathbb{F}_p)$. But q is prime, so that $\overline{P}$ is a point of order q, and hence

$$q \leq |\overline{E}(\mathbb{F}_p)| \leq p + 1 + 2\sqrt{p},$$

where the second inequality comes from Hasse's theorem (Theorem 14.12). Since $p \leq \sqrt{l}$, this implies that

$$q \leq \sqrt{l} + 1 + 2\sqrt[4]{l} = (\sqrt[4]{l} + 1)^2.$$

However, by assumption, we have

$$2q = |E_0(\mathbb{Z}/l\mathbb{Z})| \geq l + 1 - 2\sqrt{l} = (\sqrt{l} - 1)^2.$$

Combining these two inequalities, we obtain

$$\sqrt{l} - 1 \leq \sqrt{2}(\sqrt[4]{l} + 1),$$

which is easily seen to be impossible for $l > 33$. This contradiction proves the lemma. Q.E.D.

To convert this lemma into a probabilistic primality test, we need one more observation. Namely, if l is prime and $|E_0(\mathbb{Z}/l\mathbb{Z})| = 2q$, q an odd prime, then $E_0(\mathbb{Z}/l\mathbb{Z})$ must be a cyclic group, and hence exactly $q-1$ of the $2q-1$ nonidentity elements have order q. Thus, the probability that a randomly chosen $P \neq \infty$ doesn't prove primality (i.e., has order $\neq q$) is $q/(2q-1) \approx 1/2$, assuming that q is large.

Now we can state the test. Given E and l be as above, pick k randomly chosen points $P_1, \ldots, P_k$ from $E_0(\mathbb{Z}/l\mathbb{Z})$, and then compute $qP_1, \ldots, qP_k$. If any one of these is defined and equals ∞, then by the above lemma, we have a proof of primality. If none of $qP_1, \ldots, qP_k$ satisfy this condition, then either l is composite, or l is prime and we were unlucky. To see how unlucky, suppose that l were prime. Then our test fails only if all of $P_1, \ldots, P_k$ have order $\neq q$. By the above paragraph, the probability of this happening is

$$\left(\frac{q}{2q-1}\right)^k \approx \frac{1}{2^k}.$$

So we can't guarantee a proof of primality, but we have to be mighty unlucky not to find one.

This test depended on the assumptions (i) and (ii) above. The first assumption is quite reasonable, since by Hasse's theorem it holds if l is prime. So if (i) fails, we have a proof of compositeness. But the second assumption, that $|E_0(\mathbb{Z}/l\mathbb{Z})|$ is twice a prime, is very special, and certainly fails for most elliptic curves. An added difficulty is that $|E_0(\mathbb{Z}/l\mathbb{Z})|$ is a very large number (by (i), it has the same order of magnitude as l). Thus, even if $|E_0(\mathbb{Z}/l\mathbb{Z})| = 2q$ were twice a prime, we'd be unlikely to know it, since we'd have to prove that q, a number roughly the size of $l/2$, is also prime.

To overcome these problems, Goldwasser and Kilian used two ideas. The first idea is quite simple:

(14.24)
Choose *lots* of elliptic curves E over $\mathbb{Z}/l\mathbb{Z}$ at random.
If we get one where $|E_0(\mathbb{Z}/l\mathbb{Z})| = 2q$, q a probable prime,
then use the above special test to check for primality.

Notice the word "probable." Using known probabilistic compositeness tests (described in Wagon [**128**]), one can efficiently reduce to the case where $|E_0(\mathbb{Z}/l\mathbb{Z})|$

is of the form $2q$, where q is *probably* prime. If the special test succeeds, we have proved that l is prime, provided that q is prime. Then the second idea is

(14.25) Make the above process recursive.

This means proving q is prime by applying the special test to an elliptic curve over $\mathbb{Z}/q\mathbb{Z}$ of order $2q'$, q' a probable prime. In this way the primality of q' implies the primality of q. Since each iteration reduces the size by a factor of 2 (i.e., q is about the size of $l/2$, q' is about the size of $q/2$, etc.), it follows that in $O(\log l)$ steps the numbers will get small enough that primality can be verified easily.

The algorithm contained in (14.24) and (14.25) is the heart of the Goldwasser–Kilian primality test (see their article [**55**] for a fuller discussion). The key unanswered question concerns (14.24): when l is prime, how many elliptic curves do we have to choose before finding one where $|E(\mathbb{Z}/l\mathbb{Z})|$ is twice a prime? The following result of Lenstra plays a crucial role:

THEOREM 14.26. *Let l be a prime, and let*

$$S = \{2q : q \text{ prime},\ l+1-\sqrt{l} \le 2q \le l+1+\sqrt{l}\}.$$

Then there is a constant $c_1 > 0$, independent of l and S, such that the number of elliptic curves E over $\mathbb{F}_l$ satisfying $|E(\mathbb{F}_l)| \in S$ is at least

$$c_1 \cdot \frac{(|S|-2)\sqrt{l}(l-1)}{\log l}.$$

REMARK. Notice that the elliptic curves described in this theorem satisfy

$$l+1-\sqrt{l} \le |E(\mathbb{F}_l)| \le l+1+\sqrt{l},$$

which is more restrictive than the bound given by Hasse's theorem. The proof below will explain the reason for this.

PROOF. Given $2q \in S$, write $2q = l+1-a$. Then we proved in Theorem 14.18 that the number of curves with $2q = l+1-a$ points is $((l-1)/2)H(a^2-4l)$, where $H(a^2-4l)$ is the Hurwitz class number defined earlier. Using classically known bounds on class numbers, Lenstrat proved in [**94**] that for $2q \in S$, with at most two exceptions, there is the estimate

$$H(a^2-4l) \ge c \cdot \frac{\sqrt{|a^2-4l|}}{\log l}$$

where c is a constant independent of the discriminant (see [**94**, Proposition 1.8]). We are assuming that $|a| \le \sqrt{l}$, which implies $\sqrt{|a^2-4l|} \ge \sqrt{3l}$, and consequently

$$\frac{l-1}{2}H(a^2-4l) \ge c_1 \cdot \frac{\sqrt{l}(l-1)}{\log l},$$

where $c_1 = \sqrt{3}c/2$. The theorem follows immediately. Q.E.D.

By this theorem, we are reduced to knowing the number of primes in the interval $[(l+1)/2-\sqrt{l}/2, (l+1)/2+\sqrt{l}/2]$. By the Prime Number Theorem, the probability that a number in the interval $[0, N]$ is prime is $1/\log N$. It is conjectured that this holds for intervals of shorter length. Applied to the above, we get the following conjecture:

CONJECTURE 14.27. *There is a constant $c_2 > 0$ such that, for all sufficiently large primes l, the number of primes in the interval $[(l+1)/2-\sqrt{l}/2,\ (l+1)/2+\sqrt{l}/2]$ is at least*

$$c_2 \cdot \frac{\sqrt{l}}{\log l}.$$

If this conjecture were true, then Theorem 14.26 would imply that when l is large, there is a constant c_3, independent of l, such that at least

$$c_3 \cdot \frac{l(l-1)}{(\log l)^2}$$

elliptic curves E over $\mathbb{Z}/l\mathbb{Z}$ have order $|E(\mathbb{F}_l)| = 2q$ for some prime q (see Exercise 14.21). Since there are $l(l-1)$ elliptic curves over $\mathbb{F}_l$, it follows that there is a probability of at least

$$c_3/(\log l)^2 \tag{14.28}$$

that $|E(\mathbb{F}_l)|$ has the desired order.

Now we can explain how many curves need to be chosen in (14.24). Namely, pick an integer k, and pick $k(\log l)^2/c_3$ randomly chosen elliptic curves over $\mathbb{Z}/l\mathbb{Z}$. If l were prime, could all these curves fail to have order twice a prime? By (14.28), the probability of this happening is less than

$$\left(1 - \frac{c_3}{(\log l)^2}\right)^{k(\log l)^2/c_3} \approx \frac{1}{e^k}.$$

It remains to give a run time analysis of the Goldwasser–Kilian Test. For (14.24), we need to pick $O((\log l)^2)$ curves and count the number of points on each one. By an algorithm of Schoof (see Morain [**99**, §5.5]), it takes $O((\log l)^8)$ to count the points on each curve. Once a curve with $|E_0(\mathbb{Z}/l\mathbb{Z})| = 2q$ is found, we then need to pick points $P \in E_0(\mathbb{Z}/l\mathbb{Z})$ and compute qP. These operations are bounded by $O((\log l)^8)$ (see Goldwasser and Kilian [**55**, §4.3]), and thus the run time of (14.24) is $O((\log l)^{10})$. By (14.25), we have to iterate this $O(\log l)$ times, so that the run time of the whole algorithm is $O((\log l)^{11})$.

The above analysis is predicated on Conjecture 14.27, which may be very difficult to prove (or even false!). But now comes the final ingredient: using known results about the distribution of primes, Goldwasser and Kilian were able to prove that their algorithm terminates with a run time of $O(k^{11})$ for at least

$$(1 - O(2^{-k^{1/\log\log k}})) \times 100\%$$

of the prime inputs of length k (see [**55**, Theorem 3]). Thus the Goldwasser–Kilian Test is *almost* a polynomial time probabilistic primality test!

In practice, the implementation of the Goldwasser–Kilian Test is more complicated than the algorithm sketched above. The main difference is that the order $|E_0(\mathbb{Z}/l\mathbb{Z})|$ is allowed to be of the form mq, where m may be bigger than 2 but is still small compared to q. This means that fewer elliptic curves must be tried before finding a suitable one, and thus the algorithm runs faster. For the details of how this is done, see Goldwasser and Kilian [**55**, §4.4] or Morain [**99**, §§2.2.2 and 7.7].

The most "expensive" part of the Goldwasser–Kilian Test is the $O((\log l)^8)$ spent counting the points on a given elliptic curve. So rather than starting with E and then computing $|E_0(\mathbb{Z}/l\mathbb{Z})|$ the hard way, why not use the theory developed earlier to *predict* the order? This is the basis of the Goldwasser–Kilian–Atkin Test, which we will discuss next.

The wonderful thing about this test is that it brings us back to our topic of primes of the form $x^2 + ny^2$. To see why, let l be a prime, and let n be a positive integer such that l can be written as

$$l = a^2 + nb^2, \qquad a, b \in \mathbb{Z}.$$

We will use this information to produce an elliptic curve over $\mathbb{F}_l$ with $l + 1 - 2a$ points on it. The basic idea is to use the characterization of primes of the form $x^2 + ny^2$ proved in §13:

$$l = x^2 + ny^2 \Longrightarrow \begin{cases} (-n/l) = 1 \text{ and } H_{-4n}(X) \equiv 0 \bmod l \\ \text{has an integer solution,} \end{cases}$$

where $H_{-4n}(X)$ is the class equation for discriminant $-4n$ (see Theorem 13.23). Thus $l = a^2 + nb^2$ gives us a solution j of the congruence $H_{-4n}(X) \equiv 0 \bmod p$, and for simplicity, we will suppose that $j \not\equiv 0, 1728 \bmod l$. Define $k \in \mathbb{F}_l$ to be the congruence class

$$k = \left[\frac{27j}{j - 1728}\right]$$

and then consider the two elliptic curves

$$\begin{aligned} y^2 &= 4x^3 - k\,x - k \\ y^2 &= 4x^3 - c^2k\,x - c^3k, \end{aligned} \tag{14.29}$$

where $c \in \mathbb{F}_l$ is a nonsquare. We have the following result:

PROPOSITION 14.30. *Of the two elliptic curves over* $\mathbb{F}_l$ *defined in* (14.29), *one has order* $l + 1 - 2a$, *and the other has order* $l + 1 + 2a$, *where* $l = a^2 + nb^2$.

PROOF. Let L be the ring class field of the order $\mathcal{O} = \mathbb{Z}[\sqrt{-n}]$, and let $H_{-4n}(X) = \prod_{i=1}^{h}(X - j(\mathfrak{a}_i))$ be the class equation. If $\mathfrak{P}$ is prime in L containing l, then the isomorphism $\mathcal{O}_L/\mathfrak{P} \simeq \mathbb{Z}/l\mathbb{Z} = \mathbb{F}_l$ shows that our solution j of $H_{-4n}(X) \equiv 0 \bmod l$ satisfies $j \equiv j(\mathfrak{a}_i) \bmod \mathfrak{P}$ for some i. It follows that the curves (14.29) are members of the corresponding collection $\overline{E}_c$ constructed in the proof of Theorem 14.18, and our proposition then follows immediately since $l = \pi\overline{\pi}$ in $\mathcal{O}$, where $\pi = a + b\sqrt{-n}$. Q.E.D.

The curves (14.29) don't make sense when $j \equiv 0, 1728 \bmod l$, but the proof of Theorem 14.18 makes it clear how to proceed in these cases.

We can now sketch the Goldwasser–Kilian–Atkin Test. Given a potential prime l, one searches for the smallest n with l of the form $a^2 + nb^2$. Once we succeed, we check if either $l + 1 \pm 2a$ is twice a probable prime q. If not, we look for the next n with $l = a^2 + nb^2$. We continue this until $l + 1 \pm 2a$ has the right form, and then we apply the special primality test embodied in Lemma 14.23, using the two curves given in (14.29). In this way, we can prove that l is prime, provided that q is prime. Then, as in the regular Goldwasser–Kilian Test, we make the whole process recursive.

In practice, the implementation of the Goldwasser–Kilian–Atkin Test improves the run time by allowing the order $l+1\pm 2a$ to be more complicated than just twice a prime. The complete description of an implementation can be found in Morain's article [**99**]. See also Atkin and Morain [**3**].

For our purposes, this test is wonderful because it relates so nicely to our problem of when a prime is of the form x^2+ny^2. But from a practical point of view, the situation is less than ideal, for the test requires knowing $H_{-4n}(X)$, a polynomial with notoriously large coefficients. So in implementing the Goldwasser–Kilian–Atkin Test, one of the main goals is to avoid computing the full class equation. Different authors have taken different approaches to this problem, but the basic idea in each case is to use the Weber functions $\mathfrak{f}(\tau)$, $\mathfrak{f}_1(\tau)$ and $\mathfrak{f}_2(\tau)$ from §12. In [**99**, §6.2], Morain uses formulas of Weber, such as the one quoted in §12

$$\mathfrak{f}(\sqrt{-105})^6 = \sqrt{2}^{-13}(1+\sqrt{3})^3(1+\sqrt{5})^3(\sqrt{3}+\sqrt{7})^3(\sqrt{5}+\sqrt{7}),$$

to determine a root of $H_{-4n}(X)$ modulo l when n is one of Euler's convenient numbers (as defined in §3). Another approach, suggested by Kaltofen, Valente and Yui [**80**], is to use the methods of Kaltofen and Yui [**81**] to compute the minimal polynomials of the Weber functions. To see the potential savings, consider the case $n = 14$. We proved in §12 that

$$\mathfrak{f}_1(\sqrt{-14})^2 = \frac{\sqrt{2}+1+\sqrt{2\sqrt{2}-1}}{\sqrt{2}}$$

$$j(\sqrt{-14}) = \left(\mathfrak{f}_1(\sqrt{-14})^{16} + \frac{16}{\mathfrak{f}_1(\sqrt{-14})^8}\right)^3.$$

It is clear which one has the simpler minimal polynomial! The papers by Kaltofen, Valente and Yui [**80**] and Morain [**99**] give more details on the various implementations of the Goldwasser–Kilian–Atkin Test. See also the paper [**3**] by Atkin and Morain.

Primality testing is a good place to end this section, for primes are the basis of all number theory. We began in §1 with concrete questions concerning $p = x^2+y^2$, x^2+2y^2 and x^2+3y^2, and followed the general question of x^2+ny^2 through various wonderful areas of number theory. The theory of §8 was rather abstract, and even the ring class fields of §9 were not very intuitive. Complex multiplication helped bring these ideas down to earth, and now elliptic curves bring us back to the question of proving that a given number is prime. Fermat and Euler would have loved it.

E. Exercises

14.1. Let K be a field, and let $\mathbb{P}^2(K)$ be the projective plane over K, which is the set $K^3 - \{0\}/\sim$, where we set $(\lambda x, \lambda y, \lambda z) \sim (x, y, z)$ for all $\lambda \in K^*$.

(a) Show that the map $(x, y) \mapsto (x, y, 1)$ defines an injection $K^2 \to \mathbb{P}^2(K)$ and that the complement $\mathbb{P}^2(K) - K^2$ consists of those points with $z = 0$ (this is called the *line at infinity*).

(b) Given an elliptic curve E over K defined by the Weierstrass equation $y^2 = 4x^3 - g_2x - g_3$ we form the equation

$$y^2z = 4x^3 - g_2xz^2 - g_3z^3,$$

which is a homogeneous equation of degree 3. Then we define

$$\tilde{E}(K) = \{(x, y, z) \in \mathbb{P}^2(K) : y^2z = 4x^3 - g_2xz^2 - g_3z^3\}.$$

To relate this to $E(K)$, show that

$$\tilde{E}(K) = \{(x, y, 1) \in \mathbb{P}^2(K) : y^2 = 4x^3 - g_2x - g_3\} \cup \{0, 1, 0\}.$$

Thus the projective solutions consist of the solutions of the affine equation together with one point at infinity, (0,1,0). This is the point denoted ∞ in the text.

14.2. Let $L \subset \mathbb{C}$ be a lattice, and let $y^2 = 4x^3 - g_2(L)x - g_3(L)$ be the corresponding elliptic curve. Then show that the map $z \mapsto (\wp(z), \wp'(z))$ induces a bijection

$$(\mathbb{C} - L)/L \xrightarrow{\sim} E(\mathbb{C}) - \{\infty\}.$$

Hint: use Lemma 10.4 and part (b) of Exercise 10.14. Note also that $\wp'(z)$ is an odd function.

14.3. Prove Proposition 14.3.

14.4. In this exercise we will study elliptic curves with the same j-invariant.

(a) Prove Propositions 14.4 and 14.5.

(b) Consider the elliptic curves $y^2 = 4x^3 - g_3$, where g_3 is any nonzero integer. These curves all have j-invariant 0, so that they are all isomorphic over $\mathbb{C}$. Show that over $\mathbb{Q}$, these curves break up into infinitely many isomorphism classes.

14.5. In this exercise we will study the addition and duplication laws of $\wp'(z)$.

(a) Use formula (14.6) and the addition law for $\wp(z+w)$ (see Theorem 10.1) to conjecture and prove an addition law for $\wp'(z+w)$.

(b) Use formula (14.7) and the duplication law for $\wp(2z)$ (see (10.13)) to conjecture and prove a duplication law for $\wp'(2z)$.

14.6. If L and L' are lattices and $\alpha L \subset L'$, where $\alpha \neq 0$, show that the kernel of the map $\alpha : \mathbb{C}/L \to \mathbb{C}/L'$ is isomorphic to $L'/\alpha L$.

14.7. Complete the proof (begun in (14.10)) that

$$(x, y) \longmapsto \left(-\frac{2x^2+4x+9}{4(x+2)}, -\frac{1}{\sqrt{-2}}\frac{2x^2+8x-1}{4(x+2)^2}y \right)$$

defines the isogeny of $y^2 = 4x^3 - 30x - 28$ given by complex multiplication by $\sqrt{-2}$. Hint: use the discussion surrounding (10.21) and (10.22). Note that the $\wp$-function in (10.22) is $\wp(z; \lambda L)$ for the lattice λL in the display following (14.10)).

14.8. Let E be an elliptic curve over the finite field $\mathbb{F}_q$, and for any extension $\mathbb{F}_q \subset L$, define $Frob_q : E(L) \to E(L)$ by $Frob_q(x, y) = (x^q, y^q)$.

(a) Show that $Frob_q$ is a group homomorphism.

(b) Show that $Frob_q$ is not of the form $(R(x), (1/\alpha)R'(x)y)$ for any rational function $R(x)$.

14.9. Formulate and prove a version of Proposition 14.9 that applies to lattices L and L' such that $\alpha L \subset L'$ for some $\alpha \in \mathbb{C}^*$.

14.10. Use Theorem 11.23 to prove Proposition 14.11.

14.11. If E is an elliptic curve over $\mathbb{F}_q$, then prove that $|E(\mathbb{F}_q)| \leq 2q + 1$. Hint: given x, how many y's can satisfy $y^2 = 4x^3 - g_2x - g_3$?

14.12. If E is an elliptic curve over $\mathbb{F}_q$, then show that

$$E(\mathbb{F}_q) = \ker(1 - Frob_q : E(\overline{\mathbb{F}}_q) \to E(\overline{\mathbb{F}}_q)),$$

where $\overline{\mathbb{F}}_q$ is the algebraic closure of $\mathbb{F}_q$. Hint: for $x \in \overline{\mathbb{F}}_q$, recall that $x \in \mathbb{F}_q$ if and only if $x^q = x$.

14.13. This exercise is concerned with the relation between Gauss' claim (14.17) and Theorem 14.16.

(a) Verify that $(x, y) \mapsto (3x/(1+y), 9(1-y)/(1+y))$ transforms the curve $x^3 = y^3 + 1$ into the elliptic curve E defined by $y^2 = 4x^3 - 27$.

(b) The projective version of part (a) is given by $(x, y, z) \mapsto (3x, 9(z-y), z+y)$. Check that $(0, -1)$ on $x^3 = y^3 + 1$ is the only point that maps to $\infty = (0, 1, 0)$ on E.

(c) Check that $x^3 = y^3 + 1$ has three points at infinity. Hint: remember that $p \equiv 1 \bmod 3$.

(d) Show that E has complex multiplication by $\mathbb{Z}[\omega]$, $\omega = e^{2\pi i/3}$. Hint: see Exercise 10.17.

14.14. The last entry in Gauss' mathematical diary says that for a prime $p \equiv 1 \bmod 4$,

$$\begin{aligned} &\text{If } p = a^2 + b^2 \text{ and } a + bi \text{ is primary, then } N = p - 2a - 3, \text{where} \\ &N \text{ is the number of solutions modulo } p \text{ of } x^2 + y^2 + x^2y^2 = 1 \bmod p. \end{aligned}$$

Show that this is a special case of Theorem 14.16. Hint: use the change of variables $(x, y) \mapsto ((1+x)/2(1-x), (1+x^2)y/(1-x)^2)$ to transform the curve $x^2+y^2+x^2y^2 = 1$ into the elliptic curve $y^2 = 4x^3 + x$. See the discussion surrounding (4.24) for more details and references.

14.15. Let $\mathcal{O}_a$ be the order defined in the proof of Theorem 14.18. Prove that $\mathcal{O}_a$ has discriminant $a^2 - 4p$ and p does not divide the conductor. Hint: Exercise 7.12.

14.16. Let E be an elliptic curve over a field K, and assume that its j-invariant j is different from 0 and 1728. Then define $k \in K$ to be the number

$$k = \frac{27j}{j - 1728}.$$

Show that the Weierstrass equation for E can be written in the form

$$y^2 = 4x^3 - c^2k\,x - c^3k$$

for a unique $c \in K^*$. Hint: $c = g_3/g_2$.

14.17. Let $\mathcal{O}$ be an order in an imaginary quadratic field, and let p be a prime not dividing the conductor of $\mathcal{O}$. If $\pi \in \mathcal{O}$ satisfies $N(\pi) = p$, then prove that all solutions $\alpha \in \mathcal{O}$ of $N(\alpha) = p$ are given by $\alpha = \epsilon\pi$ or $\epsilon\overline{\pi}$ for $\epsilon \in \mathcal{O}^*$. Hint: this can be proved using unique factorization of ideals prime to the conductor (see Exercise 7.26).

14.18. Let $\overline{E}_c$ be one of the collections of elliptic curve over $\mathbb{F}_p$ which appear in the proof of Theorem 14.18, and let j denote their common j-invariant. By Exercise 14.16, note that $\overline{E}_c$ consists of *all* elliptic curves over $\mathbb{F}_p$ with this j-invariant.

(a) If $j \neq 0, 1728$, show that the curves break up into two isomorphism classes, each consisting of $(p-1)/2$ curves. Hint: consider the subgroup of squares in $\mathbb{F}_p^*$.

(b) If $j = 1728$ and $p \equiv 1 \bmod 4$, then show that there are four isomorphism classes, each consisting of $(p-1)/4$ curves.

(c) If $j = 0$ and $p \equiv 1 \bmod 3$, then show that there are six isomorphism classes, each consisting of $(p-1)/6$ curves.

14.19. In this exercise, we will sketch two proofs that there are $q(q-1)$ elliptic curves over the finite field $\mathbb{F}_q$. As usual, $q = p^a$, $p > 3$.

(a) Adapt the proof of Exercise 14.16 to show that there are q possible j-invariants for elliptic curves over $\mathbb{F}_q$, and show that there are $q-1$ curves with a given j-invariant. This gives $q(q-1)$ elliptic curves.

(b) A second way to prove the formula is to show that there are exactly q solutions $(g_2, g_3) \in \mathbb{F}_q^2$ of the equation $g_2^3 - 27g_3^2 = 0$. We can write this as $(g_2/3)^3 = g_3^2$, and after excluding the trivial solution (0,0) we need to study solutions of $u^3 = v^2$ in the group $\mathbb{F}_q^*$. So prove the following general fact: if G is a finite Abelian group and $a, b \in \mathbb{Z}$ are relatively prime, then the equation $u^a = v^b$ has exactly $|G|$ solutions in $G \times G$.

14.20. Let $\mathcal{O}'$ be an order in an imaginary quadratic field. Given a integer m which isn't a perfect square, show that

$$|\{\alpha \in \mathcal{O}' : N(\alpha) = m\}| = 2 \sum_{0 \le |a| \le 2\sqrt{m}} \chi(a),$$

where $\chi(a)$ is defined by

$$\chi(a) = \begin{cases} 1 & \text{if } \mathcal{O}' \text{ contains a root of } x^2 - ax + m \\ 0 & \text{otherwise.} \end{cases}$$

14.21. Use Theorem 14.26 to show that when Conjecture 14.27 is true, there is a constant $c_3 > 0$ such that for all sufficiently large primes l, there are at least

$$c_3 \cdot \frac{l(l-1)}{(\log l)^2}$$

elliptic curves E over $\mathbb{F}_l$ with $|E(\mathbb{F}_l)|$ twice a prime.

14.22. Let $y^2 = 4x^3 - g_2x - g_3$ be an elliptic curve E over a field K of characteristic $\neq 2, 3$.

(a) Use (14.7) to show that the elements of order 2 of $E(K)$ are the points $(a, 0)$, where $a \in K$ is a root of $4x^3 - g_2x - g_3$.

(b) Assume that $g_3 = 0$. If $(A, B) \in E(K)$ with $B \neq 0$ and $(x, y) = 2(A, B)$, then use (14.7) to show that

$$x = \left(\frac{4A^2 + g_2}{4B}\right)^2.$$

14.23. Let E be the elliptic curve defined by $y^2 = 4x^3 - 48x$. The discriminant of E is $\Delta = (-48)^3 = -2^{12}3^3$, so E has good reduction modulo primes > 3. Note also that E has complex multiplication by $\mathbb{Z}[i]$ and that $P = (-2, 8)$ is a point of $E(\mathbb{Q})$. In this exercise, we will assume that $q = 2^k - 1$ is prime, where $k \geq 3$. Let $\overline{E}$ be the reduction of E modulo q. Note that $q \equiv 7 \bmod 24$.

(a) Show that $\overline{E}$ is supersingular. Hint: if not, use Theorem 14.14 to show that $Frob_q \in \mathbb{Z}[i]$. Then consider degrees.

(b) By part (a) and Proposition 14.15, we have $|\overline{E}(\mathbb{F}_q)| = q + 1 = 2^k$. Then use Exercise 14.22 to show that $(0, 0)$ is the unique element of order 2 of $\overline{E}(\mathbb{F}_q)$ and conclude that $\overline{E}(\mathbb{F}_q)$ is cyclic. Hint: compute $(12/q)$.

(c) Prove that $(-2, 8) \in \overline{E}(\mathbb{F}_q)$ generates $\overline{E}(\mathbb{F}_q)$. Hint: assume not and use Exercise 14.22 to show that -2 is a square in $\mathbb{F}_q$. Then compute $(-2/q)$.

14.24. Consider the Mersenne number $q = 2^k - 1$, k an odd prime. We give a method due to Gross [**59**] to test whether q is prime using the elliptic curve E from Exercise 14.23. Let $\widetilde{E}$ be the reduction of E over the ring $\mathbb{Z}/q\mathbb{Z}$. It is easy to see that $\widetilde{E}$ is an elliptic curve over $\mathbb{Z}/q\mathbb{Z}$. Note also that $(-2,8) \in \widetilde{E}(\mathbb{Z}/q\mathbb{Z})$. Now the method: set $P_1 = (-2,8)$ and successively compute $P_\ell = 2P_{\ell-1}$ in $\widetilde{E}(\mathbb{Z}/q\mathbb{Z})$ for $\ell \geq 2$. Stop when one of the following scenarios occurs:

A: P_ℓ is defined for all $\ell \leq k-1$ and $P_{k-1} = (0,0)$.
B: P_ℓ is defined for all $\ell \leq k-1$ and $P_{k-1} \neq (0,0)$.
C: P_ℓ is undefined for some $\ell \leq k-1$ (see the discussion following (14.22)).

(a) Show that q cannot be prime in scenario B or C. Hint: Exercise 14.23. Also note that in a cyclic group of order 2^k, an element a generates if and only if $2^{k-1}a$ is the unique element of order 2.

(b) It remains to show that q is prime in scenario A. Following Gross [**59**], let p be a prime divisor of q and let $\overline{E}$ be the reduction of E over $\mathbb{F}_p$.
   (i) Show that $(-2,8) \in \overline{E}(\mathbb{F}_p)$ has order $q+1$. Thus $|\overline{E}(\mathbb{F}_p)| \geq q+1$.
   (ii) Apply the Hasse bound (Theorem 14.12 to $\overline{E}$ to show that $p = q$.

In [**59**], Gross uses the elliptic curve $Y^2 = X^3 - 12X$, which is isomorphic to the above curve E via $(X,Y) = (x, y/2)$. Gross also gives a more elementary version of the above method that highlights the relation with the classical Lucas–Lehmer Test for primality of Mersenne numbers.

15. Shimura Reciprocity

One step in the proof of Theorem 12.24 from §12 is the formula

$$\sigma\big(\mathfrak{f}_1(\sqrt{-m})^6\big) = \mathfrak{f}_1(\gamma\sqrt{-m})^6, \tag{15.1}$$

where $m \equiv 6 \bmod 8$, $\mathfrak{f}_1$ is one of the Weber functions, σ is in a certain Galois group, and $\gamma = \left(\begin{smallmatrix} 1 & -2 \\ 1 & -1 \end{smallmatrix}\right)$. As we remarked after the proof, (15.1) shows how the Galois action on a special value of a modular function is given by an element of $\mathrm{SL}(2,\mathbb{Z})$. We also noted that this was a special case of the *Shimura Reciprocity Theorem.*

In this section, we will state versions of Shimura Reciprocity that apply to ring class fields and what we call *extended ring class fields.* We will then re-prove some of the results from §12 using Shimura Reciprocity. Our exposition is based on the work of Peter Stevenhagen and Alice Gee [**53, 54, 126**] and Bumkyo Cho [**22**].

This section uses some mathematical tools not encountered in previous sections, including the p-adic integers $\mathbb{Z}_p$ and tensor products. Some proofs will be deferred until the end of the section, where we will also encounter the idelic version of class field theory mentioned briefly in §8.

A. Modular Functions

Earlier in the book, we used modular functions for the congruence subgroup $\Gamma_0(m)$. For Shimura Reciprocity, we instead use the congruence subgroup

$$\Gamma(m) = \{\gamma \in \mathrm{SL}(2,\mathbb{Z}) : \gamma \equiv I \bmod m\},$$

where I is the 2×2 identity matrix. Thus $\Gamma(m)$ is the kernel of the reduction map $\mathrm{SL}(2,\mathbb{Z}) \to \mathrm{SL}(2,\mathbb{Z}/m\mathbb{Z})$.

A *modular function of level m* is a function f defined on the upper half plane $\mathfrak{h}$, except for isolated singularities, which satisfies the following three conditions:

(1) f is meromorphic on $\mathfrak{h}$.
(2) f is invariant under $\Gamma(m)$.
(3) f is meromorphic at the cusps.

These conditions are similar to the corresponding definition for $\Gamma_0(m)$ from §11. To explain (iii), note that $f(\gamma(\tau+m)) = f(\gamma\tau)$ for $\gamma \in \mathrm{SL}(2,\mathbb{Z})$ since $\left(\begin{smallmatrix} 1 & m \\ 0 & 1 \end{smallmatrix}\right) \in \Gamma(m)$ and $\Gamma(m)$ is normal in $\mathrm{SL}(2,\mathbb{Z})$. As in §11, this gives the q-expansion

$$f(\gamma\tau) = \sum_{n=-\infty}^{\infty} a_n\, q^{n/m}, \quad q = e^{2\pi i\tau}.$$

Then "meromorphic at the cusps" means that for all $\gamma \in \mathrm{SL}(2,\mathbb{Z})$, this expansion has only finitely many terms with negative exponents.

Rather than work with all modular functions of level m, we will work with the set F_m of all modular functions of level m whose q-expansion at every cusp has coefficients in $\mathbb{Q}(\zeta_m)$, $\zeta_m = e^{2\pi i/m}$. In Exercise 15.1 we will show that F_m is a field and that $\mathsf{F}_1 = \mathbb{Q}(j)$. (The definition of F_m given here differs from the definition found in Lang [**90**, p. 66] or Shimura [**115**, §6.2]. Shimura proves the equivalence of the two definitions in [**115**, Prop. 6.9].)

For an example of a modular function, consider the Weber functions $\mathfrak{f}_1(\tau)$ from §12. In Proposition 12.25 we showed that $\mathfrak{f}_1(8\tau)^6$ is a modular function for $\Gamma_0(32)$. Here is a similar result for $\mathfrak{f}_1(\tau)^2$ and $\mathfrak{f}_1(\tau)^6$:

PROPOSITION 15.2. *$\mathfrak{f}_1(\tau)^2 \in \mathsf{F}_{24}$ and $\mathfrak{f}_1(\tau)^6 \in \mathsf{F}_8$.*

PROOF. Recall the group $\Gamma_0(2)^t = \{\left(\begin{smallmatrix} a & b \\ c & d \end{smallmatrix}\right) : b \equiv 0 \bmod 2\}$ introduced in the proof of Proposition 12.25. In (12.26), we showed how $\mathfrak{f}_1(\tau)^6$ transforms under elements of $\Gamma_0(2)^t$. For $\mathfrak{f}_1(\tau)^2$, we need the formula

$$\mathfrak{f}_1(\gamma\tau)^2 = \zeta_{12}^{-\frac{1}{2}ab+c(d(1-a^2)-a)}\,\mathfrak{f}_1(\tau)^2 \tag{15.3}$$

when $\gamma = \left(\begin{smallmatrix} a & b \\ c & d \end{smallmatrix}\right) \in \Gamma_0(2)^t$. This follows from the transformation law for the Dedekind η-function given in Schertz [**111**]. See Exercise 15.2 for the details.

The formula (15.3) makes it easy to see that $\mathfrak{f}_1(\tau)^2$ is invariant under $\Gamma(24)$. To complete the proof of $\mathfrak{f}_1(\tau)^2 \in \mathsf{F}_{24}$, we need to study what happens at the cusps. We will do this in Exercise 15.2 using the methods of the proof of Proposition 12.25.

In Exercise 15.2 we will show that $\mathfrak{f}_1(\tau)^6 \in \mathsf{F}_8$ by a similar argument. Q.E.D.

Given $\gamma \in \mathrm{SL}(2,\mathbb{Z})$, the map $f(\tau) \mapsto f^\gamma(\tau) = f(\gamma\tau)$ is easily seen to be an automorphism of F_m which is the identity on $\mathsf{F}_1(\zeta_m)$. Since every $f \in \mathsf{F}_m$ is invariant under $\pm\Gamma(m)$, we get a homomorphism

$$\mathrm{SL}(2,\mathbb{Z}/m\mathbb{Z})/\{\pm I\} \longrightarrow \mathrm{Gal}(\mathsf{F}_m/\mathsf{F}_1(\zeta_m)). \tag{15.4}$$

This sets the stage for the following well-known result (see Lang [**90**, p. 66]):

THEOREM 15.5. *F_m is a Galois extension of $\mathsf{F}_1(\zeta_m)$ and the above homomorphism is an isomorphism, i.e.,*

$$\mathrm{SL}(2,\mathbb{Z}/m\mathbb{Z})/\{\pm I\} \simeq \mathrm{Gal}(\mathsf{F}_m/\mathsf{F}_1(\zeta_m)). \qquad \text{Q.E.D.}$$

We regard this Galois group as acting *geometrically* on modular functions. For an example, recall from §12 that $\mathfrak{f}_2(\tau) = \mathfrak{f}_1(-1/\tau) = \mathfrak{f}_1(S\tau) = \mathfrak{f}_1^S(\tau)$, where $S = \left(\begin{smallmatrix} 0 & -1 \\ 1 & 0 \end{smallmatrix}\right)$. It follows from Proposition 15.2 that $\mathfrak{f}_2(\tau)^2 \in \mathsf{F}_{24}$ and $\mathfrak{f}_2(\tau)^6 \in \mathsf{F}_8$.

We can also compute the Galois group of F_m over F_1. An integer d relatively prime to m gives $\left(\begin{smallmatrix} 1 & 0 \\ 0 & d \end{smallmatrix}\right) \in \mathrm{GL}(2, \mathbb{Z}/m\mathbb{Z})$. This matrix acts *arithmetically* on the coefficients of the q-expansion of $f \in \mathsf{F}_m$ via the Galois automorphism that takes ζ_m to ζ_m^d (so $\left(\begin{smallmatrix} 1 & 0 \\ 0 & d \end{smallmatrix}\right)$ acts trivially on f when f has a rational q-expansion).

Then, as described in Lang [**90**, p. 66], we have the following result:

THEOREM 15.6. *F_m is a Galois extension of F_1 with Galois group*

$$\mathrm{GL}(2, \mathbb{Z}/m\mathbb{Z})/\{\pm I\} \simeq \mathrm{Gal}(\mathsf{F}_m/\mathsf{F}_1),$$

where the isomorphism combines the geometric action of $\mathrm{SL}(2, \mathbb{Z}/m\mathbb{Z})/\{\pm I\}$ *from* (15.4) *with the arithmetic action of* $\left(\begin{smallmatrix} 1 & 0 \\ 0 & d \end{smallmatrix}\right) \in \mathrm{GL}(2, \mathbb{Z}/m\mathbb{Z})$ *described above.* Q.E.D.

For Shimura Reciprocity, we need to consider modular functions of all levels. Define the *field of modular functions* to be the union

$$\mathsf{F} = \bigcup_{m=1}^{\infty} \mathsf{F}_m.$$

The Galois group $\mathrm{Gal}(\mathsf{F}/\mathsf{F}_1)$ consists of automorphisms of F that restrict to the identity on F_1, though this group is no longer finite. Describing $\mathrm{Gal}(\mathsf{F}/\mathsf{F}_1)$ involves letting $m \to \infty$ in the isomorphism $\mathrm{GL}(2, \mathbb{Z}/m\mathbb{Z})/\{\pm I\} \simeq \mathrm{Gal}(\mathsf{F}_m/\mathsf{F}_1)$ of Theorem 15.6. Doing so will require some new ideas.

First observe that when $m \mid n$, the surjection $\mathbb{Z}/n\mathbb{Z} \to \mathbb{Z}/m\mathbb{Z}$ and the inclusion $\mathsf{F}_m \subset \mathsf{F}_n$ induce the vertical maps in the commutative diagram

$$\begin{array}{ccc} \mathrm{GL}(2, \mathbb{Z}/n\mathbb{Z})/\{\pm I\} & \xrightarrow{\sim} & \mathrm{Gal}(\mathsf{F}_n/\mathsf{F}_1) \\ \downarrow & & \downarrow \\ \mathrm{GL}(2, \mathbb{Z}/m\mathbb{Z})/\{\pm I\} & \xrightarrow{\sim} & \mathrm{Gal}(\mathsf{F}_m/\mathsf{F}_1), \end{array}$$

where the horizontal isomorphisms are from Theorem 15.6. Since F_n and F_m are Galois over F_1, the restriction maps from $\mathrm{Gal}(\mathsf{F}/\mathsf{F}_1)$ to $\mathrm{Gal}(\mathsf{F}_n/\mathsf{F}_1)$ and $\mathrm{Gal}(\mathsf{F}_m/\mathsf{F}_1)$ can be added on the right to yield the enlarged commutative diagram

$$\begin{array}{ccccc} \mathrm{GL}(2, \mathbb{Z}/n\mathbb{Z})/\{\pm I\} & \xrightarrow{\sim} & \mathrm{Gal}(\mathsf{F}_n/\mathsf{F}_1) & \longleftarrow & \mathrm{Gal}(\mathsf{F}/\mathsf{F}_1) \\ \downarrow & & \downarrow & \swarrow & \\ \mathrm{GL}(2, \mathbb{Z}/m\mathbb{Z})/\{\pm I\} & \xrightarrow{\sim} & \mathrm{Gal}(\mathsf{F}_m/\mathsf{F}_1). & & \end{array} \tag{15.7}$$

To enlarge the other side of the diagram in a similar way, we need the *ring of p-adic integers* $\mathbb{Z}_p$, p prime. Here are some key properties of $\mathbb{Z}_p$:

(i) The only ideals of $\mathbb{Z}_p$ are $\{0\}$ and $p^\ell\mathbb{Z}_p$ for $\ell \geq 0$.
(ii) $\mathbb{Z} \subset \mathbb{Z}_p$ induces an isomorphism $\mathbb{Z}/p^\ell\mathbb{Z} \simeq \mathbb{Z}_p/p^\ell\mathbb{Z}_p$ for all $\ell \geq 0$.
(iii) Given p-adic integers x_ℓ for $\ell \geq 0$ such that $x_k \equiv x_\ell \bmod p^\ell\mathbb{Z}_p$ whenever $k \geq \ell$, there is a unique $x \in \mathbb{Z}_p$ such that $x \equiv x_\ell \bmod p^\ell\mathbb{Z}_p$ for all $\ell \geq 0$.

See Gouvêa [**56**] for more on the p-adic integers. Property (iii) captures the idea that $\mathbb{Z}_p$ is the *completion* of $\mathbb{Z}$ at the prime ideal $p\mathbb{Z}$.

Using $\mathbb{Z}_p$ for all primes p, we define the ring $\widehat{\mathbb{Z}}$ by

$$\widehat{\mathbb{Z}} = \prod_p \mathbb{Z}_p.$$

For positive integers m, the natural map $\mathbb{Z} \to \widehat{\mathbb{Z}}$ induces isomorphisms $\mathbb{Z}/m\mathbb{Z} \simeq \widehat{\mathbb{Z}}/m\widehat{\mathbb{Z}}$ (see Exercise 15.3) that fit together in a commutative diagram

$$\begin{array}{ccccc} \widehat{\mathbb{Z}} & \longrightarrow & \widehat{\mathbb{Z}}/n\widehat{\mathbb{Z}} & \xleftarrow{\sim} & \mathbb{Z}/n\mathbb{Z} \\ & \searrow & \downarrow & & \downarrow \\ & & \widehat{\mathbb{Z}}/m\widehat{\mathbb{Z}} & \xleftarrow{\sim} & \mathbb{Z}/m\mathbb{Z} \end{array}$$

when $m \mid n$. It follows that (15.7) can be enlarged to the commutative diagram

$$\begin{array}{ccccccc} \mathrm{GL}(2,\widehat{\mathbb{Z}})/\{\pm I\} & \longrightarrow & \mathrm{GL}(2,\mathbb{Z}/n\mathbb{Z})/\{\pm I\} & \xrightarrow{\sim} & \mathrm{Gal}(\mathsf{F}_n/\mathsf{F}_1) & \longleftarrow & \mathrm{Gal}(\mathsf{F}/\mathsf{F}_1) \\ & \searrow & \downarrow & & \downarrow & \swarrow & \\ & & \mathrm{GL}(2,\mathbb{Z}/m\mathbb{Z})/\{\pm I\} & \xrightarrow{\sim} & \mathrm{Gal}(\mathsf{F}_m/\mathsf{F}_1). & & \end{array}$$

PROPOSITION 15.8. *There is an isomorphism* $\mathrm{GL}(2,\widehat{\mathbb{Z}})/\{\pm I\} \simeq \mathrm{Gal}(\mathsf{F}/\mathsf{F}_1)$ *that is compatible with the above commutative diagram.*

PROOF. The diagram shows that an element of $\mathrm{GL}(2,\widehat{\mathbb{Z}})/\{\pm I\}$ gives elements of $\mathrm{Gal}(\mathsf{F}_m/\mathsf{F}_1)$ that are compatible with restriction. Since $\mathsf{F} = \bigcup_m \mathsf{F}_m$, this gives a unique element of $\mathrm{Gal}(\mathsf{F}/\mathsf{F}_1)$. The resulting map $\mathrm{GL}(2,\widehat{\mathbb{Z}})/\{\pm I\} \to \mathrm{Gal}(\mathsf{F}/\mathsf{F}_1)$ is clearly a homomorphism that is compatible with the above commutative diagram. Injectivity of the map boils down to the *uniqueness* of x in property (iii) of the p-adic integers, and surjectivity similarly reduces to the *existence* of x in property (iii) (see Exercise 15.4 for the details). Q.E.D.

For readers familiar with inverse limits and the Galois theory of infinite algebraic extensions, the above argument can summarized by noting that

$$\widehat{\mathbb{Z}} = \varprojlim_m \mathbb{Z}/m\mathbb{Z},$$

which in turn implies that

$$\mathrm{Gal}(\mathsf{F}/\mathsf{F}_1) = \varprojlim_m \mathrm{Gal}(\mathsf{F}_m/\mathsf{F}_1) \simeq \varprojlim_m \mathrm{GL}(2,\mathbb{Z}/m\mathbb{Z})/\{\pm I\} \simeq \mathrm{GL}(2,\widehat{\mathbb{Z}})/\{\pm I\}.$$

Inverse limits and the version of Galois theory used here are covered in Dummit and Foote [**40**]. See Milne [**98**] for a complete, self-contained exposition.

We also need the group $\mathrm{Gal}(\mathsf{F}/\mathbb{Q})$ of all automorphisms of F. The group

$$\mathrm{GL}(2,\mathbb{Q})^+ = \{\gamma \in \mathrm{GL}(2,\mathbb{Q}) : \det(\gamma) > 0\}$$

acts naturally on F, since given $f \in \mathsf{F}$ and $\gamma \in \mathrm{GL}(2,\mathbb{Q})^+$, the transformed function $f^\gamma(\tau) = f(\gamma\tau)$ is still a modular function, but possibly of a different level. To see why, first note that multiplying γ by an integer does not change its action on $\mathfrak{h}$. Hence we may assume that γ has integer entries. Let $N = \det(\gamma)$, which is positive since $\gamma \in \mathrm{GL}(2,\mathbb{Q})^+$. In Exercise 15.5, we will show that

$$\gamma\Gamma(mN)\gamma^{-1} \subseteq \Gamma(m).$$

This makes it easy to see that if $f \in \mathsf{F}_m$, then f^γ is invariant under $\Gamma(mN)$. We will show that $f^\gamma \in \mathsf{F}_{mN}$ in Exercise 15.5.

To combine the actions of $\mathrm{GL}(2,\widehat{\mathbb{Z}})$ and $\mathrm{GL}(2,\mathbb{Q})^+$ on F, we use the ring

$$\widehat{\mathbb{Q}} = \mathbb{Q} \otimes_{\mathbb{Z}} \widehat{\mathbb{Z}}$$

(a good reference for tensor products is Dummit and Foote [**40**]) and the obvious maps

$$\begin{aligned} \mathrm{GL}(2,\widehat{\mathbb{Z}}) &\longrightarrow \mathrm{GL}(2,\widehat{\mathbb{Q}}) \\ \mathrm{GL}(2,\mathbb{Q})^+ &\longrightarrow \mathrm{GL}(2,\widehat{\mathbb{Q}}). \end{aligned}$$

We now have everything we need to describe $\mathrm{Gal}(\mathsf{F}/\mathbb{Q})$:

THEOREM 15.9.

(i) *Every element* $\gamma \in \mathrm{GL}(2,\widehat{\mathbb{Q}})$ *can be written as* $\gamma = uv$ *for* $u \in \mathrm{GL}(2,\widehat{\mathbb{Z}})$ *and* $v \in \mathrm{GL}(2,\mathbb{Q})^+$.

(ii) $f \mapsto f^\gamma = (f^u)^v$ *gives a well-defined action of* $\mathrm{GL}(2,\widehat{\mathbb{Q}})$ *on the field* F *such that the subgroup* $\mathbb{Q}^*I \subset \mathrm{GL}(2,\widehat{\mathbb{Q}})$ *acts trivially on* F.

(iii) *This action induces an isomorphism*

$$\mathrm{GL}(2,\widehat{\mathbb{Q}})/\mathbb{Q}^*I \simeq \mathrm{Gal}(\mathsf{F}/\mathbb{Q}).$$

PROOF. See Lang [**90**, §7.2, Theorem 6]. Q.E.D.

REMARK. The decomposition $\gamma = uv$ from part (i) of the theorem is not unique since $\mathrm{GL}(2,\widehat{\mathbb{Z}}) \cap \mathrm{GL}(2,\mathbb{Q})^+ = \mathrm{SL}(2,\mathbb{Z})$. But the action defined in part (ii) can be shown to be well-defined. Also, we can think of $\mathrm{GL}(2,\widehat{\mathbb{Q}})$ as a product as follows. Let $\mathbb{Q}_p = \mathbb{Q} \otimes_{\mathbb{Z}} \mathbb{Z}_p$ be the *field of* p*-adic numbers*. Then $\widehat{\mathbb{Z}} = \prod_p \mathbb{Z}_p$ induces an injective map

$$\mathrm{GL}(2,\widehat{\mathbb{Q}}) \longrightarrow \prod_p \mathrm{GL}(2,\mathbb{Q}_p) \tag{15.10}$$

whose image consists of all $\gamma = (\gamma_p) \in \prod_p \mathrm{GL}(2,\mathbb{Q}_p)$ such that $\gamma_p \in \mathrm{GL}(2,\mathbb{Z}_p)$ for all but finitely many p (see Exercise 15.6).

B. The Shimura Reciprocity Theorem

Theorem 15.9 describes the *modular action* of $\mathrm{GL}(2,\widehat{\mathbb{Q}})$ on F. This action is the first key player in Shimura Reciprocity. The second key player comes from the *Galois action* of $\mathrm{Gal}(K^{ab}/K)$ on the *maximal Abelian extension* K^{ab} of an imaginary quadratic field K. One can think of K^{ab} as the union of all subfields of $\mathbb{C}$ that are finite Abelian Galois extensions of K. Given $\tau_0 \in K \cap \mathfrak{h}$, the basic idea of Shimura Reciprocity is that if $f(\tau_0)$ is defined for a modular function $f \in \mathsf{F}$, then $f(\tau_0)$ lies in K^{ab}, and the Galois action on $f(\tau_0)$ can be computed in terms of the modular action on f.

The link between these actions is described using a suitable idele group. We follow Stevenhagen [**126**, §3] and define the *ring of* K*-adeles* to be

$$\widehat{K} = K \otimes_{\mathbb{Z}} \widehat{\mathbb{Z}}.$$

Then the *group of* K*-ideles* is the group of units $\widehat{K}^*$ of $\widehat{K}$. The key point is that an idele $x \in \widehat{K}^*$ gives an automorphism $\sigma_x \in \mathrm{Gal}(K^{ab}/K)$ (for the Galois action on $f(\tau_0)$) and a matrix $g_{\tau_0}(x) \in \mathrm{GL}(2,\widehat{\mathbb{Q}})$ (for the modular action on f). These are defined as follows.

The matrix $g_{\tau_0}(x) \in \mathrm{GL}(2, \widehat{\mathbb{Q}})$ is straightforward to describe. First observe that $\{\tau_0, 1\}$ is a basis of K over $\mathbb{Q}$ since $\tau_0 \in K \cap \mathfrak{h}$, Then

$$\widehat{K} = K \otimes_{\mathbb{Z}} \widehat{\mathbb{Z}} \simeq (K \otimes_{\mathbb{Q}} \mathbb{Q}) \otimes_{\mathbb{Z}} \widehat{\mathbb{Z}} \simeq K \otimes_{\mathbb{Q}} (\mathbb{Q} \otimes_{Z} \widehat{\mathbb{Z}}) \simeq K \otimes_{\mathbb{Q}} \widehat{\mathbb{Q}}$$

shows that $\{\tau_0, 1\}$ is a basis of $\widehat{K}$ as a free $\widehat{\mathbb{Q}}$-module of rank 2. Multiplication by x give an invertible map $m_x : \widehat{K} \to \widehat{K}$ since $x \in \widehat{K}^*$. We define $g_{\tau_0}(x)$ to be the transpose of the 2×2 matrix representing m_x with respect to the basis $\{\tau_0, 1\}$. Note that $g_{\tau_0}(x) \in \mathrm{GL}(2, \widehat{\mathbb{Q}})$ since m_x is an isomorphism.

For $x \in \widehat{K}^*$, the description of $\sigma_x \in \mathrm{Gal}(K^{ab}/K)$ takes more work. In what follows, we give the basic setup and defer many details until the end of the section. Recall from §8 that for a modulus $\mathfrak{m}$, the Artin map for the *ray class field* $K_{\mathfrak{m}}$ induces an isomorphism

$$I_K(\mathfrak{m})/P_{K,1}(\mathfrak{m}) \simeq \mathrm{Gal}(K_{\mathfrak{m}}/K). \tag{15.11}$$

Since K is imaginary quadratic, a modulus is simply an ideal in $\mathcal{O}_K$. We will focus on moduli of the form $\mathfrak{m} = m\mathcal{O}_K$ for positive integers m and write $I_K(m)$ instead of $I_K(m\mathcal{O}_K)$. Similarly, the corresponding ray class fields will be written K_m. In Exercise 15.7, we will see that $K_m \subseteq K_n$ when $m \mid n$.

By Exercise 11.23, every finite Abelian extension of K lies in some ray class field K_m, so that $K^{ab} = \bigcup_m K_m$. It follows that constructing $\sigma_x \in \mathrm{Gal}(K^{ab}/K)$ is equivalent to constructing $\sigma_{x,m} \in \mathrm{Gal}(K_m/K)$ for all m such that

$$\sigma_{x,n}|_{K_m} = \sigma_{x,m}$$

whenever $m \mid n$ (see Exercise 15.7). By (15.11) with $\mathfrak{m} = m\mathcal{O}_K$, the construction σ_x reduces to the construction of compatible elements of $I_K(m)/P_{K,1}(m)$.

To see how $x \in \widehat{K}^*$ gives a generalized ideal class in $I_K(m)/P_{K,1}(m)$, we need to explain how the adele ring $\widehat{K} = K \otimes_{\mathbb{Z}} \widehat{\mathbb{Z}}$ and the idele group $\widehat{K}^* = (K \otimes_{\mathbb{Z}} \widehat{\mathbb{Z}})^*$ interact with the product $\widehat{\mathbb{Z}} = \prod_p \mathbb{Z}_p$. For the adeles, we have an injection

$$\widehat{K} \hookrightarrow \prod_p (K \otimes_{\mathbb{Z}} \mathbb{Z}_p), \tag{15.12}$$

so that $x = (x_p)$ for $x_p \in K \otimes_{\mathbb{Z}} \mathbb{Z}_p$. Also, if we define $\widehat{\mathcal{O}}_K = \mathcal{O}_K \otimes_{\mathbb{Z}} \widehat{\mathbb{Z}}$, then $\widehat{\mathbb{Z}} = \prod_p \mathbb{Z}_p$ induces an isomorphism

$$\widehat{\mathcal{O}}_K \simeq \prod_p (\mathcal{O}_K \otimes_{\mathbb{Z}} \mathbb{Z}_p)$$

since $\mathcal{O}_K$ is a free $\mathbb{Z}$-module of rank 2. This isomorphism is clearly compatible with (15.12). By Exercise 15.8, $(x_p) \in \prod_p (K \otimes_{\mathbb{Z}} \mathbb{Z}_p)$ is an adele, i.e., comes from some $x \in \widehat{K}$ via (15.12), if and only if $x_p \in \mathcal{O}_K \otimes_{\mathbb{Z}} \mathbb{Z}_p$ for all but finitely many p.

Turning to ideles, we similarly have

$$\widehat{K}^* \hookrightarrow \prod_p (K \otimes_{\mathbb{Z}} \mathbb{Z}_p)^* \quad \text{and} \quad \widehat{\mathcal{O}}_K^* \simeq \prod_p (\mathcal{O}_K \otimes_{\mathbb{Z}} \mathbb{Z}_p)^*.$$

By Exercise 15.8, $(x_p) \in \prod_p (K \otimes_{\mathbb{Z}} \mathbb{Z}_p)$ is an idele if and only if $x_p \in (\mathcal{O}_K \otimes_{\mathbb{Z}} \mathbb{Z}_p)^*$ for all but finitely many p. Finally, we note that the map $\beta \in K \mapsto \beta \otimes 1 \in \widehat{K} = K \otimes_{\mathbb{Z}} \widehat{\mathbb{Z}}$ induces injections

$$K \hookrightarrow \widehat{K} \quad \text{and} \quad K^* \hookrightarrow \widehat{K}^*.$$

The next step is to create a fractional ideal using an idele $x \in \widehat{K}^*$. The idea is simple: $x\widehat{\mathcal{O}}_K \subset \widehat{K}$ is a module over $\widehat{\mathcal{O}}_K$, and since the inclusion $K \hookrightarrow \widehat{K}$ takes $\mathcal{O}_K$ into $\widehat{\mathcal{O}}_K$, the intersection

$$x\mathcal{O}_K = K \cap x\widehat{\mathcal{O}}_K \subset K$$

is an $\mathcal{O}_K$-module contained in K. The following lemma (whose proof is deferred until the end of the section) guarantees that $x\mathcal{O}_K$ is a fractional $\mathcal{O}_K$-ideal in K:

LEMMA 15.13. *Let $\widehat{\mathcal{O}}_K$, $\widehat{K}$ and $\widehat{K}^*$ be as above.*

(i) *If $x \in \widehat{K}^*$, then $x\mathcal{O}_K = K \cap x\widehat{\mathcal{O}}_K$ is a fractional $\mathcal{O}_K$-ideal in K.*
(ii) *The map $\widehat{K}^* \to I_K$ defined by $x \mapsto x\mathcal{O}_K$ is a surjective homomorphism with kernel $\widehat{\mathcal{O}}_K^*$, where I_K is the group of fractional $\mathcal{O}_K$-ideals in K.*
(iii) *If $\mathfrak{b}$ is a fractional $\mathcal{O}_K$-ideal in K, then $\mathfrak{b} = K \cap \mathfrak{b}\widehat{\mathcal{O}}_K$.*

Thus our idele $x \in \widehat{K}^*$ gives the fractional ideal $x\mathcal{O}_K$. But to get a class in $I_K(m)/P_{K,1}(m)$, we need a fractional ideal that is relatively prime to m. For this purpose, we have the following factorization result whose proof will be given at the end of the section. For a positive integer m, define

$$J_m^1 = \{y \in \widehat{\mathcal{O}}_K^* : y \equiv 1 \bmod m\widehat{\mathcal{O}}_K\}.$$

Note that J_m^1 is a subgroup of $\widehat{\mathcal{O}}_K^*$. Then:

LEMMA 15.14. *Fix an integer $m > 0$ and an idele $x \in \widehat{K}^*$.*

(i) *There is a factorization $x = \beta y z$ where $\beta \in K^*$, $y \in J_m^1$, and $z = (z_p) \in \widehat{K}^*$ satisfies $z_p = 1$ for all $p \mid m$.*
(ii) *Given $x = \beta y z$ as in* (i)*, we have $z\mathcal{O}_K \in I_K(m)$, and the class $[z\mathcal{O}_K] \in I_K(m)/P_{K,1}(m)$ depends only on x.*
(iii) *Given $x \in \widehat{\mathcal{O}}_K^*$ and $\alpha \in \mathcal{O}_K$ such that $\alpha x \equiv 1 \bmod m\widehat{\mathcal{O}}_K$, x can be written $x = \beta y z$ as in* (i) *with $\beta = \alpha^{-1}$ and $z\mathcal{O}_K = \alpha\mathcal{O}_K$.*

In what follows, the factorization $x = \beta y z$ as in part (i) of Lemma 15.14 will be called an *m-factorization* of x. The lemma implies that we get a well-defined element

$$\sigma_{x,m} \in \mathrm{Gal}(K_m/K) \simeq I_K(m)/P_{K,1}(m)$$

by requiring that $\sigma_{x,m} \mapsto [z\mathcal{O}_K]$ when $x = \beta y z$ is an m-factorization. Note that if we have an n-factorization $x = \beta y z$ for n and $m \mid n$, then $x = \beta y z$ is also an m-factorization. Thus $\sigma_{x,m}$ and $\sigma_{x,n}$ are compatible when $m \mid n$ (see Exercise 15.9 for the details). As explained above, this gives $\sigma_x \in \mathrm{Gal}(K^{ab}/K)$. The map

$$\widehat{K}^* \longrightarrow \mathrm{Gal}(K^{ab}/K)$$

given by $x \mapsto \sigma_x$ is called the *Artin map*. We will see later that the Artin map is surjective, so that the σ_x for $x \in \widehat{K}^*$ gives the full Galois action of $\mathrm{Gal}(K^{ab}/K)$. Exercise 15.9 will show that K^* is contained in the kernel of the Artin map.

REMARK. This definition of the Artin map looks very different from the usual definition given in the idelic class field theory, which uses the Artin symbol from local class field theory. Proposition 15.40 below will explain the relation between the two definitions.

When we combine the Galois action $x \mapsto \sigma_x$ with the modular action $x \mapsto g_{\tau_0}(x)$ defined earlier, we are ready to state the Shimura Reciprocity Theorem. Here is the version due to Stevenhagen and Gee in [**53, 54, 126**]. For more classical versions, see Lang [**90**, Ch. 11] and Shimura [**115**, Thm. 6.31].

THEOREM 15.15 (Shimura Reciprocity). *Let K be an imaginary quadratic field and fix $\tau_0 \in K \cap \mathfrak{h}$. Then $f(\tau_0) \in K^{ab}$ for any modular function $f \in \mathsf{F}$ such that $f(\tau_0)$ is defined. Furthermore, if an idele $x \in \widehat{K}^*$ maps to $\sigma_x \in \mathrm{Gal}(K^{ab}/K)$ and $g_{\tau_0}(x) \in \mathrm{GL}(2, \widehat{\mathbb{Q}})$ as above, then*

$$\sigma_x(f(\tau_0)) = f^{g_{\tau_0}(x^{-1})}(\tau_0). \qquad \text{Q.E.D.}$$

Shimura Reciprocity thus consists of two miracles: first, that $f(\tau_0) \in K^{ab}$, and second, that given an idele $x \in \widehat{K}^*$, there is $g_{\tau_0}(x^{-1}) \in \mathrm{GL}(2, \widehat{\mathbb{Q}})$ such that the Galois action of σ_x on $f(\tau_0) \in K^{ab}$ is given by the modular action of $g_{\tau_0}(x^{-1})$ on f, followed by evaluation at τ_0.

C. Extended Ring Class Fields

Let $\mathcal{O}$ be an order of conductor f in an imaginary quadratic field K. In §9, we introduced the ring class field of $\mathcal{O}$, denoted here by $L_{\mathcal{O}}$. This field is determined via the Existence Theorem of class field theory (Theorem 8.6) by the subgroup $P_{K,\mathbb{Z}}(f) \subset I_K(f)$ generated by principal ideals $\alpha\mathcal{O}_K \in I_K(f)$ where $\alpha \equiv a \bmod f\mathcal{O}_K$ for some $a \in \mathbb{Z}$ (see Proposition 7.22). This implies that

$$\mathrm{Gal}(L_{\mathcal{O}}/K) \simeq I_K(f)/P_{K,\mathbb{Z}}(f) \simeq C(\mathcal{O}),$$

where $C(\mathcal{O})$ is the class group and the final isomorphism is from Proposition 7.22.

The goal of this section is to describe some interesting Abelian extensions of $L_{\mathcal{O}}$ indexed by positive integers m. Following Cho [**22**], we define

$$P_{K,\mathbb{Z},m}(fm) \subset I_K(fm)$$

to be the subgroup generated by the principal ideals $\alpha\mathcal{O}_K \in I_K(fm)$ where $\alpha \in \mathcal{O}_K$ satisfies

$$\alpha \equiv a \bmod fm\mathcal{O}_K \text{ for some } a \in \mathbb{Z} \text{ with } a \equiv 1 \bmod m.$$

Since this subgroup obviously contains $P_{K,1}(fm)$, the Existence Theorem mentioned above gives an extension $L_{\mathcal{O},m}$ of K with Galois group

$$\mathrm{Gal}(L_{\mathcal{O},m}/K) \simeq I_K(fm)/P_{K,\mathbb{Z},m}(fm). \tag{15.16}$$

We call $L_{\mathcal{O},m}$ the *extended ring class field of level m*. Note that $L_{\mathcal{O}_K,m}$ is the ray class field of modulus m, and for an order $\mathcal{O}$ of conductor f, $L_{\mathcal{O},m}$ is contained in the ray class field K_{fm} of modulus fm.

The field $L_{\mathcal{O},m}$ is related to the ring class field $L_{\mathcal{O}}$ as follows:

LEMMA 15.17. *$L_{\mathcal{O},m}$ is a Galois extension of $L_{\mathcal{O}}$, and there is an exact sequence*

$$\mathcal{O}^* \longrightarrow (\mathcal{O}/m\mathcal{O})^* \longrightarrow \mathrm{Gal}(L_{\mathcal{O},m}/L_{\mathcal{O}}) \longrightarrow 1.$$

PROOF. First observe that $I_K(fm) \subset I_K(f)$ induces an inclusion

$$I_K(fm)/(I_K(fm) \cap P_{K,\mathbb{Z}}(f)) \hookrightarrow I_K(f)/P_{K,\mathbb{Z}}(f)$$

which is actually an isomorphism (see Exercise 15.10). Hence

$$\mathrm{Gal}(L_{\mathcal{O}}/K) \simeq I_K(fm)/(I_K(fm) \cap P_{K,\mathbb{Z}}(f)).$$

This shows that the congruence subgroup for $L_{\mathcal{O}}$ for the modulus fm is given by $I_K(fm) \cap P_{K,\mathbb{Z}}(f)$. By Corollary 8.7, the inclusions

$$P_K(fm) \subset P_{K,\mathbb{Z},m}(fm) \subset I_K(fm) \cap P_{K,\mathbb{Z}}(f) \subset I_K(fm)$$

give the inclusion $L_{\mathcal{O}} \subset L_{\mathcal{O},m}$. Then the above isomorphisms for $\mathrm{Gal}(L_{\mathcal{O}}/K)$ and $\mathrm{Gal}(L_{\mathcal{O},m}/K)$ imply that

$$\begin{aligned}\mathrm{Gal}(L_{\mathcal{O},m}/L_{\mathcal{O}}) &= \ker\big(\mathrm{Gal}(L_{\mathcal{O},m}/K) \to \mathrm{Gal}(L_{\mathcal{O}}/K)\big)\\ &\simeq (I_K(fm) \cap P_{K,\mathbb{Z}}(f))/P_{K,\mathbb{Z},m}(fm).\end{aligned}$$

Hence it suffices to show that we have an exact sequence

$$\mathcal{O}^* \longrightarrow (\mathcal{O}/m\mathcal{O})^* \longrightarrow (I_K(fm) \cap P_{K,\mathbb{Z}}(f))/P_{K,\mathbb{Z},m}(fm) \longrightarrow 1. \tag{15.18}$$

An elementary argument (see Exercise 15.10) shows that any class in $(\mathcal{O}/m\mathcal{O})^*$ can be represented by $\alpha \in \mathcal{O}$ relatively prime to fm. Such an α gives the ideal $\alpha\mathcal{O}_K \in I_K(fm)$. To show that $\alpha\mathcal{O}_K \in P_{K,\mathbb{Z}}(f)$, we follow the proof of Proposition 7.22. Write $\mathcal{O}_K = [1, w_K]$, so that $\mathcal{O} = [1, fw_K]$. Then $\alpha = a + bfw_K$ for $a, b \in \mathbb{Z}$, and $\alpha \equiv a \bmod f\mathcal{O}_K$ follows. This shows that $\alpha\mathcal{O}_K \in I_K(fm) \cap P_{K,\mathbb{Z}}(f)$.

We next observe that $\alpha\mathcal{O}_K \in P_{K,\mathbb{Z},m}(fm)$ when α is relatively prime to fm and satisfies $\alpha \equiv 1 \bmod m\mathcal{O}$. This is easy to see, since the congruence gives $\alpha = 1 + m\beta$, and then writing $\beta = c + dfw_K$ for $c, d \in \mathbb{Z}$ shows that $\alpha \equiv 1 + mc \bmod fm\mathcal{O}_K$. It follows that $\alpha\mathcal{O}_K \in P_{K,\mathbb{Z},m}(fm)$.

Now suppose that $\alpha \equiv \beta \bmod m\mathcal{O}$ with α, β relatively prime to fm. Then we can find $\delta \in \mathcal{O}$ with $\beta\delta \equiv 1 \bmod fm\mathcal{O}$. This implies $\alpha\delta \equiv \beta\delta \equiv 1 \bmod m\mathcal{O}$. By the previous paragraph, we obtain

$$\begin{aligned}\alpha\mathcal{O}_K \cdot \delta\mathcal{O}_K &= \alpha\delta\mathcal{O}_K \in P_{K,\mathbb{Z},m}(fm)\\ \beta\mathcal{O}_K \cdot \delta\mathcal{O}_K &= \beta\delta\mathcal{O}_K \in P_{K,\mathbb{Z},m}(fm).\end{aligned}$$

If follows that $\alpha\mathcal{O}_K$ and $\beta\mathcal{O}_K$ are equal modulo $P_{K,\mathbb{Z},m}(fm)$.

This gives a well-defined map $(\mathcal{O}/m\mathcal{O})^* \to (I_K(fm) \cap P_{K,\mathbb{Z}}(f))/P_{K,\mathbb{Z},m}(fm)$. For surjectivity, note that $I_K(fm) \cap P_{K,\mathbb{Z}}(f) \subset I_K(fm)$ is generated by the principal ideals $\alpha\mathcal{O}_K \in I_K(fm)$ where $\alpha \in \mathcal{O}_K$ satisfies

$$\alpha \equiv a \bmod f\mathcal{O}_K \text{ for some } a \in \mathbb{Z}$$

(see Exercise 15.10). This congruence implies $\alpha \in \mathcal{O}$, and surjectivity follows.

To prove that (15.18) is exact at $(\mathcal{O}/m\mathcal{O})^*$, assume that $\alpha \in \mathcal{O}$ maps to $\alpha\mathcal{O}_K \in P_{K,\mathbb{Z},m}(fm)$. Then $\alpha\mathcal{O}_K = (\beta\mathcal{O}_K)(\gamma\mathcal{O}_K)^{-1}$, where $\beta, \gamma \in \mathcal{O}_K$ satisfy $\beta \equiv a \bmod fm\mathcal{O}_K$ with $a \equiv 1 \bmod m$ and $\gamma \equiv b \bmod fm\mathcal{O}_K$ with $b \equiv 1 \bmod m$. Since $f\mathcal{O}_K \subset \mathcal{O}$, we have $\beta, \gamma \in \mathcal{O}$ and $\beta \equiv \gamma \equiv 1 \bmod m\mathcal{O}$. Also, $\alpha\gamma\mathcal{O} = (\alpha\mathcal{O})(\gamma\mathcal{O}) = \beta\mathcal{O}$ by Proposition 7.20 since α and β are relatively prime to f. Hence $\alpha\gamma = u\beta$ for some $u \in \mathcal{O}^*$. This implies that $\alpha \equiv u \bmod m\mathcal{O}$, and exactness follows. Q.E.D.

A pleasant surprise is that for the order $\mathcal{O} = \mathbb{Z}[\sqrt{-n}]$, $p = x^2 + ny^2$ has an interesting relation to extended ring class fields, as proved by Cho [**22**]:

THEOREM 15.19. *Let n and m be positive integers. Then there is a monic irreducible polynomial $f_{n,m}(x) \in \mathbb{Z}[x]$ such that if an odd prime p divides neither nm nor the discriminant of $f_{n,m}(x)$, then*

$$\left\{\begin{matrix} p = x^2 + ny^2 \text{ such that} \\ x \equiv 1 \bmod m,\ y \equiv 0 \bmod m\end{matrix}\right\} \iff \left\{\begin{matrix}(-n/p) = 1 \text{ and } f_{n,m}(x) \equiv 0 \bmod p \\ \text{has an integer solution}\end{matrix}\right\}.$$

Furthermore, $f_{n,m}(x)$ may be taken to be the minimal polynomial of a real algebraic integer α for which $L = K(\alpha)$ is the level m extended ring class field of the order $\mathbb{Z}[\sqrt{-n}]$ in the imaginary quadratic field $K = \mathbb{Q}(\sqrt{-n})$.

PROOF. The proof is similar to the argument used to prove Theorems 9.2 and 9.4. The conductor f of $\mathcal{O} = \mathbb{Z}[\sqrt{-n}]$ is determined by the equation $-4n = f^2 d_K$. Now let p be an odd prime not dividing nm. Then p is unramified in $L_{\mathcal{O},m}$. Furthermore,

$$\begin{aligned} & p = x^2 + ny^2 \text{ with } x \equiv 1 \bmod m,\ y \equiv 0 \bmod m \\ \iff & p\mathcal{O}_K = \mathfrak{p}\bar{\mathfrak{p}},\ \mathfrak{p} = (x + \sqrt{-n}y)\mathcal{O}_K,\ x \equiv 1 \bmod m,\ y \equiv 0 \bmod m \\ \iff & p\mathcal{O}_K = \mathfrak{p}\bar{\mathfrak{p}},\ \mathfrak{p} = \alpha\mathcal{O}_K,\ \alpha \in \mathcal{O},\ \alpha \equiv 1 \bmod m\mathcal{O}. \end{aligned}$$

Since p is relatively prime to fm, the proof of Lemma 15.17 shows that the last condition is equivalent to $\mathfrak{p} \in P_{K,\mathbb{Z},m}(fm)$. From here, the rest of the proof follows as in the proofs of Theorems 9.2 and 9.4. We leave the details as Exercise 15.12. See also Cho [**22**, Theorem 1]. Q.E.D.

Our final task is to relate $L_{\mathcal{O},m}$ to the Artin map $\widehat{K}^* \to \mathrm{Gal}(K^{ab}/K)$ from the discussion leading up to Theorem 15.15. For an order $\mathcal{O}$ of K, define $\widehat{\mathcal{O}} = \mathcal{O} \otimes_{\mathbb{Z}} \widehat{\mathbb{Z}}$ with group of units $\widehat{\mathcal{O}}^*$. Note that $\widehat{\mathcal{O}}^*$ is a subgroup of $\widehat{K}^*$.

LEMMA 15.20. *For the order $\mathcal{O} \subset \mathcal{O}_K$ of conductor f, define*

$$J^1_{\mathcal{O},m} = \{x \in \widehat{\mathcal{O}}^* : x \equiv 1 \bmod m\widehat{\mathcal{O}}\} \subset \widehat{\mathcal{O}}^*.$$

Then there is a commutative diagram

$$\begin{array}{ccccccccc} 1 & \longrightarrow & K^* J^1_{\mathcal{O},m} & \longrightarrow & \widehat{K}^* & \longrightarrow & I_K(fm)/P_{K,\mathbb{Z},m}(fm) & \longrightarrow & 1 \\ & & \downarrow & & \downarrow \scriptstyle\text{Artin} & & \downarrow \scriptstyle\simeq & & \\ 1 & \longrightarrow & \mathrm{Gal}(K^{ab}/L_{\mathcal{O},m}) & \longrightarrow & \mathrm{Gal}(K^{ab}/K) & \longrightarrow & \mathrm{Gal}(L_{\mathcal{O},m}/K) & \longrightarrow & 1 \end{array}$$

with exact rows. Furthermore, the first two downward arrows are surjections and the third is the isomorphism (15.16).

The proof will be given at the end of the section.

D. Shimura Reciprocity for Extended Ring Class Fields

Let K be an imaginary quadratic field and fix a point $\tau_0 \in K \cap \mathfrak{h}$. Then τ_0 is a root of $ax^2 + bx + c$, where $a, b, c \in \mathbb{Z}$ are relatively prime with $a > 0$. Recall from Lemma 7.5 that the lattice $[1, \tau_0]$ is a proper ideal for the order $\mathcal{O} = [1, a\tau_0]$.

Given a modular function $f \in \mathsf{F}$ that is defined at τ_0, Shimura Reciprocity implies that $f(\tau_0)$ lies in some Abelian extension of K. We saw examples in §12 where $f(\tau_0)$ lay in the ring class field $L_{\mathcal{O}}$. This is not true in general, but we still have the following very nice consequence of Shimura Reciprocity:

THEOREM 15.21. *Fix τ_0 and $\mathcal{O}$ as above and assume that $f(\tau_0)$ is defined for a modular function $f \in \mathsf{F}_m$. Then $f(\tau_0) \in L_{\mathcal{O},m}$.*

PROOF. Since $f(\tau_0) \in K^{ab}$ by Shimura Reciprocity, $f(\tau_0) \in L_{\mathcal{O},m}$ once we prove that $f(\tau_0)$ is fixed by $\mathrm{Gal}(K^{ab}/L_{\mathcal{O},m})$. By Lemma 15.20, it suffices to show that $\sigma_x(f(\tau_0)) = f(\tau_0)$ when $x \in K^* J^1_{\mathcal{O},m}$, and since K^* is in the kernel of the Artin map by Exercise 15.9, we can assume that $x \in J^1_{\mathcal{O},m}$,

Observe that $x \in \widehat{\mathcal{O}}^*$ by the definition of $J^1_{\mathcal{O},m}$. Also recall that $a\tau_0^2+b\tau_0+c=0$ where $a,b,c \in \mathbb{Z}$ satisfy $\gcd(a,b,c)=1$ and $a>0$. Furthermore, $\mathcal{O}=[1,a\tau_0]$, and $[1,\tau_0]$ is an $\mathcal{O}$-module since it is a proper fractional $\mathcal{O}$-ideal. It follows that $[1,\tau_0]\otimes_{\mathbb{Z}}\widehat{\mathbb{Z}}$ is an $\widehat{\mathcal{O}}$-module, so that multiplication by x is an automorphism of $[1,\tau_0]\otimes_{\mathbb{Z}}\widehat{\mathbb{Z}}$. This is also a rank 2 free $\widehat{\mathbb{Z}}$-module, so that the transpose of the matrix representing multiplication by x on $[1,\tau_0]\otimes_{\mathbb{Z}}\widehat{\mathbb{Z}}$ relative to the basis $\{\tau_0,1\}$ lies in $\mathrm{GL}(2,\widehat{\mathbb{Z}})$. By the description of $g_{\tau_0}(x)$ given in the paragraph before (15.11), it follows that $g_{\tau_0}(x)\in \mathrm{GL}(2,\widehat{\mathbb{Z}})$.

Then $x \in J^1_{\mathcal{O},m}$ implies that $x \equiv 1 \bmod m\widehat{\mathcal{O}}$, which makes it easy to see that $g_{\tau_0}(x)\equiv I \bmod m$ (see Exercise 15.13). Thus $g_{\tau_0}(x^{-1})\equiv I \bmod m$ as well. Since $f \in \mathsf{F}_m$ is invariant under such matrices, we obtain

$$\sigma_x(f(\tau_0)) = f^{g_{\tau_0}(x^{-1})}(\tau_0) = f(\tau_0),$$

where the first equality uses Shimura Reciprocity (Theorem 15.15). As noted at the beginning of the proof, this implies that $f(\tau_0)\in L_{\mathcal{O},m}$. Q.E.D.

We note that a stronger result holds, namely

$$L_{\mathcal{O},m} = K(f(\tau_0) : f \in \mathsf{F}_m \text{ is defined at } \tau_0).$$

See Cho [**22**, Theorem 4] for a proof.

Here is an example from [**22**] that illustrates Theorems 15.19 and 15.21. Let $K=\mathbb{Q}(i)$ and $\mathcal{O}=\mathcal{O}_K=\mathbb{Z}[i]$. Then $L_{\mathcal{O}}=K$ since $h(\mathcal{O})=1$. It follows from Lemma 15.17 that $L_{\mathcal{O},5}$ has degree 4 over K.

We can apply Theorem 15.19 to any element of F_5. For example, consider the Rogers–Ramanujan continued fraction

$$r(\tau) = \cfrac{q^{1/5}}{1+\cfrac{q}{1+\cfrac{q^2}{1+\cdots}}}.$$

As explained by Duke [**39**], $r(\tau)$ is invariant under $\Gamma(5)$, and it also has a rational q-expansion because of the amazing identity proved by Rogers:

$$r(\tau) = q^{1/5}\prod_{n=1}^{\infty}(1-q^n)^{\left(\frac{5}{n}\right)},$$

where $(5/n)$ is built from the Legendre and Kronecker symbols in the usual way. The transformation formulas for $r(\tau+1)$ and $r(-1/\tau)$ from [**39**, Proposition 2] imply that $r(\tau)\in \mathsf{F}_5$, and then applying Theorem 15.21 with $\tau_0=i$ gives $r(i)\in L_{\mathcal{O},5}$.

In his first letter to Hardy in 1913, Ramanujan gives the value

$$r(i) = \cfrac{e^{-2\pi/5}}{1+\cfrac{e^{-2\pi}}{1+\cfrac{e^{-4\pi}}{1+\cdots}}} = \sqrt{\frac{5+\sqrt{5}}{2}} - \frac{\sqrt{5}+1}{2}$$

for the continued fraction. This is a root of the quartic $x^4+2x^3-6x^2-2x+1$, which is irreducible over K. Since $[L_{\mathcal{O},5}:K]=4$, it follows that

$$L_{\mathcal{O},5} = K(r(i)).$$

Since the discriminant of $x^4+2x^3-6x^2-2x+1$ is $32000 = 2^8 5^3$, Theorem 15.19 implies that for any prime $p > 5$, we have

$$\left\{\begin{matrix} p = x^2+y^2 \text{ with} \\ x \equiv 1 \bmod 5,\ y \equiv 0 \bmod 5 \end{matrix}\right\} \iff \left\{\begin{matrix} p \equiv 1 \bmod 4 \text{ and} \\ x^4+2x^3-6x^2-2x+1 \equiv 0 \bmod p \\ \text{has an integer solution} \end{matrix}\right\}.$$

See Exercise 15.26 for a different approach to this example.

Now suppose that we are in the general situation of Theorem 15.21 and want to know whether or not $f(\tau_0) \in L_{\mathcal{O},m}$ lies in the smaller ring class field $L_{\mathcal{O}}$. Hence we need to understand how the Galois group $\mathrm{Gal}(L_{\mathcal{O},m}/L_{\mathcal{O}})$ acts on $f(\tau_0)$. With a little work, Shimura Reciprocity will describe this action explicitly.

Recall from Lemma 15.17 that we have the exact sequence

$$\mathcal{O}^* \longrightarrow (\mathcal{O}/m\mathcal{O})^* \longrightarrow \mathrm{Gal}(L_{\mathcal{O},m}/L_{\mathcal{O}}) \longrightarrow 1.$$

Thus $u \in (\mathcal{O}/m\mathcal{O})^*$ gives $\sigma_u \in \mathrm{Gal}(L_{\mathcal{O},m}/L_{\mathcal{O}})$. But u also gives a matrix since $[1,\tau_0]$ is an $\mathcal{O}$-module, so that $[1,\tau_0]\otimes_{\mathbb{Z}}\mathbb{Z}/m\mathbb{Z}$ is a module over $\mathcal{O}\otimes_{\mathbb{Z}}\mathbb{Z}/m\mathbb{Z} = \mathcal{O}/m\mathcal{O}$. Multiplication by $u \in (\mathcal{O}/m\mathcal{O})^*$ gives an automorphism m_u of $[1,\tau_0]\otimes_{\mathbb{Z}}\mathbb{Z}/m\mathbb{Z}$. This is a free $\mathbb{Z}/m\mathbb{Z}$-module of rank 2 with $\{\tau_0, 1\}$ as basis. Define $\overline{g}_{\tau_0}(u) \in \mathrm{GL}(2,\mathbb{Z}/m\mathbb{Z})$ to be the transpose of the matrix representing m_u with respect to this basis.

We now state a version of Shimura Reciprocity taken from Stevenhagen [**126**] that is optimized for the extended ring class field $L_{\mathcal{O},m}$:

THEOREM 15.22 (Shimura Reciprocity for $L_{\mathcal{O},m}$). *Fix $f \in \mathsf{F}_m$ and let τ_0 and $\mathcal{O}$ be as in Theorem 15.21. Assume that $f(\tau_0)$ is defined. If $u \in (\mathcal{O}/m\mathcal{O})^*$ gives $\sigma_u \in \mathrm{Gal}(L_{\mathcal{O},m}/L_{\mathcal{O}})$ and $\overline{g}_{\tau_0}(u) \in \mathrm{GL}(2,\mathbb{Z}/m\mathbb{Z})$ as above, then*

$$\sigma_u(f(\tau_0)) = f^{\overline{g}_{\tau_0}(u)}(\tau_0).$$

PROOF. Shimura Reciprocity uses $g_{\tau_0}(x^{-1})$ in Theorem 15.15, yet the above equation uses $\overline{g}_{\tau_0}(u)$ without the inverse. This is unexpected, but once we construct an idele x with $\sigma_x|_{L_{\mathcal{O},m}} = \sigma_u$ and compute $g_{\tau_0}(x)$, everything will become clear.

In the proof of Lemma 15.17, we showed the following:

(i) $(I_K(fm) \cap P_{K,\mathbb{Z}}(f))/P_{K,\mathbb{Z},m}(fm) \simeq \mathrm{Gal}(L_{\mathcal{O},m}/L_{\mathcal{O}})$ via the Artin map.
(ii) We can write $u = [\alpha] \in (\mathcal{O}/m\mathcal{O})^*$ where $\alpha \in \mathcal{O}$ is relatively prime to fm and f is the conductor of $\mathcal{O}$.
(iii) $\alpha\mathcal{O}_K \in I_K(fm) \cap P_{K,\mathbb{Z}}(f)$.

It follows that σ_u is the image of $\alpha\mathcal{O}_K \in I_K(fm) \cap P_{K,\mathbb{Z}}(f)$ under the Artin map for $I_K(fm)$.

Now define the idele $x = (x_p) \in \widehat{K}^*$ as follows:

$$x_p = \begin{cases} 1 & p \nmid fm \\ \alpha^{-1} & p \mid fm, \end{cases}$$

where α^{-1} denotes the element of $K \otimes_{\mathbb{Z}} \mathbb{Z}_p$ given by $\alpha^{-1} \otimes 1$. We claim that $\sigma_x|_{K_{fm}} \in \mathrm{Gal}(K_{fm}/K) \simeq I_K(fm)/P_{K,1}(fm)$ maps to the class of $\alpha\mathcal{O}_K \in I_K(fm)$. Since $L_{\mathcal{O},m} \subset K_{fm}$, this claim and the definition of σ_u will imply that $\sigma_x|_{L_{\mathcal{O},m}} = \sigma_u$.

To compute $\sigma_x|_{K_{fm}}$, we follow the procedure described in the discussion before Shimura Reciprocity (Theorem 15.15). Recall that if $x = \beta y z$ is an fm-factorization as in Lemma 15.14, then $\sigma_x|_{K_{fm}}$ is given by the class of $z\mathcal{O}_K$.

The above definition of x makes it easy to see that $\alpha x \equiv 1 \bmod fm\widehat{\mathcal{O}}_K$. Then part (iii) of Lemma 15.14, applied with fm in place of m, tells us that $z\mathcal{O}_K = \alpha\mathcal{O}_K$. This proves our claim that $\sigma_x|_{K_{fm}}$ maps to the class of $\alpha\mathcal{O}_K$. As noted above, we obtain $\sigma_x|_{L_{\mathcal{O},m}} = \sigma_u$.

If $f \in \mathsf{F}_m$ and $f(\tau_0)$ is defined, then $f(\tau_0) \in L_{\mathcal{O},m}$ by Theorem 15.21, so that by Shimura Reciprocity, we have

$$\sigma_u(f(\tau_0)) = \sigma_x(f(\tau_0)) = f^{g_{\tau_0}(x^{-1})}(\tau_0).$$

However, $f \in \mathsf{F}_m$ implies that $f^{g_{\tau_0}(x^{-1})}$ depends only on $g_{\tau_0}(x^{-1})$ modulo m. The definition of x implies that $x^{-1} = (x_p^{-1})$, where

$$x_p^{-1} = \begin{cases} 1 & p \nmid fm \\ \alpha & p \mid fm. \end{cases}$$

Recall that $g_{\tau_0}(x^{-1})$ is the transpose of the matrix representing multiplication by x^{-1} on $\widehat{K} = K \otimes_{\mathbb{Z}} \widehat{\mathbb{Z}}$ relative to the basis $\{\tau_0, 1\}$. Since $[1, \tau_0]$ is an $\mathcal{O}$-module, $[1, \tau_0] \otimes_{\mathbb{Z}} \widehat{\mathbb{Z}}$ is a module over $\widehat{\mathcal{O}} = \mathcal{O} \otimes_{\mathbb{Z}} \widehat{\mathbb{Z}}$. Then $x^{-1} \in \widehat{\mathcal{O}}_K^*$ implies that $g_{\tau_0}(x^{-1})$ is the transpose of the matrix representing multiplication by x^{-1} on $[1, \tau_0] \otimes_{\mathbb{Z}} \widehat{\mathbb{Z}}$ relative to the basis $\{\tau_0, 1\}$. Using the map $\widehat{\mathbb{Z}} \to \mathbb{Z}/m\mathbb{Z}$ from Exercise 15.3 and the above formula for x^{-1}, multiplication by x^{-1} reduces modulo m to multiplication by $[\alpha] = u$ on $[1, \tau_0] \otimes_{\mathbb{Z}} \mathbb{Z}/m\mathbb{Z}$. It follows immediately that $g_{\tau_0}(x^{-1})$ reduces mod m to $\overline{g}_{\tau_0}(u)$. Thus

$$\sigma_u(f(\tau_0)) = \sigma_x(f(\tau_0)) = f^{g_{\tau_0}(x^{-1})}(\tau_0) = f^{\overline{g}_{\tau_0}(u)}(\tau_0),$$

completing the proof of Theorem 15.22. Q.E.D.

Here is an application that shows how easy Theorem 15.22 is to use:

THEOREM 15.23. *If* $m \equiv 6 \bmod 8$, *then* $\mathfrak{f}_1(\sqrt{-m})^6 \in L_{\mathcal{O}}$ *for* $\mathcal{O} = \mathbb{Z}[\sqrt{-m}]$.

PROOF. This was proved earlier in part (i) of Theorem 12.24 via an ad-hoc argument using $\Gamma_0(32)$. Here we use $\Gamma(8)$ and Shimura Reciprocity with $\tau_0 = \sqrt{-m}$.

For simplicity, set $L = L_{\mathcal{O}}$ and $L_8 = L_{\mathcal{O},8}$. Since $\mathfrak{f}_1(\tau)^6 \in \mathsf{F}_8$ by Proposition 15.2, we have $\mathfrak{f}_1(\sqrt{-m})^6 \in L_8$ by Theorem 15.21. We also have the exact sequence

$$\mathcal{O}^* \to (\mathcal{O}/8\mathcal{O})^* \to \mathrm{Gal}(L_8/L) \to 1$$

from Lemma 15.17. One can check that $\mathcal{O}^* = \{\pm 1\}$, and $m \equiv 6 \bmod 8$ implies that $(\mathcal{O}/8\mathcal{O})^*$ is generated by -1, 3 and $1 + \sqrt{-m}$ (see Exercise 15.14). Thus $\mathrm{Gal}(L_8/L)$ is generated by σ_3 and $\sigma_{1+\sqrt{-m}}$. By Shimura Reciprocity, the theorem will follow once we prove that $\mathfrak{f}_1(\tau)^6$ is invariant under the matrices $\overline{g}_{\sqrt{-m}}(3)$ and $\overline{g}_{\sqrt{-m}}(1 + \sqrt{-m})$ from Theorem 15.22.

This is easy, since 3 and $1 + \sqrt{-m}$ give the matrices

$$\begin{pmatrix} 3 & 0 \\ 0 & 3 \end{pmatrix}, \begin{pmatrix} 1 & -6 \\ 1 & 1 \end{pmatrix} \in \mathrm{GL}(2, \mathbb{Z}/8\mathbb{Z}).$$

The matrix for 3 is obvious, and the matrix for $1 + \sqrt{-m}$ comes from the computation

$$\begin{aligned} (1 + \sqrt{-m}) \cdot \sqrt{-m} &= 1 \cdot \sqrt{-m} - m \cdot 1 = 1 \cdot \sqrt{-m} - 6 \cdot 1 \\ (1 + \sqrt{-m}) \cdot \quad 1 \quad &= 1 \cdot \sqrt{-m} + 1 \cdot 1, \end{aligned}$$

where the first line uses $m \equiv 6 \bmod 8$ (remember that $\overline{g}_{\tau_0}(1+\sqrt{-m})$ uses the transpose of the matrix representing multiplication by $1+\sqrt{-m}$).

The matrix for 3 lies in $\mathrm{SL}(2,\mathbb{Z}/8\mathbb{Z})$, so we lift it to $\gamma = \left(\begin{smallmatrix} a & b \\ c & d\end{smallmatrix}\right) \in \mathrm{SL}(2,\mathbb{Z})$ and compute $\mathfrak{f}_1(\gamma\tau)^6$. Cubing the transformation law (15.3) for $\mathfrak{f}_1(\tau)^2$ implies that

$$\mathfrak{f}_1(\gamma\tau)^6 = i^{-\frac{1}{2}ab+c(d(1-a^2)-a)}\,\mathfrak{f}_1(\tau)^6$$

when b is even. In this equation, we need to use the lift $\gamma = \left(\begin{smallmatrix} a & b \\ c & d\end{smallmatrix}\right) \in \mathrm{SL}(2,\mathbb{Z})$ of $\left(\begin{smallmatrix} 3 & 0 \\ 0 & 3\end{smallmatrix}\right) \in \mathrm{SL}(2,\mathbb{Z}/8\mathbb{Z})$ on the left-hand side of this equation in order for $\gamma\tau$ to make sense. But on the right-hand side, changing the integers a, b, c, d by multiples of 8 doesn't affect the power of i, so we don't actually need to lift—we can use the matrix $\left(\begin{smallmatrix} 3 & 0 \\ 0 & 3\end{smallmatrix}\right)$ in the right-hand side. One computes the power of i to be $i^0 = 1$, so that $\mathfrak{f}_1(\tau)^6$ is invariant under $\overline{g}_{\sqrt{-m}}(3)$.

The matrix for $1+\sqrt{-m}$ does not lie in $\mathrm{SL}(2,\mathbb{Z}/8\mathbb{Z})$, so we factor it as

$$\begin{pmatrix} 1 & -6 \\ 1 & 1\end{pmatrix} = \begin{pmatrix} 1 & 6 \\ 1 & -1\end{pmatrix}\begin{pmatrix} 1 & 0 \\ 0 & -1\end{pmatrix}, \quad \begin{pmatrix} 1 & 6 \\ 1 & -1\end{pmatrix} \in \mathrm{SL}(2,\mathbb{Z}/8\mathbb{Z}).$$

Note that $\left(\begin{smallmatrix} 1 & 0 \\ 0 & -1\end{smallmatrix}\right) \in \mathrm{GL}(2,\mathbb{Z})$ acts trivially on $\mathfrak{f}_1(\tau)^6$ since its q-expansion has rational coefficients. Arguing as for 3 allows us to use $\left(\begin{smallmatrix} 1 & 6 \\ 1 & -1\end{smallmatrix}\right) \in \mathrm{SL}(2,\mathbb{Z}/8\mathbb{Z})$ in the right-hand side. For this matrix, the power of i is $i^{-\frac{1}{2}6-1} = 1$. Hence $\mathfrak{f}_1(\tau)^6$ is also invariant under $\overline{g}_{\sqrt{-m}}(1+\sqrt{-m})$. This completes the proof of the theorem. Q.E.D.

As noted in the proof of part (i) of Theorem 12.24, once we have $\mathfrak{f}_1(\sqrt{-m})^6 \in L_{\mathcal{O}}$, we get the equality

$$L_{\mathcal{O}} = K(\mathfrak{f}_1(\sqrt{-m})^2) \text{ when } m \equiv 6 \bmod 8 \text{ and } 3 \nmid m.$$

Exercise 15.15 will use Shimura Reciprocity to prove part (ii) of Theorem 12.24, which asserts that $L_{\mathcal{O}} = K(\mathfrak{f}(\sqrt{-m})^2)$ when $m \equiv 3 \bmod 4$ and $3 \nmid m$.

The papers [**53**] by Gee and [**111**] by Schertz give numerous examples of values of modular functions that lie in ring class fields. See also the papers [**30**] by Cox, McKay and Stevenhagen and [**64**] by Hajir and Rodríguez-Villegas.

E. Shimura Reciprocity for Ring Class Fields

By Theorem 11.1, the j-invariant of an order $\mathcal{O}$ in an imaginary quadratic field K gives the ring class field $L_{\mathcal{O}} = K(j(\mathcal{O}))$. We also saw in Corollary 11.37 that given a proper $\mathcal{O}$-ideal $\mathfrak{a}$, there is a unique $\sigma_{\mathfrak{a}} \in \mathrm{Gal}(L_{\mathcal{O}}/K)$ such that

$$\sigma_{\mathfrak{a}}(j(\mathfrak{b})) = j(\overline{\mathfrak{a}}\mathfrak{b})$$

for any proper $\mathcal{O}$-ideal $\mathfrak{b}$, and furthermore, the map $\mathfrak{a} \mapsto \sigma_{\mathfrak{a}}$ induces the isomorphism $C(\mathcal{O}) \simeq \mathrm{Gal}(L_{\mathcal{O}}/K)$ given by class field theory.

We now tackle the general problem of describing $\sigma_{\mathfrak{a}}(f(\tau_0))$ when f is a modular function such that $f(\tau_0) \in L_{\mathcal{O}}$. Before explaining how this works, let's recall a particular case studied in §12. Suppose $K = \mathbb{Q}(\sqrt{-14})$ and $\mathcal{O} = \mathcal{O}_K$, so that $L = L_{\mathcal{O}}$ is the Hilbert class field of K. The element of order 2 in $\mathrm{Gal}(L/K) \simeq C(\mathcal{O}) \simeq \mathbb{Z}/4\mathbb{Z}$ is $\sigma = \sigma_{\mathfrak{a}}$ for $\mathfrak{a} = [2, \sqrt{-14}]$. In our computation of $j(\sqrt{-14})$ in §12, one of the key steps was to prove that

$$\sigma(\mathfrak{f}_1(\sqrt{-14})^2) = \mathfrak{f}_2(\sqrt{-14}/2)^2.$$

The proof given in the discussion following (12.30) was an ad-hoc argument using the ring class field of $\mathcal{O}' = [1, 4\sqrt{-14}]$. One goal of this section is to give a better proof that uses Shimura Reciprocity to describe the Galois action on $L_{\mathcal{O}}$.

It turns out to be more convenient to describe $\sigma_{\overline{\mathfrak{a}}}(f(\tau_0))$. Written this way, the above formula for the Galois action on $j(\mathfrak{b})$ becomes

$$\sigma_{\overline{\mathfrak{a}}}(j(\mathfrak{b})) = j(\mathfrak{a}\mathfrak{b}). \tag{15.24}$$

Theorem 15.27 below will generalize this by showing how to compute $\sigma_{\overline{\mathfrak{a}}}(f(\tau_0))$ when $f(\tau_0) \in L_{\mathcal{O}}$.

Let τ_0 be chosen so that $\mathcal{O} = [1, \tau_0]$. This case occurs often in practice. Turning to the proper $\mathcal{O}$-ideal $\mathfrak{a}$, we will assume that

$$\mathfrak{a} = \mathsf{a}[1, \tau_1], \text{ where } \begin{cases} \tau_1 \in \mathfrak{h} \text{ is a root of } \mathsf{a}x^2 + \mathsf{b}x + \mathsf{c} \\ \text{for } \mathsf{a}, \mathsf{b}, \mathsf{c} \in \mathbb{Z} \text{ relatively prime, } \mathsf{a} > 0. \end{cases} \tag{15.25}$$

Lemma 7.5 implies that $\mathcal{O} = [1, \mathsf{a}\tau_1]$. Note that $N(\mathfrak{a}) = \mathsf{a}$, so that $\mathfrak{a}^{-1} = (1/\mathsf{a})\overline{\mathfrak{a}}$. We will also assume that

$$\mathfrak{a} \text{ is relatively prime to the conductor of } \mathcal{O}. \tag{15.26}$$

Since $N(\mathfrak{a}) = \mathsf{a}$, Lemma 7.18 tells us that (15.26) holds if and only if a is relatively prime to the conductor. Theorem 7.7 and Proposition 7.22 guarantee that ideals satisfying (15.25) and (15.26) represent all ideal classes in $C(\mathcal{O})$. Note also that (15.26) is automatic when $\mathcal{O} = \mathcal{O}_K$.

In this situation, we get the following version of Shimura Reciprocity due to Stevenhagen and Gee (see [**53, 54, 126**]):

THEOREM 15.27. *Fix τ_0 with $\mathcal{O} = [1, \tau_0]$ and let $\mathfrak{a}$ and τ_1 be as in* (15.25) *and* (15.26). *If $f \in \mathsf{F}$ satisfies $f(\tau_0) \in L_{\mathcal{O}}$, then there is an explicitly computable $u \in \mathrm{GL}(2, \widehat{\mathbb{Z}})$ such that*

$$\sigma_{\overline{\mathfrak{a}}}(f(\tau_0)) = f^u(\tau_1).$$

For example, when $\mathcal{O} = [1, \tau_0]$ and $\mathfrak{a} = \mathsf{a}[1, \tau_1]$ is relatively prime to the conductor, applying Theorem 15.27 to $f = j$ implies that

$$\sigma_{\overline{\mathfrak{a}}}(j(\tau_0)) = j(\tau_1)$$

since j is invariant under $\mathrm{GL}(2, \widehat{\mathbb{Z}})$ by Theorem 15.6. Thus

$$\sigma_{\overline{\mathfrak{a}}}(j(\mathcal{O})) = j(\mathfrak{a}),$$

exactly as predicted by (15.24). In fact, it is now easy to prove (15.24) in general (see Exercise 15.16).

PROOF OF THEOREM 15.27. The strategy will be to use $\mathfrak{a}$ to construct an idele $x \in \widehat{K}^*$ in Lemma 15.28 below. From x we get the matrix $g_{\tau_0}(x) \in \mathrm{GL}(2, \widehat{\mathbb{Q}})$ that appears in Shimura Reciprocity. By Theorem 15.9, we can write $g_{\tau_0}(x) = uv$ with $u \in \mathrm{GL}(2, \widehat{\mathbb{Z}})$ and $v \in \mathrm{GL}(2, \mathbb{Q})^+$. The u from this decomposition is the u that appears in Theorem 15.27. Once we prove that $\sigma_x|_{L_{\mathcal{O}}} = \sigma_{\mathfrak{a}}$, the theorem will follow easily from Shimura Reciprocity.

In the discussion preceding Lemma 15.20, we introduced $\widehat{\mathcal{O}} = \mathcal{O} \otimes_{\mathbb{Z}} \widehat{\mathbb{Z}} \simeq \prod_p (\mathcal{O} \otimes_{\mathbb{Z}} \mathbb{Z}_p)$, where the last isomorphism follows since $\mathcal{O}$ is a free $\mathbb{Z}$-module of rank 2. We call $\widehat{\mathcal{O}}$ the *ring of $\mathcal{O}$-adeles*, and its group of units $\widehat{\mathcal{O}}^*$ is the *group of $\mathcal{O}$-ideles*. Following Gee [**53**], we use $\mathfrak{a}$ to define an idele $x \in \widehat{K}^*$ as follows:

LEMMA 15.28. *Given a proper $\mathcal{O}$-ideal $\mathfrak{a} = \mathsf{a}[1, \tau_1]$ as above, define $x = (x_p) \in \widehat{K}$ by*

$$x_p = \begin{cases} \mathsf{a} & p \nmid \mathsf{a} \\ \mathsf{a}\tau_1 & p \mid \mathsf{a} \text{ and } p \nmid \mathsf{c} \\ \mathsf{a}(\tau_1 - 1) & p \mid \mathsf{a} \text{ and } p \mid \mathsf{c}. \end{cases}$$

Then $x \in \widehat{K}^*$ and $\mathfrak{a} \otimes_{\mathbb{Z}} \widehat{\mathbb{Z}} = x\widehat{\mathcal{O}}$. Furthermore, x is unique up to multiplication by an $\mathcal{O}$-idele in $\widehat{\mathcal{O}}^*$.

PROOF. Since $\mathfrak{a}$ is a free $\mathbb{Z}$-module of rank 2, we have

$$\mathfrak{a} \otimes_{\mathbb{Z}} \widehat{\mathbb{Z}} = \prod_p (\mathfrak{a} \otimes_{\mathbb{Z}} \mathbb{Z}_p).$$

For x_p as defined above, we claim that

$$\mathfrak{a} \otimes_{\mathbb{Z}} \mathbb{Z}_p = x_p(\mathcal{O} \otimes_{\mathbb{Z}} \mathbb{Z}_p).$$

Recall that by (15.25), $\mathfrak{a} = [\mathsf{a}, \mathsf{a}\tau_1]$, where τ_1 is a root of $\mathsf{a}x^2 + \mathsf{b}x + \mathsf{c}$. We can also assume $\mathcal{O} = [1, \mathsf{a}\tau_1]$ by the discussion following (15.25).

When $p \nmid \mathsf{a}$, a is invertible in $\mathbb{Z}_p$, so that $x_p(\mathcal{O} \otimes_{\mathbb{Z}} \mathbb{Z}_p) = \mathsf{a}(\mathcal{O} \otimes_{\mathbb{Z}} \mathbb{Z}_p) = \mathcal{O} \otimes_{\mathbb{Z}} \mathbb{Z}_p$. Since $\mathsf{a} \in \mathfrak{a}$, we also have $\mathfrak{a} \otimes_{\mathbb{Z}} \mathbb{Z}_p = \mathcal{O} \otimes_{\mathbb{Z}} \mathbb{Z}_p$, so the claim holds for $p \nmid \mathsf{a}$.

When $p \mid \mathsf{a}$ and $p \nmid \mathsf{c}$, we have

$$x_p\mathcal{O} = \mathsf{a}\tau_1\mathcal{O} = \mathsf{a}\tau_1[1, \mathsf{a}\tau_1] = [\mathsf{a}\tau_1, \mathsf{a}(\mathsf{a}\tau_1^2)] = [\mathsf{a}\tau_1, \mathsf{a}(-\mathsf{b}\tau_1 - \mathsf{c})] = [\mathsf{a}\tau_1, \mathsf{ac}].$$

Since c is invertible in $\mathbb{Z}_p$, $x_p(\mathcal{O} \otimes_{\mathbb{Z}} \mathbb{Z}_p) = [\mathsf{a}\tau_1, \mathsf{ac}] \otimes_{\mathbb{Z}} \mathbb{Z}_p = [\mathsf{a}\tau_1, \mathsf{a}] \otimes_{\mathbb{Z}} \mathbb{Z}_p = \mathfrak{a} \otimes_{\mathbb{Z}} \mathbb{Z}_p$. Finally, when $p \mid \mathsf{a}$ and $p \mid \mathsf{c}$, we must have $p \nmid \mathsf{b}$. We leave it as Exercise 15.17 to show that

$$x_p\mathcal{O} = \mathsf{a}(\tau_1 - 1)\mathcal{O} = [\mathsf{a}(\tau_1 - 1), \mathsf{a}(\mathsf{a} + \mathsf{b} + \mathsf{c})].$$

Since $\mathsf{a} + \mathsf{b} + \mathsf{c}$ is invertible in $\mathbb{Z}_p$, we obtain

$$\begin{aligned} x_p(\mathcal{O} \otimes_{\mathbb{Z}} \mathbb{Z}_p) &= [\mathsf{a}(\tau_1 - 1), \mathsf{a}(\mathsf{a} + \mathsf{b} + \mathsf{c})] \otimes_{\mathbb{Z}} \mathbb{Z}_p = [\mathsf{a}(\tau_1 - 1), \mathsf{a}] \otimes_{\mathbb{Z}} \mathbb{Z}_p \\ &= [\mathsf{a}\tau_1, \mathsf{a}] \otimes_{\mathbb{Z}} \mathbb{Z}_p = \mathfrak{a} \otimes_{\mathbb{Z}} \mathbb{Z}_p. \end{aligned}$$

Thus the claim is true for all p, which proves that $x = (x_p)$ satisfies $\mathfrak{a} \otimes_{\mathbb{Z}} \widehat{\mathbb{Z}} = x\widehat{\mathcal{O}}$.

The remaining assertions of the lemma will be proved in Exercise 15.17. Q.E.D.

In Lemma 15.28, x is computed using τ_1, while $g_{\tau_0}(x)$ uses τ_0. Since $\mathcal{O} = [1, \mathsf{a}\tau_1] = [1, \tau_0]$, there is an integer ℓ such that $\mathsf{a}\tau_1 = \tau_0 + \ell$ (see Exercise 15.18). Then the matrix

$$v = \begin{pmatrix} 1 & \ell \\ 0 & \mathsf{a} \end{pmatrix} \in \mathrm{GL}(2, \mathbb{Q})^+$$

has the property that

$$v\tau_0 = \frac{\tau_0 + \ell}{\mathsf{a}} = \frac{\mathsf{a}\tau_1}{\mathsf{a}} = \tau_1.$$

We claim that

$$u = g_{\tau_0}(x)\, v^{-1} \in \mathrm{GL}(2, \widehat{\mathbb{Z}}). \tag{15.29}$$

Once we prove this, we will have the decomposition $g_{\tau_0}(x) = uv$, and (15.29) then gives an explicit formula for the u that appears in the Theorem 15.27.

To prove (15.29), consider the $\widehat{\mathbb{Q}}$-linear map $m_x : \widehat{K} \to \widehat{K}$ that comes from multiplication by x. This map has two properties:

(i) $m_x(\widehat{\mathcal{O}}) = x\widehat{\mathcal{O}} = \mathfrak{a} \otimes_{\mathbb{Z}} \widehat{\mathbb{Z}}$.

(ii) The matrix of m_x with respect to the $\widehat{\mathbb{Q}}$-basis $\{\tau_0, 1\}$ of $\widehat{K}$ is the transpose of the matrix $g_{\tau_0}(x)$.

Now consider the $\widehat{\mathbb{Q}}$-linear map $T : \widehat{K} \to \widehat{K}$ defined by $T(\tau_0) = \mathsf{a}\tau_1 = \tau_0 + \ell$ and $T(1) = \mathsf{a}$. This map has two properties:

(i) $T(\widehat{\mathcal{O}}) = \mathfrak{a} \otimes_{\mathbb{Z}} \widehat{\mathbb{Z}}$.

(ii) The matrix of T with respect to the $\widehat{\mathbb{Q}}$-basis $\{\tau_0, 1\}$ of $\widehat{K}$ is the transpose of the matrix v.

Since m_x and T are automorphisms of $\widehat{K}$ and map $\widehat{\mathcal{O}}$ to the same thing, it follows that $T^{-1} \circ m_x : \widehat{\mathcal{O}} \to \widehat{\mathcal{O}}$ is an isomorphism. But $\widehat{\mathcal{O}}$ is a free $\widehat{\mathbb{Z}}$-module of rank 2, so the matrix of this isomorphism with respect to the basis $\{\tau_0, 1\}$ lies in $\mathrm{GL}(2, \widehat{\mathbb{Z}})$. This matrix is the transpose of $g_{\tau_0}(x)\, v^{-1}$, and (15.29) follows.

Now suppose that

$$\sigma_x|_{L_{\mathcal{O}}} = \sigma_{\mathfrak{a}}. \tag{15.30}$$

Then

$$\sigma_{\overline{\mathfrak{a}}}(f(\tau_0)) = \sigma_{x^{-1}}(f(\tau_0)) = f^{g_{\tau_0}(x)}(\tau_0) = f^{uv}(\tau_0) = f^u(f^v(\tau_0)) = f^u(v\tau_0) = f^u(\tau_1).$$

Here, the first equality follows from $\mathfrak{a}^{-1} = (1/\mathsf{a})\overline{\mathfrak{a}}$ and the fact that $\sigma_{\overline{\mathfrak{a}}}$ depends only on the class of $\overline{\mathfrak{a}}$ in $C(\mathcal{O})$. Also, the second equality is Shimura Reciprocity for x^{-1}, and the fourth equality uses modular action given in Theorem 15.9. This will prove the theorem.

It remains to prove (15.30). Let f be the conductor of $\mathcal{O}$. (The symbol f is also used for the modular function $f \in \mathsf{F}$. However, for the rest of this proof, f will refer exclusively to the conductor.) From the discussion surrounding (15.26), we can assume that $\mathfrak{a}$ is relatively prime to f and $\gcd(\mathsf{a}, f) = 1$.

To define σ_x, we first construct an f-factorization of x as in Lemma 15.14. This is simple to do: $x = \mathsf{a}yz$, where $y = 1$ and $z = \mathsf{a}^{-1}x$. The condition of the lemma for y is obviously satisfied, and since $\gcd(\mathsf{a}, f) = 1$ and $x_p = \mathsf{a}$ when $p \nmid a$,

$$p \mid f \implies p \nmid \mathsf{a} \implies z_p = \mathsf{a}^{-1}x_p = \mathsf{a}^{-1}\mathsf{a} = 1,$$

so the condition for z is also satisfied. Then the ideal $z\mathcal{O}_K$ is relatively prime to f, and the restriction $\sigma_x|_{K_f}$ of σ_x to the ray class field K_f is defined to be the image of the class $[z\mathcal{O}_K]$ via the isomorphism

$$I_K(f)/P_{K,1}(f) \simeq \mathrm{Gal}(K_f/K)$$

induced by the Artin map. It follows that $\sigma_x|_{L_{\mathcal{O}}}$ is the image of the class $[z\mathcal{O}_K]$ via the isomorphism

$$I_K(f)/P_{K,\mathbb{Z}}(f) \simeq \mathrm{Gal}(L_{\mathcal{O}}/K)$$

induced by the Artin map, as explained in §9. However, when we combine this isomorphism with Proposition 7.22, we get

$$C(\mathcal{O}) \simeq I_K(f)/P_{K,\mathbb{Z}}(f) \simeq \mathrm{Gal}(L_{\mathcal{O}}/K),$$

where for an $\mathcal{O}$-ideal $\mathfrak{b}$ relatively prime to f, the first isomorphism takes $[\mathfrak{b}]$ to $[\mathfrak{b}\mathcal{O}_K]$. Since $\mathfrak{a}$ is relatively prime to f, $\sigma_{\mathfrak{a}}$ is determined by $[\mathfrak{a}\mathcal{O}_K] \in I_K(f)/P_{K,\mathbb{Z}}(f)$. Yet we just noted that $\sigma_x|_{L_{\mathcal{O}}}$ is determined by $[z\mathcal{O}_K] \in I_K(f)/P_{K,\mathbb{Z}}(f)$. Thus (15.30) reduces to showing that $[z\mathcal{O}_K] = [\mathfrak{a}\mathcal{O}_K]$.

We will study $z\mathcal{O}_K$ and $\mathfrak{a}\mathcal{O}_K$ separately, beginning with $z\mathcal{O}_K$. Note that $\mathsf{a}\mathcal{O}_K \in P_{K,\mathbb{Z}}(f)$ since $\gcd(\mathsf{a}, f) = 1$. Then $x = \mathsf{a}z$ implies that

$$[z\mathcal{O}_K] = [(\mathsf{a}\mathcal{O}_K)(z\mathcal{O}_K)] = [\mathsf{a}z\mathcal{O}_K] = [x\mathcal{O}_K]$$

in $I_K(f)/P_{K,\mathbb{Z}}(f)$, where the second equality uses Lemma 15.13. Turning to $\mathfrak{a}\mathcal{O}_K$, recall that $\mathfrak{a} \otimes_{\mathbb{Z}} \widehat{\mathbb{Z}} = x\widehat{\mathcal{O}}$ by Lemma 15.28. By Exercise 15.18, $\mathfrak{a} \otimes_{\mathbb{Z}} \widehat{\mathbb{Z}} = \mathfrak{a}\widehat{\mathcal{O}}$, so that $\mathfrak{a}\widehat{\mathcal{O}} = x\widehat{\mathcal{O}}$, which in turn implies that $(\mathfrak{a}\mathcal{O}_K)\widehat{\mathcal{O}}_K = x\widehat{\mathcal{O}}_K$. Thus

$$\mathfrak{a}\mathcal{O}_K = K \cap ((\mathfrak{a}\mathcal{O}_K)\widehat{\mathcal{O}}_K) = K \cap x\widehat{\mathcal{O}}_K = x\mathcal{O}_K,$$

where the first equality follows from Lemma 15.13 applied to $\mathfrak{a}\mathcal{O}_K$ and the last equality is the definition of $x\mathcal{O}_K$. So $\mathfrak{a}\mathcal{O}_K = x\mathcal{O}_K$, which when combined with the above equation for $[z\mathcal{O}_K]$, proves that $[z\mathcal{O}_K] = [\mathfrak{a}\mathcal{O}_K]$. Thus (15.30) holds and Theorem 15.27 is proved. Q.E.D.

Given $\mathfrak{a} = [\mathsf{a}, \mathsf{a}\tau_1]$ and $\mathcal{O} = [1, \tau_0]$, the version of Shimura Reciprocity presented in Theorem 15.27 is very explicit. Here is a nice example of how it works:

THEOREM 15.31. *Let L be the Hilbert class field of $K = \mathbb{Q}(\sqrt{-14})$ and let σ be the element of order* 2 *in* $\mathrm{Gal}(L/K) \simeq \mathbb{Z}/4\mathbb{Z}$. *Then $\mathfrak{f}_1(\sqrt{-14})^2 \in L$ and*

$$\sigma(\mathfrak{f}_1(\sqrt{-14})^2) = \mathfrak{f}_2(\sqrt{-14}/2)^2.$$

PROOF. Since L is the ring class field of $\mathcal{O}_K = [1, \sqrt{-14}]$, $\mathfrak{f}_1(\sqrt{-14})^6 \in L$ by Theorem 15.23. We first show that $\sigma(\mathfrak{f}_1(\sqrt{-14})^6) = \mathfrak{f}_2(\sqrt{-14}/2)^6$.

The element of order 2 in $C(\mathcal{O}_K) \simeq C(-56) \simeq \mathbb{Z}/4\mathbb{Z}$ is represented by the ideal $\mathfrak{a} = [2, \sqrt{-14}]$. This follows, for example, from the reduced forms listed in (2.14). Hence the automorphism σ in the statement of the theorem is $\sigma = \sigma_{\mathfrak{a}} = \sigma_{\overline{\mathfrak{a}}}$.

We have $\tau_0 = \sqrt{-14}$ and $\tau_1 = \sqrt{-14}/2$, so that

$$v = \begin{pmatrix} 1 & 0 \\ 0 & 2 \end{pmatrix}.$$

Since τ_1 is a root of $2x^2 + 7$, the idele $x = (x_p)$ from Lemma 15.28 is

$$x_p = \begin{cases} 2 & p \neq 2 \\ \sqrt{-14} & p = 2. \end{cases}$$

When computing $g_{\tau_0}(x) \in \mathrm{GL}(2, \widehat{\mathbb{Q}})$, it is most convenient to work one prime at a time using the injective map from (15.10)

$$\mathrm{GL}(2, \widehat{\mathbb{Q}}) \longrightarrow \prod_p \mathrm{GL}(2, \mathbb{Q}_p),$$

where $g_{\tau_0}(x)$ maps to an element of the product denoted $(g_{\tau_0}(x_p))$. Thus $g_{\tau_0}(x_p)$ is the transpose of the matrix of multiplication by x_p with respect to the basis $\{\tau_0, 1\}$. Since $\tau_0 = \sqrt{-14}$, one easily computes that

$$g_{\tau_0}(x_p) = \begin{cases} \begin{pmatrix} 2 & 0 \\ 0 & 2 \end{pmatrix} = \begin{pmatrix} 2 & 0 \\ 0 & 1 \end{pmatrix} \begin{pmatrix} 1 & 0 \\ 0 & 2 \end{pmatrix} & p \neq 2 \\[2ex] \begin{pmatrix} 0 & -14 \\ 1 & 0 \end{pmatrix} = \begin{pmatrix} 0 & -7 \\ 1 & 0 \end{pmatrix} \begin{pmatrix} 1 & 0 \\ 0 & 2 \end{pmatrix} & p = 2. \end{cases}$$

Then $u = g_{\tau_0}(x)v^{-1}$ is given by $u = (u_p) \in \mathrm{GL}(2, \widehat{\mathbb{Z}}) = \prod_p \mathrm{GL}(2, \mathbb{Z}_p)$, where

$$u_p = \begin{cases} \begin{pmatrix} 2 & 0 \\ 0 & 1 \end{pmatrix} & p \neq 2 \\ \begin{pmatrix} 0 & -7 \\ 1 & 0 \end{pmatrix} & p = 2. \end{cases} \tag{15.32}$$

With this value of u, Theorem 15.27 implies that

$$\sigma(\mathfrak{f}_1(\sqrt{-14})^6) = \mathfrak{f}_1^u(\sqrt{-14}/2)^6. \tag{15.33}$$

It remains to compute $\mathfrak{f}_1^u(\tau)^6$. The key point is that $\mathfrak{f}_1(\tau)^6 \in \mathsf{F}_8$ by Proposition 15.2, so it is invariant under $\Gamma(8)$. Hence we only need u modulo 8 in order to compute $\mathfrak{f}_1^u(\tau)^6$. In particular, we can ignore all primes $\neq 2$. By (15.32), this means replacing u with

$$\begin{pmatrix} 0 & -7 \\ 1 & 0 \end{pmatrix} = \begin{pmatrix} 0 & 1 \\ 1 & 0 \end{pmatrix} \in \mathrm{GL}(2, \mathbb{Z}/8\mathbb{Z}).$$

To compute how this matrix acts on $\mathfrak{f}_1(\tau)^6$, we use the methods from the proof of Theorem 15.23. Write

$$\begin{pmatrix} 0 & 1 \\ 1 & 0 \end{pmatrix} = \begin{pmatrix} 0 & -1 \\ 1 & 0 \end{pmatrix} \begin{pmatrix} 1 & 0 \\ 0 & -1 \end{pmatrix}, \quad \begin{pmatrix} 0 & -1 \\ 1 & 0 \end{pmatrix} \in \mathrm{SL}(2, \mathbb{Z}/8\mathbb{Z}).$$

The second matrix in the product gives the Galois action on the coefficients of the q-expansion of $\mathfrak{f}_1(\tau)^6$, which is trivial the coefficients are rational. The first matrix is the standard generator $S = \left(\begin{smallmatrix} 0 & -1 \\ 1 & 0 \end{smallmatrix}\right)$ of $\mathrm{SL}(2, \mathbb{Z})$. Since $\mathfrak{f}_1(-1/\tau) = \mathfrak{f}_2(\tau)$ by Corollary 12.19, we see that

$$\mathfrak{f}_1^u(\tau)^6 = \mathfrak{f}_1^S(\tau)^6 = \mathfrak{f}_2(\tau)^6.$$

Combining this with (15.33), we obtain

$$\sigma(\mathfrak{f}_1(\sqrt{-14})^6) = \mathfrak{f}_1^u(\sqrt{-14}/2)^6 = \mathfrak{f}_1^S(\sqrt{-14}/2)^6 = \mathfrak{f}_2(\sqrt{-14}/2)^6. \tag{15.34}$$

In the proof of of Theorem 12.24, we showed that $\mathfrak{f}_1(\sqrt{-14})^2 \in L$ follows from $\mathfrak{f}_1(\sqrt{-14})^6 \in L$, and then the discussion surrounding (12.32) shows how the theorem follows from (15.34). This completes the proof. Q.E.D.

The situation encountered in Theorem 15.31 is especially simple because we only had to consider $p = 2$ since $\mathfrak{f}_1(\tau)^6 \in \mathsf{F}_8$. When working with $f \in \mathsf{F}_m$, one needs to consider all primes p dividing m and use the Chinese Remainder Theorem to combine the corresponding u_p's into a matrix $u_m \in \mathrm{GL}(2, \mathbb{Z}/m\mathbb{Z})$. Once we have u_m, the next step is to write it as a product $u_m = \widetilde{\gamma}\left(\begin{smallmatrix} 1 & 0 \\ 0 & d \end{smallmatrix}\right)$ with $\widetilde{\gamma} \in \mathrm{SL}(2, \mathbb{Z}/m\mathbb{Z})$. Systematic methods for lifting $\widetilde{\gamma}$ to $\gamma \in \mathrm{SL}(2, \mathbb{Z})$ are described in Gee [**53**], and then Shimura Reciprocity reduces to

$$\sigma(f(\tau_0)) = f^u(\tau_1) = f^\gamma(\tau_1)$$

when the q-expansion of f is rational. The papers [**53, 54**] by Gee and Stevenhagen give numerous applications of this approach to Shimura Reciprocity.

F. Class Field Theory

As we approach the end of §15, two tasks remain: explain how the K-adeles $\widehat{K}$ and K-ideles $\widehat{K}^*$ relate to the usual adeles and ideles of class field theory, and use tools from algebraic number theory and class field theory to prove some earlier results whose proofs were deferred until now.

For an imaginary quadratic field K, the idele group $\mathbf{I}_K$ defined at the end of §8 is the *restricted product*

$$\mathbf{I}_K = \prod_{\mathfrak{p}}{}^{*} K_{\mathfrak{p}}^* = \mathbb{C}^* \times \prod_{\mathfrak{p}\text{ finite}}{}^{*} K_{\mathfrak{p}}^*$$

since K has only one infinite prime, which is complex. Recall that $\prod_{\mathfrak{p}}^*$ means that $x = (x_{\mathfrak{p}}) \in \prod_{\mathfrak{p}} K_{\mathfrak{p}}^*$ lies in $\mathbf{I}_K$ if and only if $x_{\mathfrak{p}} \in \mathcal{O}_{K_{\mathfrak{p}}}^*$ for all but finitely many $\mathfrak{p}$. Here, $K_{\mathfrak{p}}^* = \mathcal{O}_{K_{\mathfrak{p}}}^* = \mathbb{C}^*$ when $\mathfrak{p}$ is infinite, and for a finite prime $\mathfrak{p}$, $K_{\mathfrak{p}}$ and $\mathcal{O}_{K_{\mathfrak{p}}}$ are the completions of K and $\mathcal{O}_K$ at the prime ideal $\mathfrak{p}$.

In a similar way, we have the *adele ring*

$$\mathbf{A}_K = \mathbb{C} \times \prod_{\mathfrak{p}\text{ finite}}{}^{*} K_{\mathfrak{p}}.$$

where $\prod_{\mathfrak{p}}^*$ now means that for $\mathfrak{p}$ finite, $x_{\mathfrak{p}} \in \mathcal{O}_{K_{\mathfrak{p}}}$ for all but finitely many $\mathfrak{p}$. Adeles were not mentioned in §8 but will play a useful role in what follows. Note that $\mathbf{I}_K$ is the group of units of $\mathbf{A}_K$, i.e., $\mathbf{I}_K = \mathbf{A}_K^*$. The completions used in the definitions of $\mathbf{A}_K$ and $\mathbf{I}_K$ are described most books on algebraic number theory and class field theory, including [**11, 57, 78, 89, 100, 113**].

In the imaginary quadratic case, the $\mathbb{C}^*$ factor has no influence on the class field theory. For this reason, we use the *finite adele ring* and *finite idele group*, which are the restricted products

$$\mathbf{A}_K^{\mathsf{fin}} = \prod_{\mathfrak{p}\text{ finite}}{}^{*} K_{\mathfrak{p}} \quad\text{and}\quad \mathbf{I}_K^{\mathsf{fin}} = \prod_{\mathfrak{p}\text{ finite}}{}^{*} K_{\mathfrak{p}}^*,$$

where we now use only finite primes $\mathfrak{p} \subset \mathcal{O}_K$. Thus

$$\mathbf{A}_K = \mathbb{C}^* \times \mathbf{A}_K^{\mathsf{fin}} \quad\text{and}\quad \mathbf{I}_K = \mathbb{C}^* \times \mathbf{I}_K^{\mathsf{fin}}.$$

For the rest of the section, $\mathfrak{p}$ will always denote a finite prime. With this convention,

$$\mathbf{A}_K^{\mathsf{fin}} = \prod_{\mathfrak{p}}{}^{*} K_{\mathfrak{p}} \quad\text{and}\quad \mathbf{I}_K^{\mathsf{fin}} = \prod_{\mathfrak{p}}{}^{*} K_{\mathfrak{p}}^*.$$

We relate $\mathbf{A}_K^{\mathsf{fin}}$ and $\mathbf{I}_K^{\mathsf{fin}}$ to $\widehat{K}$ and $\widehat{K}^*$ as follows. For an ordinary prime $p \in \mathbb{Z}$, we have compatible isomorphisms

$$K \otimes_{\mathbb{Z}} \mathbb{Z}_p \simeq K \otimes_{\mathbb{Q}} \mathbb{Q}_p \simeq \prod_{p \in \mathfrak{p}} K_{\mathfrak{p}} \quad\text{and}\quad \mathcal{O}_K \otimes_{\mathbb{Z}} \mathbb{Z}_p \simeq \prod_{p\in\mathfrak{p}} \mathcal{O}_{K_{\mathfrak{p}}}, \tag{15.35}$$

where $\mathbb{Q}_p$ is the field of p-adic numbers, and $K \otimes_{\mathbb{Z}} \mathbb{Z}_p \simeq K \otimes_{\mathbb{Q}} \mathbb{Q}_p$ follows from $\mathbb{Q}_p = \mathbb{Q} \otimes_{\mathbb{Z}} \mathbb{Z}_p$. The other isomorphisms in (15.35) are proved in Theorem 1 and Proposition 4 of Serre [**113**, Chapter II]. Hence we have isomorphisms

$$\mathbf{A}_K^{\mathsf{fin}} = \prod_{\mathfrak{p}}{}^{*} K_{\mathfrak{p}} \simeq \prod_{p}{}^{*} (K \otimes_{\mathbb{Z}} \mathbb{Z}_p) \quad\text{and}\quad \mathbf{I}_K^{\mathsf{fin}} = \prod_{\mathfrak{p}}{}^{*} K_{\mathfrak{p}}^* \simeq \prod_{p}{}^{*} (K \otimes_{\mathbb{Z}} \mathbb{Z}_p)^*,$$

where $\prod_p^*$ means that x_p is in $\mathcal{O}_K \otimes_{\mathbb{Z}} \mathbb{Z}_p$ (on the right) or in $(\mathcal{O}_K \otimes_{\mathbb{Z}} \mathbb{Z}_p)^*$ (on the left) for all but finitely many p. However, we know from Exercise 15.8 that the natural injection $\widehat{K} \to \prod_p (K \otimes_{\mathbb{Z}} \mathbb{Z}_p)$ induces isomorphisms

$$\widehat{K} \simeq \prod_p{}^*(K \otimes_{\mathbb{Z}} \mathbb{Z}_p) \quad\text{and}\quad \widehat{K}^* \simeq \prod_p{}^*(K \otimes_{\mathbb{Z}} \mathbb{Z}_p)^*.$$

It follows that we have isomorphisms

$$\widehat{K} \simeq \mathbf{A}_K^{\mathsf{fin}} \text{ and } \widehat{K}^* \simeq \mathbf{I}_K^{\mathsf{fin}}. \tag{15.36}$$

This explains why elements of $\widehat{K}$ and $\widehat{K}^*$ are called adeles and ideles, though strictly speaking, they should be called *finite adeles* and *finite ideles*.

The idele groups $\widehat{K}^*$ and $\mathbf{I}_K^{\mathsf{fin}}$ both have Artin maps that are compatible with (15.36), as we will see in Proposition 15.40 below. The proof uses ideas from the proofs of Lemmas 15.13 and 15.14, so we first prove these lemmas.

Proof of Lemma 15.13. We need to show the following:

(i) If $x \in \widehat{K}^*$, then $x\mathcal{O}_K = K \cap x\widehat{\mathcal{O}}_K$ is a fractional $\mathcal{O}_K$-ideal.
(ii) $x \mapsto x\mathcal{O}_K$ defines a surjective homomorphism $\widehat{K}^* \to I_K$ with kernel $\widehat{\mathcal{O}}_K^*$.
(iii) If $\mathfrak{a}$ is a fractional $\mathcal{O}_K$-ideal, then $\mathfrak{a} = K \cap \mathfrak{a}\widehat{\mathcal{O}}_K$.

The proof will use the isomorphism $\widehat{K} \simeq \mathbf{A}_K^{\mathsf{fin}} = \prod_{\mathfrak{p}}^* K_{\mathfrak{p}}$ from (15.36). We will also use some properties of $\mathcal{O}_{K_\mathfrak{p}}$ and the $\mathfrak{p}$-adic valuation $\nu_\mathfrak{p} : K_\mathfrak{p}^* \to \mathbb{Z}$ on $K_\mathfrak{p}$:

(1) $\mathfrak{p}\mathcal{O}_{K_\mathfrak{p}}$ is the only nonzero prime ideal of $\mathcal{O}_{K_\mathfrak{p}}$.
(2) If $u \in K_\mathfrak{p}^*$, then $u\mathcal{O}_{K_\mathfrak{p}} = \mathfrak{p}^{\nu_\mathfrak{p}(u)}\mathcal{O}_{K_\mathfrak{p}}$.
(3) If $u \in K^*$, then $u\mathcal{O}_K = \prod_\mathfrak{p} \mathfrak{p}^{\nu_\mathfrak{p}(u)}$.

Property (1) has the consequence that if $\mathfrak{a} = \prod_\mathfrak{p} \mathfrak{p}^{n_\mathfrak{p}}$ is a fractional $\mathcal{O}_K$-ideal, then $\mathfrak{a}\mathcal{O}_{K_\mathfrak{p}} = \mathfrak{p}^{n_\mathfrak{p}}\mathcal{O}_{K_\mathfrak{p}}$ for all $\mathfrak{p}$. Also note that $\nu_\mathfrak{p} : K_\mathfrak{p}^* \to \mathbb{Z}$ is a homomorphism.

We first prove (iii). One can check that (15.36) is compatible with the inclusions $K \hookrightarrow \widehat{K}^*$ and $K \hookrightarrow \mathbf{A}_K^{\mathsf{fin}}$. Given a fractional $\mathcal{O}_K$-ideal $\mathfrak{a}$, it follows that $\mathfrak{a}\widehat{\mathcal{O}}_K$ maps to $\mathfrak{a}\prod_\mathfrak{p} \mathcal{O}_{K_\mathfrak{p}}$. Hence

$$K \cap \mathfrak{a}\widehat{\mathcal{O}}_K = K \cap (\mathfrak{a}\prod_\mathfrak{p} \mathcal{O}_{K_\mathfrak{p}}),$$

If $\mathfrak{a} = \prod_\mathfrak{p} \mathfrak{p}^{n_\mathfrak{p}}$ is the prime factorization of $\mathfrak{a}$, then the consequence of property (1) noted above implies that $\mathfrak{a}\mathcal{O}_{K_\mathfrak{p}} = \mathfrak{p}^{n_\mathfrak{p}}\mathcal{O}_{K_\mathfrak{p}}$. Thus

$$K \cap \mathfrak{a}\widehat{\mathcal{O}}_K = K \cap \prod_\mathfrak{p} \mathfrak{p}^{n_\mathfrak{p}}\mathcal{O}_{K_\mathfrak{p}},$$

so that we are reduced to proving

$$\mathfrak{a} = K \cap \prod_\mathfrak{p} \mathfrak{p}^{n_\mathfrak{p}}\mathcal{O}_{K_\mathfrak{p}}. \tag{15.37}$$

If $\beta \in \mathfrak{a} = \prod_\mathfrak{p} \mathfrak{p}^{n_\mathfrak{p}}$, then $\beta \in K$ and $\beta \in \mathfrak{p}^{n_\mathfrak{p}} \subset \mathfrak{p}^{n_\mathfrak{p}}\mathcal{O}_{K_\mathfrak{p}}$ for all $\mathfrak{p}$, so β is in the intersection on the right. For the opposite inclusion, take β in the intersection on the right. Then $\beta \in K$ and $\beta \in \mathfrak{p}^{n_\mathfrak{p}}\mathcal{O}_{K_\mathfrak{p}}$ for all $\mathfrak{p}$. Since $\beta\mathcal{O}_{K_\mathfrak{p}} = \mathfrak{p}^{\nu_\mathfrak{p}(\beta)}\mathcal{O}_{K_\mathfrak{p}}$ by property (2), we have $\nu_\mathfrak{p}(\beta) \geq n_\mathfrak{p}$ for all $\mathfrak{p}$. But $\beta\mathcal{O}_K = \prod_\mathfrak{p} \mathfrak{p}^{\nu_\mathfrak{p}(\beta)}$ by property (3), so the inequality just proved implies that $\beta \in \beta\mathcal{O}_K \subset \prod_\mathfrak{p} \mathfrak{p}^{n_\mathfrak{p}} = \mathfrak{a}$.

To prove (i), write (15.36) as $x = (x_p) \mapsto x^* = (x^*_{\mathfrak{p}})$. Then $x\widehat{\mathcal{O}}_K \subset \widehat{K}$ maps to

$$x^* \prod_{\mathfrak{p}} \mathcal{O}_{K_\mathfrak{p}} = \prod_{\mathfrak{p}} x^*_{\mathfrak{p}} \mathcal{O}_{K_\mathfrak{p}} = \prod_{\mathfrak{p}} \mathfrak{p}^{\nu_\mathfrak{p}(x^*_\mathfrak{p})} \mathcal{O}_{K_\mathfrak{p}},$$

where the second equality uses property (2). Note that $\nu_\mathfrak{p}(x^*_\mathfrak{p}) = 0$ for all but finitely many $\mathfrak{p}$ since $x \in \widehat{K}^*$. Since $x \mapsto x^*$ is compatible with $K \hookrightarrow \widehat{K}^*$ and $K \hookrightarrow \mathbf{A}_K^{\mathrm{fin}}$, we have

$$x\mathcal{O}_K = K \cap x\widehat{\mathcal{O}}_K = K \cap \prod_{\mathfrak{p}} \mathfrak{p}^{\nu_\mathfrak{p}(x^*_\mathfrak{p})} \mathcal{O}_{K_\mathfrak{p}},$$

where the first equality is the definition of $x\mathcal{O}_K$. Applying (15.37) to the fractional $\mathcal{O}_K$-ideal $\mathfrak{a} = \prod_\mathfrak{p} \mathfrak{p}^{\nu_\mathfrak{p}(x^*_\mathfrak{p})}$, we conclude that $x\mathcal{O}_K = \mathfrak{a}$. This shows that $x\mathcal{O}_K$ is a fractional $\mathcal{O}_K$-ideal, and (i) is proved.

Finally, for (ii), first note that the above formula for $x\mathcal{O}_K$ and (15.37) applied to $\mathfrak{a} = \prod_\mathfrak{p} \mathfrak{p}^{\nu_\mathfrak{p}(x^*_\mathfrak{p})}$ imply that

$$x\mathcal{O}_K = K \cap \prod_{\mathfrak{p}} \mathfrak{p}^{\nu_\mathfrak{p}(x^*_\mathfrak{p})} \mathcal{O}_{K_\mathfrak{p}} = \prod_{\mathfrak{p}} \mathfrak{p}^{\nu_\mathfrak{p}(x^*_\mathfrak{p})}. \tag{15.38}$$

Since $x \mapsto x^*$ and $\nu_\mathfrak{p}$ are homomorphisms, it follows that $x \mapsto x\mathcal{O}_K$ defines a homomorphism $\widehat{K}^* \to I_K$. Furthermore, let $\mathfrak{a} = \prod_\mathfrak{p} \mathfrak{p}^{n_\mathfrak{p}}$ be any fractional $\mathcal{O}_K$-ideal and pick $\pi_\mathfrak{p} \in \mathfrak{p} - \mathfrak{p}^2$ for every $\mathfrak{p}$, so that $\nu_\mathfrak{p}(\pi_\mathfrak{p}) = 1$. By (15.36), there is $x \in \widehat{K}^*$ such that $x \mapsto (x^*_\mathfrak{p}) = (\pi_\mathfrak{p}^{n_\mathfrak{p}})$, and $x\mathcal{O}_K = \mathfrak{a}$ follows easily from (15.38). Thus $\widehat{K}^* \to I_K$ is surjective. To determine the kernel, note that if $x \mapsto x^* = (x^*_\mathfrak{p})$, then

$$\begin{aligned} x\mathcal{O}_K = \mathcal{O}_K &\iff \prod_{\mathfrak{p}} \mathfrak{p}^{\nu_\mathfrak{p}(x^*_\mathfrak{p})} = \mathcal{O}_K \iff \nu_\mathfrak{p}(x^*_\mathfrak{p}) = 0 \text{ for all } \mathfrak{p} \\ &\iff x^*_\mathfrak{p} \in \mathcal{O}^*_{K_\mathfrak{p}} \text{ for all } \mathfrak{p} \iff x \in \widehat{\mathcal{O}}^*_K, \end{aligned}$$

where on the top line, the first equivalence uses (15.38), and on the bottom line, the second equivalence follows because (15.36) induces an isomorphism $\widehat{\mathcal{O}}^*_K \simeq \prod_\mathfrak{p} \mathcal{O}^*_{K_\mathfrak{p}}$ via (15.35). Q.E.D.

We next turn to the proof of Lemma 15.14. We will use the following version of the Approximation Theorem that is specially adapted to $\widehat{K}^*$:

LEMMA 15.39. *Given $x = (x_p) \in \widehat{K}^*$ and a positive integer $m = \prod_{p|m} p^{n_p}$, there is $\alpha \in K^*$ such that $\alpha x_p \in 1 + p^{n_p}(\mathcal{O}_K \otimes_\mathbb{Z} \mathbb{Z}_p)$ for all $p \mid m$.*

PROOF. The Approximation Theorem (see, for example, Theorem 1.1 from Janusz [**78**, Chapter IV]) tells us that given an $\mathcal{O}_K$-ideal $\mathfrak{m} = \prod_{\mathfrak{p}|\mathfrak{m}} \mathfrak{p}^{n_\mathfrak{p}}$ and $u_\mathfrak{p} \in K$ for $\mathfrak{p} \mid \mathfrak{m}$, there is $\alpha \in K$ such that $\alpha - u_\mathfrak{p} \in \mathfrak{p}^{n_\mathfrak{p}}$ for all $\mathfrak{p} \mid \mathfrak{m}$. We need the closely related result that given $\mathfrak{m}$ and $u_\mathfrak{p} \in K^*_\mathfrak{p}$ for $\mathfrak{p} \mid \mathfrak{m}$, there is $\alpha \in K^*$ such that $\alpha u_\mathfrak{p} \in 1 + \mathfrak{p}^{n_\mathfrak{p}} \mathcal{O}_{K_\mathfrak{p}}$ for all $\mathfrak{p} \mid \mathfrak{m}$. This will be proved in Exercise 15.19.

Suppose $x_p \mapsto (x^*_\mathfrak{p})_{p \in \mathfrak{p}}$ under the isomorphism $K \otimes_\mathbb{Z} \mathbb{Z}_p \simeq \prod_{p \in \mathfrak{p}} K_\mathfrak{p}$ from (15.35). When we write $m\mathcal{O}_K = \prod_{\mathfrak{p}|m\mathcal{O}_K} \mathfrak{p}^{n_\mathfrak{p}}$, the result stated in the previous paragraph gives $\alpha \in K^*$ such that $\alpha x^*_\mathfrak{p} \in 1 + \mathfrak{p}^{n_\mathfrak{p}} \mathcal{O}_{K_\mathfrak{p}}$ for all $\mathfrak{p} \mid m\mathcal{O}_K$. Then

$$\alpha x_p \longmapsto (\alpha x^*_\mathfrak{p})_{p \in \mathfrak{p}} \in \prod_{p \in \mathfrak{p}} (1 + \mathfrak{p}^{n_\mathfrak{p}} \mathcal{O}_{K_\mathfrak{p}})$$

via (15.35). Since $p^{n_p}\mathcal{O}_K = \prod_{p\in\mathfrak{p}} \mathfrak{p}^{n_\mathfrak{p}}$, (15.35) induces a bijection

$$1 + p^{n_p}(\mathcal{O}_K \otimes_\mathbb{Z} \mathbb{Z}_p) \simeq \prod_{p\in\mathfrak{p}} (1 + \mathfrak{p}^{n_\mathfrak{p}} \mathcal{O}_{K_\mathfrak{p}}),$$

which implies that $\alpha x_p \in 1 + p^{n_p}(\mathcal{O}_K \otimes_\mathbb{Z} \mathbb{Z}_p)$, as desired. Q.E.D.

PROOF OF LEMMA 15.14. Fix $m > 0$. We need to prove the following:

(i) An idele $x = (x_p) \in \widehat{K}^*$ has an m-factorization $x = \beta yz$ with $\beta \in K^*$, $y \in J_m^1$ and $z = (z_p) \in \widehat{K}^*$ with $z_p = 1$ when $p \mid m$. Recall that $J_m^1 = \{y \in \widehat{\mathcal{O}}_K^* : y \equiv 1 \bmod m\widehat{\mathcal{O}}_K\}$.

(ii) For $x = \beta yz$ as above, $[z\mathcal{O}_K] \in I_K(m)/P_{K,1}(m)$ depends only on x.

(iii) Given $x \in \widehat{\mathcal{O}}_K^*$ and $\alpha \in \mathcal{O}_K$ with $\alpha x \equiv 1 \bmod m\widehat{\mathcal{O}}_K$, there is an m-factorization $x = \alpha^{-1}yz$ with $z\mathcal{O}_K = \alpha\mathcal{O}_K$.

For (i), we adapt the proof of Proposition (8.1) of Neukirch [**100**, Chapter IV] to $\widehat{K}^*$. Let $m = \prod_{p\mid m} p^{n_p}$. By the approximation result proved in Lemma 15.39, there is $\alpha \in K$ such that $\alpha x_p \in 1+p^{n_p}(\mathcal{O}_K \otimes_\mathbb{Z} \mathbb{Z}_p)$ when $p \mid m$. Now define $y = (y_p)$ and $z = (z_p)$ by

$$y_p = \begin{cases} \alpha x_p & p \mid m \\ 1 & p \nmid m \end{cases} \qquad z_p = \begin{cases} 1 & p \mid m \\ \alpha x_p & p \nmid m. \end{cases}$$

Then $y_p \in 1 + p^{n_p}(\mathcal{O}_K \otimes_\mathbb{Z} \mathbb{Z}_p)$ when $p \mid m$, so that $y \in \widehat{\mathcal{O}}_K^*$ and $y \equiv 1 \bmod m\widehat{\mathcal{O}}_K$. It is easy to see that $z \in \widehat{K}^*$, and we also have $z_p = 1$ when $p \mid m$. Finally, the definitions of y and z imply that $yz = \alpha x$, so that setting $\beta = \alpha^{-1}$ gives the m-factorization $x = \beta yz$.

For (ii), note that by Lemma 15.13, an m-factorization $x = \beta yz$ gives the fractional $\mathcal{O}_K$-ideal $z\mathcal{O}_K$. To show that $z\mathcal{O}_K \in I_K(m)$, let $z \mapsto z^* = (z_\mathfrak{p}^*) \in \mathbf{I}_K^{\mathrm{fin}}$ via (15.36). Then

$$z\mathcal{O}_K = \prod_\mathfrak{p} \mathfrak{p}^{\nu_\mathfrak{p}(z_\mathfrak{p}^*)}$$

by (15.38) with $x = z$. Since $z_p = 1$ when $p \mid m$, we have $z_\mathfrak{p}^* = 1$ and hence $\nu_\mathfrak{p}(z_\mathfrak{p}^*) = 0$ when $\mathfrak{p} \mid m\mathcal{O}_K$. It follows immediately that $z\mathcal{O}_K \in I_K(m)$.

To finish (ii), we need to show that the class $[z\mathcal{O}_K] \in I_K(m)/P_{K,1}(m)$ does not depend on which m-factorization we use. Consider m-factorizations $x = \beta yz = \tilde\beta\tilde y\tilde z$. In Exercise 15.20 we will see that $\tilde\beta\beta^{-1}\mathcal{O}_K \in P_{K,1}(m)$, and we also know that y, $\tilde y \in \widehat{\mathcal{O}}_K^*$. Working in $I_K(m)/P_{K,1}(m)$ and using Lemma 15.13, it follows that

$$\begin{aligned}[\tilde z\mathcal{O}_K] &= [\tilde y\tilde z\mathcal{O}_K] = [\tilde\beta^{-1}x\mathcal{O}_K] = [\tilde\beta\beta^{-1}\mathcal{O}_K][\tilde\beta^{-1}x\mathcal{O}_K] \\ &= [\beta^{-1}x\mathcal{O}_K] = [yz\mathcal{O}_K] = [z\mathcal{O}_K].\end{aligned}$$

Finally, to prove (iii), write $m = \prod_{p\mid m} p^{n_p}$ and note that $\alpha x \equiv 1 \bmod m\widehat{\mathcal{O}}_K$ implies that $\alpha x_p \in 1 + p^{n_p}(\mathcal{O}_K \otimes_\mathbb{Z} \mathbb{Z}_p)$ when $p \mid m$, exactly as in Lemma 15.39. It follows that the construction given in the proof of (i) yields $x = \beta yz$ with $\beta = \alpha^{-1}$. Then $\alpha x = yz$, so that

$$z\mathcal{O}_K = yz\mathcal{O}_K = \alpha x\mathcal{O}_K = \alpha\mathcal{O}_K,$$

where the first and third equalities follow since $x, y \in \widehat{\mathcal{O}}_K^*$ imply that $x\mathcal{O}_K = y\mathcal{O}_K = \mathcal{O}_K$ by Lemma 15.13. Q.E.D.

Our next task is to bring Artin maps into the picture. In the discussion leading up to Shimura Reciprocity, we showed that $x \in \widehat{K}^*$ gives $\sigma_x \in \mathrm{Gal}(K^{ab}/K)$, and the resulting map

$$\widehat{K}^* \longrightarrow \mathrm{Gal}(K^{ab}/K)$$

was called the *Artin map*. There also an Artin map $\Phi_K : \mathbf{I}_K \to \mathrm{Gal}(K^{ab}/K)$ in the idelic version of class field theory. It is well-known that Φ_K is surjective (see [**57**, **98**]). Furthermore, when we write $\mathbf{I}_K = \mathbb{C}^* \times \mathbf{I}_K^{\mathsf{fin}}$, Φ_K is trivial on the $\mathbb{C}^*$ factor (see [**57**, **98**]). It follows that the map from $\mathbf{I}_K^{\mathsf{fin}}$ to $\mathrm{Gal}(K^{ab}/K)$ is surjective. By abuse of notation, we will call this the *Artin map* and write it as

$$\Phi_K : \mathbf{I}_K^{\mathsf{fin}} \longrightarrow \mathrm{Gal}(K^{ab}/K)$$

The two Artin maps relate nicely as follows:

PROPOSITION 15.40. *The isomorphism $\widehat{K}^* \simeq \mathbf{I}_K^{\mathsf{fin}}$ from* (15.36) *is compatible with the Artin maps described above, i.e., there is a commutative diagram*

$$\begin{array}{ccc} \widehat{K}^* & \xrightarrow{\ \sim\ } & \mathbf{I}_K^{\mathsf{fin}} \\ & {\scriptstyle \text{Artin}} \searrow & \downarrow {\scriptstyle \Phi_K} \\ & & \mathrm{Gal}(K^{ab}/K) \end{array}$$

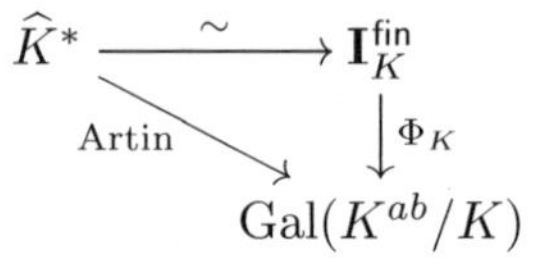

PROOF. Let K_m be the ray class field for modulus $m\mathcal{O}_K$. Since $K^{ab} = \bigcup_m K_m$, we need only show that

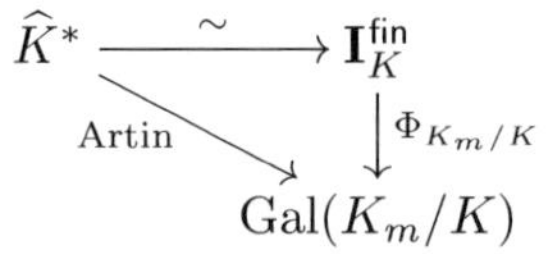

$$\begin{array}{ccc} \widehat{K}^* & \xrightarrow{\ \sim\ } & \mathbf{I}_K^{\mathsf{fin}} \\ & {\scriptstyle \text{Artin}} \searrow & \downarrow {\scriptstyle \Phi_{K_m/K}} \\ & & \mathrm{Gal}(K_m/K) \end{array}$$

commutes for every m. Recall from the discussion following Lemma 15.14 that the Artin map takes $x \in \widehat{K}^*$ to $\sigma_x \in \mathrm{Gal}(K^{ab}/K)$, where $\sigma_x|_{K_m}$ is the Artin map from Theorem 8.2 applied to $z\mathcal{O}_K \in I_K(m)$ for an m-factorization $x = \beta y z$. Thus, ignoring the dotted arrow for now, we need to show that

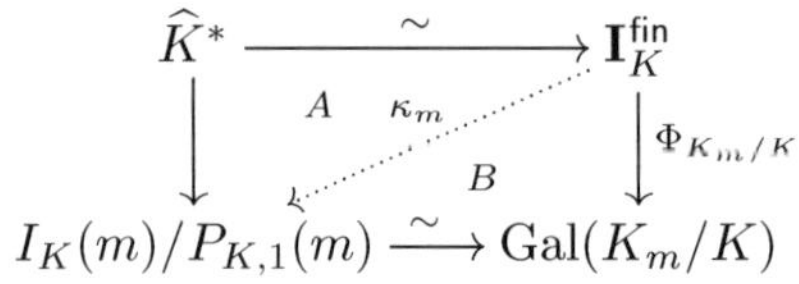

$$\begin{array}{ccc} \widehat{K}^* & \xrightarrow{\ \sim\ } & \mathbf{I}_K^{\mathsf{fin}} \\ \downarrow & A \quad \kappa_m \swarrow \quad B & \downarrow {\scriptstyle \Phi_{K_m/K}} \\ I_K(m)/P_{K,1}(m) & \xrightarrow{\ \sim\ } & \mathrm{Gal}(K_m/K) \end{array}$$

is a commutative diagram, where the vertical map on the left takes $x = \beta y z$ to the class of $z\mathcal{O}_K$. Commutativity will be proved using the dotted map κ_m constructed by Neukirch in Proposition (8.1) of [**100**, Chapter IV]. His definition of κ_m implies that triangle B in the diagram commutes by Theorem (8.2) of [**100**, Chapter IV].

The proposition will follow once we show that triangle A also commutes. To do this, we compare Neukirch's approach to our definition of the Artin map. As in the proof of Lemma 15.13, denote the isomorphism $\widehat{K}^* \simeq \mathbf{I}_K^{\mathsf{fin}}$ by $x \mapsto x^*$. Now fix $x \in \widehat{K}^*$ and $m > 0$, and let $x = \beta y z$ be an m-factorization. Then $x = \beta y z$ maps to $x^* = \beta y^* z^*$, and in Exercise 15.21, we will see that

$$y^* \in \prod_{\mathfrak{p}} \mathcal{O}_{K_\mathfrak{p}}^*, \ y_\mathfrak{p}^* \equiv 1 \bmod \mathfrak{p}^{n_\mathfrak{p}} \mathcal{O}_{K_\mathfrak{p}} \text{ when } \mathfrak{p} \mid m\mathcal{O}_K \ \text{ and } \ z_\mathfrak{p}^* = 1 \text{ when } \mathfrak{p} \mid m\mathcal{O}_K,$$

where $m\mathcal{O}_K = \prod_{\mathfrak{p}} \mathfrak{p}^{n_\mathfrak{p}}$. Neukirch defines $\kappa_m(x^*)$ to be the class of $z^*\mathcal{O}_K = \prod_{\mathfrak{p}} \mathfrak{p}^{\nu_\mathfrak{p}(z^*_\mathfrak{p})}$ in $I_K(m)/P_{K,1}(m)$. Then $z^*\mathcal{O}_K = z\mathcal{O}_K$ by (15.38), and it follows that triangle A of the diagram commutes. The proof is now complete. Q.E.D.

Our final task is to prove Lemma 15.20.

PROOF OF LEMMA 15.20. Recall that $J^1_{\mathcal{O},m} = \{x \in \widehat{\mathcal{O}}^* : x \equiv 1 \bmod m\widehat{\mathcal{O}}\}$. Then we need to prove that there is a commutative diagram

$$\begin{array}{ccccccc} 1 \longrightarrow & K^*J^1_{\mathcal{O},m} & \longrightarrow & \widehat{K}^* & \longrightarrow & I_K(fm)/P_{K,\mathbb{Z},m}(fm) & \to 1 \\ & \downarrow & & \downarrow \text{Artin} & & \downarrow \simeq & \\ 1 \to & \mathrm{Gal}(K^{ab}/L_{\mathcal{O},m}) & \to & \mathrm{Gal}(K^{ab}/K) & \longrightarrow & \mathrm{Gal}(L_{\mathcal{O},m}/K) & \longrightarrow 1 \end{array}$$

with exact rows, where the first two downward arrows are surjections and the third is the isomorphism (15.16).

Some parts of the proof are easy. The bottom row is exact by Galois theory, and the vertical isomorphism on the right follows from the construction of $L_{\mathcal{O},m}$. Also, as noted earlier in the section, the Artin map $\mathbf{I}_K^{\mathsf{fin}} \to \mathrm{Gal}(K^{ab}/K)$ is surjective, so the Artin map $\widehat{K}^* \to \mathrm{Gal}(K^{ab}/K)$ is surjective by Proposition 15.40. For the square on the right, commutativity is also easy and is left as Exercise 15.22.

Let H be the kernel of $\widehat{K}^* \to I_K(fm)/P_{K,\mathbb{Z},m}(fm)$ and consider the diagram

$$\begin{array}{ccccccc} 1 \longrightarrow & H & \longrightarrow & \widehat{K}^* & \longrightarrow & I_K(fm)/P_{K,\mathbb{Z},m}(fm) & \to 1 \\ & \downarrow & & \downarrow \text{Artin} & & \downarrow \simeq & \\ 1 \to & \mathrm{Gal}(K^{ab}/L_{\mathcal{O},m}) & \to & \mathrm{Gal}(K^{ab}/K) & \longrightarrow & \mathrm{Gal}(L_{\mathcal{O},m}/K) & \longrightarrow 1 \end{array}$$

where the first vertical map is the restriction of the second to H. Then the top row is exact and the square on the left commutes. By what we have proved so far, it follows that the diagram commutes and has exact rows. Furthermore, an easy diagram chase show that the first vertical map is surjective. So the proof of Lemma 15.20 reduces to showing that $H = K^*J^1_{\mathcal{O},m}$.

To prove $K^*J^1_{\mathcal{O},m} \subset H$, it suffices to show that $J^1_{\mathcal{O},m} \subset H$ since $K^* \subset H$ by Exercise 15.9. Take $x \in J^1_{\mathcal{O},m}$ and note that $x \in \widehat{\mathcal{O}}^* \subset \widehat{\mathcal{O}}^*_K$. By Exercise 15.23, $\mathcal{O}_K \subset \widehat{\mathcal{O}}_K$ induces an isomorphism

$$(\mathcal{O}_K/fm\mathcal{O}_K)^* \simeq (\widehat{\mathcal{O}}_K/fm\widehat{\mathcal{O}}_K)^*,$$

so there is $\alpha \in \mathcal{O}_K$ such that

$$\alpha x \equiv 1 \bmod fm\widehat{\mathcal{O}}_K. \tag{15.41}$$

By part (iii) of Lemma 15.14, x has an fm-factorization with $z\mathcal{O}_K = \alpha\mathcal{O}_K$.

Thus we need to prove that $\alpha\mathcal{O}_K \in P_{K,\mathbb{Z},m}(fm)$. By definition, $x \in J^1_{\mathcal{O},m}$ means that $x \in \widehat{\mathcal{O}}^*$ and $x \equiv 1 \bmod m\widehat{\mathcal{O}}$. Thus $x = 1 + m\gamma$ for $\gamma \in \widehat{\mathcal{O}}$. However, $\mathcal{O} \subset \mathcal{O}_K$ has conductor f, so that $\mathcal{O}_K = [1, w_K]$ and $\mathcal{O} = [1, fw_K]$ by Lemma 7.2. Since $\widehat{\mathcal{O}} = \mathcal{O} \otimes_{\mathbb{Z}} \widehat{\mathbb{Z}}$, there are $u, v \in \widehat{\mathbb{Z}}$ such that

$$x = 1 + m\gamma = 1 + m(u + fw_Kv) = (1 + mu) + fmw_Kv.$$

Hence

$$x \equiv 1 + mu \bmod fm\widehat{\mathcal{O}}_K. \tag{15.42}$$

By Exercise 15.23, (15.41) and (15.42) imply that $\alpha\mathcal{O}_K \in P_{K,\mathbb{Z},m}(fm)$. Since $z\mathcal{O}_K = \alpha\mathcal{O}_K$, we have $x \in H$. This completes the proof of $K^* J^1_{\widehat{\mathcal{O}},m} \subset H$.

To show that $H \subset K^* J^1_{\widehat{\mathcal{O}},m}$, assume for the moment that the following is true:

$$\text{(15.43)}\qquad \begin{array}{l}\text{Every element of } P_{K,\mathbb{Z},m}(fm) \text{ can be written as } \tilde{z}\mathcal{O}_K \\ \text{for an } fm\text{-factorization } \tilde{x} = \tilde{\beta}\tilde{y}\tilde{z} \text{ of some } \tilde{x} \in J^1_{\widehat{\mathcal{O}},m}.\end{array}$$

Now take $x \in H$. Then $z\mathcal{O}_K \in P_{K,\mathbb{Z},m}(fm)$ for some fm-factorization $x = \beta y z$. By (15.43), there is $\tilde{x} \in J^1_{\widehat{\mathcal{O}},m}$ and an fm-factorization $\tilde{x} = \tilde{\beta}\tilde{y}\tilde{z}$ such that $\tilde{z}\mathcal{O}_K = z\mathcal{O}_K$. By Lemma 15.13, this implies that $z = u\tilde{z}$ for $u \in \widehat{\mathcal{O}}_K^*$. Also, $u_p = 1$ for $p \mid fm$ since the same is true for z_p and $\tilde{z}_p$. This easily implies that $u \equiv 1 \bmod fm\widehat{\mathcal{O}}_K$, so that $u \in J^1_{fm}$. Hence

$$\begin{aligned} x = \beta y z = \beta y(u\tilde{z}) &= \beta(yu)\tilde{z} = \beta(yu)(\tilde{\beta}^{-1}\tilde{y}^{-1}\tilde{x}) \\ &= (\beta\tilde{\beta}^{-1})(yu\tilde{y}^{-1})\tilde{x} \in K^* J^1_{fm} J^1_{\widehat{\mathcal{O}},m}. \end{aligned}$$

But $J^1_{fm} \subset J^1_{\widehat{\mathcal{O}},m}$ by Exercise 15.23, so that $x \in K^* J^1_{\widehat{\mathcal{O}},m}$. Thus, once we prove (15.43), we will have $H \subset K^* J^1_{\widehat{\mathcal{O}},m}$ and the proof of the lemma will be complete.

To prove (15.43), let $\alpha\mathcal{O}_K$ be a generator of $P_{K,\mathbb{Z},m}(fm)$. Then $\alpha \in \mathcal{O}_K$ is relatively prime to fm and satisfies $\alpha \equiv a \bmod fm\mathcal{O}_K$ where $a \in \mathbb{Z}$ and $a \equiv 1 \bmod m$. These conditions imply that $\gcd(a, fm) = 1$. Thus $a \in \mathbb{Z}_p^*$ whenever $p \mid fm$. Now define the idele $\tilde{x} = (\tilde{x}_p)$ by

$$\tilde{x}_p = \begin{cases} 1 & p \nmid fm \\ a^{-1} & p \mid fm. \end{cases}$$

We first show that $\tilde{x} \in J^1_{\widehat{\mathcal{O}},m}$, i.e., that $\tilde{x} \in \widehat{\mathcal{O}}^*$ and $\tilde{x} \equiv 1 \bmod m\widehat{\mathcal{O}}$. Since $a^{-1} \in \mathbb{Z}_p^*$ when $p \mid fm$, it follows that $\tilde{x} \in \widehat{\mathbb{Z}}^*$. Furthermore, $a \equiv 1 \bmod m$ implies that $a^{-1} \equiv 1 \bmod m\mathbb{Z}_p$ when $p \mid fm$ (remember that m is invertible in $\mathbb{Z}_p$ when $p \mid fm$ but $p \nmid m$). Thus $\tilde{x} \equiv 1 \bmod m\widehat{\mathbb{Z}}$. Since $\widehat{\mathcal{O}} = \mathcal{O} \otimes_{\mathbb{Z}} \widehat{\mathbb{Z}}$, it follows that $\tilde{x} \in \widehat{\mathcal{O}}^*$ and $\tilde{x} \equiv 1 \bmod m\widehat{\mathcal{O}}$. Hence $\tilde{x} \in J^1_{\widehat{\mathcal{O}},m}$.

The next step is to find an fm-factorization of $\tilde{x} = \tilde{\beta}\tilde{y}\tilde{z}$ such that $\tilde{z}\mathcal{O}_K = \alpha\mathcal{O}_K$. The definition of $\tilde{x}$ makes this easy. First note that

$$p \mid fm \implies \alpha\tilde{x}_p \equiv \alpha a^{-1} \equiv a a^{-1} \equiv 1 \bmod fm(\mathcal{O}_K \otimes_{\mathbb{Z}} \mathbb{Z}_p)$$

since $\alpha \equiv a \bmod fm\mathcal{O}_K$ and $\tilde{x}_p = a^{-1}$ in the case. Furthermore, since $\alpha \in \mathcal{O}_K$ and $fm \in \mathbb{Z}_p^*$ when $p \nmid fm$, we also have

$$p \nmid fm \implies \alpha\tilde{x}_p \equiv \alpha \equiv 1 \bmod fm(\mathcal{O}_K \otimes_{\mathbb{Z}} \mathbb{Z}_p)$$

since $\tilde{x}_p = 1$ in this case. This implies that $\alpha\tilde{x} \equiv 1 \bmod fm\widehat{\mathcal{O}}_K$, so by part (iii) of Lemma 15.14, $\tilde{x}$ has an fm-factorization with $\tilde{z}\mathcal{O}_K = \alpha\mathcal{O}_K$. Since $\alpha\mathcal{O}_K$ was an arbitrary generator of $P_{K,\mathbb{Z},m}(fm)$, it follows that $\tilde{x} \in J^1_{\widehat{\mathcal{O}},m}$ can be chosen so that $\tilde{z}\mathcal{O}_K$ is an arbitrary element of $P_{K,\mathbb{Z},m}(fm)$. This proves (15.43) and completes the proof of Lemma 15.20. Q.E.D.

The treatment of Shimura Reciprocity given in this section focused on what was needed for §12 and does not give the full story of this remarkable result. The reader should consult references [**22, 30, 53, 54, 64, 111, 126**] already mentioned, together with some of the *Further Readings for Chapter* 4 listed in the *Further Reading* section that follows the References to get a better idea of how Shimura

Reciprocity is used in practice. Applications are also given in the books by Lang [**90**] and Shimura [**115**]. In spite of our many omissions, we nevertheless hope to have convinced you that Shimura Reciprocity is both powerful and surprisingly easy to use.

G. Exercises

15.1. Prove that F_m is a field and show that $\mathbb{Q}(j)$ is the field F_1 defined in the text. Hint: use Theorem 11.9 and Exercise 11.12.

15.2. Here are some details from the proof of Proposition 15.2.

(a) The Dedekind η-function $\eta(\tau)$ has the following transformation law. Suppose that $\gamma = \left(\begin{smallmatrix} a & b \\ c & d \end{smallmatrix}\right) \in \mathrm{SL}(2,\mathbb{Z})$ with $c \geq 0$ and $d > 0$ if $c = 0$. Then Schertz [**111**, Proposition 2] states that

$$\eta(\gamma\tau) = \epsilon(\gamma)\sqrt{c\tau + d}\,\eta(\tau),$$

where the square root is chosen so that $\mathrm{Re}(\sqrt{c\tau + d}) > 0$, and

$$\epsilon(\gamma) = \left(\frac{a}{c_1}\right)\zeta_{24}^{ba + c(d(1-a^2)-a) + 3(a-1)c_1 + \lambda\frac{3}{2}(a^2-1)}.$$

Here, (a/c_1) is the Legendre symbol, and c_1 and λ are determined by

$$c > 0\colon\ c = 2^\lambda c_1,\ \ c_1 \text{ odd}$$
$$c = 0\colon\ c_1 = \lambda = 1.$$

The paper [**64**] by Hajir and Rodríguez-Villegas explores a more conceptual way of thinking about $\epsilon(\gamma)$.

Use the transformation law for $\eta(\tau)$ to prove (15.3) for $\mathfrak{f}_1(\tau)^2$ when $\gamma \in \Gamma_0(2)^t$. Hint: $\mathfrak{f}_1(\tau) = \eta(\tau/2)/\eta(\tau)$. Also, if $\gamma \in \mathrm{SL}(2,\mathbb{Z})$, then the transformation law for η applies to either γ or $-\gamma$.

(b) Use part (a) to show that $\mathfrak{f}_1(\tau)^2$ is invariant under $\Gamma(24)$.

(c) Show that $\mathfrak{f}_1(\tau)^2 \in \mathsf{F}_{24}$. Hint: analyze the q-expansions using the method of Proposition 12.25.

(d) Prove that $\mathfrak{f}_1(\tau)^6 \in \mathsf{F}_8$. Note that part (b) of Exercise 12.23 asked you to prove that $\mathfrak{f}_1(\tau)^6$ is invariant under the group $\tilde{\Gamma}(8)$ from Exercise 12.21 using (12.26)). Note also that $\Gamma(8) \subseteq \tilde{\Gamma}(8)$.

15.3. As in the text, let $\widehat{\mathbb{Z}} = \prod_p \mathbb{Z}_p$.

(a) The inclusion $\mathbb{Z} \subset \widehat{\mathbb{Z}}$ induces a map $\mathbb{Z}/m\mathbb{Z} \to \widehat{\mathbb{Z}}/m\widehat{\mathbb{Z}}$. Prove that this map is an isomorphism. Hint: use the prime factorization of m and the Chinese Remainder Theorem. Property (ii) of $\mathbb{Z}_p$ will be useful.

(b) The isomorphism of part (a) gives maps $\widehat{\mathbb{Z}} \to \mathbb{Z}/m\mathbb{Z}$ for all integers $m > 0$. When $m \mid n$, prove that these maps are compatible with the natural map $\mathbb{Z}/n\mathbb{Z} \to \mathbb{Z}/m\mathbb{Z}$.

15.4. Consider the map $\mathrm{GL}(2,\widehat{\mathbb{Z}})/\{\pm I\} \to \mathrm{Gal}(\mathsf{F}/\mathsf{F}_1)$ constructed in the proof of Proposition 15.8. The goal of this exercise is to prove that the map is a bijection using property (iii) of $\mathbb{Z}_p$ and the isomorphism

$$\mathrm{GL}(2,\widehat{\mathbb{Z}}) \simeq \prod_p \mathrm{GL}(2,\mathbb{Z}_p)$$

induced by $\widehat{\mathbb{Z}} = \prod_p \mathbb{Z}_p$. So $\gamma \in \mathrm{GL}(2,\widehat{\mathbb{Z}})$ can be written $\gamma = (\gamma_p)$, $\gamma_p \in \mathrm{GL}(2,\mathbb{Z}_p)$.

(a) Show that the map is injective. Hint: suppose $\gamma \in \mathrm{GL}(2,\widehat{\mathbb{Z}})$ satisfies $\gamma \equiv \pm I \bmod m$ for all $m > 0$, where *a priori* the $\pm$ depends on m. Show that in fact the $\pm$ is independent of m. Then consider the entries of γ_p mod p^ℓ and use the uniqueness of x in of property (iii).

(b) For surjectivity, note that $\sigma \in \mathrm{Gal}(\mathsf{F}/\mathsf{F}_1)$ gives a compatible family of elements $h_m \in \mathrm{GL}(2,\mathbb{Z}/m\mathbb{Z})/\{\pm I\}$ via diagram (15.7). Prove that this family can be lifted to a compatible family $g_m \in \mathrm{GL}(2,\mathbb{Z}/m\mathbb{Z})$. Hint: fix $m_0 > 2$ and let $g_{m_0} \in \mathrm{GL}(2,\mathbb{Z}/m_0\mathbb{Z})$ be a lifting of h_{m_0}. Then construct compatible liftings $g_{mm_0} \in \mathrm{GL}(2,\mathbb{Z}/mm_0\mathbb{Z})$ for all $m > 0$, and define g_m to be the image of g_{mm_0} in $\mathrm{GL}(2,\mathbb{Z}/m\mathbb{Z})$.

(c) Use property (iii) of $\mathbb{Z}_p$ to construct $\gamma \in \mathrm{GL}(2,\widehat{\mathbb{Z}})$ that maps to σ. Hint: first construct $\gamma_p \in \mathrm{GL}(2,\mathbb{Z}_p)$ that maps to $g_{p^\ell} \in \mathrm{GL}(2,\mathbb{Z}/p^\ell\mathbb{Z})$ for all ℓ.

15.5. We will study the action of $\mathrm{GL}(2,\mathbb{Q})^+$ on F.

(a) Assume that $\gamma \in \mathrm{GL}(2,\mathbb{Q})^+$ has integer entries and set $N = \det(\gamma)$. Prove that $\gamma\Gamma(mN)\gamma^{-1} \subseteq \Gamma(m)$. Hint: write an element of $\Gamma(mN)$ as $I + mN\delta$, where δ has integer entries.

(b) Use part (a) to prove that f^γ is invariant under $\Gamma(mN)$ when $f \in \mathsf{F}_m$.

(c) Prove that $f^\gamma \in \mathsf{F}_{mN}$ when $f \in \mathsf{F}_m$. Hint: for the cusps, take $\delta \in \mathrm{SL}(2,\mathbb{Z})$ and consider $f^\gamma(\delta\tau) = f(\gamma\delta\tau)$. By Exercise 7.15, there is $B \in \mathrm{GL}(2,\mathbb{Z})$ such that $B\gamma\delta = \left(\begin{smallmatrix} a & b \\ 0 & d \end{smallmatrix}\right)$. Explain why B can be chosen to lie in $\mathrm{SL}(2,\mathbb{Z})$ with $a, d > 0$. Then use the expansion of $f(B^{-1}\tau)$ in powers of $q^{1/m}$ to show that $f^\gamma(\delta\tau)$ has a suitable expansion in powers of $q^{1/mN}$.

15.6. Prove that the image of (15.10) consists of all $\gamma = (\gamma_p) \in \prod_p \mathrm{GL}(2,\mathbb{Q}_p)$ such that $\gamma_p \in \mathrm{GL}(2,\mathbb{Z}_p)$ for all but finitely many p.

15.7. Consider the ray class fields $K_{\mathfrak{m}}$ and the Galois group $\mathrm{Gal}(K^{ab}/K)$. As in the text, we set $K_m = K_{m\mathcal{O}_K}$ when m is a positive integer.

(a) Prove that if $\mathfrak{m} \mid \mathfrak{n}$, then $K_{\mathfrak{m}} \subset K_{\mathfrak{n}}$. Hint: the Artin map $\Phi_{\mathfrak{m}} : I_K(\mathfrak{m}) \to \mathrm{Gal}(K_{\mathfrak{m}}/K)$ has kernel $P_{K,1}(\mathfrak{m})$. Now use Corollary 8.7 and Exercise 8.4.

(b) When $\mathfrak{m} \mid \mathfrak{n}$, prove that there is a commutative diagram

$$\begin{array}{ccc}
I_K(\mathfrak{n})/P_{K,1}(\mathfrak{n}) & \longrightarrow & I_K(\mathfrak{n})/(I_K(\mathfrak{n}) \cap P_{K,1}(\mathfrak{m})) \\
 & & \downarrow \simeq \\
\text{Artin} \downarrow \simeq & & I_K(\mathfrak{m})/P_{K,1}(\mathfrak{m}) \\
 & & \text{Artin} \downarrow \simeq \\
\mathrm{Gal}(K_{\mathfrak{n}}/K) & \xrightarrow{\text{res}} & \mathrm{Gal}(K_{\mathfrak{m}}/K)
\end{array}$$

Hint: use part (a) and remember that the Artin map is compatible with restriction.

(c) For $\sigma \in \mathrm{Gal}(K^{ab}/K)$, define $\sigma_m = \sigma|_{K_m}$. Part (a) implies that $K_m \subset K_n$ when $m \mid n$, and it follows that $\sigma_n|_{K_m} = \sigma_m$ when $m \mid n$. Prove that the converse works: given $\sigma_m \in \mathrm{Gal}(K_m/K)$ such that $\sigma_n|_{K_m} = \sigma_m$ when $m \mid n$, there is a unique $\sigma \in \mathrm{Gal}(K^{ab}/K)$ with $\sigma_m = \sigma|_{K_m}$ for all m. Hint: the only subtle point is proving that $\sigma_n|_{K_m} = \sigma_m$ when $K_m \subset K_n$, which might happen even if $m \mid n$ fails. Here, σ_{mn} will be useful.

15.8. Consider the map $\widehat{K} \to \prod_p (K \otimes_{\mathbb{Z}} \mathbb{Z}_p)$ constructed in the text.

(a) Prove that the map is injective.

(b) Define the *restricted product* $\prod_p^* (K \otimes_{\mathbb{Z}} \mathbb{Z}_p)$ to consist of all $x = (x_p) \in \prod_p (K \otimes_{\mathbb{Z}} \mathbb{Z}_p)$ such that $x_p \in \mathcal{O}_K \otimes_{\mathbb{Z}} \mathbb{Z}_p$ for all but finitely many p. Prove that the image of $\widehat{K} \to \prod_p (K \otimes_{\mathbb{Z}} \mathbb{Z}_p)$ lies in $\prod_p^* (K \otimes_{\mathbb{Z}} \mathbb{Z}_p)^*$. Hint: first show that if $\alpha \in \mathcal{O}_K$ is nonzero, then $\alpha \otimes 1 \in \mathcal{O}_K \otimes_{\mathbb{Z}} \mathbb{Z}_p$ for all but finitely many p. Remember that $\alpha\overline{\alpha} \in \mathbb{Z} - \{0\}$ since $\alpha \neq 0$.

(c) Prove that the image of $\widehat{K} \to \prod_p (K \otimes_{\mathbb{Z}} \mathbb{Z}_p)$ equals $\prod_p^* (K \otimes_{\mathbb{Z}} \mathbb{Z}_p)$. Hint: take $x = (x_p) \in \prod_p (K \otimes_{\mathbb{Z}} \mathbb{Z}_p)$ with $S = \{p : x_p \notin \mathcal{O}_K \otimes_{\mathbb{Z}} \mathbb{Z}_p\}$ finite. For $p \in S$ pick $n_p \in \mathbb{Z}$ with $p^{n_p} x_p \in \mathcal{O}_K \otimes_{\mathbb{Z}} \mathbb{Z}_p$ and consider $x \prod_{p \in S} p^{n_p}$. Remember $\widehat{\mathcal{O}}_K = \prod_p (\mathcal{O}_K \otimes_{\mathbb{Z}} \mathbb{Z}_p)$ since $\mathcal{O}_K$ is a free $\mathbb{Z}$-module of rank 2.

(d) Conclude that $\widehat{K} \hookrightarrow \prod_p (K \otimes_{\mathbb{Z}} \mathbb{Z}_p)$ induces $\widehat{\mathcal{O}}_K^* \simeq \prod_p (\mathcal{O}_K \otimes_{\mathbb{Z}} \mathbb{Z}_p)^*$ and $\widehat{K}^* \simeq \prod_p^* (K \otimes_{\mathbb{Z}} \mathbb{Z}_p)^*$, where the restricted product $\prod_p^*$ now means that $x_p \in (\mathcal{O}_K \otimes_{\mathbb{Z}} \mathbb{Z}_p)^*$ for all but finitely many p.

15.9. This exercise will supply some details about the Artin map defined in the discussion preceding Shimura Reciprocity (Theorem 15.15).

(a) Show that $\sigma_{x,n}|_{K_m} = \sigma_{x,m}$ when $m \mid n$. Hint: show that an n-factorization of x from Lemma 15.14 for n is also an m-factorization. Thus $\sigma_{x,n}$ and $\sigma_{x,m}$ both use the same $z\mathcal{O}_K$. Part (b) of Exercise 15.7 will be useful.

(b) Prove that $K^* \subset \widehat{K}^*$ is in the kernel of the Artin map $\widehat{K}^* \to \mathrm{Gal}(K^{ab}/K)$. Hint: $\beta \in K^*$ gives the idele x where $x_p = \beta$ for every p. Given $m > 0$, what is the m-factorization of x?

15.10. This exercise is concerned with the proof of Lemma 15.17.

(a) Show that that the map $I_K(fm) \to C(\mathcal{O})$ defined by $\mathfrak{a} \mapsto [\mathfrak{a} \cap \mathcal{O}]$ is onto. Hint: use Corollary 7.17 to show that any class in $C(\mathcal{O})$ can be written $[\mathfrak{b}]$ for some $\mathcal{O}$-ideal $\mathfrak{b}$ with norm relatively prime to fm. Then let $\mathfrak{a} = \mathfrak{b}\mathcal{O}_K$ and use Proposition 7.20.

(b) Take $\alpha = a + bfw_K \in \mathcal{O}$ relatively prime to m. Let $d = \gcd(a, m)$ and write $a = a_0 d$, $m = m_0 d$ with $\gcd(a_0, m_0) = 1$. By Exercise 15.11 below, there is an integer ℓ such that $a_0 + m_0\ell$ is relatively prime to f. Show that $\alpha + m\ell$ is relatively prime to f and hence relatively prime to fm.

(c) Prove that $I_K(fm) \cap P_{K,\mathbb{Z}}(f)$ is generated by principal ideals $\alpha\mathcal{O}_K$ where $\alpha \in \mathcal{O}_K$ satisfies $\alpha \equiv a \bmod f\mathcal{O}_K$ for $a \in \mathbb{Z}$ and α relatively prime to fm. Hint: the intersection consists of fractional ideals of the form $\mathfrak{a}\mathfrak{b}^{-1} = \alpha\mathcal{O}_K(\beta\mathcal{O}_K)^{-1}$ where $\mathfrak{a}, \mathfrak{b} \subset \mathcal{O}_K$ are relatively prime to fm and $\alpha\mathcal{O}_K, \beta\mathcal{O}_K \subset \mathcal{O}_K$ are generators of $P_{K,\mathbb{Z}}(f)$. Note also that $\mathfrak{a}\mathfrak{b}^{-1} = \mathfrak{a}\overline{\mathfrak{b}}(N(\mathfrak{b})\mathcal{O}_K)^{-1}$.

15.11. Let a_0, m_0 be relatively prime integers and let M be any positive integer. Prove that there is an integer ℓ such that $\gcd(a_0 + m_0\ell, M) = 1$. Hint: let $p_1, \ldots, p_s$ be the distinct primes dividing M but not m_0. Use the Chinese Remainder Theorem to find $x \in \mathbb{Z}$ with $x \equiv a_0 \bmod m_0$ and $x \equiv 1 \bmod p_1 \cdots p_s$.

15.12. Complete the proof of Theorem 15.19. Hint: can you explain why a prime ideal $\mathfrak{p} \in I_K(fm)$ splits completely in $L_{\mathcal{O},m}$ if and only if $\mathfrak{p} \in P_{K,\mathbb{Z},m}(fm)$?

15.13. In the proof of Theorem 15.21, we showed that $x \in \widehat{\mathcal{O}}^*$ implies that $g_{\tau_0}(x) \in \mathrm{GL}(2, \widehat{\mathbb{Z}})$. Prove that $x \equiv 1 \bmod m\widehat{\mathcal{O}}$ implies that $g_{\tau_0}(x) \equiv I \bmod m$.

15.14. Let $m \in \mathbb{Z}$ be positive with $m \equiv 6 \bmod 8$. Set $\mathcal{O} = \mathbb{Z}[\sqrt{-m}] = [1, \sqrt{-m}]$.

(a) Show that $\mathcal{O}^* = \{\pm 1\}$.

(b) Show that $|(\mathcal{O}/8\mathcal{O})^*| = 32$. Hint: set $\alpha = \sqrt{-m}$ and use $m \equiv 6 \bmod 8$ to show $\alpha^2 \equiv 2 \bmod 8\mathcal{O}$. Then prove that $a + b\alpha \in \mathcal{O}$ gives an element of $(\mathcal{O}/8\mathcal{O})^*$ if and only if a is odd.

(c) Show that $(\mathcal{O}/8\mathcal{O})^* \simeq \mathbb{Z}/2\mathbb{Z} \times \mathbb{Z}/2\mathbb{Z} \times \mathbb{Z}/8\mathbb{Z}$, with generators given by the classes of $-1, 3, 1 + \alpha$.

15.15. Prove that $L_\mathcal{O} = K(\mathfrak{f}(\sqrt{-m})^2)$ when $m \equiv 3 \bmod 4$ and $3 \nmid m$. Hint: first use the methods of Theorem 15.23 to show that $\mathfrak{f}(\sqrt{-m}))^6 \in L_\mathcal{O}$. Then explain why this gives the desired result. Use Corollary 12.19 to relate $\mathfrak{f}(\tau)$ to $\mathfrak{f}_1(\tau)$. Do the cases $m \equiv 3 \bmod 8$ and $m \equiv 7 \bmod 8$ separately.

15.16. This exercise will show that (15.24) follows from Shimura Reciprocity.

(a) In the text, we used Shimura Reciprocity to show that $\sigma_{\overline{\mathfrak{a}}}(j(\mathcal{O})) = j(\mathfrak{a})$ when $\mathfrak{a}$ is an $\mathcal{O}$-ideal relatively prime to the conductor. Prove that this equality holds for any proper $\mathcal{O}$-ideal $\mathfrak{a}$. Hint: what do $\sigma_{\overline{\mathfrak{a}}}$ and $j(\mathfrak{a})$ depend on? Then use Proposition 7.22.

(b) Prove (15.24) by applying part (a) with the ideals $\mathfrak{b}$ and $\mathfrak{ab}$ in place of $\mathfrak{a}$.

15.17. Here we will complete the proof of Lemma 15.28.

(a) Prove that $\mathsf{a}(\tau_1 - 1)\mathcal{O} = [\mathsf{a}(\tau_1 - 1), \mathsf{a}(\mathsf{a} + \mathsf{b} + \mathsf{c})]$.

(b) The definition of $x = (x_p)$ in the lemma implies that $x \in \prod_p (\mathcal{O} \otimes_\mathbb{Z} \mathbb{Z}_p)$. Prove that $x \in \widehat{K}^*$. Hint: use Exercise 15.8.

(c) Prove that x is unique up to multiplication by an element of $\widehat{\mathcal{O}}^*$. Hint: use part (b).

15.18. This exercise will supply some details needed in the proof of Theorem 15.27.

(a) In the proof of Theorem 15.27, show that $\mathsf{a}\tau_1 = \tau_0 + \ell$ for some integer ℓ. Hint: use $[1, \mathsf{a}\tau_1] = [1, \tau_0]$ and remember that $\mathsf{a}\tau_1, \tau_0 \in \mathfrak{h}$.

(b) Given a proper ideal $\mathfrak{a}$ in an order $\mathcal{O}$, there are two methods to create an ideal in $\widehat{\mathcal{O}}$. First, using $\widehat{\mathcal{O}} = \mathcal{O} \otimes_\mathbb{Z} \widehat{\mathbb{Z}}$, we get $\mathfrak{a} \otimes_\mathbb{Z} \widehat{\mathbb{Z}}$ as in Lemma 15.28, and second, using the embedding $\mathcal{O} \hookrightarrow \widehat{\mathcal{O}}$ defined by $\alpha \mapsto \alpha \otimes 1$, we get $\mathfrak{a}\widehat{\mathcal{O}}$. Prove that both methods give the same ideal, i.e, $\mathcal{O} \otimes_\mathbb{Z} \widehat{\mathbb{Z}} = \mathfrak{a}\widehat{\mathcal{O}}$.

15.19. The Approximation Theorem mentioned in the proof of Lemma 15.39 states that given $\mathfrak{m} = \prod_{\mathfrak{p}|\mathfrak{m}} \mathfrak{p}^{n_\mathfrak{p}} \subset \mathcal{O}_K$ and $v_\mathfrak{p} \in K$ for all $\mathfrak{p} \mid \mathfrak{m}$, there is $\alpha \in K$ such that $\alpha - v_\mathfrak{p} \in \mathfrak{p}^{n_\mathfrak{p}}$ for all $\mathfrak{p} \mid \mathfrak{m}$. This goal of this exercise is to prove the related result stated in the proof of the lemma. We will use two additional properties of $\mathcal{O}_{K_\mathfrak{p}}$ not mentioned in the text:

(1) For all $\ell \geq 0$, $\mathcal{O}_K \subset \mathcal{O}_{K_\mathfrak{p}}$ induces an isomorphism $\mathcal{O}_K/\mathfrak{p}^\ell \simeq \mathcal{O}_{K_\mathfrak{p}}/\mathfrak{p}^\ell \mathcal{O}_{K_\mathfrak{p}}$.

(2) Fix $\pi_\mathfrak{p} \in \mathfrak{p} - \mathfrak{p}^2$. Then $\mathfrak{p}^\ell \mathcal{O}_{K_\mathfrak{p}} = \pi_\mathfrak{p}^\ell \mathcal{O}_{K_\mathfrak{p}}$ for any $\ell \in \mathbb{Z}$.

Now take $u_\mathfrak{p} \in K_\mathfrak{p}^*$ for all $\mathfrak{p} \mid \mathfrak{m}$ and let M be a positive integer.

(a) Prove that there is $v_\mathfrak{p} \in K$ such that $u_\mathfrak{p}^{-1} - v_\mathfrak{p} \in \mathfrak{p}^{M+n_\mathfrak{p}} \mathcal{O}_{K_\mathfrak{p}}$ when $\mathfrak{p} \mid \mathfrak{m}$. Hint: explain why $\pi_\mathfrak{p}^{\nu_\mathfrak{p}(u_\mathfrak{p})} u_\mathfrak{p}^{-1} \in \mathcal{O}_{K_\mathfrak{p}}$. Here, $\nu_\mathfrak{p}$ is the $\mathfrak{p}$-adic valuation from the proof of Lemma 15.13.

(b) By the Approximation Theorem, there is $\alpha \in K$ such that $\alpha - v_\mathfrak{p} \in \mathfrak{p}^{M+n_\mathfrak{p}}$ for all $\mathfrak{p} \mid \mathfrak{m}$. Show that with the right choice of M, this implies that

$\alpha u_\mathfrak{p} \in 1 + \mathfrak{p}^{n_\mathfrak{p}}\mathcal{O}_{K_\mathfrak{p}}$ for all $\mathfrak{p} \mid \mathfrak{m}$. Thus α has the property stated in the proof of Lemma 15.39. Hint: remember that $\nu_\mathfrak{p}(u_\mathfrak{p})$ might be negative.

15.20. Here are some details from the proof of Lemma 15.14.

(a) Let $m = \prod_{p|m} p^{n_p}$. Prove that

$$P_{K,1}(m) = \{\delta\mathcal{O}_K : \gamma \in K^*,\ \delta \in 1 + p^{n_p}(\mathcal{O}_K \otimes_\mathbb{Z} \mathbb{Z}_p) \text{ for all } p \mid m\}.$$

Hint: as defined in §8, $P_{K,1}(m)$ is generated by $\alpha\mathcal{O}_K$ for $\alpha \in \mathcal{O}_K$ with $\alpha \equiv 1 \bmod m\mathcal{O}_K$. Use the exact sequence from part (a) of Exercise 8.1.

(b) Given m-factorizations $x = \beta yz = \tilde\beta\tilde y\tilde z$, prove that $\tilde\beta\beta^{-1}\mathcal{O}_K \in P_{K,1}(m)$.

15.21. As in the proof of Proposition 15.40, an m-factorization $x = \beta yz$ maps to $x^* = \beta y^* z^*$. Let $m\mathcal{O}_K = \prod_\mathfrak{p} \mathfrak{p}^{n_\mathfrak{p}}$.

(a) Prove that $y^* \in \prod_\mathfrak{p} \mathcal{O}_{K_\mathfrak{p}}^*$ and that $y_p^* \equiv 1 \bmod \mathfrak{p}^{n_\mathfrak{p}}\mathcal{O}_{K_\mathfrak{p}}$ when $\mathfrak{p} \mid m\mathcal{O}_K$.

(b) Prove that $z_\mathfrak{p}^* = 1$ when $\mathfrak{p} \mid m\mathcal{O}_K$.

15.22. In the diagram in Lemma 15.20, prove that the square on the right commutes. Hint: consider the diagram

$$\begin{array}{ccccc}
\widehat{K}^* & \longrightarrow & I_K(fm)/P_{K,1}(fm) & \longrightarrow & I_K(fm)/P_{K,\mathbb{Z},m}(fm) \\
\big\downarrow{\scriptstyle \text{Artin}} & & \big\downarrow{\scriptstyle \simeq} & & \big\downarrow{\scriptstyle \simeq} \\
\mathrm{Gal}(K^{ab}/K) & \longrightarrow & \mathrm{Gal}(K_{fm}/K) & \longrightarrow & \mathrm{Gal}(L_{\mathcal{O},m}/K)
\end{array}$$

15.23. Lemma 15.20 says that $K^* J^1_{\mathcal{O},m}$ is the kernel of $\widehat{K}^* \to I_K(fm)/P_{K,\mathbb{Z},m}(fm)$. Here are some details needed for the proof.

(a) Show that for any positive integer m, the injection $\mathcal{O}_K \hookrightarrow \widehat{\mathcal{O}}_K$ induces a ring isomorphism $\mathcal{O}_K/m\mathcal{O}_K \simeq \widehat{\mathcal{O}}_K/m\widehat{\mathcal{O}}_K$.

(b) Use (15.41) and (15.42) to prove that $\alpha\mathcal{O}_K \in P_{K,\mathbb{Z},m}(fm)$. Hint: first show that $u \equiv c \bmod f\widehat{\mathbb{Z}}$ for some $c \in \mathbb{Z}$, and then find $a \in \mathbb{Z}$ with $a(1 + mc) \equiv 1 \bmod fm$.

(c) Prove that $J^1_{fm} \subset J^1_{\mathcal{O},m}$.

15.24. Prove that $\zeta_m \in L_{\mathcal{O},m}$. Hint: $\zeta_m \in \mathsf{F}_m$.

15.25. Let $\omega = \zeta_3$ and set $K = \mathbb{Q}(\omega)$ and $\mathcal{O} = \mathcal{O}_K = \mathbb{Z}[\omega]$.

(a) Use Lemma 15.17 to show that $[L_{\mathcal{O},5} : K] = 4$.

(b) Use part (a) and the previous exercise to show that

$$L_{\mathcal{O},5} = K(\zeta_5) = \mathbb{Q}(\omega, \zeta_5) = \mathbb{Q}(\zeta_{15}).$$

(c) Theorem 15.21 implies that $r(\omega) \in L_{\mathcal{O},5}$, where $r(\tau) \in \mathsf{F}_5$ is the continued fraction function from the example following Theorem 15.21. In his so-called "lost" notebook, Ramanujan states that

$$\zeta_{10} r(\omega) = \frac{\sqrt{30 + 6\sqrt{5}} - 3 - \sqrt{5}}{4}$$

(see Duke [**39**, (2.4)]). Prove directly, without using Theorem 15.21, that this lies in $\mathbb{Q}(\omega, \zeta_5)$. Hint: $\zeta_{10} = -\zeta_5^3$, $\zeta_5 = \frac{-1+\sqrt{5}}{4} + \frac{i}{2}\sqrt{\frac{5+\sqrt{5}}{2}}$, and $\sqrt{30 + 6\sqrt{5}} = 2\sqrt{-3} \cdot i\sqrt{\frac{5+\sqrt{5}}{2}}$.

15.26. Let $K = \mathbb{Q}(i)$ and $\mathcal{O} = \mathbb{Z}[i]$. The example following Theorem 15.21 characterized $p = x^2 + y^2$ with $x \equiv 1 \bmod 5$, $y \equiv 0 \bmod 5$ using the extended ring class field $L_{\mathcal{O},5}$. Here we explore this example from some other points of view.

(a) Adapt the method of Exercise 15.25 to show that

$$L_{\mathcal{O},5} = K(\zeta_5) = \mathbb{Q}(i, \zeta_5) = \mathbb{Q}(\zeta_{20}).$$

(b) In the example following Theorem 15.21, the solution of $p = x^2 + y^2$ with $x \equiv 1 \bmod 5$ and $y \equiv 0 \bmod 5$ involved the quartic $x^4 + 2x^3 - 6x^2 + 1$. Use part (a) to show that we can use $x^4 + x^3 + x^2 + x + 1$ instead.

(c) Show that for $p > 5$, $x^4 + x^3 + x^2 + x + 1 \equiv 0 \bmod p$ has a solution if and only if $p \equiv 1 \bmod 5$, and conclude that for a prime p,

$$p = x^2 + y^2,\ x \equiv 1 \bmod 5,\ y \equiv 0 \bmod 5 \iff p \equiv 1 \bmod 20.$$

(d) The proof of Theorem 15.21 and the solution to Exercise 15.12 imply that

$$p = x^2 + y^2,\ x \equiv 1 \bmod 5,\ y \equiv 0 \bmod 5 \iff p \text{ splits completely in } L_{\mathcal{O},5}.$$

Use this together with Corollary 5.21 and the Artin map (8.3) to prove the equivalence of part (c).

(e) Give an elementary proof of the equivalence of part (c).

Solutions

Roger Lipsett and David Cox

I first encountered this book about twenty years ago and was intrigued by the subject matter, but I only recently decided to study the text in detail. I tend to learn material best by doing the exercises, and I tend to get the most out of exercises by writing down the solutions. The result is this chapter, which contains complete solutions for every exercise in the book. While most of the exercises are straightforward, some are very challenging, and a few (primarily in §15) require mathematics not covered in the text.

There are few books that entail the historical study of a single mathematical problem, particularly a problem such as this one which turns out to be unexpectedly rich. While I learned class field theory in graduate school, it was not until I read this book that I understood how powerful it can be. This book has also given me a new understanding of the depth of influence of mathematicians such as Gauss on today's mathematics.

To get the most out of the book, David Cox and I strongly urge you to work hard on an exercise before reading the solution. The more active you can be in thinking about a problem, the better. And if you do get stuck, try reading just part of the solution and see if you can finish it yourself. Of course, there are often many ways to solve a problem; the solution that appears here is just one particular approach.

David and I hope you find these solutions useful and clear. The goal is to help you deepen your understanding of the wonderful mathematics explored in *Primes of the Form* $x^2 + ny^2$.

I am grateful to David for the time and effort he willingly spent considering and responding to my innumerable questions, and for completing and correcting a number of solutions. His encouragement made it possible for me to complete this chapter.

March 2022 ROGER LIPSETT

Solutions to Exercises in §1

1.1. (a) We compute as follows:

$$\begin{aligned}(xz \pm nyw)^2 + n(xw \mp yz)^2 &= x^2z^2 \pm 2nxyzw + n^2y^2w^2 + nx^2w^2\\ &\quad \mp 2nxwyz + ny^2z^2\\ &= x^2z^2 + n^2y^2w^2 + nx^2w^2 + ny^2z^2\\ (x^2+ny^2)(z^2+nw^2) &= x^2z^2 + nx^2w^2 + ny^2z^2 + n^2y^2w^2.\end{aligned}$$

Hence the right-hand sides are equal, proving (1.6). Setting $n = 1$ gives (1.3).

(b) The identity

$$(ax^2 + cy^2)(az^2 + cw^2) = (axz \pm cyw)^2 + ac(xw \mp yz)^2$$

is proved by a method similar to part (a), expanding both sides.

1.2. (a) We use induction on k. It is true for $k = 1$. Assume true for $1, 2, \ldots, k-1$. Then

$$\Delta^k f(x) = \Delta(\Delta^{k-1} f(x)) = (\Delta^{k-1} f(x))(x+1) - (\Delta^{k-1} f(x))(x),$$

and the result follows via the inductive hypotheses on $(\Delta^{k-1} f(x))(x)$.

(b) Note that $\Delta(f+g) = \Delta f + \Delta g$, and that for $c \in \mathbb{R}$, $\Delta(cf) = c\Delta f$ and $\Delta c = 0$. Also note that if $f = a_d x^d + \sum_0^{d-1} a_i x^i$ has degree d, then $\Delta f = f(x+1) - f(x)$ has degree $d-1$ with leading coefficient da_d, since

$$\begin{aligned}\Delta f(x) &= f(x+1) - f(x)\\ &= a_d(x+1)^d + a_{d-1}(x+1)^{d-1} - a_d x^d - a_{d-1}x^{d-1} + \text{ lower terms}\\ &= a_d x^d + a_d d x^{d-1} + a_{d-1}x^{d-1} - a_d x^d - a_{d-1}x^{d-1} + \text{ lower terms}\\ &= a_d d x^{d-1} + \text{ lower terms}.\end{aligned}$$

Therefore if f has degree less than d, then $\Delta^d f = 0$. Thus if f is monic of degree d, then by induction on d (and since $\Delta x = 1$ to start the induction)

$$\Delta^d f = \Delta^d(x^d + a_{d-1}x^{d-1} + \cdots) = \Delta^d x^d = \Delta^{d-1}(dx^{d-1} + \cdots) = d\Delta^{d-1}x^{d-1} = d!$$

(c) Suppose $f(x) \equiv 0 \bmod p$ for all x. Then by part (a), $\Delta^d f(x)$, being a linear combination of $f(x), f(x+1), \ldots, f(x+d)$, is divisible by p. But by part (b), $\Delta^d f(x) = d!$, and $d < p$, a contradiction.

1.3. (a) The version of Lemma 1.4 for $x^2 + ny^2$ goes as follows.

LEMMA 1.4. *Suppose that $N = a^2 + nb^2$ with* $\gcd(a,b) = 1$, *and that $q = x^2 + ny^2$ is a prime divisor of N with $q \nmid n$. Then $N/q = c^2 + nd^2$ for* $\gcd(c,d) = 1$.

PROOF. Since $q \mid N$, we have also $q \mid x^2 N - a^2 q$, but

$$x^2N - a^2q = x^2(a^2 + nb^2) - a^2(x^2 + ny^2) = nx^2b^2 - na^2y^2 = n(xb - ay)(xb + ay)$$

Since $q \nmid n$, it follows that $q \mid xb - ay$ or $q \mid xb + ay$. By choosing the sign of a appropriately, we may assume $q \mid xb - ay$, say $xb - ay = dq$. But then

$$\begin{aligned}(a + ndy)y &= ay + ndy^2 = xb - dq + ndy^2 = xb - d(x^2 + ny^2) + ndy^2\\ &= xb - dx^2 = x(b - dx).\end{aligned}$$

Since q is prime, it follows that $\gcd(x,y) = 1$, so that $x \mid a + ndy$, say $a + ndy = cx$. Then $cxy = (a + ndy)y = x(b - dx)$, so

$$(*) \qquad a = cx - ndy, \; b = dx + cy.$$

Thus

(**) $N = a^2 + nb^2 = (cx - ndy)^2 + n(dx + cy)^2 = (x^2 + ny^2)(c^2 + nd^2) = q(c^2 + nd^2)$

by part (a) of Exercise 1.1. Dividing both sides of (**) by q gives $N/q = c^2 + nd^2$. Finally, since $\gcd(a, b) = 1$, (*) proves that $\gcd(c, d) = 1$. Q.E.D.

(b) The assumption that q was prime in part (a) was used in two places. First, when $q \mid n(xb - ay)(xb + ay)$ we inferred that $q \mid xb - ay$ or $q \mid xb + ay$. Second, if $q = x^2 + ny^2$ we concluded that $\gcd(x, y) = 1$. When $q = 4$ and $n = 3$, we have $4 = 1^2 + 3 \cdot 1^2$, so we can choose $x = y = 1$, and then indeed $\gcd(x, y) = 1$. Further, since $\gcd(q, n) = 1$, then $q \mid n(xb - ay)(xb + ay) = n(b - a)(b + a)$ implies that $q \mid (b-a)(b+a)$. If q divides neither factor, then $b-a \equiv b+a \equiv 2 \bmod 4$ since $q = 4$. But then $2a \equiv 0 \bmod 4$, and hence a is even. It follows that b is also even since $b+a$ is. This is a contradiction, since we assumed that $\gcd(a, b) = 1$. Therefore $q \mid xb - ay$ or $q \mid xb + ay$.

We conclude that the proof from part (a) can be made to work for the case of $q = 4$, $n = 3$.

1.4. (a) We prove $p \mid a^2 + 2b^2$ with $\gcd(a, b) = 1 \Rightarrow p = x^2 + 2y^2$ as follows.

Choose the smallest p for which this does not hold, and note that the theorem holds for $p = 2, 3$. Then $p \mid a^2 + 2b^2$, $\gcd(a, b) = 1$. We may modify a, b by multiples of p so that $|a|, |b| < \frac{p}{2}$ and still $p \mid a^2 + 2b^2$. Then we have $a^2 + 2b^2 < \frac{p^2}{4} + 2\frac{p^2}{4} < p^2$. If now $\gcd(a, b) = d > 1$, note that $d < p$ since $a^2 + 2b^2 < p^2$, so dividing a, b by d, we get

$$N_0 = a_0^2 + 2b_0^2, \ \gcd(a_0, b_0) = 1, \ p \mid N_0, \ N_0 < p^2.$$

Note that any prime factor $q \neq p$ of N_0 is less than p, since $qp \leq N_0 < p^2$. So write

$$N_0 = q_1 q_2 \cdots q_r p, \ q_i < p, \ q_i \text{ prime}$$
$$N_i = N_{i-1}/q_i$$

Note that if $q_i = x_i^2 + 2y_i^2$ and $N_{i-1} = a_{i-1}^2 + 2b_{i-1}^2$, then by part (a) of Exercise 1.3, we have $N_i = a_i^2 + 2b_i^2$. But $N_r = p$, so if each $q_i = x_i^2 + 2y_i^2$, then $p = a_r^2 + 2b_r^2$. But we chose p to not be so representable, so one of the q_i is not so representable. This is a contradiction, since p was the smallest such prime.

(b) We prove that $p \mid a^2 + 3b^2$ with p an odd prime, $\gcd(a, b) = 1 \Rightarrow p = x^2 + 3y^2$ as follows.

Note that $4 = 1^2 + 3 \cdot 1^2$, and that by part (b) of Exercise 1.3, if $4k = x^2 + 3y^2$, then k can be so expressed as well. Assume the result false, and let p be the smallest odd prime for which it does not hold. Choose N_0 as above. If N_0 is even, it must be $\equiv 0 \bmod 4$ by congruence considerations, so $N_0/4 = a_0^2 + 3b_0^2$ as well. Continue in this fashion until factors of 4 have been removed; then N_0 is odd and the proof proceeds as in part (a).

1.5. By Fermat's Little Theorem, $x^{p-1} \equiv 1 \bmod p$ for $p \nmid x$. Substituting $3k$ for $p - 1$ and rearranging gives $x^{3k} - 1 \equiv 0 \bmod p$ when $p \nmid x$. But then

(*) $$4(x^{3k} - 1) = (x^k - 1)((2x^k + 1)^2 + 3) \equiv 0 \bmod p.$$

Since $x^k \equiv 1 \bmod p$ has at most $k < p$ solutions modulo p, we can choose x such that $x^k - 1$ is not divisible by p. Then (*) shows that $(2x^k + 1)^2 + 3 \equiv 0 \bmod p$ and thus $(-3/p) = 1$.

1.6. We prove Lemma 1.7 as follows.

$\Leftarrow$: This is easy: $(-n/p) = 1 \Rightarrow x^2 \equiv -n \bmod p \Rightarrow x^2 + n \cdot 1^2 \equiv 0 \bmod p \Rightarrow p \mid x^2 + n \cdot 1^2$.

$\Rightarrow$: Assume $x^2 + ny^2 \equiv 0 \bmod p$ and $\gcd(x, y) = 1$. Then $p \mid y$ implies $p \mid x$, impossible by $\gcd(x, y) = 1$. Thus y is invertible modulo p, say $yz \equiv 1 \bmod p$. Multiplying both sides of $x^2 + ny^2 \equiv 0 \bmod p$ by z^2 gives $(xz)^2 + n \equiv 0 \bmod p$, or $(xz)^2 \equiv -n \bmod p$, so $(-n/p) = 1$.

1.7. If $(p^*/q) = (q/p)$, then

$$\left(\frac{p}{q}\right)\left(\frac{q}{p}\right) = \left(\frac{p}{q}\right)\left(\frac{p^*}{q}\right) = \left(\frac{p}{q}\right)^2\left(\frac{(-1)^{\frac{p-1}{2}}}{q}\right) = \left(\frac{(-1)^{\frac{p-1}{2}}}{q}\right) = \begin{cases} 1 & p \text{ or } q \equiv 1 \bmod 4 \\ -1 & p \equiv q \equiv 3 \bmod 4, \end{cases}$$

which is equivalent to the standard statement of quadratic reciprocity.

In the other direction, assume quadratic reciprocity, i.e.,

$$\left(\frac{p}{q}\right)\left(\frac{q}{p}\right) = (-1)^{\frac{p-1}{2}\frac{q-1}{2}}.$$

Then

$$\begin{aligned}\left(\frac{p^*}{q}\right)\left(\frac{q}{p}\right) &= \left(\frac{(-1)^{\frac{p-1}{2}}}{q}\right)\left(\frac{p}{q}\right)\left(\frac{q}{p}\right) \\ &= \left(\frac{(-1)^{\frac{p-1}{2}}}{q}\right)(-1)^{\frac{p-1}{2}\frac{q-1}{2}} = \begin{cases} 1\cdot 1 = 1 & p \text{ or } q \equiv 1 \bmod 4 \\ -1\cdot -1 = 1 & p \equiv q \equiv 3 \bmod 4. \end{cases}\end{aligned}$$

Thus $(p^*/q)(q/p) = 1$, which implies that $(p^*/q) = (q/p)$.

1.8. $\Leftarrow$: If $p \equiv \beta^2 \bmod 4q$ with β odd, then $p \equiv 1 \bmod 4$. Thus $p^* = p$. Also $p \equiv \beta^2 \bmod q$, so that $1 = (p/q) = (p^*/q)$.
If $p \equiv -\beta^2 \bmod 4q$ with β odd, then $p \equiv 3 \bmod 4$, so that $p^* = -p$. Also $p \equiv -\beta^2 \bmod q$, so that $-p \equiv \beta^2 \bmod q$ and thus $1 = (-p/q) = (p^*/q)$.

$\Rightarrow$: Assume $(p^*/q) = 1$. If $p \equiv 1 \bmod 4$, then $p^* = p$, so that $(p/q) = 1$ and $p \equiv \beta^2 \bmod q$ for some β. By adding q to β if necessary, we may ensure that β is odd, so that $p \equiv \beta^2 \bmod 4$. It follows that $p \equiv \beta^2 \bmod 4q$.
If $p \equiv 3 \bmod 4$, then $p^* = -p$, and $(-p/q) = (p^*/q) = 1$, so that $-p \equiv \beta^2 \bmod q$. Again by adding q to β if necessary, we may ensure that β is odd. Then $-p \equiv \beta^2 \bmod 4$, so that $p \equiv -\beta^2 \bmod 4q$.

1.9. (a) Note that $p \equiv 1 \bmod 3 \iff (p/3) = 1$, so that

$$(*) \qquad p \mid x^2 + 3y^2,\ \gcd(x, y) = 1 \iff p \equiv 1 \bmod 3$$

is equivalent to

$$p \mid x^2 + 3y^2,\ \gcd(x, y) = 1 \iff \left(\frac{p}{3}\right) = 1.$$

By Lemma 1.7, for $p \neq 3$

$$p \mid x^2 + 3y^2,\ \gcd(x, y) = 1 \iff \left(\frac{-3}{p}\right) = 1,$$

and thus $(*)$ is indeed equivalent to $(-3/p) = (p/3)$.

(b) For the first part, first note that $p \equiv 1 \bmod 4 \iff (-1)^{(p-1)/2} = 1$, so that the statement

$$p \mid x^2 + y^2,\ \gcd(x, y) = 1 \iff p \equiv 1 \bmod 4$$

is equivalent to

$$p \mid x^2 + y^2,\ \gcd(x, y) = 1 \iff (-1)^{(p-1)/2} = 1.$$

By Lemma 1.7

$$p \mid x^2 + y^2,\ \gcd(x, y) = 1 \iff \left(\frac{-1}{p}\right) = 1,$$

and thus the statement is equivalent to

$$(-1)^{(p-1)/2} = 1 \iff \left(\frac{-1}{p}\right) = 1.$$

This is clearly equivalent to $(-1/p) = (-1)^{(p-1)/2}$, and we are done.

For the second part, by Lemma 1.7

$$p \mid x^2 + 2y^2,\ \gcd(x, y) = 1 \iff \left(\frac{-2}{p}\right) = 1$$

and thus the statement

$$p \mid x^2 + 2y^2,\ \gcd(x, y) = 1 \iff p \equiv 1, 3 \bmod 8$$

is equivalent to

$$p \equiv 1, 3 \bmod 8 \iff \left(\frac{-2}{p}\right) = 1.$$

However, one checks easily that $p \equiv 1, 3 \bmod 8 \iff (-1)^{(p^2-1)/8} = (-1)^{(p-1)/2}$, and by (1.11), $(-2/p) = (-1/p)(2/p) = (-1)^{(p-1)/2}(2/p)$, so that

$$\left(\frac{-2}{p}\right) = 1 \iff \left(\frac{2}{p}\right) = (-1)^{(p-1)/2}.$$

Putting this all together, the statement is equivalent to

$$(-1)^{(p^2-1)/8} = (-1)^{(p-1)/2} \iff \left(\frac{2}{p}\right) = (-1)^{(p-1)/2},$$

which in turn is equivalent to $(2/p) = (-1)^{(p^2-1)/8}$.

1.10. (a) We have

$$\left(\frac{M}{m}\right) = \prod \left(\frac{M}{p_i}\right) = \prod \left(\frac{N}{p_i}\right) = \left(\frac{N}{m}\right)$$

since $M \equiv N \bmod m \Rightarrow M \equiv N \bmod p_i$.

(b) For the first line of (1.15), note that (1.11) implies $(MN/p_i) = (M/p_i)(N/p_i)$ for each p_i. For the second line of (1.15), note that $n = p_1 \cdots p_r$ and $m = q_1 \cdots q_s$ imply $nm = p_1 \cdots p_r q_1 \cdots q_s$.

(c) If m is odd, then $(-1/m) = \prod_{i=1}^r (-1/p_i) = \prod_{i=1}^r (-1)^{(p_i-1)/2} = (-1)^{\sum_{i=1}^r (p_i-1)/2}$. Each p_i is odd. But for any r, s odd,

$$\begin{aligned} \frac{rs-1}{2} \equiv 0 \bmod 2 &\iff rs - 1 \equiv 0 \bmod 4 \iff rs \equiv 1 \bmod 4 \\ &\iff r \equiv s \bmod 4 \iff \frac{r-1}{2} \equiv \frac{s-1}{2} \bmod 2 \\ &\iff \frac{r-1}{2} + \frac{s-1}{2} \equiv 0 \bmod 2, \end{aligned}$$

so that $(rs-1)/2 \equiv (r-1)/2 + (s-1)/2 \bmod 2$. This generalizes to the finite product $m = p_1 \cdots p_r$, with the result that

$$\left(\frac{-1}{m}\right) = (-1)^{\sum_{i=1}^r (p_i-1)/2} = (-1)^{(\prod_{i=1}^r p_i - 1)/2} = (-1)^{(m-1)/2},$$

as desired.

The rule for $(-2/m)$ follows similarly, using the identity

$$\frac{r^2s^2-1}{8} \equiv \frac{r^2-1}{8} + \frac{s^2-1}{8} \bmod 2.$$

To prove this identity, note that if r, s are odd, then $r^2 - 1 \equiv s^2 - 1 \equiv 0 \bmod 8$, so that

$$(r^2-1)(s^2-1) \equiv 0 \bmod 16.$$

But

$$(r^2-1)(s^2-1) = r^2s^2 - r^2 - s^2 + 1 = (r^2s^2-1) - ((r^2-1)+(s^2-1)) \equiv 0 \bmod 16.$$

Divide both sides of the final equivalence by 8 and rearrange, giving

$$\frac{r^2s^2-1}{8} = \frac{r^2-1}{8} + \frac{s^2-1}{8} \bmod 2,$$

proving the identity.

(d) If M is a quadratic residue for $m = \prod_{i=1}^{r} p_i$, then $(M/p_i) = 1$ for each i and thus $(M/m) = \prod_{i=1}^{r}(M/p_i) = 1$. Now consider $M = 2$, $m = 15$. In this case, 2 is not a residue modulo 15 since it is not a residue modulo 3, but $(2/15) = (2/3)(2/5) = -1 \cdot -1 = 1$.

1.11. First, if $D > 0$ and $D \equiv 0 \bmod 4$, write $D = 4^r M$ where M is odd. Then as in the text

$$\left(\frac{D}{m}\right) = \left(\frac{4^rM}{m}\right) = \left(\frac{4^r}{m}\right)\left(\frac{M}{m}\right) = \left(\frac{M}{m}\right) = (-1)^{(M-1)(m-1)/4}\left(\frac{m}{M}\right)$$
$$\left(\frac{D}{n}\right) = \left(\frac{4^rM}{n}\right) = \left(\frac{4^r}{n}\right)\left(\frac{M}{n}\right) = \left(\frac{M}{n}\right) = (-1)^{(M-1)(n-1)/4}\left(\frac{n}{M}\right).$$

Then $(m/M) = (n/M)$ since $n \equiv m \bmod M$. Further, $D \equiv 0 \bmod 4$, so that $m \equiv n \bmod D$ implies $m - 1 \equiv n - 1 \bmod 4$, and therefore the signs are the same.

If $D < 0$ and $D \equiv 0 \bmod 4$, then

$$\left(\frac{D}{m}\right) = \left(\frac{-1}{m}\right)\left(\frac{-D}{m}\right), \quad \left(\frac{D}{n}\right) = \left(\frac{-1}{n}\right)\left(\frac{-D}{n}\right).$$

Since $m \equiv n \bmod D$, we have also $m \equiv n \bmod 4$, so $(-1/m) = (-1/n)$. Further, since $-D > 0$ and $-D \equiv 0 \bmod 4$, it follows from the above that $(-D/m) = (-D/n)$.

If $D < 0$ and $D \equiv 1 \bmod 4$, then

$$\left(\frac{D}{m}\right) = \left(\frac{-1}{m}\right)\left(\frac{-D}{m}\right) = (-1)^{(m-1)/2}\left(\frac{-D}{m}\right) = (-1)^{(m-1)/2}(-1)^{(-D-1)(m-1)/4}\left(\frac{m}{-D}\right)$$
$$\left(\frac{D}{n}\right) = \left(\frac{-1}{n}\right)\left(\frac{-D}{n}\right) = (-1)^{(n-1)/2}\left(\frac{-D}{n}\right) = (-1)^{(n-1)/2}(-1)^{(-D-1)(n-1)/4}\left(\frac{n}{-D}\right).$$

Note that since $m \equiv n \bmod D$, we have $(m/-D) = (n/-D)$, so we only need to prove that the signs are equal. But $D \equiv 1 \bmod 4$ implies that $-D - 1 \equiv 2 \bmod 4$, so that $(-1)^{(-D-1)/2} = -1$. Then

$$\begin{aligned}
(-1)^{(m-1)/2}(-1)^{(-D-1)(m-1)/4} &= (-1)^{(m-1)/2}\left((-1)^{(-D-1)/2}\right)^{(m-1)/2} \\
&= \left((-1)^{(m-1)/2}\right)^2 = 1 \\
(-1)^{(n-1)/2}(-1)^{(-D-1)(n-1)/4} &= (-1)^{(n-1)/2}\left((-1)^{(-D-1)/2}\right)^{(n-1)/2} \\
&= \left((-1)^{(n-1)/2}\right)^2 = 1
\end{aligned}$$

and we are done.

1.12. (a) Any element $[a]$ of $(\mathbb{Z}/D\mathbb{Z})^*$ clearly includes a positive integer k. Then either k or $k+|D|$ is odd since $\gcd(k, D) = 1$, so that $[a]$ includes an odd positive integer m. Define $\chi([a]) = \chi([m]) = (D/m)$. By (1.17), if $[m] = [n]$ for m, n odd and positive, then $(D/m) = (D/n)$, so that χ is well-defined, and χ preserves multiplication by (1.15).

(b) Consider the four cases separately:

- $D > 0$, $D \equiv 0 \bmod 4$. In this case, $D-1$ is odd and positive, so that

$$\chi([-1]) = \chi([D-1]) = \left(\frac{D}{D-1}\right) = \left(\frac{1}{D-1}\right) = 1$$

- $D > 0$, $D \equiv 1 \bmod 4$. In this case, $2D-1$ is odd and positive, and then

$$\begin{aligned}\chi([-1]) = \chi([2D-1]) &= \left(\frac{D}{2D-1}\right) = \left(\frac{4D}{2D-1}\right) = \left(\frac{2}{2D-1}\right)\left(\frac{2D}{2D-1}\right)\\ &= \left(\frac{2}{2D-1}\right)\left(\frac{1}{2D-1}\right) = \left(\frac{2}{2D-1}\right).\end{aligned}$$

  But $D \equiv 1 \bmod 4 \Rightarrow 2D \equiv 2 \bmod 8 \Rightarrow 2D-1 \equiv 1 \bmod 8 \Rightarrow (2/2D-1) = 1$.

- $D < 0$, $D \equiv 0 \bmod 4$. Here $-1-D$ is odd and positive, so we get

$$\begin{aligned}\chi([-1]) = \chi([-1-D]) &= \left(\frac{D}{-1-D}\right) = \left(\frac{-1}{-1-D}\right)\left(\frac{-D}{-1-D}\right)\\ &= \left(\frac{-1}{-1-D}\right)\left(\frac{1}{-1-D}\right) = \left(\frac{-1}{-1-D}\right),\end{aligned}$$

  which is -1 since $-1-D \equiv 3 \bmod 4$.

- $D < 0$, $D \equiv 1 \bmod 4$. In this case, $-1-2D$ is odd and positive, so

$$\begin{aligned}\chi([-1]) = \chi([-1-2D]) &= \left(\frac{D}{-1-2D}\right) = \left(\frac{4D}{-1-2D}\right)\\ &= \left(\frac{-2}{-1-2D}\right)\left(\frac{-2D}{-1-2D}\right) = \left(\frac{-2}{-1-2D}\right)\left(\frac{1}{-1-2D}\right)\\ &= \left(\frac{-2}{-1-2D}\right) = \left(\frac{-1}{-1-2D}\right)\left(\frac{2}{-1-2D}\right)\end{aligned}$$

  But $D \equiv 1 \bmod 4 \Rightarrow -1-2D \equiv 1 \bmod 4$, so

$$\left(\frac{-1}{-1-2D}\right) = 1, \quad \left(\frac{2}{-1-2D}\right) = -1.$$

(c) Consider separately $D > 0$ and $D < 0$. If $D > 0$, then

$$\begin{aligned}\chi([2]) = \chi([D+2]) &= \left(\frac{D}{D+2}\right) = (-1)^{(D-1)(D+1)/4}\left(\frac{D+2}{D}\right)\\ &= (-1)^{(D^2-1)/4}\left(\frac{2}{D}\right) = \left(\frac{2}{D}\right) = (-1)^{(D^2-1)/8}.\end{aligned}$$

If $D < 0$, then $2-D > 0$ and $2-D \equiv 1 \bmod 4$, so that

$$\begin{aligned}\chi([2]) = \chi([2-D]) &= \left(\frac{D}{2-D}\right) = \left(\frac{-1}{2-D}\right)\left(\frac{-D}{2-D}\right)\\ &= 1\cdot\left(\frac{2-D}{-D}\right) = \left(\frac{2}{-D}\right) = (-1)^{((-D)^2-1)/8} = (-1)^{(D^2-1)/8}.\end{aligned}$$

Thus in either case, $\chi([2]) = (-1)^{(D^2-1)/8}$, so that

$$\chi([2]) = \begin{cases} 1 & D \equiv 1 \bmod 8 \\ -1 & D \equiv 5 \bmod 8. \end{cases}$$

1.13. (a) If $D \equiv 0, 1 \bmod 4$, write χ_D for the homomorphism guaranteed by Lemma 1.14. We first claim that if q is an odd prime and χ_{q^*} is nontrivial, we have $(q^*/p) = (p/q)$ for any odd prime $p \neq q$. To prove the claim, note that both $(q^*/\cdot) = \chi_{q^*}$ and $(\cdot/q)$ are nontrivial homomorphisms from the cyclic group $(\mathbb{Z}/q\mathbb{Z})^*$ to $\{\pm 1\}$, so they must both be -1 on a generator and thus are the same map.

Now, let p and q be distinct odd primes. If at least one of them is $\equiv 3 \bmod 4$, assume without loss of generality that $q \equiv 3 \bmod 4$. Then $q^* < 0$, so that χ_{q^*} is nontrivial by part (b) of Exercise 1.12 and we are done. Otherwise, $p \equiv q \equiv 1 \bmod 4$, and $q^* = q$. If $(p/q) \neq (q/p)$, then exactly one of them is -1, say (q/p), so that $\chi_{q^*} = \chi_q$ is nontrivial and thus

$$-1 = \left(\frac{q}{p}\right) = \left(\frac{q^*}{p}\right) = \left(\frac{p}{q}\right)$$

by the claim above, which is a contradiction.

(b) With $D = -4$, we get a homomorphism $\chi : \{[1], [3]\} = (\mathbb{Z}/4\mathbb{Z})^* \to \{\pm 1\}$ defined by

$$\chi([p]) = \left(\frac{-4}{p}\right) = \left(\frac{-1}{p}\right).$$

Since $D < 0$, $\chi([3]) = -1$ by part (b) of Exercise 1.12, so the homomorphism is nontrivial, and in particular

$$\left(\frac{-1}{p}\right) = \chi([p]) = \begin{cases} 1 & p \equiv 1 \bmod 4 \\ -1 & p \equiv 3 \bmod 4, \end{cases}$$

and $(-1/p) = (-1)^{(p-1)/2}$ follows easily.

With $D = 8$, we get a homomorphism $\chi : \{[1], [3], [5], [7]\} = (\mathbb{Z}/8\mathbb{Z})^* \to \{\pm 1\}$ defined by

$$\chi([p]) = \left(\frac{8}{p}\right) = \left(\frac{2}{p}\right).$$

Since 2 is not a square modulo 3 or 5 but is a square modulo 7, this homomorphism is nontrivial, and we see that

$$\left(\frac{2}{p}\right) = \chi([p]) = \begin{cases} 1 & p \equiv 1, 7 \bmod 8 \\ -1 & p \equiv 3, 5 \bmod 8, \end{cases}$$

which gives $(2/p) = (-1)^{(p^2-1)/8}$.

1.14. Consider an odd prime p not dividing n. Since $n \equiv 3 \bmod 4$, Lemma 1.14 with $D = -n$ gives a homomorphism $\chi : (\mathbb{Z}/n\mathbb{Z})^* \to \{\pm 1\}$ such that $\chi([p]) = (-n/p)$. Combining this with Lemma 1.7, we obtain

$$p \mid x^2 + ny^2,\ \gcd(x, y) = 1 \iff \left(\frac{-n}{p}\right) = 1 \iff [p] \in \ker(\chi).$$

Then we are done since $\ker(\chi)$ can be expressed in terms of congruence classes modulo n.

1.15. Note that $(-21/p) = (-1/p)(3/p)(7/p)$. By quadratic reciprocity,

$$\left(\frac{3}{p}\right) = (-1)^{(p-1)/2}\left(\frac{p}{3}\right),\quad \left(\frac{7}{p}\right) = (-1)^{(p-1)/2}\left(\frac{p}{7}\right),$$

so that $(-21/p) = (-1/p)(p/3)(p/7)$. We must determine for which residue classes modulo 84 this product is 1. Since

$$\left(\frac{p}{3}\right) = 1 \iff p \equiv 1 \bmod 3$$
$$\left(\frac{p}{7}\right) = 1 \iff p \equiv 1, 2, 4 \bmod 7,$$

we get the following table:

p mod 84	p mod 4	p mod 3	p mod 7	$(-1/p)$	$(p/3)$	$(p/7)$	$(-21/p)$
1	1	1	1	1	1	1	1
5	1	2	5	1	-1	-1	1
11	3	2	4	-1	-1	1	1
13	1	1	6	1	1	-1	-1
17	1	2	3	1	-1	-1	1
19	3	1	5	-1	1	-1	1
23	3	2	2	-1	-1	1	1
25	1	1	4	1	1	1	1
29	1	2	1	1	-1	1	-1
31	3	1	3	-1	1	-1	1
37	1	1	2	1	1	1	1
41	1	2	6	1	-1	-1	1
43	3	1	1	-1	1	1	-1
47	3	2	6	-1	-1	-1	-1
53	1	2	4	1	-1	1	-1
55	3	1	6	-1	1	-1	1
59	3	2	3	-1	-1	-1	-1
61	1	1	5	1	1	-1	-1
65	1	2	2	1	-1	1	-1
67	3	1	4	-1	1	1	-1
71	3	2	1	-1	-1	1	1
73	1	1	3	1	1	-1	-1
79	3	1	2	-1	1	1	-1
83	3	2	6	-1	-1	-1	-1

Since $p \mid x^2 + 21y^2 \iff (-21/p) = 1$, we see that the last column of the table implies that $p \mid x^2 + 21y^2 \iff p \equiv 1, 5, 11, 17, 19, 23, 25, 31, 37, 41, 55, 71 \bmod 84$.

1.16. We have $N = 4q > 0$, so to show that (i) and (ii) characterize $\ker \chi$, we need to prove that exactly one index 2 subgroup of $(\mathbb{Z}/4q\mathbb{Z})^*$ contains -1.

Since $\mathbb{Z}/4q\mathbb{Z} \cong \mathbb{Z}/4\mathbb{Z} \times \mathbb{Z}/q\mathbb{Z}$ and q is odd, we have $(\mathbb{Z}/4q\mathbb{Z})^* \cong (\mathbb{Z}/4\mathbb{Z})^* \times (\mathbb{Z}/q\mathbb{Z})^*$. Choose a generator a for $(\mathbb{Z}/q\mathbb{Z})^*$; then $(\mathbb{Z}/4\mathbb{Z})^* \times (\mathbb{Z}/q\mathbb{Z})^*$ is generated by $\{(-1, 1),\ (1, a)\}$. (Note that, as in the text, congruence classes are being written informally for clarity, so we write -1 rather than $[-1]$.) It follows that there are exactly three nontrivial homomorphisms from $(\mathbb{Z}/4\mathbb{Z})^* \times (\mathbb{Z}/q\mathbb{Z})^* \to \{\pm 1\}$:

$$\begin{aligned} \varphi_1 &: \ (-1, 1) \longmapsto 1, \qquad (1, a) \longmapsto -1, \\ \varphi_2 &: \ (-1, 1) \longmapsto -1, \quad (1, a) \longmapsto 1, \\ \varphi_3 &: \ (-1, 1) \longmapsto -1, \quad (1, a) \longmapsto -1. \end{aligned}$$

The kernels of $\varphi_1, \varphi_2, \varphi_3$ thus give all three subgroups of index 2. Note that $-1 \in (\mathbb{Z}/4q\mathbb{Z})^*$ corresponds to $(-1, -1) \in (\mathbb{Z}/4\mathbb{Z})^* \times (\mathbb{Z}/q\mathbb{Z})^*$, so we want to show that $(-1, -1)$ is in exactly one of these kernels. But

$$\varphi_i(-1,-1) = \varphi_i\big(-1, a^{(q-1)/2}\big) = \varphi_i(-1,1)\varphi_i(1,a)^{(q-1)/2} = \begin{cases} (-1)^{(q-1)/2}, & i=1 \\ -1, & i=2 \\ -(-1)^{(q-1)/2}, & i=3. \end{cases}$$

Clearly $(-1,-1) \notin \ker(\varphi_i)$ when $i=2$, and $(-1,-1) \in \ker(\varphi_i)$ for exactly one of $i=1$ and $i=3$. It follows that -1 is in exactly one index 2 subgroup of $(\mathbb{Z}/4q\mathbb{Z})^*$, proving that χ is uniquely characterized by (i) and (ii) in the discussion following Lemma 1.14.

Solutions to Exercises in §2

2.1. Suppose $m = f(x,y) = ax^2 + bxy + cy^2$. Let $d = \gcd(x,y)$ and $m' = m/d^2$. Then

$$m' = \frac{m}{d^2} = \frac{1}{d^2}(ax^2 + bxy + cy^2) = a\left(\frac{x}{d}\right)^2 + b\left(\frac{x}{d}\right)\left(\frac{y}{d}\right) + c\left(\frac{y}{d}\right)^2.$$

Since x/d and y/d are integers with $\gcd(x/d, y/d) = 1$, this shows that m' is an integer that is properly represented by $f(x,y)$.

2.2. We start by developing some notation to be used in this exercise and in Exercise 2.3. If $g(x,y) = ax^2 + bxy + cy^2$, define the matrix of $g(x,y)$ to be $M_g = \left(\begin{smallmatrix} 2a & b \\ b & 2c \end{smallmatrix}\right)$. Then

$$2g(x,y) = (x \;\; y)\, M_g \begin{pmatrix} x \\ y \end{pmatrix}.$$

Now suppose that $f(x,y) = g(px+qy, rx+sy)$ with $ps - qr = \pm 1$, and define $P = \left(\begin{smallmatrix} p & q \\ r & s \end{smallmatrix}\right) \in \mathrm{GL}(2,\mathbb{Z})$. Note that if $P \in \mathrm{SL}(2,\mathbb{Z})$ then the equivalence is proper. Then

$$\begin{aligned} 2f(x,y) &= 2g(px+qy, rx+sy) \\ &= (px+qy \;\; rx+sy)\, M_g \begin{pmatrix} px+qy \\ rx+sy \end{pmatrix} \\ &= (x \;\; y) \begin{pmatrix} p & r \\ q & s \end{pmatrix} M_g \begin{pmatrix} p & q \\ r & s \end{pmatrix} \begin{pmatrix} x \\ y \end{pmatrix}. \end{aligned}$$

Therefore

$$M_f = \begin{pmatrix} p & r \\ q & s \end{pmatrix} M_g \begin{pmatrix} p & q \\ r & s \end{pmatrix} = P^T M_g P.$$

(a) $f(x,y)$ is equivalent to itself via $P = I$ (the identity matrix), which is in $\mathrm{SL}(2,\mathbb{Z})$. If $M_f = P^T M_g P$, then $M_g = \left(P^{-1}\right)^T M_f P^{-1}$, and $P^{-1} \in \mathrm{SL}(2,\mathbb{Z})$ if P is. Transitivity is similar.

(b) From the discussion above, if $P, Q \in \mathrm{GL}(2,\mathbb{Z}) - SL(2,\mathbb{Z})$, then $M_f = Q^T P^T M_g PQ$ is a proper equivalence between $f(x,y)$ and $g(x,y)$, since $\det(PQ) = 1$. Thus applying two successive improper equivalences gives a proper equivalence. Now assume that $f(x,y) = 2x^2 + xy + 3y^2$ is improperly equivalent to itself. Apply that improper equivalence, and then apply the improper equivalence $(x,y) \mapsto (x,-y)$. The result is a proper equivalence

$$f(x,y) \xrightarrow{\text{improper}} f(x,y) \xrightarrow{\text{improper}} 2x^2 - xy + 3y^2.$$

But this is impossible since both $f(x,y) = 2x^2 + xy + 3y^2$ and $2x^2 - xy + 3y^2$ are reduced, so cannot be properly equivalent.

(c) If $f(x,y) = m$ and $g(x,y)$ is equivalent to $f(x,y)$, then we can write $m = f(x,y) = g(px+qy, rx+sy)$ for some $P = \left(\begin{smallmatrix} p & q \\ r & s \end{smallmatrix}\right) \in \mathrm{GL}(2,\mathbb{Z})$. Since P is invertible, we see that $f(x,y)$ and $g(x,y)$ represent the same numbers. If the representation by $f(x,y)$ is improper, then $\gcd(x,y) > 1$ and thus $\gcd(px+qy, rx+sy) > 1$. It follows that the representation by $g(x,y)$ is also improper. Therefore if the representation by $g(x,y)$ is proper, so is the representation by $f(x,y)$. Since equivalence is an equivalence relation, we are done.

(d) Suppose that $f(x,y) = ax^2 + bxy + cy^2$ is equivalent to a primitive form $g(x,y) = a'x^2 + b'xy + c'y^2$, and that $\gcd(a,b,c) = d$. Then every number represented by $f(x,y)$ is a multiple of d, so that by part (c) every number represented by $g(x,y)$ is as well. In particular, $g(1,0) = a'$ and $g(0,1) = c'$ are multiples of d, as is $g(1,1) = a' + b' + c'$. It follows that b' is also a multiple of d. Since $g(x,y)$ is primitive, $d = 1$ so that $f(x,y)$ is primitive as well.

2.3. From Exercise 2.2, $D = -\det(M_f)$ and $D' = -\det(M_g)$. But

$$\begin{aligned} D = -\det(M_f) &= -\det\begin{pmatrix} p & r \\ q & s \end{pmatrix} \det(M_g) \det\begin{pmatrix} p & q \\ r & s \end{pmatrix} \\ &= -(ps - qr)^2 \det(M_g) = (ps - qr)^2 D'. \end{aligned}$$

2.4. (a) Let $f(x,y) = ax^2 + bxy + cy^2$. If $a = 0$, then $D = b^2 > 0$, so in particular $b \neq 0$. Then $f(x,1) = bx + c$ clearly takes both positive and negative values. If $a < 0$, showing that $-f(x,y)$ represents both positive and negative numbers proves the same for $f(x,y)$. So assume $a > 0$. Then $f(1,0) = a > 0$, so that $f(x,y)$ represents positive numbers. For negative numbers, since $a > 0$, (2.4) shows that it suffices to find $x, y > 0$ such that $(2ax + by)^2 < Dy^2$. For example, take $(x,y) = (-b, 2a)$. Then $2ax + by = 0$, so that $y = 2a \neq 0$ and $D > 0$ show that $0 < Dy^2$.

(b) Since $D < 0$, the right-hand side of (2.4) is a sum of positive multiples of squares, so it is nonnegative. Further, $D = b^2 - 4ac < 0$ implies that $a \neq 0$, so that in fact the right-hand side of (2.4) is positive if either x or y is nonzero. Thus $4af(x,y)$ is positive for $(x,y) \neq (0,0)$, so that $f(x,y)$ is either always positive if $a > 0$, and always negative if $a < 0$.

2.5. A version of Corollary 2.6 for arbitrary discriminants goes as follows.

Corollary 2.6. *Let $D \equiv 0, 1 \bmod 4$ be an integer and let p be an odd prime not dividing D. Then $(D/p) = 1$ if and only if p is properly represented by a primitive form of discriminant D.*

Proof. This follows from Lemma 2.5 since $(D/p) = 1$ if and only if D is a quadratic residue modulo p. Q.E.D.

2.6. Apply $x \mapsto -y$, $y \mapsto x$, which is of determinant 1. Then

$$126(-y)^2 + 74(-y)x + 13x^2 = 13x^2 - 74xy + 126y^2.$$

Now apply $x \mapsto x + 3y$, $y \mapsto y$ to obtain the reduced form

$$13(x+3y)^2 - 74(x+3y)y + 126y^2 = 13x^2 + 4xy + 21y^2.$$

2.7. First observe that $ax^2 + bxy + cy^2$ and $ax^2 - bxy + cy^2$ represent the same values, since we can substitute $-x$ for x in the first to get the second. So after substituting $-b$ for b in the original form if necessary, we can write $f(x,y) = ax^2 - bxy + cy^2$ where $b \geq 0$. If x and y have the same sign, then $f(x,-y) \geq f(x,y)$, so that the minimum will occur when x and y have the same sign. So assume without loss of generality that x and y are both nonnegative. If $x \geq y \geq 0$, then

$$\begin{aligned} f(x,y) &= ax^2 - bxy + cy^2 = ax^2 - |b|xy + cy^2 \\ &\geq ax^2 - |b|x^2 + cy^2 = (a - |b|)x^2 + cy^2 \geq (a - |b| + c)y^2 \\ &= (a - |b| + c)\min(x^2, y^2) \end{aligned}$$

since $x^2 \geq xy \geq y^2$ and $a - |b| \geq 0$. Similarly, when $y \geq x \geq 0$, we get

$$\begin{aligned} f(x,y) &= ax^2 - bxy + cy^2 = ax^2 - |b|xy + cy^2 \\ &\geq ax^2 - |b|y^2 + cy^2 = ax^2 + (c - |b|)y^2 \geq (a - |b| + c)x^2 \\ &= (a - |b| + c)\min(x^2, y^2) \end{aligned}$$

since $y^2 \geq xy \geq x^2 =$ and $c - |b| \geq 0$.

2.8. (a) Both $\Leftarrow$ implications are obvious. For the converse, first suppose that $f(x,y) = a$ and $\gcd(x,y) = 1$. If $xy \neq 0$, then we have (using (2.9) and (2.10))

$$(*) \qquad f(x,y) \geq (a - |b| + c)\min(x^2, y^2) > c\min(x^2, y^2) \geq c,$$

so that $f(x,y) > c > a$, contradicting $f(x,y) = a$. Thus either x or y must be zero. If $x = 0$ and $y \neq 0$, then $f(0,y) = cy^2 \geq c > a$, again contradicting $f(x,y) = a$. Finally, if $y = 0$ and $x \neq 0$, then $f(x,0) = ax^2$, so that we must have $x = \pm 1$. It follows that $(x,y) = \pm(1,0)$.

Similarly, suppose that $f(x,y) = c$ with $\gcd(x,y) = 1$. If $xy \neq 0$, then (*) shows that $f(x,y) > c$, contradicting $f(x,y) = c$. Thus either x or y must be zero. If $y = 0$ and $x \neq 0$, then $f(x,0) = ax^2$. But then $a < c$ implies that $|x| > 1$, which contradicts $\gcd(x,y) = 1$. Finally, if $x = 0$ and $y \neq 0$, then $f(0,y) = cy^2$, so that we must have $y = \pm 1$. It follows that $(x,y) = \pm(0,1)$.

(b) We will use the fact (proved in Exercise 2.7) that (2.9) holds even in the exceptional cases $|b| = a$ and $a = c$.

If $|b| = a$, we claim that

$$(*) \qquad \begin{aligned} f(x,y) &= a,\ \gcd(x,y) = 1 &&\iff (x,y) = \pm(1,0) \\ f(x,y) &= c,\ \gcd(x,y) = 1 &&\iff (x,y) = \pm(0,1) \text{ or } (x,y) = (\mp 1, \pm 1). \end{aligned}$$

Since the form is reduced, we have $b \geq 0$, so that $a = b$. Again both $\Leftarrow$ implications are clear. For the converse, first suppose $f(x,y) = a$ with $\gcd(x,y) = 1$. If $xy \neq 0$, then by (2.9),

$$f(x,y) \geq (a - |b| + c)\min(x^2, y^2) \geq a - |b| + c = c > a.$$

This contradicts $f(x,y) = a$, so that either x or y must be zero. If $x = 0$ and $y \neq 0$, then $f(0,y) = cy^2 \geq c > a$, again a contradiction. Finally, if $y = 0$ and $x \neq 0$, then $f(x,0) = ax^2$, so that $f(x,0) = a$ forces $x = \pm 1$. Thus in this case $(x,y) = \pm(1,0)$.

Similarly, suppose $f(x,y) = c$ with $\gcd(x,y) = 1$. Since $a = b$, $f(x,y) = c$ means that

$$(**) \qquad ax^2 + axy + (cy^2 - c) = 0.$$

The discriminant of this quadratic in x is

$$a^2y^2 - 4a(cy^2 - c) = a^2y^2 - 4ac(y^2 - 1) < a^2y^2 - 4a^2(y^2 - 1) = 4a^2 - 3a^2y^2.$$

This expression is negative unless $y = 0$ or $y = \pm 1$. If $y = 0$, then (**) gives $ax^2 = c$, forcing $|x| > 1$ and thus contradicting $\gcd(x,y) = \gcd(x,0) = 1$. With $y = \pm 1$, (**) gives $ax^2 \pm ax = 0$, so that $x = 0$ or $x = \mp 1$. Thus $(x,y) = \pm(0,1)$ or $(\mp 1, \pm 1)$.

If $a = c$, we claim that

$$(***) \qquad f(x,y) = a = c,\ \gcd(x,y) = 1 \iff (x,y) = (\pm 1, 0) \text{ or } (0, \pm 1).$$

Again $\Leftarrow$ is clear. For $\Rightarrow$, if $xy \neq 0$, note first that we must have $|b| < a$ in order for f to be primitive, so that

$$f(x,y) \geq (a - |b| + c)\min(x^2, y^2) \geq a - |b| + c > c = a,$$

and we do not get equality. Thus either x or y must be zero, and it is clear that the other must be ± 1 for equality to hold.

We now complete the proof of Theorem 2.8. Suppose $g(x, y)$ is reduced and properly equivalent to $f(x, y) = ax^2 + bxy + cy^2$, where either $a = |b|$ or $a = c$. Then they represent the same numbers, so in particular the smallest number each represents is the same. Therefore $g(x, y)$ has the same leading coefficient a as $f(x, y)$ does, and we can write $g(x, y) = ax^2 + b'xy + c'y^2$ with $|b'| \le a \le c'$. Write

$$g(x, y) = f(px + qy, rx + sy),\ ps - qr = 1.$$

Suppose first that $|b| = a < c$. Then $a = g(1, 0) = f(p, r)$, so that by (*) we see that $p = \pm 1$ and $r = 0$. Then $ps - qr = 1$ implies that $s = p$, so that $\left(\begin{smallmatrix} p & q \\ r & s \end{smallmatrix}\right) = \pm\left(\begin{smallmatrix} 1 & \alpha \\ 0 & 1 \end{smallmatrix}\right)$ for some $\alpha \in \mathbb{Z}$. It follows after a computation that $b' = a \pm 2a\alpha$. Since $|b'| \le a$, we have $\alpha = 0$, so that $\left(\begin{smallmatrix} p & q \\ r & s \end{smallmatrix}\right) = \pm\left(\begin{smallmatrix} 1 & 0 \\ 0 & 1 \end{smallmatrix}\right)$, showing that $f(x, y) = g(x, y)$.

Alternatively, assume $|b| < a = c$. Then (***) implies that

$$a = g(1, 0) = f(p, r),\ \text{so that}\ p = \pm 1,\ r = 0\ \text{or}\ p = 0,\ r = \pm 1$$
$$c = g(0, 1) = f(q, s),\ \text{so that}\ q = \pm 1,\ s = 0\ \text{or}\ q = 0,\ s = \pm 1.$$

Since $ps - qr = 1$, we see that

$$\begin{pmatrix} p & q \\ r & s \end{pmatrix} = \begin{pmatrix} \pm 1 & 0 \\ 0 & \pm 1 \end{pmatrix} \text{ or } \begin{pmatrix} 0 & \pm 1 \\ \mp 1 & 0 \end{pmatrix}$$

In the second case, we get $g(x, y) = f(\pm y, \mp x) = ax^2 - bxy + ay^2$, which is not reduced since $b > 0$ and therefore $-b < 0$. Thus the transition matrix is the identity or its negation, so that $g(x, y) = f(x, y)$.

2.9. (a) Here is *Mathematica* code for computing all reduced forms of discriminant $D < 0$ where $D \equiv 0, 1 \bmod 4$:

```
reducedForms[
  disc : d_Integer /; (d < 0 && (Mod[d, 4] == 0 || Mod[d, 4] == 1))] :=
  Module[{list},
    list = Select[
      Flatten[Table[{a, b, (b^2 - disc)/(4 a)},
       {a, 1, Floor[Sqrt[-disc/3]]},
       {b, Select[Range[-a + 1, a], Mod[#^2 - disc, 4 a] == 0 &]}], 1],
      GCD[Sequence@@#] == 1 && #[[1]] <= #[[3]] &&
        (#[[2]] >= 0 || #[[1]] < #[[3]]) &];
    #.{x^2, x y, y^2} & /@ list]
```

For example, issuing the command

```
In[1]:= reducedForms[-108]
```

in *Mathematica* gives the output (assuming that `x` and `y` are not defined):

```
Out[1]:= {x^2 + 27y^2, 4x^2 - 2xy + 7y^2, 4x^2 + 2xy + 7y^2}
```

This makes it easy to verify the entries in table (2.14).

(b) $D = -3$: We have $a \le \sqrt{\frac{-D}{3}} = 1$, so that $a = 1$. Since the discriminant is odd, b must be odd, so that $b = \pm 1$. It follows that $c = 1$ by computing the discriminant. So the possibilities are $x^2 \pm xy + y^2$. But $x^2 - xy + y^2$ is not reduced ($a = |b|$, $b < 0$), so there is only one reduced form.

$D = -15$: We have $a \le \sqrt{\frac{-D}{3}} = \sqrt{5} < 3$, so that $a = 1$ or $a = 2$. Since D is odd, we must have $b = \pm 1$; computing discriminants gives $x^2 \pm xy + 4y^2$ and

$2x^2 \pm xy + 2y^2$. But $x^2 - xy + 4y^2$ ($a = |b|$, $b < 0$) and $2x^2 - xy + 2y^2$ ($a = c$, $b < 0$) are not reduced, so there are two reduced forms.

$D = -24$: We have $a \leq \sqrt{\frac{-D}{3}} = \sqrt{8} < 3$, so that $a = 1$ or $a = 2$. Since D is even, so is b. Hence we get the forms $x^2 + 6y^2$ and $2x^2 + 3y^2$, since $a = b = 2$ gives $c = \frac{b^2 - D}{4a} = \frac{7}{2} \notin \mathbb{Z}$. So there are two reduced forms.

$D = -31$: We have $a \leq \sqrt{\frac{31}{3}} = \sqrt{10} < 4$, so that $a = 1$, $a = 2$, or $a = 3$. Since D is odd, so is b. For $a = 1$, we get $x^2 \pm xy + 8y^2$, but $x^2 - xy + 8y^2$ ($a = |b|$, $b < 0$) is not reduced. For $a = 2$ we must have $b = 1$, and we get the reduced forms $2x^2 \pm xy + 4y^2$. For $a = 3$, we have $b = 1$ or $b = 3$; neither of these values of b yields a form with integral coefficients. So there are three reduced forms: $x^2 + xy + 8y^2$ and $2x^2 \pm xy + 4y^2$.

$D = -52$: We have $a \leq \sqrt{\frac{52}{3}} < \sqrt{18} < 5$, so that $a = 1$, $a = 2$, $a = 3$, or $a = 4$. Since D is even, so is b. For $a = 1$, we get only $x^2 + 13y^2$. For $a = 2$, $b = 0$ does not yield an integral form, while $b = 2$ yields $2x^2 \pm 2xy + 7y^2$, but $2x^2 - 2xy + 7y^2$ ($a = |b|$, $b < 0$) is not reduced. For $a = 3$, again neither possibility for b yields an integral form. Finally, for $a = 4$, none of the possible three values $b = 0$, $b = 1$, or $b = 2$ yields an integral form. So there are two reduced forms: $x^2 + 13y^2$ and $2x^2 + 2xy + 7y^2$.

2.10. (a) The proof is similar to the proof for negative discriminants. Given a form of discriminant $D > 0$, D not a perfect square, pick a form $f(x, y) = ax^2 + bxy + cy^2$ properly equivalent to it and minimizing $|b|$. If $|a| < |b|$, then f is properly equivalent to

$$g(x, y) = f(x + my, y) = ax^2 + (2am + b)xy + c'y^2,$$

where $c' = am^2 + bm + c$. Since $|a| < |b|$ and $a \neq 0$, we can choose $m \in \mathbb{Z}$ so that $|2am + b| < |b|$, contradicting our choice of b. So we must have $|b| \leq |a|$. In a similar manner we can also show $|b| \leq |c|$. If $|a| \leq |c|$, we are done; otherwise, exchange the outer coefficients by the proper equivalence $(x, y) \mapsto (-y, x)$, giving the new form $cx^2 - bxy + ay^2$ which satisfies the required inequalities.

(b) Suppose $ax^2 + bxy + cy^2$ has discriminant $D > 0$, where D is not a perfect square and $|b| \leq |a| \leq |c|$. If a and c have the same sign, then

$$0 < D = b^2 - 4ac = b^2 - 4\,|a|\,|c| \leq a^2 - 4a^2 < 0,$$

which is impossible. Thus a and c have opposite signs, so

$$D = b^2 - 4ac = b^2 + 4\,|a|\,|c| \geq b^2 + 4\,|a|^2 \geq 4\,|a|^2.$$

Dividing through by 4 and taking square roots gives $|a| \leq \frac{\sqrt{D}}{2}$.

(c) If D as in part (b) is fixed, then part (b) implies that there are only finitely many choices for a. But since $|b| \leq |a|$, there are only finitely many choices for b given a. Finally, for any choice of a and b, $D = b^2 - 4ac$ shows that there is at most one choice for c since $a \neq 0$. So the number of forms of discriminant D satisfying $|b| \leq |a| \leq |c|$ is finite. By part (a), it follows that the class number $h(D)$ is finite.

2.11. We use quadratic reciprocity to find the relevant congruence conditions and then apply Theorem 2.16, using table (2.14) to find the reduced forms of the given discriminants.

$x^2 + y^2$: Here $D = -4$, and

$$\chi([p]) = \left(\frac{-4}{p}\right) = \left(\frac{-1}{p}\right) = \begin{cases} 1 & p \equiv 1 \bmod 4 \\ -1 & p \equiv -1 \bmod 4, \end{cases}$$

so that by Theorem 2.16, p is represented by a reduced form of discriminant -4 if and only if $p \equiv 1 \bmod 4$. But the only reduced form of discriminant -4 is $x^2 + y^2$, which proves the first of Fermat's results.

$x^2 + 2y^2$: Here $D = -8$, and

$$\chi([p]) = \left(\frac{-8}{p}\right) = \left(\frac{-1}{p}\right)\left(\frac{2}{p}\right) = (-1)^{(p-1)/2}(-1)^{(p^2-1)/8} = \begin{cases} 1 & p \equiv 1,3 \bmod 8 \\ -1 & p \equiv 5,7 \bmod 8. \end{cases}$$

By Theorem 2.16, it follows that p is represented by a reduced form of discriminant -8 if and only if $p \equiv 1,3 \bmod 8$. But the only reduced form of discriminant -8 is $x^2 + 2y^2$, which proves the second of Fermat's results.

$x^2 + 3y^2$: Note first that $3 = 0^2 + 3 \cdot 1^2$ is so representable. Now assume $p > 3$. Here $D = -12$, and

$$\chi([p]) = \left(\frac{-12}{p}\right) = \left(\frac{-1}{p}\right)\left(\frac{3}{p}\right) = (-1)^{(p-1)/2}(-1)^{(p-1)/2}\left(\frac{p}{3}\right) = \left(\frac{3}{p}\right).$$

Thus p is represented by a reduced form of discriminant -12 if and only if p is a square mod 3, i.e., $p \equiv 1 \bmod 3$. But the only reduced form of discriminant -12 is $x^2 + 3y^2$, which proves the third of Fermat's results.

$x^2 + 7y^2$: Here $D = -28$, and

$$\chi([p]) = \left(\frac{-28}{p}\right) = \left(\frac{-1}{p}\right)\left(\frac{7}{p}\right) = (-1)^{(p-1)/2}(-1)^{(p-1)/2}\left(\frac{p}{7}\right) = (-1)^{p-1}\left(\frac{p}{7}\right) = \left(\frac{p}{7}\right).$$

Thus p is represented by a reduced form of discriminant -28 if and only if $p \equiv 1,2,4 \bmod 7$. But this is the same as $p \equiv 1,15,9,23,11,25 \bmod 28$ (since the other members of those congruence classes are even, so not prime to 28). Since the only reduced form of discriminant -28 is $x^2 + 7y^2$, this proves (2.17).

2.12. (a) Let p be a prime dividing m, and suppose $p^r \mid m$, $p^{r+1} \nmid m$. Then $m = p^r k$ where $p \nmid k$. If $p^r < k$, let $a = p^r$ and $c = k$. If $k < p^r$, let $a = k$ and $c = p^r$. Then $a < c$ and $\gcd(a,c) = 1$. Furthermore, $a > 1$ since p is not a prime power.

(b) These class numbers can be computed using the computer program from Exercise 2.9. We can also do the computation by hand as follows.

Forms of discriminant -32 have $a \leq \sqrt{\frac{32}{3}} < 4$, so that $a = 1$, $a = 2$, or $a = 3$. Since the discriminant is even, b is even. We now consider three cases:

$a = 1$: b must be 0, so that $c = 8$ and we get $x^2 + 8y^2$.

$a = 2$: If $b = 0$, we get $2x^2 + 4y^2$, which is not a primitive form; if $b = \pm 2$ we do not get an integral form.

$a = 3$: If $b = 0$, we do not get an integral form. If $b = \pm 2$, then we get $3x^2 \pm 2xy + 3y^2$. Since $a = c$, only $3x^2 + 2xy + 3y^2$ is reduced.

So there are two classes of reduced forms, $x^2 + 8y^2$ and $3x^2 + 2xy + 3y^2$, and therefore $h(-32) = 2$.

Forms of discriminant -124 have $a \leq \sqrt{\frac{124}{3}} < 7$, so that $1 \leq a \leq 6$. Since the discriminant is even, b is even. We now consider six cases:

$a = 1$: b must be zero, and we get $x^2 + 31y^2$.

$a = 2$: If $b = 0$, we do not get an integral form. If $b = \pm 2$ we get $2x^2 \pm 2xy + 16y^2$, which is not a primitive form.

$a = 3$: Neither $b = 0$ nor $b = \pm 2$ yields an integral form.

$a = 4$: Neither $b = 0$ nor $b = \pm 4$ yields an integral form. When $b = \pm 2$ we get $4x^2 \pm 2xy + 8y^2$, which is not a primitive form.

$a = 5$: Neither $b = 0$ nor $b = \pm 2$ yields an integral form. When $b = \pm 4$ we get $5x^2 \pm 4xy + 7y^2$, which are both reduced.

$a = 6$: None of $b = 0$, $b = \pm 2$, $b = \pm 4$, or $b = \pm 6$ yields an integral form.

So there are three classes of reduced forms, $x^2 + 31y^2$ and $5x^2 \pm 4xy + 7y^2$, and therefore $h(-124) = 3$.

2.13. For (2.19), when $n = 5$, Table (2.14) gives two reduced forms of discriminant -20: $x^2 + 5y^2$ and $2x^2 + 2xy + 3y^2$. Then for $D = -20$, quadratic reciprocity gives

$$\chi([p]) = \left(\frac{-20}{p}\right) = \left(\frac{-1}{p}\right)\left(\frac{5}{p}\right) = (-1)^{(p-1)/2}\left(\frac{p}{5}\right).$$

This product is 1 when both factors are positive, which occurs when $p \equiv 1 \bmod 4$ and $p \equiv \pm 1 \bmod 5$, or when both factors are negative, which occurs when $p \equiv 3 \bmod 4$ and $p \equiv \pm 2 \bmod 5$. This simplifies to $p \equiv 1, 9, 3, 7 \bmod 20$. Then (2.19) follows easily from Theorem 2.16.

For the remaining discriminants, part (b) of Exercise 2.9 computes the reduced forms for each.

$D = -3$:

$$\chi([p]) = \left(\frac{-3}{p}\right) = \left(\frac{-1}{p}\right)\left(\frac{3}{p}\right) = (-1)^{(p-1)/2}(-1)^{(3-1)\cdot(p-1)/4}\left(\frac{p}{3}\right) = \left(\frac{p}{3}\right).$$

There is only one reduced form of discriminant -3, so

$$p \equiv 1 \bmod 3 \iff \left(\frac{-3}{p}\right) = 1 \iff p = x^2 + xy + y^2.$$

Note that 3 is also representable by this form.

$D = -15$: Using quadratic reciprocity and the computation for $D = -3$,

$$\chi([p]) = \left(\frac{-15}{p}\right) = \left(\frac{-1}{p}\right)\left(\frac{3}{p}\right)\left(\frac{5}{p}\right) = \left(\frac{p}{3}\right)\left(\frac{p}{5}\right).$$

So $\chi([p]) = 1$ when $p \equiv 1 \bmod 3$ and $p \equiv 1, 4 \bmod 5$ or when $p \equiv 2 \bmod 3$ and $p \equiv 2, 3 \bmod 5$. Written mod 15 we get $p \equiv 1, 2, 4, 8 \bmod 15$. There are two reduced forms of discriminant -15, so for $p \nmid 15$, we have

$$p \equiv 1, 2, 4, 8 \bmod 15 \iff \left(\frac{-15}{p}\right) = 1$$
$$\iff p = x^2 + xy + 4y^2 \text{ or } p = 2x^2 + xy + 2y^2.$$

For primes dividing 15, note that 3 and 5 are representable by the second form but not by the first.

$D = -24$: Using quadratic reciprocity and the computation from $D = -3$,

$$\chi([p]) = \left(\frac{-24}{p}\right) = \left(\frac{-1}{p}\right)\left(\frac{2}{p}\right)\left(\frac{3}{p}\right) = \left(\frac{p}{3}\right)\left(\frac{2}{p}\right) = (-1)^{(p^2-1)/8}\left(\frac{p}{3}\right).$$

So $\chi([p]) = 1$ when $p \equiv 1 \bmod 3$ and $p \equiv 1, 7 \bmod 8$ or when $p \equiv 2 \bmod 3$ and $p \equiv 3, 5 \bmod 8$. Written mod 24 we get $p \equiv 1, 5, 7, 11 \bmod 24$. There are two reduced forms of discriminant -24, so for $p \nmid 24$, we have

$$p \equiv 1, 5, 7, 11 \bmod 24 \iff \left(\frac{-24}{p}\right) = 1 \iff p = x^2 + 6y^2 \text{ or } p = 2x^2 + 3y^2.$$

Note that $p = 2$ and $p = 3$, which divide 24, are clearly representable by the second form but not the first.

$D = -31$: Quadratic reciprocity gives

$$\chi([p]) = \left(\frac{-31}{p}\right) = \left(\frac{-1}{p}\right)\left(\frac{31}{p}\right) = \left(\frac{p}{31}\right)$$

since if $p \equiv 1 \bmod 4$ then $(-1/p) = 1$ and $(31/p) = (p/31)$, while if $p \equiv 3 \bmod 4$ then $(-1/p) = -1$ and $(31/p) = -(p/31)$. So $\chi([p]) = 1$ exactly when p is a square mod 31. There are three reduced forms of discriminant -31, so that when $p \neq 31$,

$$p \equiv 1, 2, 4, 5, 7, 8, 9, 10, 14, 16, 18, 19, 20, 25, 28 \bmod 31$$

$$\iff \left(\frac{-31}{p}\right) = 1 \iff p = x^2 + xy + 8y^2 \text{ or } p = 2x^2 \pm xy + 4y^2.$$

Finally, 31 is representable by the first form ($x = 1, y = -2$) but not the second.

$D = -52$: Proceeding as above,

$$\chi([p]) = \left(\frac{-52}{p}\right) = \left(\frac{-1}{p}\right)\left(\frac{13}{p}\right) = \left(\frac{-1}{p}\right)\left(\frac{p}{13}\right).$$

This product is 1 if $p \equiv 1 \bmod 4$ and is a square mod 13, or when $p \equiv 3 \bmod 4$ and is not a square mod 13. Squares mod 13 are $1, 3, 4, 9, 10, 12$; expressing this in terms of congruence classes mod 52 gives for $p \nmid 52$,

$$p \equiv 1, 7, 9, 11, 15, 17, 19, 25, 29, 31, 47, 49 \bmod 52$$

$$\iff \left(\frac{-52}{p}\right) = 1 \iff p = x^2 + 13y^2 \text{ or } p = 2x^2 + 2xy + 7y^2.$$

Observe that $p = 13$ is representable by the first form but not the second, while $p = 2$ is representable by the second form but not the first.

2.14. For $D = -20$, $\ker(\chi)$ consists of the congruence classes of $1, 3, 7, 9 \bmod 20$. The form $x^2 + 5y^2$ clearly represents $[1]$ and $[9]$. However, $x^2 + 5y^2 \equiv 3 \bmod 20$ implies that $x^2 \equiv 3 \bmod 5$, which is impossible, so that $x^2 + 5y^2$ does not represent $[3]$. Thus $x^2 + 5y^2$ represents the subgroup $\{[1], [9]\} \subset \ker(\chi)$. The other reduced form, $2x^2 + 2xy + 3y^2$, since it represents 3 and therefore represents $[3]$, must represent the other coset of that subgroup, $\{[3], [7]\} \subset \ker(\chi)$.

For $D = -56$, $\ker(\chi)$ consists of the congruence classes of

$$1, 3, 5, 9, 13, 15, 19, 23, 25, 27, 39, 45 \bmod 56.$$

Clearly $x^2 + 14y^2$ represents $[1]$, $[9]$, and $[25]$; it also represents $[15]$, $[23]$, and $[39]$ by setting $x = 1$, 3, or 5 and $y = 1$. However, if $x^2 + 14y^2 \equiv 3 \bmod 56$, then $x^2 \equiv 3 \bmod 7$, which is impossible. Therefore $x^2 + 14y^2$ does not represent $[3]$, and therefore it represents precisely the subgroup $\{[1], [9], [15], [23], [25], 39\} \subset \ker(\chi)$. Since $2x^2 + 7y^2$ represents 9, it represents the same subgroup. Finally, since $3x^2 \pm 2xy + 5y^2$ represents 3, these forms must represent elements of the other coset of that subgroup, $\{[3], [5], [13], [19], [27], [45]\} \subset \ker(\chi)$.

2.15. For $D = -56$, quadratic reciprocity gives

$$\chi([p]) = \left(\frac{-56}{p}\right) = \left(\frac{-1}{p}\right)\left(\frac{2}{p}\right)\left(\frac{7}{p}\right) = (-1)^{(p^2-1)/8}\left(\frac{p}{7}\right).$$

This product is 1 when $p \equiv 1, 7 \bmod 8$ and $p \equiv 1, 2, 4 \bmod 7$, or when $p \equiv 3, 5 \bmod 8$ and $p \equiv 3, 5, 6 \bmod 7$. This is equivalent to $p \equiv 1, 3, 5, 9, 13, 15, 19, 25, 27, 39, 45 \bmod 56$, so these are the congruence classes mod 56 that are representable by a form of discriminant -56. Combining this with (2.21) yields (2.23).

2.16. First, since $D \equiv 1 \bmod 4$, this is an integral form. Its discriminant is

$$b^2 - 4ac = 1^2 - 4 \cdot 1 \cdot \frac{1-D}{4} = 1 - (1-D) = D.$$

Next, since $a = b = 1$, it is clear that $\gcd(a, b, c) = 1$. Further, $D < 0$ implies that $c = (1-D)/4 \geq 1$. It follows that $|b| = a \leq c$, and also $b = 1 \geq 0$ as required by (2.7). Therefore this is a reduced form.

2.17. (a) By Lemma 2.3, if $2k$ is properly represented by a form f of discriminant D, then f is equivalent to a form $2kx^2 + bxy + cy^2$, so that $b^2 - 4(2k)c = b^2 - 8kc = D$. It follows that $b^2 \equiv D \bmod 8$, so that $D \equiv 1 \bmod 8$ since we know that D is odd.

(b) From the text, we know that m is properly represented by a form of discriminant D. Since we are dealing with the case where m is even, we know that $D \equiv 1 \bmod 8$ by part (a), so $\chi([2]) = 1$ by part (c) of Exercise 1.12. Therefore, if $m = 2^k r$ where r is odd, we have

$$\chi([m]) = \chi([2^k])\chi([r]) = \chi([2])^k \chi([r]) = \chi([r]).$$

By Lemma 2.5, D is a quadratic residue modulo m, hence a quadratic residue modulo r since $r \mid m$. Applying Lemma 2.5 again, we conclude that r is properly represented by a form of discriminant D. Since r is odd, the argument in the text implies that $\chi([r]) = 1$. Combining this with the above equation proves that $[m] \in \ker(\chi)$.

(c) We start with

$$4\left(x^2 + xy + \frac{1-D}{4}y^2\right) \equiv (2x+y)^2 \bmod D,$$

which is easily seen to be true by expanding both sides. Since D is odd, 2 is invertible modulo D, say $2r \equiv 1 \bmod D$, so that

$$x^2 + xy + \frac{1-D}{4}y^2 \equiv r^2(2x+y)^2 \equiv (2rx + ry)^2 \equiv (x + ry)^2 \bmod D.$$

This shows that the principal form represents only squares modulo D. But by setting $y = 0$ and letting x vary from 1 to $D - 1$, we see that the principal form represents all the squares modulo D. Thus H is exactly the set of squares modulo D.

(d) If $f(x, y) = ax^2 + bxy + cy^2$ is a primitive form of discriminant D, it represents some m prime to D, so that it is properly equivalent to $g(x, y) = mx^2 + b'xy + c'y^2$. So we may assume that a is prime to D. Then (2.4) gives

$$4af(x, y) = (2ax + by)^2 - Dy^2 \equiv (2ax + by)^2 \bmod D \in H.$$

Since D is odd and relatively prime to a, we see that $4a$ is invertible in $(\mathbb{Z}/D\mathbb{Z})^*$, say $4at \equiv 1 \bmod D$. Then $f(x, y) \in tH$ for all x, y. Since $(2ax + by)^2$ ranges over all elements of H, it follows that the values assumed by f are exactly the coset tH.

2.18. (a) If $f(1, 0) = a$ and $f(0, 1) = c$ are not prime to p, then $f(1, 1) = a + b + c$ must be prime to p, for if it were not, then b would not be prime to p either, so the form would not be primitive.

(b) Let the prime factors of M be $p_1, p_2, \ldots, p_r$. Then using part (a), for each p_i we can choose $(x_i, y_i) = (1, 0)$, $(0, 1)$, or $(1, 1)$ such that $f(x_i, y_i) \not\equiv 0 \bmod p_i$. By the

Chinese Remainder Theorem, there are integers x and y such that $x \equiv x_i \bmod p_i$ for all i and $y \equiv y_i \bmod p_i$ for all i. Then for each i,

$$f(x,y) \equiv f(x_i, y_i) \not\equiv 0 \bmod p_i,$$

so that $f(x,y)$ is prime to M. If $\gcd(x,y) = d > 1$, then $f(x/d, y/d)$ is still prime to M.

2.19. We will use the results of part (b) of Exercise 2.9 and Exercise 2.13.

$D = -15$: By Exercise 2.17, the congruence classes representable by the principal form are the squares in $(\mathbb{Z}/15\mathbb{Z})^*$, which are [1] and [4]; the remaining classes are [2] and [8] by Exercise 2.13. Thus

$$\begin{aligned} p \equiv 1, 4 \bmod 15 &\iff p = x^2 + xy + 4y^2 \\ p \equiv 2, 8 \bmod 15 &\iff p = 2x^2 + xy + 2y^2. \end{aligned}$$

$D = -24$: By Corollary 2.27, the congruence classes representable by the principal form are $[\beta^2], [\beta^2 + 6] \subset (\mathbb{Z}/24\mathbb{Z})^*$, which consists only of [1] and [7]. The remaining classes are [5] and [11] by Exercise 2.13. Thus

$$\begin{aligned} p \equiv 1, 7 \bmod 24 &\iff p = x^2 + 6y^2 \\ p \equiv 5, 11 \bmod 24 &\iff p = 2x^2 + 3y^2. \end{aligned}$$

$D = -31$: By Exercise 2.17, the congruence classes representable by the principal form are the squares mod 31, of which there are fifteen. Note that the principal form $x^2 + xy + 8y^2$ represents $[2] \in (\mathbb{Z}/31\mathbb{Z})^*$, for example with $(x, y) = (8, 0)$. The other two forms are $2x^2 \pm xy + 4y^2$; since these forms represent 2 and therefore represent [2], all three forms have the same genus, so that

$$\begin{aligned} &p \equiv 1, 2, 4, 5, 7, 8, 9, 10, 14, 16, 18, 19, 20, 25, 28 \bmod 31 \\ &\iff p = x^2 + xy + 8y^2 \text{ or } p = 2x^2 \pm xy + 4y^2. \end{aligned}$$

$D = -52$: By Corollary 2.27, the congruence classes representable by the principal form are the squares mod 52, which are 1, 9, 17, 25, 29, and 49 (here $n = 13$, so that adding n to any of the squares gives an even number, which is not prime to 52). The remaining classes are 7, 11, 15, 19, 31, and 47 by Exercise 2.13. Thus

$$\begin{aligned} p \equiv 1, 9, 17, 25, 29, 49 \bmod 52 &\iff p = x^2 + 13y^2 \\ p \equiv 7, 11, 15, 19, 31, 47 \bmod 52 &\iff p = 2x^2 + 2xy + 7y^2. \end{aligned}$$

2.20. The version of Corollary 2.27 for $D \equiv 1 \bmod 4$, $D < 0$, goes as follows:

COROLLARY 2.27. *Let $n \equiv 3 \bmod 4$ be a positive integer and p an odd prime not dividing n. Then p is represented by a form of discriminant $-n$ in the principal genus if and only if for some integer β,*

$$p \equiv \beta^2 \bmod n.$$

PROOF. Since $D = -n \equiv 1 \bmod 4$, part (c) of Exercise 2.17 implies that the subgroup H in Lemma 2.24 consists of the squares in $(\mathbb{Z}/n\mathbb{Z})^*$. Then the corollary follows easily from part (a) of Lemma 2.24. Q.E.D.

2.21. For each discriminant, we compute the reduced forms and show that they represent different congruence classes. Lemma 2.24 then gives the result.

$p = x^2 + 6y^2$: This is the case $D = -24$, which was handled in Exercise 2.19.

$p = x^2 + 10y^2$: Here $D = -40$. To find the reduced forms, note that $a \leq \sqrt{\frac{40}{3}} < 4$, and b is even. For $a = 1$, b must be zero and we get $x^2 + 10y^2$. For $a = 2$ and $b = 0$ we get $2x^2 + 5y^2$; when $a = 2$ and $b = \pm 2$ we do not get an integral form. Since $a = 3$

does not give an integral form for either $b = 0$ or $b = 2$, there are only two reduced forms. By quadratic reciprocity,

$$\chi([p]) = \left(\frac{-40}{p}\right) = \left(\frac{-1}{p}\right)\left(\frac{2}{p}\right)\left(\frac{5}{p}\right) = (-1)^{(p-1)/2}(-1)^{(p^2-1)/8}\left(\frac{p}{5}\right).$$

Thus solutions to $\chi([p]) = 1$ are primes with

$$p \equiv 1, 7, 9, 11, 13, 19, 23, 37 \bmod 40.$$

Of these, $x^2 + 10y^2$ represents [1], [9], [11], and [19], say with $x = 1$ or 3 and $y = 0$ or 1. It does not represent [7]. To see this, suppose that $x^2 + 10y^2 \equiv 7 \bmod 40$. This implies $x^2 \equiv 7 \bmod 10$, a contradiction. The other form, $2x^2 + 5y^2$, represents [7], say with $x = y = 1$, so it must be in a different genus, and represents [7], [13], [23], and [37]. Thus

$$p = x^2 + 10y^2 \iff p \equiv 1, 9, 11, 19 \bmod 40.$$

$p = x^2 + 13y^2$**:** This is the case $D = -52$, which was handled in Exercise 2.19.

$p = x^2 + 15y^2$**:** Here $D = -60$. To find the $h(-60) = 2$ reduced forms, note that $a \le \sqrt{\frac{60}{3}} < 5$, and b is even. For $a = 1$, b must be zero and we get $x^2 + 15y^2$. For $a = 2$ and $b = 0$ we do not get an integral form; when $a = 2$ and $b = \pm 2$ we get $2x^2 \pm 2xy + 8y^2$, which is not primitive. Then $a = 3$ and $b = 0$ gives $3x^2 + 5y^2$, but $a = 3$ and $b = 2$ does not give an integral form. Finally, $a = 4$ and $b = 0$ or $b = 4$ does not give an integral form, while $a = 4$ and $b = 2$ gives $2x^2 \pm 2xy + 4y^2$, which is not primitive. Quadratic reciprocity easily implies that

$$\chi([p]) = \left(\frac{-60}{p}\right) = \left(\frac{-1}{p}\right)\left(\frac{3}{p}\right)\left(\frac{5}{p}\right) = \left(\frac{p}{3}\right)\left(\frac{p}{5}\right).$$

Thus solutions to $\chi([p]) = 1$ are primes with

$$p \equiv 1, 2, 4, 8 \bmod 15, \text{ or equivalently, } p \equiv 1, 17, 19, 23, 31, 47, 49, 53 \bmod 60.$$

Of these, by inspection $x^2 + 15y^2$ represents [1], [19], [31], and [49]. The other form, $3x^2 + 5y^2$, does not represent [1], since $3x^2 + 5y^2 \equiv 1 \bmod 60$ implies $5y^2 \equiv 1 \bmod 3$, which in turn gives $y^2 \equiv 2 \bmod 3$, a contradiction. Therefore $3x^2 + 5y^2$ must be in a different genus, and must represent [17], [23], [47], and [53]. Thus

$$p = x^2 + 10y^2 \iff p \equiv 1, 19, 31, 49 \bmod 60.$$

$p = x^2 + 21y^2$**:** Here $D = -84$. To find the $h(-84) = 4$ reduced forms, note that $a \le \sqrt{\frac{84}{3}} = \sqrt{28} < 6$, and b is even. For $a = 1$, b must be zero and we get $x^2 + 21y^2$. For $a = 2$ and $b = 0$ we do not get an integral form; when $a = 2$ and $b = \pm 2$ we get $2x^2 \pm 2xy + 11y^2$, but $2x^2 + 2xy + 11y^2$ $(a = |b|,\ b < 0)$ is not reduced. Next, $a = 3$ and $b = 0$ gives $3x^2 + 7y^2$, but $a = 3$ and $b = 2$ does not give an integral form. Then $a = 4$ does not give an integral form for $b = 0$, 2, or 4. Finally, for $a = 5$, the only integral forms are $5x^2 \pm 4xy + 5y^2$, but $5x^2 - 4xy + 5y^2$ $(a = c,\ b < 0)$ is not reduced. Solutions to $\chi([p]) = 1$ are primes p with $(-21/p) = 1$, which were computed in Exercise 1.15 to be

$$p \equiv 1, 5, 11, 17, 19, 23, 25, 31, 37, 41, 55, 71 \bmod 84.$$

Squares less than 84 that are prime to 84 are only 1 and 25; of the other squares less than 84, only $16 + n = 16 + 21 = 37$ is prime to 84. So the principal genus represents [1], [25], and [37]. Since $2x^2 + 2xy + 11y^2$ represents 11, it is not in the principal genus. Since $3x^2 + 7y^2$ represents 19 via $x = 2$, $y = 1$, it is not in the principal genus either. Finally, $5x^2 + 4xy + 5y^2$ represents 5, so it is not in the principal genus. Therefore the principal form is the only form in its genus, and thus

$$p = x^2 + 21y^2 \iff p \equiv 1, 25, 37 \bmod 84.$$

$p = x^2 + 22y^2$: Here $D = -88$. To find the $h(-88) = 2$ reduced forms, note that $a \le \sqrt{\frac{88}{3}} < 6$, and b is even. For $a = 1$, b must be zero and we get $x^2 + 22y^2$. For $a = 2$ and $b = 0$ we get $2x^2 + 11y^2$; when $a = 2$ and $b = \pm 2$ we do not get an integral form. Then $a = 3$ does not give an integral form with either $b = 0$ or $b = 2$. Also $a = 4$ does not give an integral form for $b = 0$, 2, or 4, and the same is true when $a = 5$. By quadratic reciprocity,

$$\chi([p]) = \left(\frac{-88}{p}\right) = \left(\frac{-1}{p}\right)\left(\frac{2}{p}\right)\left(\frac{11}{p}\right) = (-1)^{(p^2-1)/8}\left(\frac{p}{11}\right).$$

One can then compute that solutions to $\chi([p]) = 1$ are primes with

$$p \equiv 1, 9, 13, 15, 19, 21, 23, 25, 29, 31, 35, 43, 47, 49, 51, 61, 71, 81, 83, 85 \bmod 88.$$

Squares less than 88 that are prime to 88 are 1, 9, 25, 49, and 81; further, $n = 22$, and $1+22 = 23$, $9+22 = 31$, $25+22 = 47$, $49+22 = 71$, and $81+22 = 103 \equiv 15 \bmod 88$ are all prime to 88. So the principal genus represents the ten classes

$$[1], [9], [23], [25], [31], [47], [49], [71], [81], [103] \in (\mathbb{Z}/88\mathbb{Z})^*.$$

Now suppose that $2x^2 + 11y^2$ represented $[1]$. Then $2x^2 + 11y^2 \equiv 1 \bmod 88$ implies that $2x^2 \equiv 1 \bmod 11$, or $x^2 \equiv 6 \bmod 11$, which is impossible. Therefore $2x^2 + 7y^2$ does not represent $[1]$, so it is not in the principal genus. It follows that the principal form is the only form in its genus, so we get

$$p = x^2 + 22y^2 \iff p \equiv 1, 9, 15, 23, 25, 31, 47, 49, 71, 81 \bmod 88.$$

$p = x^2 + 30y^2$: Here $D = -120$. To find the $h(-120) = 4$ reduced forms, note that $a \le \sqrt{\frac{120}{3}} < 7$, and b is even. For $a = 1$, b must be zero and we get $x^2 + 30y^2$. For $a = 2$ and $b = 0$ we get $2x^2 + 15y^2$; when $a = 2$ and $b = \pm 2$ we do not get an integral form. $a = 3$ and $b = 0$ gives $3x^2 + 10y^2$, while $a = 3$ and $b = 2$ does not give an integral form. $a = 4$ does not give an integral form for $b = 0$, 2, or 4. $a = 5$ and $b = 0$ gives $5x^2 + 6y^2$; the other possible values for b, which are 2 and 4, do not give integral forms. When $a = 6$, $b = 0$ gives $6x^2 + 5y^2$, which is not reduced, and none of $b = 2$, 4, or 6 give integral forms. By quadratic reciprocity,

$$\chi([p]) = \left(\frac{-120}{p}\right) = \left(\frac{-1}{p}\right)\left(\frac{2}{p}\right)\left(\frac{3}{p}\right)\left(\frac{5}{p}\right) = (-1)^{(p^2-1)/8}\left(\frac{p}{3}\right)\left(\frac{p}{5}\right).$$

After some computation, one finds that solutions to $\chi([p]) = 1$ are primes with

$$p \equiv 1, 11, 13, 17, 23, 29, 31, 37, 43, 47, 49, 59, 67, 79, 101, 113 \bmod 120.$$

Squares less than 120 that are prime to 120 are 1 and 49; further, $n = 30$, and adding 30 to squares less than 120 gives only the two additional values $1 + 30 = 31$ and $49 + 30 = 79$ prime to 120. So the principal genus represents the four classes $[1]$, $[31]$, $[49]$, and $[79]$. Since the form $2x^2 + 15y^2$ represents 17, the form $3x^2 + 10y^2$ represents 13, and the form $5x^2 + 6y^2$ represents 11, none of these forms is in the principal genus. Therefore the principal form is the only form in its genus, so we get

$$p = x^2 + 30y^2 \iff p \equiv 1, 31, 49, 79 \bmod 120.$$

2.22. Expanding $(axz + bxw + byz + cyw)^2 + n(xw - yz)^2$ gives

$$a^2x^2z^2 + 2abx^2zw + 2abxyz^2 + 2acxyzw + b^2x^2w^2 + 2b^2xyzw + 2bcxyw^2 +$$
$$b^2y^2z^2 + 2bcy^2zw + c^2y^2w^2 + n(x^2w^2 - 2xyzw + y^2z^2).$$

Since $n = ac - b^2$, further expansion and simplification yields

$$a^2x^2z^2 + 2abx^2zw + 2abxyz^2 + 4b^2xyzw + 2bcxyw^2 + 2bcy^2zw + c^2y^2w^2 + acx^2w^2 + acy^2z^2.$$

One checks that $(ax^2 + 2bxy + cy^2)(az^2 + 2bzw + cw^2)$ expands to the same quantity, proving (2.31). Substituting $a = 2$, $b = 1$, $c = 3$ gives (2.30).

2.23. (a) Since $x^{4k} - 1$ is a polynomial of degree $4k < p$, we can find $0 < x < p$ with $x^{4k} - 1 \not\equiv 0 \bmod p$ (this idea is from the proof of Theorem 1.2). Since x is not divisible by p, Fermat's Little Theorem shows that the left-hand side of the identity is 0 mod p, so that

$$((x^{2k} - 1)^2 + 2x^{2k})(x^{4k} - 1) \equiv 0 \bmod p.$$

Since $x^{4k} - 1 \not\equiv 0 \bmod p$, it follows that $(x^{2k} - 1)^2 + 2x^{2k} \equiv 0 \bmod p$, so that $-2x^{2k} \equiv (x^{2k} - 1)^2 \bmod p$. Using Fermat's Little Theorem again,

$$-2 \equiv -2x^{8k} \equiv (-2x^{2k})x^{6k} \equiv (x^{2k} - 1)^2 x^{6k} \equiv (x^{5k} - x^{3k})^2 \bmod p,$$

showing that $(-2/p) = 1$.

(b) If $p \equiv 3 \bmod 8$ and $(-2/p) = -1$, then

$$-1 = \left(\frac{-2}{p}\right) = \left(\frac{-1}{p}\right)\left(\frac{2}{p}\right) = (-1)^{(p-1)/2}\left(\frac{2}{p}\right) = -\left(\frac{2}{p}\right),$$

where the last equality uses $p \equiv 3 \bmod 8$. Thus $(2/p) = 1$, and then it follows from Corollary 2.6 that p is represented by a primitive form of discriminant 8.

(c) By part (b) of Exercise 2.10, a form of discriminant 8 is properly equivalent to a reduced form $ax^2 + bxy + cy^2$ (of discriminant 8) with $|a| \le \frac{\sqrt{8}}{2} = \sqrt{2} < 2$, so that $a = \pm 1$. Also, b must be even; since $|b| \le |a|$, we get $b = 0$. Since $-4ac = 8$, we have $c = \mp 2$. So the two reduced forms of discriminant 8 are $\pm(x^2 - 2y^2)$.

(d) If $p = \pm(x^2 - 2y^2)$, then x is odd, so that $x^2 \equiv 1 \bmod 8$. If y is even, then $\pm(x^2 - 2y^2) \equiv \pm 1 \bmod 8$, which is impossible since $p \equiv 3 \bmod 8$. If y is odd, then $y^2 \equiv 1 \bmod 8$, so that $\pm(x^2 - 2y^2) \equiv \pm(1 - 2) = \mp 1 \bmod 8$, which is again impossible. This shows that the original assumption that $(-2/p) = -1$ was false, so that $(-2/p) = 1$.

2.24. (a) First, 1, p, and $-q$ are not all of the same sign. For the second condition, $-bc = pq$ is vacuously a quadratic residue mod 1. Next, $-ac = q$ is a quadratic residue mod p since we are assuming $(q/p) = 1$. Finally, $-ab = -p$ is a quadratic residue mod q since

$$\left(\frac{-p}{q}\right) = \left(\frac{-1}{q}\right)\left(\frac{p}{q}\right) = (-1)^{(q-1)/2} \cdot (-1) = (-1) \cdot (-1) = 1.$$

since $q \equiv 3 \bmod 4$ and $(p/q) = -1$ by assumption. It follows that $x^2 + py^2 - qz^2 = 0$ has a nontrivial integer solution.

(b) Choose a nontrivial solution. If x, y, and z are all even, then $\frac{x}{2}, \frac{y}{2}, \frac{z}{2}$ is also a nontrivial solution. So we may assume that x, y, and z are not all even. If x is even, then neither y nor z can be even since then the expression would be odd. However, if x is even and y, z are both odd, then mod 4 we have

$$x^2 + py^2 - qz^2 \equiv 0 + p - q \equiv 1 - 3 \equiv 2 \bmod 4,$$

so there are no such solutions. So assume x is odd. If both y and z are even, or if they are both odd, then the expression would be odd. There are two remaining cases:

$$\begin{aligned} y \text{ even, } z \text{ odd}: &\quad x^2 + py^2 - qz^2 \equiv 1 + 0 - q \equiv 1 - 3 \equiv 2 \bmod 4 \\ y \text{ odd, } z \text{ even}: &\quad x^2 + py^2 - qz^2 \equiv 1 + p - 0 \equiv 1 + 1 \equiv 2 \bmod 4. \end{aligned}$$

So there are no nontrivial solutions. This is the desired contradiction.

2.25. Since $ax^2-bxy+cy^2$ is improperly equivalent to $ax^2+bxy+cy^2$ via $(x,y)\mapsto(-x,y)$, we have

$$ax^2-bxy+cy^2 \xrightarrow{\text{improper}} ax^2+bxy+cy^2 \xrightarrow{\text{proper}} a'x^2+b'xy+c'y^2 \xrightarrow{\text{improper}} a'x^2-b'xy+c'y^2.$$

Since improper equivalences lie in $\mathrm{GL}(2,\mathbb{Z})-\mathrm{SL}(2,\mathbb{Z})$ (matrices with determinant -1) and proper equivalences lie in $\mathrm{SL}(2,\mathbb{Z})$, the composition above lies in $\mathrm{SL}(2,\mathbb{Z})$, so is a proper equivalence. The other direction is similar.

2.26. Let $f(x,y) = 14x^2+10xy+21y^2$, $g(x,y) = 9x^2+2xy+30y^2$. Note that the discriminant of these forms is -1076, so that $n = 269$. In (2.32), we have $a = 14$, $b = 5$, $a' = 9$, $b' = 1$. Since a and a' are relatively prime, we can find B such that

$$B \equiv \pm 5 \bmod 14$$
$$B \equiv \pm 1 \bmod 9.$$

Observe that 19 is a solution for $++$, 37 for $-+$, -37 for $+-$, and -19 for $--$. Then the four possible compositions $aa'x^2+2Bxy+\frac{B^2+n}{aa'}y^2$ are

$$126x^2 \pm 38xy + \frac{19^2+269}{126}y^2 = 126x^2 \pm 38xy + 5y^2$$
$$126x^2 \pm 74xy + \frac{37^2+269}{126}y^2 = 126x^2 \pm 74xy + 13y^2.$$

We compute the corresponding reduced forms as follows:

$126x^2+74xy+13y^2$: By Exercise 2.6, this form is properly equivalent to $13x^2+4xy+21y^2$, which is reduced.

$126x^2-74xy+13y^2$: Using Exercise 2.25 and the reduced form properly equivalent to $126x^2+74xy+13y^2$ above, this form is properly equivalent to $13x^2-4xy+21y^2$, also reduced.

$126x^2+38xy+5y^2$: First apply $(x,y)\mapsto(-y,x)$:

$$126(-y)^2+38(-y)(x)+5x^2 = 5x^2-38xy+126y^2.$$

Now apply $(x,y)\mapsto(x+4y,y)$ to get a reduced form:

$$5(x+4y)^2-38(x+4y)y+126y^2 = 5x^2+2xy+54y^2.$$

$126x^2-38xy+5y^2$: By Exercise 2.25 and the reduced form properly equivalent to $126x^2+38xy+5y^2$ above, this form is properly equivalent to $5x^2-2xy+54y^2$, also reduced.

The four reduced forms are different, so all four forms lie in different equivalence classes.

2.27. (a) First note that the representation of p must be proper, since if x and y have any common factor $d>1$, then $f(x,y)$ (or $g(x,y)$) would be divisible by d^2. By Lemma 2.3, we may assume

$$f(x,y) = px^2+bxy+cy^2, \quad g(x,y) = px^2+b'xy+c'y^2.$$

Since the discriminants of these two forms are equal, we have $b^2-4pc = (b')^2-4pc'$, so that $b^2\equiv(b')^2 \bmod 4p$. There are now two cases to consider:

p odd: Here, $b^2\equiv(b')^2 \bmod 4$, so that b and b' have the same parity, and $b^2\equiv(b')^2 \bmod p$, so that $b\equiv\pm b' \bmod p$. Since $px^2+b'xy+c'y^2$ is equivalent to $px^2-b'xy+c'y^2$ via $(x,y)\mapsto(-x,y)$, we may assume $b\equiv b' \bmod p$. Parity implies that $b\equiv b' \bmod 2p$, so that $b' = b+2kp$ for some $k\in\mathbb{Z}$. Then

$$\begin{aligned} f(x+ky,y) &= p(x+ky)^2+b(x+ky)y+cy^2 \\ &= px^2+(b+2kp)xy+(pk^2+bk+c)y^2 \\ &= px^2+b'xy+(pk^2+bk+c)y^2. \end{aligned}$$

This has the same discriminant as $g(x,y)$, and $f(x+ky,y) = g(x,y)$ follows. Thus $f(x,y)$ and $g(x,y)$ are equivalent.

$p = 2$: Here, $b^2 \equiv (b')^2 \bmod 8$, which easily implies $b \equiv b' \bmod 4$ or $b \equiv -b' \bmod 4$. Using the equivalence $(x,y) \mapsto (-x,y)$, we may assume $b \equiv b' \bmod 4$, so that $b' = b + 4k$ for some $k \in \mathbb{Z}$. Then

$$\begin{aligned} f(x+ky,y) &= 2(x+ky)^2 + b(x+ky)y + cy^2 \\ &= 2x^2 + (b+4k)xy + (2k^2+bk+c)y^2 \\ &= 2x^2 + b'xy + (2k^2+bk+c)y^2. \end{aligned}$$

As above, this implies $f(x+ky,y) = g(x,y)$, so $f(x,y)$ and $g(x,y)$ are equivalent.

(b) In general, if two forms $h(x,y)$ and $k(x,y)$ are equivalent, then either $h(x,y)$ or its opposite is properly equivalent to $k(x,y)$. By (a), $f(x,y) = x^2 + ny^2$ and $g(x,y)$ are equivalent. Since $f(x,y)$ is its own opposite, it follows that $f(x,y)$ is properly equivalent to $g(x,y)$. Since $g(x,y)$ is reduced, the two forms must be equal.

Solutions to Exercises in §3

3.1. (a) For the first formula, use $x = z = 1$ and $y = w = 0$; this gives

(*) $$f(1,0)g(1,0) = F(a_1,a_2), \quad \text{or} \quad aa' = Aa_1^2 + Ba_1a_2 + Ca_2^2.$$

For the second, use $x = w = 1$ and $y = z = 0$; this gives

(*) $$f(1,0)g(0,1) = F(b_1,b_2), \quad \text{or} \quad ac' = Ab_1^2 + Bb_1b_2 + Cb_2^2.$$

For the last, set $x = w = z = 1$ and $y = 0$, giving $f(1,0)g(1,1) = F(a_1+b_1, a_2+b_2)$. Now,

$$\begin{aligned} &F(a_1+b_1,a_2+b_2) \\ &\quad = A(a_1+b_1)^2 + B(a_1+b_1)(a_2+b_2) + C(a_2+b_2)^2 \\ &\quad = (Aa_1^2 + Ba_1a_2 + Ca_2^2) + (2Aa_1b_1 + B(a_1b_2+a_2b_1) + 2Ca_2b_2) \\ &\qquad\quad + (Ab_1^2 + Bb_1b_2 + Cb_2^2) \\ &\quad = aa' + (2Aa_1b_1 + B(a_1b_2+a_2b_1) + 2Ca_2b_2) + ac' \end{aligned}$$

using the equations (*). Setting this expression equal to

$$f(1,0)g(1,1) = a(a'+b'+c') = aa' + ab' + ac'$$

gives

(**) $$ab' = 2Aa_1b_1 + B(a_1b_2+a_2b_1) + 2Ca_2b_2.$$

(b) We have $a^2(b'^2 - 4a'c') = (ab')^2 - 4(aa')(ac')$. Substituting using (*) and (**) from part (a) gives

$$\begin{aligned} a^2(b'^2 - 4a'c') &= (ab')^2 - 4(aa')(ac') \\ &= (2Aa_1b_1 + B(a_1b_2+a_2b_1) + 2Ca_2b_2)^2 \\ &\qquad - 4(Aa_1^2 + Ba_1a_2 + Ca_2^2)(Ab_1^2 + Bb_1b_2 + Cb_2^2) \\ &= 4ACa_1a_2b_1b_2 + B^2a_1^2b_2^2 - 2B^2a_1a_2b_1b_2 \\ &\qquad + B^2a_2^2b_1^2 - 4ACa_1^2b_2^2 - 4ACa_2^2b_1^2 \\ &= (a_1b_2 - a_2b_1)^2(B^2 - 4AC). \end{aligned}$$

Since all three forms have discriminant $D \neq 0$, we may divide through by $b'^2 - 4a'c' = B^2 - 4AC$ to get $a^2 = (a_1b_2 - a_2b_1)^2$, so that $a = \pm(a_1b_2 - a_2b_1)$.

(c) As in part (a),

$$(*) \qquad f(1,0)g(1,0) = F(a_1, a_2), \text{ or } a'a = Aa_1^2 + Ba_1a_2 + Ca_2^2.$$

Set $x = w = 0$ and $y = z = 1$; this gives

$$(*) \qquad f(0,1)g(1,0) = F(c_1, c_2), \text{ or } a'c = Ac_1^2 + Bc_1c_2 + Cc_2^2.$$

Finally, set $x = y = z = 1$ and $w = 0$, giving $f(1,1)g(1,0) = F(a_1 + c_1, a_2 + c_2)$. Expanding the right-hand side, we get

$$\begin{aligned} F(a_1 + c_1, a_2 + c_2) &= A(a_1 + c_1)^2 + B(a_1 + c_1)(a_2 + c_2) + C(a_2 + c_2)^2 \\ &\quad + (Ac_1^2 + Bc_1c_2 + Cc_2^2) \\ &= a'a + (2Aa_1c_1 + B(a_1c_2 + a_2c_1) + 2Ca_2c_2) + a'c \end{aligned}$$

using the equalities (*). As in part (a), since $f(1,1)g(1,0) = a'(a + b + c) = a'a + a'b + a'c$, we get

$$(**) \qquad a'b = 2Aa_1c_1 + B(a_1c_2 + a_2c_1) + 2Ca_2c_2.$$

Since $a'^2(b^2 - 4ac) = (a'b)^2 - 4(a'a)(a'c)$, substitute using (*) and (**) and simplify, giving

$$a'^2(b^2 - 4ac) = (a_1c_2 - a_2c_1)^2(B^2 - 4AC).$$

Finally $b^2 - 4ac = B^2 - 4AC = D \neq 0$, so that $a'^2 = (a_1c_2 - a_2c_1)^2$ and therefore $a' = \pm(a_1c_2 - a_2c_1)$.

3.2. For (2.30),

$$a_1b_2 - a_2b_1 = 2 \cdot 1 - 0 \cdot 1 = 2 = f(1,0), \quad a_1c_2 - a_2c_1 = 2 \cdot (-1) - 0 \cdot 1 = -2 = -g(1,0).$$

For (2.31),

$$a_1b_2 - a_2b_1 = a \cdot 1 - 0 \cdot b = a = f(1,0), \quad a_1c_2 - a_2c_1 = a \cdot (-1) - 0 \cdot b = -a = -g(1,0).$$

So neither composition is direct.

3.3. Since $\gcd(p_1, p_2, \ldots, p_r, m) = 1$, choose integers $a, a_1, a_2, \ldots, a_r$ such that $am + \sum_{i=1}^r a_ip_i = 1$. This implies that $\sum_{i=1}^r a_ip_1 \equiv 1 \bmod m$.

First assume that $p_iq_j \equiv p_jq_i \bmod m$ for all i, j. Let $B \equiv \sum_{i=1}^r a_iq_i \bmod m$. Then

$$p_kB = \sum_{i=1}^r a_ip_kq_i \equiv \sum_{i=1}^r a_ip_iq_k = q_k \sum_{i=1}^r a_ip_i \equiv q_k \bmod m.$$

This solution is unique; for suppose that C is another solution. Then for $1 \leq i \leq r$, $p_i(B - C) \equiv 0 \bmod m$, so that $m \mid p_i(B - C)$. But the divisibility condition on the p_i and m then ensures that $m \mid B - C$, so that $B \equiv C \bmod m$.

Conversely, suppose that the congruences have a unique solution B. Given $1 \leq i, j \leq r$, we have $p_iB \equiv q_i \bmod m$ and $p_jB \equiv q_j \bmod m$. Multiply the first of these by p_j and the second by p_i, giving

$$p_jp_iB \equiv p_jq_i \bmod m, \; p_ip_jB \equiv p_iq_j \bmod m.$$

Combining these congruences, we get

$$p_iq_j \equiv p_ip_jB \equiv p_jp_iB \equiv p_jq_i \bmod m$$

as desired.

3.4. Set

$$\begin{aligned} p_1 &= a, \quad q_1 = ab' \\ p_2 &= a', \quad q_2 = a'b \\ p_3 &= (b+b')/2, \quad q_3 = (bb'+D)/2, \end{aligned}$$

and let $m = 2aa'$. For $i = 1$, $j = 2$, we have $a'(ab') \equiv a(a'b) \bmod 2aa'$ since $b' \equiv b \bmod 2$. For $i = 1$, $j = 3$,

$$a\left(\frac{bb'+D}{2}\right) = \frac{abb' + ab'^2 - 4aa'c'}{2} \equiv \frac{abb' + ab'^2}{2} \equiv \left(\frac{b+b'}{2}\right) ab' \bmod 2aa'.$$

Finally, for $i = 2$, $j = 3$,

$$a'\left(\frac{bb'+D}{2}\right) = \frac{a'bb' + a'b^2 - 4aa'c}{2} \equiv \frac{a'bb' + a'b^2}{2} \equiv \left(\frac{b+b'}{2}\right) a'b \bmod 2aa'.$$

3.5. (a) For $f(x,y)$, since $B \equiv b \bmod 2aa'$ write $B = b + 2aa'r = b + 2ak$. Then

$$\begin{aligned} f(x+ky, y) &= a(x+ky)^2 + b(x+ky)y + cy^2 \\ &= ax^2 + (b+2ak)xy + (bk + ak^2 + c)y^2 \\ &= ax^2 + (b+2ak)xy + \frac{(b+2ak)^2 - (b^2 - 4ac)}{4a} y^2 \\ &= ax^2 + Bxy + a'Cy^2, \end{aligned}$$

where the final equality follows from the definition of C. Hence the two forms are equivalent. The equivalence is proper since the transformation is $x \mapsto x + ky$, $y \mapsto y$, and $1 \cdot 1 - 0 \cdot k = 1$. The argument for $g(x,y)$ is essentially identical, using the fact that $B \equiv b' \bmod 2aa'$, say $B = b' + 2aa'\tilde{r} = b' + 2a'\tilde{k}$.

(b) Following the hint, we have

$$\begin{aligned} aa'X + \frac{B+\sqrt{D}}{2}Y &= aa'(xz - Cyw) + \frac{B+\sqrt{D}}{2}(axw + a'yz + Byw) \\ &= aa'xz + a\frac{B+\sqrt{D}}{2}xw + a'\frac{B+\sqrt{D}}{2}yz - aa'Cyw + \frac{B(B+\sqrt{D})}{2}yw \\ &= aa'xz + a\frac{B+\sqrt{D}}{2}xw + a'\frac{B+\sqrt{D}}{2}yz + \frac{2B^2 + 2B\sqrt{D} - 4aa'C}{4}yw \\ &= aa'xz + a\frac{B+\sqrt{D}}{2}xw + a'\frac{B+\sqrt{D}}{2}yz + \frac{2B^2 + 2B\sqrt{D} - (B^2 - D)}{4}yw \\ &= aa'xz + a\frac{B+\sqrt{D}}{2}xw + a'\frac{B+\sqrt{D}}{2}yz + \frac{B^2 + 2B\sqrt{D} + D}{4}yw \\ &= \left(ax + \frac{B+\sqrt{D}}{2}y\right)\left(a'z + \frac{B+\sqrt{D}}{2}w\right). \end{aligned}$$

Similarly,

$$aa'X + \frac{B-\sqrt{D}}{2}Y = \left(ax + \frac{B-\sqrt{D}}{2}y\right)\left(a'z + \frac{B-\sqrt{D}}{2}w\right).$$

Multiplying the two equations gives on the left

$$\begin{aligned}\Big(aa'X + \frac{B+\sqrt{D}}{2}Y\Big)&\Big(aa'X + \frac{B-\sqrt{D}}{2}Y\Big)\\ &= a^2a'^2X^2 + aa'X\frac{B-\sqrt{D}}{2}Y + aa'X\frac{B+\sqrt{D}}{2}Y + \frac{B^2-D}{4}Y^2\\ &= a^2a'^2X^2 + aa'BXY + \frac{4aa'C}{4}Y^2\\ &= aa'(aa'X^2 + BXY + CY^2),\end{aligned}$$

and on the right

$$\begin{aligned}\Big(a^2x^2 + aBxy + \frac{B^2-D}{4}y^2\Big)&\Big(a'^2z^2 + a'Bzw + \frac{B^2-D}{4}w^2\Big)\\ &= aa'(ax^2 + Bxy + a'Cy^2)(a'z^2 + Bzw + aCw^2).\end{aligned}$$

Since $aa' \neq 0$, setting these two expressions equal and dividing through by aa' gives

$$(ax^2 + Bxy + a'Cy^2)(a'z^2 + Bzw + aCw^2) = aa'X^2 + BXY + CY^2$$

as desired. To see that the composition is direct, note that

$$\begin{aligned}B_1(x, y; z, w) &= 1xz + 0xw + 0yz - Cyw,\\ B_2(x, y; z, w) &= 0xz + axw + a'yz + Byw\end{aligned}$$

so that

$$a_1b_2 - a_2b_1 = a = f(1, 0), \qquad a_1c_2 - a_2c_1 = a' = g(1, 0).$$

(c) Suppose that $\tilde{h}(x, y) = h(\alpha x + \beta y, \gamma x + \delta y)$ where $\alpha\delta - \beta\gamma = 1$, and that $G(x, y)$ is a direct composition of $h(x, y) = ax^2 + bxy + cy^2$ and $k(x, y)$. Then

$$\begin{aligned}\tilde{h}(x, y)k(x, y) &= h(\alpha x + \beta y, \gamma x + \delta y)k(z, w)\\ &= G(B_1(\alpha x + \beta y, \gamma x + \delta y; z, w), B_2(\alpha x + \beta y, \gamma x + \delta y; z, w)).\end{aligned}$$

Expanding the B_i gives

$$\begin{aligned}B_i(&\alpha x+\beta y,\gamma x+\delta y; z, w)\\ &= a_i(\alpha x+\beta y)z+b_i(\alpha x+\beta y)w+c_i(\gamma x+\delta y)z+d_i(\gamma x+\delta y)w\\ &= (a_i\alpha+c_i\gamma)xz+(b_i\alpha+d_i\gamma)xw+(a_i\beta+c_i\delta)yz+(b_i\beta+d_i\delta)yw,\end{aligned}$$

so that $G(x, y)$ is a composition of $\tilde{h}(x, y)$ and $k(x, y)$. (Note that the a_i, b_i, c_i, and d_i are independent of α, β, γ, and δ since they are determined by the composition of $h(x, y)$ and $k(x, y)$.) To see that this composition is direct, note first that

$$\begin{aligned}(a_1\alpha + c_1\gamma)(a_2\beta + c_2\delta) &- (a_2\alpha + c_2\gamma)(a_1\beta + c_1\delta)\\ &= a_1c_2\alpha\delta + a_2c_1\beta\gamma - a_2c_1\alpha\delta - a_1c_2\beta\gamma\\ &= (a_1c_2 - a_2c_1)(\alpha\delta - \beta\gamma)\\ &= a_1c_2 - a_2c_1 = k(1, 0)\end{aligned}$$

since $G(x, y)$ is a direct composition of $h(x, y)$ and $k(x, y)$. Turning to $\tilde{h}(1, 0)$, we follow the hint and use (3.1), which gives

$$(*)\qquad \pm\tilde{h}(1, 0) = (a_1\alpha + c_1\gamma)(b_2\alpha + d_2\gamma) - (a_2\alpha + c_2\gamma)(b_1\alpha + d_1\gamma).$$

Once we show that the sign is +, we will have the desired direct composition.

To study the sign in (*), we first expand the right-hand side to obtain

$$(a_1b_2 - a_2b_1)\alpha^2 + (a_1d_2 + b_2c_1 - a_2d_1 - b_1c_2)\alpha\gamma + (c_1d_2 - c_2d_1)\gamma^2.$$

However, $a_1b_2 - a_2b_1 = h(1,0) = a$ since $G(x,y)$ is a direct composition of $h(x,y)$ and $k(x,y)$. For simplicity, set $q = a_1d_2 + b_2c_1 - a_2d_1 - b_1c_2$ and $d = c_1d_2 - c_2d_1$. Then we can write (*) as

$$\pm\tilde{h}(1,0) = a\alpha^2 + q\alpha\gamma + d\gamma^2,$$

and since

$$\tilde{h}(1,0) = h(\alpha,\gamma) = a\alpha^2 + b\alpha\gamma + c\gamma^2,$$

we see that (*) becomes

$$(**) \qquad \pm(a\alpha^2 + b\alpha\gamma + c\gamma^2) = a\alpha^2 + q\alpha\gamma + d\gamma^2.$$

The key point is that this is true for *any* $\tilde{h}(x,y) = h(\alpha x+\beta y, \gamma x+\delta y)$ with $\alpha\delta-\beta\gamma = 1$, though the sign may depend on $\alpha,\beta,\gamma,\delta$. Since q and d depend only on the a_i, b_i, c_i, and d_i, which are independent of α, β, γ, and δ, our strategy will be to prove that $b = q$ and $c = d$ using specific values for α, β, γ, and δ. We then use these equalities to show that the sign is always + in (*) and (**).

First observe that if we set $\alpha = \delta = 1$ and $\beta = 0$, then for any $\gamma \in \mathbb{Z}$, we have $\alpha\delta - \beta\gamma = 1\cdot 1 - 0\cdot\gamma = 1$. Hence, by (**), there is $\varepsilon(\gamma) \in \{\pm 1\}$ such that

$$\varepsilon(\gamma)(a + b\gamma + c\gamma^2) = a + q\gamma + d\gamma^2$$

for all $\gamma \in \mathbb{Z}$. In particular, there is $\varepsilon \in \{\pm 1\}$ such that

$$\varepsilon(a\alpha^2 + b\alpha\gamma + c\gamma^2) = a\alpha^2 + q\alpha\gamma + d\gamma^2$$

for infinitely many $\gamma \in \mathbb{Z}$. When we rewrite this equation as

$$(\varepsilon a - a) + (\varepsilon b - q)\gamma + (\varepsilon c - d)\gamma^2 = 0,$$

we get a quadratic polynomial in γ with infinitely many roots. Hence its coefficients must vanish, so that

$$\varepsilon a - a = 0,\ \varepsilon b - q = 0,\ \varepsilon c - d = 0.$$

The first equation implies $\varepsilon = 1$ since $a \neq 0$, and then the second and third equations give the desired equalities $b = q$ and $c = d$.

Now consider an arbitrary $\alpha,\beta,\gamma,\delta$ with $\alpha\delta - \beta\gamma = 1$. Using $b = q$ and $c = d$, (**) becomes

$$\pm(a\alpha^2 + b\alpha\gamma + c\gamma^2) = a\alpha^2 + b\alpha\gamma + c\gamma^2.$$

However, $h(\alpha,\gamma) = a\alpha^2 + b\alpha\gamma + c\gamma^2 > 0$ since $(\alpha,\gamma) \neq (0,0)$. It follows that the sign in (**) and hence in (*) must be +. This completes the proof that $G(x,y)$ is a direct composition of $\tilde{h}(x,y)$ and $k(x,y)$.

An almost identical argument shows that $G(x,y)$ is a direct composition of $h(x,y)$ and $\tilde{k}(x,y)$ when $\tilde{k}(x,y)$ is properly equivalent to $k(x,y)$. Since we just proved that $G(x,y)$ is a direct composition of $\tilde{h}(x,y)$ and $k(x,y)$, the previous sentence implies that $G(x,y)$ is also a direct composition of $\tilde{h}(x,y)$ and $\tilde{k}(x,y)$.

(d) Use part (a) to find forms $f'(x,y) = ax^2 + Bxy + a'Cy^2$ and $g'(x,y) = a'x^2 + Bxy + aCy^2$ properly equivalent to $f(x,y)$ and $g(x,y)$ respectively; then the form $aa'X^2 + BXY + CY^2$ is a direct composition of $f'(x,y)$ and $g'(x,y)$ by part (b). Thus by part (c), it is a direct composition of $f(x,y)$ and $g(x,y)$.

3.6. (a) Since $\gcd(a,a') = 1$, we have $\gcd(a,a',(2b+2b')/2) = 1$ as well, so we can use Lemma 3.2 to choose B such that $B \equiv 2b \bmod 2a$, $B \equiv 2b' \bmod 2a'$, and $B^2 \equiv -4n \bmod 4aa'$, where $-4n$ is the discriminant of $f(x,y)$ (or $g(x,y)$). Then the Dirichlet composition of $f(x,y)$ and $g(x,y)$ is

$$aa'x^2 + Bxy + \frac{B^2+4n}{4aa'}y^2.$$

Now, B is even, say $B = 2B_1$, so that $B_1 \equiv b \bmod a$ and $B_1 \equiv b' \bmod a'$. The Dirichlet composition becomes

$$aa'x^2 + 2B_1xy + \frac{(2B_1)^2 + 4n}{4aa'}y^2 = aa'x^2 + 2B_1xy + \frac{B_1^2 + n}{aa'}y^2,$$

which is the Legendre composition with both signs positive.

(b) By the previous exercise, the Dirichlet composition is the direct composition. We wish to solve $B \equiv 10 \bmod 14$, $B \equiv 2 \bmod 9$, and $B^2 \equiv -1076 \bmod 504$; one solution is $B = 38$. Then the Dirichlet composition is

$$126x^2 + 38xy + \frac{1444 + 1076}{4 \cdot 126}y^2 = 126x^2 + 38xy + 5y^2.$$

3.7. If b is even, say $b = 2B$, apply the transformation $x \mapsto -y$, $y \mapsto x + By$; this is proper, and gives

$$ac(-y)^2 + 2B(-y)(x + By) + (x + By)^2 = x^2 + (ac - B^2)y^2$$

which is the principal form of discriminant $-4(ac - B^2) = 4B^2 - 4ac = b^2 - 4ac$. If b is odd, say $b = 2B + 1$, apply the transformation $x \mapsto -y$, $y \mapsto x + (B+1)y$; this is proper, and gives

$$ac(-y)^2 + (2B+1)(-y)(x + (B+1)y) + (x + (B+1)y)^2 = x^2 + xy + (ac - B - B^2)y^2,$$

which is again the principal form of discriminant $1 - 4(ac - B - B^2) = -4ac + (2B+1)^2 = b^2 - 4ac$.

3.8. (a) Given a form $f(x,y)$, let $\mathcal{L}_f$ be the Lagrangian class of $f(x,y)$, and $\mathcal{S}_f$ be the union of the class of $f(x,y)$ and the class of its opposite. Since every form is improperly equivalent to its opposite via $x \mapsto x$, $y \mapsto -y$, it is clear that $\mathcal{S}_f \subset \mathcal{L}_f$. For the reverse, suppose $g(x,y) \in \mathcal{L}_f$. If $g(x,y)$ is properly equivalent to $f(x,y)$, then clearly $g(x,y) \in \mathcal{S}_f$. Otherwise, $g(x,y)$ is improperly equivalent to $f(x,y)$. Since $f(x,y)$ is improperly equivalent to its opposite, and since the composition of two improper equivalences is a proper equivalence (see the solution to Exercise 2.2), it follows that $g(x,y)$ is properly equivalent to the opposite of $f(x,y)$, so that again $g(x,y) \in \mathcal{S}_f$.

(b) **(i) ⇒ (ii):** If the Lagrangian class and the class of $f(x,y)$ are equal, then the opposite of $f(x,y)$ is contained in the class of $f(x,y)$ since it is contained in the Lagrangian class of $f(x,y)$. It follows that $f(x,y)$ is properly equivalent to its opposite.

(ii) ⇒ (iii): Since $f(x,y)$ is always properly equivalent to itself, we need only show that (ii) implies that $f(x,y)$ is improperly equivalent to itself. Note that $f(x,y)$ is always improperly equivalent to its opposite via $x \mapsto x$, $y \mapsto -y$. Writing $f^{opp}(x,y)$ for the opposite of $f(x,y)$, we have

$$f(x,y) \xrightarrow{\text{(ii)}} f^{opp}(x,y) \xrightarrow{\text{improper}} f(x,y),$$

showing that $f(x,y)$ is improperly equivalent to itself, since the composition of a proper and an improper equivalence is improper.

(iii) ⇒ (iv): If f is improperly equivalent to itself, then since it is also improperly equivalent to its opposite, it must be properly equivalent to its opposite. By Theorem 3.9, the class of f must be its own inverse, so that the class of f has order 1 or 2 in $C(D)$.

(iv) ⇒ (i): Write $[f]$ for the class of a form $f(x,y)$ in the class group, and $f^{opp}(x,y)$ for the opposite of $f(x,y)$. By Theorem 3.9, $[f]^{-1} = [f^{opp}]$. But if $[f]$ has order ≤ 2 in the class group, then $[f]^{-1} = [f]$. Combining these two statements shows that then $[f] = [f^{opp}]$, so that f is properly equivalent to its opposite. Since by part (a) the Lagrangian class of $f(x,y)$ is the union of the class of $f(x,y)$ and the class of $f^{opp}(x,y)$, it follows that the Lagrangian class of $f(x,y)$ equals the class of $f(x,y)$.

3.9. (a) First, $[a]\cdot[b]$ gives the Lagrangian classes $[ab]$, $[a^{-1}b^{-1}]$, $[ab^{-1}]$, and $[a^{-1}b]$ since $a \sim a^{-1}$ and $b \sim b^{-1}$. However,

$$[ab] = [(ab)^{-1}] = [b^{-1}a^{-1}] = [a^{-1}b^{-1}]$$
$$[ab^{-1}] = [(ab^{-1})^{-1}] = [ba^{-1}] = [a^{-1}b],$$

where in each case the final equality holds since G is Abelian. Therefore $[a]\cdot[b]$ gives at most two distinct Lagrangian classes, $[ab]$ and $[ab^{-1}]$, proving the first statement. For the second statement, notice that $[a]\cdot[b]$ takes on only one value precisely when $[ab] = [ab^{-1}]$, which means that $ab = ab^{-1}$ or $ab = a^{-1}b$; this happens if and only if $b = b^{-1}$ or $a = a^{-1}$, i.e., if and only if either a or b has order ≤ 2 in G.

(b) Let the cyclic group G of order 8 be generated by g; then the elements of $G/\sim$ are $\{1\}$, $\{g, g^7\}$, $\{g^2, g^6\}$, $\{g^3, g^5\}$, and $\{g^4\}$. Define a correspondence with the notation in (2.33) as follows:

$$A \leftrightarrow \{1\}, \quad B \leftrightarrow \{g^4\}, \quad C \leftrightarrow \{g^2, g^6\}, \quad D \leftrightarrow \{g, g^7\}, \quad E \leftrightarrow \{g^3, g^5\}.$$

Then the multiplication in $G/\sim$ is identical to that in (2.34).

(c) Since the inverse of the class of a form in $C(D)$ is the class of its opposite, identifying a class with its inverse is the same as identifying forms with their opposites, which is to say considering Lagrangian classes instead of classes.

3.10. We start by carefully doing the case in the text, which was $D = -4n$, $n \equiv 1 \bmod 4$. We wish to count reduced forms $ax^2 + 2bxy + cy^2$ with $2b = 0$, $a = 2b$, or $a = c$. As noted in the text, there are 2^{r-1} reduced forms with $2b = 0$, corresponding to factorizations of $n = ac$ with $a < c$.

Next, we will show that there are also 2^{r-1} reduced forms with $a = 2b$ or $a = c$; in fact, we claim that such forms are in a one-to-one correspondence with factorizations $n = bk$ with $\gcd(b,k) = 1$ and $b < k$. Choose such a factorization, and, as in the text, let $c = (b+k)/2$, so that $k = 2c - b$. Note that $\gcd(b,k) = 1$ implies that $\gcd(b, 2c) = 1$ and thus $\gcd(b,c) = 1$. Also note that $n \equiv 1 \bmod 4$ implies that $b \equiv k \bmod 4$ so that $c = (b+k)/2$ is odd.

Now consider the form $2bx^2 + 2bxy + cy^2$. This form has discriminant $(2b)^2 - 4(2b)c = 4b^2 - 8bc = -4b(2c-b) = -4n$, and is primitive since c is odd and $\gcd(b,c) = 1$. If $2b < c$, it is reduced. We cannot have $2b = c$ since then $n = bk = b(2c-b) = b(4b-b) = 3b^2$, but $n \equiv 1 \bmod 4$. The final possibility is $2b > c$. Here the form is not reduced, but it is properly equivalent via $(x,y) \mapsto (-y, x+y)$ to

$$2b(-y)^2 + 2b(-y)(x+y) + c(x+y)^2 = cx^2 + 2(c-b)xy + cy^2$$

which is a form with $a = c$, reduced since $c > b > 0$ and $\gcd(c, 2c-2b) = \gcd(c, 2b) = 1$.

Finally, note that any reduced form $cx^2 + 2b'xy + cy^2$ is properly equivalent to a non-reduced form with $a = 2b$, using the inverse transformation $(x,y) \mapsto (x+y, -x)$. So we can count reduced forms of these two types by counting factorizations $n = bk$ with $\gcd(b,k) = 1$ and $b < k$, proving the claim. As in the case of $2b = 0$, there are 2^{r-1} such factorizations.

Thus altogether, in the case $D = -4n$, $n \equiv 1 \bmod 4$, we get $2^{r-1} + 2^{r-1} = 2^r$ such forms of order ≤ 2 in the class group $C(D)$, which verifies that in this case $\mu = r + 1$.

For the remaining cases where $D = -4n$, again we write a reduced form as $ax^2+2bxy+cy^2$; again we wish to count forms with $2b = 0$, $a = 2b$ or $a = c$. This gives three types of forms:

(i) When $b = 0$, we get $ax^2 + cy^2$ where $n = ac$, $\gcd(a, c) = 1$, $a < c$.

(ii) When $a = 2b$, we get $2bx^2 + 2bxy + cy^2$, so that

$$D = -4n = (2b)^2 - 4 \cdot 2b \cdot c = -8bc + 4b^2 = -4(2bc - b^2).$$

Thus $n = 2bc - b^2$.

(iii) When $a = c$, we get $ax^2 + 2bxy + ay^2$, $\gcd(a, 2b) = 1$, $0 < 2b < a$, so that

$$D = -4n = 4b^2 - 4a^2.$$

Thus $n = a^2 - b^2$.

If $D = -4n$, $n \equiv 2 \bmod 4$, then n has $r + 1$ distinct prime factors, and with $\gcd(a, c) = 1$ and $a < c$ we get 2^r reduced forms of type (i) by the same analysis as in the previous case. There are no forms of type (ii) since $n = 2bc - b^2 = b(2c - b)$, and b and $2c - b$ are either both odd or both even, contradicting $n \equiv 2 \bmod 4$. There are no forms of type (iii) since $n = a^2 - b^2 \equiv 2 \bmod 4$ is impossible. So in this case there are precisely 2^r elements of order ≤ 2, so that again $\mu = r + 1$.

If $D = -4n$, $n \equiv 3 \bmod 4$, then n has r distinct (odd) prime factors, and with $\gcd(a, c) = 1$ and $a < c$ we get 2^{r-1} reduced forms of type (i). There are no forms of type (ii), since $n = 2bc - b^2 \equiv 3 \bmod 4$ implies that b is odd and c is even, impossible since the form would not be primitive. There are no forms of type (iii) since $\gcd(a, 2b) = 1$ shows that a is odd, and then $n = a^2 - b^2 \equiv 3 \bmod 4$ is impossible. So in this case we get 2^{r-1} elements of order ≤ 2, so that $\mu = r$.

If $D = -4n$, $n \equiv 4 \bmod 8$, then n has $r + 1$ distinct prime factors, so there are 2^r reduced forms of type (i). There are no forms of type (ii), since we would have $n = 2bc - b^2 = b(2c - b) \equiv 4 \bmod 8$. Since the form is primitive, c is odd, and then this congruence has no solutions. There are no forms of type (iii) since $\gcd(a, 2b) = 1$ shows that a is odd, and then $n = a^2 - b^2 \equiv 4 \bmod 8$ has no solutions. So in this case we get 2^r elements of order ≤ 2, so that $\mu = r + 1$.

If $D = -4n$, $n \equiv 0 \bmod 8$, then n has $r + 1$ distinct prime factors, so there are 2^r reduced forms of type (i). To count forms of types (ii) and (iii), we study the situation in a manner similar to the case $D = -4n$, $n \equiv 1 \bmod 4$ considered above. Write $n = bk$ where $\gcd(b, k) = 2$. There are 2^{r+1} such factorizations, exactly half of which have $0 < b < k$. Let $c = \frac{b+k}{2}$, and consider the form $2bx^2 + 2bxy + cy^2$. Then c is odd and is prime to b since b and k share no odd factors.

- If $2b < c$, then the form is reduced, of type (ii).
- If $2b > c$, then as in the text the form is equivalent to the reduced form $cx^2 + 2(c - b)xy + cy^2$, of type (iii).

So there are again 2^r such reduced forms of types (ii) and (iii), and altogether there are therefore 2^{r+1} reduced forms of this type, so that $\mu = r + 2$.

Finally, if $D \equiv 1 \bmod 4$, $D < 0$, let $n = -D$. Then forms of order two in the class group are of the form $ax^2+axy+cy^2$ or $ax^2+bxy+ay^2$, $b \neq 0$. (Note that ax^2+cy^2 is impossible because its discriminant is even, while D is odd.) Since n has r prime factors, there are 2^r factorizations $n = bk$ where $\gcd(b, k) = 1$; of these, half, or 2^{r-1}, satisfy $b < k$. Since $n \equiv 3 \bmod 4$, it follows that $b + k \equiv 0 \bmod 4$; let $c = (b + k)/4$. Then

- If $b < c$, then $bx^2+bxy+cy^2$ is a reduced form of discriminant $b^2-4bc = b(b-4c) = D$.
- If $b > c$, then $bx^2 + bxy + cy^2$ is properly equivalent, via the transformation $x \mapsto -y$, $y \mapsto x + y$, to the reduced form $cx^2 + (2c - b)xy + cy^2$.

So there are 2^{r-1} elements of order ≤ 2, and $\mu = r$.

In all cases, an analysis similar to that done in the $D = -4n$, $n \equiv 1 \bmod 4$ case at the beginning of the solution shows that these methods generate all the forms of this type.

3.11. (a) The homomorphism $\pi : (\mathbb{Z}/p^m\mathbb{Z})^* \to \{\pm 1\}$ given by $a \mapsto (a/p)$ maps squares to 1. It follows immediately that $\ker(\pi)$ contains the subgroup S of squares modulo p^m. Equality will follow once we prove that both subgroups have index 2 in $(\mathbb{Z}/p^m\mathbb{Z})^*$. Since π is clearly surjective, the isomorphism

$$(\mathbb{Z}/p^m\mathbb{Z})^*/\ker(\pi) \simeq \{\pm 1\}.$$

shows that $\ker(\pi)$ has index 2 in $(\mathbb{Z}/p^m\mathbb{Z})^*$. The subgroup S is the image of the squaring map $(\mathbb{Z}/p^m\mathbb{Z})^* \to (\mathbb{Z}/p^m\mathbb{Z})^*$. To study the kernel, suppose that $u^2 \equiv 1 \bmod p^m$, so that p^m divides $u^2 - 1 = (u-1)(u+1)$. If both factors are multiples of p, then $p \mid (u+1) - (u-1) = 2$, impossible since p is odd. It follows that $p^m \mid u - 1$ or $p^m \mid u + 1$, so that $u \equiv \pm 1 \bmod p^m$. The kernel thus has order 2, proving that S has index 2 in $(\mathbb{Z}/p^m\mathbb{Z})^*$.

(b) If $D = -4n$, let $n = 2^a m$, m odd; then the isomorphism

$$\varphi : (\mathbb{Z}/D\mathbb{Z})^* \simeq (\mathbb{Z}/2^{a+2}\mathbb{Z})^* \times (\mathbb{Z}/m\mathbb{Z})^*$$

in the statement of the exercise is defined by

$$\varphi(x) = ([x]_{2^{a+2}}, [x]_m).$$

Note that φ^{-1} is given by the Chinese Remainder Theorem. Let $H \subset \ker(\chi)$ be the subgroup of values represented by the principal form $x^2 + ny^2$; that is, the congruence classes represented by the principal genus. By Corollary 2.27,

$$H = \{[\beta^2]_D, [\beta^2 + n]_D : \beta \in \mathbb{Z}\} \cap (\mathbb{Z}/D\mathbb{Z})^* \subset (\mathbb{Z}/D\mathbb{Z})^*.$$

Note that $m \mid n \mid D$, and that $\gcd(\beta, D) = 1$ implies that $\gcd(\beta, n) = \gcd(\beta, m) = 1$.

(i) Let P be the projection of $\varphi(H)$ on the first factor, so that $P \subset (\mathbb{Z}/2^{a+2}\mathbb{Z})^*$. We want to show that $H_1 = P$. From the definition of H_1, it is obvious that $H_1 \subset P$.

For the opposite inclusion, let $x = \beta^2$ or $\beta^2 + n$ be relatively prime to D. Then $([x]_{2^{a+2}}, [\beta^2]_m)$ is the corresponding element of $\varphi(H)$, so that $[x]_{2^{a+2}}$ represents an arbitrary element of P. We need to show that $([x]_{2^{a+2}}, 1)$ lies in $\varphi(H)$. For $x = \beta^2$ or $\beta^2 + n$, observe that β is invertible modulo m since x is prime to D and $n \mid m \mid D$. Choose ρ with $\beta\rho \equiv 1 \bmod m$. Then by the Chinese Remainder Theorem, we can find γ with $\gamma \equiv 1 \bmod 2^{a+2}$ and $\gamma \equiv \rho \bmod m$. Thus $x\gamma^2 = (\beta\gamma)^2$ or $x\gamma^2 = (\beta\gamma)^2 + n\gamma^2$, and

$$\varphi(x\gamma^2) = ([x\gamma^2]_{2^{a+2}}, [\beta\gamma]_m^2) = ([x\gamma^2]_{2^{a+2}}, [\beta\rho]_m^2) = ([x]_{2^{a+2}}, 1)$$

by the choice of γ and ρ. Thus $([x]_{2^{a+2}}, 1) \in \varphi(H)$, so that $[x]_{2^{a+2}} \in H_1$.

(ii) If $x \in H$, then $x \equiv \beta^2$ or $x \equiv \beta^2 + n \bmod D$, $\beta \in (\mathbb{Z}/D\mathbb{Z})^*$, and as in part (i), $\gcd(\beta, m) = 1$ and $\varphi(x) = ([x]_{2^{a+2}}, [\beta^2]_m)$. Since $[x_{2^{a+2}}] \in H_1$ by part (i), this shows that $\varphi(x) \in H_1 \times (\mathbb{Z}/m\mathbb{Z})^{*2}$.

For the reverse, it suffices to show that $H_1 \times \{1\} \subset \varphi(H)$ and $\{1\} \times (\mathbb{Z}/m\mathbb{Z})^{*2} \subset \varphi(H)$. The first inclusion follows from the definition. For the second, choose $(1, \alpha^2) \in \{1\} \times (\mathbb{Z}/m\mathbb{Z})^{*2}$, and let $\delta = \varphi^{-1}((1, \alpha)) \in (\mathbb{Z}/D\mathbb{Z})^*$. It follows that $\varphi(\delta^2) = (1, \alpha^2)$, proving the second inclusion.

(iii) By part (i), H_1 consists of $\{[\beta^2]_{2^{a+2}}, [\beta^2 + n]_{2^{a+2}}\}$, so it suffices to show that $[\beta^2 + n]_{2^{a+2}}$ is a square in $(\mathbb{Z}/2^{a+2}m\mathbb{Z})^*$. Choose $x \in H$ with $x \equiv \beta^2 + n \bmod D$, and recall that $\gcd(\beta, n) = 1$. Therefore, since $n = 2^a m$ is even, β must be odd. It follows that $\gcd(\beta, D) = \gcd(\beta, -4n) = 1$. Pick an integer u with $u\beta \equiv 1 \bmod 2^{a+2}$. Then

$$(\beta + 2^{a-1}mu)^2 \equiv \beta^2 + 2^a m + 2^{2a-2}m^2u^2 \equiv \beta^2 + n \bmod 2^{a+2}$$

since $2^a m = n$ and $2a - 2 \geq a + 2$ for $a \geq 4$.

(iv) First, since all the characters induce surjective homomorphisms to $\{\pm 1\}$, it follows by the Chinese Remainder Theorem, as in the text, that Ψ is surjective to $\{\pm 1\}^\mu$. Let $p_1, \ldots, p_r$ be the odd primes dividing m (recall that m is odd and $D = -2^{a+2}m$), and set

$$X_2(b) = \big(\chi_1(b), \ldots, \chi_r(b)\big) = \left(\left(\frac{b}{p_1}\right), \ldots, \left(\frac{b}{p_r}\right)\right) \in \{\pm 1\}^r$$

for b relatively prime to $p_1, \ldots, p_r$. Then we can think of Ψ as a product map via the commutative diagram

$$\begin{array}{ccc}(\mathbb{Z}/D\mathbb{Z})^* = (\mathbb{Z}/2^{a+2}\mathbb{Z})^* \times (\mathbb{Z}/m\mathbb{Z})^* = & (\mathbb{Z}/2^{a+2}\mathbb{Z})^* \times \prod_{i=1}^r (\mathbb{Z}/p_i^{m_i}\mathbb{Z})^* \\ \quad\searrow^{\Psi} & \downarrow X_1 \times X_2 \\ & \{\pm 1\}^{\mu - r} \times \{\pm 1\}^r\end{array}$$

where X_2 is as above and X_1 is determined by the assigned characters involving δ, ϵ in the table preceding Lemma 3.17. It follows that $\ker(\Psi)$ can be identified with $\ker(X_1) \times \ker(X_2)$. Using part (a) of the exercise, we can thus write

$$\ker(\Psi) = \ker(X_1) \times (\mathbb{Z}/m\mathbb{Z})^{*2}.$$

The goal is to prove that $\ker(\Psi) = H$. Since $H = H_1 \times (\mathbb{Z}/m\mathbb{Z})^{*2}$ by part (ii), we need only prove $H_1 = \ker(X_1)$. Since we know from part (i) that H_1 is the projection of H onto $(\mathbb{Z}/2^{a+2}\mathbb{Z})^*$, it suffices to prove that

(*) $$H_1 = \big\{[\beta^2]_{2^{a+2}}, [\beta^2 + n]_{2^{a+2}} \mid \beta \in \mathbb{Z}\big\} \cap (\mathbb{Z}/2^{a+2}\mathbb{Z})^* \quad \text{is equal to} \quad \ker(X_1).$$

Suppose $a = 0$, so that n is odd. If $n \equiv 1 \bmod 4$, then $X_1 = \delta$, so $\ker(X_1) = \{1\}$. In H_1, the first possibility $[\beta^2]_4$ requires that β be odd, so that $\beta^2 = 1 \bmod 4$; the second requires that β be even, so $\beta^2 + n \equiv 1 \bmod 4$. Thus (*) holds in this case. On the other hand, if $n \equiv 3 \bmod 4$, then X_1 is trivial, so $\ker(X_1) = (\mathbb{Z}/4\mathbb{Z})^*$. But for $\beta = 2$, we have $\beta^2 + n \equiv 3 \bmod 4$, so that $H_1 = (\mathbb{Z}/4\mathbb{Z})^*$. Again, (*) holds.

If $a = 1$, then $n = 2m$, m odd. If $m \equiv 1 \bmod 4$, then $n \equiv 2 \bmod 8$ and $X_1 = \delta\epsilon$. Writing $(\mathbb{Z}/8\mathbb{Z})^* = \{1, 3, 5, 7\}$, one computes that $\ker(X_1) = \{1, 3\}$. But

$$H_1 = \{[\beta^2]_8, [\beta^2 + n]_8\} \cap (\mathbb{Z}/8\mathbb{Z})^* = \{[\beta^2]_8, [\beta^2 + 2]_8\} \cap (\mathbb{Z}/8\mathbb{Z})^*$$

which is easily seen to equal $\{1, 3\}$ since β must be odd in both cases. Thus (*) holds. On the other hand, if $m \equiv 3 \bmod 4$, then $n \equiv 6 \bmod 8$ and $X_1 = \epsilon$, which has $\ker(X_1) = \{1, 7\}$. Here,

$$H_1 = \{[\beta^2]_8, [\beta^2 + n]_8\} \cap (\mathbb{Z}/8\mathbb{Z})^* = \{[\beta^2]_8, [\beta^2 + 6]_8\} \cap (\mathbb{Z}/8\mathbb{Z})^*$$

which equals $\{1, 7\}$ since β is odd in both cases, and again (*) holds.

Next, if $a = 2$, then $n = 4m$, m odd and $n \equiv 4 \bmod 8$, so that $X_1 = \delta$. Writing $(\mathbb{Z}/16\mathbb{Z})^* = \{\pm 1, \pm 3, \pm 5, \pm 7\}$, one gets $\ker(X_1) = \{1, -3, 5, -7\} = \{1, 5, 9, 13\}$. If $m \equiv 1 \bmod 4$, then

$$H_1 = \{[\beta^2]_{16}, [\beta^2 + n]_{16}\} \cap (\mathbb{Z}/16\mathbb{Z})^* = \{[\beta^2]_{16}, [\beta^2 + 4]_{16} \mid \beta \text{ odd}\}.$$

Since the odd squares mod 16 are 1 and 9, one sees that H_1 consists of $1, 9$ (from β^2) and $1+4=5, 9+4=13$ (from β^2+4). Thus $H_1 = \{1, 9, 5, 13\}$ and (*) holds. On the other hand, if $m \equiv 3 \bmod 4$, then

$$H_1 = \{[\beta^2]_{16}, [\beta^2+n]_{16}\} \cap (\mathbb{Z}/16\mathbb{Z})^* = \{[\beta^2]_{16}, [\beta^2+12]_{16} \mid \beta \text{ odd}\}.$$

Thus H_1 consists of $1, 9$ (from β^2) and $1+12 = 13, 9+12 = 21 = 5$ (from β^2+12). Again, (*) holds.

Before treating the cases $a = 3$ and $a \geq 4$ separately, we compute $\ker(X_1)$ when $a \geq 3$. Here $n \equiv 0 \bmod 8$ since $n = 2^a m$, m odd, so that $X_1 = (\delta, \epsilon)$. We claim that $\ker(X_1)$ is the subgroup S of squares in $(\mathbb{Z}/2^{a+2}\mathbb{Z})^*$. Since $X_1 : (\mathbb{Z}/2^{a+2}\mathbb{Z})^* \to \{\pm 1\}^2$ is onto and maps squares to squares, $\ker(X_1)$ has index 4 in $(\mathbb{Z}/2^{a+2}\mathbb{Z})^*$ and contains S. Equality will follow once we show that S also has index 4. Since S is the image of the squaring map from $(\mathbb{Z}/2^{a+2}\mathbb{Z})^*$ to itself, it suffices to show that the kernel of the squaring map has order 4. So suppose $u^2 \equiv 1 \bmod 2^{a+2}$. Then $2^{a+2} \mid (u+1)(u-1)$, and since the difference of the factors is 2, 2^{a+1} must divide one of the factors. Suppose $2^{a+1} \mid u-1$. Then $u = 1 + k2^{a+1}$ for some $k \in \mathbb{Z}$. Then $u \equiv 1 \bmod 2^{a+2}$ (k even) or $u \equiv 1 + 2^{a+1} \bmod 2^{a+2}$ (k odd). The argument for $2^{a+1} \mid u+1$ is similar, with the result that

$$u^2 \equiv 1 \bmod 2^{a+2} \implies u \equiv \pm 1, \pm 1 + 2^{a+1} \bmod 2^{a+2}.$$

It follows that $\ker(X_1)$ is the subgroup S of squares in $(\mathbb{Z}/2^{a+2}\mathbb{Z})^*$.

If $a \geq 4$, then from part (iii), $H_1 = S$, and (*) holds. Finally, if $a = 3$, then $n = 8m$, m odd, and $H_1 = \{[\beta^2]_{32}, [\beta^2+8m]_{32}\} \cap (\mathbb{Z}/32\mathbb{Z})^*$. The first possibility gives $(\mathbb{Z}/2^{a+2}\mathbb{Z})^{*2}$, so $\ker(X_1) \subseteq H_1$ by the above paragraph. For the opposite inclusion, note that any $[b]_{32} \in H_1$ satisfies $b \equiv \beta^2 \bmod 8 \equiv 1 \bmod 8$, which easily implies $\delta(b) = \epsilon(b) = 1$. Thus $[b]_{32} \in \ker(X_1)$ and (*) holds, completing the proof of Lemma 3.17.

3.12. (a) Restate Lemma 2.24 to remove the requirement that $D \equiv 0, 1 \bmod 4$ be negative, and replace part (i) by "(i) The values in $(\mathbb{Z}/D\mathbb{Z})^*$ represented by the principal form of discriminant D form a subgroup $H \subset (\mathbb{Z}/D\mathbb{Z})^*$". Since quadratic reciprocity was used in the proof in the text only to show that $H \subset \ker(\chi)$, the proofs of (i) and (ii) then become independent of quadratic reciprocity. Additionally, those proofs, including the proof of Lemma 2.25 in Exercise 2.18, do not depend on the sign of D. Thus forms in the same genus map to the same coset of H under the induced map

$$\Phi' : C \to (\mathbb{Z}/D\mathbb{Z})^*/H.$$

It follows that $\ker(\Phi')$ is the principal genus, and $\Phi'(C)$ is the set of genera.

(b) As noted, H contains all the squares in $(\mathbb{Z}/D\mathbb{Z})^*$, so that $C^2 \subset \ker(\Phi')$. Then by Lemma 3.17 (which holds for positive D as well, see part (a) of Exercise 3.13),

$$(\mathbb{Z}/D\mathbb{Z})^*/H \simeq \{\pm 1\}^\mu,$$

and since $\Phi' : C \to (\mathbb{Z}/D\mathbb{Z})^*/H$ is a homomorphism, $C^2 \subset \ker(\Phi')$. By the same argument as in the text, $|C/C^2| = 2^{\mu-1}$ using Proposition 3.11, so that $|C/\ker(\Phi')|$, which is the number of genera, is at most $2^{\mu-1}$.

3.13. (a) The definitions of the assigned characters do not depend on the sign of D, so that (3.16) is still a homomorphism. The proof for $D \equiv 1 \bmod 4$ (which includes Exercises 2.17 and 3.11) also does not depend on the sign of D.

For $D = 4n$, the version of Lemma 2.24 proved in Exercise 3.12 immediately implies the following analog of Theorem 2.26:

THEOREM. *Let $D \equiv 0, 1 \bmod 4$ be positive, and let $H \subset (\mathbb{Z}/D\mathbb{Z})^*$ be as in the revised Lemma 2.24. If H' is a coset of H in $(\mathbb{Z}/D\mathbb{Z})^*$ and p is an odd prime not dividing D, then $[p] \in H'$ if and only if p is represented by a reduced form of discriminant D in the genus of H'.*

The principal form of discriminant $D = 4n$ is $x^2 - ny^2$, and $x^2 - ny^2$ is congruent to either x^2 or $x^2 - n$ modulo $4n$, depending on whether y is even or odd. This implies the following version of Corollary 2.27:

COROLLARY. *If $n > 0$ and p is an odd prime not dividing n, then p is represented by a form of discriminant $4n$ in the principal genus if and only if $p \equiv \beta^2$ or $\beta^2 - n \bmod 4n$.*

The proof in Exercise 3.11 then goes through essentially unchanged.

(b) Assume $(p^*/q) = 1$. By Lemma 2.5, this implies that q is properly represented by a form $f(x, y)$ of discriminant p^*. Since q is prime, $\mu = 1$, so that by Exercise 3.12 there is a single genus, which is necessarily the principal genus. Thus $f(x, y)$ is in the principal genus. Since $\mu = 1$, there is only one assigned character $(\cdot/p)$. This character takes the value 1 on any number represented by a form in the principal genus, so it follows that $(q/p) = 1$, as desired.

(c) Here we are assuming that $p \equiv 1 \bmod 4$ or $q \equiv 1 \bmod 4$. We first claim that

$$(*)\qquad \left(\frac{p^*}{q}\right) = \left(\frac{p}{q}\right) \text{ and } \left(\frac{q^*}{p}\right) = \left(\frac{q}{p}\right).$$

The first equality is obvious if $p \equiv 1 \bmod 4$. If $p \equiv 3 \bmod 4$, then $p^* = -p$ and (by our assumption) $q \equiv 1 \bmod 4$. Thus

$$\left(\frac{p^*}{q}\right) = \left(\frac{-p}{q}\right) = \left(\frac{-1}{q}\right)\left(\frac{p}{q}\right) = \left(\frac{p}{q}\right),$$

where the third equal sign uses $q \equiv 1 \bmod 4$. The second equality of (*) follows by interchanging the roles of p and q.

It is now easy to prove the contrapositive of $(q/p) = 1 \Rightarrow (p^*/q) = 1$:

$$\left(\frac{p^*}{q}\right) = -1 \implies \left(\frac{p}{q}\right) = -1 \implies \left(\frac{q^*}{p}\right) = -1 \implies \left(\frac{q}{p}\right) = -1,$$

where the first and third implications use (*) and the second uses the contrapositive of part (b) with p and q interchanged.

(d) Following the extended hint, since $f(x, y) = px^2 + pxy + ((p - q)/4)y^2$ and $-f(x, y)$ both have discriminant pq and pq has two odd prime factors with $pq \equiv 1 \bmod 4$, the assigned characters are

$$\chi_1(a) = \left(\frac{a}{p}\right), \quad \chi_2(a) = \left(\frac{a}{q}\right).$$

Since $f(x, y)$ represents $p - q$, its complete character $(\chi_1(p - q), \chi_2(p - q))$ is

$$\left(\left(\frac{p-q}{p}\right), \left(\frac{p-q}{q}\right)\right) = \left(\left(\frac{-q}{p}\right), \left(\frac{p}{q}\right)\right) = \left(-\left(\frac{q}{p}\right), \left(\frac{p}{q}\right)\right)$$

where the final equality uses $(-q/p) = (-1/p)(q/p) = -(q/p)$ since $p \equiv 3 \bmod 4$. Similarly, $-f(x, y)$ represents $q - p$, so its complete character $(\chi_1(q - p), \chi_2(q - p))$ is

$$\left(\left(\frac{q-p}{p}\right), \left(\frac{q-p}{q}\right)\right) = \left(\left(\frac{q}{p}\right), \left(\frac{-p}{q}\right)\right) = \left(\left(\frac{q}{p}\right), -\left(\frac{p}{q}\right)\right)$$

where the final equality uses $q \equiv 3 \bmod 4$.

These computations show that the complete characters differ, so that $f(x,y)$ and $-f(x,y)$ are not in the same genus. Since there are at most two genera, one of them must lie in the principal genus. If $f(x,y)$ is in the principal genus, then

$$\left(-\left(\frac{q}{p}\right), \left(\frac{p}{q}\right)\right) = (1,1),$$

which is impossible since we are assuming $(q/p) = 1$. Hence $-f(x,y)$ is in the principal genus, which implies

$$\left(\left(\frac{q}{p}\right), -\left(\frac{p}{q}\right)\right) = (1,1).$$

Thus $(-p/q) = 1$, which is what we wanted to prove.

It follows that $(q/p) = 1 \Rightarrow (p^*/q) = 1$ when $p \equiv q \equiv 3 \bmod 4$. Combining this with part (c), we see that this implication holds for all odd primes $p \neq q$. Then $(q/p) = 1 \Leftrightarrow (p^*/q) = 1$ by part (b), which completes the proof of quadratic reciprocity.

(e) If $p \equiv 1 \bmod 8$, the form $2x^2 + xy + \frac{1-p}{8}y^2$ has discriminant p, so $\mu = 1$, the only assigned character is $(\cdot/p)$, and there is only one genus. The form properly represents 2, so that $(2/p) = 1$. A similar argument works when $p \equiv 7 \bmod 8$, where $2x^2 + xy + \frac{1+p}{8}y^2$ has discriminant $-p$; again $\mu = 1$ and the only assigned character is $(\cdot/p)$. It follows that $p \equiv 1, 7 \bmod 8 \Rightarrow (2/p) = 1$.

Conversely, suppose that $(2/p) = 1$. Lemma 2.5 implies that p is properly represented by a form of discriminant 8. The only assigned character is ϵ since $8 = -4(-2)$ and $n = -2 \equiv 6 \bmod 8$. Furthermore, $\mu = 1$, so there is only the principal genus by part (b) of Exercise 3.12. This implies that $\epsilon(p) = (-1)^{(p^2-1)/8} = 1$, i.e., $p \equiv 1, 7 \bmod 8$.

3.14. Since $-164 = -4 \cdot 41$ and $41 \equiv 1 \bmod 4$, the assigned characters are

$$\chi_1(a) = \left(\frac{a}{41}\right), \quad \delta(a) = (-1)^{(a-1)/2}.$$

The complete characters of the eight reduced forms are:

$$\begin{aligned}
x^2 + 41y^2 &: \ \chi_1(1) = 1, && \delta(1) = 1 && (1,1)\\
2x^2 + 2xy + 21y^2 &: \ \chi_1(21) = \left(\frac{21}{41}\right) = \left(\frac{3}{41}\right)\left(\frac{7}{41}\right) = 1, && \delta(21) = 1 && (1,1)\\
3x^2 \pm 2xy + 14y^2 &: \ \chi_1(3) = \left(\frac{3}{41}\right) = -1, && \delta(3) = -1 && (-1,-1)\\
5x^2 \pm 4xy + 9y^2 &: \ \chi_1(5) = \left(\frac{5}{41}\right) = 1, && \delta(5) = 1 && (1,1)\\
6x^2 \pm 2xy + 7y^2 &: \ \chi_1(7) = \left(\frac{7}{41}\right) = -1, && \delta(7) = -1 && (-1,-1).
\end{aligned}$$

There are two genera, each consisting of four forms:

$$\{x^2 + 41y^2,\ 2x^2 + 2xy + 21y^2,\ 5x^2 \pm 4xy + 9y^2\},$$
$$\{3x^2 \pm 2xy + 14y^2,\ 6x^2 \pm 2xy + 7y^2\}.$$

3.15. Note that equivalence over a ring is an equivalence relation (see Exercise 2.2).

(vi) $\Rightarrow$ (iii): Given m, choose a matrix in $\mathrm{GL}(2,\mathbb{Q})$ with $f(x,y) = g(px+qy, rx+sy)$ with entries whose denominators are prime to m. Then the matrix may be written

$$\begin{pmatrix} p & q \\ r & s \end{pmatrix} = \frac{1}{d}\begin{pmatrix} p' & q' \\ r' & s' \end{pmatrix}$$

where d is the least common multiple of the denominators and p', q', r', s' are integers. Since $f(x, y)$ and $g(x, y)$ have the same discriminant, we must have $ps - qr = \pm 1$. Thus $p's' - q'r' = \pm d^2$, which is prime to m, so that $\left(\begin{smallmatrix} p' & q' \\ r' & s' \end{smallmatrix}\right)$ is invertible modulo m. Furthermore, if $ed \equiv 1 \bmod m$, then $e\left(\begin{smallmatrix} p' & q' \\ r' & s' \end{smallmatrix}\right)$ is invertible modulo m and takes $g(x, y)$ to $f(x, y)$.

(iii) ⇒ (ii): Given m, choose $\left(\begin{smallmatrix} p & q \\ r & s \end{smallmatrix}\right) \in \mathrm{GL}(2, (\mathbb{Z}/m\mathbb{Z})^*)$ with $f(x, y) = g(px + qy, rx + sy)$. If $f(x, y)$ represents a congruence class $[a]$, then

$$[a] = f(x, y) = g(px + qy, rx + sy),$$

so that $g(x, y)$ represents $[a]$ as well. Since equivalence modulo m is an equivalence relation, we are done.

(ii) ⇒ (i): Take $m = D$ in (ii).

(vi) ⇒ (v): Take $m = 2D$ in (vi).

(v) ⇒ (i): (v) implies that they are equivalent over $\mathbb{Q}$ via a matrix whose denominators are prime to D; then the argument for (vi) ⇒ (iii) ⇒ (ii) ⇒ (i) with $m = D$ gives the desired result.

3.16. The two forms are rationally equivalent using the matrix

$$\begin{pmatrix} 0 & -3 \\ \frac{1}{3} & 0 \end{pmatrix}$$

since under this transformation, $x^2 + 18y^2$ becomes

$$(-3y)^2 + 18\left(\frac{1}{3}x\right)^2 = 2x^2 + 9y^2.$$

However, the forms do not represent the same equivalence classes modulo 72, since if they did, they would also represent the same equivalence classes modulo 3, and reducing modulo 3 the two forms become x^2 and $2x^2$; only one of these represents $p \equiv 2 \bmod 3$.

3.17. (a) Suppose $g(x, y) = f(px + qy, rx + sy)$ via a matrix in $\mathrm{GL}(2, \mathbb{Q})$. Let d be the least common multiple of the denominators of the matrix entries, and $dp = p'$ etc. where p', q', r', $s' \in \mathbb{Z}$. Then

$$d^2 g(x, y) = f(p'x + q'y, r'x + s'y)$$

and we are done.

(b) By Exercise 2.1, since $f(x, y)$ represents $d^2 m$, there are integers d', m', with $d'^2 m' = d^2 m$ and $f(x, y)$ properly represents m'. Dividing out by a suitable common factor, we can assume that $\gcd(d', d) = 1$. We first show that $\gcd(m', D) = 1$. Suppose p is a prime dividing m' and D. Then $p \nmid m$ since $\gcd(m, D) = 1$, so that $d'^2 m' = d^2 m$ implies that $p \mid d^2$ and hence $p^2 \mid d^2 m = d'^2 m'$. But $\gcd(d', d) = 1$, showing that $p^2 \mid m'$. Next, since $f(x, y)$ represents m' properly, Lemma 2.3 implies that $f(x, y)$ is properly equivalent to $m'x^2 + bxy + cy^2$. Thus $D = b^2 - 4m'c$, and then $p \mid D$ and $p \mid m'$ imply that $p \mid b^2$. Thus $p^2 \mid b^2$, and since we know $p^2 \mid m'$, we see that $p^2 \mid D$. This is impossible when $p > 2$ since D is a field discriminant. It remains to see what happens when $p = 2$. Here, $D = 4k$, where k squarefree with $k \not\equiv 1 \bmod 4$. Also, b is even, say $b = 2b'$. Then

$$4k = D = b^2 - 4m'c = 4b'^2 - 4m'c \implies k = b'^2 - m'c.$$

Since m' is divisible by $p^2 = 4$, it follows that $k \equiv b'^2 \equiv 0, 1 \bmod 4$. This contradicts the above conditions on k and completes the proof that $\gcd(m', D) = 1$.

We thus have $d'^2 m' = d^2 m$ with $\gcd(d', d) = \gcd(m', D) = 1$. We claim that this forces $\gcd(d, D) = 1$. To see why, suppose that a prime p divides d and D. Then

$d'^2m' = d^2m$ and $\gcd(d', d) = 1$ imply that $p \mid m'$, which contradicts $\gcd(m', D) = 1$. Thus we can find e relatively prime to D such that $ed \equiv 1 \bmod D$. We saw above that $f(x, y)$ represents d^2m, say $f(u, v) = d^2m$. Then $f(eu, ev) = e^2d^2m \equiv m \bmod D$ shows that $f(x, y)$ represents the congruence class of m. It follows that $f(x, y)$ and $g(x, y)$ lie in the same genus.

3.18. (a) Since f is in the principal genus, it is rationally equivalent to $x^2 + ny^2$ via a matrix whose denominators are prime to $2D$. Thus if d is the least common multiple of the denominators, the matrix is $\frac{1}{d}\left(\begin{smallmatrix} p & q \\ r & s \end{smallmatrix}\right)$ with $p, q, r, s \in \mathbb{Z}$ and $\gcd(d, 2D) = 1$, so that

$$x^2 + ny^2 = \frac{1}{d^2} f(px + qy, rx + sy) \implies d^2x^2 + d^2ny^2 = f(px + qy, rx + sy).$$

Setting $(x, y) = (1, 0)$ gives $d^2 = f(p, r)$. Setting $e = \gcd(p, r)$, we may rewrite this as $d^2 = e^2 f(p/e, r/e)$. Then $a^2 = d^2/e^2$ is properly represented by $f(x, y)$, and $\gcd(a, D) = 1$ since $a \mid d$. Finally, a is odd since d is.

(b) Since $f(x, y)$ represents a^2 properly and $\gcd(a^2, D) = 1$, we may assume that $f(x, y) = a^2x^2 + 2bxy + cy^2$ by Lemma 2.3. We have $D = -4n = 4b^2 - 4a^2c$, so that $n = a^2c - b^2$. If $\gcd(a, b) = q > 1$, then $q^2 \mid n$, contrary to the assumption that n is squarefree. Thus $\gcd(a, b) = 1$, and since a is odd by construction, we have $\gcd(a, 2b) = 1$. Finally, we see that $g(x, y) = ax^2 + 2bxy + acy^2$ has discriminant D as well, and is a primitive form.

(c) To compute the Dirichlet composition of $g(x, y)$ with itself, first compute B such that

$$B \equiv 2b \bmod 2a, \quad B^2 \equiv 4b^2 - 4a^2c \bmod 4a^2.$$

This has the solution $B = 2b$. Then the Dirichlet composition is

$$F(x, y) = a^2x^2 + 2bxy + \frac{4b^2 - (4b^2 - 4a^2c)}{4a^2} y^2 = a^2x^2 + 2bxy + cy^2,$$

which is $f(x, y)$.

3.19. We show that (i) is equivalent to each of the others.

(i) ⇔ (ii): The principal genus is $C(D)^2$. If (i) holds, then, first, $C(D)^2$ consists only of the principal class, so that every element of $C(D)$ has order at most two, and second, the number of classes equals the number of genera, which is $2^{\mu-1}$. Thus $C(D) \simeq \{\pm 1\}^{\mu-1}$. Conversely, if $C(D) \simeq \{\pm 1\}^{\mu-1}$, then the square of each class is the principal class, so that $C(D)^2$ contains only the principal class. But then the principal genus consists of a single class and thus by Corollary 3.14, each genus consists of a single class.

(i) ⇔ (iii): If (i) holds, then every genus consists of a single class by Corollary 3.14. Thus every genus consists of properly equivalent forms. Since properly equivalent forms are equivalent, it follows that every genus consists of equivalent forms, proving (iii). For (iii) ⇒ (i), note that the only reduced form properly or improperly equivalent to the principal form is the principal form itself, so if the principal genus consists of equivalent forms, it must consist of a single class. Then by Corollary 3.14, every genus consists of a single class.

3.20. (a) Consider first the case where $m = p^k$ is a power of an odd prime. Since the kernel of $(\cdot/p) : (\mathbb{Z}/p^k\mathbb{Z})^* \to \{\pm 1\}$ is the subgroup of squares by part (a) of Exercise 3.11, we have

$$x^2 \equiv -n \bmod p^k \text{ has a solution} \iff \left(\frac{-n}{p}\right) = 1.$$

Thus, $(-n/p) = -1$ implies that the number of solutions is $0 = 1 + (-n/p)$. Furthermore, when $(-n/p) = 1$, we show that there are $2 = 1 + (-n/p)$ solutions as follows. Suppose x and y are solutions. Then $x^2 \equiv y^2 \equiv -n \bmod p^k$, so that $p^k \mid (x+y)(x-y)$. If p divided both factors, then $p \mid 2x$, so that $p \mid x$, which is impossible since $\gcd(p,n) = 1$. Thus p^k divides either $x+y$ or $x-y$, so that either $x \equiv y \bmod p^k$ or $x \equiv -y \bmod p^k$, proving that there are precisely two solutions in $(\mathbb{Z}/p^k\mathbb{Z})^*$. In summary, the number of solutions of the congruence $x^2 \equiv -n \bmod p^k$ is given by

$$1 + \left(\frac{-n}{p}\right) = \prod_{p \mid p^k} \left(1 + \left(\frac{-n}{p}\right)\right).$$

Hence the formula holds in the prime power case.

Finally, if $m = \prod_{i=1}^r p_i^{k_i}$ where the p_i are distinct odd primes and $k_1 \geq 1$ are integers, then $x^2 \equiv -n \bmod m$ has a solution if and only if it has a solution modulo $p_i^{k_i}$ for each i. But in this case an r-tuple of solutions $(b_1, \ldots, b_r)$ lifts, using the Chinese Remainder Theorem, to a unique solution modulo m. Thus the number of solutions modulo m is the product of the number of solutions modulo $p_i^{k_i}$ for each i, giving

$$\prod_{p \mid m} \left(1 + \left(\frac{-n}{p}\right)\right)$$

as desired.

(b) The form $g(x,y)$ has discriminant $4b^2 - 4mc = -4n$, or $b^2 = mc - n$. Then if $\gcd(b,m) > 1$ it follows that $\gcd(n,m) > 1$, a contradiction. Thus for any such form, $[b] \in (\mathbb{Z}/m\mathbb{Z})^*$ and hence $g(x,y) \mapsto [b]$ is well-defined.

To see that the correspondence is one-to-one, suppose $mx^2 + 2bxy + cy^2$ and $mx^2 + 2b'xy + c'y^2$ both map to $[b] \in (\mathbb{Z}/m\mathbb{Z})^*$ with $0 \leq b, b' < m$. Then $b \equiv b' \bmod m$ and therefore $b = b'$. Since $-4n = 4b^2 - 4mc = 4b'^2 - 4m'c'$, we see also that $c = c'$ so that the forms are equal.

It remains to show that any solution $[b]$ modulo m of $x^2 \equiv -n \bmod m$ is the image of some $g(x,y)$. Suppose $b^2 \equiv -n \bmod m$, so that $b^2 + n = km$ for some integer k. We may choose b to satisfy $0 \leq b < m$. Then $mx^2 + 2bxy + ky^2$ has discriminant $-4n$, and its image in $(\mathbb{Z}/m\mathbb{Z})^*$ is $[b]$.

(c) Since $f(x,y)$ has discriminant $-4n$, we can write $f(x,y) = Ax^2 + 2Bxy + Cy^2$. Then

$$\begin{aligned} g(x,y) &= mx^2 + (2km + 2Aur_0 + 2B(vr_0 + us_0) + 2Cvs_0)xy + f(r,s)y^2 \\ &= mx^2 + 2(km + Aur_0 + B(vr_0 + us_0) + Cvs_0)xy + f(r,s)y^2. \end{aligned}$$

But there is a unique value of k such that $0 \leq km + Aur_0 + B(vr_0 + us_0) + Cvs_0 < m$. This form is $g_{u,v}(x,y)$.

(d) Suppose $g(x,y) = mx^2 + 2bxy + cy^2$ is a form as in part (b). We need to find a form $f(x,y)$ and a proper representation $f(u,v) = m$ such that the form $g_{u,v}(x,y)$ constructed in part (c) is equal to $g(x,y)$. This is immediate, since $f(x,y) = mx^2 + 2bxy + cy^2$ represents m primitively via $f(1,0) = m$. Following part (c), choose $r_0 = v = 0$, $s_0 = u = 1$, so that $r = k$ and $s = 1$; then

$$f(x + ky, y) = mx^2 + 2(b + km)xy + (c + 2bk + k^2m)y^2.$$

Then $k = 0$ is the unique k found in part (c), so that $g_{1,0}(x,y) = g(x,y)$.

(e) If $f(x,y) = Ax^2 + 2Bxy + Cy^2$, then define the matrix of $f(x,y)$ to be $M_f = \left(\begin{smallmatrix} A & B \\ B & C \end{smallmatrix}\right)$. This matrix satisfies

$$f(x,y) = (x \;\; y)\, M_f \begin{pmatrix} x \\ y \end{pmatrix},$$

and it follows without difficulty that the matrices of $g_{u,v}$ and $g_{u',v'}$ are

$$M_{u,v} = \begin{pmatrix} u & v \\ r & s \end{pmatrix} M_f \begin{pmatrix} u & r \\ v & s \end{pmatrix}$$
$$M_{u',v} = \begin{pmatrix} u' & v' \\ r' & s' \end{pmatrix} M_f \begin{pmatrix} u' & r' \\ v' & s' \end{pmatrix}$$

Setting these two equal gives, after some rearrangement,

$$M_f = \begin{pmatrix} \alpha & \beta \\ \gamma & \delta \end{pmatrix} M_f \begin{pmatrix} \alpha & \gamma \\ \beta & \delta \end{pmatrix}$$

with $\alpha\delta - \beta\gamma = 1$. Thus $f(\alpha x + \gamma y, \beta x + \delta y) = f(x, y)$. This gives

$$f(x, y) = f(\alpha x + \gamma y, \beta x + \delta y) = f(\alpha, \beta)x^2 + Qxy + f(\gamma, \delta)y^2,$$

so that $f(\alpha, \beta) = A$ and $f(\gamma, \delta) = C$.

We are assuming $f(x, y)$ is reduced. If $A = C$, then from part (b) of Exercise 2.8, we have $(\alpha, \beta) = \pm(1, 0)$ or $\pm(0, 1)$ and $(\gamma, \delta) = \pm(1, 0)$ or $\pm(0, 1)$. Since $\alpha\delta - \beta\gamma = 1$, we get the following possibilities:

$$\begin{pmatrix} \alpha & \beta \\ \gamma & \delta \end{pmatrix} = \pm\begin{pmatrix} 1 & 0 \\ 0 & 1 \end{pmatrix}, \ \pm\begin{pmatrix} 0 & 1 \\ -1 & 0 \end{pmatrix}.$$

The second of these, however, maps $Ax^2 + 2Bxy + Cy^2$ to $Ax^2 - 2Bxy + Cy^2$. Since these two must be equal, we get $B = 0$, and since $f(x, y)$ is reduced, we have $A = C = 1$. But this case is impossible since then $f(x, y)$ has discriminant $-4n = -4$ and we are assuming $n > 1$. So in this case $\left(\begin{smallmatrix} \alpha & \beta \\ \gamma & \delta \end{smallmatrix}\right) = \pm\left(\begin{smallmatrix} 1 & 0 \\ 0 & 1 \end{smallmatrix}\right)$.

Next, if $A = 2B$, so that $f(x, y) = Ax^2 + Axy + Cy^2$, then part (b) of Exercise 2.8 gives

$$(\alpha, \beta) = \pm(1, 0), \ (\gamma, \delta) = \pm(0, 1) \text{ or } \pm(1, -1).$$

Since $\alpha\delta - \beta\gamma = 1$, we get the following possibilities:

$$\begin{pmatrix} \alpha & \beta \\ \gamma & \delta \end{pmatrix} = \pm\begin{pmatrix} 1 & 0 \\ 0 & 1 \end{pmatrix}, \ \pm\begin{pmatrix} 1 & 0 \\ -1 & 1 \end{pmatrix}.$$

However, the second of these matrices maps $Ax^2 + Axy + Cy^2$ to $Cx^2 + (A - 2C)xy + Cy^2$. Since the two must be equal, we get $A = A - 2C$, so that $C = 0$, which is impossible. So in this case as well we get $\left(\begin{smallmatrix} \alpha & \beta \\ \gamma & \delta \end{smallmatrix}\right) = \pm\left(\begin{smallmatrix} 1 & 0 \\ 0 & 1 \end{smallmatrix}\right)$.

Finally, if $|2B| < A < C$, we can apply (2.11) to get $(\alpha, \beta) = \pm(1, 0)$ and $(\gamma, \delta) = \pm(0, 1)$. Again $\alpha\delta - \beta\gamma = 1$ forces $\left(\begin{smallmatrix} \alpha & \beta \\ \gamma & \delta \end{smallmatrix}\right) = \pm\left(\begin{smallmatrix} 1 & 0 \\ 0 & 1 \end{smallmatrix}\right)$.

(f) Write $J = \left(\begin{smallmatrix} \alpha & \beta \\ \gamma & \delta \end{smallmatrix}\right)$. If $J = \pm I$, then $\left(\begin{smallmatrix} u' & v' \\ r' & s' \end{smallmatrix}\right)^{-1} = \pm\left(\begin{smallmatrix} u & v \\ r & s \end{smallmatrix}\right)^{-1}$, so that $(u', v') = \pm(u, v)$. Conversely, if $(u', v') = \pm(u, v)$, then the uniqueness condition on k together with the fact that the determinant of each matrix is one forces $r' = \pm r$ and $s' = \pm s$, so that $J = \pm I$.

Since the map in part (d) is surjective from proper representations of m by a form of the required discriminant to the $g_{u,v}$, this shows that it is exactly two-to-one; since the $g_{u,v}(x, y)$ are mapped bijectively onto the solutions of $x^2 \equiv -n \bmod m$ by part (b), we get an extra factor of two in part (a), so that the number of ways of representing m properly by a form of discriminant $-4n$ is as asserted.

3.21. (a) Since m and m^3 have the same prime factors, and m is odd, setting $n = 2$ in Lemma 3.25 shows that the number of proper representations of either by a reduced form of discriminant -8 is

$$2\prod_{p|m}\left(1 + \left(\frac{-2}{p}\right)\right).$$

Then we are done since the only reduced form of discriminant -8 is $x^2 + 2y^2$.

(b) It is a representation, since

$$(a^3 - 6ab^2)^2 + 2(3a^2b - 2b^3)^2 = a^6 + 6a^4b^2 + 12a^2b^4 + 8b^6 = (a^2 + 2b^2)^3 = m^3.$$

Note that a is odd, so that the same is true for $a^3 - 6ab^2$. Also note that 3 cannot be a common divisor of $a^3 - 6ab^2$ and $3a^2b - 2b^3$ since $\gcd(a, b) = 1$. Now suppose a prime $p > 3$ divides $a^3 - 6ab^2 = a(a^2 - 6b^2)$ and $3a^2b - 2b^3 = b(3a^2 - 2b^2)$. Then again using $\gcd(a, b) = 1$, we have

$$p \mid a,\ p \mid b(3a^2 - 2b^2) \Rightarrow p \mid a,\ p \mid 3a^2 - 2b^2 \Rightarrow p \mid a,\ p \mid b$$
$$p \mid b,\ p \mid a(a^2 - 6b^2) \Rightarrow p \mid b,\ p \mid a^2 - 6b^2 \Rightarrow p \mid b,\ p \mid a,$$

both of which are impossible since $\gcd(a, b) = 1$ and $p > 3$. This implies that $p \nmid a$ and $p \nmid b$, so that $p \mid a^2 - 6b^2$ and $p \mid 3a^2 - 2b^2$, which implies

$$p \mid (3a^2 - 2b^2) - 3(a^2 - 6b^2) = 16b^2, \quad p \mid 3(3a^2 - 2b^2) - (a^2 - 6b^2) = 8a^2,$$

which is impossible since $p > 3$ and $p \nmid a$. Thus $\gcd(a^3 - 6ab^2, 3a^2b - 2b^3) = 1$ and the representation is proper.

(c) Suppose that (a, b) and (a', b') have the same image under the given map. Then from the identity shown in the hint, we have $(a + b\sqrt{-2})^3 = (a' + b'\sqrt{-2})^3$, so that $a + b\sqrt{-2}$ and $a' + b'\sqrt{-2}$ differ by a factor of a cube root of unity. This factor must be 1 since 1 is the only cube root of unity lying in $\mathbb{Q}(\sqrt{-2})$.

(d) As (a, b) ranges over all possible proper representations of m, the image of the map in part (c) produces the same number of proper solutions of $m^3 = x^2+2y^2$. Since the two equations have the same number of proper solutions, each solution for $m^3 = x^2 + 2y^2$ must arise from a proper representation of m.

3.22. Suppose $x^3 = y^2 + 2 = y^2 + 2 \cdot 1^2$. If x is even, then $x^3 \equiv 0 \bmod 8$. Then y must be even, but $y^2 \equiv 0, 4 \bmod 8$, so that $y^2 + 2 \not\equiv 0 \bmod 8$. Thus x must be odd. Then from the previous exercise, since the solution $(y, 1)$ arises from a solution (a, b) of $x = y^2 + 2z^2$, we have

$$y = a^3 - 6ab^2, \quad 1 = 3a^2b - 2b^3 = b(3a^2 - 2b^2).$$

The second equation forces $b = \pm 1$, so that $1 = \pm(3a^2 - 2)$, giving $3a^2 = 2 \pm 1$, which implies $a = \pm 1$. Thus $y = \pm 1 \mp 6 = \pm 5$, and then $x^3 = y^2 + 2 = 27$ so that $x = 3$. The only solutions are $(3, \pm 5)$, as claimed by Fermat.

3.23. By Lemma 3.25, the number of proper representations (i.e. representations) of p by a form of discriminant $-4n$ is

$$2\prod_{p|p}\left(1 + \left(\frac{-n}{p}\right)\right) = 2\left(1 + \left(\frac{-n}{p}\right)\right).$$

Since p has a representation, we must have $(-n/p) = 1$, so p has four representations. Since it is represented by $x^2 + ny^2$, the four solutions must be $(\pm a)^2 + n(\pm b)^2$ for some integers a and b. Only one of these has $x, y \geq 0$.

3.24. (a) We want to prove the formula

$$\left(\frac{P}{Q}\right)\left(\frac{Q}{P}\right) = (-1)^{(P-1)(Q-1)/4 + (\operatorname{sgn}(P)-1)(\operatorname{sgn}(Q)-1)/4}$$

where P and Q are odd and relatively prime, but may be negative.

Observe that if $P > 0$ or $Q > 0$ then $(\operatorname{sgn}(P) - 1)(\operatorname{sgn}(Q) - 1)/4 = 0$, while if $P < 0$ and $Q < 0$, then $(\operatorname{sgn}(P) - 1)(\operatorname{sgn}(Q) - 1)/4 = -1$. So to prove the desired formula, it suffices to show that

$$(*)\qquad \left(\frac{P}{Q}\right)\left(\frac{Q}{P}\right) = \begin{cases} (-1)^{(P-1)(Q-1)/4} & P > 0 \text{ or } Q > 0 \\ (-1)^{(P-1)(Q-1)/4-1} & P < 0 \text{ and } Q < 0. \end{cases}$$

Multiply both sides of the third identity of (1.16) by (m/M), giving

$$\left(\frac{M}{m}\right)\left(\frac{m}{M}\right) = (-1)^{(M-1)(m-1)/4}\left(\frac{m}{M}\right)^2 = (-1)^{(M-1)(m-1)/4}.$$

This holds if $m > 0$ is odd and relatively prime to $M > 0$. Writing P for M and Q for m, we see that if $P > 0$ and $Q > 0$, we get

$$(**)\qquad \left(\frac{P}{Q}\right)\left(\frac{Q}{P}\right) = (-1)^{(P-1)(Q-1)/4},$$

so that (*) holds when $P > 0$ and $Q > 0$. Note that this shows that (*) holds even in the exceptional case where $P = 1$ or $Q = 1$ since (**) holds in this case, where both sides are 1.

Next suppose that $P < 0$ and $Q > 0$. Then (**) gives

$$\left(\frac{-P}{Q}\right)\left(\frac{Q}{-P}\right) = (-1)^{(-P-1)(Q-1)/4},$$

even if $P = -1$. So when using this formula below, we will not explicitly differentiate the cases where $P = -1$. Then

$$\begin{aligned}\left(\frac{P}{Q}\right)\left(\frac{Q}{P}\right) = \left(\frac{-(-P)}{Q}\right)\left(\frac{Q}{-P}\right) &= \left(\frac{-1}{Q}\right)\left(\frac{-P}{Q}\right)\left(\frac{Q}{-P}\right) \\ &= (-1)^{(Q-1)/2}(-1)^{(-P-1)(Q-1)/4} \\ &= (-1)^{((-P-1)(Q-1)+2(Q-1))/4} \\ &= (-1)^{-(P-1)(Q-1)/4} \\ &= (-1)^{(P-1)(Q-1)/4}\end{aligned}$$

as desired. If $Q < 0$, $P > 0$, exchange P and Q and use the previous case. This proves (*) when P and Q have different signs.

Finally, if $Q, P < 0$, then

$$\begin{aligned}\left(\frac{P}{Q}\right)\left(\frac{Q}{P}\right) = \left(\frac{P}{-Q}\right)\left(\frac{Q}{-P}\right) &= \left(\frac{-(-P)}{-Q}\right)\left(\frac{-(-Q)}{-P}\right) \\ &= \left(\frac{-1}{-Q}\right)\left(\frac{-1}{-P}\right)\left(\frac{-P}{-Q}\right)\left(\frac{-Q}{-P}\right) \\ &= (-1)^{(-Q-1)/2}(-1)^{(-P-1)/2}(-1)^{(-P-1)(-Q-1)/4} \\ &= (-1)^{(-Q-1)/2+(-P-1)/2+(-P-1)(-Q-1)/4}.\end{aligned}$$

It remains to simplify the exponent:

$$\begin{aligned}\frac{-Q-1}{2} + \frac{-P-1}{2} + \frac{(-P-1)(-Q-1)}{4} &= \frac{-2Q-2-2P-2+PQ+P+Q+1}{4} \\ &= \frac{PQ-P-Q+1-4}{4} \\ &= \frac{(P-1)(Q-1)}{4} - 1\end{aligned}$$

as required by (*).

(b) Gauss' ten cases assume that P and Q are relatively prime odd integers. They are:

	P	Q	$(-1)^{(P-1)(Q-1)/4}$	$(-1)^{(\mathrm{sgn}(P)-1)(\mathrm{sgn}(Q)-1)/4}$
1.	$\equiv 1 \bmod 4, > 0$	$\equiv 1 \bmod 4, > 0$	1	1
2.	$\equiv 1 \bmod 4, > 0$	$\equiv 3 \bmod 4, < 0$	1	1
3.	$\equiv 1 \bmod 4, > 0$	$\equiv 3 \bmod 4, > 0$	1	1
4.	$\equiv 1 \bmod 4, > 0$	$\equiv 1 \bmod 4, < 0$	1	1
5.	$\equiv 3 \bmod 4, < 0$	$\equiv 3 \bmod 4, < 0$	-1	-1
6.	$\equiv 3 \bmod 4, > 0$	$\equiv 1 \bmod 4, < 0$	1	1
7.	$\equiv 3 \bmod 4, < 0$	$\equiv 3 \bmod 4, > 0$	-1	1
8.	$\equiv 3 \bmod 4, < 0$	$\equiv 1 \bmod 4, < 0$	1	-1
9.	$\equiv 3 \bmod 4, > 0$	$\equiv 3 \bmod 4, > 0$	-1	1
10.	$\equiv 1 \bmod 4, < 0$	$\equiv 1 \bmod 4, < 0$	1	-1

(If you want to compare this to the table in [**51**, §133], you should note that, for example, Gauss' notation "$-A$" means that the number is the negation of a positive integer which is congruent to 1 modulo 4, so that it is negative, and congruent to 3 modulo 4.)

For cases 1 through 6, Gauss claims that $(P/Q)(Q/P) = 1$, while for cases 7 through 10, he claims that $(P/Q)(Q/P) = -1$. From the table above, this is clear, using the formula from part (a).

(c) Let P be an odd integer. For $P > 1$ these are just the Jacobi supplementary laws. If $P = 1$ both left-hand sides and both right-hand sides are 1, so the equations hold. If $P = -1$, the left-hand sides are both 1, and the right-hand sides are $-(-1)^1 = 1$ and $(-1)^{(1^2-1)/8} = 1$ respectively. Finally, for $P < -1$, we get

$$\begin{aligned}\left(\frac{-1}{P}\right) = \left(\frac{-1}{-P}\right) &= (-1)^{(-P-1)/2} = (-1)^{-P}(-1)^{(P-1)/2}\\ &= -(-1)^{(P-1)/2} = \mathrm{sgn}(P)(-1)^{(P-1)/2}\\ \left(\frac{2}{P}\right) = \left(\frac{2}{-P}\right) &= (-1)^{((-P)^2-1)/8} = (-1)^{(P^2-1)/8},\end{aligned}$$

where $(-1)^{-P} = -1$ since P is odd.

3.25. (a) By the structure theorem for Abelian groups, $C(-4p) \simeq H \times G$, where $|H| = 2^a$, $a \geq 0$, and $|G|$ is odd. Since $p \equiv 1 \bmod 4$, we have $\mu = r + 1 = 2$, so there are exactly two elements of order ≤ 2 (Proposition 3.11). These elements must be (e, e) and $(b, e) \in H \times G$ for some nonidentity $b \in H$, so that $|H| > 1$. Further, this shows that H has a unique subgroup of order two and therefore is cyclic. So

$$C(-4p) \simeq (\mathbb{Z}/2^a\mathbb{Z}) \times G$$

where $a \geq 1$ and $|G|$ is odd. It follows that $2 \mid h(-4p)$.

(b) The discriminant of $f(x, y)$ is

$$2^2 - 4 \cdot 2 \cdot \frac{p+1}{2} = 4 - 4(p+1) = -4p.$$

Since $p \equiv 1 \bmod 4$, the table preceding (3.16) shows that the relevant characters are

$$\chi_1(a) = \left(\frac{a}{p}\right), \quad \delta(a) = (-1)^{(a-1)/2}.$$

Since $x^2 + py^2$ represents 1, its complete character is $(\chi_1(1), \delta(1)) = (1, 1)$.

The form $f(x, y) = 2x^2 + 2xy + ((p+1)/2)y^2$ represents $a = (p+1)/2$ with $(x, y) = (0, 1)$. Note that $a \equiv 1 \bmod 4$ since $p \equiv 1 \bmod 8$. This implies $\delta(a) = 1$, and

$p \equiv 1 \bmod 8$ implies $(2/p) = 1$. Then

$$\chi_1(a) = \left(\frac{a}{p}\right) = \left(\frac{2}{p}\right)\left(\frac{a}{p}\right) = \left(\frac{2a}{p}\right) = \left(\frac{p+1}{p}\right) = \left(\frac{1}{p}\right) = 1.$$

Thus the complete character of this form is also $(1, 1)$, so it is in the principal genus.

(c) The form $f(x, y) = 2x^2 + 2xy + ((p + 1)/2)y^2$ is reduced, so that its class has order ≤ 2 by Lemma 3.10. That order must be exactly 2 since the form is reduced and different from the principal form $x^2 + py^2$.

Since by part (b) the form $f(x, y)$ is in the principal genus, Theorem 3.15 implies that its class is the square of some class in $C(-4p)$. That class has order 4 since the class of $f(x, y)$ has order 2. Then $4 \mid h(-4p)$ by Lagrange's Theorem.

Solutions to Exercises in §4

4.1. To prove that these are subrings of $\mathbb{C}$ it suffices to show that they are closed under subtraction and multiplication, and that 1, the multiplicative identity of $\mathbb{C}$, lies in both rings. Choose $a, b, c, d \in \mathbb{Z}$. Then

$$\begin{aligned}(a + b\omega) - (c + d\omega) &= (a - b) + (c - d)\omega \in \mathbb{Z}[\omega]\\ (a + bi) - (c + di) &= (a - b) + (c - d)i \in \mathbb{Z}[i]\\ (a + b\omega)(c + d\omega) &= ac + (ad + bc)\omega + bd\omega^2 = ac - bd + (ad + bc - bd)\omega \in \mathbb{Z}[\omega]\\ (a + bi)(c + di) &= ac + (ad + bc)i + bdi^2 = (ac - bd) + (ad + bc)i \in \mathbb{Z}[i],\end{aligned}$$

where the third line uses $\omega^2 = -1 - \omega$. Since $1 = 1 + 0\omega = 1 + 0i$, we conclude that $\mathbb{Z}[\omega]$ and $\mathbb{Z}[i]$ are both subrings of $\mathbb{C}$.

For the second part, we have

$$\omega^2 = \left(\frac{-1 + \sqrt{-3}}{2}\right)^2 = \frac{1 - 2\sqrt{-3} + (-3)}{4} = \frac{-2 - 2\sqrt{-3}}{4} = \frac{-1 - \sqrt{-3}}{2} = \overline{\omega}.$$

Then for $a, b \in \mathbb{Z}$,

$$\overline{a + b\omega} = a + b\overline{\omega} = a + b\omega^2 = a + b(-1 - \omega) = (a - b) - b\omega \in \mathbb{Z}[\omega],$$

so $\mathbb{Z}[\omega]$ is closed under conjugation. Since $\overline{a + bi} = a - bi$, the same holds for $\mathbb{Z}[i]$.

4.2. (a) From the previous solution, $\overline{\omega} = \omega^2$, so that $\overline{r + s\omega} = r + s\omega^2$. Then

$$N(r + s\omega) = (r + s\omega)(r + s\omega^2) = r^2 + rs(\omega + \omega^2) + s^2\omega^3 = r^2 - rs + s^2,$$

since $\omega + \omega^2 = -1$ and $\omega^3 = 1$.

(b) Let $u, v \in \mathbb{Z}[\omega]$. Then

$$N(uv) = uv \cdot \overline{uv} = uv \cdot \overline{u}\,\overline{v} = (u\overline{u})(v\overline{v}) = N(u)N(v),$$

where the second equality uses properties of complex conjugation and third uses the associativity and commutativity of multiplication.

4.3. Let $R = \mathbb{C}[x, y]$ and $I = \{f(x, y) \in R : f(0, 0) = 0\}$. To prove that I is an ideal, first note $0 \in I$. If $f(x, y), g(x, y) \in I$, then $(f + g)(0, 0) = f(0, 0) + g(0, 0) = 0 + 0 = 0$, so $f(x, y) + g(x, y) \in I$. Furthermore, if $f(x, y) \in I$, $g(x, y) \in R$, then $(fg)(0, 0) = f(0, 0) \cdot g(0, 0) = 0 \cdot g(0, 0) = 0$, so $f(x, y)g(x, y) \in I$. Hence I is an ideal of R.

Suppose I is principal, say $I = g(x, y)R$. Since $x, y \in I$, both x and y must be multiples of $g(x, y)$. Both x and y are irreducible by degree considerations: if $x = rs$ or $y = rs$, then either r or s must be a nonzero constant, so is a unit in R. Since $1 \notin I$ it follows that $g(x, y)$ cannot be a unit. Thus there are units u, v such that $x = ug(x, y)$ and $y = vg(x, y)$, which implies that $x = ug(x,y) = uv^{-1}vg(x,y) = (uv^{-1})y$, so that x and y are associates. But this is not the case in $\mathbb{C}[x,y]$. Thus I is not a principal ideal.

4.4. We must prove that statements (i), (ii), (iii), and (iv) just before Lemma 4.5 are equivalent. We do so as follows:

(iv) $\Rightarrow$ (iii)**:** It is a standard fact from abstract algebra that in a commutative ring, all maximal ideals are prime.

(iii) $\Rightarrow$ (ii)**:** Suppose that αR is prime and that $\alpha \mid \beta\gamma$; then $\beta\gamma \in \alpha R$, so that $\beta \in \alpha R$ or $\gamma \in \alpha R$. But this means that either $\alpha \mid \beta$ or $\alpha \mid \gamma$, proving that α is prime.

(ii) $\Rightarrow$ (i)**:** Suppose that α is prime and that $\alpha = \beta\gamma$. Then $\alpha \mid \beta\gamma$, so that $\alpha \mid \beta$ (say). Thus $\beta = \alpha\mu$, so that $\alpha = \beta\gamma = \alpha\mu\gamma$. Since $\alpha \neq 0$ and R is a domain, it follows that $\mu\gamma = 1$, so that γ is a unit.

(i) $\Rightarrow$ (iv)**:** Suppose α is irreducible and $\alpha R \subset J \subset R$ for some ideal J. Then $J = \beta R$ since R is a PID. But $\alpha \in \alpha R \subset \beta R$, so that $\alpha = \beta\gamma$. Since α is irreducible, either β or γ is a unit. In the first case, $J = \beta R = R$; in the second, $J = \beta R = \alpha R$. Since α is a nonunit, it follows that $\alpha R \neq R$, so that αR is a maximal ideal.

4.5. (i) If $\alpha \in \mathbb{Z}[\omega]$ is a unit, say $\alpha\beta = 1$, then $N(\alpha)N(\beta) = N(\alpha\beta) = N(1) = 1$; since norms are nonnegative integers, $N(\alpha) = 1$. Conversely, suppose $N(\alpha) = 1$; then $1 = N(\alpha) = \alpha\overline{\alpha}$. Since $\overline{\alpha} \in \mathbb{Z}[\omega]$ by Exercise 4.1, we conclude that α is a unit.

(ii) A computation shows that $\pm1, \pm\omega, \pm\omega^2$ all have norm 1, so are all units. It remains to show that they are the only units. By (2.4), $4(a^2 - ab + b^2) = (2a - b)^2 + 3b^2$. If $\alpha = a + b\omega$ is a unit in $\mathbb{Z}[\omega]$, then $N(\alpha) = a^2 - ab + b^2 = 1$ by (i), so that $(2a - b)^2 + 3b^2 = 4$. If $|b| \geq 2$, then this equation implies

$$4 = (2a - b)^2 + 3b^2 \geq 0 + 3 \cdot 4 = 12,$$

which is clearly impossible. Hence $b = 0$ or $b = \pm1$. Then:

- If $b = 0$, then $a = \pm1$, giving $\alpha = \pm1$.
- If $b = 1$, then $(2a - 1)^2 + 3 = 4$, so that $2a - 1 = \pm1$ and thus $a = 0$ or $a = 1$. This gives $\alpha = \omega$ and $\alpha = 1 + \omega = -\omega^2$.
- If $b = -1$, then $(2a + 1)^2 + 3 = 4$, so that $2a + 1 = \pm1$ and thus $a = 0$ or $a = -1$. This gives $\alpha = -\omega$ and $\alpha = -1 - \omega = \omega^2$.

4.6. (a) A unit $u \in \mathbb{Z}[\sqrt{-3}]$ satisfies $uv = 1$ for some $v \in \mathbb{Z}[\sqrt{-3}]$. Since $\mathbb{Z}[\sqrt{-3}] \subset \mathbb{Z}[\omega]$, part (i) of Lemma 4.5 implies that $N(u) = 1$. Writing $u = a + b\sqrt{-3}$, $a, b \in \mathbb{Z}$, we obtain

$$1 = N(u) = u\overline{u} = (a + b\sqrt{-3})(a - b\sqrt{-3}) = a^2 + 3b^2,$$

so that $a = \pm1$ and $b = 0$. Hence the only units are ±1.

(b) Since the only units in $\mathbb{Z}[\sqrt{-3}]$ are ±1 by part (a), it follows immediately that no two of 2, $1 + \sqrt{-3}$, and $1 - \sqrt{-3}$ are associates. It remains to prove that each is irreducible. Part (a) shows that for $a + b\sqrt{-3} \in \mathbb{Z}[\omega]$, the norm function is

$$N(a + b\sqrt{-3}) = a^2 + 3b^2.$$

Thus $N(2) = N(1 \pm \sqrt{-3}) = 4$, so if any of these are reducible, they must factor into elements of norm 2. But clearly $a^2 + 3b^2 \neq 2$ for integers a and b, so there are no elements of norm 2 in $\mathbb{Z}[\sqrt{-3}]$. This proves that all three are irreducible.

It follows immediately that $\mathbb{Z}[\sqrt{-3}]$ is not a UFD since $4 = 2 \cdot 2 = (1 + \sqrt{-3})(1 - \sqrt{-3})$ has two distinct factorizations into irreducibles.

(c) Let $\mathfrak{a} \subset \mathbb{Z}[\sqrt{-3}]$ be the ideal generated by 2 and $1 + \sqrt{-3}$. Assume that this ideal is principal, say $\mathfrak{a} = \alpha\mathbb{Z}[\sqrt{-3}]$. Then $2 = \alpha\beta$, $1 + \sqrt{-3} = \alpha\gamma$ for some $\beta, \gamma \in \mathbb{Z}[\sqrt{-3}]$.

Since 2 and $1+\sqrt{-3}$ are irreducible by part (b) and the units in $\mathbb{Z}[\sqrt{-3}]$ are ± 1 by part (a), it follows that one of α, β is ± 1 and one of α, γ is ± 1. If $\alpha \neq \pm 1$, then one concludes easily that 2 and $1+\sqrt{3}$ are associates, which is impossible by part (b). On other hand, if $\alpha = \pm 1$, then $1 \in \mathfrak{a} = \alpha\mathbb{Z}[\sqrt{-3}]$, which would imply

$$1 = 2(a+b\sqrt{-3}) + (1+\sqrt{-3})(c+d\sqrt{-3}) = (2a+c-3d) + (2b+c+d)\sqrt{-3}$$

for some $a, b, c, d \in \mathbb{Z}$. Then $2b+c+d=0$, which implies that c and d have the same parity, but this means that $2a+c-3d$ is even, which is impossible. Thus $1 \notin \mathfrak{a}$. It follows that $\mathfrak{a}$ is not principal, so that $\mathbb{Z}[\sqrt{-3}]$ is not a PID.

4.7. (a) If $p = \pi\overline{\pi}$ where $\pi = a + b\omega$ and $\overline{\pi} = a - b\omega$, then $p = a^2 - ab + b^2$, so that neither a nor b is zero. From Lemma 4.5, the units in $\mathbb{Z}[\omega]$ are ± 1, $\pm\omega$, and $\pm\omega^2$. Now compute the product of each unit with $\overline{\pi}$:

$$\begin{aligned}
1 \cdot \overline{\pi} &= a - b\omega\\
-1 \cdot \overline{\pi} &= -a + b\omega\\
\pm\omega \cdot \overline{\pi} &= \pm\omega(a - b\omega) = \pm(a\omega - b\omega^2) = \pm(a\omega - b(-1-\omega)) = \pm(b + (a+b)\omega)\\
\pm\omega^2 \cdot \overline{\pi} &= \pm\omega^2(a - b\omega) = \pm(a\omega^2 - b\omega^3) = \pm(-b + a(-1-\omega)) = \mp(a + b + a\omega).
\end{aligned}$$

We want to show that none of these is equal to π. The first cannot be, since $b \neq 0$. Similarly, the second cannot equal π since $a \neq 0$. For the rest,

$$\begin{aligned}
a + b\omega = b + (a+b)\omega \ &\Rightarrow\ b = a + b \ \Rightarrow\ a = 0\\
a + b\omega = -b - (a+b)\omega \ &\Rightarrow\ a = -b \text{ and } b = -a - b \Rightarrow\ b = 0\\
a + b\omega = a + b + a\omega \ &\Rightarrow\ a = a + b \ \Rightarrow b = 0\\
a + b\omega = -a - b - a\omega \ &\Rightarrow\ a = -a - b \text{ and } b = -a \ \ \Rightarrow a = 0.
\end{aligned}$$

These all produce contradictions since $a \neq 0$ and $b \neq 0$, so π and $\overline{\pi}$ are nonassociate.

(b) If $p \equiv 2 \bmod 3$, the version of quadratic reciprocity given in (1.12) implies that $(-3/p) = (p/3) = (2/3) = -1$. By Lemma 2.5, it follows that p is not represented by a form of discriminant -3; in particular, $p \neq a^2 - ab + b^2 = N(a + b\omega)$ for $a + b\omega \in \mathbb{Z}[\omega]$. But if $p = \alpha\beta$ in $\mathbb{Z}[\omega]$, then $N(p) = p^2 = N(\alpha)N(\beta)$; since neither norm can be p, one must be 1, so that either α or β is a unit by Lemma 4.5. It follows that p is irreducible in $\mathbb{Z}[\omega]$, hence prime since $\mathbb{Z}[\omega]$ is a UFD.

4.8. Since $\pi\mathbb{Z}[\omega]$ is maximal in $\mathbb{Z}[\omega]$, we see that $\mathbb{Z}[\omega]/\pi\mathbb{Z}[\omega]$ is in fact a finite field, not just a ring, with $N(\pi)$ elements. The proof of Proposition 4.7 shows that $N(\pi) = p$ or p^2 and furthermore that the former occurs when $p = 3$ or $p \equiv 1 \bmod 3$ and the latter occurs when $p \equiv 2 \bmod 3$.

Thus $p = 3$ or $p \equiv 1 \bmod 3$ implies $N(\pi) = p$ and hence $\mathbb{Z}[\omega]/\pi\mathbb{Z}[\omega] \simeq \mathbb{F}_p \simeq \mathbb{Z}/p\mathbb{Z}$. Also, $p \equiv 2 \bmod 3$ implies $N(\pi) = p^2$ and hence $\mathbb{Z}[\omega]/\pi\mathbb{Z}[\omega] \simeq \mathbb{F}_{p^2}$, as desired.

In the last case, where $p \equiv 2 \bmod 3$, $p = \pi$ and we can consider $\mathbb{Z}/p\mathbb{Z}$ as a subfield of $\mathbb{Z}[\omega]/\pi\mathbb{Z}[\omega]$ under the natural inclusion map $\mathbb{Z} \hookrightarrow \mathbb{Z}[\omega]$. This proves the last statement, since finite fields of order p^k have a unique subfield isomorphic to $\mathbb{Z}/p\mathbb{Z}$.

4.9. (a) If π is not associate to $1 - \omega$, then π is (associate to) a factor of an integral prime $p \equiv 1 \bmod 3$, or else π is (associate to) an integral prime $p \equiv 2 \bmod 3$, by Proposition 4.7. In the first case, $N(\pi) = p \equiv 1 \bmod 3$, while in the second case, $N(\pi) = p^2 \equiv 1 \bmod 3$. In either case, $3 \mid N(\pi) - 1$.

(b) If $1 \equiv \omega^2 \bmod \pi$, then $\omega \equiv \omega^3 = 1 \bmod \pi$. If $\omega \equiv \omega^2 \bmod \pi$, then $\omega \cdot \omega^2 \equiv \omega^2 \cdot \omega^2 \bmod \pi$, or $1 \equiv \omega \bmod \pi$. Thus if any two of 1, ω, and ω^2 are congruent modulo π, then $1 \equiv \omega \bmod \pi$. But $1 \equiv \omega \bmod \pi$ means that $1 - \omega \mid \pi$. Since π is prime, it is irreducible, so that $1 - \omega$ and π must be associates since $1 - \omega$ is also prime by Proposition 4.7. This contradicts our assumption on π, so that 1, ω, and ω^2 are incongruent modulo π.

4.10. (a) Assuming π is prime and not associate to $1-\omega$, then using (4.10),

$$\left(\frac{\alpha\beta}{\pi}\right)_3 \equiv (\alpha\beta)^{(N(\pi)-1)/3} \equiv \alpha^{(N(\pi)-1)/3}\beta^{(N(\pi)-1)/3} \equiv \left(\frac{\alpha}{\pi}\right)_3\left(\frac{\beta}{\pi}\right)_3 \mod \pi.$$

Since $(\alpha\beta/\pi)_3$, $(\alpha/\pi)_3$, and $(\beta/\pi)_3$ are all cube roots of unity, and no two different cube roots of unity are congruent modulo π by part (b) of Exercise 4.9, the two sides must be equal.

(b) If $\alpha \equiv \beta \mod \pi$, then $\alpha^{(N(\pi)-1)/3} \equiv \beta^{(N(\pi)-1)/3} \mod \pi$. It follows that $(\alpha/\pi)_3 \equiv (\beta/\pi)_3 \mod \pi$ and again we conclude equality since no two different cube roots of unity are congruent modulo π.

4.11. The first equivalence follows from (4.10) since the cube roots of unity are incongruent modulo π. For the second, if $\beta^3 \equiv \alpha \mod \pi$, then

$$\alpha^{(N(\pi)-1)/3} \equiv \left(\beta^3\right)^{(N(\pi)-1)/3} = \beta^{N(\pi)-1}.$$

But $\pi \nmid \alpha$, so that $\pi \nmid \beta$. Thus by Corollary 4.9, $\beta^{N(\pi)-1} \equiv 1 \mod \pi$ since π is prime. Conversely, assume that $\alpha^{(N(\pi)-1)/3} \equiv 1 \mod \pi$ and that π is not an associate of $1-\omega$. Since we are assuming that $(\mathbb{Z}[\omega]/\pi\mathbb{Z}[\omega])^*$ is cyclic, choose $\gamma \in \mathbb{Z}[\omega]$ such that $[\gamma]$ generates $(\mathbb{Z}[\omega]/\pi\mathbb{Z}[\omega])^*$. Then $[\alpha] = [\gamma]^k$ for some k, so that $\alpha \equiv \gamma^k \mod \pi$. It follows that

$$\alpha^{(N(\pi)-1)/3} \equiv \gamma^{k(N(\pi)-1)/3} \equiv 1 \mod \pi.$$

By Lemma 4.8, $|(\mathbb{Z}[\omega]/\pi\mathbb{Z}[\omega])^*| = N(\pi)-1$, so that $N(\pi)-1 \mid k(N(\pi)-1)/3$, which implies that $3 \mid k$. Let $\beta = \gamma^{k/3}$; then $\beta^3 \equiv \gamma^k \equiv \alpha \mod \pi$.

4.12. If $\pi = a + b\omega$ with a and b not both zero, then

$$\begin{aligned}\pm\pi &= \pm a \pm b\omega\\ \pm\omega\pi &= \pm a\omega \pm b\omega^2 = \mp b \pm (a-b)\omega\\ \pm\omega^2\pi &= \pm a\omega^2 \pm b\omega^3 = \pm(b-a) \mp a\omega.\end{aligned}$$

In determining which of these associates are primary, we need only consider a and b modulo 3, so we need only consider

$$(a,b) \equiv \pm(1,-1),\ \pm(1,1),\ \pm(0,1),\ \pm(1,0) \mod 3.$$

If $(a,b) \equiv \pm(1,-1) \mod 3$, then $\pm\pi \equiv \pm 1 \mp \omega \mod 3$, so that π is associate to $1-\omega$, impossible by assumption. Next suppose $(a,b) \equiv \pm(1,1) \mod 3$. Then $\pm\pi$ are not primary since $\pm b \not\equiv 0 \mod 3$. Further, $\pm\omega^2\pi$ are not primary since $\mp a \not\equiv 0 \mod 3$. Finally, $\pm\omega\pi$ are both primary since $\mp b \equiv \pm 1 \mod 3$ and $a - b \equiv 0 \mod 3$. If $(a,b) \equiv (0,\pm 1) \mod 3$, a similar analysis shows that only $\pm\omega^2\pi$ are primary. Finally, if $(a,b) \equiv (\pm 1, 0) \mod 3$, a similar analysis shows that only $\pm\pi$ are primary.

4.13. As noted in the discussion before (4.13), since π is not associate to $1-\omega$, we may assume that π is primary and $\pi \equiv -1 \mod 3$. Then $\pi = -1 + 3m + 3n\omega$, so that

$$\begin{aligned}\frac{N(\pi)-1}{3} &= \frac{1}{3}((3m-1)^2 - (3m-1)(3n) + (3n)^2 - 1)\\ &= 3m^2 + 3n^2 + n - 2m - 3mn \equiv n - 2m \equiv n + m \mod 3.\end{aligned}$$

Then by (4.13),

$$\left(\frac{\omega}{\pi}\right)_3 \equiv \omega^{3m^2+3n^2+n-2m-3mn} \equiv \omega^{n+m} \mod \pi.$$

and $(\omega/\pi)_3 = \omega^{n+m}$ follows since cube roots of unity are incongruent modulo π.

4.14. A nontrivial element in the kernel of the cubing map from $(\mathbb{Z}/p\mathbb{Z})^*$ to itself has order 3 in $(\mathbb{Z}/p\mathbb{Z})^*$, which is impossible since $(\mathbb{Z}/p\mathbb{Z})^*$ has order $p-1$ and $3 \nmid p-1$. Thus the kernel is trivial, so that the cubing map is one-to-one. It is then onto since it maps $(\mathbb{Z}/p\mathbb{Z})^*$ to itself. Thus every element of $(\mathbb{Z}/p\mathbb{Z})^*$ is a cube.

4.15. (a) The proof of Theorem 4.15 shows that $4p = (2r-3b)^2 + 27b^2$ for integers r and b. Note that since p is prime, r cannot be divisible by 3. If $r \equiv 1 \bmod 3$, let $a = 3b - 2r$; otherwise, let $a = 2r - 3b$. This gives $4p = a^2 + 27b^2$ with $a \equiv 1 \bmod 3$. Let

$$\pi = \frac{a+3\sqrt{-3}\,b}{2} = \frac{a+3b}{2} + 3b \cdot \frac{-1+\sqrt{-3}}{2} = \frac{a+3b}{2} + 3b\omega = a' + 3b\omega.$$

Note that a and b have the same parity, and hence a and $3b$ do as well. Thus a' is an integer and $\pi \in \mathbb{Z}[\omega]$. We also have

$$\begin{aligned} N(\pi) = a'^2 - 3a'b + (3b)^2 &= \left(\frac{a+3b}{2}\right)^2 - 3b \cdot \frac{a+3b}{2} + 9b^2 \\ &= \frac{a^2+6ab+9b^2}{4} - \frac{6ab-18b^2}{4} + \frac{36b^2}{4} = \frac{a^2+27b^2}{4} = p, \end{aligned}$$

so that π is prime by Lemma 4.6. Finally, $2a' = a + 3b \equiv a \equiv 1 \bmod 3$, which implies that $a' \equiv -1 \bmod 3$. It follows that $\pi = a' + 3b\omega \equiv -1 \bmod 3$ is primary.

(b) Note that

$$\pi = a' + 3b\omega = \frac{a+3b}{2} + 3b\omega = -1 + \frac{a+3b+2}{2} + 3b\omega.$$

Next observe that $(a+3b+2)/2$ is a multiple of 3 since $a' \equiv -1 \bmod 3$. Then using (4.13) and the fact that $\omega = \omega^4$, we get

$$\begin{aligned} \left(\frac{\omega}{\pi}\right)_3 &= \omega^{(a+3b+2)/6+b} = \omega^{(a+2+9b)/6} = \omega^{4(a+2+9b)/6} = \omega^{2(a+2)/3+6b} = \omega^{2(a+2)/3} \\ \left(\frac{1-\omega}{\pi}\right)_3 &= \omega^{2(a+3b+2)/6} = \omega^{(a+3b+2)/3} = \omega^{(a+2)/3+b}. \end{aligned}$$

(c) Since $3 = -\omega^2(1-\omega)^2$, we have

$$\left(\frac{3}{\pi}\right)_3 = \left(\frac{-1}{\pi}\right)_3 \left(\frac{\omega}{\pi}\right)_3^2 \left(\frac{1-\omega}{\pi}\right)_3^2 = \omega^{4(a+2)/3}\omega^{2(a+2)/3+2b} = \omega^{2(a+2)+2b} = \omega^{2b}$$

since $(-1/\pi)_3 = ((-1)^3/\pi)_3 = 1$ and $a+2$ is a multiple of 3.

(d) If $4p = x^2 + 243y^2$, then x cannot be a multiple of 3 since otherwise $9 \mid p$. Therefore $x^2 + 243y^2 \equiv 1 \bmod 3$, so that $p \equiv 1 \bmod 3$. Then as in part (a), $4p = x^2 + 27(3y)^2$, and it then follows from part (c) that

$$\left(\frac{3}{\pi}\right)_3 = \omega^{2\cdot 3y} = 1,$$

so that 3 is a cubic residue modulo p by (4.14). For the other direction, if $p \equiv 1 \bmod 3$ then (from the proof of Theorem 4.15) there are integers a, b such that

$$4p = (2a-3b)^2 + 27b^2.$$

It suffices to show that b is divisible by 3. But 3 is a cubic residue modulo p, so $1 = (3/\pi)_3 = \omega^{2b}$ (by (4.14) and part (c)) and thus $b \equiv 0 \bmod 3$.

4.16. (a) This is very similar to the proof that $\mathbb{Z}[\omega]$ is a Euclidean domain. We want to show that if $\alpha, \beta \in \mathbb{Z}[i]$, $\beta \neq 0$, then there are $\gamma, \delta \in \mathbb{Z}[i]$ such that $\alpha = \gamma\beta + \delta$ with $N(\delta) < N(\beta)$.

We first extend the norm function on $\mathbb{Z}[i]$ to $\mathbb{Q}(i)$ by setting $N(\alpha) = \alpha\overline{\alpha}$ for $\alpha \in \mathbb{Q}(i)$. Thus if $\alpha = a + bi$, we have $N(\alpha) = a^2 + b^2$. As in part (b) of Exercise 4.2, the norm function is multiplicative on $\mathbb{Q}(i)$ by the properties of complex conjugation. Now choose $\alpha, \beta \in \mathbb{Z}[i] \subset \mathbb{Q}(i)$ with $\beta \neq 0$; then $\alpha/\beta \in \mathbb{Q}(i)$. Let $\alpha/\beta = r + si$ for $r, s \in \mathbb{Q}$. Choose $r_1, s_1 \in \mathbb{Z}$ such that $|r - r_1| \leq 1/2$ and $|s - s_1| \leq 1/2$, and set $\gamma = r_1 + s_1 i$ and $\delta = \alpha - \gamma\beta$. Note that $\gamma, \delta \in \mathbb{Z}[i]$.

Then $\alpha = \gamma\beta + \delta$ by the definition of δ. It remains to show that $N(\delta) < N(\beta)$. So let $\epsilon = \alpha/\beta - \gamma = (r + si) - (r_1 + s_1 i) = (r - r_1) + (s - s_1)i$. Then

$$\delta = \alpha - \gamma\beta = \beta(\alpha/\beta - \gamma) = \beta\epsilon.$$

Since the norm is multiplicative and $N(\beta) \neq 0$, it suffices to show that $N(\epsilon) < 1$. But

$$N(\epsilon) = N((r - r_1) + (s - s_1)i) = (r - r_1)^2 + (s - s_1)^2 \leq \frac{1}{4} + \frac{1}{4} < 1.$$

(b) For Lemma 4.5, let $\alpha \in \mathbb{Z}[i]$ be a unit, say $\alpha\beta = 1$ for some $\beta \in \mathbb{Z}[i]$. Then $N(\alpha)N(\beta) = N(\alpha\beta) = N(1) = 1$; since norms are nonnegative integers, $N(\alpha) = 1$. Conversely, suppose $N(\alpha) = 1$. Then $1 = N(\alpha) = \alpha\overline{\alpha}$, so that α is a unit since $\alpha \in \mathbb{Z}[i]$ implies $\overline{\alpha} \in \mathbb{Z}[i]$. Clearly, ± 1 and $\pm i$ are units. To see that there are no others, let $\alpha = a + bi$ be a unit, so that $1 = N(\alpha) = a^2 + b^2$, This forces $a = \pm 1$, $b = 0$ or $a = 0$, $b = \pm 1$, so $\alpha = \pm 1$ or $\pm i$.

For Lemma 4.6, we imitate the proof given for $\mathbb{Z}[\omega]$. Suppose $\alpha \in \mathbb{Z}[i]$ and $N(\alpha)$ is a prime in $\mathbb{Z}$. We wish to show that α is prime in $\mathbb{Z}[i]$. Since $\mathbb{Z}[i]$ is a PID, it suffices to show that α is irreducible in $\mathbb{Z}[i]$. So suppose that $\alpha = \beta\gamma$, $\beta, \gamma \in \mathbb{Z}[i]$. Taking norms gives $N(\alpha) = N(\beta\gamma) = N(\beta)N(\gamma)$. Since $N(\alpha)$ is prime, either $N(\beta) = 1$ or $N(\gamma) = 1$, so that β or γ is a unit by the preceding paragraph.

(c) (i) $1 + i$ is prime since $N(1 + i) = 2$ is a prime, and

$$i^3(1 + i)^2 = -i(1 + 2i + i^2) = 2.$$

(ii) If $p \equiv 1 \bmod 4$, then $p = a^2 + b^2$ for some integers a and b, so that $p = N(a+bi)$ and thus $p = (a + bi)(a - bi)$. Also, $\pi = a + bi$ and $\overline{\pi} = a - bi$ are prime since they have prime norms. It remains to prove that π and $\overline{\pi}$ are not associates. However,

$$\begin{aligned}
a + bi &= a - bi \ \Rightarrow\ b = 0\\
a + bi &= -(a - bi) = -a + bi \ \Rightarrow\ a = 0\\
a + bi &= i(a - bi) = b + ai \ \Rightarrow\ a = b\\
a + bi &= -i(a - bi) = -b - ai \ \Rightarrow\ a = -b.
\end{aligned}$$

The first two are impossible since $p = a^2 + b^2$ implies $a, b \neq 0$, and the last two imply $p^2 = a^2 + b^2 = 2a^2$, impossible since p is odd.

(iii) If $p \equiv 3 \bmod 4$, $p \neq a^2 + b^2$ for any a and b, so $p \neq N(\alpha)$ for any $\alpha \in \mathbb{Z}[i]$. If $p = \pi\rho$ in $\mathbb{Z}[i]$, then $N(p) = p^2 = N(\pi)N(\rho)$. Since $N(\pi) \neq p$ and $N(\rho) \neq p$, one of these must equal 1. Hence either π or ρ is a unit, which proves that p is irreducible, hence prime.

For the final assertion of Proposition 4.18, let α be any prime of $\mathbb{Z}[i]$. Then $N(\alpha) = \alpha\overline{\alpha} \in \mathbb{Z}$, so it factors into a product of integer primes. By (i)–(iii), each of these integer primes is a product of primes of $\mathbb{Z}[i]$ of the forms in (i)–(iii); since α is prime and $\mathbb{Z}[i]$ is a UFD, α is associate to one of these primes.

(d) The analog of Lemma 4.8 is

LEMMA 4.8. *If π is a prime of $\mathbb{Z}[i]$, then the quotient field $\mathbb{Z}[i]/\pi\mathbb{Z}[i]$ is a finite field with $N(\pi)$ elements. Furthermore, $N(\pi) = p$ or p^2 for some integer prime p, and:*

(i) *If $p = 2$ or $p \equiv 1 \bmod 4$, then $N(\pi) = p$ and $\mathbb{Z}/p\mathbb{Z} \simeq \mathbb{Z}[i]/\pi\mathbb{Z}[i]$.*

(ii) *If $p \equiv 3 \bmod 4$, then $N(\pi) = p^2$ and $\mathbb{Z}/p\mathbb{Z}$ is the unique subfield of order p of the field $\mathbb{Z}[i]/\pi\mathbb{Z}[i]$ of p^2 elements.*

PROOF. Since π is prime and $\mathbb{Z}[i]$ is a PID, $\pi\mathbb{Z}[i]$ is maximal by Exercise 4.4, so that $\mathbb{Z}[i]/\pi\mathbb{Z}[i]$ is a field with $N(\pi)$ elements. Further, $N(\pi) = p$ or p^2 by part (c).

If $p = 2$, then π is an associate of $1 + i$, so we may as well assume $\pi = 1 + i$. Thus by part (c), $N(\pi) = 2$ and $\mathbb{Z}[i]/\pi\mathbb{Z}[i] \simeq \mathbb{Z}/2\mathbb{Z}$. If $p \equiv 1 \bmod 4$, then by part (c), $N(\pi) = p$ and $\mathbb{Z}[i]/\pi\mathbb{Z}[i] \simeq \mathbb{Z}/p\mathbb{Z}$. Finally, if $p \equiv 3 \bmod 4$, then part (c) shows that $N(\pi) = p^2$, so that $\mathbb{Z}[i]/\pi\mathbb{Z}[i] \simeq \mathbb{F}_{p^2}$, a finite field with p^2 elements.

In the last case, where $p \equiv 3 \bmod 4$, $p = \pi$ and we can consider $\mathbb{Z}/p\mathbb{Z}$ as a subfield of $\mathbb{Z}[i]/\pi\mathbb{Z}[i]$ under the natural inclusion map $\mathbb{Z} \hookrightarrow \mathbb{Z}[i]$. This proves the last statement, since finite fields of order p^k have a unique subfield congruent to $\mathbb{Z}/p\mathbb{Z}$. Q.E.D.

(e) From part (d), $(\mathbb{Z}[i]/\pi\mathbb{Z}[i])^*$ is a finite group with $N(\pi) - 1$ elements. Since π does not divide a, we have $[a] \in (\mathbb{Z}[i]/\pi\mathbb{Z}[i])^*$. Then $[a]^{N(\pi)-1} = 1 \in (\mathbb{Z}[i]/\pi\mathbb{Z}[i])^*$ by Lagrange's Theorem, so that $a^{N(\pi)-1} \equiv 1 \bmod \pi$.

4.17. If π is not associate to $1 + i$, then by the version of Lemma 4.8 in part (d) of Exercise 4.16, there is an odd prime p such that $N(\pi) = p$ (if $p \equiv 1 \bmod 4$) or p^2 (if $p \equiv 3 \bmod 4$). In the former case, $N(\pi) - 1 = p - 1 \equiv 0 \bmod 4$, and in the latter case, $N(\pi) - 1 = p^2 - 1 \equiv 0 \bmod 4$. Thus $4 \mid N(\pi) - 1$.

If $\pm 1 \equiv \pm i \bmod \pi$, then $\pm 1 \pm i \in \pi\mathbb{Z}[i]$. But $\pm 1 \pm i$ are all associate to $1 + i$, so that $1 + i \in \pi\mathbb{Z}[i]$, a contradiction since π is a prime that is not associate to $1 + i$. If $1 \equiv -1 \bmod \pi$, then $\pi \mid 1 - (-1) = 2$. This implies that $N(\pi) \mid 4$, clearly impossible since $N(\pi) = p$ or p^2 is odd. Finally, if $i \equiv -i \bmod \pi$, then $\pi \mid 2i$. As in the previous case, this implies $N(\pi) \mid 4$, which is a contradiction.

4.18. Recall that we are assuming throughout that π is not associate to $1 + i$, so that $N(\pi) = p$ or p^2 for some odd prime p.

(a) First note that $(N(\pi) - 1)/4 \in \mathbb{Z}$ by the previous exercise. Then $\alpha^{(N(\pi)-1)/4}$ is a root of $x^4 - 1$ modulo π by (4.19). However,

$$x^4 - 1 \equiv (x - 1)(x + 1)(x - i)(x + i) \bmod \pi,$$

where π is prime. Since $\mathbb{Z}[i]/\pi\mathbb{Z}[i]$ is a field and hence an integral domain, $\alpha^{(N(\pi)-1)/4}$ is congruent to one of ± 1, $\pm i$. By Exercise 4.17, these are incongruent modulo π.

(b) Using (4.20),

$$\left(\frac{\alpha\beta}{\pi}\right)_4 \equiv (\alpha\beta)^{(N(\pi)-1)/4} \equiv \alpha^{(N(\pi)-1)/4}\beta^{(N(\pi)-1)/4} \equiv \left(\frac{\alpha}{\pi}\right)_4\left(\frac{\beta}{\pi}\right)_4 \bmod \pi.$$

Since $(\alpha\beta/\pi)_4$, $(\alpha/\pi)_4$, and $(\beta/\pi)_4$ are all fourth roots of unity, and no two different fourth roots of unity are congruent modulo π by the previous exercise, the two sides must be equal. Next, if $\alpha \equiv \beta \bmod \pi$, then $\alpha^{(N(\pi)-1)/4} \equiv \beta^{(N(\pi)-1)/4} \bmod \pi$, so that $(\alpha/\pi)_4 \equiv (\beta/\pi)_4 \bmod \pi$ and again we conclude equality since no two different fourth roots of unity are congruent modulo π.

(c) By (4.20), $(\alpha/\pi)_4 = 1$ is equivalent to saying that $\alpha^{(N(\pi)-1)/4} \equiv 1 \bmod \pi$ since $\pm 1, \pm i$ are incongruent modulo π. Hence it suffices to show that

$$\alpha^{(N(\pi)-1)/4} \equiv 1 \bmod \pi \iff x^4 \equiv \alpha \bmod \pi \text{ is solvable in } \mathbb{Z}[i].$$

The proof is similar to that in Exercise 4.11. Suppose $\pi \nmid \alpha$. If $\beta \in \mathbb{Z}[i]$ with $\beta^4 \equiv \alpha \bmod \pi$, then

$$\alpha^{(N(\pi)-1)/4} \equiv \left(\beta^4\right)^{(N(\pi)-1)/4} = \beta^{N(\pi)-1} \equiv 1 \bmod \pi$$

by (4.19) since $\pi \nmid \alpha$ and therefore $\pi \nmid \beta$. Conversely, assume $\alpha^{(N(\pi)-1)/4} \equiv 1 \bmod \pi$. By part (d) of Exercise 4.16, $\mathbb{Z}[i]/\pi\mathbb{Z}[i]$ is a finite field, so that as noted in the text, $(\mathbb{Z}[i]/\pi\mathbb{Z}[i])^*$ is cyclic and $|(\mathbb{Z}[i]/\pi\mathbb{Z}[i])^*| = N(\pi) - 1$. Choose $\gamma \in \mathbb{Z}[i]$ such that $[\gamma]$ generates $(\mathbb{Z}[i]/\pi\mathbb{Z}[i])^*$. Then $[\alpha] = [\gamma]^k$ for some k, so that $\alpha \equiv \gamma^k \bmod \pi$. It follows that

$$\alpha^{(N(\pi)-1)/4} \equiv \gamma^{k(N(\pi)-1)/4} \equiv 1 \bmod \pi.$$

So we must have $N(\pi) - 1 \mid k(N(\pi) - 1)/4$, whence $4 \mid k$. Let $\beta = \gamma^{k/4}$; then $\beta^4 \equiv \gamma^k \equiv \alpha \bmod \pi$. This proves that $x^4 \equiv \alpha \bmod \pi$ is solvable in $\mathbb{Z}[i]$, as desired.

4.19. (a) In this case, $N(\pi) = p$, so squaring both sides of (4.20) gives

$$\left(\frac{a}{\pi}\right)_4^2 \equiv a^{(N(\pi)-1)/2} = a^{(p-1)/2} \equiv \left(\frac{a}{p}\right) \bmod \pi,$$

where the final congruence follows from $a^{(p-1)/2} \equiv (a/p) \bmod p$ since $\pi \mid p$. Both sides are units in $\mathbb{Z}[i]$ and hence are equal by Exercise 4.17. The second part follows easily.

(b) By multiplicativity of $(\cdot/\pi)_4$, the map $(\mathbb{Z}[i]/\pi\mathbb{Z}[i])^* \to \{\pm 1, \pm i\}$ defined by

$$a \longmapsto \left(\frac{a}{\pi}\right)_4$$

is a group homomorphism. The kernel has order $|(\mathbb{Z}[i]/\pi\mathbb{Z}[i])^*|/4 = (N(\pi) - 1)/4$, as we now prove. By part (c) of Exercise 4.18, the kernel of $(\cdot/\pi)_4$ is the subgroup of fourth powers, which is the image of the fourth power map $(\mathbb{Z}[i]/\pi\mathbb{Z}[i])^* \to (\mathbb{Z}[i]/\pi\mathbb{Z}[i])^*$. The kernel of that fourth power map consists of the four classes represented by $\pm 1, \pm i$ (see the solution to part (a) of Exercise 4.18). It follows that the subgroup of fourth powers, hence the kernel of $(\cdot/\pi)_4$, has index 4 in $(\mathbb{Z}[i]/\pi\mathbb{Z}[i])^*$, as claimed. This shows that $(\cdot/\pi)_4$ is onto and therefore that the fibers of each element in $\{\pm 1, \pm i\}$ have the same size, proving that $(\cdot/\pi)_4$ divides $(\mathbb{Z}[i]/\pi\mathbb{Z}[i])^*$ into four equal classes.

Now assume $p \equiv 1 \bmod 4$, so that $(\mathbb{Z}[i]/\pi\mathbb{Z}[i])^*$ can be identified with $(\mathbb{Z}/p\mathbb{Z})^*$. Then $(\cdot/\pi)_4 : (\mathbb{Z}/p\mathbb{Z})^* \to \{\pm 1, \pm i\}$ divides $(\mathbb{Z}/p\mathbb{Z})^*$ into four equal classes as follows:

- The inverse image of 1 is the kernel, i.e., the subgroup of fourth powers, or more colloquially, the set of biquadratic residues.
- By part (a), the inverse image of -1 consists of quadratic residues that are not biquadratic residues, since the latter comprise the inverse image of 1.
- The remaining two classes are the inverse images of i and $-i$. However,

$$\left(\frac{a}{\pi}\right)_4 = \pm i \iff \left(\frac{a}{\pi}\right)_4^2 = -1 \iff \left(\frac{a}{p}\right) = -1$$
$$\iff a \text{ is not a quadratic residue,}$$

where the second equivalence follows from part (a).

4.20. (a) By (4.20), since $p = \pi$ is prime in $\mathbb{Z}[i]$,

$$\left(\frac{a}{p}\right)_4 \equiv a^{(p^2-1)/4} \bmod p.$$

By Fermat's Little Theorem for $a \in \mathbb{Z}$, since $p + 1 \equiv 0 \bmod 4$ and $p \nmid a$,

$$a^{(p^2-1)/4} = \left(a^{p-1}\right)^{(p+1)/4} \equiv 1 \bmod p$$

and we are done.

(b) Every biquadratic residue is a quadratic residue, so it suffices to show that every quadratic residue is a biquadratic residue. Let R be the multiplicative group of quadratic residues in $(\mathbb{Z}/p\mathbb{Z})^*$ and consider the squaring map $\varphi : R \to R$, which is a group homomorphism. We claim that φ is an injection. If $x \in \ker(\varphi)$, then $x^2 \equiv 1 \bmod p$, so that $x \equiv \pm 1 \bmod p$. However, -1 is not a quadratic residue modulo p since $p \equiv 3 \bmod 4$, so that $-1 \notin R$. Since $x \in R$, $x \equiv \pm 1 \bmod p$ implies $x \equiv 1 \bmod p$, proving that φ is injective, as claimed. Then φ is onto because R is finite. Thus $R = \operatorname{im}(\varphi)$ consists wholly of biquadratic residues, and we are done.

4.21. We first claim that a prime $a + bi$ with a odd and b even is a primary prime if and only if $a + b \equiv 1 \bmod 4$. To see this, note that $a + bi \equiv 1 \bmod 2 + 2i$ means that for $c, d \in \mathbb{Z}$

$$a + bi = (2 + 2i)(c + di) + 1 = (2c - 2d + 1) + (2c + 2d)i.$$

Solving for c and d gives

$$c = \frac{a + b - 1}{4}, \; d = \frac{-a + b + 1}{4}.$$

Then c and d are integers if and only if $a + b \equiv 1 \bmod 4$, proving the claim.

Now suppose that $\pi = a + bi$ is prime. Since it is not associate to $1 + i$, we have $N(\pi) = p$ or p^2 for an odd prime p. Thus $N(\pi) = a^2 + b^2$ is odd, so that a and b have opposite parity; after multiplying by the unit i and relabeling if necessary, we may assume that a is odd and b is even. The associates of π are $a + bi$, $-a - bi$, $i(a + bi) = -b + ai$, and $b - ai$. The latter two cannot be primary since b is even, so it remains to show that exactly one of $a + bi$ and $-a - bi$ is primary. But a odd and b even implies that exactly one of $a + b$ and $-a - b$ is congruent to 1 modulo 4, and the above claim finishes the proof.

4.22. Since π is primary, the analysis of Exercise 4.21 shows that $\pi = a + bi$, where a is odd, b is even, and $a + b \equiv 1 \bmod 4$. Then $b = 1 - a + 4k$ for some integer k. Thus

$$\begin{aligned} \frac{N(\pi) - 1}{4} &= \frac{1}{4}(a^2 + b^2 - 1) = \frac{1}{4}\left(a^2 + (1 - a + 4k)^2 - 1\right) \\ &= \frac{1}{4}\left(a^2 + (1 - a)^2 + 2(1 - a)4k + (4k)^2 - 1\right) \\ &= \frac{a^2 - a}{2} + 4\left(\frac{1 - a}{2}k + k^2\right). \end{aligned}$$

Note that $(1 - a)/2 \in \mathbb{Z}$ since a is odd. Since $i^4 = 1$, (4.20) implies that

$$\left(\frac{i}{\pi}\right)_4 \equiv i^{(N(\pi)-1)/4} = i^{(a^2-a)/2} \bmod \pi.$$

and then $(i/\pi)_4 = i^{(a^2-a)/2}$ by Exercise 4.17. However, a is odd, so that $a^2 \equiv 1 \bmod 8$, giving finally

$$\left(\frac{i}{\pi}\right)_4 = i^{(1-a)/2} = i^{-(a-1)/2}.$$

4.23. Since $2 = i^3(1+i)^2$ and $\pi = a + bi$ is primary, (4.22) implies

$$\left(\frac{2}{\pi}\right)_4 = \left(\frac{i}{\pi}\right)_4^3 \left(\frac{1+i}{\pi}\right)_4^2 = i^{3(-(a-1)/2)} i^{2(a-b-1-b^2)/4} = i^{-3(a-1)/2+(a-b-1-b^2)/2}.$$

To show that this equals $i^{ab/2}$, it suffices to show that the difference of the exponents is a multiple of 4. A simple computation shows that

$$-\frac{3(a-1)}{2} + \frac{a-b-1-b^2}{2} - \frac{ab}{2} = \frac{1}{2}(b+2)(1-a-b).$$

Since b is even, $b+2$ is even, and from the solution of Exercise 4.21 we have $a+b \equiv 1 \bmod 4$, so that $1-a-b \equiv 0 \bmod 4$. The result follows.

4.24. (a) Since $p \equiv 1 \bmod 4$, we know that $p = a^2 + b^2$, where we may assume that a is odd and positive. Since $\gcd(a,p) = 1$, quadratic reciprocity for the Jacobi symbol (see (1.16)) gives

$$\left(\frac{a}{p}\right) = \left(\frac{p}{a}\right) = \left(\frac{a^2+b^2}{a}\right) = \left(\frac{b^2}{a}\right) = 1.$$

(b) When $p = a^2 + b^2$ with $a > 0$ odd, note that b must be even, and we can also assume $b > 0$. Then $a+b$ is odd and positive, and prime to p, so again using quadratic reciprocity for the Jacobi symbol,

$$\left(\frac{a+b}{p}\right) = \left(\frac{p}{a+b}\right) = \left(\frac{4p}{a+b}\right) = \left(\frac{2}{a+b}\right)\left(\frac{2p}{a+b}\right).$$

As noted in the statement of the exercise, $2p = (a+b)^2 + (a-b)^2$. Thus

$$\begin{aligned}\left(\frac{a+b}{p}\right) &= \left(\frac{2}{a+b}\right)\left(\frac{(a+b)^2+(a-b)^2}{a+b}\right)\\ &= \left(\frac{2}{a+b}\right)\left(\frac{(a-b)^2}{a+b}\right) = \left(\frac{2}{a+b}\right) = (-1)^{((a+b)^2-1)/8},\end{aligned}$$

where the last equality is the supplementary law for the Jacobi symbol (see (1.16)).

(c) Since $p \equiv 1 \bmod 4$, we have $N(\pi) = p = a^2 + b^2$. Also, $(i/\pi)_4 \equiv i^{(N(\pi)-1)/4} \bmod \pi$ by (4.20), which by Exercise 4.17 implies $(i/\pi)_4 = i^{(N(\pi)-1)/4}$. Then

$$\begin{aligned}\left(\frac{i}{\pi}\right)_4 i^{ab/2} &= i^{(N(\pi)-1)/4} i^{ab/2} = i^{(a^2+b^2-1)/4+(ab)/2} = i^{((a+b)^2-1)/4}\\ &= (-1)^{((a+b)^2-1)/8} = \left(\frac{a+b}{p}\right),\end{aligned}$$

where the final equality is the result from part (b).

(d) (i) Note that $p = a^2 + b^2$ implies $(a+b)^2 \equiv 2ab \bmod p$. Since $p \equiv 1 \bmod 4$, raising both sides of to the $(p-1)/4$ power gives the desired result.

(ii) Since the congruence in part (i) holds modulo $p = \pi\overline{\pi}$, it also holds modulo π, so that by (4.20),

$$\left(\frac{2ab}{\pi}\right)_4 \equiv (2ab)^{(N(\pi)-1)/4} = (2ab)^{(p-1)/4} \equiv (a+b)^{(p-1)/2} \bmod \pi.$$

For the same reason, $(a+b/p) \equiv (a+b)^{(p-1)/2} \bmod p$ gives a congruence modulo π, with the result that $(2ab/\pi)_4 \equiv (a+b/p) \bmod \pi$. Since the fourth roots of unity are incongruent modulo π by Exercise 4.17, we conclude that

$$\left(\frac{2ab}{\pi}\right)_4 = \left(\frac{a+b}{p}\right).$$

(e) Note that $2ab-2a^2i = 2a(b-ai) = 2a(a+bi)(-i) = -2ai\pi$, so that $2ab \equiv 2a^2 i \bmod \pi$. It follows that

$$\left(\frac{2ab}{\pi}\right)_4 = \left(\frac{2a^2 i}{\pi}\right)_4 = \left(\frac{a^2}{\pi}\right)_4 \left(\frac{2i}{\pi}\right)_4.$$

By Exercise 4.19 and part (a), $(a^2/\pi)_4 = (a/\pi)_4^2 = (a/p) = 1$, so that

$$\left(\frac{2ab}{\pi}\right)_4 = \left(\frac{2i}{\pi}\right)_4.$$

(f) Combining parts (c), (d)(ii), and (e), we have

$$\left(\frac{i}{\pi}\right)_4 i^{ab/2} \overset{\text{(c)}}{=} \left(\frac{a+b}{p}\right) \overset{\text{(d)(ii)}}{=} \left(\frac{2ab}{\pi}\right)_4 \overset{\text{(e)}}{=} \left(\frac{2i}{\pi}\right)_4 = \left(\frac{i}{\pi}\right)_4 \left(\frac{2}{\pi}\right)_4.$$

Dividing by $(i/\pi)_4$ gives the desired equality $(2/\pi)_4 = i^{ab/2}$.

4.25. (a) If π is primary, then as in the solution to Exercise 4.21, $\pi = a + bi$, where a is odd, b is even, and $a + b \equiv 1 \bmod 4$. If $a \equiv 1 \bmod 4$, then $a + b \equiv 1 \bmod 4$ implies $b \equiv 0 \bmod 4$, and $\pi \equiv 1 \bmod 4$ follows. On the other hand, if $a \equiv 3 \bmod 4$, then $a + b \equiv 1 \bmod 4$ implies $b \equiv 2 \bmod 4$, and $\pi \equiv 3 + 2i \bmod 4$ follows.

(b) Let

$$\mu = (-1)^{(N(\pi)-1)(N(\theta)-1)/16} = (-1)^{(N(\pi)-1)/4\cdot(N(\theta)-1)/4}.$$

Comparing Theorem 4.21 to Gauss' statement of biquadratic reciprocity, proving that they are equivalent amounts to showing that

$$\begin{aligned} &\pi \equiv 1 \bmod 4 \text{ or } \theta \equiv 1 \bmod 4 \iff \mu = 1 \\ &\pi, \theta \equiv 3 \bmod 4 \iff \mu = -1. \end{aligned}$$

It suffices to prove the first equivalence. For this, first note that $\pi \equiv 1 \bmod 4$ implies $\pi = 1 + 4\delta$ for $\delta \in \mathbb{Z}[i]$, so that

$$N(\pi) - 1 = (1 + 4\delta)(1 + 4\overline{\delta}) - 1 = 4(\delta + \overline{\delta}) + 16N(\delta).$$

Since $\delta \in \mathbb{Z}[i]$ implies $\delta + \overline{\delta} \in 2\mathbb{Z}$, we see that $(N(\pi) - 1)/4$ is even. On the other hand, suppose $\pi \equiv 3 \bmod 4$. Then $\pi = 3 + 2i + 4\delta$ for some $\delta \in \mathbb{Z}[i]$, so that

$$\begin{aligned} N(\pi) - 1 &= (3 + 2i + 4\delta)(3 - 2i + 4\overline{\delta}) - 1 \\ &= (3 + 2i)(3 - 2i) + (3 + 2i)4\overline{\delta} + (3 - 2i)4\delta + (4\delta)(4\overline{\delta}) - 1 \\ &= 4 + 12(\delta + \overline{\delta}) - 8(\delta - \overline{\delta}) + 16N(\delta). \end{aligned}$$

As above, $\delta + \overline{\delta}$ is even, so that $(N(\pi) - 1)/4$ is odd. These considerations imply that

$$\pi \equiv 1 \bmod 4 \iff (N(\pi) - 1)/4 \text{ is even.}$$

Thus

$$\begin{aligned} \mu = 1 &\iff (N(\pi) - 1)/4 \cdot (N(\theta) - 1)/4 \text{ is even} \\ &\iff (N(\pi) - 1)/4 \text{ or } N(\theta) - 1)/4 \text{ is even} \\ &\iff \pi \equiv 1 \bmod 4 \text{ or } \theta \equiv 1 \bmod 4. \end{aligned}$$

4.26. If 2 is a biquadratic residue, then it is surely a quadratic residue. But for p an odd prime, $(2/p) = 1$ only for $p \equiv \pm 1 \bmod 8$.

4.27. As in the statement of the exercise, write $p = a^2 + 2b^2$. Then a, b are nonzero, so we can assume $a, b > 0$.

(a) If $p \equiv 1 \bmod 8$, then $N(\pi) = p$, so that by (4.20),

$$\left(\frac{-1}{\pi}\right)_4 \equiv (-1)^{(N(\pi)-1)/4} = (-1)^{(p-1)/4} = 1 \bmod \pi,$$

which implies $(-1/\pi)_4 = 1$ since $\pm 1, \pm i$ are incongruent modulo π by Exercise 4.17.

(b) Note that a is odd and $\gcd(a,p) = 1$. Since $p \equiv 1 \bmod 4$, quadratic reciprocity for the Jacobi symbol implies

$$\left(\frac{a}{p}\right) = \left(\frac{p}{a}\right) = \left(\frac{a^2+2b^2}{a}\right) = \left(\frac{2b^2}{a}\right) = \left(\frac{2}{a}\right)\left(\frac{b^2}{a}\right) = \left(\frac{2}{a}\right) = (-1)^{(a^2-1)/8}.$$

(c) Writing $b = 2^m c$ with c odd gives $p = a^2 + 2(2^m c)^2 = a^2 + 2^{2m+1}c^2$. Recall that $(2/p) = 1$ since $p \equiv 1 \bmod 8$, and so by quadratic reciprocity for the Jacobi symbol,

$$\left(\frac{b}{p}\right) = \left(\frac{2^m}{p}\right)\left(\frac{c}{p}\right) = \left(\frac{c}{p}\right) = \left(\frac{p}{c}\right) = \left(\frac{a^2+2^{2m+1}c^2}{c}\right) = \left(\frac{a^2}{c}\right) = 1.$$

(d) Since $(b/\pi)_4^2 = (b/p) = 1$ by Exercise 4.19 and part (c) above, and $(-1/\pi)_4 = 1$ from part (a), we get

$$\left(\frac{2}{\pi}\right)_4 = \left(\frac{-1}{\pi}\right)_4\left(\frac{b^2}{\pi}\right)_4\left(\frac{2}{\pi}\right)_4 = \left(\frac{-2b^2}{\pi}\right)_4 = \left(\frac{p-2b^2}{\pi}\right)_4 = \left(\frac{a^2}{\pi}\right)_4 = \left(\frac{a}{p}\right),$$

where we use $\pi \mid p$ in the third equality, $p = a^2 + 2b^2$ in the fourth, and Exercise 4.19 in the last.

From parts (b) and (d), we conclude that

$$\left(\frac{2}{\pi}\right)_4 = (-1)^{(a^2-1)/8},$$

which gives Gauss' result that 2 is a biquadratic residue modulo $p = a^2 + 2b^2$ if and only if $a \equiv \pm 1 \bmod 8$. Note by the way that the representation $p = a^2 + 2b^2$ is unique for $a, b \geq 0$ by Lemma 3.25.

4.28. Since $\mu_j = \mu g^{ej}$, expanding the right-hand side gives:

$$\begin{aligned}\sum_{j=1}^{f}(f, \lambda + \mu_j) &= \sum_{j=1}^{f}\sum_{k=1}^{f} \zeta_p^{(\lambda+\mu_j)g^{ek}} = \sum_{j=1}^{f}\sum_{k=1}^{f} \zeta_p^{\lambda g^{ek}} \zeta_p^{\mu_j g^{ek}} \\ &= \sum_{j=1}^{f}\sum_{k=1}^{f} \zeta_p^{\lambda g^{ek}} \zeta_p^{\mu g^{ej} g^{ek}} = \sum_{j=1}^{f}\sum_{k=1}^{f} \zeta_p^{\lambda g^{ek}} \zeta_p^{\mu g^{e(j+k)}} = \sum_{k=1}^{f}\sum_{j=1}^{f} \zeta_p^{\lambda g^{ek}} \zeta_p^{\mu g^{e(j+k)}}.\end{aligned}$$

For fixed k, as j varies from 1 to f, $j + k$ ranges over a full set of congruence classes modulo f, so the above sum is the same as

$$\sum_{k=1}^{f}\sum_{j=1}^{f} \zeta_p^{\lambda g^{ek}} \zeta_p^{\mu g^{ej}} = \left(\sum_{k=1}^{f} \zeta_p^{\lambda g^{ek}}\right)\left(\sum_{j=1}^{f} \zeta_p^{\mu g^{ej}}\right) = (f, \lambda)(f, \mu).$$

4.29. (a) We first show that any pair (m, n) with $0 \leq m, n \leq f - 1$ such that $1 + g^{3m} \equiv g^{3n} \bmod p$ gives nine distinct solutions

$$(x, y) = (g^{n+kf}, g^{m+\ell f}), \quad k, \ell = 0, 1, 2,$$

to $x^3 - y^3 \equiv 1 \bmod p$. To see this, note that $g^{p-1} \equiv 1 \bmod p$ by Fermat's Little Theorem. Since $3f = p - 1$, we get

$$g^{3(n+kf)} - g^{3(m+\ell f)} = g^{3n+k(p-1)} - g^{3m+\ell(p-1)} \equiv g^{3n} - g^{3m} \equiv 1 \bmod p.$$

Conversely, suppose $x^3 - y^3 \equiv 1 \bmod p$ and $x, y \not\equiv 0 \bmod p$. Since g is a primitive root modulo p, we can choose integers $0 \leq u, v < p - 1$ such that $x \equiv g^u \bmod p$ and

$y \equiv g^v \bmod p$. Dividing by f gives $u = n + fk$, $v = m + \ell k$ with $0 \le n, m \le f - 1$, and $k, \ell = 0, 1, 2$ since $0 \le u, v < p - 1 = 3f$. The pair (m, n) is in the set counted by (00) and (x, y) is one of the nine solutions associated to (m, n). Finally, there are three solutions to the equation when $x \equiv 0 \bmod p$ and three when $y \equiv 0 \bmod p$, namely the cube roots of unity in $(\mathbb{Z}/p\mathbb{Z})^*$ (see Exercise 4.14). So altogether there are $9(00) + 6$ solutions modulo p to the congruence $x^3 - y^3 \equiv 1 \bmod p$.

(b) From Exercise 4.28, we have

$$(*) \qquad (f, 1) \cdot (f, 1) = (f, 1) \cdot \sum_{j=1}^{f} \zeta_p^{g^{3j}} = (f, 1 + g^3) + (f, 1 + g^6) + \cdots + (f, 1 + g^{3f}).$$

Suppose $1 + g^{3j} \equiv 1 + g^{3k} \bmod p$. Then $g^{3(j-k)} \equiv 1 \bmod p$, so that $j - k$ is a multiple of f and thus $j = k$. So the $1 + g^{3j}$ are all distinct modulo p.

Note that since $p - 1 = 3f$, the period $(f, \lambda) = \sum_{j=0}^{f-1} \zeta_p^{\lambda g^{3j}}$ satisfies

$$(**) \qquad (f, \lambda) = (f, g^{3i}\lambda) \text{ for any } i.$$

Also observe that f is even since p is odd. When $j = f/2$, $1 + g^{3j} = 0$ since $g^{3j} = g^{(p-1)/2} = -1$. Thus $(f, 1 + g^{3f/2}) = (f, 0)$. For all other j, since $(\mathbb{Z}/p\mathbb{Z})^*$ is cyclic with generator g, $1 + g^{3j} \equiv g^{3i+k} \bmod p$ for some $0 \le i \le f - 1$, $k \in \{0, 1, 2\}$, so that

$$(f, 1 + g^{3j}) = (f, g^{3i+k}) = (f, g^{3i}g^k) = (f, g^k),$$

where the final equality uses (**). There are (00) values of j corresponding to $k = 0$, since that means $1 + g^{3j} \equiv g^{3i} \bmod p$, and in this case $(f, g^k) = (f, 1)$. Similarly, there are (01) values of j corresponding to $k = 1$, since that means $1 + g^{3j} \equiv g^{3i+1} \bmod p$, and in this case $(f, g^k) = (f, g)$. Finally, there are (02) values of j corresponding to $k = 2$, since that means $1 + g^{3j} \equiv g^{3i+2} \bmod p$, and in this case $(f, g^k) = (f, g^2)$. Thus, in the right-hand side of (*), we see that $(f, 0) = f$ occurs exactly once and (f, g^k) occurs exactly $(0k)$ times for $k = 0, 1, 2$, proving that

$$\begin{aligned}(f, 1) \cdot (f, 1) &= (f, 0) + (00)(f, 1) + (01)(f, g) + (02)(f, g^2) \\ &= f + (00)(f, 1) + (01)(f, g) + (02)(f, g^2).\end{aligned}$$

Since there are f summands in (*),

$$1 + (00) + (01) + (02) = f, \text{ i.e., } (00) + (01) + (02) = f - 1.$$

Similarly,

$$(f, 1) \cdot (f, g) = (f, 1) \cdot \sum_{j=1}^{f} \zeta_p^{g^{3j+1}} = (f, 1 + g^4) + (f, 1 + g^7) + \cdots + (f, g^{3f+1}).$$

As above, each $1 + g^{3j+1}$ is congruent modulo p to g^{3i+k} for $k \in \{0, 1, 2\}$; there is no exponent equal to $p - 1$, so there is no $(f, 0)$ term. Then using the same analysis as above, we see that the sum is equal to

$$(f, 1) \cdot (f, g) = (10)(f, 1) + (11)(f, g) + (12)(f, g^2),$$

and the number of summands is

$$(10) + (11) + (12) = f.$$

(c) Using Exercise 4.28 as we did in (*) gives

$$(f, g) \cdot (f, 1) = (f, g + g^3) + (f, g + g^6) + \cdots + (f, g + g^{3f}).$$

which can be written

$$(f, g) \cdot (f, 1) = (f, g(1 + g^2)) + \cdots + (f, g(1 + g^{3(f-2)+2})) + (f, g(1 + g^{3(f-1)+2})).$$

As above, each element of the form $1+g^{3m+2}$ is congruent modulo p to one of g^{3n}, g^{3n+1}, or g^{3n+2}, and $(f, g\cdot g^{3n+j}) = (f, g^{j+1})$ by (**). So we get

$$(f,g)\cdot(f,1) = (22)(f,1) + (20)(f,g) + (21)(f,g^2).$$

Now, $\zeta_p, \zeta_p^2, \ldots, \zeta_p^{p-1}$ are linearly independent over $\mathbb{Q}$. To see this, note that writing $0 = \sum_{i=1}^{p-1} a_i\zeta_p^i = \zeta_p\big(\sum_{i=1}^{p-1} a_i\zeta_p^{i-1}\big)$ produces a polynomial $\sum_{i=1}^{p-1} a_i x^{i-1}$ of degree $p-2$ with ζ_p as a root, which is a contradiction since the minimal polynomial of ζ_p over $\mathbb{Q}$ has degree $p-1$. This proves the desired linear independence, which implies that (f,g), (f,g^2), and $(f,1)$ are also linearly independent over $\mathbb{Q}$. Comparing this to the result of part (b) shows that $(22)=(10)$, $(20)=(11)$, and $(21)=(12)$.

(d) We first prove $(10)=(01)$. As noted in the solution to part (b), $g^{3f/2} \equiv -1 \bmod p$. Now suppose that (m,n) is a pair counted by (10). Then $1+g^{3m+1} \equiv g^{3n} \bmod p$ and $0 \le m,n \le f-1$. Define a pair (n',m') as follows:

$$\begin{aligned} n + f/2 &\equiv n' \bmod f,\ 0 \le n' \le f-1\\ m + f/2 &\equiv m' \bmod f,\ 0 \le m' \le f-1. \end{aligned}$$

We prove that (n',m') is counted by (01) as follows. From $1+g^{3m+1} \equiv g^{3n} \bmod p$ we get $1-g^{3n} \equiv -g^{3m+1} \bmod p$. However,

$$-g^{3n} \equiv g^{3f/2}g^{3n} = g^{3(n+f/2)} \equiv g^{3n'} \bmod p$$

since $3(n+f/2) \equiv 3n' \bmod p$ by the definition of n' (remember that $p-1=3f$). In a similar way we get $-g^{3m+1} \equiv g^{3m'+1} \bmod p$, so that

$$1+g^{3n'} \equiv 1-g^{3n} \equiv -g^{3m+1} \equiv g^{3m'+1} \bmod p,$$

proving that (n',m') is indeed counted by (01). The same construction applies to pairs counted by (01), which shows that the map sending $(m,n) \mapsto (n',m')$ is a bijection, proving that $(10)=(01)$. A similar proof shows that $(20)=(02)$. Finally, combining

$$(00)+(01)+(02)+1 = (12)+(10)+(20) = f$$

from part (b) with $(10)=(01)$, $(20)=(02)$ and the equalities proved in part (c), we get $(00)+1=(12)$, giving the desired formulas for α, β and γ.

(e) The product is an integer since these three are roots of an irreducible integer cubic. Write $u=(f,1)$, $u'=(f,g)$, and $u''=(f,g^2)$. In this notation, the formulas from parts (b) and (d) can be written

$$\begin{aligned} u^2 &= f + (\alpha-1)u + \beta u' + \gamma u''\\ uu' &= \beta u + \gamma u' + \alpha u''. \end{aligned}$$

Furthermore, a computation similar to part (b) shows that

$$u'u'' = \alpha u + \beta u' + \gamma u''.$$

Since $u+u'+u''$ is the sum of the nontrivial pth-roots of unity, we have $u+u'+u'' = -1$. Furthermore, the solution to part (c) shows that u, u', u'' are linearly independent over $\mathbb{Q}$. Since their sum is -1, it follows that $1, u, u'$ are also linearly independent over $\mathbb{Q}$. In terms of $1, u, u'$, the above formulas become

$$\begin{aligned} u^2 &= f - \gamma + (\alpha - 1 - \gamma)u + (\beta-\gamma)u'\\ uu' &= -\alpha + (\beta-\alpha)u + (\gamma-\alpha)u'\\ u'u'' &= -\gamma + (\alpha-\gamma)u + (\beta-\gamma)u'. \end{aligned}$$

Then

$$\begin{aligned} uu'u'' &= u(u'u'') \\ &= u\big(-\gamma + (\alpha-\gamma)u + (\beta-\gamma)u'\big) = -\gamma u + (\alpha-\gamma)u^2 + (\beta-\gamma)uu' \\ &= -\gamma u + (\alpha-\gamma)\big(f - \gamma + (\alpha-1-\gamma)u + (\beta-\gamma)u'\big) \\ &\qquad + (\beta-\gamma)\big(-\alpha + (\beta-\alpha)u + (\gamma-\alpha)u'\big) \\ &= (\cdots)\cdot 1 + \Big(-\gamma + (\alpha-\gamma)(\alpha-1-\gamma) + (\beta-\gamma)(\beta-\alpha)\Big)\cdot u + (\cdots)\cdot u' \end{aligned}$$

The quantities inside the parentheses are integers, and since $uu'u''$ is an integer, linear independence implies that the coefficients of u and u' must vanish. For the coefficient of u, this implies

$$-\gamma + (\alpha^2 - \alpha - \alpha\gamma - \gamma\alpha + \gamma + \gamma^2) + (\beta^2 - \beta\gamma - \beta\alpha + \gamma\alpha) = 0.$$

which easily implies the desired equation

$$\alpha^2 + \beta^2 + \gamma^2 - \alpha = \alpha\beta + \beta\gamma + \alpha\gamma.$$

(f) Expanding $(6\alpha - 3\beta - 3\gamma - 2)^2 + 27(\beta-\gamma)^2$ gives

$$\begin{aligned} &36\alpha^2 + 9\beta^2 + 9\gamma^2 + 4 - 36\alpha\beta - 36\alpha\gamma - 24\alpha + 18\beta\gamma + 12\beta + 12\gamma + 27(\beta^2 - 2\beta\gamma + \gamma^2) \\ &\quad = 36(\alpha^2 + \beta^2 + \gamma^2 - \alpha) + 12(\alpha + \beta + \gamma) + 4 - 36(\alpha\beta + \alpha\gamma + \beta\gamma). \end{aligned}$$

By part (e), we can substitute $\alpha\beta + \beta\gamma + \alpha\gamma$ for $\alpha^2 + \beta^2 + \gamma^2 - \alpha$, which gives the desired answer $12(\alpha + \beta + \gamma) + 4$.

(g) Since $p - 1 = 3f$, we have $p = 3f + 1 = 3(\alpha + \beta + \gamma) + 1$, so that by part (f),

$$4p = 12(\alpha + \beta + \gamma) + 4 = (6\alpha - 3\beta - 3\gamma - 2)^2 + 27(\beta - \gamma)^2 = a^2 + 27b^2,$$

where $a = 6\alpha - 3\beta - 3\gamma - 2$ and $b = \beta - \gamma$.

(h) First,

$$a = 6\alpha - 3\beta - 3\gamma - 2 = 9\alpha - 3\alpha - 3\beta - 3\gamma - 2 = 9\alpha - 3(\alpha + \beta + \gamma) - 2.$$

Since $\alpha + \beta + \gamma = f$ and $p = 3f - 1$, this simplifies to

$$a = 9\alpha - 3(\alpha + \beta + \gamma) - 2 = 9\alpha - 3f - 2 = 9\alpha - p - 1.$$

Finally, $N = 9(00) + 6$ from part (a), and $\alpha = (00) + 1$ from part (d), so that

$$a = 9\alpha - p - 1 = 9(00) + 9 - p - 1 = (9(00) + 6) - p + 2 = N - p + 2,$$

so that $N = p + a - 2$.

Solutions to Exercises in §5

5.1. (a) Suppose that $x^n + a_1x^{n-1} + \cdot + a_n$ is the minimal polynomial for $\alpha \in \mathfrak{a}$ over $\mathbb{Q}$; then the $a_i \in \mathbb{Z}$ since $\alpha \in \mathcal{O}_K$. Note that $a_n \neq 0$ since the polynomial is irreducible and $\alpha \neq 0$. Thus

$$\alpha^n + a_1\alpha^{n-1} + \cdots + a_{n-1}\alpha + a_n = 0,$$

so that

$$a_n = -\left(\alpha^n + a_1\alpha^{n-1} + \cdots + a_{n-1}\alpha\right) = -\alpha\left(\alpha^{n-1} + a_1\alpha^{n-2} + \cdots + a_{n-1}\right) \in \mathfrak{a}$$

since $\mathfrak{a}$ is an ideal.

(b) Suppose that $[K : \mathbb{Q}] = d$. Then $\mathcal{O}_K \simeq \mathbb{Z}^d$ as a $\mathbb{Z}$-module by Proposition 5.3. Using part (a), choose $m \neq 0$ in $\mathbb{Z} \cap \mathfrak{a}$. Then $\mathcal{O}_K/m\mathcal{O}_K \simeq (\mathbb{Z}/m\mathbb{Z})^d$ as a $\mathbb{Z}$-module, so it is finite. Since $m \in \mathfrak{a}$, there is a natural surjection $\mathcal{O}_K/m\mathcal{O}_K \to \mathcal{O}_K/\mathfrak{a}$, which is therefore finite as well.

(c) If $\mathfrak{a}$ is a nonzero ideal of $\mathcal{O}_K$, it is a subgroup of a free Abelian group of finite rank, so is itself free Abelian of finite rank. Since $\mathcal{O}_K/\mathfrak{a}$ is finite, the rank of $\mathfrak{a}$ must be the same as the rank of $\mathcal{O}_K$, which is $[K:\mathbb{Q}]$.

(d) $\mathcal{O}_K/\mathfrak{a}_i$ is finite for every $i \geq 1$ by part (b). Additionally, for each $i \geq 1$, $\mathcal{O}_K/\mathfrak{a}_{i+1}$ is a quotient of $\mathcal{O}_K/\mathfrak{a}_i$, so that $|\mathcal{O}_K/\mathfrak{a}_i| \geq |\mathcal{O}_K/\mathfrak{a}_{i+1}|$. Thus there must be some n such that $|\mathcal{O}_K/\mathfrak{a}_n| = |\mathcal{O}_K/\mathfrak{a}_{n+1}| = \cdots$. It follows that for every $i \geq n$, the surjection $\mathcal{O}_K/\mathfrak{a}_i \to \mathcal{O}_K/\mathfrak{a}_{i+1}$ is an isomorphism. Since $\mathfrak{a}_i \subset \mathfrak{a}_{i+1}$, this implies that $\mathfrak{a}_i = \mathfrak{a}_{i+1}$, and $\mathfrak{a}_n = \mathfrak{a}_{n+1} = \cdots$ follows.

(e) Since $\mathfrak{a}$ is prime, $\mathcal{O}_K/\mathfrak{a}$ is an integral domain, finite by part (b) since $\mathfrak{a}$ is nonzero. But any finite integral domain is a field by a standard result from abstract algebra. Thus $\mathfrak{a}$ is a maximal ideal.

5.2. (a) We first observe that ideals of $\mathcal{O}_K$ are simply $\mathcal{O}_K$-modules contained in $\mathcal{O}_K$ where the action of $\mathcal{O}_K$ is given by multiplication.

Now suppose $\mathfrak{a}$ is a fractional ideal of K. Since K is the field of fractions of $\mathcal{O}_K$, we can write the generators of $\mathfrak{a}$ as α_i/β_i, $\alpha_i, \beta_i \in \mathcal{O}_K$, $\beta_i \neq 0$, $1 \leq i \leq r$. Let $\beta = \prod_{i=1}^r \beta_i \neq 0$. Then

$$\beta\mathfrak{a} = \beta\Big\{\sum_{i=1}^r \gamma_i \frac{\alpha_i}{\beta_i} \mid \gamma_i \in \mathcal{O}_K,\ 1 \leq i \leq r\Big\} = \Big\{\sum_{i=1}^r \gamma_i \alpha_i \frac{\beta}{\beta_i} \mid \gamma_i \in \mathcal{O}_K,\ 1 \leq i \leq r\Big\}.$$

Since $\alpha_i(\beta/\beta_i) \in \mathcal{O}_K$ for all i by the definition of β, it follows that $\beta\mathfrak{a}$ is a submodule of $\mathcal{O}_K$ generated by the $\alpha_i(\beta/\beta_i)$, so that it is an ideal of $\mathcal{O}_K$, which we denote $\mathfrak{b}$. Then $\beta\mathfrak{a} = \mathfrak{b}$ implies $\mathfrak{a} = \alpha\mathfrak{b}$ for $\alpha = 1/\beta$. For the converse, if $\mathfrak{b}$ is an ideal of $\mathcal{O}_K$, then $\mathfrak{b}$, and therefore $\alpha\mathfrak{b}$, is an $\mathcal{O}_K$-module. Finally, $\mathfrak{b}$ is finitely generated as a $\mathbb{Z}$-module by part (c) of Exercise 5.1. Hence $\mathfrak{b}$ is also finitely generated as an $\mathcal{O}_K$-module, and then the same is true for $\alpha\mathfrak{b}$.

(b) By part (a), a fractional ideal $\mathfrak{a}$ can be written $\mathfrak{a} = \alpha\mathfrak{b}$, where $\alpha \in K$ and $\mathfrak{b} \subset \mathcal{O}_K$ is an ideal. Note that α and $\mathfrak{b}$ are nonzero since fractional ideals are nonzero by definition. By part (c) of Exercise 5.1, there is a $\mathbb{Z}$-module isomorphism $\mathfrak{b} \simeq \mathbb{Z}^d$, where $d = [K:\mathbb{Q}]$. Since $\alpha \neq 0$, we also have the $\mathbb{Z}$-module isomorphism $\mathfrak{a} \simeq \mathfrak{b}$ defined by $\alpha u \mapsto u$. Composing these gives a $\mathbb{Z}$-module isomorphism $\mathfrak{a} \simeq \mathbb{Z}^d$, so that $\mathfrak{a}$ is a free $\mathbb{Z}$-module of rank $d = [K:\mathbb{Q}]$.

(c) Suppose $\alpha\mathfrak{a}$ and $\beta\mathfrak{b}$ are two fractional ideals, where $\mathfrak{a}, \mathfrak{b}$ are ideals of $\mathcal{O}_K$. Let $\mathfrak{c} = \mathfrak{a}\mathfrak{b}$, so that $\mathfrak{c}$ is an $\mathcal{O}_K$-ideal. Then $(\alpha\mathfrak{a})(\beta\mathfrak{b}) = (\alpha\beta)\mathfrak{a}\mathfrak{b} = (\alpha\beta)\mathfrak{c}$, which is a fractional ideal by part (a).

5.3. (a) First note that $\sigma : L \to L$ induces an automorphism $\sigma : \mathcal{O}_L \to \mathcal{O}_L$ that takes $\mathfrak{P}$ to $\sigma(\mathfrak{P})$ and is the identity on $\mathcal{O}_K$. This gives an isomorphism of finite fields $\mathcal{O}_L/\mathfrak{P} \simeq \mathcal{O}_L/\sigma(\mathfrak{P})$ which is the identity on the subfield $\mathcal{O}_K/\mathfrak{p}$. Thus

$$f_{\mathfrak{P}|\mathfrak{p}} = [\mathcal{O}_L/\mathfrak{P} : \mathcal{O}_K/\mathfrak{p}] = [\mathcal{O}_L/\sigma(\mathfrak{P}) : \mathcal{O}_K/\mathfrak{p}] = f_{\sigma(\mathfrak{P})|\mathfrak{p}}.$$

To prove $e_{\mathfrak{P}|\mathfrak{p}} = e_{\sigma(\mathfrak{P})|\mathfrak{p}}$, let the primes of L lying over $\mathfrak{p}$ be $\mathfrak{P}_1, \ldots, \mathfrak{P}_r$ with $\mathfrak{P} = \mathfrak{P}_1$. For simplicity, write $e_i = e_{\mathfrak{P}_i|\mathfrak{p}}$. Since σ is the identity on K and $\mathfrak{p}$ is an ideal of $\mathcal{O}_K$, we have

$$\prod_{i=1}^r \mathfrak{P}_i^{e_i} = \mathfrak{p}\mathcal{O}_L = \sigma(\mathfrak{p}\mathcal{O}_L) = \prod_{i=1}^r \sigma(\mathfrak{P}_i)^{e_i}.$$

By unique factorization of ideals, $\sigma(\mathfrak{P}) = \sigma(\mathfrak{P}_1) = \mathfrak{P}_j$ for some j, and it follows that $e_1 = e_j$. Then

$$e_{\sigma(\mathfrak{P})|\mathfrak{p}} = e_{\mathfrak{P}_j|\mathfrak{p}} = e_j = e_1 = e_{\mathfrak{P}_1|\mathfrak{p}} = e_{\mathfrak{P}|\mathfrak{p}}.$$

(b) By part (i) of Theorem 5.9, $\mathrm{Gal}(L/K)$ acts transitively on the primes $\mathfrak{P}_1, \ldots, \mathfrak{P}_r$ of L lying over $\mathfrak{p}$. By part (a) of this exercise, it follows that the $\mathfrak{P}_i$ all have the same ramification index e and inertial degree f over $\mathfrak{p}$. Then using Theorem 5.8,

$$n = [L : K] = \sum_{i=1}^{g} e_i f_i = \sum_{i=1}^{g} ef = efg.$$

5.4. (a) If $\sigma \in I_{\mathfrak{P}}$ and $x \in \mathfrak{P}$, then $\sigma(x) \equiv x \equiv 0 \bmod \mathfrak{P}$, so that $\sigma(x) \in \mathfrak{P}$. Thus $\sigma(\mathfrak{P}) \subset \mathfrak{P}$. But $\mathfrak{P}$ and hence $\sigma(\mathfrak{P})$ are maximal ideals, so that $\sigma(\mathfrak{P}) = \mathfrak{P}$. Hence $\sigma \in D_{\mathfrak{P}}$.

(b) If $\sigma \in D_{\mathfrak{P}}$, then $\sigma(\mathfrak{P}) = \mathfrak{P}$, so σ induces an automorphism $\tilde{\sigma}$ of $\mathcal{O}_L/\mathfrak{P}$. Now, σ is the identity on $\mathcal{O}_K$, and since $\mathfrak{p} \subset \mathfrak{P}$, σ preserves $\mathfrak{p}$, so that $\tilde{\sigma}$ restricts to the identity automorphism of $\mathcal{O}_K/\mathfrak{p}$.

(c) Take $\sigma \in D_{\mathfrak{P}}$. Since $\tilde{\sigma}(\alpha + \mathfrak{P}) = \sigma(\alpha) + \mathfrak{P}$ for $\alpha \in \mathcal{O}_L$, we have

$$\begin{aligned} \sigma \in I_{\mathfrak{P}} &\iff \sigma(\alpha) \equiv \alpha \bmod \mathfrak{P} \text{ for all } \alpha \iff \sigma(\alpha) + \mathfrak{P} = \alpha + \mathfrak{P} \text{ for all } \alpha \\ &\iff \tilde{\sigma}(\alpha + \mathfrak{P}) = \alpha + \mathfrak{P} \text{ for all } \alpha \iff \tilde{\sigma} \text{ is the identity on } \mathcal{O}_L/\mathfrak{P}. \end{aligned}$$

5.5. Assuming that part (ii) of Proposition 5.11 holds, the prime factors of $\mathfrak{P}_i$ of $\mathfrak{p}\mathcal{O}_L$ are distinct by hypothesis, so that $\mathfrak{p}$ is unramified in L and part (i) holds.

Now we assume that part (ii) holds and prove part (iii). Note that $\deg(f) = [L : K]$. First if $\mathfrak{p}$ splits completely in L, then $g = [L : K] = \deg(f)$, so that the factors $f_1(x), \ldots, f_g(x)$ are linear. Thus $f(x) \equiv 0 \bmod \mathfrak{p}$ has a solution in $\mathcal{O}_K$. Conversely, if $f(x) \equiv 0 \bmod \mathfrak{p}$ has a solution in $\mathcal{O}_K$, then since $f_1(x), \ldots, f_g(x)$ are irreducible, one of them must be linear. Part (ii) together with part (e) of Exercise 5.6 implies that the $f_i(x)$ all have the same degree. Thus all factors are linear, so that again $g = [L : K] = \deg(f)$. This implies that $e = f = 1$, which shows that $\mathfrak{p}$ splits completely in L.

5.6. Suppose $f(x) \equiv f_1(x) \cdots f_g(x) \bmod \mathfrak{p}$ with $f_i(x)$ irreducible modulo $\mathfrak{p}$. Since $f(x)$ is separable modulo $\mathfrak{p}$, the factors are distinct.

(a) Since $f(\alpha) = 0$, it follows that $f_1(\alpha) \cdots f_g(\alpha) \in \mathfrak{p}$. Since $\mathfrak{p}\mathcal{O}_L \subset \mathfrak{P}$, it follows that $\prod_{i=1}^{g} f_i(\alpha) \in \mathfrak{P}$. But $\mathfrak{P}$ is prime, so that $f_i(\alpha) \in \mathfrak{P}$ for some i; assume after renumbering that $i = 1$.

(b) Let $[\alpha] = \alpha + \mathfrak{P} \in \mathcal{O}_L/\mathfrak{P}$. Then $f_1(\alpha) \in \mathfrak{P}$ implies $f_1([\alpha]) = 0$. Since $f_1(x)$ is irreducible modulo $\mathfrak{p}$, $f(x_1)$ modulo $\mathfrak{p}$ is the minimal polynomial for $[\alpha] \in \mathcal{O}_L/\mathfrak{P}$ over $\mathcal{O}_K/\mathfrak{p}$, so its degree is at most $f = [\mathcal{O}_L/\mathfrak{P} : \mathcal{O}_K/\mathfrak{p}]$.

(c) First note that since α is an integral primitive element of L, its minimal polynomial $f(x)$ splits completely in $\mathcal{O}_L$ as

$$(*) \qquad f(x) = \prod_{\sigma \in \mathrm{Gal}(L/K)} (x - \sigma(\alpha)).$$

Separability over $\mathcal{O}_K/\mathfrak{p}$ implies separability over $\mathcal{O}_L/\mathfrak{P}$, and reducing (*) modulo $\mathfrak{P}$ shows that the $\sigma(\alpha)$ for $\sigma \in \mathrm{Gal}(L/K)$ are distinct modulo $\mathfrak{P}$.

Then $f_1(\alpha) \in \mathfrak{P}$ implies $f_1(\sigma(\alpha)) = \sigma(f_1(\alpha)) \in \mathfrak{P}$ for all $\sigma \in D_{\mathfrak{P}}$. Thus $\{\sigma(\alpha)\}_{\sigma \in D_{\mathfrak{P}}}$ are roots of $f_1(x)$ modulo $\mathfrak{P}$. Since they are distinct modulo $\mathfrak{P}$, it follows that $f_1(x)$ has at least $|D_{\mathfrak{P}}| = ef$ distinct roots. Thus $\deg(f_1(x)) \geq |D_{\mathfrak{P}}| = ef$.

(d) Since $f \geq \deg(f_1(x)) \geq ef$, we get $e = 1$ and $f = \deg(f_1(x))$. Thus $\mathfrak{p}$ is unramified in L.

(e) Since L/K is Galois, all primes lying over $\mathfrak{p}$ are unramified with inertial degree f. It remains to show that the number of such primes is g and that each $f_i(\alpha)$ lies in a different prime over $\mathfrak{p}$.

We first claim that $f_i(\alpha)$ lies in some prime $\mathfrak{P}'$ of $\mathcal{O}_L$ containing $\mathfrak{p}$. For a fixed prime $\mathfrak{P}$ over $\mathfrak{p}$, reducing (*) modulo $\mathfrak{P}$ gives

$$f(x) \equiv f_1(x)\cdots f_g(x) \equiv \prod_{\sigma\in \mathrm{Gal}(L/K)} (x-\sigma(\alpha)) \bmod \mathfrak{P}.$$

It follows that $x-\sigma(\alpha)$ is a factor of $f_i(x)$ modulo $\mathfrak{P}$ for some $\sigma \in \mathrm{Gal}(L/K)$, which implies $f_i(\sigma(\alpha)) \in \mathfrak{P}$. Since $f_i(\sigma(\alpha)) = \sigma(f_i(\alpha))$, we obtain $f_i(\alpha) \in \mathfrak{P}' = \sigma^{-1}(\mathfrak{P})$.

This argument has two nice consequences. First, from $f_i(\alpha) \in \mathfrak{P}'$, the arguments of parts (b), (c) and (d) imply that $\deg(f_i(x)) = f$, so that all of the $f_i(x)$ have the same degree f. Second, since $\mathrm{Gal}(L/K)$ acts transitively on the primes of $\mathcal{O}_L$ over $\mathfrak{p}$, given any $\mathfrak{P}'$ over $\mathfrak{p}$, there is $\sigma \in \mathrm{Gal}(L/K)$ such that $\mathfrak{P}' = \sigma^{-1}(\mathfrak{P})$. Then the argument of the previous paragraph implies that $f_i(\alpha) \in \mathfrak{P}'$ for at least one i.

It remains to prove that $f_i(\alpha) \in \mathfrak{P}'$ for precisely one i. Suppose that $f_i(\alpha), f_j(\alpha) \in \mathfrak{P}'$. Then $f_i([\alpha]) = f_j([\alpha]) = 0$ in $\mathcal{O}_K/\mathfrak{P}'$, where $[\alpha] = \alpha + \mathfrak{P}'$. As noted in part (b), $f_i(x)$ and $f_j(x)$ modulo $\mathfrak{p}$ must both be the minimal polynomial of $[\alpha]$ over $\mathcal{O}_K/\mathfrak{p}$, so $i = j$ since the $f_i(x)$ are distinct modulo $\mathfrak{p}$ since $f(x)$ is separable modulo $\mathfrak{p}$.

Thus all of the $f_i(x)$ have degree f and correspond to the distinct primes of $\mathcal{O}_L$ lying over $\mathfrak{p}$, so that $\mathfrak{p}\mathcal{O}_L = \mathfrak{P}_1\cdots\mathfrak{P}_g$. By renumbering if necessary, we may assume $f_i(\alpha) \in \mathfrak{P}_i$.

(f) First suppose we have ideals $\mathfrak{a}$ and $\mathfrak{b}$ in $\mathcal{O}_L$. Then $\mathfrak{a} \subset \mathfrak{b}$ implies $\mathfrak{c} = \mathfrak{b}^{-1}\mathfrak{a} \subset \mathfrak{b}^{-1}\mathfrak{b} = \mathcal{O}_L$ is an ideal of $\mathcal{O}_L$ satisfying $\mathfrak{a} = \mathfrak{b}\mathfrak{c}$. Conversely, if $\mathfrak{a} = \mathfrak{b}\mathfrak{c}$, then $\mathfrak{a} \subset \mathfrak{b}$ follows since a product of two ideals is contained in each one (this is a standard fact from abstract algebra).

Let $I_i = \mathfrak{p}\mathcal{O}_L + f_i(\alpha)\mathcal{O}_L$, $1 \le i \le g$, and note that $\mathfrak{p}\mathcal{O}_L \subset I_i \subset \mathfrak{P}_i$ for each i. By the above paragraph, there are ideals $\mathfrak{c}$ and $\mathfrak{d}$ of $\mathcal{O}_L$ such that

$$\mathfrak{p}\mathcal{O}_L = I_i\mathfrak{c} \ \text{ and } \ I_i = \mathfrak{P}_i\mathfrak{d}.$$

These equations imply $\mathfrak{p}\mathcal{O}_L = \mathfrak{P}_i\mathfrak{d}\mathfrak{c}$. By the unique factorization $\mathfrak{p}\mathcal{O}_L = \mathfrak{P}_1\cdots\mathfrak{P}_g$ from part (e), the only possible nontrivial factors of $\mathfrak{d}$ are $\mathfrak{P}_j$ for $j \neq i$. Suppose $\mathfrak{P}_j$ is a factor of $\mathfrak{d}$. Then it is also a factor of I_i, which implies $I_i \subset \mathfrak{P}_j$. By the definition of I_i, this gives $f_i(\alpha) \in \mathfrak{P}_j$, yet we also have $f_j(\alpha) \in \mathfrak{P}_j$. The proof of part (e) given above shows that this can't happen since $i \neq j$. Hence $\mathfrak{d}$ is the trivial ideal, which proves that $I_i = \mathfrak{P}_i$.

5.7. (a) The definitions give

$$\begin{aligned} T(\alpha+\beta) &= (\alpha+\beta) + (\alpha+\beta)' = \alpha+\beta+\alpha'+\beta' \\ &= \alpha+\alpha'+\beta+\beta' = T(\alpha)+T(\beta) \\ N(\alpha\beta) &= (\alpha\beta)(\alpha\beta)' = \alpha\beta\alpha'\beta' = (\alpha\alpha')(\beta\beta') = N(\alpha)N(\beta). \end{aligned}$$

(b) With $\alpha = r + s\sqrt{N}$, if $T(\alpha),\, N(\alpha) \in \mathbb{Z}$, then

$$f(x) = x^2 - T(\alpha)x + N(\alpha) = x^2 - (\alpha+\alpha')x + \alpha\alpha' = (x-\alpha)(x-\alpha')$$

by the definition of $T(\alpha)$ and $N(\alpha)$. Thus the roots of $f(x)$, which is a monic integer polynomial, are α and α', showing that $\alpha, \alpha' \in \mathcal{O}_K$.

Conversely, if $\alpha \in \mathcal{O}_K$, then its minimal polynomial is $x^2 + cx + d \in \mathbb{Z}[x]$ and its other root is α'. Thus $x^2 + cx + d = (x-\alpha)(x-\alpha') = x^2 - (\alpha+\alpha') + \alpha\alpha'$, so that $-c = \alpha + \alpha' = T(\alpha)$ and $d = \alpha\alpha' = N(\alpha)$ are both integers.

(c) Suppose $\alpha \in \mathcal{O}_K$ and write $\alpha = r + s\sqrt{N}$ with $r, s \in \mathbb{Q}$. Then $T(\alpha) = 2r \in \mathbb{Z}$ implies that $r \in \frac{1}{2}\mathbb{Z}$, and then $N(\alpha) = r^2 - s^2 N \in \mathbb{Z}$ means either that $r, s \in \mathbb{Z}$ or that r and s are both half-integers. If they are both half-integers, say

$$r = \frac{2a+1}{2}, \quad s = \frac{2b+1}{2},$$

then

$$r^2 - s^2 N = a^2 + a - b^2 - b + \frac{1-N}{4}.$$

This is an integer if and only if $N \equiv 1 \bmod 4$. It follows that for $N \not\equiv 1 \bmod 4$, $r + s\sqrt{N} \in \mathcal{O}_K$ if and only if $r, s \in \mathbb{Z}$, so that $\mathcal{O}_K = \mathbb{Z}[\sqrt{N}]$, while if $N \equiv 1 \bmod 4$, both r and s may be integers or both may be half-integers, so that $\mathcal{O}_K = \mathbb{Z}\Big[\frac{1+\sqrt{N}}{2}\Big]$.

(d) If $N \not\equiv 1 \bmod 4$, then $d_K = 4N$, so that

$$\mathbb{Z}\left[\frac{d_K + \sqrt{d_K}}{2}\right] = \mathbb{Z}\left[\frac{4N + 2\sqrt{N}}{2}\right] = \mathbb{Z}[2N + \sqrt{N}] = \mathbb{Z}[\sqrt{N}],$$

while if $N \equiv 1 \bmod 4$, then $d_K = N$ is odd, say $d_K = 2a + 1$, and then

$$\mathbb{Z}\left[\frac{d_K + \sqrt{d_K}}{2}\right] = \mathbb{Z}\left[\frac{2a + 1 + \sqrt{N}}{2}\right] = \mathbb{Z}\left[a + \frac{1 + \sqrt{N}}{2}\right] = \mathbb{Z}\left[\frac{1 + \sqrt{N}}{2}\right].$$

5.8. If n is not squarefree, say $n = a^2 m$, $a > 1$, m squarefree, then $K = \mathbb{Q}(\sqrt{-m})$, so that $d_K = -m$ or $-4m$, neither of which equals $-4n$. Additionally, $\mathcal{O}_K \neq \mathbb{Z}[\sqrt{-n}]$ since $\sqrt{-m} \in \mathcal{O}_K$ but $\sqrt{-m} \notin \mathbb{Z}[\sqrt{-n}]$, and n does not satisfy (5.2), so none of the three statements holds.

Otherwise, assume that n is squarefree. Then $d_K = -4n$ if and only if $-n \not\equiv 1 \bmod 4$ by (5.12), which is equivalent to $\mathcal{O}_K = \mathbb{Z}[\sqrt{-n}]$ by (5.13). Further, $-n \not\equiv 1 \bmod 4$ is the same as $n \not\equiv 3 \bmod 4$, so that $d_K = -4n$ if and only if n satisfies (5.2).

5.9. (a) If $\alpha = r + s\sqrt{N}$ is a unit, say $\alpha\beta = 1$, then $N(\alpha)N(\beta) = N(\alpha\beta) = N(1) = 1$. But since $N < 0$, it follows that $N(\alpha) = r^2 - s^2 N \geq 0$. Therefore $N(\alpha) = N(\beta) = 1$. Conversely, suppose $N(\alpha) = 1$; then $\alpha\alpha' = N(\alpha) = 1$, which shows that α is a unit since $\alpha' \in \mathcal{O}_K$. (See Exercise 4.5 and part (b) of Exercise 4.16 for comparison.)

(b) Suppose $\alpha \in \mathcal{O}_K$ is a unit. If $N \not\equiv 1 \bmod 4$, then $\alpha = m + n\sqrt{N}$, $m, n \in \mathbb{Z}$. Thus $N(\alpha) = m^2 - n^2 N = m^2 + n^2(-N) = 1$. If $N < -1$, then $-N > 1$. Then n must be zero, and the only solutions are $m = \pm 1$, $n = 0$. If $N = -1$, we get the case of $\mathbb{Q}(i)$ that was dealt with in Exercise 4.5.

If $N \equiv 1 \bmod 4$, then $\alpha = (m + n\sqrt{N})/2$ with $m, n \in \mathbb{Z}$, $m \equiv n \bmod 2$. Then

$$N(\alpha) = \frac{m^2 - n^2 N}{4} = 1,$$

so we must have $m^2 - n^2 N = m^2 + n^2(-N) = 4$. If $N < -4$, then $-N > 4$. As above, this forces $n = 0$ and $m = \pm 1$. If $N \geq -4$, then $N = -3$ since $N \equiv 1 \bmod 4$ is negative. Hence we get the case of $\mathbb{Q}(\omega)$ that was dealt with in Exercise 4.16.

Thus unless $\mathbb{Q}(\sqrt{N}) = \mathbb{Q}(i)$ or $\mathbb{Q}(\omega)$, the only units in $\mathcal{O}_{\mathbb{Q}(\sqrt{N})}$ for $N < 0$ are ± 1.

5.10. (a) If $2 \mid d_K$, then $d_K = 4N$ since $d_K \equiv 0, 1 \bmod 4$. If N is odd, let $\mathfrak{p} = 2\mathcal{O}_K + (1 + \sqrt{N})\mathcal{O}_K$. Then as in the proof of Proposition 5.16,

$$\begin{aligned} \mathfrak{p}^2 &= \left(2\mathcal{O}_K + (1 + \sqrt{N})\mathcal{O}_K\right)^2 = 2^2\mathcal{O}_K + 2(1 + \sqrt{N})\mathcal{O}_K + (1 + \sqrt{N})^2\mathcal{O}_K \\ &= 4\mathcal{O}_K + (2 + 2\sqrt{N})\mathcal{O}_K + (N + 1 + 2\sqrt{N})\mathcal{O}_K. \end{aligned}$$

This ideal is contained in $2\mathcal{O}_K$ since each generator is a multiple of 2 (N is odd). For the reverse inclusion, note that $N \equiv 3 \bmod 4$ (since otherwise $d_K = N$), say $N = 4m + 3$. Then

$$\begin{aligned}(N + 1 + 2\sqrt{N}) - (2 + 2\sqrt{N}) - 4m &= (4m + 3) + 1 + 2\sqrt{N} - 2 - 2\sqrt{N} - 4m \\ &= 2 \in \mathfrak{p}^2.\end{aligned}$$

Thus $2\mathcal{O}_K \subset \mathfrak{p}^2$ and therefore $2\mathcal{O}_K = \mathfrak{p}^2$; since $1 - \sqrt{N} = 2 - (1 + \sqrt{N})$ we also see that $\mathfrak{p} = \mathfrak{p}'$.

If N is even, let $\mathfrak{p} = 2\mathcal{O}_K + \sqrt{N}\mathcal{O}_K$. Then as above,

$$\mathfrak{p}^2 = (2\mathcal{O}_K + \sqrt{N}\mathcal{O}_K)^2 = 4\mathcal{O}_K + 2\sqrt{N}\mathcal{O}_K + N\mathcal{O}_K.$$

It follows easily that $\mathfrak{p}^2 \subset 2\mathcal{O}_K$. Since N is squarefree and even, we have $N \equiv 2 \bmod 4$, so that $\gcd(4, N) = 2$. Then

$$2 \in 4\mathbb{Z} + N\mathbb{Z} \subset 4\mathcal{O}_K + 4\sqrt{N}\mathcal{O}_K + N\mathcal{O}_K = \mathfrak{p}^2,$$

and again $2\mathcal{O}_K = \mathfrak{p}^2$. Since $-\sqrt{N} \in \mathfrak{p}$, we also have $\mathfrak{p} = \mathfrak{p}'$.

In both cases, since $2\mathcal{O}_K = \mathfrak{p}^2$, any prime in the factorization of $\mathfrak{p}$ must appear to an even power in $2\mathcal{O}_K$, so that $e \geq 2$. But then $efg = 2$ shows that $e = 2$ and $f = g = 1$ so that $\mathfrak{p}$ is prime.

(b) If $2 \nmid d_K$, then $d_K = N \equiv 1 \bmod 4$ and $K = \mathbb{Q}(\sqrt{d_K}) = \mathbb{Q}\big(\frac{1+\sqrt{d_K}}{2}\big)$. The minimal polynomial of $\alpha = (1 + \sqrt{d_K})/2$ over $\mathbb{Q}$ is $f(x) = x^2 - x + (1 - d_K)/4$, separable modulo 2 since its discriminant is d_K and $2 \nmid d_K$. Thus we can apply Proposition 5.11 to determine the factorization of $2\mathcal{O}_K$.

Since $f(x)$ is quadratic, it is either irreducible or a product of two distinct linear factors modulo 2. Then Proposition 5.11 implies that $2\mathcal{O}_K$ is either prime or splits completely, and the latter occurs if and only if $f(x) \equiv 0 \bmod 2$ has an integer solution. One computes that $f(0) \equiv f(1) \equiv (1 - d_K)/4 \bmod 2$, so that

$$\begin{aligned}2\mathcal{O}_K \text{ splits completely} &\iff \frac{1 - d_K}{4} \equiv 0 \bmod 2 \\ &\iff d_K \equiv 1 \bmod 8 \iff \left(\frac{d_K}{2}\right) = 1.\end{aligned}$$

This then implies that $2\mathcal{O}_K$ is prime in $\mathcal{O}_K$ if and only if $d_K \equiv 5 \bmod 8$, i.e., if $(d_K/2) = -1$. Furthermore, when $2\mathcal{O}_K$ splits completely as a product $\mathfrak{p}_1\mathfrak{p}_2$, then by the same argument as in the text, since $\mathrm{Gal}(K/\mathbb{Q})$ acts transitively on the primes over 2, we must have $\mathfrak{p}_2 = \mathfrak{p}_1'$.

(c) Suppose $\mathfrak{p}$ is a nonzero prime ideal of $\mathcal{O}_K$. Part (a) of Exercise 5.1 implies that $\mathfrak{p}$ contains a nonzero integer m, which can be assumed to be positive. Then m is a product of integer primes, and for each prime p in this factorization, $p\mathcal{O}_K$ is a product of primes described in parts (i)–(iii) of Proposition 5.16. Therefore $m\mathcal{O}_K$ is also a product of primes described in the proposition. However, $m \in \mathfrak{p}$ implies $m\mathcal{O}_K \subset \mathfrak{p}$, so that $\mathfrak{p}$ is one of the prime factors of $m\mathcal{O}_K$ by the final assertion of Corollary 5.6. This proves that Proposition 5.16 describes all prime ideals of $\mathcal{O}_K$.

5.11. (a) $[\mathcal{O}_K/\mathfrak{p} : \mathbb{Z}/p\mathbb{Z}] = f$ by the definition of f as the inertial degree, and $|\mathbb{Z}/p\mathbb{Z}| = p$, so that $|\mathcal{O}_K/\mathfrak{p}| = p^f$.

(b) Note that $N(\mathfrak{p}) = |\mathcal{O}_K/\mathfrak{p}| = p^f$ by part (a). If $p \mid d_K$, then Proposition 5.16 shows that $p\mathcal{O}_K = \mathfrak{p}^2$. Then $efg = 2$ with $e = 2$ gives $f = 1$, so $N(\mathfrak{p}) = p$. If $p \nmid d_K$, then by the proposition, $p = \mathfrak{p}\mathfrak{p}'$, $\mathfrak{p} \neq \mathfrak{p}'$ or $p\mathcal{O}_K$ is prime. In the former case, p splits completely, so that $g = 2$. Then $efg = 2$ implies $e = f = 1$ and $N(\mathfrak{p}) = p$. In the latter case, p remains prime in $\mathcal{O}_K$. Then $e = g = 1$ implies $f = 2$ and $N(\mathfrak{p}) = p^2$.

5.12. (a) If $\alpha \in \mathcal{O}_L$ and $\sigma \in \mathrm{Gal}(L/K)$, then

$$\left(\frac{L/K}{\mathfrak{P}}\right)(\sigma^{-1}\alpha) \equiv (\sigma^{-1}\alpha)^{N(\mathfrak{p})} \equiv \sigma^{-1}(\alpha^{N(\mathfrak{p})}) \bmod \mathfrak{P}.$$

Now apply σ to both sides, giving

$$\sigma\left(\frac{L/K}{\mathfrak{P}}\right)\sigma^{-1}\alpha \equiv \sigma\sigma^{-1}(\alpha^{N(\mathfrak{p})}) \equiv \alpha^{N(\mathfrak{p})} \bmod \sigma(\mathfrak{P}).$$

But $((L/K)/\sigma(\mathfrak{P}))(\alpha) \equiv \alpha^{N(\mathfrak{p})} \bmod \sigma(\mathfrak{P})$ by definition of the Artin symbol. Since the Artin symbol for $\sigma(\mathfrak{P})$ is unique, we get

$$\left(\frac{L/K}{\sigma(\mathfrak{P})}\right) = \sigma\left(\frac{L/K}{\mathfrak{P}}\right)\sigma^{-1}.$$

(b) If $\sigma \in \mathrm{Gal}(L/K)$, then since σ permutes the primes lying over $\mathfrak{p}$, part (a) shows that conjugating $((L/K)/\mathfrak{P})$ by σ gives the Artin symbol $((L/K)/\mathfrak{P}')$ for another prime $\mathfrak{P}' = \sigma(\mathfrak{P})$ over $\mathfrak{p}$. Since the action of $\mathrm{Gal}(L/K)$ on the primes over $\mathfrak{p}$ is transitive, all members of

$$\left\{\left(\frac{L/K}{\mathfrak{P}}\right) \,\middle|\, \mathfrak{P} \text{ is a prime of } L \text{ lying over } \mathfrak{p}\right\}$$

are conjugates of one another. Hence this set is a conjugacy class.

5.13. (a) Since $n \notin \mathfrak{p}$, $(x^n - 1)' = nx^{n-1}$ is nonzero modulo $\mathfrak{p}$ and is relatively prime to $x^n - 1$ over $\mathcal{O}_K/\mathfrak{p}$ since zero is not a root of the latter. Thus

$$x^n - 1 = (x-1)(x-\zeta)\cdots(x-\zeta^{n-1})$$

is separable modulo $\mathfrak{p}$, so its roots in K, namely $1, \zeta, \ldots, \zeta^{n-1}$, are distinct mod $\mathfrak{p}$.

(b) By part (a), $1, \zeta, \ldots, \zeta^{n-1}$ are distinct modulo $\mathfrak{p}$. They are closed under multiplication and hence form a subgroup of $(\mathcal{O}_K/\mathfrak{p})^*$ of order n. Since $|(\mathcal{O}_K/\mathfrak{p})^*| = N(\mathfrak{p}) - 1$, Lagrange's Theorem implies that $n \mid N(\mathfrak{p}) - 1$.

(c) Since $(\mathcal{O}_K/\mathfrak{p})^*$ has order $N(\mathfrak{p}) - 1$, $a^{N(\mathfrak{p})-1} \equiv 1 \bmod \mathfrak{p}$, so that $a^{(N(\mathfrak{p})-1)/n}$ is a root of $x^n - 1$ modulo $\mathfrak{p}$. Since $\mathcal{O}_K/\mathfrak{p}$ is a field, the only roots of $x^n - 1 \equiv 0 \bmod \mathfrak{p}$ are $1, \zeta, \ldots, \zeta^{n-1}$ modulo $\mathfrak{p}$. Thus $a^{(N(\mathfrak{p})-1)/n}$ is congruent modulo $\mathfrak{p}$ to an nth root of unity, necessarily unique by part (a).

(d) This is similar to Exercise 4.11. If a is an nth power residue modulo $\mathfrak{p}$, then we can write $a \equiv b^n \bmod \mathfrak{p}$. Because $a \notin \mathfrak{p}$, it follows that $b \not\equiv 0 \bmod \mathfrak{p}$ since $b^n \equiv a \not\equiv 0 \bmod \mathfrak{p}$. Therefore

$$\left(\frac{a}{\mathfrak{p}}\right)_n \equiv a^{(N(\mathfrak{p})-1)/n} \equiv (b^n)^{(N(\mathfrak{p})-1)/n} \equiv b^{N(\mathfrak{p})-1} \equiv 1 \bmod \mathfrak{p}$$

since $(\mathcal{O}_K/\mathfrak{p})^*$ has order $N(\mathfrak{p}) - 1$. Part (a) then shows that $(a/\mathfrak{p})_n = 1$. Conversely, suppose that $(a/\mathfrak{p})_n = 1$. Then $a^{(N(\mathfrak{p})-1)/n} \equiv 1 \bmod \mathfrak{p}$. Let $g \in \mathcal{O}_K$ be such that $[g]$ is a generator for $(\mathcal{O}_K/\mathfrak{p})^*$, and choose $k \in \mathbb{Z}$ such that $[a] = [g]^k$. Then $a \equiv g^k \bmod \mathfrak{p}$, and

$$a^{(N(\mathfrak{p})-1)/n} \equiv g^{k(N(\mathfrak{p})-1)/n} \equiv 1 \bmod \mathfrak{p}.$$

Then we must have $N(\mathfrak{p}) - 1 \mid k(N(\mathfrak{p}) - 1)/n$, and thus $n \mid k$. Let $b = g^{k/n}$; then $b^n = g^k \equiv a \bmod \mathfrak{p}$.

5.14. (a) Since $a, n \notin \mathfrak{p}$, $(x^n - a)' = nx^{n-1}$ is nonzero modulo $\mathfrak{p}$, and $x^n - a$ has a nonzero constant term modulo $\mathfrak{p}$. Thus they have no common factors, so $x^n - a$ is separable modulo $\mathfrak{p}$. The minimal polynomial $f(x)$ of $\sqrt[n]{a}$ over K divides $x^n - a$, so $f(x)$ is also separable modulo $\mathfrak{p}$. By Proposition 5.11, it follows that $\mathfrak{p}$ is unramified in L.

(b) By (5.20),

$$\left(\frac{L/K}{\mathfrak{p}}\right)(\sqrt[n]{a}) \equiv (\sqrt[n]{a})^{N(\mathfrak{p})} \equiv (\sqrt[n]{a})^{N(\mathfrak{p})-1} \cdot \sqrt[n]{a} \bmod \mathfrak{p}$$
$$\equiv a^{(N(\mathfrak{p})-1)/n} \cdot \sqrt[n]{a} \equiv \left(\frac{a}{\mathfrak{p}}\right)_n \sqrt[n]{a} \bmod \mathfrak{p}.$$

Since $\mathrm{Gal}(L/K)$ permutes the roots of $x^n - a = (x - \sqrt[n]{a})(x - \zeta\sqrt[n]{a})\cdots(x - \zeta^{n-1}\sqrt[n]{a})$, $((L/K)/\mathfrak{p})(\sqrt[n]{a})$ is equal to $\sqrt[n]{a}$ times an nth root of unity; since these are all incongruent modulo $\mathfrak{p}$, we get

$$\left(\frac{L/K}{\mathfrak{p}}\right)(\sqrt[n]{a}) = \left(\frac{a}{\mathfrak{p}}\right)_n \sqrt[n]{a}.$$

5.15. (a) Suppose $\mathfrak{p} = \prod_{i=1}^{g_1} \mathfrak{P}_i^{e_{\mathfrak{P}_i|\mathfrak{p}}}$ in M, and for simplicity that $\mathfrak{P} = \mathfrak{P}_1$. Then $\mathfrak{P} = \prod_{i=1}^{g_2} (\mathfrak{P}'_i)^{e_{\mathfrak{P}'_i|\mathfrak{P}}}$ in L, and we assume that $\mathfrak{P}' = \mathfrak{P}'_1$. Then the factorization of $\mathfrak{p}$ in L contains the factor

$$(\mathfrak{P}'^{e_{\mathfrak{P}'|\mathfrak{P}}})^{e_{\mathfrak{P}|\mathfrak{p}}} = (\mathfrak{P}')^{e_{\mathfrak{P}'|\mathfrak{P}} e_{\mathfrak{P}|\mathfrak{p}}}$$

as well as the factor $(\mathfrak{P}')^{e_{\mathfrak{P}'|\mathfrak{p}}}$, so that by unique factorization,

$$e_{\mathfrak{P}'|\mathfrak{p}} = e_{\mathfrak{P}'|\mathfrak{P}} e_{\mathfrak{P}|\mathfrak{p}}.$$

(b) From part (a), if $\mathfrak{P}$ is any prime of M lying over $\mathfrak{p}$, and $\mathfrak{P}'$ any prime of L lying over $\mathfrak{p}$, then

$$e_{\mathfrak{P}'|\mathfrak{p}} = e_{\mathfrak{P}'|\mathfrak{P}} e_{\mathfrak{P}|\mathfrak{p}}.$$

For $\Rightarrow$, if $\mathfrak{p}$ is unramified in L, then $e_{\mathfrak{P}'|\mathfrak{p}} = 1$ for every $\mathfrak{P}'$ lying over $\mathfrak{p}$, showing that $e_{\mathfrak{P}'|\mathfrak{P}} = e_{\mathfrak{P}|\mathfrak{p}} = 1$ for each such prime. But as $\mathfrak{P}'$ varies over all primes of L over $\mathfrak{p}$, $\mathfrak{P}$ varies over all primes of M over $\mathfrak{p}$. This proves both statements.

For $\Leftarrow$, choose a prime $\mathfrak{P}'$ of L lying over $\mathfrak{p}$, and let $\mathfrak{P} = \mathfrak{P}' \cap M$. Then $\mathfrak{P}$ lies over $\mathfrak{p}$ as well, so that by the assumptions, $e_{\mathfrak{P}|\mathfrak{p}} = e_{\mathfrak{P}'|\mathfrak{P}} = 1$, so their product is $e_{\mathfrak{P}'|\mathfrak{p}} = 1$. Since $\mathfrak{P}'$ was arbitrary, this shows that $\mathfrak{p}$ is unramified in L.

(c) L is unramified over K if and only if every prime $\mathfrak{p}$ of K is unramified in L, which by part (b) happens if and only if $\mathfrak{p}$ is unramified in M and every prime of M lying over $\mathfrak{p}$ is unramified in L. Since every prime of M lies over some prime of K, it follows (letting $\mathfrak{p}$ vary over all primes of K) that this holds if and only if M is unramified over K and L is unramified over M.

5.16. Since the Artin map is a homomorphism, it suffices to show that if $\mathfrak{p}$ is prime in K, then

$$\left(\frac{M/K}{\mathfrak{p}}\right) = r \circ \left(\frac{L/K}{\mathfrak{p}}\right).$$

Given $\mathfrak{p}$ prime in $\mathcal{O}_K$, fix $\mathfrak{P}_M \subset \mathcal{O}_M$ and $\mathfrak{P}_L \subset \mathcal{O}_L$ with $\mathfrak{P}_L$ lying over $\mathfrak{P}_M$ and $\mathfrak{P}_M$ lying over $\mathfrak{p}$. Since r is the restriction map, then for any $\sigma \in \mathrm{Gal}(L/K)$ and $\alpha \in M$, we have $\sigma(\alpha) = (r \circ \sigma)(\alpha)$. Then if $\alpha \in \mathcal{O}_M$,

$$\left(r \circ \left(\frac{L/K}{\mathfrak{p}}\right)\right)(\alpha) = \left(\frac{L/K}{\mathfrak{p}}\right)(\alpha) \equiv \alpha^{N(\mathfrak{p})} \bmod \mathfrak{P}_L.$$

Both sides of this congruence lie in $\mathcal{O}_M$, and since $\mathfrak{P}_M = \mathfrak{P}_L \cap \mathcal{O}_M$, it follows that

$$\left(r \circ \left(\frac{L/K}{\mathfrak{p}}\right)\right)(\alpha) \equiv \alpha^{N(\mathfrak{p})} \bmod \mathfrak{P}_M$$

for all $\alpha \in \mathcal{O}_M$. Uniqueness of the Artin symbol then shows that

$$\left(\frac{M/K}{\mathfrak{p}}\right) = r \circ \left(\frac{L/K}{\mathfrak{p}}\right).$$

5.17. Let L be the Hilbert class field of K. Any unramified Abelian extension of K is a subfield of L by Theorem 5.18, and any subfield of L is unramified over K by part (b) of Exercise 5.15. Thus there is a one-to-one correspondence between unramified Abelian extensions of K and subfields of L.

By Galois theory, subfields of L correspond one-to-one to subgroups $\mathrm{Gal}(L/K)$, and we have $C(\mathcal{O}_K) \simeq \mathrm{Gal}(L/K)$ by Theorem 5.23, where the isomorphism is induced by the Artin map. Therefore unramified Abelian extensions of K, which are the subfields of L, correspond one-to-one to subgroups of $C(\mathcal{O}_K)$. This proves the first statement in the corollary.

For the second statement, consider an unramified Abelian extension $K \subset M$, which gives an intermediate field $K \subset M \subset L$. Since M is Galois over K, $\mathrm{Gal}(M/K) \simeq \mathrm{Gal}(L/K)/\mathrm{Gal}(L/M)$ by Galois theory. Combining this with both Theorem 5.23 and Exercise 5.16 gives a commutative diagram

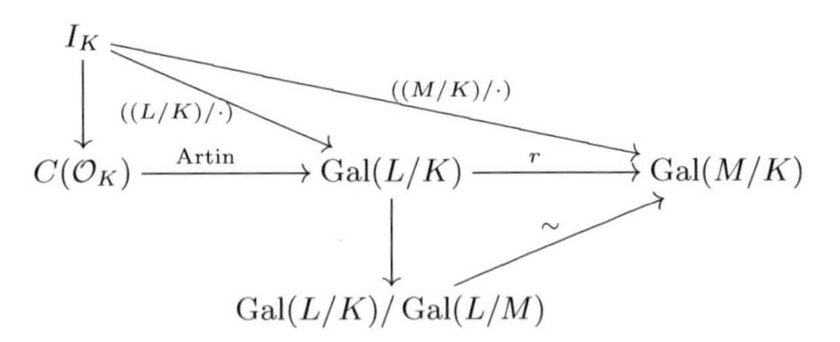

Now let H be the preimage in $C(\mathcal{O}_K)$ of $\mathrm{Gal}(L/M) \subset \mathrm{Gal}(L/K)$ under the Artin map. We can expand the above commutative diagram to obtain

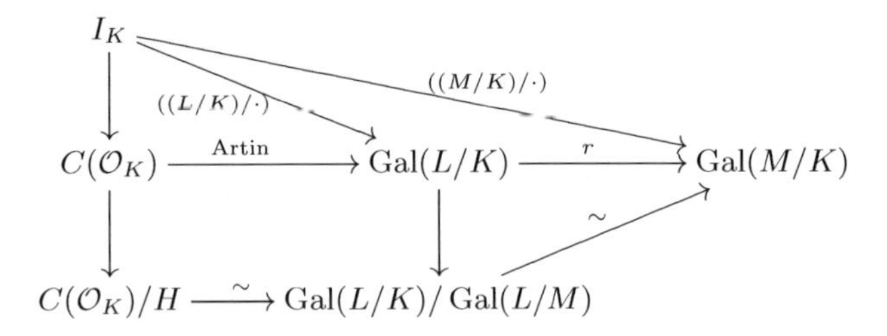

Composing gives an isomorphism $C(\mathcal{O}_K)/H \xrightarrow{\sim} \mathrm{Gal}(M/K)$ that comes from the restriction of the Artin map for the Hilbert class field.

5.18. Suppose $\mathfrak{P}_L$ is a prime factor of $\mathfrak{p}\mathcal{O}_L$, and let $\mathfrak{P}_M = \mathfrak{P}_L \cap \mathcal{O}_M$. By Exercise 5.15, we have

$$e_{\mathfrak{P}_L|\mathfrak{p}} = e_{\mathfrak{P}_L|\mathfrak{P}_M} e_{\mathfrak{P}_M|\mathfrak{p}}.$$

The inertial degrees are also multiplicative:

$$f_{\mathfrak{P}_L|\mathfrak{p}} = [\mathcal{O}_L/\mathfrak{P}_L : \mathcal{O}_K/\mathfrak{p}] = [\mathcal{O}_L/\mathfrak{P}_L : \mathcal{O}_M/\mathfrak{P}_M][\mathcal{O}_M/\mathfrak{P}_M : \mathcal{O}_K/\mathfrak{p}] = f_{\mathfrak{P}_L|\mathfrak{P}_M} f_{\mathfrak{P}_M|\mathfrak{p}}.$$

If $\mathfrak{p} \subset \mathcal{O}_K$ splits completely in L, then there are $g = [L : K]$ factors and $e = f = 1$. This means that $e_{\mathfrak{P}_L|\mathfrak{p}} = f_{\mathfrak{P}_L|\mathfrak{p}} = 1$. Hence

$$\begin{aligned} e_{\mathfrak{P}_L|\mathfrak{P}_M} &= f_{\mathfrak{P}_L|\mathfrak{P}_M} = 1 \\ e_{\mathfrak{P}_M|\mathfrak{p}} &= f_{\mathfrak{P}_M|\mathfrak{p}} = 1. \end{aligned}$$

This holds for all primes of L and M lying over $\mathfrak{p}$. Therefore $\mathfrak{p}$ splits completely in M and each prime of $\mathcal{O}_M$ lying over $\mathfrak{p}$ splits completely in L.

Conversely, suppose $\mathfrak{p}$ splits completely in M and $\mathfrak{P}_M$ is a prime lying over $\mathfrak{p}$ that splits completely in L. Choose some $\mathfrak{P}_L$ lying over $\mathfrak{P}_M$. Since $\mathfrak{p}$ splits completely in M, $e_{\mathfrak{P}_M|\mathfrak{p}} = f_{\mathfrak{P}_M|\mathfrak{p}} = 1$; since $\mathfrak{P}_M$ splits completely in L, $e_{\mathfrak{P}_L|\mathfrak{P}_M} = f_{\mathfrak{P}_L|\mathfrak{P}_M} = 1$. Then multiplicativity of e and f in towers shows that

$$e_{\mathfrak{P}_L|\mathfrak{p}} = f_{\mathfrak{P}_L|\mathfrak{p}} = 1,$$

Since L/K is Galois, every prime $\mathfrak{P}$ of L lying over $\mathfrak{p}$ satisfies $e_{\mathfrak{P}|\mathfrak{p}} = e_{\mathfrak{P}_L|\mathfrak{p}} = 1$ and $f_{\mathfrak{P}|\mathfrak{p}} = f_{\mathfrak{P}_L|\mathfrak{p}} = 1$, Hence $\mathfrak{p}$ splits completely in L.

5.19. (a) Recall that in this book, number fields are all subfields of $\mathbb{C}$. Suppose $L \subset \mathbb{C}$ is Galois over $\mathbb{Q}$. Let $f(x) \in \mathbb{Q}[x]$ be the minimal polynomial over $\mathbb{Q}$ of $\alpha \in L$. Then $f(x)$ splits completely over L since L is Galois over $\mathbb{Q}$. Since $f(\alpha) = 0$ implies $f(\overline{\alpha}) = 0$, it follows that $\overline{\alpha} \in L$. Thus $\tau(L) \subset L$, and equality follows since both fields have the same degree over $\mathbb{Q}$,

Conversely, suppose $\tau(L) = L$, so that $\tau \in \mathrm{Gal}(L/\mathbb{Q})$. Since L is Galois over K, we have $|\mathrm{Gal}(L/K)| = [L : K]$, and since $L/\mathbb{Q}$ is finite, we know that $|\mathrm{Gal}(L/\mathbb{Q})|$ divides $[L : \mathbb{Q}] = 2[L : K] = 2\,|\mathrm{Gal}(L/K)|$. Since $|\mathrm{Gal}(L/K)|$ divides $|\mathrm{Gal}(L/\mathbb{Q})|$, it follows that $|\mathrm{Gal}(L/\mathbb{Q})| = [L : \mathbb{Q}]$ or $[L : K]$, and the latter implies $\mathrm{Gal}(L/\mathbb{Q}) = \mathrm{Gal}(L/K)$. This case can't occur since $\tau \in \mathrm{Gal}(L/\mathbb{Q})$ is not the identity on K. Hence we must have $|\mathrm{Gal}(L/\mathbb{Q})| = [L : \mathbb{Q}]$, which by Galois theory implies that L is Galois over $\mathbb{Q}$.

(b) (i) Since $H = \{e, \tau\} \subset \mathrm{Gal}(L/\mathbb{Q})$ has fixed field $L \cap \mathbb{R}$ and $|H| = 2$, Galois theory for the Galois extension $\mathbb{Q} \subset L$ implies $[L : L \cap \mathbb{R}] = 2$. Thus

$$[L \cap \mathbb{R} : \mathbb{Q}] = [L : \mathbb{Q}]/2 = [L : K].$$

(ii) As noted in the proof of part (i), $[L : L \cap \mathbb{R}] = 2$. For $\alpha \in L \cap \mathbb{R}$, we compute $[L : \mathbb{Q}(\alpha)]$ in two different ways. First, $\mathbb{Q}(\alpha) \subset L \cap \mathbb{R} \subset L$ implies

$$(*) \qquad [L : \mathbb{Q}(\alpha)] = [L : L \cap \mathbb{R}][L \cap \mathbb{R} : \mathbb{Q}(\alpha)] = 2[L \cap \mathbb{R} : \mathbb{Q}(\alpha)].$$

Next, $K = \mathbb{Q}(\sqrt{N})$, $N < 0$, so that $K(\alpha) = \mathbb{Q}(\sqrt{N}, \alpha)$, which is an extension of degree 2 of $\mathbb{Q}(\alpha)$ since $x^2 - N$ has no real roots, making it irreducible over $\mathbb{Q}(\alpha) \subset \mathbb{R}$. In other words, $[K(\alpha) : \mathbb{Q}(\alpha)] = 2$. Since $\mathbb{Q}(\alpha) \subset K(\alpha) \subset L$, we get

$$[L : \mathbb{Q}(\alpha)] = [L : K(\alpha)][K(\alpha) : \mathbb{Q}(\alpha)] = 2[L : K(\alpha)].$$

Comparing this to (*) gives $[L \cap \mathbb{R} : \mathbb{Q}(\alpha)] = [L : K(\alpha)]$. It follows immediately that $L \cap \mathbb{R} = \mathbb{Q}(\alpha) \iff L = K(\alpha)$.

5.20. Write $K = \mathbb{Q}(\sqrt{-19})$. Since $-19 \equiv 1 \bmod 4$, $\mathcal{O}_K = \mathbb{Z}[(1 + \sqrt{-19})/2]$. If $C(\mathcal{O}_K)$ is trivial, then $\mathcal{O}_K$ is a PID and thus a UFD.

Any reduced form $ax^2 + bxy + cy^2$ of discriminant $D = -19$ must have $a \leq \sqrt{-D/3} \approx 2.5$. Additionally, b must be odd since D is, so $|b| \leq a$ forces $b = \pm 1$. Then $a \neq 2$, since $a = 2$ implies $b^2 - 4ac = 1 - 8c \neq -19$. Thus $a = 1$, so that (since $a = |b|$) $b = 1$ as well and the only reduced form is $x^2 + xy + 5y^2$. Thus $h(-19) = 1$, so that $|C(\mathcal{O}_K)| = h(-19) = 1$ by Theorem 5.30.

5.21. (a) $\mathbb{Z}[\sqrt{-2}]$ is the ring of integers $\mathcal{O}_K$ of $K = \mathbb{Q}(\sqrt{-2})$ by (5.13). By (2.14), the only reduced form of discriminant -8 is $x^2 + 2y^2$, so that $|C(\mathcal{O}_K)| = h(-8) = 1$ by Theorem 5.30. It follows that $\mathbb{Z}[\sqrt{-2}]$ is a PID and therefore a UFD.

(b) $N(\sqrt{-2}) = N(0 + 1\sqrt{-2}) = 0^2 + 1^2 \cdot 2 = 2$, which is prime, so $\sqrt{-2}$ is prime as well.

(c) Since $\mathbb{Z}[\sqrt{-2}]$ is a UFD, a and b each factor uniquely as a unit times a product of distinct primes to various exponents. Then $\gcd(a, b) = 1$ implies that the factorizations have no primes in common, so that each prime in each factorization must appear with an exponent that is a multiple of 3. Since the units ± 1 are also cubes $(1 = 1^3, -1 = (-1)^3)$, it follows that both a and b are cubes in $\mathbb{Z}[\sqrt{-2}]$.

5.22. Suppose $x^3 = y^2 + 2$, $x, y \in \mathbb{Z}$.

(a) Let $d = \gcd(y + \sqrt{-2}, y - \sqrt{-2})$. Then d divides their difference $2\sqrt{-2} = -(\sqrt{-2}\,)^3$. Since $\sqrt{-2}$ is prime by part (b) of Exercise 5.21, either d is a unit or d is associate to $(\sqrt{-2})^i$, $1 \leq i \leq 3$. Suppose the second case holds; then $\sqrt{-2}$ divides $y + \sqrt{-2}$, giving

$$y + \sqrt{-2} = (a + b\sqrt{-2})\sqrt{-2} = -2b + a\sqrt{-2}, \quad a, b \in \mathbb{Z}.$$

Then y must be even, so that $x^3 \equiv 2 \bmod 4$, which is impossible. Thus d is a unit, and $y + \sqrt{-2}$ and $y - \sqrt{-2}$ are relatively prime.

(b) From part (c) of Exercise 5.21, both $y + \sqrt{-2}$ and $y - \sqrt{-2}$ must be cubes in $\mathbb{Z}[\sqrt{-2}]$. If

$$y + \sqrt{-2} = (a + b\sqrt{-2}\,)^3 = (a^3 - 6ab^2) + (3a^2 b - 2b^3)\sqrt{-2}, \quad a, b \in \mathbb{Z},$$

then $3a^2b - 2b^3 = b(3a^2 - 2b^2) = 1$, so that $b = \pm 1$ and therefore $3a^2 - 2b^2 = 3a^2 - 2 = b = \pm 1$. Then $3a^2 = b + 2 = 1$ or 3. The former is impossible, so $b = 1$ and $a = \pm 1$. But then

$$y = a^3 - 6ab^2 = \pm 1 \mp 6 = \pm 5,$$

so that $x^3 = 27$ and $x = 3$. Thus the only integer solutions are $(x, y) = (3, \pm 5)$.

5.23. The analog of Theorem 5.26 is:

THEOREM 5.26. *Let L be the Hilbert class field of $K = \mathbb{Q}(\sqrt{-n})$. Assume that $n > 0$ is squarefree with $n \equiv 3 \bmod 4$, so that $\mathcal{O}_K = \mathbb{Z}[(1 + \sqrt{-n})/2]$. If p is an odd prime not dividing n, then*

$$p = x^2 + xy + \frac{n+1}{4}y^2 \iff p \text{ splits completely in } L.$$

PROOF. If p is an odd prime not dividing n, then it does not divide $d_K = -n$, so that p is unramified in K. We claim that

$$\begin{aligned} p = x^2 + xy + \frac{n+1}{4}y^2 &\iff p\mathcal{O}_K = \mathfrak{p}\overline{\mathfrak{p}},\ \mathfrak{p} \neq \overline{\mathfrak{p}},\ \mathfrak{p} \text{ is principal in } \mathcal{O}_K \\ &\iff p\mathcal{O}_K = \mathfrak{p}\overline{\mathfrak{p}},\ \mathfrak{p} \neq \overline{\mathfrak{p}},\ \mathfrak{p} \text{ splits completely in } L \\ &\iff p \text{ splits completely in } L. \end{aligned}$$

For the first equivalence, suppose $p = x^2 + xy + ((n+1)/4)y^2 = (x + wy)(x + \overline{w}y)$, where $w = (1 + \sqrt{-n})/2$. Let $\mathfrak{p} = (x + wy)\mathcal{O}_K$. Then $\mathfrak{p}$ is prime since it has prime norm, so that $\mathfrak{p}\overline{\mathfrak{p}}$ must be the unique factorization of $p\mathcal{O}_K$ in K. Since p is unramified in K we must have $\mathfrak{p} \neq \overline{\mathfrak{p}}$, and $\mathfrak{p}$ is principal by construction. For the converse, if $p\mathcal{O}_K = \mathfrak{p}\overline{\mathfrak{p}}$ with $\mathfrak{p}$ principal, then $\mathfrak{p} = (x + wy)\mathcal{O}_K$, and thus

$$p\mathcal{O}_K = (x + wy)(x + \overline{w}y)\mathcal{O}_K = \Big(x^2 + xy + \frac{n+1}{4}y^2\Big)\mathcal{O}_K.$$

It follows that $x^2 + xy + ((n+1)/4)y^2 = pu$ for $u \in \mathcal{O}_K$ a unit. Thus pu is positive and real, so that $u = 1$ and then $x^2 + xy + ((n+1)/4)y^2 = p$.

The proofs of the remaining two equivalences are identical to the proof of Theorem 5.26 in the text. Q.E.D.

The analog of Theorem 5.1 is:

THEOREM 5.1. *Let $n > 0$ be a squarefree integer with $n \equiv 3 \bmod 4$. Then there is a monic irreducible polynomial $f_n(x) \in \mathbb{Z}[x]$ of degree $h(-n)$ such that if an odd prime p divides neither n nor the discriminant of $f_n(x)$, then*

$$p = x^2 + xy + \frac{1+n}{4}y^2 \iff \begin{cases} (-n/p) = 1 \text{ and } f_n(x) \equiv 0 \bmod p \\ \text{has an integer solution.} \end{cases}$$

Furthermore, $f_n(x)$ may be taken to be the minimal polynomial of a real algebraic integer α for which $L = K(\alpha)$ is the Hilbert class field of $K = \mathbb{Q}(\sqrt{-n})$.

PROOF. The proof is very similar to the proof of Theorem 5.1 in the text. Since L is Galois over $\mathbb{Q}$ by Lemma 5.28, Proposition 5.29 shows that $L = K(\alpha)$ for some real algebraic integer $\alpha \in L$. Let $f_n(x)$ be the monic minimal polynomial of α, and let p be an odd prime dividing neither n nor the discriminant of $f_n(x)$. Then Theorem 5.26 from this exercise together with Proposition 5.29 show that

$$p = x^2 + xy + \frac{1+n}{4}y^2 \iff \begin{cases} (-n/p) = 1 \text{ and } f_n(x) \equiv 0 \bmod p \\ \text{has an integer solution.} \end{cases}$$

(Note that $d_K = -n$ in this case.)

Finally, using Galois theory and Theorem 5.23, $\deg(f_n(x)) = [L : K] = |\mathrm{Gal}(L/K)| = |C(\mathcal{O}_K)|$. But $C(\mathcal{O}_K) \simeq C(d_K) = C(-n)$. Thus $f_n(x)$ has degree $|h(-n)|$ and we are done. Q.E.D.

5.24. With $D = b^2 - 4c$, the roots of $x^4 + bx^2 + c$ are

$$r_1 = \sqrt{\frac{-b+\sqrt{D}}{2}},\ r_2 = \sqrt{\frac{-b-\sqrt{D}}{2}},\ r_3 = -\sqrt{\frac{-b+\sqrt{D}}{2}} = -r_1,\ r_4 = -\sqrt{\frac{-b-\sqrt{D}}{2}} = -r_2,$$

and the discriminant of the quartic is

$$\begin{aligned}\Delta &= \prod_{1\le i<j\le 4} (r_i - r_j)^2 \\ &= (r_1-r_2)^2(r_1-r_3)^2(r_1-r_4)^2(r_2-r_3)^2(r_2-r_4)^2(r_3-r_4)^2 \\ &= (r_1-r_2)^2\ \ (2r_1)^2\ \ (r_1+r_2)^2(r_2+r_1)^2\ \ (2r_2)^2\ \ (-r_1+r_2)^2 \\ &= 2^4 r_1^2 r_2^2 (r_1-r_2)^4 (r_1+r_2)^4 = 2^4 r_1^2 r_2^2 (r_1^2 - r_2^2)^4.\end{aligned}$$

However,

$$r_1^2 r_2^2 = \left(\frac{-b+\sqrt{D}}{2}\right)\left(\frac{-b-\sqrt{D}}{2}\right) = \frac{b^2-D}{4} = \frac{b^2-(b^2-4c)}{4} = c$$

$$r_1^2 - r_2^2 = \left(\frac{-b+\sqrt{D}}{2}\right) - \left(\frac{-b-\sqrt{D}}{2}\right) = \sqrt{D} = \sqrt{b^2-4c}.$$

Then the above formula for the discriminant simplifies to

$$\Delta = 2^4 r_1^2 r_2^2 (r_1^2 - r_2^2)^4 = 2^4 c \left(\sqrt{b^2-4c}\right)^4 = 2^4 c (b^2-4c)^2.$$

5.25. (a) The discriminant is $d_K = -68$, and the reduced forms of this discriminant are

$$x^2 + 17y^2, \quad 2x^2 + 2xy + 9y^2, \quad 3x^2 - 2xy + 6y^2, \quad 3x^2 + 2xy + 6y^2.$$

Since the last two forms are inverses of each other, they each have order four, so the form class group is isomorphic to $\mathbb{Z}/4\mathbb{Z}$. Then Theorem 5.30 shows that $C(\mathcal{O}) \simeq \mathbb{Z}/4\mathbb{Z}$ as well.

(b) We first show that $L = K(\alpha)$, $\alpha = \sqrt{(1+\sqrt{17})/2}$, is a Galois extension of degree 4 of $K = \mathbb{Q}(\sqrt{-17})$. One checks that α is a root of $f(x) = x^4 - x^2 - 4$ and hence is an algebraic integer. Also note that $\alpha^2 = (1+\sqrt{17})/2$, so that $\mathbb{Q}(\sqrt{17}) \subset \mathbb{Q}(\alpha)$. There are various ways to show that $f(x)$ is irreducible over $\mathbb{Q}$ (use a computer, or note that $\mathbb{Q}(\sqrt{17}) \subset \mathbb{Q}(\alpha)$ and show explicitly that $\alpha \notin \mathbb{Q}(\sqrt{17})$ since we know $\mathcal{O}_{\mathbb{Q}(\sqrt{17})}$). Thus $[\mathbb{Q}(\alpha):\mathbb{Q}] = 4$. Since $\sqrt{-17} = i\sqrt{17}$ and $\sqrt{17} \in \mathbb{Q}(\alpha)$, we obtain

$$L = K(\alpha) = \mathbb{Q}(\sqrt{-17}, \alpha) = \mathbb{Q}(i, \alpha).$$

Thus L has degree 2 over $\mathbb{Q}(\alpha)$ since $\alpha \in \mathbb{R}$, so that

$$[L:\mathbb{Q}] = [\mathbb{Q}(i,\alpha):\mathbb{Q}(\alpha)][\mathbb{Q}(\alpha):\mathbb{Q}] = 8.$$

Since $[K:\mathbb{Q}] = 2$, it follows immediately that $[L:K] = 4$ and then that L/K is Galois.

We next claim that $K(\alpha)$ is Galois over $\mathbb{Q}$. Let $\beta = \sqrt{(\sqrt{17}-1)/2}$. An easy computation shows that $\alpha\beta = 2$, so that $\beta = 2/\alpha \in K(\alpha)$. Since the roots of $g(x) = (x^2+1)(x^4-x^2-4)$ are

$$\pm i,\ \pm\alpha,\ \pm i\beta,$$

it follows that $K(\alpha) = \mathbb{Q}(i,\alpha) = \mathbb{Q}(\pm i, \pm\alpha, \pm i\beta)$ is the splitting field of $g(x)$ over $\mathbb{Q}$, proving our claim. We conclude that $L = K(\alpha)$ is Galois of degree 4 over K, so in particular $\mathrm{Gal}(L/K)$ is Abelian.

Our final claim is that L is unramified over K. As noted above, $\sqrt{17} \in L$. Let $K_1 = K(\sqrt{17})$; then K_1 lies strictly between K and L, and thus $[L:K_1] = [K_1:K] = 2$. Note $K_1 = K(\sqrt{17}) = K(i)$ since $K = \mathbb{Q}(\sqrt{-17})$.

Choose $\mathfrak{p} \subset \mathcal{O}_K$ prime, and first suppose that $2 \notin \mathfrak{p}$. Considering $K_1 = K(i) = K(\sqrt{-1})$, $u = -1 \notin \mathfrak{p}$ since otherwise $\mathfrak{p} = \mathcal{O}_K$. So $\mathfrak{p}$ is unramified in K_1 by part (i) of Lemma 5.32. Next suppose that $2 \in \mathfrak{p}$, and consider $K_1 = K(\sqrt{17})$; then $u = 17 \notin \mathfrak{p}$ since otherwise $\gcd(2,17) = 1 \in \mathfrak{p}$. Further, $17 = 1^2 - 4\cdot(-4)$, so that by part (ii) of Lemma 5.32, $\mathfrak{p}$ is again unramified in K_1. Thus $K \subset K_1$ is unramified.

For the extension $K_1 \subset L$, we have $L = K_1(\sqrt{\mu})$, $\mu = (1+\sqrt{17})/2$. Let $\mu' = (1-\sqrt{17})/2$. Then $\sqrt{\mu}\sqrt{\mu'} = \sqrt{\mu\mu'} = \sqrt{-4} = 2i \in K_1 = K(i)$, so that

$$L = K_1(\sqrt{\mu}) = K_1(\sqrt{\mu'})$$

since $\sqrt{\mu'} = 2i/\sqrt{\mu} \in K_1(\sqrt{\mu})$ and $\sqrt{\mu} = 2i/\sqrt{\mu'} \in K_1(\sqrt{\mu'})$. Now let $\mathfrak{p}$ be prime in K_1. Since $\mu + \mu' = 1$, either μ or μ' is not in $\mathfrak{p}$. So if $2 \notin \mathfrak{p}$, $\mathfrak{p}$ is unramified by part (i) of Lemma 5.32. If $2 \in \mathfrak{p}$, note that both μ and μ' satisfy $x = x^2 - 4$, and that $\mu^2, \mu'^2 \in \mathcal{O}_K$. Since either μ or μ' is not in $\mathfrak{p}$, we see that the conditions of part (ii) of Lemma 5.32 are satisfied, so that again $\mathfrak{p}$ is unramified.

The above paragraphs, combined with part (c) of Exercise 5.15, imply that L is an unramified Abelian extension of K of degree 4. Hence L is contained in the maximal unramified Abelian extension of K, namely its Hilbert class field. Since $h(-68) = 4$, the Hilbert class field also has degree 4 over K by Theorem 5.23. Since it contains L, they are equal, which proves that $L = K(\alpha)$ is the Hilbert class field of $K = \mathbb{Q}(\sqrt{-17})$.

5.26. By Exercise 5.25, $L = K(\alpha)$, $\alpha = \sqrt{(1+\sqrt{17})/2}$, is the Hilbert class field of $K = \mathbb{Q}(\sqrt{-17})$. The solution shows α is a real algebraic integer with minimal polynomial $f_{17}(x) = x^4 - x^2 - 4$. By Exercise 5.24, this polynomial has discriminant $2^4(-4)(1+4\cdot4)^2 = -2^6\cdot17^2$, so the only excluded primes are 2 and 17. By Theorem 5.1, we thus get

THEOREM. *If* $p \neq 17$ *is an odd prime, then*

$$p = x^2 + 17y^2 \iff \begin{cases} (-17/p) = 1 \text{ and } x^4 - x^2 - 4 \equiv 0 \bmod p \\ \text{has an integer solution.} \end{cases}$$

The only primes under 400 so representable are

$$53 = 6^2 + 17 \cdot 1^2, \quad 149 = 9^2 + 17 \cdot 2^2, \quad 157 = 2^2 + 17 \cdot 3^2, \quad 281 = 3^2 + 17 \cdot 4^2,$$
$$293 = 15^2 + 17 \cdot 2^2, \quad 349 = 14^2 + 17 \cdot 3^2, \quad 353 = 9^2 + 17 \cdot 4^2.$$

Solutions to Exercises in §6

6.1. Although the proof of Proposition 3.11 applies only to negative discriminants, the integer μ defined in the proposition makes sense for all discriminants. Let r be the number of odd primes dividing a field discriminant d_K. If $d_K \equiv 1 \bmod 4$, then since d_K is odd, Proposition 3.11 gives $\mu = r$. If $d_K \equiv 0 \bmod 4$, then we can write $d_K = -4n$ for some squarefree integer $-n \not\equiv 1 \bmod 4$, i.e., $n \not\equiv 3 \bmod 4$. We also know that $n \not\equiv 0 \bmod 4$ since n is squarefree. Thus $n \equiv 1, 2 \bmod 4$, so that $\mu = r + 1$ by the table in Proposition 3.11. This is the number of prime divisors of d_K since d_K is even.

6.2. We first recall the following standard result from group theory: if G is a group, and N a subgroup of G, then $[G, G] \subset N \Leftrightarrow N$ is a normal subgroup of G and G/N is Abelian. Here, $[G, G]$ is the subgroup of G generated by the commutators $[g, h] = g^{-1}h^{-1}gh$ for all $g, h \in G$.

We apply this result to the case where $G = \mathrm{Gal}(L/K)$ for a Galois extension $K \subset L$. Note that L is Galois over M with Galois group $H = \mathrm{Gal}(L/M)$. If $[G, G] \subset H$, then by the above, H is normal in G and G/H is Abelian. Thus by Galois theory, M is Galois over K with Galois group

$$\mathrm{Gal}(M/K) \simeq \mathrm{Gal}(L/K)/\mathrm{Gal}(L/M) = G/H,$$

so that M is Abelian over K. Conversely, if M is Abelian over K, then $\mathrm{Gal}(M/K) \simeq G/H$ is Abelian, so that $[G, G] \subset H = \mathrm{Gal}(L/M)$.

6.3. Note that since L/K is an unramified Abelian extension, $((L/K)/\mathfrak{p})$ is defined, and is the same as $((L/K)/\mathfrak{P})$ for any $\mathfrak{P}$ lying over $\mathfrak{p}$. By part (a) of Exercise 5.12, we have

$$\left(\frac{L/K}{\tau(\mathfrak{P})}\right) = \tau\left(\frac{L/K}{\mathfrak{P}}\right)\tau^{-1}.$$

Since $\tau(\mathfrak{P})$ lies over $\tau(\mathfrak{p})$, we conclude that

$$\left(\frac{L/K}{\tau(\mathfrak{p})}\right) = \tau\left(\frac{L/K}{\mathfrak{p}}\right)\tau^{-1}.$$

6.4. Suppose M is an unramified Abelian extension of K; then $\mathbb{Q} \subset K \subset M \subset L$, where L is the Hilbert class field of K. As noted in the text, L is a Galois extension of $\mathbb{Q}$, and $\mathrm{Gal}(L/\mathbb{Q}) \simeq \mathrm{Gal}(L/K) \rtimes \mathbb{Z}/2\mathbb{Z}$, where $\mathbb{Z}/2\mathbb{Z}$ acts on $\mathrm{Gal}(L/K)$ by conjugation.

It suffices to show that $\mathrm{Gal}(L/M)$ is normal in $\mathrm{Gal}(L/\mathbb{Q})$. Since $\mathrm{Gal}(L/K)$ is Abelian, elements of $\mathrm{Gal}(L/M)$ commute with elements of $\mathrm{Gal}(L/K)$. It remains to analyze conjugation with τ. Since conjugation by τ acts by sending each element of $C(\mathcal{O}_K)$ to its inverse (see the discussion following (6.4)), it also sends each element of $\mathrm{Gal}(L/K)$, and thus of $\mathrm{Gal}(L/M)$, to its inverse. So if $\sigma \in \mathrm{Gal}(L/M)$, then $\tau\sigma\tau^{-1} = \sigma^{-1} \in \mathrm{Gal}(L/M)$. It follows that $\mathrm{Gal}(L/M)$ is normal in $\mathrm{Gal}(L/\mathbb{Q})$, so that M is Galois over $\mathbb{Q}$.

6.5. As in the text, we identity $G = \mathrm{Gal}(L/\mathbb{Q})$ with $C(\mathcal{O}_K) \rtimes \mathbb{Z}/2\mathbb{Z}$, and the first isomorphism follows immediately. Now consider the map

$$(*) \qquad C(\mathcal{O}_K) \rtimes \mathbb{Z}/2\mathbb{Z} \longrightarrow (C(\mathcal{O}_K)/C(\mathcal{O}_K)^2) \times \mathbb{Z}/2\mathbb{Z}$$

defined by $([\mathfrak{a}], \gamma) \mapsto ([\mathfrak{a}] \bmod C(\mathcal{O}_K)^2, \gamma)$, where $[\mathfrak{a}]$ denotes the class of $\mathfrak{a}$. In the semidirect product $C(\mathcal{O}_K) \rtimes \mathbb{Z}/2\mathbb{Z}$, the action of the nontrivial element of $\mathbb{Z}/2\mathbb{Z}$ on $C(\mathcal{O}_K)$ is given by $[\mathfrak{a}] \mapsto [\overline{\mathfrak{a}}] = [\mathfrak{a}^{-1}]$. Hence it acts trivially on $C(\mathcal{O}_K)/C(\mathcal{O}_K)^2$ because $\mathfrak{a} \equiv \mathfrak{a}^{-1} \bmod C(\mathcal{O}_K)^2$. It follows easily that (*) is a group homomorphism. It is clearly onto, and its kernel is $C(\mathcal{O}_K)^2 \times \{0\}$, so that (*) induces an isomorphism

$$(C(\mathcal{O}_K) \rtimes \mathbb{Z}/2\mathbb{Z})/C(\mathcal{O}_K)^2 \simeq (C(\mathcal{O}_K)/C(\mathcal{O}_K)^2) \times \mathbb{Z}/2\mathbb{Z}.$$

6.6. (a) Since $L, M \subset LM$, this follows immediately from part (b) of Exercise 5.15.

(b) Let $\mathfrak{p}$ be a prime of K, and let $\mathfrak{P}$ be a prime of LM lying over $\mathfrak{p}$ with ramification degree e. Let $\mathfrak{P}_L = \mathfrak{P} \cap \mathcal{O}_L$; then $\mathfrak{P}_L$ lies over $\mathfrak{p}$ and under $\mathfrak{P}$. Choose $\sigma \in I_{\mathfrak{P}}$, so σ induces the identity map on $\mathcal{O}_{LM}/\mathfrak{P}$. Then for $\alpha \in \mathcal{O}_L$, $\sigma(\alpha) \equiv \alpha \bmod \mathfrak{P}$ implies $\sigma|_L(\alpha) \equiv \alpha \bmod \mathfrak{P}_L$, so that $\sigma|_L \in I_{\mathfrak{P}_L}$. But L is unramified over K, so $I_{\mathfrak{P}_L}$ is trivial by Proposition 5.10, and thus $\sigma|_L$ is the identity map. Similar considerations hold for $\sigma|_M$. Finally, since the map

$$\mathrm{Gal}(LM/K) \hookrightarrow \mathrm{Gal}(L/K) \times \mathrm{Gal}(M/K)$$

is an injection given by restriction in each factor, σ must be the identity as well. Thus $e = |I_{\mathfrak{P}}| = 1$ and $\mathfrak{p}$ is unramified in LM.

(c) By Exercise 5.16, the restriction map $\mathrm{Gal}(LM/K) \to \mathrm{Gal}(L/K)$ maps $((LM/K)/\mathfrak{p})$ to $((L/K)/\mathfrak{p})$, and similarly for $\mathrm{Gal}(M/K)$.

(d) Applying Corollary 5.21 to the Abelian extension $K \subset LM$, we see that $\mathfrak{p}$ splits completely in LM if and only if $((LM/K)/\mathfrak{p}) = 1$, with similar statements for L and M. By part (c), the map

$$\mathrm{Gal}(LM/K) \longrightarrow \mathrm{Gal}(L/K) \times \mathrm{Gal}(M/K)$$

takes $((LM/K)/\mathfrak{p})$ to $(((L/K)/\mathfrak{p}), ((M/K)/\mathfrak{p}))$. Part (ii) of Lemma 6.7 shows that this map is injective, and it follows that $\mathfrak{p}$ splits completely in LM if and only if it splits completely in L and M.

6.7. The next three exercises prove Lemma 6.8.

(a) Since $(1+i)^2 = 2i$ and $\alpha = (1+i)\sqrt{2m}/2$, we have $\alpha^2 = (1+i)^2 \cdot 2m/4 = 2i \cdot m/2 = im$. Thus α is a root of $x^4 + m^2$ and hence an algebraic integer. It lies in $K = \mathbb{Q}(i, \sqrt{2m})$, so that $\alpha \in \mathcal{O}_K$.

(b) The definition of $\mathfrak{P}$ implies that $\mathfrak{P}^2$ is the ideal generated by $(1+i)^2 = 2i$, $(1+i)(1+\alpha)$ and $(1+\alpha)^2$. We first show that $\mathfrak{P}^2$ is generated by $2i$, $1+i$ and $\sqrt{2m}$. Let $\mathfrak{I}$ be the ideal generated by these elements. To show that $\mathfrak{P}^2 \subset \mathfrak{I}$, first note that $2i$ and $(1+i)(1+\alpha)$ lie in $\mathfrak{I}$. Furthermore, writing $m = 2k+1$ for $k \in \mathbb{Z}$ (m is odd), we have

$$(*) \qquad \begin{aligned}(1+\alpha)^2 &= 1 + 2\alpha + \alpha^2 = 1 + (1+i)\sqrt{2m} + im = 1 + (1+i)\sqrt{2m} + i(2k+1)\\ &= k \cdot 2i + (1 + \sqrt{2m})(1+i) \in \mathfrak{I}.\end{aligned}$$

Proving $\mathfrak{I} \subset \mathfrak{P}^2$ takes a bit more work. Obviously $2i \in \mathfrak{P}^2$. Then observe that $(1+i)\alpha = (1+i)^2\sqrt{2m}/2 = 2i \cdot \sqrt{2m}/2 = i\sqrt{2m}$. Using this and (*), we obtain

$$\begin{aligned}(1+i)(1+\alpha) &= (1+i) + (1+i)\alpha = 1 \cdot (1+i) + i \cdot \sqrt{2m}\\ (1+\alpha)^2 - k \cdot 2i &= (1+\sqrt{2m})(1+i) = 1 \cdot (1+i) + (1+i) \cdot \sqrt{2m}.\end{aligned}$$

Since the left-hand side of each equation lies in $\mathfrak{P}^2$, subtracting the first equation from the second shows that $\sqrt{2m} \in \mathfrak{P}^2$, and then this and the first equation show that $1+i \in \mathfrak{P}^2$.

Hence we have proved that $\mathfrak{P}^2$ is generated by $2i$, $1+i$ and $\sqrt{2m}$. It follows that $\mathfrak{P}^4$ is generated by

$$(2i)^2 = -4,\ (1+i)^2 = 2i,\ \sqrt{2m}^{\,2} = 2m,$$
$$2i \cdot (1+i),\ 2i \cdot \sqrt{2m},\ (1+i) \cdot \sqrt{2m} = 2\alpha.$$

All generators are multiples of 2, so that $\mathfrak{P}^4 \subset 2\mathcal{O}_K$. However, $2 = -i(1+i)^2 \in \mathfrak{P}^4$, so that $2\mathcal{O}_K \subset \mathfrak{P}^4$, and equality follows. But $[K:\mathbb{Q}] = 4$ and K is Galois over $\mathbb{Q}$ (it is a compositum of Galois extensions), so $\mathfrak{P}$ must be prime since $efg = 4$.

6.8. (a) We have the following diagram of fields:

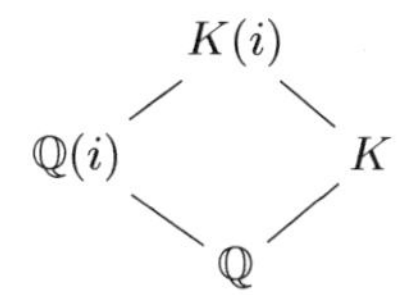

Note that 2 is unramified in K by Corollary 5.17 since $2 \nmid d_K \equiv 1 \bmod 4$. If $K \subset K(i)$ is unramified, then 2 is unramified in $K(i)$ by part (b) of Exercise 5.15. But $2 \mid -4 = d_{\mathbb{Q}(i)}$, so that 2 ramifies in $\mathbb{Q}(i)$ by Corollary 5.17 and hence must ramify in $K(i)$, a contradiction. Thus $K \subset K(i)$ is ramified when $d_K \equiv 1 \bmod 4$.

(b) Note that $d_K \equiv 0 \bmod 8$ implies that $K = \mathbb{Q}(\sqrt{2m})$ where m is odd and squarefree. Since $2 \mid d_K$, 2 is ramified in K, and we must have $e_{\mathfrak{P}_K|2\mathbb{Z}} = 2$ where $\mathfrak{P}_K$ is the unique prime of K lying over $2\mathbb{Z}$ since $[K:\mathbb{Q}] = 2 = efg$. From Exercise 6.7, we know that $2\mathcal{O}_{K(i)} = \mathfrak{P}^4$ where $\mathfrak{P}$ is prime in $K(i) = \mathbb{Q}(i, \sqrt{2m})$. Then $\mathfrak{P}$ must lie over $\mathfrak{P}_K$, and

$$4 = e_{\mathfrak{P}|2\mathbb{Z}} = e_{\mathfrak{P}|\mathfrak{P}_K} e_{\mathfrak{P}_K|2\mathbb{Z}} = 2e_{\mathfrak{P}|\mathfrak{P}_K},$$

showing that $e_{\mathfrak{P}|\mathfrak{P}_K} = 2$ and therefore $K \subset K(i)$ is ramified.

If d_K is odd, then $d_K \equiv 1 \bmod 4$, so that $K \subset K(i)$ is ramified by part (a). Otherwise, $d_K = 4N$, $N \equiv 2, 3 \bmod 4$, so that $d_K \equiv 8, 12 \bmod 16$. If $d_K \equiv 8 \bmod 16$ then $K \subset K(i)$ is ramified by part (b). Thus if $K \subset K(i)$ is unramified, we must have $d_K \equiv 12 \bmod 16$ (so that $K = \sqrt{N}$, $N \equiv 3 \bmod 4$).

6.9. (a) Assume $a \mid d_K$ and $a \equiv 1 \bmod 4$ and let $\mathfrak{p}$ be a prime of K. First suppose $2 \in \mathfrak{p}$. Then $a \notin \mathfrak{p}$ since otherwise $1 = \gcd(a, 2) \in \mathfrak{p}$. Since $a \equiv 1 \bmod 4$, we have $a = a^2 - 4c$ for some c. So by part (ii) of Lemma 5.32, $\mathfrak{p}$ is unramified in $K(\sqrt{a})$. Next suppose $2 \notin \mathfrak{p}$. Since $a \mid d_K$, $d_K = ab$. Then $K(\sqrt{b}) = K(\sqrt{a})$, since $\sqrt{a}\sqrt{b} = \sqrt{d_K}$ and $\sqrt{d_K} \in K \subset K(\sqrt{a}), K(\sqrt{b})$. Since d_K is squarefree except for a possible factor of 4, we have $\gcd(a, b) = 1$ or 2, so that $a\mathcal{O}_K + b\mathcal{O}_K = \mathcal{O}_K$ or $2\mathcal{O}_K$. Since $2 \notin \mathfrak{p}$, we see in either case that either a or b is not in $\mathfrak{p}$. So by part (i) of Lemma 5.32, since either $2a$ or $2b \notin \mathfrak{p}$, $\mathfrak{p}$ is unramified in $K(\sqrt{a})$.

(b) Now assume $K \subset K(\sqrt{a})$ is unramified with a squarefree. Consider the diagram

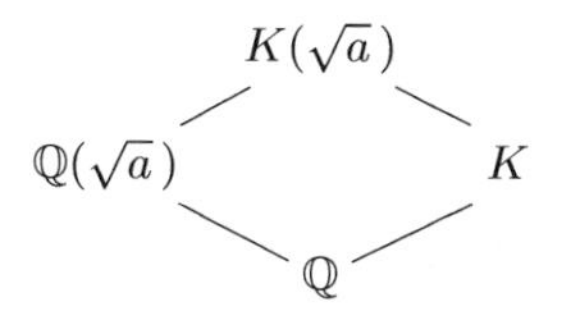

If $a \nmid d_K$, then since a is squarefree we can choose a prime p such that $p \mid a$ and $p \nmid d_K$, so that p is unramified in K, but is ramified in $\mathbb{Q}(\sqrt{a})$. The latter implies that p is ramified in $K(\sqrt{a})$, so that $K \subset K(\sqrt{a})$ must ramify at some prime over p. This contradiction proves that $a \mid d_K$.

(c) If d_K is even, Proposition 5.16 shows that $2\mathcal{O}_K = \mathfrak{p}^2$ for some prime $\mathfrak{p}$. Since $K \subset L = K(\sqrt{a})$ is assumed unramified, $\mathfrak{p}$ does not ramify in L. Let $\mathfrak{P}$ lie over $\mathfrak{p}$; then $e_{\mathfrak{P}|2\mathbb{Z}} = 2$. It follows that $|I_{\mathfrak{P}}(L/\mathbb{Q})| = e_{\mathfrak{P}|2\mathbb{Z}} = 2$ by Proposition 5.10, so that the fixed field K' of $I_{\mathfrak{P}}(L/\mathbb{Q})$ is a field lying between $\mathbb{Q}$ and L with $[K:\mathbb{Q}] = 2$. Since $\mathrm{Gal}(L/K') = I_{\mathfrak{P}}(L/\mathbb{Q})$, we get

$$\begin{aligned} I_{\mathfrak{P}}(L/K') &= \{\sigma \in \mathrm{Gal}(L/K') \mid \sigma(\alpha) \equiv \alpha \bmod \mathfrak{P} \text{ for all } \alpha \in \mathcal{O}_L\} \\ &= \{\sigma \in I_{\mathfrak{P}}(L/\mathbb{Q}) \mid \sigma(\alpha) \equiv \alpha \bmod \mathfrak{P} \text{ for all } \alpha \in \mathcal{O}_L\} = I_{\mathfrak{P}}(L/\mathbb{Q}). \end{aligned}$$

Therefore $e_{\mathfrak{P}|\mathfrak{P}\cap\mathcal{O}_{K'}} = e_{\mathfrak{P}|2\mathbb{Z}} = 2$, and then part (a) of Exercise 5.15 implies that $e_{\mathfrak{P}\cap\mathcal{O}_{K'}|2\mathbb{Z}} = 1$, so that 2 is unramified in K'. Thus $d_{K'}$ is odd, so that $K' = \mathbb{Q}(\sqrt{a'})$ for a' odd. But then $\mathbb{Q}(\sqrt{d_K}, \sqrt{a'}) \subset L$, so they are equal by degree considerations. Thus $L = K(\sqrt{a'})$, and $a' \mid d_K$ by part (b).

(d) Here we assume that $K \subset K(\sqrt{a})$ is unramified for $a \mid d_K$ odd.

(i) If $a \equiv 3 \bmod 4$, then $-a \equiv 1 \bmod 4$; since $-a \mid d_K$, $K \subset K(\sqrt{-a})$ is unramified by part (a). By Lemma 6.7, their compositum $K(\sqrt{a}, \sqrt{-a}) = K(i, \sqrt{a})$ is unramified over K. But $K \subset K(i) \subset K(i, \sqrt{a})$, so that $K(i)$ must be unramified over K. Then Exercise 6.8 implies that $d_K \equiv 12 \bmod 16$.

(ii) If $a \equiv 3 \bmod 4$ and $d_K \equiv 12 \bmod 16$, let $b = d_K/4a$. Then $K(\sqrt{a}) = K(\sqrt{b})$ since $\sqrt{a}\sqrt{b} = 2\sqrt{d_K}$. But

$$12 \equiv d_K = 4ab \equiv 12b \bmod 16.$$

where the final congruence follows from $a \equiv 3 \bmod 4$. This easily implies $b \equiv 1 \bmod 4$.

Part (a) proves $\Leftarrow$ in Lemma 6.8. For $\Rightarrow$, parts (b) and (c) show that we can choose $a \mid d_K$ and odd, and then part (d) allows us to choose an equivalent $b \equiv 1 \bmod 4$ dividing d_K if $a \equiv 3 \bmod 4$.

6.10. Choose a $(\mathbb{Z}/2\mathbb{Z})$-basis $\sigma_1, \ldots, \sigma_m$ for $\mathrm{Gal}(M/\mathbb{Q})$, and let H_i be the subgroup generated by $\sigma_1, \ldots, \sigma_{i-1}, \sigma_{i+1}, \ldots, \sigma_m$. Since $|H_i| = 2^{m-1}$, its fixed field M_i has degree 2 over $\mathbb{Q}$, and thus $M_i = \mathbb{Q}(\sqrt{a_i})$ for some squarefree $a_i \in \mathbb{Z}$. We claim that $M = \mathbb{Q}(\sqrt{a_1}, \ldots, \sqrt{a_m})$. To prove this, suppose that $\sigma \in \mathrm{Gal}(M/\mathbb{Q})$ is the identity on $\mathbb{Q}(\sqrt{a_1}, \ldots, \sqrt{a_m})$. Then it is the identity on M_i, hence $\sigma \in \mathrm{Gal}(M/M_i) = H_i$. This is true for all i, so that $\sigma \in H_1 \cap \cdots \cap H_m = \{1\}$. By Galois theory, it follows that $M = \mathbb{Q}(\sqrt{a_i}, \ldots, \sqrt{a_m})$.

6.11. Since $d_K = -4n$, we know that $n \equiv 1 \bmod 4$ if n is odd and $n \equiv 2, 6 \bmod 8$ if n is even. Also, since n is squarefree, $n = 2^\epsilon \prod_{i=1}^r p_i$, $\epsilon \in \{0, 1\}$, p_i distinct. If $n \equiv 1 \bmod 4$ or $2 \bmod 8$, then an even number of the p_i are $3 \bmod 4$; otherwise an odd number are, so that

$$n = \begin{cases} \prod_{i=1}^r p_i^* & n \equiv 1 \bmod 4 \\ -2\prod_{i=1}^r p_i^* & n \equiv 6 \bmod 8 \\ 2\prod_{i=1}^r p_i^* & n \equiv 2 \bmod 8. \end{cases}$$

It follows that

$$\sqrt{d_K} = \sqrt{-4n} = 2\sqrt{-n} = \begin{cases} 2i\prod_{i=1}^r \sqrt{p_i^*} & n \equiv 1 \bmod 4 \\ 2\sqrt{2}\prod_{i=1}^r \sqrt{p_i^*} & n \equiv 6 \bmod 8 \\ 2\sqrt{-2}\prod_{i=1}^r \sqrt{p_i^*} & n \equiv 2 \bmod 8. \end{cases}$$

Comparing this to $M = \mathbb{Q}(\sqrt{d_K}, \sqrt{p_1^*}, \ldots, \sqrt{p_r^*})$ gives (6.9).

6.12. We assume $d_K \equiv 1 \bmod 4$, $d_K < 0$. Then $\mu = r$ and $|\mathrm{Gal}(M/K)| = [M : K] = 2^{\mu-1} = 2^{r-1}$. Since M is the compositum of $K_i = K(\sqrt{p_i^*})$ for $1 \le i \le r$, we get an injection

$$A : C(\mathcal{O}_K)/C(\mathcal{O}_K)^2 \hookrightarrow \{\pm 1\}^r,$$

where the image has index 2. (The map is not onto since $\prod_{i=1}^r \sqrt{p_i^*} = \sqrt{d_K} \in K$.)

Since $d_K \equiv 1 \bmod 4$, the genus theory developed in §3 gives $\mu = r$ assigned characters $\chi_1, \ldots, \chi_r$, with $\chi_i(a) = (a/p_i)$. By Lemma 3.20, the genus of a form representing any a prime to d_K is determined by $(\chi_1(a), \ldots, \chi_r(a))$, so we get an injective map

$$G : C(d_K)/P \longrightarrow \{\pm 1\}^r.$$

Note that this implies that $C(d_K)^2 \subset P$ since everything in the image is its own inverse. We construct the diagram

$$(*) \qquad \begin{array}{ccc} C(d_K)/C(d_K)^2 & \xrightarrow{\ \alpha\ } & C(d_K)/P \\ \downarrow \wr & & \downarrow G \\ C(\mathcal{O}_K)/C(\mathcal{O}_K)^2 & \underset{A}{\hookrightarrow} & \{\pm 1\}^r \end{array}$$

where α is the natural surjection. We claim that this diagram commutes: the image of the class of a form $f(x, y) = ax^2 + bxy + cy^2$ of discriminant d_K, with a relatively prime to d_K, maps by $G \circ \alpha$ to

$$(\chi_1(a), \ldots, \chi_r(a)) = \left(\left(\frac{a}{p_1}\right), \ldots, \left(\frac{a}{p_r}\right)\right).$$

Under the isomorphism of Theorem 5.30, the class of the form $f(x, y)$ in $C(d_K)$ maps to the class of the ideal $\mathfrak{a} = [a, (b + \sqrt{d_K})/2]$ in $C(\mathcal{O}_K)$. If we show that the natural map (6.17)

$$\mathbb{Z}/a\mathbb{Z} \longrightarrow \mathcal{O}_K/\mathfrak{a}$$

is an isomorphism, then $N(\mathfrak{a}) = a$ and we again get (6.18), giving commutativity of the diagram. (Note that we are assuming here that $d_K \equiv 1 \bmod 4$. The the proof that this map is an isomorphism when $d_K \equiv 0 \bmod 4$ is handled in Exercise 6.14.) To see that this map is indeed an isomorphism, note that

$$\mathcal{O}_K = \mathbb{Z}[(1 + \sqrt{d_K})/2], \quad \mathfrak{a} = [a, (-b + \sqrt{d_K})/2].$$

Let $\rho : \mathbb{Z} \hookrightarrow \mathcal{O}_K \to \mathcal{O}_K/\mathfrak{a}$ be the natural map, and write $w = (1 + \sqrt{d_K})/2$. Note that $k = (b + 1)/2 \in \mathbb{Z}$ since b is odd. Then

$$w = (1 + \sqrt{d_K})/2 = (b + 1)/2 + (-b + \sqrt{d_K})/2 \equiv k \bmod \mathfrak{a},$$

so that $\alpha = r + sw \in \mathcal{O}_K$ satisfies $\alpha \equiv r + sk \bmod \mathfrak{a}$. Thus ρ is surjective. Next suppose $m \in \ker(\rho)$, so that $m \in \mathfrak{a}$. Then $m = r + sw = ra + s(-b + \sqrt{d_K})/2$ for some $r, s \in \mathbb{Z}$. Since $1, \sqrt{d_K}$ are linearly independent over $\mathbb{Q}$, this forces $s = 0$, so that $m = ra$ and thus $m \in a\mathbb{Z}$. So $\ker(\rho) \subset a\mathbb{Z}$. Since $\rho(a) = 0$ because $a \in \mathfrak{a}$, it follows that $\ker(\rho) = a\mathbb{Z}$ and therefore the map $\mathbb{Z}/a\mathbb{Z} \to \mathcal{O}_K/\mathfrak{a}$ is an isomorphism. This completes the proof that (*) is commutative.

Now the proof is easy. Commutativity implies that $G \circ \alpha$ is injective, hence α is injective. From here, the rest of the proof follows exactly as in the text.

6.13. Suppose $K = \mathbb{Q}(\sqrt{-n})$, $n \not\equiv 3 \bmod 4$ and squarefree. Consider a form $f(x, y) = ax^2 + bxy + cy^2$ of discriminant $-4n$ where $n = 2^\alpha p_1 p_2 \cdots p_r$, p_i distinct odd primes,

$\alpha \in \{0,1\}$. We may assume $\gcd(a,4n)=1$ since $f(x,y)$ properly represents some integer prime to $d_K=-4n$. Under the composite map

$$C(-4n) \longrightarrow C(\mathcal{O}_K) \longrightarrow \mathrm{Gal}(M/K) \longrightarrow \mathrm{Gal}(M/\mathbb{Q}) \longrightarrow \{\pm 1\}^{\mu},$$

the class of $f(x,y)$ maps to the class of $\mathfrak{a}=[a,(-b+\sqrt{d_K})/2]$, and the map to $\mathrm{Gal}(M/K)$ is then given by the Artin map $((M/K)/\cdot)$. The complete character of $f(x,y)$ is

$$\left(\chi_0(a), \left(\frac{a}{p_1}\right), \dots, \left(\frac{a}{p_r}\right)\right),$$

where χ_0 is one of ϵ, δ, or $\epsilon\delta$ depending on the congruence class of n modulo 8. So to prove that the composite map above is the same as the complete character, it suffices to show that evaluating the Artin map on each generator of $M/\mathbb{Q}$ gives the same result as evaluating the characters. We already know that

$$\left(\frac{M/K}{\mathfrak{a}}\right)(\sqrt{p_i^*}) = \left(\frac{N(\mathfrak{a})}{p}\right)\sqrt{p_i^*}$$

by (6.12) and (6.13), so it suffices to show that

$$\left(\frac{M/K}{\mathfrak{a}}\right)(g) = \chi_0(N(\mathfrak{a}))g$$

where $g=i$, $\sqrt{2}$, or $\sqrt{-2}$ is the remaining generator of $M/\mathbb{Q}$. As in (6.13) we can reduce to the case where $\mathfrak{p}$ is a prime of K that, as in the discussion preceding (6.12), is prime to $2d_K$. Let $\mathfrak{P}$ be a fixed prime lying over $\mathfrak{p}$. There are three cases to consider, depending on whether $n \equiv 1 \bmod 4$, $n \equiv 2 \bmod 8$ or $n \equiv 6 \bmod 8$.

Suppose first $n \equiv 1 \bmod 4$, so that $\alpha = 0$, $g=i$. From (6.9),

$$M = \mathbb{Q}(i, \sqrt{p_1^*}, \dots, \sqrt{p_r^*}).$$

Then

$$\left(\frac{M/K}{\mathfrak{p}}\right)(i) \equiv i^{N(\mathfrak{p})} \equiv i^{N(\mathfrak{p})-1} i \equiv (i^2)^{(N(\mathfrak{p})-1)/2} i \equiv (-1)^{(N(\mathfrak{p})-1)/2} i \equiv \delta(N(\mathfrak{p}))i \bmod \mathfrak{P}.$$

From here, equivalence modulo $\mathfrak{P}$ implies equality because $1 \not\equiv -1 \bmod \mathfrak{P}$ since $\mathfrak{p}$ is prime to $2d_K$.

If $n \equiv 6 \bmod 8$, then $g=\sqrt{2}$, and we want to show that $((M/K)/\mathfrak{p})(\sqrt{2}) = \epsilon(N(\mathfrak{p}))\sqrt{2} = (2/N(\mathfrak{p}))\sqrt{2}$. In this case

$$M = \mathbb{Q}(\sqrt{2}, \sqrt{p_1^*}, \dots, \sqrt{p_r^*}).$$

Then

$$\left(\frac{M/K}{\mathfrak{p}}\right)(\sqrt{2}) \equiv \sqrt{2}^{N(\mathfrak{p})} \equiv 2^{(N(\mathfrak{p})-1)/2}\sqrt{2} \bmod \mathfrak{P}.$$

If $N(\mathfrak{p}) = p$, the above gives

$$\left(\frac{M/K}{\mathfrak{p}}\right)(\sqrt{2}) \equiv 2^{(p-1)/2}\sqrt{2} \equiv \left(\frac{2}{p}\right)\sqrt{2} \bmod \mathfrak{P}$$

as desired. If $N(\mathfrak{p}) = p^2$, then

$$2^{(p^2-1)/2} = \left(2^{(p+1)/2}\right)^{p-1} \equiv 1 \bmod p, \qquad \epsilon(N(\mathfrak{p})) = \epsilon(p^2) = \epsilon(p)^2 = 1,$$

so the two are again equal.

Finally, if $n \equiv 2 \bmod 8$, then $g = i\sqrt{2}$, and we want to show that

$$\left(\frac{M/K}{\mathfrak{p}}\right)(i\sqrt{2}) = \delta\epsilon(N(\mathfrak{p}))i\sqrt{2} = \left(\frac{-2}{N(\mathfrak{p})}\right)i\sqrt{2}.$$

Since $\mathfrak{p}$ is prime to $2d_K$, we get

$$\left(\frac{M/K}{\mathfrak{p}}\right)(i\sqrt{2}) \equiv (i\sqrt{2})^{N(\mathfrak{p})} \equiv i^{N(\mathfrak{p})-1} 2^{(N(\mathfrak{p})-1)/2} i\sqrt{2} \bmod \mathfrak{P}.$$

Arguing as above shows that $i^{N(\mathfrak{p})-1} = \delta(N(\mathfrak{p}))$ and $2^{(N(\mathfrak{p})-1)/2} \equiv \epsilon(N(\mathfrak{p}))$ mod p. Thus

$$i^{N(\mathfrak{p})-1} 2^{(N(\mathfrak{p})-1)/2} i\sqrt{2} \equiv \delta(N(\mathfrak{p}))\epsilon(N(\mathfrak{p}))i\sqrt{2} \equiv \delta\epsilon(N(\mathfrak{p}))i\sqrt{2} \bmod \mathfrak{P},$$

as desired.

6.14. Here $d_K = -4n$ and $f(x,y) = ax^2 + 2bxy + cx^2$ is a form of discriminant $-4n$. We assume that a is coprime to d_K. Then

$$\mathfrak{a} = [a, (-2b + \sqrt{d_K})/2] = [a, -b + \sqrt{-n}] \subset \mathcal{O}_K.$$

Let $\rho : \mathbb{Z} \hookrightarrow \mathcal{O}_K \to \mathcal{O}_K/\mathfrak{a}$ be the natural map, and note that by (5.13) $\mathcal{O}_K = \mathbb{Z}[\sqrt{-n}]$. Then ρ is surjective, since if $r + s\sqrt{-n} \in \mathcal{O}_K$, $r, s \in \mathbb{Z}$,

$$r + s\sqrt{-n} \equiv (r + s\sqrt{-n}\,) - s(-b + \sqrt{-n}\,) \equiv r + sb \bmod \mathfrak{a}.$$

Then $r + sb \in \mathbb{Z}$ maps to $[r + s\sqrt{-n}] \in \mathcal{O}_K/\mathfrak{a}$.

Next, if $\rho(m) = 0$, then $m \in \mathfrak{a}$, i.e., $m = ra + s(-b + \sqrt{-n})$ for some $r, s \in \mathbb{Z}$. Since $1, \sqrt{-n}$ are linearly independent over $\mathbb{Q}$, this forces $s = 0$, so that $m = ra$ and thus $m \in a\mathbb{Z}$. So $\ker(\rho) \subset a\mathbb{Z}$. Since $\rho(a) = 0$ because $a \in \mathfrak{a}$, it follows that $\ker(\rho) = a\mathbb{Z}$ and therefore $\mathbb{Z}/a\mathbb{Z} \to \mathcal{O}_K/\mathfrak{a}$ is an isomorphism.

6.15. (a) Let L be the Hilbert class field of K and $M \subset L$ be its genus field. Let $G = \mathrm{Gal}(L/\mathbb{Q})$; then $\mathrm{Gal}(L/M) = [G, G]$. Thus

$$M = L \iff [G, G] = \{1\}.$$

In the proof of Theorem 6.1, we identified G with $C(\mathcal{O}_K) \rtimes \mathbb{Z}/2\mathbb{Z}$ and showed that $[G, G] = C(\mathcal{O}_K)^2$. Since $C(\mathcal{O}_K) \simeq C(d_K)$, we obtain

$$[G, G] = \{1\} \iff C(d_K)^2 = \{1\}.$$

Since the principal genus equals $C(d_K)^2$ by Theorem 3.15, it follows that

$$C(d_K)^2 = \{1\} \iff \text{the principal genus consists of only one class.}$$

Since every genus is a coset of the principal genus, every genus consists of the same number of classes. Hence the above equivalences imply that $M = L$ if and only if every genus consists of a single class.

(b) These are the n's on Gauss's list from §3 that are squarefree and are $\not\equiv 3$ mod 4. Thus we immediately conclude that the Hilbert class field and the genus field coincide when $d_K = -4n$ for the following 34 values of n:

1, 2, 5, 6, 10, 13, 22, 37, 58, 21, 30, 33, 42, 57, 70, 78, 85, 93, 102, 130, 133, 177, 190, 253, 105, 165, 210, 273, 330, 345, 357, 385, 462, 1365.

6.16. Since 6 is a convenient number, the genus field of $K = \mathbb{Q}(\sqrt{-6})$ is its Hilbert class field. But $d_K = -24$ has only one odd prime divisor, $p_1 = 3$, so the Hilbert class field is

$$L = K(\sqrt{p_1^*}) = K(\sqrt{-3}) = \mathbb{Q}(\sqrt{-6}, \sqrt{-3}) = \mathbb{Q}(\sqrt{-6}, \sqrt{2}).$$

Since 10 is a convenient number, the genus field of $K = \mathbb{Q}(\sqrt{-10})$ is its Hilbert class field. But $d_K = -40$ has only one odd prime divisor, $p_1 = 5$, so the Hilbert class field is

$$L = K(\sqrt{p_1^*}) = K(\sqrt{5}) = \mathbb{Q}(\sqrt{-10}, \sqrt{5}) = \mathbb{Q}(\sqrt{-10}, \sqrt{-2}).$$

(For both of these, see also (6.9)).

These arguments do not apply to $K = \mathbb{Q}(\sqrt{-35})$ since $d_K = -35$. However, $d_K = -35$ has only two reduced forms, $x^2 + xy + 9y^2$ and $3x^2 + xy + 3y^2$, and they represent different primes. To see this, note that $x^2 + xy + 9y^2$ represents 1, but since

$$3x^2 + xy + 3y^2 = 3(x+y)^2 - 5xy \equiv 3(x+y)^2 \equiv 0, 2, 3 \bmod 5,$$

it follows that $3x^2+xy+3y^2$ does not represent [1]. Thus each genus for this discriminant has exactly one form, so again the Hilbert class field and the genus field coincide. For $p_1 = 5$, $p_1^* = 5$, while for $p_2 = 7$, $p_2^* = -7$. Thus

$$L = K(\sqrt{5}, \sqrt{-7}) = \mathbb{Q}(\sqrt{-35}, \sqrt{5}, \sqrt{-7}) = \mathbb{Q}(\sqrt{5}, \sqrt{-7}).$$

6.17. As shown in the text, the Hilbert class field L is a degree 2 extension of the genus field $M = K(\sqrt{2}) = \mathbb{Q}(\sqrt{-14}, \sqrt{2})$. Since $[M \cap \mathbb{R} : \mathbb{Q}] = [M : K] = 2$ by part (b) of Exercise 5.19 and $\sqrt{2} \in M$, we see that $M \cap \mathbb{R} = \mathbb{Q}(\sqrt{2})$. Exercise 5.19 also implies $[L \cap \mathbb{R} : \mathbb{Q}] = [L : K] = 4$, so that $L \cap \mathbb{R}$ is a quadratic extension of $\mathbb{Q}(\sqrt{2})$. Thus $L \cap \mathbb{R} = \mathbb{Q}(\sqrt{2})(\sqrt{u}) = \mathbb{Q}(\sqrt{2}, \sqrt{u})$ for some $u \in \mathbb{Q}(\sqrt{2})$. Writing $u = a + b\sqrt{2}$ for $a, b \in \mathbb{Q}$, we can replace u with $m^2 u$ for any positive integer m and get the same extension of $\mathbb{Q}(\sqrt{2})$. Thus we may assume $a, b \in \mathbb{Z}$, and of course $u = a + b\sqrt{2} > 0$ since $\sqrt{u} \in L \cap \mathbb{R}$.

It remains to prove that $L = K(\sqrt{u})$. We know from §3 that $\mathrm{Gal}(L/K)$ is cyclic of order 4, so M is the unique field strictly between K and L. Since $K(\sqrt{u})$ is strictly bigger than the imaginary quadratic field K, we see that $K(\sqrt{u})$ equals either M or L. The previous paragraph showed that $\sqrt{u} \notin \mathbb{Q}(\sqrt{2}) = M \cap \mathbb{R}$, and since $\sqrt{u} \in \mathbb{R}$, it follows that $\sqrt{u} \notin M$ and therefore $L = K(\sqrt{u})$.

6.18. (a) We know from §3 that $\mathrm{Gal}(L/K)$ is cyclic of order 4, say generated by σ. As noted in the second paragraph of the solution to Exercise 6.4, complex conjugation $\tau \in \mathrm{Gal}(L/\mathbb{Q})$ satisfies $\tau\sigma\tau = \sigma^{-1}$. These properties show that $\mathrm{Gal}(L/\mathbb{Q})$ is isomorphic to the dihedral group of order 8.

Next note that $\sigma(u) = \sigma(a + b\sqrt{2}) = a \pm b\sqrt{2}$. But $\sigma(u) = u$ would imply $u \in K$, which implies $[K(\sqrt{u}) : K] \leq 2$, impossible since $[K(\sqrt{u}) : K] = [L : K] = 4$. Thus $\sigma(u) = u' = a - b\sqrt{2} \neq u$. So in particular, $b \neq 0$ and $\sigma(\sqrt{2}) = -\sqrt{2}$. Then

$$\sigma(\sqrt{u})^2 = \sigma(u) = u' = (\sqrt{u'})^2,$$

so that $\sigma(\sqrt{u}) = \pm\sqrt{u'}$. We can pick a choice of square root so that $\sigma(\sqrt{u}) = \sqrt{u'}$.

Let $M = K(\sqrt{2})$ be the genus field of K as in Exercise 6.17. One easily sees that $\mathrm{Gal}(L/M) = \{1, \sigma^2\}$. Since $L = M(\sqrt{u})$ and $u \in M$, it follows that $\sigma^2(\sqrt{u}) = -\sqrt{u}$. Finally, $\tau\sigma = \sigma^{-1}\tau = \sigma^3\tau$ since conjugation with τ sends each element to its inverse, Since $u > 0$, we have $\sqrt{u} \in \mathbb{R}$ so that $\tau(\sqrt{u}) = \sqrt{u}$. Then

$$\tau(\sqrt{u'}) = \tau\sigma(\sqrt{u}) = \sigma^3\tau(\sqrt{u}) = \sigma^3(\sqrt{u}) = \sigma(\sigma^2(\sqrt{u})) = \sigma(-\sqrt{u}) = -\sqrt{u'}.$$

(b) First observe that σ fixes $\sqrt{-14} \in K$. We saw in part (a) that $\sigma(\sqrt{2}) = -\sqrt{2}$. Then $\sqrt{-14} = \sqrt{2}\sqrt{-7}$ implies that $\sigma(\sqrt{-7}) = -\sqrt{-7}$. Since we also have $\tau(\sqrt{-7}) = -\sqrt{-7}$, it follows that σ^2 and $\sigma\tau$ fix $\sqrt{-7}$.

The description of $\mathrm{Gal}(L/\mathbb{Q})$ given in part (a) implies that σ^2 and $\sigma\tau$ generate the subgroup $\{1, \sigma^2, \sigma\tau, \tau\sigma\}$, Hence its fixed field has degree 2 over $\mathbb{Q}$. The fixed field contains $\mathbb{Q}(\sqrt{-7})$ by the previous paragraph, so that the fixed field equals $\mathbb{Q}(\sqrt{-7})$.

(c) From part (a), we have

$$\sigma^2(\sqrt{u'}) = \sigma^2(\sigma(\sqrt{u})) = \sigma(\sigma^2(\sqrt{u})) = \sigma(-\sqrt{u}) = -\sqrt{u'}$$
$$\sigma(\sqrt{u'}) = \sigma(\sigma(\sqrt{u}) = \sigma^2(\sqrt{u}) = -\sqrt{u}.$$

Thus

$$\sigma^2(\sqrt{uu'}) = \sigma^2(\sqrt{u})\sigma^2(\sqrt{u'}) = (-\sqrt{u})(-\sqrt{u'}) = \sqrt{uu'}$$
$$\sigma\tau(\sqrt{uu'}) = \sigma\tau(\sqrt{u})\sigma\tau(\sqrt{u'}) = \sigma(\sqrt{u})\sigma(-\sqrt{u'}) = \sqrt{u'}\sqrt{u} = \sqrt{uu'}.$$

Then $\sqrt{uu'} \in \mathbb{Q}(\sqrt{-7})$ by part (b). Since $\sqrt{u}$ and $\sqrt{u'}$ are both algebraic integers, $\sqrt{uu'} \in \mathcal{O}_{\mathbb{Q}(\sqrt{-7})} = \mathbb{Z}[(1+\sqrt{-7})/2]$, so $\sqrt{uu'} = a + b(1+\sqrt{-7})/2 = (a+b/2) + b\sqrt{-7}/2$,

$a, b \in \mathbb{Z}$. Since $\tau(\sqrt{uu'}) = -\sqrt{uu'}$ and τ is complex conjugation, it follows that $a + b/2 = 0$, so that $m = b/2 \in \mathbb{Z}$ and thus $\sqrt{uu'} = m\sqrt{-7}$, $m \in \mathbb{Z}$.

(d) Let $\pi = 2\sqrt{2} - 1$, so that $\pi' = -2\sqrt{2} - 1$. Then $N(\pi) = \pi\pi' = -7$, so π is prime. By part (c), $uu' = -7m^2 = \pi\pi' m^2$. Since $\mathbb{Z}[\sqrt{2}]$ is a UFD, either $\pi \mid u$ or $\pi \mid u'$. Reversing u and u' if necessary, we may assume $\pi \mid u$, so that $u = \pi\alpha$ for some $\alpha \in \mathbb{Z}[\sqrt{2}]$. Then

$$-7N(\alpha) = N(\pi)N(\alpha) = N(\pi\alpha) = N(u) = uu' = -7m^2,$$

from which we conclude that $N(\alpha) = m^2$.

(e) Write α as a product of primes (and possibly a unit). Since u was chosen such that $L = K(\sqrt{u})$, we may assume that u is squarefree in $\mathbb{Z}[\sqrt{2}]$. Thus each prime in the factorization of α appears to the first power; that is, $\alpha = \epsilon \prod_{i=1}^{r} \pi_i$ where the π_i are distinct primes of $\mathbb{Z}[\sqrt{2}]$ and ϵ is a unit in $\mathbb{Z}[\sqrt{2}]$. Integer primes factor as follows in the UFD $\mathbb{Z}[\sqrt{2}]$:

$$\begin{aligned} p = 2: &\quad 2 = \sqrt{2}^2,\ \sqrt{2} \text{ prime in } \mathbb{Z}[\sqrt{2}] \\ p \equiv 1, 7 \bmod 8: &\quad p = \varpi\varpi',\ \varpi, \varpi' \text{ prime in } \mathbb{Z}[\sqrt{2}],\ \varpi \neq \text{unit} \cdot \varpi' \\ p \equiv 3, 5 \bmod 8: &\quad p \text{ is prime in } \mathbb{Z}[\sqrt{2}], \end{aligned}$$

and any prime of $\mathbb{Z}[\sqrt{2}]$ is a unit times a prime on this list. Note also that $N(\sqrt{2}) = -2$ and $N(\varpi) = N(\varpi') = \pm p$. The above factorization of α implies

$$m^2 = \pm \prod_{i=1}^{r} N(\pi_i)$$

None of the π_i can be $\sqrt{2}$, since otherwise 2 would appear only to the first power on the right-hand side of the above equation. Furthermore, primes ϖ or ϖ' from the above list (when $p \equiv 1, 7 \bmod 8$) must occur in pairs in the factorization of α, again because $N(\alpha)$ is a square, and their product is $\varpi\varpi' = p$. Thus α is a unit ϵ times an integer n and $u = \epsilon\pi n$. Note that n is odd since $\sqrt{2}$ is not in the factorization. Additionally, since u is squarefree, $7 \nmid n$, since $7 = -\pi\pi'$ and then u would not be squarefree. The result follows: $u = \epsilon\pi n$, where ϵ is a unit and n is squarefree and prime to 14.

(f) In $\mathcal{O}_L$, u is a square, so that $u\mathcal{O}_L = \pi\mathcal{O}_L \cdot n\mathcal{O}_L$ is a square. Since $\gcd(n, 14) = 1$ and $N(\pi) = -7$, π and n share no prime factors in $\mathcal{O}_L$, so that $n\mathcal{O}_L$ is a square. Since n is squarefree, this means that every prime dividing n must ramify in L. But the only primes ramifying in L are those ramifying in $K = \mathbb{Q}(\sqrt{-14})$ (since L is the Hilbert class field of K), so only 2 and 7 can ramify in L. Thus $n = \pm 1$.

(g) From part (f), $u = \epsilon\pi$ where $\epsilon = \pm(\sqrt{2} - 1)^\ell$ is a unit. The sign is positive since u, $(\sqrt{2} - 1)^\ell$, and π are all positive. Next,

$$N(u) = N(\epsilon)N(\pi) = (-1)^\ell \cdot (-7),$$

and then $N(u) = -7m^2 < 0$ from part (d) implies that ℓ must be even, so that $u = (\sqrt{2} - 1)^{2k}\pi$. Since we are adjoining $\sqrt{u}$, we can ignore the square $(\sqrt{2} - 1)^{2k}$, giving $u = \pi$. Then $L = K(\sqrt{\pi}) = K(\sqrt{2\sqrt{2} - 1})$ is the Hilbert class field.

6.19. Set $K = \mathbb{Q}(\sqrt{-17})$ and let L be the Hilbert class field of K. The genus field, by (6.9), is $M = K(i) = K(\sqrt{17})$. Then L is a degree 2 extension of $M = \mathbb{Q}(i, \sqrt{17})$. From here, we follow the argument in to the solution of Exercise 6.17, replacing $\sqrt{2}$ with $\sqrt{17}$, to conclude that $L = K(\sqrt{u})$ where $u = a + b\sqrt{17} > 0$ for $a, b \in \mathbb{Z}$.

From part (a) of Exercise 5.25, we know that $\mathrm{Gal}(L/K) \simeq \mathbb{Z}/4\mathbb{Z}$; let $\sigma \in \mathrm{Gal}(L/K)$ be a generator. The argument of part (a) of Exercise 6.18 adapts without change to show that $\mathrm{Gal}(L/\mathbb{Q})$ is a dihedral group of order 8. If we set $u' = a - b\sqrt{17}$, then the solution from part (a) adapts word-for-word to show that $\sigma(u) = u'$ and that $b \neq 0$ and $\sigma(\sqrt{17}) = -\sqrt{17}$. Following part (a), we also obtain

$$\sigma(\sqrt{u}) = \sqrt{u'}, \quad \sigma^2(\sqrt{u}) = -\sqrt{u}, \quad \tau(\sqrt{u'}) = -\sqrt{u'}.$$

Next observe that σ fixes $\sqrt{-17} \in K$. We saw above that $\sigma(\sqrt{17}) = -\sqrt{17}$. Then $\sqrt{-17} = i\sqrt{17}$ implies that $\sigma(i) = -i$. Since we also have $\tau(i) = -i$, it follows that σ^2 and $\sigma\tau$ fix i. As in the solution to part (b) of Exercise 6.18, σ^2 and $\sigma\tau$ generate the subgroup $\{1, \sigma^2, \sigma\tau, \tau\sigma\}$, so again its fixed field has degree 2 over $\mathbb{Q}$. The fixed field contains $\mathbb{Q}(i)$, so they are equal.

Now, arguing as in part (c) of Exercise 6.18, one computes that $\sqrt{uu'}$ is fixed by σ^2 and $\sigma\tau$ and hence lies in the fixed field $\mathbb{Q}(i)$. Since u and u' are algebraic integers, we have $\sqrt{uu'} \in \mathbb{Z}[i]$, so $\sqrt{uu'} = a + mi$ with $a, m \in \mathbb{Z}$. Then $\tau(\sqrt{uu'}) = -\sqrt{uu'}$ implies $a = 0$, so that $\sqrt{uu'} = mi$, $m \in \mathbb{Z}$.

As in part (d) of Exercise 6.18, we want to write $uu' = -m^2$ as $uu' = \pi\pi' m^2$ where π is prime, so we want to find π with $N(\pi) = -1$. One choice is $\pi = 4 + \sqrt{17}$. This is a unit in $\mathcal{O}_{\mathbb{Q}(\sqrt{17})}$, so that (after reversing the roles of u and u' if necessary) we can write $u = \pi\alpha$ for $\alpha \in \mathcal{O}_{\mathbb{Q}(\sqrt{17})}$. Then

$$-m^2 = uu' = N(u) = N(\pi)N(\alpha) = -N(\alpha),$$

so that $u = \pi\alpha$ with $N(\alpha) = m^2$.

As stated in the exercise, $\mathcal{O}_{\mathbb{Q}(\sqrt{17})} = \mathbb{Z}[(1+\sqrt{17})/2]$ is a UFD. We can assume that u and hence α are squarefree, so that α is a product of primes (and possibly a unit), where each prime appears to the first power. Thus $\alpha = \epsilon \prod_{i=1}^{r} \pi_i$, where ϵ is a unit and the π_i are distinct primes of $\mathbb{Z}[(1+\sqrt{17})/2]$. However:

> $\sqrt{17}$ is prime in $\mathbb{Z}[(1+\sqrt{17})/2]$, and if $p \neq 17$ is an integer prime, then either:
>
> p is prime in $\mathbb{Z}[(1+\sqrt{17})/2]$, or
>
> $p = \varpi\varpi'$, ϖ, ϖ' prime in $\mathbb{Z}[(1+\sqrt{17})/2]$, $\varpi \neq \text{unit} \cdot \varpi'$.

Furthermore, any prime of $\mathbb{Z}[(1+\sqrt{17})/2]$ is a unit times one of these primes. Note also that $N(\sqrt{17}) = -17$ and $N(\varpi) = N(\varpi') = \pm p$.

Then the argument of part (e) shows that $\sqrt{17}$ cannot occur in the factorization of α and that if p is not prime in $\mathbb{Z}[(1+\sqrt{17})/2]$, then both ϖ and ϖ' in the above list must occur since $N(\alpha)$ is a square. Thus α is a unit times an integer n. Since α is squarefree and not divisible by $\sqrt{17}$, n cannot be divisible by 17. Thus $u = \epsilon\pi n$, where ϵ is a unit and n is squarefree and prime to 17.

In $\mathcal{O}_L$, u is a square. Since $u = \epsilon\pi n$ and $\epsilon\pi$ is a unit, we see that $u\mathcal{O}_L = n\mathcal{O}_L$ is a square. But the only primes ramifying in L are those ramifying in $K = \mathbb{Q}(\sqrt{-17})$ (since L is unramified over K), so only 2 and 17 can ramify in L since $d_K = -4 \cdot 17$. We know $17 \nmid n$. It follows that $n = \pm 1$ or ± 2, so that $u = \epsilon\pi$ or $\epsilon 2\pi$ where ϵ is a unit in $\mathbb{Z}[(1+\sqrt{17})/2]$. Since $\pi = 4 + \sqrt{17}$, the description of the units given in the exercise implies that

$$u = \pm\pi^{\ell} \text{ or } \pm 2\pi^{\ell}, \quad \ell \in \mathbb{Z}.$$

The sign is positive since $u > 0$, and ℓ is odd since $N(u)$ and $N(\pi)$ are negative. Setting $\ell = 2k+1$, we have $u = (\pi^k)^2\pi$ or $(\pi^k)^2 2\pi$. Since we are adjoining $\sqrt{u}$, we can ignore the square factor, so that the Hilbert class field of $K = \mathbb{Q}(\sqrt{-17})$ is either

$$K(\sqrt{\pi}) = K\big(\sqrt{4+\sqrt{17}}\big) \text{ or } K(\sqrt{2\pi}) = K\big(\sqrt{8+2\sqrt{17}}\big).$$

To show that the first is correct, we use Theorem 5.1. The minimal polynomial of $\sqrt{8+2\sqrt{17}}$ is $(x^2-8)^2-68$, which has discriminant $-2^{10}\cdot 17$ by Exercise 5.24. So if $\sqrt{8+2\sqrt{17}}$ gave the Hilbert class field, then Theorem 5.1 would imply that for an odd prime $p \neq 17$,

$$p = x^2 + 17y^2 \iff \begin{cases} (-17/p) = 1 \text{ and } (x^2-8)^2 \equiv 68 \bmod p \\ \text{has an integer solution.} \end{cases}$$

However, for the prime $53 = 6^2 + 17\cdot 1^2$, the congruence $(x^2-8)^2 \equiv 68 \bmod 53$ has no integer solutions. It follows that $K\big(\sqrt{4+\sqrt{17}}\big)$ is the Hilbert class field of $K = \mathbb{Q}(\sqrt{-17})$.

6.20. (a) The reduced forms of discriminant -55 are

$$x^2+xy+14y^2, \quad 2x^2-xy+7y^2, \quad 2x^2+xy+7y^2, \quad 4x^2+3xy+4y^2.$$

Since the middle two are inverses of each other, this group must be $\mathbb{Z}/4\mathbb{Z}$, so that $C(\mathcal{O}_K) \simeq C(d_K) \simeq \mathbb{Z}/4\mathbb{Z}$.

(b) If L is the Hilbert class field of K, then $[L:K]=4$ since $\mathrm{Gal}(L/K) \simeq C(\mathcal{O}_K)$. The genus field M of K is

$$M = K(\sqrt{5}, \sqrt{-11}) = K(\sqrt{5}) = K(\sqrt{-11}).$$

As before, we have $L = K(\sqrt{u})$, where $u = a + b\sqrt{5} > 0$, $a, b \in \mathbb{Z}$. Let $u' = a - b\sqrt{5}$. The proof that $\mathrm{Gal}(L/\mathbb{Q})$ is dihedral is identical in spirit to that given in part (a) of Exercise 6.18. Then arguing as in Exercises 6.18 and 6.19, one finds that

$$\sigma(u) = u', \quad \sigma(\sqrt{5}) = -\sqrt{5}$$
$$\sigma(\sqrt{u}) = \sqrt{u'}, \quad \sigma^2(\sqrt{u}) = -\sqrt{u}, \quad \tau(\sqrt{u}) = \sqrt{u}, \quad \tau(\sqrt{u'}) = -\sqrt{u'}.$$

Since σ fixes $\sqrt{-55} = \sqrt{5}\sqrt{-11} \in K$, it follows that $\sigma(\sqrt{-11}) = -\sqrt{-11}$. Since we also have $\tau(\sqrt{-11}) = -\sqrt{-11}$, we see as in the previous exercises that $\mathbb{Q}(\sqrt{-11})$ is the fixed field of σ^2 and $\sigma\tau$.

Since $\sqrt{uu'}$ is fixed by both σ^2 and $\sigma\tau$ as before, we see that $\sqrt{uu'} \in \mathbb{Z}[(1+\sqrt{-11})/2]$, so that $\sqrt{uu'} = a + b(1+\sqrt{-11})/2 = (a+b/2) + b\sqrt{-11}/2$, $a, b \in \mathbb{Z}$. Since $\tau(\sqrt{uu'}) = -\sqrt{uu'}$ and τ is complex conjugation, we get $a + b/2 = 0$, so that $m = b/2 \in \mathbb{Z}$ and then $\sqrt{uu'} = m\sqrt{-11}$, $m \in \mathbb{Z}$.

If $N(\cdot)$ is the norm on $\mathbb{Q}(\sqrt{5})$, then $N(u) = uu' = -11m^2$. Following the previous two exercises, we should find $\pi \in \mathbb{Z}[(1+\sqrt{5})/2]$ such that $N(\pi) = -11$ and then show that we can take u to be π. Such a π is $\pi = 3 + 2\sqrt{5}$. But rather than follow the approach of Exercises 6.18 and 6.19, we now use the hint to apply the methods of Proposition 5.31 to show that $L' = K(\sqrt{\pi})$ is the Hilbert class field of K.

For $\pi = 3+2\sqrt{5}$, note that $\sqrt{\pi}$ is a root of $x^4 - 6x^2 - 11$. One can check (by computer or by hand) that this polynomial is irreducible over $\mathbb{Q}$, so that $[\mathbb{Q}(\sqrt{\pi}):\mathbb{Q}] = 4$. Then $L' = K(\sqrt{\pi}) = \mathbb{Q}(\sqrt{\pi}, \sqrt{-55})$ has degree 8 over $\mathbb{Q}$. Hence $[L':K] = 4$, so that we need only show that this extension is unramified. Recall that genus field is

$$M = K(\sqrt{5}) = K(\pi) \subset K(\sqrt{\pi}) = L'.$$

Then M is unramified over K since M is a subfield of the Hilbert class field L. So we only need to show that quadratic extension $M \subset L'$ is unramified. We will use Lemma 5.32.

Let $\pi' = 3 - 2\sqrt{5}$. Since $N(\pi) = \pi\pi' = -11$, we have $\sqrt{\pi\pi'} = \sqrt{-11} \in M$ and $\sqrt{\pi} \in L'$, also $\sqrt{\pi'} \in L'$, so that

$$L' = K(\sqrt{\pi}) = K(\sqrt{\pi'}).$$

Choose $\mathfrak{p}$ prime in M. Suppose first $2 \notin \mathfrak{p}$. If π or π' is not in $\mathfrak{p}$, then $\mathfrak{p}$ is unramified in L by part (i) of Lemma 5.32. If both π and π' are in $\mathfrak{p}$, then $\pi + \pi' = 6$ and $\sqrt{5}(\pi - \pi') = 20$ lie in $\mathfrak{p}$, which in turn implies that $2 = \gcd(6, 20) \in \mathfrak{p}$, a contradiction. Next suppose that $2 \in \mathfrak{p}$. Then $u \notin \mathfrak{p}$ since otherwise $3 = \pi - 2\sqrt{5} \in \mathfrak{p}$ and then $1 \in \mathfrak{p}$. But then

$$\pi = \left(1 + 2\sqrt{5}\right)^2 - 4\left(\frac{9}{2} + \frac{1}{2}\sqrt{5}\right)$$

and thus $\mathfrak{p}$ is unramified in L' by part (ii) of Lemma 5.32.

(c) The minimal polynomial for $\sqrt{\pi} = \sqrt{3 + 2\sqrt{5}}$ is $x^4 - 6x^2 - 11 = (x^2 - 3)^2 - 20$. By Exercise 5.24, the discriminant of this quartic is $-2^{12} \cdot 5^2 \cdot 11$. Since $D = -55$ is negative, squarefree, and satisfies $D \equiv 1 \bmod 4$, Exercise 5.23 gives the following theorem:

THEOREM. *If $p \neq 5, 11$ is an odd prime, then*

$$p = x^2 + xy + 14y^2 \iff \begin{cases} (-55/p) = 1 \text{ and } (x^2 - 3)^2 \equiv 20 \bmod p \\ \text{has an integer solution.} \end{cases}$$

Following the hint, we now show that $x^2 + xy + 14y^2$ and $x^2 + 55y^2$ represent the same odd numbers. Let $m \in \mathbb{Z}$ be odd. First suppose that $m = X^2 + 55Y^2$ with $X, Y \in \mathbb{Z}$ and set $x = X - Y$ and $y = 2Y$. Then

$$\begin{aligned} x^2 + xy + 14y^2 &= (X - Y)^2 + (X - Y)2Y + 14(2Y)^2 \\ &= X^2 - 2XY + Y^2 + 2XY - 2Y^2 + 56Y^2 = X^2 + 55Y^2 = m. \end{aligned}$$

Conversely, suppose that $m = x^2 + xy + 14y^2$ with $x, y \in \mathbb{Z}$. Since m is odd, one sees that y must be even. Setting $X = x + y/2$ and $Y = y/2$ gives

$$\begin{aligned} X^2 + 55Y^2 &= (x + \tfrac{y}{2})^2 + 55(\tfrac{y}{2})^2 = x^2 + xy + \tfrac{1}{4}y^2 + \tfrac{55}{4}y^2 \\ &= x^2 + xy + 14y^2 = m. \end{aligned}$$

Thus the above theorem can be restated as:

THEOREM. *If $p \neq 5, 11$ is an odd prime, then*

$$p = x^2 + 55y^2 \iff \begin{cases} (-55/p) = 1 \text{ and } (x^2 - 3)^2 \equiv 20 \bmod p \\ \text{has an integer solution.} \end{cases}$$

Solutions to Exercises in §7

7.1. (a) First note that M is torsion-free since it is a subset of $\mathbb{C}$. The structure theorem for finitely generated Abelian groups (a standard fact from algebra) then implies that $M \simeq \mathbb{Z}^d$ for some integer d. Hence M is a free $\mathbb{Z}$-module.

(b) Suppose M has a $\mathbb{Z}$-basis $\omega_1, \ldots, \omega_d$. Then the ω_i are linearly independent over $\mathbb{Z}$, so clearing denominators in a $\mathbb{Q}$-linear relation shows that they are linearly independent over $\mathbb{Q}$. Since $[K : \mathbb{Q}] = n$ and the $\omega_i \in M \subset K$, we must have $d \leq n$. Thus M has rank $\leq n$.

If M has rank n, then by the above, $\omega_1, \ldots, \omega_n$ form a $\mathbb{Q}$-basis for K. Conversely, if M contains a $\mathbb{Q}$-basis $\omega_1, \ldots, \omega_n$ of K, then the ω_i are linearly independent over $\mathbb{Z}$, so the rank of M is at least n and therefore must equal n.

7.2. More generally, we claim the following: Let R be a subring of a field K, and $\alpha \in K$. If there is a nonzero finitely generated R-module $M \subset K$ with $\alpha M \subset M$, then α is integral over R. This will prove the desired result, since for any $\alpha \in \mathcal{O}$, we choose $R = \mathbb{Z}$ and $M = \mathcal{O}$. Then $\alpha\mathcal{O} \subset \mathcal{O}$, so that α is integral over $\mathbb{Z}$ and therefore lies in $\mathcal{O}_K$.

To prove the claim, choose nonzero generators $\omega_1, \ldots, \omega_m$ of M as an R-module, and write

$$\alpha\omega_i = \sum_{j=1}^{m} r_{ij}\omega_j, \quad r_{ij} \in R,\ 1 \le i, j \le m.$$

Consider the $m \times m$ matrix $A = (r_{ij})$ and let I_m be the $m \times m$ identity matrix. Then $\alpha I_m - A$ annihilates the nonzero column vector $(\omega_1, \ldots, \omega_m)^T$, so its determinant is zero. This gives a monic polynomial in $R[X]$ satisfied by α. Hence α is integral over R, proving the claim.

7.3. (a) If $\{\alpha, \beta\}$ and $\{\gamma, \delta\}$ are two integral bases, the change of basis matrix is an integral matrix

$$A = \begin{pmatrix} p & q \\ r & s \end{pmatrix},$$

and $(\gamma, \delta) = (\alpha, \beta)\, A$. Then $(\alpha, \beta) = (\gamma, \delta)\, A^{-1}$; since both bases are integral, A^{-1} must be an integral matrix, so that $\det(A) = \pm 1$. This implies

$$\left(\det\begin{pmatrix} \gamma & \delta \\ \gamma' & \delta' \end{pmatrix}\right)^2 = \left(\det\begin{pmatrix} \alpha & \beta \\ \alpha' & \beta' \end{pmatrix}\det(A)\right)^2 = \left(\pm\det\begin{pmatrix} \alpha & \beta \\ \alpha' & \beta' \end{pmatrix}\right)^2 = D.$$

So D is independent of the particular choice of integral basis.

(b) The discriminant of the basis $[1, fw_K]$ is

$$D = \left(\det\begin{pmatrix} 1 & fw_K \\ 1 & fw_K' \end{pmatrix}\right)^2 = f^2(w_K' - w_K)^2.$$

With $w_K = (d_K + \sqrt{d_K})/2$, we have $w_K' = (d_K - \sqrt{d_K})/2$, so that

$$D = f^2(w_K' - w_K)^2 = f^2(-\sqrt{d_K})^2 = f^2 d_K.$$

(c) Suppose $D \equiv 0, 1 \bmod 4$ is the discriminant of an order in a quadratic field K of discriminant d_K. Any order in K has discriminant $f^2 d_K$ by part (b), and any order $\mathcal{O}$ with that discriminant is, from Lemma 7.2, given by $\mathcal{O} = \mathbb{Z} + f\mathcal{O}_K$, so is uniquely determined by f and thus by the discriminant.

(d) Write $D = f^2 d$ where $d \ne 0, 1$ is squarefree and let $K = \mathbb{Q}(\sqrt{d})$. If $d \equiv 1 \bmod 4$, then K has discriminant $d_K = d$ and $D = f^2 d = f^2 d_K$ is the discriminant of the order of conductor f in $\mathcal{O}_K$. If $d \not\equiv 1 \bmod 4$, then $d_K = 4d$. If f were odd, then $d \equiv f^2 d = D \equiv 0, 1 \bmod 4$ would imply $d \equiv 0 \bmod 4$, which is impossible since d is squarefree. Thus f is even, so that $D = (f/2)^2(4d) = (f/2)^2 d_K$ is the discriminant of the order of conductor $f/2$ in $\mathcal{O}_K$.

7.4. (a) Let $\mathcal{O} = \mathbb{Z} + f\mathcal{O}_K = [1, fw_K]$, and choose $\alpha \ne 0$ in an ideal $\mathfrak{a} \subset \mathcal{O}$. Since $\mathbb{Z}$ and $\mathcal{O}_K$ are both stable under the action of $\mathrm{Gal}(K/\mathbb{Q})$, it follows that $\alpha' \in \mathcal{O}$, where $\alpha \mapsto \alpha'$ is the nontrivial element of $\mathrm{Gal}(K/\mathbb{Q})$. Then $m = N(\alpha) = \alpha\alpha' \ne 0$ is in $\mathbb{Q}$ and is an algebraic integer since α and α' are. Thus m is an integer. Finally, since $\mathfrak{a}$ is an $\mathcal{O}$-ideal and $\alpha \in \mathfrak{a}$, it follows that $m = \alpha\alpha' \in \mathfrak{a}$.

(b) Since $\mathcal{O} \simeq \mathbb{Z}^2$ and the nonzero ideal $\mathfrak{a} \subset \mathcal{O}$ contains a nonzero integer m, we are done by the argument used in the solution to part (b) of Exercise 5.1.

(c) If $\mathfrak{p}$ is a prime $\mathcal{O}$-ideal, then $\mathfrak{p}$ is maximal by part (b) and the argument used in the solution to part (e) of Exercise 5.1.

(d) We need to prove that any ascending chain of ideals in $\mathcal{O}$ eventually stabilizes. This follows from part (b) and the same argument used in the solution to part (d) of Exercise 5.1.

7.5. Note that for any nonzero ideal $\mathfrak{a}$ of an order $\mathcal{O}$ we have $S = \{\beta \in K \mid \beta\mathfrak{a} \subset \mathfrak{a}\} \subset \mathcal{O}_K$. To see this, apply the claim from the solution to Exercise 7.2 to the case $R = \mathbb{Z}$, $M = \mathfrak{a}$. For $K = \mathbb{Q}(\sqrt{-3})$, consider the order $\mathcal{O} = \mathbb{Z}[\sqrt{-3}]$. Note that $\mathcal{O} \neq \mathcal{O}_K$ since $(1+\sqrt{-3})/2$ lies in $\mathcal{O}_K$ but not in $\mathcal{O}$. Now let $\mathfrak{a}$ be the $\mathcal{O}$-ideal generated by 2 and $1+\sqrt{-3}$. To show that $S = \mathcal{O}_K = [1, (1+\sqrt{-3})/2]$, it suffices to show that $1 \in S$ and $(1+\sqrt{-3})/2 \in S$. Clearly $1 \in S$. To show that $(1+\sqrt{-3})/2 \in S$, we must show that its product with each generator of $\mathfrak{a}$ lies in $\mathfrak{a}$:

$$2\left(\frac{1+\sqrt{-3}}{2}\right) = 1+\sqrt{-3} \in \mathfrak{a}$$

$$(1+\sqrt{-3})\left(\frac{1+\sqrt{-3}}{2}\right) = \frac{-2+2\sqrt{3}}{2} = -1+\sqrt{3} = -2+(1+\sqrt{-3}) \in \mathfrak{a}.$$

Thus $(1+\sqrt{-3})/2 \in S$; it follows that $\mathcal{O}_K \subset S$, so that $S = \mathcal{O}_K$.

7.6. (a) Let $S = \{\beta \in K \mid \beta\mathfrak{a} \subset \mathfrak{a}\}$. An $\mathcal{O}$-ideal $\mathfrak{a}$ satisfies $\mathcal{O}\mathfrak{a} \subset \mathfrak{a}$, so that $\mathcal{O} \subset S$. If $\mathfrak{a}$ is principal and nonzero, write $\mathfrak{a} = \alpha\mathcal{O}$ for $\alpha \neq 0$. If $\beta \in K$ lies in S, then $\beta\alpha\mathcal{O} = \beta\mathfrak{a} \subset \mathfrak{a} = \alpha\mathcal{O}$, so that for some $\gamma \in \mathcal{O}$ we have $\beta\alpha = \gamma\alpha$. Since $\alpha \neq 0$, this implies $\beta = \gamma$, so that $\beta \in \mathcal{O}$ and then $S = \mathcal{O}$.

(b) By the solution to Exercise 7.5, $S = \{\beta \in K \mid \beta\mathfrak{a} \subset \mathfrak{a}\} \subset \mathcal{O}_K$ for a nonzero $\mathcal{O}_K$-ideal $\mathfrak{a}$. But any $\beta \in \mathcal{O}_K$ satisfies $\beta\mathfrak{a} \subset \mathfrak{a}$ precisely because $\mathfrak{a}$ is an $\mathcal{O}_K$-ideal. Hence $S = \mathcal{O}_K$.

7.7. For $\Leftarrow$, first note that an $\mathcal{O}$-ideal $\mathfrak{a}$ is a subgroup of the free Abelian group $\mathcal{O}$ of finite rank, so that $\mathfrak{a}$ is also a free Abelian group of finite rank. It follows an any $\mathcal{O}$-ideal $\mathfrak{a}$ is finitely generated, so that $\mathfrak{a}$ is a finitely generated $\mathcal{O}$-module. Then $\alpha\mathfrak{a}$ is also finitely generated for any $\alpha \in K$.

For $\Rightarrow$, suppose $\mathfrak{b}$ is generated as an $\mathcal{O}$-module by $\beta_1, \ldots, \beta_r \in K$. Since K is the field of fractions of $\mathcal{O}$, we may choose $\gamma \in \mathcal{O} - \{0\}$ such that $\gamma\beta_i \in \mathcal{O}$ for all i. Let $\mathfrak{a}$ be the $\mathcal{O}$-ideal generated by $\gamma\beta_i$ for $1 \leq i \leq r$. Then $\mathfrak{a} = \gamma\mathfrak{b}$, so that $\mathfrak{b} = \alpha\mathfrak{a}$ for $\alpha = 1/\gamma \in K$.

7.8. If $\mathfrak{a}$ is a nonzero fractional $\mathcal{O}$-ideal, then by the previous exercise, $\mathfrak{a} = \alpha\mathfrak{a}'$, where $\mathfrak{a}'$ is a nonzero $\mathcal{O}$-ideal. In the solution to the previous exercise, we noted that $\mathfrak{a}'$ is free Abelian of finite rank. Since $\mathcal{O} \simeq \mathbb{Z}^2$ and $\mathcal{O}/\mathfrak{a}'$ is finite by part (b) of Exercise 7.4, it follows that $\mathfrak{a}'$ has rank 2. Then the same is true for $\mathfrak{a} = \alpha\mathfrak{a}'$ since $\alpha \neq 0$.

7.9. (a) Since $\mathcal{O} = \mathbb{Z}[\sqrt{-3}]$ has conductor 2, its discriminant is -12. By (2.14), the only reduced form of discriminant -12 is $x^2 + 3y^2$, so $C(-12) \simeq \{1\}$, which implies that $C(\mathcal{O}) \simeq \{1\}$ by Theorem 7.7.

(b) See the discussion following Corollary 4.4 for the definition of a UFD. Assume that proper $\mathcal{O}$-ideals have unique factorization. By part (a), all proper $\mathcal{O}$-ideals are principal, so we are assuming that principal $\mathcal{O}$-ideals have unique factorization. We first show that if $\pi \in \mathcal{O}$ is irreducible, then $\pi\mathcal{O}$ has no factorization into nontrivial principal $\mathcal{O}$-ideals. To see why, suppose that $\pi\mathcal{O} = (\alpha\mathcal{O})(\beta\mathcal{O}) = \alpha\beta\mathcal{O}$, so that $\pi = \epsilon\alpha\beta = (\epsilon\alpha)\beta$ for $\epsilon \in \mathcal{O}^*$. Since π is irreducible, $\epsilon\alpha$ or β must be a unit, so that either $\alpha\mathcal{O} = \epsilon\alpha\mathcal{O}$ or $\beta\mathcal{O}$ is equal to $\mathcal{O}$.

Using the norm function on $\mathcal{O}$, one easily sees that a nonunit $a \in \mathcal{O}$ can be factored into irreducibles. If a has two such factorizations, say

$$a = p_1 \cdots p_m = q_1 \cdots q_n,$$

then

$$a\mathcal{O} = p_1\mathcal{O} \cdots p_m\mathcal{O} = q_1\mathcal{O} \cdots q_n\mathcal{O}.$$

Since the p_i are irreducible, the previous paragraph shows that the factorization on the left cannot be refined further, and the same is true for the factorization on the right. By unique factorization of principal ideals, these two factorizations are the same, so that $m = n$ and (renumbering if necessary) $p_i\mathcal{O} = q_i\mathcal{O}$, $1 \leq i \leq m$. Hence $p_i = \epsilon_i q_i$, $1 \leq i \leq m$, where $\epsilon_i \in \mathcal{O}^*$, so factorization is unique up to multiplication by units. Thus $\mathcal{O}$ is a UFD.

(c) Suppose $1 + \sqrt{-3} = \alpha\beta$, where $\alpha, \beta \in \mathbb{Z}[\sqrt{-3}]$. Taking norms gives

$$4 = 1^2 + 3 \cdot 1^2 = N(1 + \sqrt{-3}) = N(\alpha)N(\beta)$$

Since $2 = N(r + s\sqrt{-3}) = r^2 + 3s^2$ has no integer solutions, the above equation implies $N(\alpha) = 1$ or $N(\beta) = 1$, so that $\alpha = \pm 1$ or $\beta = \pm 1$ are units. Hence $1 + \sqrt{-3}$ is irreducible in $\mathbb{Z}[\sqrt{-3}]$. Since $N(2) = N(1 - \sqrt{-3}) = 4 = N(1 + \sqrt{-3})$, the same argument shows that 2 and $1 - \sqrt{-3}$ are also irreducible.

7.10. Let $\mathfrak{a}$ and $\mathfrak{b}$ be invertible fractional $\mathcal{O}$-ideals. We first show that $\mathfrak{ab}$ is a fractional ideal. By Exercise 7.7, $\mathfrak{a} = \alpha\mathfrak{a}'$ and $\mathfrak{b} = \beta\mathfrak{b}'$ for $\mathcal{O}$-ideals $\mathfrak{a}', \mathfrak{b}'$ and $\alpha, \beta \in K^*$. Then $\mathfrak{ab} = (\alpha\beta)\mathfrak{a}'\mathfrak{b}'$, which by Exercise 7.7 is again a fractional ideal since $\mathfrak{a}'\mathfrak{b}'$ is an $\mathcal{O}$-ideal and $\alpha\beta \neq 0$.

Since $\mathfrak{a}, \mathfrak{b}$ are invertible, Proposition 7.4 implies that there are fractional ideals $\mathfrak{a}^{-1}, \mathfrak{b}^{-1}$ such that $\mathfrak{a}\mathfrak{a}^{-1} = \mathfrak{b}\mathfrak{b}^{-1} = \mathcal{O}$. Then

$$(\mathfrak{ab})(\mathfrak{a}^{-1}\mathfrak{b}^{-1}) = (\mathfrak{a}\mathfrak{a}^{-1})(\mathfrak{b}\mathfrak{b}^{-1}) = \mathcal{O},$$

which proves that $\mathfrak{ab}$ is invertible. Finally, the fractional ideal $\mathfrak{a}^{-1}$ is invertible because $\mathfrak{a}^{-1}\mathfrak{a} = \mathfrak{a}\mathfrak{a}^{-1} = \mathcal{O}$.

7.11. We start with

$$\frac{p\tau + q}{r\tau + s} = \frac{(p\tau + q)(r\overline{\tau} + s)}{(r\tau + s)(r\overline{\tau} + s)} = \frac{(p\tau + q)(r\overline{\tau} + s)}{(r\tau + s)\overline{(r\tau + s)}} = \frac{pr\tau\overline{\tau} + ps\tau + qr\overline{\tau} + qs}{|r\tau + s|^2}.$$

The first and last terms in the numerator are real, so letting $\tau = a + bi$, $a, b \in \mathbb{R}$, we get

$$\begin{aligned} \operatorname{Im}\left(\frac{p\tau + q}{r\tau + s}\right) &= \frac{\operatorname{Im}(ps\tau + qr\overline{\tau})}{|r\tau + s|^2} = \frac{psb - qrb}{|r\tau + s|^2} \\ &= \frac{(ps - qr)b}{|r\tau + s|^2} = \det\begin{pmatrix} p & q \\ r & s \end{pmatrix} |r\tau + s|^{-2} \operatorname{Im}(\tau). \end{aligned}$$

7.12. (a) Note that $f(x, y)$ having negative discriminant is equivalent to $\tau \notin \mathbb{R}$.

The solution to part (b) of Exercise 2.4 shows that $f(x, y)$ is positive definite when $a > 0$ and it has negative discriminant. Conversely, if $f(x, y)$ is positive definite, then $f(1, 0) = a > 0$. If $\tau \in \mathbb{R}$, then $f(x, y)$ has positive discriminant and hence is is not positive definite by the solution to part (a) of Exercise 2.4, a contradiction. Thus $a > 0$ and $\tau \notin \mathbb{R}$.

(b) The discriminant of $\mathcal{O} = [1, a\tau]$ is

$$\left(\det\begin{pmatrix} 1 & a\tau \\ 1 & a\overline{\tau} \end{pmatrix}\right)^2 = (a(\overline{\tau} - \tau))^2.$$

Since $\tau + \overline{\tau} = -b/a$ and $\tau\overline{\tau} = c/a$, we get

$$(a(\overline{\tau} - \tau))^2 = a^2(\tau^2 - 2\tau\overline{\tau} + \overline{\tau}^2) = a^2((\tau + \overline{\tau})^2 - 4\tau\overline{\tau})$$
$$= a^2\left(\frac{b^2}{a^2} - 4\frac{c}{a}\right) = b^2 - 4ac,$$

which is the discriminant of $f(x, y)$.

(c) If $f(x, y) = ax^2 + bxy + cy^2$ is primitive and positive definite, then $x^2 + (b/a)x + (c/a)$ is the monic minimal polynomial of τ over $\mathbb{Q}$, which is unique. Since $\gcd(a, b, c) = 1$ and $a > 0$, the minimal polynomial determines the form $f(x, y) = ax^2 + bxy + cy^2$ uniquely. So a primitive positive definite form $g(x, y)$ with the same root τ must equal $f(x, y)$.

7.13. Suppose $\gcd(a, a', B) = d$. Then from $B \equiv b \bmod 2a$ and $B \equiv b' \bmod 2a'$ we see that $d \mid b$, $d \mid b'$ and therefore $d \mid b + b'$. It follows that $d = 1$ or $d = 2$ since otherwise either d or $\frac{d}{2}$ would divide all of a, a', and $(b + b')/2$.

We now show that $d = 2$ leads to a contradiction. Assume that $d = 2$. Then a, a', b, b' are all even. Since $D = b^2 - 4ac = b'^2 - 4a'c'$, we get $(b - b')(b + b') = 4(ac - a'c') \equiv 0 \bmod 8$ because a and a' are even. Hence one of the factors on the left is divisible by 4. If $b - b' \equiv 0 \bmod 4$, then since b' is even, $b - b' + 2b' = b + b' \equiv 0 \bmod 4$ as well. Thus in any event $b + b' \equiv 0 \bmod 4$, so that 2 divides $(b + b')/2$. Therefore $\gcd(a, a', (b + b')/2) \geq 2$, which is a contradiction. It follows that $d = \gcd(a, a', B) = 1$.

7.14. (a) Choose $\alpha = a + bu \in \mathcal{O}$, $\alpha \neq 0$, $a, b \in \mathbb{Z}$. Then $u \in \mathcal{O}_K$ since $\mathcal{O} = [1, u]$ is an order. Let $x^2 + sx + t$ be its minimal polynomial, and let $\alpha \mapsto \alpha'$ denote the nontrivial automorphism of K. Then $u + u' = -s$ and $uu' = N(u) = t$, so that

$$N(\alpha) = \alpha\alpha' = (a + bu)(a + bu') = a^2 + ab(u + u') + b^2uu' = a^2 - abs + b^2t.$$

Next,

$$\alpha u = (a + bu)u = au + bu^2 = au + b(-su - t) = -bt + (a - bs)u.$$

Comparing this to $\alpha u = c + du$ gives $c = -bt$ and $d = a - bs$, so

$$ad - bc = a(a - bs) - b(-bt) = a^2 - abs + b^2t = N(\alpha).$$

Since $N(\alpha) \neq 0$, we see that $ad - bc \neq 0$ as well.

(b) From Exercise 7.15, $N(\alpha\mathcal{O}) = |\mathcal{O}/\alpha\mathcal{O}| = |ad - bc| = |N(\alpha)|$. Note that this proves part (i) of Lemma 7.14, since there $\mathcal{O}$ is imaginary quadratic, so that norms are positive and $|N(\alpha)| = N(\alpha)$.

7.15. (a) Since $c = 0$, we have $0 \neq ad - bc = ad$, so that $a, d \neq 0$. For $re_1 + se_2 \in M$, use the division algorithm to write

$$s = w_2d + v_2, \quad 0 \leq v_2 < |d|$$
$$r - w_2b = w_1a + v_1, \quad 0 \leq v_1 < |a|.$$

Then

$$re_1 + se_2 = (w_1a + w_2b + v_1)e_1 + (w_2d + v_2)e_2$$
$$= v_1e_1 + v_2e_2 + w_1ae_1 + w_2(be_1 + de_2).$$

Since (under the assumption that $c = 0$) ae_1 and $be_1 + de_2$ are both in MA, it follows that every element of M is congruent modulo MA to $v_1e_1 + v_2e_2$ with $0 \leq v_1 < |a|$, $0 \leq v_2 < |d|$. Further, these ad elements are incongruent modulo MA, so that $|M/MA| = ad = |\det(A)|$.

(b) Since $|\det(B)| = \pm 1$, B is invertible; it follows that right multiplication by B is an automorphism of M. Then $MA \simeq MBA$, so we get an isomorphism

$$M/MA \simeq M/MBA.$$

Further, $|\det(BA)| = |\det(B)|\,|\det(A)| = |\det(A)|$. Thus

$$|M/MA| = |\det(A)| \iff |M/MBA| = |\det(BA)|\,.$$

(c) If $A = \left(\begin{smallmatrix} a & b \\ c & d \end{smallmatrix}\right)$ with $c \neq 0$, write $a = pc + q$, $p, q \in \mathbb{Z}$, $0 \leq q < |c|$. Let

$$B_1 = \begin{pmatrix} 0 & 1 \\ 1 & -p \end{pmatrix}.$$

Then $\det(B_1) = -1$ so that $B_1 \in \mathrm{GL}(2, \mathbb{Z})$, and

$$B_1 A = \begin{pmatrix} c & * \\ q & * \end{pmatrix}.$$

Then the lower left entry of $B_1 A$ is smaller in absolute value than the lower left entry of A. Continuing this process, we can multiply by appropriate elements of $\mathrm{GL}(2, \mathbb{Z})$ until the lower left entry is zero. The product of all these transformation matrices is the matrix B, which is of determinant ± 1.

(d) Given $A = \left(\begin{smallmatrix} a & b \\ c & d \end{smallmatrix}\right)$ with $ad - bc \neq 0$ choose $B \in \mathrm{GL}(2, \mathbb{Z})$ using part (c) so that BA has lower left entry zero. Then by parts (b) and (a),

$$|M/MA| = |M/MBA| = |\det(BA)| = |\det(B)\det(A)| = |\pm\det(A)| = |\det(A)|\,.$$

7.16. (a) By part (iii) of Lemma 7.14, $\mathfrak{a}\overline{\mathfrak{a}} = N(\mathfrak{a})\mathcal{O}$ is the principal ideal generated by $N(\mathfrak{a})$. Since $C(\mathcal{O}) = I(\mathcal{O})/P(\mathcal{O})$, it follows that the class of $\mathfrak{a}\overline{\mathfrak{a}}$ is the identity in $C(\mathcal{O})$. Thus the inverse of the class of $\mathfrak{a}$ is the class of $\overline{\mathfrak{a}}$.

(b) Let $f(x, y) = ax^2 + bxy + cy^2$, and let $g(x, y) = ax^2 - bxy + cy^2$ be its opposite. Under the isomorphism given in part (ii) of Theorem 7.7,

$$f(x, y) \longmapsto \mathfrak{a} = [a, (-b + \sqrt{D})/2]$$
$$g(x, y) \longmapsto [a, (b + \sqrt{D})/2] = [a, (-b - \sqrt{D})/2] = \overline{\mathfrak{a}}.$$

Thus $g(x, y)$ maps to $\overline{\mathfrak{a}}$ under that isomorphism. Since $[f(x, y)]^{-1} = [g(x, y)]$ in the form class group, it follows that $[\mathfrak{a}]^{-1} = [\overline{\mathfrak{a}}]$ in $C(\mathcal{O})$.

7.17. (a) As in the discussion following Lemma 7.5, $\mathfrak{a} = [\alpha, \beta] = \alpha[1, \tau]$ for $\tau = \beta/\alpha$; let $ax^2 + bx + c$ with $\gcd(a, b, c) = 1$ and $a > 0$ be the minimal polynomial of τ. Using (7.16),

$$f(x, y) = \frac{N(\alpha x - \beta y)}{N(\mathfrak{a})} = \frac{N(\alpha) N(x - \tau y)}{N(\alpha)/a} = a N(x - \tau y) = a(x - \tau y)(x - \overline{\tau} y)$$
$$= ax^2 + bxy + cy^2,$$

where the last line follows since τ and $\overline{\tau}$ are the roots of $ax^2 + bx + c$. After interchanging τ and $\overline{\tau}$ if necessary, we may assume that $\mathrm{Im}(\tau) > 0$. Lemma 7.5 shows that $\mathcal{O} = [1, a\tau]$, and then part (b) of Exercise 7.12 implies that the discriminant $b^2 - 4ac$ of $f(x, y)$ equals the discriminant D of $\mathcal{O}$.

To show that $\mathfrak{a} = [\alpha, \beta] \mapsto N(\alpha x - \beta y)/N(\mathfrak{a})$ gives a well-defined map $C(\mathcal{O}) \to C(D)$, suppose that proper $\mathcal{O}$-ideals $\mathfrak{a} = [\alpha, \beta]$ and $\mathfrak{a}' = [\alpha', \beta']$ give the same class in $C(\mathcal{O})$. As above, we may assume that the corresponding τ and τ' have positive imaginary parts. We need to show that the associated forms $f(x, y)$ and $g(x, y)$ are properly equivalent. Note that τ and τ' are the roots of these forms. Since

$$\mathfrak{a} = \alpha[1, \tau] \text{ and } \mathfrak{a}' = \alpha'[1, \tau']$$

and $\mathfrak{a}, \mathfrak{a}'$ give the same class in $C(\mathcal{O})$, the same is true for $[1, \tau]$ and $[1, \tau']$. Thus $[1, \tau] = \lambda[1, \tau']$ for some $\lambda \in K^*$, which by (7.8) implies that $f(x, y)$ and $g(x, y)$ are properly equivalent. This together with the fact that every class in $C(\mathcal{O})$ is represented by a proper $\mathcal{O}$-ideal show that we indeed get a well-defined map $C(\mathcal{O}) \to C(D)$.

Finally, observe that applying the map of Theorem 7.7 to the above form $f(x, y)$ yields the $\mathcal{O}$-ideal $\mathfrak{b} = [a, (-b + \sqrt{D})/2]$. Since $D = b^2 - 4ac$, the quadratic formula implies that

$$\mathfrak{b} = [a, a\tau] = a[1, \tau] = a[1, \beta/\alpha] = (a/\alpha)[\alpha, \beta] = (a/\alpha)\mathfrak{a}.$$

Thus $\mathfrak{a}$ gives the same class as $\mathfrak{b}$, and since the map $C(D) \to C(\mathcal{O})$ of Theorem 7.7 is an isomorphism, it follows that $\mathfrak{a} = [\alpha, \beta] \mapsto N(\alpha x - \beta y)/N(\mathfrak{a})$ gives the desired inverse.

(b) Let $K = \mathbb{Q}(\sqrt{-5})$, with $\mathcal{O}_K = \mathbb{Z}[\sqrt{-5}]$. Note that $d_K = -20$ since $-5 \not\equiv 1 \bmod 4$.

Not injective: The forms $2x^2 + 2xy + 3y^2$ and $2x^2 - 2xy + 3y^2$ of discriminant -20 map to the proper $\mathcal{O}_K$-ideals $[2, -1 + \sqrt{-5}]$ and $[2, 1 + \sqrt{-5}]$. But these two ideals are equal (*not* just in the same class), since $1 + \sqrt{-5} = 2 + (-1 + \sqrt{-5})$. Thus the map in Theorem 7.7 is not injective as a map from forms to ideals.

Not surjective: Consider the ideal $2\mathcal{O}_K$. A form $f(x, y)$ of discriminant -20 has even middle coefficient, so $f(x, y) = ax^2 + 2b'xy + cy^2$ for $a, b', c \in \mathbb{Z}$. This form maps to the ideal $[a, (-2b' + \sqrt{-20})/2] = [a, -b' + \sqrt{-5}]$. Since $-b' + \sqrt{-5} \notin 2\mathcal{O}_K = [2, 2\sqrt{-5}]$, we see that $f(x, y)$ cannot map to the ideal $2\mathcal{O}_K$. Thus the map in Theorem 7.7 is not surjective as a map from forms to ideals.

7.18. (a) The norm function is $N(x + y\sqrt{3}) = x^2 - 3y^2$, so we try $\left|x^2 - 3y^2\right|$ as our Euclidean norm. Let $x = a + b\sqrt{3}$, $y = c + d\sqrt{3}$ with $a, b, c, d \in \mathbb{Z}$ be elements of $\mathbb{Z}[\sqrt{3}]$ with $y \neq 0$. Then

$$\begin{aligned}\frac{x}{y} = \frac{a + b\sqrt{3}}{c + d\sqrt{3}} &= \frac{(a + b\sqrt{3})(c - d\sqrt{3})}{(c + d\sqrt{3})(c - d\sqrt{3})} \\ &= \frac{(ac - 3bd) + (bc - ad)\sqrt{3}}{c^2 - 3d^2} = r + s\sqrt{3}, \ r, s \in \mathbb{Q}.\end{aligned}$$

Let n, m be integers closest to the rational numbers r, s, respectively. Then we have $|r - n| \leq \frac{1}{2}$ and $|s - m| \leq \frac{1}{2}$. Write $t = (r - n) + (s - m)\sqrt{3}$. Then

$$t = r + s\sqrt{3} - (n + m\sqrt{3}) = \frac{x}{y} - (n + m\sqrt{3}),$$

so that $yt = x - (n + m\sqrt{3})y \in \mathbb{Z}[\sqrt{3}]$. Thus $x = (n + m\sqrt{3})y + yt$ with $n + m\sqrt{3} \in \mathbb{Z}[\sqrt{3}]$ and $yt \in \mathbb{Z}[\sqrt{3}]$. Finally,

$$N(yt) = N(y)N(t) = N(y)\left|(r - n)^2 - 3(s - m)^2\right|.$$

Since $(r - n)^2$ and $3(s - m)^2$ are both nonnegative, we see that

$$\left|(r - n)^2 - 3(s - m)^2\right| \leq \max((r - n)^2, 3(s - m)^2) \leq \frac{3}{4},$$

so that $N(yt) \leq \frac{3}{4}N(y) < N(y)$. This shows that $\mathcal{O}_K$ has a Euclidean norm. Thus $\mathcal{O}_K$ is Euclidean and hence a PID, so that $C(\mathcal{O}_K) \simeq \{1\}$.

(b) Suppose that $x^2 - 3y^2$ and $-x^2 + 3y^2$ were equivalent, say

$$\begin{aligned}-x^2 + 3y^2 &= (px + qy)^2 - 3(rx + sy)^2 \\ &= (p^2 - 3r^2)x^2 + (2pq - 6rs)xy + (q^2 - 3s^2)y^2.\end{aligned}$$

Then $p^2 - 3r^2 = -1$, which implies $p^2 \equiv -1 \equiv 2 \bmod 3$. This has no solutions, so there are at least two classes of forms of discriminant 12, and thus $C(d_K) = C(12)$ is nontrivial.

7.19. (a) Since K is a real field, $\alpha'\beta - \beta'\alpha \in \mathbb{R}$. If $\alpha'\beta = \alpha\beta'$, then since $(\alpha\beta')' = \alpha'\beta$, it must be the case that $\alpha\beta' \in \mathbb{Q}$ and hence is a nonzero integer n since α, β' are algebraic integers. Similarly, $N(\beta)$ is a nonzero integer m. Then

$$\alpha\beta' = n \quad\Longrightarrow\quad \alpha\beta'\beta = n\beta \quad\Longrightarrow\quad \alpha N(\beta) = \beta n \quad\Longrightarrow\quad \alpha m = \beta n,$$

By Exercise 7.8, $\mathfrak{a} = [\alpha, \beta]$ is a $\mathbb{Z}$-module of rank 2, which is impossible if $\alpha m = \beta n$. Hence we must have $\alpha'\beta - \beta'\alpha \neq 0$.

(b) Computing $\tilde{\alpha}'\tilde{\beta} - \tilde{\alpha}\tilde{\beta}'$, we get

$$\begin{aligned}
\tilde{\alpha}'\tilde{\beta} - \tilde{\alpha}\tilde{\beta}' &= (p\alpha' + q\beta')(r\alpha + s\beta) - (p\alpha + q\beta)(r\alpha' + s\beta') \\
&= pr\alpha'\alpha + ps\alpha'\beta + qr\beta'\alpha + qs\beta'\beta - pr\alpha\alpha' - ps\alpha\beta' - qr\beta\alpha' - qs\beta\beta' \\
&= ps(\alpha'\beta - \alpha\beta') - qr(\alpha'\beta - \alpha\beta') = (ps - qr)(\alpha'\beta - \alpha\beta') \\
&= \det\begin{pmatrix} p & q \\ r & s \end{pmatrix}(\alpha'\beta - \alpha\beta').
\end{aligned}$$

This shows that the two bases have the same orientation if and only if the transition matrix has determinant 1, i.e., if and only if it lies in $\mathrm{SL}(2, \mathbb{Z})$.

7.20. (a) The proof that $N(\mathfrak{a}) = |N(\alpha)|/a$ is similar to the proof of (7.16): first prove that $N(\alpha\mathfrak{a}) = |N(\alpha)|\, N(\mathfrak{a})$ using the same method as in the text, but using the fact (from Exercise 7.14) that $N(\alpha\mathcal{O}) = |N(\alpha)|$. Having established that, again we get $\mathcal{O} = [1, a\tau]$ and $N(a[1,\tau]) = a$ since $a > 0$. Now, $a \cdot \mathfrak{a} = \alpha \cdot a[1,\tau]$, so that

$$N(\mathfrak{a}) = \frac{N(\alpha \cdot a[1,\tau])}{N(a)} = \frac{|N(\alpha)|\, N(a[1,\tau])}{a^2} = \frac{|N(\alpha)|\, a}{a^2} = \frac{|N(\alpha)|}{a}.$$

(b) Expanding the definition of $f(x, y)$ and using part (a), we get

$$\begin{aligned}
f(x,y) &= \frac{N(\alpha x - \beta y)}{N(\mathfrak{a})} = \frac{N(\alpha)N(x - \tau y)}{|N(\alpha)|/a} = \frac{N(\alpha)}{|N(\alpha)|} aN(x - \tau y) \\
&= \mathrm{sgn}(N(\alpha)) aN(x - \tau y) = \mathrm{sgn}(N(\alpha)) a(x - \tau y)(x - \tau' y).
\end{aligned}$$

Since $\tau\tau' = \frac{c}{a}$, $\tau + \tau' = -\frac{b}{a}$, this gives

$$f(x,y) = \mathrm{sgn}(N(\alpha))(ax^2 + bxy + cy^2).$$

(c) As in Exercise 7.12, $[\alpha, \beta] = \alpha[1, \tau]$ is a proper ideal for $\mathcal{O} = [1, a\tau]$. The computation in part (b) of Exercise 7.12 does not in fact depend on whether $f(x, y)$ is positive definite or not, so that the discriminant of $\mathcal{O}$ is $b^2 - 4ac$, the discriminant of $f(x, y)$.

We therefore get a map ϕ from proper ideals in an order $\mathcal{O}$ of discriminant D with a given basis to primitive forms of discriminant D.

7.21. (a) If $\mathfrak{a} = [\alpha, \beta] = [\tilde{\alpha}, \tilde{\beta}]$, write $\tilde{\alpha} = p\alpha + q\beta$, $\tilde{\beta} = r\alpha + s\beta$. If both bases are positively oriented, then $ps - qr = 1$ by Exercise 7.19. Now we compute:

$$\begin{aligned}
g(x,y) &= \frac{N(\tilde{\alpha}x - \tilde{\beta}y)}{N(\mathfrak{a})} \\
&= \frac{N((p\alpha + q\beta)x - (r\alpha + s\beta)y)}{N(\mathfrak{a})} \\
&= \frac{N((px - ry)\alpha - (-qx + sy)\beta)}{N(\mathfrak{a})} = f(px - ry, -qx + sy).
\end{aligned}$$

Thus $g(x,y)$ is equivalent to $f(x,y)$ via the matrix $\left(\begin{smallmatrix} p & -r \\ -q & s \end{smallmatrix}\right)$ of determinant $ps-qr=1$. Hence the equivalence is proper. It follows that for an $\mathcal{O}$-ideal $\mathfrak{a}$, all positively oriented bases give properly equivalent forms of discriminant D, so that $\mathfrak{a}$ determines a well-defined class in $C(D)$.

To complete part (a), suppose that $g(x,y)$ is properly equivalent to $f(x,y)$ via a matrix $\left(\begin{smallmatrix} p & r \\ q & s \end{smallmatrix}\right) \in \mathrm{SL}(2,\mathbb{Z})$. Define a basis $\tilde{\alpha}$, $\tilde{\beta}$ for $\mathfrak{a}$ by

$$\tilde{\alpha} = p\alpha - q\beta, \qquad \tilde{\beta} = -r\alpha + s\beta.$$

Since the determinant of the corresponding matrix is 1, $[\tilde{\alpha}, \tilde{\beta}]$ is also a positively oriented basis of $\mathfrak{a}$, Then

$$\begin{aligned}\frac{N(\tilde{\alpha}x - \tilde{\beta}y)}{N(\mathfrak{a})} &= \frac{N((p\alpha - q\beta)x - (-r\alpha + s\beta)y)}{N(\mathfrak{a})} = \frac{N((px+ry)\alpha - (qx+sy)\beta)}{N(\mathfrak{a})} \\ &= f(px+ry, qx+sy) = g(x,y).\end{aligned}$$

(b) Recall that $\mathrm{sgn} : \mathbb{R}^* \to \{\pm 1\}$ is the group homomorphism defined by $\mathrm{sgn}(r) = |r|/r$. In this notation, part (a) of Exercise 7.19 defines $\mathrm{sgn}(\alpha,\beta)$ to be $\mathrm{sgn}(\alpha'\beta - \alpha\beta')$. Given $\lambda \in K$ with $N(\lambda) > 0$, we compute:

$$\begin{aligned}\mathrm{sgn}(\lambda\alpha, \lambda\beta) &= \mathrm{sgn}((\lambda'\alpha')(\lambda\beta) - (\lambda\alpha)(\lambda'\beta')) = \mathrm{sgn}(\lambda\lambda'(\alpha'\beta - \alpha\beta')) \\ &= \mathrm{sgn}(\lambda\lambda')\mathrm{sgn}((\alpha'\beta - \alpha\beta')) = \mathrm{sgn}(N(\lambda))\mathrm{sgn}(\alpha,\beta) = \mathrm{sgn}(\alpha,\beta) > 0\end{aligned}$$

since $N(\lambda) = \lambda\lambda' > 0$. Thus, if $N(\lambda) > 0$ and α, β are positively oriented, then the same is true for $\lambda\alpha, \lambda\beta$.

Now let $\mathfrak{a} = [\alpha, \beta]$ be a proper ideal in $\mathcal{O}$, where we assume that α, β are positively oriented. By part (a), $\mathfrak{a}$ gives the class in $C(D)$ represented by

$$f(x,y) = \frac{N(\alpha x - \beta y)}{N(\mathfrak{a})}.$$

If $N(\lambda) > 0$, then the previous paragraph shows that $\lambda\alpha, \lambda\beta$ are positively oriented, so that $\lambda\mathfrak{a} = [\lambda\alpha, \lambda\beta]$ gives the form

$$\frac{N(\lambda\alpha x - \lambda\beta y)}{N(\lambda\mathfrak{a})} = \frac{N(\lambda)N(\alpha x - \beta y)}{|N(\lambda)|\, N(\mathfrak{a})} = \frac{N(\alpha x - \beta y)}{N(\mathfrak{a})} = f(x,y).$$

It follows that the forms produced by all positively oriented bases of $\lambda\mathfrak{a}$ produce forms properly equivalent to $f(x,y)$ by part (a). Thus $\lambda\mathfrak{a}$ gives the same class of forms as does $\mathfrak{a}$.

(c) From part (b), multiplying a basis by a positive norm element doesn't change the class of the resulting form; interpreting that in light of the map ϕ from Exercise 7.20 and part (a) gives a well-defined map

$$\phi : I(\mathcal{O})/P^+(\mathcal{O}) = C^+(\mathcal{O}) \to C(D).$$

Now suppose that $\mathfrak{a}, \tilde{\mathfrak{a}}$ are $\mathcal{O}$-ideals that give the same class in $C(D)$. Choose positively oriented bases $\mathfrak{a} = [\alpha, \beta]$ and $\tilde{\mathfrak{a}} = [\tilde{\alpha}, \tilde{\beta}]$ that give the same form; this is possible by part (a) and Exercise 7.19.

(i) From Exercise 7.20, $[\alpha, \beta]$ produces the form $\mathrm{sgn}(N(\alpha))(ax^2 + bxy + cy^2)$ with $a > 0$, while $[\tilde{\alpha}, \tilde{\beta}]$ gives the form $\mathrm{sgn}(N(\tilde{\alpha}))(a'x^2 + b'xy + c'y^2)$ with $a' > 0$. Since these two forms are equal and $\mathrm{sgn}(a) = \mathrm{sgn}(a') = 1$, it follows that $\mathrm{sgn}(N(\tilde{\alpha})) = \mathrm{sgn}(N(\alpha))$, so that $N(\alpha\tilde{\alpha})$ is positive. Then since $N(\alpha^2) = N(\alpha)^2$ is also positive, part (b) shows that $\alpha\tilde{\alpha}\mathfrak{a}$ and $\alpha^2\tilde{\mathfrak{a}}$ give the same class in $C(D)$ as well. These are the ideals $[\alpha^2\tilde{\alpha}, \alpha\beta\tilde{\alpha}]$ and $[\alpha^2\tilde{\alpha}, \alpha^2\tilde{\beta}]$. After renaming, then, we may assume that the two ideals are $\mathfrak{a} = [\alpha, \beta]$ and $\tilde{\mathfrak{a}} = [\alpha, \tilde{\beta}]$.

(ii) Let $\tau = \beta/\alpha$ and $\tilde{\tau} = \tilde{\beta}/\alpha$. Since

$$f(x,y) = \frac{N(\alpha x - \beta y)}{N(\mathfrak{a})} = \frac{N(\alpha x - \tilde{\beta} y)}{N(\tilde{\mathfrak{a}})},$$

we see that $f(\tau,1) = f(\tilde{\tau},1) = 0$. Thus $\tilde{\tau} = \tau$ or $\tilde{\tau} = \tau'$. If $\tilde{\tau} = \tau'$, then $\tau' = \tilde{\beta}/\alpha$, so that $\tilde{\beta} = \alpha\tau'$. Then

$$\operatorname{sgn}(\alpha, \tilde{\beta}) = \operatorname{sgn}(\alpha, \alpha\tau') = \operatorname{sgn}(\alpha'(\alpha\tau') - \alpha(\alpha\tau')') = \operatorname{sgn}(\alpha\alpha'(\tau' - \tau)).$$

However, we also have $\beta = \alpha\tau$, so that a similar computation gives

$$\operatorname{sgn}(\alpha, \beta) = \operatorname{sgn}(\alpha, \alpha\tau) = \operatorname{sgn}(\alpha'(\alpha\tau) - \alpha(\alpha\tau')) = \operatorname{sgn}(\alpha\alpha'(\tau - \tau')).$$

It follows that $\operatorname{sgn}(\alpha, \tilde{\beta}) = -\operatorname{sgn}(\alpha, \beta) = -1$. This contradicts the fact that $[\alpha, \tilde{\beta}]$ is a positively oriented basis, so that $\tilde{\tau} = \tau$ and thus $\tilde{\beta} = \beta$. So the map is injective.

(d) For the primitive form $f(x,y) = ax^2 + bxy + cy^2$ of discriminant $D = b^2 - 4ac$, τ is a root of $ax^2 + bx + c = 0$ and hence lies in $K = \mathbb{Q}(\sqrt{D})$ by the quadratic formula. Then $a\tau \in \mathcal{O}_K$ since it is a root of the monic integer polynomial $x^2 + bx + ac$. Applying Lemma 7.5, we conclude that $[1, \tau]$ is a proper fractional ideal for the order $[1, a\tau]$, which has discriminant

$$\left(\det\begin{pmatrix} 1 & a\tau \\ 1 & a\tau' \end{pmatrix}\right)^2 = a^2(\tau' - \tau)^2 = b^2 - 4ac = D.$$

The third equality follows from the argument used in the solution to part (b) of Exercise 7.12. Then $\mathcal{O} = [1, a\tau]$ is the order of discriminant D in K. This works no matter which root of $ax^2 + bx + c$ is chosen to be τ.

In the first case, where $a > 0$, we choose the root with $\operatorname{sgn}(1, \tau) = 1$ and set

$$\mathfrak{a} = [a, a\tau] = a[1, \tau].$$

This is a proper ideal in $\mathcal{O} = [1, a\tau]$ by the previous paragraph. Also, $\mathfrak{a} = [a, a\tau] \subset [1, a\tau] = \mathcal{O}$ clearly has index a, so that

$$N(\mathfrak{a}) = |\mathcal{O}/\mathfrak{a}| = a.$$

Since a is an integer, one sees easily that $\operatorname{sgn}(a, a\tau) = \operatorname{sgn}(1, \tau) = 1$. Thus $[a, a\tau]$ is positively oriented, so that $\mathfrak{a}$ gives the quadratic form

$$\begin{aligned} \frac{N(ax - a\tau y)}{N(\mathfrak{a})} &= \frac{1}{a}(ax - a\tau y)(ax - a\tau' y) \\ &= a(x - \tau y)(x - \tau' y) = ax^2 + bxy + cy^2 = f(x,y), \end{aligned}$$

where the third equality follows since τ, τ' are the roots of $ax^2 + bx + c$.

In the second case, where $a < 0$, we choose the root with $\operatorname{sgn}(1, \tau) = -1$ and set $\mathfrak{a} = \sqrt{D}[a, a\tau]$. By definition, D is the square of an element of $\mathcal{O}$, so that $\sqrt{D} \in \mathcal{O}$. Since $\operatorname{sgn}(1, \tau) = -1$, it follows that $\tau - \tau' < 0$, and thus

$$\begin{aligned} (\sqrt{D}a)'(\sqrt{D}a\tau) - (\sqrt{D}a)(\sqrt{D}a\tau)' &= (-\sqrt{D}a)(\sqrt{D}a\tau) - (\sqrt{D}a)(-\sqrt{D}a\tau') \\ &= -Da^2(\tau - \tau') > 0, \end{aligned}$$

so that $[\sqrt{D}a, \sqrt{D}a\tau]$ is a positively oriented basis for the $\mathcal{O}$-ideal $\mathfrak{a} = [\sqrt{D}a, \sqrt{D}a\tau]$. We compute $N(\mathfrak{a})$ using the solution of part (a) of Exercise 7.20, which implies that

$$N(\mathfrak{a}) = \frac{|N(\sqrt{D}a)|}{-a} = \frac{Da^2}{-a} = -Da,$$

where $-a$ is the denominator because τ is a root of $-ax^2 - bx - c$ with $-a > 0$, as required by Exercise 7.20. Also remember that $D > 0$ since K is a real quadratic field. Then the associated quadratic form is

$$\frac{N(\sqrt{D}ax - \sqrt{D}a\tau y)}{N(\mathfrak{a})} = \frac{-a^2D(x-\tau y)(x-\tau' y)}{-Da} = a(x-\tau y)(x-\tau' y) = f(x,y).$$

7.22. (a) This is similar to part (a) of Exercise 7.21. Suppose that $\mathfrak{a} = [\alpha, \beta] = [\tilde{\alpha}, \tilde{\beta}]$ are two bases for $\mathfrak{a}$ (not necessarily positively oriented), and write $\tilde{\alpha} = p\alpha + q\beta$, $\tilde{\beta} = r\alpha + s\beta$, with $ps - qr = \pm 1$. From the computation in part (b) of Exercise 7.19, we see that

$$\operatorname{sgn}(\tilde{\alpha}, \tilde{\beta}) = \det \begin{pmatrix} p & q \\ r & s \end{pmatrix} \operatorname{sgn}(\alpha, \beta).$$

Then the form produced by $[\tilde{\alpha}, \tilde{\beta}]$ is

$$\begin{aligned} g(x,y) = \operatorname{sgn}(\tilde{\alpha}, \tilde{\beta}) \frac{N(\tilde{\alpha}x - \tilde{\beta}y)}{N(\mathfrak{a})} &= \operatorname{sgn}(\tilde{\alpha}, \tilde{\beta}) \frac{N((p\alpha + q\beta)x - (r\alpha + s\beta)y)}{N(\mathfrak{a})} \\ &= \operatorname{sgn}(\tilde{\alpha}, \tilde{\beta}) \frac{N((px - ry)\alpha - (-qx + sy)\beta)}{N(\mathfrak{a})} \\ &= \frac{\operatorname{sgn}(\tilde{\alpha}, \tilde{\beta})}{\operatorname{sgn}(\alpha, \beta)} f(px - ry, -qx + sy) \\ &= \det \begin{pmatrix} p & q \\ r & s \end{pmatrix} f(px - ry, -qx + sy) \\ &= \det \begin{pmatrix} p & -q \\ -r & s \end{pmatrix} f(px - ry, -qx + sy), \end{aligned}$$

showing that $g(x,y)$ is signed equivalent to $f(x,y)$. Thus any two bases for $\mathfrak{a}$ produce signed equivalent forms. To show that all signed equivalent forms arise in this way, suppose that $g(x,y)$ is signed equivalent to $f(x,y)$ via a matrix $\left(\begin{smallmatrix} p & q \\ r & s \end{smallmatrix}\right) \in \mathrm{GL}(2,\mathbb{Z})$. Define a basis $\tilde{\alpha}, \tilde{\beta}$ for $\mathfrak{a}$ by

$$\tilde{\alpha} = p\alpha - q\beta, \qquad \tilde{\beta} = -r\alpha + s\beta.$$

Then this basis produces the form

$$\begin{aligned} \operatorname{sgn}(\tilde{\alpha}, \tilde{\beta}) \frac{N(\tilde{\alpha}x - \tilde{\beta}y)}{N(\mathfrak{a})} &= \operatorname{sgn}(\tilde{\alpha}, \tilde{\beta}) \frac{N((p\alpha - q\beta)x - (-r\alpha + s\beta)y)}{N(\mathfrak{a})} \\ &= \operatorname{sgn}(\tilde{\alpha}, \tilde{\beta}) \frac{N((px + ry)\alpha - (qx + sy)\beta)}{N(\mathfrak{a})} \\ &= \det \begin{pmatrix} p & q \\ r & s \end{pmatrix} \operatorname{sgn}(\alpha, \beta) \frac{N((px + ry)\alpha - (qx + sy)\beta)}{N(\mathfrak{a})} \\ &= \det \begin{pmatrix} p & q \\ r & s \end{pmatrix} f(px + ry, qx + ry) = g(x,y). \end{aligned}$$

(b) For an $\mathcal{O}$-ideal $\mathfrak{a}$ with basis $[\alpha, \beta]$, denote by $f_{\alpha,\beta}(x,y)$ the form produced by the transformation in part (a). Note that while the form actually depends on the basis chosen, part (a) shows that the signed equivalence class does not.

To show that the map from $C(\mathcal{O}) \to C_s(D)$ is well-defined, choose an $\mathcal{O}$-ideal $\mathfrak{a} = [\alpha, \beta]$. If $\mathfrak{a}'$ is an $\mathcal{O}$-ideal in the same class, then $\mathfrak{a}' = \lambda\mathfrak{a} = [\lambda\alpha, \lambda\beta]$ for some $\lambda \in K^*$.

As in part (b) of Exercise 7.21, $\operatorname{sgn}(\lambda\alpha,\lambda\beta)=\operatorname{sgn}(N(\lambda))\operatorname{sgn}(\alpha,\beta)$. We get

$$\begin{aligned}f_{\lambda\alpha,\lambda\beta}(x,y)&=\operatorname{sgn}(\lambda\alpha,\lambda\beta)\frac{N(\lambda\alpha x-\lambda\beta y)}{N(\lambda\mathfrak{a})}\\&=\operatorname{sgn}(N(\lambda))\operatorname{sgn}(\alpha,\beta)\frac{N(\lambda)N(\alpha x-\beta y)}{|N(\lambda)|\,N(\mathfrak{a})}\\&=\operatorname{sgn}(\alpha,\beta)\frac{N(\alpha x-\beta y)}{N(\mathfrak{a})}=f_{\alpha,\beta}(x,y).\end{aligned}$$

Thus the forms associated with these bases are identical, so the same form class is associated with $\mathfrak{a}$ and $\mathfrak{a}'$, proving that the map is well-defined.

Next, we show that the induced map $C(\mathcal{O})\to C_s(D)$ is injective. To do this we must show that if $\mathfrak{a}$ and $\tilde{\mathfrak{a}}$ are $\mathcal{O}$-ideals that map to the same signed class, then they are in the same class in $C(\mathcal{O})$. Note that since the map is well-defined, two bases $[\alpha,\beta]$ and $[\beta,\alpha]$ of an ideal produce signed equivalent forms. So we may assume that we have chosen positively oriented bases $[\alpha,\beta]$ and $[\tilde{\alpha},\tilde{\beta}]$ that map to the same signed class. But then

$$f_{\alpha,\beta}(x,y)=\frac{N(\alpha x-\beta y)}{N(\mathfrak{a})},\qquad f_{\tilde{\alpha},\tilde{\beta}}(x,y)=\frac{N(\tilde{\alpha}x-\tilde{\beta}y)}{N(\tilde{\mathfrak{a}})},$$

and we know from Exercise 7.21 that these two forms are in the same (unsigned) class in $C(D)$ since both bases are positively oriented. Then the injectivity proved in Exercise 7.21 implies that $\mathfrak{a}$ and $\tilde{\mathfrak{a}}$ lie in the same class in $C(\mathcal{O})$. This proves that $C(\mathcal{O})\to C_s(D)$ is injective.

For surjectivity, given a form $f(x,y)=ax^2+bxy+cy^2$ of discriminant D, define the same $\mathcal{O}$-ideal $\mathfrak{a}$ and basis for that ideal as in part (d) of Exercise 7.21, and recall that the basis was positively oriented in both cases (depending on $a>0$ or $a<0$). Then

$$f_{\alpha,\beta}(x,y)=\frac{N(\alpha x-\beta y)}{N(\mathfrak{a})}.$$

But Exercises 7.20 and 7.21 show that this form is precisely $f(x,y)$, proving surjectivity.

7.23. (a) (i) If K is imaginary quadratic, then the norm of any nonzero element is positive since the norm form is positive definite. Thus $P^+(\mathcal{O})=P(\mathcal{O})$. It follows that $C^+(\mathcal{O})=I(\mathcal{O})/P^+(\mathcal{O})=I(\mathcal{O})/P(\mathcal{O})=C(\mathcal{O})$.

(ii) A negative definite form can be written $f(x,y)=ax^2+bxy+cy^2$ where $a<0$ and $b^2-4ac<0$. Thus $c<0$ as well. Let $M=\left(\begin{smallmatrix}p&q\\r&s\end{smallmatrix}\right)=\left(\begin{smallmatrix}0&1\\1&0\end{smallmatrix}\right)$; then

$$\det(M)f(px+qy,rx+sy)=-(cx^2+bxy+ay^2)=-cx^2-bxy-ay^2,$$

which is positive definite since its determinant $(-b)^2-4(-a)(-c)=b^2-4ac$ is negative, and $-c>0$. Thus any form in $C_s(D)$ is signed equivalent to a positive definite form.

Now define a map $C(D)\to C_s(D)$ by $[f(x,y)]\mapsto[f(x,y)]_s$, where by $[f(x,y)]_s$ we mean the signed equivalence class of $f(x,y)$. This map is well-defined, since if $f(x,y)$ and $g(x,y)$ are properly equivalent, they are also signed equivalent. To see that it is surjective, suppose $[g(x,y)]_s\in C_s(D)$. Then $g(x,y)$ is signed equivalent to a positive definite form $f(x,y)$ by the above paragraph, so that $[f(x,y)]\mapsto[f(x,y)]_s=[g(x,y)]_s$. To show that the map is injective, suppose that $f(x,y)$ and $f'(x,y)$ are positive definite forms that both map to $[g(x,y)]_s$ in $C_s(D)$. Then $f(x,y)$ and $f'(x,y)$ must be signed equivalent. But since they are both positive definite, the equivalence matrix has determinant 1, so it lies in $\mathrm{SL}(2,\mathbb{Z})$ and therefore $f(x,y)$ and $f'(x,y)$ are properly equivalent. Thus $[f(x,y)]=[f'(x,y)]$, proving injectivity.

(b) (i) The natural surjection $C^+(\mathcal{O}) = I(\mathcal{O})/P^+(\mathcal{O}) \to I(\mathcal{O})/P(\mathcal{O}) = C(\mathcal{O})$ is induced by the inclusion $P^+(\mathcal{O}) \subset P(\mathcal{O})$ (recall that $P^+(\mathcal{O})$ consists of principal ideals generated by elements of positive norm, while $P(\mathcal{O})$ consists of principal ideals generated by all nonzero elements of $\mathcal{O}$). The natural map $C(D) \to C_s(D)$ given by $[f(x,y)] \mapsto [f(x,y)]_s$ is well-defined since if $f(x,y)$ is properly equivalent to $g(x,y)$, then $f(x,y)$ is also signed equivalent to $g(x,y)$ because the proper equivalence matrix is in $\mathrm{SL}(2,\mathbb{Z}) \subset \mathrm{GL}(2,\mathbb{Z})$. It is a surjection since if $[f(x,y)]_s \in C_s(D)$, then $[f(x,y)] \in C(D)$ maps to $[f(x,y)]_s$.

If $[\alpha,\beta]$ is a positively oriented basis for $\mathfrak{a}$, then

$$\begin{array}{ccccc} C^+(\mathcal{O}) & \xrightarrow{\sim} & C(D) & \longrightarrow & C_s(D) \\ [\alpha,\beta] & \longmapsto & \left[\frac{N(\alpha x-\beta y)}{N(\mathfrak{a})}\right] & \longmapsto & \left[\frac{N(\alpha x-\beta y)}{N(\mathfrak{a})}\right]_s, \end{array}$$

and

$$\begin{array}{ccccc} C^+(\mathcal{O}) & \longrightarrow & C(\mathcal{O}) & \xrightarrow{\sim} & C_s(D) \\ [\alpha,\beta] & \longmapsto & [\alpha,\beta] & \longmapsto & \left[\frac{N(\alpha x-\beta y)}{N(\mathfrak{a})}\right]_s \end{array}$$

since $\mathrm{sgn}(\alpha,\beta) = 1$. Hence the diagram in the statement of the exercise commutes.

(ii) The kernel of $C^+(\mathcal{O}) \to C(\mathcal{O})$ consists of classes $[\mathfrak{a}] \in I(\mathcal{O})/P^+(\mathcal{O}) = C^+(\mathcal{O})$ that map to the zero element of $I(\mathcal{O})/P(\mathcal{O}) = C(\mathcal{O})$. This happens if and only if $\mathfrak{a} \in P(\mathcal{O})$. Therefore the kernel can be identified with $P(\mathcal{O})/P^+(\mathcal{O})$.

Next observe that $P^+(\mathcal{O}) \cup \sqrt{d_K}\,P^+(\mathcal{O})$ consists of principal fractional ideals and hence is contained in $P(\mathcal{O})$. For the opposite inclusion, take $\lambda\mathcal{O} \in P(\mathcal{O})$. If $N(\lambda) > 0$, then $\lambda\mathcal{O} \in P^+(\mathcal{O})$. Otherwise, if $N(\lambda) < 0$, then we can write $\lambda\mathcal{O} = \sqrt{d_K}\big((\lambda/\sqrt{d_K})\mathcal{O}\big)$. Since K is real quadratic, it follows that $N(\sqrt{d_K}) = -d_K < 0$, so that $N(\lambda/\sqrt{d_K}) > 0$. Therefore $(\lambda/\sqrt{d_K})\mathcal{O} \in P^+(\mathcal{O})$, so that $\lambda\mathcal{O} \in \sqrt{d_K}P^+(\mathcal{O})$.

Cosets are either equal or disjoint. When $P^+(\mathcal{O}) \neq \sqrt{d_K}\,P^+(\mathcal{O})$, there are precisely 2 cosets by the previous paragraph, so that $\left|C^+(\mathcal{O})\right| / |C(\mathcal{O})| = 2$. Since $\mathcal{O} \in P^+(\mathcal{O})$,

$$\begin{aligned} \left|C^+(\mathcal{O})\right| / |C(\mathcal{O})| = 1 &\iff P^+(\mathcal{O}) = \sqrt{d_K}\,P^+(\mathcal{O}) \\ &\iff \mathcal{O} \in \sqrt{d_K}\,P^+(\mathcal{O}) \\ &\iff \mathcal{O} = \lambda\sqrt{d_K}\mathcal{O},\ \lambda \in K^*,\ N(\lambda) > 0. \end{aligned}$$

If the equality on the last line occurs, then $\lambda\sqrt{d_K}$ is a unit whose norm is $N(\lambda)N(\sqrt{d_K}) = -d_K N(\lambda) < 0$. Thus $\epsilon = \lambda\sqrt{d_K}$ has norm -1 since units have norm ± 1, proving that $\mathcal{O}$ has a unit of norm -1. Conversely, if $\mathcal{O}$ has such a unit ϵ, then $\lambda = \epsilon\sqrt{d_k}$ has positive norm and $\mathcal{O} = \lambda\sqrt{d_K}\mathcal{O}$ where $N(\lambda) > 0$. Thus $\left|C^+(\mathcal{O})\right| / |C(\mathcal{O})| = 1$ if and only if $\mathcal{O}$ has a unit of norm -1, and the desired formula for $\left|C^+(\mathcal{O})\right| / |C(\mathcal{O})|$ follows easily.

(iii) Since the diagram in part (i) commutes and the top and bottom maps are isomorphisms, the result is an immediate consequence of part (ii).

7.24. To find the inverse of $C^+(\mathcal{O}) \xrightarrow{\sim} C(D)$, take a form $f(x,y) = ax^2 + bxy + cy^2$ of discriminant D. It suffices to construct an $\mathcal{O}$-ideal $\mathfrak{a}$ and a positively oriented basis of $\mathfrak{a}$ that maps to $f(x,y)$ under the map defined in Exercise 7.21. For if we do that, then we map the class of f in $C(D)$ to the class of $\mathfrak{a}$ in $C^+(\mathcal{O})$, and the two maps are then inverse bijections.

Given $f(x, y)$, the procedure described in part (d) of Exercise 7.21 constructs an $\mathcal{O}$-ideal $\mathfrak{a}$ with a positively oriented basis that maps to $f(x, y)$. This gives the desired inverse map.

To find the inverse of $C(\mathcal{O}) \xrightarrow{\sim} C_s(D)$, take a form $f(x, y)$ of discriminant D. As above, it suffices to construct an $\mathcal{O}$-ideal $\mathfrak{a}$ that, with some basis, maps back under the map defined in Exercise 7.22, to a form signed equivalent to $f(x, y)$. For if we do that, then we map the class of f in $C_s(D)$ to the class of $\mathfrak{a}$ in $C(\mathcal{O})$, and the two maps are then inverse bijections.

Given $f(x, y)$, the proof of surjectivity in the solution to Exercise 7.22 shows that again the procedure described in part (d) of Exercise 7.21 constructs an $\mathcal{O}$-ideal $\mathfrak{a}$ and a basis that maps to $f(x, y)$. Again, we get the desired inverse.

7.25. By definition, $I_K(f)$ (resp. $I(\mathcal{O}, f)$) is the subgroup of I_K (resp. $I(\mathcal{O})$) generated by $\mathcal{O}_K$-ideals (resp. $\mathcal{O}$-ideals) prime to f. To extend the map ϕ to $\overline{\phi} : I_K(f) \to I(\mathcal{O}, f)$, take an element of $I_K(f)$ and write it as $\mathfrak{a}\mathfrak{b}^{-1}$ for $\mathcal{O}_K$-ideals $\mathfrak{a}, \mathfrak{b}$ prime to f. Then define

$$\overline{\phi}(\mathfrak{a}\mathfrak{b}^{-1}) = \phi(\mathfrak{a})\phi(\mathfrak{b})^{-1}.$$

To check that $\overline{\phi}$ is well-defined, note that

$$\begin{aligned} \mathfrak{a}\mathfrak{b}^{-1} = \mathfrak{c}\mathfrak{d}^{-1} \ &\Rightarrow \ \mathfrak{a}\mathfrak{d} = \mathfrak{b}\mathfrak{c} \ \Rightarrow \ \phi(\mathfrak{a}\mathfrak{d}) = \phi(\mathfrak{b}\mathfrak{c}) \ \Rightarrow \ \phi(\mathfrak{a})\phi(\mathfrak{d}) = \phi(\mathfrak{b})\phi(\mathfrak{c}) \\ &\Rightarrow \phi(\mathfrak{a})\phi(\mathfrak{b})^{-1} = \phi(\mathfrak{c})\phi(\mathfrak{d})^{-1}, \end{aligned}$$

where the first line uses the fact that ϕ preserves multiplication. To show that $\overline{\phi}$ is a group homomorphism, note that

$$\begin{aligned} \overline{\phi}(\mathfrak{a}\mathfrak{b}^{-1} \cdot \mathfrak{c}\mathfrak{d}^{-1}) &= \overline{\phi}((\mathfrak{a}\mathfrak{c})(\mathfrak{b}\mathfrak{d})^{-1}) = \phi(\mathfrak{a}\mathfrak{c})\phi(\mathfrak{b}\mathfrak{d})^{-1} = \phi(\mathfrak{a})\phi(\mathfrak{c})\left(\phi(\mathfrak{b})\phi(\mathfrak{d})\right)^{-1} \\ &= \phi(\mathfrak{a})\phi(\mathfrak{b})^{-1} \cdot \phi(\mathfrak{c})\phi(\mathfrak{d})^{-1} = \overline{\phi}(\mathfrak{a}\mathfrak{b}^{-1}) \cdot \overline{\phi}(\mathfrak{c}\mathfrak{d}^{-1}), \end{aligned}$$

where we use the definition of $\overline{\phi}$ and the fact that ϕ preserves multiplication.

Injective: Suppose that $\overline{\phi}(\mathfrak{a}\mathfrak{b}^{-1}) = \mathcal{O}$ (the identity of $I(\mathcal{O}, f)$). By the definition of $\overline{\phi}$, this means $\phi(\mathfrak{a})\phi(\mathfrak{b})^{-1} = \mathcal{O}$, which implies $\phi(\mathfrak{a}) = \phi(\mathfrak{b})$. Since ϕ is injective, we get $\mathfrak{a} = \mathfrak{b}$, which implies $\mathfrak{a}\mathfrak{b}^{-1} = \mathcal{O}_K$, and injectivity is proved.

Surjective: Take $\tilde{\mathfrak{a}}\tilde{\mathfrak{b}}^{-1} \in I(\mathcal{O}, f)$. Since ϕ is surjective, there are $\mathcal{O}_K$-ideals $\mathfrak{a}$ and $\mathfrak{b}$ prime to f with $\phi(\mathfrak{a}) = \tilde{\mathfrak{a}}$ and $\phi(\mathfrak{b}) = \tilde{\mathfrak{b}}$. Then $\overline{\phi}(\mathfrak{a}\mathfrak{b}^{-1}) = \phi(\mathfrak{a})\phi(\mathfrak{b})^{-1} = \tilde{\mathfrak{a}}\tilde{\mathfrak{b}}^{-1}$, proving surjectivity.

7.26. (a) Since $\mathfrak{a} = (\mathfrak{a}\mathcal{O}_K) \cap \mathcal{O}$ by (7.21), we get a natural injection $\mathcal{O}/\mathfrak{a} \to \mathcal{O}_K/\mathfrak{a}\mathcal{O}_K$. By part (ii) of Proposition 7.20, the $\mathcal{O}$-ideal $\mathfrak{a}$ and the $\mathcal{O}_K$-ideal $\mathfrak{a}\mathcal{O}_K$ have the same norm, i.e., $|\mathcal{O}/\mathfrak{a}| = |\mathcal{O}_K/\mathfrak{a}\mathcal{O}_K|$. Since both are finite, the injection is an isomorphism $\mathcal{O}/\mathfrak{a} \simeq \mathcal{O}_K/\mathfrak{a}\mathcal{O}_K$. Thus one of these rings is an integral domain if and only if the other is, so that $\mathfrak{a}$ is a prime $\mathcal{O}$-ideal if and only if $\mathfrak{a}\mathcal{O}_K$ is a prime $\mathcal{O}_K$-ideal.

(b) If $\mathfrak{a}$ is an $\mathcal{O}$-ideal relatively prime to f, then $\mathfrak{a}\mathcal{O}_K$ factors uniquely into a product of primes, $\mathfrak{a}\mathcal{O}_K = \prod_{i=1}^r \mathfrak{p}_i$. Each of these primes is prime to f since $\mathfrak{a}\mathcal{O}_K$ is. Since the map $I_K(f) \to I(\mathcal{O}, f)$ from part (iii) of Proposition 7.20, which is given by $\mathfrak{a} \mapsto \mathfrak{a} \cap \mathcal{O}$, is a group isomorphism, it follows that $\mathfrak{a} = \prod_{i=1}^r (\mathfrak{p}_i \cap \mathcal{O})$. Since each $\mathfrak{p}_i$ is prime in $\mathcal{O}_K$, each $\mathfrak{p}_i \cap \mathcal{O}$ is prime by standard facts from abstract algebra, and each is prime to f by part (i) of Proposition 7.20. Uniqueness follows from uniqueness of the factorization of $\mathfrak{a}\mathcal{O}_K$.

7.27. Since $\alpha \equiv \beta \bmod m\mathcal{O}_K$, we have $\alpha = \beta + m\gamma$, $\gamma \in \mathcal{O}_K$. Then $\alpha' = \beta' + m\gamma'$, where the prime $'$ denotes the nontrivial automorphism of K. Now we compute:

$$\begin{aligned} N(\alpha) = \alpha\alpha' &= (\beta + m\gamma)(\beta' + m\gamma') = \beta\beta' + m(\gamma\beta' + \gamma'\beta + m\gamma\gamma') \\ (*) \qquad &= N(\beta) + m(\gamma\beta' + \gamma'\beta + m\gamma\gamma'). \end{aligned}$$

Since $\beta, \gamma \in \mathcal{O}_K$, $\mathrm{Tr}(\gamma\beta') = \gamma\beta' + (\gamma\beta')' = \gamma\beta' + \gamma'\beta$ and $N(\gamma) = \gamma\gamma'$ are integers, so that the quantity in parentheses in (*) is an integer. Thus $N(\alpha) \equiv N(\beta) \bmod m$.

7.28. Denote by π the natural map $(\mathcal{O}_K/\mathfrak{p}^n)^* \to (\mathcal{O}_K/\mathfrak{p}^{n-1})^*$; then we want to show that

$$1 \longrightarrow \mathcal{O}_K/\mathfrak{p} \xrightarrow{\ \phi\ } (\mathcal{O}_K/\mathfrak{p}^n)^* \xrightarrow{\ \pi\ } (\mathcal{O}_K/\mathfrak{p}^{n-1})^* \longrightarrow 1$$

is exact for $n \geq 2$.

(a) To prove that π is surjective, choose $[\alpha] \in (\mathcal{O}_K/\mathfrak{p}^{n-1})^*$ for $\alpha \in \mathcal{O}_K$, and choose $\beta \in \mathcal{O}_K$ such that $[\alpha\beta] = 1$, i.e., such that $\alpha\beta = 1 + \gamma$, $\gamma \in \mathfrak{p}^{n-1}$. Then

$$\alpha\beta(1-\gamma) = 1 - \gamma^2 \equiv 1 \bmod \mathfrak{p}^n$$

since $\gamma^2 \in \mathfrak{p}^{2(n-1)} \subset \mathfrak{p}^n$ for $n \geq 2$. It follows that $\alpha \in (\mathcal{O}_K/\mathfrak{p}^n)^*$, and clearly $\pi([\alpha]) = [\alpha]$.

(b) To determine the kernel of π, choose some fixed $u \in \mathfrak{p}^{n-1} - \mathfrak{p}^n$.

(i) For $\alpha \in \mathcal{O}_K$, note that $(1+\alpha u)(1-\alpha u) = 1 - \alpha^2 u^2$; since $u \in \mathfrak{p}^{n-1}$, we have $u^2 \in \mathfrak{p}^{2n-2} \subset \mathfrak{p}^n$ because $n \geq 2$. Therefore $1 + \alpha u$ is invertible modulo $\mathfrak{p}^n$, so that $[1+\alpha u] \in (\mathcal{O}_K/\mathfrak{p}^n)^*$.

(ii) Consider the map $\phi : \mathcal{O}_K/\mathfrak{p} \to (\mathcal{O}_K/\mathfrak{p}^n)^*$ defined by $[\alpha] \mapsto [1+\alpha u]$. This map has image in $(\mathcal{O}_K/\mathfrak{p}^n)^*$ by part (i). To see that it is well-defined, suppose that $[\alpha] = [\beta] \in \mathcal{O}_K/\mathfrak{p}$; we need to show that $[1+\alpha u] = [1+\beta u] \in (\mathcal{O}_K/\mathfrak{p}^n)^*$. But $(1+\alpha u) - (1+\beta u) = (\alpha - \beta)u \in \mathfrak{p}^n$ since $\alpha \equiv \beta \bmod \mathfrak{p}$ and $u \in \mathfrak{p}^{n-1}$. Thus ϕ is well-defined. Further, it is a homomorphism, since in $(\mathcal{O}_K/\mathfrak{p}^n)^*$

$$\begin{aligned}\phi([\alpha])\phi([\beta]) &= [(1+\alpha u)][(1+\beta u)] = [1 + (\alpha+\beta)u + \alpha\beta u^2]\\ &= [1 + (\alpha+\beta)u] = \phi([\alpha+\beta]),\end{aligned}$$

since $\alpha\beta u^2 \in \mathfrak{p}^n$ when $n \geq 2$. To see that ϕ is injective, suppose $\phi([\alpha]) = [1+\alpha u] = [1]$. Then $\alpha u \in \mathfrak{p}^n$. If $\alpha \notin \mathfrak{p}$, then since $\mathfrak{p}$ is maximal, $\alpha\mathcal{O}_K + \mathfrak{p} = \mathcal{O}_K$, so we can choose $r \in \mathcal{O}_K$ and $x \in \mathfrak{p}$ such that $r\alpha + x = 1$, giving $u = r\alpha u + xu$. But both αu and xu lie in $\mathfrak{p}^n$, contradicting the choice of u. This proves that $\alpha \in \mathfrak{p}$ and therefore that ϕ is injective.

It remains to show exactness at $(\mathcal{O}_K/\mathfrak{p}^n)^*$. First, $\mathrm{im}(\phi) \subset \ker(\pi)$, since $\alpha u \in \mathfrak{p}^{n-1}$ for any $\alpha \in \mathcal{O}_K$. In order to prove that $\ker(\pi) \subset \mathrm{Im}(\phi)$, we need the following fact:

$$\mathfrak{p}^n + u\mathcal{O}_K = \mathfrak{p}^{n-1}.$$

To prove this, let $\mathfrak{a} = \mathfrak{p}^n + u\mathcal{O}_K$. Then $\mathfrak{p}^n \subset \mathfrak{a} \subset \mathfrak{p}^{n-1}$, where the second inclusion follows from $u \in \mathfrak{p}^{n-1}$. Applying the first paragraph of the solution to part (f) of Exercise 5.6, we get $\mathfrak{p}^n = \mathfrak{a}\mathfrak{b}$ and $\mathfrak{a} = \mathfrak{p}^{n-1}\mathfrak{c}$ for some ideals $\mathfrak{b}$, $\mathfrak{c}$. This gives $\mathfrak{p}^n = \mathfrak{p}^{n-1}\mathfrak{b}\mathfrak{c}$, and unique factorization implies that $\mathfrak{p} = \mathfrak{b}\mathfrak{c}$. It follows that $\mathfrak{c} = \mathfrak{p}$ or $\mathfrak{c} = \mathcal{O}_K$. But $\mathfrak{c} = \mathfrak{p}$ is impossible since that gives $\mathfrak{a} = \mathfrak{p}^n$, so that $u \in \mathfrak{p}^n$, which is not the case. Therefore $\mathfrak{c} = \mathcal{O}_K$ and $\mathfrak{a} = \mathfrak{p}^{n-1}$.

Now suppose $\beta \in \ker(\pi)$; then $\beta \equiv 1 \bmod \mathfrak{p}^{n-1}$, so that $\beta = 1 + \gamma$ for some $\gamma \in \mathfrak{p}^{n-1}$. Since $\mathfrak{p}^{n-1} = \mathfrak{p}^n + u\mathcal{O}_K$, we can write $\gamma = \delta + \alpha u$ for $\delta \in \mathfrak{p}^n$ and $\alpha \in \mathcal{O}_K$. Then $\gamma \equiv \alpha u \bmod \mathfrak{p}^n$, so that $\beta = 1 + \gamma \equiv 1 + \alpha u \bmod \mathfrak{p}^n$. Hence $[\beta] = [1+\alpha u] \in \mathrm{Im}(\phi)$, and exactness follows.

7.29. (a) We begin with a version of the Chinese Remainder Theorem: if $\mathcal{O}_K$-ideals $\mathfrak{a}$ and $\mathfrak{b}$ are relatively prime (i.e. have no common factors), then the map

$$\phi : \mathcal{O}_K/\mathfrak{a}\mathfrak{b} \longrightarrow \mathcal{O}_K/\mathfrak{a} \times \mathcal{O}_K/\mathfrak{b}, \quad [\alpha] \longmapsto (\alpha \bmod \mathfrak{a}, \alpha \bmod \mathfrak{b}).$$

is an isomorphism. This map is well-defined since $\mathfrak{a}\mathfrak{b} \subset \mathfrak{a}$ and $\mathfrak{a}\mathfrak{b} \subset \mathfrak{b}$. It is also surjective. To see why, note that any prime ideal containing $\mathfrak{a} + \mathfrak{b}$ would contain $\mathfrak{a}$ and $\mathfrak{b}$ and hence divide both by Corollary 5.6. This is impossible since $\mathfrak{a}$ and $\mathfrak{b}$ are relatively prime, so that $\mathfrak{a} + \mathfrak{b} = \mathcal{O}_K$. Thus $1 = u + v$ for some $u \in \mathfrak{a}$ and $v \in \mathfrak{b}$. Given $([\alpha], [\beta]) \in \mathcal{O}_K/\mathfrak{a} \times \mathcal{O}_K/\mathfrak{b}$, one easily checks that $\phi([v\alpha + u\beta]) = ([\alpha], [\beta])$, so

ϕ is surjective. Since $N(\mathfrak{a}\mathfrak{b}) = N(\mathfrak{a})N(\mathfrak{b})$ by Lemma 7.14, ϕ is a surjection between finite sets of the same cardinality, so that ϕ is an isomorphism.

By an easy induction, the result of the previous paragraph extends to the case of mutually relatively prime ideals $\mathfrak{a}_1, \ldots, \mathfrak{a}_r$. In particular, if $\mathfrak{a}$ has a prime factorization $\prod_{i=1}^r \mathfrak{p}_i^{n_i}$ for distinct prime ideals $\mathfrak{p}_i$, then the $\mathfrak{p}_i^{n_i}$ are mutually relatively prime for $i = 1, \ldots, r$, so that the map

$$\phi : \mathcal{O}_K/\mathfrak{a} \longrightarrow \prod_{i=1}^r (\mathcal{O}_K/\mathfrak{p}_i^{n_i}), \quad [\alpha] \longmapsto (\alpha \bmod \mathfrak{p}_1^{n_1}, \ldots, \alpha \bmod \mathfrak{p}_r^{n_r})$$

is an isomorphism.

(b) This is a computation. By the isomorphism of part (a),

$$|(\mathcal{O}_K/\mathfrak{a})^*| = \left| \prod_{i=1}^r (\mathcal{O}_K/\mathfrak{p}_i^{n_i})^* \right| = \prod_{i=1}^r |(\mathcal{O}_K/\mathfrak{p}_i^{n_i})^*| = \prod_{i=1}^r N(\mathfrak{p}_i)^{n_i-1}(N(\mathfrak{p}_i) - 1)$$

where the last equality is from Exercise 7.28. Using Lemma 7.14, this becomes

$$\begin{aligned} &= \prod_{i=1}^r N(\mathfrak{p}_i)^{n_i} \left(\frac{N(\mathfrak{p}_i) - 1}{N(\mathfrak{p}_i)} \right) = \prod_{i=1}^r N\left(\mathfrak{p}_i^{n_i}\right) \prod_{i=1}^r \left(1 - \frac{1}{N(\mathfrak{p}_i)}\right) \\ &= N(\mathfrak{a}) \prod_{\mathfrak{p} \mid \mathfrak{a}} \left(1 - \frac{1}{N(\mathfrak{p})}\right). \end{aligned}$$

(c) If $m = \prod_{i=1}^r p_i^{n_i}$, where the p_i are distinct primes, then the $p_i^{n_i}\mathcal{O}_K$ are mutually relatively prime, which by the solution to part (a) implies that

$$\mathcal{O}_K/m\mathcal{O}_K \simeq \prod_{i=1}^r \mathcal{O}_K/p_i^{n_i}\mathcal{O}_K,$$

so that

$$|(\mathcal{O}_K/m\mathcal{O}_K)^*| = \left| \prod_{i=1}^r (\mathcal{O}_K/p_i^{n_i}\mathcal{O}_K)^* \right|$$

Furthermore, by part (b),

$$|(\mathcal{O}_K/p_i^{n_i}\mathcal{O}_K)^*| = N(p_i^{n_i}) \prod_{\mathfrak{p} \mid p_i^{n_i}\mathcal{O}_K} \left(1 - \frac{1}{N(\mathfrak{p})}\right) = p_i^{2n_i} \prod_{\mathfrak{p} \mid p_i\mathcal{O}_K} \left(1 - \frac{1}{N(\mathfrak{p})}\right)$$

($\mathfrak{p} \mid p_i^{n_i}\mathcal{O}_K$ if and only if $\mathfrak{p} \mid p_i\mathcal{O}_K$ since $\mathfrak{p}$ is prime). For p an integer prime, the ideal $p\mathcal{O}_K$ factors as follows:

$$p\mathcal{O}_K = \begin{cases} \mathfrak{p}^2 & p \mid d_K, \quad N(\mathfrak{p}) = p \\ \mathfrak{p}\mathfrak{p}' & (d_K/p) = 1, \quad N(\mathfrak{p}) = N(\mathfrak{p}') = p \\ p\mathcal{O}_K & (d_K/p) = -1, \quad N(p\mathcal{O}_K) = p^2. \end{cases}$$

Then:

- If $p \mid d_K$, then $p\mathcal{O}_K$ is divisible by only one prime of $\mathcal{O}_K$, of norm p, and $(d_K/p) = 0$, so that

$$|(\mathcal{O}_K/p^n\mathcal{O}_K)^*| = p^{2n}\left(1 - \frac{1}{p}\right) = p^{2n}\left(1 - \frac{1}{p}\right)\left(1 - \left(\frac{d_K}{p}\right)\frac{1}{p}\right).$$

- If $(d_K/p) = 1$, then $p\mathcal{O}_K$ is divisible by two primes of $\mathcal{O}_K$, each of norm p, so we get

$$\begin{aligned}|(\mathcal{O}_K/p^n\mathcal{O}_K)^*| &= p^{2n}\left(1-\frac{1}{p}\right)^2 = p^{2n}\left(1-\frac{1}{p}\right)\left(1-\frac{1}{p}\right)\\ &= p^{2n}\left(1-\frac{1}{p}\right)\left(1-\left(\frac{d_K}{p}\right)\frac{1}{p}\right).\end{aligned}$$

- If $(d_K/p) = -1$, then $p\mathcal{O}_K$ is divisible by one prime of $\mathcal{O}_K$, of norm p^2, so we get

$$\begin{aligned}|(\mathcal{O}_K/p^n\mathcal{O}_K)^*| &= p^{2n}\left(1-\frac{1}{p^2}\right) = p^{2n}\left(1-\frac{1}{p}\right)\left(1+\frac{1}{p}\right)\\ &= p^{2n}\left(1-\frac{1}{p}\right)\left(1-\left(\frac{d_K}{p}\right)\frac{1}{p}\right).\end{aligned}$$

The result follows since $|(\mathcal{O}_K/m\mathcal{O}_K)^*| = \left|\prod_{i=1}^r(\mathcal{O}_K/p_i^{n_i}\mathcal{O}_K)^*\right|$. Note that if we define

$$\phi_K(f) = f\prod_{p\mid f}\left(1-\left(\frac{d_K}{p}\right)\frac{1}{p}\right),$$

then the result can be written $|(\mathcal{O}_K/f\mathcal{O}_K)^*| = \phi(f)\phi_K(f)$, where ϕ is the Euler ϕ-function.

7.30. (a) We want to show that

$$1 \to \{\pm 1\} \xrightarrow{\iota} (\mathbb{Z}/f\mathbb{Z})^* \times \mathcal{O}_K^* \xrightarrow{\psi} (\mathcal{O}_K/f\mathcal{O}_K)^* \xrightarrow{\phi} I_K(f)\cap P_K/P_{K,\mathbb{Z}}(f) \to 1$$

is exact, where ι and ψ are the obvious maps given by

$$\iota(u) = ([u],u),\ \ \psi([m],u) = [mu].$$

Clearly ι is injective, and ϕ is surjective from the proof of (7.27). Further, $\psi\iota(-1) = \psi([-1],-1) = [1]$, so that $\mathrm{im}(\iota) \subset \ker(\psi)$. Next, if $([m],u) \in \ker(\psi)$, then $mu \equiv 1 \bmod f\mathcal{O}_K$. If $u = \pm 1$, then $m \equiv \pm 1 \bmod f\mathcal{O}_K$, which implies $m \equiv u \equiv \pm 1 \bmod f$. It follows that $([m],u)$ equals either $([1],1) = \iota(1)$ or $([-1],-1) = \iota(-1)$, proving that $([m],u) \in \mathrm{im}(\iota)$.

It remains to consider what happens when $u \neq \pm 1$. Since K is imaginary quadratic, this means that $K = \mathbb{Q}(i)$ and $u = \pm i$ or $K = \mathbb{Q}(\omega)$ and $u = \pm\omega, \pm\omega^2$. The description of $\mathcal{O}_K$ for these fields implies that $\mathcal{O}_K = \mathbb{Z}[u]$. Then $mu \equiv 1 \bmod f\mathcal{O}_K$ implies

$$mu = 1 + f(a+bu) = (1+fa) + fbu, \quad a,b \in \mathbb{Z}.$$

This implies $m = fb$, which contradicts $[m] \in (\mathbb{Z}/f\mathbb{Z})^*$. Hence $u \neq \pm 1$ cannot occur, and exactness at $(\mathbb{Z}/f\mathbb{Z})^* \times \mathcal{O}_K^*$ follows.

Next, $\mathrm{im}(\psi) \subset \ker(\phi)$ by definition of $P_{K,\mathbb{Z}}(f)$. Now choose $\alpha \in \mathcal{O}_K$ with $[\alpha] \in \ker(\phi)$. Then $\alpha\mathcal{O}_K \in P_{K,\mathbb{Z}}(f)$, so that $\alpha\mathcal{O}_K = (\beta\mathcal{O}_K)(\gamma\mathcal{O}_K)^{-1}$ with $\beta \equiv b \bmod f\mathcal{O}_K$ and $\gamma \equiv c \bmod f\mathcal{O}_K$ for some $[b]$, $[c] \in (\mathbb{Z}/f\mathbb{Z})^*$. Thus $\alpha = \epsilon\beta\gamma^{-1}$ for some $\epsilon \in \mathcal{O}_K^*$, and

$$\psi([b][c]^{-1},\epsilon) = [\epsilon bc^{-1}] = [\epsilon\beta\gamma^{-1}] = [\alpha] \in (\mathcal{O}_K/f\mathcal{O}_K)^*.$$

This shows exactness at $(\mathcal{O}_K/f\mathcal{O}_K)^*$ and completes the proof.

(b) First, if $\mathcal{O} = \mathcal{O}_K$, then $f = 1$ and the product over $p \mid f$ is vacuous, so the formula becomes

$$h(\mathcal{O}_K) = \frac{h(\mathcal{O}_K)\cdot 1}{[\mathcal{O}_K^* : \mathcal{O}_K^*]},$$

which clearly holds. So assume $\mathcal{O} \subsetneq \mathcal{O}_K$ and thus $f > 1$. Since $|(\mathbb{Z}/f\mathbb{Z})^*| = \phi(f)$ and $|(\mathcal{O}_K/f\mathcal{O}_K)^*| = \phi(f)\phi_K(f)$, where $\phi_K(f)$ is the function defined at the end of the solution to Exercise 7.29, the exact sequence from part (a) and (7.26) imply that

$$\begin{aligned}\frac{h(\mathcal{O})}{h(\mathcal{O}_K)} &= |(I_K(f) \cap P_K)/P_{K,\mathbb{Z}}(f)| = \frac{|(\mathcal{O}_K/f\mathcal{O}_K)^*| \cdot |\{\pm 1\}|}{|(\mathbb{Z}/f\mathbb{Z})^* \times \mathcal{O}_K^*|}\\ &= \frac{\phi(f)\phi_K(f)\,|\{\pm 1\}|}{\phi(f)\,|\mathcal{O}_K^*|} = \frac{\phi_K(f)\,|\{\pm 1\}|}{|\mathcal{O}_K^*|}.\end{aligned}$$

But $\mathcal{O} \subsetneq \mathcal{O}_K$ implies that $\mathcal{O}^* = \{\pm 1\}$ and we get

$$\frac{h(\mathcal{O})}{h(\mathcal{O}_K)} = \frac{\phi_K(f)\,|\mathcal{O}^*|}{|\mathcal{O}_K^*|} = \frac{\phi_K(f)}{[\mathcal{O}_K^* : \mathcal{O}^*]} = \frac{f}{[\mathcal{O}_K^* : \mathcal{O}^*]} \prod_{p \mid f} \left(1 - \left(\frac{d_K}{p}\right)\frac{1}{p}\right).$$

7.31. Let $\mathcal{O}$ have conductor f, so that $D = f^2 d_K$ and $\mathcal{O}'$ has conductor mf. Using Theorem 7.24 gives

$$\begin{aligned}h(m^2D) = h(\mathcal{O}') &= \frac{h(\mathcal{O}_K)mf}{[\mathcal{O}_K^* : \mathcal{O}'^*]} \prod_{p \mid mf} \left(1 - \left(\frac{d_K}{p}\right)\frac{1}{p}\right)\\ h(D) = h(\mathcal{O}) &= \frac{h(\mathcal{O}_K)f}{[\mathcal{O}_K^* : \mathcal{O}^*]} \prod_{p \mid f} \left(1 - \left(\frac{d_K}{p}\right)\frac{1}{p}\right),\end{aligned}$$

so that

$$\begin{aligned}\frac{h(m^2D)}{h(D)} &= \frac{m}{[\mathcal{O}^* : \mathcal{O}'^*]} \cdot \frac{\prod_{p \mid mf}\left(1 - (d_K/p)\frac{1}{p}\right)}{\prod_{p \mid f}\left(1 - (d_K/p)\frac{1}{p}\right)}\\ &= \frac{m}{[\mathcal{O}^* : \mathcal{O}'^*]} \prod_{\substack{p \mid m\\ p \nmid f}} \left(1 - \left(\frac{d_K}{p}\right)\frac{1}{p}\right).\end{aligned}$$

Now, $(f^2/p) = 0$ if $p \mid f$ and $(f^2/p) = 1$ if $p \nmid f$, so the product above can be rewritten

$$\begin{aligned}\frac{m}{[\mathcal{O}^* : \mathcal{O}'^*]} \prod_{\substack{p \mid m\\ p \nmid f}} \left(1 - \left(\frac{d_k}{p}\right)\frac{1}{p}\right) &= \frac{m}{[\mathcal{O}^* : \mathcal{O}'^*]} \prod_{p \mid m} \left(1 - \left(\frac{f^2}{p}\right)\left(\frac{d_k}{p}\right)\frac{1}{p}\right)\\ &= \frac{m}{[\mathcal{O}^* : \mathcal{O}'^*]} \prod_{p \mid m} \left(1 - \left(\frac{D}{p}\right)\frac{1}{p}\right),\end{aligned}$$

as desired.

7.32. (a) The inequality $1 - [2\sqrt{p}]/(p+1) \geq 1/2$ is equivalent to $p + 1 \geq 2[2\sqrt{p}]$. Note that $2[2\sqrt{p}] \leq 4\sqrt{p}$, and solving $p + 1 \geq 4\sqrt{p}$ gives $p \geq 7 + 4\sqrt{3} \approx 13.9$. So the given statement holds for $p \geq 17$. Checking 11 and 13 separately, we get

$$\begin{aligned}1 - \frac{[2\sqrt{11}]}{12} &= 1 - \frac{[2 \cdot 3.317\ldots]}{12} = 1 - \frac{6}{12} = \frac{1}{2} \geq \frac{1}{2}\\ 1 - \frac{[2\sqrt{13}]}{14} &= 1 - \frac{[2 \cdot 3.606\ldots]}{14} = 1 - \frac{7}{14} = \frac{1}{2} \geq \frac{1}{2},\end{aligned}$$

and we are done.

(b) From Theorem 6.1 (and Exercise 6.1), the number of genera of primitive positive definite forms of discriminant d_K is $\left|C(d_K)/C(d_K)^2\right| = 2^{\mu-1}$, where μ is the number of distinct primes dividing d_K. It follows that $2^{\mu-1} \mid h = h(d_K)$, so that $\nu_2(h) \geq \mu - 1$ and thus $\mu \leq \nu_2(h) + 1$.

Let $f(p) = 1 - [2\sqrt{p}]/(p+1)$. Then $f(p) < 1/2$ for $p < 11$, and part (a) shows that $f(p) \geq 1/2$ if $p > 11$. Thus the product will be minimized if it includes as many as possible of the primes less than 11. So if d_K is divisible by at least $\mu \geq 4$ distinct primes (even or odd), the smallest that the product can be is if it includes the primes 2, 3, 5, and 7, in which case the product is (using the result of part (a))

$$f(2)f(3)f(5)f(7)\prod_{\substack{p|d_k\\ p\geq 11}} f(p) \geq \frac{1}{3}\cdot\frac{1}{4}\cdot\frac{1}{3}\cdot\frac{3}{8}\cdot\frac{1}{2^{\mu-4}}$$
$$\geq \frac{1}{3}\cdot\frac{1}{4}\cdot\frac{1}{3}\cdot\frac{3}{8}\cdot\frac{1}{2^{\nu_2(h)-3}} \geq \frac{1}{3\cdot 2^{\nu_2(h)+2}}.$$

If d_K is prime, the empty product is 1, and the inequality is satisfied. If d_K is the product of 2 or 3 primes, then $\nu_2(h) \geq 1$ or 2 respectively. The smallest possible values for the product for 2 or 3 primes are

$$\frac{1}{4}\cdot\frac{1}{3} = \frac{1}{12} \geq \frac{1}{24} = \frac{1}{3\cdot 2^{1+2}} \geq \frac{1}{3\cdot 2^{\nu_2(h)+2}}$$
$$\frac{1}{4}\cdot\frac{1}{3}\cdot\frac{1}{3} = \frac{1}{36} \geq \frac{1}{48} = \frac{1}{3\cdot 2^{2+2}} \geq \frac{1}{3\cdot 2^{\nu_2(h)+2}}.$$

(c) This is a straightforward computation. The inequality (∗) given in the exercise implies

$$\log|d_K| < 55h\left(\prod_{\substack{p|d_K\\ p<d_K}}\left(1 - \frac{[2\sqrt{p}]}{p+1}\right)\right)^{-1}.$$

Applying the result of part (b) gives

$$\log|d_K| < 55h\cdot 3\cdot 2^{\nu_2(h)+2} = 165\cdot 2^{\nu_2(h)+2}h.$$

Now exponentiate both sides to get the desired inequality.

(d) Let $\mathcal{O}$ be the order of discriminant $D = f^2 d_K$. Using Theorem 7.24, if $f = \prod_{i=1}^r p_i^{n_i}$, then

$$h(D) = \frac{h(d_K)f}{[\mathcal{O}_K^* : \mathcal{O}^*]}\prod_{p|f}\left(1 - \left(\frac{d_K}{p}\right)\frac{1}{p}\right)$$

By genus theory, $h(D) \geq 2^{\mu-1}$, so if $h(D) = h$ we get $\mu \leq 1 + \log(h)/\log(2)$. By Proposition 3.11, the number of odd primes dividing D is at most μ, so that the total number of primes (odd or even) dividing D is at most $t = \mu + 1 \leq 2 + \log(h)/\log(2)$. But then

$$h(D) = \frac{h(d_K)f}{[\mathcal{O}_K^* : \mathcal{O}^*]}\prod_{p|f}\left(1 - \left(\frac{d_K}{p}\right)\frac{1}{p}\right) \geq \frac{h(d_K)f}{[\mathcal{O}_K^* : \mathcal{O}^*]}\prod_{i=1}^{t}\left(1 - \frac{1}{2}\right)$$
$$= \frac{h(d_K)f}{[\mathcal{O}_K^* : \mathcal{O}^*]}\left(\frac{1}{2}\right)^{2+\log(h)/\log(2)} \geq \frac{h(d_K)f}{3}\left(\frac{1}{2}\right)^{2+\log(h)/\log(2)},$$

where the last inequality on the first line follows since $(d_K/p)/p \leq 1/2$ for all $p \mid f$. For fixed h, $h(D) \leq h$ places upper bounds on both $h(d_K)$ and f. There are only finitely many choices for K by (c), so there are only a finite number of orders with $h(D) \leq h$.

7.33. Part (i) of Theorem 7.30 says that the only d_K with $h(\mathcal{O}_K) = h(d_K) = 1$ are

$$d_K = -3,\ -4,\ -7,\ -8,\ -11,\ -19,\ -43,\ -67,\ -163.$$

If $D = f^2 d_K$, and $\mathcal{O} \subset \mathcal{O}_K$ is the corresponding order, we have

$$\frac{h(D)}{h(d_K)} = \frac{f}{[\mathcal{O}_K^* : \mathcal{O}^*]}\prod_{p|f}\left(1 - \left(\frac{d_K}{p}\right)\frac{1}{p}\right).$$

We want to find orders with $h(D) = 1$, so we must have $h(d_K) = 1$ as well since $h(d_K)$ divides $h(D)$. Thus the list above can be used to determine the possibilities for d_K. Clearly if $f = 1$, then $h(D) = h(d_K) = 1$. So assume $f > 1$, and let $f = \prod_{i=1}^r p_i^{n_i}$ where the p_i are distinct primes. Then we want to solve

$$1 = \frac{f}{[\mathcal{O}_K^* : \mathcal{O}^*]} \prod_{p|f} \left(1 - \left(\frac{d_K}{p}\right)\frac{1}{p}\right) = \frac{P}{[\mathcal{O}_K^* : \mathcal{O}^*]}, \quad P = \prod_{i=1}^r p_i^{n_i - 1} \prod_{p|f} \left(p - \left(\frac{d_K}{p}\right)\right).$$

Examining the possible discriminants d_K, we find the following:

- If $d_K \neq -3, -4$, then $[\mathcal{O}_K^* : \mathcal{O}^*] = 1$ and we want to find all f such that P equals 1. Since the two products making up P are positive integers, each product must be 1. The second product can be 1 only if $f = p = 2$ and $(d_K/2) = 1$, since $p - (d_K/p) \geq p - 1$. This implies that $d_K \equiv 1 \bmod 8$. The only d_K in the list that is 1 modulo 8 is -7, so we get an order of conductor 2 in $\mathcal{O}_K$, with discriminant $D = -28$.
- If $d_K = -3$, then $[\mathcal{O}_K^* : \mathcal{O}^*] = 3$ and we want to find f such that P equals 3. From the first product, the only possible square factor of f would be 3, but then the second product would be at least 2. Therefore f is squarefree. Further, no prime greater than 3 can appear as a factor of f, since the second product would be at least $p - 1 > 3$. Thus $f = 2$, $f = 3$, or $f = 6$. If $f = 2$, we need $(d_K/2) = -1$, so $d_K \equiv 5 \bmod 8$; this holds for $d_K = -3$, so we get $D = -12$. If $f = 3$, we need $(d_K/3) = 0$, which holds for $d_K = -3$, and we get $D = -3^2 \cdot 3 = -27$. Finally, if $f = 6$, we need $(d_K/2) = 0$ and $(d_K/3) = 0$; but $(d_K/2) = 0$ does not hold for $d_K = -3$. So in summary, the D with class number one that are multiples of $d_K = -3$ are $D = -12$ and $D = -27$.
- If $d_K = -4$, then $[\mathcal{O}_K^* : \mathcal{O}^*] = 2$ and we want find f such that P equals 2. There are three possibilities for f. First, we can have $f = 2$, with $(d_K/2) = 0$, giving $D = -2^2 \cdot 4 = -16$. Second, $f = 3$, with $(d_K/3) = 1$; but $(-4/3) = -1$. Finally, $f = 6$, with $(d_K/3) = (d_K/2) = 0$, but this is false for $d_K = -4$. So the only additional order is the order of discriminant -16.

Thus the orders with discriminant $D \equiv 0, 1 \bmod 4$, $D < 0$, with $h(D) = 1$ are the orders with discriminant

$$D = -3,\ -4,\ -7,\ -8,\ -11,\ -12,\ -16,\ -19,\ -27,\ -28,\ -43,\ -67,\ -163.$$

Solutions to Exercises in §8

8.1. (a) We first show that the map $\phi : (\mathcal{O}_K/\mathfrak{m})^* \to I_K(\mathfrak{m}) \cap P_K/P_{K,1}(\mathfrak{m})$, where $[\alpha] \mapsto [\alpha\mathcal{O}_K]$, is well-defined. If $[\alpha] \in (\mathcal{O}_K/\mathfrak{m})^*$, then $\alpha\mathcal{O}_K$ is prime to $\mathfrak{m}$, so that $\alpha\mathcal{O}_K \in I_K(\mathfrak{m}) \cap P_K$. Next, if $[\alpha] = [\beta] \in (\mathcal{O}_K/\mathfrak{m})^*$, choose $u \in \mathcal{O}_K$ with $u\alpha \equiv u\beta \equiv 1 \bmod \mathfrak{m}$; then $u\alpha\mathcal{O}_K$ and $u\beta\mathcal{O}_K$ are in $P_{K,1}(\mathfrak{m})$. But

$$\alpha\mathcal{O}_K \cdot u\beta\mathcal{O}_K = \beta\mathcal{O}_K \cdot u\alpha\mathcal{O}_K,$$

so that $\alpha\mathcal{O}_K$ and $\beta\mathcal{O}_K$ are in the same coset of $P_{K,1}(\mathfrak{m})$, which shows that the map is well-defined.

It remains to show that the sequence

$$\mathcal{O}_K^* \xrightarrow{\ \iota\ } (\mathcal{O}_K/\mathfrak{m})^* \xrightarrow{\ \phi\ } I_K(\mathfrak{m}) \cap P_K/P_{K,1}(\mathfrak{m}) \longrightarrow 1$$

is exact, where $\iota(\alpha) = [\alpha]$. To prove that ϕ is surjective, we follow the proof of Theorem 7.24. Choose an element $\alpha\mathcal{O}_K \in I_K(\mathfrak{m}) \cap P_K$; then $\alpha\mathcal{O}_K = \mathfrak{a}\mathfrak{b}^{-1}$, where $\alpha \in K^*$ and $\mathfrak{a}, \mathfrak{b}$ are $\mathcal{O}_K$-ideals prime to $\mathfrak{m}$. Let $n = N(\mathfrak{b})$; then $\mathfrak{b}\overline{\mathfrak{b}} = n\mathcal{O}_K$ implies that $\overline{\mathfrak{b}} = n\mathfrak{b}^{-1}$. Further, n is prime to $\mathfrak{m}$, so that $rn \equiv 1 \bmod \mathfrak{m}$ for some $r \in \mathcal{O}_K$ prime to $\mathfrak{m}$. Thus

$$rn\alpha\mathcal{O}_K = rn\mathfrak{a}\mathfrak{b}^{-1} = r\mathfrak{a}\overline{\mathfrak{b}} \subset \mathcal{O}_K,$$

so that $rn\alpha \in \mathcal{O}_K$. Since r, $\mathfrak{a}$, and $\mathfrak{b}$ are all prime to $\mathfrak{m}$, so is $rn\alpha$. By construction $rn\mathcal{O}_K \in P_{K,1}(\mathfrak{m})$, so we get finally $[\alpha\mathcal{O}_K] = [rn\alpha\mathcal{O}_K] = \phi([rn\alpha])$.

Next, if $u \in \mathcal{O}_K^*$, then

$$\phi \circ \iota(u) = \phi([u]) = [u\mathcal{O}_K] = [\mathcal{O}_K],$$

where the last equality follows since u is a unit. This shows that $\mathrm{im}(\iota) \subset \ker(\phi)$. For the other inclusion, let $[\alpha] \in \ker(\phi)$. Then $\alpha\mathcal{O}_K \in P_{K,1}(\mathfrak{m})$, so that $\alpha\mathcal{O}_K = \beta\mathcal{O}_K \cdot \gamma^{-1}\mathcal{O}_K$ where $\beta, \gamma \equiv 1 \bmod \mathfrak{m}$, and therefore $\alpha = u\beta\gamma^{-1}$ for some unit $u \in \mathcal{O}_K^*$. Since $\alpha \in \mathcal{O}_K$ it follows that $\beta\gamma^{-1} \in \mathcal{O}_K$, and thus modulo $\mathfrak{m}$, $[\alpha] = [u\beta\gamma^{-1}] = [u][\beta\gamma^{-1}] = [u]$, so that $\phi(u) = [\alpha]$.

From the exact sequence, it follows that

$$I_K(\mathfrak{m}) \cap P_K/P_{K,1}(\mathfrak{m}) \simeq (\mathcal{O}_K/\mathfrak{m})^*/\,\mathrm{im}(\iota),$$

so that $I_K(\mathfrak{m}) \cap P_K/P_{K,1}(\mathfrak{m})$ is isomorphic to a quotient of a finite group, hence finite.

(b) Since $I_K(\mathfrak{m}) \subset I_K$, $P_{K,1}(\mathfrak{m}) \subset P_K$, and $P_{K,1}(\mathfrak{m}) \subset I_K(\mathfrak{m}) \cap P_K$, we have, analogously to (7.25), the sequence

$$0 \longrightarrow I_K(\mathfrak{m}) \cap P_K/P_{K,1}(\mathfrak{m}) \longrightarrow I_K(\mathfrak{m})/P_{K,1}(\mathfrak{m}) \longrightarrow I_K/P_K.$$

We claim that this is exact. The middle arrow is induced by the inclusion, so is injective. For exactness at $I_K(\mathfrak{m})/P_{K,1}(\mathfrak{m})$, note that $\mathfrak{a} \in I_K(\mathfrak{m})$ lies in P_K if and only if $\mathfrak{a} \in I_K(\mathfrak{m}) \cap P_K$. This completes the proof of exactness. Since $I_K(\mathfrak{m}) \cap P_K/P_{K,1}(\mathfrak{m})$ is finite by part (a), and $I_K/P_K \simeq C(\mathcal{O}_K)$ is finite, it follows that $I_K(\mathfrak{m})/P_{K,1}(\mathfrak{m})$ is finite as well.

8.2. (a) Let $p \nmid m$ be a prime and let $f(x) = x^m - 1$. Then $f'(x) = mx^{m-1}$ is not identically zero modulo p, and its only root is $x = 0$ (with multiplicity $m - 1$). Since 0 is not a root of $f(x)$, we see that $f(x)$ and $f'(x)$ have no common roots. Therefore $f(x)$ is separable modulo p, so it has distinct roots modulo p. Since $f(\zeta_m) = 0$, the minimal polynomial for ζ_m (which is the mth cyclotomic polynomial) must divide $f(x)$, so that it too has distinct roots modulo p and thus is separable modulo p. It follows from part (i) of Proposition 5.11 that p is unramified in $\mathbb{Q}(\zeta_m)$. Therefore all ramified primes of $\mathbb{Q}(\zeta_m)$ divide m, so that $\Phi_{m\infty}$ is defined.

(b) Let $p \nmid m$ be prime, and let $\mathfrak{p}$ be a prime of $\mathbb{Q}(\zeta_m)$ containing p, i.e., containing the prime ideal $p\mathbb{Z}$. Since p is unramified in $\mathbb{Q}(\zeta_m)$ by part (a), Lemma 5.19 gives

$$\left(\frac{\mathbb{Q}(\zeta_m)/\mathbb{Q}}{\mathfrak{p}}\right)(\alpha) \equiv \alpha^{N(p\mathbb{Z})} \equiv \alpha^p \bmod \mathfrak{p}$$

for all $\alpha \in \mathcal{O}_{\mathbb{Q}(\zeta_m)}$. Let $\sigma = ((\mathbb{Q}(\zeta_m)/\mathbb{Q})/\mathfrak{p}) \in \mathrm{Gal}(\mathbb{Q}(\zeta_m)/\mathbb{Q})$. Since $\sigma(\zeta_m) = \zeta_m^d$ where $\gcd(d, m) = 1$, we obtain

$$\zeta_m^d = \sigma(\zeta_m) \equiv \zeta_m^p \bmod \mathfrak{p}.$$

The mth roots of unity are distinct mod $\mathfrak{p}$ by part (a), so that $\zeta_m^d = \zeta_m^p$. It follows that σ is σ_p, the automorphism of $\mathbb{Q}(\zeta_m)$ determined by $\sigma_p(\zeta_m) = \zeta_m^p$. Hence

$$\Phi_{m\infty}(p\mathbb{Z}) = \left(\frac{\mathbb{Q}(\zeta_m)/\mathbb{Q}}{\mathfrak{p}}\right) = \sigma_p.$$

After identifying $\mathrm{Gal}(\mathbb{Q}(\zeta_m)/\mathbb{Q})$ with $(\mathbb{Z}/m\mathbb{Z})^*$, we get $\Phi_{m\infty}(p\mathbb{Z}) = [p] \in (\mathbb{Z}/m\mathbb{Z})^*$. Then take a fractional ideal $(a/b)\mathbb{Z}$ where $\gcd(a, m) = \gcd(b, m) = 1$ and write a and b as products of primes not dividing m. Since $\Phi_{m\infty}$ is a homomorphism, it follows that

$$\Phi_{m\infty}\left(\frac{a}{b}\mathbb{Z}\right) = \Phi_{m\infty}(a\mathbb{Z})\Phi_{m\infty}(b\mathbb{Z})^{-1} = [a][b]^{-1} \in (\mathbb{Z}/m\mathbb{Z})^*.$$

(c) To prove $\ker(\Phi_{m\infty}) \subset P_{\mathbb{Q},1}(m\infty)$, take $(a/b)\mathbb{Z} \in \ker(\Phi_{m\infty})$. We may assume (by using $-(a/b)\mathbb{Z}$ if necessary) that $a, b > 0$. Then by part (b), $[a][b]^{-1} = [1]$ in $(\mathbb{Z}/m\mathbb{Z})^*$, so that $a \equiv b \bmod m$. Choosing a positive integer c such that $bc \equiv 1 \bmod m$, we obtain

$$\left(\frac{a}{b}\mathbb{Z}\right) = \left(\frac{ac}{bc}\mathbb{Z}\right),$$

and ac, $bc \equiv 1 \bmod m$, $ac > 0$, $bc > 0$ and therefore $(a/b)\mathbb{Z} \in P_{\mathbb{Q},1}(m\infty)$. The other inclusion is clear since if $(a/b)\mathbb{Z} \in P_{\mathbb{Q},1}(m\infty)$ with $a, b \equiv 1 \bmod m$, then $[a][b]^{-1} = [1]$ in $(\mathbb{Z}/m\mathbb{Z})^*$. Hence $(a/b)\mathbb{Z} \in \ker(\Phi_{m\infty})$ by part (b).

8.3. (a) Let $K = \mathbb{Q}(\zeta_m + \zeta_m^{-1})$. Note that $\zeta_m^{-1} = \overline{\zeta_m}$, so that K is a real field. Since ζ_m is a root of the quadratic $x^2 - (\zeta_m + \zeta_m^{-1})x + 1 \in K[x]$, it follows that $[\mathbb{Q}(\zeta_m) : K] = 2$ since $\mathbb{Q}(\zeta_m)$ is not a real field for $m > 2$. Thus K is the maximal real field contained in $\mathbb{Q}(\zeta_m)$; it follows that $K = \mathbb{Q}(\zeta_m + \zeta_m^{-1}) = \mathbb{R} \cap \mathbb{Q}(\zeta_m)$.

Next, $\zeta_m = e^{2\pi i/m} = \cos(2\pi/m) + i\sin(2\pi/m)$, so that $\zeta_m + \zeta_m^{-1} = 2\cos(2\pi/m)$. Thus

$$\mathbb{R} \cap \mathbb{Q}(\zeta_m) = \mathbb{Q}(\zeta_m + \zeta_m^{-1}) = \mathbb{Q}(2\cos(2\pi/m)) = \mathbb{Q}(\cos(2\pi/m)).$$

Finally, the fact that $[\mathbb{Q}(\zeta_m) : \mathbb{Q}] = \phi(m)$ together with

$$[\mathbb{Q}(\zeta_m) : \mathbb{Q}] = [\mathbb{Q}(\zeta_m) : \mathbb{Q}(\cos(2\pi/m))][\mathbb{Q}(\cos(2\pi/m)) : \mathbb{Q}] = 2[\mathbb{Q}(\cos(2\pi/m)) : \mathbb{Q}]$$

shows that $[\mathbb{Q}(\cos(2\pi/m)) : \mathbb{Q}] = (1/2)\phi(m)$.

(b) As in part (a), let $K = \mathbb{Q}(\cos(2\pi/m)) = \mathbb{Q}(\zeta_m + \zeta_m^{-1})$. From Exercise 5.16,

$$\left(\frac{K/\mathbb{Q}}{p}\right) = r \circ \left(\frac{\mathbb{Q}(\zeta_m)/\mathbb{Q}}{p}\right),$$

where $r : \mathrm{Gal}(\mathbb{Q}(\zeta_m)/\mathbb{Q}) \to \mathrm{Gal}(K/\mathbb{Q})$ is the restriction map. The kernel of r is $\mathrm{Gal}(\mathbb{Q}(\zeta_m)/K)$, which consists solely of the identity and complex conjugation since $[\mathbb{Q}(\zeta_m) : K] = 2$. Since $\overline{\zeta_m} = \zeta_m^{-1}$, conjugation maps to $[-1] \in (\mathbb{Z}/m\mathbb{Z})^*$ under the isomorphism $\mathrm{Gal}(\mathbb{Q}(\zeta_m)/\mathbb{Q}) \simeq (\mathbb{Z}/m\mathbb{Z})^*$. Thus the restriction map r induces an isomorphism

$$\mathrm{Gal}(K/\mathbb{Q}) \simeq \mathrm{Gal}(\mathbb{Q}(\zeta_m)/\mathbb{Q})/\,\mathrm{Gal}(\mathbb{Q}(\zeta_m)/K) \simeq (\mathbb{Z}/m\mathbb{Z})^*/\{[\pm 1]\}.$$

Also observe that $I_{\mathbb{Q}}(m) = I_{\mathbb{Q}}(m\infty)$. Now consider the following diagram, which commutes by Exercise 5.16:

$$\begin{array}{ccccc} I_{\mathbb{Q}}(m) = I_{\mathbb{Q}}(m\infty) & \xrightarrow{\Phi_{m\infty}} & \mathrm{Gal}(\mathbb{Q}(\zeta_m)/\mathbb{Q}) & \xrightarrow{\sim} & (\mathbb{Z}/m\mathbb{Z})^* \\ & \searrow^{\Phi_m} & \downarrow r & & \downarrow \\ & & \mathrm{Gal}(K/\mathbb{Q}) & \xrightarrow{\sim} & (\mathbb{Z}/m\mathbb{Z})^*/\{[\pm 1]\} \end{array}$$

This diagram together with the previous exercise shows that Φ_m can be identified with the map that sends $(a/b)\mathbb{Z}$, for a, b relatively prime to m, to $[a][b]^{-1} \bmod [\pm 1]$ in $(\mathbb{Z}/m\mathbb{Z})^*/\{[\pm 1]\}$.

(c) We first show that $\ker(\Phi_m) \subset P_{\mathbb{Q},1}(m)$. Suppose $\mathfrak{a} = (a/b)\mathbb{Z}$. By part (b), $\mathfrak{a} \in \ker(\Phi_m)$ if and only if $[a][b]^{-1} = [\pm 1]$ if and only if $\pm a \equiv b \bmod m$. If $-a \equiv b \bmod m$, we note that $\mathfrak{a} = (a/b)\mathbb{Z} = (-a/b)\mathbb{Z}$, so replacing a by $-a$ we may assume that $a \equiv b \bmod m$. Choose c with $ac \equiv bc \equiv 1 \bmod m$; then $\mathfrak{a} = (a/b)\mathbb{Z} = (ac/bc)\mathbb{Z} \in P_{\mathbb{Q},1}(m)$.

For the other inclusion, choose $(a/b)\mathbb{Z} \in P_{\mathbb{Q},1}(m)$ with $a, b \equiv 1 \bmod m$. Then $[a] = [b] = [1]$ in $(\mathbb{Z}/m\mathbb{Z})^*$, so that $\Phi_m((a/b)\mathbb{Z}) = [a][b]^{-1} \bmod [\pm 1] = [1] \bmod [\pm 1]$, proving that $(a/b)\mathbb{Z} \in \ker(\Phi_m)$.

8.4. Let L/K be an Abelian extension and $\mathfrak{m}$ be a modulus divisible by all primes of K that ramify in L, so that $\Phi_{\mathfrak{m}}$ is defined. Assume $\mathfrak{m} \mid \mathfrak{n}$. Since any ideal prime to $\mathfrak{n}$ is also prime to $\mathfrak{m}$, we see that $I_K(\mathfrak{n}) \subset I_K(\mathfrak{m})$, so we may define

$$(*) \qquad \Phi_{\mathfrak{n}} : I_K(\mathfrak{n}) \longrightarrow \mathrm{Gal}(L/K)$$

by restriction. It follows that $\ker(\Phi_{\mathfrak{n}}) = \ker(\Phi_{\mathfrak{m}}) \cap I_K(\mathfrak{n})$. However, if $\alpha \in \mathcal{O}_K$ satisfies $\alpha \equiv 1 \bmod \mathfrak{n}_0$ and $\sigma(\alpha) > 0$ for every infinite real prime of $\mathfrak{n}$, then $\alpha \equiv 1 \bmod \mathfrak{m}_0$, and since the infinite real primes of $\mathfrak{m}$ are a subset of those of $\mathfrak{n}$, also $\sigma(\alpha) > 0$ for every infinite real prime of $\mathfrak{m}$. Thus $\alpha\mathcal{O}_K \in P_{K,1}(\mathfrak{m})$. Since $P_{K,1}(\mathfrak{n})$ is generated by the $\mathcal{O}_K$-ideals $\alpha\mathcal{O}_K$, we obtain $P_{K,1}(\mathfrak{n}) \subset P_{K,1}(\mathfrak{m})$.

Now assume $P_{K,1}(\mathfrak{m}) \subset \ker(\Phi_{\mathfrak{m}})$. Then the previous paragraph implies

$$P_{K,1}(\mathfrak{n}) \subset P_{K,1}(\mathfrak{m}) \cap I_K(\mathfrak{n}) \subset \ker(\Phi_{\mathfrak{m}}) \cap I_K(\mathfrak{n}) = \ker(\Phi_{\mathfrak{n}}),$$

as desired, where the first inclusion follows from $P_{K,1}(\mathfrak{n}) \subset P_{K,1}(\mathfrak{m})$ and $P_{K,1}(\mathfrak{n}) \subset I_K(\mathfrak{n})$ and the second follows since $P_{K,1}(\mathfrak{m}) \subset \ker(\Phi_{\mathfrak{m}})$.

For the final assertion of the exercise, suppose that $\mathrm{Gal}(L/K)$ is a generalized ideal class group for $\mathfrak{m}$, i.e., $\mathrm{Gal}(L/K) \simeq I_K(\mathfrak{m})/\ker(\Phi_{\mathfrak{m}})$ with $P_{K,1}(\mathfrak{m}) \subset \ker(\Phi_{\mathfrak{m}})$. Since $\mathfrak{m} \mid \mathfrak{n}$ and all ramified primes of K divide $\mathfrak{m}$, they also divide $\mathfrak{n}$. Then this implies that the Artin map $\Phi_{\mathfrak{n}} : I_K(\mathfrak{n}) \to \mathrm{Gal}(L/K)$ is onto by Theorem 8.2. Hence $\mathrm{Gal}(L/K) \simeq I_K(\mathfrak{n})/\ker(\Phi_{\mathfrak{n}})$. Since $P_{K,1}(\mathfrak{m}) \subset \ker(\Phi_{\mathfrak{m}})$, the above proof shows that $P_{K,1}(\mathfrak{n}) \subset \ker(\Phi_{\mathfrak{n}})$, so that $\mathrm{Gal}(L/K)$ is a generalized ideal class group for $\mathfrak{n}$.

8.5. If $m \le 2$, then $\mathbb{Q}(\zeta_m) = \mathbb{Q}$, so that $\mathfrak{f}(\mathbb{Q}/\mathbb{Q}) = 1$ since no primes of $\mathbb{Q}$ ramify in $\mathbb{Q}$. Otherwise, $m > 2$. By part (c) of Exercise 8.2, $\ker(\Phi_{m\infty}) = P_{\mathbb{Q},1}(m\infty)$ is a congruence subgroup for $m\infty$, so that by part (ii) of Theorem 8.5, $\mathfrak{f} = \mathfrak{f}(\mathbb{Q}(\zeta_m)/\mathbb{Q}) \mid m\infty$ and thus for some $n \mid m$, $\mathfrak{f} = n$ or $n\infty$. Since ∞ ramifies in the non-real field $\mathbb{Q}(\zeta_m)$, we must have $\mathfrak{f} = n\infty$, $n \mid m$ by part (i) of Theorem 8.5.

By part (c) of Exercise 8.2, $P_{K,1}(n\infty) = \ker(\Phi_{\mathbb{Q}(\zeta_n)/\mathbb{Q},n\infty})$. Further, since $n\infty$ is the conductor for $\mathbb{Q}(\zeta_m)/\mathbb{Q}$, part (ii) of Theorem 8.5 shows that $\ker(\Phi_{\mathbb{Q}(\zeta_m)/\mathbb{Q},n\infty})$ is a congruence subgroup for $n\infty$. It follows that

$$P_{K,1}(n\infty) = \ker(\Phi_{\mathbb{Q}(\zeta_n)/\mathbb{Q},n\infty}) \subset \ker(\Phi_{\mathbb{Q}(\zeta_m)/\mathbb{Q},n\infty}),$$

so that by Corollary 8.7, $\mathbb{Q}(\zeta_m) \subset \mathbb{Q}(\zeta_n)$. It follows that $\phi(m) = [\mathbb{Q}(\zeta_m) : \mathbb{Q}]$ divides $\phi(n) = [\mathbb{Q}(\zeta_n) : \mathbb{Q}]$. But $n \mid m$ implies that $\phi(n) \le \phi(m)$, so that $\phi(m) = \phi(n)$. Using the standard formula for the ϕ-function, we see that $n = m$, except when $m = 2n$, n odd, in which case $n = \frac{m}{2}$. Thus

$$\mathfrak{f}(\mathbb{Q}(\zeta_m)/\mathbb{Q}) = n\infty = \begin{cases} 1 & m \le 2 \\ (m/2)\infty & m = 2n, n > 1 \text{ odd} \\ m\infty & \text{otherwise.} \end{cases}$$

8.6. (a) Let L/K be an Abelian extension, and write $\mathfrak{f} = \mathfrak{f}(L/K)$. If $\mathfrak{m}$ is any modulus for which $L \subset K_{\mathfrak{m}}$, then Corollary 8.7 shows that $\ker(\Phi_{L/K,\mathfrak{m}})$ is a congruence subgroup for $\mathfrak{m}$. Then part (ii) of Theorem 8.5 shows that $\mathfrak{f} \mid \mathfrak{m}$. Thus $\mathfrak{f}$ is a common divisor of all moduli $\mathfrak{m}$ with $L \subset K_{\mathfrak{m}}$. But $\ker(\Phi_{L/K,\mathfrak{f}})$ is a congruence subgroup for $\mathfrak{f}$ by Theorem 8.5, so using Corollary 8.7 again with the data

$$P_{K,1}(\mathfrak{f}) = \ker(\Phi_{K_{\mathfrak{f}}/K,\mathfrak{f}}) \subset \ker(\Phi_{L/K,\mathfrak{f}})$$

shows that $L \subset K_{\mathfrak{f}}$ as well, showing that $\mathfrak{f}$ is indeed the greatest common divisor.

(b) By (8.4) or Exercise 8.2, the Artin map $\Phi_{m\infty} : I_{\mathbb{Q}}(m\infty) \to \mathrm{Gal}(\mathbb{Q}(\zeta_m)/\mathbb{Q})$ has the property that $\ker(\Phi_{m\infty}) = P_{\mathbb{Q},1}(m\infty)$. It follows that $\mathbb{Q}(\zeta_m) = \mathbb{Q}_{m\infty}$, the ray class field of $\mathbb{Q}$ for the modulus $m\infty$. Then $L \subset \mathbb{Q}(\zeta_m)$ implies, using part (a), that

$\mathfrak{f}(L/\mathbb{Q}) \mid m\infty$. Since m is the smallest positive integer such that $L \subset \mathbb{Q}(\zeta_m)$, we see that $\mathfrak{f}(L/\mathbb{Q}) = m$ or $m\infty$. If L is not a real field, then ∞ ramifies in L, so that $\mathfrak{f}(L/\mathbb{Q}) = m\infty$. If L is a real field, then $L \subset \mathbb{R} \cap \mathbb{Q}(\zeta_m)$ which, by Exercise 8.3, is the ray class field $\mathbb{Q}_m$. Then $\mathfrak{f}(L/\mathbb{Q}) = m$.

8.7. (a) Suppose $K \subset L$ is Abelian Galois and that $\mathrm{Gal}(L/K)$ is a generalized ideal class group for the modulus $\mathfrak{m}$ of K. This means that $\ker(\Phi_{L/K,\mathfrak{m}})$ is a congruence subgroup for $\mathfrak{m}$. Let $K \subset M \subset L$ be an intermediate field. Note that M/K is Galois since $\mathrm{Gal}(L/K)$ is Abelian. Since $\Phi_{M/K,\mathfrak{m}} = r \circ \Phi_{L/K,\mathfrak{m}}$ where r is the restriction map, we have $P_{K,1}(\mathfrak{m}) \subset \ker(\Phi_{L/K,\mathfrak{m}}) \subset \ker(\Phi_{M/K,\mathfrak{m}})$. Thus $\mathrm{Gal}(M/K)$ is a generalized ideal class group for $\mathfrak{m}$.

(b) If K ramifies only at p, then by Corollary 5.17 the only prime dividing d_K must be p, so that $K = \mathbb{Q}(\sqrt{\pm p})$. Suppose first $p \equiv 1 \bmod 4$. Then $K = \mathbb{Q}(\sqrt{p})$, since $d_{\mathbb{Q}(\sqrt{-p})} = -4p$, which is divisible by 2. Thus $p = p^*$ and $K = \mathbb{Q}(\sqrt{p}) = \mathbb{Q}(\sqrt{p^*})$. If instead $p \equiv 3 \bmod 4$, we must have $K = \mathbb{Q}(\sqrt{-p})$ for the same reason. Then $-p = p^*$ and $K = \mathbb{Q}(\sqrt{-p}) = \mathbb{Q}(\sqrt{p^*})$.

(c) Consider the homomorphism $(\mathbb{Z}/2p\mathbb{Z})^* \to (\mathbb{Z}/p\mathbb{Z})^*$ defined by $[a]_{2p} \mapsto [a]_p$. To see that this map is surjective, choose $[a] \in (\mathbb{Z}/p\mathbb{Z})^*$. Then either a or $a+p$ is odd since p is odd, so that one of these represents a congruence class in $(\mathbb{Z}/2p\mathbb{Z})^*$ that maps to $[a]$. Next, note that $|(\mathbb{Z}/2p\mathbb{Z})^*| = |(\mathbb{Z}/p\mathbb{Z})^*|$ since $\phi(2p) = \phi(p)$ when p is odd. Thus the homomorphism is an isomorphism, and its inverse provides the required isomorphism $(\mathbb{Z}/p\mathbb{Z})^* \to (\mathbb{Z}/2p\mathbb{Z})^*$. For the second map, define

$$\varphi : (\mathbb{Z}/2p\mathbb{Z})^* \longrightarrow I_{\mathbb{Q}}(2p\infty)/P_{\mathbb{Q},1}(2p\infty), \quad \varphi([a]) = [a\mathbb{Z}] \text{ when } a > 0.$$

We first show φ is well-defined. Choose $a \equiv b \bmod 2p$ with $a, b > 0$ prime to $2p$. We want to show that $[a\mathbb{Z}] = [b\mathbb{Z}]$ in $I_{\mathbb{Q}}(2p\infty)/P_{\mathbb{Q},1}(2p\infty)$, so we must show that $[(a/b)\mathbb{Z}] = [\mathbb{Z}]$. Choose some $c > 0$ such that $ac \equiv bc \equiv 1 \bmod 2p$; then $(ac/bc)\mathbb{Z} \in P_{K,1}(2p\infty)$ implies

$$\left[\frac{a}{b}\mathbb{Z}\right] = \left[\frac{ac}{bc}\mathbb{Z}\right] = [\mathbb{Z}].$$

To show φ is surjective, choose $(a/b)\mathbb{Z} \in I_{\mathbb{Q}}(2p\infty)$; we may assume that $a, b > 0$ without changing the ideal. Choose $c > 0$ with $bc \equiv 1 \bmod 2p$; then since $a, b, c > 0$ and $bc\mathbb{Z} \in P_{K,1}(2p\infty)$, we have

$$\varphi([ac]) = [ac\mathbb{Z}] = \left[\frac{ac}{bc}\mathbb{Z}\right] = \left[\frac{a}{b}\mathbb{Z}\right].$$

Finally, for injectivity, if $\varphi([a]) = [\mathbb{Z}]$ with $a > 0$, then $a\mathbb{Z} \in P_{K,1}(2p\infty)$. This means that $a\mathbb{Z} = (b\mathbb{Z})(c\mathbb{Z})^{-1}$ for some b, $c > 0$ with $b \equiv c \equiv 1 \bmod 2p$. It follows that $ac\mathbb{Z} = b\mathbb{Z}$, so that $ac = \pm b$. But a, b, and c are all positive, forcing $ac = b$. Since $b \equiv c \equiv 1 \bmod 2p$, also $a \equiv 1 \bmod 2p$ and thus $[a] = [1] \in (\mathbb{Z}/2p\mathbb{Z})^*$.

8.8. (a) By Exercise 8.2, the kernel of the Artin map $I_{\mathbb{Q}}(8\infty) \to \mathrm{Gal}(\mathbb{Q}(\zeta_8)/\mathbb{Q}) \simeq (\mathbb{Z}/8\mathbb{Z})^*$ is $P_{\mathbb{Q},1}(8\infty)$. Thus we have isomorphisms

$$I_{\mathbb{Q}}(8\infty)/P_{\mathbb{Q},1}(8\infty) \simeq \mathrm{Gal}(\mathbb{Q}(\zeta_8)/\mathbb{Q}) \simeq (\mathbb{Z}/8\mathbb{Z})^*,$$

where $[(a/b)\mathbb{Z}] \in I_{\mathbb{Q}}(8\infty)/P_{\mathbb{Q},1}(8\infty)$, for odd $a, b > 0$, maps to $[a][b]^{-1} \in (\mathbb{Z}/8\mathbb{Z})^*$. It follows that $I_{\mathbb{Q}}(8\infty)/P_{\mathbb{Q},1}(8\infty) = \{[\mathbb{Z}], [3\mathbb{Z}], [5\mathbb{Z}], [7\mathbb{Z}]\}$.

(b) The group $H = \{\mathbb{Z}, 7\mathbb{Z}\}P_{\mathbb{Q},1}(8\infty)$ contains $P_{\mathbb{Q},1}(8\infty)$ as a subgroup of index 2. Thus H is a congruence subgroup for 8∞, which by the existence theorem corresponds to an extension F of $\mathbb{Q}$ with

$$\mathrm{Gal}(F/\mathbb{Q}) \simeq I_{\mathbb{Q}}(8\infty)/H.$$

The quotient group on the right has order 2, so that F is a quadratic extension of $\mathbb{Q}$. Furthermore, $\mathbb{Q}(\zeta_8)$ corresponds to the congruence subgroup $P_{\mathbb{Q},1}(8\infty)$, and the inclusion $P_{\mathbb{Q},1}(8\infty) \subseteq H$ implies that $F \subset \mathbb{Q}(\zeta_8)$ by Corollary 8.7. Then we have a commutative diagram

$$(*)\qquad \begin{array}{ccccc} \mathrm{Gal}(\mathbb{Q}(\zeta_8)/F) & \simeq & H/P_{\mathbb{Q},1}(8\infty) = \{[\mathbb{Z}], [7\mathbb{Z}]\} & \simeq & \{[1],[7]\} = \{[1],[-1]\} \\ \cap & & \cap & & \cap \\ \mathrm{Gal}(\mathbb{Q}(\zeta_8)/\mathbb{Q}) & \simeq & I_{\mathbb{Q}}(8\infty)/P_{\mathbb{Q},1}(8\infty) & \simeq & (\mathbb{Z}/8\mathbb{Z})^*. \end{array}$$

Notice that $[-1] \in (\mathbb{Z}/8\mathbb{Z})^*$ corresponds to complex conjugation $\tau \in \mathrm{Gal}(\mathbb{Q}(\zeta_8)/\mathbb{Q})$ since $\tau(\zeta_8) = \overline{\zeta_8} = \zeta_8^{-1}$. By (*), F is fixed by τ, so that $F \subset \mathbb{R}$. The only finite prime that can ramify in F is 2, so that by Corollary 5.17, $d_F = -4$ or ± 8, corresponding respectively to $\mathbb{Q}(i)$, $\mathbb{Q}(\sqrt{2})$, and $\mathbb{Q}(\sqrt{-2})$. Therefore F must be $\mathbb{Q}(\sqrt{2})$.

(c) Since 8∞ is divisible by all primes containing $n\alpha = 4$ and $\ker(\Phi_{\mathbb{Q}(\zeta_8)/\mathbb{Q},8\infty}) = P_{\mathbb{Q},1}(8\infty)$ is a congruence subgroup for 8∞, by Weak Reciprocity with $n = 2$ we get a commutative diagram

$$\begin{array}{ccccccc} (\mathbb{Z}/8\mathbb{Z})^* & \xrightarrow{\simeq} & I_{\mathbb{Q}}(8\infty)/P_{\mathbb{Q},1}(8\infty) & \xrightarrow{\simeq} & \mathrm{Gal}(\mathbb{Q}(\zeta_8)/\mathbb{Q}) & \twoheadrightarrow & \mathrm{Gal}(\mathbb{Q}(\sqrt{2})/\mathbb{Q}) \\ & & & \searrow^{(2/\cdot)} & & & \downarrow \simeq \\ & & & & & & \mu_2 = \{\pm 1\} \end{array}$$

where $(2/\cdot)$ is a homomorphism. The kernel of the composite map $(\mathbb{Z}/8\mathbb{Z})^* \to \mathrm{Gal}(\mathbb{Q}(\sqrt{2})/\mathbb{Q})$ maps isomorphically to the kernel of $\mathrm{Gal}(\mathbb{Q}(\zeta_8)/\mathbb{Q}) \to \mathrm{Gal}(\mathbb{Q}(\sqrt{2})/\mathbb{Q})$, which is $\mathrm{Gal}(\mathbb{Q}(\zeta_8)/\mathbb{Q}(\sqrt{2}))$. By (*), it follows that the kernel of the composite map is $\{[\pm 1]\} \subset (\mathbb{Z}/8\mathbb{Z})^*$, so the kernel of the map $(\mathbb{Z}/8\mathbb{Z})^* \to \{\pm 1\}$ is also $\{[\pm 1]\}$.

(d) Denote by χ the surjective homomorphism $(\mathbb{Z}/8\mathbb{Z})^* \to I_{\mathbb{Q}}(8\infty)/P_{\mathbb{Q},1}(8\infty) \to \{\pm 1\}$ from part (c). Since $\ker(\chi) = \{[\pm 1]\} \subset (\mathbb{Z}/8\mathbb{Z})^*$, we have

$$\chi(p) = \begin{cases} 1 & p \equiv 1, 7 \bmod 8 \\ -1 & p \equiv 3, 5 \bmod 8, \end{cases}$$

so that $(2/p) = \chi(p) = (-1)^{(p^2-1)/8}$.

8.9. (a) We show inductively that for $n \geq 4$, we can find u_{n+1} such that

$$\alpha \equiv u_{n+1}^3 \bmod \lambda^{n+1}\mathcal{O}_\lambda, \ u_{n+1} \in \mathcal{O}_\lambda^*, \text{ and } u_{n+1} \equiv u_n \bmod \lambda^{n-2}\mathcal{O}_\lambda.$$

Let $u_4 = 1$ and suppose $\alpha \equiv u_n^3 \bmod \lambda^n \mathcal{O}_\lambda$ with $u_n \in \mathcal{O}_\lambda^*$. Write

$$\alpha = u_n^3 + r\lambda^n, \ r \in \mathcal{O}_\lambda$$

and set $u_{n+1} = u_n + a_n\lambda^{n-2}$, where $a_n = -ru_n^{-2}\omega^{-2}$. Note that $a_n \in \mathcal{O}_\lambda$ since we assume $u_n \in \mathcal{O}_\lambda^*$. Then since $3 = -\omega^2(1-\omega)^2 = -\omega^2\lambda^2$,

$$\begin{aligned} u_{n+1}^3 = (u_n + a_n\lambda^{n-2})^3 &= u_n^3 + 3a_nu_n^2\lambda^{n-2} + 3a_n^2u_n\lambda^{2n-2} + a_n^3\lambda^{3n-6} \\ &= u_n^3 - a_nu_n^2\omega^2\lambda^n - a_n^2u_n\omega^2\lambda^{2n} + a_n^3\lambda^{3n-6} \\ &= u_n^3 + r\lambda^n - a_n^2u_n\omega^2\lambda^{2n} + a_n^3\lambda^{3n-6} \\ &= \alpha - a_n^2u_n\omega^2\lambda^{2n} + a_n^3\lambda^{3n-6}. \end{aligned}$$

With $n \geq 4$, we have both $2n \geq n+1$ and $3n-6 \geq n+1$, so that

$$u_{n+1}^3 \equiv \alpha \bmod \lambda^{n+1}\mathcal{O}_\lambda,$$

as desired. Since u_n is a unit and $a_n \in \mathcal{O}_\lambda$, it follows that $u_{n+1} \equiv u_n \bmod \lambda^{n-2}\mathcal{O}_K$ and that u_{n+1} is a unit. This completes the induction. Then $u = \lim_{n\to\infty} u_n$ exists in $\mathcal{O}_\lambda$ and satisfies $\alpha = u^3$. Furthermore, u is a unit since $u \equiv u_4 = 1 \bmod \lambda^2\mathcal{O}_\lambda$.

(b) $\alpha \in \mathcal{O}_\lambda^*$ and $\alpha \equiv \alpha' \bmod \lambda^4\mathcal{O}_\lambda$ together imply that $\alpha'\alpha^{-1} \equiv 1 \bmod \lambda^4\mathcal{O}_\lambda$, so that by (a), $\alpha'\alpha^{-1} = u^3$ for some $u \in \mathcal{O}_\lambda$. Thus $\alpha' = \alpha(\alpha'\alpha^{-1}) = \alpha u^3$, so by property (i) and the fact that this Hilbert symbol is a cube root of unity,

$$\left(\frac{\alpha',\beta}{\lambda}\right)_3 = \left(\frac{\alpha u^3,\beta}{\lambda}\right)_3 = \left(\frac{\alpha,\beta}{\lambda}\right)_3\left(\frac{u^3,\beta}{\lambda}\right)_3 = \left(\frac{\alpha,\beta}{\lambda}\right)_3\left(\frac{u,\beta}{\lambda}\right)_3^3 = \left(\frac{\alpha,\beta}{\lambda}\right)_3.$$

(c) Since $\alpha \equiv 1 \bmod \lambda^2\mathcal{O}_\lambda$, we may write $\alpha = 1+a\lambda^2$ for $a \in \mathcal{O}_\lambda$. Since $\beta \equiv 1 \bmod \lambda^2\mathcal{O}_\lambda$, we have $a\beta - a = a(\beta-1) \equiv 0 \bmod \lambda^2\mathcal{O}_\lambda$ so that $a\beta\lambda^2 \equiv a\lambda^2 \bmod \lambda^4\mathcal{O}_\lambda$. This implies $1 + a\beta\lambda^2 \equiv 1 + a\lambda^2 \bmod \lambda^4\mathcal{O}_\lambda$. Then by property (v) and part (b) we have

$$1 = \left(\frac{1+a\beta\lambda^2, 1-(1+a\beta\lambda^2)}{\lambda}\right)_3 = \left(\frac{1+a\lambda^2, -a\beta\lambda^2}{\lambda}\right)_3 = \left(\frac{\alpha, -a\beta\lambda^2}{\lambda}\right)_3.$$

Now use $-a\lambda^2 = 1-\alpha$ and properties (ii) and (v):

$$= \left(\frac{\alpha, \beta(1-\alpha)}{\lambda}\right)_3 = \left(\frac{\alpha,\beta}{\lambda}\right)_3\left(\frac{\alpha,1-\alpha}{\lambda}\right)_3 = \left(\frac{\alpha,\beta}{\lambda}\right)_3.$$

8.10. (a) By part (c) of Exercise 5.1, a nonzero ideal of $\mathcal{O}_K$ is a free $\mathbb{Z}$-module of rank $[K:\mathbb{Q}]$, and therefore is generated by $[K:\mathbb{Q}]$ elements of $\mathcal{O}_K$. Since $\mathcal{O}_K$ is countable, it follows that $\mathcal{O}_K$ has only countably many ideals, so we can order them as $\mathcal{A} = \{\mathfrak{a}_1, \mathfrak{a}_2, \dots\}$. Let $\chi_{\mathcal{S}} : \mathcal{A} \to \{0,1\}$ be defined by $\chi_{\mathcal{S}}(\mathfrak{a}_i) = 1$ if $\mathfrak{a}_i \in \mathcal{S}$ and $\chi_{\mathcal{S}}(\mathfrak{a}_i) = 0$ otherwise. Then we are given that

$$\sum_{i=1}^{\infty} \left|N(\mathfrak{a}_i)^{-s}\right|$$

converges, so that

$$\sum_{\mathfrak{p}\in\mathcal{S}} \left|N(\mathfrak{p})^{-s}\right| = \sum_{i=1}^{\infty} \chi_{\mathcal{S}}(\mathfrak{a}_i)\left|N(\mathfrak{a}_i)^{-s}\right|$$

converges as well by the comparison test.

(b) We first remark that in the formula for $\delta(\mathcal{S})$, all terms in the numerator are positive, and the denominator is positive for s near 1.

For part (ii), note that $\mathcal{S} \subset \mathcal{T}$ implies

$$\frac{\sum_{\mathfrak{p}\in\mathcal{S}} N(\mathfrak{p})^{-s}}{-\log(s-1)} \le \frac{\sum_{\mathfrak{p}\in\mathcal{T}} N(\mathfrak{p})^{-s}}{-\log(s-1)}$$

for s near 1 by the remark above. Then $\delta(\mathcal{S}) \le \delta(\mathcal{T})$ since limits preserve weak inequalities.

For (iii), the remark above implies that $\delta(\mathcal{S}) \ge 0$. Further, $\delta(\mathcal{S}) \le 1$ by (i) and (ii).

Part (iv) follows from

$$\begin{aligned}\delta(\mathcal{S}\cup\mathcal{T}) &= \lim_{s\to1+}\frac{\sum_{\mathfrak{p}\in\mathcal{S}\cup\mathcal{T}} N(\mathfrak{p})^{-s}}{-\log(s-1)} = \lim_{s\to1+}\left(\frac{\sum_{\mathfrak{p}\in\mathcal{S}} N(\mathfrak{p})^{-s}}{-\log(s-1)} + \frac{\sum_{\mathfrak{p}\in\mathcal{T}} N(\mathfrak{p})^{-s}}{-\log(s-1)}\right)\\ &= \lim_{s\to1+}\frac{\sum_{\mathfrak{p}\in\mathcal{S}} N(\mathfrak{p})^{-s}}{-\log(s-1)} + \lim_{s\to1+}\frac{\sum_{\mathfrak{p}\in\mathcal{T}} N(\mathfrak{p})^{-s}}{-\log(s-1)} = \delta(\mathcal{S}) + \delta(\mathcal{T}),\end{aligned}$$

where the second equality on the first line uses $\mathcal{S}\cap\mathcal{T} = \emptyset$.

For part (v), note that if $\mathcal{S}$ is finite, then the numerator in the definition of $\delta(\mathcal{S})$ is bounded while the denominator grows without bound. Hence the limit is zero.

Finally, for part (vi), the hypothesis on $\mathcal{S}$ and $\mathcal{T}$ implies that $\mathcal{S}-\mathcal{T}$ and $\mathcal{T}-\mathcal{S}$ are both finite. Applying parts (iv) and (v) to the disjoint unions

$$\mathcal{S} = (\mathcal{S}\cap\mathcal{T}) \cup (\mathcal{S}-\mathcal{T}), \quad \mathcal{T} = (\mathcal{T}\cap\mathcal{S}) \cup (\mathcal{T}-\mathcal{S})$$

implies that $\delta(\mathcal{S}) = \delta(\mathcal{S} \cap \mathcal{T}) = \delta(\mathcal{T} \cap \mathcal{S}) = \delta(\mathcal{T})$.

8.11. Write $K = \mathbb{Q}(\zeta_m)$. Then $\mathrm{Gal}(K/\mathbb{Q}) \simeq (\mathbb{Z}/m\mathbb{Z})^*$, and the elements are σ_k where $\sigma_k(\zeta_m) = \zeta_m^k$ for $\gcd(k, m) = 1$, $0 \le k < m$. Since $\mathrm{Gal}(K/\mathbb{Q})$ is Abelian, the conjugacy class $\langle \sigma_k \rangle$ consists of just σ_k. Then Theorem 8.17 implies that the set

$$\mathcal{S}_k = \{p \in \mathcal{P}_{\mathbb{Q}} : p \text{ is unramified in } K \text{ and } ((K/\mathbb{Q})/p) = \sigma_k\}$$

has Dirichlet density $\delta(\mathcal{S}_k) = 1/\phi(m)$, However, (8.3) implies that $((K/\mathbb{Q})/p) = \sigma_p$, and $\sigma_p = \sigma_k$ if and only if $p \equiv k \bmod m$. Such primes are relatively prime to m and hence unramified in K. It follows that the above set $\mathcal{S}_k$ simplifies to

$$\mathcal{S}_k = \{p \in \mathcal{P}_{\mathbb{Q}} : p \equiv k \bmod m\}.$$

Since $\delta(\mathcal{S}_k) = 1/\phi(m)$ is positive, $\mathcal{S}_k$ must be infinite.

8.12. (a) It is always the case that $\mathcal{S}_{M/K} \subset \tilde{\mathcal{S}}_{M/K}$, since if $\mathfrak{p}$ splits completely then it is unramified and $f_{\mathfrak{P}|\mathfrak{p}} = 1$ for each $\mathfrak{P}$ over $\mathfrak{p}$. For the reverse, if $\mathfrak{p} \in \tilde{\mathcal{S}}_{M/K}$, then for some $\mathfrak{P}$ lying over $\mathfrak{p}$, $f_{\mathfrak{P}|\mathfrak{p}} = 1$. If M/K is Galois, then $f = 1$ for every prime lying over $\mathfrak{p}$; since $\mathfrak{p}$ is unramified, it splits completely in M and therefore $\tilde{\mathcal{S}}_{M/K} \subset \mathcal{S}_{M/K}$.

(b) Suppose $\mathfrak{p} \in \tilde{\mathcal{S}}_{M/K}$ is unramified in M and that $\mathfrak{P} \subset \mathcal{O}_M$ lying over $\mathfrak{p}$ is such that $f_{\mathfrak{P}|\mathfrak{p}} = 1$. Let $\mathfrak{P}' = \mathfrak{P} \cap \mathcal{O}_L$; then also $f_{\mathfrak{P}'|\mathfrak{p}} = 1$. Since $\mathfrak{P}$ is unramified over $\mathfrak{p}$, so is $\mathfrak{P}'$; since L/K is Galois, all primes of L lying over $\mathfrak{p}$ have residue degree 1 and are unramified, so that $\mathfrak{p}$ splits completely in L and thus $\mathfrak{p} \in \mathcal{S}_{L/K}$.

(c) Suppose $\mathfrak{a} \in \mathcal{S}_{M/K}$, so that $\mathfrak{p}$ splits completely in M. Choose any prime $\mathfrak{P}'$ of L lying over $\mathfrak{p}$, and let $\mathfrak{P}$ be any prime of M lying over $\mathfrak{P}'$. Then $1 = f_{\mathfrak{P}|\mathfrak{p}} = f_{\mathfrak{P}|\mathfrak{P}'} f_{\mathfrak{P}'|\mathfrak{p}}$ and $1 = e_{\mathfrak{P}|\mathfrak{p}} = e_{\mathfrak{P}|\mathfrak{P}'} e_{\mathfrak{P}'|\mathfrak{p}}$, so that $\mathfrak{P}'$ is unramified with residue degree 1 over $\mathfrak{p}$. This holds for each prime of L lying over $\mathfrak{p}$, so that $\mathfrak{p} \in \mathcal{S}_{L/K}$.

8.13. (a) Note that N is Galois over $N_{\mathfrak{P}}$ since N is Galois over K, and $[N : N_{\mathfrak{P}}] = |D_{\mathfrak{P}}| = ef$. Let $\mathfrak{P}' = \mathfrak{P} \cap \mathcal{O}_{N_{\mathfrak{P}}}$ be the prime of $N_{\mathfrak{P}}$ under $\mathfrak{P}$.

Then $D_{\mathfrak{P}} = \mathrm{Gal}(N/N_{\mathfrak{P}})$ acts transitively on the set of primes lying over $\mathfrak{P}'$, which includes $\mathfrak{P}$. But $D_{\mathfrak{P}}$ fixes $\mathfrak{P}$ by definition, so that $\mathfrak{P}$ is the only prime lying over $\mathfrak{P}'$, and thus $g_{\mathfrak{P}|\mathfrak{P}'} = 1$. Now

$$e_{\mathfrak{P}|\mathfrak{P}'} f_{\mathfrak{P}|\mathfrak{P}'} = e_{\mathfrak{P}|\mathfrak{P}'} f_{\mathfrak{P}|\mathfrak{P}'} g_{\mathfrak{P}|\mathfrak{P}'} = [N : N_{\mathfrak{P}}] = ef = e_{\mathfrak{P}|\mathfrak{P}'} f_{\mathfrak{P}|\mathfrak{P}'} e_{\mathfrak{P}'|\mathfrak{p}} f_{\mathfrak{P}'|\mathfrak{p}},$$

where the last equality uses Exercise 5.15. Thus $e_{\mathfrak{P}'|\mathfrak{p}} f_{\mathfrak{P}'|\mathfrak{p}} = 1$ and we are done.

(b) For $\Leftarrow$, since $\mathfrak{P}$ lies over $\mathfrak{P}_M$, so does $\mathfrak{P}'$. Therefore $e_{\mathfrak{P}'|\mathfrak{p}} = e_{\mathfrak{P}'|\mathfrak{P}_M} e_{\mathfrak{P}_M|\mathfrak{p}}$. But $e_{\mathfrak{P}'|\mathfrak{p}} = 1$ by part (a), so that $e_{\mathfrak{P}_M|\mathfrak{p}} = 1$. A similar argument shows that $f_{\mathfrak{P}_M|\mathfrak{p}} = 1$. For $\Rightarrow$, clearly $D_{N/M,\mathfrak{P}} = D_{N/K,\mathfrak{P}} \cap \mathrm{Gal}(N/M)$, so that by the Fundamental Theorem of Galois Theory, the fixed field of $D_{N/M,\mathfrak{P}}$ is the compositum of the two fixed fields, or $N_{\mathfrak{P}} M$. Since

$$e_{\mathfrak{P}|\mathfrak{p}} = e_{\mathfrak{P}|\mathfrak{P}_M} e_{\mathfrak{P}_M|\mathfrak{p}} = e_{\mathfrak{P}|\mathfrak{P}_M}$$
$$f_{\mathfrak{P}|\mathfrak{p}} = f_{\mathfrak{P}|\mathfrak{P}_M} f_{\mathfrak{P}_M|\mathfrak{p}} = f_{\mathfrak{P}|\mathfrak{P}_M},$$

and $[N : N_{\mathfrak{P}} M] = |D_{N/M,\mathfrak{P}}| = e_{\mathfrak{P}|\mathfrak{P}_M} f_{\mathfrak{P}|\mathfrak{P}_M}$ since $N_{\mathfrak{P}} M$ is the fixed field of $D_{N/M,\mathfrak{P}}$, we have finally

$$[N : N_{\mathfrak{P}} M] = e_{\mathfrak{P}|\mathfrak{P}_M} f_{\mathfrak{P}|\mathfrak{P}_M} = e_{\mathfrak{P}|\mathfrak{p}} f_{\mathfrak{P}|\mathfrak{p}} = [N : N_{\mathfrak{P}}],$$

so that $N_{\mathfrak{P}} M = N_{\mathfrak{P}}$ and thus $M \subset N_{\mathfrak{P}}$.

8.14. As the hint suggests, let N be a Galois extension of K containing both L and M, let $\mathfrak{P}$ be a prime of N lying over $\mathfrak{p}$, and construct the intermediate field $N_{\mathfrak{P}}$. Let $\mathfrak{P}_L = \mathfrak{P} \cap \mathcal{O}_L$. Then since $\mathfrak{p}$ splits completely in L, we have $e_{\mathfrak{P}_L|\mathfrak{p}} = f_{\mathfrak{P}_L|\mathfrak{p}} = 1$, so that by part (b) of Exercise 8.13, $L \subset N_{\mathfrak{P}}$. Similarly, $M \subset N_{\mathfrak{P}}$, so that $LM \subset N_{\mathfrak{P}}$. Applying part (b) of Exercise 8.13 again shows that $e_{\mathfrak{P}_{LM}|\mathfrak{p}} = f_{\mathfrak{P}_{LM}|\mathfrak{p}} = 1$, where $\mathfrak{P}_{LM} = \mathfrak{P} \cap \mathcal{O}_{LM}$. Thus $\mathfrak{p}$ splits completely in LM.

8.15. (a) If $\mathfrak{p} \in \mathcal{S}_{L/K}$, then $\mathfrak{p}\mathcal{O}_L = \prod_{i=1}^r \mathfrak{P}_i$ with the $\mathfrak{P}_i$ distinct primes of L with $f_{\mathfrak{P}_i|\mathfrak{p}} = 1$. But then

$$\mathfrak{p}\mathcal{O}_L = \sigma(\mathfrak{p}\mathcal{O}_L) = \sigma\Big(\prod_{i=1}^r \mathfrak{P}_i\Big) = \prod_{i=1}^r \sigma(\mathfrak{P}_i).$$

The $\sigma(\mathfrak{P}_i)$ are distinct since σ is injective, and they are prime since the $\mathfrak{P}_i$ are. So $\mathfrak{p}$ splits completely in $\sigma(L)$.

(b) Part (a) shows that $\mathfrak{p}$ splits completely in each $\sigma(L)$. Extending Exercise 8.14 to any finite number of fields, this shows that $\mathfrak{p}$ splits completely in the compositum of the $\sigma(L)$. It is easy to see that the Galois closure L' of L is the compositum of its $\mathbb{C}$-embeddings; therefore $\mathfrak{p}$ splits completely in L'.

By the discussion in the statement of the exercise, this proves that $\mathcal{S}_{L/K} = \mathcal{S}_{L'/K}$.

8.16. First note that $\Rightarrow$ is part (a) of Exercise 8.12. For $\Leftarrow$, assume that $\tilde{\mathcal{S}}_{M/K} = \mathcal{S}_{M/K}$ and let M' be the Galois closure of M over K. It suffices to show $M' \subset M$. By part (b) of Proposition 8.20, $M' \subset M$ if and only if $\tilde{\mathcal{S}}_{M/K} \dot{\subset} \mathcal{S}_{M'/K}$. But

$$\tilde{\mathcal{S}}_{M/K} = \mathcal{S}_{M/K} = \mathcal{S}_{M'/K}$$

where the first equality is our assumption and the second follows from Exercise 8.15. Thus $M' \subset M$ so that M is Galois over K.

8.17. (a) Given $\epsilon_i \in \{\pm 1\}$, the hint gives $\alpha \in K^*$ with $\sigma_i(\epsilon_i\alpha) > 0$ for all i. Multiplying α by a suitable positive integer gives an element of $\mathcal{O}_K$ and does not change the sign of $\sigma_i(\epsilon_i\alpha)$. So we may assume $\alpha \in \mathcal{O}_K$. Then pick a nonzero element $\beta \in \mathfrak{m}_0$ and set $\lambda = 1 + d\alpha\beta^2$, where d is a positive integer chosen so that $|\sigma_i(d\alpha\beta^2)| > 1$ for all i. Since $\beta \in \mathfrak{m}_0$ and $\alpha \in \mathcal{O}_K$, it follows that $\lambda \equiv 1 \bmod \mathfrak{m}_0$. Since $\epsilon_i = \pm 1$ for each i, we have

$$\sigma_i(\epsilon_i\lambda) = \sigma_i(\epsilon_i + d(\epsilon_i\alpha)\beta^2) = \epsilon_i + d\sigma_i(\epsilon_i\alpha)\sigma_i(\beta)^2 \geq -1 + d\sigma_i(\epsilon_i\alpha)\sigma_i(\beta)^2.$$

The choice of α and d implies that $d\sigma_i(\epsilon_i\alpha)\sigma_i(\beta)^2 > 1$, so that $\sigma_i(\epsilon_i\lambda) > 0$ as desired.

(b) Take $a, b \in \mathcal{O}_K$ relatively prime to $\mathfrak{m}_0$, $a \equiv b \bmod \mathfrak{m}_0$, and $\sigma_i(a/b) > 0$ for all i. We must show that $(a/b)\mathcal{O}_K \in P_{K,1}(\mathfrak{m})$. Since a is prime to $\mathfrak{m}_0$, there is some $c \in \mathcal{O}_K$ with $ac \equiv 1 \bmod \mathfrak{m}_0$, which implies $bc \equiv 1 \bmod \mathfrak{m}_0$. Since $(a/b)\mathcal{O}_K = (ac)/(bc)\mathcal{O}_K$, replacing a, b with ac, bc gives $a \equiv b \equiv 1 \bmod \mathfrak{m}_0$. For the infinite primes, pick $\epsilon_i \in \{\pm 1\}$ such that $\sigma_i(\epsilon_i a) > 0$ for all i. By part (a), there is $\lambda \in \mathcal{O}_K$ with $\lambda \equiv 1 \bmod \mathfrak{m}_0$ and $\sigma_i(\epsilon_i\lambda) > 0$ for all i. Then $a\lambda \equiv 1 \bmod \mathfrak{m}_0$, and for each i,

$$\sigma_i(a\lambda) = \sigma_i\big((\epsilon_i a)(\epsilon_i\lambda)\big) = \sigma_i(\epsilon_i a)\sigma_i(\epsilon_i\lambda) > 0.$$

Finally, we show that $b\lambda$ has the same properties. Since $b \equiv 1 \bmod \mathfrak{m}_0$ and $\lambda \equiv 1 \bmod \mathfrak{m}_0$, we have $b\lambda \equiv 1 \bmod \mathfrak{m}_0$. Also, $0 < \sigma_i(a/b) = \sigma_i((\epsilon_i a)/(\epsilon_i b))$ implies that $\sigma_i(\epsilon_i b) > 0$ since ϵ_i was chosen so that $\sigma_i(\epsilon_i a) > 0$. Then the above proof that $\sigma_i(a\lambda) > 0$ adapts to show that $\sigma_i(b\lambda) > 0$ for all i. Replacing a, b with $a\lambda, b\lambda$, we see that $(a/b)\mathcal{O}_K \in P_{K,1}(\mathfrak{m})$.

Solutions to Exercises in §9

9.1. By Theorem 8.10 and the discussion preceding it, the Hilbert class field of K is the Abelian extension determined by $C(\mathcal{O}_K) = I_K/P_K$, while the ring class field of the maximal order $\mathcal{O}_K$ of conductor $f = 1$ is the Abelian extension of K determined by $I_K(1)/P_{K,\mathbb{Z}}(1)$. By the Existence Theorem, it suffices to show that $I_K(1) = I_K$ and $P_{K.\mathbb{Z}}(1) = P_K$. But $I_K(1)$ is generated by ideals with norms prime to 1, so it is clearly equal to I_K. Similarly, $P_{K,\mathbb{Z}}(1)$ is generated by all nonzero $\alpha\mathcal{O}_K$ for $\alpha \in \mathcal{O}_K$ such that $\alpha \equiv a \bmod \mathcal{O}_K$ for $a \in \mathbb{Z}$ with $\gcd(a, 1) = 1$. But any nonzero $\alpha \in \mathcal{O}_K$ satisfies this condition.

9.2. (a) Suppose $f\alpha \in \mathfrak{m} = f\mathcal{O}_K$ for $\alpha \in \mathcal{O}_K$. Then $\tau(f\alpha) = f\overline{\alpha} \in f\mathcal{O}_K$ since $\mathcal{O}_K$ is closed under complex conjugation. So $\tau(\mathfrak{m}) \subset \mathfrak{m}$; since τ is an involution, it follows that $\tau(\mathfrak{m}) = \mathfrak{m}$.

For the second part, if we show $\tau(P_{K,\mathbb{Z}}(f)) \subset P_{K,\mathbb{Z}}(f)$, then again since τ is an involution we can conclude that $\tau(P_{K,\mathbb{Z}}(f)) = P_{K,\mathbb{Z}}(f)$. Since $P_{K,\mathbb{Z}}(f)$ is generated by principal ideals $\alpha\mathcal{O}_K$ where $\alpha \in \mathcal{O}_K$ satisfies $\alpha \equiv a \bmod f\mathcal{O}_K$, $a \in \mathbb{Z}$ with $\gcd(a, f) = 1$, it suffices to show that $\tau(\alpha\mathcal{O}_K) \in P_{K,\mathbb{Z}}(f)$ for α of this form. So choose such an α. Since $\alpha \equiv a \bmod f\mathcal{O}_K$, the previous paragraph implies that $\overline{\alpha} - a = \overline{\alpha - a} \in f\mathcal{O}_K$, so that $\overline{\alpha} \equiv a \bmod f\mathcal{O}_K$. Using the first paragraph again, we see that $\overline{\alpha\mathcal{O}_K} = \overline{\alpha}\mathcal{O}_K = \tau(\alpha\mathcal{O}_K) \in P_{K,\mathbb{Z}}(f)$ and we are done.

(b) Note that $\tau : L \xrightarrow{\sim} \tau(L)$, so that $\sigma \mapsto \tau\sigma\tau^{-1}$ induces an isomorphism

$$\mathrm{Gal}(L/K) \xrightarrow{\sim} \mathrm{Gal}(\tau(L)/K).$$

Also, τ induces an isomorphism $\mathcal{O}_K/\mathfrak{p} \simeq \mathcal{O}_{\tau(L)}/\tau(\mathfrak{p})$, so that $N(\mathfrak{p}) = N(\tau(\mathfrak{p}))$.

Now let $\mathfrak{p} \in I_K(\mathfrak{m})$ and let $\mathfrak{P}$ be a prime of L over $\mathfrak{p}$. Then for $\alpha \in \mathcal{O}_{\tau(L)}$, (5.20) implies

$$\begin{aligned}\left(\tau\left(\frac{L/K}{\mathfrak{P}}\right)\tau^{-1}\right)(\alpha) = \tau\left(\left(\frac{L/K}{\mathfrak{P}}\right)(\tau^{-1}(\alpha))\right) &\equiv \tau\left((\tau^{-1}(\alpha))^{N(\mathfrak{p})}\right) \bmod \tau(\mathfrak{P})\\ &\equiv \left(\tau\tau^{-1}(\alpha)\right)^{N(\mathfrak{p})} \equiv \alpha^{N(\mathfrak{p})} \equiv \alpha^{N(\tau(\mathfrak{p}))} \bmod \tau(\mathfrak{P}).\end{aligned}$$

However, for $\alpha \in \mathcal{O}_{\tau(L)}$, the Artin symbol $((\tau(L)/K)/\tau(\mathfrak{P}))$ satisfies

$$\left(\frac{\tau(L)/K}{\tau(\mathfrak{P})}\right)(\alpha) \equiv \alpha^{N(\tau(\mathfrak{p}))} \bmod \tau(\mathfrak{P})$$

By uniqueness of the Artin symbol, it follows that

$$\tau\left(\frac{(L/K)}{\mathfrak{P}}\right)\tau^{-1} = \left(\frac{(\tau(L)/K)}{\tau(\mathfrak{P})}\right).$$

The extensions are Abelian, so

$$\left(\frac{L/K}{\mathfrak{P}}\right) = \left(\frac{L/K}{\mathfrak{p}}\right), \quad \left(\frac{\tau(L)/K}{\tau(\mathfrak{P})}\right) = \left(\frac{\tau(L)/K}{\tau(\mathfrak{p})}\right).$$

Hence

$$\tau\left(\frac{L/K}{\mathfrak{p}}\right)\tau^{-1} = \left(\frac{\tau(L)/K}{\tau(\mathfrak{p})}\right).$$

Then $((\tau(L)/K)/\tau(\mathfrak{p})) = 1$ if and only if $((L/K)/\mathfrak{p}) = 1$. Thus $\tau(\mathfrak{p}) \in \ker(\Phi_{\tau(L)/K,\mathfrak{m}})$ if and only if $\mathfrak{p} \in \ker(\Phi_{L/K,\mathfrak{m}})$, and therefore

$$\ker(\Phi_{\tau(L)/K,\mathfrak{m}}) = \tau(\ker(\Phi_{L/K,\mathfrak{m}})).$$

(c) Note that $\tau : L \xrightarrow{\sim} \tau(L)$, so that $\sigma \mapsto \tau\sigma\tau^{-1}$ induces an isomorphism

$$\mathrm{Gal}(L/K) \xrightarrow{\sim} \mathrm{Gal}(\tau(L)/K).$$

Let $\mathfrak{P}$ be a prime of L over $\mathfrak{p}$, so that $\tau(\mathfrak{P})$ is a prime of $\tau(L)$ over $\tau(\mathfrak{p}) = \mathfrak{p}$. The extensions are Abelian, so

$$\left(\frac{L/K}{\mathfrak{P}}\right) = \left(\frac{L/K}{\mathfrak{p}}\right), \quad \left(\frac{\tau(L)/K}{\tau(\mathfrak{P})}\right) = \left(\frac{\tau(L)/K}{\mathfrak{p}}\right).$$

Then for $\alpha \in \mathcal{O}_{\tau(L)}$, (5.20) implies

$$\left(\tau\left(\frac{L/K}{\mathfrak{p}}\right)\tau^{-1}\right)(\alpha) \equiv \tau\left(\left(\frac{L/K}{\mathfrak{p}}\right)(\tau^{-1}(\alpha))\right) \equiv \tau\left((\tau^{-1}(\alpha))^{N(\mathfrak{p})}\right) \mod \tau(\mathfrak{P})$$
$$\equiv \left(\tau\tau^{-1}(\alpha)\right)^{N(\mathfrak{p})} \equiv \alpha^{N(\mathfrak{p})} \mod \tau(\mathfrak{P}).$$

By uniqueness of the Artin symbol, it follows that $\tau((L/K)/\mathfrak{p})\tau^{-1} = ((\tau(L)/K)/\mathfrak{p})$. But then by (6.4)

$$\left(\frac{\tau(L)/K}{\mathfrak{p}}\right) = \tau\left(\frac{L/K}{\mathfrak{p}}\right)\tau^{-1} = \left(\frac{L/K}{\tau(\mathfrak{p})}\right),$$

so that $((\tau(L)/K)/\mathfrak{p}) = 1$ if and only if $((L/K)/\tau(\mathfrak{p})) = 1$. Hence $\mathfrak{p} \in \ker(\Phi_{\tau(L)/K,\mathfrak{m}})$ if and only if $\tau(\mathfrak{p}) \in \ker(\Phi_{L/K,\mathfrak{m}})$. But since τ is an involution, $\tau(\mathfrak{p}) \in \ker(\Phi_{L/K,\mathfrak{m}})$ if and only if $\mathfrak{p} \in \tau(\ker(\Phi_{L/K,\mathfrak{m}}))$. Thus $\ker(\Phi_{\tau(L)/K,\mathfrak{m}}) = \tau(\ker(\Phi_{L/K,\mathfrak{m}}))$.

(d) Combine parts (a) and (b) to get

$$\ker(\Phi_{\tau(L)/K,\mathfrak{m}}) \overset{(b)}{=} \tau(\ker(\Phi_{L/K,\mathfrak{m}})) = \tau(P_{K,\mathbb{Z}}(f)) \overset{(a)}{=} P_{K,\mathbb{Z}}(f) = \ker(\Phi_{L/K,\mathfrak{m}}).$$

9.3. Let $D \equiv 1 \mod 4$ be negative. Then $D = -n$, where $n > 0$ and $n \equiv 3 \mod 4$, The analog of Theorem 9.4 is

THEOREM 9.4. *Let $n > 0$ be an integer, $n \equiv 3 \mod 4$, and let L be the ring class field of the order of discriminant $-n$ in the imaginary quadratic field $K = \mathbb{Q}(\sqrt{-n})$. If p is an odd prime not dividing n, then*

$$p = x^2 + xy + \frac{1+n}{4}y^2 \iff p \text{ splits completely in } L.$$

PROOF. The proof is very similar to the proof of Theorem 9.4 in the text. Let $\mathcal{O}$ be the order of discriminant $-n$. Then $-n = f^2 d_K$, where f is the conductor of $\mathcal{O}$. Let p be an odd prime not dividing n. Then $p \nmid d_K$, which implies that p is unramified in K. We prove the same equivalences as in Theorem 9.4:

$$\begin{aligned}
p = x^2 + xy + \frac{n+1}{4}y^2 &\iff p\mathcal{O}_K = \mathfrak{p}\overline{\mathfrak{p}},\ \mathfrak{p} \neq \overline{\mathfrak{p}},\text{ and } \mathfrak{p} \text{ is principal in } \mathcal{O}_K\\
&\iff p\mathcal{O}_K = \mathfrak{p}\overline{\mathfrak{p}},\ \mathfrak{p} \neq \overline{\mathfrak{p}},\text{ and } \mathfrak{p} \in P_{K,\mathbb{Z}}(f)\\
&\iff p\mathcal{O}_K = \mathfrak{p}\overline{\mathfrak{p}},\ \mathfrak{p} \neq \overline{\mathfrak{p}},\text{ and } ((L/K)/\mathfrak{p}) = 1\\
&\iff p\mathcal{O}_K = \mathfrak{p}\overline{\mathfrak{p}},\ \mathfrak{p} \neq \overline{\mathfrak{p}},\text{ and } \mathfrak{p} \text{ splits completely in } L\\
&\iff p \text{ splits completely in } L.
\end{aligned}$$

Let $w = (1+\sqrt{-n})/2$. For the first equivalence, take $p = x^2 + xy + \frac{n+1}{4}y^2 = (x+wy)(x+\overline{w}y)$. Set $\mathfrak{p} = (x+wy)\mathcal{O}_K$; this ideal is prime since it has prime norm. Then $\mathfrak{p}\overline{\mathfrak{p}}$ must be the unique factorization of $p\mathcal{O}_K$ in K. Since $\mathfrak{p}$ is unramified in K we must have $\mathfrak{p} \neq \overline{\mathfrak{p}}$, and $\mathfrak{p}$ is principal by construction. For the converse, if $p\mathcal{O}_K = \mathfrak{p}\overline{\mathfrak{p}}$ with $\mathfrak{p}$ principal, then $\mathfrak{p} = (x+wy)\mathcal{O}_K$ for some $x, y \in \mathbb{Z}$, and thus

$$p\mathcal{O}_K = (x+wy)(x+\overline{w}y)\mathcal{O}_K = \left(x^2 + xy + \frac{1+n}{4}y^2\right)\mathcal{O}_K.$$

It follows that $x^2 + xy + \frac{1+n}{4}y^2 = pu$ for $u \in \mathcal{O}_K$ a unit. Thus pu is positive and real, so that $u = 1$ and then $x^2 + xy + \frac{1+n}{4}y^2 = p$. The proofs of the remaining equivalences are identical to the proof in Theorem 9.4 in the text. Q.E.D.

The analog of Theorem 9.2 is

THEOREM 9.2. *Let $n > 0$ be an integer, $n \equiv 3 \bmod 4$. Then there is a monic irreducible polynomial $f_n(x) \in \mathbb{Z}[x]$ of degree $h(-n)$ such that if an odd prime p divides neither n nor the discriminant of $f_n(x)$, then*

$$p = x^2 + xy + \frac{1+n}{4}y^2 \iff \begin{cases} (-n/p) = 1 \text{ and } f_n(x) \equiv 0 \bmod p \\ \text{has an integer solution.} \end{cases}$$

Furthermore, $f_n(x)$ may be taken to be the minimal polynomial of a real algebraic integer α for which $L = K(\alpha)$ is the ring class field of the order of discriminant $-n$ in the imaginary quadratic field $K = \mathbb{Q}(\sqrt{-n})$.

Finally, if $f_n(x)$ is any monic integer polynomial of degree $h(-n)$ for which the above equivalence holds, then $f_n(x)$ is irreducible over $\mathbb{Z}$ and is the minimal polynomial of a primitive element of the ring class field L described above.

PROOF. The proof is very similar to the proof of Theorem 9.2 in the text. Since L is Galois over $\mathbb{Q}$ by Lemma 5.28, Proposition 5.29 shows that $L = K(\alpha)$ for some real algebraic integer $\alpha \in L$. Let $f_n(x)$ be the monic minimal polynomial of α, and let p be an odd prime dividing neither n nor the discriminant of $f_n(x)$. Then the above version of Theorem 9.4 together with Proposition 5.29 and the fact that $(-n/p) = (d_K/p)$ show that

$$p = x^2 + xy + \frac{1+n}{4}y^2 \iff \begin{cases} (-n/p) = 1 \text{ and } f_n(x) \equiv 0 \bmod p \\ \text{has an integer solution.} \end{cases}$$

Using Galois theory and Theorem 5.23, we have $\deg(f_n(x)) = [L : K] = |\mathrm{Gal}(L/K)| = |C(\mathcal{O})|$. But $C(\mathcal{O}) \simeq C(-n)$. Thus $\deg(f_n(x)) = h(-n)$.

For the final assertion of the theorem, the proof is the same as the in the text, with one modification: replace $x^2 + ny^2$ and $-4n$ with $x^2 + xy + \frac{1+n}{4}y^2$ and $-n$ respectively. Q.E.D.

9.4. If $u_1 = u_2 = 0$, then

$$\begin{pmatrix} u_0 \\ 0 \\ 0 \end{pmatrix} = \begin{pmatrix} 1 & 1 & 1 \\ 1 & \omega & \omega^2 \\ 1 & \omega^2 & \omega \end{pmatrix} \begin{pmatrix} \alpha \\ \sigma^{-1}\alpha \\ \sigma^{-2}\alpha \end{pmatrix},$$

so that by Cramer's rule

$$\alpha = \frac{\det \begin{pmatrix} u_0 & 1 & 1 \\ 0 & \omega & \omega^2 \\ 0 & \omega^2 & \omega \end{pmatrix}}{\det \begin{pmatrix} 1 & 1 & 1 \\ 1 & \omega & \omega^2 \\ 1 & \omega^2 & \omega \end{pmatrix}} = \frac{u_0(\omega^2 - \omega^4)}{\det \begin{pmatrix} 1 & 1 & 1 \\ 1 & \omega & \omega^2 \\ 1 & \omega^2 & \omega^4 \end{pmatrix}} = \frac{u_0(\omega^2 - \omega)}{(1-\omega)(1-\omega^2)(\omega - \omega^2)} = \frac{u_0}{3},$$

where the second and third equalities use $\omega^3 = 1$, the third also uses the standard formula for a Vandermonde determinant, and the fourth uses $x^2 + x + 1 = (x - \omega)(x - \omega^2)$ with $x = 1$. Then $u_0 \in K$ implies $\alpha \in K$, as claimed.

9.5. Consider the following diagram of fields:

$$\begin{array}{ccc} & L = K(\sqrt[3]{m}) & \\ \diagup & & \diagdown \\ K' = \mathbb{Q}(\sqrt[3]{m}) & & K = \mathbb{Q}(\sqrt{-n}) \\ \diagdown & & \diagup \\ & \mathbb{Q} & \end{array}$$

Let $\mathfrak{p}$ be a prime of K dividing m and let $\mathfrak{p} \cap \mathbb{Z} = p\mathbb{Z}$. Then $p \mid m$, so $m = p^e m'$ where $p \nmid m'$. Note that $e = 1$ or 2 since m is cubefree.

We claim that $p\mathcal{O}_{K'} = \mathfrak{a}^3$ for some ideal $\mathfrak{a}$ of K'. To see this, start with $m\mathcal{O}_{K'} = (\sqrt[3]{m}\mathcal{O}_{K'})^3$. Then $p^e\mathcal{O}_{K'} \cdot m'\mathcal{O}_{K'} = (\sqrt[3]{m}\mathcal{O}_{K'})^3$. Since the ideals $p^e\mathcal{O}_{K'}$ and $m'\mathcal{O}_{K'}$ are relatively prime, unique factorization of ideals in $\mathcal{O}_{K'}$ implies that $p^e\mathcal{O}_{K'}$ is the cube of an $\mathcal{O}_{K'}$-ideal, and since $e = 1$ or 2, $p\mathcal{O}_{K'}$ is the cube of an $\mathcal{O}_{K'}$-ideal $\mathfrak{a}$, again by unique factorization.

We next claim that the ideal $\mathfrak{a}$ must be prime. Since $\mathcal{O}_{K'}$ is a free $\mathbb{Z}$-module of rank 3 by Proposition 5.3, it follows that $N(p\mathcal{O}_{K'}) = p^3$. But then

$$p^3 = N(p\mathcal{O}_{K'}) = N(\mathfrak{a}^3) = N(\mathfrak{a})^3,$$

so that $N(\mathfrak{a}) = p$ and therefore $\mathfrak{a}$ is prime.

Now let $\mathfrak{P}$ be a prime of L lying over $\mathfrak{p}$. Then $\mathfrak{P}$ lies over p and hence $\mathfrak{P} \cap \mathcal{O}_{K'} = \mathfrak{a}$ since $\mathfrak{a}$ is the only prime of K' lying over p. Applying Exercise 5.15 twice, we get

$$e_{\mathfrak{P}|\mathfrak{p}} e_{\mathfrak{p}|p\mathbb{Z}} = e_{\mathfrak{P}|p\mathbb{Z}} = e_{\mathfrak{P}|\mathfrak{a}} e_{\mathfrak{a}|p\mathbb{Z}} = e_{\mathfrak{P}|\mathfrak{a}} \cdot 3.$$

Thus $3 \mid e_{\mathfrak{P}|\mathfrak{p}}$ or $3 \mid e_{\mathfrak{p}|p\mathbb{Z}}$. The latter can't occur since $e_{\mathfrak{p}|p\mathbb{Z}} \leq 2$ (K is a quadratic extension of $\mathbb{Q}$), so that $3 \mid e_{\mathfrak{P}|\mathfrak{p}}$. Hence $\mathfrak{p}$ ramifies in L.

9.6. The only cubefree integers whose only prime factors are 2 or 3 are 2, 3, 4, 9, 6, 12, 18, and 36. If a and b are positive integers such that $ab = m^3$ for some $m \in \mathbb{Z}$, then $\sqrt[3]{a} = m/\sqrt[3]{b} \in K(\sqrt[3]{b})$ and $\sqrt[3]{b} = m/\sqrt[3]{a} \in K(\sqrt[3]{a})$ imply that $K(\sqrt[3]{a}) = K(\sqrt[3]{b})$. Thus

$$2 \cdot 4 = 2^3 \Rightarrow K(\sqrt[3]{2}) = K(\sqrt[3]{4}), \qquad 3 \cdot 9 = 3^3 \Rightarrow K(\sqrt[3]{3}) = K(\sqrt[3]{9})$$
$$6 \cdot 36 = 6^3 \Rightarrow K(\sqrt[3]{6}) = K(\sqrt[3]{36}), \quad 12 \cdot 18 = 6^3 \Rightarrow K(\sqrt[3]{12}) = K(\sqrt[3]{18}).$$

So the only four distinct cubic fields in that list are $K(\sqrt[3]{2})$, $K(\sqrt[3]{3})$, $K(\sqrt[3]{6})$, and $K(\sqrt[3]{12})$.

9.7. The roots of $x^3 - a$ are $\sqrt[3]{a}$, $\omega\sqrt[3]{a}$, and $\omega^2\sqrt[3]{a}$, so the discriminant is

$$(\sqrt[3]{a} - \omega\sqrt[3]{a})^2(\sqrt[3]{a} - \omega^2\sqrt[3]{a})^2(\omega\sqrt[3]{a} - \omega^2\sqrt[3]{a})^2 = a^2((1-\omega)(1-\omega^2)(\omega-\omega^2))^2.$$

Note that $(1-\omega^2)(\omega-\omega^2) = (1-\omega^2)\omega(1-\omega) = (\omega-\omega^3)(1-\omega) = -(1-\omega)^2$, where the final equality uses $\omega^3 = 1$. Thus the discriminant is $a^2(1-\omega)^6$. However, $3 = -\omega^2(1-\omega)^2$ by Proposition 4.7. Cubing each side gives $27 = -(1-\omega)^6$, so that the discriminant of $x^3 - a$ is $a^2(1-\omega)^6 = -27a^2$.

9.8. Suppose that the ring class field of $\mathbb{Z}[\sqrt{-27}]$ is $K(\sqrt[3]{a})$ for $a = 3$, 6 or 12. Then Theorem 9.2 would hold for $f_{27}(x) = x^3 - a$, which has discriminant $-27a^3$ by the previous exercise. The prime 31 divides neither 27 nor $-27a^3$, so since $31 = 2^2 + 27 \cdot 1^2$, it would follow from Theorem 9.2 that $x^3 - a$ has a solution modulo 31; i.e., that a is a cube modulo 31. But the cubes modulo 31 are

$$0,\ 1,\ 2,\ 4,\ 8,\ 15,\ 16,\ 23,\ 27,\ 29,\ 30,$$

Since 3, 6 and 12 do not appear on this list, the ring class field of $\mathbb{Z}[\sqrt{-27}]$ cannot equal $K(\sqrt[3]{a})$ for $a = 3$, 6 or 12.

9.9. Let L be the ring class field of the order $\mathbb{Z}[\sqrt{-64}] \subset \mathcal{O}_K$ for $K = \mathbb{Q}(\sqrt{-64}) = \mathbb{Q}(i)$. Note that $\mathcal{O}_K = \mathbb{Z}[i]$ and that the discriminant of $\mathbb{Z}[\sqrt{-64}]$ is $-4 \cdot 64$. There are four reduced forms of discriminant $-4 \cdot 64$:

$$x^2 + 64y^2,\ 4x^2 + 4xy + 17y^2,\ 5x^2 - 2xy + 13y^2,\ 5x^2 + 2xy + 13y^2,$$

so that $[L : K] = h(-4 \cdot 64) = 4$. Thus L is a degree four extension of K. Further, since $5x^2 - 2xy + 13y^2$ and $5x^2 + 2xy + 13y^2$ are inverses of each other, $\mathrm{Gal}(L/K) \simeq \mathbb{Z}/4\mathbb{Z}$. Also, by Lemma 9.3, L is Galois over $\mathbb{Q}$, and its Galois group is $\mathbb{Z}/4\mathbb{Z} \rtimes \mathbb{Z}/2\mathbb{Z}$ with $\mathbb{Z}/2\mathbb{Z}$

acting nontrivially; this group is the dihedral group of order 8. Finally, $\mathbb{Z}[\sqrt{-64}] = \mathbb{Z}[8i]$ is an order of conductor 8, so that L corresponds to a generalized ideal class group for the modulus $8\mathcal{O}_K$, and the primes of K ramifying in L must divide the modulus. To show that only one field satisfies these criteria, we need the analog of Lemma 9.6, which is:

LEMMA. *If M is a quartic extension of $K = \mathbb{Q}(i)$ with $\mathrm{Gal}(M/\mathbb{Q})$ the dihedral group of order 8, then $M = K(\sqrt[4]{m})$ for some nonsquare, fourth-power free integer m.*

PROOF. Since M is Galois over $\mathbb{Q}$, complex conjugation τ is in $\mathrm{Gal}(M/\mathbb{Q})$. Let σ be a generator of $\mathrm{Gal}(M/K) \simeq \mathbb{Z}/4\mathbb{Z}$. Since $\mathrm{Gal}(M/\mathbb{Q})$ is the dihedral group of order 8, we have $\tau\sigma\tau = \sigma^3 = \sigma^{-1}$. Let α be a real algebraic integer such that $M = K(\alpha)$, and define $u_j \in M$ by

$$u_j = \alpha + i^j\sigma^{-1}(\alpha) + i^{2j}\sigma^{-2}(\alpha) + i^{3j}\sigma^{-3}(\alpha), \quad i = 0, 1, 2, 3.$$

The u_i are all algebraic integers, and since $\sigma(i) = i$ because $i \in K$, we have

$$\begin{aligned}\sigma(u_j) &= \sigma(\alpha) + i^j\alpha + i^{2j}\sigma^{-1}(\alpha) + i^{3j}\sigma^{-2}(\alpha)\\ &= i^j\left(\alpha + i^j\sigma^{-1}(\alpha) + i^{2j}\sigma^{-2}(\alpha) + i^{3j}\sigma^{-3}(\alpha)\right) = i^j u_j.\end{aligned}$$

Further, since α is real, and using $\sigma\tau = \tau\sigma^{-1}$, $\sigma^4 = 1$, and $\tau(i) = -i = i^3$, we have

$$\begin{aligned}\tau(u_j) &= \tau(\alpha) + \tau(i^j)\tau(\sigma^{-1}(\alpha)) + \tau(i^{2j})\tau(\sigma^{-2}(\alpha)) + \tau(i^{3j})\tau(\sigma^{-3}(\alpha))\\ &= \alpha + \tau(i^j)\sigma(\tau\alpha) + \tau(i^{2j})\sigma^2(\tau\alpha)) + (\tau(i^{3j})\sigma^3(\tau\alpha))\\ &= \alpha + i^{3j}\sigma(\alpha) + i^{6j}\sigma^2(\alpha) + i^{9j}\sigma^3(\alpha)\\ &= \alpha + i^{3j}\sigma^{-3}(\alpha) + i^{2j}\sigma^{-2}(\alpha) + i^j\sigma^{-1}(\alpha) = u_j.\end{aligned}$$

It follows that the u_j are all real. Further, u_0 is fixed by σ, so that $u_0 \in K$. Finally, for $j = 1, 2, 3$, we have

$$\sigma(u_j^4) = \sigma(u_j)^4 = i^{4j}u_j^4 = u_j^4,$$

so that $u_j^4 \in K \cap \mathbb{R} = \mathbb{Q}$; since they are algebraic integers, they are all integers. If $u_1 \neq 0$, we claim that $M = K(u_1)$. Well, $[M : K] = 4$, so if $M \neq K(u_1)$ then either $u_1 \in K$ or $u_1^2 \in K$. In the first case, since u_1 is real and an algebraic integer, it would be an integer, contradicting $\sigma(u_1) = iu_1$ with $u_1 \neq 0$. If $u_1^2 \in K$, a similar argument shows that u_1^2 is an integer, but $\sigma(u_1^2) = -u_1^2$. This proves the claim. So if $u_1 \neq 0$, then $M = K(u_1)$; set $u_1^4 = m$; then $M = K(\sqrt[4]{m})$. The same argument applies if $u_2 \neq 0$ or if $u_3 \neq 0$. Lastly, if $u_1 = u_2 = u_3 = 0$, then

$$\begin{pmatrix} u_0\\ 0\\ 0\\ 0\end{pmatrix} = \begin{pmatrix} 1 & 1 & 1 & 1\\ 1 & i & -1 & -i\\ 1 & -1 & 1 & -1\\ 1 & -i & -1 & i\end{pmatrix}\begin{pmatrix}\alpha\\ \sigma^{-1}\alpha\\ \sigma^{-2}\alpha\\ \sigma^{-3}\alpha\end{pmatrix},$$

so that by Cramer's rule

$$\alpha = \frac{\det\begin{pmatrix} u_0 & 1 & 1 & 1\\ 0 & i & -1 & -i\\ 0 & -1 & 1 & -1\\ 0 & -i & -1 & i\end{pmatrix}}{\det\begin{pmatrix} 1 & 1 & 1 & 1\\ 1 & i & -1 & -i\\ 1 & -1 & 1 & -1\\ 1 & -i & -1 & i\end{pmatrix}} = \frac{u_0\det\begin{pmatrix} i & -1 & -i\\ -1 & 1 & -1\\ -i & -1 & i\end{pmatrix}}{\det\begin{pmatrix} 1 & 1 & 1 & 1\\ 1 & i & -1 & -i\\ 1 & -1 & 1 & -1\\ 1 & -i & -1 & i\end{pmatrix}} = u_0 \cdot \beta, \quad \beta \in \mathbb{Q}(i).$$

The denominator is nonzero since it is a Vandermonde determinant with unequal rows. Since $u_0 \in K$, it follows that $\alpha \in K$. This contradicts $M = K(\alpha)$, so that one of u_1, u_2, or u_3 must be nonzero. Thus $M = K(\sqrt[4]{m})$, and we may take m to be fourth-power free. Since $[M : K] = 4$, m is not a square. Q.E.D.

By the lemma, $L = K(\sqrt[4]{m})$ for some nonsquare fourth-power free nonsquare integer m. Then an argument similar to the solution of Exercise 9.5 shows that any prime of $\mathcal{O}_K$ dividing m ramifies in L. But all ramified primes divide the modulus $8\mathcal{O}_K$, so they must divide $2\mathcal{O}_K$. It follows that 2 is the only integer prime that can divide m; since m is nonsquare and is fourth power free, we have $L = K(\sqrt[4]{2})$ or $L = K(\sqrt[4]{8})$. But since $2 \cdot 8 = 2^4$, arguing as in the solution to Exercise 9.6 shows that these two fields are equal, so the ring class field is $L = K(\sqrt[4]{2})$.

9.10. (a) The discriminant of $\mathcal{O} = \mathbb{Z}[9\omega]$ is $9^2 \cdot (-3) = -243$; there are only three classes of reduced forms with this discriminant:

$$x^2 + xy + 61y^2,\ 7x^2 - 3xy + 9y^2,\ 7x^2 + 3xy + 9y^2.$$

Thus $[L : K] = h(-243) = 3$, so L is a cubic extension of K. Lemma 9.3 then implies that L is Galois over $\mathbb{Q}$ with Galois group S_3. By Lemma 9.6, $L = K(\sqrt[3]{m})$ for some cubefree positive integer m. Any prime of K that ramifies in L must divide $9\mathcal{O}_K$, since $\mathcal{O}$ has conductor 3, so that 3 is the only rational prime that can ramify in L. Since m is cubefree and any rational prime dividing m ramifies in L, we must have $L = K(\sqrt[3]{3})$ or $L = K(\sqrt[3]{9})$. But these two fields are equal since $3 \cdot 9 = 3^3$, so that $L = K(\sqrt[3]{3})$.

(b) The minimal polynomial of $\sqrt[3]{3}$ is $x^3 - 3$, which by Exercise 9.7 has discriminant $-27 \cdot 3 = -3^4$. Hence the version of Theorem 9.2 given in Exercise 9.3 shows that $p = x^2 + xy + 61y^2$ for $p \geq 5$ if and only if $(-3/p) = 1$ and $x^3 - 3 = 0 \bmod 3$ has a solution. But

$$\left(\frac{-3}{p}\right) = \left(\frac{-1}{p}\right)\left(\frac{3}{p}\right) = \begin{cases} 1 & p \equiv 1 \bmod 3 \\ -1 & p \equiv -1 \bmod 3, \end{cases}$$

so this is the same as saying that for a prime $p \geq 5$, $p = x^2 + xy + 61y^2$ if and only if $p \equiv 1 \bmod 3$ and 3 is a cubic residue modulo p.

(c) Note that

$$(*) \qquad (2x + y)^2 + 243y^2 = 4x^2 + 4xy + 244y^2 = 4(x^2 + xy + 61y^2).$$

This shows that if p can be represented by $x^2+xy+61y^2$, then $4p$ can be represented as $x^2 + 243y^2$. Conversely, if $4p = u^2 + 243v^2$ for integers u and v, then u and v have the same parity. Thus $x = (u - v)/2$ and $y = v$ are integers that satisfy $u = 2x + y$ and $v = y$. Then $4p = (2x + y)^2 + 243y^2$ and (*) show that p is represented by $x^2 + xy + 61y^2$. This equivalence and part (b) give the desired result.

9.11. From Exercise 9.9, the ring class field of $\mathbb{Z}[\sqrt{-64}]$ is $L = K(\sqrt[4]{2})$, where $K = \mathbb{Q}(i)$. Since $\sqrt[4]{2}$ is a real algebraic integer, we may take $f_{64} = x^4 - 2$. Then Theorem 9.2 provides the statement in part (ii) of Theorem 9.8 except that we must check that $p \equiv 1 \bmod 4$ is equivalent to $(-64/p) = 1$. But this is clear, since

$$\left(\frac{-64}{p}\right) = \left(\frac{-1}{p}\right) = \begin{cases} 1 & p \equiv 1 \bmod 4 \\ -1 & p \equiv 3 \bmod 4. \end{cases}$$

Finally, the discriminant of $x^4 + bx^2 + c$ is $2^4c(b^2 - 4c)^2$ by Exercise 5.24, so that $x^4 - 2$ has discriminant $2^4(-2)(0^2 - 4 \cdot (-2))^2 = -2^{11}$. Thus the only excluded prime is 2.

9.12. (a) Since $N(1+3\omega)=7$ and $N(2)=4$, (4.10) gives the following congruences in $\mathcal{O}_K=\mathbb{Z}[\omega]$:

$$\left(\frac{2}{\theta}\right)_3=\left(\frac{2}{1+3\omega}\right)_3\equiv 2^{(N(1+3\omega)-1)/3}\equiv 4 \bmod 1+3\omega$$
$$\left(\frac{\theta}{2}\right)_3=\left(\frac{1+3\omega}{2}\right)_3\equiv (1+3\omega)^{(N(2)-1)/3}\equiv 1+3\omega\equiv 1+\omega \bmod 2.$$

To compute $4 \bmod 1+3\omega$, first note that $N(\theta)=1^2-1\cdot 3+3^2=7$. It follows that $(\mathbb{Z}/7\mathbb{Z})^*\simeq(\mathcal{O}_K/\theta\mathcal{O}_K)^*$. Since $2^3=8\equiv 1 \bmod 7$, it follows that 2 is also a root of $x^3\equiv 1 \bmod \theta$. Since $2\not\equiv 1 \bmod \theta$, we must have $2\equiv\omega \bmod \theta$ or $2\equiv\omega^2 \bmod \theta$. However,

$$\frac{2-\omega^2}{1+3\omega}=\frac{(2-\omega^2)(1+3\omega^2)}{N(1+3\omega)}=\frac{2+5\omega^2-3\omega^4}{7}=\frac{-3-8\omega}{7}\notin\mathcal{O}_K$$

using $\omega^4=\omega$ and $\omega^2=-1-\omega$. Thus $2\equiv\omega \bmod \theta$, so that $4\equiv\omega^2 \bmod \theta$. It follows that

$$\left(\frac{2}{\theta}\right)_3=\omega^2.$$

To compute $1+\omega \bmod 2$, note that since $1+\omega+\omega^2=0$, we have

$$1+\omega=-\omega^2\equiv\omega^2 \bmod 2,$$

so that

$$\left(\frac{\theta}{2}\right)_3\equiv\omega^2 \bmod 2, \text{ hence } \left(\frac{\theta}{2}\right)_3=\omega^2.$$

Thus the two are equal.

(b) Let $K=\mathbb{Q}(i)$; then from Exercise 9.9, $L=K(\sqrt[4]{2})$ is the ring class field of the order $\mathbb{Z}[8i]$ of conductor 8, so it corresponds to a (congruence) subgroup of $I_K(8)$ containing $P_{K,1}(8)$. Thus the conductor $\mathfrak{f}(L/K)$ divides $8\mathcal{O}_K$. From Weak Reciprocity, we get a well-defined homomorphism

$$\left(\frac{2}{\cdot}\right)_4 : I_K(8)/P_{K,1}(8)\longrightarrow\mu_4,$$

which is surjective since $\mathrm{Gal}(L/K)\simeq\mathbb{Z}/4\mathbb{Z}$ whose generator σ is determined by $\sigma(\sqrt[4]{2})=i\sqrt[4]{2}$. The map sending $\alpha\in\mathcal{O}_K$ to the principal ideal $\alpha\mathcal{O}_K$ induces a homomorphism $(\mathcal{O}_K/8\mathcal{O}_K)^*\to I_K(8)/P_{K,1}(8)$, well-defined by the argument following (7.26). Composing these maps gives a homomorphism

$$\left(\frac{2}{\cdot}\right)_4 : (\mathcal{O}_K/8\mathcal{O}_K)^*\longrightarrow\mu_4.$$

Elements of $(\mathcal{O}_K/8\mathcal{O}_K)^*$ are of the form $a+bi$ where $a\not\equiv b \bmod 2$. One computes that $(\mathcal{O}_K/8\mathcal{O}_K)^*\simeq\mathbb{Z}/4\mathbb{Z}\times\mathbb{Z}/4\mathbb{Z}\times\mathbb{Z}/2\mathbb{Z}$. The elements corresponding to primary primes must have a odd and b even, which gives a subgroup of index 2, of order 16 ($\simeq\mathbb{Z}/4\mathbb{Z}\times\mathbb{Z}/2\mathbb{Z}\times\mathbb{Z}/2\mathbb{Z}$), generated by (for example) the three primes $3+2i$, $-1+4i$, and $-5+4i$. It suffices to show that the formula holds for these generators. We compute as follows:

$$\left(\frac{2}{3+2i}\right)_4\equiv 2^{(N(3+2i)-1)/4}\equiv 8 \bmod 3+2i$$
$$\left(\frac{2}{-1+4i}\right)_4\equiv 2^{(N(-1+4i)-1)/4}\equiv 2^4 \bmod -1+4i$$
$$\left(\frac{2}{-5+4i}\right)_4\equiv 2^{(N(-5+4i)-1)/4}\equiv 2^{10} \bmod -5+4i.$$

For the first of these, note that $N(3+2i) = 13$, so that $(\mathbb{Z}/13\mathbb{Z})^* \simeq (\mathcal{O}_K/(3+2i)\mathcal{O}_K)^*$. Since $8^2 = 64 \equiv -1 \bmod 13$, it follows that 8 must be congruent modulo $3+2i$ to a primitive fourth root of unity, so that $8 \equiv i \bmod 3+2i$ or $8 \equiv -i \bmod 3+2i$. However,

$$\frac{8-i}{3+2i} = \frac{(8-i)(3-2i)}{N(3+2i)} = \frac{26-19i}{13} \notin \mathcal{O}_K,$$

so that $8 \equiv -i \bmod 3+2i$ and therefore

$$\left(\frac{2}{3+2i}\right)_4 = -i = i^{3\cdot 2/2}$$

as required. Next, $N(-1+4i) = 17$, so that $(\mathbb{Z}/17\mathbb{Z})^* \simeq (\mathcal{O}_K/(-1+4i)\mathcal{O}_K)^*$. Since $2^4 = 16 \equiv -1 \bmod 17$, it follows that $2^4 \equiv -1 \bmod -1+4i$ as well, so that

$$\left(\frac{2}{-1+4i}\right)_4 = -1 = i^{-1\cdot 4/2}$$

as required. Finally, $N(-5+4i) = 41$, so that $(\mathbb{Z}/41\mathbb{Z})^* \simeq (\mathcal{O}_K/(-5+4i)\mathcal{O}_K)^*$. Since $2^{10} = 24\cdot 41 + 40$, we see that $2^{10} \equiv -1 \bmod 41$ and thus $2^{10} \equiv -1 \bmod -5+4i$ as well. Therefore

$$\left(\frac{2}{-5+4i}\right)_4 = -1 = i^{-5\cdot 4/2}$$

as required.

9.13. (a) From (4.10),

$$\left(\frac{\omega}{\pi}\right)_3 \equiv \omega^{(N(\pi)-1)/3} \bmod \pi,$$

so we want to prove that

$$\frac{N(\pi)-1}{3} \equiv \frac{2(a+2)}{3} \bmod 3,$$

or that $N(\pi) - 1 \equiv 2(a+2) \bmod 9$. Since $[K : \mathbb{Q}] = 2$ and $p \equiv 1 \bmod 3$ splits, we have $N(\pi) = p$. Note that $4p = a^2 + 27b^2$, so that $4p \equiv a^2 \bmod 9$ and thus $p \equiv -2a^2 \bmod 9$. Using this and $-1-4 \equiv 4 \bmod 9$, we get

$$\begin{aligned} N(\pi) - 1 - 2(a+2) &\equiv p - 1 - 2a - 4 \equiv -2a^2 - 2a + 4 \\ &\equiv -2(a-1)(a+2) \equiv 0 \bmod 9, \end{aligned}$$

where the last congruence follows since $a \equiv 1 \bmod 3$ implies that both $a-1$ and $a+2$ are multiples of 3. Therefore, $N(\pi) - 1 \equiv 2(a+2) \bmod 9$ and we are done.

(b) Arguing as in Theorem 9.9, we know from Exercise 9.10 that $L = K(\sqrt[3]{3})$ is the ring class field of the order of $\mathcal{O}_K$ of conductor 9, so it corresponds to a subgroup of $I_K(9)$ and the conductor $\mathfrak{f}$ divides $9\mathcal{O}_K$. From Weak Reciprocity we get a homomorphism (a surjection since $\mathrm{Gal}(L/K) \simeq \mathbb{Z}/3\mathbb{Z}$)

$$(\mathcal{O}_K/9\mathcal{O}_K)^* \longrightarrow I_K(9)/P_{K,1}(9) \longrightarrow \mu_3$$

which is the Legendre symbol $(3/\cdot)_3$. Note that $\pi = \frac{1}{2}(a + \sqrt{-27}\,b) = \frac{a+3b}{2} + 3b\omega$; since $a \equiv 1 \bmod 3$, it follows that $\frac{a+3b}{2} \equiv -1 \bmod 3$. Of the 54 elements of

$$(\mathcal{O}_K/9\mathcal{O}_K)^* \simeq \mathbb{Z}/6\mathbb{Z} \times \mathbb{Z}/6\mathbb{Z} \times \mathbb{Z}/3\mathbb{Z},$$

nine are of the form $c + 3d\omega$ with $c \equiv -1 \bmod 3$. These elements are generated by (say) $2+3\omega$ and $5+3\omega$, so it suffices to show that the formula holds for these generators.

- Since $N(2+3\omega) = 7$, we get

$$\left(\frac{3}{2+3\omega}\right)_3 = 3^{(N(2+3\omega)-1)/3} \equiv 3^2 \equiv 9 \bmod 2+3\omega.$$

From $N(2+3\omega)=7$ it follows that $(\mathbb{Z}/7\mathbb{Z})^* \simeq (\mathcal{O}_K/(2+3\omega)\mathcal{O}_K)^*$. Since $9^3 = 729 \equiv 1 \bmod 7$, it follows that 9 must be congruent modulo $2+3\omega$ to a cube root of unity. Since $9 \not\equiv 1 \bmod 2+3\omega$, we have $9 \equiv \omega$ or $9 \equiv \omega^2$. But

$$\frac{9-\omega}{2+3\omega} = \frac{(9-\omega)(2+3\omega^2)}{N(2+3\omega)} = \frac{18-2\omega+27\omega^2-3\omega^3}{7} = \frac{-12-29\omega}{7} \notin \mathcal{O}_K$$

using $\omega^3 = 1$ and $\omega^2 = -1-\omega$. Thus $9 \equiv \omega^2 \bmod 2+3\omega$, and since $b=1$ here, we get

$$9 \equiv \omega^2 = \omega^{2b} \bmod 2+3\omega.$$

- Since $N(5+3\omega) = 19$, we have

$$\left(\frac{3}{5+3\omega}\right)_3 = 3^{(N(5+3\omega)-1)/3} \equiv 3^6 \equiv 729 \bmod 5+3\omega.$$

From $N(5+3\omega) = 19$ we also see that $(\mathbb{Z}/19\mathbb{Z})^* \simeq (\mathcal{O}_K/(5+3\omega)\mathcal{O}_K)^*$. Since $729^3 = 3^{18} \equiv 1 \bmod 19$ by Fermat's Little Theorem, it follows that 729 must be congruent modulo $5+3\omega$ to a cube root of unity. Since $729 \not\equiv 1 \bmod 5+3\omega$, we have $729 \equiv \omega$ or $729 \equiv \omega^2$. But

$$\begin{aligned}\frac{729-\omega}{5+3\omega} &= \frac{(729-\omega)(5+3\omega^2)}{N(5+3\omega)} = \frac{3645-5\omega+2187\omega^2-3\omega^3}{19}\\ &= \frac{1455-2192\omega}{19} \notin \mathcal{O}_K\end{aligned}$$

using $\omega^3 = 1$ and $\omega^2 = -1-\omega$. Thus $729 \equiv \omega^2 \bmod 5+3\omega$, and since $b=1$ here, we get

$$729 \equiv \omega^2 = \omega^{2b} \bmod 5+3\omega.$$

(c) Using $3 = -\omega^2(1-\omega)^2$, we obtain

$$\left(\frac{3}{\pi}\right)_3 = \left(\frac{-\omega^2(1-\omega)^2}{\pi}\right)_3 = \left(\frac{-1}{\pi}\right)_3 \left(\frac{\omega}{\pi}\right)_3^2 \left(\frac{1-\omega}{\pi}\right)_3^2.$$

We know that $(-1/\pi)_3 \equiv (-1)^{(p-1)/3} \bmod \pi$. But $p \equiv 1 \bmod 3$ is odd, so $p \equiv 1 \bmod 6$. Hence $(-1/\pi)_3 = 1$. Using parts (a) and (b) we then get

$$\omega^{2b} = \left(\frac{3}{\pi}\right)_3 = 1 \cdot \omega^{4(a+2)/3} \cdot \left(\frac{1-\omega}{\pi}\right)_3^2 \implies \left(\frac{1-\omega}{\pi}\right)_3^2 = \omega^{2b-(4(a+2)/3)},$$

so that since $-2 \cdot \frac{a+2}{3} \equiv \frac{a+2}{3} \bmod 3$,

$$\left(\frac{1-\omega}{\pi}\right)_3 = \omega^{b-2(a+2)/3} = \omega^{b+(a+2)/3}.$$

(d) Note that with $\pi = \frac{a+3b}{2} + 3b\omega$ and $a \equiv 1 \bmod 3$, we have $\frac{a+3b}{2} \equiv -1 \bmod 3$. Hence π is in the form $-1+3m+3n\omega$ required by (4.13) with

$$m = \frac{\frac{a+3b}{2}+1}{3} = \frac{a+3b+2}{6}, \qquad 2m = \frac{a+3b+2}{3} = b + \frac{a+2}{3}, \qquad n = b.$$

Thus from part (c)

$$\left(\frac{1-\omega}{\pi}\right)_3 = \omega^{b+(a+2)/3} = \omega^{2m}.$$

For the first formula of 4.13, we want to show that $m+n \equiv \frac{2(a+2)}{3} \bmod 3$:

$$m+n-\frac{2(a+2)}{3} = \frac{a+3b+2}{6} + b - \frac{2(a+2)}{3} = \frac{-3a+9b-6}{6} = \frac{-a+3b-2}{2}.$$

Since $a \equiv 1 \bmod 3$, the integer $\frac{-a+3b-2}{2}$ is divisible by 3, so that from (a)

$$\left(\frac{\omega}{\pi}\right)_3 = \omega^{2(a+2)/3} = \omega^{m+n}.$$

9.14. Using (9.14), we see that proving $\mathcal{S} \dot{\subset} \mathcal{S}'$ is the same as proving that

$$\{p \text{ prime}: \ p \nmid f, \ p = \mathfrak{p}\overline{\mathfrak{p}}, \ \mathfrak{p} \in [\mathfrak{a}_0\mathcal{O}_K]\} \dot{\subset} \mathcal{S}'.$$

Choose p prime with $p \nmid f$ and $p = \mathfrak{p}\overline{\mathfrak{p}} = N(\mathfrak{p})$ for $\mathfrak{p} \in [\mathfrak{a}_0\mathcal{O}_K] \in I_K(f)/P_{K,\mathbb{Z}}(f) \simeq C(\mathcal{O})$. Under the isomorphism $C(\mathcal{O}) \simeq \mathrm{Gal}(L/K)$, $[\mathfrak{a}_0]$ maps to $\sigma_0 \in \mathrm{Gal}(L/K)$.

Assume (discarding a finite number of primes) that p is unramified in L. Choose a prime ideal $\mathfrak{P} \in \mathcal{O}_L$ lying over $\mathfrak{p}$. Since $N(\mathfrak{p}) = p$, equation (5.20) implies that the Artin symbol $((L/K)/\mathfrak{P})$ satisfies

$$\left(\frac{L/K}{\mathfrak{P}}\right)(\alpha) \equiv \alpha^{N(\mathfrak{p})} \equiv \alpha^p \bmod \mathfrak{P}$$

for all $\alpha \in \mathcal{O}_L$. Note that $((L/K)/\mathfrak{P}) = \sigma_0$ since $\mathfrak{p} \in [\mathfrak{a}_0\mathcal{O}_K]$. In the same way, the Artin symbol $((L/\mathbb{Q})/\mathfrak{P})$ satisfies

$$\left(\frac{L/\mathbb{Q}}{\mathfrak{P}}\right)(\alpha) \equiv \alpha^{N_{\mathbb{Q}/\mathbb{Q}}(p)} \equiv \alpha^p \bmod \mathfrak{P},$$

so that $((L/\mathbb{Q})/\mathfrak{P}) = ((L/K)/\mathfrak{P}) = \sigma_0$, where we regard σ_0 as an element of both $\mathrm{Gal}(L/\mathbb{Q})$ and $\mathrm{Gal}(L/K)$. Replacing $\mathfrak{P}$ with $\sigma(\mathfrak{P})$ for $\sigma \in \mathrm{Gal}(L/\mathbb{Q})$, Corollary 5.21 gives elements of the conjugacy class of σ_0, so that

$$\left(\frac{L/\mathbb{Q}}{p}\right) = \langle \sigma_0 \rangle,$$

proving that $p \in \mathcal{S}'$. Thus $\mathcal{S} \dot{\subset} \mathcal{S}'$.

9.15. This is a general fact about groups of this form:

THEOREM. *Suppose H is Abelian and $G = H \rtimes \mathbb{Z}/2\mathbb{Z}$ where $1 \in \mathbb{Z}/2\mathbb{Z}$ acts on H via $1 \cdot h = h^{-1}$. Then the conjugacy class in G of an element $h \in H$ is $\{h, h^{-1}\}$, where we identify $h \in H$ with its image $(h, 0) \in G$.*

PROOF. Each element $g \in G$ can be written uniquely as (k, μ), $k \in H$, $\mu \in \mathbb{Z}/2\mathbb{Z}$. Recall that multiplication in $H \rtimes \mathbb{Z}/2\mathbb{Z}$ is defined by

$$(h_1, \mu_1) \cdot (h_2, \mu_2) = (h_1(\mu_1 \cdot h_2), \mu_1 + \mu_2).$$

Note that $(k, 0)^{-1} = (k^{-1}, 0)$, and since $(k,1)(k,1) = (k(1 \cdot k), 1+1) = (kk^{-1}, 0) = (1, 0)$, we have $(k,1)^{-1} = (k,1)$. Then if $h = (h, 0) \in H \subset G$ and $k \in H$, since H is Abelian we have

$$(k,0)h(k,0)^{-1} = (k,0)(h,0)(k^{-1},0) = (kh,0)(k^{-1},0) = (khk^{-1},0) = (h,0) = h$$
$$(k,1)h(k,1)^{-1} = (k,1)(h,0)(k,1) = (kh^{-1},1)(k,1) = (kh^{-1}k^{-1}, 1+1) = (h^{-1},0) = h^{-1}.$$

Hence the conjugacy class of h is $\{h, h^{-1}\}$. Q.E.D.

9.16. (a) By (9.14), $\mathcal{S} \doteq \{p \text{ prime} : p \nmid f, p = N(\mathfrak{p}), \mathfrak{p} \text{ prime}, \mathfrak{p} \in [\mathfrak{a}_0\mathcal{O}_K]\}$. We may assume, after discarding the finite number of primes that divide D (the discriminant of ax^2+bx+c), that p does not ramify in K, so that either p is prime in K or $p = \mathfrak{p}\overline{\mathfrak{p}}$. Since $p = N(\mathfrak{p})$, we have the second case, and thus

$$\mathcal{S} \doteq \{p \text{ prime} : p \nmid f, \ p = \mathfrak{p}\overline{\mathfrak{p}}, \ \mathfrak{p} \in [\mathfrak{a}_0\mathcal{O}_K]\}.$$

(b) Suppose that $[\mathfrak{a}_0\mathcal{O}_K]$ maps to $\sigma \in \mathrm{Gal}(L/K)$ under the Artin map. Then since only a finite number of primes ramify in L,

$$\begin{aligned} \mathcal{S}'' &= \{\mathfrak{p} \in \mathcal{P}_K : \mathfrak{p} \in [\mathfrak{a}_0\mathcal{O}_k]\} \\ &\doteq \{\mathfrak{p} \in \mathcal{P}_K : \mathfrak{p} \text{ unramified in } L, \ \mathfrak{p} \in [\mathfrak{a}_0\mathcal{O}_k]\} \\ &= \{\mathfrak{p} \in \mathcal{P}_K : \mathfrak{p} \text{ unramified in } L, \ ((L/K)/\mathfrak{p}) = \langle\sigma\rangle\}. \end{aligned}$$

But L/K is Abelian, so that $|\langle\sigma\rangle| = 1$. By the Čebotarev density theorem we get

$$\delta(\mathcal{S}'') = \frac{|\langle\sigma\rangle|}{[L:K]} = \frac{1}{h(D)}.$$

By (8.16), $\delta(\mathcal{S}'' \cap \mathcal{P}_{K,1}) = \delta(\mathcal{S}'')$, which proves the result.

(c) From part (a),

$$\mathcal{S} = \{p \text{ prime} : p = \mathfrak{p}\overline{\mathfrak{p}},\ \mathfrak{p} \in [\mathfrak{a}_0\mathcal{O}_K]\},$$

where $\mathfrak{a}_0$ is a proper $\mathcal{O}$-ideal and $[\mathfrak{a}_0\mathcal{O}_K] \in C(\mathcal{O}_K)$.

Consider the map $\varphi : \mathcal{S}'' \cap \mathcal{P}_{K,1} \to \mathcal{S}$ defined by

$$\varphi(\mathfrak{p}) = N(\mathfrak{p}),$$

where we restrict the map by removing the finite set of primes of $\mathcal{O}_K$ that lie over ramified rational primes; this will not change the density of $\mathcal{S}'' \cap \mathcal{P}_{K,1}$. To see that the image of this map indeed lies in $\mathcal{S}$, note first that the image of $\mathfrak{p}$ is prime by definition of $\mathcal{P}_{K,1}$. Next, since $\mathfrak{p}$ has prime norm and p is unramified, $p = N(\mathfrak{p})$ must split in K so that $p = \mathfrak{p}\overline{\mathfrak{p}}$. Therefore the map is well-defined, and it is clearly surjective.

We want to study the inverse image of an arbitrary $p \in \mathcal{S}$. The two primes above p, which are $\mathfrak{p}$ and $\overline{\mathfrak{p}}$, are inverses in $C(\mathcal{O}_K)$, and $\mathfrak{p} \in [\mathfrak{a}_0\mathcal{O}_k]$. So if $[\mathfrak{a}_0\mathcal{O}_K]$ has order ≤ 2, it is its own inverse, so that $\overline{\mathfrak{p}} \in [\mathfrak{a}_0\mathcal{O}_K]$ as well and $\varphi^{-1}(p) = \{\mathfrak{p}, \overline{\mathfrak{p}}\}$. If $[\mathfrak{a}_0\mathcal{O}_K]$ has order greater than 2, then $\overline{\mathfrak{p}} \notin [\mathfrak{a}_0\mathcal{O}_K]$ and thus $\overline{\mathfrak{p}} \notin \mathcal{S}'' \cap \mathcal{P}_{K,1}$. It follows that $\varphi^{-1}(p) = \{\mathfrak{p}\}$. This holds for each $p \in \mathcal{S}$, so φ is two-to-one if $[\mathfrak{a}_0\mathcal{O}_K]$ has order at most 2, and one-to-one otherwise. Then using part (b),

$$\delta(\mathcal{S}) = \begin{cases} \frac{1}{2}\delta(\mathcal{S}'' \cap \mathcal{P}_{K,1}) = \frac{1}{2h(D)} & [\mathfrak{a}_0\mathcal{O}_K] \text{ has order} \leq 2 \\ \delta(\mathcal{S}'' \cap \mathcal{P}_{K,1}) = \frac{1}{h(D)} & \text{otherwise.} \end{cases}$$

Finally, the order of $[\mathfrak{a}_0\mathcal{O}_K]$ in the class group $C(\mathcal{O}_K)$ is equal to the order of $[\mathfrak{a}_0]$ in $C(\mathcal{O})$, which is also the order of $ax^2 + bxy + cy^2$ in $C(D)$. This finishes the proof of Theorem 9.12 since a form has order at most two if and only if it is properly equivalent to its opposite.

9.17. (a) Of the $h = h(D)$ reduced forms of discriminant D, suppose that r of them are properly equivalent to their opposites, i.e., have order ≤ 2 in the class group. Being reduced, these r reduced forms all have middle coefficient $b \geq 0$. The $h - r$ remaining forms are not properly equivalent to their opposites, so there are $\frac{h-r}{2}$ pairs of consisting of a reduced form and its opposite. In each pair, the middle coefficients are nonzero and have opposite sign. Picking the ones with $b > 0$, we get an additional $\frac{h-r}{2}$ reduced forms with $b \geq 0$. These, together with the r forms described above, give all reduced forms with $b \geq 0$. By Theorem 9.12, the Dirichlet densities of these $r + \frac{h-r}{2}$ forms add up to

$$r \cdot \frac{1}{2h} + \frac{h-r}{2} \cdot \frac{1}{h} = \frac{h}{2h} = \frac{1}{2}.$$

(b) If p is represented by a reduced form of discriminant D, then it is represented by a reduced form with $b \geq 0$. This is the case since $ax^2 + bxy + cy^2$ and $ax^2 - bxy + cy^2$ are equivalent (though not necessarily properly equivalent). Then applying part (c) of Exercise 2.2 shows that they represent the same numbers. It follows that

$$\mathcal{S} = \{p \text{ prime} : p \nmid D,\ p \text{ represented by some form from part (a)}\}$$

is the same as the set of all primes represented by a reduced form of discriminant D. By Lemma 2.5, if p is a prime not dividing D (thus excluding a finite number of primes), then p is represented by a form of discriminant D if and only if $(D/p) = 1$, so that $p \in \mathcal{S}$ if and only if $(D/p) = 1$. By Exercise 5.22, $(D/p) = ((\mathbb{Q}(\sqrt{D})/\mathbb{Q})/p)$, and then the Čebotarev density theorem shows that

$$\delta(S) = \frac{|\langle 1 \rangle|}{[\mathbb{Q}(\sqrt{D}) : \mathbb{Q}]} = \frac{1}{2}$$

9.18. Suppose as in Lemma 9.3 that L is the ring class field of some order $\mathcal{O}$ in an imaginary quadratic field K. Let M be an intermediate field, so that $\mathbb{Q} \subset K \subset M \subset L$.

Since L is Abelian over K, we know that M is Galois and Abelian over K. Exercise 9.15 shows that the conjugacy class of an element $\sigma \in \mathrm{Gal}(L/M) \subset \mathrm{Gal}(L/K)$ in $\mathrm{Gal}(L/\mathbb{Q})$ is $\{\sigma, \sigma^{-1}\} \subset \mathrm{Gal}(L/M)$, so that $\mathrm{Gal}(L/M)$ is a normal subgroup of $\mathrm{Gal}(L/\mathbb{Q})$. It follows that M is Galois over $\mathbb{Q}$. Finally, the action of the nontrivial element of $\mathbb{Z}/2\mathbb{Z}$ on $\mathrm{Gal}(M/K)$ is induced from its action on $\mathrm{Gal}(L/K)$, which by Lemma 9.3 is given by $\sigma \mapsto \sigma^{-1}$. It follows that M is a generalized dihedral extension of $\mathbb{Q}$, as desired.

9.19. (a) Since $\mathcal{O}_1$ has finite index in $\mathcal{O}_K$, the inclusions $\mathcal{O}_1 \subset \mathcal{O}_2 \subset \mathcal{O}_K$ imply that

$$[\mathcal{O}_K : \mathcal{O}_2][\mathcal{O}_2 : \mathcal{O}_1] = [\mathcal{O}_K : \mathcal{O}_1].$$

If $f_i = [\mathcal{O}_K : \mathcal{O}_i]$ is the conductor of $\mathcal{O}_i$, then the above equation implies that $f_2 \mid f_1$. The ring class field L_i is the Abelian extension of K determined by the congruence subgroup $\ker(\Phi_{L_i/K, f_i\mathcal{O}_K}) = P_{K,\mathbb{Z}}(f_i)$ for the modulus $f_i\mathcal{O}_K$.

Since $f_2\mathcal{O}_K \mid f_1\mathcal{O}_K$, Exercise 8.4 implies that L_2 is a generalized ideal class group for $f_1\mathcal{O}_K$ with congruence subgroup

$$\ker(\Phi_{L_2/K, f_1\mathcal{O}_K}) = \ker(\Phi_{L_2/K, f_2\mathcal{O}_K}) \cap I_K(f_1\mathcal{O}_K) = P_{K,\mathbb{Z}}(f_2) \cap I_K(f_1\mathcal{O}_K).$$

By Corollary 8.7, the desired inclusion $L_2 \subset L_1$ will follow once we prove that

$$(*) \qquad P_{K,\mathbb{Z}}(f_1) \subset P_{K,\mathbb{Z}}(f_2) \cap I_K(f_1\mathcal{O}_K).$$

To prove this, take a generator $\alpha\mathcal{O}_K$ of $P_{K,\mathbb{Z}}(f_1)$, so that $\alpha \equiv a \bmod f_1\mathcal{O}_K$ for some $a \in \mathbb{Z}$ with $\gcd(a, f_1) = 1$. Then $\alpha\mathcal{O}_K$ is relatively prime to $f_1\mathcal{O}_K$, so that $\alpha\mathcal{O}_K \in I_K(f_1\mathcal{O}_K)$. Furthermore, $f_2 \mid f_1$ implies $\alpha \equiv a \bmod f_2\mathcal{O}_K$ and $\gcd(a, f_2) = 1$, so that $\alpha\mathcal{O}_K \in P_{K,\mathbb{Z}}(f_2)$, and (*) follows.

(b) Let f_i be the conductor of $\mathcal{O}_i$. In part (a), we proved that $\mathcal{O}_1 \subset \mathcal{O}_2$ implies $f_2 \mid f_1$. For the converse, note that $\mathcal{O}_i = [1, f_i w_K]$ by Lemma 7.2. Then $f_2 \mid f_1$ implies that $f_1 w_k \in [1, f_2 w_k] = \mathcal{O}_2$, and $\mathcal{O}_1 \subset \mathcal{O}_2$ follows easily. This proves the desired equivalence. The final assertion, that $f_2 \mid f_1 \Rightarrow L_2 \subset L_1$, was proved in part (a).

(c) Let $\mathcal{O}$ be the order of conductor f in $\mathcal{O}_K$. Since $\mathcal{O} \subset \mathcal{O}_K$, part (a) implies that the ring class field L_2 of $\mathcal{O}_K$, which is the Hilbert class field of K by Exercise 9.1, is contained in the ring class field L_1 of $\mathcal{O}$. Since the discriminant of $\mathcal{O}$ is $f^2 d_K$, $L_2 \subset L_1$ implies that

$$h(d_K) = [L_2 : K] \bigm| [L_1 : K] = h(f^2 d_K).$$

9.20. Write $\mathcal{O}_f$ and $\mathcal{O}_{f'}$ for the orders of conductor f and f' respectively in $\mathcal{O}_K$. Then by part (b) of Exercise 9.19, $f' \mid f$ implies that $L' \subset L$.

Since 2 splits completely in K, we know that K is neither $\mathbb{Q}(\sqrt{-3})$ nor $\mathbb{Q}(i)$, so that $\mathcal{O}_K^* = \mathcal{O}_f^* = \mathcal{O}_{f'}^* = \{\pm 1\}$. Additionally, we have $(d_K/2) = 1$ by Corollary 5.17. Using the class number formula of Theorem 7.24 together with $f = 2f'$ and f' odd, we obtain

$$\begin{aligned} h(\mathcal{O}_f) &= \frac{h(\mathcal{O}_K)2f'}{[\mathcal{O}_K^* : \mathcal{O}_f^*]} \prod_{p \mid f} \left(1 - \left(\frac{d_K}{p}\right)\frac{1}{p}\right) \\ &= h(\mathcal{O}_K)2f'\left(1 - \left(\frac{d_K}{2}\right)\frac{1}{2}\right) \prod_{p \mid f'} \left(1 - \left(\frac{d_K}{p}\right)\frac{1}{p}\right) \\ &= \frac{h(\mathcal{O}_K)f'}{[\mathcal{O}_K^* : \mathcal{O}_{f'}^*]} \prod_{p \mid f'} \left(1 - \left(\frac{d_K}{p}\right)\frac{1}{p}\right) \\ &= h(\mathcal{O}_{f'}), \end{aligned}$$

where the third equality uses $(d_K/2) = 1$. Since $[L : K] = h(\mathcal{O}_f) = h(\mathcal{O}_{f'}) = [L' : K]$ and $L' \subset L$, we get $L' = L$. Thus the two ring class fields coincide and therefore

$$\mathfrak{f}(L/K) = \mathfrak{f}(L'/K).$$

9.21. (a) Since $\ker(\Phi_{L/K, f\mathcal{O}_K}) = \ker(\Phi_{f\mathcal{O}_K})$ is a congruence subgroup for the modulus $f\mathcal{O}_K$, we have $\mathfrak{f}(L/K) \mid f\mathcal{O}_K$ by Theorem 8.5. Since $\mathfrak{f}(L/K) \neq f\mathcal{O}_K$, there is at least one prime ideal $\mathfrak{p}$ whose exponent in the factorization of $f\mathcal{O}_K$ exceeds its exponent in the factorization of $\mathfrak{f}(L/K)$. Let $\mathfrak{m} = \mathfrak{p}^{-1} f\mathcal{O}_K$ be the product of the remaining factors of $f\mathcal{O}_K$; then indeed $f\mathcal{O}_K = \mathfrak{p}\mathfrak{m}$ and $\mathfrak{f}(L/K) \mid \mathfrak{m}$.

(b) Since $\mathfrak{f}(L/K) \mid \mathfrak{m} \mid f\mathcal{O}_K$, Theorem 8.5 implies that $\ker(\Phi_{\mathfrak{m}})$ is a congruence subgroup for $\mathfrak{m}$, i.e., $P_{K,1}(\mathfrak{m}) \subset \ker(\Phi_{\mathfrak{m}})$. However, since $\mathfrak{m} \mid f\mathcal{O}_K$ we have $\ker(\Phi_{f\mathcal{O}_K}) = \ker(\Phi_{\mathfrak{m}}) \cap I_K(f\mathcal{O}_K)$. Thus

$$P_{K,1}(\mathfrak{m}) \cap I_K(f\mathcal{O}_K) \subset \ker(\Phi_{\mathfrak{m}}) \cap I_K(f\mathcal{O}_K) = \ker(\Phi_{f\mathcal{O}_K}) = P_{K,\mathbb{Z}}(f),$$

where the last equality follows from the definition of the ring class field L of the order of conductor f in $\mathcal{O}_K$.

(c) Take $[\alpha] \in \ker(\pi)$. Then $\alpha \in \mathcal{O}_K$ is relatively prime to f and satisfies $\alpha \equiv 1 \bmod \mathfrak{m}$. It follows that $\alpha\mathcal{O}_K \in I_K(f) \cap P_{K,1}(\mathfrak{m})$, which by part (b) implies that $\alpha\mathcal{O}_K \in P_{K,\mathbb{Z}}(f)$. Thus

$$(*) \qquad \alpha\mathcal{O}_K = (\gamma\mathcal{O}_K)(\delta\mathcal{O}_K)^{-1}$$

where $\gamma, \delta \in \mathcal{O}_K$ satisfy $\gamma \equiv c \bmod f\mathcal{O}_K$ and $\delta \equiv d \bmod f\mathcal{O}_K$ for integers c and d satisfying $\gcd(c, f) = \gcd(d, f) = 1$. Pick $r \in \mathbb{Z}$ such that $rd \equiv 1 \bmod f$. Then $r\gamma \equiv rc \bmod f\mathcal{O}_K$ and $r\delta \equiv rd \equiv 1 \bmod f\mathcal{O}_K$. So in (*), we can replace γ, δ with $r\gamma, r\delta$. This allows us to assume that $\delta \equiv 1 \bmod f\mathcal{O}_K$ in (*).

From (*), we obtain $\alpha\delta\mathcal{O}_K = \gamma\mathcal{O}_K$, so that $\alpha\delta = u\gamma$ for some unit $u \in \mathcal{O}_K^*$. Then in $(\mathcal{O}_K/f\mathcal{O}_K)^*$, we have

$$[\alpha] = [\alpha\delta] = [u\gamma] = [u][\gamma] = [u][c] = [u]\beta([c]) \in \mathcal{O}_K^* \cdot \operatorname{im}(\beta),$$

where the first equality uses $\delta \equiv 1 \bmod f\mathcal{O}_K$ and the fourth uses $\gamma \equiv c \bmod f\mathcal{O}_K$. The final equality follows since $c \in \mathbb{Z}$ is relatively prime to f.

9.22. (a) With $f\mathcal{O}_K = \mathfrak{p}\mathfrak{m}$, let $\mathfrak{p}_i$, $i = 1, \ldots, k$ be the prime factors of $\mathfrak{m}$ other than (possibly) $\mathfrak{p}$. Then Exercise 7.29 implies

$$|(\mathcal{O}_K/\mathfrak{m})^*| = \begin{cases} N(\mathfrak{m})\left(1 - \dfrac{1}{N(\mathfrak{p})}\right)\displaystyle\prod_{i=1}^{k}\left(1 - \dfrac{1}{N(\mathfrak{p}_i)}\right), & \mathfrak{p} \mid \mathfrak{m} \\ N(\mathfrak{m})\displaystyle\prod_{i=1}^{k}\left(1 - \dfrac{1}{N(\mathfrak{p}_i)}\right), & \mathfrak{p} \nmid \mathfrak{m}. \end{cases}$$

and

$$
\begin{aligned}
|(\mathcal{O}_K/f\mathcal{O}_K)^*| &= N(f\mathcal{O}_K)\left(1-\frac{1}{N(\mathfrak{p})}\right)\prod_{i=1}^{k}\left(1-\frac{1}{N(\mathfrak{p}_i)}\right)\\
&= N(\mathfrak{p})N(\mathfrak{m})\left(1-\frac{1}{N(\mathfrak{p})}\right)\prod_{i=1}^{k}\left(1-\frac{1}{N(\mathfrak{p}_i)}\right).
\end{aligned}
$$

Dividing the second equation by the first and noting that π is surjective, we get

$$
|\ker(\pi)| = \frac{|(\mathcal{O}_K/f\mathcal{O}_K)^*|}{|(\mathcal{O}_K/\mathfrak{m})^*|} = \begin{cases} N(\mathfrak{p}), & \mathfrak{p} \mid \mathfrak{m} \\ N(\mathfrak{p})\left(1-\frac{1}{N(\mathfrak{p})}\right) = N(\mathfrak{p})-1, & \mathfrak{p} \nmid \mathfrak{m}. \end{cases}
$$

(b) (i) Since $N(\mathfrak{p}) = p$, either p splits in K, in which case $p\mathcal{O}_K = \mathfrak{p}\overline{\mathfrak{p}}$, or it is ramified. If $\mathfrak{p}$ is ramified, then $\mathfrak{p}\overline{\mathfrak{p}} = N(\mathfrak{p})\mathcal{O}_K = p\mathcal{O}_K = \mathfrak{p}^2$, so that $\mathfrak{p} = \overline{\mathfrak{p}}$. In either case $p\mathcal{O}_K = \mathfrak{p}\overline{\mathfrak{p}}$. Now $f\mathcal{O}_K = \mathfrak{p}\mathfrak{m}$ gives after taking norms

$$
pN(\mathfrak{m}) = N(\mathfrak{p})N(\mathfrak{m}) = N(\mathfrak{p}\mathfrak{m}) = N(f\mathcal{O}_K) = f^2,
$$

where the second equality follows from Lemma 7.14. Thus $p \mid f$, say $f = pm$. Then $m\mathfrak{p}\overline{\mathfrak{p}} = m(p\mathcal{O}_K) = f\mathcal{O}_K = \mathfrak{p}\mathfrak{m}$, so that $\mathfrak{m} = m\overline{\mathfrak{p}}$.

(ii) An element of $\ker(\pi) \cap \operatorname{im}(\beta)$ is of the form $[a]$ for $a \in \mathbb{Z}$ relatively prime to f with $a \equiv 1 \bmod \mathfrak{m}$. By (i), this means that $a-1 \in \mathfrak{m} = m\overline{\mathfrak{p}}$, so that $(a-1)/m \in \overline{\mathfrak{p}} \subset \mathcal{O}_K$. It follows that $(a-1)/m \in \mathbb{Q}$ is an algebraic integer and is therefore an integer. Thus $(a-1)/m = k \in \mathbb{Z}$, and then $a-1 = km \in m\overline{\mathfrak{p}}$ implies that $k \in \overline{\mathfrak{p}}$. Since k is an integer, we see that $p \mid k$, so that $f = pm \mid km = a-1$. Finally, $f \mid a-1$ shows that $[a] = [1] \in (\mathcal{O}_K/f\mathcal{O}_K)^*$, so that $\ker(\pi) \cap \operatorname{im}(\beta) = \{[1]\}$.

(iii) Since $\ker(\pi) \cap \operatorname{im}(\beta) = \{[1]\}$, the inclusion $\ker(\pi) \subset \operatorname{im}(\beta)$ implies $\ker(\pi) = \{[1]\}$. Since $\mathfrak{p}$ is prime, part (a) implies that $\mathfrak{p} \nmid \mathfrak{m}$ and $N(\mathfrak{p}) = 2$, so that $p = 2$, and then by (i), $f = 2m$. Further, since $\mathfrak{m} = m\overline{\mathfrak{p}}$ and $\mathfrak{p} \nmid \mathfrak{m}$, we see that $\mathfrak{p} \neq \overline{\mathfrak{p}}$ and $\mathfrak{p} \nmid m\mathcal{O}_K$. The first of these shows that 2 splits completely in K, and the second shows that m is odd, contradicting the assumption on f and completing the proof when $N(\mathfrak{p}) = p$.

(c) When $N(\mathfrak{p}) = p^2$, p remains prime in K, so that $\mathfrak{p} = p\mathcal{O}_K$ and $f\mathcal{O}_K = p\mathfrak{m}$. Therefore $f = pm$ for some integer m. It follows that $\mathfrak{m} = m\mathcal{O}_K$.

(i) In the commutative diagram

$$
\begin{array}{ccc}
(\mathbb{Z}/f\mathbb{Z})^* & \xrightarrow{\beta} & (\mathcal{O}_K/f\mathcal{O}_K)^* \\
{\scriptstyle\theta}\downarrow & & \downarrow{\scriptstyle\pi} \\
(\mathbb{Z}/m\mathbb{Z})^* & \longrightarrow & (\mathcal{O}_K/m\mathcal{O}_K)^*
\end{array}
$$

the bottom horizontal map is clearly injective, so that $\ker(\theta) = \ker(\pi \circ \beta)$. However, the latter maps isomorphically via β to $\ker(\pi) \cap \operatorname{im}(\beta)$ since β is injective, so we get the desired isomorphism $\ker(\pi) \cap \operatorname{im}(\beta) \simeq \ker(\theta)$.

(ii) The inclusion $\ker(\pi) \subset \operatorname{im}(\beta)$ implies $\ker(\pi) \cap \operatorname{im}(\beta) = \ker(\pi) \simeq \ker(\theta)$. We have

$$
|\ker(\theta)| = \frac{\phi(f)}{\phi(m)} = \begin{cases} p, & p \mid m \\ p-1, & p \nmid m. \end{cases}
$$

But part (a) together with $\ker(\pi) \simeq \ker(\theta)$ and $N(\mathfrak{p}) = p^2$ then implies that $p^2 = p$ or $p^2 - 1 = p - 1$, both of which are impossible.

The contradictions in parts (b) and (c) show that $\mathfrak{f}(L/K) = f\mathcal{O}_K$ in this situation.

9.23. (a) If $\mathcal{O}_K^* = \{\pm 1\}$, then $K \neq \mathbb{Q}(i)$ and $K \neq \mathbb{Q}(\sqrt{-3})$, so we are in the last two cases of the formula from Exercise 9.20. Exercise 9.22 shows that $\mathfrak{f}(L/K) = f\mathcal{O}_K$ except possibly in the case where $f = 2f'$, f' odd, and 2 splits completely in K. Exercise 9.20 shows that in this event, $\mathfrak{f}(L/K) = \mathfrak{f}(L'/K)$, where L' is the ring class field of f'. Since f' is odd, using Exercise 9.22 again shows that

$$\mathfrak{f}(L/K) = \mathfrak{f}(L'/K) = f'\mathcal{O}_K = (f/2)\mathcal{O}_K.$$

(b) For the orders $\mathcal{O}$ of conductor $f = 2$ or $f = 3$ in $\mathbb{Q}(\sqrt{-3})$, and $f = 2$ in $\mathbb{Q}(i)$, the class number formula of Theorem 7.24 gives $h(\mathcal{O}) = 1$; thus the ring class field is $L = K$ and the conductor is $\mathfrak{f}(L/K) = \mathcal{O}_K$. So we now assume that $\mathcal{O}$ is an order of conductor f other than those above in one of these two fields, L is the ring class field for $\mathcal{O}$, and $\mathfrak{f}(L/K) \neq f\mathcal{O}_K$. The goal is to derive a contradiction. Note that since 2 does not split completely in either $\mathbb{Q}(\sqrt{-3})$ or in $\mathbb{Q}(i)$, the third case of the formula in Exercise 9.20 is excluded.

Note that except for parts (b)(iii) and (c)(ii) of Exercise 9.22, neither Exercises 9.21 nor 9.22 depend on the size of $\mathcal{O}_K^*$ and thus still hold for $\mathbb{Q}(i)$ and $\mathbb{Q}(\omega)$; we will make use of those results. As in Exercise 9.21, we write $f\mathcal{O}_K = \mathfrak{p}\mathfrak{m}$ with $\mathfrak{p}$ prime and $\mathfrak{f}(L/K) \mid \mathfrak{m}$. Then $\ker(\pi) \subset \mathcal{O}_K^* \cdot \operatorname{im}(\beta)$ by part (c) of Exercise 9.21. Furthermore, if $\ker(\pi) \subset \operatorname{im}(\beta)$, then the arguments of parts (b) and (c) of Exercise 9.22 lead to a contradiction. Hence we may assume that $\ker(\pi) \not\subset \operatorname{im}(\beta)$.

Let p be the integer prime contained in $\mathfrak{p}$. Then $N(\mathfrak{p}) = p$ or p^2, so by Lemma 7.14,

$$p \mid N(\mathfrak{p})N(\mathfrak{m}) = N(\mathfrak{p}\mathfrak{m}) = N(f\mathcal{O}_K) = f^2.$$

Thus $p \mid f$. Write $f = pm$ for some integer m, and note that m is positive since f and p are.

Let $[\alpha]$ be an element of $\ker(\pi)$ not contained in $\operatorname{im}(\beta)$. Since $f\mathcal{O}_K = \mathfrak{p}\mathfrak{m}$ implies $m\mathcal{O}_K \mid \mathfrak{m}$, we have $\alpha \equiv 1 \bmod m\mathcal{O}_K$. Also, $[\alpha] \in \ker(\pi) \subset \mathcal{O}_K^* \cdot \operatorname{im}(\beta)$, so there is $c \in \mathbb{Z}$ relatively prime to f and a unit $u \in \mathcal{O}_K^*$ such that $[\alpha] = [uc]$ in $(\mathcal{O}_K/f\mathcal{O}_K)^*$. Note that $[\alpha] \notin \operatorname{im}(\beta)$ implies $u \neq \pm 1$. Thus

$$\alpha \equiv uc \bmod f\mathcal{O}_K.$$

where $u = \pm i$ when $K = \mathbb{Q}(i)$ or $u = \pm\omega$ or $\pm\omega^2$ when $K = \mathbb{Q}(\omega)$. All of these give $\mathcal{O}_K = \mathbb{Z}[u]$. Since $\alpha \equiv 1 \bmod m\mathcal{O}_K$, the above congruence implies $1 \equiv uc \bmod m\mathcal{O}_K$, so that

$$1 = uc + (a + bu)m = am + (c + bm)u, \quad a, b \in \mathbb{Z}.$$

Thus $1 = am$; since m is positive we get $m = 1$ and therefore $f = p$.

To study the case when $f = p$, we begin with the following exact sequence built from the inclusion $\ker(\pi) \subset \mathcal{O}_K^* \cdot \operatorname{im}(\beta)$:

$$1 \longrightarrow \ker(\pi) \cap \operatorname{im}(\beta) \longrightarrow \ker(\pi) \longrightarrow (\mathcal{O}_K^* \cdot \operatorname{im}(\beta))/\operatorname{im}(\beta).$$

Since $(\mathcal{O}_K^* \cdot \operatorname{im}(\beta))/\operatorname{im}(\beta) \simeq \mathcal{O}_K^*/(\mathcal{O}_K^* \cap \operatorname{im}(\beta))$, we get an exact sequence

$$(*) \qquad 1 \longrightarrow \ker(\pi) \cap \operatorname{im}(\beta) \longrightarrow \ker(\pi) \longrightarrow \mathcal{O}_K^*/(\mathcal{O}_K^* \cap \operatorname{im}(\beta)).$$

Let $A \subset \mathcal{O}_K^*/(\mathcal{O}_K^* \cap \operatorname{im}(\beta))$ be the image of the last map of (*), and note that $[\pm 1] \in \operatorname{im}(\beta)$. It follows that if $K = \mathbb{Q}(i)$, where $\mathcal{O}_K^* = \{\pm 1, \pm i\}$, then $|A| = 1$ or 2. Similarly, if $K = \mathbb{Q}(\omega)$, where $\mathcal{O}_K^* = \{\pm 1, \pm\omega, \pm\omega^2\}$, then $|A| = 1$ or 3. Furthermore, $\ker(\pi) \not\subset \operatorname{im}(\beta)$ implies $|A| > 1$. There are now two cases to consider.

Case 1: $N(\mathfrak{p}) = p$. Part (b)(ii) of Exercise 9.22 shows that $\ker(\pi) \cap \operatorname{im}(\beta) = \{[1]\}$. The exact sequence (*) then implies that $\ker(\pi) \simeq A$, so that $|\ker(\pi)| \leq 3$. Since $p - 1 = N(\mathfrak{p}) - 1 \leq |\ker(\pi)|$ by part (a) of Exercise 9.22, we have $p - 1 \leq 3$, so that $f = p = 2$ or 3. But $f = 2$ is excluded for both $\mathbb{Q}(i)$ and $\mathbb{Q}(\omega)$, and $f = 3$ is excluded for $\mathbb{Q}(\omega)$. Finally, $f = p = 3$ cannot occur for $\mathbb{Q}(i)$ since $\mathbb{Q}(i)$ has no ideals of norm 3 (i.e., $3 = x^2 + y^2 = N((x+iy)\mathcal{O}_K)$ has no solutions).

Case 2: $N(\mathfrak{p}) = p^2$. Then $\mathfrak{p} = p\mathcal{O}_K$, so p is prime in $\mathcal{O}_K$. By part (a) of Exercise 9.22,

$$|\ker(\pi)| = N(\mathfrak{p}) - 1 = p^2 - 1,$$

and by the formula for $|\ker(\theta)|$ from the solution to part (c) of Exercise 9.22,

$$|\ker(\theta)| = p - 1$$

since $m = 1$. When we combine these with $\ker(\pi) \cap \operatorname{im}(\beta) \simeq \ker(\theta)$ from part (c)(i) of Exercise 9.22, the exact sequence (*) implies that

$$|A| = \frac{|\ker(\pi)|}{|\ker(\theta)|} = \frac{p^2-1}{p-1} = p+1.$$

Recall that $|A| > 1$. Then:

- If $K = \mathbb{Q}(i)$, then $|A| > 1$ implies $|A| = 2$, so that $2 = p+1$, which is impossible.
- If $K = \mathbb{Q}(\omega)$, then $|A| > 1$ implies $|A| = 3$, so that $3 = p+1$, which implies $p = 2$, so that $f = 2$, which is an excluded case.

9.24. For the first part, choose $D \equiv 1 \bmod 8$, $D < 0$ and squarefree, and let $K = \mathbb{Q}(\sqrt{D})$. Then 2 splits completely in K since $(d_K/2) = (D/2) = 1$. So for any $f \equiv 2 \bmod 4$, applying the formula from Exercise 9.20 to the ring class field L corresponding to the order of conductor f in K gives $\mathfrak{f}(L/K) = (f/2)\mathcal{O}_K$, which is not equal to $f\mathcal{O}_K$. This gives infinitely many examples.

For the second part, let K and f be as above. Write $f = 2f'$ where f' is odd, and let L' be the ring class field corresponding to the order of conductor f'. Then Exercise 9.20 shows that $L' = L$. If the converse of part (b) of Exercise 9.19 held, then $L_2 \subset L_1$ would imply $f_2 \mid f_1$. When applied to $L \subset L'$, this would imply $f \mid f'$, which is impossible since $f = 2f'$.

Solutions to Exercises in §10

10.1. (a) For (x, y) satisfying $x^2 + y^2 = 1$, $(x, y) \mapsto |x\omega_1 + y\omega_2|$ is a continuous function on a compact set, so it achieves its minimum at some point (x_0, y_0) on the unit circle. Thus $M = |x_0\omega_1 + y_0\omega_2| > 0$ since $x_0\omega_1 + y_0\omega_2 \neq 0$ by the linear independence of ω_1, ω_2. Now take $x, y \in \mathbb{R}$. Since $|x\omega_1 + y\omega_2| \geq M\sqrt{x^2+y^2}$ holds when $(x, y) = (0, 0)$, we may assume $(x, y) \neq (0, 0)$. Then

$$|x\omega_1 + y\omega_2| = \sqrt{x^2+y^2}\left|\frac{x}{\sqrt{x^2+y^2}}\omega_1 + \frac{y}{\sqrt{x^2+y^2}}\omega_2\right| \geq M\sqrt{x^2+y^2},$$

where the final inequality follows from the definition of M since $\left(\frac{x}{\sqrt{x^2+y^2}}, \frac{y}{\sqrt{x^2+y^2}}\right)$ lies on the unit circle.

(b) Convert to polar coordinates, so that $r = \sqrt{x^2+y^2}$, and use a in place of r in the exponent to avoid confusion. Then

$$\iint_{x^2+y^2 \geq 1} (x^2+y^2)^{-a/2}\,dx\,dy = \int_0^{2\pi}\int_1^{\infty} r^{-a} \cdot r\,dr\,d\theta = \frac{2\pi}{2-a} r^{2-a}\Big|_1^{\infty} = \frac{2\pi}{a-2}$$

since $a > 2$.

(c) We consider the points $(m,n) \in \mathbb{Z}^2 - \{(0,0)\}$ lying along one of the axes, the points $(m,n) = (\pm 1, \pm 1)$, and the remaining points separately. For the first set of points, the sum is

$$2\sum_{m=1}^{\infty} |m|^{-r} + 2\sum_{n=1}^{\infty} |n|^{-r},$$

which converges since $r > 2$. The second set of points is finite, so the associated sum is finite. Now consider the remaining points (m,n) in the first quadrant, which are described by $m \geq 1, n \geq 1$ with m or $n > 1$. Since $(x^2+y^2)^{-r/2}$ is decreasing in $|x|$ and $|y|$, it follows that the minimum value of $(x^2+y^2)^{-r/2}$ on the square $[m-1,m] \times [n-1,n]$ occurs at (m,n), so that

$$\sum_{\substack{m\geq 1, n\geq 1\\ m \text{ or } n>1}} (m^2+n^2)^{-r/2} \leq \sum_{\substack{m\geq 1, n\geq 1\\ m \text{ or } n>1}} \int_{m-1}^{m}\int_{n-1}^{n} (x^2+y^2)^{-r/2}\, dx\, dy.$$

Since each integrand is positive, we can exchange the summation and integration signs, giving

$$\sum_{\substack{m\geq 1, n\geq 1\\ m \text{ or } n>1}} (m^2+n^2)^{-r/2} \leq \iint_S (x^2+y^2)^{-r/2}\, dx\, dy \leq \iint_{\substack{x^2+y^2\geq 1\\ x,y\geq 0}} (x^2+y^2)^{-r/2}\, dx\, dy$$

where S is the region in the first quadrant excluding the unit square $[0,1] \times [0,1]$. Since the integral on the right converges by part (b), the sum is finite. The remaining points in the other quadrants may be treated similarly. Thus all three sums are finite; since all consist of positive terms, we see that the original series converges as well.

10.2. (a) Take $z \notin L$. Then $\wp'(z)$ is defined, and $\omega - z \neq 0$ for $\omega \in L$. Note that there are only a finite number of $\omega \in L$ with $|\omega| < 2|z|$, so it suffices to show that $-2\sum_{\omega\in L}(z-\omega)^{-3}$ converges for $|\omega| \geq 2|z|$. But in this case

$$|z-\omega| \geq |\omega| - |z| \geq |\omega| - \frac{1}{2}|\omega| = \frac{1}{2}|\omega|,$$

so that

$$\sum_{\substack{\omega\in L\\ |\omega|\geq 2|z|}} \left|\frac{1}{(z-\omega)^3}\right| \leq \sum_{\substack{\omega\in L\\ |\omega|\geq 2|z|}} \frac{8}{|\omega|^3}.$$

The sum on the right converges by Lemma 10.2, so the sum on the left converges by the comparison test.

(b) The series for $\wp'(z)$ gives

$$\wp'(z+\omega) = -2\sum_{\tau\in L} \frac{1}{((z+\omega)-\tau))^3} = -2\sum_{\tau\in L}\frac{1}{(z-(\tau-\omega))^3}$$
$$= -2\sum_{\tau\in L}\frac{1}{(z-\tau)^3} = \wp'(z),$$

where the first equality on the second line follows from the fact that $\omega \in L$, so that as τ ranges over L, so does $\tau - \omega$. Note that we can rearrange the terms in the sum since the series converges absolutely by part (a).

10.3. The power series $(1-x)^{-1} = \sum_{n=0}^{\infty} x^n = 1 + \sum_{n=1}^{\infty} x^n$ converges absolutely for $|x| < 1$, so taking the derivative of both sides gives

$$(1-x)^{-2} = \frac{d}{dx}(1-x)^{-1} = \sum_{n=1}^{\infty} nx^{n-1} = \sum_{n=0}^{\infty}(n+1)x^n.$$

Finally, the derivative of a power series converges on the interior of the region of convergence of the original series, so this expansion is also valid for $|x| < 1$.

10.4. By Lemma 10.3, the first few terms of the expansion of $\wp(z)$ are

$$\wp(z) = \frac{1}{z^2} + 3G_4(L)z^2 + 5G_6(L)z^4 + \cdots = \frac{1}{z^2} + z^2\big(3G_4(L) + 5G_6(L)z^2 + \cdots\big).$$

Cubing each side gives

$$\begin{aligned}\wp(z)^3 &= \frac{1}{z^6} + 3\cdot\frac{1}{z^4}\cdot z^2\big(3G_4(L)+5G_6(L)z^2+\cdots\big) + 3\cdot\frac{1}{z^2}\cdot z^4\big(3G_4(L)+5G_6(L)z^2+\cdots\big)^2 + \cdots\\ &= \frac{1}{z^6} + \frac{9G_4(L)}{z^2} + 15G_6(L) + \cdots,\end{aligned}$$

where again the omitted terms all have positive degree in z. This proves the first formula.

For the second formula, the series for $\wp'(z)$ given in the text begins with

$$\wp'(z) = \frac{-2}{z^3} + 6G_4(L)z + 20G_6(L)z^3 + \cdots = \frac{-2}{z^3} + z\big(6G_4(L) + 20G_6(L)z^2 + \cdots\big).$$

Squaring each side gives

$$\begin{aligned}\wp'(z)^2 &= \frac{4}{z^6} + 2\cdot\frac{-2}{z^3}\cdot z\big(6G_4(L) + 20G_6(L)z^2 + \cdots\big) + \cdots\\ &= \frac{1}{z^6} - \frac{24G_4(L)}{z^2} - 80G_6(L) + \cdots,\end{aligned}$$

where again the omitted terms in the second line all have positive degree in z.

10.5. Since $f(z)$ is holomorphic, $|f(z)|$ is continuous and hence bounded on the compact region $\mathbf{P} = \{s\omega_1 + t\omega_2 : 0 \le s,t \le 1\}$, say by M. Then given any $z \in \mathbb{C}$, we can use Exercise 10.6 to find $z' \in \mathbf{P}$ with $z \equiv z' \bmod L$. It follows that $|f(z)| = |f(z')| \le M$ since $f(z)$ is periodic modulo L. It follows that $f(z)$ is a bounded entire function and is therefore constant by Liouville's Theorem.

10.6. Finding $0 \le s,t \le 1$ such that $z \equiv z' = \alpha + s\omega_1 + t\omega_2 \bmod L$ is the same as finding s,t such that $z - \alpha \equiv s\omega_1 + t\omega_2 \bmod L$. Since $\{\omega_1,\omega_2\}$ is an $\mathbb{R}$-basis for $\mathbb{C}$, we can write $z - \alpha = a\omega_1 + b\omega_2$, $a,b \in \mathbb{R}$. Let s and t be the fractional parts of a and b respectively, so that $a\omega_1 + b\omega_2 = (m+s)\omega_1 + (n+t)\omega_2$ for $m,n \in \mathbb{Z}$, $0 \le s,t < 1$. Thus

$$z - \alpha = a\omega_1 + b\omega_2 = (m\omega_1 + n\omega_2) + (s\omega_1 + t\omega_2) \equiv s\omega_1 + t\omega_2 \bmod L.$$

10.7. (a) As in the proof of Theorem 10.1, the power series expansion of $\wp(z)$ about $z = -w$ is

$$\wp(z) = \wp(w) - \wp'(w)(z+w) + \tfrac{1}{2}\wp''(w)(z+w)^2 + \cdots,$$

where here and below, $\cdots$ refers to higher powers of $z+w$. Differentiating gives

$$\wp'(z) = -\wp'(w) + \wp''(w)(z+w) + \cdots,$$

so that

$$\begin{aligned}\wp(z) - \wp(w) &= -\wp'(w)(z+w) + \tfrac{1}{2}\wp''(w)(z+w)^2 + \cdots\\ \wp'(z) - \wp'(w) &= -2\wp'(w) + \wp''(w)(z+w) + \cdots.\end{aligned}$$

We are assuming $2w \notin L$, so that Corollary 10.5 assures us that $\wp'(w) \neq 0$, and we can write

$$\begin{aligned}\frac{1}{4}\left(\frac{\wp'(z) - \wp'(w)}{\wp(z) - \wp(w)}\right)^2 &= \frac{1}{4}\left(\frac{-2\wp'(w) + \wp''(w)(z+w) + \cdots}{-\wp'(w)(z+w) + \frac{1}{2}\wp''(w)(z+w)^2 + \cdots}\right)^2\\ &= \frac{1}{4}\left(\frac{2}{z+w}\cdot\frac{1 - \frac{1}{2}(\wp''(w)/\wp'(w))(z+w) + \cdots}{1 - \frac{1}{2}(\wp''(w)/\wp'(w))(z+w) + \cdots}\right)^2.\end{aligned}$$

However, there is a constant $A \in \mathbb{C}$ such that

$$\frac{1 - \frac{1}{2}(\wp''(w)/\wp'(w))(z+w) + \cdots}{1 - \frac{1}{2}(\wp''(w)/\wp'(w))(z+w) + \cdots} = 1 + A(z+w)^2 + \cdots$$

because the numerator and denominator have the same constant and linear terms. Thus

$$\frac{1}{4}\left(\frac{\wp'(z) - \wp'(w)}{\wp(z) - \wp(w)}\right)^2 = \frac{1}{(z+w)^2}\left(1 + A(z+w)^2 + \cdots\right)^2 = \frac{1}{(z+w)^2} + g(z+w),$$

where $g(z+w)$ is holomorphic at $z = -w$. Also, by Lemma 10.3,

$$\wp(z+w) = \frac{1}{(z+w)^2} + h(z+w).$$

where $h(z+w)$ is holomorphic at $z = -w$. Then

$$\begin{aligned} G(z) &= \wp(z+w) + \wp(z) + \wp(w) - \frac{1}{4}\left(\frac{\wp'(z) - \wp'(w)}{\wp(z) - \wp(w)}\right)^2 \\ &= \frac{1}{(z+w)^2} + h(z+w) + \wp(z) + \wp(w) - \left(\frac{1}{(z+w)^2} + g(z+w)\right) \\ &= h(z+w) + \wp(z) + \wp(w) - g(z+w). \end{aligned}$$

The final line is holomorphic at $z = -w$, so that $G(-w)$ is defined.

(b) The proof given in the text shows that the addition law

$$(*) \qquad \wp(z+w) = -\wp(z) + \wp(w) + \frac{1}{4}\left(\frac{\wp'(z) - \wp'(w)}{\wp(z) - \wp(w)}\right)^2$$

holds when $z, w, z+w \notin L$ and $2w \notin L$. To complete the proof, we need to show that (*) holds when $2w \in L$. For this purpose, fix $z, w \in \mathbb{C}$ with $z, w, z+w \notin L$ and $2w \in L$.

A first observation is that $\wp(z) \neq \wp(w)$ in this case, since equality implies $z \equiv \pm w \bmod L$ by Lemma 10.4. But $z \equiv -w \bmod L$ implies $z + w \in L$, which is a contradiction, and since $2w \in L$, $z \equiv w \bmod L$ implies $z + w \equiv 2w \equiv 0 \bmod L$, so that $z + w \in L$, again a contradiction. Thus the denominator in (*) is nonzero.

The set $L \cup (L+z) \cup (L-z) \cup \frac{1}{2}L$ is countable, so we can choose a sequence $w_i \to w$, with $w_i, z \pm w_i, 2w_i \notin L$ for all i. Then the addition law (*) holds for z and w_i, so that

$$\wp(z+w_i) = -\wp(z) + \wp(w_i) + \frac{1}{4}\left(\frac{\wp'(z) - \wp'(w_i)}{\wp(z) - \wp(w_i)}\right)^2$$

for all i. Letting $i \to \infty$, note that our assumptions on w_i and w imply three things: first, that

$$\lim_{i\to\infty} \wp(z+w_i) = \wp(z+w), \ \lim_{i\to\infty} \wp(w_i) = \wp(w), \ \lim_{i\to\infty} \wp'(w_i) = \wp'(w);$$

second, that the denominator $\wp(z) - \wp(w_i)$ never vanishes; and third, that these quantities are all finite. Then taking the limit as $i \to \infty$ of the above formula for $\wp(z+w_i)$ gives (*). Thus the addition law holds when $2w \in L$.

10.8. (a) This follows from Vieta's formulas that express the coefficients of a polynomial as the elementary symmetric functions of its roots. If e_1, e_2, and e_3 are the roots of

$$(*) \qquad x^3 - \frac{g_2}{4}x - \frac{g_3}{4},$$

then the coefficient of x^2, which is zero, is the negative of the root sum, showing that $e_1 + e_2 + e_3 = 0$. Furthermore, the coefficient of x is

$$-\frac{g_2}{4} = e_1e_2 + e_1e_3 + e_2e_3,$$

and the constant term is the negative of the root product, so that

$$\frac{g_3}{4} = e_1e_2e_3.$$

(b) Since $e_3 = -e_1 - e_2$, the discriminant of (*) is

$$(**) \qquad \begin{aligned}(e_1-e_2)^2(e_1-e_3)^2(e_2-e_3)^2 &= ((e_1-e_2)(2e_1+e_2)(e_1+2e_2))^2\\ &= (2e_1^3+3e_1^2e_2-3e_1e_2^2-2e_2^3)^2.\end{aligned}$$

Applying $e_3 = -e_1 - e_2$ to the formulas for $-g_2/4$ and $g_3/4$ from part (a) gives

$$\begin{aligned}\frac{g_2}{4} &= -\big(e_1e_2 + e_1(-e_1-e_2) + e_2(-e_1-e_2)\big) = e_1^2 + e_1e_2 + e_2^2\\ \frac{g_3}{4} &= e_1e_2(-e_1-e_2) = -(e_1^2e_2 + e_1e_2^2).\end{aligned}$$

Now use these to evaluate $\frac{1}{16}\left(g_2^3 - 27g_3^2\right)$:

$$\begin{aligned}\frac{1}{16}\left(g_2^3-27g_3^2\right) &= 4\left(\frac{g_2}{4}\right)^3 - 27\left(\frac{g_3}{4}\right)^2\\ &= 4(e_1^2+e_1e_2+e_2^2)^3 - 27(-e_1^2e_2-e_1e_2^2)^2\\ &= 4e_1^6 + 12e_1^5e_2 + 24e_1^4e_2^2 + 28e_1^3e_2^3 + 24e_1^2e_2^4 + 12e_1e_2^5 + 4e_2^6\\ &\qquad - 27e_1^4e_2^2 - 54e_1^3e_2^3 - 27e_1^2e_2^4\\ &= 4e_1^6 + 12e_1^5e_2 - 3e_1^4e_2^2 - 26e_1^3e_2^3 - 3e_1^2e_2^4 + 12e_1e_2^5 + 4e_2^6\\ &= (2e_1^3+3e_1^2e_2-3e_1e_2^2-2e_2^3)^2.\end{aligned}$$

Comparing this to (**), this expression equals the discriminant, and we are done.

10.9. Assume $j(L) = j(L')$. We know that $g_2(L')$ and $g_3(L')$ cannot both be zero since $\Delta(L) \neq 0$. If $g_2(L') = 0$, then $j(L') = j(L) = 0$. It follows that $g_2(L) = 0$ as well. Since $g_3(L), g_3(L') \neq 0$, there is some $\lambda \in \mathbb{C}$ such that $g_3(L') = \lambda^{-6}g_3(L)$. This λ also satisfies $g_2(L') = \lambda^{-4}g_2(L)$ since both sides are zero. If on the other hand $g_3(L') = 0$, then $j(L') = j(L) = 1728$, which implies in turn that $g_3(L) = 0$. Since $g_2(L), g_2(L') \neq 0$, there is some $\lambda \in \mathbb{C}$ such that $g_2(L') = \lambda^{-4}g_2(L)$. This λ also satisfies $g_3(L') = \lambda^{-6}g_3(L)$ since both sides are zero. So in either case we can find a nonzero $\lambda \in \mathbb{C}$ such that (10.11) holds.

10.10. (a) Differentiating $\wp'(z)^2 = 4\wp(z)^3 - g_2(L)\wp(z) - g_3(L)$ gives

$$2\wp'(z)\wp''(z) = 12\wp(z)^2\wp'(z) - g_2(L)\wp'(z),$$

which implies

$$\wp''(z) = 6\wp(z)^2 - \frac{1}{2}g_2(L)$$

since $\wp'(z)$ has only isolated zeros and poles.

(b) With $\wp(z) = z^{-2} + \sum_{n=1}^{\infty} a_n z^{2n}$, differentiating gives

$$\begin{aligned}\wp'(z) &= -2z^{-3} + \sum_{n=1}^{\infty} 2na_n z^{2n-1}\\ \wp''(z) &= 6z^{-4} + \sum_{n=1}^{\infty} 2n(2n-1)a_n z^{2n-2}.\end{aligned}$$

Next,

$$\wp(z)^2 = \Big(z^{-2} + \sum_{n=1}^{\infty} a_n z^{2n}\Big)^2 = z^{-4} + \sum_{n=1}^{\infty} 2a_n z^{2n-2} + \sum_{n=1}^{\infty}\sum_{i=1}^{n-1} a_i z^{2i} a_{n-i} z^{2(n-i)}$$
$$= z^{-4} + \sum_{n=1}^{\infty} 2a_n z^{2n-2} + \sum_{n=2}^{\infty}\sum_{i=1}^{n-1} a_i a_{n-i} z^{2n}$$
$$= z^{-4} + \sum_{n=1}^{\infty} 2a_n z^{2n-2} + \sum_{n=3}^{\infty}\sum_{i=1}^{n-2} a_i a_{n-1-i} z^{2n-2},$$

where the last equality results from changing the summation bounds on the double summation.

Since $g_2(L)$ is a constant, part (a) shows that for $n \geq 3$ the coefficients of z^{2n-2} in $\wp''(z)$ and $6\wp(z)^2$ are equal. Equating those coefficients in the series expansions above gives

$$2n(2n-1)a_n = 6\Big(2a_n + \sum_{i=1}^{n-2} a_i a_{n-i-1}\Big), \quad n \geq 3.$$

10.11. (a) We can write $\mathcal{O} = [1, w]$ where $w \in \mathbb{C} - \mathbb{R}$. By Exercise 7.8, $\mathfrak{a} = [\alpha, \beta]$ is a $\mathbb{Z}$-module of rank 2, so in particular $\alpha, \beta \neq 0$. If α and β were linearly dependent over $\mathbb{R}$, then $\beta = \lambda\alpha$ for some nonzero $\lambda \in \mathbb{R}$, and then $\mathfrak{a} = [\alpha, \beta] = [\alpha, \lambda\alpha] \subset \alpha\mathbb{R}$. Since $\alpha \in \mathfrak{a}$ and $\mathfrak{a}$ is an $\mathcal{O}$-ideal, it follows that $w\alpha \in \mathfrak{a}$. This implies that $w\alpha \in \alpha\mathbb{R}$, so that $w \in \mathbb{R}$ since $\alpha \neq 0$. This contradicts $w \in \mathbb{C} - \mathbb{R}$.

(b) This is an easy consequence of Lemma 7.5. If $L = [\alpha, \beta]$ is a lattice contained in K, then $L = \alpha[1, \tau]$ where $\tau = \beta/\alpha \in K$. Furthermore, $\tau \notin \mathbb{R}$ since α, β are linearly independent over $\mathbb{R}$. Then $[K : \mathbb{Q}] = 2$ implies that $K = \mathbb{Q}(\tau)$, so Lemma 7.5 does the trick: $[1, \tau]$ is a proper fractional ideal for the order $\mathcal{O} = [1, a\tau]$, where $ax^2 + bx + c$, $\gcd(a, b, c) = 1$, is the minimal polynomial for τ. The same is true for $L = \alpha[1, \tau]$ since $\alpha \neq 0$ is in K.

10.12. (a) This is an easy consequence of Lemma 10.17. First, $\wp(nz)$ is meromorphic since $\wp(z)$ is, and since $nL \subset L$ for $n \in \mathbb{Z}$, we have $\wp(n(z+\omega)) = \wp(nz+n\omega) = \wp(nz)$ for all $\omega \in L$. Thus $\wp(nz)$ is an elliptic function for L. Furthermore, $\wp(n(-z)) = \wp(-nz) = \wp(nz)$ since $\wp(z)$ is even. Thus $\wp(nz)$ is even and hence is a rational function of $\wp(z)$ by Lemma 10.17.

Alternatively, we can prove the result by induction on $n \geq 2$. To do this, first observe that the text gives an explicit formula for $\wp(2z)$ as a rational function of $\wp(z)$. Now assume that $\wp(nz)$ is a rational function of $\wp(z)$ for some $n \geq 2$. Then the addition formula gives

$$(*) \qquad \wp((n+1)z) = \wp(nz) + \wp(z) - \frac{1}{4}\left(\frac{\wp'(nz) - \wp'(z)}{\wp(nz) - \wp(z)}\right)^2.$$

Except for $(\wp'(nz) - \wp'(z))^2$, it is clear that this expression is a rational function of $\wp(z)$ using the inductive hypothesis. But when we write $\wp(nz) = R(\wp(z))$ for some rational function $R(x) \in \mathbb{C}(x)$, differentiating with respect to z gives

$$\wp'(nz) \cdot n = R'(\wp(z)) \cdot \wp'(z), \;\; R'(x) \in \mathbb{C}(x).$$

Then

$$(\wp'(nz) - \wp'(z))^2 = (\tfrac{1}{n}R'(\wp(z))\wp'(z) - \wp'(z))^2 = \wp'(z)^2(\tfrac{1}{n}R'(\wp(z)) - 1)^2$$
$$= (4\wp(z)^3 - g_2(L)\wp(z) - g_3(L))(\tfrac{1}{n}R'(\wp(z)) - 1)^2,$$

where the second line uses part (ii) of Theorem 10.1. Substituting this into (*) expresses $\wp((n+1)z)$ as a rational function of $\wp(z)$ and completes the inductive step.

(b) The proof is very similar to the proof of Theorem 10.14. Just as in Theorem 10.14, we have

$$\wp(nz) = \frac{A(\wp(z))}{B(\wp(z))}$$

where $A(x)$ and $B(x)$ are polynomials, and $\deg(A(x)) = \deg(B(x)) + 1$ by consideration of the double pole at the origin. We want to show that $\deg(A(x)) = n^2$. As in the proof of Theorem 10.14, choose $z \in \mathbb{C}$ with $2z \notin (1/n)L$. Then $\deg(A(x) - \wp(nz)B(x)) = \deg(A(x))$, and by Exercise 10.14 we can choose z so that it has distinct roots.

We have $[L : nL] = [(1/n)L : L] = n^2$; let $\{w_i\}_{1\leq i\leq n^2}$ be coset representatives of L in $(1/n)L$. If we can show that the $\wp(z + w_i)$ are distinct and give all roots of $A(x) - \wp(nz)B(x)$, then $\deg(A(x)) = n^2$ and we are done. To see that they are all distinct, suppose that $\wp(z+w_i) = \wp(z+w_j)$, $i \neq j$. Then $z+w_i \equiv \pm(z+w_j) \bmod L$ by Lemma 10.4. If plus, then $w_i \equiv w_j \bmod L$, which contradicts the fact that $i \neq j$ since the w_i are distinct coset representatives. If minus, then $2z \equiv -w_i - w_j \bmod L$, which implies that $2z \in (1/n)L$ since $w_i, w_j \in (1/n)L$, a contradiction. Thus the $\wp(z + w_i)$ are distinct.

Next, $A(\wp(z + w_i)) = \wp(n(z + w_i))B(\wp(z + w_i))$. Then $w_i \in (1/n)L$ implies that $nw_i \in L$, so that $n(z + w_i) \equiv nz \bmod L$ and therefore $\wp(n(z + w_i)) = \wp(nz)$. Thus the $\wp(z + w_i)$ are roots of $A(x) - \wp(nz)B(x)$. Finally, suppose u is a root of $A(x) - \wp(nz)B(x)$. Then if $B(u) = 0$, we also have $A(u) = 0$, which contradicts the assumption that A and B are relatively prime. Thus $B(u) \neq 0$, and Exercise 10.14 gives $u = \wp(w)$ for some $w \in \mathbb{C}$. This gives

$$\wp(nz) = \frac{A(u)}{B(u)} = \frac{A(\wp(w))}{B(\wp(w))} = \wp(nw),$$

so that $nw \equiv \pm nz \bmod L$. Changing w to $-w$ if necessary (recall that $\wp$ is even, so $u = \wp(w) = \wp(-w)$), we may assume $w \equiv z \bmod (1/n)L$, so that $w \equiv z + w_i \bmod L$ for some i. Thus $u = \wp(w) = \wp(z + w_i)$ is one of the roots above, completing the proof.

10.13. (a) Let $f(z)$ be even, elliptic, and holomorphic on $\mathbb{C} - L$. We claim that there is a polynomial $A(x) \in \mathbb{C}[x]$ such that $f(z) - A(\wp(z))$ is holomorphic at the origin. If not, choose $A(x)$ such that $f(z) - A(\wp(z))$ has a pole of minimal order at the origin, necessarily even since $f(z)$ and $\wp(z)$ are even functions. Thus

$$f(z) - A(\wp(z)) = \frac{a}{z^{2M}} + \text{higher order terms}, \quad a \neq 0, M > 0.$$

But $\wp(z) = z^{-2} + \text{higher order terms}$, so that $\wp(z)^M = z^{-2M} + \text{higher order terms}$. Therefore $f(z) - A(\wp(z)) - a\wp(z)^M$ has a pole of order $< 2M$ at the origin. This contradicts the minimality of $A(x)$ and proves the claim.

Once $f(z) - A(\wp(z))$ is holomorphic at $z = 0$, it is holomorphic at all points of L by periodicity. Since $f(z)$ is holomorphic on $\mathbb{C} - L$ by assumption, we see that the elliptic function $f(z) - A(\wp(z))$ is holomorphic on all of $\mathbb{C}$ and therefore is constant by Exercise 10.5. It follows that $f(z)$ is a polynomial in $\wp(z)$.

(b) We have $(\wp(z) - \wp(w))^m f(z) = \left(\frac{\wp(z)-\wp(w)}{z-w}\right)^m g(z)$, where $g(z) = (z - w)^m f(z)$ is holomorphic at $z = w$ since $f(z)$ has a pole of order m at $z = w$. Since $\wp(z)$ is holomorphic in a neighborhood of $w \notin L$, we get

$$\lim_{z\to w} \frac{\wp(z) - \wp(w)}{z - w} = \wp'(w),$$

so that $(\wp(z) - \wp(w))^m f(z)$ is holomorphic at w.

(c) Since $f(z)$ is meromorphic, it has only a finite number of poles in the compact region $\mathbf{P}$. Let the poles lying in $\mathbf{P} - L$ be at $w_1, \ldots, w_k$, of orders $m_1, \ldots, m_k$, and if two differ by an element of L (which can happen for poles lying on the boundary of $\mathbf{P}$), discard one and relabel accordingly. In this way, every pole of $f(z)$ on $\mathbb{C} - L$ is congruent modulo L to a unique w_i. By part (b), it follows that

$$g(z) = f(z) \prod_{i=1}^{k} (\wp(z) - \wp(w_i))^{m_i}$$

is holomorphic at $w_1, \ldots w_k$, so it is holomorphic in $\mathbb{C} - L$ by periodicity. By part (a), there is a polynomial $A(x)$ with $g(z) = A(\wp(z))$. Let $B(x) = \prod_{i=1}^{k} (\wp(x) - \wp(w_i))^{m_i}$. Then we get

$$f(z) = \frac{A(\wp(z))}{B(\wp(z))}.$$

(d) Write

$$f(z) = \Big(\frac{f(z) + f(-z)}{2}\Big) + \Big(\frac{f(z) - f(-z)}{2\wp'(z)}\Big)\wp'(z).$$

Both quantities in parentheses are elliptic functions. The first is an even function, so it is a rational function of $\wp(z)$ by part (c). The second is also even since its numerator and denominator are both odd, so it is also a rational function of $\wp(z)$. The result follows.

10.14. (a) Suppose w is a multiple root of $A(x) - \lambda B(x)$. Then w is also a root of $(A(x) - \lambda B(x))' = A'(x) - \lambda B'(x)$. It follows that $A(w) = \lambda B(w)$ and $A'(w) = \lambda B'(w)$, so that

$$A(w)B'(w) - A'(w)B(w) = (\lambda B(w))B'(w) - B'(w)(\lambda B(w)) = 0.$$

Hence w is also a root of $A(x)B'(x) - A'(x)B(x)$. If there are an infinite number of such w, then $A(x)B'(x) - A'(x)B(x) = 0$ identically. It follows that $A(x) \mid A'(x)B(x)$; since $\gcd(A(x), B(x)) = 1$, this implies $A(x) \mid A'(x)$, which is impossible by degree considerations since we are in characteristic 0. This contradiction proves part (a).

(b) Let $f(z) = \wp(z) - u$, and choose a parallelogram $\mathbf{P} = \{s\omega_1 + t\omega_2 \mid \delta \leq s, t \leq \delta + 1\}$ with $-1 < \delta < 0$ such that $f(z)$ has no poles or zeros on the boundary Γ of $\mathbf{P}$; this is possible since $f(z)$ is meromorphic. If Z (resp. P) is the number of zeros (resp. poles) of $f(z)$ in $\mathbf{P}$, then

$$\frac{1}{2\pi i} \int_\Gamma \frac{f'(z)}{f(z)}\, dz = Z - P.$$

Both $f(z)$ and $f'(z)$ are periodic, so that the integrals on opposite sides of the bounding rectangle cancel, giving $Z = P$. However, $f(z)$ has at least one pole in $\mathbf{P}$, at the origin, which lies in $\mathbf{P}$ since $\delta < 0 < \delta + 1$. Therefore $f(z) = \wp(z) - u$ also has at least one zero in $\mathbf{P}$, and thus there is some $w \in \mathbf{P}$ such that $\wp(w) = u$.

10.15. If the ideals $\mathfrak{a}$ and $\mathfrak{b}$ are in the same class in $C(\mathcal{O})$, then $\mathfrak{b} = k\mathfrak{a}$ for some $k \in K$. Writing $\mathfrak{a} = [\alpha_1, \alpha_2]$ and $\mathfrak{b} = [\beta_1, \beta_2]$, this becomes $[\beta_1, \beta_2] = k[\alpha_1, \alpha_2]$, which shows that $\mathfrak{a}$ and $\mathfrak{b}$ are homothetic as lattices. Conversely, if $\mathfrak{b}$ is homothetic to $\mathfrak{a}$, then $[\beta_1, \beta_2] = \lambda[\alpha_1, \alpha_2]$ for some $\lambda \in \mathbb{C}$. This implies $\lambda\alpha_1 \in [\beta_1, \beta_2] \subset K$, and since $\alpha_1 \neq 0$ is in K, we get $\lambda \in K$, so that $\mathfrak{a}$ and $\mathfrak{b} = \lambda\mathfrak{a}$ give the same class in $C(\mathcal{O})$.

10.16. (a) If a lattice L has complex multiplication by $\sqrt{-5}$, then by Theorem 10.14 there is an imaginary quadratic field K and an order $\mathcal{O} \subset \mathcal{O}_K$ with $\sqrt{-5} \in \mathcal{O}$ such that L is homothetic to a proper fractional $\mathcal{O}$-ideal. Since $\sqrt{-5} \in K - \mathbb{Q}$ and

$[K:\mathbb{Q}]=2$, it follows that $K=\mathbb{Q}(\sqrt{-5})$. The only order in $\mathbb{Q}(\sqrt{-5})$ containing $\sqrt{-5}$ is $\mathcal{O}_K=\mathbb{Z}[\sqrt{-5}]$, of discriminant -20, and the reduced forms of discriminant -20 are x^2+5y^2 and $2x^2+2xy+3y^2$. By Theorem 7.7, these forms correspond to the proper $\mathcal{O}_K$-ideals

$$x^2+5y^2 \longmapsto \left[1,\frac{\sqrt{-20}}{2}\right]=[1,\sqrt{-5}]$$
$$2x^2+2xy+3y^2 \longmapsto \left[2,\frac{-2+\sqrt{-20}}{2}\right]=[2,-1+\sqrt{-5}]=[2,1+\sqrt{-5}],$$

whose ideal classes are the two elements of the class group. So by Corollary 10.20, these are the only homothety classes of lattices with complex multiplication by $\sqrt{-5}$.

(b) Arguing as in part (a), if a lattice L has complex multiplication by $\sqrt{-14}$, then by Theorem 10.14 there is an imaginary quadratic field K and an order $\mathcal{O}\subset\mathcal{O}_K$ with $\sqrt{-14}\in\mathcal{O}$ such that L is homothetic to a proper fractional $\mathcal{O}$-ideal. Since $\sqrt{-14}\in K-\mathbb{Q}$ and $[K:\mathbb{Q}]=2$, it follows that $K=\mathbb{Q}(\sqrt{-14})$. The only order in $\mathbb{Q}(\sqrt{-14})$ that contains $\sqrt{-14}$ is $\mathcal{O}_K=\mathbb{Z}[\sqrt{-14}]$, of discriminant $d_K=-56$, and we get

$$x^2+14y^2 \longmapsto \left[1,\frac{\sqrt{-56}}{2}\right]=[1,\sqrt{-14}]$$
$$2x^2+7y^2 \longmapsto \left[2,\frac{\sqrt{-56}}{2}\right]=[2,\sqrt{-14}]$$
$$3x^2-2xy+5y^2 \longmapsto \left[3,\frac{2+\sqrt{-56}}{2}\right]=[3,1+\sqrt{-14}]$$
$$3x^2+2xy+5y^2 \longmapsto \left[3,\frac{-2+\sqrt{-56}}{2}\right]=[3,-1+\sqrt{-14}]$$

by the example following Theorem 5.30. By Corollary 10.20, these are the only homothety classes of lattices with complex multiplication by $\sqrt{-14}$.

(c) Fix $\alpha\in\mathcal{O}_K-\mathbb{Z}$. We first determine which orders $\mathcal{O}\subset\mathcal{O}_K$ contain α. Since $[K:\mathbb{Q}]=2$ and $\alpha\notin\mathbb{Q}$, it follows that $K=\mathbb{Q}(\alpha)$, which implies that $\mathbb{Z}[\alpha]\subset\mathcal{O}_K$ is an order. By definition, $\mathbb{Z}[\alpha]$ has conductor $[\mathcal{O}_K:\mathbb{Z}[\alpha]]$. If $\mathcal{O}$ is an order in K of conductor f, then

$$(*)\qquad \alpha\in\mathcal{O} \iff \mathbb{Z}[\alpha]\subset\mathcal{O} \iff f \mid [\mathcal{O}_K:\mathbb{Z}[\alpha]],$$

where the first equivalence is obvious ($\mathcal{O}$ is a ring containing 1), and the second follows from part (b) of Exercise 9.19.

Now we turn to lattices. Let $[L]$ denote the homothety class of a lattice $L\subset\mathbb{C}$. The goal of the problem is to determine the cardinality of the set

$$A=\{[L]:\alpha L\subset L\}.$$

The proof of (ii) $\Rightarrow$ (iii) of Theorem 10.14 shows that up to homothety, $L=[1,\tau]$ where $\mathbb{Q}(\tau)$ is imaginary quadratic. Then $\alpha L\subset L$ implies $\alpha\in\mathbb{Q}(\tau)$, so that $L\subset\mathbb{Q}(\tau)=\mathbb{Q}(\alpha)=K$. Thus we may express the above set as

$$A=\{[L]:\alpha L\subset L\subset K\}.$$

Theorem 10.14 also tells us that L is a proper ideal for an order $\mathcal{O}\subset\mathcal{O}_K$, and then $\alpha L\subset L$ implies that $\alpha\in\mathcal{O}$ since $\mathcal{O}$ is the full set of complex multiplications of L. Thus we may write A as a disjoint union

$$A=\bigcup_{\alpha\in\mathcal{O}}\{[L]:L\text{ is a proper ideal for }\mathcal{O}\}.$$

This is a finite union by (*), so that by Corollary 10.20, its cardinality is

$$|A| = \sum_{\alpha\in\mathcal{O}} |\{[L] : L \text{ is a proper ideal for } \mathcal{O}\}| = \sum_{\alpha\in\mathcal{O}} h(\mathcal{O}).$$

An order $\mathcal{O}$ is uniquely determined by its conductor f. Then $h(\mathcal{O}) = h(f^2 d_K)$ by (7.3) and Theorem 7.7, and (*) implies that $\alpha \in \mathcal{O}$ if and only if $f \mid [\mathcal{O}_K : \mathbb{Z}[\alpha]]$. Thus

$$|A| = \sum_{f\mid[\mathcal{O}_K:\mathbb{Z}[\alpha]]} h(f^2 d_K).$$

10.17. Since $\omega L = \omega[1,\omega] = [\omega,\omega^2] = [\omega,-\omega-1] = [\omega,\omega+1] = [1,\omega] = L$, we have

$$g_2(L) = g_2(\omega L) = \omega^{-4} g_2(L) = \omega^2 g_2(L)$$

by (10.11). This forces $g_2(L) = 0$ and thus $j(\omega) = j(L) = 0$.

10.18. We start with

$$\wp(z) = \frac{1}{z^2} + \sum_{n=1}^{\infty}(2n+1)G_{2n+2}(L)z^{2n}.$$

We know that $g_2 = 60G_4$ and $g_3 = 140G_6$, so the first four terms on the right-hand side are

$$\frac{1}{z^2} + 3G_4z^2 + 5G_6z^4 + 7G_8z^6 = \frac{1}{z^2} + \frac{g_2}{20}z^2 + \frac{g_3}{28}z^4 + 7G_8z^6.$$

It remains to compute $7G_8$. Let $a_n = (2n+1)G_{2n+2}$ as in the proof of Lemma 10.12. Then $a_1 = 3G_4$ and $a_3 = 7G_8$. However, Exercise 10.10 with $n = 3$ gives

$$30a_3 = 6(2a_3 + a_1a_1) \implies 18a_3 = 6a_1^2 \implies 18(7G_8) = 6(3G_4)^2 = 6\Big(3\frac{g_2}{60}\Big)^2,$$

from which we obtain

$$7G_8 = \frac{1}{3}\Big(\frac{g_2}{20}\Big)^2 = \frac{g_2^2}{1200},$$

as desired.

10.19. (a) If L is a lattice with $g_2(L) = 0$, then $j(L) = 0$. Since $j([1,\omega]) = 0$ from Exercise 10.17, it follows from Theorem 10.9 that L and $[1,\omega]$ are homothetic.

(b) If L is a lattice with $g_3(L) = 0$, then $j(L) = 1728$. But $j([1,i]) = 1728$ by the computation following Corollary 10.20, so again Theorem 10.9 shows that L and $[1,i]$ are homothetic.

(c) Given $g_2g_3 \neq 0$, we solve the system

$$\begin{aligned} 20g &= \lambda^{-4}g_2 \\ 28g &= \lambda^{-6}g_3 \end{aligned}$$

by dividing the first equation by the second and solving for λ, giving

$$\lambda = \sqrt{\frac{20g_3}{28g_2}} = \sqrt{\frac{5g_3}{7g_2}} \neq 0, \quad g = \frac{1}{20}\Big(\sqrt{\frac{5g_3}{7g_2}}\Big)^{-4} g_2 = \frac{1}{20}\frac{(7g_2)^2}{(5g_3)^2}g_2 = \frac{49g_2^3}{500g_3^2}.$$

(d) The first equality results from simply adding the two series expansions. For the second,

$$\begin{aligned}
\Big(-\frac{3g}{2}z^2+\frac{9g}{2}z^4-\frac{5g^2}{2}z^6+\cdots\Big)^{-1} &= \Big(-\frac{3gz^2}{2}\Big(1-3z^2+\frac{5g}{3}z^4+\cdots\Big)\Big)^{-1}\\
&= \frac{-2}{3gz^2}\Big(1-z^2\Big[3-\frac{5g}{3}z^2+\cdots\Big]\Big)^{-1}\\
&= \frac{-2}{3gz^2}\Big(1+z^2\Big[3-\frac{5g}{3}z^2+\cdots\Big]+z^4\Big[3-\frac{5g}{3}z^2+\cdots\Big]^2+\cdots\Big)\\
&= \frac{-2}{3gz^2}\Big(1+3z^2-\frac{5g}{3}z^4+\cdots+9z^4+\cdots\Big)\\
&= \frac{-2}{3gz^2}-\frac{2}{g}-\frac{2}{3g}\Big(9-\frac{5g}{3}\Big)z^2+\cdots.
\end{aligned}$$

10.20. Write $w=\frac{1+\sqrt{-7}}{2}$. We have $N(w)=w\overline{w}=2$, so that by Theorem 10.14, $\wp(wz)$ is a rational function of $\wp(z)$ with quadratic numerator and linear denominator. Thus

$$\wp(wz)=a\wp(z)+b+\frac{1}{c\wp(z)+d},\quad a,b,c,d\in\mathbb{C}.$$

The first terms of the Laurent expansion of $\wp(z)$ are

$$\wp(z)=\frac{1}{z^2}+\frac{g_2}{20}z^2+\frac{g_3}{28}z^4+\frac{g_2^2}{1200}z^6+\cdots,$$

and $g_2g_3\neq 0$ since $L=[1,w]$ is homothetic to neither $[1,\omega]$ nor $[1,i]$. After replacing L by a homothetic lattice, we may assume that $g_2=20g$, $g_3=28g$; making these replacements gives

$$\wp(z)=\frac{1}{z^2}+gz^2+gz^4+\frac{g^2}{3}z^6+\cdots,$$

so that

$$\wp(wz)=\frac{1}{w^2z^2}+gw^2z^2+gw^4z^4+\frac{g^2}{3}w^6z^6+\cdots.$$

As in the text, $\wp(wz)-a\wp(z)-b$ must vanish at 0, so that $a=\frac{1}{w^2}$ and $b=0$. For the moment, assume that there are constants $A,B,C\in\mathbb{C}$ such that

$$(*)\quad \wp(wz)-\frac{1}{w^2}\wp(z)=\frac{z^2}{A}\big(1+Bz^2+Cgz^4+\cdots\big)=\frac{z^2}{A}\big(1+z^2[B+Cgz^2+\cdots]\big).$$

Then

$$\begin{aligned}
\Big(\wp(wz)-\frac{1}{w^2}\wp(z)\Big)^{-1} &= \frac{A}{z^2}\Big(1-z^2[B+Cgz^2+\cdots]+z^4[B+Cgz^2+\cdots]^2+\cdots\Big)\\
&= \frac{A}{z^2}\Big(1-Bz^2+(-Cg+B^2)z^4+\cdots\Big)\\
&= \frac{A}{z^2}-AB+A(-Cg+B^2)z^2+\cdots.
\end{aligned}$$

But this equals $c\wp(z)+d=\frac{c}{z^2}+cgz^2+\cdots+d$ for $c,d\in\mathbb{C}$. Comparing the coefficients of z^{-2} shows that $c=A$, and then comparing the coefficients of z^2 gives the equation

$$Ag=cg=A\big(-Cg+B^2\big),$$

which implies $g=-Cg+B^2$. Thus

$$g=\frac{B^2}{1+C}.$$

The next step is to determine A, B, C in (*). The negative powers of z in $\wp(wz) - \frac{1}{w^2}\wp(z)$ cancel, so that the above expansions of $\wp(z)$ and $\wp(wz)$ give

$$\begin{aligned}
\wp(wz) - \frac{1}{w^2}\wp(z) &= gw^2z^2 + gw^4z^4 + \frac{g^2}{3}z^6 + \cdots - \Big(gw^{-2}z^2 + gw^{-2}z^4 + \frac{g^2}{3}w^6z^6 + \cdots\Big)\\
&= g(w^2 - w^{-2})z^2 + g(w^4 - w^{-2})z^4 + \frac{g^2}{3}(w^6 - w^{-2})z^6 + \cdots\\
&= gz^2\Big(\frac{w^4-1}{w^2} + \frac{w^6-1}{w^2}z^2 + \frac{g(w^8-1)}{3w^2}z^4 \cdots\Big)\\
&= \frac{g(w^4-1)z^2}{w^2}\Big(1 + \frac{w^6-1}{w^4-1}z^2 + \frac{g(w^8-1)}{3(w^4-1)}z^4 + \cdots\Big).
\end{aligned}$$

It follows that (*) holds with

$$A = \frac{w^2}{g(w^4-1)},\quad B = \frac{w^6-1}{w^4-1} = \frac{w^4+w^2+1}{w^2+1},\quad C = \frac{w^8-1}{3(w^4-1)} = \frac{w^4+1}{3}.$$

Then

$$g = \frac{B^2}{1+C} = \frac{\Big(\dfrac{w^4+w^2+1}{w^2+1}\Big)^2}{1+\dfrac{w^4+1}{3}} = \frac{3(w^4+w^2+1)^2}{(w^2+1)^2(w^4+4)}.$$

To simplify the numerator and denominator, note that $w^2 - w + 2 = 0$ implies $w^2 = w - 2$, $w - w^2 = 2$, and $w^4 = 2 - 3w$, so that

$$\begin{aligned}
w^4 + w^2 + 1 &= (2-3w) + (w-2) + 1 = 1 - 2w = -\sqrt{-7}\\
(w^2+1)^2(w^4+4) &= (w-1)^2(5-3w) = (w^2-2w+1)(6-3w) = (-1-w)(6-3w)\\
&= -3(1+w)(2-w) = -3(2+w-w^2) = -3\cdot 4 = -12.
\end{aligned}$$

Thus finally

$$g = \frac{3(w^4+w^2+1)^2}{(w^2+1)^2(w^4+4)} = \frac{3(-7)}{-12} = \frac{7}{4},$$

so that

$$j\Big(\frac{1+\sqrt{-7}}{2}\Big) = 1728\frac{g_2^3}{g_2^3 - 27g_3^2} = 1728\frac{\big(20\cdot\frac{7}{4}\big)^3}{\big(20\cdot\frac{7}{4}\big)^3 - 27\big(28\cdot\frac{7}{4}\big)^2} = -3375 = (-15)^3.$$

Note that the formula

$$g = \frac{3(w^4+w^2+1)^2}{(w^2+1)^2(w^4+4)}$$

derived above uses only $N(w) = 2$. Hence it applies when $w = \sqrt{-2}$, and for this value of w, an easy computation gives $g = \frac{27}{8}$, as in the text.

Solutions to Exercises in §11

11.1. (a) Since $\overline{L} = \{\overline{\omega} : \omega \in L\}$, it follows that $\omega \in L$ if and only if $\overline{\omega} \in \overline{L}$. Thus

$$\overline{g_2(L)} = 60\sum_{\omega\in L-\{0\}} \overline{\frac{1}{\omega^4}} = 60\sum_{\omega\in L-\{0\}} \frac{1}{\overline{\omega}^4} = 60\sum_{\omega\in \overline{L}-\{0\}} \frac{1}{\omega^4} = g_2(\overline{L})$$

$$\overline{g_3(L)} = 140\sum_{\omega\in L-\{0\}} \overline{\frac{1}{\omega^6}} = 140\sum_{\omega\in L-\{0\}} \frac{1}{\overline{\omega}^6} = 140\sum_{\omega\in \overline{L}-\{0\}} \frac{1}{\omega^6} = g_3(\overline{L}),$$

where the first equality in each line follows since complex conjugation is continuous. It follows immediately that $j(\overline{L}) = \overline{j(L)}$.

(b) We have the following equivalences:

$$j(\mathfrak{a}) \text{ is real} \iff j(\mathfrak{a}) = \overline{j(\mathfrak{a})} \iff j(\mathfrak{a}) = j(\overline{\mathfrak{a}}) \iff \mathfrak{a} \text{ and } \overline{\mathfrak{a}} \text{ are homothetic}$$
$$\iff \mathfrak{a} \text{ and } \overline{\mathfrak{a}} \text{ are in the same class in } C(\mathcal{O}),$$

where the second equivalence uses part (a), the third uses Theorem 10.9, and the fourth uses Corollary 10.20. The discussion following Lemma 7.14 shows that in $C(\mathcal{O})$, $[\overline{\mathfrak{a}}]$ is the inverse of $[\mathfrak{a}]$, so the last line above can be rephrased as saying that $[\mathfrak{a}]$ equals its inverse in $C(\mathcal{O})$, i.e., that $[\mathfrak{a}]$ has order ≤ 2 in $C(\mathcal{O})$.

For the final statement, part (b) implies that $j(\mathcal{O})$ is real since $[\mathcal{O}]$ has order 1 in $C(\mathcal{O})$.

11.2. Using (7.10) (proved in Exercise 7.11),

$$\operatorname{Im}(\gamma\tau) = \operatorname{Im}\left(\frac{a\tau + b}{c\tau + d}\right) = \det\begin{pmatrix} a & b \\ c & d \end{pmatrix} |c\tau + d|^{-2} \operatorname{Im}(\tau).$$

The determinant is $+1$ since $\gamma \in \mathrm{SL}(2,\mathbb{Z})$, and $\operatorname{Im}(\tau) > 0$ since $\tau \in \mathfrak{h}$. Therefore the right-hand side is positive, so the left-hand side is positive as well, showing that $\gamma\tau \in \mathfrak{h}$.

11.3. (a) For $\tau = a + bi$ we have $b^2y^2 = \operatorname{Im}(\tau)^2 y^2 \geq \epsilon^2 y^2$. Thus if $|x + ay| \geq (\epsilon/2)\,|x|$,

$$|x + y\tau|^2 = (x + ay)^2 + b^2y^2 \geq \frac{\epsilon^2}{4}x^2 + \epsilon^2 y^2 \geq \frac{\epsilon^2}{4}(x^2 + y^2).$$

(b) Using $\epsilon < 1$, $|x + ay| < (\epsilon/2)\,|x|$, $|A + B| \geq |A| - |B|$, and $|a| \leq 1/2$, we obtain

$$\frac{1}{2}\,|x| \geq \frac{\epsilon}{2}\,|x| > |x + ay| \geq |x| - |a|\,|y| \geq |x| - \frac{1}{2}\,|y|,$$

which implies $(1/2)\,|y| > (1/2)\,|x|$. Hence $|y| > |x|$ when $|x + ay| < (\epsilon/2)\,|x|$.

(c) Since $(x + ay)^2 \geq 0$, and $|b| \geq \epsilon$ by assumption, we get

$$|x + y\tau|^2 = (x + ay)^2 + b^2y^2 \geq b^2y^2 \geq \epsilon^2y^2 \geq \frac{\epsilon^2}{4}(2y^2) \geq \frac{\epsilon^2}{4}(x^2 + y^2),$$

where the final inequality follows since $|y| > |x|$ by part (b).

11.4. (a) First, $f(x,y)$ is $\mathbb{R}^+$-equivalent to itself via the identity matrix, so that $\mathbb{R}^+$-equivalence is reflexive. If $f(x,y)$ is $\mathbb{R}^+$-equivalent to $g(x,y)$, say $f(x,y) = \lambda g(px + qy, rx + sy)$ with $ps - rq = 1$, then

$$\begin{aligned}
&\frac{1}{\lambda} f(sx - qy, -rx + py)\\
&\qquad = \frac{1}{\lambda} \cdot \lambda g(p(sx - qy) + q(-rx + py), r(sx - qy) + s(-rx + py))\\
&\qquad = g((ps - qr)x, (ps - rq)y)\\
&\qquad = g(x,y).
\end{aligned}$$

We have $1/\lambda > 0$ since $\lambda > 0$, and $sp - (-q)(-r) = ps - rq = 1$, so this shows that $g(x,y)$ is $\mathbb{R}^+$-equivalent to $f(x,y)$, proving that $\mathbb{R}^+$-equivalence is symmetric. Finally, if

$$\begin{aligned}
f(x,y) &= \lambda g(px + qy, rx + sy), \quad ps - rq = 1\\
g(x,y) &= \rho h(ax + by, cx + dy), \quad ad - bc = 1,
\end{aligned}$$

then

$$\begin{aligned}
f(x,y) &= \lambda\rho \cdot h(a(px + qy) + b(rx + sy), c(px + qy) + d(rx + sy))\\
&= \lambda\rho \cdot h((ap + br)x + (aq + bs)y, (cp + dr)x + (cq + ds)y),
\end{aligned}$$

and

$$(ap + br)(cq + ds) - (aq + bs)(cp + dr) = (ad - bc)(ps - rq) = 1,$$

so that $\mathbb{R}^+$-equivalence is transitive.

(b) The proof that every form is properly equivalent to a reduced form is identical to the first part of the proof of Theorem 2.8, except that we now have to show that $f(x,y)$ is properly equivalent to a form whose middle coefficient has minimal absolute value. In §2, this was obvious since the coefficients were integers, but here, they can be real numbers.

Let $f(x,y) = ax^2 + bxy + cy^2$, $a,b,c \in \mathbb{R}$, be positive definite of discriminant D. By (2.4), $f(x,y) \geq \frac{|D|}{4a}y^2$, and similarly $f(x,y) \geq \frac{|D|}{4c}x^2$. Thus, given any bound $M > 0$, there are only finitely many $(m,n) \in \mathbb{Z}^2$ such that $f(m,n) \leq M$. In particular, $f(m,n)$ has a minimum nonzero value A for $(m,n) \in \mathbb{Z}^2 - \{(0,0)\}$.

Now take $\left(\begin{smallmatrix} p & q \\ r & s \end{smallmatrix}\right) \in \mathrm{SL}(2,\mathbb{Z})$ and set

$$f(px+qy, rx+sy) = f(p,q)x^2 + b'xy + f(r,s)y^2 = a'x^2 + b'xy + c'y^2.$$

Assume $|b'| \leq |b|$. Since proper equivalence preserves the discriminant, we get

$$D + 4a'c' = b'^2 \leq b^2 = D + 4ac \implies a'c' \leq ac.$$

Using $a' = f(p,q) \geq A$ and $c' = f(r,s) \geq A$ from the previous paragraph, it follows that

$$f(p,q) = a' \leq \frac{ac}{c'} \leq \frac{ac}{A} \quad \text{and} \quad f(r,s) = c' \leq \frac{ac}{a'} \leq \frac{ac}{A}.$$

The previous paragraph also shows that only finitely many $\left(\begin{smallmatrix} p & q \\ r & s \end{smallmatrix}\right) \in \mathrm{SL}(2,\mathbb{Z})$ satisfy these inequalities. Hence $|b'| > |b|$ with at most finitely many exceptions. This proves the existence of a minimal value of $|b|$.

The uniqueness part of the proof is similar to Theorem 2.8 as well; start by assuming that $g(x,y)$ and $f(x,y)$ are $\mathbb{R}^+$-equivalent reduced forms, and normalize $g(x,y)$ by multiplying by an appropriate positive real so that the coefficients of x^2 in $f(x,y)$ and $g(x,y)$ are the same.

(c) Since $f(x,y)$ is positive definite, its discriminant is negative. It follows that $f(x,1) = ax^2 + bx + c$, which has the same discriminant, has two complex conjugate roots. Let τ be the one that lies in $\mathfrak{h}$. Then

$$f(x,y) = a(x-\tau y)(x-\overline{\tau}y) = a(x-\tau y)(\overline{x-\tau y}) = a\,|x-\tau y|^2.$$

From this expansion we see that

$$f(x,y) = ax^2 - a(\tau+\overline{\tau})xy + a\tau\overline{\tau}y^2 = ax^2 + bxy + cy^2.$$

Since $\tau + \overline{\tau} = 2\,\mathrm{Re}(\tau)$ and $\tau\overline{\tau} = |\tau|^2$, we get

$$b = -2a\,\mathrm{Re}(\tau), \quad c = a\,|\tau|^2.$$

For uniqueness, if $\mu \in \mathfrak{h}$ satisfies $f(x,y) = a\,|x-\mu y|^2 = a(x-\mu y)(x-\overline{\mu}y)$, then μ and $\overline{\mu}$ are the roots of $f(x,1)$, so that $\mu = \tau$ since they both lie in $\mathfrak{h}$.

(d) As in (7.8), suppose

$$f(x,y) = \lambda g(px+qy, rx+sy), \quad \lambda > 0,\ \gamma = \begin{pmatrix} p & q \\ r & s \end{pmatrix} \in \mathrm{SL}(2,\mathbb{Z}).$$

Let τ (resp. τ') be the root of $f(x,y)$ (resp. $g(x,y)$). Then

$$(*) \quad 0 = f(\tau,1) = \lambda g(p\tau+q, r\tau+s) = \lambda(r\tau+s)^2 g\Big(\frac{p\tau+q}{r\tau+s}, 1\Big) = \lambda(r\tau+s)^2 g(\gamma\tau, 1).$$

It follows that $g(\gamma\tau,1) = 0$, so that $\gamma\tau = \tau'$ or $\overline{\tau'}$. But by Exercise 11.2, $\tau \in \mathfrak{h}$ implies $\gamma\tau \in \mathfrak{h}$ since $\gamma \in \mathrm{SL}(2,\mathbb{Z})$. Thus $\gamma\tau = \tau'$ and thus τ and τ' are $\mathrm{SL}(2,\mathbb{Z})$-equivalent. Conversely, suppose that the roots τ and τ' of $f(x,y)$ and $g(x,y)$ are $\mathrm{SL}(2,\mathbb{Z})$-equivalent, say $\tau' = \frac{p\tau+q}{r\tau+s}$. If we set $f'(x,y) = g(p\tau+q, r\tau+s)$, then arguing as in (*), one sees that $g(\tau',1) = 0$ implies $f'(\tau,1) = 0$, so that τ is the root of $f'(x, y)$. By the uniqueness proved in part (c), it follows that $f(x, y)$ is a constant multiple of $f'(x, y)$ and hence is $\mathbb{R}^+$-equivalent to $g(x, y)$.

(e) Suppose $f(x, y)$ is a positive definite form with root τ. Then using parts (b) and (c),

$$|\mathrm{Re}(\tau)| \le \frac{1}{2} \iff \frac{|b|}{2a} \le \frac{1}{2} \iff |b| \le a$$
$$|\tau| \ge 1 \iff |\tau|^2 = \frac{c}{a} \ge 1 \iff a \le c.$$

It follows that $|b| \le a \le c$ if and only if $|\mathrm{Re}(\tau)| \le 1/2$ and $|\tau| \ge 1$. For the two boundary conditions, note that

$$a = |b| \text{ or } a = c \iff |\mathrm{Re}(\tau)| = \frac{1}{2} \text{ or } |\tau| = 1.$$

Under these conditions, $f(x, y)$ is reduced if and only if $b \ge 0$. But $b = -2a\,\mathrm{Re}(\tau)$, and $a > 0$, so that $f(x, y)$ is reduced if and only if $\mathrm{Re}(\tau) \le 0$, which means that $\tau \in F$.

(f) If $\tau \in \mathfrak{h}$, then $f(x, y) = |x - \tau y|^2$ is positive definite with τ as root. Let $g(x, y)$ be an $\mathbb{R}^+$-equivalent reduced positive definite form. Then from part (b), $g(x, y)$ is unique up to a constant multiple, so it has a well-defined root $\tau' \in \mathfrak{h}$. Parts (d) and (e) show that $\tau' \in F$ and that τ and τ' are $\mathrm{SL}(2, \mathbb{Z})$-equivalent. This shows that every $\tau \in \mathfrak{h}$ is $\mathrm{SL}(2, \mathbb{Z})$-equivalent to some point of F.

To prove uniqueness, it suffices to show that no two distinct points of F are $\mathrm{SL}(2, \mathbb{Z})$-equivalent. So suppose τ, $\tau' \in F$ are $\mathrm{SL}(2, \mathbb{Z})$-equivalent. Then from part (d), the positive definite forms $f(x, y)$ and $g(x, y)$ with these roots are $\mathbb{R}^+$-equivalent, and from part (e), they are both reduced. But then from part (b), $f(x, y)$ and $g(x, y)$ are constant multiples of one another. It follows that their roots are equal, so that $\tau = \tau'$.

11.5. (a) (i) By (7.10), with $\gamma \in \mathrm{SL}(2, \mathbb{Z})$,

$$\mathrm{Im}(\gamma\tau) = \det(\gamma)\,|c\tau + d|^{-2}\,\mathrm{Im}(\tau),$$

so that since $\det(\gamma) = 1$, the inequalities $\mathrm{Im}(\tau) \le 1/\epsilon$ and $\mathrm{Im}(\gamma\tau) \ge \epsilon$ imply that

$$|c\tau + d|^2 = \frac{\mathrm{Im}(\tau)}{\mathrm{Im}(\gamma\tau)} \le \frac{1/\epsilon}{\epsilon} = \frac{1}{\epsilon^2}.$$

Then $|c\tau + d| \le 1/\epsilon$ follows immediately.

(ii) From part (i),

$$(*) \qquad \frac{1}{\epsilon^2} \ge |c\tau + d|^2 = |(c\,\mathrm{Re}(\tau) + d) + c\,\mathrm{Im}(\tau)|^2 = (c\,\mathrm{Re}(\tau) + d)^2 + (c\,\mathrm{Im}(\tau))^2.$$

From (*) we see that $|c\,\mathrm{Im}(\tau)| \le 1/\epsilon$. Since $\mathrm{Im}(\tau) \ge \epsilon > 0$, we obtain

$$|c| \le \frac{1}{\epsilon\,\mathrm{Im}(\tau)} \le \frac{1}{\epsilon^2}.$$

From (*) we also get $|c\,\mathrm{Re}(\tau) + d| \le 1/\epsilon$. Using the bound on $|c|$ just proved and the bound on $|\mathrm{Re}(\tau)|$ from the definition of C, we see that

$$|d| - \frac{M}{\epsilon^2} \le |d| - |c\,\mathrm{Re}(\tau)| \le |d + c\,\mathrm{Re}(\tau)| \le \frac{1}{\epsilon},$$

so that $|d| \le (\epsilon + M)/\epsilon^2$.

(iii) Since $\gamma \in \Delta(C)$, we can choose $\tau \in C$ with $\gamma\tau \in C$. Then $\gamma^{-1}(\gamma\tau) = \tau \in C$, so that $\gamma^{-1} \in \Delta(C)$. But

$$\gamma = \begin{pmatrix} a & b \\ c & d \end{pmatrix} \implies \gamma^{-1} = \begin{pmatrix} d & -b \\ -c & a \end{pmatrix}.$$

Since $\gamma^{-1} \in C$, part (ii) shows that $|a| \leq (\epsilon + M)/\epsilon^2$.

(iv) The definition of the action of γ gives $|a\tau + b| = |c\tau + d|\,|\gamma\tau|$, and therefore

$$|b| - |a\tau| \leq |a\tau + b| = |c\tau + d|\,|\gamma\tau| \quad \Rightarrow \quad |b| \leq |c\tau + d|\,|\gamma\tau| + |a\tau|.$$

But by parts (ii) and (iii), $|a|$, $|c|$, and $|d|$ are all bounded in terms of M and ϵ, as are $|\tau|$ and $|\gamma\tau|$, so that $|b|$ is bounded in terms of M and ϵ as well.

(b) Since $\overline{U}$ is compact and contained in $\mathfrak{h}$, it is contained in some rectangle $C \subset \mathfrak{h}$ of the form in part (a). By part (a), $\Delta(C) = \{\gamma \in \mathrm{SL}(2,\mathbb{Z}) \mid \gamma(C) \cap C \neq \emptyset\}$ is finite, so that $\{\gamma \in \mathrm{SL}(2,\mathbb{Z}) \mid \gamma(U) \cap U \neq \emptyset\} \subset \Delta(C)$ must be finite as well.

To see that this implies Lemma 11.5, choose a bounded open set W containing both τ and τ' such that $\overline{W} \subset \mathfrak{h}$. Then W is a neighborhood of both τ and τ', and by the previous paragraph, the set $\{\gamma \in \mathrm{SL}(2,\mathbb{Z}) : \gamma(W) \cap W \neq \emptyset\}$ is finite.

(c) By part (b), there is a neighborhood W of τ such that $\mathcal{S} = \{\gamma \in \mathrm{SL}(2,\mathbb{Z}) : \gamma(W) \cap W \neq \emptyset\}$ is finite. For $\gamma \in \mathcal{S}$ with $\gamma\tau \neq \tau$, pick a neighborhood U_γ of τ such that $\gamma(U_\gamma) \cap U_\gamma = \emptyset$. Then $U = W \cap \bigcap_{\gamma \in \mathcal{S}, \gamma\tau \neq \tau} U_\gamma$ is open since $\mathcal{S}$ is finite. Note also that $\tau \in U \subset W$.

Now take any $\gamma \in \mathrm{SL}(2,\mathbb{Z})$. If $\gamma(U) \cap U \neq \emptyset$, clearly $\gamma(W) \cap W \neq \emptyset$, so that $\gamma \in \mathcal{S}$. Then $\gamma\tau = \gamma$, since otherwise $\gamma(U) \cap U \subset \gamma(U_\gamma) \cap U_\gamma = \emptyset$. Conversely, if $\gamma\tau = \tau$, then $\tau \in U$ implies $\tau \in \gamma(U) \cap U$, hence $\gamma(U) \cap U \neq \emptyset$.

11.6. (a) If $c = 0$, then $\gamma = \left(\begin{smallmatrix} a & b \\ 0 & d \end{smallmatrix}\right)$. Since $\det(\gamma) = 1$, it follows that $a = d = \pm 1$, so that

$$\gamma = \begin{pmatrix} \pm 1 & b \\ 0 & \pm 1 \end{pmatrix} \quad \Longrightarrow \quad \gamma\tau = \frac{\pm\tau + b}{\pm 1} = \tau \pm b.$$

But $b \neq 0$ since $\gamma \neq \pm I$, so $\gamma\tau \neq \tau$. This contradicts our assumption that $\gamma\tau = \tau$, so $c \neq 0$.

(b) Suppose $\mathcal{O}^* \neq \{\pm 1\}$. Any unit in $\mathcal{O}$ is also a unit in $\mathcal{O}_K$, so that $K = \mathbb{Q}(i)$ or $K = \mathbb{Q}(\omega)$ by Exercise 5.9. If $K = \mathbb{Q}(i)$, then $\pm i \in \mathcal{O}$ implies $\mathcal{O} = \mathbb{Z}[i] = \mathcal{O}_K$. If $K = \mathbb{Q}(\omega)$, then $\pm\omega \in \mathcal{O}$ implies $\mathcal{O} = \mathbb{Z}[\omega] = \mathcal{O}_K$, and similarly $\pm\omega^2 = \pm(-1-\omega) \in \mathcal{O}$ implies that $\mathcal{O} = \mathbb{Z}[-1-\omega] = \mathbb{Z}[\omega] = \mathcal{O}_K$.

(c) If $\gamma = \left(\begin{smallmatrix} a & b \\ c & d \end{smallmatrix}\right)$ fixes i, then $i = \frac{ai+b}{ci+d}$, so that $di - c = ai + b$. Thus $a = d$ and $c = -b$, hence

$$\gamma = \begin{pmatrix} a & b \\ -b & a \end{pmatrix}.$$

Then $1 = \det(\gamma) = a^2 + b^2$, so since a and b are integers, either $a = \pm 1$, $b = 0$ or $a = 0$, $b = \pm 1$, and thus

$$\gamma = \pm\begin{pmatrix} 1 & 0 \\ 0 & 1 \end{pmatrix},\ \pm\begin{pmatrix} 0 & 1 \\ -1 & 0 \end{pmatrix}.$$

(d) Since $j(\tau) = 1728\frac{g_2(\tau)^3}{\Delta(\tau)}$, $g_2(\omega) = 0$, and $\Delta(\tau)$ never vanishes, it follows that ω is a triple root of $j(\tau)$, so that $j'(\omega) = j''(\omega) = 0$. Choose a neighborhood U of ω satisfying the conditions of Corollary 11.6 with respect to ω. Suppose $j'''(\omega) = 0$; then ω is a root of multiplicity 4 of $j(\tau)$, so that there are four distinct points $\tau_i \in U$, $i = 0, 1, 2, 3$ such that $j(\tau_i) = w$ where $w \in \mathbb{C}$ is close to zero. Since the $j(\tau_i)$ are equal, part (ii) of Theorem 11.2 shows that $\tau_i = \gamma_i\tau_0$, $i = 1, 2, 3$ for some $\gamma_i \in \mathrm{SL}(2,\mathbb{Z})$. It follows that $\pm I$, $\pm\gamma_1$, $\pm\gamma_2$, and $\pm\gamma_3$ are distinct elements of $\mathrm{SL}(2,\mathbb{Z})$. By Corollary 11.6, all eight of these elements fix ω since the τ_i all lie in U. But if $\gamma = \left(\begin{smallmatrix} a & b \\ c & d \end{smallmatrix}\right) \in \mathrm{SL}(2,\mathbb{Z})$ fixes ω, then

$$\omega = \frac{a\omega + b}{c\omega + d}, \quad a, b, c, d \in \mathbb{Z},\ ad - bc = 1.$$

Simplifying, we get

$$a\omega + b = \omega(c\omega + d) = c\omega^2 + d\omega = c(-1-\omega) + d\omega = -c + (d-c)\omega,$$

so that $c = -b$, and then $a = d - c = d + b$, so that $d = a - b$. Then

$$\gamma = \begin{pmatrix} a & b \\ -b & a-b \end{pmatrix}.$$

Then $\det(\gamma) = a(a-b) + b^2 = a^2 - ab + b^2 = 1$. Finally, $a^2 - ab + b^2$ is a positive definite form, and it takes the value 1 for elements $a + b\omega \in \mathbb{Q}(\omega)$ of norm 1; there are only six of these (the units of $\mathbb{Z}[\omega]$). This is a contradiction, and therefore $j'''(\omega) \neq 0$.

11.7. (a) If $f(\tau)$ is a modular function with a pole of order m at i, we can write $f(\tau)$ in the form $f(\tau) = g(\tau)/(\tau - i)^m$, where $g(\tau)$ is holomorphic and nonvanishing at i. Then for any $\gamma \in \mathrm{SL}(2,\mathbb{Z})$,

$$f(\tau) = f(\gamma\tau) = \frac{g(\gamma\tau)}{(\gamma\tau - i)^m}.$$

Take $\gamma = \left(\begin{smallmatrix} 0 & 1 \\ -1 & 0 \end{smallmatrix}\right)$, so that $\gamma\tau = -1/\tau$. Then

$$f(\tau) = \frac{g(\gamma\tau)}{(\gamma\tau - i)^m} = \frac{g(\gamma\tau)}{(-1/\tau + 1/i)^m} = \frac{g(\gamma\tau)}{(1/(\tau i)^m)(\tau - i)^m} = \frac{(\tau i)^m g(\gamma\tau)}{(\tau - i)^m}.$$

But also $f(\tau) = g(\tau)/(\tau - i)^m$, so equating these expressions gives $g(\tau) = (\tau i)^m g(\gamma\tau)$. Evaluating at $\tau = i$, and noting that γ fixes i, gives

$$g(i) = (-1)^m g(i).$$

Since $g(i) \neq 0$, m must be even.

(b) Since $f(\tau)$ is a modular function with a pole of order m at ω, we can write $f(\tau) = g(\tau)/(\tau - \omega)^m$, where $g(\tau)$ is holomorphic and nonvanishing at ω. Then for $\gamma \in \mathrm{SL}(2,\mathbb{Z})$,

$$f(\tau) = f(\gamma\tau) = \frac{g(\gamma\tau)}{(\gamma\tau - \omega)^m}.$$

Take $\gamma = \left(\begin{smallmatrix} 1 & 1 \\ -1 & 0 \end{smallmatrix}\right)$, so that $\gamma\tau = -(1/\tau) - 1$. Then

$$f(\tau) = \frac{g(\gamma\tau)}{(\gamma\tau - \omega)^m} = \frac{g(\gamma\tau)}{\left(-\frac{1}{\tau} - (\omega + 1)\right)^m} = \frac{g(\gamma\tau)}{\left(-\frac{1}{\tau} + \omega^2\right)^m} = \frac{g(\gamma\tau)}{\left(-\frac{1}{\tau} + \frac{1}{\omega}\right)^m},$$

and simplifying the denominator gives

$$= \frac{g(\gamma\tau)}{(1/(\tau\omega)^m)(\tau - \omega)^m} = \frac{(\tau\omega)^m g(\gamma\tau)}{(\tau - \omega)^m}.$$

But also $f(\tau) = g(\tau)/(\tau - \omega)^m$, so that $g(\tau) = (\tau\omega)^m g(\gamma\tau)$. An easy computation shows that γ fixes ω, so evaluating this equation at $\tau = \omega$ gives

$$g(\omega) = (\omega^2)^m g(\omega).$$

Since $g(\omega) \neq 0$, we get $\omega^{2m} = 1$, so that m must be divisible by three.

11.8. (a) To prove $\subset$, take $\gamma = \left(\begin{smallmatrix} a & b \\ mc & d \end{smallmatrix}\right) \in \Gamma_0(m)$ and set $\gamma' = \sigma_0\gamma\sigma_0^{-1}$. Then

$$\gamma' = \begin{pmatrix} m & 0 \\ 0 & 1 \end{pmatrix}\begin{pmatrix} a & b \\ mc & d \end{pmatrix}\begin{pmatrix} 1/m & 0 \\ 0 & 1 \end{pmatrix} = \begin{pmatrix} ma & mb \\ mc & d \end{pmatrix}\begin{pmatrix} 1/m & 0 \\ 0 & 1 \end{pmatrix} = \begin{pmatrix} a & mb \\ c & d \end{pmatrix} \in \mathrm{SL}(2,\mathbb{Z}),$$

so that $\gamma = \sigma_0^{-1}\gamma'\sigma_0 \in (\sigma_0^{-1}\,\mathrm{SL}(2,\mathbb{Z})\sigma_0) \cap \mathrm{SL}(2,\mathbb{Z})$. For the opposite inclusion, take $\gamma = \left(\begin{smallmatrix} a & b \\ c & d \end{smallmatrix}\right)$ in this intersection. So $\gamma \in \mathrm{SL}(2,\mathbb{Z})$, and

$$\sigma_0\gamma\sigma_0^{-1} = \begin{pmatrix} a & mb \\ c/m & d \end{pmatrix} \in \mathrm{SL}(2,\mathbb{Z}).$$

Thus c/m is integral, which implies that $\gamma \in \Gamma_0(m)$.

(b) For $\sigma \in C(m)$, set

$$S_\sigma = (\sigma_0^{-1}\,\mathrm{SL}(2,\mathbb{Z})\sigma) \cap \mathrm{SL}(2,\mathbb{Z}).$$

To prove that S_σ is a coset of $\Gamma_0(m)$, we first show that $S_\sigma \neq \emptyset$. Write $\sigma = \left(\begin{smallmatrix} a & b \\ 0 & d \end{smallmatrix}\right)$, where $ad = m$ and $\gcd(a,b,d) = 1$. Suppose for the moment that we can find $\mu \in \mathbb{Z}$ such that

(*) $$\gcd(d, a\mu - b) = 1.$$

Then let

$$\gamma_1 = \begin{pmatrix} d & a\mu - b \\ u & v \end{pmatrix},$$

where $u, v \in \mathbb{Z}$ satisfy $vd - u(a\mu - b) = 1$. The integers u, v exist by (*), and the equation they satisfy guarantees that $\gamma_1 \in \mathrm{SL}(2,\mathbb{Z})$. Using $ad = m$, one computes that

$$\gamma_0 = \sigma_0^{-1}\gamma_1\,\sigma = \begin{pmatrix} 1 & \mu \\ ua & vd - ub \end{pmatrix} \in \mathrm{SL}(2,\mathbb{Z})$$

Hence $\gamma_0 \in S_\sigma$, proving that $S_\sigma \neq \emptyset$.

It remains to prove (*). Consider all divisors ℓ of d such that $\gcd(\ell, b) = 1$, and let μ be their least common multiple, which also divides d and is relatively prime to b. Suppose a prime p divides d and $a\mu - b$. If $p \mid b$, then $p \mid a\mu$. But $p \nmid a$ since $\gcd(a,b,d) = 1$, and $p \nmid \mu$ since $\gcd(\mu, b) = 1$, so $p \nmid a\mu$. This contradiction shows that $p \nmid b$, so $\gcd(p, b) = 1$. Then $p \mid \mu$ by the definition of μ. Combining this with $p \mid a\mu - b$ shows that $p \mid b$, again a contradiction. We conclude that $\gcd(d, a\mu - b) = 1$, proving (*).

We prove that $S_\sigma = \Gamma_0(m)\gamma_0$ as follows. Recall from above that $\gamma_0 = \sigma_0^{-1}\gamma_1\,\sigma \in S_\sigma$ with $\gamma_1 \in \mathrm{SL}(2,\mathbb{Z})$. Then by part (a),

$$\Gamma_0(m)\gamma_0 \subset (\sigma_0^{-1}\,\mathrm{SL}(2,\mathbb{Z})\sigma_0)(\sigma_0^{-1}\gamma_1\,\sigma) = \sigma_0^{-1}\,\mathrm{SL}(2,\mathbb{Z})\gamma_1\,\sigma \subset \sigma_0^{-1}\,\mathrm{SL}(2,\mathbb{Z})\,\sigma,$$

where the last inclusion uses $\gamma_1 \in \mathrm{SL}(2,\mathbb{Z})$. Since $\gamma_0 \in \mathrm{SL}(2,\mathbb{Z})$, we also have $\Gamma_0(m)\,\gamma_0 \subset \mathrm{SL}(2,\mathbb{Z})$, so that

$$\Gamma_0(m)\,\gamma_0 \subset (\sigma_0^{-1}\,\mathrm{SL}(2,\mathbb{Z})\,\sigma) \cap \mathrm{SL}(2,\mathbb{Z}) = S_\sigma.$$

For the opposite inclusion, note that

$$\begin{aligned} S_\sigma\gamma_0^{-1} \subset (\sigma_0^{-1}\,\mathrm{SL}(2,\mathbb{Z})\,\sigma)\gamma_0^{-1} &= (\sigma_0^{-1}\,\mathrm{SL}(2,\mathbb{Z})\,\sigma)(\sigma_0^{-1}\gamma_1\,\sigma)^{-1} \\ &= (\sigma_0^{-1}\,\mathrm{SL}(2,\mathbb{Z})\,\sigma)(\sigma^{-1}\gamma_1^{-1}\,\sigma_0) \\ &= \sigma_0^{-1}(\mathrm{SL}(2,\mathbb{Z})\,\gamma_1^{-1})\,\sigma_0 \subset \sigma_0^{-1}\,\mathrm{SL}(2,\mathbb{Z})\,\sigma_0 \end{aligned}$$

since $\gamma_1 \in \mathrm{SL}(2,\mathbb{Z})$. But we also have $S_\sigma\gamma_0^{-1} \subset \mathrm{SL}(2,\mathbb{Z})$ since $\gamma_0 \in \mathrm{SL}(2,\mathbb{Z})$, so that by part (a),

$$S_\sigma\gamma_0^{-1} \subset (\sigma_0^{-1}\,\mathrm{SL}(2,\mathbb{Z})\,\sigma_0) \cap \mathrm{SL}(2,\mathbb{Z}) = \Gamma_0(m).$$

This implies that $S_\sigma \subset \Gamma_0(m)\,\gamma_0$ and completes the proof that $S_\sigma = \Gamma_0(m)\,\gamma_0$.

(c) To see that different elements of $C(m)$ produce different cosets of $\Gamma_0(m)$, choose $\sigma, \tau \in C(m)$ and suppose that $S_\sigma = S_\tau$. By part (a), an element of S_σ is an element of $\mathrm{SL}(2,\mathbb{Z})$ of the form $\sigma_0^{-1}\gamma_1\sigma$ for some $\gamma_1 \in \mathrm{SL}(2,\mathbb{Z})$. Since it is also in S_τ, we have

$$\sigma_0^{-1}\gamma_1\sigma = \sigma_0^{-1}\gamma_2\tau$$

for some $\gamma_2 \in \mathrm{SL}(2,\mathbb{Z})$. Thus $\tau = \gamma\sigma$ for $\gamma = \gamma_2^{-1}\gamma_1 \in \mathrm{SL}(2,\mathbb{Z})$. If

$$\gamma = \begin{pmatrix} p & q \\ r & s \end{pmatrix} \in \mathrm{SL}(2,\mathbb{Z}), \quad \sigma = \begin{pmatrix} a & b \\ 0 & d \end{pmatrix} \in C(m), \quad \tau = \begin{pmatrix} e & f \\ 0 & h \end{pmatrix} \in C(m),$$

then $\gamma\sigma = \tau$ means that

$$\begin{pmatrix} ap & bp+dq \\ ar & br+ds \end{pmatrix} = \begin{pmatrix} e & f \\ 0 & h \end{pmatrix}.$$

Since a is nonzero, this forces $r = 0$. But $\gamma \in \mathrm{SL}(2,\mathbb{Z})$, so that $p = s = \pm 1$. Since $e = ap$ is positive and a is positive, p must be positive as well, so that $p = s = 1$. But then

$$\begin{pmatrix} a & b+qd \\ 0 & d \end{pmatrix} = \tau \in C(m).$$

Thus $0 \le b+qd < d$, and $0 \le b < d$ since $\sigma \in C(m)$; this forces $q = 0$, so that $\sigma = \tau$.

Finally, we want to prove that every coset of $\Gamma_0(m)$ in $\mathrm{SL}(2,\mathbb{Z})$ is one of the S_σ's. This will follow if we can show that any $\gamma \in \mathrm{SL}(2,\mathbb{Z})$ lies in some S_σ, i.e., that γ can be written as $\sigma_0^{-1}\gamma'\sigma$ for some $\gamma' \in \mathrm{SL}(2,\mathbb{Z})$ and some $\sigma \in C(m)$. Equivalently, it suffices to show that $\sigma_0\gamma = \gamma'\sigma$. For this purpose, write $\sigma_0\gamma = \left(\begin{smallmatrix} a & b \\ c & d \end{smallmatrix}\right)$, and note that $\det(\sigma_0\gamma) = m$. Choose relatively prime integers x and y with $xa + yc = 0$, and then choose relatively prime integers u, v with $uy - vx = 1$. Thus $\left(\begin{smallmatrix} u & v \\ x & y \end{smallmatrix}\right) \in \mathrm{SL}(2,\mathbb{Z})$, and

$$\begin{pmatrix} u & v \\ x & y \end{pmatrix}\begin{pmatrix} a & b \\ c & d \end{pmatrix} = \begin{pmatrix} ua+vc & ub+vd \\ xa+yc & xb+yd \end{pmatrix} = \begin{pmatrix} a' & b' \\ 0 & d' \end{pmatrix},$$

with $a'd' = m$. Premultiplying if necessary by $-I$, we may assume that $a' > 0$, $d' > 0$. Now choose integers q, r such that $b' = d'q + r$, $0 \le r < d'$; we get

$$\underbrace{\begin{pmatrix} 1 & -q \\ 0 & 1 \end{pmatrix}\begin{pmatrix} u & v \\ x & y \end{pmatrix}}_{\gamma'}\underbrace{\begin{pmatrix} a & b \\ c & d \end{pmatrix}}_{\sigma_0\gamma} = \begin{pmatrix} 1 & -q \\ 0 & 1 \end{pmatrix}\begin{pmatrix} a' & b' \\ 0 & d' \end{pmatrix} = \underbrace{\begin{pmatrix} a' & r \\ 0 & d' \end{pmatrix}}_{\sigma}.$$

Then $\gamma' \in \mathrm{SL}(2,\mathbb{Z})$, so multiplying this equation on the left by $(\gamma')^{-1}$ shows that

$$\sigma_0\gamma = (\gamma')^{-1}\sigma.$$

The entries of σ_0 have no common divisor > 1, and since $\gamma, \gamma' \in \mathrm{SL}(2,\mathbb{Z})$, the same is true for σ. It follows that $\sigma \in C(m)$. Thus $\sigma_0\gamma = (\gamma')^{-1}\sigma$ for $\gamma' \in \mathrm{SL}(2,\mathbb{Z})$ and $\sigma \in C(m)$. This proves that γ is in the coset S_σ.

11.9. (a) Note that $a = m/d$. We say that $0 \le b < \gcd(a,d)$ is *admissible* if $\gcd(a,b,d) = 1$. This happens if and only if b is prime to $\gcd(a,d)$, so that there are $\phi(\gcd(a,d))$ admissible b's. If $b \ge \gcd(a,d)$ is such that $\gcd(a,b,d) = 1$, then also $\gcd(a, b - \gcd(a,d), d) = 1$, so such a b must be congruent to an admissible b. Each admissible b is congruent to $d/\gcd(a,d)$ numbers less than d, so that for each choice of $d \mid m$ we get

$$\frac{d}{\gcd(d,a)}\phi(\gcd(d,a)) = \frac{d}{\gcd(d,m/d)}\phi(\gcd(d,m/d))$$

possibilities for b. Thus

$$\Psi(m) = |C(m)| = \sum_{d\mid m} \frac{d}{\gcd(d,m/d)}\phi(\gcd(d,m/d)) = \sum_{d\mid m} \frac{d\phi(\gcd(d,m/d))}{\gcd(d,m/d)}.$$

(b) For any m_1 and m_2,

$$\begin{aligned}\Psi(m_1)\Psi(m_2) &= \Bigg(\sum_{d_1\mid m_1} \frac{d_1\phi(\gcd(d_1,m_1/d_1))}{\gcd(d_1,m_1/d_1)}\Bigg)\Bigg(\sum_{d_2\mid m_2} \frac{d_2\phi(\gcd(d_2,m_2/d_2))}{\gcd(d_2,m_2/d_2)}\Bigg) \\ &= \sum_{\substack{d_1\mid m_1 \\ d_2\mid m_2}} \frac{d_1 d_2}{\gcd(d_1,m_1/d_1)\gcd(d_2,m_2/d_2)}\phi(\gcd(d_1,m_1/d_1))\phi(\gcd(d_2,m_2/d_2)).\end{aligned}$$

If $\gcd(m_1, m_2) = 1$, then the sum is the same as the sum over all $d_1 d_2 \mid m_1 m_2$. Further, ϕ is multiplicative, and clearly the $\gcd(d_i, m_i/d_i)$ are relatively prime since the m_i are. The sum thus becomes

$$\begin{aligned}\Psi(m_1)\Psi(m_2) &= \sum_{d_1 d_2 \mid m_1 m_2} \frac{d_1 d_2}{\gcd(d_1 d_2, (m_1 m_2)/(d_1 d_2))}\phi(\gcd(d_1 d_2, (m_1 m_2)/(d_1 d_2))) \\ &= \sum_{d \mid m_1 m_2} \frac{d}{\gcd(d, m_1 m_2/d)}\phi(\gcd(d, m_1 m_2/d)) = \Psi(m_1 m_2).\end{aligned}$$

(c) From part (a),

$$\begin{aligned}\Psi(p^r) &= \sum_{i=0}^{r} \frac{p^i}{\gcd(p^i, p^{r-i})}\phi(\gcd(p^i, p^{r-i})) \\ &= \frac{p^0}{\gcd(p^0, p^r)} + \frac{p^r}{\gcd(p^r, p^0)} + \sum_{i=1}^{r-1} \frac{p^i}{\gcd(p^i, p^{r-i})}\phi(\gcd(p^i, p^{r-i})) \\ &= 1 + p^r + \sum_{i=1}^{[r/2]} \frac{p^i}{\gcd(p^i, p^{r-i})}\phi(\gcd(p^i, p^{r-i})) \\ &\quad + \sum_{i=[r/2]+1}^{r-1} \frac{p^i}{\gcd(p^i, p^{r-i})}\phi(\gcd(p^i, p^{r-i})),\end{aligned}$$

where $[\ ]$ is the greatest integer function. For the first sum, $1 \leq i \leq r - i$, so that

$$\frac{p^i}{\gcd(p^i, p^{r-i})}\phi(\gcd(p^i, p^{r-i})) = 1 \cdot \phi(p^i) = p^{i-1}(p-1),$$

while for the second, $r - i < i \leq r - 1$, so that

$$\frac{p^i}{\gcd(p^i, p^{r-i})}\phi(\gcd(p^i, p^{r-i})) = p^{2i-r}\phi(p^{r-i}) = p^{2i-r}p^{r-i-1}(p-1) = p^{i-1}(p-1).$$

Thus $\Psi(p^r) = 1 + p^r + \sum_{i=1}^{r-1} p^{i-1}(p-1) = 1 + p^r + (p-1) \cdot \frac{p^{r-1}-1}{p-1} = p^r + p^{r-1}$.

(d) From (b) and (c), if $m = \prod_{i=1}^{k} p_i^{r_i}$, where the p_i are distinct primes and each $r_i > 0$, then

$$\begin{aligned}\Psi(m) &= \prod_{i=1}^{k} \Psi(p_i^{r_i}) = \prod_{i=1}^{k} (p_i^{r_i} + p_i^{r_i - 1}) = \prod_{i=1}^{k} p_i^{r_i} \cdot \prod_{i=1}^{k} \left(1 + \frac{1}{p_i}\right) \\ &= m \prod_{i=1}^{k} \left(1 + \frac{1}{p_i}\right) = m \prod_{p \mid m} \left(1 + \frac{1}{p}\right).\end{aligned}$$

11.10. We have defined

$$\Phi_m(X, j(\tau)) = \prod_{i=1}^{|C(m)|} (X - j(m\gamma_i \tau)), \quad \mathcal{F}_m = \mathbb{C}(j(\tau), j(m\tau)),$$

where the γ_i are a set of coset representatives for $\Gamma_0(m)$ in $\mathrm{SL}(2, \mathbb{Z})$ with $\gamma_1 = I$. Then the introduction to the exercise notes that $\Phi_m(X, j(\tau))$ is a polynomial in $\mathbb{C}(j(\tau))[X]$ with $j(m\tau)$ a root. It follows that $[\mathcal{F}_m : \mathbb{C}(j(\tau))]$ is the degree of the minimal polynomial of $j(m\tau)$ over $\mathbb{C}(j(\tau))$ and thus has degree at most $\deg(\Phi_m(X, j(\tau))) = \Psi(m) = |C(m)|$.

(a) Fix $\gamma \in \mathrm{SL}(2, \mathbb{Z})$. First note that the poles of $f(\gamma\tau)$ correspond one-to-one to the poles of $f(\tau)$. Since γ and γ^{-1} act continuously on $\mathfrak{h}$, the isolated poles of $f(\tau)$ remain isolated in $f(\gamma\tau)$. It follows that $f(\gamma\tau)$ is meromorphic on $\mathfrak{h}$ and therefore lies in $\mathcal{F}$. It is clear that $f(\tau) \mapsto f(\gamma\tau)$ is a field homomorphism. It is nonzero since γ is an automorphism of $\mathfrak{h}$, and therefore is an embedding of fields. It is the identity on $\mathbb{C}(j(\tau))$ since $j(\tau)$ is $\mathrm{SL}(2, \mathbb{Z})$-invariant.

(b) Suppose the embeddings $f(\tau) \mapsto f(\gamma_i\tau)$ and $f(\tau) \mapsto f(\gamma_j\tau)$ are the same. Then applying these embeddings to $j(m\tau)$ we get $j(m\gamma_i\tau) = j(m\gamma_j\tau)$. From (11.12), $j(m\gamma_i\tau) = j(\sigma_i\tau)$ and $j(m\gamma_j\tau) = j(\sigma_j\tau)$ for some $\sigma_i, \sigma_j \in C(m)$, so that the q-expansions of $j(\sigma_i\tau)$ and $j(\sigma_j\tau)$ are equal. Looking at (11.13), since $a > 0$, the term with negative exponent shows us that the values of a are the same for σ_i and σ_j, which implies that their d's are also the same since $ad = m$. The numerator of that term forces them to have the same b as well since $\zeta_m^{-ab} = \zeta_d^{-b}$ and $0 \le b < d$. Therefore $\sigma_i = \sigma_j$, so that $\gamma_i = \gamma_j$ and thus $i = j$. So the embeddings are distinct on coset representatives. It follows that there are at least $|C(m)|$ distinct embeddings, so that $[\mathcal{F}_m : C(j(\tau))] \ge \Psi(m)$.

It follows that $[\mathcal{F}_m : C(j(\tau))] = \Psi(m)$, proving part (ii) of Theorem 11.18.

11.11. The coefficients of $G(X,\tau)$ are polynomials in the $f(\gamma_i\tau)$ and $j(m\gamma_i\tau)$; since each of these is meromorphic on $\mathfrak{h}$, so are their sums and products. To show invariance, note that each coefficient is a sum of terms and that the sum is invariant under a permutation of the γ_i. Then if $\gamma \in \mathrm{SL}(2,\mathbb{Z})$, the cosets $\Gamma_0(m)\gamma_i\gamma$ are a permutation of the $\Gamma_0(m)\gamma_i$. Since both $j(m\tau)$ and $f(\tau)$ are invariant under $\Gamma_0(m)$, we see that the coefficient is unchanged under the action of γ. Thus the coefficients are $\mathrm{SL}(2,\mathbb{Z})$-invariant. Finally, since the q-expansions of the $j(m\gamma_i\tau)$ and of $f(\tau)$ have only finitely many nonzero negative exponents, so does each coefficient. Thus the coefficients are modular functions for $\mathrm{SL}(2,\mathbb{Z})$.

11.12. If the q-expansion of $f(\tau)$ has $M < 0$, then $-M > 0$, so that $f(\tau)$ is holomorphic on $\mathfrak{h} \cup \{\infty\}$ and vanishes at ∞. By the proof of Lemma 11.10, it follows that $f(\tau)$ is identically zero. So we may assume that $M \ge 0$, and we prove by induction on M that there is $B(x) \in A[x]$ such that $f(\tau) = B(j(\tau))$. If $M = 0$, then $f(\tau)$ is holomorphic on $\mathfrak{h} \cup \{\infty\}$ and hence is constant. Then $f(\tau) = a_0 \in A$. Now assume the result is true for $M-1$ and $f(\tau) = a_{-M}q^{-M} + \cdots$. Since $j(\tau) = 1/q + \cdots$ has integer coefficients and A is an additive subgroup, it follows that $a_{-M}j(\tau)^M = a_{-M}q^{-M} + \cdots$ also has coefficients in A. Furthermore, in the q-expansion

$$g(\tau) = f(\tau) - a_{-M}j(\tau)^M = (a_{-M}q^{-M} + \cdots) - (a_{-M}q^{-M} + \cdots) = \sum_{n=-M+1} b_n q^n,$$

we have $b_n \in A$ for $n \le 0$ since A is an additive subgroup. By the inductive hypothesis, $g(\tau) = C(j(\tau))$ for some $C \in A[x]$. Then we are done since $f(\tau) = a_{-M}j(\tau)^M + C(j(\tau))$ and $a_{-M}x^M + C(x) \in A[x]$.

11.13. (a) Since $x^{p-1} + \cdots + x + 1 = \frac{x^p-1}{x-1}$, the roots of the left-hand side are the pth roots of unity other than one, so they are ζ_p^i, $i = 1, \ldots, p-1$. Then

$$x^{p-1} + \cdots + x + 1 = (x - \zeta_p)(x - \zeta_p^2)\cdots(x - \zeta_p^{p-1}).$$

Now set $x = 1$.

(b) Let $G = \mathrm{Gal}(\mathbb{Q}(\zeta_p)/\mathbb{Q})$. Then $N(\alpha) = \prod_{\sigma\in G}\sigma(\alpha)$ is symmetric in the $\sigma(\alpha)$. Since $\tau \in G$ permutes the $\sigma(\alpha)$, it follows that τ fixes $N(\alpha)$. Hence $N(\alpha) \in \mathbb{Q}$. But α is an algebraic integer, so that the same is true for each $\sigma(\alpha)$. Thus $N(\alpha)$ is an algebraic integer in $\mathbb{Q}$, hence $N(\alpha) \in \mathbb{Z}$. To see that $N(\alpha)N(\beta) = N(\alpha\beta)$, note that

$$N(\alpha\beta) = \prod_{\sigma\in G}\sigma(\alpha\beta) = \prod_{\sigma\in G}\sigma(\alpha)\sigma(\beta) = \prod_{\sigma\in G}\sigma(\alpha)\prod_{\sigma\in G}\sigma(\beta) = N(\alpha)N(\beta).$$

Finally, the conjugates of $1 - \zeta_p$ are $1 - \zeta_p^i$ for $1 \le i \le p-1$. So the product of the conjugates, which is $N(1 - \zeta_p)$, is equal to p by part (a).

(c) For $a \in \mathbb{Z}$, we have $N(a) = a^{p-1}$. Then $a = (1 - \zeta_p)\alpha$, $\alpha \in \mathbb{Z}[\zeta_p]$, implies that

$$a^{p-1} = N(a) = N((1 - \zeta_p)\alpha) = N(1 - \zeta_p)N(\alpha) = pN(\alpha)$$

by part (b). Since $N(\alpha) \in \mathbb{Z}$ by part (b), a^{p-1} is divisible by p, so that a is as well.

11.14. With $\sigma_0 = \left(\begin{smallmatrix} 1 & 0 \\ 0 & p \end{smallmatrix}\right)$, then $j(\tau) = 1/q + \sum_{n=0}^{\infty} c_n q^n$ with $q = e^{2\pi i \tau}$ implies that

$$j(\sigma_0 \tau) = j(\tau/p) = \frac{1}{q^{1/p}} + \sum_{n=0}^{\infty} c_n (q^{1/p})^n.$$

Now take pth powers of both sides. All cross-product terms in the expansion of the sum on the right-hand side have coefficients divisible by p. Thus

$$j(\sigma_0 \tau)^p = \left(\frac{1}{q^{1/p}} + \sum_{n=0}^{\infty} c_n (q^{1/p})^n \right)^p \equiv \frac{1}{q} + \sum_{n=0}^{\infty} c_n^p q^n \bmod p.$$

However, $c_n^p \equiv c_n \bmod p$ by Fermat's Little Theorem, so that finally

$$j(\sigma_0 \tau)^p \equiv \frac{1}{q} + \sum_{n=0}^{\infty} c_n q^n = j(\tau) \bmod p.$$

11.15. Consider the coefficient $g(\tau)$ of some power of X in $f(X, j(\tau))$. It is a polynomial in $j(\tau)$, so is a holomorphic modular function. By assumption, its q-expansion lies in $p\mathbb{Z}((q))$, so the coefficients of its q-expansion are integral and lie in $p\mathbb{Z}$, which is an additive subgroup of $\mathbb{Z}$. It follows from the Hasse q-expansion principle (Exercise 11.12) that $g(\tau)$ is a polynomial in $j(\tau)$ with integral coefficients in $p\mathbb{Z}$, and therefore that $f(X,Y)$ is a polynomial with integral coefficients in $p\mathbb{Z}$, i.e., that $f(X,Y) \in p\mathbb{Z}[X,Y]$.

11.16. (a) First assume that G contains a subgroup isomorphic to $(\mathbb{Z}/d\mathbb{Z})^2$ for some $d > 1$. Since every subgroup of a cyclic group is cyclic, it follows that G is not cyclic. Conversely, assume that G is not cyclic. Decompose G into a direct product of prime power subgroups, $G = \prod_{i=1}^{k} \mathbb{Z}/p_i^{r_i}\mathbb{Z}$, $r_i \geq 1$. Then we must have $p_i = p_j$ for at least one pair $i \neq j$, since otherwise G would be cyclic. However, $\mathbb{Z}/p_i^{r_i}\mathbb{Z}$ has a subgroup isomorphic to $\mathbb{Z}/p_i\mathbb{Z}$, and similarly for $\mathbb{Z}/p_j^{r_j}\mathbb{Z}$. Then G contains a subgroup isomorphic to $(\mathbb{Z}/d\mathbb{Z})^2$ for $d = p_i = p_j$.

(b) Using the hint, suppose $d > 1$ is a common divisor of the entries of A, and let $A = dA'$. Then A' is an integer matrix as well, and $\det(A') \neq 0$ since $\det(A) \neq 0$. Then we have $AM = dA'M \subset A'M \subset M$, which gives

$$A'M/dA'M = A'M/AM \hookrightarrow M/AM.$$

But $A'M$ has finite index in M since $\det(A') \neq 0$. Then $M = \mathbb{Z}^2$ implies $A'M \simeq \mathbb{Z}^2$, so that $A'M/dA'M \simeq \mathbb{Z}^2/d\mathbb{Z}^2 \simeq (\mathbb{Z}/d\mathbb{Z})^2$. By part (a), M/AM is not cyclic.

(c) Since M/AM is not cyclic, it contains a subgroup isomorphic to $(\mathbb{Z}/d\mathbb{Z})^2$ for some $d > 1$, so that there is some M' with $AM \subset M' \subset M$ and $M'/AM \simeq (\mathbb{Z}/d\mathbb{Z})^2$. Thus $dM' \subset AM$, and since $M'/dM' \simeq (\mathbb{Z}/d\mathbb{Z})^2$, we must have $AM = dM'$, so that every element of AM is divisible by d. If we write $A = \left(\begin{smallmatrix} p & q \\ r & s \end{smallmatrix}\right)$, then $M = \mathbb{Z}^2$ implies that $AM = \{(pm + qn, rm + sn) : m, n \in \mathbb{Z}\}$. Since the elements of AM are all divisible by d, it follows easily that d divides the entries of A.

11.17. (a) If L' has index N in $[1, \tau]$, then $1 \in [1, \tau]$ implies that $N = N \cdot 1 \in L'$. Thus L' contains some positive integer, so that $L' \cap \mathbb{Z} \subset \mathbb{Z}$ is a nontrivial subgroup. It follows that $L' \cap \mathbb{Z} = d\mathbb{Z}$ for some $d > 0$, and this d is therefore the smallest positive integer in $L' \cap \mathbb{Z}$. Since L' has rank 2, the quotient $L'/d\mathbb{Z}$ has rank 1. We claim that $L'/d\mathbb{Z}$ is torsion-free. To see why, take $[\alpha] \in L'/d\mathbb{Z}$ and suppose that $m[\alpha] = [m\alpha] = [0]$ for some $m \neq 0$ in $\mathbb{Z}$. Write $\alpha = a + b\tau$ for $a, b \in \mathbb{Z}$. Then $m\alpha = ma\tau + mb \in d\mathbb{Z}$, so that $a = 0$. Then $\alpha = b \in L' \cap \mathbb{Z} = d\mathbb{Z}$, which implies $[\alpha] = [0]$ in $L'/d\mathbb{Z}$. Thus $L'/d\mathbb{Z}$ is torsion-free.

Since it has rank 1 and is torsion-free, $L'/d\mathbb{Z} \simeq \mathbb{Z}$. Let $\alpha \in L'$ map to a generator of $L'/d\mathbb{Z}$. It follows easily that $L' = d\mathbb{Z} + \alpha\mathbb{Z} = [d, \alpha]$, and since $L' \subset [1, \tau]$, we can write $\alpha = a + b\tau$ for $a, b \in \mathbb{Z}$. This proves that $L' = [d, a + b\tau]$ for $a, b \in \mathbb{Z}$.

(b) Let

$$\sigma = \begin{pmatrix} a & b \\ 0 & d \end{pmatrix}, \ \sigma' = \begin{pmatrix} a' & b' \\ 0 & d' \end{pmatrix} \in C(m),$$

and observe that $d[1, \sigma\tau] = d[1, (a\tau+b)/d] = [d, a\tau+b]$. Similarly, we have $d'[1, \sigma'\tau] = [d', a'\tau + b']$. Also, for $r, s \in \mathbb{Z}$, $rd + s(a\tau + b) \in \mathbb{Z}$ implies $s = 0$, so that d is the smallest positive integer in $d[1, \sigma\tau]$. Similarly, d' is the smallest positive integer in $d'[1, \sigma'\tau]$.

Now suppose that $d[1, \sigma\tau] = d'[1, \sigma'\tau]$. By the previous paragraph, it follows immediately that $d = d'$. Then $ad = m = a'd'$ implies $a = a'$. Thus $[d, a\tau + b] = [d, a\tau + b']$. In particular,

$$a\tau + b = rd + s(a\tau + b'), \quad r, s \in \mathbb{Z}.$$

This implies $a\tau = sa\tau$, so $s = 1$ since $a \neq 0$. Thus $b = rd + b'$, which forces $r = 0$ since $0 \leq b, b' < d$. It follows that $b = b'$ and $\sigma = \sigma'$.

11.18. Let $L = [1, \tau]$ and $j(L) = j(\tau)$. Then by (11.15) and Lemma 11.24,

(*) the roots of $\Phi_m(X, j(L))$ are $j(L')$ for cyclic sublattices $L' \subset L$ of index m.

If $L' \subset L$ is such a lattice, then $(u, v) = (j(L'), j(L))$ satisfies

$$\Phi_m(u, v) = \Phi_m(j(L'), j(L)) = 0$$

by (*). This proves one direction of Theorem 11.23. Conversely, suppose $\Phi_m(u, v) = 0$. Since $j : \mathfrak{h} \to \mathbb{C}$ is surjective by Theorem 11.2, choose $\tau \in \mathfrak{h}$ with $j(\tau) = v$, and let $L = [1, \tau]$. Then $j(L) = v$, and since $\Phi_m(u, j(L)) = \Phi_m(u, v) = 0$, we know from (*) that u is the j-invariant of some cyclic sublattice of L of index m. This completes the proof of Theorem 11.23.

11.19. As an Abelian group, $\mathfrak{b} \simeq \mathbb{Z}^2$, say $\mathfrak{b} = [\alpha, \beta]$. Let $\mathfrak{a}$ be a proper $\mathcal{O}$-ideal that is not primitive, say $\mathfrak{a} = d\mathfrak{a}'$ for some $d > 1$ and some proper $\mathcal{O}$-ideal $\mathfrak{a}'$. Then $\mathfrak{a}'\mathfrak{b} \subset \mathfrak{b}$ is generated by two elements $p\alpha + q\beta$, $r\alpha + s\beta$ for $a, b, c, d \in \mathbb{Z}$. Setting $A = d\left(\begin{smallmatrix} p & q \\ r & s \end{smallmatrix}\right)$ gives $A\mathfrak{b} = d\mathfrak{a}'\mathfrak{b} = \mathfrak{a}\mathfrak{b}$. Since $d > 1$ divides the entries of A, $\mathfrak{b}/\mathfrak{a}\mathfrak{b}$ is not cyclic by part (b) of Exercise 11.16.

11.20. Let $\mathcal{O} = [1, fw_K]$, where $w_K = \frac{d_K + \sqrt{d_K}}{2}$. Then $\alpha = fw_K$ is a primitive element of $\mathcal{O}$, since if $fw_k = d(m + nfw_k)$ with $d > 1$ and $m, n \in \mathbb{Z}$, then $m = 0$ and $dn = 1$, a contradiction. The norm of α is

$$N(\alpha) = N(f)N(w_K) = f^2 w_K \overline{w_K} = f^2 \cdot \frac{d_K + \sqrt{d_K}}{2} \cdot \frac{d_K - \sqrt{d_K}}{2} = f^2 \cdot \frac{d_K(d_K - 1)}{4}.$$

At least one of $|d_K|$ and $|d_K - 1|$ is a nonsquare, and since they are relatively prime, their product cannot be a square. It follows that $N(\alpha)$ is not a perfect square.

11.21. (a) Let $n = [L : \mathbb{Q}]$, $m = [L : K]$, and $\ell = [K : \mathbb{Q}]$, so that $n = m\ell$. Then by Proposition 5.3, $\mathcal{O}_L$ is a free $\mathbb{Z}$-module of rank n and $\mathcal{O}_K$ is a free $\mathbb{Z}$-module of rank ℓ. By part (b) of Exercise 7.1, $\mathcal{O}_K$ contains a $\mathbb{Q}$-basis $\{\beta_1, \dots, \beta_\ell\}$ of K over $\mathbb{Q}$. Furthermore, $L = K(\alpha)$ implies that $\{1, \alpha, \alpha^2, \dots, \alpha^{m-1}\}$ is a basis for L over K. Then $\{\beta_i\alpha^j : 1 \leq i \leq \ell, 0 \leq j \leq m - 1\}$ is a basis of L over $\mathbb{Q}$ that lies in $\mathcal{O}_K[\alpha]$. However, $\alpha \in \mathcal{O}_L$ implies that $\mathcal{O}_K[\alpha] \subset \mathcal{O}_L$. Since $\mathcal{O}_L$ is a free $\mathbb{Z}$-module of finite rank, the same is true for $\mathcal{O}_K[\alpha]$. Thus $\mathcal{O}_K[\alpha] \subset L$ is a finitely generated $\mathbb{Z}$-module that contains a $\mathbb{Q}$-basis of L. By part (b) of Exercise 7.1, $\mathcal{O}_K[\alpha]$ has rank n. Since $\mathcal{O}_L$ has the same rank, it follows that $\mathcal{O}_K[\alpha]$ has finite index in $\mathcal{O}_L$.

(b) Let $N = [\mathcal{O}_L : \mathcal{O}_K[\alpha]]$. If $\gcd(p, N) = 1$, then $|\mathcal{O}_L/\mathfrak{P}| = p^f$ shows that multiplication by N induces an isomorphism of $\mathcal{O}_L/\mathfrak{P}$. If $\beta \in \mathcal{O}_L$, choose $\beta' \in \mathcal{O}_L$ with $N\beta' \equiv \beta \bmod \mathfrak{P}$. Since $N\mathcal{O}_L \subset \mathcal{O}_K[\alpha]$, we see that $N\beta' \in \mathcal{O}_K[\alpha]$, and then

$$\beta^p \equiv (N\beta')^p \equiv N\beta' \equiv \beta \bmod \mathfrak{P},$$

where $(N\beta')^p \equiv N\beta' \bmod \mathfrak{P}$ by the hypothesis of part (b) and $N^p \equiv N \bmod \mathfrak{P}$ by Fermat's Little Theorem, applicable since $\gcd(p, N) = 1$.

11.22. When $\mathfrak{p}$ is a prime of $\mathcal{O}_K$ relatively prime to the conductor f of $\mathcal{O}$, Theorem 11.36 tells us that $((L/K)/\mathfrak{p})(j(\mathfrak{b})) = j(\overline{\mathfrak{p} \cap \mathcal{O}}\mathfrak{b})$. Let $\mathfrak{a}$ be an $\mathcal{O}$-ideal relatively prime to f. By Proposition 7.20, $\mathfrak{a}\mathcal{O}_K$ is an $\mathcal{O}_K$-ideal relatively prime to f. Write $\mathfrak{a}\mathcal{O}_K = \mathfrak{p}_1 \cdots \mathfrak{p}_r$, where $\mathfrak{p}_i$ is prime in $\mathcal{O}_K$ and relatively prime to f. As in the discussion preceding Theorem 5.23,

$$\begin{aligned}\left(\frac{L/K}{\mathfrak{a}\mathcal{O}_K}\right)(j(\mathfrak{b})) &= \left(\frac{L/K}{\mathfrak{p}_1}\right)\cdots\left(\frac{L/K}{\mathfrak{p}_r}\right)(j(\mathfrak{b}))\\ &= j(\overline{(\mathfrak{p}_1 \cap \mathcal{O})}\cdots\overline{(\mathfrak{p}_r \cap \mathcal{O})}\mathfrak{b}) = j(\overline{(\mathfrak{p}_1 \cap \mathcal{O})\cdots(\mathfrak{p}_r \cap \mathcal{O})}\mathfrak{b}).\end{aligned}$$

The proof of Proposition 7.20 shows that $\mathfrak{c} \mapsto \mathfrak{c} \cap \mathcal{O}$ preserves multiplication for $\mathcal{O}_K$-ideals relatively prime to f. It follows that $(\mathfrak{p}_1 \cap \mathcal{O}) \cdots (\mathfrak{p}_r \cap \mathcal{O}) = (\mathfrak{p}_1 \cdots \mathfrak{p}_r) \cap \mathcal{O} = \mathfrak{a}\mathcal{O}_K \cap \mathcal{O} = \mathfrak{a}$, where the last equality uses (7.21). Hence the above formula simplifies to

$$(*) \qquad \left(\frac{L/K}{\mathfrak{a}\mathcal{O}_K}\right)(j(\mathfrak{b})) = j(\overline{\mathfrak{a}}\mathfrak{b}).$$

It follows that $\sigma_\mathfrak{a} = ((L/K)/\mathfrak{a}\mathcal{O}_K) \in \mathrm{Gal}(L/K)$ is well-defined, and that $\sigma_\mathfrak{a}(j(\mathfrak{b})) = j(\overline{\mathfrak{a}}\mathfrak{b})$, which proves the first assertion of Corollary 11.37. Furthermore, we have isomorphisms

$$C(\mathcal{O}) \simeq I(\mathcal{O}, f)/P(\mathcal{O}, f) \simeq I_K(f)/P_K(f) \simeq \mathrm{Gal}(L/K),$$

where the final map is the Artin map. By (*), this map sends the class of $\mathfrak{a}$ to $\sigma_\mathfrak{a}$. This proves the second assertion of the corollary.

11.23. Let L be an Abelian extension of K. By Theorem 8.2, there is a modulus $\mathfrak{m}$ such that $\mathrm{Gal}(L/K)$ is a generalized congruence subgroup for $\mathfrak{m}$. We claim that if $\mathfrak{n}$ is a modulus with $\mathfrak{m} \mid \mathfrak{n}$, then L is contained in the ray class field $K_\mathfrak{n}$. To prove this, note that by Exercise 8.4, $\mathrm{Gal}(L/K)$ is a generalized congruence subgroup for $\mathfrak{n}$, which means that L corresponds to a subgroup of $I_K(\mathfrak{n})$ containing $P_{K,1}(\mathfrak{n})$. Since the ray class field $K_\mathfrak{n}$ corresponds to $P_{K,1}(\mathfrak{n})$, the inclusion $L \subset K_\mathfrak{n}$ follows from Corollary 8.7.

Since K is imaginary quadratic, a modulus is simply an ideal of $\mathcal{O}_K$. Let $N = N(\mathfrak{m})$, so that $\mathfrak{m}\overline{\mathfrak{m}} = N\mathcal{O}_K$. Then $\mathfrak{m} \mid N\mathcal{O}_K$ and the previous paragraph imply that $L \subset K_{N\mathcal{O}_K}$.

11.24. In §10, we saw that $\wp(\lambda z; \lambda L) = \lambda^{-2}\wp(z; L)$, and by (10.10),

$$(*) \qquad g_2(\lambda L) = \lambda^{-4} g_2(L), \quad g_3(\lambda L) = \lambda^{-6} g_3(L).$$

Thus

$$\begin{aligned} g_2(\lambda L)^2 \wp(\lambda z; \lambda L)^2 &= \lambda^{-12} g_2(L)^2 \wp(z; L)^2\\ g_3(\lambda L)\wp(\lambda z; \lambda L)^3 &= \lambda^{-12} g_3(L)\wp(z; L)^3\\ g_2(\lambda L) g_3(\lambda L)\wp(\lambda z; \lambda L) &= \lambda^{-12} g_2(L) g_3(L)\wp(z; L).\end{aligned}$$

By (*), we also have $\Delta(\lambda L) = g_3(\lambda L)^2 - 27g_2(\lambda L)^3 = \lambda^{-12}\Delta(L)$. Then dividing the above equations by $\Delta(\lambda L)$ yields

$$\frac{g_2(\lambda L)^2\wp(\lambda z;\lambda L)^2}{\Delta(\lambda L)} = \frac{g_2(L)^2\wp(z;L)^2}{\Delta(L)}$$
$$\frac{g_3(\lambda L)\wp(\lambda z;\lambda L)^3}{\Delta(\lambda L)} = \frac{g_3(L)\wp(z;L)^3}{\Delta(L)}$$
$$\frac{g_2(\lambda L)g_3(\lambda L)\wp(\lambda z;\lambda L)}{\Delta(\lambda L)} = \frac{g_2(L)g_3(L)\wp(z;L)}{\Delta(L)}.$$

By the definition of the Weber function, we obtain $h(\lambda z;\lambda L) = h(z;L)$ in all three cases.

Solutions to Exercises in §12

12.1. Suppose $\tau = ai$ for $a > 0$ in $\mathbb{R}$. First note that $[1, ai] = [1, -ai] = [1, \overline{ai}] = \overline{[1, ai]}$. Then, using Exercise 11.1, we obtain

$$g_2(ai) = g_2([1, ai]) = g_2(\overline{[1, ai]}) = \overline{g_2([1, ai])} = \overline{g_2(ai)}.$$

Thus $g_2(\tau)$ is real-valued on the positive imaginary axis. The same argument holds for $g_3(\tau)$, and this implies that $\Delta(\tau)$ is real-valued there as well.

12.2. (a) One can define a single-valued holomorphic function $H(z) = \sqrt[m]{z}$ in a small neighborhood U of $1 \in \mathbb{C}$ such that $H(1) = 1$. Thus $H(z)^m = z$ for $z \in U$. Since $F(0) = 1$ and $F(q)$ is continuous at $0 \in \mathbb{C}$, there is a neighborhood V of 0 such that $F(V) \subset U$. Then the composition $G = H \circ F$ is defined and holomorphic on V and satisfies

$$G(q)^m = H(F(q))^m = F(q)$$

for $q \in V$. Also, $G(0) = H(F(0)) = H(1) = 1$.

(b) Write $G(q) = 1 + \sum_{n=1}^{\infty} b_n q^n$. Since $F(q) = 1 + \sum_{n=1}^{\infty} a_n q^n$, the equation $F(q) = G(q)^m$ implies $1 + \sum_{n=1}^{\infty} a_n q^n = (1 + \sum_{n=1}^{\infty} b_n q^n)^m$. Comparing the coefficients of q^n on each side, we obtain

$$a_n = \sum_{\substack{i_1+\cdots+i_m=n \\ 0\le i_1,\ldots,i_m\le n}} b_{i_1}\cdots b_{i_m}$$

for $n \geq 0$, where $a_0 = b_0 = 1$. When $n \geq 1$, the only way b_n can appear in the sum on the right is if $i_j = n$ for some j and $i_k = 0$ for $k \neq j$. Thus

$$(*) \qquad a_n = mb_n + \sum_{\substack{i_1+\cdots+i_m=n \\ 0\le i_1,\ldots,i_m< n}} b_{i_1}\cdots b_{i_m}.$$

The a_n are all rational by assumption. By (*), $a_1 = mb_1$, so that b_1 is rational. Furthermore, if $b_1, \ldots, b_{n-1}$ are rational, then (*) shows that b_n is also rational. By complete induction, it follows that a_n is rational for all $n \geq 1$.

12.3. (a) Note that

$$S\begin{pmatrix} 0 & b \\ c & d \end{pmatrix} = \begin{pmatrix} -c & -d \\ 0 & b \end{pmatrix},$$

so that it suffices to show that matrices in $\mathrm{SL}(2,\mathbb{Z})$ of the form $\left(\begin{smallmatrix} a & b \\ 0 & d \end{smallmatrix}\right)$ lie in Γ. Given such a matrix, we must have $a = d = 1$ or $a = d = -1$. Since $S^2 = -I$, we may, after multiplying by S^2 if necessary, assume that $a = d = 1$. Then we are done since

$$\begin{pmatrix} 1 & b \\ 0 & 1 \end{pmatrix} = T^b \in \Gamma.$$

(b) Choose $\gamma_0 \in \mathrm{SL}(2,\mathbb{Z})$, and choose $\gamma \in \Gamma$ such that

$$\gamma\gamma_0 = \begin{pmatrix} a & b \\ c & d \end{pmatrix}$$

has minimal $|c|$.

(i) If $a = 0$ or $c = 0$, then it is immediate from part (a) that $\gamma\gamma_0 \in \Gamma$, so that $\gamma_0 \in \Gamma$.

(ii) Now assume for purposes of contradiction that $c \neq 0$. Choose a γ giving the minimal $|a|$ from among those that give the minimal $|c|$. Since

$$S\gamma\gamma_0 = \begin{pmatrix} -c & * \\ a & * \end{pmatrix} \in \Gamma,$$

minimality of $|c|$ shows that $|a| \geq |c|$. Now,

$$T^{\pm 1}\gamma\gamma_0 = \begin{pmatrix} a \pm c & * \\ c & * \end{pmatrix} \in \Gamma.$$

We can write $a \pm c = a \pm |c|$ (possibly interchanging the plus and minus signs). If $a > 0$, then $|c| \leq |a|$ shows that $a > a - |c| \geq 0$, contradicting minimality of $|a|$. If $a < 0$, then $a < a + |c| \leq 0$, which again contradicts minimality of $|a|$. This contradiction shows that $c = 0$, which implies by (i) that $\gamma_0 \in \Gamma$.

(c) In all of this, γ_0 was chosen to be an arbitrary element of $\mathrm{SL}(2,\mathbb{Z})$, and part (b) showed that by multiplying it by an appropriate element of Γ, we get a matrix with lower left entry equal to 0. This matrix is in Γ by part (a). It follows that $\gamma_0 \in \Gamma$ and we are done.

12.4. We begin with two observations. First, the groups $\Gamma_0(2)$, $\Gamma_0(2)^t$ and $\Gamma(2)$ have the property that if $\left(\begin{smallmatrix} a & b \\ c & d \end{smallmatrix}\right)$ lies in any of them, then a is odd. This follows because $ad - bc = 1$ and at least one of b, c is even. Second, if $x, y \in \mathbb{R}$, then

$$(*) \qquad |x \pm 2y| \geq |x| \implies y = 0 \text{ or } |y| \geq |x|.$$

Here and below, $|x \pm 2y| \geq |x|$ is short for $|x + 2y| \geq |x|$ and $|x - 2y| \geq |x|$. To prove (*), first suppose $x, y \geq 0$. Then the inequality with the minus sign implies that either $x - 2y \geq x$ or $-x + 2y \geq x$. The former implies $y \leq 0$, hence $y = 0$, and the latter implies $y \geq x$, hence $|y| \geq |x|$. When $x \geq 0$ and $y \leq 0$, the inequality with the plus sign implies that either $x + 2y \geq x$ or $-x - 2y \geq x$, which again implies $y = 0$ or $|y| \geq |x|$. The other cases are handled similarly, and (*) is proved.

(a) Let $\Gamma \subset \mathrm{SL}(2,\mathbb{Z})$ be the subgroup generated by $-I$, A^2 and B. Note that $\Gamma \subset \Gamma_0(2)$. We first claim that if $\gamma = \left(\begin{smallmatrix} a & b \\ 0 & d \end{smallmatrix}\right) \in \mathrm{SL}(2,\mathbb{Z})$, then $\gamma \in \Gamma$. To see this, note that $a = d = \pm 1$. If $a = d = 1$, then $\gamma = B^b \in \Gamma$, while if $a = d = -1$, then $\gamma = (-I)B^{-b} \in \Gamma$. This proves the claim.

Now, pick $\gamma_0 \in \Gamma_0(2)$. Choose $\gamma \in \Gamma$ such that $\gamma\gamma_0 = \left(\begin{smallmatrix} a & b \\ c & d \end{smallmatrix}\right) \in \Gamma_0(2)$ has minimal $|c|$ and, from among those with that minimal $|c|$, also has minimal $|a|$. We know that c is even since $\gamma\gamma_0 \in \Gamma_0(2)$.

Assume $c \neq 0$. If $|a| > |c|$, then either $|a + c| < |a|$ or $|a - c| < |a|$. But

$$B^{\pm 1}\begin{pmatrix} a & b \\ c & d \end{pmatrix} = \begin{pmatrix} a \pm c & * \\ c & * \end{pmatrix},$$

contradicting the minimality of $|a|$. Further, $|a| \neq |c|$ since a is odd and c is even. Thus $|a| < |c|$. Now

$$A^{\pm 2}\begin{pmatrix} a & b \\ c & d \end{pmatrix} = \begin{pmatrix} a & * \\ c \pm 2a & * \end{pmatrix}.$$

By the minimality of c, $|c \pm 2a| \geq |c|$, which by (*) forces $a = 0$ since $|a| < |c|$. But we know that a is odd. This contradiction shows that $c = 0$. It follows that $\gamma\gamma_0 \in \Gamma$, so that $\gamma_0 \in \Gamma$ and we are done.

(b) Since $(A^2)^t = B^2$, $B^t = A$ and $(-I)^t = -I$, it is immediate from part (a) that $-I$, A and B^2 generate the transpose $\Gamma_0(2)^t$.

(c) Let $\Gamma \subset \mathrm{SL}(2,\mathbb{Z})$ be the the subgroup generated by $-I$, A^2 and B^2. Note that $\Gamma \subset \Gamma(2)$.

We first prove that any matrix of the form $\left(\begin{smallmatrix} a & b \\ 0 & d \end{smallmatrix}\right) \in \Gamma(2)$ lies in Γ. To see this, note that $a = d = \pm 1$. If $a = d = 1$, then $\gamma = B^b \in \Gamma$, while if $a = d = -1$, then $\gamma = (-I)B^{-b} \in \Gamma$ since b is even for matrices in $\Gamma(2)$. This proves the claim.

Now, pick $\gamma_0 \in \Gamma(2)$. Choose $\gamma \in \Gamma$ such that $\gamma\gamma_0 = \left(\begin{smallmatrix} a & b \\ c & d \end{smallmatrix}\right) \in \Gamma(2)$ has minimal $|c|$ and, from among those with that minimal $|c|$, also has minimal $|a|$. Note also that a is odd and c is even since $\gamma\gamma_0 \in \Gamma(2)$.

Assume $c \neq 0$. Then

$$B^{\pm 2}\begin{pmatrix} a & b \\ c & d \end{pmatrix} = \begin{pmatrix} a \pm 2c & * \\ c & * \end{pmatrix},$$

The right-hand side has the minimal c, so by the minimality of $|a|$, $|a \pm 2c| \geq |a|$. Since $c \neq 0$, we get $|c| \geq |a|$ by (*). But we also have

$$A^{\pm 2}\begin{pmatrix} a & b \\ c & d \end{pmatrix} = \begin{pmatrix} a & * \\ c \pm 2a & * \end{pmatrix}.$$

which by the minimality of c implies $|c \pm 2a| \geq |c|$. By (*), either $a = 0$ or $|a| \geq |c|$. The former cannot occur since a is odd, and the latter implies $|a| = |c|$ since we proved above that $|c| \geq |a|$. But $|a| = |c|$ is impossible since a is odd and c is even. This contradiction shows that $c = 0$. Above, we showed that any $\left(\begin{smallmatrix} a & b \\ 0 & d \end{smallmatrix}\right) \in \Gamma(2)$ lies in Γ, and from here it follows easily that $\gamma_0 \in \Gamma$. This completes the proof that $\Gamma = \Gamma(2)$.

12.5. (a) If $\gamma = \left(\begin{smallmatrix} a & b \\ c & d \end{smallmatrix}\right) \in \mathrm{SL}(2,\mathbb{Z})$, let $\rho(\gamma) = ac - ab + a^2cd - cd$ be the exponent of ζ_3 in (12.6).

We first show that (12.6) holds for S, S^{-1}, T, T^{-1}. To begin, $\rho(S) = \rho(S^{-1}) = 0$ since $a = d = 0$, and $\rho(T) = -1$, $\rho(T^{-1}) = 1$ since all terms except for ab vanish. Thus, for S and T, according to (12.6), we should have

$$\gamma_2(-1/\tau) = \gamma_2(S\tau) = \zeta_3^0\gamma_2(\tau) = \gamma_2(\tau)$$
$$\gamma_2(\tau + 1) = \gamma_2(T\tau) = \zeta_3^{-1}\gamma_2(\tau),$$

and indeed both of these are true by (12.4). Next, (12.6) holds for S^{-1} since $S^{-1}\tau = S\tau$ and $\rho(S^{-1}) = \rho(S)$. Finally, $\gamma_2(\tau) = \gamma_2(TT^{-1}\tau) = \zeta_3^{-1}\gamma_2(T^{-1}\tau)$, so that $\gamma_2(T^{-1}\tau) = \zeta_3\gamma_2(\tau)$. Since $\rho(T) = 1$, this shows that (12.6) holds for T^{-1}.

We next claim that

(*) $$\delta \in \{S^{\pm 1}, T^{\pm 1}\},\ \gamma \in \mathrm{SL}(2,\mathbb{Z}) \implies \rho(\delta) + \rho(\gamma) \equiv \rho(\delta\gamma) \bmod 3.$$

This is proved as follows. Fix $\gamma = \left(\begin{smallmatrix} a & b \\ c & d \end{smallmatrix}\right) \in \mathrm{SL}(2,\mathbb{Z})$. Then:

- For $\delta = S$, we have

$$S\gamma = \begin{pmatrix} -c & -d \\ a & b \end{pmatrix}.$$

  Then

$$\rho(\gamma) = ac - ab + a^2cd - cd, \qquad \rho(S\gamma) = -ac - cd + c^2ab - ab.$$

  It follows that $\rho(\gamma) - \rho(S\gamma) = 2ac + ac(ad - bc) = 3ac \equiv 0 \bmod 3$. This and $\rho(S) = 0$ show that (*) holds for $\delta = S$.

- When $\delta = S^{-1}$, what we just proved implies
$$\rho(\gamma) = \rho((SS^{-1})\gamma) = \rho(S(S^{-1}\gamma)) \equiv \rho(S) + \rho(S^{-1}\gamma) \bmod 3.$$
Thus $\rho(S^{-1}\gamma) \equiv -\rho(S) + \rho(\gamma) \bmod 3$, and since $-\rho(S) = \rho(S^{-1})$ (both are zero), we get $\rho(S^{-1}\gamma) \equiv \rho(S^{-1}) + \rho(\gamma) \bmod 3$. Thus (*) holds for $\delta = S^{-1}$.

- Next, suppose $\delta = T$. Then
$$T\gamma = \begin{pmatrix} a+c & b+d \\ c & d \end{pmatrix},$$
so that
$$\begin{aligned}\rho(T\gamma) &= (a+c)c - (a+c)(b+d) + (a+c)^2cd - cd \\ &= \rho(\gamma) + c^2 - ad - bc - cd + 2ac^2d + c^3d\end{aligned}$$
Since $ad - bc = 1$, we can substitute $bc + 1$ for ad:
$$\begin{aligned}&= \rho(\gamma) + c^2 - (bc+1) - bc - cd + 2c^2(bc+1) + c^3d \\ &= \rho(\gamma) - 1 + c(3c - 2b - d + 2c^2b + c^2d) \\ &\equiv \rho(\gamma) - 1 + c(-2b - d + 2c^2b + c^2d) \bmod 3 \\ &\equiv \rho(\gamma) - 1 + c(c-1)(c+1)(2b+d) \bmod 3 \\ &\equiv \rho(\gamma) - 1 \bmod 3,\end{aligned}$$
where the last line follows since one of $c-1$, c and $c+1$ is divisible by 3. Then (*) holds for $\delta = T$ since $\rho(T) = -1$.

- When $\delta = T^{-1}$, note that $-\rho(T) = -(-1) = 1 = \rho(T^{-1})$. Then the argument for $\delta = S^{-1}$ implies that (*) holds for $\delta = T^{-1}$.

Since $\mathrm{SL}(2,\mathbb{Z})$ is generated by S and T by Exercise 12.3, it suffices by induction to show that if (12.6) holds for γ, then it holds for $\delta\gamma$ when $\delta \in \{S^{\pm 1}, T^{\pm 1}\}$. This is now easy:
$$\begin{aligned}\gamma_2((\delta\gamma)\tau) &= \gamma_2(\delta(\gamma\tau)) = \zeta_3^{\rho(\delta)}\gamma_2(\gamma\tau) = \zeta_3^{\rho(\delta)}\zeta_3^{\rho(\gamma)}\gamma_2(\tau) \\ &= \zeta_3^{\rho(\delta)+\rho(\gamma)}\gamma_2(\tau) = \zeta_3^{\rho(\delta\gamma)}\gamma_2(\tau),\end{aligned}$$
where the first line follows since δ and γ satisfy (12.6) and the second line uses (*).

(b) Since $ac - ab + a^2cd - cd \equiv 0 \bmod 3$ if $b \equiv c \equiv 0 \bmod 3$, we have $\zeta_3^{\rho(\gamma)} = 1$ for $\gamma \in \tilde{\Gamma}(3)$. Then (12.6) implies that $\gamma_2(\tau)$ is invariant under $\tilde{\Gamma}(3)$.

(c) If $\gamma = \left(\begin{smallmatrix} a & b \\ c & d \end{smallmatrix}\right) \in \Gamma_0(9)$, then
$$\begin{pmatrix} a & b \\ c & d \end{pmatrix} = \begin{pmatrix} 1/3 & 0 \\ 0 & 1 \end{pmatrix}\begin{pmatrix} a & 3b \\ c/3 & d \end{pmatrix}\begin{pmatrix} 3 & 0 \\ 0 & 1 \end{pmatrix}.$$
Since $c \equiv 0 \bmod 9$ we see that $c/3$ is an integer and divisible by 3, so the middle matrix on the right-hand side lies in $\tilde{\Gamma}(3)$. Conversely, if $\left(\begin{smallmatrix} a & b \\ c & d \end{smallmatrix}\right) \in \tilde{\Gamma}(3)$ with $b \equiv c \equiv 0 \bmod 3$ then
$$(*)\qquad \begin{pmatrix} 1/3 & 0 \\ 0 & 1 \end{pmatrix}\begin{pmatrix} a & b \\ c & d \end{pmatrix}\begin{pmatrix} 3 & 0 \\ 0 & 1 \end{pmatrix} = \begin{pmatrix} a & b/3 \\ 3c & d \end{pmatrix},$$
which lies in $\Gamma_0(9)$ since $b/3$ is an integer and $3c \equiv 0 \bmod 9$. These computations prove the first assertion of part (c).

For the second assertion, take $\gamma \in \Gamma_0(9)$. By what we just proved, $\gamma = \left(\begin{smallmatrix} 1/3 & 0 \\ 0 & 1 \end{smallmatrix}\right)\tilde{\gamma}\left(\begin{smallmatrix} 3 & 0 \\ 0 & 1 \end{smallmatrix}\right)$ with $\tilde{\gamma} \in \tilde{\Gamma}(3)$. Thus $\left(\begin{smallmatrix} 3 & 0 \\ 0 & 1 \end{smallmatrix}\right)\gamma = \tilde{\gamma}\left(\begin{smallmatrix} 3 & 0 \\ 0 & 1 \end{smallmatrix}\right)$, which when applied to $\tau \in \mathfrak{h}$ implies that
$$3\gamma(\tau) = \tilde{\gamma}(3\tau).$$

Hence

$$\gamma_2(3\gamma(\tau)) = \gamma_2(\tilde{\gamma}(3\tau)) = \gamma_2(3\tau),$$

where the last equality follows from part (b). Thus $\gamma_2(3\tau)$ is invariant under $\Gamma_0(9)$.

(d) Let

$$\mathcal{S} = \left\{ \begin{pmatrix} a & b \\ c & d \end{pmatrix} \in \mathrm{SL}(2,\mathbb{Z}) \mid a \equiv d \equiv 0 \bmod 3 \text{ or } b \equiv c \bmod 3 \right\},$$

and let $\Gamma \subset \mathrm{SL}(2,\mathbb{Z})$ be the subgroup consisting of all elements of $\mathrm{SL}(2,\mathbb{Z})$ that fix $\gamma_2(\tau)$.

We first show $\mathcal{S} \subset \Gamma$. For each $\gamma = \left(\begin{smallmatrix} a & b \\ c & d \end{smallmatrix}\right) \in S$. we want to show that

$$ac - ab + a^2cd - cd \equiv 0 \bmod 3.$$

The congruence clearly holds if $a \equiv d \equiv 0 \bmod 3$. If $b \equiv c \bmod 3$, then

$$ac - ab + a^2cd - cd \equiv a^2cd - cd \equiv (a^2 - 1)cd \bmod 3.$$

If $b \equiv c \equiv 0 \bmod 3$, then this expression is $0 \bmod 3$. Otherwise, $bc \equiv 1 \bmod 3$, and then $ad - bc = 1$ implies $ad \equiv 2 \bmod 3$. Thus a is not divisible by 3, so that $a^2 - 1 \equiv 0 \bmod 3$. Again the expression above is divisible by 3, completing the proof that $\mathcal{S} \subset \Gamma$.

Next we must prove $\Gamma \subset \mathcal{S}$. That is, we must show that

$$ac - ab + a^2cd - cd \equiv 0 \bmod 3 \implies a \equiv d \equiv 0 \bmod 3 \text{ or } b \equiv c \bmod 3.$$

We start with

$$(*) \qquad ac - ab + a^2cd - cd = a(c - b) + (a^2 - 1)cd \equiv 0 \bmod 3.$$

If $a \equiv 0 \bmod 3$, then modulo 3, (*) becomes $2cd \equiv 0 \bmod 3$, so that either c or d must be divisible by 3. But c cannot be since $ad - bc = 1$, so that $a \equiv d \equiv 0 \bmod 3$. Otherwise, if $a \not\equiv 0 \bmod 3$, then $a^2 - 1 \equiv 0 \bmod 3$, so that (*) gives $a(c-b) \equiv 0 \bmod 3$, which implies $b \equiv c \bmod 3$. Thus $\gamma \in \mathcal{S}$.

12.6. If $f(\tau)$ is identically zero, the result clearly holds. Otherwise, as in the text, $G(j(m\tau), j(\tau))$ is a polynomial in $j(m\tau)$ and $j(\tau)$, say

$$G(X, Y) = \sum_{i=0}^{M} \sum_{k=0}^{N} a_{ik} X^i Y^k$$

with $a_{ik} \in \mathbb{C}$ not all zero, and by (12.8) we have

$$G(j(m\tau), j(\tau)) = f(\tau) \frac{\partial \Phi_m}{\partial X}(j(m\tau), j(\tau)).$$

Thus

$$\sum_{i=0}^{M} \sum_{k=0}^{N} a_{ik} j(m\tau)^i j(\tau)^k = f(\tau) \frac{\partial \Phi_m}{\partial X}(j(m\tau), j(\tau))$$

Using the q-expansions of $f(\tau)$, $j(\tau)$, and $j(m\tau)$, we get an infinite system of inhomogeneous linear equations with the a_{ik}'s as unknowns. This system of equations has rational coefficients since the q-expansions have rational coefficients and since coefficients of the partial derivative are also rational by part (i) of Theorem 11.18. Since the system has a solution over $\mathbb{C}$, it must also have a solution over $\mathbb{Q}$. Thus we can assume $G(X, Y) \in \mathbb{Q}(X, Y)$. As noted in the text, this implies that $f(\tau_0) \in \mathbb{Q}(j(m\tau_0), j(\tau_0))$ when the partial derivative doesn't vanish at τ_0.

12.7. Note that $3 \nmid D$ implies that $3 \nmid m$. If $D \equiv 0 \bmod 4$, then $\tau_0 = \sqrt{-m}$, and the minimal polynomial for $\tau_0/3$ is $9x^2 + m$. Since $3 \nmid m$, this polynomial has relatively prime integral coefficients, so Lemma 7.5 shows that $[1, \tau_0/3]$ is a proper fractional ideal for $[1, 9(\tau_0/3)] = [1, 3\tau_0]$. If $D \equiv 1 \bmod 4$, then $\tau_0 = \frac{3+\sqrt{-m}}{2}$, and $36x^2 - 36x + m + 9$

has $\tau_0/3$ as a root. However, since $m \equiv -1 \bmod 4$, $m+9$ is a multiple of 4, so that $9x^2 - 9x + \frac{m+9}{4}$, which also has $\tau_0/3$ as a root, has integer coefficients. Furthermore, since $3 \nmid m$, the coefficients are relatively prime, so by Lemma 7.5, $[1, \tau_0/3]$ is again a proper fractional ideal for $[1, 9(\tau_0/3)] = [1, 3\tau_0]$.

12.8. With $\mathcal{O} = \mathbb{Z}[i]$, we have (in the notation of Lemma 12.11) $\alpha = i$, and since the discriminant of $K = \mathbb{Q}(i)$ is -4, we have $\tau_0 = i$ as well. The proof of Lemma 12.11 with $m = 9$, $s = 3$, and $\alpha = \tau_0 = i$ goes through in this case to show that if the partial vanishes, then there is a unit $\lambda \in \mathcal{O}$ with

$$\lambda[1, 9i/3] = [d, ai/3 + b], \quad \begin{pmatrix} a & b \\ 0 & d \end{pmatrix} \in C(9).$$

Further, the proof of the lemma shows that $\lambda = \pm 1$ leads to a contradiction, so in our case we must have $\lambda = \pm i$. But then

$$\lambda[1, 9i/3] = \pm i[1, 3i] = [\pm 3, \pm i] = [3, i],$$

so that we have $[3, i] = [d, ai/3 + b]$. Since $a \neq 0$ it follows that $d = 3$; then $ad = 9$ shows that $a = 3$ as well and therefore $[3, i] = [3, i + b]$. This shows that b is a multiple of 3; however, $0 \leq b < d = 3$, so that $b = 0$. But then $\gcd(a, b, d) = \gcd(3, 0, 3) = 3$, which is impossible since elements of $C(9)$ satisfy $\gcd(a, b, d) = 1$.

12.9. (a) With $q = e^{2\pi i \tau}$, we have

$$\begin{aligned}
\eta((\tau+1)/2) &= \left(e^{2\pi i(\tau+1)/2}\right)^{1/24} \prod_{n=1}^{\infty} \left(1 - \left(e^{2\pi i(\tau+1)/2}\right)^n\right) \\
&= \left(e^{2\pi i \tau}\right)^{1/48} \left(e^{\pi i}\right)^{1/24} \prod_{n=1}^{\infty} \left(1 - e^{i\pi n\tau} e^{i\pi n}\right) \\
&= q^{1/48} \zeta_{48} \prod_{n=1}^{\infty} \left(1 - (-1)^n q^{n/2}\right) \\
&= q^{1/48} \zeta_{48} \prod_{n=1}^{\infty} (1 - q^n) \prod_{n=1}^{\infty} \left(1 + q^{n-1/2}\right),
\end{aligned}$$

where in the last line, the first product consists of factors of $\prod_{n=1}^{\infty} \left(1 - (-1)^n q^{n/2}\right)$ with n even and the second product consists of factors with n odd. Thus

$$\begin{aligned}
\mathfrak{f}(\tau) = \zeta_{48}^{-1} \frac{\eta((\tau+1)/2)}{\eta(\tau)} &= \zeta_{48}^{-1} \frac{q^{1/48} \zeta_{48} \prod_{n=1}^{\infty} (1 - q^n) \prod_{n=1}^{\infty} \left(1 + q^{n-1/2}\right)}{q^{1/24} \prod_{n=1}^{\infty} (1 - q^n)} \\
&= q^{-1/48} \prod_{n=1}^{\infty} \left(1 + q^{n-1/2}\right).
\end{aligned}$$

Next,

$$\begin{aligned}
\eta(\tau/2) &= \left(e^{2\pi i \tau/2}\right)^{1/24} \prod_{n=1}^{\infty} \left(1 - \left(e^{2\pi i \tau/2}\right)^n\right) \\
&= \left(e^{2\pi i \tau}\right)^{1/48} \prod_{n=1}^{\infty} \left(1 - e^{2\pi i n\tau/2}\right) \\
&= q^{1/48} \prod_{n=1}^{\infty} (1 - q^n) \prod_{n=1}^{\infty} \left(1 - q^{n-1/2}\right).
\end{aligned}$$

where the last line follows as in the analysis of $\eta((\tau+1)/2)$. Thus

$$\mathfrak{f}_1(\tau) = \frac{\eta(\tau/2)}{\eta(\tau)} = \frac{q^{1/48}\prod_{n=1}^{\infty}(1-q^n)\prod_{n=1}^{\infty}(1-q^{n-1/2})}{q^{1/24}\prod_{n=1}^{\infty}(1-q^n)}$$
$$= q^{-1/48}\prod_{n=1}^{\infty}(1-q^{n-1/2}).$$

Finally,

$$\eta(2\tau) = \left(e^{2\pi i 2\tau}\right)^{1/24}\prod_{n=1}^{\infty}(1-e^{2\pi i n 2\tau}) = q^{1/12}\prod_{n=1}^{\infty}(1-(q^n)^2)$$
$$= q^{1/12}\prod_{n=1}^{\infty}(1-q^n)\prod_{n=1}^{\infty}(1+q^n).$$

Thus

$$\mathfrak{f}_2(\tau) = \sqrt{2}\,\frac{\eta(2\tau)}{\eta(\tau)} = \sqrt{2}\,\frac{q^{1/12}\prod_{n=1}^{\infty}(1-q^n)\prod_{n=1}^{\infty}(1+q^n)}{q^{1/24}\prod_{n=1}^{\infty}(1-q^n)}$$
$$= \sqrt{2}\,q^{1/24}\prod_{n=1}^{\infty}(1+q^n).$$

(b) Using the formulas for $\mathfrak{f}(\tau)$, $\mathfrak{f}_1(\tau)$, $\mathfrak{f}_2(\tau)$ proved in part (a), we obtain

$$\mathfrak{f}(\tau)\mathfrak{f}_1(\tau) = q^{-1/48}\prod_{n=1}^{\infty}(1+q^{n-1/2})\cdot q^{-1/48}\prod_{n=1}^{\infty}(1-q^{n-1/2})$$
$$= q^{-1/24}\prod_{n=1}^{\infty}(1-q^{2n-1})$$
$$\eta(\tau)\mathfrak{f}_2(\tau) = q^{1/24}\prod_{n=1}^{\infty}(1-q^n)\cdot\sqrt{2}q^{1/24}\prod_{n=1}^{\infty}(1+q^n)$$
$$= q^{1/12}\sqrt{2}\prod_{n=1}^{\infty}(1-q^{2n}).$$

Then

$$\eta(\tau)\mathfrak{f}(\tau)\mathfrak{f}_1(\tau)\mathfrak{f}_2(\tau) = q^{-1/24}\prod_{n=1}^{\infty}(1-q^{2n-1})\cdot q^{1/12}\sqrt{2}\prod_{n=1}^{\infty}(1-q^{2n})$$
$$= q^{1/24}\sqrt{2}\prod_{n=1}^{\infty}(1-q^n)$$
$$= \sqrt{2}\,\eta(\tau),$$

and dividing both sides by $\eta(\tau)$ gives the result.

(c) From the definitions,

$$\mathfrak{f}_1(2\tau)\,\mathfrak{f}_2(\tau) = \frac{\eta(\tau)}{\eta(2\tau)}\cdot\sqrt{2}\,\frac{\eta(2\tau)}{\eta(\tau)} = \sqrt{2}.$$

12.10. (a) Since

$$\ln\sigma(z) = \ln z + \sum_{\omega\in L-\{0\}}\left(\ln\left(1-\frac{z}{\omega}\right)+\frac{z}{\omega}+\frac{z^2}{2\omega^2}\right),$$

we have

$$\frac{\sigma'(z)}{\sigma(z)} = (\ln \sigma(z))' = \frac{1}{z} + \sum_{\omega \in L-\{0\}} \Big(\frac{-1/\omega}{1-\frac{z}{\omega}} + \frac{1}{\omega} + \frac{z}{\omega^2} \Big) = \frac{1}{z} + \sum_{\omega \in L-\{0\}} \Big(\frac{1}{z-\omega} + \frac{1}{\omega} + \frac{z}{\omega^2} \Big).$$

(b) Since

$$\zeta'(z) = -\frac{1}{z^2} + \sum_{\omega \in L-\{0\}} \Big(-\frac{1}{(z-\omega)^2} + \frac{1}{\omega^2} \Big),$$

it is immediate that $\wp(z) = -\zeta'(z)$.

(c) Pick $-1 < \gamma < 0$, and let Γ be the boundary of the rectangle $\mathbf{P} = \{s + t\tau \mid \gamma \leq s, t \leq \gamma + 1\}$. The only pole of $\zeta(z)$ in $\mathbf{P}$ is 0, so that by part (a) and residue theory, $\int_\Gamma \zeta(z)\, dz = 2\pi i$ when Γ is oriented counterclockwise. Let C_i, $i = 1, 2, 3, 4$, be the bottom, right, top, and left boundaries respectively, with orientations induced from the orientation of Γ. Then

$$2\pi i = \int_\Gamma \zeta(z)\, dz = \int_{C_1} \zeta(z)\, dz + \int_{C_3} \zeta(z)\, dz + \int_{C_2} \zeta(z)\, dz + \int_{C_4} \zeta(z)\, dz.$$

However, using $\eta_1 = \zeta(z+\tau) - \zeta(z)$, we have

$$\begin{aligned}
\int_{C_1} \zeta(z)\, dz + \int_{C_3} \zeta(z)\, dz &= \int_\gamma^{\gamma+1} \zeta(s + \gamma\tau)\, ds - \int_\gamma^{\gamma+1} \zeta(s + (\gamma+1)\tau)\, ds \\
&= \int_\gamma^{\gamma+1} \zeta(s + \gamma\tau)\, ds - \int_\gamma^{\gamma+1} \zeta(s + \gamma\tau) + \eta_1\, ds \\
&= -\int_\gamma^{\gamma+1} \eta_1\, ds = -\eta_1.
\end{aligned}$$

Similarly, $\eta_2 = \zeta(z+1) - \zeta(z)$ gives

$$\begin{aligned}
\int_{C_2} \zeta(z)\, dz + \int_{C_4} \zeta(z)\, dz &= \int_\gamma^{\gamma+1} \zeta(\gamma + 1 + t\tau)\, \tau dt - \int_\gamma^{\gamma+1} \zeta(\gamma + t\tau)\, \tau dt \\
&= \int_\gamma^{\gamma+1} (\zeta(\gamma + t\tau) + \eta_2)\, \tau dt - \int_\gamma^{\gamma+1} \zeta(\gamma + t\tau)\, \tau dt \\
&= \int_\gamma^{\gamma+1} \eta_2\, \tau dt = \eta_2 \tau.
\end{aligned}$$

Combining these results, we get $2\pi i = \eta_2 \tau - \eta_1$.

(d) (i) Using $\zeta(z) = \sigma'(z)/\sigma(z)$ and $\eta_1 = \zeta(z+\tau) - \zeta(z)$, we obtain

$$\begin{aligned}
\frac{d}{dz} \frac{\sigma(z+\tau)}{\sigma(z)} &= \frac{\sigma(z)\sigma'(z+\tau) - \sigma'(z)\sigma(z+\tau)}{\sigma(z)^2} \\
&= \frac{\sigma(z)\sigma(z+\tau)\zeta(z+\tau) - \sigma(z)\sigma(z+\tau)\zeta(z)}{\sigma(z)^2} \\
&= \frac{\sigma(z+\tau)(\zeta(z) + \eta_1) - \sigma(z+\tau)\zeta(z)}{\sigma(z)} = \eta_1 \frac{\sigma(z+\tau)}{\sigma(z)}.
\end{aligned}$$

Thus $f(z) = \frac{\sigma(z+\tau)}{\sigma(z)}$ satisfies the differential equation $f'(z) = \eta_1 f(z)$, so that $f(z) = \frac{\sigma(z+\tau)}{\sigma(z)} = Ce^{\eta_1 z}$ for some constant C that depends on τ. It follows that

(*) $$\sigma(z+\tau) = Ce^{\eta_1 z} \sigma(z).$$

(ii) Setting $z = -\tau/2$ and using the fact that σ is an odd function, (*) gives

$$\sigma(\tau/2) = Ce^{-\eta_1 \tau/2} \sigma(-\tau/2) = -Ce^{-\eta_1 \tau/2} \sigma(\tau/2).$$

Thus $-Ce^{-\eta_1\tau/2} = 1$, so that $C = -e^{\eta_1\tau/2}$. Then (*) can be written

$$\sigma(z+\tau) = -e^{\eta_1\tau/2}e^{\eta_1 z}\sigma(z) = -e^{\eta_1(z+\tau/2)}\sigma(z).$$

(iii) Using $\zeta(z) = \sigma'(z)/\sigma(z)$ and $\eta_2 = \zeta(z+1) - \zeta(z)$, we obtain

$$\begin{aligned}\frac{d}{dz}\frac{\sigma(z+1)}{\sigma(z)} &= \frac{\sigma(z)\sigma'(z+1) - \sigma'(z)\sigma(z+1)}{\sigma(z)^2}\\ &= \frac{\sigma(z)\sigma(z+1)\zeta(z+1) - \sigma(z)\sigma(z+1)\zeta(z)}{\sigma(z)^2}\\ &= \frac{\sigma(z+1)(\zeta(z)+\eta_2) - \sigma(z+1)\zeta(z)}{\sigma(z)} = \eta_2\frac{\sigma(z+1)}{\sigma(z)}.\end{aligned}$$

Thus $f(z) = \frac{\sigma(z+1)}{\sigma(z)}$ satisfies the differential equation $f'(z) = \eta_2 f(z)$, so that $f(z) = \frac{\sigma(z+1)}{\sigma(z)} = Ce^{\eta_2 z}$. It follows that

$$\sigma(z+1) = Ce^{\eta_2 z}\sigma(z).$$

Evaluating this at $z = -1/2$, and using the fact that σ is an odd function, the above equation gives

$$\sigma(1/2) = Ce^{-\eta_2/2}\sigma(-1/2) = -Ce^{-\eta_2/2}\sigma(1/2).$$

Thus $-Ce^{-\eta_2/2} = 1$, so that $C = -e^{\eta_2/2}$. Finally,

$$\sigma(z+1) = -e^{\eta_2/2}e^{\eta_2 z}\sigma(z) = -e^{\eta_2(z+1/2)}\sigma(z).$$

12.11. (a) Since $\sigma(z)$ is odd, it is easy to show that $f(z) = -\frac{\sigma(z+w)\sigma(z-w)}{\sigma(z)^2\sigma(w)^2}$ is even:

$$f(-z) = -\frac{\sigma(-z+w)\sigma(-z-w)}{\sigma(-z)^2\sigma(-w)^2} = -\frac{\sigma(z-w)\sigma(z+w)}{\sigma(z)^2\sigma(w)^2} = f(z).$$

To see that $f(z)$ is periodic, it suffices to show that $f(z+1) = f(z)$ and $f(z+\tau) = f(z)$. Using Exercise 12.10,

$$\begin{aligned}f(z+1) &= -\frac{\sigma(z+w+1)\sigma(z-w+1)}{\sigma(z+1)^2\sigma(w)^2}\\ &= -\frac{e^{\eta_2(z+w+1/2)}\sigma(z+w)e^{\eta_2(z-w+1/2)}\sigma(z-w)}{e^{2\eta_2(z+1/2)}\sigma(z)^2\sigma(w)^2}\\ &= -\frac{\sigma(z+w)\sigma(z-w)}{\sigma(z)^2\sigma(w)^2} = f(z)\\ f(z+\tau) &= -\frac{\sigma(z+w+\tau)\sigma(z-w+\tau)}{\sigma(z+\tau)^2\sigma(w)^2}\\ &= -\frac{e^{\eta_2(z+w+\tau/2)}\sigma(z+w)e^{\eta_2(z-w+\tau/2)}\sigma(z-w)}{e^{2\eta_2(z+\tau/2)}\sigma(z)^2\sigma(w)^2}\\ &= -\frac{\sigma(z+w)\sigma(z-w)}{\sigma(z)^2\sigma(w)^2} = f(z)\end{aligned}$$

Finally, since $\sigma(z)$ is meromorphic, so is $f(z)$. So $f(z)$ is an even elliptic function for L, and thus is a rational function in $\wp(z;L)$.

(b) The only zeros of $\sigma(z)$ are zeros of order 1 at all lattice points, and since $w \notin L$, the denominator of $f(z)$ never vanishes on $\mathbb{C} - L$. Further, $\sigma(z)$ has no poles on $\mathbb{C} - L$. It follows that $f(z)$ is holomorphic on $\mathbb{C} - L$.

For the second part, we have

$$f(z) = \frac{1}{\sigma^2(z)}\Big(-\frac{\sigma(z+w)\sigma(z-w)}{\sigma^2(w)}\Big)$$
$$= \frac{1}{z^2}\Big(\prod_{\omega\in L-\{0\}}\Big(1-\frac{z}{\omega}\Big)e^{z/\omega+(1/2)(z/\omega)^2}\Big)^{-1}\Big(-\frac{\sigma(z+w)\sigma(z-w)}{\sigma(w)^2}\Big)$$

The two factors in parentheses are holomorphic near zero, so their Laurent expansions at $z=0$ have only terms with nonnegative exponents, and their constant terms are found by setting $z=0$. The first of these evaluates to 1, and the second is

$$-\frac{\sigma(0+w)\sigma(0-w)}{\sigma(w)^2} = -\frac{-\sigma(w)^2}{\sigma(w)^2} = 1.$$

Thus the Laurent expansion for $f(z)$ at $z=0$ starts with $1/z^2$.

(c) Since $f(z)$ is an even elliptic function holomorphic on $\mathbb{C}-L$, part (a) of Exercise 10.13 shows that $f(z)$ is a polynomial in $\wp(z)$. Since the Laurent expansions of both $f(z)$ and $\wp(z)$ start with $1/z^2$, the solution of the exercise shows that $f(z)=\wp(z)+C$ for some constant C. To find C, note that $\sigma(0)=0$ by the definition of $\sigma(z)$ given in Exercise 12.10. Then

$$\wp(w)+C = f(w) = -\frac{\sigma(w+w)\sigma(w-w)}{\sigma(w)^2\sigma(w)^2} = -\frac{\sigma(2w)\sigma(0)}{\sigma(w)^4} = 0.$$

Thus $C=-\wp(w)$, so that

$$\wp(z)-\wp(w) = -\frac{\sigma(z+w)\sigma(z-w)}{\sigma(z)^2\sigma(w)^2}.$$

12.12. For the first equality, we use Exercise 12.11 and part (d) of Exercise 12.10 to obtain

$$e_2-e_1 = \wp\Big(\frac{1}{2}\Big)-\wp\Big(\frac{\tau}{2}\Big) = -\frac{\sigma\left(\frac{\tau+1}{2}\right)\sigma\left(\frac{1-\tau}{2}\right)}{\sigma^2\left(\frac{1}{2}\right)\sigma^2\left(\frac{\tau}{2}\right)} = -\frac{\sigma\left(\frac{\tau+1}{2}\right)\sigma\left(-\frac{1+\tau}{2}+1\right)}{\sigma^2\left(\frac{1}{2}\right)\sigma^2\left(\frac{\tau}{2}\right)}$$
$$= e^{\eta_2(-(1+\tau)/2+1/2)}\frac{\sigma\left(\frac{\tau+1}{2}\right)\sigma\left(-\frac{1+\tau}{2}\right)}{\sigma^2\left(\frac{1}{2}\right)\sigma^2\left(\frac{\tau}{2}\right)}$$
$$= -e^{-\eta_2\tau/2}\frac{\sigma^2\left(\frac{\tau+1}{2}\right)}{\sigma^2\left(\frac{1}{2}\right)\sigma^2\left(\frac{\tau}{2}\right)},$$

where the last line follows since $\sigma(z)$ is odd. Similarly,

$$e_2-e_3 = \wp\Big(\frac{1}{2}\Big)-\wp\Big(\frac{\tau+1}{2}\Big) = -\frac{\sigma\left(1+\frac{\tau}{2}\right)\sigma\left(-\frac{\tau}{2}\right)}{\sigma^2\left(\frac{1}{2}\right)\sigma^2\left(\frac{\tau+1}{2}\right)} = e^{\eta_2(\tau+1)/2}\frac{\sigma\left(\frac{\tau}{2}\right)\sigma\left(-\frac{\tau}{2}\right)}{\sigma^2\left(\frac{1}{2}\right)\sigma^2\left(\frac{\tau+1}{2}\right)}$$
$$= -e^{\eta_2(\tau+1)/2}\frac{\sigma^2\left(\frac{\tau}{2}\right)}{\sigma^2\left(\frac{1}{2}\right)\sigma^2\left(\frac{\tau+1}{2}\right)}$$
$$e_3-e_1 = \wp\Big(\frac{\tau+1}{2}\Big)-\wp\Big(\frac{\tau}{2}\Big) = -\frac{\sigma\left(\tau+\frac{1}{2}\right)\sigma\left(\frac{1}{2}\right)}{\sigma^2\left(\frac{\tau+1}{2}\right)\sigma^2\left(\frac{\tau}{2}\right)} = e^{\eta_1(\tau+1)/2}\frac{\sigma^2\left(\frac{1}{2}\right)}{\sigma^2\left(\frac{\tau+1}{2}\right)\sigma^2\left(\frac{\tau}{2}\right)}.$$

12.13. (a) The definition of $\sigma(z)$ shows that its zeros of occur at the points of L, and are all order one zeros. For $f(z)$ as defined in the exercise, the zeros occur when

- $q_z^{1/2} = q_z^{-1/2}$, which means $e^{\pi i z} = e^{-\pi i z}$, so $e^{2\pi i z}=1$. Thus $z\in\mathbb{Z}\subset L=[1,\tau]$.
- $q_\tau^n q_z^{\pm 1} = 1$, which means that $e^{2\pi i n\tau \pm 2\pi i z} = e^{2\pi i(n\tau\pm z)} = 1$, so that $n\tau\pm z\in\mathbb{Z}$ and therefore $z\pm n\tau\in\mathbb{Z}$. This implies that $z=m\pm n\tau$ for $m,n\in\mathbb{Z}$, so that $z\in L$.

Conversely, let $z = m + n\tau \in L$. If $n = 0$, then $q_z^{1/2} - q_z^{-1/2} = (-1)^m - (-1)^{-m} = 0$, and if $n > 0$, then $q_\tau^n/q_z = 1$; and if $n < 0$ then $q_\tau^{-n}q_z = 1$. These are all clearly zeros of order one. It follows that $\sigma(z)/f(z)$ is holomorphic on $\mathbb{C} - L$.

(b) We use formulas for $\sigma(z+1)$ and $\sigma(z+\tau)$ from part (d) of Exercise 12.10. For $f(z+1)$,

$$\begin{aligned} f(z+1) &= \frac{1}{2\pi i} e^{\eta_2(z+1)^2/2}\left(e^{\pi i(z+1)} - e^{-\pi i(z+1)}\right)\prod_{n=1}^{\infty}\frac{\left(1-q_\tau^n e^{2\pi i(z+1)}\right)\left(1-q_\tau^n e^{-2\pi i(z+1)}\right)}{(1-q_\tau^n)^2} \\ &= e^{\eta_2(2z+1)/2}e^{\pi i}\cdot\frac{1}{2\pi i}e^{\eta_2 z^2/2}\left(q_z^{1/2} - q_z^{-1/2}\right)\prod_{n=1}^{\infty}\frac{\left(1-q_\tau^n q_z e^{2\pi i}\right)\left(1-q_\tau^n q_z^{-1} e^{-2\pi i}\right)}{(1-q_\tau^n)^2} \\ &= -e^{\eta_2(z+1/2)}f(z) \end{aligned}$$

Since $\sigma(z)$ transforms the same way, we get $\sigma(z+1)/f(z+1) = \sigma(z)/f(z)$. For $f(z+\tau)$,

$$\begin{aligned} f(z+\tau) &= \frac{1}{2\pi i} e^{\eta_2(z+\tau)^2/2}\left(e^{\pi i(z+\tau)} - e^{-\pi i(z+\tau)}\right)\prod_{n=1}^{\infty}\frac{\left(1-q_\tau^n e^{2\pi i(z+\tau)}\right)\left(1-q_\tau^n e^{-2\pi i(z+\tau)}\right)}{(1-q_\tau^n)^2} \\ &= e^{\eta_2(2z\tau+\tau^2)/2}\frac{1}{2\pi i}e^{\eta_2 z^2/2}\left(q_\tau^{1/2}q_z^{1/2} - q_\tau^{-1/2}q_z^{-1/2}\right)\prod_{n=1}^{\infty}\frac{\left(1-q_\tau^{n+1}q_z\right)\left(1-q_\tau^{n-1}/q_z\right)}{(1-q_\tau^n)^2} \end{aligned}$$

Using the Legendre relation $\eta_2\tau - \eta_1 = 2\pi i$, we have

$$\begin{aligned} e^{\eta_2(2z\tau+\tau^2)/2} &= e^{\eta_2\tau(2z+\tau)/2} = e^{(2\pi i+\eta_1)(z+\tau/2)} \\ &= e^{\eta_1(z+\tau/2)}e^{2\pi i(z+\tau/2)} = e^{\eta_1(z+\tau/2)}q_z q_\tau^{1/2}. \end{aligned}$$

Furthermore, rearranging factors in the product gives

$$\prod_{n=1}^{\infty}\frac{\left(1-q_\tau^{n+1}q_z\right)\left(1-q_\tau^{n-1}/q_z\right)}{(1-q_\tau^n)^2} = \frac{1-1/q_z}{1-q_\tau q_z}\prod_{n=1}^{\infty}\frac{\left(1-q_\tau^n q_z\right)\left(1-q_\tau^n/q_z\right)}{(1-q_\tau^n)^2}.$$

Combining all this, we obtain

$$f(z+\tau) = q_z q_\tau^{1/2}\frac{1-1/q_z}{1-q_\tau q_z}\,\frac{q_\tau^{1/2}q_z^{1/2} - q_\tau^{-1/2}q_z^{-1/2}}{q_z^{1/2} - q_z^{-1/2}}\cdot e^{\eta_1(z+\tau/2)}f(z).$$

It remains to simplify the first factor:

$$\begin{aligned} q_z q_\tau^{1/2}\frac{1-1/q_z}{1-q_\tau q_z}\,\frac{q_\tau^{1/2}q_z^{1/2} - q_\tau^{-1/2}q_z^{-1/2}}{q_z^{1/2} - q_z^{-1/2}} &= \frac{(q_z-1)(q_\tau q_z^{1/2} - q_z^{-1/2})}{(1-q_\tau q_z)(q_z^{1/2} - q_z^{-1/2})} \\ &= \frac{(q_z-1)(q_\tau q_z^{1/2} - q_z^{-1/2})}{(1-q_\tau q_z)(q_z^{1/2} - q_z^{-1/2})}\cdot\frac{q_z^{1/2}}{q_z^{1/2}} \\ &= \frac{(q_z-1)(q_\tau q_z - 1)}{(1-q_\tau q_z)(q_z-1)} = -1. \end{aligned}$$

Thus

$$f(z+\tau) = -e^{\eta_1\left(z+\frac{\tau}{2}\right)}f(z).$$

Since $\sigma(z)$ transforms the same way, we get $\sigma(z+\tau)/f(z+\tau) = \sigma(z)/f(z)$.

(c) Both $\sigma(z)$ and $f(z)$ have zeros of order one at 0, so $\sigma(z)/f(z)$ is defined at 0. Also,

$$\frac{\sigma(z)}{f(z)} = 2\pi i\,\frac{z}{q_z^{1/2} - q_z^{-1/2}}\cdot\frac{\displaystyle\prod_{\omega\in L-\{0\}}\left(1-\frac{z}{\omega}\right)e^{z/\omega+(z/\omega)^2/2}}{\displaystyle e^{\eta_2 z^2/2}\prod_{n=1}^{\infty}\frac{(1-q_\tau^n q_z)(1-q_\tau^n/q_z)}{(1-q_\tau^n)^2}}$$

The numerator and denominator of the big fraction both evaluate to 1 at $z = 0$ (note that $q_z = 1$ at $z = 0$), and by L'Hospital's rule

$$2\pi i \lim_{z\to 0} \frac{z}{e^{\pi i z} - e^{-\pi i z}} = 2\pi i \lim_{z\to 0} \frac{1}{\pi i(e^{\pi i z} + e^{-\pi i z})} = 1.$$

Thus $\sigma(z)/f(z)$ takes the value 1 at $z = 0$.

(d) Since $\sigma(z)/f(z)$ is holomorphic and elliptic, it is constant by Exercise 10.5; since it takes the value 1 at $z = 0$, we get $\sigma(z) = f(z)$, as desired.

12.14. (a) At $z = \frac{1}{2}$, $q_z = e^{i\pi}$. Writing q for q_τ and using the product expansion of $\mathfrak{f}_2(\tau)$ from (12.15) together with the product expansion of $\eta(\tau)$, we obtain

$$\begin{aligned}\sigma\Big(\frac{1}{2}\Big) &= \frac{1}{2\pi i} e^{\eta_2/8}\big(e^{i\pi/2} - e^{-i\pi/2}\big) \prod_{n=1}^{\infty} \frac{(1+q^n)(1+q^n)}{(1-q^n)^2}\\ &= \frac{1}{2\pi i} e^{\eta_2/8}(2i)\frac{q^{-1/12}\mathfrak{f}_2(\tau)^2/2}{q^{-1/12}\eta(\tau)^2},\end{aligned}$$

and thus

$$\boxed{\sigma\left(\frac{1}{2}\right) = \frac{1}{2\pi} e^{\eta_2/8}\frac{\mathfrak{f}_2(\tau)^2}{\eta(\tau)^2}.}$$

At $z = \frac{\tau}{2}$, $q_z = e^{2\pi i\tau/2} = q_\tau^{1/2}$. When we write q for q_τ, we obtain

$$\sigma\left(\frac{\tau}{2}\right) = \frac{1}{2\pi i} e^{\eta_2\tau^2/8}(q^{1/4} - q^{-1/4}) \prod_{n=1}^{\infty} \frac{(1-q^{n+1/2})(1-q^{n-1/2})}{(1-q^n)^2}.$$

Rearranging terms in the product gives

$$\begin{aligned}\sigma\left(\frac{\tau}{2}\right) &= \frac{1}{2\pi i} e^{\eta_2\tau^2/8}\frac{q^{1/4} - q^{-1/4}}{1-q^{1/2}} \prod_{n=1}^{\infty} \frac{(1-q^{n-1/2})(1-q^{n-1/2})}{(1-q^n)^2}\\ &= \frac{1}{2\pi i} e^{\eta_2\tau^2/8}\cdot \frac{q^{1/4} - q^{-1/4}}{1-q^{1/2}} \frac{q^{1/24}\,\mathfrak{f}_1(\tau)^2}{q^{-1/12}\eta(\tau)^2}.\end{aligned}$$

where the second line uses the product expansion of $\mathfrak{f}_1(\tau)$ from (12.15) together with the product expansion of $\eta(\tau)$. Since

$$\frac{q^{1/4} - q^{-1/4}}{1-q^{1/2}} \cdot \frac{q^{1/24}}{q^{-1/12}} = q^{1/8}\cdot \frac{q^{1/4} - q^{-1/4}}{1-q^{1/2}} \cdot \frac{q^{1/4}}{q^{1/4}} = q^{-1/8}\frac{q^{1/2}-1}{1-q^{1/2}} = -q^{-1/8},$$

we get

$$\boxed{\sigma\left(\frac{\tau}{2}\right) = -\frac{1}{2\pi i} e^{\eta_2\tau^2/8} q^{-1/8}\frac{\mathfrak{f}_1(\tau)^2}{\eta(\tau)^2} = \frac{i}{2\pi} e^{\eta_2\tau^2/8} q^{-1/8}\frac{\mathfrak{f}_1(\tau)^2}{\eta(\tau)^2}}.$$

Finally, at $z = \frac{\tau+1}{2}$, $q_z = e^{2\pi i(\tau+1)/2} = q_\tau^{1/2}e^{\pi i} = -q_\tau^{1/2}$. We again write q for q_τ to obtain

$$\begin{aligned}\sigma\Big(\frac{\tau+1}{2}\Big) &= \frac{1}{2\pi i} e^{\eta_2(\tau+1)^2/8}\big(q^{1/4}e^{\pi i/2} - q^{-1/4}e^{-\pi i/2}\big) \prod_{n=1}^{\infty} \frac{(1-(-q^{n+1/2}))(1-(-q^{n-1/2}))}{(1-q^n)^2}\\ &= \frac{1}{2\pi i} e^{\eta_2(\tau+1)^2/8}\cdot i(q^{1/4} + q^{-1/4}) \prod_{n=1}^{\infty} \frac{(1+q^{n+1/2})(1+q^{n-1/2})}{(1-q^n)^2}\\ &= \frac{1}{2\pi} e^{\eta_2(\tau+1)^2/8}\frac{q^{1/4} + q^{-1/4}}{1+q^{1/2}} \prod_{n=1}^{\infty} \frac{(1+q^{n-1/2})(1+q^{n-1/2})}{(1-q^n)^2}\\ &= \frac{1}{2\pi} e^{\eta_2(\tau+1)^2/8}\frac{q^{1/24}(q^{1/4} + q^{-1/4})}{q^{-1/12}(1+q^{1/2})}\frac{\mathfrak{f}(\tau)^2}{\eta(\tau)^2},\end{aligned}$$

where we again use the product expansion of $\mathfrak{f}(\tau)$ from (12.15) together with the product expansion of $\eta(\tau)$. Since

$$\frac{q^{1/24}(q^{1/4}+q^{-1/4})}{q^{-1/12}(1+q^{1/2})} = \frac{q^{1/24}q^{-1/4}(q^{1/2}+1)}{q^{-1/12}(1+q^{1/2})} = q^{1/24-1/4+1/12} = q^{-1/8},$$

we get finally

$$\boxed{\sigma\left(\frac{\tau+1}{2}\right) = \frac{1}{2\pi}e^{\eta_2(\tau+1)^2/8}q^{-1/8}\frac{\mathfrak{f}(\tau)^2}{\eta(\tau)^2}}.$$

(b) Exercise 12.12 expresses e_2-e_1, e_2-e_3, e_3-e_1 in terms of $\sigma(\frac{1}{2})$, $\sigma(\frac{\tau}{2})$, $\sigma(\frac{\tau+1}{2})$ and various exponentials. We now combine these formulas with part (a). For e_2-e_1, this gives

$$\begin{aligned} e_2-e_1 &= -e^{-\eta_2\tau/2}\frac{(1/(4\pi^2))e^{\eta_2(\tau+1)^2/4}q^{-1/4}\frac{\mathfrak{f}(\tau)^4}{\eta(\tau)^4}}{(1/(4\pi^2))e^{\eta_2/4}\frac{\mathfrak{f}_2(\tau)^4}{\eta(\tau)^4}\cdot(-1/(4\pi^2))e^{\eta_2\tau^2/4}q^{-1/4}\frac{\mathfrak{f}_1(\tau)^4}{\eta(\tau)^4}} \\ &= 4\pi^2\frac{\eta(\tau)^4\mathfrak{f}(\tau)^4}{\mathfrak{f}_1(\tau)^4\mathfrak{f}_2(\tau)^4}. \end{aligned}$$

Since $\mathfrak{f}_1(\tau)\mathfrak{f}_2(\tau) = \frac{\sqrt{2}}{\mathfrak{f}(\tau)}$ by (12.16), this simplifies to

$$e_2-e_1 = 4\pi^2\frac{\eta(\tau)^4\mathfrak{f}(\tau)^4}{4/\mathfrak{f}(\tau)^4} = \pi^2\eta(\tau)^4\mathfrak{f}(\tau)^8.$$

Next, for e_2-e_3, we get

$$\begin{aligned} e_2-e_3 &= -e^{\eta_2(\tau+1)/2}\frac{(-1/(4\pi^2))e^{\eta_2\tau^2/4}q^{-1/4}\frac{\mathfrak{f}_1(\tau)^4}{\eta(\tau)^4}}{(1/(4\pi^2))e^{\eta_2/4}\frac{\mathfrak{f}_2(\tau)^4}{\eta(\tau)^4}\cdot(1/4\pi^2)e^{\eta_2(\tau+1)^2/4}q^{-1/4}\frac{\mathfrak{f}(\tau)^4}{\eta(\tau)^4}} \\ &= 4\pi^2\frac{\eta(\tau)^4\mathfrak{f}_1(\tau)^4}{\mathfrak{f}(\tau)^4\mathfrak{f}_2(\tau)^4}. \end{aligned}$$

Applying (12.16) again, this simplifies to

$$e_2-e_3 = 4\pi^2\frac{\eta(t)^4\mathfrak{f}_1(\tau)^4}{4/\mathfrak{f}_1(\tau)^4} = \pi^2\eta(\tau)^4\mathfrak{f}_1(\tau)^8.$$

Finally, for e_3-e_1, we get

$$\begin{aligned} e_3-e_1 &= e^{\eta_1(\tau+1)/2}\frac{(1/(4\pi^2))e^{\eta_2/4}\frac{\mathfrak{f}_2(\tau)^4}{\eta(\tau)^4}}{(1/4\pi^2)e^{\eta_2(\tau+1)^2/4}q^{-1/4}\frac{\mathfrak{f}(\tau)^4}{\eta(\tau)^4}\cdot(-1/(4\pi^2))e^{\eta_2\tau^2/4}q^{-1/4}\frac{\mathfrak{f}_1(\tau)^4}{\eta(\tau)^4}} \\ &= -4\pi^2e^{(\eta_1-\eta_2\tau)(\tau+1)/2}q^{1/2}\frac{\eta(\tau)^4\mathfrak{f}_2(\tau)^4}{\mathfrak{f}(\tau)^4\mathfrak{f}_1(\tau)^4}. \end{aligned}$$

Using the Legendre relation and 12.16 gives

$$\begin{aligned} e_3-e_1 &= -4\pi^2e^{-2\pi i(\tau+1)/2}q^{1/2}\frac{\eta(\tau)^4\mathfrak{f}_2(\tau)^4}{4/\mathfrak{f}_2(\tau)^4} \\ &= -\pi^2e^{-\pi i\tau}e^{-\pi i}q^{1/2}\eta(\tau)^4\mathfrak{f}_2(\tau)^8 = \pi^2\eta(\tau)^4\mathfrak{f}_2(\tau)^8. \end{aligned}$$

12.15. (a) First note that

$$(*)\quad \begin{aligned} 4((e_2-e_1)^2-(e_2-e_3)(e_3-e_1)) &= 4(e_2^2-2e_1e_2+e_1^2-e_2e_3+e_1e_2+e_3^2-e_1e_3) \\ &= -4(e_1e_2+e_1e_3+e_2e_3)+4(e_1^2+e_2^2+e_3^2). \end{aligned}$$

Since $0=(e_1+e_2+e_3)^2 = e_1^2+e_2^2+e_3^2+2(e_1e_2+e_1e_3+e_2e_3)$, this simplifies to

$$-4(e_1e_2+e_1e_3+e_2e_3)-8(e_1e_2+e_1e_3+e_2e_3) = -12(e_1e_2+e_1e_3+e_2e_3).$$

Since $g_2(\tau) = -4(e_1e_2 + e_1e_3 + e_2e_3)$, we obtain the desired formula

$$(**) \qquad 3g_2(\tau) = 4((e_2 - e_1)^2 - (e_2 - e_3)(e_3 - e_1)).$$

(b) By (*), $4((e_2 - e_1)^2 - (e_2 - e_3)(e_3 - e_1))$ is symmetric in the e_i. Therefore permuting the e_i cyclically in (**) gives two additional formulas:

$$3g_2(\tau) = 4((e_3 - e_2)^2 - (e_3 - e_1)(e_1 - e_2))$$
$$3g_2(\tau) = 4((e_1 - e_3)^2 - (e_1 - e_2)(e_2 - e_3)).$$

Substituting the formulas from Exercise 12.14 and using (12.16) gives

$$\begin{aligned}
3g_2(\tau) &= 4(\pi^4\eta(\tau)^8\mathfrak{f}_1(\tau)^{16} - (\pi^2\eta(\tau)^4\mathfrak{f}_2(\tau)^8)(-\pi^2\eta(\tau)^4\mathfrak{f}(\tau)^8))\\
&= 4\pi^4\eta(\tau)^8(\mathfrak{f}_1(\tau)^{16} + \mathfrak{f}(\tau)^8\mathfrak{f}_2(\tau)^8)\\
&= 4\pi^4\eta(\tau)^8\left(\mathfrak{f}_1(\tau)^{16} + \frac{16}{\mathfrak{f}_1(\tau)^8}\right)\\
&= 4\pi^4\eta(\tau)^8\frac{\mathfrak{f}_1(\tau)^{24} + 16}{\mathfrak{f}_1(\tau)^8}\\
3g_2(\tau) &= 4(\pi^4\eta(\tau)^8\mathfrak{f}_2(\tau)^{16} - (-\pi^2\eta(\tau)^4\mathfrak{f}(\tau)^8)(\pi^2\eta(\tau)^4\mathfrak{f}_1(\tau)^8))\\
&= 4\pi^4\eta(\tau)^8(\mathfrak{f}_2(\tau)^{16} + \mathfrak{f}(\tau)^8\mathfrak{f}_1(\tau)^8)\\
&= 4\pi^4\eta(\tau)^8\left(\mathfrak{f}_2(\tau)^{16} + \frac{16}{\mathfrak{f}_2(\tau)^8}\right)\\
&= 4\pi^4\eta(\tau)^8\frac{\mathfrak{f}_2(\tau)^{24} + 16}{\mathfrak{f}_2(\tau)^8}.
\end{aligned}$$

Then substituting these formulas into the equation $\gamma_2(\tau) = \frac{3g_2(\tau)}{4\pi^4\eta(\tau)^8}$ from the proof of Theorem 12.17 gives

$$\gamma_2(\tau) = \frac{\mathfrak{f}_1(\tau)^{24} + 16}{\mathfrak{f}_1(\tau)^8} = \frac{\mathfrak{f}_2(\tau)^{24} + 16}{\mathfrak{f}_2(\tau)^8}.$$

12.16. Multiplying together the three formulas for $\gamma_2(\tau) = \sqrt[3]{j(\tau)}$ from Theorem 12.17 and using (12.16) gives

$$\begin{aligned}
j(\tau) &= \frac{\mathfrak{f}(\tau)^{24}\mathfrak{f}_1(\tau)^{24}\mathfrak{f}_2(\tau)^{24} + 16P(\mathfrak{f}(\tau), \mathfrak{f}_1(\tau), \mathfrak{f}_2(\tau))}{\mathfrak{f}(\tau)^8\mathfrak{f}_1(\tau)^8\mathfrak{f}_2(\tau)^8}\\
&= \frac{2^{12} + 16P(\mathfrak{f}(\tau), \mathfrak{f}_1(\tau), \mathfrak{f}_2(\tau))}{16}\\
&= 256 + P(\mathfrak{f}(\tau), \mathfrak{f}_1(\tau), \mathfrak{f}_2(\tau)),
\end{aligned}$$

where $P(X, Y, Z)$ is a polynomial with integer coefficients. Substituting the q-expansions of the three functions, which all clearly have integer coefficients, we see that the q-expansion of $j(\tau)$ has integer coefficients as well.

12.17. The behavior of the Weber functions under $\tau \mapsto \tau + 1$ follows from their definitions and $\eta(\tau + 1) = \zeta_{24}\eta(\tau)$:

$$\begin{aligned}
\mathfrak{f}(\tau + 1) &= \zeta_{48}^{-1}\frac{\eta((\tau + 1 + 1)/2)}{\eta(\tau + 1)} = \zeta_{48}^{-1}\frac{\eta(\tau/2 + 1)}{\eta(\tau + 1)} = \zeta_{48}^{-1}\frac{\zeta_{24}\eta(\tau/2)}{\zeta_{24}\eta(\tau)} = \zeta_{48}^{-1}\mathfrak{f}_1(\tau)\\
\mathfrak{f}_1(\tau + 1) &= \frac{\eta((\tau + 1)/2)}{\eta(\tau + 1)} = \frac{\eta((\tau + 1)/2)}{\zeta_{24}\eta(\tau)} = \zeta_{48}^{-1}\left(\zeta_{48}^{-1}\frac{\eta((\tau + 1)/2)}{\eta(\tau)}\right) = \zeta_{48}^{-1}\mathfrak{f}(\tau)\\
\mathfrak{f}_2(\tau + 1) &= \sqrt{2}\frac{\eta(2(\tau + 1))}{\eta(\tau + 1)} = \sqrt{2}\frac{\eta(2\tau + 2)}{\eta(\tau + 1)} = \sqrt{2}\frac{\zeta_{24}^2\eta(2\tau)}{\zeta_{24}\eta(\tau)} = \zeta_{24}\mathfrak{f}_2(\tau)
\end{aligned}$$

For the second set of transformations, we use $\eta(-1/\tau) = \sqrt{-i\tau}\,\eta(\tau)$ to obtain

$$\mathfrak{f}_1(-1/\tau) = \frac{\eta((-1/\tau)/2)}{\eta(-1/\tau)} = \frac{\eta(-1/2\tau)}{\eta(-1/\tau)} = \frac{\sqrt{-2i\tau}\,\eta(2\tau)}{\sqrt{-i\tau}\,\eta(\tau)} = \sqrt{2}\,\frac{\eta(2\tau)}{\eta(\tau)} = \mathfrak{f}_2(\tau).$$

Substituting $-1/\tau$ for τ in this transformation gives

$$\mathfrak{f}_2(-1/\tau) = \mathfrak{f}_1(\tau).$$

Finally, combining the formulas for $\mathfrak{f}_1(-1/\tau)$ and $\mathfrak{f}_2(-1/\tau)$ with (12.16) gives

$$\mathfrak{f}(-1/\tau) = \frac{\sqrt{2}}{\mathfrak{f}_1(-1/\tau)\mathfrak{f}_2(-1/\tau)} = \frac{\sqrt{2}}{\mathfrak{f}_2(\tau)\mathfrak{f}_1(\tau)} = \mathfrak{f}(\tau).$$

12.18. The text explains how to estimate $\gamma_2(\tau_0)$ numerically when τ_0 is as defined in Theorem 12.2 and the class number is 1. In (12.21) and the discussion that followed, we derived the formulas

$$\gamma_2(\sqrt{-m}) = [\![256q^{2/3} + q^{-1/3}]\!], \quad d_K \text{ even}$$
$$\gamma_2((3+\sqrt{-m})/2) = [\![256q^{1/3} - q^{-1/6}]\!], \quad d_K \text{ odd},$$

where $[\![\,\cdot\,]\!]$ is the nearest integer function. Applying these to the ten class number 1 fields, we get:

d_K	τ_0	$\gamma_2(\tau_0)$ (est.)	$\gamma_2(\tau_0)$	$j(\tau_0)$
-4	i	11.8820	$12 = 2^2 \cdot 3$	12^3
-7	$\frac{1}{2}(3+i\sqrt{7})$	-14.9647	$-15 = -3 \cdot 5$	-15^3
-8	$i\sqrt{2}$	19.9984	$20 = 2^2 \cdot 5$	20^3
-11	$\frac{1}{2}(3+i\sqrt{11})$	-31.9922	$-32 = -2^5$	-32^3
-16	$2i$	66.	$66 = 2 \cdot 3 \cdot 11$	66^3
-19	$\frac{1}{2}(3+i\sqrt{19})$	-95.9991	$-96 = -2^5 \cdot 3$	-96^3
-28	$i\sqrt{7}$	255.	$255 = 3 \cdot 5 \cdot 17$	255^3
-43	$\frac{1}{2}(3+i\sqrt{43})$	$-960.$	$-960 = -2^6 \cdot 3 \cdot 5$	-960^3
-67	$\frac{1}{2}(3+i\sqrt{67})$	$-5280.$	$-5280 = -2^5 \cdot 3 \cdot 5 \cdot 11$	-5280^3
-163	$\frac{1}{2}(3+i\sqrt{163})$	$-640320.$	$-640320 = -2^6 \cdot 3 \cdot 5 \cdot 23 \cdot 29$	-640320^3

12.19. Using part (ii) of Theorem 6.1, the genus field of $\mathbb{Q}(\sqrt{-105})$ is seen to be $M = K(\sqrt{-3}, \sqrt{5}, \sqrt{-7})$. The discriminant is $d_K = -4 \cdot 105 \equiv 0 \bmod 4$, so by part (iii) of Theorem 6.1, the number of genera of primitive positive definite forms is $2^{4-1} = 8$. However, there are also eight primitive positive definite forms of that discriminant:

$$x^2 + 105y^2, \quad 3x^2 + 35y^2, \quad 5x^2 + 21y^2, \quad 7x^2 + 15y^2,$$
$$2x^2 + 2xy + 53y^2, \quad 6x^2 + 6xy + 19y^2, \quad 10x^2 + 10xy + 13y^2, \quad 11x^2 + 8xy + 11y^2.$$

Thus each genus consists of a single form, so by (i) $\Leftrightarrow$ (v) of Theorem 3.22, the class number is $h(d_K) = 2^{\mu-1}$, where μ is the number determined in Proposition 3.11. To calculate μ, we have $D = d_K = -4 \cdot 105$, and $105 \equiv 1 \bmod 4$, so that from Proposition 3.11, $\mu = r + 1$ where r is the number of odd primes dividing $105 = 3 \cdot 5 \cdot 7$. Thus $\mu = 3 + 1 = 4$, and $h(d_K) = 2^3 = 8$.

However, $|h(d_K)| = |C(\mathcal{O}_K)| = [L : K]$ where L is the Hilbert class field of K. Since $[M : K] = 8$ and $M \subset L$, it follows that $M = L$ so the genus field is in this case the Hilbert class field. (For a different proof of this equality, see Exercise 6.15.)

For the second part, note that the formula for the genus field given in (6.9) implies that $M = \mathbb{Q}(i, \sqrt{-3}, \sqrt{5}, \sqrt{-7}) = \mathbb{Q}(i, \sqrt{3}, \sqrt{5}, \sqrt{7})$. Since $L = M$, it follows that $L = L_0(i)$ for $L_0 = \mathbb{Q}(\sqrt{3}, \sqrt{5}, \sqrt{7})$. Hence $[L : L_0] = 2$, and since $L_0 \subset L \cap \mathbb{R} \subset L$, we see that $L \cap \mathbb{R}$ is either L or L_0. The former is impossible since $i \in L$. It follows that $L_0 = L \cap \mathbb{R}$ is the maximal real subfield of L.

12.20. (a) Using Corollary 12.19,

$$\begin{aligned}\mathfrak{f}_1(U\tau) &= \mathfrak{f}_1\Big(\frac{\tau}{\tau+1}\Big) = \mathfrak{f}_1\Big(-\frac{1}{\tau+1}+1\Big) = \zeta_{48}^{-1}\mathfrak{f}\Big(-\frac{1}{\tau+1}\Big)\\ &= \zeta_{48}^{-1}\mathfrak{f}(\tau+1) = \zeta_{48}^{-2}\mathfrak{f}_1(\tau)\\ \mathfrak{f}_1(V\tau) &= \mathfrak{f}_1(\tau+2) = \zeta_{48}^{-1}\mathfrak{f}(\tau+1) = \zeta_{48}^{-2}\mathfrak{f}_1(\tau)\end{aligned}$$

Taking sixth powers, we get

$$\mathfrak{f}_1(U\tau)^6 = \mathfrak{f}_1(V\tau)^6 = \zeta_{48}^{-12}\mathfrak{f}_1(\tau)^6 = -i\,\mathfrak{f}_1(\tau)^6.$$

(b) If $\gamma = \left(\begin{smallmatrix} a & b\\ c & d\end{smallmatrix}\right) \in \Gamma_0(2)^t$, write $\rho(\gamma) = -ac - \frac{1}{2}bd + \frac{1}{2}b^2c$ for the exponent of i in (12.26). Note that $\mathfrak{f}_1(-I\tau)^6 = \mathfrak{f}_1(\tau)^6$ and that $\rho(-I) = 0$ since $b = c = 0$, so that (12.26) holds for $-I$. One sees easily that $\rho(U) = \rho(V) = -1$, so that part(a) shows that (12.26) also holds for U and V.

Also, part (a) implies that $\mathfrak{f}_1(\tau)^6 = \mathfrak{f}_1(UU^{-1}\tau)^6 = -i\mathfrak{f}_1(U^{-1}\tau)^6$, so that $\mathfrak{f}_1(U^{-1}\tau)^6 = i\mathfrak{f}_1(\tau)^6$, and similarly, $\mathfrak{f}_1(V^{-1}\tau)^6 = i\mathfrak{f}_1(\tau)^6$. One computes that $\rho(U^{-1}) = \rho(V^{-1}) = 1$, so that (12.26) holds for U^{-1} and V^{-1}.

We next claim that

$$(*)\qquad \delta \in \{-I, U^{\pm 1}, V^{\pm 1}\},\ \gamma \in \Gamma_0(2)^t \implies \rho(\delta) + \rho(\gamma) \equiv \rho(\delta\gamma) \bmod 4.$$

(Note that $(-I)^{-1} = -I$.). This is proved as follows. Fix $\gamma = \left(\begin{smallmatrix} a & b\\ c & d\end{smallmatrix}\right) \in \Gamma_0(2)^t$. Then:

- For $\delta = -I$, since b is even, we have $\frac{1}{2}b^2c \equiv -\frac{1}{2}b^2c \bmod 4$, so that

$$\begin{aligned}\rho(\gamma) &= -ac - \frac{1}{2}bd + \frac{1}{2}b^2c \equiv -ac - \frac{1}{2}bd - \frac{1}{2}b^2c \bmod 4.\\ \rho(-I\gamma) &= -(-a)(-c) - \frac{1}{2}(-b)(-d) + \frac{1}{2}(-b)^2(-c),\end{aligned}$$

  and $\rho(\gamma) \equiv \rho(-I\gamma) \bmod 4$ follows. Since $\rho(-I) = 0$, we see that (*) holds for $\delta = -I$.

- Next, for $\delta = U$, we have $U\gamma = \left(\begin{smallmatrix} a & b\\ a+c & b+d\end{smallmatrix}\right)$, so that

$$\rho(U\gamma) = -a(a+c) - \frac{1}{2}b(b+d) + \frac{1}{2}b^2(a+c) = \rho(\gamma) + \frac{1}{2}b^2(a-1) - a^2.$$

  Since b is even and a is odd, $\frac{1}{2}b^2(a-1) \equiv 0 \bmod 4$ and $-a^2 \equiv -1 \bmod 4$, so that $\rho(U\gamma) \equiv \rho(\gamma) - 1 \bmod 4$. Then $\rho(U) = -1$ shows that (*) holds for $\delta = U$.

- Now suppose $\delta = U^{-1}$. Since $\rho(U^{-1}) = 1 = -(-1) = -\rho(U)$, the argument for $\delta = S^{-1}$ in the solution to part (a) of Exercise 12.5 shows that (*) holds for $\delta = U^{-1}$.

- For $\delta = V$, we have $V\gamma = \left(\begin{smallmatrix} a+2c & b+2d\\ c & d\end{smallmatrix}\right)$, so that

$$\begin{aligned}\rho(V\gamma) &= -(a+2c)c - \frac{1}{2}(b+2d)d + \frac{1}{2}(b+2d)^2c\\ &= \rho(\gamma) - 2c^2 - d^2 + 2bcd + 2cd^2.\end{aligned}$$

  But b is even and d is odd, so that

$$-2c^2 - d^2 + 2bcd + 2cd^2 \equiv 2c(d-c) - d^2 \equiv 2c(d-c) - 1 \bmod 4.$$

  Since d is odd, either c or $d - c$ is even and thus $2c(d-c) \equiv 0 \bmod 4$, so that in the end, $\rho(V\gamma) \equiv \rho(\gamma) - 1 \bmod 4$. Then $\rho(V) = -1$ shows that (*) holds for $\delta = V$.

- Finally, when $\delta = V^{-1}$, note that $\rho(V^{-1}) = 1 = -(-1) = -\rho(V)$, so that as for U^{-1}, (*) holds for $\delta = V^{-1}$ by the argument for $\delta = S^{-1}$ in the solution given in part (a) of Exercise 12.5.

Since $\Gamma(2)$ is generated by $-I$, U and V, it suffices by induction to show that if (12.26) holds for γ, then it holds for $\delta\gamma$ when $\delta \in \{-I, U^{\pm 1}, V^{\pm 1}\}$. This is now easy:

$$\begin{aligned}\mathfrak{f}_1((\delta\gamma)\tau)^6 &= \mathfrak{f}_1(\delta(\gamma\tau))^6 = i^{\rho(\delta)}\mathfrak{f}_1(\gamma\tau)^6 = i^{\rho(\delta)}i^{\rho(\gamma)}\mathfrak{f}_1(\tau)^6\\ &= i^{\rho(\delta)+\rho(\gamma)}\mathfrak{f}_1(\tau)^6 = i^{\rho(\delta\gamma)}\mathfrak{f}_1(\tau)^6,\end{aligned}$$

where the first line follows since δ and γ satisfy (12.26) and the second line uses (*).

12.21. (a) Note that $\gamma \in \Gamma_0(2)^t - \Gamma(2)$ can be written $\gamma = \left(\begin{smallmatrix} a & 2b \\ 2c+1 & d\end{smallmatrix}\right)$, where a is odd since $\det(\gamma) = 1$. However, one computes that

$$\begin{pmatrix} a & 2b \\ 2c+1 & d\end{pmatrix} = U\begin{pmatrix} a & 2b \\ 2c+1-a & d-2b\end{pmatrix},$$

where the last matrix above is in $\Gamma(2)$ since a is odd.. Thus every element of $\Gamma_0(2)^t$ is either of the form $I\Gamma(2)$ or $U\Gamma(2)$. Since $U \notin \Gamma(2)$, these are distinct coset representatives.

(b) Choose

$$A = \begin{pmatrix} a & b \\ c & d\end{pmatrix} \in \mathrm{SL}(2,\mathbb{Z}), \qquad M = \begin{pmatrix} p & 8q \\ 8r & s\end{pmatrix} \in \tilde{\Gamma}(8).$$

Then

$$AMA^{-1} = \begin{pmatrix} * & 8(a^2q - b^2r) + ab(s-p) \\ 8(d^2r - c^2q) + cd(p-s) & *\end{pmatrix}.$$

But p and s are odd, and $ps \equiv 1 \bmod 8$ since $\det(M) = 1$. It follows that $s \equiv p \bmod 8$, so $AMA^{-1} \in \tilde{\Gamma}(8)$.

For the second part, note that since $\tilde{\Gamma}(8)$ is normal in $\mathrm{SL}(2,\mathbb{Z})$, it is also normal in $\Gamma(2)$, so that the quotient is defined. Now,

$$[U^2, V] = \begin{pmatrix} 1 & 0 \\ 2 & 1\end{pmatrix}\begin{pmatrix} 1 & 2 \\ 0 & 1\end{pmatrix}\begin{pmatrix} 1 & -2 \\ 0 & 1\end{pmatrix}\begin{pmatrix} 1 & 0 \\ -2 & 1\end{pmatrix} = \begin{pmatrix} -3 & 8 \\ -8 & 21\end{pmatrix} \in \tilde{\Gamma}(8).$$

Therefore U^2 and V commute in $\Gamma(2)/\tilde{\Gamma}(8)$. Since $-I$ commutes with everything and $-I$, U^2 and V generate $\Gamma(2)$ by Exercise 12.4, the commutator subgroup of $\Gamma(2)$ is contained in $\tilde{\Gamma}(8)$, so that the quotient is Abelian.

(c) (i) Part (a) of Exercise 12.20 showed that (12.26) holds for U and V, so that

$$\begin{aligned}\mathfrak{f}_1(V^{b_i}\tau)^6 &= (-i)^{b_i}\mathfrak{f}_1(\tau)^6 = i^{-b_i}\mathfrak{f}_1(\tau)^6\\ \mathfrak{f}_1(U^{2a_i}\tau)^6 &= (-i)^{2a_i}\mathfrak{f}_1(\tau)^6 = i^{-2a_i}\mathfrak{f}_1(\tau)^6,\end{aligned}$$

so that $\mathfrak{f}_1(\gamma\tau)^6 = i^{-\sum_{i=1}^s 2a_i - \sum_{i=1}^s b_i}\mathfrak{f}_1(\tau)^6 = i^{-2A-B}\mathfrak{f}_1(\tau)^6$.

(ii) By the analysis in part (b), we can rearrange U^2's and V's in any element of $\Gamma(2)$ modulo $\tilde{\Gamma}(8)$, so that $\gamma \equiv U^{2A}V^B \bmod \tilde{\Gamma}(8)$. Computing the right-hand side, we find

$$\gamma = \begin{pmatrix} a & b \\ c & d\end{pmatrix} \equiv \begin{pmatrix} 1 & 0 \\ 2A & 1\end{pmatrix}\begin{pmatrix} 1 & 2B \\ 0 & 1\end{pmatrix} \equiv \begin{pmatrix} 1 & 2B \\ 2A & 1+4AB\end{pmatrix} \bmod \tilde{\Gamma}(8).$$

(iii) We compute

$$\begin{pmatrix} a & b \\ c & d\end{pmatrix}^{-1}\begin{pmatrix} 1 & 2B \\ 2A & 1+4AB\end{pmatrix} = \begin{pmatrix} * & -4bAB + 2Bd - b \\ 2aA - c & *\end{pmatrix}.$$

From part (ii), this matrix is in $\tilde{\Gamma}(8)$. However, $a^2 \equiv d^2 \equiv 1 \bmod 8$ since a and d are both odd. Also, b is even since $\gamma \in \Gamma(2)$. But then $8 \mid -4bAB + 2Bd - b$ implies

$$2Bd \equiv b \bmod 8 \ \Rightarrow\ 2Bd^2 \equiv bd \bmod 8 \ \Rightarrow\ 2B \equiv bd \bmod 8.$$

Similarly, $8 \mid 2aA - c$ implies

$$2aA \equiv c \bmod 8 \ \Rightarrow\ 2a^2A \equiv ac \bmod 8 \ \Rightarrow\ 2A \equiv ac \bmod 8.$$

(iv) Since $2A \equiv ac \bmod 8$ and $2B \equiv bd \bmod 8$, it follows that $i^{2A} = i^{ac}$ and that $i^B = i^{(1/2)bd}$ (note again that b is even, so that $(1/2)bd \in \mathbb{Z}$). Then from the previous two parts, we see that $\mathfrak{f}_1(\gamma\tau)^6 = i^{-2A-B}\mathfrak{f}_1(\tau)^6 = i^{-ac-(1/2)bd}\mathfrak{f}_1(\tau)^6$ for $\gamma \in \Gamma(2)$.

(d) Let $\gamma = \left(\begin{smallmatrix} a & b \\ c & d \end{smallmatrix}\right) \in \Gamma_0(2)^t - \Gamma(2)$. Then by the solution to part (a), $\gamma = U\tilde{\gamma}$ for

$$\tilde{\gamma} = U^{-1}\gamma = \begin{pmatrix} a & b \\ -a+c & -b+d \end{pmatrix} \in \Gamma(2).$$

Using part (a) of Exercise 12.20 and part (iv) of part (c) of this exercise, we get

$$\begin{aligned} \mathfrak{f}_1(\gamma\tau)^6 &= \mathfrak{f}_1(U\tilde{\gamma}\tau)^6 = -i\mathfrak{f}_1(\tilde{\gamma}\tau)^6 \\ &= i^{-1}i^{-a(-a+c)-(1/2)b(-b+d)}\mathfrak{f}_1(\tau)^6 \\ &= i^{-1+a^2-ac+(1/2)b^2-(1/2)bd}\mathfrak{f}_1(\tau)^6. \end{aligned}$$

Since a is odd, $a^2 - 1 \equiv 0 \bmod 4$, so this simplifies to

$$= i^{-ac-(1/2)bd+(1/2)b^2}\mathfrak{f}_1(\tau)^6.$$

(e) If $\gamma = \left(\begin{smallmatrix} a & b \\ c & d \end{smallmatrix}\right) \in \Gamma(2) \subset \Gamma_0(2)^t$, then $b \equiv c \equiv 0 \bmod 2$, so that $\frac{1}{2}b^2c \equiv 0 \bmod 4$. Otherwise, $\gamma \in \Gamma_0(2)^t - \Gamma(2)$, so that b is even but c is odd, and therefore $\frac{1}{2}b^2c \equiv \frac{1}{2}b^2 \bmod 4$. Using this and applying it to the formulas in parts (c) and (d) gives

$$\mathfrak{f}_1(\gamma\tau)^6 = i^{-ac-(1/2)bc+(1/2)b^2c}\mathfrak{f}_1(\tau)^6, \quad \gamma \in \Gamma_0(2)^t.$$

12.22. Let $K = \mathbb{Q}(\sqrt{-m})$, where $m > 0$ and $m \equiv 6 \bmod 8$. Let $\mathcal{O} = [1, \sqrt{-m}]$ and $\mathcal{O}' = [1, 4\sqrt{-m}]$.

(a) Let $\tau_0 = \sqrt{-m}/8$ and note that $K = \mathbb{Q}(\tau_0)$. Then τ_0 satisfies $64x^2 + m = 0$, and since $m \equiv 6 \bmod 8$, its minimal polynomial with relatively prime coefficients is $32x^2 + (m/2)$. By Lemma 7.5, $[1, \sqrt{-m}/8]$ is a proper fractional ideal for $[1, 32\sqrt{-m}/8] = [1, 4\sqrt{-m}] = \mathcal{O}'$. Hence the same is true for $\mathfrak{b} = 8[1, \sqrt{-m}/8] = [8, \sqrt{-m}]$.

Next, let $\tau_0 = (1 + \sqrt{-m})/4$ and $K = \mathbb{Q}(\tau_0)$. Since $m + 1$ is odd, the minimal polynomial of τ_0 with relatively prime coefficients is $16x^2 - 8x + m + 1 = 0$. Then Lemma 7.5 implies that $[1, (1 + \sqrt{-m})/4]$ is a proper fractional ideal for

$$[1, 16(1 + \sqrt{-m})/4] = [1, 4 + 4\sqrt{-m}] = [1, 4\sqrt{-m}] = \mathcal{O}'.$$

The same is true for $\mathfrak{a} = 4[1, (1 + \sqrt{-m})/4] = [4, 1 + \sqrt{-m}]$.

(b) The natural map $C(\mathcal{O}') \to C(\mathcal{O})$ sends $[\mathfrak{a}] \mapsto [\mathfrak{a}\mathcal{O}]$, and since $m + 1$ is odd,

$$\begin{aligned} \mathfrak{a}\mathcal{O} = [4, 1 + \sqrt{-m}][1, \sqrt{-m}] &= [4, 4\sqrt{-m}, 1 + \sqrt{-m}, -m + \sqrt{-m}] \\ &= [4, 1 + \sqrt{-m}, 1 + m] \\ &= [1 + \sqrt{-m}, 1] = \mathcal{O}, \end{aligned}$$

so that $[\mathfrak{a}]$ is in the kernel. If L and L' are the ring class fields of $\mathcal{O}$ and $\mathcal{O}'$, the text shows that $[L' : L] = 4$, so that the restriction map $\mathrm{Gal}(L'/K) \to \mathrm{Gal}(L/K)$ has kernel of order 4. Since the map $C(\mathcal{O}') \to C(\mathcal{O})$ induces this restriction map, it follows that the order of $[\mathfrak{a}]$ is 1, 2, or 4. But

$$\begin{aligned} \mathfrak{a}^2 &= [4, 1 + \sqrt{-m}][4, 1 + \sqrt{-m}] = [16, 4 + 4\sqrt{-m}, 1 - m + 2\sqrt{-m}] \\ &= [16, 4 - 2(1 - m), 1 - m + 2\sqrt{-m}] \\ &= [16, 2 + 2m, 1 - m + 2\sqrt{-m}]. \end{aligned}$$

Since $m \equiv 6 \bmod 8$, $2+2m \equiv 14 \bmod 16$, so this is the same as $[2, 1-m+2\sqrt{-m}] = [2, 1+2\sqrt{-m}]$. Suppose $\mathfrak{a}^2 = [2, 1+2\sqrt{-m}]$ is principal, say $\mathfrak{a}^2 = \lambda\mathcal{O}'$ for some λ. Then $1 \in \mathcal{O}'$ and $\mathfrak{a}^2 \subset \mathcal{O}$ imply that $\lambda = a + b\sqrt{-m}$ for $a, b \in \mathbb{Z}$. Then $2 \in \lambda\mathcal{O}'$ implies $2 = (a+b\sqrt{-m})(c+4d\sqrt{-m})$ for $c, d \in \mathbb{Z}$. Taking norms, we get

$$4 = N(2) = N(a+b\sqrt{-m})N(c+4d\sqrt{-m}) = (a^2+b^2m)(c^2+16d^2m).$$

Since $a, b, c, d \in \mathbb{Z}$ and $m \equiv 6 \bmod 8$ implies $m \geq 6$, one sees that $\lambda = a+b\sqrt{-m} = \pm 1$ or ± 2. Thus $\mathfrak{a}^2 = [2, 1+2\sqrt{-m}] = \mathcal{O}'$ or $2\mathcal{O}'$, both of which are impossible since $1+2\sqrt{-m} \notin \mathcal{O}' = [1, 4\sqrt{-m}]$. Thus $\mathfrak{a}$ has order 4 in $C(\mathcal{O}')$.

(c) We have $\overline{\mathfrak{a}} = [4, 1-\sqrt{m}] = [4, -1+\sqrt{-m}]$. Multiplying gives

$$\begin{aligned}\overline{\mathfrak{a}}\mathfrak{b} &= [4, -1+\sqrt{-m}][8, \sqrt{-m}]\\ &= [32, 4\sqrt{-m}, -8+8\sqrt{-m}, -m-\sqrt{-m}]\\ &= [32, 4\sqrt{-m}, -8, -m-\sqrt{-m}]\\ &= [8, 4\sqrt{-m}, -m-\sqrt{-m}].\end{aligned}$$

Since $-m \equiv 2 \bmod 8$, this is the same as

$$= [8, 4\sqrt{-m}, 2-\sqrt{-m}] = [8, 2-\sqrt{-m}] = [8, -2+\sqrt{-m}].$$

12.23. (a) $\mathfrak{f}(\sqrt{-m})^2$ is an algebraic integer for the same reason as in Theorem 12.24. Since $3 \nmid m$ and $m \equiv 3 \bmod 4$, in this case $\tau_0 = (3+\sqrt{-m})/2$ and $L = K(\gamma_2(\tau_0)) = K(j(\tau_0))$ by Theorem 12.13. From the discussion following the table of j-invariants in (12.20), we have

$$\gamma_2(\tau_0) = \frac{256 - \mathfrak{f}(\sqrt{-m})^{24}}{\mathfrak{f}(\sqrt{-m})^{16}} \in L,$$

so that $L \subset K(\mathfrak{f}(\sqrt{-m})^2)$. Suppose $\mathfrak{f}(\sqrt{-m})^6 \in L$. Then $\mathfrak{f}(\sqrt{-m})^{24} \in L$, which by the above formula for $\gamma_2(\tau_0)$ implies that $\mathfrak{f}(\sqrt{-m})^{16} \in L$. Since $\mathfrak{f}(\sqrt{-m})^6 \in L$, it follows easily that $\mathfrak{f}(\sqrt{-m})^2 \in L$ and thus $K(\mathfrak{f}(\sqrt{-m})^2) \subset L$. Therefore $\mathfrak{f}(\sqrt{-m})^6 \in L$ implies that $L = K(\mathfrak{f}(\sqrt{-m})^2)$.

(b) Let $\tilde{\Gamma}(8) \subset \mathrm{SL}(2, \mathbb{Z})$ be as in Exercise 12.21. We first show that $\mathfrak{f}_1(\tau)^6$ is $\tilde{\Gamma}(8)$-invariant. Take $\gamma = \left(\begin{smallmatrix} a & b \\ c & d \end{smallmatrix}\right) \in \tilde{\Gamma}(8)$. Since $\tilde{\Gamma}(8) \subset \Gamma_0(2)^t$, (12.26) implies

$$\mathfrak{f}_1(\gamma\tau)^6 = i^{-ac-(1/2)bd-(1/2)bc^2}\mathfrak{f}_1(\tau)^6.$$

Since $b \equiv c \equiv 0 \bmod 8$, the exponent is a multiple of four, and invariance under $\tilde{\Gamma}(8)$ follows.

We next show that $\mathfrak{f}(\tau)$ is also $\tilde{\Gamma}(8)$-invariant. Choose $\gamma \in \tilde{\Gamma}(8)$. By part (b) of Exercise 12.21, $\tilde{\Gamma}(8)$ is normal in $\mathrm{SL}(2, \mathbb{Z})$. This implies that $\gamma\tau + 1 = T\gamma\tau = \tilde{\gamma}T\tau$ for some $\tilde{\gamma} \in \tilde{\Gamma}(8)$. From here, we use Corollary 12.19 to obtain

$$\mathfrak{f}(\gamma\tau)^6 = \zeta_8\mathfrak{f}_1(\gamma\tau+1)^6 = \zeta_8\mathfrak{f}_1(\tilde{\gamma}T\tau)^6 = \zeta_8\mathfrak{f}_1(T\tau)^6 = \zeta_8\mathfrak{f}_1(\tau+1)^6 = \mathfrak{f}(\tau)^6.$$

To show that $\mathfrak{f}(8\tau)^6$ is a modular function for $\Gamma_0(64)$, we note first that it is meromorphic on $\mathfrak{h}$ since $\mathfrak{f}(\tau)$ is. Next, to show it is $\Gamma_0(64)$-invariant, let $\gamma = \left(\begin{smallmatrix} a & b \\ 64c & d \end{smallmatrix}\right) \in \Gamma_0(64)$. Then

$$8\gamma\tau = 8\begin{pmatrix} a & b \\ 64c & d \end{pmatrix}\tau = \begin{pmatrix} a & 8b \\ 8c & d \end{pmatrix}8\tau,$$

The matrix on the right is in $\tilde{\Gamma}(8)$ and $\mathfrak{f}(\tau)^6$ is $\tilde{\Gamma}(8)$-invariant. Hence $\mathfrak{f}(8\gamma\tau)^6 = \mathfrak{f}(8\tau)$, which proves invariance under $\Gamma_0(64)$.

Finally, the proof of part (i) of Theorem 12.24 shows that $\mathfrak{f}_1(8\tau)^6$ is meromorphic at the cusps. The argument remains valid when $\mathfrak{f}_1$ is replaced with $\mathfrak{f}$, which shows that

$\mathfrak{f}(8\tau)^6$ is meromorphic at the cusps. We conclude that $\mathfrak{f}(8\tau)^6$ is a modular function for $\Gamma_0(64)$.

(c) Since $m \equiv 3 \bmod 4$ and $3 \nmid m$, we know that $m \neq 1, 3$, so that $\mathcal{O}^* = \{\pm 1\}$ for $\mathcal{O} = [1, \sqrt{-m}]$. Then $T(\sqrt{-m}) = 0$ and $N(\sqrt{-m}) = m$, so choosing $s = 8$, $m = 64$ (this is a different m than that in $\sqrt{-m}$) in Lemma 12.11 gives

$$\frac{\partial \Phi_{64}}{\partial X}(j(8\sqrt{-m}), j(\sqrt{-m}/8)) = \frac{\partial \Phi_{64}}{\partial X}(j([1, 8\sqrt{-m}]), j([8, \sqrt{-m}])) \neq 0.$$

Then by Proposition 12.7, $\mathfrak{f}(8\tau)^6 \in \mathbb{Q}(j(\tau), j(64\tau))$, which we can write as

$$\mathfrak{f}(\tau)^6 \in \mathbb{Q}(j(\tau/8), j(8\tau)) = \mathbb{Q}(j([1, \tau/8]), j([1, 8\tau])) = \mathbb{Q}(j([8, \tau]), j([1, 8\tau])),$$

so that $\mathfrak{f}(\tau)^6 = S(j([8, \tau]), j([1, 8\tau]))$ for some $S(X, Y) \in \mathbb{Q}(X, Y)$. Set $\tau = \sqrt{-m}$. Using the fact that the partial derivative does not vanish, Proposition 12.7 implies that

$$\mathfrak{f}(\sqrt{-m})^6 = S(j([8, \sqrt{-m}]), j([1, 8\sqrt{-m}])), \quad S[X, Y] \in \mathbb{Q}(X, Y).$$

(d) Let $\mathcal{O}' = [1, 8\sqrt{-m}]$ and $\mathfrak{a}' = \left[1, \frac{2+\sqrt{-m}}{8}\right] = [1, \tau_0]$. Then the minimal polynomial for τ_0 is $64x^2 - 32x + m + 4$, whose coefficients are relatively prime since m is odd. Lemma 7.5 shows that $\mathfrak{a}'$ is a proper fractional ideal for $[1, 64\tau_0] = [1, 16 + 8\sqrt{-m}] = [1, 8\sqrt{-m}] = \mathcal{O}'$, so that the same is true for $\mathfrak{a} = 8\mathfrak{a}'$. For $\mathfrak{b}' = [1, \sqrt{-m}/8]$, the minimal polynomial of $\sqrt{-m}/8$ is $64x^2 + m$, which has relatively prime coefficients since m is odd. Lemma 7.5 shows that $\mathfrak{b}'$ is a proper fractional ideal in $[1, 64(\sqrt{-m}/8)] = [1, 8\sqrt{-m}] = \mathcal{O}'$, so that the same is true for $\mathfrak{b} = 8\mathfrak{b}'$. Note that $\mathfrak{a}$ and $\mathfrak{b}$ are still fractional ideals; neither lies in $\mathcal{O}'$.

The ring class field L' of $\mathcal{O}'$ is generated over K by $j(\mathfrak{a})$ where $\mathfrak{a}$ is any proper fractional ideal of $\mathcal{O}'$, by Theorem 11.1. Therefore both $j(\mathfrak{b}) = j([8, \sqrt{-m}])$ and $j(\mathcal{O}') = j([1, 8\sqrt{-m}])$ lie in L', so that by part (c), so does $\mathfrak{f}(\sqrt{-m})^6$.

(e) The discriminant of $\mathcal{O}$ is $D = -4m = 4d_K$, and the discriminant of $\mathcal{O}'$ is $-256m = 64D = 8^2 D$. Then Corollary 7.28 shows that $h(\mathcal{O}') = 8h(\mathcal{O})$ since $(D/2) = 0$. Since $K \subset L \subset L'$, $[L : K] = h(\mathcal{O})$, and $[L' : K] = h(\mathcal{O}')$, this shows that $[L' : L] = 8$.

For the second part, we first show that $\mathfrak{a}^4 = \mathfrak{b}^2 = \mathcal{O}'$. For $\mathfrak{a} = [8, 2 + \sqrt{-m}]$, we have

$$\begin{aligned}
\mathfrak{a}^2 &= [64, 16 + 8\sqrt{-m}, 4 - m + 4\sqrt{-m}] \\
&= [64, 16 + 8\sqrt{-m} - 2(4 - m + 4\sqrt{-m}), 4 - m + 4\sqrt{-m}] \\
&= [64, 8 + 2m, 4 - m + 4\sqrt{-m}] \\
&= [2, 4 - m + 4\sqrt{-m}] = [2, -m + 4\sqrt{-m}] = [2, 1 + 4\sqrt{-m}].
\end{aligned}$$

The fourth line follows from the third since is m odd. Then

$$\begin{aligned}
\mathfrak{a}^4 &= [2, 1 + 4\sqrt{-m}]^2 = [4, 2 + 8\sqrt{-m}, 1 - 16m + 8\sqrt{-m}] \\
&= [4, 2 + 8\sqrt{-m}, 1 + 8\sqrt{-m}] \\
&= [4, 2 + 8\sqrt{-m}, (1 + 8\sqrt{-m}) - (2 + 8\sqrt{-m})] \\
&= [4, 2 + 8\sqrt{-m}, -1] \\
&= [1, 2 + 8\sqrt{-m}] = [1, 8\sqrt{-m}] = \mathcal{O}'.
\end{aligned}$$

For $\mathfrak{b} = [8, \sqrt{-m}]$, the computation is easier:

$$\mathfrak{b}^2 = [64, 8\sqrt{-m}, -m] = [1, 8\sqrt{-m}] = \mathcal{O}'$$

where the second equality again follows since $-m$ is odd.

To show that $[\mathfrak{a}]$ and $[\mathfrak{b}]$ generate a subgroup of $C(\mathcal{O}')$ of order 8, we need to show that certain ideals are not principal. We will use the following fact:

(*) If $\mathfrak{c}$ is a principal fractional $\mathcal{O}'$-ideal with $8 \in \mathfrak{c} \subset \mathcal{O}$, then $\mathfrak{c} \subset \mathcal{O}'$.

To prove this, assume $\mathfrak{c} = \lambda\mathcal{O}'$ for some λ. Then $1 \in \mathcal{O}'$ and $\mathfrak{c} \subset \mathcal{O}$ imply that $\lambda = a + b\sqrt{-m}$ for $a, b \in \mathbb{Z}$. Then $8 \in \lambda\mathcal{O}'$ implies $8 = (a+b\sqrt{-m})(c+8d\sqrt{-m})$ for $c, d \in \mathbb{Z}$. Taking norms, we get

$$64 = N(8) = N(a+b\sqrt{-m})N(c+8d\sqrt{-m}) = (a^2+b^2m)(c^2+64d^2m).$$

Since $a, b, c, d \in \mathbb{Z}$ and $m \equiv 3 \bmod 4$ implies $m \geq 3$, the second factor is > 64 when $d \neq 0$. Thus $d = 0$, which easily implies $b = 0$. Then λ is a nonzero integer, so $\mathfrak{c} = \lambda\mathcal{O}' \subset \mathcal{O}'$, proving (*).

To show that $[\mathfrak{a}]$ has order 4 in $C(\mathcal{O}')$, we need to show that $\mathfrak{a}^2 = [2, 1+4\sqrt{-m}]$ is not principal. If it is, then $8 \in \mathfrak{a}^2 \subset \mathcal{O}$ and (*) imply that $\mathfrak{a}^2 \subset \mathcal{O}'$. This is impossible since $1+4\sqrt{-m} \in \mathfrak{a}^2$ yet $1+4\sqrt{-m} \notin \mathcal{O}' = [1, 8\sqrt{-m}]$. Thus $\mathfrak{a}$ has order 4 in $C(\mathcal{O}')$.

To show that $[\mathfrak{b}]$ has order 2 in $C(\mathcal{O}')$, we need to show that $\mathfrak{b} = [8, \sqrt{-m}]$ is not principal. If it is, then $8 \in \mathfrak{b} \subset \mathcal{O}$ and (*) imply that $\mathfrak{b} \subset \mathcal{O}'$. This is impossible since $\sqrt{-m} \in \mathfrak{b}$ yet $\sqrt{-m} \notin \mathcal{O}' = [1, 8\sqrt{-m}]$. This proves that $\mathfrak{b}$ has order 2 in $C(\mathcal{O}')$.

Finally, we need to show that $[\mathfrak{a}]^2 \neq [\mathfrak{b}]$ in $\mathbb{C}(\mathcal{O}')$. Suppose $\mathfrak{b} = \lambda\mathfrak{a}^2$ for some $\lambda \neq 0$. Since $\mathfrak{a}^4 = \mathcal{O}'$, we get $\mathfrak{a}^2\mathfrak{b} = \lambda\mathcal{O}'$. Using the above computation of $\mathfrak{a}^2$, we have

$$\begin{aligned}\mathfrak{a}^2\mathfrak{b} &= [2, 1+4\sqrt{-m}][8, \sqrt{-m}] = [16, 8+32\sqrt{-m}, 2\sqrt{-m}, -4m+\sqrt{-m}]\\ &= [16, 8, 2\sqrt{-m}, -4m+\sqrt{-m}] = [8, 2\sqrt{-m}, 4+\sqrt{-m}]\\ &= [8, 2\sqrt{-m} - 2(4+\sqrt{-m}), 4+\sqrt{-m}] = [8, -8, 4+\sqrt{-m}]\\ &= [8, 4+\sqrt{-m}],\end{aligned}$$

where the second line follows since m is odd. Thus $\mathfrak{a}^2\mathfrak{b} = [8, 4+\sqrt{-m}] = \lambda\mathcal{O}'$. Then $8 \in \mathfrak{a}^2\mathfrak{b} \subset \mathcal{O}$ and (*) imply that $\mathfrak{a}^2\mathfrak{b} \subset \mathcal{O}'$. This is impossible since $4+\sqrt{-m} \in \mathfrak{a}^2\mathfrak{b}$ yet $4+\sqrt{-m} \notin \mathcal{O}' = [1, 8\sqrt{-m}]$. Thus $[\mathfrak{a}]^2 \neq [\mathfrak{b}]$ in $C(\mathcal{O}')$.

It follows that $[\mathfrak{a}]$ and $[\mathfrak{b}]$ generate a subgroup of $\mathrm{Gal}(L'/K)$ of order 8 under the isomorphism $C(\mathcal{O}') \to \mathrm{Gal}(L'/K)$. The map $C(\mathcal{O}') \to C(\mathcal{O})$ is the map $[\mathfrak{c}] \mapsto [\mathfrak{c}\mathcal{O}]$ for an ideal class $[\mathfrak{c}] \in C(\mathcal{O}')$, and

$$\begin{aligned}[8, 2+\sqrt{-m}][1, \sqrt{-m}] &= [8, 8\sqrt{-m}, 2+\sqrt{-m}, -m+2\sqrt{-m}]\\ &= [8, 8\sqrt{-m}, m+2, 2+\sqrt{-m}] = [1, 8\sqrt{-m}, \sqrt{-m}] = \mathcal{O}\\ [8, \sqrt{-m}][1, \sqrt{-m}] &= [8, \sqrt{-m}, -m, 8\sqrt{-m}] = \mathcal{O}\end{aligned}$$

since m is odd, so $[\mathfrak{a}]$ and $[\mathfrak{b}]$ are in the kernel of the map $C(\mathcal{O}') \to C(\mathcal{O})$, which corresponds to the quotient (restriction) map $\mathrm{Gal}(L'/K) \to \mathrm{Gal}(L/K)$. Since $[L' : L] = 8$, they generate the kernel, which is isomorphic to $\mathrm{Gal}(L'/L)$. Let the corresponding generators be σ_1 and σ_2.

(f) Recall from parts (d) and (e) that $\mathfrak{a} = [8, 2+\sqrt{-m}]$ and $\mathfrak{b} = [8, \sqrt{-m}]$. Then using Corollary 11.37 we observe that $\sigma_1 \in \mathrm{Gal}(L'/L)$ is the map $j(\mathfrak{c}) \mapsto j(\bar{\mathfrak{a}}\mathfrak{c})$, and $\sigma_2 \in \mathrm{Gal}(L'/L)$ is the map $j(\mathfrak{c}) \mapsto j(\bar{\mathfrak{b}}\mathfrak{c})$. Since $\mathfrak{f}(\sqrt{-m})^6$ is a rational function in $j([8, \sqrt{-m}])$ and $j([1, 8\sqrt{-m}])$, we must evaluate $\bar{\mathfrak{a}}\mathfrak{c}$ and $\bar{\mathfrak{b}}\mathfrak{c}$ for $\mathfrak{c} = [8, \sqrt{-m}]$ and for

$\mathfrak{c} = [1, 8\sqrt{-m}]$. Using the fact that $m \equiv 3 \bmod 4$, we have

$$\begin{aligned}
\overline{\mathfrak{a}}[8, \sqrt{-m}] &= [8, 2-\sqrt{-m}][8, \sqrt{-m}] = [64, 8\sqrt{-m}, 16 - 8\sqrt{-m}, m + 2\sqrt{-m}]\\
&= [16, 8\sqrt{-m}, m + 2\sqrt{-m}] = [16, -4m, m + 2\sqrt{-m}]\\
&= [4, m + 2\sqrt{-m}] = [4, 3 + 2\sqrt{-m}]\\
\overline{\mathfrak{a}}[1, 8\sqrt{-m}] &= [8, 2-\sqrt{-m}]\mathcal{O}' = [8, 2-\sqrt{-m}] = [8, -2+\sqrt{-m}] = [8, 6+\sqrt{-m}]\\
\overline{\mathfrak{b}}[8, \sqrt{-m}] &= [8, -\sqrt{-m}][8, \sqrt{-m}] = \mathfrak{b}^2 = \mathcal{O}' = [1, 8\sqrt{-m}] \text{ (computed earlier)}\\
\overline{\mathfrak{b}}[1, 8\sqrt{-m}] &= [8, -\sqrt{-m}]\mathcal{O}' = [8, \sqrt{-m}].
\end{aligned}$$

Then part (c) implies that

$$\begin{aligned}
\sigma_1(\mathfrak{f}(\sqrt{-m})^6) &= S(j(\overline{\mathfrak{a}}[8, \sqrt{-m}]), j(\overline{\mathfrak{a}}[1, 8\sqrt{-m}]))\\
&= S(j([4, 3+2\sqrt{-m}]), j([8, 6+\sqrt{-m}]))\\
\sigma_2(\mathfrak{f}(\sqrt{-m})^6) &= S(j(\overline{\mathfrak{b}}[8, \sqrt{-m}]), j(\overline{\mathfrak{b}}[1, 8\sqrt{-m}]))\\
&= S(j([1, 8\sqrt{-m}], j([8, \sqrt{-m}]).
\end{aligned}$$

(g) We want to evaluate the behavior of $\mathfrak{f}(\tau)^6 = S(j([8,\tau]), j([1, 8\tau]))$ under $\gamma_1 = \left(\begin{smallmatrix} 2 & 11 \\ 1 & 6 \end{smallmatrix}\right)$ and $\gamma_2 = \left(\begin{smallmatrix} 0 & -1 \\ 1 & 0 \end{smallmatrix}\right)$. Computing the effects of these matrices on τ and using $\sim$ to indicate homotheties, we get

$$\begin{aligned}
[8, \gamma_1\tau] &= \left[8, \frac{2\tau+11}{\tau+6}\right] \sim [8\tau + 48, 2\tau + 11] = [4, 2\tau+11] = [4, 2\tau+3]\\
[1, 8\gamma_1\tau] &= \left[1, 8\frac{2\tau+11}{\tau+6}\right] \sim [\tau+6, 16\tau+88] = [\tau+6, 8]\\
[8, \gamma_2\tau] &= \left[8, \frac{-1}{\tau}\right] \sim [1, 8\tau]\\
[1, 8\gamma_2\tau] &= \left[1, 8\frac{-1}{\tau}\right] \sim [8, \tau].
\end{aligned}$$

By part (c), $\mathfrak{f}(\tau)^6 = S(j([8,\tau]), j([1,8\tau]))$, so that $\mathfrak{f}(\gamma_i\tau)^6 = S(j([8, \gamma_i\tau]), j([1, 8\gamma_i\tau]))$. Since j is homothety invariant, the above equations give the desired formula for $\mathfrak{f}(\gamma_i\tau)^6$.

(h) Parts (f) and (g) imply that $\sigma_i(\mathfrak{f}(\sqrt{-m})^6) = \mathfrak{f}(\gamma_i\sqrt{-m})^6$ for $i = 1, 2$. By Corollary 12.19, $\mathfrak{f}(-1/\tau) = \mathfrak{f}(\tau)$ and $\mathfrak{f}(\tau + n) = \zeta_{48}^{-n}\mathfrak{f}(\tau)$ for $n \in \mathbb{Z}$. Then

$$\begin{aligned}
\mathfrak{f}(\gamma_1\tau)^6 &= \mathfrak{f}\Big(\frac{2\tau+11}{\tau+6}\Big)^6 = \mathfrak{f}\Big(2 - \frac{1}{\tau+6}\Big)^6 = \Big(\zeta_{48}^{-2}\mathfrak{f}\Big(-\frac{1}{\tau+6}\Big)\Big)^6\\
&= \big(\zeta_{48}^{-2}\mathfrak{f}(\tau+6)\big)^6 = \big(\zeta_{48}^{-8}\mathfrak{f}(\tau)\big)^6 = \mathfrak{f}(\tau)^6\\
\mathfrak{f}(\gamma_2\tau)^6 &= \mathfrak{f}\Big(\frac{-1}{\tau}\Big)^6 = \mathfrak{f}(\tau)^6.
\end{aligned}$$

If follows immediately that $\sigma_i(\mathfrak{f}(\sqrt{-m})^6) = \mathfrak{f}(\gamma_i\sqrt{-m})^6 = \mathfrak{f}(\sqrt{-m})^6$, so that $\mathfrak{f}(\sqrt{-m})^6$ is fixed by σ_1 and σ_2. This shows that $\mathfrak{f}(\sqrt{-m})^6 \in L$ and thus $L = L'$.

12.24. The map $C(\mathcal{O}') \to C(\mathcal{O})$ is $[\mathfrak{a}] \mapsto [\mathfrak{a}\mathcal{O}]$, where $\mathcal{O} = [1, \sqrt{-14}]$. Hence the class of $\mathfrak{b} = [8, \sqrt{-14}]$ maps to the class of

$$[8, \sqrt{-14}][1, \sqrt{-14}] = [8, \sqrt{-14}, 8\sqrt{-14}, -14] = [2, \sqrt{-14}].$$

There are four classes of reduced forms with discriminant -56:

$$x^2 + 14y^2, \quad 2x^2 + 7y^2, \quad 3x^2 - 2xy + 5y^2, \quad 3x^2 + 2xy + 5y^2.$$

The first of these is the identity in the form class group, so has order 1. The two forms $3x^2 \pm 2xy + 5y^2$ are inverses of each other, so $2x^2 + 7y^2$ is the only element whose class has order 2. Under the isomorphism from Theorem 7.7, $[2, \sqrt{-14}] = \big[2, \frac{\sqrt{-56}}{2}\big]$ maps to

$2x^2 + 7y^2$. It follows that the class of $\mathfrak{b} = [8, \sqrt{-14}]$ maps to the unique element of order 2 in $C(\mathcal{O})$.

12.25. The reduced forms of these two discriminants are:

$$\begin{aligned} -4 \cdot 46: &\quad x^2 + 46y^2,\ 2x^2 + 23y^2,\ 5x^2 - 4xy + 10y^2,\ 5x^2 + 4xy + 10y^2 \\ -4 \cdot 142: &\quad x^2 + 142y^2,\ 2x^2 + 71y^2,\ 11x^2 - 2xy + 13y^2, 11x^2 + 2xy + 13y^2, \end{aligned}$$

so that the Hilbert class fields each have Galois groups over K cyclic of order 4. Then if one substitutes $m = 46 = 2 \cdot 23$ or $m = 142 = 2 \cdot 71$ for $14 = 2 \cdot 7$ in the computation of $j(\sqrt{-14})$ in the text and assumes that $\mathrm{Gal}(L/K)$ is cyclic of order 4, the argument goes through, with $\alpha = \mathfrak{f}_1(\sqrt{-m})^2$, up through the equation

$$\frac{2}{\alpha} = \mathfrak{f}_2(\sqrt{-m}/2)^2 = 2q^{1/12} \prod_{n=1}^{\infty} (1 + q^n)^2,$$

where $q = e^{-\pi\sqrt{m}}$.

In particular, we know that in both cases, the genus field is $K(\sqrt{2})$ and $\alpha + 2/\alpha = a + b\sqrt{2}$ for positive integers a and b. So we can determine $\alpha + 2/\alpha$ by computing a sufficiently accurate numerical approximation.

Again using the analysis from the class number 1 case (see the material between (12.22) and (12.23)), the q-expansion of $\mathfrak{f}_2(\tau)$ gives

$$2q^{1/12} < \frac{2}{\alpha} = \mathfrak{f}_2(\sqrt{-m}/2)^2 < 2q^{1/12}e^{2.004q}.$$

Therefore also

$$q^{-1/12}e^{-2.004q} < \alpha < q^{-1/12},$$

so that

$$q^{-1/12}e^{-2.004q} + 2q^{1/12} < \alpha + \frac{2}{\alpha} < q^{-1/12} + 2q^{1/12}e^{2.004q}.$$

The difference in the upper and lower bounds is then

$$E = 2q^{1/12}(e^{2.004q} - 1) + q^{-1/12}(1 - e^{-2.004q}).$$

With the inequality $1 - e^{-x} < \frac{x}{1-x}$, we get

$$E < 2q^{1/12}(e^{2.004q} - 1) + q^{-1/12}\frac{2.004q}{1 - 2.004q} < 10^{-4},$$

where the final inequality uses the fact that $q < e^{-4\pi}$ in both cases. Thus

$$\alpha + \frac{2}{\alpha} \approx q^{-1/12} + 2q^{1/12}$$

with an error of at most 10^{-4}.

For $m = 46$, we have $q = e^{-\pi\sqrt{46}}$, while for $m = 142$, $q = e^{-\pi\sqrt{142}}$, so that

$$\begin{aligned} m = 46: \quad \alpha + \frac{2}{\alpha} &\approx e^{-\pi\sqrt{46}/12} + 2e^{\pi\sqrt{46}/12} \approx 5.9039 + 2 \cdot 0.1640 \\ &\approx 6.2426 \approx 2 + 3\sqrt{2} \\ m = 142: \quad \alpha + \frac{2}{\alpha} &\approx e^{-\pi\sqrt{142}/12} + 2e^{\pi\sqrt{142}/12} \approx 22.6396 + 2 \cdot 0.04417 \\ &\approx 22.7279 \approx 10 + 9\sqrt{2}. \end{aligned}$$

For $m = 46$, $\alpha + 2/\alpha = 2 + 3\sqrt{2}$ gives $\alpha^2 - (2 + 3\sqrt{2})\alpha + 2 = 0$, so that

$$\alpha = \frac{2 + 3\sqrt{2} \pm \sqrt{14 + 12\sqrt{2}}}{2} = \frac{3 + \sqrt{2} \pm \sqrt{7 + 6\sqrt{2}}}{\sqrt{2}}.$$

Taking again the larger root, we get

$$\begin{aligned}\gamma_2(\sqrt{-46}) &= \mathfrak{f}_1(\sqrt{-46})^{16} + \frac{16}{\mathfrak{f}_1(\sqrt{-46})^8} = \alpha^8 + \frac{16}{\alpha^4} = \alpha^8 + \Big(\frac{2}{\alpha}\Big)^4 \\ &= \left(\frac{3+\sqrt{2}+\sqrt{7+6\sqrt{2}}}{\sqrt{2}}\right)^8 + \left(\frac{3+\sqrt{2}-\sqrt{7+6\sqrt{2}}}{\sqrt{2}}\right)^4 \\ &= 6\big(61553 + 43524\sqrt{2} + 15615\sqrt{7+6\sqrt{2}} + 11043\sqrt{2}\sqrt{7+6\sqrt{2}}\,\big).\end{aligned}$$

Cubing this, we get

$$j(\sqrt{-46}) = 216\big(61553 + 43524\sqrt{2} + 15615\sqrt{7+6\sqrt{2}} + 11043\sqrt{2}\sqrt{7+6\sqrt{2}}\,\big)^3.$$

Thus $L = K\big(\sqrt{7+6\sqrt{2}}\big)$ is the Hilbert class field of $K = \mathbb{Q}\left(\sqrt{-46}\right)$.

For $m = 142$, $\alpha + 2/\alpha = 10 + 9\sqrt{2}$ gives $\alpha^2 - (10+9\sqrt{2})\alpha + 2 = 0$, so that

$$\alpha = \frac{10+9\sqrt{2}+\sqrt{254+180\sqrt{2}}}{2} = \frac{9+5\sqrt{2}\pm\sqrt{127+90\sqrt{2}}}{\sqrt{2}}.$$

Using the larger root for α,

$$\begin{aligned}\gamma_2(\sqrt{-142}) &= \mathfrak{f}_1(\sqrt{-142})^{16} + \frac{16}{\mathfrak{f}_1(\sqrt{-142})^8} = \alpha^8 + \frac{16}{\alpha^4} = \alpha^8 + \Big(\frac{2}{\alpha}\Big)^4 \\ &= \left(\frac{9+5\sqrt{2}+\sqrt{127+90\sqrt{2}}}{\sqrt{2}}\right)^8 + \left(\frac{9+5\sqrt{2}-\sqrt{127+90\sqrt{2}}}{\sqrt{2}}\right)^4 \\ &= 30\big(575131981 + 406679724\sqrt{2} + (36066897 + 25503147\sqrt{2})\sqrt{127+90\sqrt{2}}\big).\end{aligned}$$

Cubing this, we see that $j(\sqrt{-142})$ equals

$$27000\big(575131981 + 406679724\sqrt{2} + (36066897 + 25503147\sqrt{2})\sqrt{127+90\sqrt{2}}\,\big)^3.$$

Thus $L = K\big(\sqrt{127+90\sqrt{2}}\big)$ is the Hilbert class field of $K = \mathbb{Q}(\sqrt{-142})$.

12.26. (a) If $2(b^2 - 4a) = (2b - a^2)^2$, then a must be even since the left hand side is even. Then modulo 4 the equation reads $2b^2 \equiv (2b)^2 = 4b^2 \bmod 4$, so that b must be even as well.

(b) Evaluating $2X(X^3+1) - Y^2$ for $X = -a/2$ and $Y = (b-a^2)/2$, we get

$$\begin{aligned}2X(X^3+1) - Y^2 &= 2\Big(-\frac{a}{2}\Big)\Big(-\frac{a^3}{8}+1\Big) - \Big(\frac{b-a^2}{2}\Big)^2 \\ &= \frac{1}{8}\left(a(a^3-8) - 2(b^2 - 2a^2b + a^4)\right) \\ &= \frac{1}{8}(-a^4 - 8a - 2b^2 + 4a^2b).\end{aligned}$$

However, we are assuming that a and b satisfy the equation

$$\begin{aligned}0 &= 2(b^2-4a) - (2b-a^2)^2 = 2b^2 - 8a - (4b^2 - 4a^2b + a^4) \\ &= -a^4 - 8a - 2b^2 + 4a^2b.\end{aligned}$$

Comparing this to the above equation shows that $2X(X^3+1) = Y^2$. Note also that $X = -a/2$ and $Y = (b-a^2)/2$ are integers since a and b are even by part (a).

12.27. (a) First, we rewrite the equation as $X^3 = -Z^2 - 1 = -Z^2 + i^2 = (i+Z)(i-Z)$ and let $d = \gcd(i+Z, i-Z)$ in $\mathbb{Z}[i]$, where Z is an integer. Then d divides their sum, $2i$, so that either d is a unit or d is a power of $1+i$ (which is prime). In the latter case, we have $1+i \mid Z+i$, so that for some integers a, b

$$Z + i = (1+i)(a+bi) = (a-b) + (a+b)i.$$

This implies that $a = \frac{Z+1}{2}$, so that Z must be odd, and therefore $X^3 + 1 \equiv 3 \bmod 4$. But this is impossible. So $i + Z$ and $i - Z$ are relatively prime, and therefore each of them must be a cube since $\mathbb{Z}[i]$ is a PID and every unit in $\mathbb{Z}[i]$ is a cube. Thus

$$Z + i = (a + bi)^3 = (a^3 - 3ab^2) + (3a^2b - b^3)i,$$

so that $3a^2b - b^3 = b(3a^2 - b^2) = 1$. It follows that $3a^2 - b^2 = b = \pm 1$. Then $b^2 = 1$ so that $3a^2 - 1 = b$, which we write as $3a^2 = b + 1$. Hence $3a^2 = 0$ or 2; the latter is impossible. It follows that $a = 0$ and $b = -1$, so that $Z + i = (-i)^3 = i$. Thus $Z = 0$, so that $X^3 = -Z^2 - 1 = -1$. Since X is an integer, we must have $X = -1$, so that $(-1, 0)$ is the only solution.

(b) Over $\mathbb{Z}[\omega]$, consider the factorization

$$W^6 + 1 = (W^2 + 1)(W^4 - W^2 + 1) = (W^2 + 1)(W^2 + \omega)(W^2 + \omega^2).$$

We claim that $W^2 + 1$ is prime to both $W^2 + \omega$ and $W^2 + \omega^2$. To see this, first suppose that $d \mid W^2 + 1$ and $d \mid W^2 + \omega$ for $d \in \mathbb{Z}[\omega]$. Then $d \mid 1 - \omega$. Since $1 - \omega$ is prime and therefore irreducible, it follows that either $d = 1 - \omega$ or d is a unit. But $1 - \omega \mid W^2 + 1$ implies that $3 \mid W^2 + 1$, which is impossible. Thus d is a unit, proving that $W^2 + 1$ is prime to $W^2 + \omega$. Similarly, if $d \mid W^2 + 1$ and $d \mid W^2 + \omega^2$, then $d \mid 1 - \omega^2 = (1 - \omega)(1 + \omega)$. The latter factor is a unit, so that d is a unit or $d = 1 - \omega$, and again d cannot be $1 - \omega$ since then $3 \mid W^2 + 1$. This proves the claim. So $W^2 + 1$ is prime to each of the other two factors, which implies that in $\mathbb{Z}$, $W^2 + 1$ is relatively prime to $W^4 - W^2 + 1$. Since $W^2 + 1$ and $W^4 - W^2 + 1$ are positive with product $2Z^2$, we see that $W^2 + 1$ is either a square or twice a square. If $W^2 + 1$ is a square, it is straightforward to prove that $W = 0$. This implies that $1 = W^6 + 1 = 2Z^2$, which is impossible. Thus $W^2 + 1$ is twice a square, so that $W^4 - W^2 + 1$ is a square. Set $n = W^2$, so $n \geq 0$ and $n^2 - n + 1$ is a square. However,

$$n > 1 \implies (n-1)^2 < n^2 - n + 1 < n^2 \implies n^2 - n + 1 \text{ is not a square.}$$

We conclude that $n = 0$ or $n = 1$. But $n = 0$ implies $W = 0$, which we already know to be impossible. Thus $W^2 = n = 1$, so that $W = \pm 1$. Then $2 = W^6 + 1 = 2Z^2$, which implies that $Z = \pm 1$. This completes the proof that the only integral solutions are $(W, Z) = (\pm 1, \pm 1)$.

(c) Here, we write the equation as $X^3 = -2Z^2 - 1$. In $\mathbb{Z}[\sqrt{-2}]$, the right-hand side factors to give $X^3 = (Z\sqrt{-2} + 1)(Z\sqrt{-2} - 1)$. If $d = \gcd(Z\sqrt{-2} + 1, Z\sqrt{-2} - 1)$, then d divides their difference, 2, so that d is a unit or d is a power of $\sqrt{-2}$, which is prime. In the latter case, $\sqrt{-2} \mid Z\sqrt{-2} + 1$, so that

$$1 + Z\sqrt{-2} = (a + b\sqrt{-2})\sqrt{-2} = -2b + a\sqrt{-2},$$

which is impossible. Thus the two factors are relatively prime. Since their product is a cube, part (c) of Exercise 5.21 shows that each of them is a cube in $\mathbb{Z}[\sqrt{-2}]$. Writing the first factor as a cube we get

$$1 + Z\sqrt{-2} = (a + b\sqrt{-2})^3 = (a^3 - 6ab^2) + (3a^2b - 2b^3)\sqrt{-2},$$

so that $a^3 - 6ab^2 = a(a^2 - 6b^2) = 1$. Thus $a^2 - 6b^2 = a = \pm 1$, which implies $1 - 6b^2 = a$. Then $6b^2 = 1 - a$, which is either 0 or 2. But 2 is impossible, so that $b = 0$. Then $Z = 3a^2b - 2b^3 = 0$, which implies $X^3 = -2Z^2 - 1 = -1$. Thus $X = -1$. We conclude that $(X, Z) = (-1, 0)$ is the only integer solution of $X^3 + 1 = -2Z^2$.

12.28. (a) Let p be prime. If $p \mid b$ and $p \mid c^2 - 3bc + 3b^2$, then $p \mid c^2$, which is impossible since $\gcd(b, c) = 1$. Similarly, if $p \mid c$ and $p \mid c^2 - 3bc + 3b^2$, then $p \mid 3b^2$. This is impossible since $p \nmid 3$ and $\gcd(b, c) = 1$. Thus b, c, and $c^2 - 3bc + 3b^2$ are pairwise relatively prime and positive. So if their product is a perfect square, each of them is also a perfect square.

In particular, $c^2 - 3bc + 3b^2 = k^2$ for some positive integer k. Choose relatively prime positive integers $m, n \in \mathbb{Z}$ such that

$$\frac{m}{n} = \frac{k+c}{b}.$$

This implies $\frac{m}{n}b - c = k$, so that $\left(\frac{m}{n}b - c\right)^2 = k^2 = c^2 - 3bc + 3b^2$. Expanding gives $\frac{m^2}{n^2}b^2 - \frac{2m}{n}bc + c^2 = c^2 - 3bc + 3b^2$, so that $\frac{m^2}{n^2}b^2 - \frac{2m}{n}bc = 3b^2 - 3bc$. Removing a factor of b gives $\frac{m^2}{n^2}b - \frac{2m}{n}c = 3b - 3c$. Next, we clear fractions and collect terms to obtain

$$(*) \qquad (3n^2 - m^2)b = (3n^2 - 2mn)c \implies \frac{b}{c} = \frac{2mn - 3n^2}{m^2 - 3n^2}.$$

We next show that $2mn - 3n^2,\ m^2 - 3n^2 > 0$. These have the same sign by (*) since $b/c > 0$. So it suffices to show $2mn - 3n^2 = n(2m - 3n) > 0$. Since $n > 0$, we are reduced to showing $2m - 3n > 0$. By the definition of m and n, we obtain

$$2m - 3n > 0 \iff \frac{m}{n} > \frac{3}{2} \iff \frac{k+c}{b} > \frac{3}{2} \iff k > \frac{3}{2}b - c.$$

The last inequality is true because

$$k = \sqrt{c^2 - 3bc + 3b^2} = \sqrt{(c - \tfrac{3}{2}b)^2 + \tfrac{3}{4}b^2} > \left|c - \tfrac{3}{2}b\right| \geq \tfrac{3}{2}b - c.$$

(b) This is the case where $3 \nmid m$.

(i) We first claim that $\gcd(2mn - 3n^2, m^2 - 3n^2) = 1$. Suppose that $p \mid 2mn - 3n^2$ and $p \mid m^2 - 3n^2$ for some prime p. If $p \mid m$, then note that $p \neq 3$ since $3 \nmid m$. But then $p \mid m^2 - 3n^2$ and $p \mid m$ implies $p \mid 3n^2$, contradicting $\gcd(m, n) = 1$. Thus $p \nmid m$. We also have $p \mid (m^2 - 3n^2) - (2mn - 3n^2) = m(m - 2n)$, so $p \nmid m$ implies that $m \equiv 2n \bmod p$. Since $p \mid 2mn - 3n^2$, we get

$$0 \equiv 2mn - 3n^2 \equiv 2(2n)n - 3n^2 \equiv n^2 \bmod p,$$

hence $p \mid n$. Then $m \equiv 2n \equiv 0 \bmod p$, which contradicts $p \nmid m$ and proves the claim.

In part (a), (*) is an equality of fractions with positive numerator and denominator. Furthermore, b and c are assumed to be relatively prime, and we just proved the same for $2mn - 3n^2$ and $m^2 - 3n^2$. It follows that $b = 2mn - 3n^2$ and $c = m^2 - 3n^2$.

(ii) Since c is a square, $c = m^2 - 3n^2$ implies that $m^2 - 3n^2 = r^2$ for $r \in \mathbb{Z}$. Then

$$m^2 - 3n^2 = r^2 = (\pm r)^2 = \left(\frac{m \pm r}{n}n - m\right)^2.$$

But $m + r$ and $m - r$ cannot both be multiples of 3, since then $3 \mid 2m$, yet we are assuming $3 \nmid m$. So choose $p = m \pm r$, $q = n$ with $3 \nmid p$. Then $m^2 - 3n^2 = \left(\frac{p}{q}n - m\right)^2$. For the second part,

$$\frac{p^2 + 3q^2}{2pq} = \frac{(m \pm \sqrt{m^2 - 3n^2})^2 + 3n^2}{2(m \pm \sqrt{m^2 - 3n^2})n} = \frac{2m^2 \pm 2m\sqrt{m^2 - 3n^2}}{2mn \pm 2n\sqrt{m^2 - 3n^2}} = \frac{m}{n}.$$

Notice that $m^2 - 3n^2 = r^2$ implies $p = m \pm r > 0$. Finally, replacing p, q with $p/d, q/d$ for $d = \gcd(p, q)$ allows us to assume that $\gcd(p, q) = 1$ with $p, q > 0$ and $3 \nmid p$.

(iii) Using (ii) and $b = 2mn - 3n^2$ from (i),

$$\frac{b}{n^2} = \frac{2mn - 3n^2}{n^2} = 2 \cdot \frac{m}{n} - 3$$
$$= \frac{2p^2 + 6q^2}{2pq} - 3 = \frac{2p^2 - 6pq + 6q^2}{2pq} = \frac{p^2 - 3pq + 3q^2}{pq}.$$

Clearing fractions, we have

$$b(pq)^2 = n^2 pq(p^2 - 3pq + 3q^2),$$

and since b is a square, it follows that $pq(p^2 - 3pq + 3q^2)$ is a square as well. If $p = q$, then $m/n = 2$, so that $m = 2$ and $n = 1$, and then $b = 2mn - 3n^2 = 1$ and $c = m^2 - 3n^2 = 1$, which contradicts $b \neq c$. Thus $p \neq q$.

(iv) The definition of p and q in (ii) shows that $q \mid n$. But the definition of m and n in (i) shows that $n \mid b$. Hence $q \mid b$, so that $q < b$ unless $q = n = b$. But in this case we have $b = 2mn - 3n^2 = b(2m - 3b)$, so that $2m = 3b + 1$, and then

$$c = m^2 - 3n^2 = \frac{(3b+1)^2}{4} - 3b^2 = \frac{-3b^2 + 6b + 1}{4}.$$

The only positive b for which this fraction is a positive integer is $b = 1$, so that $c = 1$, which as in (iii) contradicts $b \neq c$. Thus $q < b$ and we have found a smaller solution than the original one.

(c) This is the case where $3 \mid m$, say $m = 3k$. Since $\gcd(m, n) = 1$, we see that $3 \nmid n$ and $\gcd(k, n) = 1$. Part (a) gives

$$(**) \qquad \frac{b}{c} = \frac{2mn - 3n^2}{m^2 - 3n^2} = \frac{6kn - 3n^2}{9k^2 - 3n^2} = \frac{2nk - n^2}{3k^2 - n^2}.$$

(i) We claim that $\gcd(2nk - n^2, 3k^2 - n^2) = 1$. Suppose that $p \mid 2nk - n^2$ and $p \mid 3k^2 - n^2$ for some prime p. If $p \mid k$, then $p \mid 3k^2 - n^2$ implies $p \mid n^2$, contradicting $\gcd(k, n) = 1$. Thus $p \nmid k$. We also have $p \mid (3k^2 - n^2) - (2kn - n^2) = k(3k - 2n)$, so $p \nmid k$ implies that $3k \equiv 2n \bmod p$. Since $p \mid 3k^2 - n^2$, we get

$$0 \equiv 3(3k^2 - n^2) \equiv (3k)^2 - 3n^2 \equiv (2n)^2 - 3n^2 \equiv n^2 \bmod p,$$

hence $p \mid n$. Then $3k \equiv 2n \equiv 0 \bmod p$. Note that $p \neq 3$ since $3 \nmid n$. Thus $p \mid k$, which contradicts $\gcd(k, n) = 1$. The claim follows.

Consider the equality of fractions given by (**). We know that $\gcd(b, c) = 1$, and we just showed that $\gcd(2nk - n^2, 3k^2 - n^2) = 1$. Furthermore, $b, c > 0$, and by part (a), we also know that $2mn - 3n^2$, $m^2 - 3n^2 > 0$. Since $m = 3k$, this easily implies that $2kn - n^2$, $3k^2 - n^2 > 0$. As in part (b), it follows that $b = 2kn - n^2$ and $c = 3k^2 - n^2$.

(ii) Since $3 \nmid n$, we have $n^2 \equiv 1 \bmod 3$. Then $c = 3k^2 - n^2 \equiv -1 \bmod 3$, which is impossible since c is a square. Hence there are no solutions when $3 \mid m$.

The above analysis shows that all solutions must satisfy $3 \nmid m$. Given any solution, part (b)(iv) shows that we can find a solution with a smaller b. By infinite descent, no solutions exist.

12.29. (a) Since

$$X^3 + 1 = \left(\frac{a}{b}\right)^3 + 1 = \frac{a^3}{b^3} + 1 = Z^2,$$

clearing fractions gives $a^3 + b^3 = b^3 Z^2$, so that $b(a^3 + b^3) = b^4 Z^2 = (b^2 Z)^2$. We know that $Z \in \mathbb{Q}$, so that $b^2 Z$ is a rational root of $x^2 - b(a^3 + b^3)$. Thus $b^2 Z$ is also an algebraic integer, which implies that $b^2 Z \in \mathbb{Z}$. Hence $b(a^3+b^3) = (b^2Z)^2$ is a perfect square. Let $c = a+b$. Note that $a/b \neq -1, 0$ means that $a \neq 0$ and $c \neq 0$. Then

$$\begin{aligned} bc(c^2 - 3bc + 3b^2) &= b(a+b)\big((a+b)^2 - 3(a+b)b + 3b^2\big) = b(a+b)(a^2 - ab + b^2) \\ &= b(a^3 + b^3) \end{aligned}$$

is a perfect square. Next, b and c are relatively prime since $\gcd(b,c) = \gcd(b, a+b) = \gcd(a,b) = 1$. In addition, $a \neq 0$ implies that $c = a + b \neq b$.

Finally, we know that $b > 0$. Then $c > 0$ as well since $bc(c^2 - 3bc + 3b^2)$ is a square and $c^2 - 3bc + 3b^2 = (c - \frac{3}{2}b)^2 + \frac{3}{4}b^2 > 0$.

(b) From Exercise 12.28, it follows that $3 \mid c$ since $b \neq c$, so that $c = 3d$. Then $\gcd(b,c) = 1$ implies $\gcd(b,d) = 1$ and $3 \nmid b$. Then

$$bc(c^2 - 3bc + 3b^2) = 3bd(9d^2 - 9bd + 3b^2) = 3^2 db(b^2 - 3bd + 3d^2),$$

so that $db(b^2 - 3bd + 3d^2)$ is a perfect square with $\gcd(b,d) = 1$. Then Exercise 12.28 shows that either $3 \mid b$ or $b = d$. But $3 \mid b$ contradicts $3 \nmid b$. Thus $b = d$, and so $b = d = 1$ since $\gcd(b,d) = 1$. Hence $c = 3d = 3$, so that $a = 2$ (since $c = a + b$), contradicting the assumption that $a/b \neq 2$.

This shows that if $X^3 + 1 = Z^2$, then $X = \frac{a}{b} = -1$, 0 or 2, so that the only solutions are $(-1,0)$, $(0,\pm 1)$ and $(2,\pm 3)$.

12.30. Since $j(\mathcal{O}_K) = j([1, w_K])$ and $j(\mathcal{O}_{K'}) = j([1, w_{K'}])$, if these two are equal, then $[1, w_K]$ and $[1, w_{K'}]$ are homothetic by Theorem 10.9. This means $\lambda[1, w_K] = [\lambda, \lambda w_K] = [1, w_{K'}]$ for some nonzero complex number λ. Thus $\lambda = a + bw_{K'}$ and $\lambda w_K = c + dw_{K'}$ for integers a, b, c, d. But then

$$w_K = \frac{\lambda w_K}{\lambda} = \frac{c + dw_{K'}}{a + bw_{K'}} \in K',$$

so that $w_K \in K'$, forcing $K = K'$ since $K = \mathbb{Q}(w_K)$.

12.31. (a) For P, first note that

$$\begin{aligned} q &= e^{-\pi\sqrt{2/7}} = e^{2\pi i \sqrt{-1/14}} = e^{2\pi i(-1/\sqrt{-14})} \\ q^7 &= e^{-7\pi\sqrt{2/7}} = e^{2\pi i \sqrt{-14}/2}. \end{aligned}$$

Since $f(-q) = q^{-1/24}\eta(\tau)$, we get

$$\begin{aligned} P = \frac{f^2(-q)}{q^{1/2} f^2(-q^7)} &= \frac{q^{-1/12}\eta^2(-1/\sqrt{-14})}{q^{1/2}q^{-7/12}\eta^2(\sqrt{-14}/2)} = \Big(\frac{\eta(-1/\sqrt{-14})}{\eta(\sqrt{-14}/2)}\Big)^2 \\ &= \Big(\frac{\sqrt{-i\sqrt{-14}}\,\eta(\sqrt{-14})}{\eta(\sqrt{-14}/2)}\Big)^2 = \sqrt{14}\Big(\frac{\eta(\sqrt{-14})}{\eta(\sqrt{-14}/2)}\Big)^2 = \frac{\sqrt{14}}{\alpha}. \end{aligned}$$

Here, the second line uses Corollary 12.19 and

$$(*) \qquad \alpha = \mathfrak{f}_1(\sqrt{-14})^2 = \Big(\frac{\eta(\sqrt{-14}/2))}{\eta(\sqrt{-14})}\Big)^2,$$

where the second equality follows from the definition of $\mathfrak{f}_1(\tau)$. Turning to Q, we have

$$\begin{aligned} q^2 &= e^{-2\pi\sqrt{2/7}} = e^{2\pi i\sqrt{-2/7}} = e^{2\pi i(-1/\sqrt{-7/2})} \\ q^{14} &= e^{-14\pi\sqrt{2/7}} = e^{2\pi i\sqrt{-14}}. \end{aligned}$$

Thus

$$Q = \frac{f^2(-q^2)}{qf^2(-q^{14})} = \frac{q^{-1/6}\eta^2(-1/\sqrt{-7/2})}{q\cdot q^{-7/6}\eta^2(\sqrt{-14})} = \Big(\frac{\eta(-1/\sqrt{-7/2})}{\eta(\sqrt{-14})^2}\Big)^2$$

$$= \left(\frac{\sqrt{-i\sqrt{-7/2}}\,\eta(\sqrt{-7/2})}{\eta(\sqrt{-14})}\right)^2 = \sqrt{\frac{7}{2}}\Big(\frac{\eta(\sqrt{-14}/2))}{\eta(\sqrt{-14})}\Big)^2 = \sqrt{\frac{7}{2}}\,\alpha,$$

where the second line again uses Corollary 12.19 and (*). It follows immediately that $PQ = 7$ and $Q/P = \alpha^2/2$.

(b) Set $a = \alpha^2/2 = Q/P$ and $b = 1/a = 2/\alpha^2 = P/Q$. Since $PQ = 7$, Ramanujan's modular equation gives

$$14 = PQ + \frac{49}{PQ} = \Big(\frac{Q}{P}\Big)^3 - 8\frac{Q}{P} - 8\frac{P}{Q} + \Big(\frac{P}{Q}\Big)^3 = a^3 - 8a - 8b + b^3.$$

However, $ab = 1$ implies $(a+b)^3 = a^3 + 3a^2b + 3ab^2 + b^3 = a^3 + 3a + 3b + b^3$, so that

$$(a+b)^3 - 11(a+b) = a^3 + 3a + 3b + b^3 - 11a - 11b = a^3 - 8a - 8b + b^3 = 14.$$

It follows that $\beta = a + b = \alpha^2/2 + 2/\alpha^2$ is a root of the cubic $x^3 - 11x - 14 = (x^2 - 2x - 7)(x+2) = 0$. The roots are -2 and $1 \pm 2\sqrt{2}$. But β is positive since α is real, so that $\beta = \alpha^2/2 + 2/\alpha^2 = 1 + 2\sqrt{2}$. Clearing denominators gives the equation

$$\alpha^4 - 2(1+2\sqrt{2})\alpha^2 + 4 = 0.$$

To get from here to the formula for α given in the text, note that the above equation factors

$$\big(\alpha^2 - (2+\sqrt{2})\alpha + 2\big)\big(\alpha^2 + (2+\sqrt{2})\alpha + 2\big) = 0.$$

The second factor cannot vanish since $\alpha > 0$. Thus $\alpha^2 - (2+\sqrt{2})\alpha + 2 = 0$, which as in the text gives

$$\alpha = \mathfrak{f}_1(\sqrt{-14})^2 = \frac{\sqrt{2}+1+\sqrt{2\sqrt{2}-1}}{\sqrt{2}}.$$

Solutions to Exercises in §13

13.1. Let L be the ring class field of $\mathcal{O}$. From Corollary 11.37, given a proper fractional ideal $\mathfrak{a}$ of $\mathcal{O}$, there is $\sigma_{\mathfrak{a}} \in \mathrm{Gal}(L/K)$ such that $\sigma_{\mathfrak{a}}(j(\mathfrak{b})) = j(\bar{\mathfrak{a}}\mathfrak{b})$, and $\mathfrak{a} \mapsto \sigma_{\mathfrak{a}}$ induces an isomorphism from $C(\mathcal{O})$ to $\mathrm{Gal}(L/K)$. Since $L = K(j(\mathcal{O}))$, Galois theory tells us that the minimal polynomial $H_{\mathcal{O}}(X)$ of $j(\mathcal{O})$ has degree $h = |C(\mathcal{O})| = |\mathrm{Gal}(L/K)|$ and that the roots of $H_{\mathcal{O}}(X)$ are all of the form $\sigma_{\mathfrak{a}}(j(\mathcal{O})) = j(\bar{\mathfrak{a}})$.

Let $\mathfrak{a}_1, \ldots, \mathfrak{a}_h$ be a complete set of ideal class representatives. Applying the previous paragraph to $\mathfrak{a} = \bar{\mathfrak{a}}_i$, it follows that $j(\mathfrak{a}_i)$ is a root of $H_{\mathcal{O}}(X)$. Since the $\mathfrak{a}_i$ were chosen so that no two are in the same class in $C(\mathcal{O})$, no two are homothetic, and therefore all of the $j(\mathfrak{a}_i)$ are distinct by Theorem 10.9. Thus $H_{\mathcal{O}}(X)$ is divisible by $\prod_{i=1}^h (X - j(\mathfrak{a}_i))$. Since both polynomials are monic of degree h, they must be equal.

13.2. (a) Let K be a quadratic field of discriminant $d_K < 0$. Since $\mathcal{O} \subset \mathcal{O}_K$, it suffices to show that there are only finitely many $\alpha \in \mathcal{O}_K$ such that $N(\alpha) = m$. By (5.13), $\mathcal{O}_K = [1, \tau_0]$, where

$$\tau_0 = \begin{cases} \sqrt{-n} & d_K = -4n \equiv 0 \bmod 4 \\ \frac{1+\sqrt{-n}}{2} & d_K = -n \equiv 1 \bmod 4. \end{cases}$$

Thus $\alpha = a + b\tau_0 \in \mathcal{O}_K$ has norm $f(a,b) = N(a + b\tau_0)$, which equals $a^2 + nb^2$ (when $d_K = -4n$) or $a^2 + ab + \frac{1+n}{4}b^2$ (when $d_K = -n$). Note that $f(a,b)$ is the principal form of discriminant d_K. From (2.9), $f(a,b) \geq C\min(a^2, b^2)$ when $ab \neq 0$, where

$$C = \begin{cases} n+1 & d_K = -4n \\ \frac{1+n}{4} & d_K = -n. \end{cases}$$

Suppose $0 < a^2 \leq b^2$ and $f(a,b) = m$. Then $m \geq Ca^2$, so only finitely many a's can occur. For each such a, at most two b's can satisfy the quadratic equation $f(a,b) = m$, giving a finite number of solutions in this case. The argument is similar when $0 < b^2 \leq a^2$. Finally, when $ab = 0$, $f(a,b) = m$ has at most four solutions. Thus only a finite number of integers a and b can satisfy $f(a,b) = m$, and the finiteness of $r(\mathcal{O}, m)$ follows.

(b) Using the discussion following (2.9), we see that if $d_K = -4n$ with $n > m$, then the two smallest values represented by the principal form are 1 and $n > m$, so that there are no elements of norm m since $m > 1$. By a similar argument, if $d_K = -n$ with $(1+n)/4 > m$, then there are no elements of norm m. Thus in these two cases, $r(\mathcal{O}, m) = 0$ for any order in the field.

Since m is fixed, this leaves only a finite number of fields to consider. Let K be one of these fields. Then part (a) shows that K has only a finite number of primitive elements with norm m, so it suffices to show that each of these can lie in only a finite number of orders. With $\mathcal{O}_K = [1, \tau_0]$ as in part (a), a primitive element of norm $m > 1$ is of the form $\alpha = a + b\tau_0$ with $b \neq 0$. The order of conductor f is $\mathcal{O} = [1, f\tau_0]$, and if $\alpha \in \mathcal{O}$, then we must have $f \mid b$. Since only finitely many b's can occur, the same is true for the number of conductors f. So α lies in only a finite number of orders of K and we are done.

13.3. (a) Observe that $F(X,0) = X^3$ has a root of multiplicity 3 at 0, but $F(X,X) = 2X^3 + X^2 = X^2(X+1)$ has a root of multiplicity 2 at 0.

(b) Since both $F(X,X)$ and $F(X,X_0)$ have a root at X_0, there are positive integers m, n such that

$$F(X,X) = (X - X_0)^m P(X)$$
$$F(X,X_0) = (X - X_0)^n Q(X)$$

for $P(X), Q(X) \in \mathbb{C}[X]$ with $P(X_0) \neq 0$ and $Q(X_0) \neq 0$. Then

$$\lim_{X \to X_0} \frac{F(X,X)}{F(X,X_0)} = \lim_{X \to X_0} \frac{(X - X_0)^m P(X)}{(X - X_0)^n Q(X)} = \frac{P(X_0)}{Q(X_0)} \lim_{X \to X_0} (X - X_0)^{m-n}.$$

It follows immediately that the limit exists and is nonzero if and only if $m = n$, in which case the orders of the zero of $F(X,X)$ and of $F(X,X_0)$ at X_0 are equal.

13.4. (a) Since $\tilde{\sigma}(\tau_0) = \tau_0$, solving $\frac{a\tau_0 + b}{c\tau_0 + d} = \tau_0$ shows that τ_0 satisfies the equation

$$(*) \qquad c\tau_0^2 + (d - a)\tau_0 - b = 0.$$

If $c = 0$, then $(d-a)\tau_0 - b = 0$ and $\tau_0 \in \mathfrak{h}$ imply that $d = a$ and $b = 0$, so that $\tilde{\sigma} = aI$. Since $\tilde{\sigma}$ has relatively prime entries, we see that $a = \pm 1$, so $\det(\tilde{\sigma}) = 1$. But $\det(\tilde{\sigma}) = m > 1$. This contradiction shows that $c \neq 0$.

(b) From the discussion in the text, if (13.7) failed to hold for $j(\tau_0) = 1728$, we would have $m^2 = (ci + d)^4 = (d^2 - c^2)^2 - 4c^2d^2 + 4(d^2 - c^2)cdi$. Since the imaginary part on the right hand side must be zero, we have either $d = \pm c$, $c = 0$, or $d = 0$. But $d = \pm c$ is impossible since then $m^2 = -4c^2d^2$ with $m > 1$. Part (a) shows that $c \neq 0$, so we must have $d = 0$.

Since $\tilde{\sigma}$ fixes $\tau_0 = i$, (*) implies that $(d-a)i-(b+c)=0$, so that $a=d$ and $b=-c$. Since $d=0$, we have $a=d=0$, so that $\tilde{\sigma} = \left(\begin{smallmatrix} 0 & -c \\ c & 0 \end{smallmatrix}\right)$. But as noted above, $\tilde{\sigma}$ has relatively prime entries, so that $c=\pm 1$ and hence $\det(\tilde{\sigma})=1$. Yet $\det(\tilde{\sigma})=m>1$. Thus (13.7) holds when $j(\tau_0)=1728$.

(c) If $j(\tau_0)=0$, we can assume $\tau_0=\omega=e^{2\pi i/3}$. Then Theorem 11.2 implies that ω is a triple root of j, so that $k=3$ and we must show that

$$\tilde{\sigma}'(\omega)^3 = \frac{m^3}{(c\omega+d)^6} \neq 1.$$

Suppose to the contrary that $m^3=(c\omega+d)^6$ for $c,d\in\mathbb{Z}$. Expanding $(c\omega+d)^6$ produces

$$\begin{aligned}(c\omega+d)^6 &= (6cd^5-15c^2d^4+15c^4d^2-6c^5d)\omega+N\\ &= 3cd(2d-c)(d-2c)(d+c)(d-c)\omega+N\end{aligned}$$

for some integer N. If $m^3=(c\omega+d)^6$, then the coefficient of ω is zero, so one of the factors in the coefficient of ω must vanish. We know that $c=0$ is impossible. If $2d=c$, $d=2c$, or $d=-c$, then one computes that

$$(c\omega+d)^6 = \begin{cases} d^6(2\omega+1)^6 = -27d^6 & \text{if } 2d=c\\ c^6(\omega+2)^6 = -27c^6 & \text{if } d=2c\\ c^6(\omega-1)^6 = -27c^6 & \text{if } d=-c,\end{cases}$$

where we use $2\omega+1=\sqrt{-3}$, $\omega+2=\omega^2(\omega-1)$, and $(\omega-1)^2=-3\omega$. Thus we have $(c\omega+d)^6\le 0$ in these cases, which contradicts $m^3>0$. It remains to consider what happens when $d=c$ or $d=0$. Since $\tilde{\sigma}=\left(\begin{smallmatrix} a & b \\ c & d\end{smallmatrix}\right)$ fixes ω, we get

$$\frac{a\omega+b}{c\omega+d}=\omega \implies a\omega+b=c\omega^2+d\omega=(d-c)\omega-c,$$

where the last equality uses $\omega^2=-1-\omega$. It follows that $a=d-c$ and $b=-c$. When $d=c$ or $d=0$, the formulas for a and b imply that

$$\tilde{\sigma}=\begin{pmatrix} a & b\\ c & d\end{pmatrix} = \begin{cases} c\left(\begin{smallmatrix} 0 & -1\\ 1 & 1\end{smallmatrix}\right) & \text{if } d=c\\ c\left(\begin{smallmatrix} -1 & -1\\ 1 & 0\end{smallmatrix}\right) & \text{if } d=0.\end{cases}$$

Since $\tilde{\sigma}$ has relatively prime entries, we must have $c=\pm1$. Then $\det(\tilde{\sigma})=1$, which contradicts $\det(\tilde{\sigma})=m>1$. Thus $\tilde{\sigma}'(\omega)^3\neq 1$ and (13.7) is proved.

13.5. To see that the map is surjective, choose $\sigma=\left(\begin{smallmatrix} a & b\\ 0 & d\end{smallmatrix}\right)\in C(m)$ with $j(\sigma\tau_0)=j(\tau_0)$. Then $d[1,\sigma\tau_0]$ is a cyclic sublattice of $[1,\tau_0]$ of index m by Lemma 11.24, and

$$j(d[1,\sigma\tau_0]) = j([1,\sigma\tau_0]) = j([1,\tau_0]),$$

so that $d[1,\sigma\tau]$ is homothetic to $[1,\tau_0]$. Therefore by Corollary 11.27 there is a primitive $\alpha\in\mathbb{C}$ of norm m with $d[1,\sigma\tau_0]=\alpha[1,\tau_0]$. Then $\alpha\in d[1,\sigma\tau_0]=[d,a\tau_0+b]$, so that $\alpha\in\mathcal{O}$ and maps to σ.

For injectivity, suppose $\alpha,\beta\in\mathcal{O}$ both map to $\sigma\in C(m)$. Then we have $\alpha\mathcal{O}=d[1,\sigma\tau_0]$ and $\beta\mathcal{O}=d[1,\sigma\tau_0]$, so that $\alpha\mathcal{O}=\beta\mathcal{O}$. This implies that α and β are associates and hence give the same class in $\mathcal{A}$.

13.6. (a) Substitute the q-expansion of $j(\tau)=\frac{1}{q}+\cdots$ for X in $\Phi_m(X,X)$. If $\Phi_m(X,X)$ has degree N, it is clear that the most negative power of q in the result will be $-N$.

(b) By (11.19),

$$j(\tau)-j(\sigma\tau) = q^{-1}-\zeta_m^{-ab}q^{-a/d}+\sum_{n=0}^{\infty} d_n(q^{1/m})^n,\ d_n\in\mathbb{C}.$$

Then $a < d$ implies that $a/d < 1$ so that the q-expansion is $q^{-1} - \zeta_m^{-ab} q^{-a/d} + \cdots$. Next, $a > d$ means that $a/d > 1$, so that the q-expansion is $-\zeta_m^{-ab} q^{-a/d} + q^{-1} + \cdots$. Finally, if $a = d$, so that $a/d = 1$, the q-expansion is $(1 - \zeta_m^{-ab})q^{-1} + \cdots$.

(c) By (11.15), $\Phi_m(j(\tau), j(\tau)) = \prod_{\sigma \in C(m)} (j(\tau) - j(\sigma\tau))$. We will compute the number of possible $\sigma \in C(m)$ with $a < d$, $a > d$, and $a = d$ and apply parts (a) and (b). When $a < d$, the first term in the q-expansion of $j(\tau) - j(\sigma\tau)$ is q^{-1}, so by part (a) of Exercise 11.9, the absolute value of the largest negative exponent of the product of those terms is

$$\sum_{\substack{a|m \\ a<\sqrt{m}}} \frac{d}{e}\phi(e)$$

with $e = \gcd(a, d)$. When $a > d$, the first term in the q-expansion is $-\zeta_m^{-ab} q^{-a/d}$, so the the absolute value of the largest negative exponent of the product of those terms is

$$\frac{a}{d} \sum_{\substack{a|m \\ a>\sqrt{m}}} \frac{d}{e}\phi(e) = \sum_{\substack{a|m \\ a>\sqrt{m}}} \frac{a}{d} \cdot \frac{d}{e}\phi(e).$$

Finally, if m is a perfect square and $a = d$, the first term in the q-expansion is $(1 - \zeta_m^{-ab})q^{-1}$, and there are $\phi(d) = \phi(\sqrt{m})$ such terms, one for each possible b. Since $\phi(\sqrt{m}) = 0$ if m is not a perfect square, we get

$$N = \sum_{\substack{a|m \\ a<\sqrt{m}}} \frac{d}{e}\phi(e) + \sum_{\substack{a|m \\ a>\sqrt{m}}} \frac{a}{d} \cdot \frac{d}{e}\phi(e) + \phi(\sqrt{m}).$$

(d) Using the fact that $ad = m$, the second sum is

$$\sum_{\substack{a|m \\ a>\sqrt{m}}} \frac{a}{d} \cdot \frac{d}{e}\phi(e) = \sum_{\substack{a|m \\ a>\sqrt{m}}} \frac{a}{e}\phi(e) = \sum_{\substack{d|m \\ d<\sqrt{m}}} \frac{a}{e}\phi(e).$$

Reversing the roles of a and d shows that this is the same as the first sum, proving Proposition 13.8.

13.7. (a) Let $K = \mathbb{Q}(\sqrt{-n})$, $n > 0$ squarefree, and suppose that $\alpha \in \mathcal{O}_K$ has norm 3. If $n \equiv 1, 2 \bmod 4$, so that $d_K = -4n$, then $\alpha = a + b\sqrt{-n}$ and $N(\alpha) = a^2 + nb^2 = 3$, forcing $n \leq 2$. There are no solutions when $n = 1$. When $n = 2$ we get $a^2 + 2b^2 = 3$, which has solutions $(a, b) = \pm(1, \pm 1)$. Thus $\alpha = \pm(1 \pm \sqrt{-2})$, and since we count α's modulo units, we get $r(-8, 3) = 2$. These values of α lie in no order properly contained in $\mathcal{O}_{\mathbb{Q}(\sqrt{-2})}$, and we conclude that if $\mathcal{O}$ is the order of discriminant D in $\mathbb{Q}(\sqrt{-n})$, $n \equiv 1, 2 \bmod 4$, then $r(D, 3) = 2$ for $D = -8$ and $r(D, 3) = 0$ otherwise.

The other case is $n \equiv 3 \bmod 4$, $d_K = -n$. Let $\tau_0 = (-1 + \sqrt{-n})/2$. Then $\mathcal{O}_K = [1, \tau_0]$ by (5.13). If $\alpha = a + b\tau_0 \in \mathcal{O}_K$, then

$$N(\alpha) = (a + b\tau_0)(a + b\overline{\tau_0}) = a^2 - ab + \frac{n+1}{4}b^2.$$

From (2.9), $a^2 - ab + \frac{n+1}{4}b^2 \geq \frac{n+1}{4}\min(a^2, b^2)$ if $ab \neq 0$, and in this case the inequality holds when $a = 0$ as well. Note that $b \neq 0$ since α must be primitive. So we want $\frac{n+1}{4} \leq 3$ and thus $n = 3$, 7, or 11.

For $n = 3$, note that $\tau_0 = \omega$. The equation $N(\alpha) = a^2 - ab + b^2 = 3$ has solutions $(a, b) = \pm(2, 1)$, $\pm(1, 2)$, and $\pm(1, -1)$. Hence $\alpha = a + b\omega$ equals one of $\pm(2 + \omega)$, $\pm(1 + 2\omega)$ or $\pm(1 - \omega)$. Since

$$2 + \omega = (-\omega)(1 + 2\omega) = (-\omega^2)(1 - \omega),$$

these are all associates, so $r(-3,3) = 1$. Now consider the order $\mathcal{O}$ of conductor f in $\mathcal{O}_K$. Then $\mathcal{O} = [1, f\omega]$. When $f \geq 3$, none of the above six α's lie in $\mathcal{O}$. When $f = 2$, we have $\alpha = \pm(1+2\omega) = \pm\sqrt{-3} \in \mathcal{O}$. Since the only units in $\mathcal{O} = [1, 2\omega]$ are ± 1, it follows that $r(-12,3) = 1$.

For $n = 7$, the equation $N(\alpha) = a^2 - ab + 2b^2 = 3$ has no solutions: $3 = a^2 - ab + 2b^2 \geq 2\min(a^2, b^2)$ forces $|a|, |b| \leq 1$. Neither can be zero since $a^2 = 3$ and $2b^2 = 3$ have no solutions. If they both have absolute value 1, then $a^2 - ab + 2b^2 = 2$ or 4. So $r(-7,3) = 0$, and the same holds for all other orders in K.

Finally, when $n = 11$, we want solutions to $N(\alpha) = a^2 - ab + 3b^2 = 3$. Again we must have $|a|, |b| \leq 1$, and the solutions are $(a,b) = \pm(1,1)$ and $\pm(0,1)$, so that $\alpha = \pm(1+\tau_0)$ or $\alpha = \pm\tau_0$. Modulo units (± 1 in this case), we get $r(-11,3) = 2$. None of these solutions lie in any order of conductor $f > 1$ in $\mathcal{O}_K$, so that all other orders in K have $r = 0$.

The factorization of $\Phi_3(X,X)$ given in the problem statement follows immediately from Theorem 13.4.

(b) For $d_K = -4n$, $N(a + b\sqrt{-n}) = x^2 + ny^2 = 5$ gives $n = 1$, 2, or 5. For $n = 1$, solving $x^2 + y^2 = 5$ gives two equivalence classes, represented by $1 - 2i$ and $1 + 2i$, which lie both in $[1, i]$ and $[1, 2i]$ of discriminants -4 and -16. When $n = 2$ there are no solutions to $x^2 + 2y^2 = 5$ since $(-2/5) = -1$. When $n = 5$, the only solutions to $x^2 + 5y^2 = 5$ are $(0, \pm 1)$, which are associates. Thus the only solution up to units is $\sqrt{-5}$, which lies only in $[1, \sqrt{-5}]$. So we get three factors, $H_{-4}(X)^2$, $H_{-16}(X)^2$, and $H_{-20}(X)$.

For $d_K = -n$, $\tau_0 = (-1+\sqrt{-n})/2$, we want to solve $N(a + b\tau_0) = a^2 - ab + \frac{n+1}{4}b^2 = 5$. Here, $\frac{n+1}{4} \leq 5$ and $n \equiv 3 \bmod 4$ imply that $n = 3, 7, 11, 15$, or 19. The cases $n = 3$, $n = 7$, and $n = 15$ are easily seen to have no solutions by writing the equation as

$$(2a+b)^2 + nb^2 = 20, \quad n = 3, 7, 15.$$

When $n = 11$, the equation $a^2 - ab + 3b^2 = 5$ has solutions $(a,b) = \pm(2,1)$ and $\pm(1,-1)$, corresponding to $\alpha = \pm(2+\tau_0)$ and $\pm(1-\tau_0)$. The units are ± 1, so that $r(-11,5) = 2$. Note also that none of these α's lie in an order of conductor $f > 1$. Finally, for $n = 19$, $a^2 - ab + 5b^2 = 5$ has the solutions $(a,b) = \pm(0,1)$ and $\pm(1,1)$, corresponding to $\alpha = \pm\tau_0$ and $\pm(1+\tau_0)$, so again $r(-19,5) = 2$, and again, none of these α's lie in an order of conductor $f > 1$. So finally,

$$\Phi_5(X,X) = \pm H_{-4}(X)^2 H_{-16}(X)^2 H_{-11}(X)^2 H_{-19}(X)^2 H_{-20}(X).$$

And in fact retrieving $\Phi_5(X,X)$ online and factoring it gives

$$\Phi_5(X,X) = -(x - 12^3)^2(x - 66^3)^2(x + 32^3)^2(x + 96^3)^2(x^2 - 1264000x - 681472000)$$

where $j(\sqrt{-5}) = 320(1975 + 884\sqrt{5})$ is one of the roots of the quadratic, and the factors correspond to the factorization above. Note this is consistent with Proposition 13.11.

13.8. (a) Fix $m > 1$ with $r(-3,m) = 1$. This means that $N(a + b\omega) = a^2 - ab + b^2 = m$ has a unique solution modulo units in $\mathbb{Z}[\omega]$. Let $\alpha = a + b\omega$ be a solution. Note that $\gcd(a,b) = 1$ since α is primitive, and $ab \neq 0$ since otherwise $(a,b) = (0, \pm 1)$ or $(\pm 1, 0)$, which imply that $N(\alpha) = 1$, yet $m > 1$.

Then observe that $\overline{\alpha} = a + b\overline{\omega} = a + b\omega^2 = a - b - b\omega$ also has norm m. Thus $a - b - b\omega$ is a unit times $a + b\omega$. This leads to several cases:

$$a - b - b\omega = \pm(a + b\omega) \Rightarrow a - b = \pm a \Rightarrow b = 2a \text{ or } b = 0$$
$$a - b - b\omega = \pm\omega(a + b\omega) = \pm(-b + (a - b)\omega) \Rightarrow a - b = \pm b \Rightarrow a = 2b \text{ or } a = 0$$
$$a - b - b\omega = \pm\omega^2(a + b\omega) = \pm(b - a - a\omega) \Rightarrow b = \pm a.$$

Using $\gcd(a, b) = 1$ and recalling that $ab \neq 0$, we see that (a, b) is one of $\pm(1, 2)$, $\pm(2, 1)$, $\pm(1, 1)$, or $\pm(1, -1)$. The corresponding α's have norm 3 or 1, so that $m = 3$ or 1. Since $m > 1$, we conclude that $r(-3, m) = 1$ implies $m = 3$. The converse follows from part (a) of Exercise 13.7.

(b) Note first that $r(-4, 2) = 1$ since the solutions to $a^2 + b^2 = 2$ are $(a, b) = (\pm 1, \pm 1)$, and $\pm 1 \pm i$ are all associates in $\mathbb{Z}[i]$. Now let $a+bi$ be primitive of norm m. As in part (a), we have $\gcd(a, b) = 1$ and $ab \neq 0$. Since $a-bi$ also has norm m, $a+bi$ and $a-bi$ are associates. Since $ab \neq 0$, the only possibility is that $a - bi = \pm i(a + bi) = \pm(-b + ai)$, so that $a = \pm b$. Since $\gcd(a, b) = 1$, this implies that $a, b = \pm 1$ and thus $m = 2$.

13.9. Suppose $\alpha = a + b\sqrt{-m} \in \mathbb{Z}[\sqrt{-m}] = [1, \sqrt{-m}]$ is primitive of norm $N(\alpha) = a^2 + mb^2 = (m + 1)/4$. If $b = 0$, then $\gcd(a, b) = 1$ forces $a = \pm 1$ and then $N(\alpha) = (\pm 1)^2 + m0^2 = 1$, yet $m > 3$ implies that $N(\alpha) = (m + 1)/4 > 1$. Hence $b \neq 0$, so $b^2 \geq 1$. This implies $N(\alpha) = a^2 + mb^2 \geq m > (m+1)/4$ since $m > 3$. Thus there are no elements of norm $(m + 1)/4$ in $\mathbb{Z}[\sqrt{-m}]$.

13.10. (a) For the discriminant $D = -56 = -4 \cdot 14$, we have $m = 14$, $m \not\equiv 3 \bmod 4$ and $m \neq 3k^2$. By Proposition 13.11, $H_{-56}(X) = \Phi_{14,1}(X, X)$. So after factoring $\Phi_{14}(X, X)$ into irreducibles, $H_{-56}(X)$ is the only factor that appears with an exponent of 1.

(b) For $D = -11$ or $D = -44 = -4 \cdot 11$, we have $m = 11$, so that $m \equiv 3 \bmod 8$ and $m \neq 3k^2$. Thus by Proposition 13.11, $H_{-11}(X)H_{-44}(X) = \Phi_{11,1}(X, X)$. But $-11 \equiv 5 \bmod 8$, so $(-11/2) = -1$ and thus $h(-44) = 3h(-11)$ by Corollary 7.28 with $m = 2$. For any discriminant D, we know that $H_D(X)$ has degree $h(D)$ by Proposition 13.2. Thus $\deg(H_{-11}(X)) = h(-11) < h(-44) = \deg(H_{-44}(X))$. Therefore, of the two factors that appear in $\Phi_{11}(X, X)$ with exponent one, $H_{-11}(X)$ is the one of smaller degree and $H_{-44}(X)$ is the one of larger degree.

(c) With $D = -7$ or $D = -28 = -4 \cdot 7$, we apply Proposition 13.11 with $m = 7$, so that $m \equiv 7 \bmod 8$ and $m \neq 3k^2$. This gives $H_{-7}(X)H_{-28}(X) = \Phi_{7,1}(X, X)$. However, since $D \equiv 1 \bmod 8$, $(D/2) = 1$ and thus $h(-28) = h(-7)$ by Corollary 7.28. But $H_{-7}(X)$ divides $\Phi_2(X, X)$ since

$$N\Big(\frac{1 + \sqrt{-7}}{2}\Big) = 0^2 - 0 \cdot 1 + \frac{7 + 1}{4} \cdot 1^2 = 2$$

and this element is primitive. Additionally, $H_{-28}(X)$ does not divide $\Phi_2(X, X)$ since by Exercise 13.9 there are no elements of norm 2 in the order $\mathbb{Z}[\sqrt{-7}]$ of discriminant -28. So once $\Phi_7(X, X)$ and $\Phi_2(X, X)$ are factored into irreducibles, $H_{-7}(X)$ is the factor of exponent one of $\Phi_7(X, X)$ dividing $\Phi_2(X, X)$ and $H_{-28}(X)$ is the other one.

13.11. Write $g(\tau) = \prod_{i=1}^{|C(m)|} f(\gamma_i \tau)$.

(a) When $\rho, \gamma \in \mathrm{SL}(2, \mathbb{Z})$ are in the same $\Gamma_0(m)$-coset, the $\Gamma_0(m)$-invariance of $f(\tau)$ implies

$$f(\rho\tau) = f((\rho\gamma^{-1})\gamma\tau) = f(\gamma\tau),$$

so that $f(\rho\tau)$ depends only on the $\Gamma_0(m)$-coset of ρ in $\mathrm{SL}(2, \mathbb{Z})$. Next, since the γ_i are coset representatives of $\Gamma_0(m)$ in $\mathrm{SL}(2, \mathbb{Z})$, and any $\gamma \in \mathrm{SL}(2, \mathbb{Z})$ permutes the $\Gamma_0(m)$

cosets of $\mathrm{SL}(2,\mathbb{Z})$, it follows that $\gamma_i' = \gamma_i\gamma$, $i = 1,\ldots,|C(m)|$, also give a complete set of coset representatives. Then we have

$$g(\gamma\tau) = \prod_{i=1}^{|C(m)|} f(\gamma_i\gamma\tau) = \prod_{i=1}^{|C(m)|} f(\gamma_i'\tau) = \prod_{i=1}^{|C(m)|} f(\gamma_i\tau) = g(\tau).$$

Thus $g(\tau)$ is $\mathrm{SL}(2,\mathbb{Z})$-invariant. Clearly $g(\tau)$ is meromorphic on $\mathfrak{h}$ since $f(\tau)$ is. Also, since $f(\tau)$ vanishes at the cusps, each $f(\gamma_i\tau)$ has a Laurent expansion in positive powers of $q^{1/m}$. So the same is true for their product $g(\tau)$. Being $\mathrm{SL}(2,\mathbb{Z})$-invariant, the product $g(\tau)$ is in fact an expansion in positive powers of q. Thus $g(\tau)$ is holomorphic at ∞ and vanishes there.

(b) If $f(\tau)$ is holomorphic on $\mathfrak{h}$, then $g(\tau)$ is as well, so that by Lemma 11.10, it is a constant, and since g vanishes at ∞, we see that $g(\tau)$ is identically zero. This means that $f(\gamma_i\tau)$ is identically zero for some i. Then $f(\tau) = f((\gamma_i\gamma_i^{-1})\tau) = f(\gamma_i(\gamma_i^{-1}\tau)) = 0$. Thus $f(\tau)$ is identically zero.

13.12. First expand the infinite product in the formula for $\Delta(\tau)$ to obtain

$$\Delta(\tau) = (2\pi)^{12} q\prod_{i=1}^{\infty}(1-q^n)^{24} = (2\pi)^{12}q(1+qh(q)),$$

where $h(q)$ is holomorphic at $q=0$ and its power series expansion about $q=0$ has integer coefficients. Since $qh(q)$ maps the origin to itself, $|qh(q)| < 1$ for $|q|$ sufficiently small. Using $(1+z)^{-1} = \sum_{n=0}^{\infty}(-1)^n z^n$ for $|z|<1$, we see that

$$\frac{1}{\Delta(\tau)} = \frac{1}{(2\pi)^{12}q(1+qh(q))} = \frac{1}{(2\pi)^{12}q}\Big(\sum_{n=0}^{\infty}(-1)^n q^n h^n(q)\Big).$$

The series inside the parentheses converges for q near 0 and has integer coefficients. Using the formula for $g_2(\tau)$ given in the exercise, we obtain

$$\begin{aligned} j(\tau) = 1728\,\frac{g_2(\tau)^3}{\Delta(\tau)} &= 1728\Big(\frac{(2\pi)^4}{12}\Big)^3\Big(1+240\sum_{n=1}^{\infty}\sigma_3(n)q^n\Big)^3\cdot\frac{1}{\Delta(\tau)}\\ &= (2\pi)^{12}\Big(1+240\sum_{n=1}^{\infty}\sigma_3(n)q^n\Big)^3\cdot\frac{1}{(2\pi)^{12}q}\Big(\sum_{n=0}^{\infty}(-1)^n q^n h^n(q)\Big)\\ &= \frac{1}{q}\Big(1+240\sum_{n=1}^{\infty}\sigma_3(n)q^n\Big)^3\Big(\sum_{n=0}^{\infty}(-1)^n q^n h^n(q)\Big).\end{aligned}$$

The two series in parentheses have integer coefficients, so the same is true for product of the cube of the first times the second. We conclude that $j(\tau)$ has an integral q-expansion.

13.13. Let $d_i = d_{K_i}$ for $i=1,2$ and $w_i = \left|\mathcal{O}_{K_1}^*\right|$. As usual, $\gcd(d_1,d_2)=1$.

(a) If $d_1, d_2 < -4$, then $w_i = 2$ for $i=1,2$. Thus we need to show that

$$J(d_1,d_2) = \prod_{i=1}^{h_1}\prod_{j=1}^{h_2}(j(\mathfrak{a}_i) - j(\mathfrak{b}_j)) \tag{*}$$

is an integer. Let $L_i = \mathbb{Q}(j(\mathcal{O}_{K_i}))$ be the Hilbert class field of K_i for $i=1,2$, and let $L = L_1L_2$ be the compositum. Since the L_i are Galois over $\mathbb{Q}$ by Lemma 5.28, the same is true for L. Take $\sigma\in\mathrm{Gal}(L/\mathbb{Q})$. Since $j(\mathfrak{a}_i)$ is a root of $H_{d_1}(X)\in\mathbb{Z}[X]$, it follows from Proposition 13.2 that $\sigma(j(\mathfrak{a}_i)) = j(\mathfrak{a}_{i'})$ for some i', and similarly, $\sigma(j(\mathfrak{b}_j)) = j(\mathfrak{b}_{j'})$ for some j'. Thus σ permutes the factors of the product that defines $J(d_1,d_2)$, so that $J(d_1,d_2)$ is fixed by $\mathrm{Gal}(L/\mathbb{Q})$. Thus $J(d_1,d_2)\in\mathbb{Q}$, and since $j(\mathfrak{a}_i)$ and $j(\mathfrak{b}_j)$ are algebraic integers, the same is true for $J(d_1,d_2)$, and $J(d_1,d_2)\in\mathbb{Z}$ follows.

(b) If $d_1, d_2 < -4$, then part (a) shows that in fact $J(d_1, d_2)$ is an integer, so that $J(d_1, d_2)^2$ is as well. The remaining cases are when one of the d_i (or both) are -3 or -4. When $d_i = -3$, we have $K_i = \mathbb{Q}(\omega)$ and $w_i = 6$, and when $d_i = -4$, we have $K_i = \mathbb{Q}(i)$ and $w_i = 4$. Also recall that $h(-3) = h(-4) = 1$, $j(\omega) = 0$, and $j(i) = 1728$. For simplicity, write $P(d_1, d_2)$ for the product in the definition of $J(d_1, d_2)$, so that $J(d_1, d_2) = P(d_1, d_2)^{4/(w_1 w_2)}$. Then we have three cases:

- $(d_1, d_2) = (-4, d)$ or $(d, -4)$ for $d < -4$. Here $4/(w_1 w_2) = 4/(4 \cdot 2) = 1/2$. Then $J(d_1, d_2)^2 = (P(d_1, d_2)^{1/2})^2 = P(d_1, d_2)$, which is an integer by the argument of part (a).
- $(d_1, d_2) = (-3, d)$ or $(d, -3)$ for $d < -4$. Here $4/(w_1 w_2) = 4/(6 \cdot 2) = 1/3$. Then

$$J(-3, d) = \left(\prod_{i=1}^{h} (j(\omega) - j(\mathfrak{b}_i))\right)^{1/3} = \left(\prod_{i=1}^{h_2} (-j(\mathfrak{b}_i))\right)^{1/3} = N(-j(\mathcal{O}_K))^{1/3}.$$

  But $3 \nmid d$ since $\gcd(-3, d) = 1$, so by Theorem 12.2, $\gamma_2(\tau_0)$ lies in $K(j(\mathcal{O}_K)) = K(j(\tau_0))$. Then $-j(\mathcal{O}_K) = -j(\tau_0) = -\gamma_2(\tau_0)^3 = (-\gamma_2(\tau_0))^3$, so that

$$J(-3, d) = N((-\gamma_2(\tau_0))^3)^{1/3} = \left(N(-\gamma_2(\tau_0))^3\right)^{1/3} = N(-\gamma_2(\tau_0)).$$

  This lies in $\mathbb{Z}$ since $-\gamma_2(\tau_0)$ is an algebraic integer. Similarly, $J(d, -3) = N(\gamma_2(\tau_0))$ is in $\mathbb{Z}$. Then $J(-3, d)^2$ and $J(d, -3)^2$ are also both integers.
- $(d_1, d_2) = (-3, -4)$ or $(-4, -3)$. Here $4/(w_1 w_2) = 4/(4 \cdot 6) = 1/6$. Then

$$J(-3, -4)^2 = \left((0 - 1728)^{1/6}\right)^2 = \left((-12)^3\right)^{1/3} = -12.$$

  A similar argument shows that $J(-4, -3)^2 = 12$.

13.14. (a) If $(d_1 d_2/p) = 1$, then $p \nmid d_1 d_2$ and $(d_1/p) = (d_2/p) = \pm 1$, so the two are equal and $\epsilon(p)$ is well-defined. If $(d_1 d_2/p) = 0$, then since d_1 and d_2 are relatively prime, exactly one of (d_1/p) and (d_2/p) is zero so again $\epsilon(p)$ is well-defined.

(b) We assume $p \mid (d_1 d_2 - x^2)/4$. If $p = 2$, then $d_1 d_2 \equiv x^2 \bmod 8$, so $d_1 d_2 \equiv 0, 1, 4 \bmod 8$. Thus $(d_1 d_2/2) = 0$ or 1 and hence $(d_1 d_2/2) \neq -1$. If $p \neq 2$, then $p \mid (d_1 d_2 - x^2)/4$ implies that $d_1 d_2 \equiv x^2 \bmod p$. Thus $(d_1 d_2/p) = 0$ if $p \mid d_1 d_2$ and $(d_1 d_2/p) = 1$ if $p \nmid d_1 d_2$. This implies $(d_1 d_2/p) \neq -1$.

(c) Suppose $\epsilon(p) = -1$. If $p \mid d_1 d_2$, say $p \mid d_1$, then $(d_1/p) = 0$ and $\epsilon(p) = (d_2/p) = -1$, so neither Legendre symbol is 1. Otherwise, $p \nmid d_1 d_2$, and since $\epsilon(p)$ is defined, we must have $(d_1/p) = (d_2/p) = -1$, and again neither Legendre symbol is 1.

13.15. (a) By definition, $(D/m) = \chi([m])$ depends only on the congruence class of m modulo D. Next,

$$\left(\frac{D}{mn}\right) = \chi([mn]) = \chi([m][n]) = \chi([m])\chi([n]) = \left(\frac{D}{m}\right)\left(\frac{D}{n}\right),$$

where the third equality follows since χ is a homomorphism. Thus (D/m) is multiplicative in m. If $m > 1$ with $m = p_1^{a_1} \cdots p_r^{a_r}$, then multiplicativity in m implies that

$$\left(\frac{D}{m}\right) = \prod_{i=1}^{r} \left(\frac{D}{p_i}\right)^{a_i},$$

and from the uniqueness in Lemma 1.14, (D/p_i) is the Kronecker symbol. Therefore if m is odd and positive, (D/m) is just the Jacobi symbol.

To see that (D/m) is multiplicative in D, choose $m' > 0$ with $m \equiv m' \bmod D$, so that $(D/m) = (D/m')$. By the above, if $D = D_1 D_2$, then $(D/m') = (D_1 D_2/m')$ is the Jacobi symbol, which is multiplicative in D, so that

$$\left(\frac{D}{m}\right)=\left(\frac{D}{m'}\right)=\left(\frac{D_1D_2}{m'}\right)=\left(\frac{D_1}{m'}\right)\left(\frac{D_2}{m'}\right)=\left(\frac{D_1}{m}\right)\left(\frac{D_2}{m}\right),$$

where the first equality follows since $m \equiv m' \bmod D$, and the last equality follows similarly since $m \equiv m' \bmod D_1, D_2$.

For $(D/-1)$, part (b) of Exercise 1.12 implies that $\chi([-1]) = 1$ when $D > 0$ and -1 when $D < 0$. Then $(D/-1) = \operatorname{sgn}(D)$ follows immediately.

(b) First suppose that $m > 0$, $m \equiv 1 \bmod 4$. Then $D = \operatorname{sgn}(D)\,|D|$ and quadratic reciprocity for the Jacobi symbol (see (1.16)) imply that

$$(*)\qquad \left(\frac{D}{m}\right)=\left(\frac{\operatorname{sgn}(D)\,|D|}{m}\right)=\left(\frac{\operatorname{sgn}(D)}{m}\right)\cdot\left(\frac{|D|}{m}\right)=1\cdot\left(\frac{m}{|D|}\right)=\left(\frac{m}{|D|}\right)$$

since $m \equiv 1 \bmod 4$. For general $m \equiv 0, 1 \bmod 4$, adding a suitable multiple of $D \equiv 1 \bmod 4$ to m give $m' \equiv m \bmod D$ with $m' > 0$, $m' \equiv 1 \bmod 4$. Then

$$\left(\frac{D}{m}\right)=\left(\frac{D}{m'}\right)=\left(\frac{m'}{|D|}\right)=\left(\frac{m}{|D|}\right),$$

where the first equality uses part (a), the second uses (*), and the third follows because the Jacobi symbol $(\ell/\,|D|)$ depends only on ℓ modulo $|D|$ by Exercise 1.10.

For the second part, if $D > 0$ and $m < 0$ this follows from the above. If $D < 0$ and $m > 0$, then part (a) implies that

$$\left(\frac{m}{|D|}\right)=\left(\frac{m}{-D}\right)=\left(\frac{m}{-1}\right)\left(\frac{m}{D}\right)=\operatorname{sgn}(m)\left(\frac{m}{D}\right)=\left(\frac{m}{D}\right)$$

since $m > 0$. Thus $(D/m) = (m/\,|D|) = (m/D)$.

(c) Write $m = \prod_{i=1}^{r} p_i^{a_i}$ where the p_i are distinct primes and $a_i > 0$ are integers. Note that $\gcd(m, d_1) = 1$ implies that $\gcd(p_i, d_1) = 1$ for each i. Since $\epsilon(m)$ is defined, it follows that $\epsilon(p_i)$ is defined for each i, so that $\epsilon(p_i) = (d_1/p_i)$ since $p_i \nmid d_1$. Then

$$\epsilon(m)=\prod_{i=1}^{r}\epsilon(p_i)^{a_i}=\prod_{i=1}^{r}\left(\frac{d_1}{p_i}\right)^{a_i}=\prod_{i=1}^{r}\left(\frac{d_1}{p_i^{a_i}}\right)=\left(\frac{d_1}{m}\right)$$

by the multiplicativity of $(d_1/\cdot)$.

(d) We first explain the factorization $m = ab$ described in the statement of part (d). Let $a = \gcd(d_1, m)$. Then $m = ab$ for some $b \in \mathbb{Z}$, where $a, b > 0$. To understand $\gcd(d_1, b)$, let $p \mid d_1$ and $p \mid b$ for some prime p. Hence $p \mid a$ by the definition of a, so that $p^2 \mid ab = m$. Then $4m = d_1d_2 - x^2$ implies that $p \mid x^2$, and it follows easily that $p^2 \mid d_1d_2$. But $p \nmid d_2$ since $\gcd(d_1, d_2) = 1$, so $p^2 \mid d_1$, which impossible since d_1 is an odd field discriminant. This contradiction shows that $\gcd(d_1, b) = 1$.

(i) By part (b) of Exercise 13.14, we know that $(d_1d_2/p) \neq -1$ for any $p \mid m$, so that $\epsilon(m)$ is defined. Since $m = ab$, we have $\epsilon(m) = \epsilon(a)\epsilon(b)$, and since $\gcd(d_1, b) = 1$, we get $\epsilon(b) = (d_1/b)$ by part (c). Also, $a \mid d_1$ implies that $\gcd(a, d_2) = 1$, so that by part (c) applied to d_2, we get $\epsilon(a) = (d_2/a)$. Thus $\epsilon(m) = \epsilon(a)\epsilon(b) = (d_2/a)(d_1/b)$.

(ii) Set $\varepsilon = (-1)^{(a-1)/2}$ and write $d_1 = \varepsilon ad$. Note that $\varepsilon a \equiv d \equiv 1 \bmod 4$ since $d_1 \equiv 1 \bmod 4$. Also, $4ab = 4m = d_1d_2 - x^2 = \varepsilon add_2 - x^2$ implies that $a \mid x$ since a is squarefree. Writing $x = ay$ gives $4ab = \varepsilon add_2 - a^2y^2$, or $4b = \varepsilon dd_2 - ay^2$. It follows that $4b \equiv \varepsilon dd_2 \bmod a$, and then $\varepsilon = \pm 1$ implies that $4b \equiv \varepsilon dd_2 \bmod \varepsilon a$.

Since d_1 is odd, $(d_1/4) = (d_1/2)^2 = (\pm 1)^2 = 1$, and then by multiplicativity

$$\begin{aligned}\left(\frac{d_1}{b}\right) &= \left(\frac{d_1}{4}\right)\left(\frac{d_1}{b}\right) = \left(\frac{d_1}{4b}\right) = \left(\frac{\varepsilon a}{4b}\right)\left(\frac{d}{4b}\right) \\ &= \left(\frac{\varepsilon a}{\varepsilon d d_2}\right)\left(\frac{d}{4b}\right) = \left(\frac{\varepsilon a}{d}\right)\left(\frac{\varepsilon a}{\varepsilon d_2}\right)\left(\frac{d}{4b}\right).\end{aligned}$$

Since $\varepsilon a \equiv d \equiv 1 \bmod 4$, part (b) shows that $(\varepsilon a/d) = (d/|\varepsilon a|) = (d/a)$. Then

$$\begin{aligned}\left(\frac{\varepsilon a}{d}\right)\left(\frac{\varepsilon a}{\varepsilon d_2}\right)\left(\frac{d}{4b}\right) &= \left(\frac{d}{a}\right)\left(\frac{\varepsilon a}{\varepsilon d_2}\right)\left(\frac{d}{4b}\right) = \left(\frac{\varepsilon a}{\varepsilon d_2}\right)\left(\frac{d}{4ab}\right) \\ &= \left(\frac{\varepsilon a}{\varepsilon d_2}\right)\left(\frac{d}{-1}\right)\left(\frac{d}{-4ab}\right).\end{aligned}$$

But $-4ab = x^2 - d_1 d_2$ is a square modulo d since $d \mid d_1$, so the last factor is equal to 1. Combining this with the previous two equations and using part (a) to evaluate $(d/-1)$, we obtain

$$\left(\frac{d_1}{b}\right) = \left(\frac{\varepsilon a}{\varepsilon d_2}\right)\left(\frac{d}{-1}\right) = \left(\frac{\varepsilon a}{\varepsilon}\right)\left(\frac{\varepsilon a}{d_2}\right)\left(\frac{d}{-1}\right) = \left(\frac{\varepsilon a}{d_2}\right)\left(\frac{\varepsilon a}{\varepsilon}\right)\operatorname{sgn}(d).$$

It remains to study the last two factors. First, checking the cases $\varepsilon = 1$ and -1 separately, one sees that $(\varepsilon a/\varepsilon) = \varepsilon$ since $a > 0$. Second, $\varepsilon a d = d_1 < 0$ implies $\operatorname{sgn}(\varepsilon)\operatorname{sgn}(a)\operatorname{sgn}(d) = -1$, so that $\operatorname{sgn}(d) = -\varepsilon$ since $a > 0$. Hence

$$\left(\frac{d_1}{b}\right) = \left(\frac{\varepsilon a}{d_2}\right)\cdot\varepsilon\cdot(-\varepsilon) = -\left(\frac{\varepsilon a}{d_2}\right).$$

(iii) Combining parts (i) and (ii) gives

$$\begin{aligned}\epsilon(m) &= \left(\frac{d_2}{a}\right)\left(\frac{d_1}{b}\right) = \left(\frac{d_2}{a}\right)\left(-\left(\frac{\varepsilon a}{d_2}\right)\right) = -\left(\frac{d_2}{a}\right)\left(\frac{d_2}{|\varepsilon a|}\right) \\ &= -\left(\frac{d_2}{a}\right)\left(\frac{d_2}{a}\right) = -1,\end{aligned}$$

where we use $a > 0$ and part (b), which applies since $\varepsilon a \equiv 1 \bmod 4$ and $d_2 \equiv 0, 1 \bmod 4$.

13.16. (a) Since $\gcd(m_1, m_2) = 1$, divisors of $m_1 m_2$ are counted by divisors of m_1 times divisors of m_2, so that, using the fact that ϵ is multiplicative,

$$\begin{aligned}F(m_1 m_2) &= \prod_{\substack{n_1 n_1' = m_1 \\ n_2 n_2' = m_2 \\ n_i > 0}} (n_1 n_2)^{\epsilon(n_1' n_2')} = \prod_{\substack{n_1 n_1' = m_1 \\ n_2 n_2' = m_2 \\ n_i > 0}} \left(n_1^{\epsilon(n_1')}\right)^{\epsilon(n_2')} \left(n_2^{\epsilon(n_2')}\right)^{\epsilon(n_1')} \\ (*) \qquad &= \prod_{\substack{n_1 n_1' = m_1 \\ n_2 n_2' = m_2 \\ n_i > 0}} \left(n_1^{\epsilon(n_1')}\right)^{\epsilon(n_2')} \prod_{\substack{n_1 n_1' = m_1 \\ n_2 n_2' = m_2 \\ n_i > 0}} \left(n_2^{\epsilon(n_2')}\right)^{\epsilon(n_1')}.\end{aligned}$$

Now, the first of these products is

$$\begin{aligned}\prod_{\substack{n_1 n_1' = m_1 \\ n_2 n_2' = m_2 \\ n_i > 0}} \left(n_1^{\epsilon(n_1')}\right)^{\epsilon(n_2')} &= \prod_{\substack{n_2 n_2' = m_2 \\ n_2 > 0}} \Bigg(\prod_{\substack{n_1 n_1' = m_1 \\ n_1 > 0}} n_1^{\epsilon(n_1')}\Bigg)^{\epsilon(n_2')} \\ &= \Bigg(\prod_{\substack{n_1 n_1' = m_1 \\ n_1 > 0}} n_1^{\epsilon(n_1')}\Bigg)^{\sum_{n_2' \mid m_2,\, n_2' > 0} \epsilon(n_2')} = F(m_1)^{s(m_2)}.\end{aligned}$$

Similarly, the second product in (*) is equal to $F(m_2)^{s(m_1)}$, so that

$$F(m_1m_2) = F(m_1)^{s(m_2)}F(m_2)^{s(m_1)}$$

as desired.

(b) We first claim that s is a multiplicative function. Indeed, if $\gcd(m_1, m_2) = 1$, then

$$\begin{aligned} s(m_1m_2) &= \sum_{\substack{n_1|m_1\\ n_2|m_2\\ n_1,n_2>0}} \epsilon(n_1n_2) = \sum_{\substack{n_1|m_1\\ n_2|m_2\\ n_1,n_2>0}} \epsilon(n_1)\epsilon(n_2) \\ &= \sum_{\substack{n_1|m_1\\ n_1>0}} \epsilon(n_1) \sum_{\substack{n_2|m_2\\ n_2>0}} \epsilon(n_2) = s(m_1)s(m_2), \end{aligned}$$

proving the claim. Therefore

$$s(m) = s(p_1^{a_1})\cdots s(p_r^{a_r}) \cdot s(q_1^{b_1})\cdots s(q_s^{b_s}).$$

If p is any prime for which $\epsilon(p)$ is defined, then

$$s(p^a) = \sum_{i=0}^{a} \epsilon(p^i) = \sum_{i=0}^{a} \epsilon(p)^i.$$

It follows that

$$\epsilon(p) = -1 \quad \Longrightarrow \quad s(p^a) = \sum_{i=0}^{a}(-1)^i = \begin{cases} 1 & a \text{ even} \\ 0 & a \text{ odd} \end{cases}$$

$$\epsilon(p) = 1 \quad \Longrightarrow \quad s(p^a) = \sum_{i=0}^{a} 1^i = a+1.$$

Since $\epsilon(q_i) = 1$, it follows that

$$s(m) = s(p_1^{a_1})\cdots s(p_r^{a_r}) \cdot s(q_1^{b_1})\cdots s(q_s^{b_s}) = s(p_1^{a_1})\cdots s(p_r^{a_r}) \cdot \prod_{i=1}^{s}(b_i+1).$$

If any a_i is odd, then $s(p_i^{a_i}) = 0$ so that $s(m) = 0$, while if all the a_i are even, then each $s(p_i^{a_i}) = 1$ and then $s(m) = \prod_{i=1}^{s}(b_i+1)$.

(c) If $m = \prod_{i=1}^{k} p_i^{r_i}$, then $\epsilon(m) = \prod_{i=1}^{k} \epsilon(p_i)^{r_i}$. If $\epsilon(m) = -1$, one of the factors must be -1, so there is some i with $\epsilon(p_i) = -1$ and r_i odd, so that $\nu_{p_i}(m)$ is odd. Then $s(m) = 0$ by part (b).

(d) From the hint and the assumptions, we can write $m = p^{2a+1}q^{2b+1}m'$ where m' is a positive integer prime to p and q. By part (a),

$$\begin{aligned} F(m) &= F\left(p^{2a+1}q^{2b+1}\right)^{s(m')} F(m')^{s(p^{2a+1}q^{2b+1})} \\ &= F\left(p^{2a+1}\right)^{s(q^{2b+1})s(m')} F\left(q^{2b+1}\right)^{s(p^{2a+1})s(m')} F(m')^{s(p^{2a+1}q^{2b+1})}. \end{aligned}$$

But $s(p^{2a+1}) = s(q^{2b+1}) = s(p^{2a+1}q^{2b+1}) = 0$ by part (b), so that $F(m) = 1$.

(e) Write m as in the exercise. By the definition of $F(m)$, we have

$$F(p^{2a+1}) = \prod_{i=0}^{2a+1} (p^i)^{\epsilon(p^{2a+1-i})} = \prod_{i=0}^{2a+1} (p^i)^{(-1)^{2a+1-i}}$$

since $\epsilon(p) = -1$. The terms corresponding to even values of i evaluate to p^{-i}, while those corresponding to odd values evaluate to p^i, so this product is the same as

$$\prod_{i=0}^{a} \frac{p^{2i+1}}{p^{2i}} = p^{a+1}.$$

Now let $p_1^{a_1} \cdots p_r^{a_r} q_1^{b_1} \cdots q_s^{b_s} = m'$. Note that by assumption all the a_i are even. Then $\gcd(p, m') = 1$, so that $F(m) = F(p^{2a+1})^{s(m')} F(m')^{s(p^{2a+1})}$ by part (a). But $s(p^{2a+1}) = 0$ by part (b), and also by part (b), $s(m') = (b_1+1)\cdots(b_s+1)$. It follows that

$$F(m) = F(p^{2a+1})^{s(m')} = (F(p^{2a+1}))^{(b_1+1)\cdots(b_s+1)} = p^{(a+1)(b_1+1)\cdots(b_s+1)}.$$

This exercise shows that if $\epsilon(m) = -1$, then $F(m)$ is given by the formula in Lemma 13.26. Since $m = (d_1 d_2 - x^2)/4$ implies $\epsilon(m) = -1$ by the previous exercise, Lemma 13.26 is proved.

13.17. (a) Assume first that p is odd. By Corollary 13.25, $p \mid (d_1 d_2 - x^2)/8$ for some x. Since $d_1 d_2 - x^2 \equiv 0 \bmod 4$ and $d_1 d_2 \equiv 1 \bmod 8$, we see that x must be odd, so that $x^2 \equiv 1 \bmod 8$ and therefore $(d_1 d_2 - x^2)/8$ is an integer and divisible by p since p is odd. Thus

$$p \,\Big|\, \frac{d_1 d_2 - x^2}{8} < \frac{d_1 d_2}{8},$$

where the final strict inequality follows since x is odd and cannot be zero. It remains to consider $p = 2$. First, d_1 and d_2 are distinct since $\gcd(d_1, d_2) = 1$, and $d_1 d_2 \equiv 1 \bmod 8$ implies $d_1 \equiv d_2 \equiv 1 \bmod 4$. Thus the choices for d_1 and d_2 are $-3, -7, -11, \ldots$, which makes it easy to see that $d_1 d_2 \geq 33$. It follows that $p = 2 < 33/8 \leq d_1 d_2/8$.

(b) Note that d_1 and d_2 satisfy the conditions of part (a), so that the conclusions reached there are valid here as well. Assume first that p is an odd prime. For any x with $p \mid (d_1 d_2 - x^2)/4$, we know from part (a) that $p \mid (d_1 d_2 - x^2)/8$. If we can show that $p < (d_1 d_2 - x^2)/8$, then we can conclude that $p < (d_1 d_2 - x^2)/16 \leq d_1 d_2/16$, proving the result for odd p. So assume by way of contradiction that $p = (d_1 d_2 - x^2)/8$, i.e. that $2p = (d_1 d_2 - x^2)/4$. Then by Exercise 13.15,

$$\epsilon(2p) = \epsilon(2)\epsilon(p) = \epsilon\Big(\frac{d_1 d_2 - x^2}{4}\Big) = -1.$$

Since $d_1 \equiv d_2 \equiv 5 \bmod 8$, $\epsilon(2) = (5/2) = -1$, so that $\epsilon(p) = 1$. This implies that $(d_1/p) = 1$ or $(d_2/p) = 1$, so that by Corollary 13.25, p cannot divide $J(d_1, d_2)^2$. This contradiction proves the result for odd p. If $p = 2$, then from part (a), $d_1 d_2 \geq 33$ so that in this case as well $p = 2 < 33/16 \leq d_1 d_2/16$.

13.18. We begin with a general observation. Take $d_1 < -4$ and set $h = h(d_1)$. As noted in the proof of Corollary 13.27, $J(d_1, -3)^2 = \big(\prod_{i=1}^h j(\mathfrak{a}_i)\big)^{2/3}$. Proposition 13.2 tells us that the constant term of $H_{d_1}(X)$ is $H_{d_1}(0) = \prod_{i=1}^h (0 - j(\mathfrak{a}_i)) = (-1)^h \prod_{i=1}^4 j(\mathfrak{a}_i)$. Thus

$$H_{d_1}(0)^{2/3} = \Big((-1)^h \prod_{i=1}^4 j(\mathfrak{a}_i)\Big)^{2/3} = \Big(\prod_{i=1}^4 j(\mathfrak{a}_i)\Big)^{2/3} = J(d_1, -3)^2.$$

It follows that $H_{d_1}(0) = \pm\big(J(d_1, -3)^2\big)^{3/2}$; the reason for the $\pm$ is due to the square root. We also note that by the solution to part (b) of Exercise 13.13, $J(d_1, -3)$ is an integer.

Now let $d_1 = -56$. Since $d_1 d_2 = (-56)(-3) = 168$ is divisible by 4, $x^2 \equiv d_1 d_2 \bmod 4$ if and only if x is even. Thus Theorem 13.24 implies

$$(*) \qquad J(-56, -3)^2 = \pm \prod_{\substack{x^2 < 168 \\ x \text{ even}}} F\Big(\frac{168 - x^2}{4}\Big).$$

But $x^2 < 168$ means $|x| < 13$. Thus $x = -12, -10, \ldots, 10, 12$ and

$$\frac{d_1 d_2 - x^2}{4} = 6, 17, 26, 33, 38, 41, 42, 41, 38, 33, 26, 17, 6.$$

By Lemma 13.26, computing $F(m)$ for these values of m requires the prime factorization of m and in particular, we need to know $\epsilon(p)$ for each prime factor.

Looking at the above list reveals that the prime factors involved are thus 2, 3, 7, 11, 13, 17, 19 and 41. We know that ϵ is defined for each of these by Exercise 13.14. Computing these values, we get

$$\begin{aligned}
\epsilon(2) &= \left(\frac{-3}{2}\right) = -1\\
\epsilon(3) &= \left(\frac{-56}{3}\right) = \left(\frac{1}{3}\right) = 1\\
\epsilon(7) &= \left(\frac{-3}{7}\right) = \left(\frac{4}{7}\right) = 1\\
\epsilon(11) &= \left(\frac{-56}{11}\right) = \left(\frac{-1}{11}\right) = -1\\
\epsilon(13) &= \left(\frac{-56}{13}\right) = \left(\frac{9}{13}\right) = 1\\
\epsilon(17) &= \left(\frac{-56}{17}\right) = \left(\frac{12}{17}\right) = \left(\frac{3}{17}\right) = \left(\frac{17}{3}\right) = -1\\
\epsilon(19) &= \left(\frac{-56}{19}\right) = \left(\frac{1}{19}\right) = 1\\
\epsilon(41) &= \left(\frac{-3}{41}\right) = \left(\frac{-1}{41}\right)\left(\frac{3}{41}\right) = \left(\frac{41}{3}\right) = -1.
\end{aligned}$$

Then using Lemma 13.26,

$$\begin{gathered}
F(6) = 2^2, \quad F(17) = 17, \quad F(26) = 2^2,\\
F(33) = 11^2, \quad F(38) = 2^2, \quad F(41) = 41, \quad F(42) = 2^4.
\end{gathered}$$

Each of these values except the one corresponding to 42 appears twice in the product for $J(-56,-3)^2$, so by (*), multiplying these together with multiplicity 2 except for the term corresponding to 42 gives

$$J(-56,-3)^2 = \pm 2^{16} \cdot 17^2 \cdot 11^4 \cdot 41^2.$$

As noted above, $J(-56,-3)$ is a integer, so that sign in the above equation is $+$. It follows that the constant term of $H_{-56}(X)$ is

$$H_{-56}(0) = \pm(J(-56,-3)^2)^{3/2} = \pm(2^{16} \cdot 17^2 \cdot 11^4 \cdot 41^2)^{3/2} = \pm(2^8 \cdot 11^2 \cdot 17 \cdot 41)^3,$$

which agrees with (13.1) up to sign.

For $d_1 = -71$, the analysis is similar. Since $d_1 d_2 = (-71)(-3) = 213$ is 1 modulo 4, $x^2 \equiv d_1 d_2 \bmod 4$ if and only if x is odd. Thus Theorem 13.24 implies

$$(**) \qquad J(-71,-3)^2 = \pm \prod_{\substack{x^2<213\\ x \text{ odd}}} F\Big(\frac{213 - x^2}{4}\Big).$$

But $x^2 < 213$ means $|x| < \sqrt{213} \approx 14.6$. Thus $x = -13, -11, \ldots, 11, 13$ gives

$$\frac{d_1 d_2 - x^2}{4} = 11, 23, 33, 41, 47, 51, 53, 53, 51, 47, 41, 33, 23, 11.$$

The prime factors involved are thus 3, 11, 17, 23, 41, 47, and 53. We know that ϵ is defined for each of these by Exercise 13.14. Computing these values, we get

$$\begin{aligned}
\epsilon(3) &= \left(\frac{-71}{3}\right) = \left(\frac{1}{3}\right) = 1\\
\epsilon(11) &= \left(\frac{-3}{11}\right) = \left(\frac{-1}{11}\right)\left(\frac{3}{11}\right) = \left(\frac{11}{3}\right) = -1\\
\epsilon(17) &= \left(\frac{-71}{17}\right) = \left(\frac{-3}{17}\right) = \left(\frac{-1}{17}\right)\left(\frac{3}{17}\right) = \left(\frac{17}{3}\right) = -1\\
\epsilon(23) &= \left(\frac{-3}{23}\right) = \left(\frac{-1}{23}\right)\left(\frac{3}{23}\right) = \left(\frac{23}{3}\right) = -1\\
\epsilon(41) &= \left(\frac{-3}{41}\right) = \left(\frac{-1}{41}\right)\left(\frac{3}{41}\right) = \left(\frac{41}{3}\right) = -1\\
\epsilon(47) &= \left(\frac{-3}{47}\right) = \left(\frac{-1}{47}\right)\left(\frac{3}{47}\right) = \left(\frac{47}{3}\right) = -1\\
\epsilon(53) &= \left(\frac{-3}{53}\right) = \left(\frac{-1}{53}\right)\left(\frac{3}{53}\right) = \left(\frac{53}{3}\right) = -1.
\end{aligned}$$

Then using Lemma 13.26,

$$F(11) = 11, \quad F(23) = 23, \quad F(33) = 11^2,$$
$$F(41) = 41, \quad F(47) = 47, \quad F(51) = 17^2, \quad F(53) = 53.$$

Each of these values appears twice in the product for $J(-56,-3)^2$, so multiplying these together with multiplicity 2 gives via (**)

$$J(-71,-3)^2 = \pm 11^6 \cdot 17^4 \cdot 23^2 \cdot 41^2 \cdot 47^2 \cdot 53^2,$$

and the sign is $+$ since $J(-71,-3)$ is an integer. It follows that the constant term of $H_{-71}(X)$ is

$$\begin{aligned}
H_{-71}(0) &= \pm(J(-71,-3)^2)^{3/2} = \pm(11^6 \cdot 17^4 \cdot 23^2 \cdot 41^2 \cdot 47^2 \cdot 53^2)^{3/2}\\
&= \pm(11^3 \cdot 17^2 \cdot 23 \cdot 41 \cdot 47 \cdot 53)^3,
\end{aligned}$$

which agrees with (13.20) up to sign.

Solutions to Exercises in §14

14.1. (a) Suppose (a,b) and (c,d) are points in K^2. Then $(a,b) \mapsto (a,b,1)$ and $(c,d) \mapsto (c,d,1)$. If $(a,b,1) = (c,d,1)$ in $\mathbb{P}^2(K)$, then $(a,b,1) = \lambda(c,d,1)$ so clearly $\lambda = 1$, $a = c$, $b = d$. Thus the map is an injection. The image of the map contains none of the points $(x,y,0)$, since then we would have $(x,y,0) = \lambda(a,b,1)$ for $\lambda \in K^*$, which is impossible. Finally, if $(a,b,c) \in \mathbb{P}^2(K)$ with $c \neq 0$, then

$$\left(\frac{a}{c},\frac{b}{c}\right) \longmapsto \left(\frac{a}{c},\frac{b}{c},1\right) = \frac{1}{c}(a,b,c) \sim (a,b,c),$$

so the image of the map includes every point of $\mathbb{P}^2(K)$ except the line at infinity.

(b) Let

$$\tilde{E}(K)' = \{(x,y,1) \in \mathbb{P}^2(K) \mid y^2 = 4x^3 - g_2x - g_3\}.$$

It is clear that $\tilde{E}(K)' \cup \{(0,1,0)\} \subset \tilde{E}(K)$. For the reverse inclusion, suppose $(x,y,z) \in \tilde{E}(K)$. If $z = 0$, the equation $y^2z = 4x^3 - g_2z^2x - g_3z^3$ becomes $0 = 4x^3$, so that $x = 0$. Since y cannot then be zero, we get the solution $(0, y, 0) \sim (0, 1, 0)$. Otherwise, if $z \neq 0$, then dividing $y^2z = 4x^3 - g_2xz^2 - g_3z^3$ through by z^3 gives

$$\left(\frac{y}{z}\right)^2 = 4\left(\frac{x}{z}\right)^3 - g_2\frac{x}{z} - g_3,$$

so that $(x/z, y/z) \in E(K)$ and thus $(x/z, y/z, 1) \in \tilde{E}(K)'$. But $(x/z, y/z, 1) \sim (x, y, z)$, so that $(x, y, z) \in \tilde{E}(K)'$.

14.2. Define $\phi : \mathbb{C} - L \to E(\mathbb{C})$ by $\phi(z) = (\wp(z), \wp'(z))$. The differential equation satisfied by $\wp(z)$ and $\wp'(z)$ implies that $\phi(z) \in E(\mathbb{C})$. If $w, z \in \mathbb{C} - L$ and $w \equiv z \bmod L$, then $\wp(z) = \wp(w)$ and $\wp'(z) = \wp'(w)$ since $\wp(z)$ and $\wp'(z)$ are elliptic functions for L by Theorem 10.1 and Exercise 10.2. Thus $\phi(z) = \phi(w)$, so that ϕ may be regarded as a map $\phi : (\mathbb{C} - L)/L \to E(\mathbb{C})$. In fact the image of ϕ lies in $E(\mathbb{C}) - \{\infty\}$ since both $\wp(z)$ and $\wp'(z)$ are holomorphic away from L.

To see that ϕ is surjective, take $(u, v) \in E(\mathbb{C}) - \{\infty\}$. By part (b) of Exercise 10.14, there is $z \in \mathbb{C} - L$ with $\wp(z) = u$, so that $v^2 = 4\wp(z)^3 - g_2\wp(z) - g_3$. Comparing this to the differential equation for $\wp(z)$ and $\wp'(z)$ implies that $v = \pm\wp'(z)$. If the sign is plus, then $\phi(z) = (\wp(z), \wp'(z)) = (u, v)$. If the sign is minus, then since $\wp(z)$ is even and $\wp'(z)$ is odd, we get $\phi(-z) = (\wp(-z), \wp'(-z)) = (\wp(z), -\wp'(z)) = (u, v)$. Thus $\phi(z)$ is surjective.

For injectivity, if $(\wp(z), \wp'(z)) = (\wp(w), \wp'(w))$, then $\wp(z) = \wp(w)$ implies that $z \equiv \pm w \bmod L$ by Lemma 10.4. If $z \equiv w \bmod L$ then $z = w$ in $(\mathbb{C} - L)/L$. Otherwise, $z \equiv -w \bmod L$, so that $(\wp(w), \wp'(w)) = (\wp(-z), \wp'(-z)) = (\wp(z), -\wp'(z))$, and equality is impossible unless $\wp'(z) = 0$. But $\wp'(z) = 0$ means that $2z \in L$ by Corollary 10.5, so that $z \equiv -w \bmod L$ is the same as $z \equiv w \bmod L$ and again $z = w$ in $(\mathbb{C} - L)/L$.

14.3. The existence of L follows directly from Corollary 11.7. For uniqueness, suppose there are two such lattices L and L'. Then $j(L) = j(L')$, so L and L' are homothetic by Theorem 10.9. Suppose $L' = \lambda L$ for $\lambda \in \mathbb{C}^*$. Then the definitions of g_2 and g_3 show that

$$g_2(L) = \lambda^4 g_2(L')$$
$$g_3(L) = \lambda^6 g_3(L').$$

But $g_2(L') = g_2(L) = g_2$, $g_3(L') = g_3(L) = g_3$. If $g_2g_3 \neq 0$, then $\lambda^4 = \lambda^6 = 1$, so that $\lambda^2 = 1$. Hence $\lambda = \pm 1$, which implies $L' = \pm L = L$. It remains to study $g_2 = 0$ and $g_3 = 0$:

- If $g_2 = 0$, then $g_3 \neq 0$, so that $\lambda^6 = 1$, hence $\lambda = \pm\omega^\ell$. But $g_2 = 0$ implies that $j(L) = j(L') = 0 = j([1, \omega])$, so that $L = \lambda_0[1, \omega]$ for $\lambda_0 \in \mathbb{C}^*$. Consequently
$$L' = \lambda L = \pm\omega^\ell(\lambda_0[1, \omega]) = \lambda_0\big(\pm\,\omega^\ell[1, \omega]\big) = \lambda_0[1, \omega] = L.$$
- If $g_3 = 0$, then $g_2 \neq 0$, so that $\lambda^4 = 1$, hence $\lambda = i^\ell$. But $g_3 = 0$ implies that $j(L) = j(L') = 1728 = j([1, i])$, so that $L = \lambda_0[1, i]$ for $\lambda_0 \in \mathbb{C}^*$. Consequently
$$L' = \lambda L = i^\ell(\lambda_0[1, i]) = \lambda_0\big(i^\ell[1, i]\big) = \lambda_0[1, i] = L.$$

14.4. (a) For Proposition 14.4, note first that by Proposition 14.3, $j(E) = j(E')$ if and only if $j(L) = j(L')$. Then (ii) $\Leftrightarrow$ (iii) by Theorem 10.9. The argument for (i) $\Leftrightarrow$ (iii) is similar: if $j(E') = j(E)$, then the proof of Theorem 10.9 shows that there is some $c \in \mathbb{C}^*$ with

$$g_2' = c^4 g_2$$
$$g_3' = c^6 g_3,$$

so that E and E' are isomorphic over $\mathbb{C}$. Conversely, if E and E' are isomorphic over $\mathbb{C}$, then by definition there is $c \in \mathbb{C}$ such that the above equations hold, and it is immediate that $j(E') = j(E)$.

For Proposition 14.5, note that (i) implies (ii) since if K is algebraically closed it has no nontrivial finite extensions. It remains to prove (i). We mimic the proof of Theorem 10.9. If E and E' are isomorphic over some finite extension L of K, there is some $c \in L$ with $g_2' = c^4 g_2$ and $g_3' = c^6 g_3$. A computation shows that $j(E) = j(E')$. Conversely, suppose that $j(E) = j(E')$. First assume $g_2(E) \neq 0$ and $g_3(E) \neq 0$ and pick a finite extension L of K such that

$$c^4 = \frac{g_2(E')}{g_2(E)}$$

for some $c \in L$. Note that $g_2(E) \neq 0 \Rightarrow j(E') = j(E) \neq 0 \Rightarrow g_2(E') \neq 0$. Thus $c \in L^*$. A computation with $j(E')$ and $j(E)$ shows that

$$c^{12} = \Big(\frac{g_3(E')}{g_3(E)}\Big)^2.$$

We thus have $g_2(E') = c^4 g_2(E)$ and $g_3(E') = \pm c^6 g_3(E)$. If the sign is negative in the second equation, enlarge L to contain a square root of -1 and replace c by $\sqrt{-1}c$. Then we can assume that the sign is positive. This gives $g_3(E') = c^6 g_3(E)$, and we continue to have $g_2(E') = c^4 g_2(E)$. Thus E' and E are isomorphic over L. The proof when $g_2(E) = 0$ or $g_3(E) = 0$ follows the proof in Exercise 10.9 and is similar to the above. Note that the fact that K does not have characteristic 2 or 3 is required since otherwise $j(E)$ is identically zero.

(b) Two curves $y^2 = 4x^3 - g_3$ and $y^2 = 4x^3 - g_3'$ are isomorphic over $\mathbb{Q}$ if and only if there is some rational number c with $g_3' = c^6 g_3$. To construct infinite many isomorphism classes, consider the curves E_p defined by $y^2 = 4x^3 - p$ for $p \in \mathbb{Z}$ prime. Then E_p and E_q are isomorphic over $\mathbb{Q}$ if and only if $p = c^6 q$ for some $c \in \mathbb{Q}$ if and only if $p = q$.

14.5. (a) Set $(x_1, y_1) = (\wp(z), \wp'(z))$ and $(x_2, y_2) = (\wp(w), \wp'(w))$ in (14.6). Since we want (x_3, y_3) to be $(\wp(z+w), \wp'(z+w))$, it is natural to conjecture that

$$(*) \qquad \wp'(z+w) = -\wp'(z) - (\wp(z+w) - \wp(z))\Big(\frac{\wp'(z) - \wp'(w)}{\wp(z) - \wp(w)}\Big).$$

To prove this, differentiate the addition law for $\wp(z+w)$ with respect to z, giving

$$\begin{aligned}\wp'(z+w) &= -\wp'(z) + \frac{1}{2}\Big(\frac{\wp'(z) - \wp'(w)}{\wp(z) - \wp(w)}\Big)\Big(\frac{\wp'(z) - \wp'(w)}{\wp(z) - \wp(w)}\Big)' \\ &= -\wp'(z) + \frac{1}{2}\Big(\frac{\wp'(z) - \wp'(w)}{\wp(z) - \wp(w)}\Big)\Big(\frac{\wp''(z)(\wp(z) - \wp(w)) - \wp'(z)(\wp'(z) - \wp'(w))}{(\wp(z) - \wp(w))^2}\Big).\end{aligned}$$

We want to show this expression is equal to the right-hand side of (*), which amounts to showing that

$$-(\wp(z+w) - \wp(z)) = \frac{1}{2}\left(\frac{\wp''(z)(\wp(z) - \wp(w)) - \wp'(z)(\wp'(z) - \wp'(w))}{(\wp(z) - \wp(w))^2}\right),$$

or

$$2(\wp(z) - \wp(z+w))(\wp(z) - \wp(w))^2 = \wp''(z)(\wp(z) - \wp(w)) - \wp'(z)(\wp'(z) - \wp'(w)).$$

Evaluating the left-hand side, we start by substituting for $\wp(z+w)$ using the addition formula, giving

$$2\left(2\wp(z) + \wp(w) + \frac{1}{4}\left(\frac{\wp'(z) - \wp'(w)}{\wp(z) - \wp(w)}\right)^2\right)(\wp(z) - \wp(w))^2$$

which after simplification becomes

$$4\wp(z)^3 - 6\wp(w)\wp(z)^2 + \wp'(w)\wp'(z) - \frac{1}{2}\wp'(z)^2 + 2\wp(w)^3 - \frac{1}{2}\wp'(w)^2.$$

Now use the differential equation for $\wp(z)$ to write $\frac{1}{2}\wp'(w)^2 = 2\wp(w)^3 - \frac{1}{2}g_2\wp(w) - \frac{1}{2}g_3$, giving

$$(1)\qquad 4\wp(z)^3 - 6\wp(w)\wp(z)^2 + \wp'(w)\wp'(z) - \frac{1}{2}\wp'(z)^2 + \frac{1}{2}g_2\wp(w) + \frac{1}{2}g_3.$$

Next work on the right side. Start by substituting $6\wp(z)^2 - \frac{1}{2}g_2$ for $\wp''(z)$ (see the proof of Lemma 10.12). Simplify, giving

$$(2)\qquad 6\wp(z)^3 - 6\wp(w)\wp(z)^2 + \wp'(w)\wp'(z) - \wp'(z)^2 - \frac{1}{2}g_2\wp(z) + \frac{1}{2}g_2\wp(w).$$

Subtracting (1) from (2), it remains to see that

$$2\wp(z)^3 - \frac{1}{2}\wp'(z)^2 - \frac{1}{2}\wp(z) - \frac{1}{2}g_3 = 0.$$

But the differential equation for $\wp'(z)^2$ (see part (ii) of Theorem 10.1) shows that this is true, proving (*).

(b) We claim that

$$(**)\qquad \wp'(2z) = -\wp'(z) - (\wp(2z) - \wp(z))\Big(\frac{12\wp(z)^2 - g_2}{2\wp'(z)}\Big).$$

To prove this, fix z and take the limit of (*) as $w \to z$. This gives

$$\begin{aligned}\wp'(2z) &= \lim_{w\to z}\wp'(z+w) = \lim_{w\to z}\Big(-\wp'(z) - (\wp(z+w) - \wp(z))\cdot\frac{\wp'(z) - \wp'(w)}{\wp(z) - \wp(w)}\Big)\\ &= -\wp'(z) - (\wp(2z) - \wp(z))\lim_{w\to z}\frac{\wp'(z) - \wp'(w)}{\wp(z) - \wp(w)}\\ &= -\wp'(z) - (\wp(2z) - \wp(z))\cdot\frac{-\wp''(z)}{-\wp'(z)},\end{aligned}$$

where the final line uses L'Hospital's Rule. However, $\wp''(z) = 6\wp(z)^2 - \frac{1}{2}g_2$ by the proof of Lemma 10.12. From here, (**) follows easily.

14.6. Define $\phi : \ker(\alpha) \to L'/\alpha L$ as follows. Take $u \in \ker(\alpha)$. Writing $u = [z] \in \mathbb{C}/L$, $u \in \ker(\alpha)$ means that $\alpha(u) = [\alpha z]$ is the identity element of $\mathbb{C}/L'$, so that $\alpha z \in L'$. Then $\phi(u) = [\alpha z] \in L'/\alpha L$. To check that this is well defined, suppose $u = [z']$, so $z' = z + \omega$ for some some $\omega \in L$. Then $[\alpha z'] = [\alpha z + \alpha\omega] = [\alpha z]$ in $L'/\alpha L$ since $\alpha\omega \in \alpha L$.

Surjective: Take $[z'] \in L'/\alpha L$, where $z' \in L'$. Set $z = \alpha^{-1}z' \in \mathbb{C}$ (remember $\alpha \neq 0$). Then $\alpha z = z' \in L'$ shows that $[z] \in \ker(\alpha)$, and $\phi([z]) = [\alpha z] = [z'] \in L'/\alpha L$. Thus ϕ is surjective.

Injective: Suppose that $\phi([z]) = [\alpha z] \in L'/\alpha L$ is the zero element. Then $\alpha z \in \alpha L$, which implies $z \in L$ since $\alpha \neq 0$. Then $[z]$ is the zero element of $\ker(\alpha) \subset \mathbb{C}/L$. Thus ϕ is injective.

14.7. For the elliptic curve E defined by $y^2 = 4x^3 - 30x - 28$, we have $g_2 = 30$ and $g_3 = 28$. The corresponding lattice is $\lambda' L = \lambda'[1, \sqrt{-2}]$, as explained in the discussion following (14.10). Therefore E has the same complex multiplication as $[1, \sqrt{-2}]$ by the discussion following Corollary 10.20, so that $\mathrm{End}_{\mathbb{C}}(E) = \mathbb{Z}[\sqrt{-2}]$.

Following the same method as in the computation of $j(\sqrt{-2})$ (see (10.21)), we have

$$(*)\qquad \wp(\sqrt{-2}\,z) = a\wp(z) + b + \frac{1}{c\wp(z) + d},\quad a, b, c, d \in \mathbb{C},\quad a, c \neq 0.$$

By Exercise 10.18, the Laurent expansion of $\wp(z)$ about $z = 0$ is

$$\wp(z) = \frac{1}{z^2} + \frac{g_2}{20}z^2 + \frac{g_3}{28}z^4 + \frac{g_2^2}{1200}z^6 \cdots = \frac{1}{z^2} + \frac{3}{2}z^2 + z^4 + \frac{3}{4}z^6 + \cdots,$$

and thus

$$\wp(\sqrt{-2}\,z) = -\frac{1}{2z^2} - 3z^2 + 4z^4 - 6z^6 + \cdots.$$

Since $\wp(\sqrt{-2}\,z) - a\wp(z) - b$ must be zero at $z = 0$, we get $a = -\frac{1}{2}$, $b = 0$. Then (*) implies that $(\wp(\sqrt{-2}\,z) + (1/2)\wp(z))^{-1} = c\wp(z) + d$, which is linear in $\wp(z)$. But

$$\begin{aligned}\Big(\wp(\sqrt{-2}\,z) + \frac{1}{2}\wp(z))^{-1}\Big)^{-1} &= \Big(-\frac{9}{4}z^2 + \frac{9}{2}z^4 - \frac{45}{8}z^6 + \cdots\Big)^{-1}\\ &= -\frac{4}{9z^2} - \frac{8}{9} - \frac{2}{3}z^2 + \cdots.\end{aligned}$$

Since this must be linear in $\wp(z)$, we get

$$c\wp(z) + d = -\frac{4}{9}\wp(z) - \frac{8}{9}.$$

Then

$$\wp(\sqrt{-2}\,z) = -\frac{1}{2}\wp(z) - \frac{1}{(4/9)\wp(z) + (8/9)} = -\frac{2\wp(z)^2 + 4\wp(z) + 9}{4(\wp(z) + 2)}.$$

Differentiating,

$$\sqrt{-2}\wp'(\sqrt{-2}\,z) = -\frac{2\wp(z)^2 + 8\wp(z) - 1}{4(\wp(z) + 2)^2}\wp'(z).$$

Since the isomorphism from $\mathbb{C}/L \to E(\mathbb{C})$ is given by $z \mapsto (\wp(z), \wp'(z))$, we have

$$\sqrt{-2}\,(x, y) = \Big(-\frac{2x^2 + 4x + 9}{4(x + 2)}, -\frac{1}{\sqrt{-2}}\frac{2x^2 + 8x - 1}{4(x + 2)^2}y\Big).$$

14.8. (a) Since $E : y^2 = 4x^3 - g_2x - g_3$ is defined over $\mathbb{F}_q$, taking the qth power of each side fixes all the coefficients, so that if $(x, y) \in E(L)$, then $y^2 = 4x^3 - g_2x - g_3$ implies

$$(y^q)^2 = (y^2)^q = (4x^3 - g_2x - g_3)^q = 4^q(x^q)^3 - g_2^q x^q - g_3^q = 4(x^q)^3 - g_2(x^q) - g_3,$$

proving that $(x^q, y^q) \in E(L)$. Raising each side of (14.6) to the qth power and arguing as above shows that this map is a group homomorphism $E(L) \to E(L)$.

(b) If $(x^q, y^q) = (R(x), (1/\alpha)R'(x)y)$, then $R(x) = x^q$, so that $R'(x) = qx^{q-1} = 0$ since $\mathbb{F}_q \subset L$ and $q = p^a$ imply that L has characteristic p.

14.9. The analogous statement is:

PROPOSITION. *Assume that $\alpha \neq 0$ in $\mathbb{C}$ satisfies $\alpha L \subset L'$ and let $\alpha : E(\mathbb{C}) \to E'(\mathbb{C})$ be the isogeny corresponding to the map $\mathbb{C}/L \to \mathbb{C}/L'$ induced by multiplication by α. Then there is a rational function $R(x) \in \mathbb{C}(x)$ such that for $(x, y) \in E(\mathbb{C})$, we have*

$$\alpha(x, y) = \Big(R(x), \frac{1}{\alpha}R'(x)y\Big) = \Big(R(x), \frac{1}{\alpha}R'(x)y\Big).$$

PROOF. Let $f(z) = \wp(\alpha z; L')$. Clearly $f(z)$ is meromorphic, and if $\omega \in L$, $z \notin L$, then

$$f(z + \omega) = \wp(\alpha(z + \omega); L') = \wp(\alpha z + \alpha\omega; L') = \wp(\alpha z; L') = f(z)$$

since $\alpha\omega \in \alpha L \subset L'$. Thus $f(z)$ is an elliptic function for L, and it is even since $\wp(z)$ is even. So by Lemma 10.17, $\wp(\alpha z; L') = R(\wp(z; L))$ where $R(x) \in \mathbb{C}(x)$. Differentiating with respect to z gives $\alpha\wp'(\alpha z; L') = R'(\wp(z))\wp'(z)$. The map $\mathbb{C}/L \to \mathbb{C}/L'$ is given by $[z] \mapsto [\alpha z]$. For $(x, y) \in E(\mathbb{C})$, the solution to Exercise 14.2 shows that we can choose $z \in \mathbb{C}$ with $(\wp(z; L), \wp'(z; L) = (x, y)$. Then

$$\alpha(x, y) = (\wp(\alpha z; L'), \wp'(\alpha z; L')) = \Big(R(\wp(z; L), \frac{1}{\alpha}R'(\wp(z; L)\wp'(z; L)\Big) = \Big(R(x), \frac{1}{\alpha}R'(x)y\Big).$$

Q.E.D.

14.10. Let E and E' correspond to lattices L and L'. As explained in the text, an isogeny $\alpha : E \to E'$ corresponds to $\alpha L \subset L'$. Also note that $\ker(\alpha) \simeq L'/\alpha L$ by Exercise 14.6. It follows that the degree of α is $|\ker(\alpha)| = |L'/\alpha L|$. In other words, $\deg(\alpha) = [L' : \alpha L]$.

If $\alpha : E \to E'$ is cyclic of degree m, then $L'/\alpha L$ is cyclic of order m. Then Theorem 11.23 implies that $\Phi_m(j(\alpha L), j(L')) = \Phi_m(j(L), j(L')) = \Phi_m(j(E), j(E')) = 0$. Conversely, suppose that $\Phi_m(j(E), j(E')) = 0$. Then $\Phi_m(j(L), j(L')) = 0$. By the argument at the end of the proof of Theorem 11.23, the roots of $\Phi_m(X, j(L'))$ are the j-invariants of the cyclic sublattices $L'' \subset L'$ of index m. Thus $j(L) = j(L'')$ for some such L''. Then L is homothetic to L'', which means that $\alpha L = L'' \subset L'$ is a cyclic sublattice of index m for some α. As noted above, this gives an isogeny from E to E' of order m, which is cyclic since $\ker(\alpha) \simeq L'/\alpha L$.

14.11. Any value of x produces at most two values of y since $y^2 = 4x^3 - g_2x - g_3$ is quadratic in y. Thus the q choices for $x \in \mathbb{F}_q$ lead to at most $2q$ solutions $(x, y) \in E(\mathbb{F}_q)$. Adding the point at infinity gives $|E(\mathbb{F}_q)| \leq 2q + 1$.

14.12. For $x \in \overline{\mathbb{F}}_q$, $x^q = x$ if and only if $x \in \mathbb{F}_q$. Then $(x, y) \in \ker(1 - \mathit{Frob}_q)$ if and only if $\mathit{Frob}_q(x, y) = (x^q, y^q) = (x, y)$ if and only if $(x, y) \in E(\mathbb{F}_q)$.

14.13. (a) Substituting the given formulas into $y^2 = 4x^3 - 27$ gives

$$\Big(9 \cdot \frac{1-y}{1+y}\Big)^2 - 4\Big(\frac{3x}{1+y}\Big)^3 + 27 = \frac{81(1-y)^2(1+y) - 108x^3 + 27(1+y)^3}{(1+y)^3}$$
$$= \frac{108y^3 - 108x^3 + 108}{(1+y)^3} = \frac{108(y^3 - x^3 + 1)}{(1+y)^3}.$$

Then if (x, y) lies on the curve $x^3 = y^3 + 1$, we have $y^3 - x^3 + 1 = 0$ so that the fraction above is zero. This proves that the given formulas define a point on $y^2 = 4x^3 - 27$. For the inverse map, suppose $Y^2 = 4X^3 - 27$. Solving

$$X = \frac{3x}{1+y}, \quad Y = \frac{9(1-y)}{1+y}$$

for x and y gives $y = (9-Y)/(9+Y)$ and then $x = (X/3)(1+y) = 6X/(9+Y)$. This inverse map shows that the given formulas transform $x^3 = y^3 + 1$ into $y^2 = 4x^3 - 27$.

(b) Under the map $(x, y, z) \mapsto (3x, 9(z - y), 9(z + y))$, assume that (x, y, z) maps to $(0, 1, 0)$. Then $(3x, 9(1-y), 9(1+y)) \sim (0, 1, 0)$, which implies $3x = 0$ and $9(z+y) = 0$. Since we are working over $\mathbb{F}_p$ with $p \equiv 1 \bmod 3$, we have $x = 0$ and $y = -z$. It follows that $(x, y, z) = (0, -z, z) \sim (0, -1, 1)$, which corresponds to the solution $(0, -1)$ of $x^3 = y^3 + 1$.

(c) The points at infinity are the solutions (when $z = 0$) to $x^3 = y^3$, so we may assume $xy \neq 0$. Rescaling, we may assume $y = 1$, so we need solutions of $x^3 - 1 = (x - 1)(x^2 + x + 1) = 0$ in $\mathbb{F}_p$. One solution is clearly $x = 1$. Further, $x^2 + x + 1$ has discriminant -3; since $p \equiv 1 \bmod 3$, Exercise 1.5 shows that $(-3/p) = 1$, so that the quadratic has two roots in $\mathbb{F}_p$. Finally, $x^3 - 1$ is separable since $p \neq 3$, so the roots are distinct, and we are done.

(d) Let L be a lattice corresponding to E. Then $j(E) = 0$ since $g_2 = 0$, so that $j(L) = 0$. But $j(\omega) = j([1, \omega]) = 0$ for $\omega = e^{2\pi i/3}$ by Exercise 10.17, so L is homothetic to $[1, \omega]$ by Theorem 10.9. Since $[1, \omega]$ has complex multiplication by ω, so does L and therefore so does E.

14.14. To see that the given formulas map $x^2+y^2+x^2y^2=1$ into the curve E defined by $y^2=4x^3+x$, we compute as follows:

$$\begin{aligned} y^2-4x^3-x &= \Big(\frac{(1+x^2)y}{(1-x)^2}\Big)^2-4\Big(\frac{1+x}{2(1-x)}\Big)^3-\frac{1+x}{2(1-x)}\\ &= \frac{x^4y^2+2x^2y^2+y^2+x^4-1}{(x-1)^4}=\frac{(x^2+1)(x^2y^2+x^2+y^2-1)}{(x-1)^4}. \end{aligned}$$

If (x,y) lies on $x^2+y^2+x^2y^2=1$, the numerator vanishes, so this point indeed lies on the elliptic curve E. For the inverse map, suppose $Y^2=4X^3+X$. Solving

$$X=\frac{1+x}{2(1-x)},\qquad Y=\frac{(1+x^2)y}{(1-x)^2}$$

for x and y gives $x=(2X-1)/(2X+1)$ and then $y=Y(1-x)^2/(1+x^2)=2Y/(4X^2+1)$. This inverse map shows that $x^2+y^2+x^2y^2=1$ transforms into $y^2=4x^3+x$. (These curves are not isomorphic in the sense of algebraic geometry. This is because $x^2+y^2+x^2y^2=1$ has singular points—see below—but $y^2=4x^3+x$ doesn't. The map between $x^2+y^2+x^2y^2=1$ and $y^2=4x^3+x$ studied here is an example of a *birational transformation*.)

Since $g_3(E)=0$, we see that $j(E)=1728$, and since $j(i)=1728$, E arises from a lattice homothetic to $[1,i]$, so it has complex multiplication by $\mathbb{Z}[i]$.

Note that $x^2+y^2+x^2y^2=1$ gives the projective curve $x^2z^2+y^2z^2+x^2y^2=z^4$. Setting $z=0$ gives $x^2y^2=0$, which implies that $x=0$ or $y=0$. Thus we have two points at infinity, $(1,0,0)$ and $(0,1,0)$, each of which is a double point because of the squares in the equation $x^2y^2=0$.

Taking these four additional solutions into account, Gauss' conjecture is then equivalent to: If $p=a^2+b^2$ and $a+bi$ is primary, then $|\overline{E}(\mathbb{F}_p)|=p-2a-3+4=p-2a+1$. To compute $|\overline{E}(\mathbb{F}_p)|$ using Theorem 14.16, write $p=(a+bi)(a-bi)$ with $\pi=a+bi\equiv 1 \bmod 2+2i$, and let $L=\mathbb{Q}(i)$, $\mathfrak{P}=\pi\mathbb{Z}[i]$. The discriminant of $\overline{E}$ is $[g_2]^3-27[g_3]^2=[-27]\in\mathbb{F}_p$, which is nonzero since $p\neq 3$. Therefore E has good reduction modulo $\mathfrak{P}$, so by Theorem 14.16, $|\overline{E}(\mathbb{F}_p)|=p+1-(\pi+\overline{\pi})=p+1-2a$. The two answers agree.

14.15. As in the proof of Theorem 14.18, let π be a root of x^2-ax+p, so that $\mathcal{O}_a=\mathbb{Z}[\pi]$ is an order in the imaginary quadratic field $K(\sqrt{a^2-4p})$. By part (b) of Exercise 7.12, applied to the positive definite form $f(x,y)=x^2-axy+py^2$, the discriminant of $\mathcal{O}_a$ equals the discriminant of $f(x,y)$, which is a^2-4p.

Now consider the conductor of $\mathcal{O}_a$. If $a=0$, then $\mathcal{O}_a$ has conductor $f=1$ or $f=2$ depending on whether $p\equiv 3 \bmod 4$ or $p\equiv 1 \bmod 4$. Since $p>3$, we see that $p\nmid f$. Now assume $a\neq 0$. Since $|a|<2\sqrt{p}$, we have $|a|<p$ since $p>3$, so that $p\nmid a$. It follows that $p\nmid a^2-4p$, so that in particular p does not divide the conductor.

14.16. Suppose that the Weierstrass equation for E is $y^2=4x^3-g_2x-g_3$. Since $j\neq 0,1728$, $k=27j/(j-1728)$ is defined and nonzero, and we also know that $g_2g_3\neq 0$. Then

$$k=\frac{27j}{j-1728}=\frac{27\cdot 1728\frac{g_2^3}{g_2^3-27g_3^2}}{1728\frac{g_2^3}{g_2^3-27g_3^2}-1728}=27\frac{g_2^3}{g_2^3-(g_2^3-27g_3^2)}=\frac{g_2^3}{g_3^2}.$$

Now set $c=g_3/g_2$ and note that $c\neq 0$. Then the above equation shows that

$$k=\frac{g_2^3}{g_3^2}=\begin{cases}\dfrac{g_2^2}{g_3^2}g_2=\dfrac{1}{c^2}g_2 \implies g_2=c^2k\\[2ex] \dfrac{g_2^3}{g_3^3}g_3=\dfrac{1}{c^3}g_3 \implies g_3=c^3k.\end{cases}$$

Hence the Weierstrass equation of E becomes $y^2 = 4x^3 - c^2kx - c^3k$. Finally, c is unique because $g_2 = c^2k$ and $g_3 = c^3k$ with $g_2g_3 \neq 0$ imply that $ck \neq 0$, and then $c = (c^3k)/(c^2k) = g_3/g_2$.

14.17. Suppose $N(\pi) = p$ for $\pi \in \mathcal{O}$. By Lemma 7.14, $p = N(\pi) = N(\pi\mathcal{O}) = |\mathcal{O}/\pi\mathcal{O}|$, which implies that $\mathcal{O}/\pi\mathcal{O} \simeq \mathbb{F}_p$. This is a field, so that $\pi\mathcal{O}$ is a prime ideal in $\mathcal{O}$. Since $N(\overline{\pi}) = p$ as well, $\overline{\pi}\mathcal{O}$ is also prime in $\mathcal{O}$. This together with $p = N(\pi) = \pi\overline{\pi}$ shows that

$$p\mathcal{O} = (\pi\mathcal{O})(\overline{\pi}\mathcal{O})$$

is a factorization into prime $\mathcal{O}$-ideals. Now suppose $\alpha \in \mathcal{O}$ satisfies $N(\alpha) = p$. Arguing as above, $p\mathcal{O} = (\alpha\mathcal{O})(\overline{\alpha}\mathcal{O})$ is also a prime factorization of $p\mathcal{O}$. By assumption, p is prime to the conductor of $\mathcal{O}$, so the ideals appearing above are all prime to the conductor. Thus these factorizations are the same up to order by Exercise 7.26, so that $\alpha\mathcal{O} = \pi\mathcal{O}$ or $\overline{\pi}\mathcal{O}$. This gives the desired result that $\alpha = \epsilon\pi$ or $\epsilon\overline{\pi}$ for some $\epsilon \in \mathcal{O}^*$.

14.18. (a) Consider the curves $y^2 = 4x^3 - [k][a]^2x - [k][a]^3$ and $y^2 = 4x^3 - [k][b]^2x - [k][b]^3$ in the collection $\overline{E}_c$ and note that $[a], [b], [k] \in \mathbb{F}_p^*$. By definition, they are isomorphic over $\mathbb{F}_p$ if and only if there is $\lambda \in \mathbb{F}_p^*$ such that

$$k[a]^2 = \lambda^4[k][b]^2, \quad [k][a]^3 = \lambda^6[k][b]^3,$$

which is equivalent to $[a] = \lambda^2[b]$. It follows that $\overline{E}_a$ and $\overline{E}_b$ are isomorphic if and only if $[a]$ and $[b]$ differ by a square in $\mathbb{F}_p^*$, i.e., they are in the same coset of $(\mathbb{F}_p^*)^2$ in $\mathbb{F}_p^*$. Since p is odd, $\mathbb{F}_p^*$ is cyclic of even order, so that $(\mathbb{F}_p^*)^2$ has index 2 in $\mathbb{F}_p^*$. Thus there are two equally sized isomorphism classes of curves in the collection $\overline{E}_c$, each containing $(p-1)/2$ elements.

(b) Two curves $y^2 = 4x^3 - [r]x$ and $y^2 = 4x^3 - [s]x$ in the collection $\overline{E}_c$ are isomorphic over $\mathbb{F}_p$ if and only if there is $\lambda \in \mathbb{F}_p^*$ such that $[r] = \lambda^4[s]$, i.e., if and only if $[r]$ and $[s]$ are in the same coset of $(\mathbb{F}_p^*)^4$ in $\mathbb{F}_p$. Since $p \equiv 1 \bmod 4$, $\mathbb{F}_p^*$ is cyclic of order divisible by 4, so that $(\mathbb{F}_p^*)^4$ has index 4 in $\mathbb{F}_p^*$. Thus there are four equally sized isomorphism classes of curves in the collection $\overline{E}_c$, each containing $(p-1)/4 = (p-1)/|\mathcal{O}'^*|$ elements.

(c) Two curves $y^2 = 4x^3 - [r]$ and $y^2 = 4x^3 - [s]$ in the collection $\overline{E}_c$ are isomorphic over $\mathbb{F}_p$ if and only if there is $\lambda \in \mathbb{F}_p^*$ such that $[r] = \lambda^6[s]$, i.e., if and only if $[r]$ and $[s]$ are in the same coset of $(\mathbb{F}_p^*)^6$ in $\mathbb{F}_p$. Since $p \equiv 1 \bmod 3$ is odd, we have $p \equiv 1 \bmod 6$, so that $\mathbb{F}_p^*$ is cyclic of order divisible by 6, so that $(\mathbb{F}_p^*)^6$ has index 6 in $\mathbb{F}_p^*$. Thus there are six equally sized isomorphism classes of curves in the collection $\overline{E}_c$, each containing $(p-1)/6 = (p-1)/|\mathcal{O}'^*|$ elements.

14.19. In this exercise, $\mathbb{F}_q$ is the field with $q = p^a$ elements, where $p > 3$.

(a) By Exercise 14.16, an elliptic curve E over $\mathbb{F}_q$ with $j \neq 0, 1728$ is defined by $y^2 = 4x^3 - c^2kx - c^3k$ where $k = \frac{27j}{j-1728}$ and $c \in \mathbb{F}_q^*$ is uniquely determined. Then $E \mapsto (j, c)$ defines a map

$$\{\text{elliptic curves over } \mathbb{F}_q \text{ with } j \neq 0, 1728\} \longrightarrow (\mathbb{F}_q - \{0, 1728\}) \times \mathbb{F}_q^*.$$

which is clearly injective since j and c determine the equation of E. It is also surjective since given $(j, c) \in (\mathbb{F}_q - \{0, 1728\}) \times \mathbb{F}_q^*$, setting $k = \frac{27j}{j-1728}$ gives the equation $y^2 = 4x^3 - c^2kx - c^3k$ which is easily seen to define an elliptic curve with $j(E) = j$. Thus the above map is a bijection. Since $1728 \neq 0$ in $\mathbb{F}_q$, it follows that there are $(q-2)(q-1)$ elliptic curves over $\mathbb{F}_q$ with $j \neq 0, 1728$,

Elliptic curves E with $j(E) = 0$ are defined by $y^2 = 4x^3 - g_3$, where $g_3 \neq 0$ in $\mathbb{F}_q^*$. There are $q-1$ choices for g_3, giving $q-1$ such curves. Similarly, elliptic curves E with $j(E) = 1728$ are defined by $y^2 = 4x^3 - g_2x$, where $g_2 \neq 0$ in $\mathbb{F}_q^*$. There are $q-1$ choices for g_2, again giving $q-1$ such curves. So there are altogether $(q-2)(q-1)+(q-1)+(q-1)=q(q-1)$ elliptic curves over $\mathbb{F}_q$.

(b) We want to prove:

THEOREM. *Let G be a finite Abelian group and $a,b \in \mathbb{Z}$ relatively prime integers. Then $u^a = v^b$ has exactly $|G|$ solutions in $G \times G$.*

PROOF. Since $\gcd(a,b)=1$, we can find $m,n \in \mathbb{Z}$ with $ma+nb=1$. Consider the map $\varphi : G \to G \times G$ defined by $\varphi(g)=(g^b,g^a)$. We claim that φ is injective, with image equal to $\{(u,v)\in G\times G : u^a=v^b\}$. To see that $\varphi(g)$ is injective, suppose $\varphi(g)=(e,e)$. Then $e=g^a=g^b$, so that $g=g^{ma+nb}=(g^a)^m(g^b)^n=e$, proving injectivity. For the second part of the claim, if (u,v) is a solution to $u^a=v^b$, let $g=u^nv^m$; then using $u^a=v^b$, we obtain

$$\begin{aligned}\varphi(g) &= ((u^nv^m)^b,(u^nv^m)^a) = (u^{nb}(v^b)^m,(u^a)^nv^{ma}) = (u^{nb}(u^a)^m,(v^b)^nv^{ma})\\ &= (u^{nb+ma},v^{nb+ma}) = (u,v).\end{aligned}$$

This shows that any solution of $u^a=v^b$ lies in the image of the injective map φ, which proves the theorem. Q.E.D.

An elliptic curve over $\mathbb{F}_q$ is defined by $y^2=4x^3-g_2x-g_3$ with $g_2^3-27g_3^2\neq 0$. Thus

$$\begin{aligned}\big|\{\text{elliptic curves over } \mathbb{F}_q\}\big| &= \big|\{(g_2,g_3)\in\mathbb{F}_q^2 : g_2^3-27g_3^2\neq 0\}\big|\\ &= \big|\mathbb{F}_q^2-\{(g_2,g_3)\in\mathbb{F}_q^2 : g_2^3-27g_3^2=0\}\big|\\ &= q^2-\big|\{(g_2,g_3)\in\mathbb{F}_q^2 : g_2^3-27g_3^2=0\}\big|.\end{aligned}$$

Furthermore, since 3 is invertible in $\mathbb{F}_q$, the only solution of $g_2^3-27g_3^2=0$ with $g_2=0$ or $g_3=0$ is the trivial solution $(0,0)$. Thus

$$\big|\{(g_2,g_3)\in\mathbb{F}_q^2 : g_2^3-27g_3^2=0\}\big| = 1+\big|\{(g_2,g_3)\in(\mathbb{F}_q^*)^2 : g_2^3-27g_3^2=0\}\big|$$

Then setting $a=3^{-1}g_2$ and $b=g_3$ gives

$$a^3=b^2 \iff (3^{-1}g_2)^3=g_2^2 \iff g_2^3=27g_3^2 \iff g_2^3-27g_3^2=0,$$

so that $|\{(g_2,g_3)\in(\mathbb{F}_q^*)^2 : g_2^3-27g_3^2=0\}| = |\{(a,b)\in(\mathbb{F}_q^*)^2 : a^3=b^2\}|$, which by the theorem equals $|\mathbb{F}_q^*|=q-1$. Putting this all together gives

$$\big|\{\text{elliptic curves over } \mathbb{F}_q\}\big| = q^2-(1+(q-1)) = q^2-q = q(q-1).$$

14.20. Assume $\mathcal{O}'$ is contained in the imaginary quadratic field K. Let $\mathcal{S}=\{\alpha\in\mathcal{O}' : N(\alpha)=m\}$, and let $\mathcal{T}$ be the set of integers a with $0\le|a|\le 2\sqrt{m}$ such that $\mathcal{O}'$ contains a root of x^2-ax+m. The equality we want to prove is equivalent to the statement that $|\mathcal{S}|=2\,|\mathcal{T}|$. We will define a two-to-one map $\varphi:\mathcal{S}\to\mathcal{T}$ and show that it is surjective.

Suppose $\alpha\in\mathcal{S}$. We claim that $\alpha\neq\overline{\alpha}$. To see this, suppose that $\alpha=\overline{\alpha}$, then $\alpha\in\mathcal{O}'\cap\mathbb{R}\subset K\cap\mathbb{R}=\mathbb{Q}$ since K is imaginary quadratic. Since α is an algebraic integer, it follows that $\alpha\in\mathbb{Z}$ and therefore $m=N(\alpha)=\alpha\overline{\alpha}=\alpha^2$, contradicting the assumption that m is not a perfect square, and proving the claim. Let $a=\alpha+\overline{\alpha}$, so that α and $\overline{\alpha}$ are the two roots of x^2-ax+m. The discriminant of this quadratic is $a^2-4m<0$ since α is a nonreal root. It follows that $0\le|a|\le 2\sqrt{m}$. Since $N(\overline{\alpha})=m$, we see that $\overline{\alpha}\in\mathcal{S}$ as well. Let $\varphi(\alpha)=\varphi(\overline{\alpha})=a$. The function $\varphi(\alpha)$ thus defined is clearly two-to-one. Finally, φ is surjective, since if $a\in\mathcal{T}$, then by the definition of $\mathcal{T}$, $\mathcal{O}'$ contains a root α of x^2-ax+m, and by the construction above $\varphi(\alpha)=a$.

Since φ is two-to-one and surjective to $\mathcal{T}$, it follows that $|\mathcal{S}|=2\,|\mathcal{T}|$ as required.

14.21. Let N_l be the number of elliptic curves over $\mathbb{F}_l$ satisfying $|E(\mathbb{F}_l)|\in S$, where S is defined as in Theorem 14.26. Assuming Conjecture 14.27 holds, we also have $|S|\ge$

$c_2\sqrt{l}/\log l$, so that by Theorem 14.26,

$$\begin{aligned} N_l &\geq c_1 \frac{(|S|-2)\sqrt{l}(l-1)}{\log l} = c_1\Big(c_2 \frac{|S|\,\sqrt{l}(l-1)}{\log l} - 2\frac{\sqrt{l}(l-1)}{\log l}\Big) \\ &\geq c_1\Big(c_2\frac{l(l-1)}{(\log l)^2} - 2\frac{\sqrt{l}(l-1)}{\log l}\Big) = c_1\frac{l-1}{(\log l)^2}(c_2 l - 2\sqrt{l}\log l) \\ &= c_1\frac{l(l-1)}{(\log l)^2}\Big(c_2 - \frac{2\log l}{\sqrt{l}}\Big). \end{aligned}$$

Since $2\log l/\sqrt{l} \to 0$ as $l \to \infty$, there is l_0 such that $l \geq l_0$ implies $2\log l/\sqrt{l} \leq c_2/2$. Then $l \geq l_0$ implies

$$c_1\frac{l(l-1)}{(\log l)^2}\Big(c_2 - \frac{2\log l}{\sqrt{l}}\Big) \geq c_1\frac{l(l-1)}{(\log l)^2}\Big(c_2 - \frac{c_2}{2}\Big) = \frac{c_1 c_2}{2}\frac{l(l-1)}{(\log l)^2}.$$

These inequalities imply that $N_l \geq c_3 l(l-1)/(\log l)^2$ when $l \geq l_0$ and $c_3 = c_1c_2/2$.

14.22. (a) Suppose $(a,0) \in E(K)$. By the discussion preceding (14.7), the inverse of $(a,0)$ is $(a,-0) = (a,0)$, so that $2(a,0) = \infty$ and therefore $(a,0)$ has order two.

Conversely, suppose $(a,b) \in E(K)$ with $2(a,b) = \infty$. Then if $b \neq 0$, (14.7) gives $2(a,b) = (x,y)$ with

$$\begin{aligned} x &= -2a + \frac{1}{16}\Big(\frac{12a^2 - g_2}{b}\Big)^2 \\ y &= -b - (x-a)\frac{12a^2 - g_2}{2b}. \end{aligned}$$

If $b \neq 0$, this is not the point at infinity, since the denominators are defined. Therefore $b = 0$ and the point is of the form $(a,0)$.

We can see this over $\mathbb{C}$ as follows: if L is the lattice corresponding to E, then the points at infinity on $E(\mathbb{C})$ correspond to lattice points, and points of order 2 in $E(\mathbb{C})$ thus correspond to the half-lattice points $z = \omega_1/2$, $\omega_2/2$ and $(\omega_1+\omega_2)/2$. Further, for these points $\wp'(z) = 0$ by Corollary 10.5, so they map to the three points of the form $(a,0)$ in $E(\mathbb{C})$.

(b) Since $g_3 = 0$, E is $y^2 = 4x^3 - g_2 x$, so that (A,B) satisfies $B^2 = 4A^3 - g_2A$. By (14.7),

$$\begin{aligned} x - \Big(\frac{4A^2+g_2}{4B}\Big)^2 &= -2A + \frac{1}{16}\Big(\frac{12A^2-g_2}{B}\Big)^2 - \frac{1}{16}\Big(\frac{4A^2+g_2}{B}\Big)^2 \\ &= -2A + \frac{1}{16B^2}\big((12A^2-g_2)+(4A^2+g_2)\big)\big((12A^2-g_2)-(4A^2+g_2)\big) \\ &= -2A + \frac{1}{16B^2}16A^2(8A^2-2g_2) = -2A + \frac{2A^2}{B^2}(4A^2-g_2) \\ &= \frac{2A}{B^2}(-B^2+4A^3-g_2A) = 0. \end{aligned}$$

14.23. (a) Suppose $\overline{E}$ is ordinary. Then $\mathrm{End}_{\overline{\mathbb{F}}_q}(\overline{E})$ is an order in an imaginary quadratic field, and since reduction modulo q induces a map from $\mathbb{Z}[i] \simeq \mathrm{End}_{\mathbb{C}}(E) \to \mathrm{End}_{\overline{\mathbb{F}}_q}(\overline{E})$, the order must be contained in $\mathbb{Z}[i]$. It follows that $Frob_q \in \mathbb{Z}[i]$. The degree of $Frob_q$ is $q \equiv 7 \bmod 24$, and this degree is the norm of the corresponding element in $\mathbb{Z}[i]$. However, the norm of $a+bi$ is a^2+b^2, and the squares modulo 24 are 0, 1, 4, 9, 12, and 16; no pair of those adds to 7 modulo 24. Thus $\overline{E}$ must be supersingular.

(b) Part (a) and Proposition 14.15 imply that $\big|\overline{E}(\mathbb{F}_q)\big| = q+1 = 2^k$. The elements of $\overline{E}(\mathbb{F}_q)$ of order two are the roots of $4x^3 - 48x$ in $\mathbb{F}_q$. Those roots are $x = 0$ and possibly the two square roots of 12. Since $q \equiv 3 \bmod 4$ and $q \equiv 1 \bmod 3$, we have by quadratic reciprocity

$$\left(\frac{12}{q}\right) = \left(\frac{3}{q}\right) = -\left(\frac{q}{3}\right) = -\left(\frac{1}{3}\right) = -1,$$

so that 12 is not a square modulo q. Thus $(0,0)$ is the only point of order two. Since $\overline{E}(\mathbb{F}_q)$ is a product of groups of order a power of two, this implies that the group must be cyclic of order 2^k.

(c) If $(-2,8)$ is not a generator, then from part (a) it must be the double of some element, say $(-2,8) = 2(A,B)$. Then part (b) of Exercise 14.22 shows that -2 is a square modulo q, but, as noted in the statement of the exercise, $q \equiv 7 \bmod 8$. By quadratic reciprocity,

$$\left(\frac{-2}{q}\right) = \left(\frac{-1}{q}\right)\left(\frac{2}{q}\right) = (-1)^{(q-1)/4} \cdot (-1)^{(q^2-1)/8} = -1 \cdot 1 = -1,$$

which is a contradiction. Thus $(-2,8)$ generates $\overline{E}(\mathbb{F}_q)$.

14.24. (a) If q is prime, then $\widetilde{E}(\mathbb{Z}/q\mathbb{Z}) = \widetilde{E}(\mathbb{F}_q)$ is cyclic by the previous exercise, with $|\widetilde{E}(\mathbb{F}_q)| = 2^k = q+1$ and with $(-2,8)$ as a generator. It follows that $2^{k-1}(-2,8)$ must be the unique element $(0,0)$ of order two in the group, so that if $P_{k-1} = 2^{k-1}(-2,8)$ is defined but is not equal to $(0,0)$, then q cannot be prime. So q cannot be prime in Scenario B. Similarly, in Scenario C, some P_ℓ for $\ell \le k-1$ is undefined, so q cannot be prime as noted in the discussion following (14.22).

(b) (i) Scenario A implies that for $0 \le \ell \le k-1$, P_ℓ is defined for $\widetilde{E}$ over $\mathbb{Z}/q\mathbb{Z}$, which means that the denominators that appear in going from $P_{\ell-1}$ to P_ℓ are represented by integers prime to q. When we reduce from $\widetilde{E}$ to $\overline{E}$ over $\mathbb{Z}/p\mathbb{Z} = \mathbb{F}_p$, going from $P_{\ell-1}$ to P_ℓ uses the same denominators, but now reduced modulo p, and they are relatively prime to p since $p \mid q$. Furthermore, $P_{k-1} = (0,0)$ over $\mathbb{Z}/q\mathbb{Z}$, the same holds over $\mathbb{Z}/p\mathbb{Z} = \mathbb{F}_p$ since $p \mid q$. Thus $2^{k-1}P_1 = P_{k-1}$ is a point of order 2 in $\overline{E}(\mathbb{F}_p)$, which proves that P_1 has order $2^k = q+1$ in $\overline{E}(\mathbb{F}_p)$.

(ii) Assume $p < q$. Since q is odd, $q \ge 3p$, so that $q+1 \ge 3p+1$, which implies that $q+1 > p+1+2\sqrt{p}$ since $p > 1$. Thus $|\widetilde{E}(\mathbb{F}_p)| = q+1 > p+1+2\sqrt{p}$, contradicting the Hasse bound. Therefore $p = q$ and q is prime.

Solutions to Exercises in §15

15.1. If $f, g \in \mathsf{F}_m$, then clearly $f+g$ and fg are meromorphic and $\Gamma(m)$-invariant since f and g are. Further, at each cusp, the q-expansions of f and g have coefficients in $\mathbb{Q}(\zeta_m)$ and finitely many terms with negative exponents. Hence the same is true for the q-expansions of $f+g$ and fg. It follows that $f+g$ and fg both lie in F_m.

Next consider $1/f$ for $0 \ne f \in \mathsf{F}_m$. Clearly $1/f$ is $\Gamma(m)$-invariant since f is. Suppose the q-expansion of f at a cusp (including the cusp corresponding to $\gamma = I$) is

$$\sum_{n=-M}^{\infty} a_n q^n, \ M \in \mathbb{Z}, \ a_M \ne 0, \ a_n \in \mathbb{Q}(\zeta_m) \text{ for all } n.$$

(Note that since $f \in \mathsf{F}_m$, it has only a finite number of terms with negative exponents in its q-expansion). Since $a_{-M} \neq 0$, we can write this q-expansion as

$$\begin{aligned} a_{-M}q^{-M} + a_{-M}q^{-M} \sum_{n=-M+1}^{\infty} \frac{a_n}{a_{-M}} q^{n+M} &= a_{-M}q^{-M} + a_{-M}q^{-M} \sum_{n=1}^{\infty} \frac{a_{n-M-1}}{a_{-M}} q^n \\ &= a_{-M}q^{-M}\left(1 + q\sum_{n=0}^{\infty} \frac{a_{n-M}}{a_{-M}} q^n\right) \\ &= a_{-M}q^{-M}(1 + qh(q)). \end{aligned}$$

Observe that $h(q)$ is holomorphic at $q = 0$. Further, since $a_n \in \mathbb{Q}(\zeta_m)$ for all n, it follows that the coefficients of $h(q)$ lie in $\mathbb{Q}(\zeta_m)$ as well. Since $qh(q)$ maps the origin to itself, $|qh(q)| < 1$ for $|q|$ sufficiently small. Using $(1+z)^{-1} = \sum_{n=0}^{\infty}(-1)^n z^n$ for $|z| < 1$, we see that the q-expansion for $1/f$ near $q = 0$ is

$$\frac{1}{a_{-M}q^{-M}(1 + qh(q))} = a_{-M}^{-1}q^M\left(1 + \sum_{n=1}^{\infty}(-1)^n q^n h(q)^n\right).$$

This is meromorphic, and it has coefficients in $\mathbb{Q}(\zeta_m)$ since $h(q)$ does.

For the second part, since $j \in \mathsf{F}_1$, clearly $\mathbb{Q}(j) \subset \mathsf{F}_1$. If $f \in \mathsf{F}_1$, then by Theorem 11.9, f is a rational function in $j(\tau)$ (with coefficients in $\mathbb{C}$). Since the q-expansion of $f \in \mathsf{F}_1$ has coefficients in $\mathbb{Q}(\zeta_1) = \mathbb{Q}$, Exercise 11.12 (the Hasse expansion principle) shows that in fact f is a rational function in $j(\tau)$ with rational coefficients, so that $f \in \mathbb{Q}(j)$.

15.2. (a) Note that $\gamma\tau = (-\gamma)\tau$, and that the right-hand side of (15.3) is invariant under $\gamma \mapsto -\gamma$. Thus if we prove (15.3) for elements of $\Gamma_0(2)^t$ with $c \geq 0$, and with $d > 0$ if $c = 0$, we are done.

So choose $\gamma = \left(\begin{smallmatrix} a & b \\ c & d \end{smallmatrix}\right) \in \Gamma_0(2)^t$ with $c \geq 0$. Using the hint, write $\mathfrak{f}_1(\gamma\tau) = \dfrac{\eta((\gamma\tau)/2)}{\eta(\gamma\tau)}$. Then

$$\frac{1}{2}\gamma\tau = \frac{a\tau + b}{2(c\tau + d)} = \frac{2a(\tau/2) + b}{4c(\tau/2) + 2d} = \frac{a(\tau/2) + b/2}{2c(\tau/2) + d} = \begin{pmatrix} a & b/2 \\ 2c & d \end{pmatrix}\frac{\tau}{2} = \gamma'\frac{\tau}{2}$$

for some $\gamma' \in \mathrm{SL}(2, \mathbb{Z})$ with $2c \geq 0$. The transformation law then gives

$$\mathfrak{f}_1(\gamma\tau)^2 = \frac{\eta((\gamma\tau)/2)^2}{\eta(\gamma\tau)^2} = \frac{\eta(\gamma'(\tau/2))^2}{\eta(\gamma\tau)^2} = \frac{\epsilon(\gamma')^2(c\tau + d)\,\eta(\tau/2)^2}{\epsilon(\gamma)^2(c\tau + d)\,\eta(\tau)^2} = \frac{\epsilon(\gamma')^2\,\eta(\tau/2)^2}{\epsilon(\gamma)^2\,\eta(\tau)^2}.$$

If $c = 2^\lambda c_1$ with c_1 odd, then $2c = 2^{\lambda+1}c_1$, so that

$$\begin{aligned} \mathfrak{f}_1(\gamma\tau)^2 &= \frac{\epsilon(\gamma')^2\,\eta(\tau/2)^2}{\epsilon(\gamma)^2\,\eta(\tau)^2} \\ &= \frac{(a/c_1)\zeta_{12}^{\frac{1}{2}ba + 2c(d(1-a^2)-a) + 3(a-1)c_1 + \frac{3}{2}(\lambda+1)(a^2-1)}}{(a/c_1)\zeta_{12}^{ba + c(d(1-a^2)-a) + 3(a-1)c_1 + \frac{3}{2}\lambda(a^2-1)}} \cdot \frac{\eta(\tau/2)^2}{\eta(\tau)^2} \\ (*) \qquad &= \zeta_{12}^{-\frac{1}{2}ba + c(d(1-a^2)-a) + \frac{3}{2}(a^2-1)} \cdot \frac{\eta(\tau/2)^2}{\eta(\tau)^2}. \end{aligned}$$

Since $\gamma \in \Gamma_0(2)^t$, we know that b is even. It follows that a is odd, so that $a^2 - 1 \equiv 0 \bmod 8$ and thus $\frac{3}{2}(a^2 - 1) \equiv 0 \bmod 12$. Thus (*) reduces to

$$\mathfrak{f}_1(\gamma\tau)^2 = \zeta_{12}^{-\frac{1}{2}ba + c(d(1-a^2)-a)}\mathfrak{f}_1(\tau)^2.$$

(b) By part (a), if $\gamma = \left(\begin{smallmatrix} a & b \\ c & d \end{smallmatrix}\right) \in \Gamma_0(2)^t$ then

$$\mathfrak{f}_1(\gamma\tau)^2 = \zeta_{12}^{-\frac{1}{2}ab + c(d(1-a^2)-a)}\mathfrak{f}_1(\tau)^2,$$

so to show $\Gamma(24)$-invariance it suffices to show that the exponent is a multiple of 12 when $\gamma \in \Gamma(24)$. But $\gamma \in \Gamma(24)$ means that $a \equiv d \equiv 1 \bmod 24$ and $b \equiv c \equiv 0 \bmod 24$, so that

$$-\frac{1}{2}ab \equiv 0 \bmod 12$$
$$c(d(1-a^2)-a) \equiv 0 \bmod 24.$$

(c) Note first that by (12.15), $\mathfrak{f}_1(\tau)$ is meromorphic on $\mathfrak{h}$, so that $\mathfrak{f}_1(\tau)^2$ is as well. Part (b) showed that $\mathfrak{f}_1(\tau)^2$ is $\Gamma(24)$-invariant. To check the cusps, use the transformations in Corollary 12.19 and write $\gamma \in \mathrm{SL}(2,\mathbb{Z})$ as a product of the matrices $\left(\begin{smallmatrix}1&1\\0&1\end{smallmatrix}\right)$ and $\left(\begin{smallmatrix}0&-1\\1&0\end{smallmatrix}\right)$. We conclude that

$$\mathfrak{f}_1(\gamma\tau)^2 = \epsilon\mathfrak{f}(\tau)^2,\ \epsilon\mathfrak{f}_1(\tau)^2,\ \text{or } \epsilon\mathfrak{f}_2(\tau)^2$$

for some root of unity ϵ. This shows that $\mathfrak{f}_1(\tau)^2$ is meromorphic at the cusps since in all cases the right-hand side has a q-expansion with only finitely many nonzero coefficients of terms with negative exponents. Finally, since all coefficients in Corollary 12.19 are powers of ζ_{48}, it follows that when squared, all these coefficients are powers of ζ_{24}, and therefore $\epsilon \in \mathbb{Q}(\zeta_{24})$. Since the coefficients of the q-expansions of $\mathfrak{f}(\tau)^2$, $\mathfrak{f}_1(\tau)^2$, and $\mathfrak{f}_2(\tau)^2$ also lie in $\mathbb{Q}(\zeta_{24})$ by (12.15), it follows that the coefficients of the q-expansion of $\mathfrak{f}_1(\gamma\tau)^2$ lie in $\mathbb{Q}(\zeta_{24})$. It follows that $\mathfrak{f}_1(\tau)^2 \in \mathsf{F}_{24}$.

(d) From Exercise 12.23, $\mathfrak{f}_1(\tau)^6$ is $\tilde{\Gamma}(8)$-invariant. Since $\Gamma(8) \subset \tilde{\Gamma}(8)$, it is also $\Gamma(8)$-invariant. It is meromorphic since $\mathfrak{f}_1(\tau)$ is, and (12.26) shows that for any $\gamma \in \mathrm{SL}(2,\mathbb{Z})$, $\mathfrak{f}_1(\gamma\tau)^6 = i^n\mathfrak{f}_1(\tau)^6 = \zeta_8^{2n}\mathfrak{f}_1(\tau)^6$ where n is an integer, so that it is meromorphic at the cusps as well. Finally, since the coefficients of the q-expansion of $\mathfrak{f}_1(\tau)^6$ lie in $\mathbb{Q}(\zeta_8)$ by (12.15), the coefficients of the q-expansion of $\mathfrak{f}_1(\gamma\tau)^6 = \zeta_8^{2n}\mathfrak{f}_1(\tau)^6$ do as well, so that $\mathfrak{f}_1(\tau)^6 \in \mathsf{F}_8$.

15.3. (a) Write $m = \prod_p p^{v_p(m)}$ where p ranges over all primes (so that all but a finite number of the $v_p(m)$ are equal to 0). Under the inclusion $\mathbb{Z} \hookrightarrow \widehat{\mathbb{Z}}$, we see that $m\mathbb{Z} \hookrightarrow m\widehat{\mathbb{Z}}$, so that the composite map $\mathbb{Z} \to \widehat{\mathbb{Z}} \to \widehat{\mathbb{Z}}/m\widehat{\mathbb{Z}}$ induces a map $\mathbb{Z}/m\mathbb{Z} \to \widehat{\mathbb{Z}}/m\widehat{\mathbb{Z}}$. The Chinese Remainder Theorem gives

$$\mathbb{Z}/m\mathbb{Z} \simeq \prod_{p\mid m} \mathbb{Z}/p^{v_p(m)}\mathbb{Z}.$$

We also have

$$\widehat{\mathbb{Z}}/m\widehat{\mathbb{Z}} = \prod_p \mathbb{Z}_p \Big/ m\prod_p \mathbb{Z}_p \simeq \prod_p \mathbb{Z}_p/m\mathbb{Z}_p \simeq \prod_{p\mid m} \mathbb{Z}_p/p^{v_p(m)}\mathbb{Z}_p,$$

where the last isomorphism comes from the fact that $m\mathbb{Z}_p = \mathbb{Z}_p$ if $p \nmid m$. Since the map $\mathbb{Z}/m\mathbb{Z} \to \widehat{\mathbb{Z}}/m\widehat{\mathbb{Z}}$ is induced by the inclusion $\mathbb{Z} \hookrightarrow \mathbb{Z}_p$ on each component of $\widehat{\mathbb{Z}}$, it follows that it is given by the component-wise maps

$$\mathbb{Z}/p^{v_p(m)}\mathbb{Z} \longrightarrow \mathbb{Z}_p/p^{v_p(m)}\mathbb{Z}_p,$$

which are all isomorphisms by property (ii) of $\mathbb{Z}_p$. Thus $\mathbb{Z}/m\mathbb{Z} \to \widehat{\mathbb{Z}}/m\widehat{\mathbb{Z}}$ is an isomorphism.

(b) If $m \mid n$, then we have a diagram

$$\begin{array}{ccccc} \widehat{\mathbb{Z}} & \longrightarrow & \widehat{\mathbb{Z}}/n\widehat{\mathbb{Z}} & \xrightarrow{\sim} & \mathbb{Z}/n\mathbb{Z} \\ & \searrow & \downarrow & & \downarrow \\ & & \widehat{\mathbb{Z}}/m\widehat{\mathbb{Z}} & \xrightarrow{\sim} & \mathbb{Z}/m\mathbb{Z} \end{array}$$

The triangle on the left commutes since $n\widehat{\mathbb{Z}} \subset m\widehat{\mathbb{Z}}$. For the square on the right, the isomorphism $\widehat{\mathbb{Z}}/m\widehat{\mathbb{Z}} \simeq \mathbb{Z}/m\mathbb{Z}$ is given by

$$\widehat{\mathbb{Z}}/m\widehat{\mathbb{Z}} \simeq \prod_{p \mid m} \mathbb{Z}_p/p^{v_p(m)}\mathbb{Z}_p \simeq \prod_{p \mid m} \mathbb{Z}/p^{v_p(m)}\mathbb{Z} \simeq \mathbb{Z}/m\mathbb{Z}$$

by part (a). For $m \mid n$, the natural map $\widehat{\mathbb{Z}}/n\widehat{\mathbb{Z}} \to \widehat{\mathbb{Z}}/m\widehat{\mathbb{Z}}$ is therefore given by the natural projections

$$(*) \qquad \mathbb{Z}_p/p^{v_p(n)}\mathbb{Z}_p \longrightarrow \mathbb{Z}_p/p^{v_p(m)}\mathbb{Z}_p.$$

Since the isomorphism from $\mathbb{Z}/p^\ell\mathbb{Z}$ to $\mathbb{Z}_p/p^\ell\mathbb{Z}_p$ is induced by the inclusion $\mathbb{Z} \hookrightarrow \mathbb{Z}_p$, we see that the maps (*) are consistent with the natural projection map $\mathbb{Z}/n\mathbb{Z} \to \mathbb{Z}/m\mathbb{Z}$ given by

$$\mathbb{Z}/p^{v_p(n)}\mathbb{Z} \longrightarrow \mathbb{Z}/p^{v_p(m)}\mathbb{Z}.$$

It follows that the square on the right commutes, so the whole diagram is commutative.

15.4. (a) Choose $\gamma = \left(\begin{smallmatrix} a & b \\ c & d \end{smallmatrix}\right) \in \mathrm{GL}(2, \widehat{\mathbb{Z}})$ with $\gamma \equiv \pm I \bmod m$ for all m. We need to prove that $\gamma = \pm I$. For a fixed $m_0 > 2$, suppose that $\gamma \equiv I \bmod m_0$. We claim that $\gamma \equiv I \bmod m$ for all $m > 0$. To see why, take m and note that $\gamma \equiv \pm I \bmod mm_0$. If $-I$, then γ is congruent to both I and $-I$ modulo m_0, which is impossible since $m_0 > 2$. Thus $\gamma \equiv I \bmod mm_0$, which implies $\gamma \equiv I \bmod m$ as well. Similarly, $\gamma \equiv -I \bmod m_0$ implies that $\gamma \equiv -I \bmod m$ for all $m > 0$.

Now write $\gamma = (\gamma_p) \in \mathrm{GL}(2, \widehat{\mathbb{Z}})$, and assume for definiteness that $\gamma \equiv I \bmod m$ for all $m > 0$ (a similar proof works if $\gamma \equiv -I \bmod m$ for all m). For a given p, the map $\mathrm{GL}(2, \widehat{\mathbb{Z}}) \to \mathrm{GL}(2, \mathbb{Z}/p^\ell\mathbb{Z})$ is induced by

$$\widehat{\mathbb{Z}} \longrightarrow \widehat{\mathbb{Z}}/p^\ell\widehat{\mathbb{Z}} = \mathbb{Z}_p/p^\ell\mathbb{Z}_p \simeq \mathbb{Z}/p^\ell\mathbb{Z}.$$

Therefore the entries of γ_p satisfy $a \equiv d \equiv 1 \bmod p^\ell$ and $b \equiv c \equiv 0 \bmod p^\ell$ for all ℓ, and also clearly satisfy the compatibility condition of (iii). Since $(1, 1, 1, \ldots)$ and $(0, 0, 0, \ldots)$ satisfy those congruences as well, we must have $\gamma_p = I$ by uniqueness. Thus $\gamma = I$ and the map is injective.

(b) To simplify notation, define $G(m) = \mathrm{GL}(2, \mathbb{Z}/m\mathbb{Z})$ and $H(m) = G(m)/\{\pm I\} = \mathrm{GL}(2, \mathbb{Z}/m\mathbb{Z})/\{\pm I\}$. In order to construct a compatible lifting of the h_ℓ to g_ℓ, first fix some integer $m_0 > 2$; we will construct a compatible lifting g_{mm_0} for all multiples of m_0. Arbitrarily fix $g_{m_0} \in G(m_0)$ with $h_{m_0} = [g_{m_0}]$. For any $m > 0$, let $u \in G(mm_0)$ be any lift of h_{mm_0}. Since the h_ℓ are compatible, we see that u maps to $\pm g_{m_0} \in G(m_0)$. If $u \mapsto g_{m_0}$, then $-u \mapsto -g_{m_0}$, and vice versa. Since $m_0 > 2$, it follows that $g_{m_0} \neq -g_{m_0}$, so we can uniquely define g_{mm_0} to be the one which maps to g_{m_0}.

This gives a lifting of h_{mm_0} for all $m > 0$. We now claim that this lifting is compatible under the projections $G(nm_0) \to G(mm_0)$ for $m \mid n$. To see why, consider the commutative diagram

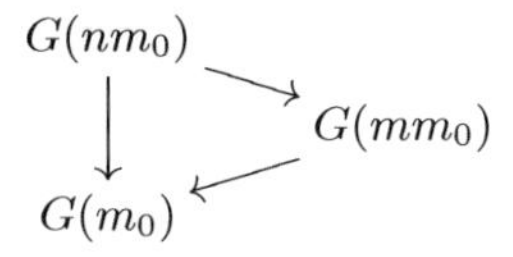

Let $u \in G(mm_0)$ be the image of g_{nm_0}. Then u lifts h_{mm_0} since g_{nm_0} lifts h_{nm_0} and the h_ℓ are compatible. Since the diagram commutes and g_{nm_0} maps to g_{m_0}, it follows that u maps to g_{m_0} as well. Since the lifting to $G(mm_0)$ constructed above was unique, we see that $u = g_{mm_0}$, proving the claim.

Now given an arbitrary integer $m > 0$, we define g_m to be the image of g_{mm_0} under the projection $G(mm_0) \to G(m)$. We will show that under this definition, g_m lifts h_m for all m, and that the resulting g_m are compatible under the projection maps.

To see that g_m lifts h_m for all $m > 0$, note that $g_{mm_0} \mapsto g_m$ implies that

$$h_{mm_0} = [g_{mm_0}] \longmapsto [g_m].$$

But the h_ℓ are compatible, so that $h_{mm_0} \mapsto h_m$, and therefore $[g_m] = h_m$ as desired.

Finally, we need to show that the g_m are compatible under the projection maps. That is, we must show that if $m \mid n$, then the natural map $G(n) \to G(m)$ maps g_n to g_m. Consider the commutative diagram

$$\begin{array}{ccc} G(nm_0) & \longrightarrow & G(mm_0) \\ \downarrow & & \downarrow \\ G(n) & \longrightarrow & G(m) \end{array}$$

By construction, $g_{nm_0} \mapsto g_n$ and $g_{mm_0} \mapsto g_m$. Since $g_{nm_0} \mapsto g_{mm_0}$ (the lifting of multiples of m_0 is a compatible lifting), it follows that $g_{nm_0} \mapsto g_{mm0} \mapsto g_m$ in the maps $G(nm_0) \to G(mm_0) \to G(m)$. Since the above diagram commutes and $g_{nm_0} \mapsto g_n$ via $G(nm_0) \to G(n)$ it must be that $g_n \mapsto g_m$, proving that the lift is compatible.

(c) Since the $h_m \in H(m)$ lift to a compatible family $g_m \in G(m)$ for all m, we can prove surjectivity as follows. For each prime p, we get a compatible sequence g_{p^ℓ} of elements of $\mathrm{GL}(2, \mathbb{Z}/p^\ell\mathbb{Z})$ for each $\ell > 0$. Then property (iii) of $\mathbb{Z}_p$ shows that these matrices determine an element $\gamma_p \in \mathrm{GL}(2, \mathbb{Z}_p)$ such that γ_p maps to g_{p^ℓ}. Doing this for each prime p, we construct an element $\gamma = (\gamma_p) \in \mathrm{GL}(2, \widehat{\mathbb{Z}})$.

For $m > 0$, we prove that γ maps to $g_m \in \mathrm{GL}(2, \mathbb{Z}/m\mathbb{Z})$ as follows. Write $m = \prod_{p \mid m} p^{v_p(m)}$ and consider the commutative diagram

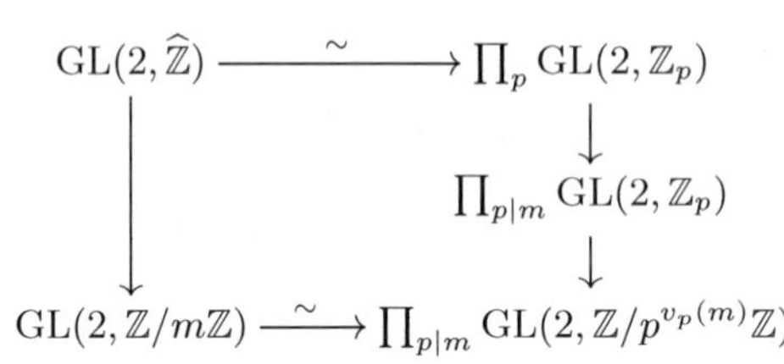

In the bottom map, $g_m \mapsto (g_{p^{v_p(m)}})_{p \mid m}$ by compatibility. Also, if we go over and then down in the diagram, the constructions of γ and γ_p imply that

$$\gamma \longmapsto (\gamma_p)_p \longmapsto (\gamma_p)_{p \mid m} \longmapsto (g_{p^{v_p(m)}})_{p \mid m}.$$

Since the diagram commutes and the bottom map is an isomorphism, it follows that $\gamma \mapsto g_m$. Thus, in the enlarged version of diagram (15.7), γ maps to $[h_m]$ and then to $\sigma|_{\mathsf{F}_m}$ for all m. Hence $\gamma \mapsto \sigma$ since $\mathsf{F} = \bigcup_m \mathsf{F}_m$.

15.5. (a) Write $\rho \in \Gamma(mN)$ as $\rho = I + mN\delta$, where δ is a matrix with integer entries; this is possible since $\rho \equiv I \bmod mN$. If $\gamma \in \mathrm{GL}(2, \mathbb{Q})^+$ with integer entries and $\det(\gamma) = N$, then $\gamma^{-1} = \frac{1}{N}\gamma'$ where γ' has integer entries. So

$$\gamma\rho\gamma^{-1} = \gamma(I + mN\delta)\gamma^{-1} = I + mN\gamma\delta\gamma^{-1} = I + m\gamma\delta\gamma'.$$

Since γ, δ, and γ' all have integer entries, this matrix lies in $\Gamma(m)$.

(b) If $\delta \in \Gamma(mN)$, then

$$(f^\gamma)^\delta(\tau) = f(\gamma\delta\tau).$$

By part (a), there is some $\rho \in \Gamma(m)$ with $\gamma\delta = \rho\gamma$, so that $f(\gamma\delta\tau) = f(\rho\gamma\tau)$. But $f \in \mathsf{F}_m$, so it is invariant under the action of ρ. Therefore $f(\rho\gamma\tau) = f(\gamma\tau) = f^\gamma(\tau)$, showing that $f^\gamma(\tau)$ is $\Gamma(mN)$-invariant.

(c) By part (b), $f^\gamma(\tau)$ is $\Gamma(mN)$-invariant. It is meromorphic since $f(\tau)$ is. To study f^γ at the cusps, take $\delta \in \mathrm{SL}(2,\mathbb{Z})$. The goal is to show that $f^\gamma(\delta\tau) = f(\gamma\delta\tau)$ has a meromorphic expansion in $q^{1/(mN)}$ with coefficients in $\mathbb{Q}(\zeta_{mN})$.

Using the solution to part (c) of Exercise 7.15, we can find $B \in \mathrm{GL}(2,\mathbb{Z})$ with $\det(B) = \pm 1$ such that

$$(*) \qquad B\gamma\delta = \begin{pmatrix} a & b \\ 0 & d \end{pmatrix}.$$

Multiplying B by $\left(\begin{smallmatrix} -1 & 0 \\ 0 & 1 \end{smallmatrix}\right)$ if $a < 0$ and by $\left(\begin{smallmatrix} 1 & 0 \\ 0 & -1 \end{smallmatrix}\right)$ if $d < 0$, one can assume that a, $d > 0$. Since the right-hand side of (*) has determinant $ad > 0$ and $\det(\gamma\delta) = N > 0$, it follows that $\det(B) > 0$, so that $B \in \mathrm{SL}(2,\mathbb{Z})$ and $ad = N$. Then

$$f^\gamma(\delta\tau) = f(\gamma\delta\tau) = f\left(B^{-1}\begin{pmatrix} a & b \\ 0 & d \end{pmatrix}\tau\right) = f(B^{-1}((a\tau+b)/d)).$$

Since $f \in \mathsf{F}_m$ and $B^{-1} \in \mathrm{SL}(2,\mathbb{Z})$, we know that $f(B^{-1}\tau)$ has a meromorphic expansion in $q^{1/m}$ with coefficients in $\mathbb{Q}(\zeta_m)$, say

$$f(B^{-1}\tau) = \sum_{j=-M}^{\infty} a_j q^{j/m},\ q = e^{2\pi i\tau},\ M \in \mathbb{Z},\ a_j \in \mathbb{Q}(\zeta_m).$$

Then the q-expansion of $f^\gamma(\delta\tau)$ is

$$\begin{aligned} f^\gamma(\delta\tau) = f(B^{-1}((a\tau+b)/d)) &= \sum_{j=-M}^{\infty} a_j e^{2\pi i j(((a\tau+b)/d)/m)} \\ &= \sum_{j=-M}^{\infty} a_j e^{2\pi i j(a\tau/(dm))} e^{jb/(dm)} \\ &= \sum_{j=-M}^{\infty} a_j \zeta_{md}^{bj} q^{ja/(md)}. \end{aligned}$$

Since $ad = N$, it follows that $\zeta_{md} \in \mathbb{Q}(\zeta_{mN})$, so that $f^\gamma(\delta\tau)$ has a q-expansion in powers of $q^{1/(mN)}$, is meromorphic, and has coefficients in $\mathbb{Q}(\zeta_{mN})$, as desired. This completes the proof that $f^\gamma \in \mathsf{F}_{mN}$.

15.6. We first represent $\widehat{\mathbb{Q}} = \mathbb{Q}\otimes\widehat{\mathbb{Z}} = \mathbb{Q}\otimes\prod_p \mathbb{Z}_p$ as the restricted product $\prod_p^* \mathbb{Q}_p$ consisting of all $(y_p) \in \prod_p \mathbb{Q}_p$ such that $y_p \in \mathbb{Z}_p$ for all but finitely many p. Clearly $\prod_p^* \mathbb{Q}_p \subset \prod_p \mathbb{Q}_p$. There is a natural map $\widehat{\mathbb{Q}} = \mathbb{Q}\otimes_{\mathbb{Z}}\widehat{\mathbb{Z}} \to \prod_p \mathbb{Q}_p$ defined by

$$\frac{r}{s}\otimes(x_p) \longmapsto (rx_p/s).$$

Since only a finite number of primes divide s, s is invertible in $\mathbb{Z}_p$ for all but a finite number of p, so that $rx_p/s \in \mathbb{Z}_p$ for all but finitely many p. This proves that the image lies in $\prod_p^* \mathbb{Q}_p$. To show that the resulting map $\widehat{\mathbb{Q}} \to \prod_p^* \mathbb{Q}_p$ is a ring isomorphism, first suppose that $(r/s)\otimes(x_p)$ maps to the zero element, then $rx_p/s = 0$ for all p. If $r = 0$, then $(r/s)\otimes(x_p) = 0\otimes(x_p) = (0)$, and if $r \neq 0$, then $x_p = 0$ for all p, so that $(r/s)\otimes(x_p) = (r/s)\otimes(0) = (0)$. Injectivity follows. Now take $(y_p) \in \prod^* \mathbb{Q}_p$ and pick a positive integer N such that $Ny_p \in \mathbb{Z}_p$ for the finitely many p's for which $y_p \notin \mathbb{Z}_p$. Then $(1/N)\otimes(Ny_p) \in \mathbb{Q}\otimes\widehat{\mathbb{Z}}$ maps to (y_p).

The ring isomorphism $\widehat{\mathbb{Q}} \simeq \prod_p^* \mathbb{Q}_p$ induces a group isomorphism $\mathrm{GL}(2, \widehat{\mathbb{Q}}) \simeq \mathrm{GL}(2, \prod_p^* \mathbb{Q}_p)$. The latter group is naturally isomorphic to $\prod_p^* \mathrm{GL}(2, \mathbb{Q}_p)$ since the data

$$\begin{pmatrix} (a_p) & (b_p) \\ (c_p) & (d_p) \end{pmatrix}, \quad (a_p), (b_p), (c_p), (d_p) \in \prod_p{}^{*} \mathbb{Q}_p, \ (a_p)(d_p) - (b_p)(c_p) \in \Big(\prod{}^{*} \mathbb{Q}_p\Big)^{*}$$

is clearly equivalent to the data

$$\begin{pmatrix} a_p & b_p \\ c_p & d_p \end{pmatrix} \in \mathrm{GL}(2, \mathbb{Q}_p) \text{ for all } p, \ \begin{pmatrix} a_p & b_p \\ c_p & d_p \end{pmatrix} \in \mathrm{GL}(2, \mathbb{Z}_p) \text{ for all but finitely many } p.$$

15.7. (a) By Exercise 8.4, since $\mathrm{Gal}(K_{\mathfrak{m}}/K)$ is a generalized ideal class group for $\mathfrak{m}$, it is also a generalized class group for $\mathfrak{n}$. This implies that

$$\ker(\Phi_{K_{\mathfrak{n}}/K,\mathfrak{n}}) = P_{K,1}(\mathfrak{n}) \subset \ker(\Phi_{K_{\mathfrak{m}}/K,\mathfrak{n}}) \subset I_K(\mathfrak{n})$$

with $I_K(\mathfrak{n})/\ker(\Phi_{K_{\mathfrak{m}}/K,\mathfrak{n}}) \simeq \mathrm{Gal}(K_{\mathfrak{m}}/K)$. Then by Corollary 8.7 $K_{\mathfrak{m}} \subset K_{\mathfrak{n}}$.

(b) Suppose $\mathfrak{m} \mid \mathfrak{n}$. Since the Artin map is compatible with restriction, the inclusion $K_{\mathfrak{m}} \subset K_{\mathfrak{n}}$ gives a commutative diagram (ignore the dotted arrow for now):

$$\begin{array}{ccc} I_K(\mathfrak{n}) & = & I_K(\mathfrak{n}) \\ & & \downarrow \\ \Big\downarrow \text{Artin} & \searrow & I_K(\mathfrak{m}) \\ & & \Big\downarrow \text{Artin} \\ \mathrm{Gal}(K_{\mathfrak{n}}/K) & \xrightarrow{\text{res}} & \mathrm{Gal}(K_{\mathfrak{m}}/K) \end{array}$$

The kernels of the Artin maps on the left and right are $P_{K,1}(\mathfrak{n})$ and $P_{K,1}(\mathfrak{m})$ respectively, and the kernel of the composed map $I_K(\mathfrak{n}) \to \mathrm{Gal}(K_{\mathfrak{m}}/K)$ on the right is $I_K(\mathfrak{n}) \cap P_{K,1}(\mathfrak{m})$. Since the Artin maps are surjective, it follows that we get a commutative diagram with isomorphisms as indicated:

$$\begin{array}{ccc} I_K(\mathfrak{n})/P_{K,1}(\mathfrak{n}) & \longrightarrow & I_K(\mathfrak{n})/I_K(\mathfrak{n}) \cap P_{K,1}(\mathfrak{m}) \\ & & \downarrow \\ \text{Artin} \Big\downarrow \simeq & & I_K(\mathfrak{m})/P_{K,1}(\mathfrak{m}) \\ & & \simeq \Big\downarrow \text{Artin} \\ \mathrm{Gal}(K_{\mathfrak{n}}/K) & \xrightarrow{\quad\text{res}\quad} & \mathrm{Gal}(K_{\mathfrak{m}}/K) \end{array}$$

It remains to show that the map

$$(*) \qquad I_K(\mathfrak{n})/(I_K(\mathfrak{n}) \cap P_{K,1}(\mathfrak{m})) \longrightarrow I_K(\mathfrak{m})/P_{K,1}(\mathfrak{m})$$

is an isomorphism. It is clearly injective, so we must prove surjectivity. Observe that from part (a), $\mathrm{Gal}(K_{\mathfrak{m}}/K)$ is a generalized ideal class group for $\mathfrak{n}$. Therefore the Artin map $I_K(\mathfrak{n}) \to \mathrm{Gal}(K_{\mathfrak{m}}/K)$, which is the dotted arrow in the first diagram, is surjective. It follows that (*) is also surjective and is therefore an isomorphism.

(c) Suppose for the moment that $\sigma_n|_{K_m} = \sigma_m$ whenever $K_m \subset K_n$. Define $\sigma \in K^{ab}$ as follows. For $x \in K^{ab}$, choose m such that $x \in K_m$ (possible since $K^{ab} = \bigcup_m K_m$). Define $\sigma(x) = \sigma_m(x)$. The compatibility condition above shows that it doesn't matter which m we choose, so that σ is well-defined. It is unique since the requirement that $\sigma|_{K_m} = \sigma_m$ makes this the only possible definition.

To complete the proof, we claim that given $\sigma_j \in \mathrm{Gal}(K_j/K)$ with $\sigma_n|_{K_m} = \sigma_m$ whenever $m \mid n$, then $\sigma_n|_{K_m} = \sigma_m$ whenever $K_m \subset K_n$. Suppose $K_m \subset K_n$. Since

$m \mid mn$, we have $\sigma_{mn}|_{K_m} = \sigma_m$. Similarly, since $n \mid mn$, we have $\sigma_{mn}|_{K_n} = \sigma_n$. Since $K_m \subset K_n \subset K_{mn}$, we get

$$\sigma_n|_{K_m} = \big(\sigma_{mn}|_{K_n}\big)|_{K_m} = \sigma_{mn}|_{K_m} = \sigma_m,$$

proving the claim.

15.8. (a) Consider the commutative diagram

$$\begin{array}{ccc} K \otimes_{\mathbb{Z}} \widehat{\mathbb{Z}} & \longrightarrow & \prod_p (K \otimes_{\mathbb{Z}} \mathbb{Z}_p) \\ \uparrow & \nearrow & \uparrow \\ \mathcal{O}_K \otimes_{\mathbb{Z}} \widehat{\mathbb{Z}} & \longrightarrow & \prod_p (\mathcal{O}_K \otimes_{\mathbb{Z}} \mathbb{Z}_p) \end{array}$$

where the vertical arrows are injections. The bottom arrow is an isomorphism since $\mathcal{O}_K$ is a free $\mathbb{Z}$-module of finite rank, and hence the dotted map is injective.

Suppose that $r \in K \otimes_{\mathbb{Z}} \widehat{\mathbb{Z}}$ maps to zero in $\prod_p (K \otimes_{\mathbb{Z}} \mathbb{Z}_p)$. Then the same is true for Nr for any positive integer N. But r is a finite sum of tensors $x_i \otimes (y_p)_i$ for $x_i \in K$ and $(y_p)_i \in \widehat{\mathbb{Z}}$. Hence we can a pick positive integer N such that $Nx_i \in \mathcal{O}_K$ for all i. Thus $Nr \in \mathcal{O}_K \otimes_{\mathbb{Z}} \widehat{\mathbb{Z}}$. Since Nr maps to zero in $\prod_p K \otimes_{\mathbb{Z}} \mathbb{Z}_p$ and the dotted map is injective, we see that $Nr = 0$ in $K \otimes_{\mathbb{Z}} \widehat{\mathbb{Z}}$. Multiplying by $1/N$ (note that $K \otimes_{\mathbb{Z}} \widehat{\mathbb{Z}}$ is a vector space over $\mathbb{Q}$) shows that $r = 0$, and injectivity follows.

(b) The map $\widehat{K} = K \otimes_{\mathbb{Z}} \widehat{\mathbb{Z}} \to \prod_p (K \otimes_{\mathbb{Z}} \mathbb{Z}_p)$ takes a tensor $x \otimes (y_p) \in K \otimes_{\mathbb{Z}} \widehat{\mathbb{Z}}$ to $(x \otimes y_p)$. Pick a positive integer $N > 0$ such that $Nx \in \mathcal{O}_K$. Since N is invertible in $\mathbb{Z}_p$ when $p \nmid N$, it follows that $x \otimes y_p = x \otimes NN^{-1} y_p = Nx \otimes N^{-1} y_p \in \mathcal{O}_K \otimes_{\mathbb{Z}} \mathbb{Z}_p$ when $p \nmid N$. It follows that $x \otimes y_p \in \mathcal{O}_K \otimes_{\mathbb{Z}} \mathbb{Z}_p$ except for a finite set of primes dividing N. Thus $(x \otimes y_p)$ lies in the restricted product $\prod_p^* (K \otimes_{\mathbb{Z}} \mathbb{Z}_p)$. Since an arbitrary element of $\widehat{K}$ is a finite sum of such tensors, part (a) gives an injective ring homomorphism

$$(*) \qquad \widehat{K} \longrightarrow \prod_p{}^* (K \otimes_{\mathbb{Z}} \mathbb{Z}_p).$$

(c) Following the hint, take $(x_p) \in \prod_p^* (K \otimes_{\mathbb{Z}} \mathbb{Z}_p)$ and let S be the finite set $\{p : x_p \notin \mathcal{O}_K \otimes_{\mathbb{Z}} \mathbb{Z}_p\}$. For $p \in S$ pick a nonnegative integer n_p with $p^{n_p} x_p \in \mathcal{O}_K \otimes_{\mathbb{Z}} \mathbb{Z}_p$ and set $N = \prod_{p \in S} p^{n_p}$. Since p^{n_p} is invertible in $\mathbb{Z}_q$ for $q \neq p$, it follows that $Nx_p \in \mathcal{O}_K \otimes_{\mathbb{Z}} \mathbb{Z}_p$ for all primes p and therefore that $(Nx_p) \in \prod_p (\mathcal{O}_K \otimes_{\mathbb{Z}} \mathbb{Z}_p)$. Since $\mathcal{O}_K$ is a free $\mathbb{Z}$-module of finite rank, we have $\widehat{\mathcal{O}}_K = \prod_p (\mathcal{O}_K \otimes_{\mathbb{Z}} \mathbb{Z}_p)$, so that there is $y \in \widehat{\mathcal{O}}_K$ that maps to (Nx_p).

Since $N \in K$ is nonzero, $N^{-1} \in K$ so that $N^{-1} \otimes (1) \in \widehat{K}$. Since the map (*) is a ring homomorphism,

$$(N^{-1} \otimes (1))y \in \widehat{K} \longmapsto (N^{-1} \otimes 1)(Nx_p) \in \prod_p{}^* (K \otimes_{\mathbb{Z}} \mathbb{Z}_p).$$

However, in $K \otimes_{\mathbb{Z}} \mathbb{Z}_p$, multiplication by the integer N is the same as multiplication by the tensor $N \otimes_{\mathbb{Z}} 1$. Thus, for each p, we have

$$(N^{-1} \otimes 1)(Nx_p) = (N^{-1} \otimes 1)(N \otimes 1) \cdot x_p = x_p.$$

Hence $(N^{-1} \otimes (1))y \in \widehat{K}$ maps to (x_p), so that the image of the map is $\prod_p^* (K \otimes_{\mathbb{Z}} \mathbb{Z}_p)$.

Note that parts (a)–(c) together show that $\widehat{K} \simeq \prod_p^* (K \otimes_{\mathbb{Z}} \mathbb{Z}_p)$.

(d) Recall that since $\mathcal{O}_K$ is a free $\mathbb{Z}$-module of finite rank, we have $\widehat{\mathcal{O}}_K \simeq \prod_p (\mathcal{O}_K \otimes_{\mathbb{Z}} \mathbb{Z}_p)$. Therefore $\widehat{\mathcal{O}}_K^* \simeq \big(\prod_p (\mathcal{O}_K \otimes_{\mathbb{Z}} \mathbb{Z}_p)\big)^* = \prod_p (\mathcal{O}_K \otimes_{\mathbb{Z}} \mathbb{Z}_p)^*$.

Finally, to see that $\widehat{K}^* \simeq \big(\prod^*(K\otimes_{\mathbb{Z}}\mathbb{Z}_p)\big)^* \simeq \prod^*(K\otimes_{\mathbb{Z}}\mathbb{Z}_p)^*$, suppose that $(x_p) \in \widehat{K}^*$. Then $(x_p) \in \widehat{K}$ and $(x_p^{-1}) \in \widehat{K}$. Thus $x_p \in \mathcal{O}_K \otimes_{\mathbb{Z}} \mathbb{Z}_p$ for all but finitely many p, and similarly for x_p^{-1}. This implies that with finitely many exceptions, x_p and x_p^{-1} both lie in $\mathcal{O}_K \otimes_{\mathbb{Z}} \mathbb{Z}_p$, which implies $x_p \in (\mathcal{O}_K \otimes_{\mathbb{Z}} \mathbb{Z}_p)^*$ for all but finitely many p. Hence $(x_p) \in \prod_p^*(K \otimes_{\mathbb{Z}} \mathbb{Z}_p)^*$.

For the opposite inclusion, take $(x_p) \in \prod_p^*(K\otimes_{\mathbb{Z}}\mathbb{Z}_p)^*$. Then $(x_p) \in \prod_p^* K\otimes_{\mathbb{Z}}\mathbb{Z}_p = \widehat{K}$. Furthermore, $x_p \in (K \otimes_{\mathbb{Z}} \mathbb{Z}_p)^*$ for all p and $x_p \in (\mathcal{O}_K \otimes_{\mathbb{Z}} \mathbb{Z}_p)^*$ for all but finitely many p. It follows that $x_p^{-1} \in K \otimes_{\mathbb{Z}} \mathbb{Z}_p$ for all p and $x_p^{-1} \in \mathcal{O}_K \otimes_{\mathbb{Z}} \mathbb{Z}_p$ for all but finitely many p, so that $(x_p^{-1}) \in \prod_p^* K \otimes_{\mathbb{Z}} \mathbb{Z}_p = \widehat{K}$. Thus $(x_p) \in \widehat{K}^*$.

15.9. (a) Suppose $x \in \widehat{K}^*$ is an idele, and $x = \beta yz$ is an n-factorization. If $m \mid n$, then $y \equiv 1 \bmod n\widehat{\mathcal{O}}_K$ implies that $y \equiv 1 \bmod m\widehat{\mathcal{O}}_K$, so that $J_n^1 \subset J_m^1$. Next, writing $z = (z_p)$, we have $z_p = 1$ for all $p \mid n$, so that $z_p = 1$ for all $p \mid m$ as well. Since $\beta \in K^*$, it follows that $x = \beta yz$ is an m-factorization of x. Then $\sigma_{x,m}$ and $\sigma_{x,n}$ are defined by

$$\begin{aligned}\sigma_{x,m} &\longmapsto [z\mathcal{O}_K] \in I_K(m)/P_{K,1}(m)\\ \sigma_{x,n} &\longmapsto [z\mathcal{O}_K] \in I_K(n)/P_{K,1}(n).\end{aligned}$$

It remains to show that these are compatible, i.e. that $\sigma_{x,n}|_{K_m} = \sigma_{x,m}$. But since $m \mid n$ implies that $K_m \subset K_n$, this follows immediately from the commutative diagram in part (b) of Exercise 15.7:

$$\begin{array}{ccc} I_K(n)/P_{K,1}(n) & \longrightarrow & I_K(m)/P_{K,1}(m) \\ \text{Artin}\big\downarrow\simeq & & \text{Artin}\big\downarrow\simeq \\ \mathrm{Gal}(K_n/K) & \xrightarrow{\text{res}} & \mathrm{Gal}(K_m/K) \end{array}$$

To see this, note that going from $\mathrm{Gal}(K_n/K)$ up, then right, then down, gives first $[z\mathcal{O}_K] \in I_K(n)/P_{K,1}(n)$, which maps to $[z\mathcal{O}_K] \in I_K(m)/P_{K,1}(m)$ since that map is induced by the inclusion $I_K(n) \hookrightarrow I_K(m)$. Then $[z\mathcal{O}_K] \in I_K(m)/P_{K,1}(m)$ maps to $\sigma_{x,m}$ by definition. However, going along the bottom gives $\sigma_{x,n} \mapsto \sigma_{x,n}|_{K_m}$. Since the diagram commutes, $\sigma_{x,n}$ and $\sigma_{x,m}$ are compatible.

(b) If $x \in K^* \subset \widehat{K}^*$, then $x = \beta yz$ is a factorization, where $\beta = x$, $y = 1 \in \widehat{\mathcal{O}}^*$, and $z = (z_p) = (1) \in \widehat{K}^*$. This factorization is an m-factorization for any $m > 0$, since $1 \in J_m^1$ for all m and $z = (1)$ satisfies $z_p = 1$ for $p \mid m$. But then $\sigma_{x,m} \mapsto [z\mathcal{O}_K] = [\mathcal{O}_K]$, so that $\sigma_{x,m}$ is the identity element of $\mathrm{Gal}(K_m/K)$. It follows that the element $\sigma_x \in \mathrm{Gal}(K^{ab}/K)$ determined by the $\sigma_{x,m}$ is the identity of $\mathrm{Gal}(K^{ab}/K)$, so that $K^* \to \widehat{K}^* \to \mathrm{Gal}(K^{ab}/K)$ is the trivial map. Thus K^* is in the kernel of the Artin map.

15.10. (a) Given a class $[\mathfrak{c}] \in C(\mathcal{O})$, use Corollary 7.17 to choose a proper $\mathcal{O}$-ideal $\mathfrak{b}$ with $[\mathfrak{c}] = [\mathfrak{b}]$ and $N(\mathfrak{b})$ prime to fm. Let $\mathfrak{a} = \mathfrak{b}\mathcal{O}_K$. By Proposition 7.20, $N(\mathfrak{a})$ is also prime to fm, so that $\mathfrak{a} \in I_K(fm)$. Using Proposition 7.20 again,

$$\mathfrak{a} \longmapsto [\mathfrak{a} \cap \mathcal{O}] = [\mathfrak{b}] = [\mathfrak{c}],$$

showing that the map is surjective.

(b) Write $\alpha = a + bfw_K$ and let $d = \gcd(a, m)$, so that $a = a_0 d$, $m = m_0 d$ with $\gcd(a_0, m_0) = 1$. Using Exercise 15.11, choose $\ell \in \mathbb{Z}$ such that $a_0 + m_0\ell$ is prime to f.

We first claim that $a + m\ell$ is relatively prime to f. Since $a + m\ell = d(a_0 + m_0\ell)$, and $\gcd(a_0 + m_0\ell, f) = 1$, it suffices to show that $\gcd(d, f) = 1$. Suppose $r = \gcd(d, f)$

with $d = d_0 r$ and $f = f_0 r$. Then

$$\alpha = a + bfw_K = da_0 + bfw_K = rd_0a_0 + rbf_0w_K = r(d_0a_0 + bf_0w_K).$$

It follows that $r \mid N(\alpha)$. However, m is relatively prime to α and thus relatively prime to $N(\alpha)$. Since $r \mid d \mid m$, we must have $r = 1$. Thus $\gcd(d, f) = 1$, proving the claim.

Next, since $a + m\ell$ and f are relatively prime to f in $\mathbb{Z}$, they are also relatively prime in $\mathcal{O}_K$. This implies that

$$\alpha + m\ell = (a + bfw_k) + m\ell = (a + m\ell) + bfw_k$$

is also relatively prime to f in $\mathcal{O}_K$. Finally, since α is relatively prime to m in $\mathcal{O}_K$ by assumption, the same holds for $\alpha + m\ell$. Thus $\alpha + m\ell$ is relatively prime to both f and m, and therefore is relatively prime to fm.

(c) Write $I = I_K(fm) \cap P_{K,\mathbb{Z}}(f)$, and let J be the subgroup generated by ideals of the form $\alpha\mathcal{O}_K \in I_K(fm)$ where

$$\alpha \in \mathcal{O}_K \text{ with } \alpha \equiv a \bmod f\mathcal{O}_K \text{ for some } a \in \mathbb{Z}.$$

We wish to show that $I = J$.

One inclusion is easy. Since $\alpha\mathcal{O}_K$ is prime to fm and $\alpha \equiv a \bmod f\mathcal{O}_K$, it follows that $\gcd(a, f) = 1$, so that $J \subset I$.

For $I \subset J$, observe that an element of I is of the form $\mathfrak{a}\mathfrak{b}^{-1} = \alpha\mathcal{O}_K(\beta\mathcal{O}_K)^{-1} = \alpha\beta^{-1}\mathcal{O}_K$, where $\mathfrak{a}, \mathfrak{b} \subset \mathcal{O}_K$ are relatively prime to fm and $\alpha, \beta \in \mathcal{O}_K$ are relatively prime to f and satisfy $\alpha \equiv a \bmod f\mathcal{O}_K$ and $\beta \equiv b \bmod f\mathcal{O}_K$ where $a, b \in \mathbb{Z}$ are each relatively prime to f. Since $\alpha\beta^{-1} = \alpha\overline{\beta}N(\beta)^{-1}$, we can also write $\mathfrak{a}\mathfrak{b}^{-1} = (\alpha\overline{\beta}\mathcal{O}_K)(N(\beta)\mathcal{O}_K)^{-1}$. Note also that $\alpha\overline{\beta} \equiv ab \bmod f\mathcal{O}_K$, so that after replacing α by $\alpha\overline{\beta}$ and β by $N(\beta)$, we can assume at the outset that β is an integer b prime to f. Since $\mathfrak{a}\mathfrak{b}^{-1}N(\mathfrak{b}) = \mathfrak{a}\overline{\mathfrak{b}}$, we get

$$\mathfrak{a}\overline{\mathfrak{b}} = \alpha b^{-1}N(\mathfrak{b})\mathcal{O}_K.$$

Since $\mathfrak{a}\overline{\mathfrak{b}} \subset \mathcal{O}_K$ is relatively prime to fm, it follows that $\alpha_0 = \alpha b^{-1}N(\mathfrak{b})$ is in $\mathcal{O}_K$ and is relatively prime to fm. If we can show that $\alpha_0 \equiv u \bmod f\mathcal{O}_K$ for $u \in \mathbb{Z}$, then $\alpha_0\mathcal{O}_K$ is a generator of J, and the same is true for $N(\beta)\mathcal{O}_K$ since $N(\mathfrak{b})$ is relatively prime to fm. Thus

$$\mathfrak{a}\mathfrak{b}^{-1} = \alpha b^{-1}\mathcal{O}_K = \underbrace{(\alpha b^{-1}N(\beta))}_{\alpha_0}N(\beta)^{-1}\mathcal{O}_K = (\alpha_0\mathcal{O}_K)(N(\beta)\mathcal{O}_K)^{-1} \in J.$$

It remains to prove that $\alpha_0 \equiv u \bmod f\mathcal{O}_K$ for $u \in \mathbb{Z}$. For simplicity, set $N = N(\mathfrak{b})$. Then we have

$$\alpha_0 = \alpha b^{-1}N = (a + \gamma f)Nb^{-1}, \ \gamma \in \mathcal{O}_K,$$

so that if we write $\gamma = s + tw_K$, then

$$\alpha_0 = (a + \gamma f)Nb^{-1} = (a + (s + tw_K)f)Nb^{-1} = (a + sf)Nb^{-1} + (tfNb^{-1})w_K.$$

Since $\alpha_0 \in \mathcal{O}_K = [1, w_K]$, we get $u = (a + sf)Nb^{-1} \in \mathbb{Z}$ and $tfNb^{-1} \in \mathbb{Z}$. But f and b are relatively prime, so the latter implies that $v = tNb^{-1} \in \mathbb{Z}$. Then

$$\alpha_0 = u + vfw_K \equiv u \bmod f\mathcal{O}_K,$$

as desired.

15.11. Let $p_1, \ldots, p_s$ be the distinct primes dividing M but not dividing m_0. Then m_0 and the p_i are pairwise relatively prime, so that by the Chinese Remainder Theorem there is some $x \in \mathbb{Z}$ such that $x \equiv a_0 \bmod m_0$, say $x = a_0 + m_0\ell$ for $\ell \in \mathbb{Z}$, and $x \equiv 1 \bmod p_i$, $i = 1, \ldots, s$. It follows that $\gcd(x, m_0) = \gcd(a_0, m_0) = 1$ and that $\gcd(x, p_i) = 1$,

$i = 1, \ldots, s$. Since the primes dividing M are the p_i together with some of those dividing m_0, we see that $\gcd(x, M) = 1$. Then $\gcd(a_0 + m_0\ell) = \gcd(x, M) = 1$.

15.12. We first show that

$$p\mathcal{O}_K = \mathfrak{p}\overline{\mathfrak{p}},\ \mathfrak{p} = \alpha\mathcal{O}_K,\ \alpha \in \mathcal{O},\ \alpha \equiv 1 \bmod m\mathcal{O} \iff p\mathcal{O}_K = \mathfrak{p}\overline{\mathfrak{p}},\ \mathfrak{p} \in P_{K,\mathbb{Z},m}(fm).$$

For $\Rightarrow$, write $\alpha = 1 + m\beta$ for $\beta \in \mathcal{O}$. Then $\beta = c + dfw_K$ for some $c, d \in \mathbb{Z}$, so that $\alpha \equiv 1 + mc \bmod fm\mathcal{O}_K$. It follows that $\mathfrak{p} = \alpha\mathcal{O}_K \in P_{K,\mathbb{Z},m}(fm)$.

For $\Leftarrow$, since $\mathfrak{p} \in P_{K,\mathbb{Z},m}(fm)$ is an $\mathcal{O}_K$-ideal, we have $\mathfrak{p} = \alpha\mathcal{O}_K$ for $\alpha \in \mathcal{O}_K$, and $\alpha \equiv a \bmod fm\mathcal{O}_K$, $a \in \mathbb{Z}$ with $a \equiv 1 \bmod m$. Write $a = 1 + rm$. Then for some $\beta \in \mathcal{O}_K$ we get

$$\alpha = a + fm\beta = 1 + rm + fm\beta \equiv 1 \bmod m\mathcal{O}$$

since $f\beta \in f\mathcal{O}_K \subset \mathcal{O}$. Thus $\alpha \equiv 1 \bmod m\mathcal{O}$, and $\alpha \in \mathcal{O}$ follows immediately.

We next show that

$$p\mathcal{O}_K = \mathfrak{p}\overline{\mathfrak{p}},\ \mathfrak{p} \in P_{K,\mathbb{Z},m}(fm) \iff p\mathcal{O}_K = \mathfrak{p}\overline{\mathfrak{p}},\ \mathfrak{p} \text{ splits completely in } L_{\mathcal{O},m}.$$

By Corollary 5.21, $\mathfrak{p}$ splits completely in $L_{\mathcal{O},m}$ if and only if $((L_{\mathcal{O},m}/K)/\mathfrak{p}) = 1$. However, $\mathrm{Gal}(L_{\mathcal{O},m}/K) \simeq I_K(fm)/P_{K,\mathbb{Z},m}(fm)$, so this happens if and only if $\mathfrak{p} \in P_{K,\mathbb{Z},m}(fm)$, proving the desired equivalence.

Finally, since $L_{\mathcal{O},m}$ is Galois over $\mathbb{Q}$, we see that

$$p\mathcal{O}_K = \mathfrak{p}\overline{\mathfrak{p}},\ \mathfrak{p} \text{ splits completely in } L_{\mathcal{O},m} \iff p \text{ splits completely in } L_{\mathcal{O},m}.$$

The rest of the proof is identical to the proof of Theorems 9.2 and 9.4.

15.13. Since $\mathcal{O} = [1, a\tau_0]$, where $ax^2 + bx + c$ is the minimal polynomial for τ_0, $x \equiv 1 \bmod m\widehat{\mathcal{O}}$ means that $x = s + ra\tau_0$ where $r \equiv 0 \bmod m$ and $s \equiv 1 \bmod m$ (here $r, s \in \widehat{\mathbb{Z}}$). Since $g_{\tau_0}(x)$ is the transpose of the matrix representing multiplication by x with respect to the basis $\{\tau_0, 1\}$, we get

$$g_{\tau_0}(x)^t \begin{pmatrix} \tau_0 \\ 1 \end{pmatrix} = (ra\tau_0 + s) \begin{pmatrix} \tau_0 \\ 1 \end{pmatrix} = \begin{pmatrix} ra\tau_0^2 + s\tau_0 \\ ra\tau_0 + s \end{pmatrix} = \begin{pmatrix} r(-b\tau_0 - c) + s\tau_0 \\ ra\tau_0 + s \end{pmatrix}$$
$$= \begin{pmatrix} (s - br)\tau_0 - cr \\ ra\tau_0 + s \end{pmatrix} = \begin{pmatrix} s - br & -cr \\ r & s \end{pmatrix} \begin{pmatrix} \tau_0 \\ 1 \end{pmatrix}.$$

Thus modulo m

$$g_{\tau_0}(x) = \begin{pmatrix} s - br & -cr \\ r & s \end{pmatrix}^t = \begin{pmatrix} s - br & r \\ -cr & s \end{pmatrix} \equiv \begin{pmatrix} 1 & 0 \\ 0 & 1 \end{pmatrix} \equiv I \bmod m.$$

Alternatively, recall that $g_{\tau_0}(x)$ is the transpose of the matrix representing multiplication by x on the free $\widehat{\mathbb{Z}}$-module $\widehat{\mathbb{Z}} + \widehat{\mathbb{Z}}\tau_0$ relative to the basis $\{\tau_0, 1\}$. This operation makes sense for any $x \in \widehat{\mathcal{O}}$ and defines a $\widehat{\mathbb{Z}}$-module homomorphism $g : \widehat{\mathcal{O}} \to M_{2\times 2}(\widehat{\mathbb{Z}})$. Applied to $x = 1 + my$ for $x \in \widehat{\mathcal{O}}^*$ and $y \in \widehat{\mathcal{O}}$, we obtain

$$g_{\tau_0}(x) = g(1 + my) = g(1) + mg(y) = I + mg(y) \equiv I \bmod m.$$

15.14. (a) Let $K = \mathbb{Q}(\sqrt{-m})$; then $\mathcal{O}^* \subset \mathcal{O}_K^*$ and it suffices to show that $K \neq \mathbb{Q}(\omega)$, $K \neq \mathbb{Q}(i)$. Note that $\mathcal{O} = [1, \sqrt{-m}]$ has discriminant $-4m$. If $K = \mathbb{Q}(\omega)$, then $-4m = f^2 d_K = -12f^2$, so that $m = 3f^2$. Then $3f^2 = m \equiv 6 \bmod 8$, which implies $f^2 \equiv 2 \bmod 8$. This is clearly impossible. If $K = \mathbb{Q}(i)$, then $-4m = f^2 d_K = -4f^2$, so that $m = f^2$. Then $f^2 = m \equiv 6 \bmod 8$, which is also impossible. Thus $\mathcal{O}^* = \{\pm 1\}$.

(b) Set $\alpha = \sqrt{-m}$ so that $\alpha^2 = -m \equiv -6 \equiv 2 \bmod 8\mathcal{O}$ since $m \equiv 6 \bmod 8$. In this notation, $8\mathcal{O} = [8, 8\alpha]$. If $a + b\alpha$ gives an element of $(\mathcal{O}/8\mathcal{O})^*$, then for some $c, d \in \mathbb{Z}$,

$$(a + b\alpha)(c + d\alpha) = ac + bd\alpha^2 + (ad + bc)\alpha \equiv ac + 2bd + (ad + bc)\alpha \equiv 1 \bmod 8\mathcal{O},$$

so that $ac + 2bd \equiv 1 \bmod 8$, which forces a to be odd. Conversely, if a is odd, then $(a + b\alpha)(a - b\alpha) = a^2 - b^2\alpha^2 \equiv a^2 - 2b^2 \equiv 1 - 2b^2 \bmod 8\mathcal{O}$. If b is odd this is $-1 \bmod 8$, while if b is even, it is $1 \bmod 8$, so in either case $[a+b\alpha]$ is a unit in $\mathcal{O}/8\mathcal{O}$. Thus $(\mathcal{O}/8\mathcal{O})^*$ consists of congruence classes of elements $a + b\alpha \in \mathcal{O}/8\mathcal{O}$ with a odd. Since $|\mathcal{O}/8\mathcal{O}| = 64$, it follows that $|(\mathcal{O}/8\mathcal{O})^*| = 32$.

(c) From the previous part, $(\mathcal{O}/8\mathcal{O})^*$ decomposes as a product of cyclic 2-groups. Now,

$$\begin{aligned}(1+\alpha)^2 &= 1 + \alpha^2 + 2\alpha \equiv 3 + 2\alpha \bmod 8\mathcal{O}\\ (1+\alpha)^4 &= (3+2\alpha)^2 = 9 + 4\alpha^2 + 12\alpha \equiv 1 + 4\alpha \bmod 8\mathcal{O}\\ (1+\alpha)^8 &= (1+4\alpha)^2 = 1 + 16\alpha^2 + 8\alpha \equiv 1 \bmod 8\mathcal{O}.\end{aligned}$$

Thus $1 + \alpha$ generates a cyclic subgroup of order 8 that has trivial intersection with the subgroup $\{\pm 1, \pm 3\}$ generated by -1 and 3. Since -1 and 3 each have order 2, we conclude that

$$(\mathcal{O}/8\mathcal{O})^* \simeq \mathbb{Z}/2\mathbb{Z} \times \mathbb{Z}/2\mathbb{Z} \times \mathbb{Z}/8\mathbb{Z},$$

where generators of the three direct factors are the classes of -1, 3, and $1 + \alpha$.

15.15. Let $\mathcal{O} = \mathbb{Z}[\sqrt{-m}]$ where $m \equiv 3 \bmod 4$, $3 \nmid m$. This implies $m \equiv 3 \bmod 8$ or $m \equiv 7 \bmod 8$. Lemma 15.17 gives an exact sequence

$$\mathcal{O}^* \longrightarrow (\mathcal{O}/8\mathcal{O})^* \longrightarrow \mathrm{Gal}(L_8/L) \longrightarrow 1.$$

Since $m > 0$, the conditions $m \equiv 3, 7 \bmod 8$ and $3 \nmid m$ imply that $m \geq 7$. It follows that $\mathcal{O}^* \subset \mathcal{O}_K^*$ must be $\{\pm 1\}$.

We first study the group $(\mathcal{O}/8\mathcal{O})^*$. Write $\alpha = \sqrt{-m}$; then $\alpha^2 = -m$, so that if $a + b\alpha$ gives an element of $(\mathcal{O}/8\mathcal{O})^*$, then for some $c, d \in \mathbb{Z}$,

$$(a + b\alpha)(c + d\alpha) = ac + bd\alpha^2 + (ad + bc)\alpha \equiv ac - mbd + (ad + bc)\alpha \equiv 1 \bmod 8\mathcal{O},$$

and therefore $ac - mbd \equiv 1 \bmod 8$ and $ad + bc \equiv 0 \bmod 8$. If a and b are both even, then $ac - mbd \equiv 1 \bmod 8$ is impossible. If a and b are both odd, then $ad + bc \equiv 0 \bmod 8$ forces c and d to have the same parity, so that $ac - mbd$ is even, and again $ac - mbd \equiv 1 \bmod 8$ is impossible. Therefore a and b must have opposite parity. Since there are 16 elements $a + b\alpha$ with a odd and b even, and also 16 elements with a even and b odd, it follows that $|(\mathcal{O}/8\mathcal{O})^*| \leq 32$.

We now consider the cases $m \equiv 3 \bmod 8$ and $m \equiv 7 \bmod 8$ separately. First suppose that $m \equiv 3 \bmod 8$.

It is easily computed that $1 + 2\alpha$ and $2 + \alpha$ generate subgroups of order four that intersect only in the identity and do not contain -1. So together with -1, they form a subgroup of $(\mathcal{O}/8\mathcal{O})^*$ of order 32, which must therefore be the complete group of units, and the classes of these three elements generate it. It follows incidentally that

$$(\mathcal{O}/8\mathcal{O})^* \simeq \mathbb{Z}/4\mathbb{Z} \times \mathbb{Z}/4\mathbb{Z} \times \mathbb{Z}/2\mathbb{Z}.$$

Recall that Corollary 12.19 gives

$$\mathfrak{f}(\tau) = \zeta_{48}\mathfrak{f}_1(\tau + 1) = \zeta_{48}\mathfrak{f}_1^T(\tau), \quad T = \begin{pmatrix} 1 & 1 \\ 0 & 1 \end{pmatrix} \in \mathrm{SL}(2, \mathbb{Z}/8\mathbb{Z}).$$

Thus

$$(*) \qquad \mathfrak{f}(\tau)^6 = \zeta_8\left(\mathfrak{f}_1(T\tau)\right)^6 = \zeta_8(\mathfrak{f}_1(\tau)^6)^T.$$

Since by Proposition 15.2, $\mathfrak{f}_1(\tau)^6 \in \mathsf{F}_8$, Theorem 15.5 implies that $\mathfrak{f}(\tau)^6 \in \mathsf{F}_8$ as well. So following Theorem 15.23, $\mathrm{Gal}(L_8/L_{\mathcal{O}})$ is generated by $\sigma_{1+2\alpha}$ and $\sigma_{2+\alpha}$. By Shimura Reciprocity, we will know that $\mathfrak{f}(\alpha)^6 \in L_{\mathcal{O}}$ if we show it is invariant under the matrices

$\overline{g}_\alpha(1+2\alpha)$ and $\overline{g}_\alpha(2+\alpha)$. Since $\mathcal{O} = [1,\alpha]$, these matrices are computed as follows (using $m \equiv 3 \bmod 8$):

$$\begin{aligned}(1+2\alpha)\cdot\alpha &= 1\cdot\alpha - 2m \equiv 1\cdot\alpha + 2\cdot 1\\ (1+2\alpha)\cdot 1 &= 2\cdot\alpha + 1\cdot 1\end{aligned}$$

$$\begin{aligned}(2+\alpha)\cdot\alpha &= 2\cdot\alpha - m\cdot 1 \equiv 2\cdot\alpha - 3\cdot 1\\ (2+\alpha)\cdot 1 &= 1\cdot\alpha + 2\cdot 1,\end{aligned}$$

so that the two matrices are

$$\overline{g}_\alpha(1+2\alpha) = \begin{pmatrix}1 & 2\\ 2 & 1\end{pmatrix} \in \mathrm{GL}(2,\mathbb{Z}/8\mathbb{Z}),$$

$$\overline{g}_\alpha(2+\alpha) = \begin{pmatrix}2 & -3\\ 1 & 2\end{pmatrix} \in \mathrm{GL}(2,\mathbb{Z}/8\mathbb{Z}).$$

These matrices do not lie in $\mathrm{SL}(2,\mathbb{Z}/8\mathbb{Z})$, but we can factor them (modulo 8) as

$$(**)\qquad \begin{aligned}\begin{pmatrix}1 & 2\\ 2 & 1\end{pmatrix} &= \begin{pmatrix}1 & 2\\ 2 & 5\end{pmatrix}\begin{pmatrix}1 & 0\\ 0 & 5\end{pmatrix}, \quad \begin{pmatrix}1 & 2\\ 2 & 5\end{pmatrix} \in \mathrm{SL}(2,\mathbb{Z}/8\mathbb{Z})\\ \begin{pmatrix}2 & -3\\ 1 & 2\end{pmatrix} &= \begin{pmatrix}2 & 3\\ 1 & -2\end{pmatrix}\begin{pmatrix}1 & 0\\ 0 & -1\end{pmatrix}, \quad \begin{pmatrix}2 & 3\\ 1 & -2\end{pmatrix} \in \mathrm{SL}(2,\mathbb{Z}/8\mathbb{Z}).\end{aligned}$$

In what follows, we denote by γ a lift to $\mathrm{SL}(2,\mathbb{Z})$ of either of the first factors on the right-hand side of (**). As in the text, since $\mathfrak{f}(\tau)$ is invariant under each of the second factor in each of these on the right-hand side of (**), we may ignore it. By (*),

$$\mathfrak{f}(\rho\tau)^6 = \zeta_8\mathfrak{f}_1^T(\rho\tau)^6 = \zeta_8\mathfrak{f}_1(T\rho\tau)^6,\ \ \rho \in \mathrm{SL}(2,\mathbb{Z}),$$

so if we show that $\mathfrak{f}_1(T\gamma\tau)^6 = \mathfrak{f}_1(T\tau)^6$, applying (*) again shows that $\mathfrak{f}(\gamma\tau)^6 = \mathfrak{f}(\tau)^6$ as desired. But

$$\mathfrak{f}_1(T\gamma\tau)^6 = \mathfrak{f}_1(T\gamma T^{-1}\cdot T\tau)^6 = \mathfrak{f}_1(T\tau)^6,$$

so it suffices to show that $\mathfrak{f}_1(\tau)^6$ is invariant under $T\gamma T^{-1}$. Under conjugation by T we get modulo 8

$$\begin{aligned}T\begin{pmatrix}1 & 2\\ 2 & 5\end{pmatrix}T^{-1} &= \begin{pmatrix}3 & 4\\ 2 & 3\end{pmatrix}\\ T\begin{pmatrix}2 & 3\\ 1 & -2\end{pmatrix}T^{-1} &= \begin{pmatrix}3 & -2\\ 1 & -3\end{pmatrix}.\end{aligned}$$

Both of these matrices lie in $\mathrm{SL}(2,\mathbb{Z}/8\mathbb{Z})$. As in the proof of Theorem 15.23, we again note that lifting these matrices to $\mathrm{SL}(2,\mathbb{Z})$ involves adding multiples of 8 to the entries, which does not change the power of i in the transformation law (15.3), so computing the relevant exponent for these matrices from (15.3) we get

$$\begin{aligned}-\frac{1}{2}\cdot 3\cdot 4 + 2(3(1-3^2)-3) &= -6-54 = -60 \equiv 0 \bmod 4\\ -\frac{1}{2}\cdot 3(-2) + 1(-3(1-3^2)-3) &= 3+21 = 24 \equiv 0 \bmod 4\end{aligned}$$

which proves $T\gamma T^{-1}$-invariance of $\mathfrak{f}_1(\tau)$ for each of the γ's. This shows that $\mathfrak{f}(\alpha)^6$ is invariant under the action of $(\mathcal{O}/8\mathcal{O})^*$, and it follows from Theorem 15.22 that $\mathfrak{f}(\alpha)^6 \in L_{\mathcal{O}}$.

To see why this implies that $L_{\mathcal{O}} = K(\mathfrak{f}(\alpha)^2)$, note that $\mathcal{O}$ has discriminant $-4m$, so that by Theorem 12.2,

$$(**)\qquad \gamma_2(\alpha) = \frac{\mathfrak{f}(\alpha)^{24}-16}{\mathfrak{f}(\alpha)^8} \in L_{\mathcal{O}}.$$

Since $\mathfrak{f}(\alpha)^6 \in L_{\mathcal{O}}$, we also have $\mathfrak{f}(\alpha)^{24} \in L_{\mathcal{O}}$. Then it follows that $\mathfrak{f}(\alpha)^8 \in L_{\mathcal{O}}$, and therefore that $\mathfrak{f}(\alpha)^2 = \mathfrak{f}(\alpha)^8/\mathfrak{f}(\alpha)^6 \in L_{\mathcal{O}}$. But cubing (**), $j(\alpha)$ is a polynomial in $\mathfrak{f}(\alpha)^2$;

since $L_{\mathcal{O}} = K(j(\alpha))$, it follows that $L_{\mathcal{O}} = K(\mathfrak{f}(\alpha)^2)$ as well. This completes the proof when $m \equiv 3 \bmod 8$, $3 \nmid m$.

We now assume $m \equiv 7 \bmod 8$, $3 \nmid m$. Recall that $|(\mathcal{O}/8\mathcal{O})^*| \leq 32$ from the discussion above.

In what follows, we write $\langle p, q\rangle$ for the subgroup of $(\mathcal{O}/8\mathcal{O})^*$ generated by the (classes of) the elements p, q of $\mathcal{O}$. Note that $(-1)^2 \equiv 1 \bmod 8\mathcal{O}$ and $3^2 \equiv 1 \bmod 8\mathcal{O}$, so that $\langle -1, 3\rangle = \{\pm 1, \pm 3\}$ is a subgroup H of $(\mathcal{O}/8\mathcal{O})^*$ of order four. Next, note that $\alpha^2 = -m \equiv -7 \equiv 1 \bmod 8\mathcal{O}$ and that $\alpha \not\equiv \pm 1, \pm 3 \bmod 8\mathcal{O}$, so that $\langle \alpha\rangle = \{1, \alpha\}$ intersects H only in the identity. It follows that $\langle -1, 3, \alpha\rangle$ is a subgroup K of $(\mathcal{O}/8\mathcal{O})^*$ of order 8, isomorphic to $(\mathbb{Z}/2\mathbb{Z})^3$, which can easily be enumerated. Finally, it is easily computed that $1 + 2\alpha$ generates a subgroup of $(\mathcal{O}/8\mathcal{O})^*$ of order four that intersects K only in the identity and therefore that $\langle -1, 3, \alpha, 1+2\alpha\rangle$ is a subgroup of $(\mathcal{O}/8\mathcal{O})^*$ of order 32, which must therefore be the complete group of units. (It follows incidentally that $(\mathcal{O}/8\mathcal{O})^* \simeq \mathbb{Z}/2\mathbb{Z} \times \mathbb{Z}/2\mathbb{Z} \times \mathbb{Z}/2\mathbb{Z} \times \mathbb{Z}/4\mathbb{Z}$.)

Again following Theorem 15.23, $\mathrm{Gal}(L_8/L_{\mathcal{O}})$ is generated by σ_3, σ_α, and $\sigma_{1+2\alpha}$. The matrices are

$$\begin{aligned} \overline{g}_\alpha(3) &= \begin{pmatrix} 3 & 0 \\ 0 & 3 \end{pmatrix} \\ \overline{g}_\alpha(\alpha) &= \begin{pmatrix} 0 & 1 \\ 1 & 0 \end{pmatrix} = \begin{pmatrix} 0 & -1 \\ 1 & 0 \end{pmatrix}\begin{pmatrix} 1 & 0 \\ 0 & -1 \end{pmatrix} \\ \overline{g}_\alpha(1+2\alpha) &= \begin{pmatrix} 1 & 2 \\ 2 & 1 \end{pmatrix} = \begin{pmatrix} 1 & 2 \\ 2 & 5 \end{pmatrix}\begin{pmatrix} 1 & 0 \\ 0 & 5 \end{pmatrix} \end{aligned} \tag{***}$$

As above, the second matrices in each of the last two act trivially on $\mathfrak{f}(\tau)$, so they may be ignored. Again we denote by γ a lift of any of the first factors on the right-hand side of (***). Conjugation by T gives modulo 8

$$\begin{aligned} T\begin{pmatrix} 3 & 0 \\ 0 & 3 \end{pmatrix}T^{-1} &= \begin{pmatrix} 3 & 0 \\ 0 & 3 \end{pmatrix} \\ T\begin{pmatrix} 1 & 0 \\ 0 & -1 \end{pmatrix}T^{-1} &= \begin{pmatrix} 1 & -2 \\ 1 & -1 \end{pmatrix} \\ T\begin{pmatrix} 1 & 2 \\ 2 & 5 \end{pmatrix}T^{-1} &= \begin{pmatrix} 3 & 4 \\ 2 & 3 \end{pmatrix}. \end{aligned}$$

Again note that as in the proof of Theorem 15.23, we do not actually need lifts γ of these matrices, but can use them as is. The powers of i associated with these matrices are

$$\begin{aligned} -\frac{1}{2}\cdot 3\cdot 0 + 0(3(1-3^2) - 3) &\equiv 0 \bmod 4 \\ -\frac{1}{2}\cdot 1(-2) + 1(-1(1-1^2) - 1) &\equiv 0 \bmod 4 \\ -\frac{1}{2}\cdot 3\cdot 4 + 2(3(1-3^2) - 3) &\equiv 0 \bmod 4, \end{aligned}$$

proving $T\gamma T^{-1}$-invariance. The rest of the proof proceeds as in the case $m \equiv 3 \bmod 8$.

15.16. (a) Let $\mathfrak{a}$ be a proper $\mathcal{O}$-ideal. By Proposition 7.22, there is a proper $\mathcal{O}$-ideal $\mathfrak{a}'$ prime to the conductor of $\mathcal{O}$ such that $[\mathfrak{a}] = [\mathfrak{a}'] \in C(\mathcal{O})$. Since the map $\mathfrak{a} \mapsto \sigma_{\mathfrak{a}}$ defines a map from $C(\mathcal{O}) \to \mathrm{Gal}(L/K)$, it follows that $\sigma_{\mathfrak{a}} = \sigma_{\mathfrak{a}'}$. Further, using Theorem 10.9 and Corollary 10.20, we know that $j(\mathfrak{a}) = j(\mathfrak{a}')$. Thus

$$\sigma_{\overline{\mathfrak{a}}}(j(\mathcal{O})) = \sigma_{\overline{\mathfrak{a}'}}(j(\mathcal{O})) = j(\mathfrak{a}') = j(\mathfrak{a}),$$

which proves that $\sigma_{\overline{\mathfrak{a}}}(j(\mathcal{O})) = j(\mathfrak{a})$ for any proper $\mathcal{O}$-ideal $\mathfrak{a}$.

(b) If $\mathfrak{a}$ and $\mathfrak{b}$ are proper $\mathcal{O}$-ideals, then so is $\mathfrak{ab}$ (use Proposition 7.4). Then by part (a),

$$\sigma_{\overline{\mathfrak{ab}}}(j(\mathcal{O})) = \sigma_{\overline{\mathfrak{a}}\overline{\mathfrak{b}}}(j(\mathcal{O})) = (\sigma_{\overline{\mathfrak{a}}} \circ \sigma_{\overline{\mathfrak{b}}})(j(\mathcal{O})) = \sigma_{\overline{\mathfrak{a}}}(\sigma_{\overline{\mathfrak{b}}}(j(\mathcal{O})) = \sigma_{\overline{\mathfrak{a}}}(j(\mathfrak{b})).$$

But again by part (a), $\sigma_{\overline{\mathfrak{ab}}}(j(\mathcal{O})) = j(\mathfrak{ab})$, so that $\sigma_{\overline{\mathfrak{a}}}(j(\mathfrak{b})) = j(\mathfrak{ab})$.

15.17. (a) Recall that $\mathcal{O} = [1, \mathsf{a}\tau_1]$ where $\tau_1 \in \mathfrak{h}$ is a root of $\mathsf{a}x^2 + \mathsf{b}x + \mathsf{c}$. Then:

$$\begin{aligned}
\mathsf{a}(\tau_1 - 1)\mathcal{O} &= \mathsf{a}(\tau_1 - 1)[1, \mathsf{a}\tau_1] = [\mathsf{a}(\tau_1 - 1), \mathsf{a} \cdot \mathsf{a}\tau_1^2 - \mathsf{a}^2\tau_1] \\
&= [\mathsf{a}(\tau_1 - 1), \mathsf{a}(-\mathsf{b}\tau_1 - \mathsf{c}) - \mathsf{a}^2\tau_1] \\
&= [\mathsf{a}(\tau_1 - 1), \mathsf{a}(-\mathsf{b}\tau_1 - \mathsf{c}) - \mathsf{a}^2\tau_1 + \mathsf{b} \cdot \mathsf{a}(\tau_1 - 1) + \mathsf{a} \cdot \mathsf{a}(\tau_1 - 1)] \\
&= [\mathsf{a}(\tau_1 - 1), -\mathsf{a}(\mathsf{a} + \mathsf{b} + \mathsf{c}) = [\mathsf{a}(\tau_1 - 1), \mathsf{a}(\mathsf{a} + \mathsf{b} + \mathsf{c})].
\end{aligned}$$

(b) Let $x = (x_p)$ be as in Lemma 15.28. Since $x_p \in K^*$, it follows that $x_p = x_p \otimes 1 \in (K \otimes_{\mathbb{Z}} \mathbb{Z}_p)^*$ for all p. Further, if $p \nmid \mathsf{a}$, then

$$x_p = \mathsf{a} = \mathsf{a} \otimes 1 = 1 \otimes \mathsf{a}.$$

But since $p \nmid \mathsf{a}$, it follows that a is invertible in $\mathbb{Z}_p$, so that $x_p \in (\mathcal{O} \otimes_{\mathbb{Z}} \mathbb{Z}_p)^*$. Since only a finite number of primes divide a, this shows that $x_p \in (\mathcal{O} \otimes_{\mathbb{Z}} \mathbb{Z}_p)^*$ for all but finitely many p. Thus $x = (x_p) \in \prod_p^*(K \otimes_{\mathbb{Z}} \mathbb{Z}_p)^*$, which by Exercise 15.8 equals $\widehat{K}^*$.

(c) Lemma 15.28 constructs x with $\mathfrak{a} \otimes \widehat{\mathbb{Z}} = x\widehat{\mathcal{O}}$, and $x \in \widehat{K}^*$ by part (b). Suppose we also have $\mathfrak{a} \otimes \widehat{\mathbb{Z}} = y\widehat{\mathcal{O}}$ for some $y \in \widehat{K}$. Then $x\widehat{\mathcal{O}} = y\widehat{\mathcal{O}}$, and multiplying by x^{-1} (this is legal since $x \in \widehat{K}^*$) gives $\widehat{\mathcal{O}} = x^{-1}y\widehat{\mathcal{O}}$. This implies both that $u = x^{-1}y \in \widehat{\mathcal{O}}$ and that $1 = x^{-1}yv = uv$ for some $v \in \widehat{\mathcal{O}}$. It follows that $u \in \widehat{\mathcal{O}}^*$, and then we are done since $y = ux$ by the definition of u.

15.18. (a) $[1, \mathsf{a}\tau_1] = [1, \tau_0]$ means that there are integers ℓ, s, p, q such that

$$\mathsf{a}\tau_1 = \ell + s\tau_0, \qquad \tau_0 = p + q\mathsf{a}\tau_1.$$

This implies that

$$\tau_0 = p + q(\ell + s\tau_0) = p + q\ell + qs\tau_0.$$

Thus $qs = 1$, so that $s = \pm 1$. This implies that $\mathsf{a}\tau_1 = \ell \pm \tau_0$. Since $\tau_0, \tau_1 \in \mathfrak{h}$ we must take the plus sign, giving $\mathsf{a}\tau_1 = \tau_0 + \ell$, $\ell \in \mathbb{Z}$.

(b) This is a matter of unwinding the definitions:

$$\mathfrak{a}\widehat{\mathcal{O}} = \mathfrak{a}(\mathcal{O} \otimes_{\mathbb{Z}} \widehat{\mathbb{Z}}) = (\mathfrak{a} \otimes_{\mathbb{Z}} \{1\})(\mathcal{O} \otimes_{\mathbb{Z}} \widehat{\mathbb{Z}}) = \mathfrak{a}\mathcal{O} \otimes_{\mathbb{Z}} \widehat{\mathbb{Z}} = \mathfrak{a} \otimes_{\mathbb{Z}} \widehat{\mathbb{Z}}.$$

15.19. (a) The choice of $\pi_\mathfrak{p}$ implies that $\nu_\mathfrak{p}(\pi_\mathfrak{p}) = 1$ and $\pi_\mathfrak{p}\mathcal{O}_{K_\mathfrak{p}} = \mathfrak{p}\mathcal{O}_{K_\mathfrak{p}}$. Thus

$$\nu_\mathfrak{p}(\pi_\mathfrak{p}^{\nu_\mathfrak{p}(u_\mathfrak{p})} u_\mathfrak{p}^{-1}) = \nu_\mathfrak{p}(u_\mathfrak{p}) + \nu_\mathfrak{p}(u_\mathfrak{p}^{-1}) = \nu(u_\mathfrak{p}u_\mathfrak{p}^{-1}) = \nu_\mathfrak{p}(1) = 0$$

since $\nu_\mathfrak{p} : K_\mathfrak{p}^* \to \mathbb{Z}$ is a homomorphism. Then $\pi_\mathfrak{p}^{\nu_\mathfrak{p}(u_\mathfrak{p})} u_\mathfrak{p}^{-1} \in \mathcal{O}_{K_\mathfrak{p}}$, so that for any $\ell \geq 0$, the isomorphism in property (1) in the statement of the exercise implies that there is $w_\mathfrak{p} \in \mathcal{O}_K$ such that

$$\pi_\mathfrak{p}^{\nu_\mathfrak{p}(u_\mathfrak{p})} u_\mathfrak{p}^{-1} - w_\mathfrak{p} \in \mathfrak{p}^\ell \mathcal{O}_{K_\mathfrak{p}}.$$

Multiplying by $\pi_\mathfrak{p}^{-\nu_\mathfrak{p}(u_\mathfrak{p})}$ gives

$$u_\mathfrak{p}^{-1} - \pi_\mathfrak{p}^{-\nu_\mathfrak{p}(u_\mathfrak{p})} w_\mathfrak{p} \in \pi_\mathfrak{p}^{-\nu_\mathfrak{p}(u_\mathfrak{p})} \mathfrak{p}^\ell \mathcal{O}_{K_\mathfrak{p}} = \mathfrak{p}^{\ell - \nu_\mathfrak{p}(u_\mathfrak{p})} \mathcal{O}_{K_\mathfrak{p}}.$$

Picking $\ell \geq M + n_\mathfrak{p} + \nu_\mathfrak{p}(u_\mathfrak{p})$ and setting $v_\mathfrak{p} = \pi_\mathfrak{p}^{-\nu_\mathfrak{p}(u_\mathfrak{p})} w_\mathfrak{p} \in K$, we obtain

$$u_\mathfrak{p}^{-1} - v_\mathfrak{p} \in \mathfrak{p}^{\ell - \nu_\mathfrak{p}(u_\mathfrak{p})} \mathcal{O}_{K_\mathfrak{p}} \subset \mathfrak{p}^{M + n_\mathfrak{p}} \mathcal{O}_{K_\mathfrak{p}}$$

since $\ell - \nu_\mathfrak{p}(u_\mathfrak{p}) \geq M + n_\mathfrak{p}$.

(b) Choose $M > 0$ such that $M + \nu_\mathfrak{p}(u_\mathfrak{p}) \geq 0$ for all $\mathfrak{p} \mid \mathfrak{m}$, and for that value of M, choose $v_\mathfrak{p}$ as in part (a) for each $\mathfrak{p} \mid \mathfrak{m}$. Then by the Approximation Theorem using the modulus $\mathfrak{m}' = \prod_{n_\mathfrak{p}>0} \mathfrak{p}^{M+n_\mathfrak{p}}$, we can find $\alpha \in K$ such that $\alpha - v_\mathfrak{p} \in \mathfrak{p}^{M+n_\mathfrak{p}} \subset \mathfrak{p}^{M+n_\mathfrak{p}}\mathcal{O}_{K_\mathfrak{p}}$ for $\mathfrak{p} \mid \mathfrak{m}$. Since $u_\mathfrak{p}^{-1} - v_\mathfrak{p} \in \mathfrak{p}^{M+n_\mathfrak{p}}\mathcal{O}_{K_\mathfrak{p}}$, we get $\alpha - u_\mathfrak{p}^{-1} \in \mathfrak{p}^{M+n_\mathfrak{p}}\mathcal{O}_{K_\mathfrak{p}}$. Multiplying by $u_\mathfrak{p}$ gives

$$\alpha u_\mathfrak{p} - 1 \in u_\mathfrak{p}\mathfrak{p}^{M+n_\mathfrak{p}}\mathcal{O}_{K_\mathfrak{p}} = \mathfrak{p}^{\nu_\mathfrak{p}(u_\mathfrak{p})}\mathfrak{p}^{M+n_\mathfrak{p}}\mathcal{O}_{K_\mathfrak{p}} = \mathfrak{p}^{M+\nu_\mathfrak{p}(u_\mathfrak{p})+n_\mathfrak{p}}\mathcal{O}_{K_\mathfrak{p}} \subset \mathfrak{p}^{n_\mathfrak{p}}\mathcal{O}_{K_\mathfrak{p}}.$$

This shows that $\alpha u_\mathfrak{p} \in 1 + \mathfrak{p}^{n_\mathfrak{p}}\mathcal{O}_{K_\mathfrak{p}}$ as desired.

15.20. (a) Let $J = \{\delta\mathcal{O}_K : \delta \in K^*,\ \delta \in 1 + p^{n_p}(\mathcal{O}_K \otimes_\mathbb{Z} \mathbb{Z}_p) \text{ for all } p \mid m\}$, where $m = \prod_{p\mid m} p^{n_p}$. Note that J is a subgroup of $I_K(m)$. We prove $P_{K,1}(m) = J$ as follows.

For the inclusion $P_{K,1}(m) \subset J$, note that $\alpha \in \mathcal{O}_K$ with $\alpha \equiv 1 \bmod m\mathcal{O}_K$ satisfies $\alpha \in 1 + p^{n_p}(\mathcal{O}_K \otimes_\mathbb{Z} \mathbb{Z}_p)$ when $p \mid m$. Thus the generators of $P_{K,1}(m)$ lie in J, and $P_{K,1}(m) \subset J$ follows.

For the opposite inclusion, take $\delta \in K^*$ with $\delta \in 1 + p^{n_p}(\mathcal{O}_K \otimes_\mathbb{Z} \mathbb{Z}_p)$ for $p \mid m$. We need to show that $\delta\mathcal{O}_K \in P_{K,1}(\mathfrak{m})$. As noted above, we have $\delta\mathcal{O}_K \in I_K(m)$. Then part (a) of Exercise 8.1 gives the exact sequence

$$\mathcal{O}_K^* \longrightarrow (\mathcal{O}_K/m\mathcal{O}_K)^* \xrightarrow{\ \phi\ } I_K(m) \cap P_K/P_{K,1}(m) \longrightarrow 1,$$

where $\phi([\alpha]) = [\alpha\mathcal{O}_K]$. Since $\delta \in P_{K,1}(m)$, we have $\delta\mathcal{O}_K \in I_K(m) \cap P_K$, so there is $\alpha \in \mathcal{O}_K$ relatively prime to m such that $\alpha\mathcal{O}_K$ and $\delta\mathcal{O}_K$ give the same class in $I_K(m) \cap P_K/P_{K,1}(m)$. Thus there are $\alpha_0, \beta_0 \in \mathcal{O}_K$ with $\alpha_0 \equiv \beta_0 \equiv 1 \bmod m\mathcal{O}_K$ such that

$$\alpha\mathcal{O}_K = \delta\mathcal{O}_K \cdot (\alpha_0/\beta_0)\mathcal{O}_K.$$

Since $\delta, \alpha_0, \beta_0$ are $\equiv 1 \bmod m\mathcal{O}_K$, the same is true for α. But $\alpha \equiv 1 \bmod m\mathcal{O}_K$ implies $[\alpha] = [1] \in (\mathcal{O}_K/m\mathcal{O}_K)^*$. Then $\phi([\alpha]) = [\alpha\mathcal{O}_K] = [\delta\mathcal{O}_K]$ implies that $\delta\mathcal{O}_K \in P_{K,1}(\mathfrak{m})$.

(b) Given m-factorizations $x = \beta yz = \tilde\beta\tilde y\tilde z$, let $\delta = \tilde\beta\beta^{-1}$. Then $\delta = y\tilde y^{-1}z\tilde z^{-1}$. Remember that $y,\ \tilde y \in \widehat{\mathcal{O}}_K$ with $y,\ \tilde y \equiv 1 \bmod m\widehat{\mathcal{O}}_K$, and we also have $z_p = \tilde z_p = 1$ when $p \mid m$. Thus $y_p\tilde y_p^{-1}z_p\tilde z_p^{-1} \in 1+p^{n_p}(\mathcal{O}_K\otimes_\mathbb{Z}\mathbb{Z}_p)$ for $p \mid m$, so that $\delta \in 1+p^{n_p}(\mathcal{O}_K\otimes_\mathbb{Z}\mathbb{Z}_p)$ for $p \mid m$. Then $\tilde\beta\beta^{-1}\mathcal{O}_K = \delta\mathcal{O}_K \in P_{K,1}(m)$ by part (a).

15.21. (a) The isomorphism $\widehat{K}^* \simeq \mathbf{I}_K^{\mathrm{fin}}$ of (15.36) takes $\widehat{\mathcal{O}}_K^*$ to $\prod_\mathfrak{p} \mathcal{O}_{K_\mathfrak{p}}^*$. Thus $y = (y_p) \in \widehat{\mathcal{O}}_K^* = \prod_p(\mathcal{O}_K \otimes_\mathbb{Z} \mathbb{Z}_p)^*$ maps to $y^* = (y_\mathfrak{p}^*) \in \prod_\mathfrak{p} \mathcal{O}_{K_\mathfrak{p}}^*$, so that $y_\mathfrak{p}^* \in \mathcal{O}_{K_\mathfrak{p}}^*$ for all $\mathfrak{p}$. Furthermore, $y \equiv 1 \bmod m\widehat{\mathcal{O}}_K$ means that $y_p \equiv 1 \bmod m(\mathcal{O}_K \otimes_\mathbb{Z} \mathbb{Z}_p)$. If we write $m = \prod_{p\mid m} p^{n_p}$ and $m\mathcal{O}_K = \prod_\mathfrak{p} \mathfrak{p}^{n_\mathfrak{p}}$, then

$$p \mid m \implies y_p \equiv 1 \bmod p^{n_p}(\mathcal{O}_K \otimes_\mathbb{Z} \mathbb{Z}_p).$$

Since (15.35) takes $p^{n_p}(\mathcal{O}_K \otimes_\mathbb{Z} \mathbb{Z}_p)$ to $\prod_{p\in\mathfrak{p}} \mathfrak{p}^{n_\mathfrak{p}}\mathcal{O}_{K_\mathfrak{p}}$ and takes y_p to $(y_\mathfrak{p}^*)_{p\in\mathfrak{p}}$, we conclude that $y_\mathfrak{p}^* \equiv 1 \bmod \mathfrak{p}^{n_\mathfrak{p}}\mathcal{O}_{K_\mathfrak{p}}$ when $\mathfrak{p} \mid m\mathcal{O}_K$.

(b) $z = (z_p) \mapsto z^* = (z_\mathfrak{p}^*)$ via (15.36), which uses $z_p \mapsto (z_\mathfrak{p}^*)_{p\in\mathfrak{p}}$ via (15.35). Thus $z_\mathfrak{p}^* = 1$ whenever $z_p = 1$ and $p \in \mathfrak{p}$. Now suppose that $\mathfrak{p} \mid m\mathcal{O}_K$. Then $m = \prod_{p\mid m} p^{n_p} \in \mathfrak{p}$, so that $p \in \mathfrak{p}$ for some $p \mid m$ since $\mathfrak{p}$ is a prime ideal. By assumption, $z_p = 1$ when $p \mid m$, so it follows that $z_\mathfrak{p}^* = 1$ when $\mathfrak{p} \mid m\mathcal{O}_K$.

15.22. In the diagram in Lemma 15.20, we need to prove that the the square on the right commutes. As suggested by the hint, it suffices to prove that we have a commutative diagram

$$\begin{array}{ccccc} \widehat{K}^* & \longrightarrow & I_K(fm)/P_{K,1}(fm) & \longrightarrow & I_K(fm)/P_{K,\mathbb{Z},m}(fm) \\ \big\downarrow \text{Artin} & & \big\downarrow \simeq & & \big\downarrow \simeq \\ \mathrm{Gal}(K^{ab}/K) & \longrightarrow & \mathrm{Gal}(K_{fm}/K) & \longrightarrow & \mathrm{Gal}(L_{\mathcal{O},m}/K) \end{array}$$

where the vertical map are Artin maps from Theorem 8.2. In this diagram, the construction of the Artin map $x \in \widehat{K}^* \mapsto \sigma_x \in \mathrm{Gal}(K^{ab}/K)$ used in Shimura Reciprocity shows that the square on the left commutes.

To prove that the square on the right commutes, we use the class field theory developed in §8. The fields K_{fm} and $L_{\mathcal{O},m}$ correspond to the congruence subgroups $P_{K,1}(fm)$ and $P_{K,\mathbb{Z},m}(fm)$ by Theorem 8.6. Furthermore, the inclusion $L_{\mathcal{O},m} \subset K_{fm}$ follows from $P_{K,1}(fm) \subset P_{K,\mathbb{Z},m}(fm)$ by Corollary 8.7, and the proof of the corollary notes that the corresponding Artin maps are compatible with restriction. It follows that the square on the right commutes.

15.23. (a) We have the standard exact sequence

$$0 \longrightarrow \mathbb{Z} \xrightarrow{m} \mathbb{Z} \longrightarrow \mathbb{Z}/m\mathbb{Z} \longrightarrow 0.$$

Tensoring with the $\mathbb{Z}$-module $\mathcal{O}_K$ is right exact, so we get the exact sequence

$$\mathcal{O}_K \xrightarrow{m} \mathcal{O}_K \longrightarrow \mathcal{O}_K \otimes_{\mathbb{Z}} \mathbb{Z}/m\mathbb{Z} \longrightarrow 0.$$

This proves that $\mathcal{O}_K/m\mathcal{O}_K \simeq \mathcal{O}_K \otimes_{\mathbb{Z}} (\mathbb{Z}/m\mathbb{Z})$ as rings. Similarly, the isomorphism $\mathbb{Z}/m\mathbb{Z} \simeq \widehat{\mathbb{Z}}/m\widehat{\mathbb{Z}}$ from Exercise 15.3 gives the exact sequence

$$0 \longrightarrow \widehat{\mathbb{Z}} \xrightarrow{m} \widehat{\mathbb{Z}} \longrightarrow \mathbb{Z}/m\mathbb{Z} \longrightarrow 0,$$

which when tensored with $\mathcal{O}_K$ gives the exact sequence

$$\mathcal{O}_K \otimes_{\mathbb{Z}} \widehat{\mathbb{Z}} \xrightarrow{m} \mathcal{O}_K \otimes_{\mathbb{Z}} \widehat{\mathbb{Z}} \longrightarrow \mathcal{O}_K \otimes_{\mathbb{Z}} (\mathbb{Z}/m\mathbb{Z}) \longrightarrow 0.$$

Thus $(\mathcal{O}_K \otimes_{\mathbb{Z}} \widehat{\mathbb{Z}})/m(\mathcal{O}_K \otimes_{\mathbb{Z}} \widehat{\mathbb{Z}}) \simeq \mathcal{O}_K \otimes_{\mathbb{Z}} (\mathbb{Z}/m\mathbb{Z})$ as rings, so that

$$\mathcal{O}_K/m\mathcal{O}_K \simeq \mathcal{O}_K \otimes_{\mathbb{Z}} (\mathbb{Z}/m\mathbb{Z}) \simeq (\mathcal{O}_K \otimes_{\mathbb{Z}} \widehat{\mathbb{Z}})/m(\mathcal{O}_K \otimes_{\mathbb{Z}} \widehat{\mathbb{Z}}) = \widehat{\mathcal{O}}_K/m\widehat{\mathcal{O}}_K,$$

where the last line uses $\widehat{\mathcal{O}}_K = \mathcal{O}_K \otimes_{\mathbb{Z}} \widehat{\mathbb{Z}}$. Since $\mathbb{Z}/m\mathbb{Z} \simeq \widehat{\mathbb{Z}}/m\widehat{\mathbb{Z}}$ is induced by the injection $\mathbb{Z} \hookrightarrow \widehat{\mathbb{Z}}$, the above isomorphism is induced by

$$\mathcal{O}_K = \mathcal{O}_K \otimes_{\mathbb{Z}} \mathbb{Z} \hookrightarrow \mathcal{O}_K \otimes_{\mathbb{Z}} \widehat{\mathbb{Z}} = \widehat{\mathcal{O}}_K.$$

(b) We have $x \subset \widehat{\mathcal{O}}_K^*$ and $\alpha \in K^*$ such that $\alpha x \equiv 1 \bmod fm\widehat{\mathcal{O}}_K$ by (15.41) and $x \equiv 1 + mu \bmod fm\widehat{\mathcal{O}}_K$ for some $u \in \widehat{\mathbb{Z}}$ by (15.42). We need to prove that $\alpha\mathcal{O}_K \in P_{K,\mathbb{Z},m}(fm)$.

Since $\mathbb{Z}/f\mathbb{Z} \simeq \widehat{\mathbb{Z}}/f\widehat{\mathbb{Z}}$ by Exercise 15.3, there is $c \in \mathbb{Z}$ such that $u \equiv c \bmod f\widehat{\mathbb{Z}}$. It follows that $x \equiv 1 + mc \bmod fm\widehat{\mathcal{O}}_K$. The idele x is invertible in $\widehat{\mathcal{O}}_K$ and hence also in $\widehat{\mathcal{O}}_K/fm\widehat{\mathcal{O}}_K$. It follows that the integer $1 + mc$ is relatively prime to fm, so that there is $a \in \mathbb{Z}$, also relatively prime to fm, such that $a(1 + mc) \equiv 1 \bmod fm$. Then

$$\alpha x \equiv 1 \equiv a(1 + mc) \equiv ax \bmod fm\widehat{\mathcal{O}}_K,$$

which implies that $\alpha \equiv a \bmod fm\widehat{\mathcal{O}}_K$ since x is invertible mod fm. Using the isomorphism of part (a) for fm, we conclude that $\alpha \equiv a \bmod fm\mathcal{O}$. Since $a(1+mc) \equiv 1 \bmod fm$ implies $a \equiv 1 \bmod m$, we get $\alpha\mathcal{O}_K \in P_{K,\mathbb{Z},m}(fm)$.

(c) Take $y \in J^1_{fm}$. By the definition given before Lemma 15.14, $y \in \widehat{\mathcal{O}}_K^*$ and $y \equiv 1 \bmod fm\widehat{\mathcal{O}}_K$. Since the conductor f of $\mathcal{O}$ satisfies $f\mathcal{O}_K \subset \mathcal{O}$, we have

$$f\widehat{\mathcal{O}}_K = f(\mathcal{O}_K \otimes_{\mathbb{Z}} \widehat{\mathbb{Z}}) = (f\mathcal{O}_K) \otimes_{\mathbb{Z}} \widehat{\mathbb{Z}} \subset \mathcal{O} \otimes_{\mathbb{Z}} \widehat{\mathbb{Z}} = \widehat{\mathcal{O}}.$$

Then $y \equiv 1 \bmod fm\widehat{\mathcal{O}}_K$ implies that

$$y \in 1 + fm\widehat{\mathcal{O}}_K \subset 1 + m\widehat{\mathcal{O}}.$$

This shows that $y \in \widehat{\mathcal{O}}$ and that $y \equiv 1 \bmod m\widehat{\mathcal{O}}$. Since J^1_{fm} is a group under multiplication, this argument applies to y^{-1}. Thus $y^{-1} \in \widehat{\mathcal{O}}$. Hence $y \in \widehat{\mathcal{O}}^*$ and $y \equiv 1 \bmod m\widehat{\mathcal{O}}$, which proves that $y \in J^1_{\mathcal{O},m}$.

15.24. Given a positive integer m, the constant function $f(\tau) = \zeta_m$ is meromorphic on $\mathfrak{h}$, invariant under $\Gamma(m)$, and meromorphic at the cusps. Furthermore, its q-expansion $f(\tau) = \zeta_m + \sum_{n=1}^{\infty} 0 \cdot q^n$ has coefficients in $\mathbb{Q}(\zeta_m)$. Thus $f(\tau) \in \mathsf{F}_m$ by the definition of F_m. If $\mathcal{O} = [1, \tau_0]$, then $f(\tau_0)$ is defined, so by Theorem 15.21, we have $f(\tau_0) = \zeta_m \in L_{\mathcal{O},m}$.

15.25. (a) With $K = \mathbb{Q}(\omega)$ and $\mathcal{O} = \mathcal{O}_K = \mathbb{Z}[\omega]$, note that $|\mathcal{O}^*| = 6$. Further, since 5 is prime in $\mathcal{O}$, $\mathcal{O}/5\mathcal{O}$ is a field of order $N(5) = 25$, so that $|(\mathcal{O}/5\mathcal{O})^*| = 24$. We next claim that the map $\mathcal{O}^* \to (\mathcal{O}/5\mathcal{O})^*$ of Lemma 15.17 is injective. If $u \in \mathcal{O}^*$ is a unit other than 1, so that $u = -1$, $\pm\omega$, or $\pm\omega^2 = \pm(-1-\omega)$, it is easy to see that $u - 1 \notin 5\mathcal{O} = [5, 5\omega]$. This shows that $[u] \neq [1]$ in $\mathcal{O}/5\mathcal{O}$, so that u is not in the kernel of $\mathcal{O}^* \to (\mathcal{O}/5\mathcal{O})^*$, proving the claim. Then Lemma 15.17 shows that $|\mathrm{Gal}(L_{\mathcal{O},5}/L_{\mathcal{O}})| = 4$, and it follows that $[L_{\mathcal{O},5} : L_{\mathcal{O}}] = 4$. However, since $\mathcal{O}$ is a PID, the Hilbert class field $L_{\mathcal{O}}$ of K must equal K, so that $[L_{\mathcal{O},5} : K] = 4$.

(b) Since $K = L_{\mathcal{O}} \subset L_{\mathcal{O},5}$ and (from Exercise 15.24) $\zeta_5 \in L_{\mathcal{O},5}$, we have $K(\zeta_5) \subset L_{\mathcal{O},5}$. Since $\omega\zeta_5 = \zeta_3\zeta_5 = e^{2\pi i(8/15)} = \zeta_{15}^8$, we have $\zeta_{15} = (\omega\zeta_5)^2 \in \mathbb{Q}(\omega, \zeta_5)$. This gives inclusions

$$\mathbb{Q}(\zeta_{15}) \subset \mathbb{Q}(\omega, \zeta_5) = K(\zeta_5) \subset L_{\mathcal{O},5}.$$

By part (a), $[L_{\mathcal{O},5} : \mathbb{Q}] = [L_{\mathcal{O},5} : K][K : \mathbb{Q}] = 4 \cdot 2 = 8$. Since $[\mathbb{Q}(\zeta_{15}) : \mathbb{Q}] = \phi(15) = 8$, where ϕ is the Euler ϕ-function, the above inclusions are all equalities.

(c) From the hint,

$$\begin{aligned}\zeta_{10} r(\omega) = \frac{\sqrt{30+6\sqrt{5}} - 3 - \sqrt{5}}{4} &= \frac{2\sqrt{-3}\, i\sqrt{\frac{5+\sqrt{5}}{2}} - 3 - \sqrt{5}}{4}\\ &= \sqrt{-3}\,\frac{i}{2}\sqrt{\frac{5+\sqrt{5}}{2}} - \frac{3+\sqrt{5}}{4}\\ &= \sqrt{-3}\,\zeta_5 - \sqrt{-3}\cdot\frac{-1+\sqrt{5}}{4} - \frac{3+\sqrt{5}}{4}.\end{aligned}$$

Clearly $\sqrt{-3} \in \mathbb{Q}(\omega, \zeta_5)$ and $\zeta_5 \in \mathbb{Q}(\omega, \zeta_5)$, so that $\zeta_{10} = -\zeta_5^3 \in \mathbb{Q}(\omega, \zeta_5)$. It remains to show that

$$\frac{-1+\sqrt{5}}{4},\ \frac{3+\sqrt{5}}{4} = 1 + \frac{-1+\sqrt{5}}{4} \in \mathbb{Q}(\omega, \zeta_5).$$

But $\frac{-1+\sqrt{5}}{2} = \zeta_5 + \overline{\zeta_5} = \zeta_5 + \zeta_5^4 \in \mathbb{Q}(\omega, \zeta_5)$, and $r(\omega) \in \mathbb{Q}(\omega, \zeta_5) = L_{\mathcal{O},5}$ follows.

15.26. (a) For $K = \mathbb{Q}(i)$ and $\mathcal{O} = \mathcal{O}_K = \mathbb{Z}[i]$, we prove that $L_{\mathcal{O},5} = K(\zeta_5)$ by a method similar to parts (a) and (b) of Exercise 15.25. Since $5 = (1+2i)(1-2i)$ in $\mathbb{Z}[i]$,

$$(\mathcal{O}/5\mathcal{O})^* \simeq (\mathcal{O}/(1+2i))^* \times (\mathcal{O}/(1-2i))^* \simeq (\mathbb{Z}/5\mathbb{Z})^* \times (\mathbb{Z}/5\mathbb{Z})^* \simeq \mathbb{Z}/4\mathbb{Z} \times \mathbb{Z}/4\mathbb{Z},$$

showing that $|(\mathcal{O}/5\mathcal{O})^*| = 16$. We next claim that the map $\mathcal{O}^* \to (\mathcal{O}/5\mathcal{O})^*$ given in Lemma 15.17 is injective. If $u \in \mathcal{O}^*$ is a unit other than 1, so that $u = -1$ or $\pm i$, it is easy to see that $u - 1 \notin 5\mathcal{O} = [5, 5i]$. This shows that $[u] \neq [1]$ in $\mathcal{O}/5\mathcal{O}$, so that u is not in the kernel of $\mathcal{O}^* \to (\mathcal{O}/5\mathcal{O})^*$, proving the claim. Since $|\mathcal{O}^*| = 4$, it follows from Lemma 15.17 that $|\mathrm{Gal}(L_{\mathcal{O},5}/L_{\mathcal{O}})| = 4$, and therefore that $[L_{\mathcal{O},5} : L_{\mathcal{O}}] = 4$. Since $\mathcal{O} = \mathcal{O}_K$ is a PID, the Hilbert class field $L_{\mathcal{O}}$ of K must equal K, and we conclude that $[L_{\mathcal{O},5} : K] = 4$ and $K = L_{\mathcal{O}} \subset L_{\mathcal{O},5}$. Further, $\zeta_5 \in L_{\mathcal{O},5}$ by Exercise 15.24, so that $K(\zeta_5) \subset L_{\mathcal{O},5}$.

Next, since $i\zeta_5 = \zeta_4\zeta_5 = e^{2\pi i(9/20)} = \zeta_{20}^9$, we have $\zeta_{20} = (i\zeta_5)^9 \in \mathbb{Q}(i, \zeta_5)$, so we have inclusions

$$\mathbb{Q}(\zeta_{20}) \subset \mathbb{Q}(i, \zeta_5) = K(\zeta_5) \subset L_{\mathcal{O},5}.$$

Then $[L_{\mathcal{O},5} : \mathbb{Q}] = [L_{\mathcal{O},5} : K][K : \mathbb{Q}] = 4 \cdot 2 = 8$, also $[\mathbb{Q}(\zeta_{20}) : \mathbb{Q}] = \phi(20) = 8$, where ϕ is the Euler ϕ-function. Hence the above inclusions are all equalities.

(b) For $K = \mathbb{Q}(i) \subset L_{\mathcal{O},5}$ with $\mathcal{O} = \mathcal{O}_K = \mathbb{Z}[i]$, we have $n = 1$ and $m = 5$ in the notation of Theorem 15.19. In the theorem, the polynomial $f_{1,5}$ is chosen to have a real root, which is not the case for $x^4 + x^3 + x^2 + x + 1$. So we adapt the proof of the theorem as follows. If p is an odd prime, then

$$\begin{aligned} & p = x^2 + y^2 \text{ with } x \equiv 1 \bmod 5,\ y \equiv 0 \bmod 5 \\ \iff & p\mathcal{O}_K = \mathfrak{p}\bar{\mathfrak{p}},\ \mathfrak{p} = \alpha\mathcal{O}_K,\ \alpha \in \mathcal{O}_K,\ \alpha \equiv 1 \bmod 5\mathcal{O}_K \\ \iff & p\mathcal{O}_K = \mathfrak{p}\bar{\mathfrak{p}} \text{ in } \mathcal{O}_K,\ \mathfrak{p} \text{ splits completely in } L_{\mathcal{O}_K,5}, \end{aligned}$$

where the first equivalence is from the proof of Theorem 15.19 and the second is from the solution to Exercise 15.12. However, since p is odd,

$$p\mathcal{O}_K = \mathfrak{p}\bar{\mathfrak{p}} \text{ in } \mathcal{O}_K \iff p \equiv 1 \bmod 4.$$

Furthermore, part (a) shows that $L_{\mathcal{O}_K,5} = K(\zeta_5)$, and also that $[K(\zeta_5) : \mathbb{Q}] = 8$. Since $[K(\zeta_5) : \mathbb{Q}] = [K(\zeta_5) : K][K : \mathbb{Q}] = [K(\zeta_5) : K] \cdot 2$, it follows that $[K(\zeta_5) : K] = 4$. Thus $f(x) = x^4 + x^3 + x^2 + x + 1$ is the minimal polynomial of ζ_5 over K, so that by Proposition 5.11,

$$\begin{aligned} & \mathfrak{p} \text{ splits completely in } K(\zeta_5) = L_{\mathcal{O}_K,5} \\ \iff & f(x) \equiv 0 \bmod \mathfrak{p} \text{ has a solution in } \mathcal{O}_K \end{aligned}$$

for any prime $\mathfrak{p}$ of $\mathcal{O}_K$ for which $f(x)$ is separable modulo $\mathfrak{p}$. The discriminant of $f(x)$ is $125 = 5^3$ (easily computed by a computer algebra system), so that this equivalence holds whenever $\mathfrak{p} \nmid 5$. Furthermore, when $p \equiv 1 \bmod 4$, the splitting $p\mathcal{O}_K = \mathfrak{p}\bar{\mathfrak{p}}$ implies that $\mathcal{O}_K/\mathfrak{p} \simeq \mathbb{Z}/p\mathbb{Z}$. Thus such $\mathfrak{p}$ satisfy

$$\begin{aligned} & f(x) \equiv 0 \bmod \mathfrak{p} \text{ has a solution in } \mathcal{O}_K \\ \iff & f(x) \equiv 0 \bmod p \text{ has an integer solution.} \end{aligned}$$

Putting all of these equivalences together, we conclude that for any prime $p > 5$,

$$\left\{ \begin{matrix} p = x^2 + y^2 \text{ with} \\ x \equiv 1 \bmod 5,\ y \equiv 0 \bmod 5 \end{matrix} \right\} \iff \left\{ \begin{matrix} p \equiv 1 \bmod 4 \text{ and} \\ x^4 + x^3 + x^2 + x + 1 \equiv 0 \bmod p \\ \text{has an integer solution} \end{matrix} \right\}.$$

(c) For the first equivalence, note that $p \equiv 1 \bmod 5$ if and only if $5 \mid p - 1 = |(\mathbb{Z}/p\mathbb{Z})^*|$ if and only if $(\mathbb{Z}/p\mathbb{Z})^*$ has an element of order 5. Since $x^5 - 1 = (x-1)(x^4+x^3+x^2+x+1)$ and $p > 5$, $(\mathbb{Z}/p\mathbb{Z})^*$ has an element of order 5 if and only if $x^4+x^3+x^2+x+1 \equiv 0 \bmod p$ has an integer solution ($p > 5$ guarantees that the solution is not $\equiv 1 \bmod p$ and hence has order 5 in $(\mathbb{Z}/p\mathbb{Z})^*$). This proves the first equivalence, and the second follows from part (b) together with the obvious equivalence

$$p \equiv 1 \bmod 4 \text{ and } p \equiv 1 \bmod 5 \iff p \equiv 1 \bmod 20.$$

(d) The proof of Theorem 15.19 and the solution to Exercise 15.12 show that

$$p = x^2 + y^2,\ x \equiv 1 \bmod 5,\ y \equiv 0 \bmod 5 \iff p \text{ splits completely in } L_{\mathcal{O},5}.$$

We know that $L_{\mathcal{O},5} = \mathbb{Q}(\zeta_{20})$ by part (a). For the modulus $\mathfrak{m} = 20\infty$, the Artin map of $\mathbb{Q} \subset \mathbb{Q}(\zeta_{20})$ takes p to $[p] \in (\mathbb{Z}/20\mathbb{Z})^*$ by (8.3), and since p splits completely in $\mathbb{Q}(\zeta_{20})$ if and only if the Artin map for p is trivial (Corollary 5.21), we see that

$$p \text{ splits completely in } L_{\mathcal{O},5} = \mathbb{Q}(\zeta_{20}) \iff p \equiv 1 \bmod 20$$

The desired result follows immediately from the two displayed equivalences.

(e) If $p = x^2 + y^2$ with $x \equiv 1 \bmod 5$, $y \equiv 0 \bmod 5$, then clearly p is odd, so $p \equiv 1 \bmod 4$ since it is a sum of squares. Furthermore, modulo 5, we have $p = x^2 + y^2 \equiv 1^2 + 0^2 = 1 \bmod 5$, and then $p \equiv 1 \bmod 20$ as in the solution to part (c). Conversely, if $p \equiv 1 \bmod 20$, then $p \equiv 1 \bmod 4$, so that $p = x^2 + y^2$ for some $x, y \in \mathbb{Z}$. We also have $p \equiv 1 \bmod 5$, which implies that $x^2, y^2 \equiv 0, 1, 4 \bmod 5$ satisfy $x^2 + y^2 \equiv 1 \bmod 5$. Switching x and y if necessary, we have $x^2 \equiv 1 \bmod 5$ and $y^2 \equiv 0 \bmod 5$. The former implies $x \equiv \pm 1 \bmod 5$, so replacing x with $-x$, we may assume $x \equiv 1 \bmod 5$. Since $y^2 \equiv 0 \bmod 5$ implies $y \equiv 0 \bmod 5$, we get the desired representation of p.

References

[1] Tom M Apostol, *Modular functions and Dirichlet series in number theory*, Graduate Texts in Mathematics, No. 41, Springer-Verlag, New York-Heidelberg, 1976. MR0422157

[2] Emil Artin, *Galois theory*, Notre Dame Mathematical Lectures, no. 2, University of Notre Dame Press, South Bend, Ind., 1959. Edited and supplemented with a section on applications by Arthur N. Milgram; Second edition, with additions and revisions; Fifth reprinting. MR0265324

[3] A. O. L. Atkin and F. Morain, *Elliptic curves and primality proving*, Math. Comp. **61** (1993), no. 203, 29–68, DOI 10.2307/2152935. MR1199989

[4] A. Baker, *Linear forms in the logarithms of algebraic numbers. I, II, III*, Mathematika **13** (1966), 204–216; ibid. 14 (1967), 102–107; ibid. 14 (1967), 220–228, DOI 10.1112/s0025579300003843. MR220680

[5] Bruce C. Berndt, *Ramanujan's notebooks. Part IV*, Springer-Verlag, New York, 1994, DOI 10.1007/978-1-4612-0879-2. MR1261634

[6] Bruce C. Berndt, Ronald J. Evans, and Kenneth S. Williams, *Gauss and Jacobi sums*, Canadian Mathematical Society Series of Monographs and Advanced Texts, John Wiley & Sons, Inc., New York, 1998. A Wiley-Interscience Publication. MR1625181

[7] W. E. H. Berwick, *Modular Invariants expressible in terms of quadratic and cubic irrationalities*, Proc. London Math. Soc. (2) **28** (1928), no. 1, 53–69, DOI 10.1112/plms/s2-28.1.53. MR1575872

[8] K. R. Biermann, E. Schuhmann, H. Wussing and O. Neumann, *Mathematisches Tagebuch 1796–1814 von Carl Friedrich Gauss*, 3rd edition, Ostwalds Klassiker **256**, Leipzig, 1981 MR656518.

[9] B. J. Birch, *Diophantine analysis and modular functions*, Algebraic Geometry (Internat. Colloq., Tata Inst. Fund. Res., Bombay, 1968), Oxford Univ. Press, London, 1969, pp. 35–42. MR0258832

[10] B. J. Birch, *Weber's class invariants*, Mathematika **16** (1969), 283–294, DOI 10.1112/S0025579300008251. MR262206

[11] Z. I. Borevich and I. R. Shafarevich, *Number theory*, Pure and Applied Mathematics, Vol. 20, Academic Press, New York-London, 1966. Translated from the Russian by Newcomb Greenleaf. MR0195803

[12] Jonathan M. Borwein and Peter B. Borwein, *Pi and the AGM*, Canadian Mathematical Society Series of Monographs and Advanced Texts, John Wiley & Sons, Inc., New York, 1987. A study in analytic number theory and computational complexity; A Wiley-Interscience Publication. MR877728

[13] W. E. Briggs, *An elementary proof of a theorem about the representation of primes by quadratic forms*, Canad. J. Math. **6** (1954), 353–363, DOI 10.4153/cjm-1954-034-0. MR63408

[14] John Brillhart and Patrick Morton, *Class numbers of quadratic fields, Hasse invariants of elliptic curves, and the supersingular polynomial*, J. Number Theory **106** (2004), no. 1, 79–111, DOI 10.1016/j.jnt.2004.01.006. MR2049594

[15] Gottfried Bruckner, *Charakterisierung der galoisschen Zahlkörper, deren zerlegte Primzahlen durch binäre quadratische Formen gegeben sind* (German), Math. Nachr. **32** (1966), 317–326, DOI 10.1002/mana.19660320604. MR217043

[16] Duncan A. Buell, *Class groups of quadratic fields I, II*, Math. Comp. **30** (1976), no. 135, 610–623, ibid **48** (1987), 85–93, DOI 10.2307/2005330. MR404205, MR0866100

[17] Walter Kaufmann-Bühler, *Gauss: A biographical study*, Springer-Verlag, Berlin-New York, 1981. MR617739

[18] J. J. Burckhardt, *Euler's work on number theory: a concordance for A. Weil's Number theory [Birkhäuser, Boston, Mass., 1984; MR0734177 (85c:01004)]*, Historia Math. **13** (1986), no. 1, 28–35, DOI 10.1016/0315-0860(86)90223-5. MR840974, MR0866100

[19] P. Bussotti, *From Fermat to Gauss: Infinite Descent and Methods of Reduction in Number Theory*, Dr. Erwin Rauner Verlag, Augsburg, 2006.

[20] Ph. Cassou-Noguès and M. J. Taylor, *Elliptic functions and rings of integers*, Progress in Mathematics, vol. 66, Birkhäuser Boston, Inc., Boston, MA, 1987. MR886887

[21] K. Chandrasekharan, *Elliptic functions*, Grundlehren der mathematischen Wissenschaften [Fundamental Principles of Mathematical Sciences], vol. 281, Springer-Verlag, Berlin, 1985, DOI 10.1007/978-3-642-52244-4. MR808396

[22] Bumkyu Cho, *Primes of the form $x^2 + ny^2$ with conditions $x \equiv 1 \bmod N$, $y \equiv 0 \bmod N$*, J. Number Theory **130** (2010), no. 4, 852–861, DOI 10.1016/j.jnt.2009.07.013. MR2600406

[23] S. Chowla, *An extension of Heilbronn's class number theorem*, Quarterly J. Math. **5** (1934), pp. 304–307.

[24] Paula Cohen, *On the coefficients of the transformation polynomials for the elliptic modular function*, Math. Proc. Cambridge Philos. Soc. **95** (1984), no. 3, 389–402, DOI 10.1017/S0305004100061697. MR755826

[25] Harvey Cohn, *A classical invitation to algebraic numbers and class fields*, Universitext, Springer-Verlag, New York-Heidelberg, 1978. With two appendices by Olga Taussky: "Artin's 1932 Göttingen lectures on class field theory" and "Connections between algebraic number theory and integral matrices". MR506156

[26] Harvey Cohn, *Advanced number theory*, Dover Books on Advanced Mathematics, Dover Publications, Inc., New York, 1980. Reprint of *A second course in number theory*, 1962. MR594936

[27] Harvey Cohn, *Introduction to the construction of class fields*, Cambridge Studies in Advanced Mathematics, vol. 6, Cambridge University Press, Cambridge, 1985. MR812270

[28] Mary Joan Collison, *The origins of the cubic and biquadratic reciprocity laws*, Arch. History Exact Sci. **17** (1977), no. 1, 63–69, DOI 10.1007/BF00348402. MR0441648

[29] David A. Cox, *Galois theory*, 2nd ed., Pure and Applied Mathematics (Hoboken), John Wiley & Sons, Inc., Hoboken, NJ, 2012, DOI 10.1002/9781118218457. MR2919975

[30] David Cox, John McKay, and Peter Stevenhagen, *Principal moduli and class fields*, Bull. London Math. Soc. **36** (2004), no. 1, 3–12, DOI 10.1112/S0024609303002583. MR2011972

[31] David A. Cox, *The arithmetic-geometric mean of Gauss*, Enseign. Math. (2) **30** (1984), no. 3-4, 275–330. MR767905

[32] M. Deuring, *Die Klassenkörper der komplexen Multiplikation* (German), Enzyklopädie der mathematischen Wissenschaften mit Einschluss ihrer Anwendungen, Band I 2, Heft 10, Teil II (Article I 2, vol. 23, B. G. Teubner Verlagsgesellschaft, Stuttgart, 1958. MR0167481

[33] Max Deuring, *Teilbarkeitseigenschaften der singulären Moduln der elliptischen Funktionen und die Diskriminante der Klassengleichung* (German), Comment. Math. Helv. **19** (1946), 74–82, DOI 10.1007/BF02565948. MR18704

[34] L. E. Dickson, *History of the Theory of Numbers*, Carnegie Institute, Washington D.C., 1919–1923. (Reprint by Chelsea, New York, 1971.) MR0245499.

[35] P. G. L. Dirichlet, *Werke*, Berlin, 1889–1897. (Reprint by Chelsea, New York, 1969.)

[36] P. G. L. Dirichlet, *Zahlentheorie*, 4th edition, Vieweg, Braunschweig, 1894.

[37] David R. Dorman, *Singular moduli, modular polynomials, and the index of the closure of $\mathbf{Z}[j(\tau)]$ in $\mathbf{Q}(j(\tau))$*, Math. Ann. **283** (1989), no. 2, 177–191, DOI 10.1007/BF01446429. MR980592

[38] David R. Dorman, *Special values of the elliptic modular function and factorization formulae*, J. Reine Angew. Math. **383** (1988), 207–220, DOI 10.1515/crll.1988.383.207. MR921991

[39] W. Duke, *Continued fractions and modular functions*, Bull. Amer. Math. Soc. (N.S.) **42** (2005), no. 2, 137–162, DOI 10.1090/S0273-0979-05-01047-5. MR2133308

[40] David S. Dummit and Richard M. Foote, *Abstract algebra*, 3rd ed., John Wiley & Sons, Inc., Hoboken, NJ, 2004. MR2286236

[41] Harold M. Edwards, *Fermat's last theorem: A genetic introduction to algebraic number theory*, Graduate Texts in Mathematics, vol. 50, Springer-Verlag, New York-Berlin, 1977. MR616635

[42] W. Ellison and F. Ellison, *Théorie des nombres*, in *Abrégé d'Histoire des Mathématiques 1700–1900*, Vol. I, ed. by J. Dieudonné, Hermann, Paris, 1978, pp. 165–334.

[43] L. Euler, *Opera Omnia*, Series prima, Vols. I–V, Teubner, Leipzig and Berlin, 1911–1944.

[44] P. Eymard and J. P. Lafon, *Le journal mathématique de Gauss*, Revue d'Histoire des Sciences **9** (1956), pp. 21–51.

[45] P. de Fermat, *Oeuvres*, Gauthier-Villars, Paris, 1891–1896.

[46] Daniel E. Flath, *Introduction to number theory*, A Wiley-Interscience Publication, John Wiley & Sons, Inc., New York, 1989. MR972739

[47] Wolfgang Franz, *Die Teilwerte der Weberschen Tau-Funktion* (German), J. Reine Angew. Math. **173** (1935), 60–64, DOI 10.1515/crll.1935.173.60. MR1581458

[48] Günther Frei, *Leonhard Euler's convenient numbers*, Math. Intelligencer **7** (1985), no. 3, 55–58, 64, DOI 10.1007/BF03025809. MR795540

[49] Günther Frei, *On the development of the genus of quadratic forms* (English, with French summary), Ann. Sci. Math. Québec **3** (1979), no. 1, 5–62. MR530752

[50] P.-H. Fuss, *Correspondance mathématique et physique de quelques célèbres géomètres du XVIIIème siècle. Tomes I, II* (French), St. Petersburg, 1843, The Sources of Science, No. 35, Johnson Reprint Corp., New York-London, 1968. Précédée d'une notice sur les travaux de Léonard Euler, tant imprimés qu'inédits et publiée sous les auspices de l'Académie Impériale des Sciences de Saint-Pétersbourg. MR0225627

[51] C. F. Gauss, *Disquisitiones Arithmeticae*, Leipzig, 1801. Republished in 1863 as Volume I of *Werke* (see [42]). French translation, *Recherches Arithmétiques*, Paris, 1807. (Reprint by Hermann, Paris, 1910.) German translation, *Untersuchungen über Höhere Arithmetik*, Berlin, 1889. (Reprint by Chelsea, New York, 1965.) English Translation, Yale, New Haven, 1966. (Reprint by Springer-Verlag, Berlin, Heidelberg, and New York, 1986,) MR0197380

[52] C. F. Gauss, *Werke*, Gottingen and Leipzig, 1863–1927.

[53] Alice Gee, *Class invariants by Shimura's reciprocity law* (English, with English and French summaries), J. Théor. Nombres Bordeaux **11** (1999), no. 1, 45–72. Les XXèmes Journées Arithmétiques (Limoges, 1997). MR1730432

[54] Alice Gee and Peter Stevenhagen, *Generating class fields using Shimura reciprocity*, Algorithmic number theory (Portland, OR, 1998), Lecture Notes in Comput. Sci., vol. 1423, Springer, Berlin, 1998, pp. 441–453, DOI 10.1007/BFb0054883. MR1726092

[55] S. Goldwasser and J. Kilian, *Almost all primes can be quickly quickly certified*, Proc. 18th Annual ACM Symp. on Theory of Computing (STOC, Berkeley), ACM Press, 1986, pp. 316–329.

[56] Fernando Q. Gouvêa, *p-adic numbers: An introduction*, 2nd ed., Universitext, Springer-Verlag, Berlin, 1997, DOI 10.1007/978-3-642-59058-0. MR1488696

[57] Georges Gras, *Class field theory: From theory to practice*, Springer Monographs in Mathematics, Springer-Verlag, Berlin, 2003. Translated from the French manuscript by Henri Cohen, DOI 10.1007/978-3-662-11323-3. MR1941965

[58] J. J. Gray, *A commentary on Gauss's mathematical diary, 1796–1814, with an English translation*, Exposition. Math. **2** (1984), no. 2, 97–130. (Gauss' diary has also been translated into French [**34**] and German [**5**].) MR783128

[59] Benedict H. Gross, *An elliptic curve test for Mersenne primes*, J. Number Theory **110** (2005), no. 1, 114–119, DOI 10.1016/j.jnt.2003.11.011. MR2114676

[60] Benedict H. Gross, *Arithmetic on elliptic curves with complex multiplication*, Lecture Notes in Mathematics, vol. 776, Springer, Berlin, 1980. With an appendix by B. Mazur. MR563921

[61] Benedict H. Gross and Don B. Zagier, *On singular moduli*, J. Reine Angew. Math. **355** (1985), 191–220. MR772491

[62] Emil Grosswald, *Representations of integers as sums of squares*, Springer-Verlag, New York, 1985, DOI 10.1007/978-1-4613-8566-0. MR803155

[63] G. H. Hardy and E. M Wright, *An introduction to the theory of numbers*, 5th ed., The Clarendon Press, Oxford University Press, New York, 1979. MR568909

[64] Farshid Hajir and Fernando Rodriguez Villegas, *Explicit elliptic units. I*, Duke Math. J. **90** (1997), no. 3, 495–521, DOI 10.1215/S0012-7094-97-09013-X. MR1480544

[65] H. Hasse, *Bericht über neuere Untersuchungen und Probleme aus der Theorie def algebraischen Zahlkörper, I, Ia and II*, Jahresber. Deutch. Math. Verein **35** (1926), pp. 1–55, **36** (1927), pp. 233–311, and Erg. Bd. **6** (1930), pp. 1–201. (Reprint by Physica-Verlag, Würzburg Vienna, 1965.) MR0195848

[66] Helmut Hasse, *Number Theory*, Springer-Verlag, Berlin, Heidelberg, and New York, 1980.

[67] Helmut Hasse, *Zur Geschlechtertheorie in quadratischen Zahlkörpern* (German), J. Math. Soc. Japan **3** (1951), 45–51, DOI 10.2969/jmsj/00310045. MR43828

[68] Kurt Heegner, *Diophantische Analysis und Modulfunktionen* (German), Math. Z. **56** (1952), 227–253, DOI 10.1007/BF01174749. MR53135

[69] Oskar Herrmann, *Über die Berechnung der Fourierkoeffizienten der Funktion $j(\tau)$* (German), J. Reine Angew. Math. **274(275)** (1975), 187–195, DOI 10.1515/crll.1975.274-275.187. MR374032

[70] I. N. Herstein, *Topics in Algebra*, 2nd edition, Wiley, New York, 1975. MR0356988

[71] C. S. Herz, *Computation of singular j-invariants, in Seminar on Complex Multiplication*, Lecture Notes in Math. **21**, Springer-Verlag, Berlin, Heidelberg, and New York, 1966, pp. VIII-1 to VIII-11.

[72] C. S. Herz, *Construction of class fields, in Seminar on Complex Multiplication*, Lecture Notes in Math. **21**, Springer-Verlag, Berlin, Heidelberg, and New York, 1966, pp. VII-1 to VII-21.
5

[73] L.-K. Hua, *Introduction to Number Theory*, Springer-Verlag, Berlin, Heidelberg, and New York, 1982.

[74] D. Husemöller, *Elliptic Curves*, Springer-Verlag, Berlin, Heidelberg, and New York, 1987.

[75] Kenneth F. Ireland and Michael I. Rosen, *A classical introduction to modern number theory*, Graduate Texts in Mathematics, vol. 84, Springer-Verlag, New York-Berlin, 1982. Revised edition of *Elements of number theory*. MR661047

[76] Makoto Ishida, *The genus fields of algebraic number fields*, Lecture Notes in Mathematics, Vol. 555, Springer-Verlag, Berlin-New York, 1976. MR0435028

[77] C. G. J. Jacobi, *Gesammelte Werke*, Vol. 6, Berlin, 1891. (Reprint by Chelsea, New York, 1969.)

[78] Gerald J. Janusz, *Algebraic number fields*, Pure and Applied Mathematics, Vol. 55, Academic Press [Harcourt Brace Jovanovich, Publishers], New York-London, 1973. MR0366864

[79] Burton W. Jones, *The Arithmetic Theory of Quadratic Forms*, Carcus Monograph Series, no. 10, Mathematical Association of America, Buffalo, N.Y., 1950. MR0037321

[80] E. Kaltofen, T. Valente and N. Yui, *An improved Las Vegas primality test*, Proc. International Symposium on Symbolic and Algebraic Computation (ISSAC '89, Portland), ACM Press, 1989, pp. 26–33.

[81] Erich Kaltofen and Noriko Yui, *Explicit construction of the Hilbert class fields of imaginary quadratic fields by integer lattice reduction*, Number theory (New York, 1989/1990), Springer, New York, 1991, pp. 149–202. MR1124640

[82] E. Kaltofen and N. Yui, *On the modular equation of order 11*, in *Third MACSYMA User's Conference, Proceedings*, General Electric, 1984, pp. 472–485.

[83] Anthony W. Knapp, *Elliptic curves*, Mathematical Notes, vol. 40, Princeton University Press, Princeton, NJ, 1992. MR1193029

[84] Neal Koblitz, *Introduction to elliptic curves and modular forms*, Graduate Texts in Mathematics, vol. 97, Springer-Verlag, New York, 1984, DOI 10.1007/978-1-4684-0255-1. MR766911

[85] L. Kronecker, *Werke*, Leipzig, 1895–1931. (Reprint by Chelsea, New York, 1968.)

[86] J. L. Lagrange, *Oeuvres*, Vol. 3, Gauthier-Villars, Paris, 1869.

[87] Edmund Landau, *Über die Klassenzahl der binären quadratischen Formen von negativer Discriminante* (German), Math. Ann. **56** (1903), no. 4, 671–676, DOI 10.1007/BF01444311. MR1511192

[88] E. Landau, *Vorlesungen über Zahlentheorie*, Hirzel, Leipzig, 1927.

[89] S. Lang, *Algebraic Number Theory*, Springer-Verlag, Berlin, Heidelberg, and New York, 1986.

[90] S. Lang, *Elliptic Functions*, 2nd edition, Springer-Verlag, Berlin, Heidelberg, and New York, 1987.

[91] A. M. Legendre, *Essai sur la Théorie des Nombres*, Paris, 1798. Third edition retitled *Théorie des Nombres*, Paris, 1830. (Reprint by Blanchard, Paris, 1955.) MR2859036

[92] A. M. Legendre, *Recherches d'analyse indéterminée*, in *Histoire de l'Académie Royale des Sciences, 1785*, Paris, 1788, pp. 465–559.

[93] Franz Lemmermeyer, *Reciprocity laws: From Euler to Eisenstein*, Springer Monographs in Mathematics, Springer-Verlag, Berlin, 2000, DOI 10.1007/978-3-662-12893-0. MR1761696

[94] H. W. Lenstra Jr., *Factoring integers with elliptic curves*, Ann. of Math. (2) **126** (1987), no. 3, 649–673, DOI 10.2307/1971363. MR916721

[95] H. W. Lenstra Jr. and Carl Pomerance, *Primality testing with Gaussian periods*, J. Eur. Math. Soc. (JEMS) **21** (2019), no. 4, 1229–1269, DOI 10.4171/JEMS/861. MR3941463

[96] Daniel A. Marcus, *Number fields*, Universitext, Springer-Verlag, New York-Heidelberg, 1977. MR0457396

[97] G. B. Mathews, *Theory of numbers*, Chelsea Publishing Co., New York, 1961. 2nd ed. MR0126402

[98] J. Milne, *Field and Galois Theory*, `https://www.jmilne.org/math/CourseNotes/ft.html`

[99] F. Morain, *Implementation of the Goldwasser-Kilian-Atkin primality testing algorithm*, Rapport de Recherche 911, INRIA-Rocquencourt, Octobre 1988.

[100] Jürgen Neukirch, *Class field theory*, Grundlehren der mathematischen Wissenschaften [Fundamental Principles of Mathematical Sciences], vol. 280, Springer-Verlag, Berlin, 1986, DOI 10.1007/978-3-642-82465-4. MR819231

[101] J. Oesterlé, *Le problème de Gauss sur le nombre de classes* (French), Enseign. Math. (2) **34** (1988), no. 1-2, 43–67. MR960192

[102] Joseph Oesterlé, *Nombres de classes des corps quadratiques imaginaires* (French), Astérisque **121-122** (1985), 309–323. Seminar Bourbaki, Vol. 1983/84. MR768967

[103] T. Ono, *Arithmetic of Algebraic Groups and its Applications*, St. Paul's International Exchange Series, Occasional Papers VI, St. Paul's University, 1986.

[104] H. L. S. Orde, *On Dirichlet's class number formula*, J. London Math. Soc. (2) **18** (1978), no. 3, 409–420, DOI 10.1112/jlms/s2-18.3.409. MR518225

[105] G. J. Rieger, *Die Zahlentheorie bei C. F. Gauss*, in *C. F. Gauss, Gedenkband Anlässlich des 100. Todestages, am 23. Februar 1955*, Teubner, Leipzig, 1957, pp. 38–77.

[106] Peter Roquette, *On class field towers*, Algebraic Number Theory (Proc. Instructional Conf., Brighton, 1965), Thompson, Washington, D.C., 1967, pp. 231–249. MR0218331

[107] K. Rubin and A. Silverberg, *Point counting on reductions of CM elliptic curves*, J. Number Theory **129** (2009), no. 12, 2903–2923, DOI 10.1016/j.jnt.2009.01.020. MR2560842

[108] Winfried Scharlau and Hans Opolka, *From Fermat to Minkowski*, Undergraduate Texts in Mathematics, Springer-Verlag, New York, 1985. Lectures on the theory of numbers and its historical development; Translated from the German by Walter K. Bühler and Gary Cornell, DOI 10.1007/978-1-4757-1867-6. MR770936

[109] Reinhard Schertz, *Complex multiplication*, New Mathematical Monographs, vol. 15, Cambridge University Press, Cambridge, 2010, DOI 10.1017/CBO9780511776892. MR2641876

[110] Reinhard Schertz, *Die singulären Werte der Weberschen Funktionen* $\mathfrak{f}, \mathfrak{f}_1, \mathfrak{f}_2, \gamma_2, \gamma_3$ (German), J. Reine Angew. Math. **286(287)** (1976), 46–74, DOI 10.1515/crll.1976.286-287.46. MR422213

[111] Reinhard Schertz, *Weber's class invariants revisited* (English, with English and French summaries), J. Théor. Nombres Bordeaux **14** (2002), no. 1, 325–343. MR1926005

[112] J.-P. Serre, *A course in arithmetic*, Graduate Texts in Mathematics, No. 7, Springer-Verlag, New York-Heidelberg, 1973. Translated from the French. MR0344216

[113] Jean-Pierre Serre, *Local fields*, Graduate Texts in Mathematics, vol. 67, Springer-Verlag, New York-Berlin, 1979. Translated from the French by Marvin Jay Greenberg. MR554237

[114] Daniel Shanks, *Class number, a theory of factorization, and genera*, 1969 Number Theory Institute (Proc. Sympos. Pure Math., Vol. XX, State Univ. New York, Stony Brook, N.Y., 1969), Amer. Math. Soc., Providence, R.I., 1971, pp. 415–440. MR0316385

[115] Goro Shimura, *Introduction to the arithmetic theory of automorphic functions*, Kanô Memorial Lectures, No. 1, Iwanami Shoten Publishers, Tokyo; Princeton University Press, Princeton, N.J., 1971. Publications of the Mathematical Society of Japan, No. 11. MR0314766

[116] Carl Ludwig Siegel, *Equivalence of quadratic forms*, Amer. J. Math. **63** (1941), 658–680, DOI 10.2307/2371381. MR5506

[117] C. L. Siegel, *Über die Classenzahl quadratischer Zahlkörper*, Acta Arithmetica **1** (1935), pp. 83–86.

[118] Joseph H. Silverman, *The arithmetic of elliptic curves*, Graduate Texts in Mathematics, vol. 106, Springer-Verlag, New York, 1986, DOI 10.1007/978-1-4757-1920-8. MR817210

[119] Joseph H. Silverman, *Advanced topics in the arithmetic of elliptic curves*, Graduate Texts in Mathematics, vol. 151, Springer-Verlag, New York, 1994, DOI 10.1007/978-1-4612-0851-8. MR1312368

[120] Joseph H. Silverman and John Tate, *Rational points on elliptic curves*, Undergraduate Texts in Mathematics, Springer-Verlag, New York, 1992, DOI 10.1007/978-1-4757-4252-7. MR1171452
Soc. **10** (1878), pp. 87–91.

[121] Henry J. Stephen Smith, *Note on a Modular Equation for the Transformation of the Third Order*, Proc. Lond. Math. Soc. **10** (1878/79), 87–91, DOI 10.1112/plms/s1-10.1.87. MR1576202

[122] Henry J. Stephen Smith, *Report on the Theory of Numbers*, Reports of the British Association, 1859–1865. (Reprint by Chelsea, New York, 1965.)

[123] H. M. Stark, *A complete determination of the complex quadratic fields of class-number one*, Michigan Math. J. **14** (1967), 1–27. MR222050

[124] H. M. Stark, *Class-numbers of complex quadratic fields*, Modular functions of one variable, I (Proc. Internat. Summer School, Univ. Antwerp, Antwerp, 1972), Springer, Berlin, 1973, pp. 153–174. Lecture Notes in Mathematics, Vol. 320. MR0344225

[125] H. M. Stark, *On the "gap" in a theorem of Heegner*, J. Number Theory **1** (1969), 16–27, DOI 10.1016/0022-314X(69)90023-7. MR241384

[126] Peter Stevenhagen, *Hilbert's 12th problem, complex multiplication and Shimura reciprocity*, Class field theory—its centenary and prospect (Tokyo, 1998), Adv. Stud. Pure Math., vol. 30, Math. Soc. Japan, Tokyo, 2001, pp. 161–176, DOI 10.2969/aspm/03010161. MR1846457

[127] S. G. Vlăduţ, *Kronecker's Jugentraum and Modular Functions*, Gordon and Breach, New York, 1991.

[128] Stan Wagon, *Primality testing*, Math. Intelligencer **8** (1986), no. 3, 58–61, DOI 10.1007/BF03025793. MR846996

[129] J. Wallis, *Opera Mathematica*, Oxford, 1695–1699. (Reprint by G. Olms, Hindesheim, New York, 1972.)

[130] H. Weber, *Beweis des Satzes, dass jede eigentlich primitive quadratische Form unendlich viele Primzahlen darzustellen fähig ist* (German), Math. Ann. **20** (1882), no. 3, 301–329, DOI 10.1007/BF01443599. MR1510171

[131] H. Weber, *Lehrbuch der Algebra*, Vol. III, 2nd edition, Vieweg, Braunschweig, 1908. (Reprint by Chelsea, New York, 1961.)

[132] H. Weber, *Zur complexen Multiplication elliptischer Functionen* (German), Math. Ann. **33** (1889), no. 3, 390–410, DOI 10.1007/BF01443968. MR1510550

[133] André Weil, *Basic Number Theory*, 3rd edition, Springer-Verlag, Berlin, Heidelberg, and New York, 1974.

[134] André Weil, *La cyclotomie jadis et naguère* (French), Enseign. Math. (2) **20** (1974), 247–263. MR441831

[135] André Weil, *Number theory: An approach through history; From Hammurapi to Legendre*, Birkhäuser Boston, Inc., Boston, MA, 1984, DOI 10.1007/978-0-8176-4571-7. MR734177

[136] André Weil, *Two lectures on number theory, past and present*, Enseign. Math. (2) **20** (1974), 87–110. MR366788

[137] P. J. Weinberger, *Exponents of the class groups of complex quadratic fields*, Acta Arith. **22** (1973), 117–124, DOI 10.4064/aa-22-2-117-124. MR313221

[138] E. T. Whittaker and G. N. Watson, *A course of modern analysis*, Cambridge Mathematical Library, Cambridge University Press, Cambridge, 1996. An introduction to the general theory of infinite processes and of analytic functions; with an account of the principal transcendental functions; Reprint of the fourth (1927) edition, DOI 10.1017/CBO9780511608759. MR1424469

[139] Noriko Yui, *Explicit form of the modular equation*, J. Reine Angew. Math. **299(300)** (1978), 185–200, DOI 10.1515/crll.1978.299-300.185. MR476642

[140] D. B. Zagier, *Zetafunktionen und quadratische Körper* (German), Hochschultext [University Textbooks], Springer-Verlag, Berlin-New York, 1981. Eine Einführung in die höhere Zahlentheorie. [An introduction to higher number theory]. MR631688

[141] D. Zagier, *L-series of elliptic curves, the Birch-Swinnerton-Dyer conjecture, and the class number problem of Gauss*, Notices Amer. Math. Soc. **31** (1984), no. 7, 739–743. MR765835

Further Reading

A. Further Reading for Chapter 1

A1. D. A. Buell, *Binary Quadratic Forms: Classical Theory and Modern Computations*, Springer-Verlag, New York, 1989.

A2. K. Burde, *Ein rationales biquadratisches Reziprozitätsgesetz*, J. Reine Angew. Math. **235** (1969), pp. 175–184.

A3. A. Schönhage, *Fast reduction and composition of binary quadratic forms*, Proc. International Symposium on Symbolic and Algebraic Computation (ISSAC '91, Bonn), ACM Press, 1991, pp. 128–133.

A4. B. K. Spearman and K. S. Williams, *Representing primes by binary quadratic forms*, Amer. Math. Monthly **99** (1992), pp. 423–426.

B. Further Reading for Chapter 2

B1. H. Cohen and P. Stevenhagen, *Computational class field theory*, in *Algorithmic Number Theory: Lattices, Number Fields, Curves and Cryptography*, Math. Sci. Res. Inst. Publ. **44**, Cambridge University Press, Cambridge, 2008, pp. 497–534.

B2. K. Conrad, *History of class field theory*, available online at `www.math.uconn.edu/~kconrad/blurbs/gradnumthy/cfthistory.pdf`.

B3. J. Lagarias, *Sets of primes determined by systems of polynomial congruences*, Illinois J. Math. **27** (1983), pp. 224–239.

B4. K. S. Williams and R. H. Hudson, *Representation of primes by the principal form of discriminant $-D$ when the classnumber $h(-D)$ is* 3, Acta Arith. **57** (1991), pp. 131–153.

C. Further Reading for Chapter 3

C1. J. V. Belding, R. Bröker, A. Enge and K. Lauter, *Computing Hilbert class polynomials*, in *Algorithmic Number Theory*, Lecture Notes in Comput. Sci. **5011**, Springer-Verlag, Berlin, 2008, pp. 282–295.

C2. B. C. Berndt and H. H. Chan, *Ramanujan and the modular j-invariant*, Canad. Math. Bull. **42** (1999), pp. 427–440.

C3. S. Chowla and M. Cowles, *On the coefficients c_n in the expansion $x\prod_1^\infty(1-x^n)^2(1-x^{11n})^2 = \sum_1^\infty c_n x^n$*, J. Reine Angew. Math. **292** (1977), pp. 115–116.

C4. H. Cohen and P. Stevenhagen, *Computational class field theory*, in *Algorithmic Number Theory: Lattices, Number Fields, Curves and Cryptography*, Math. Sci. Res. Inst. Publ. **44**, Cambridge University Press, Cambridge, 2008, pp. 497–534.

C5. A. V. Sutherland, *Computing Hilbert class polynomials with the Chinese remainder theorem*, Math. Comp. **80** (2011), pp. 501–538.

C6. N. Yui and D. Zagier, *On the singular values of Weber modular functions*, Math. Comp. **66** (1997), pp. 1645–1662.

D. Further Reading for Chapter 4

D1. A. Abatzoglou, A. Silverberg, A. V. Sutherland and A. Wong, *Deterministic elliptic curve primality proving for a special sequence of numbers*, preprint, 2012, available online at `http://arXiv.org/abs/1202.3695`.

D2. H. H. Chan, A. Gee and V. Tan, *Cubic singular moduli, Ramanujan's class invariants λ_n and the explicit Shimura reciprocity law*, Pacific J. Math. **208** (2003), pp. 23–37.

D3. B. Cho and J. K. Koo, *Construction of class fields over imaginary quadratic fields and applications*, Quarterly J. Math. **61** (2010), pp. 199–216.

D4. H. Cohen and P. Stevenhagen, *Computational class field theory*, in *Algorithmic Number Theory: Lattices, Number Fields, Curves and Cryptography*, Math. Sci. Res. Inst. Publ. **44**, Cambridge University Press, Cambridge, 2008, pp. 497–534.

D5. K. J. Hong and J. K. Koo, *Singular values of some modular functions and their applications to class fields*, Ramanujan J. **16** (2008), pp. 321–337.

D6. F. Morain, *Primality proving using elliptic curves: an update*, in *Algorithmic Number Theory* (*Portland, OR, 1998*), Lecture Notes in Comput. Sci. **1423**, Springer-Verlag, Berlin, 1998, pp. 111–127.

Index

G

H

I

J